P9-CMO-861
The Oxford dictionary of art.

THE OXFORD DICTIONARY *of* **ART**

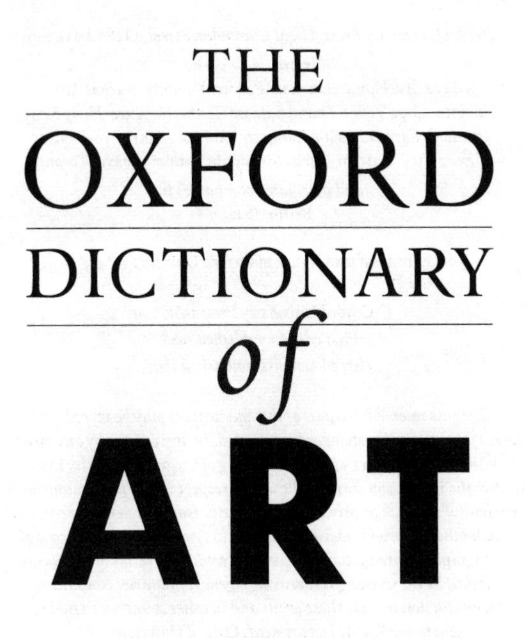

NEW EDITION

EDITED BY
IAN CHILVERS
AND
HAROLD OSBORNE

Consultant Editor
DENNIS FARR

Oxford New York
OXFORD UNIVERSITY PRESS
1997

Oxford University Press, Great Clarendon Street, Oxford OX2 6DP
Oxford New York
Athens Auckland Bangkok Bogota Bombay Buenos Aires
Calcutta Cape Town Dar es Salaam Delhi Florence Hong Kong
Istanbul Karachi Kuala Lumpur Madras Madrid Melbourne
Mexico City Nairobi Paris Singapore Taipei Tokyo Toronto
and associated companies in
Berlin Ibadan

First edition published 1988
This edition first published 1997

British Library Cataloguing in Publication Data
Data available

Library of Congress Cataloging in Publication Data
Data available

ISBN 0-19-860084-4

1 3 5 7 9 10 8 6 4 2

Typeset by Jayvee, Trivandrum, India
Printed in Great Britain
on acid-free paper by
Bookcraft (Bath) Ltd
Midsomer Norton, Somerset

Foreword

The success of *The Oxford Companion to Art*, edited by the late Harold Osborne and first published in 1970, with many subsequent reprintings, and of *The Oxford Companion to Twentieth-Century Art*, first published in 1981, also edited by Mr Osborne, has proved the demand for handy reference books on the fine arts. Another volume, *The Oxford Companion to the Decorative Arts*, which first appeared under Mr Osborne's editorship in 1975, has also fulfilled a need. Excellent as these books are, it was felt that an *Oxford Dictionary of Art* would serve a useful purpose in providing a thoroughly up-to-date work in a single volume, with the added advantage of drawing on the expertise of the many specialist contributors to the three *Companions*.

The aim has been to provide an overview of Western art forms and of individual artists from antiquity to our own day. Architecture and architects have been excluded unless they have some relevance to painting, sculpture, the graphic arts, or design. Thus, a tympanum is defined since it is often a field for sculptural decoration; but the architectural work of Michelangelo, for example, is treated in much more cursory fashion than his activities as a painter and sculptor. Oriental art has been included only in so far as it has had an influence on Western art. Ian Chilvers, the compiler and editor of this volume, has drawn on the entries in *The Oxford Companion to Art* and its twentieth-century counterpart, and to a lesser extent on the decorative arts volume; but the result is no mere conflation of selected entries from the three works. Virtually all the major entries and many lesser ones have been completely rewritten, and all have been carefully revised (and corrected) so as to take account of recent published research where appropriate. *The Oxford Dictionary of Art* also includes many artists and art terms omitted from *OCA* and *OC20CA* but which one would now expect to find in a dictionary of art. Similarly, the number of entries for noteworthy collectors, patrons, dealers, and art historians has been much expanded to include Henry Clay Frick, Paul Durand-Ruel, Kenneth Clark, Ernst Gombrich, J. Paul Getty, and Nikolaus Pevsner, among others; and the scope of entries for museums and galleries has also been enlarged.

A dictionary of this kind is not intended to be encyclopaedic, but we have aimed at comprehensiveness within the limits defined above. Like the *Companion* volumes, the *ODA* is meant for the layman who needs reliable information in an easily accessible form; it is also designed to be a handy reference book for art students and teachers. Inevitably, there are omissions, and, as with all works of this nature, some new biographical facts about particular artists will emerge after its publication. The compiler has had to weigh the evidence available where facts, or their interpretation, are still in dispute, and we have had to make decisions in the light of that evidence. We have not included a bibliography of the type found in the *Companions*, but in certain instances leading authorities have been quoted and sources given in the text of the entry concerned.

As consultant editor of *The Oxford Dictionary of Art*, I should like to pay tribute to the professionalism and skill of Ian Chilvers, collaboration with whom has been a most agreeable experience. Harold Osborne, who died 13 March 1987 at the age of 82, was a civil servant for most of his life, but in 1950 joined with Sir Herbert Read in founding the British Society of Aesthetics, whose journal he edited for many years. In retirement, he began a second career as editor of the *Companion* volumes, and he brought to this enterprise a passionate and well-informed interest in the arts. We salute his memory.

DENNIS FARR

31 October 1987

NOTE TO NEW EDITION

For this edition the text has been thoroughly revised, with numerous changes and additions, including more than a hundred new entries. The bulk of the new material deals with twentieth-century art. It is impossible in a book of this nature to provide balanced coverage of the contemporary art scene, in which reputations are often ephemeral; but the new material includes several entries on recent developments to bring the treatment more up to date. Examples are Graffiti art, New British Sculpture, Neo-Expressionism, Pomidou Centre, and Turner Prize. As in the original edition, no artist born after 1945 has an individual entry, but several younger artists are discussed within general entries and are provided with cross-references of the type 'Schnabel, Julian. See NEO-EXPRESSIONISM'.

In addition to revision and updating, there is a completely new section, containing a chronology and other reference, compiled by Caroline Juler.

Contents

Introduction

THE *Oxford Dictionary of Art* is a descendant of the three *Oxford Companions** edited by Harold Osborne, but is in effect a new book. A few of the shorter entries have been taken over more or less unchanged from the *Companions*, but most of the text has been completely rewritten, and there are also over 300 new entries (out of a total of about 3,000) on personalities and topics not covered by the *Companions*. Harold Osborne, who died in 1987, was not directly involved in the production of this *Dictionary*, but it is firmly based on the foundations laid by his books, so his name rightly appears on the title-page.

To keep the *Dictionary* within manageable bounds, certain classes of entry in the *Companions* have been omitted, particularly the long articles on the art of individual countries. Architecture also is omitted, although there are entries on individuals who were active chiefly as architects but who made significant contributions to other fields of the visual arts (Aalto, Bramante, and Brunelleschi, for example). Oriental art, too, has been almost entirely excluded, although the entry on Ukiyo-e has been retained, as the subject of Japanese prints occurs so frequently in the discussion of late 19th-century French art.

The field covered by this book is, then, Western and Western-inspired painting, sculpture, and graphic art from ancient times to the present day. The exact boundaries, however, have deliberately been kept flexible, so that usefulness to the general reader rather than adherence to a fixed scheme has been the criterion determining whether a topic or artist should be included. Thus, although an arbitrary cut-off point has been adopted for contemporary art (no artist born after 1945 is included), the starting-point at the other end of the time-scale is intentionally more vague. Detailed coverage begins with the art of ancient Greece, but entries on Altamira and Lascaux have been retained because these names are so well-known, and there are also a few entries on ancient Egyptian topics of non-specialist interest, such as Book of the Dead. Similarly, although the great majority of the artists included are painters, sculptors, draughtsmen, or engravers, there are some entries on personalities who are thought of primarily as craftsmen or designers. They have been included not only for their inherent interest, but also because they help to elucidate other entries; an example is C. R. Ashbee, included principally because of his role in the Arts and Crafts Movement.

The length of individual entries is roughly correlated to the importance of the subject, but with many qualifications, some artists' lives being much more easily summarized than others. Those who travelled a great deal, or had fingers in many pies, or who for one reason or another led especially interesting lives are likely to have longer entries than equally accomplished artists who stayed at home and devoted themselves to one speciality. It is of course tempting to write more about one's own favourites, but I hope this kind of personal bias (for or against) has intruded only rarely.

In line with the practice of the catalogues of the National Gallery in London and of the Witt Library at the Courtauld Institute of Art, artists from the Low Countries are called

* *The Oxford Companion to Art* (1970); *The Oxford Companion to the Decorative Arts* (1975); *The Oxford Companion to Twentieth-Century Art* (1981).

Netherlandish up to about 1600; they are then distinguished as either Dutch or Flemish; and after about 1830 'Flemish' becomes 'Belgian'.

There is no system of alphabetizing artists' names that will satisfy logic but not offend against usage. Thus, one says 'van Dyck' or 'van Gogh', rather than 'Dyck' or 'Gogh', but they are almost invariably indexed under D and G (as they are here) rather than under V. Cross-references are given when there is likely to be doubt about where an artist will be found, but the following general rules may be taken as guidelines. Prefixes such as 'de', 'van', and 'von' are generally ignored, but an exception is 'La' or 'Le' (thus La Tour, Georges de). There are certain names where usage goes against this principle—thus Willem de Kooning and Peter De Wint are found under D. Italian Old Masters whose names include 'da', 'del', or 'di' are usually found under their first name (Leonardo da Vinci, rather than Vinci, Leonardo da), but again usage occasionally dictates otherwise; thus Andrea del Verrocchio is found under Verrocchio, not Andrea. For purposes of alphabetization 'Mc' is treated as 'Mac' and 'St' as 'Saint'. Artists from the same family are usually covered in one composite entry, except where it seemed more reasonable to treat major and distinct personalities separately.

Artists' names are given in the form most commonly used; so various elements of full names have sometimes been dropped, and nicknames or pseudonyms are used as the heading where these are better known than the artist's real name. Thus Delacroix's Christian name is given as Eugène rather than Ferdinand-Victor-Eugène, Velázquez is called Diego Velázquez rather than Diego Rodriguez de Silva y Velázquez, and Giovanni Francesco Barbieri appears under Guercino.

Names of galleries are also sometimes given in slightly shortened form; thus the Museum Boymans-van Beuningen in Rotterdam is generally referred to as 'Boymans Museum'. When a gallery is mentioned more than once within the same entry, the town in which it is located is generally omitted after the first mention. Every attempt has been made to be fully up to date with locations of works of art, but this is sometimes no easy matter. During the course of work on this book, for example, Matisse's *Luxe, calme et volupté* moved from a private collection to the Musée National d'Art Moderne in Paris and then to the Musée d'Orsay in Paris, and Joseph Wright's *An Experiment on a Bird in the Air Pump* moved from the Tate Gallery to the National Gallery at galley proof stage. Locations are not given for prints unless they are known to be exceptionally rare (see, for example, Bramante).

Cross-references from one article to another are indicated by an asterisk (*) within the main part of the text or by the use of small capitals when the formula 'see so-and-so' is used. Names of all people who have their own entries are automatically asterisked on their first mention in another entry, but cross-references are used selectively for art media, styles, terms, etc.; and given only when further elucidation under that heading might be helpful to the reader.

Dr Johnson defined a lexicographer as 'a writer of dictionaries, a harmless drudge'. I can vouch for the drudgery, but there is also much satisfaction involved in compiling a dictionary such as this, and I should like to thank the various people who have helped to make the task often such an enjoyable one. I am most deeply indebted to Dr Dennis Farr, who read every word of the text (apart from a few late additions) and, with learning, tact, and not a little wit, made numerous corrections and a great many suggestions for improvements to both content and style. He also discussed various problems with me and picked up several points that needed attention when he looked over the galleys.

At the Oxford University Press, I am grateful to everyone who worked on the book—for their enthusiasm and encouragement no less than for their skills; most notably to Nicholas Wilson, who commissioned the book, Betty Palmer, who guided it through its early stages, and Pam Coote, who handled with aplomb all the problems involved in seeing a book such as this through the press, unruffled by errant computer typesetting equipment on one side or my innumerable requests for last-minute changes on the other.

In a more general sense, I would like to say what a comfort it is to anyone involved in the field of reference books to be able to draw on the resources and tradition of the OUP. I have been able to use the various members of the family of Oxford English Dictionaries to help with definitions, and I have taken much information from *The Dictionary of National Biography* and the various *Oxford Companions* outside the field of art. Margaret Drabble's new edition of *The Oxford Companion to English Literature*, for example, contains many excellent entries (by Helen Langdon) on artists and their relationship to literature, and I have also made frequent use of the *Oxford Companions* to *French Literature* and *German Literature*, and of *The Oxford Classical Dictionary* and *The Oxford Dictionary of the Christian Church*. The *Oxford Companions* to *Film*, *Music*, and the *Theatre* have also proved helpful when their fields overlapped with mine, and, less obviously, I have also benefited from the *Oxford Companion to Ships and the Sea* (in relation to Clarkson Stanfield) and from the *Oxford Companion to Chess* (for some information concerning a Lucas van Leyden painting and for the delightful anecdote in the Duchamp entry).

On a more personal note, I would like to thank my sister, Doreen Chilvers, for undertaking the tedious task of pasting up the slips from the three *Oxford Companions* that formed the raw material for this book, and my nephew, Gavin Chilvers, for similar help. Many friends and former colleagues have given me advice, information, or encouragement, and several have read and commented on entries in fields in which they have specialist knowledge. Among them, I would like to thank, first, Claudia Stumpf, and then in alphabetical order: Tim Ayers, Georgina Barker, Alison Bolus, Caroline Bugler, Vanessa Cawley, Caroline Christian, Sue Churchill, Alison Cole, Celia de la Hey, Vanessa Fletcher, Janet Furze, John Gaisford, John Glaves-Smith, Bina Goldman, Clive Gregory, Flavia Howard, Miranda Innes, Dr Michael Jacobs, Stuart John, Jessica Johnson, Jon Kirkwood, Anne Lyons, Margaret Mauger, Jenny Mohammadi, Sir Felix Moore, Anna Morter, Nigel O'Gorman, Alice Peebles, Maggie Ramsay, Benedict Read, John Roberts, Carolyn Rogers, Antonia Spowers, Kate Sprawson, Julie Staniland, Ruth Taylor, Jack Tresidder, Dr Malcolm Warner, Jude Welton, and Iain Zaczek. I am also grateful to the staff of the Tate Gallery Archive who allowed me access to their press cuttings files to try to ensure that the book was as up to date as reasonably possible in recording the deaths of recently deceased artists. Branches of the OUP in Australia, Canada, and the USA similarly provided up-to-date information on artists in those countries. Finally, for inspiration from afar, thanks to Deborah Lambert, Victoria Kirkham, and G.G.

IAN CHILVERS

November 1987

Abbreviations

Bib.	Bibliothèque or Biblioteca
Bib. Nat.	Bibliothèque Nationale
BL	British Library
BM	British Museum
Coll.	Collection
DNB	*The Dictionary of National Biography*
Gal.	Gallery
Inst.	Institute
Lib.	Library
Met. Mus.	Metropolitan Museum
MOMA	Museum of Modern Art
Mus.	Musée, Museo, Museum, etc.
NG	National Gallery
NPG	National Portrait Gallery
OED	*The Oxford English Dictionary*
Univ.	University
V&A	Victoria and Albert Museum

AAA. See ALLIED ARTISTS' ASSOCIATION and AMERICAN ABSTRACT ARTISTS.

Aa, Dirck van der (1731–1809). The best-known member of a family of Dutch painters from The Hague. His speciality was *grisaille decorative panels for interiors (examples in V&A, London). There was another **van der Aa** family of artists active in the 18th cent. in Leyden; most of the members were illustrators or engravers.

Aachen, Hans von (1552–1615). German painter, active in the Netherlands, Italy (1574–87), and most notably Prague, where he settled in 1596 as court painter to the emperor Rudolf II (1552–1612). On Rudolf's death he worked for the emperor Matthias (1557–1619). His paintings, featuring elegant, elongated figures, are—like those of his colleague Bartholomeas *Spranger—leading examples of the sophisticated *Mannerist art then in vogue at the courts of northern Europe, and he was particularly good with playfully erotic nudes (*The Triumph of Truth*, Alte Pinakothek, Munich, 1598). Engravings after his work gave his style wide influence and he ranks as one of the most important German artists of his time. He was married to Regina de Lassus, daughter of the composer Orlando de Lassus (1532–94), who erected his tomb in St Vitus's Cathedral in Prague.

Aalto, Alvar (1898–1976). Finnish architect, designer, sculptor, and painter. One of the most illustrious architects of the 20th cent., Aalto was also a talented abstract painter and sculptor and an important furniture designer. In 1925 he married Aino Marsio, who was his chief collaborator until her death in 1949, particularly with the *Artek* furniture (first designed for the Paimio Sanatorium in 1928), whose new methods of bending and jointing and revolutionary flowing lines became internationally popular and had a lasting influence on furniture design. During the period 1927 to 1954, but particularly during the 1930s, Aalto engaged in what have been described as artistic laboratory experiments, making both abstract reliefs and free abstract sculptures in laminated wood. These sculptural experiments, many of which were fine works of art in their own right, had the dual purpose of solving technical problems concerning the pliancy of wood in the manufacture of furniture and of developing spatial ideas which served as an inspiration for his architectural work, as in the Institute of International Education at New York (1965), where walls are conceived as abstract sculptural reliefs in wood. In the 1960s Aalto came to the fore as a monumental sculptor, working in bronze, marble, and mixed media. Outstanding among his

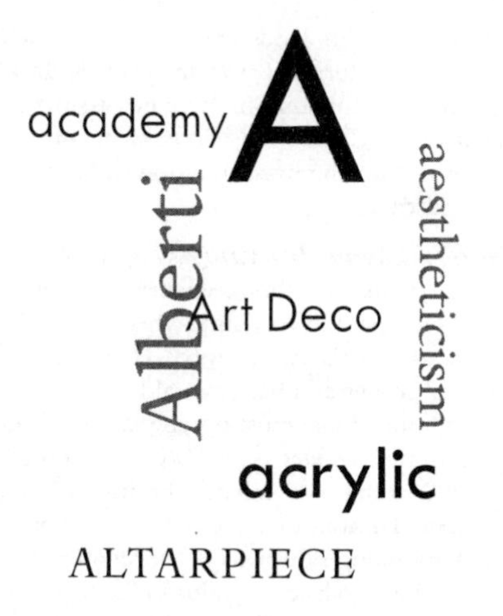

monumental works is his memorial for the Battle of Suomussalmi, a leaning bronze pillar raised 9 m. on a stone pedestal set up in the arctic wastes of the battlefield (1960). Aalto was also influential in introducing modern art—particularly the works of Alexander *Calder and Fernand *Léger, his close friends—to the Finnish public.

Aaltonen, Wäinö (1894–1966). Finnish sculptor who came to be regarded as the personification of the patriotic spirit of his country in the years following the declaration of independence from Russia (1917). The bronze monument to the runner Paavo Nurmi (1925, the best-known cast is outside the athletics stadium in Helsinki) and the bust of the composer Sibelius (1928, various casts exist) are probably his most famous portrayals of national heroes.

Abakanowicz, Magdalena (1930–). Polish sculptor. She is the pioneer and leading exponent of sculpture made from woven fabrics and has been widely imitated in Europe and the USA.

Abbate, Niccolò dell' (*c.*1510–71). Italian painter. He was trained in the tradition of his birthplace, Modena, but he developed his mature style in Bologna (1547–52) under the influence of *Correggio and *Parmigianino. There he decorated palaces, combining painted stucco with figure compositions and landscapes (the best surviving examples are at the Palazzo Pozzi, now Palazzo dell' Università). His elegant figure style was influenced particularly by Parmigianino. He was invited to France in 1552, probably at the suggestion of *Primaticcio, under whom he worked at *Fontainebleau, and he remained in France for the rest of his life. Most of his work in the palace itself has been lost, and he is now considered most important for his landscapes with figures from mythological stories (*Landscape with the Death of Eurydice*, NG, London). In these he was the direct precursor of *Claude and *Poussin and one of the sources of the long-lived tradition of French classical landscape.

Abbey, Edwin Austin (1852–1911). American painter, etcher, and book illustrator, active and highly successful in England (where he settled in 1878) as well as his own country. He specialized in historical scenes and had several large and prestigious commissions, most notably a set of murals representing *The Quest of the Holy Grail* (completed 1902) in Boston Public Library (his friend *Sargent also painted murals there) and the official painting commemorating Edward VII's coronation in 1902 (Buckingham Palace, London). Such paintings now seem rather overblown and ponderous, and he is most highly esteemed for his lively book illustrations. He was particularly prolific for *Harper's Weekly*, his association with the magazine lasting from 1870 until his death.

Abbot, Lemuel Francis (*c*.1760–1802). English portrait painter. His clientele included many naval officers and he is best known for his portrayals of Lord Nelson. He did several portraits of him with slight variations (an example is in the NPG, London). In 1798 Abbot became insane, and his unfinished works were completed by other hands.

Abildgaard, Nicolai Abraham (1743–1809). Danish *Neoclassical painter. From 1772 to 1777 he studied in Rome, where his friendship with *Fuseli helped to form his *eclectic early style. On his return to Denmark his work became more purely Classical, as is best seen in his cycles of paintings illustrating Apuleius and Terence (Statens Mus. for Kunst, Copenhagen). He became one of the leading figures in Danish art and had great influence because of his post as Director of the Copenhagen Academy (1789–91 and 1801–9), where his pupils included *Runge and *Thorvaldsen. Abildgaard occasionally worked as an architect, sculptor, and designer (most notably of some Grecian furniture for himself), and he also wrote on art. His most ambitious work, a huge decorative scheme at Christianborg Palace, was destroyed by fire in 1794.

abstract art. Art that does not depict recognizable scenes or objects, but instead is made up of forms and colours that exist for their own expressive sake. Much decorative art can thus be described as abstract, but in normal usage the term refers to 20th-cent. painting and sculpture that abandon the traditional European conception of art as the imitation of nature. Herbert *Read (*Art Now*, 1948) gave the following definition: 'in practice we call "abstract" all works of art which, though they may start from the artist's awareness of an object in the external world, proceed to make a self-consistent and independent aesthetic unity in no sense relying on an objective equivalence.' Abstract art in this sense was born and achieved its distinctive identity in the decade 1910–20 and is now regarded as the most characteristic form of 20th-cent. art. It has developed into many different movements and 'isms', but three basic tendencies are recognizable: (i) the reduction of natural appearances to radically simplified forms, exemplified in the sculpture of *Brancusi (one meaning of the verb 'abstract' is to summarize or concentrate); (ii) the construction of works of art from non-representational basic forms (often simple geometric shapes), as in Ben *Nicholson's reliefs; (iii) spontaneous, 'free' expression, as in the *Action painting of Jackson *Pollock. Many exponents of such art dislike the word 'abstract', but the alternatives they prefer, although perhaps more precise, are usually cumbersome, notably non-figurative, non-representational, and *Non-Objective.

The aesthetic premiss of abstract art—that formal qualities can be thought of as existing independently of subject-matter—existed long before the 20th cent. In 1780, in his 10th *Discourse* to the students of the Royal Academy, Sir Joshua *Reynolds advised that 'we are sure from experience that the beauty of form alone, without the assistance of any other quality, makes of itself a great work, and justly claims our esteem and admiration'; and in discussing the **Belvedere Torso* he referred to 'the perfection of this science of abstract form'. In the 19th cent. several notable writers followed this line (Maurice *Denis, for example) and many of the leading painters of the 1890s—notably the *Symbolists—stressed the expressive properties of colour, line, and shape rather than their representative function. This process was taken further by the major avant-garde movements of the first decade of the 20th cent.—especially *Cubism, *Expressionism, and *Fauvism. By 1910, then, the time was ripe for abstract art, and it developed more or less simultaneously in various countries. *Kandinsky is often cited as the first person to paint an abstract picture, but no artist can in fact be singled out for the distinction. (A work by Kandinsky known as *'First Abstract Watercolour'* (Musée National d'Art Moderne, Paris) is signed and dated 1910, but some scholars believe that it is later and was inscribed by Kandinsky several years after its execution. This kind of problem arises not only with Kandinsky: several early abstract artists were keen to stress the primacy of their ideas and were not above backdating works.) Among the other artists who produced abstract paintings at about the same early date as Kandinsky were the American Arthur *Dove and the Swiss Augusto Giacometti, cousin of Alberto *Giacometti.

The individual pioneers were soon followed by abstract groups and movements—among the first were *Orphism and *Synchromism in France. There was a particularly rich crop in Russia, with *Constructivism, *Rayonism, and *Suprematism all launched by 1915. The almost religious fervour with which some of the Russian artists pursued their ideals was matched by the members of the De *Stijl group in Holland, founded in 1917. To such artists, abstraction was not simply a matter of style, but a question of finding a visual idiom capable of expressing their most deeply felt ideas. In the period between the two world wars, the severely geometrical style of De Stijl and the technologically orientated Constructivism were the most influential currents in abstraction (they came together in the *Bauhaus), although *Surrealism also had a strong abstract element. The first international exhibition of abstract art was held in Paris in 1930 and there were many outstanding individual achievements in abstraction in this period—the sculpture of *Calder and *Hepworth, for example. However, in general figurative art was dominant, and abstract art was banned in totalitarian countries such as Germany and Russia. The second heroic period of abstract art came after the Second World War, when the enormous success of *Abstract Expressionism and its European equivalent *Art Informel made abstraction for a time virtually the dominant orthodoxy in Western art.

Abstract Expressionism. The dominant movement in American painting in the late 1940s and 1950s. It was the first major development in American art to lead rather than follow Europe, and it is often reckoned the most significant art movement anywhere since the Second World War. The energy and excitement it brought to the American art scene helped New York to replace Paris as the world capital of contemporary art, and much of the subsequent history of painting can be written in terms of reactions to it. The phrase 'Abstract Expressionism' had originally been used in 1919 to describe certain paintings by *Kandinsky, but in the context of modern American painting it was first used by the *New Yorker* art critic Robert Coates in 1945; by the end of the decade it had become part of the standard critical vocabulary. The painters embraced by the term worked mainly in New York (hence the term *New York School) and there were various ties of friendship and loose groupings among them, but they shared a similarity of outlook rather than of style—an outlook characterized by a spirit of revolt against tradition and a demand for spontaneous freedom of expression. The stylistic roots of Abstract Expressionism are complex, but despite its name it owed more to *Surrealism—with its stress on *automatism and intuition—than to *Expressionism. A direct source of inspiration came from the European Surrealists who took refuge in the USA during the Second World War. The most famous Abstract Expressionist is Jackson *Pollock, whose explosive energy best sums up the movement, but the work of other leading exponents was sometimes neither abstract (the leering *Women* of *de Kooning) nor expressionist (the serene visions of *Rothko). Even allowing for these wide differences, however, there are certain qualities that are basic to most Abstract Expressionist painting: the preference for working on a huge scale; the emphasis placed on surface qualities so that the flatness of the canvas is stressed; the adoption of an *all-over type of treatment, in which the whole area of the picture is regarded as equally important; and the glorification of the act of painting itself (see ACTION PAINTING).

Alongside de Kooning, Pollock, and Rothko, the painters who are considered central to Abstract Expressionism include *Gorky, *Gottlieb, *Guston, *Kline, *Motherwell, *Newman, and *Still. Most of them struggled for recognition early in their careers, but during the 1950s the movement became an enormous critical and financial success. It had passed its peak by 1960, but several of the major figures continued productively after this. By 1960, also, reaction against the emotionalism of the movement was under way, in the shape principally of *Pop art and *Post-Painterly Abstraction. Sculptors as well as painters were influenced by Abstract Expressionism, the leading figures including Ibram Lassaw (1913–), Seymour Lipton (1903–86), and Theodore Roszak (1907–81).

Abstraction-Création. The name taken by a group of abstract painters and sculptors formed in Paris in February 1931, following the first international exhibition of abstract art held there in 1930. It was successor to the group *Cercle et Carré, founded in 1930. The group was open to artists of all nationalities and the organization was loose, so that at one time its numbers rose to as many as 400. As indicated by its title, the association was intended to encourage so-called 'creative' abstraction, by which was meant abstract works constructed from non-figurative, usually geometrical, elements (rather than abstraction derived from natural appearances, of the kind being developed by Roger *Bissière among others in France). The association operated by arranging group exhibitions and by publishing an illustrated annual with the title *Abstraction-Création: Art non-figuratif*, which appeared from 1932 to 1936 with different editors for each issue. Within this general principle the association was extremely catholic in its outlook and embraced many kinds of non-figurative art, from the *Constructivism of *Gabo, *Pevsner, and *Lissitzky, and the *Neo-Plasticism of *Mondrian, to the expressive abstraction of *Kandinsky, and even the *biomorphic abstraction of *Arp and some forms of abstract *Surrealism. Owing, however, to the strength of the Constructivist element and the supporters of De *Stijl, the emphasis fell increasingly upon geometrical rather than expressive abstraction. After *c.*1936 the activities of the association dwindled as some of the leading Constructivists moved from France to England.

académie. A French term for a private art school, several of which flourished in Paris in the late 19th and early 20th cents. The term 'atelier libre' has also been used to refer to such establishments. Entry to the official École des *Beaux-Arts was difficult (almost impossible for foreigners, who from 1884 had to take a vicious examination in French) and teaching there was conservative, so private art schools, with their more liberal regimes, were often frequented by progressive young artists. Four of them are particularly well known. The **Académie Carrière** was opened in 1890 by Eugène Carrière (1849–1906), a painter of portraits, religious pictures, and—his speciality—scenes of motherhood. His style was characteristically misty, monochromatic, and vaguely *Symbolist. *Rodin was a great admirer of his work. There was no regular teaching at the school, though Carrière visited it once a week. It was here that *Matisse met *Derain, thus helping to form the nucleus of the future *Fauves. The **Académie Julian** was founded in 1873 by Rodolphe Julian (1839–1907), whose work as a painter is now forgotten. The school had no entrance requirements, was open from 8 a.m. to nightfall, and was soon the most popular establishment of its type. Julian opened several branches throughout Paris, one of them for women artists, and by the 1880s the student population was about 600. Although the Académie Julian became famous for the unruly behaviour of its students, it was regarded as a stepping-stone to the École des Beaux-Arts (Julian had been astute in engaging teachers from the École as visiting professors). Among the French artists who studied there were *Bonnard, *Denis, *Matisse, and *Vuillard. The list of distinguished foreign artists who studied there is very long. The **Académie Ranson** was founded in 1908 by Paul Ranson (1864–1909), who had studied at the Académie Julian. After Ranson's early death, his wife took over as director, and his friends Denis and *Sérusier were among the teachers. Among later teachers the most important was Roger *Bissière, whose style of expressive abstraction influenced many young painters in the 1930s; his pupils included *Manessier. The **Académie Suisse** was founded in about 1850 by a former artists' model called Suisse 'in an old and sordid building where a well-known dentist pulled teeth at one franc apiece . . . artists could for a small fee work from the living model without any examinations or tuition' (John *Rewald, *The History of Impressionism*). *Courbet, *Manet, and several of the *Impressionists drew at the Académie Suisse, and it was there in 1861 that Camille *Pissarro first noticed the 'strange Provençal' Paul *Cezanne, whose life drawings were ridiculed by his fellow students.

Académie Royale de Peinture et de Sculpture, Paris. See ACADEMY.

academy. Association of artists, scholars, etc., arranged in a professional institution. The original Academy was an olive grove outside Athens where Plato and his successors taught philosophy, and his school of philosophy was therefore known as 'The Academy'. In the Italian *Renaissance the word began to be applied to almost any philosophical or literary circle and was sometimes employed of groups of artists who discussed theoretical as well as practical problems—in this sense it was used somewhat ironically of *Botticelli's studio *c.*1500. Similarly the famous Accademia of *Leonardo da Vinci—the reference has come down to us in engravings of the first years of the 16th cent.—almost certainly implied no more than a group of men who discussed scientific and artistic problems with Leonardo. Another premature 'academy' is the one attributed by *Vasari to Lorenzo de' *Medici under the direction of the sculptor *Bertoldo, a suitable ancestor for the one that Vasari himself promoted in 1563; all the evidence suggests that while Lorenzo allowed artists and others easy access to his collections, there was no organized body of any kind concerned with the arts in the Florence of his day. Finally, there is an engraving of 1531 by Agostino Veneziano (1490–1540) which shows the *Accademia di Baccio Bandin in Roma in luogo detto Belvedere*. Although this portrays sculptors drawing a small statue in *Bandinelli's studio, it probably represents no more than a group of friends discussing the theory and practice of art.

The first art academy proper was set up some 30 years later when Duke Cosimo de' Medici founded the Accademia del Disegno in Florence in 1563. The prime mover was Giorgio Vasari, whose aim was to emancipate artists from control by the guilds, and to confirm the rise in social standing they had achieved during the previous hundred years. *Michelangelo, who more than any other embodied this change of status in his own person, was made one of the two heads and Duke Cosimo himself was the other. Thirty-six artist members were elected, and amateurs and theoreticians were also eligible for membership. Lectures in geometry and anatomy were planned, but there was no scheme of compulsory training to replace regular workshop practice. The Academy quickly won great international authority, but in Florence itself it soon degenerated into a sort of glorified artists' guild.

The next important step was taken in Rome, where in 1593 was founded the Accademia di S. Luca, of which Federico *Zuccaro was elected president. Though more stress was laid on practical instruction than at Florence, theoretical lectures were also prominent in the Academy's plans. However, the Academy was not at all successful in its war against the guilds until the powerful support it received from Pope Urban VIII (Maffeo *Barberini) in 1627 and 1633. Thereafter it grew in wealth and prestige. All the leading Italian and many foreign artists in Rome were members; debates on artistic policy were held; some influence over important commissions was wielded; and everything was done to make the lives of those who were not members (such as many of the Flemish freelance artists working in Rome—see FIAMMINGO) as unpleasant as possible.

The only other similar organization in Italy was the Academy established in Milan by Cardinal Federico Borromeo in 1620. But meanwhile the word was very frequently used of private institutions where artists met either in a studio or in some patron's palace to draw from life. The most famous example of this kind was organized by the *Carracci in Bologna.

In France a group of painters, moved by the same reasons of prestige as had earlier inspired the Italians, persuaded Louis XIV (1638–1715) to found the Académie Royale de Peinture et de Sculpture in 1648. Here too the guilds put up a powerful opposition, and its supremacy was not assured until *Colbert was elected Vice-Protector in 1661 and found in the Académie an instrument for imposing the official standards and principles of taste. Colbert and *Lebrun, the Director, envisaged dictatorship of the arts, and for the first time in history the expression 'academic art' acquired a precise significance. The Académie arrogated to itself a virtual monopoly of teaching and of exhibition, and by applying rigidly its own standards of membership came to wield an important economic influence on the profession of artist. For the first time an orthodoxy of artistic and aesthetic doctrine obtained official sanction. Implicit in the academic theory and teaching was the assumption that everything to do with the practice and appreciation of art, or the cultivation of taste, can be brought within the scope of rational understanding and reduced to logical precepts that can be taught and studied.

Other art academies were founded in Germany, Spain, and elsewhere between the middle and end of the 17th cent., and it has been calculated that in 1720 there were nineteen of them active in Europe. An upsurge of activity occurred in the middle of the 18th cent., and by 1790 well over a hundred art academies were flourishing throughout Europe. Among these was the *Royal Academy in London, which was founded in 1768 (previously in England there had been only private teaching academies, one of which was run by *Thornhill and then by *Hogarth). For the most part these academies were the product of a new awareness on the part of the State of the place that art might be expected to play in the life of society. The Church and the court were no longer the chief patrons of art. But with the growth of industry and commerce the economic importance attributed to good design led to official support for teaching academies. Inseparably linked to this motive was the promotion of *Neoclassicism in opposition to the surviving styles of *Baroque and *Rococo. Everywhere the academies made themselves the champions of the new return to the *antique. As far as instruction went, the copying of casts and life drawing were paramount, and Classical subjects were particularly encouraged.

There was some opposition to these bodies from the start. Towards the end of the 18th cent. French Revolutionary sentiment was especially bitter about the exclusive privileges enjoyed by members of the Académie, and many artists, with *David in the lead, demanded its dissolution. This step was taken in 1793, but after various experiments had been tried and substitute bodies set up, the Académie was reinstated in 1816 as the Académie des Beaux-Arts.

In fact the real threat to academies came rather from the *Romantic notion of the artist as a genius who produces his masterpieces by the light of inspiration which cannot be taught or subjected to rule. Opposition was accentuated by the widening breach between creative artists and the bourgeois public after aristocratic patronage declined.

Virtually all the finest and most creative artists of the 19th cent. stood outside the academies and sought alternative channels for exhibiting their works, although *Manet, for example, always craved traditional academic success. When the revaluation of 19th-cent. art took place with the final recognition of the *Impressionists this contrast was too blatant to be ignored. Academies were condemned out of hand by the adventurous, though they still retained prestige among those who were too confused by the prevailing breakdown of taste to risk supporting what they did not understand. Gradually compromises were made on both sides, facilitated by the liberalization of the academies and the increasing number of art schools which have finally broken academic monopolies. Academies now generally hold the view that they should at least insist on certain standards of sheer craftsmanship which any artist, however personal his vision, is expected to display. During the 20th cent. the pace has quickened. Most of the aesthetic movements, however revolutionary they seemed at the time, have rapidly produced their own 'academicism' as minor artists of no more than moderate endowment have aped the mannerisms of the few geniuses who have shaped each movement. Thus, the word 'academic' now almost always carries a pejorative meaning, and is associated with mediocrity and lack of originality.

academy board. A pasteboard used for painting, especially in oils, since the early 19th cent. It is made of sheets of paper sized and pressed together, and treated with a *ground consisting usually of white lead, oil, and chalk. Sometimes it is embossed with indentations—a sort of mechanical 'grain'—in imitation of canvas weave. Because it is fairly inexpensive, academy board is a popular *support with amateur painters, and it is also used by professional artists for sketches and studies.

academy figure. A careful painting or drawing (usually about half life-size) from the nude made as an exercise. The figure is usually depicted in a heroic pose, and there is a tradition of suitable postures which goes back to the *Carracci. An academy figure ascribed to *Géricault is in the National Gallery, London.

Accademia di Belli Arti, Venice. The municipal picture gallery of Venice and one of the most important collections in Italy. It was founded by decree of Napoleon in 1807 and combined the collection of the old Academy, the Galleria dei Gessi founded in 1767, with works of art from suppressed churches and monasteries. The collection has continually been enriched by additions, in particular Venetian paintings returned after the fall of Napoleon in 1815 and those ceded by Austria in 1918.

Acconci, Vito. See BODY ART.

Achilles Painter. Greek vase painter, active in Athens in the mid-5th cent. BC. He was a contemporary of *Phidias and his paintings have some of the nobility associated with the great sculptor's work. His compositions are simple (usually limited to one or two figures) and his figures are serene and graceful (*A Muse*, Staatliche Antikensammlungen, Munich).

acroterion (pl. acroteria). An ornamental block on the apex and at the lower angles of a pediment, bearing a statue or a carved finial.

acrylic. A modern synthetic paint combining some of the properties of *oils and *watercolour. Most acrylic paints are water-based, although some are oil compatible, using turpentine as a thinner. They can be used on a wide variety of surfaces to create effects ranging from thin washes to rich *impasto and with a matt or gloss finish. Thinly applied paint dries in a matter of minutes, thickly applied paint in hours—much quicker than oils. Acrylic paint first became available to artists in the 1940s in the USA and certain American painters discovered that it offered them advantages over oils. *Colour stain painters such as Helen *Frankenthaler and Morris *Louis, for example, found that they could thin the paint so that it flowed over the canvas yet still retained its full brilliance of colour. David *Hockney took up acrylic during his first visit to Los Angeles in 1963; he had earlier tried and rejected the medium, but American-manufactured acrylic was at this time far superior to that available in Britain, and he felt that the flat, bold colours helped him to capture the strong Californian light. Hockney used acrylic almost exclusively for his paintings until 1972, when he returned to oils because he now regarded their slow-drying properties as an advantage: 'you can work for days and keep altering it as well; you can scrape it off if you don't like it. Once acrylic is down you can't get it off.' Other artists have remained loyal to acrylic and it is now a serious rival to oil paint, but in the 1990s doubts were expressed about its permanence.

Action painting. A type of dynamic, impulsive painting in which the artist applies paint with energetic *gestural movements—sometimes by dribbling or splashing—and with no preconceived idea of what the picture will look like. It is sometimes used loosely as a synonym for *Abstract Expressionism, but such usage is misleading, as Action painting represents only one aspect of that movement. The term was coined by the critic Harold *Rosenberg in an article entitled 'The American Action Painters' in *Art News* in December 1952. Rosenberg regarded Action painting as a means of giving free expression to the artist's instinctive creative forces and he regarded the act of painting itself as more significant than the finished work. Although the term soon became established, many critics were unconvinced by Rosenberg's idea of the canvas being 'not a picture but an event': Mary McCarthy, for example, wrote that 'you cannot hang an event on a wall, only a picture'. Rosenberg's article did not mention individual painters, but the one who is associated above all with Action painting is Jackson *Pollock, who vividly described how he felt when working on a canvas laid on the floor: 'I feel nearer, more a part of the painting, since this way I can walk around it, work from the four sides and literally be *in* the painting . . . When I am *in* my painting, I am not aware of what I'm doing. It is only after a sort of "get acquainted" period that I see what I have been about. I have no fears about making changes, destroying the image, etc., because the painting has a life of its own. I try to let it come through.' In the work of lesser artists than Pollock, Action painting could easily degenerate into messy self-indulgence, and it came in for a good deal of mockery, especially after the British painter William Green (1934–) took to riding a bicycle over the canvas, a feat imitated by the comedian Tony Hancock in the film *The Rebel* (1961).

Adam, François-Gérard (1710–61); **Lambert-Sigisbert** (1700–59); **Nicolas-Sébastien** (1705–78). French sculptors from Nancy, brothers. All three brothers went to Rome after training with their father **Jacob-Sigisbert** (1670–1747), and on their return adapted the Roman *Baroque style to French *Rococo taste. Lambert-Sigisbert was the most distinguished member of the family. His masterpiece is the *Neptune Fountain* (1740) at Versailles, a work showing the influence of *Bernini in its exuberant movement. Nicolas-Sébastien is remembered mainly for the monument of Queen Catharina Opalinska (1749) in the church of Notre Dame de Bon Secours in Nancy. François-Gérard's best works are probably his garden statues for Frederick the Great of Prussia at Sanssouci, Potsdam. Better known than any of the three brothers is their nephew *Clodion.

Adam, Henri-Georges (1904–67), French sculptor, graphic artist, and tapestry designer. His figure sculptures were conceived in large expressive planes somewhat in the manner of *Brancusi; but he also did geometrical abstracts, influenced by *Arp, which he sometimes decorated with engraved patterns. Among his best-known works is the *Beacon of the Dead* monument at Auschwitz (1957–8). He designed tapestries for *Aubusson, as well as ones for the United Nations and the French Embassy in Washington.

Adam, Robert (1728–92). Scottish architect and designer, the outstanding member of a famous family of architects. Highly prolific and successful in both Scotland and England, he supervised the furnishing of his buildings down to the last detail, creating a distinctively elegant and highly influential style of interior decoration. In 1812 the architect Sir John Soane (1753–1837) wrote: 'the light and elegant ornaments, the varied compartments in the ceilings of Mr Adam, imitated from Ancient Works in the Baths and Villas of the Romans, were soon applied in designs for chairs, tables, carpets, and in every other species of furniture.' Adam gave work to a number of outstanding craftsmen, and Angelica *Kauffmann and her husband Antonio Zucchi (1726–95) were among the artists who painted decorative panels for his interiors (examples by Zucchi are at 20 Portman Square, London, formerly the *Courtauld Institute of Art). About 9,000 of Adam's drawings are in the Soane Museum in London.

Adami, Valerio (1935–). One of the leading Italian *Pop artists. He uses pure, unmodulated colours and depicts his subjects against a background of advertisements and posters, framing his pictures with drawings in the manner of comic-strip cartoons.

Ada School (or **Ada Group**). Term applied to a group of stylistically related Carolingian manuscripts from the Middle Rhine region, the first of which is the *Ada Gospels* (Stadtbibliothek, Trier), so called because it was made for Ada, reputedly a sister of Charlemagne (742?–814). The manuscripts are written in gold, sometimes on purple vellum, and are sumptuously *illuminated, combining *Early Christian and *Byzantine models in style and *iconography. Some *ivory carvings similar in style to the manuscripts have also been included within the group. It has been argued, however, that the name 'Ada School' is inappropriate and too restricted; that there was a more general movement in illumination which cannot be confined to one group of manuscripts. Consequently, the name is now often used within inverted commas.

Addison, Joseph (1672–1719). English essayist, poet, statesman, and critic. His papers 'On the Pleasures of the Imagination' published in *The Spectator* in 1712 were regarded by 18th-cent. writers on *aesthetics as striking out new ground and formulating problems which became the basis of aesthetic discussion well into the 19th cent. By 'Pleasures of the Imagination' Addison meant 'such as arise from visible Objects, either when we have them actually in our View, or when we call up their Ideas into our Minds by Paintings, Statues, Descriptions, or any the like Occasion'. He laid the basis for the notion of 'sensibility' and for that of an 'inner sense' of beauty which was taken up by *Hutcheson and the philosopher David Hume (1711–76). As a member of the *Kit-Cat Club Addison was painted by *Kneller.

Adler, Jankel (1895–1949). Polish painter, active in Germany (1922–33), France (1933–9), and Britain (1939–49). His work is eclectic and varied, usually with strong expressionist overtones; his best-known paintings feature Jewish life in Poland.

Aelst, Willem van (1625–*c.*1683). Dutch painter of lavish flower pieces and still lifes, a pupil of his uncle **Evert van Aelst** (1602–57). He worked in Paris, Florence, and Rome (he was court painter to Ferdinando de' *Medici, Grand Duke of Tuscany), before settling in Amsterdam in 1657. His pupils included Rachel *Ruysch.

aerial perspective. Term describing the means of producing a sense of depth in a painting by imitating the effect of atmosphere whereby objects look paler and bluer the further away they are from the viewer. Scientific analysis shows that the presence in the atmosphere of dust and large moisture particles causes some scattering of light as it passes through it. The amount of scattering depends on the wavelength (hence colour) of the light. Short wavelength (blue) light is scattered most and long wavelength (red) is scattered least. This is the reason why the sky is blue and why distant dark objects appear to lie behind a veil of blue. Distant bright objects tend to appear redder than they would be if near because some blue is lost from the light by which we see them.

The term 'aerial perspective' was invented by *Leonardo, but the device was used by Roman painters, for example at Pompeii. In the work of Italian painters of Leonardo's time, backgrounds sometimes look artificially blue, and in general aerial perspective has been more subtly exploited in northern Europe, where the atmosphere tends to be hazier. No one used it more beautifully than *Turner, in some of whose late works it is virtually the subject of the painting.

Aertsen, Pieter (1508/9–75). Netherlandish painter. He was born and died in Amsterdam but spent most of his career in Antwerp. A pioneer of still life and *genre painting, he is best known for scenes that at first glance look like pure examples of these types, but which in fact have a religious scene incorporated in them (*Butcher's Stall with the Flight into Egypt*, Univ. of Uppsala, 1551). Aertsen was the head of a long dynasty of painters, of whom the most talented was his nephew and pupil Joachim *Bueckelaer.

Aestheticism. A term applied to various exaggerations of the doctrine that art is self-sufficient and need serve no ulterior purpose, whether moral, political, or religious. Both the doctrine and its exaggerations have found expression in the phrase 'art for art's sake' (*l'art pour l'art*), which was apparently first used by the French philosopher Victor Cousin (1792–1867) in his lectures on *Le Vrai, le Beau et le Bien* (1818, first published in 1836) at the Sorbonne. In England 'art for art's sake' became the catchword of an exaggerated Aestheticism which was satirized as early as 1827 by Thomas de Quincey in his essay *On Murder considered as one of the Fine Arts*. The affected dandyism and extravagant cult of the beautiful that characterized the 'Aesthetic Movement' in late 19th-cent. England was brilliantly parodied by Gilbert and Sullivan in *Patience* (1881). In the 1870s and 1880s the hypersensibility cultivated by certain followers of the *Pre-Raphaelites obtained a sanction that was almost official in Walter *Pater, who in the conclusion to *The Renaissance* (1873) advocated a sensibility which finds the most precious moments of life in the pursuit of sensations raised to a pitch of 'poetic passion, the desire of beauty, the love of art for art's sake'. Oscar Wilde's *The Picture of Dorian Gray* (1891) expressed the same primacy for the aesthetic experience. Reaction from the tendency to regard the artist and connoisseur as specially endowed individuals whose role was to withdraw from everyday life and remain shut off in what the critic Sainte-Beuve (1804–69) first (in 1829) called the 'Ivory Tower' came from the *Arts and Crafts Movement of William *Morris and *Lethaby. *Ruskin, despite his enthusiastic worship of beauty, threw in his weight against an art which was out of touch with common life, and his controversy with *Whistler on the 'art for art's sake' doctrine has become famous. The would-be emancipation of fine art from moral standards and the common man was challenged by *Tolstoy in *What is Art?* (1898).

The exaggerated one-sidedness of the doctrine that art may have no ulterior motive, religious, political, social, or moral, hardly survived the turn of the century, though an echo of the implied emphasis may be seen in the extreme version of the

'Formalist' doctrine advocated by Clive *Bell, who maintained that the values of visual art reside solely in its formal qualities to the exclusion of subject or representation. But the more moderate form of the doctrine, in which it is held that aesthetic standards are autonomous, and that the creation and appreciation of beautiful art are 'self-rewarding' activities, has become an integral part of 20th-cent. aesthetic outlook.

aesthetics. Term defined in the *Oxford English Dictionary* as 'the philosophy or theory of taste, or of the perception of the beautiful in nature and art'. It was first used about the middle of the 18th cent. by the German philosopher Alexander Gottlieb Baumgarten (1714–62), who applied it to the theory of the liberal arts or the science of perceptible beauty. The scope and usefulness of the term have been much discussed, and in Gwilt's *Encyclopaedia of Architecture* (1842), it was still described as a 'silly pedantic term' and one of 'the useless additions to nomenclature in the arts' which had been introduced by the Germans. In the 20th cent. there is no general agreement about the scope of philosophical aesthetics, but it is understood to be wider than the theory of *fine art and to include the theory of natural beauty and non-perceptible (e.g. moral or intellectual) beauty in so far as these are thought to be susceptible of philosophical or scientific study.

Affichiste. Name (literally 'poster designer') taken by the French artists and photographers Raymond Hains (1926–) and Jacques de la Villeglé (1926–), who met in 1949 and during the early 1950s devised a technique of making *collages from fragments of torn-down posters. While Villeglé tried to organize his *affiches lacérées* (torn posters) into aesthetic structures, Hains used them to demonstrate the aesthetic bankruptcy of the advertising world.

Afro (Afro Basaldella) (1912–76). Italian painter. After passing through a *Cubist phase he evolved a personal idiom of *Lyrical Abstraction and became one of the best-known Italian painters in this style. During the 1950s he painted murals for public buildings in his home town of Udine. In 1958 he executed a mural for the Unesco Building, Paris. Afro was the brother of the sculptors Dino (1909–) and Mirko Basaldella (1910–69).

Agam, Yaacov (Jacob Gipstein) (1928–). Israeli sculptor and experimental artist, based in Paris from 1951. In 1955 Agam participated, with *Bury, *Tinguely, *Calder, and other artists, in the exhibition 'Le Mouvement' at the Denise René Gallery, considered the definitive exhibition for the *Kinetic movement, and from this time he was recognized as a pioneer in those branches of abstract art that lay stress on movement and spectator participation. Agam often uses light and sound effects in conjunction with his sculptures, and sometimes the components of his works can be rearranged by the spectator.

Agasse, Jacques-Laurent (1767–1849). Swiss-born animal painter who settled permanently in England in 1800 and became one of the principal successors to *Stubbs. He studied veterinary science in Paris as well as painting (with J.-L. *David) and his work is distinguished by anatomical accuracy as well as grace of line. Although Agasse was initially successful in England (George IV (1762–1830) was among his patrons), he died poor and virtually forgotten. The Musée d'Art et d'Histoire in his native Geneva has the best collection of his work.

Agatharcus of Samos. Greek painter of the mid or later 5th cent. BC. According to *Vitruvius he was noted for stage scenery and was a pioneer of *perspective.

Agoracritus. Greek sculptor from Paros of the second half of the 5th cent. BC, a pupil of *Phidias. His most celebrated work was a colossal marble statue of Nemesis at Rhamnus; a fragment of the head is in the British Museum, London.

Agostino di Duccio (1418–81). Florentine sculptor and architect. He was the most original if not the greatest sculptor of his time, and the only 15th-cent. sculptor born in Florence who owed little or nothing either to *Donatello or to *Ghiberti. His fresh and lively style was linear and graceful, with distinctive swirling draperies. Reliefs at Modena Cathedral executed by 1442 are accepted as his earliest work. Some have seen in them indications of a debt to Jacopo della *Quercia, and others of possible training by Luca della *Robbia. In 1446 he fled Florence after being accused of stealing silver from a church, and from *c.*1450 to 1457 he worked on the sculptural *reliefs for the Tempio Malatestiano at Rimini (see MALATESTA), which include personifications of the Trivium and Quadrivium (see LIBERAL ARTS). His other memorable large work is the series of reliefs, partly in *terracotta, on the façade of the Oratory of S. Bernadino at Perugia, on which he worked *c.*1457–61, as architect as well as sculptor. Agostino also executed several delightful reliefs of the Virgin and Child (examples in the Louvre, Paris, and V&A, London).

Agostino Veneziano. See ACADEMY.

air brush. An instrument for spraying paint or varnish by means of compressed air. It looks rather like an outsize fountain pen and is operated in a similar fashion, the pressure of the forefinger on a lever regulating the air supply. It can be controlled so as to give large areas of flat colour, delicate gradations of tone, or a fine mist. The device was invented by Charles Burdick, an American watercolour painter, who patented it in England in 1893. He used the tradename Aerograph, which for many years was used as a general term for air brushes (*Man Ray called paintings he did with an air brush 'aerographs'). In the early 20th cent. air brushes were mainly used for photographic retouching, and their principal use is now in commercial art. However, they are also used by painters such as *Hard Edge Abstractionists and *Superrealists, who require a very smooth finish.

Aken, Joseph van (*c.*1699–1749). Flemish-born painter who settled in London around 1720, and in the 1730s and 1740s became the leading specialist drapery painter of his day, working for numerous portraitists, notably *Hudson and *Ramsay, who were his executors. He also painted some *genre scenes and *conversation pieces. His brother **Alexander** (*c.*1701–57) was his assistant and also made *mezzotints. **Arnold van Aken** (d. 1735/6) is presumed to have been another brother. Almost nothing is known of him. The family name was also spelled 'Hacken'.

alabaster. Soft, semi-transparent stone (a form of gypsum) extensively used in sculpture in the later Middle Ages. Its most notable use was in small *retables, which were made in great numbers in England, many of them for export—they were sent even as far as Italy, Spain, and Russia. The best collection of such retables is in the Hildeburgh bequest at the Victoria and Albert Museum, London, and other fine examples are in the Castle Museum at Nottingham, a town that was famed for its 'alabastermen'. The production of religious images was cut off abruptly by the Reformation, but some tombs continued to be carved in alabaster until the 18th cent. Several modern sculptors, notably Henry *Moore, have used alabaster, generally for small carvings.

Albani, Cardinal Alessandro (1692–1779). Italian churchman, the leading collector and art patron in Rome in his day. He came from a distinguished family that included several cardinals and also Pope Clement XI (1649–1721), but he led a worldly life and was notorious for his lucrative dealings in the art market, not hesitating to have works heavily restored if it made them sell better. With the help of his librarian *Winckelmann, he made a fine collection of *antique sculpture, much of which is now in the Glyptothek at Munich. It was housed in the villa he had built (1755–63) by Carlo Marchionni, where *Mengs painted his famous ceiling painting *Parnassus* (1761), one of the key works of *Neoclassicism.

Albani (or **Albano**), **Francesco** (1578–1660). Bolognese painter. After a period in the studio of Denis *Calvaert and subsequently in the *Carracci academy under Lodovico Carracci, he moved to Rome (*c.*1600), where he collaborated with Annibale Carracci and *Domenichino on various decorative schemes, including work in the Palazzo Farnese. In 1616 he returned to Bologna and produced, besides altarpieces, many allegorical paintings and idyllic landscapes in a charmingly light-hearted style which proved very popular in England in the late 18th cent.

Albers, Josef (1888–1976). German-born painter, designer, writer, and teacher, who became an American citizen in 1939. After studying at the Royal School of Art, Berlin, 1913–15, he did lithographs and woodcuts in the *Expressionist manner while teaching at the School of Arts and Crafts, Essen. He then studied painting at the Munich Academy under Franz von Stuck (1863–1928), who had been a teacher of *Kandinsky and *Klee, from 1919 to 1920, when he entered the *Bauhaus and occupied himself particularly with glass pictures. He studied and taught at the Bauhaus, where his activities embraced stained glass, typography, and designing furniture and utility objects. When the Bauhaus closed in 1933 he emigrated to the USA. He was one of the first of the Bauhaus teachers to move there and one of the most energetic in propagating its ideas. From 1933 to 1949 he taught at *Black Mountain College, and from 1950 to 1959 he was head of the Department of Design at Yale University (the art gallery there has an outstanding collection of his work); he lectured at many other places. As a painter Albers was best known for his long series, begun in 1950, entitled *Homage to the Square*. The paintings in this series consisted of squares within squares of closely calculated sizes and subtly varied hues, within a narrow range of colour. The research into colour which they embodied was published in *Interaction of Color* (1963), and his disciplined ideas and techniques were influential on *Op art. Albers' wife, **Anni Albers** (1899–1994), whom he met when she was a student at the Bauhaus, was a weaver; her rectilinear designs have something of the severe economy of her husband's paintings.

Albert, Prince (Francis Charles Augustus Albert Emmanuel of Saxe-Coburg-Gotha) (1819–61). German-born Consort of Queen Victoria, whom he married in 1840, an influential figure in many fields of art. Albert was an amateur painter, designer, and architect (Osborne House, the royal residence on the Isle of Wight, is largely his creation), but his importance lies more in his roles of patron, collector, and administrator. He was a tireless committee worker on behalf of the arts; in 1841 he was appointed chairman of the Royal Commission set up 'to take into consideration the Promotion of the Fine Arts of the Country, in connection with the Rebuilding of the Houses of Parliament', and ten years later he was the inspiration behind the Great *Exhibition, held at the Crystal Palace. The profits from the Exhibition were used to endow the South Kensington (later renamed *Victoria and Albert) Museum. As a collector he was ahead of his time in his appreciation of *'primitive' artists (the *Duccio Crucifixion in the royal collection was one of his purchases). He greatly improved the administration of the art treasures at Hampton Court and Windsor Castle, and was a pioneer in the use of photography in art history. After his death, Victoria had him commemorated in numerous monuments, of which the Albert Memorial (1863–72) in Kensington Gardens, designed by Sir George Gilbert Scott (1811–78), is the most famous.

Alberti, Leon Battista (1404–72). Italian architect, sculptor, painter, and writer, the most important art theorist of the *Renaissance. Born in Genoa, the illegitimate son of an exiled Florentine merchant, he was educated at Padua and Bologna, and proved himself a precocious scholar; at the age of 20 he wrote a Latin comedy, *Philodoxeos*, which passed as an original Latin work. In about 1428 he went to Florence and became a friend of the most advanced artists—*Brunelleschi, *Donatello, *Ghiberti, Luca della *Robbia, and *Masaccio. To all these, jointly, he dedicated his first theoretical work on the arts, *Della Pittura* (1436), which contains the first description of *perspective construction. (This was first written in Latin as *De Pictura*, but Alberti translated it into Italian for Brunelleschi's benefit.) Alberti wrote on a wide variety of other topics, complementing *Della Pittura* with treatises on architecture (*De Re Aedificatoria*) and sculpture (*De Statua*). He worked on *De Re Aedificatoria* until his death and it became the first printed book on architecture in 1485: *De Statua* is generally dated to the 1460s.

Alberti was the most representative figure in the change which took place at the Renaissance from the medieval attitude to art as a symbolic expression of theological truths to the humanistic outlook, and the new ideas of scientific naturalism found their fullest expression in his writings. In the breadth of his knowledge and the rational and scientific temper of his mind he was typical of the early humanists. He thought of architecture as a civic activity and broke new ground by dealing with it in the context of unified town planning. In accordance with the new status of the artist as an exponent of a *liberal art and not a mere manual worker, Alberti emphasized in all his works the rational and scientific basis of the arts, and the necessity for the artist to have a thorough grounding in such 'sciences' as history, poetry, and mathematics. His definitions of the various arts were no longer in terms of religious function, but entirely in human terms, with a strong admiration for classical antiquity. In some contexts he defines painting in language of unqualified naturalism, but elsewhere he recognizes that the artist is bound to seek an ideal of beauty which cannot be achieved simply by accurate reproduction of nature, but involves selection. At times he identifies the beautiful with the typical in nature. In the *De Re Aedificatoria* he evolves a theory of beauty in terms of an ultimately mathematical symmetry and proportion of parts. He distinguished the rational appreciation of beauty from the vagaries of individual taste, though at times he held that beauty is what pleases the eye. While many of Alberti's ideas were of Platonic origin, often derived through *Vitruvius, he was not identified with the mystical Neo-Platonism of Marsilio Ficino (1433–99) and Pico della Mirandola (1463–94) encouraged by Lorenzo de' *Medici.

Alberti spent most of his life in Florence and Rome (he held a secretarial post in the papal court from 1432 to 1464), and he worked as an architect in both these places as well as in Mantua and Rimini. His buildings—among them the churches of S. Andrea and S. Sebastiano in Mantua, and the façade of Sta Maria Novella and the Palazzo Rucellai in Florence—are among the outstanding architectural works of the early Renaissance, but almost no trace survives of his work as a painter or sculptor. Two self-portrait plaques are attributed to him (Louvre, Paris, and NG, Washington), but no paintings are extant.

Albertina, Vienna. One of the world's most celebrated collections of Old Master drawings and prints. It was founded in 1776 by Archduke Albert of Sachsen-Teschen (1738–1822) and his wife Christina (1742–98), a daughter of the empress Maria Theresa, who devoted their lives to the collection. They found unique opportunities in the Netherlands, of which the Archduke was governor, and also through the breaking up of the great French collections shortly before the Revolution in 1789. The Albertina collection was enlarged under Albert's successors and taken over by the state in 1918. It is preserved in the archducal palace, where a changing selection of its treasures (among them a superlative collection of *Dürer's drawings that can be traced back directly to the artist's widow) is on view to the public.

Albertinelli, Mariotto (1474–1515). Florentine painter, trained by Cosimo *Rosselli, in whose studio he met Fra *Bartolommeo. The two went into partnership in 1508, but soon after this Albertinelli temporarily abandoned painting to become an inn keeper, saying (according to *Vasari) that he was fed up with criticism and wanted a 'less difficult and more cheerful craft'. Albertinelli's paintings are elegant but rather insipid. His best-known pictures are the *Visitation* (Uffizi, Florence, 1503) and an *Annunciation* (Accademia, Florence, 1510). Vasari also says he was 'a restless man, a follower of Venus, and a good liver'.

Albright, Ivan Le Lorraine (1897–1983). American painter, the son of **Adam Emery Albright** (1862–1957), a painter who studied under *Eakins. During the First World War Albright served in France as a medical draughtsman and worked with a meticulous detail and clinical precision that anticipated his later paintings, which show a morbid obsession with death and corruption: sagging, almost putrescent flesh (which he described as 'corrugated mush'), decrepit, decaying objects, and lurid lighting are typical of his work. Often it evokes a feeling of melancholy for a beauty that is past. He came from a wealthy family and his financial independence allowed him to work slowly, producing a small number of elaborate, highly finished paintings. For most of his life he lived in or near Chicago, and the city's Art Institute has the best collection of his works. It includes the painting Albright did for the Hollywood film (1943) of Oscar Wilde's *The Picture of Dorian Gray*, showing the loathsomely corrupted title figure; Albright's identical twin brother, **Malvin Marr Albright** (1897–), did the portrait of the young and beautiful Dorian for this film.

Alcamenes. Athenian sculptor of the second half of the 5th cent. BC, a pupil of *Phidias. *Pausanias' assertion that he made the sculptures of the west pediment of the temple of Zeus at Olympia is implausible, but numerous other works by him are mentioned in ancient sources, and after the end of Phidias' career in Athens, Alcamenes was probably the leading sculptor in the city. Several of his works are known from copies, and a badly mutilated marble statue in the Acropolis Museum in Athens is perhaps an original by him representing *Procne with Itys*.

Aldegrever, Heinrich (1502–c.1555). German engraver and painter. He worked mainly in Soest and was one of the leading artists in Westphalia in his day. He is usually grouped with the so-called *Little Masters, since he liked working on a rather small scale. His numerous engravings, usually of religious subjects, betray the all-pervading influence of *Dürer. Little is known of his activity as a painter.

Aldine Press. A famous press established in Venice in 1494 by one of the most distinguished scholar-printers of the 16th cent., Aldus Manutius (1449–1515). During the twenty-one years of his activity Aldus deliberately devoted his energies to the cause of scholarship and did more than any other man to facilitate the spread of the new learning among the scholars of Europe. The work by which he is most widely known is the long series of small octavo volumes of Greek and Latin classics, bearing his device of an anchor and dolphin. Aldus was the first printer to design a small book which was convenient for the student. In order to compress the texts into the limited compass of the octavo volume Aldus needed a type quite different from the roman and gothic letters then in use for large library folios. In 1501 he therefore designed a new small type, which was based on the cursive script then fashionable in Italy. It is from this type that the letter we call *italic* derives. Aldus also designed a Greek type for which the handwriting of his friend Marcus Musurus is said to have served as a model. The wide influence of the Aldine books set a fashion in Greek letters which lasted for three centuries.

Alechinsky, Pierre (1927–). Belgian painter and graphic artist. In 1947 he became a member of the association *Jeune Peinture Belge and in 1949 joined the *Cobra group. He severed his connection with Cobra in 1951 and settled in Paris, where he studied graphic techniques at Studio 17 under S. W. *Hayter. At this time he made contact with Japanese calligraphers and in 1955 he travelled to the Far East and produced a film *Calligraphie japonaise*. Alechinsky paints in a style of vigorous, even violent, expressive abstraction which has closer affinities with Nordic *Expressionism than with the Classical restraint of the School of *Paris. Residual figurative motifs remained constant to his work, and these were redolent of a turbulent fantasy, often approaching *Surrealism and also showing a strong debt to *Ensor.

Alexander Mosaic. A *mosaic from Pompeii (now in the Archaeological Mus., Naples) depicting Alexander the Great (356–323 BC) face to face with Darius III, the king of Persia, at the Battle of Issus (333 BC). It is about 5 m. long, spectacular and dynamic in composition, and one of the finest mosaics to survive (albeit substantially damaged) from the ancient world or any other period. *Pliny praises a painting of the subject by Philoxenos of Eretria, and the mosaic is probably modelled on this lost original.

Alexander Sarcophagus. A marble *sarcophagus, shaped like a chest with roofed lid, found at Sidon and now in Istanbul (Archaeological Mus.). It takes its name from the high *reliefs of the panels on the sides and ends, which show Alexander the Great lion-hunting, etc. The sculpture, good work of *c.*300 BC, is exceptional for the admirably preserved and effective colouring.

Alexandros. See VENUS DE MILO.

Algardi, Alessandro (1598–1654). Italian sculptor, born in Bologna, where he had his initial training in the *Carracci academy. He settled in Rome in 1625 and became, after *Bernini, the leading sculptor in the city. During the pontificate of Innocent X (1644–55) Bernini was out of favour and Algardi replaced him at the papal court. Three most prestigious commissions were the tomb of Leo XI (Alessandro de' *Medici) (1634–44) and the huge relief of *Pope Leo driving Attila from Rome* (1646–53), both in St Peter's, Rome, and the free-standing group of *The Decapitation of St Paul* (1641–7) in S. Paolo, Bologna. He was a prolific sculptor of portrait *busts, and these are his works that are now generally most admired—indeed he ranks as one of the greatest portrait sculptors of all time (excellent examples are in the Victoria and Albert Museum, London, and the City Art Gallery, Manchester). His style was generally more sober and Classical than Bernini's (although portraits have occasionally been disputed between them), reflecting his Bolognese upbringing, his work as a restorer of antique statuary, and his friendship with artists such as *Domenichino, *Duquesnoy, *Poussin, and *Sacchi.

Algarotti, Francesco (1712–64). The foremost Italian art critic of his day, a cosmopolitan snob of great charm, whose friendship with some of the leading men of Europe—notably the French philosopher Voltaire and Frederick the Great of Prussia—played a part in spreading Venetian culture. His writings proclaimed a watered-down version of the *Neoclassicism which was then gaining ground in Europe (though not yet in Venice). For some years he influenced the practice of his friends *Tiepolo ('restraining his wilder fantasies' as he claimed), and *Canaletto (encouraging his architectural *capricci*), as well as *Piazzetta and other Venetian painters. His real importance lies in his having helped to make acceptable the bolder ideas of other men and thus to break down the cultural isolation of Italy. He was a notable collector of paintings and drawings, and was employed by Frederick Augustus II (1696–1763), Elector of Saxony, to buy pictures for the Dresden Gemäldegalerie.

Alison, Archibald (1757–1839). Scottish aesthetician and divine. His *Essays on the Nature and Principles of Taste* (1790), published in the same year as *Kant's *Critique of Judgement*, had a revolutionary influence on English aesthetic writing and important bearings on later *Romantic theory. He held that previous thinkers had erred in supposing aesthetic enjoyment to be a simple emotion, believing himself to have shown that it is produced only when the arousal of a simple emotion or the exercise of some moral affection is followed by the 'Excitement of a peculiar exercise of the Imagination', and that the two constituent elements in the complex aesthetic emotion are always distinguishable and sometimes distinguished in our experience. The play of imagination which Alison held to be essential to aesthetic appreciation consists in the indulgence of an associative train of thought and imagery activated by, and emotionally related to, the object of appreciation.

Alken. Family of British sporting artists of Danish origin. **Samuel Alken** senior (1750–1815) did hunting and sporting landscapes in the manner of *Stubbs. He was the father of **Samuel Alken** junior (1784–1825) and of **Henry** (1785–1851), the latter of whom became one of the most prolific sporting painters and illustrators of his time. His gaunt but sprightly style was already archaic in his day, but he excelled at representing the life and movement of the hunting field and at suggesting typical aspects of the English countryside. The great popularity of coloured *aquatints after his paintings (some of them produced under the name 'Ben Tally-Ho') has persisted. He was also proficient in etching and published *The Art and Practice of Etching* (1849). His other publications included *Sporting Sketches* (1817) and *National Sports of Great Britain* (1820). He had four sons, among them **Henry Gordon Alken** (1810–92), with whose work his own is often confused.

Allan, David (1744–96). Scottish portrait and *genre painter. He spent the years 1764–*c.*1775 in

Italy, studying with Gavin *Hamilton and winning a prize for history painting at the Academy of St Luke in Rome. In 1770–80 Allan worked in London, then at the end of this period he settled in Edinburgh as a painter of portraits and *conversation pieces. When abroad he had made studies of French and Italian peasants and he painted scenes of Scottish life in a similar vein, which earned him the misleading title of 'the Scottish *Hogarth'. Such works influenced *Wilkie.

Allan, Sir William (1782–1850). Scottish historical painter, who travelled extensively in Russia, the Middle East, and elsewhere. He was elected RA in 1835 and knighted in 1842. His accuracy of detail, the heroic scale of his work, and his exotic subject matter (*The Black Dwarf*, NG, Edinburgh) satisfied the same appetite as did the novels of Sir Walter Scott (1771–1832), who was his enthusiastic supporter. With *Wilkie, he did much to establish the vogue for historical *genre painting in Scotland.

alla prima. Method of painting, primarily in oils, in which the paint is applied directly to the ground, without underpainting, and the finished surface is achieved with a single application of *pigments. *Alla prima* is Italian for 'at first'; synonymous terms are 'direct painting', 'wet on wet', and the French *au premier coup* (at first stroke). Direct painting was practised from the 17th cent. (for example by *Hals), but it was not until the middle of the 19th cent. that it became the chief method in oil painting. Its growing popularity was connected with the availability of commercial paints of a buttery consistency which after drying retained the marks of manipulation better than the earlier studio-manufactured paints, as well as with Romantic ideas about spontaneity of expression.

Allied Artists' Association (AAA). A society of British artists formed by the critic Frank Rutter (1876–1937) and artists in *Sickert's circle for the purpose of organizing annual exhibitions of independent progressive painters in the jury-free manner of the French *Salon des Indépendants. The group, which represented a reaction to the conservative values of the *New English Art Club, was formed in 1908 and held the first of its exhibitions in the Albert Hall in that year where more than 3,000 works were shown. Smaller shows were held at the Grafton galleries between 1916 and 1920. *Brancusi (1913), *Kandinsky (1909) and *Zadkine (1913) received the first British showing of their works at these exhibitions. In addition to Sickert, the Association included Spencer *Gore, Harold *Gilman, the future president of the *London Group, Lucien *Pissarro, Augustus *John, Henry *Lamb, Walter *Bayes, and Charles *Ginner. It was from the Allied Artists' Association that the *Camden Town Group emerged in 1911.

Allori, Alessandro (1535–1607). Florentine painter, the pupil and adopted son of *Bronzino. An early visit to Rome added the influence of late *Michelangelo paintings to that of his master's courtly *Mannerism. *The Pearl Fishers* (Studiolo of Francesco I, Palazzo Vecchio, Florence, *c*.1570) is generally considered his masterpiece; playful and full of artifice, it combines nude figures obviously drawn from Michelangelo with Bronzino's svelteness and enamelled colouring. His frescos include *grotesques for the *Uffizi (1580) and two large antique histories for the *Medici villa of Poggio a Caiano. His son **Cristofano** (1577–1621) was one of the leading Florentine painters of his period, working in a style that was more naturalistic and *Baroque than that of his father. He is remembered primarily for one work, *Judith with the Head of Holofernes* (*c*.1615, Pitti, Florence, and other versions), in which his *femme fatale* mistress is portrayed as Judith and he has depicted his own features in Holofernes' severed head. In the 18th and 19th cents. it was one of the most famous paintings in Italy. The Ashmolean Museum in Oxford possesses portraits by both Alessandro and Cristofano Allori.

All-over painting. A term used for a style of painting in which the whole surface of the canvas is treated in a relatively uniform manner and traditional ideas of composition—of the picture having a top, bottom, or centre—are abandoned. First used of the 'drip' paintings of Jackson *Pollock, the term has since been applied to paintings of other artists where the overall treatment of the canvas is relatively uniform, whether relying on texture or on 'scribbled' material as with Cy *Twombly or on colour as with the *Colour Field painters. The German term is *Streu-Komposition* (see also FIELD PAINTING).

Alloway, Lawrence (1926–90). British art critic and curator, active for much of his career in the USA. In the 1950s he worked at the *Institute of Contemporary Arts, London (he was director, 1957–9), and he was one of the leading figures in the *Independent Group, the cradle of British *Pop art (Alloway himself coined this term). He was also a strong supporter of American *Abstract Expressionism, and in 1961 he emigrated to the USA, settling in New York, where he became a curator at the *Guggenheim Museum and art critic for *The Nation*. His books include *American Pop Art* (1974) and *Topics in American Art since 1945* (1975).

Allston, Washington (1779–1843). American painter and writer, the most important artistic personality of the first generation of *Romanticism in the USA. Samuel Taylor Coleridge (1772–1834), who met him in Rome and whose portrait Allston painted (NPG, London), considered him 'a man of . . . high and rare genius . . . whether I contemplate him in the character of a Poet, a Painter, or a Philosophic Analyst'. Allston spent his working life in Boston apart from two lengthy visits to Europe: during the first, 1801–8, he studied under Benjamin *West at the *Royal Academy and exhibited three pictures there, subsequently visiting France with John *Vanderlyn; the second stay in England was from 1811 to 1818. Up to *c.*1818 his Romanticism expressed itself in the grandiose and dramatic, and his large canvases exploited the mysterious, monumental, and terrific aspects of nature (*The Rising of a Thunderstorm at Sea*, Mus. of Fine Arts, Boston, 1804). In his later period he was a forerunner of the subjective and visionary trend in American landscape painting, which relied more upon mood and reverie than upon observation or drama. Typical of this quietist style is the smaller, more dreamlike picture *The Moonlight Landscape* (Mus. of Fine Arts, Boston, 1819). Through his pupil Samuel F. B. *Morse this type of landscape painting became indigenous to the USA. Allston's writing included poetry, a novel, and *Lectures on Art* (posthumously published, 1850).

Alma-Tadema, Sir Lawrence (1836–1912). Dutch painter who settled in London in 1870 and took British nationality in 1873. He specialized in historical genre scenes, beginning with medieval subjects, but then—following a visit to Pompeii in 1863—turning to the ancient world. His paintings evoke a Hollywood vision of ancient Greece and Rome (and sometimes Egypt), with their sensuous depiction of beautiful women, exotic costumes, and marbled settings—*Punch* called him a 'marbellous artist'. They were enormously successful and he had a sumptuous lifestyle in his house in St John's Wood, which had previously been owned by *Tissot and which Alma-Tadema remodelled as a Roman villa. He became RA in 1879, was knighted in 1899, and received the Order of Merit in 1905. His success encouraged several imitators, including his second wife **Laura** (1852–1909), his daughter **Anna** (1865–1943), and painters such as John William Godward (1861–1922) and Edwin Long (1829–91). However, Alma-Tadema's work went completely out of favour after his death, and his reputation did not revive until the 1970s. Now he once again pleases a large public; visitor surveys at the Getty Museum, Malibu, have revealed that his *Spring* (1894) is the most popular work in the collection.

Alsloot, Denis van (*c.*1570–*c.*1627). Flemish painter of landscapes with religious and mythological figures, especially noted for scenes of processions and fêtes. In 1599 he became a master in Brussels and worked for the court there. Until *c.*1610 Alsloot painted in the manner of *Coninxloo and Jan *Brueghel, but then developed a more realistic, unromanticized style. In his earlier pictures the figures were often by Hendrick de *Clerck. His best-known works are a series, painted for the Archduchess Isabella, of the Ommeganck procession in 1615 (surviving examples are in the Prado, Madrid, and the Victoria and Albert Museum, London). He also painted several scenes connected with festivals at the Abbaye de la Cambre at Brussels. The winter landscapes are among the most attractive of Alsloot's paintings, with small brightly coloured figures set off against the light backgrounds.

Altamira. A site of palaeolithic rock painting 30 km. from Santander near the village of Santillana del Mar in northern Spain, the first prehistoric rock paintings to be discovered. The entrance to the cave was found by accident in 1869. Excavations were started by Don Marceliano de Sautuola in 1879 and during the operations the roof paintings were discovered by his infant daughter. The antiquity and authenticity of the paintings were at first denied by most prehistorians and they were repudiated at the Congress of Anthropology and Prehistoric Archaeology held at Lisbon in 1880, but the doubts were dispelled after the discoveries in 1901 of cave art near Les Eyzies by Henri Breuil (1877–1961). The cave extends for 300 m. into a limestone massif, but the paintings are in a gallery, often no more than 2 m. high, about 30 m. from the entrance. The best-preserved paintings are those on the roof, which comprise twenty-five polychrome figures of animals, mainly bison, drawn almost life-size with the contours accentuated here and there by engraving. The paintings are naturalistic in style, making full use of the irregularities of the rock, and they display a remarkable grasp of essential form and an eye for characteristic attitude and movement. The cave also contains fine engravings of animal heads. The polychrome paintings are assigned to the Upper Magdalenian period (*c.*12,000 BC) and are regarded, with those at *Lascaux, France, as being the outstanding paintings known from the prehistoric era.

altarpiece. A picture, sculpture, screen, or decorated wall standing on or behind an altar in a

Christian church. They vary enormously in size and conception, from tiny portable pictures to huge structures embracing the arts of architecture, sculpture, and painting (enormous and extraordinarily complex altarpieces are particularly common in Spanish art). They do, however, divide into two main types: the reredos, which rises from ground level behind the altar; and the retable, which stands either on the back of the altar itself or on a pedestal behind it. Many altars have both.

Altdorfer, Albrecht (*c.*1480–1538). German painter and printmaker working in Regensburg, of which town he was a citizen from 1505 onwards, the leading artist of the so-called *Danube School. It seems that Altdorfer learned his trade during the early years of the century in Austria, where he was influenced by Lucas *Cranach, with whom he shared a hitherto unusual interest in landscape. *Dürer's art too was known to him through the woodcuts and engravings. Mingled with these German impressions was a knowledge of the art of *Mantegna, perhaps through the mediation of Michael *Pacher. Yet in spite of these varied influences Altdorfer's style always remained personal. Most of his paintings are religious works, but he was one of the first artists to show an interest in landscape as an independent genre. In works such as the altar for S. Florian near Linz (1518) or the *Christ Taking Leave of His Mother* (NG, London) he achieved a wonderful unity of mood between action and landscape, and two pure landscape paintings (without any figures) by him are known (NG, London, and Alte Pinakothek, Munich). His patrons included the emperor Maximilian (1527–76) and Louis X, Duke of Bavaria (1516–45), for whom he painted the celebrated *Battle of Issus* (Alte Pinakothek, Munich, 1529), which formed part of a large series of famous battle-pieces from Classical antiquity. With its dazzling light effects, teeming figures, and brilliant colours, it is one of the finest examples of Altdorfer's rich imaginative powers. From 1526 until his death Altdorfer was employed as town architect of Regensburg. No architectural work by him is known, but his interest in architecture and his skill in handling intricate problems of *perspective are demonstrated by his *Birth of the Virgin* (Alte Pinakothek, Munich).

Altichiero (*c.*1330–*c.*1395). Italian painter. He probably came from Zevio near Verona and is sometimes considered to be the founder of the Veronese School, although the only surviving example of his work in that town is a fresco in Sta Anastasia dated from near the end of his life. Most of his surviving work is in Padua, where he had a hand in fresco cycles in the Basilica of St Anthony (between 1372 and 1379) and in the Oratory of St George (between 1377 and 1384), in the latter of which he collaborated with an artist called Avanzo, who is otherwise unknown and whose contribution to the work is uncertain. Altichiero's gravity and the solidity and voluminousness of his figures clearly reveal his debt to *Giotto's frescos in the Arena Chapel of Padua. But his pageant-like scenes with their elaborate architectural views express the taste of the late 14th cent. for *Gothic intricacy, while his naturalism in the study of plants, animals, and portraiture formed the point of departure for a new style which is reflected in *Pisanello.

Álvarez y Cubero, José (1768–1827). The leading Spanish sculptor of the *Neoclassical period, sometimes called 'the Spanish *Canova'. After studying at Granada, Madrid, and Paris he settled in Rome (1805–25), where Canova befriended him. Among his admirers were the emperor Napoleon, and Ferdinand VII (1784–1833) of Spain. He preferred Classical themes such as *Nestor and Antilochus* (Modern Art Mus., Madrid, 1818), but was also an accomplished portrait sculptor.

Amarna art. Art associated with the city of (el-) Amarna, founded by the Pharaoh Akhenaten (1379–1361 BC) when he moved his court northwards from Thebes to a virgin site. Akhenaten tried to impose a form of solar monotheism upon Egypt, and the art of the period represents the only real break in Egyptian tradition, for the rejection of the old gods and more particularly of the accepted notion of the afterlife implied a reappraisal of the basic concept of artistic representation. Although certain conventions such as the drawing of the eye remained unchanged, Amarna art was in general more naturalistic and relaxed than art of earlier and succeeding periods. There was greater play with sloping lines and curves and a marked interest in drapery and streamers, which together gave a curiously feminine allure.

The new approach to art, in some degree inspired by the king himself, could not as such survive his death and the return to orthodox religious ideology, but certain features had a lasting influence on work of later periods. Many of the treasures found in the tomb of Akhenaten's successor, Tutankhamun (*c.*1334–1325 BC), are characteristic of Amarna art.

amateur. An artist who works purely for pleasure, rather than for a livelihood. In Western art, the idea of amateur status began to have meaning only with the advent of the *Renaissance, before when the visual arts were considered mere crafts and therefore socially disreputable. The change in the intellectual and social status of the artist brought about by figures such as *Leonardo meant that even princes began to submit to the lure of the arts, not only as patrons but as practising amateurs. In the 15th cent. René d'Anjou (1434–80) gave up his dukedom to devote himself to painting and landscape gardening, and England witnessed several distinguished aristocratic amateur artists in the 17th and 18th cents., notably Sir Nathaniel *Bacon, the third Earl of *Burlington, and Prince *Rupert. With the development of *watercolour in the 18th and 19th cents.—the amateur medium *par excellence*—amateur painters proliferated all over Europe. They included eminent men such as *Goethe, and statesmen began to advocate painting as a means of relaxation—'a joyride in a paintbox' as Sir Winston Churchill, author of *Painting as a Pastime* (1948), expressed it. At the same time, sketching and watercolour became accepted 'accomplishments' for the prototypes of the elegant young ladies depicted in the novels of Jane Austen (1775–1817). Queen Victoria (1819–1901) practised both painting and etching.

Local drawing and painting societies had received parliamentary recognition before the end of the 19th cent. and increased steadily in numbers during the first half of the 20th cent. Although the phrase 'Sunday afternoon painter' has become a term of opprobrium, amateurs bulked largely in the recognized associations such as the *London Group and the National Association. Amateurs even invaded the once jealously guarded exhibitions of the *Royal Academy and, finally, the dubious economic status of the many professional artists who are not in regular employment has tended to obscure any sharp distinction between professional and amateur.

An equivalent term 'dilettante' became current in England after the foundation of the Society of *Dilettanti in 1733–4, at a time when 'amateur' had not yet been adopted and 'virtuoso' (see VIRTU) had already acquired a professional meaning. The dilettante interests himself in the arts for 'delight' as the amateur does for 'love'. Both words are apt to suggest a lack of serious aim or study, but there are amateurs who cannot be regarded as dilettanti, this term implying a degree of social distinction. In the non-dilettante category are some of the 'Sunday afternoon painters', who first appeared in France in the late 19th cent. In general these were painters who without formal training exercised a natural talent and depicted either their idiosyncratic vision (e.g. Henri *Rousseau), or the simple everyday things and events around them (e.g. Louis *Vivin). To this category belongs the venerable 'Grandma' *Moses, who took her native America by storm with scenes of remembered childhood set down on canvas with naïve but evident sincerity. From the 1930s *primitives have become a regular stock-in-trade of art dealers throughout the world.

The word 'amateur' was also used as early as 1784 for a person who had a taste for any art or craft. About 1803 it developed the specialized meaning of a person who cultivates any pursuit purely as a pastime, and in the field of sport this meaning has become formalized into the distinction between amateur and professional. But the phrase 'amateur of the arts' is still current in the sense of a person who cultivates several of the arts as a connoisseur and neither for gain nor as an executant.

In China, contrary to the position in the West, art has always been accorded a noble position, and the amateur has enjoyed a higher status than the professional. (The difference in status might be seen as analogous to the distinction between 'Gentlemen' and 'Players' that applied in English cricket until 1963.) The emperor Hui Tsung (1082–1135) was perhaps the most distinguished of all such amateurs.

amber. A fossil *resin derived from various trees and found mainly on the southern shores of the Baltic, used as an ingredient of some oil *varnishes. As it is a hard resin, difficult to dissolve, it is not suitable for varnishing modern pictures painted with *linseed oil and *turpentine, since cracks may be caused by rapid hardening over the comparatively soft layers of *pigment underneath. It also turns yellow and brown, becoming cloudy with time, and can be removed only by the strongest solvents.

Amberger, Christoph (d. 1561/2). German painter (mainly of portraits) and designer. He worked in Augsburg, which had many cultural and economic ties with Italy (he met *Titian when he visited the city in 1548), and his style emulates the grand manner of the Venetian School, paying as much attention to rich effects of dress and jewellery as to psychological subtlety. His *Charles V* (Staatliche Museen, Berlin, *c.*1532) is typical of his portrait manner. His rare figure compositions (*Virgin and Child between Saints Ulrich and Afra*, Augsburg Cathedral, 1554) are less distinguished. Amberger also designed coins, façade paintings, and statuary.

Ambrosiana, Milan. Picture gallery founded by Cardinal Federico Borromeo in 1618 as a sister institution to the already famous Ambrosian Library. It has a small but choice collection of Italian and Flemish paintings (Cardinal Borromeo's friend Jan

*Brueghel is well represented) and an important collection of drawings, including works by *Leonardo and *Raphael.

American Abstract Artists (AAA). An association of American abstract painters and sculptors formed in New York in 1936 with the aim of promoting their work and fostering understanding of it. The association held annual exhibitions (the first in 1937) and disseminated the principles of abstract art by lectures, publications, etc. The first President was Balcomb *Greene (1904–90) and among the early members were: Josef *Albers, Ad *Reinhardt, Ilya *Bolotowsky, Willem *de Kooning, Jackson *Pollock, Burgoyne *Diller, and David *Smith. In 1940 members picketed the *Museum of Modern Art, demanding that it should show American art, but by the mid-1940s abstract art had achieved recognition and the activities of the association dwindled, though it continued to stage exhibitions. In the 1950s, however, it became active again with more than 200 members.

American Scene painting. Term applied to the work of various painters who in the 1920s and 1930s depicted aspects of American life and landscape in a naturalistic, descriptive style. The term does not signify an organized movement, but rather an aspect of a broad tendency for American artists to move away from abstraction and the avant-garde in the period between the two world wars. *Burchfield and *Hopper are among the best known exponents of American Scene painting, and the *Regionalists, who were more self-conscious in their nationalism, are also embraced by the term.

Amigoni, Jacopo (*c.*1682–1752). Italian *Rococo decorative painter and portraitist. He was born in Naples, formed his style in Venice, and had an international career, working in Bavaria, England, Flanders, France, and Spain. His English sojourn lasted from 1730 to 1739 (with a break for a visit to Paris in 1736), his finest surviving work from this period being a series of four paintings on the *Story of Jupiter and Io* at Moor Park (now Moor Park Golf Club) in Hertfordshire. He was the last of the Venetian decorators to come to England, in the wake of *Pellegrini and the *Ricci, for the rise of Lord *Burlington's *Palladian school had stifled the demand for decorative painting. Amigoni, however, earned a good living with his portraits, and is said to have persuaded *Canaletto to try his fortune in England.

Amman, Jost (1539–91). Swiss engraver, born in Zurich and active mainly in Nuremberg, Germany, where he is documented from 1561. He was perhaps the most prolific book illustrator of his day and one of his pupils boasted that he produced more drawings in four years than could be carted away in a hay-wagon. On the whole his woodcuts and engravings are perhaps more important as documents of contemporary life than for their artistic value and his *Panoplia Omnium Artium* (1568) in particular contains a wealth of pictorial evidence on contemporary crafts and techniques.

Ammanati, Bartolommeo (1511–92). Florentine *Mannerist architect and sculptor, strongly influenced by *Michelangelo and *Sansovino, on whose Library in Venice he worked. From 1551 he worked with *Vasari and the architect Giacomo Vignola (1507–83) on Pope Julius III's Villa Giulia in Rome. His best-known works in Florence are the Ponte Sta Trinità (1567–70), destroyed during the Second World War, but rebuilt, and his additions to the *Pitti Palace (1558–70), including the rusticated courtyard. In sculpture his chief work is the rather ponderous fountain (1560–75) in the Piazza della Signoria, Florence, with its marble *Neptune* and bronze *Nymphs*. Ammanati beat several sculptors, including *Cellini and *Giambologna, in a competition for this commission, but the work was not well received. In old age, influenced by Counter-Reformation piety, he wrote a recantation of his secular works and destroyed some. He was married to Laura Battiferri, a poet who was the subject of a memorable portrait by *Bronzino.

Analytical Cubism. See CUBISM.

anamorphosis. Painting or drawing which is so executed as to give a distorted image of the object represented but which, if viewed from a certain point or reflected in a curved mirror, shows the object in true proportion, the purpose being to mystify or amuse. The earliest examples appear in *Leonardo's notes, and the word first appears in the 17th cent. Well-known examples of anamorphosis in painting are the portrait of Edward VI in the National Portrait Gallery, London (1546, of uncertain attribution), and the distorted skull in *Holbein's *Ambassadors* (NG, London, 1533).

Ancients. Group of English *Romantic artists active for about a decade in the 1820s and 1830s. The leading member of the group was Samuel *Palmer; others included Edward *Calvert and George *Richmond. The name derives from their passion for the medieval world, but they concentrated on pastoral subjects, sharing the mystical outlook of their inspiration, William *Blake, who himself took part in some of their activities, which included poetry-reading and music-making.

Anderson, Laurie. See PERFORMANCE ART.

Andre, Carl (1935–). American sculptor, a leading exponent of *Minimal art. He produces his works by stacking identical ready-made commercial units such as styrofoam planks, bricks, cement blocks, ceramic magnets, etc. (occasionally 'natural products' like logs or bales of hay), in simple geometrical arrangements. His most characteristic products abjure height and are arranged as horizontal configurations on the ground—'more like roads than buildings', in his own words. In Britain Andre is best known for the sensational publicity accompanying 'the Tate bricks' incident in 1976. His *Equivalent VIII* (1966) (consisting of 120 bricks arranged two deep in a rectangle) was vandalized and there was an outcry about the alleged waste of public money on its purchase by the Tate Gallery (which had occurred some four years earlier). In 1985 Andre was charged with murdering his wife (who died after falling from a window); he was acquitted at his trial.

Andrea da Firenze (Andrea Bonaiuti) (active *c.*1343–77). Florentine painter, remembered mainly for his frescos in the Spanish Chapel of Sta Maria Novella, a church of the Dominican Order. The frescos illustrate the Triumph of the Faith and the Dominican doctrine. In their descriptive detail they belong to the tradition of *Giotto, but otherwise, in the more rigid composition and impassive countenances, they mark a return to the Italo-Byzantine style of painting. Both the severity and the meticulous detail accorded with the expository style of the Dominican preaching friars. Andrea is last recorded in 1377 working on frescos of the *Life of St Raynerius* in the Campo Santo at Pisa.

Andrea del Castagno (Andrea di Bartolo di Bargilla) (*c.*1421–57). One of the most powerful Florentine painters in the generation after *Masaccio. According to tradition, in 1440 he painted frescos at the Palazzo del Podestà depicting rebels against Cosimo de' *Medici, who were sentenced to be hanged by the heels, earning him the sobriquet Andreino degli Impiccati (of the hanged men). These have been destroyed, and Andrea's earliest known surviving works are frescos in the church of Sta Zaccaria in Venice (1442), painted in collaboration with an obscure artist called Francesco da Faenza. By 1444 he was back in Florence, designing a stained-glass window for the cathedral, and soon after he began his greatest work, a series of frescos on Christ's Passion for the monastery of Sta Apollonia (now a Castagno museum), dominated by one of the most celebrated of all portrayals of *The Last Supper*. In their emotional vigour and sinewy realism these paintings have been regarded as the pictorial equivalent of the sculpture of *Donatello, but they also have something of Masaccio's monumentality. *Vasari said of him 'He excels in showing movement in his figures and the disquieting expression of their faces . . . the vigorous drawing of which emphasizes their gravity.' Andrea's other noteworthy works in Florence include the *St Julian* (SS. Annunziata, 1454–5), a frescoed equestrian portrait—*Niccolò da Tolentino*—a pendant to *Uccello's earlier fresco of Sir John Hawkwood in the cathedral, and an extraordinarily intense pair of altar frescos for SS. Annunziata (*c.*1455).

Vasari started the rumour that Castagno murdered his friend *Domenico Veneziano, and it was not until the 19th cent. that it was discovered that Castagno had died early of the plague and that Domenico in fact had outlived him. The story, however, makes it easy to believe that the intensity of his work reflected a fierce temperament.

Andrea del Sarto (1486–1530). Florentine painter. The epithet 'del sarto' (of the tailor) is derived from his father's profession; his real name was Andrea d'Agnolo di Francesco. After an apprenticeship under *Piero di Cosimo he soon absorbed the style of poise and beauty developed by Fra *Bartolommeo and *Raphael in Florence during the first decade of the 16th cent., and following the departures of *Leonardo, Raphael, and *Michelangelo (all of whom had left Florence by 1509) he became established with Bartolommeo as the outstanding painter of the city. Apart from a visit to Fontainebleau in 1518–19 to work for Francis I, Andrea was based in Florence all his life, although he probably visited Rome soon after his return from France, and made short visits elsewhere. He excelled as a fresco decorator (there are outstanding examples in Florence in SS. Annunziata and the Chiostro dello Scalzo), and he also painted superb altarpieces (*The Madonna of the Harpies*, Uffizi, Florence, 1517) and portraits (*A Young Man*, NG, London).

The reputation of Andrea del Sarto was largely made and marred by *Vasari, who said that his works were 'faultless' but represented him as a weakling completely under the thumb of his wicked wife. In Robert Browning's poem on the painter (1855) and in a psychoanalytic essay by Freud's disciple Ernest Jones (1913) attempts are made to link a supposed lack of vigour in his mellifluous art with these traits of character. This, however, is hardly just and a good deal of Vasari's account of Andrea's private life has been shown to be factually inaccurate (the scandal-mongering is mainly in the 1550 edition of his book and was suppressed in the 1568 edition). Andrea has suffered from being the contemporary of such giants as Michelangelo and Raphael, but he undoubtedly ranks as one of the greatest masters of his time. In grandeur and gracefulness he approaches Raphael, and he had a feel for colour and atmosphere that was unrivalled among Florentine painters of his period. He also numbers among the finest draughtsmen of the Renaissance (the best collection of his drawings is in the Uffizi). His compositional mastery was well brought out by *Wölfflin (*Classic Art*, ch. vi). More recent criticism has concentrated on those features of his art which foreshadow the *Mannerist experiments of his great pupils *Pontormo and *Rosso Fiorentino. The many other artists who trained in his busy workshop include *Salviati and Vasari.

Andrea del Verrocchio. See VERROCCHIO.

Andrea di Bartolo. See BARTOLO DI FREDI.

Andrews, Michael (1928–95). British painter. He studied at the *Slade School under *Coldstream, 1949–53. A slow, fastidious worker, he concentrated on ambitious figure compositions, subtly handled and often with an underlying emotional tension. He shunned publicity and was little known to the public until an *Arts Council exhibition of his work in 1980, after which he achieved a considerable reputation. In the mid-1980s he had a change of direction with a series of huge, brilliantly coloured landscapes featuring Ayers Rock in Australia. See also SCHOOL OF LONDON.

Andriessen, Jurriaen (1742–1819). Dutch artist whose unsentimental paintings and prints depicted scenes of everyday life. He was director of a private drawing academy in Amsterdam from 1799 until his death, and had great influence on early 19th-cent. Dutch artists. *Troostwijck was his most important pupil.

Angelico, Fra (Guido di Pietro) (*c*.1395–1455). Florentine painter, a Dominican friar. Although in popular tradition he has been seen as, in *Ruskin's words, 'not an artist properly so-called but an inspired saint', Angelico was in fact a highly professional artist. The Dominican Order discouraged individual inspiration and sought to maintain the tradition of art in the service of religion, yet he remained in touch with the most advanced developments in contemporary Florentine art and in later life travelled extensively for prestigious commissions. Angelico probably began his career as a manuscript *illuminator, and his early paintings are strongly influenced by *International Gothic. But even in the most lavishly decorative of them all—the *Annunciation* in the Diocesan Museum in Cortona—*Masaccio's influence is evident in the insistent perspective of the architecture.

For most of his career Angelico was based in S. Domenico in Fiesole (he became Prior there in 1450), but his most famous works were painted at S. Marco in Florence (now an Angelico museum), a Sylvestrine monastery which was taken over by his Order in 1436. He and his assistants painted about fifty frescos in the friary (*c*.1438–45) that are at once the expression of and a guide to the spiritual life and disciplined devotion of the community. Many of the frescos are in the friars' cells and were intended as aids to devotion; with their immaculate colouring, their economy in drawing and composition, and their freedom from the accidents of time and place, they attain a sense of blissful serenity.

In the last decade of his life Angelico also worked in Orvieto and Perugia, and most importantly in Rome, where he frescoed the private chapel of Pope Nicholas V in the Vatican with *Scenes from the Lives of SS. Stephen and Lawrence* (1447–50). These differ considerably from the S. Marco frescos, with new emphasis on the story and on circumstantial detail, which brings Angelico more clearly into the mainstream of 15th-cent. Italian fresco painting.

Angelico died in Rome and was buried in the church of S. Maria sopra Minerva, where his tombstone still exists. His most important pupil was Benozzo *Gozzoli and he had considerable influence on Italian painting. He painted numerous altarpieces as well as frescos, several outstanding examples being in the S. Marco museum. His particular grace and sweetness stimulated the school of Perugia, and Fra *Bartolommeo, who followed him into the Convento di S. Marco in 1500, had something of his restraint and grandeur.

*Vasari, who referred to Fra Giovanni as 'a simple and most holy man', popularized the use of the name Angelico for him, but he says it is the name by which he was always known, and it was certainly used as early as 1469. The painter has long been called 'Beato Angelico' (the Blessed Angelico), but his beatification was not made official until 1984.

Angry Penguins. An Australian avant-garde journal (1940–6) devoted to arts and letters, published first in Adelaide and then in Melbourne. It encouraged and provided a focus for a group of young painters who worked in an *Expressionist vein and attempted to create an authentic Australian art free from European influences; among them were Arthur *Boyd and Sidney *Nolan.

They were opposed by a group of *Social Realist painters, among them Noel Counihan (1913–86), and the debate between the two factions in the pages of *Angry Penguins* helped to make Melbourne a lively artistic centre in the early 1940s.

Anguier, François (*c.*1604–69) and **Michel** (*c.*1613–86). French sculptors, brothers, who stood apart from the mainstream which in the middle of the 17th cent. was dominated by Jacques *Sarrazin. They went to Rome about 1641 and joined the studio of *Algardi. On their return to France (François in 1643, Michel in 1651), the brothers collaborated on the tomb of Henry de Montmorency in the chapel of the Lycée at Moulins (1648–52), which reveals the new Roman influence they introduced into France. Later the two brothers worked mainly apart, Michel having the more interesting career. His work includes the decoration of the interior of the church of the Val-de-Grâce, Paris (1662–7), and the Nativity group in St Roch, Paris (1665).

Anguisciola, Sofonisba (*c.*1530–1625). Italian portrait painter, one of six painter-sisters from Cremona. She was the first woman artist to achieve international renown, being called to Spain by Philip II (1527–98) and visited by van *Dyck in Genoa in 1623, when she was in her nineties. Her self-portraits and portraits of her family are considered her finest works; they are somewhat stiff, but can have great charm.

Angus, Rita (1908–70). New Zealand painter, mainly of portraits and landscapes. She was considered one of the leading figures in New Zealand art, particularly in the 1940s. Working in both oils and watercolours, she painted in a forthright, brightly coloured style. She also painted under her married name, Rita Cook.

Animal style. Term applied to a distinctive tradition of art in portable goods which developed from the 7th cent. BC and was once spread, with local variations, among the mounted nomads of Europe and northern Asia right across the steppes from Hungary to the Gobi desert, and occasionally even beyond. The style was first studied in modern times when gold plaques representing animals were collected from western Siberia under Peter the Great (1672–1725) of Russia, but it did not become well known until the barrows of south Russia began to be opened during the 18th cent. and excavated more scientifically during the 19th cent. Its origins are disputed and the whole subject is full of pitfalls.

Anker, Albert (1831–1910). Swiss painter. He worked in a style of sentimental naturalism, and was once very popular, his chief title to fame being the favour his works found with Adolf Hitler.

Annigoni, Pietro (1910–88). Italian painter (and occasional sculptor), the only artist of his time to become internationally famous as a society and state portraitist. The turning-point in his career was a commission from the Worshipful Company of Fishmongers to paint a portrait of Queen Elizabeth II (1954–5); it was reproduced endlessly, notably on the postage stamps and banknotes of various countries, and the jacket blurb of Annigoni's autobiography (*An Artist's Life,* 1977) claims that it made him 'the most famous artist in the world—not excluding even *Picasso'. Subsequently he painted many other celebrity sitters, including several other members of the British royal family, Presidents Kennedy and Johnson, and Pope John XXIII. In style and technique he based himself on the masters of the Italian Renaissance, placing great stress on draughtsmanship and often working in *tempera. Characteristically his work was very smoothly finished and detailed, melodramatic in lighting, and often rather melancholy in mood. Annigoni also painted religious works (including frescos in Italian churches) and ambitious allegorical scenes, and he regarded these as more important than his portraits. Critics often dismissed his work as portentously inflated and tasteless, but his royal portraits particularly were highly popular with the general public: more than 200,000 people went to see his second portrait of the Queen (NPG, London) during the fortnight when it was first exhibited in 1970.

Anquetin, Louis (1861–1932). French painter, designer, and writer. He was a friend of van *Gogh and *Toulouse-Lautrec (fellow-students in *Cormon's studio), and with *Bernard he was one of the pioneers of *cloisonnism. After about 1890, however, Anquetin's work was much more traditional, and he became interested in research into the techniques of the Old Masters, notably *Rubens, on whom he wrote a book (1924). He was a prolific draughtsman and made tapestry cartoons for the Beauvais and Gobelins factories.

Anrep, Boris (1883–1969). Russian-born mosaicist and painter, active mainly in England. He organized the Russian section of Roger *Fry's second *Post-Impressionist exhibition in 1912 and after the First World War, in which he served in the Russian Army, he settled in London, although he also spent a good deal of time in Paris. Anrep was deeply interested in Byzantine art and came to specialize in mosaic pavements, the best-known examples being in the Tate Gallery, representing William Blake's *Proverbs* (1923), and the National Gallery—four floors on and around the main staircase, executed between 1926 and 1952. The subjects are *The Awakening of the Muses*, *The Modern Virtues*, *The Labours of Life*, and *The Pleasures of Life*; portraits of many well-known contemporaries are incorporated in them, for example the philosopher Bertrand Russell representing 'Lucidity' and the film actress Greta Garbo as Melpomene, the Muse of Tragedy. They were paid for by Samuel *Courtauld and other benefactors.

Anshutz, Thomas Pollock. See HENRI.

Antal, Frederick (1887–1954). Hungarian-born art historian who settled in England in 1933 and became a British citizen in 1946. He was deeply interested in historical methodology, and is regarded as the leading exponent of an approach that attempted to apply the Marxist interpretation of history to the arts. His views are most fully expressed in *Florentine Painting and its Social Background* (1948), in which he argues that developments in style and subject matter were directly influenced by social and political changes. The book was widely regarded as an important contribution to the study of Italian Renaissance art, but it was also attacked as being over-rigid in the way it linked artistic phenomena with social and economic causes. Antal's other books include the posthumously published *Hogarth and his Place in European Art* (1962), in which he applied his methods more subtly and flexibly, revealing his fascination with *Hogarth as an expression of English middle-class morality and culture. A collection of his articles, *Classicism and Romanticism, with Other Studies in Art History* (1966), includes 'Remarks on the Method of Art History', which is a statement of his own credo. Antal never held a regular teaching post in England (he occasionally lectured at the *Courtauld Institute), but he was an influential figure.

antefix. An architectural ornament, often of *terracotta, placed at the eaves or along the cornice of a Classical building to mask the end of each ridge of tiling.

Antelami, Benedetto (active late 12th cent.). Italian *Romanesque sculptor, the most notable figure in the history of Italian sculpture before Nicola *Pisano. His name first appears on a marble panel representing the *Descent from the Cross* (1178) in Parma Cathedral. He is chiefly known for his *reliefs on the doors of the Baptistery at Parma, which was begun in 1196, and it has been suggested that he may have overseen the whole structure as architect as well as sculptor. On stylistic grounds he has also been credited with a hand in the sculptural decorations of Fidenza Cathedral (formerly Borgo San Donnino) and of S. Andrea at Vercelli. Little or nothing is known about Antelami's history, but certain affinities of his work with Provence (particularly at S. Gilles and Arles) have given rise to the belief that he may have been trained there. His elongated figures, his compact compositions, and skilful use of drapery folds to emphasize design give his work a gravity and dramatic expressiveness hitherto unknown to north Italian sculpture. His work had great influence in Italy.

Antenor. Athenian sculptor active in the late 6th cent. BC. In antiquity he was famous for his bronze group of the *Tyrannicides* (now lost), which was carried off to Susa by Xerxes in 480 BC and restored by Alexander (356–323 BC) or one of his successors. An impressively solid marble **kore* from the Acropolis at Athens may belong to a base signed by Antenor; and by comparison of style some archaeologists attribute to him the pedimental sculpture of the late *Archaic temple of Apollo at Delphi.

Antes, Horst (1936–). German painter, sculptor, and graphic artist. He is indebted to the German *Expressionist tradition, and is best known for his 'Gnome' paintings, clumsy, heavily stylized figures, usually represented in profile, with eyes set one above the other and over-large, hulking hands and feet.

anthemion. A Greek decorative motif with radiating plant forms of honeysuckle or of lotus and palmette pattern. The term is sometimes applied to decorative patterns which incorporate only one of these motifs.

Anthonisz., Cornelis (*c.*1499–after 1556). Netherlandish painter, etcher, and designer of woodcuts. He was active in Amsterdam and in 1544 made a large map of that city, now in the Weigh House there. He worked as a cartographer in the service of the emperor Charles V and also painted group portraits, most notably *The Civic Guard* (Historical Mus., Amsterdam, 1533), an early example of a type for which *Hals later became renowned.

Antico (Pier Jacopo Alari Bonacolsi) (*c.*1460–1528). Italian sculptor, goldsmith, and medallist. His nickname derived from his classically inspired statuettes, which won him great popularity (he also restored ancient sculpture). He visited Rome in the 1490s, but worked mainly in and around his native Mantua, particularly for the *Gonzaga family and Isabella d'*Este.

Antinous. A representation in sculpture of the beautiful youth of this name who was a favourite of the emperor Hadrian. After Antinous was drowned while accompanying Hadrian up the Nile in AD 130 his name became surrounded by romantic legend, and the grief-stricken emperor commemorated him in lavish fashion. He founded a city called Antinoöpolis in Egypt, erected temples in his memory, and had him honoured in festivals. Antinous was frequently represented in sculpture, sometimes as Apollo or Dionysus, and several examples survive, although the identification is not always certain, and the title 'Antinous' has sometimes been given loosely to similar figures of beautiful and graceful youths. Particularly famous were the Belvedere Antinous (Vatican Museums), which was regarded as one of the standards of male beauty (for *Bernini's views on it, see ANTIQUE), and a relief (Villa Albani, Rome) excavated at Hadrian's Villa at Tivoli in 1735. It was one of the greatest treasures of Cardinal *Albani and was regarded by his librarian *Winckelmann as one of the peaks of ancient art.

antiphonal (or **antiphonary**). A liturgical book containing those parts of the Mass or Divine Office which are sung antiphonally (i.e. in responses) by the choir. The *Gradual*, which has only the hymns sung after the Epistle, often forms part of it. Although antiphonals were in common use since Early Christian days, their *illumination and decoration seems to have begun only *c.* AD 1000. Appropriate scenes from the Bible or the lives of the saints were chosen to illustrate the text. During the Middle Ages rich pictorial cycles were preferred, but from the 14th cent. onward words and music took precedence over pictures, the latter being confined to historiated initials, while scrolls and *grotesques framed the page.

Antipodeans. The name adopted by a group of Australian painters (Arthur *Boyd was the best known) who held an exhibition in Melbourne in 1959; the catalogue contained a manifesto of their aims and ideas, attacking abstraction and championing figurative art.

antique, the. The physical remains of the Greek and Roman world, or more particularly the remains of antique sculpture, which have been for later artists an inspiration, a challenge, and a canon of perfection. Such remains have never been totally absent and seldom totally disregarded. Memories of classical ornament or drapery forms recur throughout the Middle Ages, and occasionally, as in the 13th-cent. *Visitation* group at Reims Cathedral, a true classic dignity was attained. The revival of interest in classical antiquity at the Italian *Renaissance of the 15th cent. was not a new or unheralded phenomenon, but it was in 15th-cent. Italy that the recovery of the classical antique as a Golden Age in the past became a deliberate ideal. Artists and scholars found in classical sculpture a key to reality, awakening their awareness of the beauty of the human body and its expressive potentialities. *Ghiberti's writings testify to his admiration for antique statues and *cameos found in his lifetime which, as we can infer, he eagerly collected. We know from a letter by Poggio Bracciolini (1380–1459) that *Donatello 'highly praised' antique marbles, and his *David* would be unthinkable without a close study of the antique. In the latter half of the 15th cent. architects such as Giuliano da Sangallo (*c.*1443–1516) vied with the humanists in searching for and recording classical remains and inscriptions, and even such a conservative Sienese artist as *Neroccio is documented as the owner of an antique marble head and plaster casts of Apollo.

When Filippino *Lippi went to Rome he carried with him a letter of introduction by Lorenzo de' *Medici asking one of the cardinals to show the artist his collection of antiques. The role which Lorenzo's own collection in the gardens of S. Marco played in the formation of artists such as *Bertoldo and of *Michelangelo has been vividly described by *Vasari, whose account, however, may be coloured by his own academic bias.

Vasari, indeed, attributed the attainment of perfection by the generation of *Leonardo, Michelangelo, and *Raphael in no small measure to the discovery of the famous antiques of the Vatican collection, notably the **Apollo Belvedere* and the **Laocoön* (found 1506). There can be no doubt that the challenge of these late masterpieces did have its effect on the artists of the day, though it was mainly the next generation, particularly the apprentices from northern Europe, who systematically drew after the antique. *Heemskerck's sketch-books are characteristic of this trend. Towards the middle of the 16th cent. the role of the antique in the curriculum of artists became firmly established. Vasari advises the study from statues, simply because these do not move and are easier to draw from than the live model; but he also warns (in the 'Life' of *Mantegna) against transferring the habits thus acquired to the rendering of life. Giovanni Armenini in his *De' Veri Precetti della Pittura* (1587) already gives a list of 'canonic' antiques, including the famous **Belvedere Torso*, and the group of the river-god Nile (Vatican), which achieved a genuinely 'classical' authority and were carried by means of engravings, casts, and copies into every artist's studio.

In 17th-cent. Europe every artist had to define his attitude towards the canons of the antique, and the philosophical justification for this dependence on antique models was given in *Bellori's famous oration, *Idea*, where he claimed in the ancient statuary a revelation of an absolute beauty that had been discovered once and for all (see IDEAL). To the followers of the academic doctrine each of the great antiques, to which now had to be added the **Farnese Hercules*, the **Borghese Warrior*, the **Medici Venus*, and the **Barberini Faun*, represented a type of physique that could serve as a permanent standard for the artist. Nor was antique influence confined to those artists whose work was most obviously Classical (for example, *Poussin). *Rubens warned painters (in a letter printed by *de Piles) against the harsh manner that can result from too much attention to the antique, but he was a great student of Classical art himself; and *Bernini, when he addressed the Academy in Paris in 1666, said, 'In my early youth I drew a great deal from Classical figures, and when I was in difficulties with my first statue I turned to the **Antinous* as to the oracle.'

Reverence for the antique was given a new lease of life when the *Neoclassical movement reacted against the frivolities of *Rococo fashions. The coldness of the marble seemed to symbolize the aloofness of Stoic grandeur. Art was purged of its sensuality and superficial prettiness, and antique statues were imitated in a cold and often mechanical manner. In opposition to earlier ideas, *Winckelmann preached the belief that Classical artists had deliberately avoided representing extreme passions, and he regarded the antique less as a source of expressive formulas than as a model of noble restraint. During the later 18th cent., however, travellers began to discover Greece, and the Roman copies of Greek originals which had hitherto supplied the main knowledge of Classical art were supplanted by the fresher sculptures of 5th-cent. Greece and the virile forms of *Archaic and *Minoan art.

The authority of the antique declined with the onset of *Romanticism, with its stress on self-expression, but its influence has still continued. Drawing from casts of antique sculpture remained a part of most official art training into the 20th cent., and *Picasso, for example, often drew on Classical art as a source of inspiration; in particular, his 'Neoclassical' paintings of the 1920s owed much to visits to the Archaeological Museum in Naples.

Antolínez, José (1635–75). Spanish painter, born and active in Madrid. Like *Murillo in Seville, he had a penchant for paintings of the *Immaculate Conception*, and his colourful, sweet style has much in common with Murillo's. His temperament was anything but sweet, however, for he was renowned for his arrogance and died of wounds received in a duel.

Antonello da Messina (*c.*1430–79). Italian painter, from Messina in Sicily, a pioneer of *oil painting in Italy. According to *Vasari he was a pupil of Jan van *Eyck and brought the 'secret' of oil painting to Italy, but he is most unlikely to have visited northern Europe. He probably acquired his knowledge of northern techniques either in Naples, then artistically dominated by the Netherlands, or by contact with Flemish artists, perhaps in Milan. Early examples from the many years he worked in his home town of Messina (*c.*1455–75) are a *Crucifixion* (Museum, Sibiu, Romania), in which the drama is played out against a background of the local landscape in the Flemish manner, and a *St Jerome* (NG, London).

At the same time, perhaps inspired by the visit to Sicily of the Italian sculptor Francesco *Laurana, he was developing a style based on simplified, rounded, and sculptural modelling. In another *Crucifixion* (NG, London) the composition is Italian but the treatment of light and atmosphere is Flemish. This was painted (1475) just before a visit to Venice (1475/6). The principal work he executed in Venice itself was the S. Cassiano altarpiece, of which two fragments only remain, both in the Kunsthistorisches Museum, Vienna. Its arrangement was perhaps derived from Giovanni *Bellini's lost altarpiece for SS. Giovanni e Paolo; but it was itself influential, for several younger Venetian artists borrowed directly from it and Bellini himself admired the modelling of its figures. Antonello in turn softened his modelling under Bellini's influence, as we see in the *St Sebastian* in the Gemäldegalerie, Dresden.

Antonello's *bust portraits—in three-quarter view, of Flemish type—also enjoyed a notable vogue in Venice: their expressions were more lively than in the portraits by *Memlinc then being imported and, like Antonello's religious works, they show a remarkable ability to combine Northern particularity of detail with the Italian tradition of grandeur and clarity of form. (The example in the National Gallery, London, is sometimes considered a self-portrait.)

Antonio de Comontes. See BORGOÑA.

An Túr Gloine. See PURSER.

Anuszkiewicz, Richard. See OP ART.

Apelles. Greek painter active in the 4th cent. BC, born at Colophon in Asia Minor. Apelles was reckoned in antiquity to be the greatest of Greek painters, but none of his work remains. His technical treatises on painting also are lost and more anecdotes survive about him than useful information. He was court painter to Philip of Macedon (382–336 BC) and his son Alexander the Great, and one of the most famous stories about Apelles tells how Alexander gave him his mistress Pancaspe after the artist had fallen in love with her while painting her in the nude. Apelles was said to excel all other painters in grace and was also agreed in antiquity to have been a master of composition and of *chiaroscuro. The names of about 30 of his works are recorded by ancient sources. Among his subjects were portraits of Alexander the Great (particularly famous was one for the temple of Artemis at Ephesus), *Aphrodite Anadyomene* (Venus rising from the sea) made for the temple of Asclepius at Cos, brought to Rome by Augustus, and set up in the temple of Caesar, and *Calumny*.

Descriptions of his work by classical authors were well known during the *Renaissance and inspired several major artists to attempt to emulate them. *Botticelli made a painting and *Mantegna a drawing of *Calumny* (Uffizi, Florence, and BM, London, respectively), and *Titian did an *Aphrodite Anadyomene* (NG, Edinburgh, on loan from Ellesmere collection). A small but majestic painting of *Zeus Enthroned* in the House of the Vettii at Pompeii may well be a reminiscence of Apelles' picture of Alexander holding a thunderbolt.

Aphrodite of Cnidus. Statue by *Praxiteles, made for the city of Cnidus in Asia Minor. It is now lost, but it was his most famous work in antiquity, and was the ancestress of the modern female nude—the first life-size statue showing the goddess completely naked. Several Roman copies survive (for example, in the Vatican): they show the goddess in a gently twisted pose, with right hand casually masking the pudenda and left hand dropping her robe over an urn. The statue was placed in an open shrine so it could be seen from all four sides, each view being equally admired. *Pliny thought it was the finest statue in the world, and *Lucian wrote of 'the smile playing gently over her parted lips' and of 'the melting gaze of the eyes with their bright and joyous expression'.

Apollinaire, Guillaume (1880–1918). French poet and art critic. Apollinaire was 'the most influential art critic writing in France during the decade before the outbreak of war in 1914 . . . a cardinal figure in creating the artistic climate of Paris early in this century—a climate in which anything and everything was thought possible' (John Golding, *Visions of the Modern*, 1994). He was among the first to acclaim *Picasso (in 1905), praised *Matisse in 1907 and *Braque in 1908, and led the way in doing honour to the talent of Henri *Rousseau. One of the earliest champions of the *Cubist movement, he published *Les peintres cubistes* in 1913, and helped to found the offshoot group *Section d'Or. Apollinaire was also a friend of the *Futurists and composed one of the Futurist Manifestos. He coined the word 'surrealist' in 1917 to describe his own play *Les Mamelles de Tirésias* and through André *Breton he influenced the views of the *Surrealist school. His early death was hastened by wounds sustained fighting in the trenches in the First World War.

Apollo Belvedere. The most famous of the many Classical representations of the Greek god Apollo, discovered towards the end of the 15th cent. and regarded for centuries afterwards as one of the supreme masterpieces of world art and the absolute standard for male beauty. The statue (Vatican) is a marble copy from the Roman period of a *Classical or *Hellenistic Greek bronze. *Leochares has been proposed as the sculptor of the lost original. It was often copied or adapted, sometimes with curious effect, as when *Reynolds painted his *Commodore Keppel* (National Maritime Mus., Greenwich, 1753–4) in the posture of the statue but in 18th-cent. dress. *Winckelmann's rapturous description of the *Apollo Belvedere* enshrined it as one of the models of *Neoclassicism, but following the revelation of the Parthenon sculptures (see ELGIN MARBLES) and the aesthetic discovery of *Archaic Greek sculpture, it fell from grace, seeming cold and academic to many critics. Whereas to Winckelmann it appeared 'the highest ideal of art among all the works of antiquity', to Kenneth *Clark it seemed that 'in no other famous work of art are idea and execution more distressingly divorced'.

Apollodorus. Athenian painter of the 5th cent. BC. *Pliny records the tradition that he was called *Sciagraphus* because he was the first to model his figures in light and shade, a notable step forward in pictorial illusion which was carried still further by *Zeuxis.

Apollonio di Giovanni. See CASSONE.

Apollonius. See BELVEDERE TORSO.

Apoxyomenos. See LYSIPPUS.

Appel, Karel (1921–). Dutch abstract painter, sculptor, graphic artist, and ceramicist, regarded as the most powerful of the post-war generation of Dutch artists. In 1948 he helped to found Reflex, the experimental group of Dutch and Belgian artists from which the *Cobra group sprang. He moved to Paris in 1950 and with the help of the critic Michel Tapié he began during the 1950s to gain an international reputation, spending much of his time in the USA.

In his Cobra period Appel was among the most energetic exponents of expressive abstraction, anticipating *Art Informel. His work, executed in very thick impasto and violent colours, had the restless, agitated character of northern *Expressionism rather than the more restrained classicism of the French: in the words of Herbert *Read, looking at his pictures one has the impression 'of a spiritual tornado that has left these images of its passage'. Latterly he has often painted with a rather smoother facture and in a style somewhat closer to *Hard Edge. But his abstract images retained suggestions of human masks, animal or fantasy figures, which have often been regarded as fraught with terror as well as a childlike naïvety. His works are included in many public collections.

Appiani, Andrea (1754–1817). Italian painter. He was the leading Italian painter of the *Neoclassical period, but more on account of the lack of native competition than because of the quality of his work, which generally looks like a rather tired imitation of *David. He was painter to Napoleon I (in his capacity as King of Italy) and was a favourite of Pope Pius VI (1717–99). Most of his career was spent in his native Milan, where he did large decorative schemes at Sta Maria presso S. Celso (1792–8) and the Palazzo Reale (1808). He also painted numerous portraits which are generally more accomplished than his decorative work.

applied art. Term describing the design or decoration of functional objects so as to make them aesthetically pleasing. It is used in distinction to *fine art, although there is often no clear dividing line between the two.

applied work (or **appliqué**). A form of embroidery in which bold designs are produced by sewing shaped pieces of cloth or other material on to a contrasting ground. The term is sometimes transferred to applied decoration in leatherwork, silverwork, etc.

Apt, Ulrich the Elder (d. 1532). German painter, the best-known member of a family of artists. He worked in Augsburg, where he is recorded from 1481, painting altarpieces, portraits, and in 1516 a series of wall-paintings in Augsburg Town Hall in collaboration with his pupil Jörg *Breu.

aquarelle. The French term for true *watercolour painting in transparent washes of colour, as distinct from the opaque method known as *gouache.

aquatint. A method of *etching producing finely granulated tonal areas rather than lines. The aquatint ground is normally formed by shaking on to the copper plate a dust of finely powdered *resin, which is subsequently fixed to the plate by heating. When the plate is bitten, the acid attacks the surface through the reticulation of the hardened resin, causing the copper to be pitted all over with a texture of little dots which will eventually print as a finely speckled grey tone. The coarseness or fineness of the texture depends on the size and quality of the grains of resin forming the acid resist, while the depth of tone is regulated, as in ordinary etching, by the length of time the plate is exposed to the acid. The artist forms his design and varies his tones by 'stopping out', that is to say by painting over with a varnish those areas of the plate which he wishes to protect from the acid. In this way white-line effects on grey or black are commonly obtained.

Among several variants of this procedure, sugar aquatint, a method dating from the 18th cent. and used by *Gainsborough but later neglected, has been revived in the 20th cent. It enables the artist to overcome to some degree the indirectness of the aquatint medium; for instead of making his design by the negative means of 'stopping out' his whites, he is able to establish his dark areas directly by drawing on the copper plate with a solution of black watercolour and sugar, applied either by pen or by brush. The plate is then varnished and immersed in water. While under water the sugar begins to swell and, lifting off the plate, bursts through the varnish, leaving the bare copper beneath, the result being a copper plate covered with varnish except where the drawing was made with sugar solution. An aquatint ground is next laid over the whole surface and the plate bitten in a slow acid. By this means the drawing alone is etched in an aquatint texture, for the rest of the plate, being protected by the varnish, remains unbitten. Finally both ground and varnish are cleaned off and the plate is printed, showing the design as a dark tone on a white background.

Attempts at printing from a granulated, tonal surface date back to the 17th cent., but a satisfactory system was not evolved until the 18th, by Jean Baptiste Le Prince (1733–81), whose first plates date from 1768. In England, a country always partial to tonal methods of print-making, aquatint soon became popular. The principal English pioneer was Paul *Sandby, who around 1775 invented the spirit-ground method whereby resin dissolved in alcohol is poured over the plate, the evaporation of the spirit leaving the resin to crystallize so as to produce a reticulated effect. Aquatint, both hand-coloured and printed in colour by means of a separate plate for each colour, was much used in England in the late 18th and early 19th cents. for interpreting the luminosity and transparency of the washes in water-colour drawings.

Until modern times few great artists used the medium. The exception was *Goya, whose favourite medium was a combination in varying proportions of etching and aquatint. Goya's expressive use of aquatint was not emulated until late in the 19th cent., when *Degas and Camille *Pissarro made a number of highly original prints in which aquatint was combined with etching and other *intaglio techniques. Between the two world wars the establishment of 'Atelier 17', S. W. *Hayter's workshop of experimental engraving, introduced many artists to the possibilities of aquatint as an element in composite prints of an abstract nature. Several of the greatest masters of the School of *Paris have used sugar aquatint with outstanding success, notably *Picasso, *Rouault, and André *Masson.

arabesque. A scrolling or interlacing plant form, the most typical motif of Islamic ornament. In the late Middle Ages such patterns were in Europe termed 'moresques', but in the 16th cent. the word 'arabesques' came into use when Europeans began to be interested in the art of the Muslim world. But the motif itself is much older, for it is found in *Hellenistic art, notably in Asia Minor. Its Islamic form first appeared *c.* AD 1000 and thereafter it became the stock-in-trade of the Muslim artist, for whom it had a particular appeal since—in theory at least—his religion precluded the representing of living creatures. The term is applied also by extension to the combinations of flowing lines interwoven with flowers and fruit and fanciful figures used by *Renaissance decorators.

Ara Pacis. State monument, the 'Altar of Peace', set up by the first Roman emperor, Augustus, on the Campus Martius at Rome (13–9 BC) to commemorate his victorious return from Spain and Gaul. It consists of an altar on a podium, enclosed by screen walls and approached by a stairway. The lavish sculptural decoration ranks among the finest products of Roman art, and the reliefs representing the ceremonial procession at the dedication of the altar are the first in Western art that can strictly be called documentary, that is, showing identifiable individuals in a contemporary event. Other panels contain allegorical figures, such as Italia flanked by the personifications of the ocean and inland waters, and scenes from Roman mythology and legend. This combination of realistic representation with symbolism and allegory set a tradition which had a long history in Western commemorative sculpture. Following excavations in the 1930s, the *Ara Pacis* has been reconstructed on a covered site slightly to the north of its original situation, where the Palazzo Fiano now stands. The sculptural decoration is substantially complete, although fragments are dispersed in various European museums.

Archaic art. Term applied to Greek art in the period before the *Classical period, roughly from about 650 BC until about 480 BC (the date of the Persian sack of Athens). The Archaic period is marked by the development of the life-size stone statue (the **kouros* and **kore*) and by the change from *black-figure vase-painting to *red-figure. Corinth was the leading centre of early Archaic art, gradually giving way to Athens.

Archaic smile. Conventional smiling expression, often seen in Greek statues of the *Archaic period, especially during the second quarter of the 6th cent. BC. A smile was suggested by drawing the mouth upwards in a clear, flat curve applied to the surface of the face. Various and conflicting theories have been offered to explain the convention, which has been described by John Boardman as an expression of 'strained cheerfulness'. According to one view it may have originated in the technical difficulty of fitting the curved mouth into the block-like form which the early sculptors of **kouroi* gave to the head: certainly as the century progressed the increasing use of chisels led to a disappearance of the block-like character and the archaic smile gave place to a straighter, graver, and more serious or almost sulky expression of the mouth. Other writers have suggested that the smile was intended to express a state of health and well-being. A similar shift from a patterned smile to a more naturalistic gravity may be seen in some *Gothic sculpture, e.g. figures carved early and late in the 13th cent. at Bamberg Cathedral.

archaism. The taste for and imitation of earlier, especially *primitive, styles. Quintilian (1st cent. AD) mentions collectors who preferred Archaic Greek art to that of later periods, but a general interest in the 'purity' and 'simplicity' of early styles began in Europe only in the 18th cent. in conjunction with the Greek and *Gothic Revivals. The archaisms of the German *Nazarenes and the English *Pre-Raphaelites were matched by a renewed interest in the products of the 'childhood of art', works which were then prized for their 'charmingly naïve' character. In modern movements this taste for the naïve and archaic has been swallowed up in a general reassessment of the elemental and the primitive in the history of art.

Archipenko, Alexander (1887–1964). Russian-born sculptor who took American nationality in 1928. In 1908 he moved from Moscow to Paris, where he played an active part in the development of the *Cubist movement. In sculptures such as *Walking Woman* (Denver Art Mus., 1912), he analysed the human figure into geometrical forms and opened parts of it up with holes and concavities to create a contrast of solid and void, issuing in a new idiom in modern sculpture. In 1913 he had a one-man exhibition at the *Sturm gallery and he taught in Berlin between 1921 and 1923, when he settled in the USA. He taught in various places, but principally in New York, where he opened his own school of sculpture in 1939. Archipenko pursued an independent course, and was inventive in technique. He created sculpto-painting, in which forms project from and develop a painted background, and he was a pioneer in the revival of *polychromy in sculpture. In 1924 he invented the *Archipentura*, an attempt to make movable paintings, and from *c.*1946 he experimented with 'light' sculpture, making structures of Plexiglas lit from within. Archipenko had a considerable influence on the course of sculpture both in Europe and in America, particularly in the use of new materials and in pointing a course away from the sculpture of solid form towards one of space and light.

Arcimboldo, Giuseppe (1527–93). Milanese painter famous for his *grotesque symbolical compositions of fruits or animals, landscapes or implements arranged into human forms. He began his career as a designer of stained-glass windows for Milan Cathedral, but from 1562 he worked in Vienna and Prague as court painter to the emperors Ferdinand I (1503–64), Maximilian II (1527–76), and Rudolf II (1552–1612). He returned to Milan in 1587. His paintings, though much imitated, were generally regarded as curiosities in very poor taste until the *Surrealists revived interest in 'visual punning'.

Ardizzone, Edward (1900–79). British artist, best known as an illustrator, particularly of children's books, which he also wrote. He was a war artist in the Second World War.

Arentsz., Arent (called Cabel) (*c.*1585–1635). Dutch painter of winter and summer activities on the sea-shore and canals of Holland, active mainly in and around Amsterdam. His works are sometimes difficult to distinguish from Hendrick *Avercamp's.

Arman (Armand Fernandez) (1928–). French-born artist who became an American citizen in 1972. In 1957, with his friend Yves *Klein, he decided to be known by his first name only, and the form 'Arman' was adopted in 1958 as a result of a printer's error on the cover of a catalogue. He moved to New York in 1963. Arman is best known for his assemblages of junk material, ranging from modest collections to household debris (*Accumulation of Sliced Teapots*, Walker Art Center, Minneapolis, 1964) to towers of crushed automobiles encased in concrete.

armature. A framework or skeleton round which a figure of clay, plaster, or other similar material can be modelled. The term is also applied to the iron framework of *stained-glass windows.

Armitage, Kenneth (1916–). British sculptor. He first exhibited in 1952, and very rapidly established an international reputation. His characteristic work began in the mid-1940s, when he destroyed his pre-war carvings and began to model in plaster round an *armature. Though not naturalistic, his sculpture was figurative and humanistic, capturing typical gesture and attitude, as for example in *People in the Wind* (Tate, London, 1951). At the same time all his work gave great prominence to the qualities of the material—usually at this period bronze—in which the sculptures were cast. From the mid-1950s his figures became more impersonal in character and often larger in scale, and a new phase began in 1969, when he combined sculpture and drawing in figures of wood, plaster, and paper. In the 1970s he returned to bronze and began to explore non-human subject-matter.

Armory Show. An art exhibition (officially entitled the International Exhibition of Modern Art) held in New York, 17 February–15 March 1913, at the Sixty-ninth Regiment Armory. The initiative came from a group of artists, several of them from the circle of Robert *Henri, who in 1911 formed the Association of American Painters and Sculptors to organize the show. The breadth of conception with which the project was envisaged and conceived was largely due to the president, Arthur B. *Davies, whose enthusiasm for presenting a comprehensive picture of current European artistic movements largely overshadowed the original idea of an American exhibition. The Armory Show was both a mammoth exhibition in sheer quantity (estimated at 1,600 works) and a daring presentation of new and still controversial art. It was in effect two exhibitions in one. The American portion gave a cross-section of contemporary US art heavily weighted in favour of the younger and more radical groups. The foreign section, which was the real core of the exhibition and became the focus of controversy, traced the evolution of modern art, showing works by *Goya, *Delacroix, *Courbet, and the *Impressionists and *Post-Impressionists, as well as leading contemporary artists such as *Duchamp and *Kandinsky.

From New York the show went to Chicago (Art Institute) and Boston (Copley Hall). It was estimated that over a quarter of a million visitors paid to see it, and its impact was enormous. There was a good deal of ridicule and indignation (directed particularly at Duchamp's *Nude Descending a Staircase*), but there were also many positive reviews. It created a climate more favourable to experimentation and had a profound effect on many young American artists. Several important patrons and collectors made their first tentative purchases of modern art at the show. It has therefore become a commonplace to speak of the Armory Show as the beginning of an interest in progressive art in the USA.

Arnolfo di Cambio (d. probably 1302, certainly before 1310). Italian sculptor and architect. He is first mentioned in 1265 as Nicola *Pisano's assistant on the pulpit for Siena Cathedral. In 1277 he went to Rome, where he was in the service of Charles of Anjou (1227–85): his portrait of Charles (Capitoline Mus., Rome) is one of the earliest portrait-statues since the ancient world. Also in Rome, he designed and made a tomb, now dismembered, and a bust of Boniface VIII (Grotte Vaticane). His most important surviving tomb, however, is that of Cardinal de Braye (d. 1282), in S. Domenico at Orvieto, which set the type of wall tomb for more than a century. The most famous piece of sculpture attributed to him is the bronze statue of St Peter in St Peter's, Rome.

As an architect, Arnolfo evidently had a great reputation in his day. A document of 1300 praises him as the designer of Florence Cathedral (begun 1294–6), and describes him as 'more famous and skilled in the building of churches than anyone else in the region'. His plans for the cathedral were altered after his death, and the façade he began was dismantled in 1587; his sculptural decoration from the façade is now mainly in the cathedral museum. No other buildings are documented as being by Arnolfo, but several other important Florentine buildings, including Sta Croce and the Palazzo Vecchio, have been attributed to him, notably by *Vasari. If they really are to be credited to Arnolfo, he must rank as one of the greatest architects of the Middle Ages, as well as a distinguished sculptor.

Armstrong, Elizabeth. See NEWLYN SCHOOL.

Arnold, Ann and **Graham.** See BLAKE, PETER.

Arp, Jean (or **Hans**) (1886–1966). French sculptor, painter, and poet who was involved with several of the most important movements in European art in the first half of the 20th cent. He was born in Strasburg, then under German rule, and spoke French and German with equal ease (he wrote poetry in both languages). Before the First World War he came into contact with the *Blaue Reiter group in Munich and participated in their second exhibition (1912), where he met Robert *Delaunay. During the war he met Max *Ernst in Cologne, and was a member of a circle in Paris which included *Modigliani, the poet Max Jacob (1876–1944), *Apollinaire, and *Picasso. In 1915 he met Sophie *Taeuber (whom he married in 1922) at Zurich and collaborated with her in experiments with cut-paper compositions and *collages. He helped to found the *Dada movement and made illustrations for Dada publications 1916–19, during which years he also made his first abstract polychrome *relief carvings in wood (*Dada Relief*, Kunsthaus, Basle, 1916). In 1919–20 he worked with Max Ernst at Cologne and met *Schwitters in Berlin. During the 1920s he settled at Meudon near Paris and was associated with the *Surrealist movement, participating in the first Surrealist exhibition in 1925. He joined *Cercle et Carré in 1930 and was a founder member of *Abstraction-Création in 1931. In the 1930s he turned to sculpture, and produced what are now his most familiar and distinctive works—sensuous abstract pieces that convey a suggestion of organic forms and growth without reproducing actual plant or animal shapes (*Growth*, Guggenheim Mus., New York, 1938). During the 1940s he lived first at Grasse with Sophie Taeuber, Sonia *Delaunay, and Alberto *Magnelli, then in Switzerland. He returned to Meudon in 1946. In his final years he won honours and prestigious public commissions, including a relief for the Unesco building in Paris (1958), but he did not seriously add to his earlier achievements.

Arpino, Cavaliere d'. See CESARI, GIUSEPPE.

Arras tapestries. Tapestries produced at Arras in northern France. They are first mentioned in 1313 and their pre-eminent reputation during the later Middle Ages is attested by the fact that the name of the town passed into several European languages as a generic term for tapestry hangings (Polonius is hiding 'behind the arras' when he is killed by Hamlet). But they were already overshadowed by the productions of Tournai when the capture of Arras and the dispersal of its inhabitants by Louis XI (1423–83) brought the industry to an end there in

1477. Subsequent attempts to re-establish it bore little fruit. A number of tapestries have been ascribed to Arras looms. Those in Tournai Cathedral depicting the lives of SS. Piat and Eleuthère formerly bore the signature of Pierrot Feré of Arras and the date 1402.

Arretine ware. A class of Roman pottery characterized by its red, glossy surface and named after its most important centre of production, Arretium (Arezzo) in Italy. Fragments found in the 13th cent. were very much admired by contemporary artists.

arriccio (or **arricciato**). In *fresco painting, the rough coating of lime and sand plaster to which the final layer of plaster forming the painting surface (the **intonaco*) is applied.

Ars Moriendi ('The Art of Dying'). A popular moral treatise, intended particularly for the spiritual edification of the dying, issued as a *block book *c.*1466. The illustrations, enjoining the dying to save their souls by repenting of their sins and affirming their Christian faith, appeared in both woodcut and engraved form, the latter being in the manner of the *Master E. S. They were a popular source for painters; the deathbed scene in *Bosch's *The Seven Deadly Sins and the Four Last Things* (Prado, Madrid, *c.*1475), for example, derives from the similar scene in the block book.

Art Autre. A name for expressive or non-geometrical abstraction, virtually synonymous with *Art Informel. The expression originated with the critic Michel Tapié, who in 1952 published a book entitled *Un art autre* and in 1960 *Morphologie autre*. In using the phrase 'art autre' (other art), Tapié claimed that post-war art showed a complete break with the past. Following suggestions in *Kandinsky, he argued that expressive, non-geometrical abstract art is a method of discovery and of communicating intuitive awareness of the fundamental nature of reality. He cited among others *Dubuffet, *Mathieu, *Wols, and *Matta as representatives of *Art Autre*.

Art Brut. Term coined by Jean *Dubuffet for the art produced by people outside the established art world, people such as solitaries, the maladjusted, patients in psychiatric hospitals, prisoners, and fringe-dwellers of all kinds. Dubuffet claimed that such art—'springing from pure invention and in no way based, as cultural art constantly is, on chameleon- or parrot-like processes'—is evidence of a power of originality that all people possess but which in most has been stifled by educational training and social constraints. In 1945 he began to make a collection of works free from cultural norms and fashionable or traditional trends in art, and in 1972 his collection, by then numbering more than 5,000 items, was presented to the city of Lausanne, where it was inaugurated at the Château de Beaulieu in 1976. Although nearly half the collection was produced by patients, usually schizophrenics, in psychiatric hospitals, Dubuffet repudiated the concept of psychiatric art, claiming that 'there is no art of the insane any more than there is an art of dyspeptics or an art of people with knee complaints'. He also distinguished Art Brut from *naïve art on the more dubious ground that the naïve or 'primitive' painters remain within the mainstream of painting proper, hoping for public if not official recognition, whereas the *Art Brut* artists create their works for their own use as a kind of private theatre.

Art Deco. The most fashionable style of design and interior decoration in the 1920s and 1930s in Europe and the USA. Art Deco owes something to *Art Nouveau, but its characteristic shapes are geometric or stylized rather than organic. The style takes its name from the Exposition Internationale des Arts Décoratifs et Industriels Modernes, held in Paris in 1925. Initially it was a luxury style, with costly materials such as ivory, jade, and lacquer much in evidence, but partly because of the effects of the Depression it also found expression in materials that could be easily and economically mass produced. Although the term is not often applied to painting or sculpture, the Art Deco style is clearly reflected in the streamlined forms of certain artists of the period, for example the painter Tamara de *Lempicka and the sculptor Paul *Manship.

Arteaga, Sebastián de. See JUÁREZ.

Arte Povera (or **Art Povera**). Term (Italian for 'poor' or 'impoverished art') coined by the Italian art critic Germano Celant to unite certain aspects of *Conceptual, *Minimal, and *Performance art. Celant, who organized an exhibition of Arte Povera at the Museo Civico, Turin, in 1970 and edited a book on the subject (*Arte Povera: conceptual, actual or impossible art?*, 1969), hoped that the use of 'worthless' materials such as soil, twigs, and newspapers and the avoidance of the traditional idea of art as a collectable 'product' would undermine the art world's commercialism. However, dealers have shown that even this kind of art can be commercially exploited. Among the artists whom Celant embraced by the term were Carl *Andre, Joseph *Beuys, and Walter *de Maria.

art for art's sake. See AESTHETICISM.

Arthois, Jacques d' (1613–86). Flemish landscape painter active in Brussels. He specialized in large wooded landscapes, with figures that were often added by other artists, notably *Teniers the Younger. Few dated works exist, the development of his style is not easily followed, and the work of his brother, **Nicolas**, and his son, **Jean-Baptiste**, is sometimes indistinguishable from his. D'Arthois led an unstable life, being imprisoned for debt, and dying in poverty despite his successful career. Paintings from his busy studio were often used to decorate churches; examples are in Brussels Cathedral.

Art Informel. Term coined by the French critic Michel Tapié in his book *Un art autre* (1952) to describe a type of spontaneous abstract painting prominent among European artists in the 1940s and 1950s, roughly equivalent to 'Abstract Expressionism' in the USA. The terms Art Informel and *Art Autre are often used more or less synonymously with *Tachisme, although some critics regard Tachisme as a more clearly defined style, representing one aspect of the broader trend. In English the term 'Informalism' is sometimes used to cover the same style or tendency, but the word 'informel' (which Tapié himself devised) might be translated as 'without form' rather than 'informal'.

art mobilier. Term in common use for small movable works of art. Examples are the figurines, engraved stones, bone carvings, and miscellaneous objects found in caves, rock shelters, and other prehistoric sites, and the art objects of nomadic tribes.

Art Nouveau. Decorative style flourishing in most of western Europe and the USA from about 1890 to the First World War. It was a deliberate attempt to create a new style in reaction from the academic 'historicism' of much 19th-cent. art, its most characteristic theme being the use of sinuous asymmetrical lines based on plant forms. Primarily an art of ornament, it had its most typical manifestations in the practical and applied arts, *art mobilier, *graphic work, and illustration. The style takes its name from a gallery called L'Art Nouveau opened in Paris in 1895 by the art dealer Siegfried (not Samuel) Bing, a leading propagandist for modern design. However, the roots of the style were in England, where the *Arts and Crafts Movement had established a tradition of vitality in the applied arts, and it spread to the Continent chiefly from London. In France, indeed, Art Nouveau is sometimes known by the name 'Modern Style', reflecting its English origin. In Germany the style was called Jugendstil (a name connected with the popular review *Die Jugend* founded in 1896); in Austria it was called Sezessionstil (see SEZESSION); in Italy Stile Liberty after the Regent Street store which had played so large a part in the dissemination of designs; and in Spain Modernista. The style was truly international, the most celebrated exponents ranging from *Beardsley in England to *Mucha, a Czech whose most characteristic work was done in Paris, and *Tiffany in the USA. Leading Art Nouveau architects included Charles Rennie *Mackintosh in Scotland, Antonio Gaudí in Spain, and Victor *Horta in Belgium. Although not primarily associated with painting and sculpture, the influence of Art Nouveau can clearly be seen in these fields, for example in the work of Alfred *Gilbert and Jan *Toorop. The style nowhere survived the outbreak of the First World War to any significant extent.

Art Students League of New York. An art school established in 1875 when that of the *National Academy of Design temporarily closed. Unlike the Academy School (which reopened in 1877), the Art Students League had no entrance requirements and no set course. Its more progressive methods soon attracted many students—at the turn of the century the enrolment stood at nearly a thousand. By this time it was the most important art school in the country. The teachers have included *Chase, *Eakins, *Henri, *Saint-Gaudens, and *Sloan; and the students have included many of the most illustrious names in 20th-cent. American art.

arts, classification of. See APPLIED ART, FINE ARTS, DECORATIVE ARTS, LIBERAL ARTS.

Arts and Crafts Movement. English social and aesthetic movement of the latter half of the 19th cent. that aimed to reassert the importance of craftsmanship in the face of increasing mechanization and mass production. The name derives from the Arts and Crafts Exhibition Society founded in 1888, but the movement had its basis in the ideas of *Pugin and *Ruskin, the most eloquent and influential of the writers who deplored the aesthetic as well as the social effects of industrialization and nostalgically longed for the standards of craftsmanship of the medieval guilds. Ruskin himself may have had some idea of passing from theory to actual social organization and teaching at the time when he planned the Guild of St George in 1871, but it was left to the businesslike genius of William *Morris to translate Ruskin's ideas into practical activity. Morris set about the re-creation of hand industry in a machine age, producing hand-printed, hand-woven, hand-dyed textiles, printed books, wallpaper, furniture, and so forth. Aesthetically his work was highly successful, but his ideal of producing art for the masses failed for the simple reason that his products were necessarily expensive. Nevertheless, his ideas had great influence on craftsmen, teachers, and propagandists such as Walter *Crane, W. R. *Lethaby, and C. R. *Ashbee, bearing fruit in organizations such as the Art Workers' Guild and the Guild of Handicrafts, founded in 1884 and 1888 respectively. In the 1890s the Arts and Crafts Movement was connected with the International style of *Art Nouveau. It spread abroad in the early years of this century, being particularly successful in Germany and the Low Countries, Austria, where it led to the establishment of the *Wiener Werkstätte, and Scandinavia, where it is still influential. In Germany it came to terms with the machine and may be considered an ancestor of the *Bauhaus. The movement died out, or rather was transformed by the acceptance of modern industrial methods. It has, however, left a legacy in the persistence of studio weavers and potters, in organizations such as the Craft Centre, in the manual education both of children and of adults, which continues to be valued, at any rate for therapeutic reasons, and by some who look to it for relief from the prefabricated character of modern civilization.

Arts Council of Great Britain. Independent government body established in 1946 for the purpose of developing greater knowledge, understanding, and practice of the arts. It has a large collection of British art, but this does not have a permanent home, and its main activity in connection with the visual arts consists of the organization and circulation of exhibitions, many of which are accompanied by scholarly catalogues. The Arts Council has two galleries in London used for such exhibitions: the Hayward Gallery, part of the South Bank arts complex alongside the Thames—a large, rather grim concrete building opened in 1968; and the much smaller and prettier Serpentine Gallery in Kensington Gardens, which was originally an Edwardian tea-house and was opened as a gallery in 1970. Many Arts Council exhibitions travel to other venues in Britain.

Arundel, Thomas Howard, 2nd Earl of (1586–1646). English collector and patron of the arts, who gave the primary impetus to the great interest in the arts at Charles I's court. Apart from Charles I, he was the greatest English collecter of his time. His knowledge of art was based partly on extensive travels in Europe, his most important journey being made with Inigo *Jones in 1613–14, when they carried out archaeological investigations in Rome. His agents sought out antiquities that he imported from all over Europe and even the Levant. He also patronized living artists, notably *Rubens and van *Dyck (both of whom painted him), and he brought *Hollar to England. Of the Old Masters, he collected especially works by *Holbein and *Dürer. His great collections were gradually dispersed after his death, but much of his collection of classical sculpture is in the *Ashmolean Museum, Oxford.

Asam, Cosmas Damian (1686–1739) and **Egid Quirin** (1692–1750). Bavarian architects and decorators, brothers. They studied in Rome (1711–14) and developed further the dramatic effects of light and illusionism with which Italian *Baroque artists, notably *Bernini and *Pozzo, had experimented. Both men worked as architects, but Cosmas Damian was also a prolific fresco painter, and Egid Quirin was a sculptor and *stuccoist. They worked best as a team, and their ecclesiastical buildings were the supreme expression of the Bavarian delight in decorative display; architecture, painting, and sculpture unite to set a scene in which light and colour are the chief actors. The best known of their churches is that of St John Nepomuk, Munich (1733–46). The brothers themselves paid for the building (which was attached to Egid Quirin's house), and it is often referred to simply as the 'Asamkirche'.

Ashbee, Charles Robert (1863–1942). English architect and designer, a leading advocate of the principles which inspired the *Arts and Crafts Movement. As well as being the architect of some of the finest small houses of the time (good examples are in Cheyne Walk, London), Ashbee was also a designer of metalwork and jewellery, a poet, and essayist. In 1888 he founded the Guild of Handicraft, which moved from London to Chipping Campden in Gloucestershire in 1902, and in 1898 he founded the Essex House Press, one of the many private presses inspired by William *Morris's *Kelmscott Press. Ashbee published a pamphlet *Should We Stop Teaching Art?* in 1911 and in this expressed a change in outlook that perhaps owed something to his meeting with Frank Lloyd *Wright in 1900. He abandoned his advocacy of the artist-craftsman, and argued that the machine is the vital instrument of contemporary civilization and that it is by the correct use of the machine that the ideals of the Arts and Crafts Movement are to be promoted.

Ash-can School. A term (first used in print in 1934) loosely applied to a group of American painters active from about 1908 until the First World War, in reference to the everyday urban subject-matter they favoured. The painters embraced by the term were inspired largely by Robert *Henri, and the four central figures had been members of the group The *Eight, which he had founded. They were *Glackens, *Luks, *Shinn, and *Sloan. All four had been artist-reporters on the *Philadelphia Press*. At a time when the camera was little used for press work the job of making rapid sketches on the spot for subsequent publication demanded a quick eye and rapid pencil and an exact memory for detail, and encouraged an interest in scenes of everyday life. In style and technique the artists of the Ash-can School are now seen to have differed less from contemporary academic painting than they themselves believed. Although they often painted slum life and outcasts, they were interested more in the picturesque aspects of these subjects than in the social issues they raised. *Bellows and *Hopper are among the other artists associated with the group.

Ashendene Press. A private printing press conducted from 1894 to 1935 by C. H. St John Hornby, first at Ashendene, Herts., and from 1901 at Shelley House, Chelsea. Among the artists connected with the press were Eric *Gill, who designed initial letters, and the illustrators Charles Gere (1869–1957) (*Dante*, 1909) and Gwen Raverat (1885–1957) (*Daphnis and Chloe*, 1933).

Ashmolean Museum, Oxford. The most important of the museums belonging to the University of Oxford. The nucleus of the Ashmolean collection was formed by the 'closet of rarities' (a collection of curiosities rather than works of art) assembled by the traveller and gardener John Tradescant (1608–62) and given in 1659 to the virtuoso (see VIRTU) Elias Ashmole (1617–92), who offered it to Oxford University in 1675. The original Ashmolean Museum was built by the University to the design of the mason and sculptor Thomas Wood (1643/4–95) to house this collection and was opened in 1683—the first public museum in Great Britain. The new building in Beaumont Street designed by Charles Robert Cockerell (1788–1863) was opened in 1845, and this was enlarged by the C. D. E. Fortnum extension in 1894. In 1899 the designation 'Ashmolean Museum' was transferred to the new building and the original museum of Thomas Wood became known as the 'Old Ashmolean Building'. It now houses the Museum of the History of Science.

The collections of the Ashmolean are large and varied. It is particularly rich in works from the ancient world (highlights including material from the excavations of Sir Arthur Evans (1851–1941) at Knossos (see MINOAN ART) and Classical marbles from the collection formed by Thomas *Arundel in the 17th cent.) and in Italian *Renaissance paintings (several of the finest given in 1850 by William Fox-Strangways, who formed his collection whilst on diplomatic service in Italy). Also outstanding are the collections of coins and of Old Master drawings, the latter including a superlative representation of *Michelangelo and *Raphael from the collection of Sir Thomas *Lawrence.

Aspertini, Amico (*c*.1475–1552). Italian *Mannerist painter from Bologna, where he was a pupil of *Francia. *Vasari describes him as having an eccentric personality and this comes out in his paintings, which are often bizarre in expression: *The Holy Family with Saints* (St Nicolas aux Champs, Paris) is described by S. J. Freedberg in *Painting in Italy 1500–1600* (1971) as 'suggesting to the spectator the image of what he would expect from a demented Michelangelo'. Aspertini was in Rome 1500–3 and his sketchbooks of Roman remains (BM, London) are important sources of information about contemporary knowledge of the *antique.

Asselyn, Jan (*c*. 1615–52). Dutch painter, mainly of landscapes. He studied in Italy in the 1630s and 1640s and came to specialize in real and imaginary scenes of the Roman Campagna, his glowing light effects owing much to *Claude Lorraine. His most famous painting, however, is not a landscape, but *The Threatened Swan* (Rijksmuseum, Amsterdam), an unusual work showing a bird defending its nest against a dog that is said to be an allegory of Dutch nationalism. *Rembrandt, who was Asselyn's friend, etched his portrait. Because of a crippled hand he was nicknamed 'Crabbetje' (Little Crab).

assemblage. Term coined in 1953 by Jean *Dubuffet to describe works of art made from fragments of natural or preformed materials, such as household debris. Some critics maintain that the term should apply only to three-dimensional found material and not to *collage, but it is not usually employed with any precision and has been used to embrace *photomontage at one extreme and room *environments at the other. It gained wide currency with an exhibition called 'The Art of Assemblage' staged at the Museum of Modern Art, New York, in 1961.

Ast, Balthasar van der (1593/4–1657). Dutch still-life painter, the brother-in-law of Ambrosius *Bosschaert the Elder, with whom he trained in Middelburgh. He worked in Utrecht before settling in Delft in 1632. His touch was less exquisite than Bosschaert's, but his range was wider, his paintings often including fruit and shells as well as flowers. Jan Davidsz. de *Heem was his pupil in Utrecht.

atelier. French term for an artist's workshop or studio. The term 'ateliers libres' was sometimes applied to the private *académies that became centres for avant-garde art in 19th-cent. Paris.

Atlan, Jean-Michel (1913–60). French painter and poet, born in Algeria, of Jewish-Berber stock, and active mainly in Paris. He turned to abstraction *c.*1947 and developed a characteristic style of rhythmical forms in deep, rich colours painted in a mixture of oils and pastel, often enhanced by thick black outlines. These paintings belong to the category known by the general terms *Abstract Expressionism in America and *Tachisme in France. He himself said: 'For me a picture cannot be the result of a preconceived idea, the part played by chance [*aventure*] is too important and moreover it is this part played by chance which is decisive in creation. At the outset there is a rhythm which tends to develop itself; it is the perception of this rhythm which is fundamental and it is on its development that the vital quality of the work depends.' (*Art d'Aujourd'hui*, 1953.)

atlas (pl. atlantes). A sculpted male figure used in place of a column or other supporting feature in architecture. In Greek mythology, Atlas was the giant who supported the sky, and atlantes are often depicted so as to suggest the strain of carrying a huge weight on the shoulders. The female equivalent—the *caryatid—is, in contrast, usually shown standing serenely upright. 'Telamon' (pl. 'telamones') is an alternative term for atlas.

atmospheric perspective. An alternative term for *aerial perspective.

attributes. Objects habitually associated with a person (real or imaginary) by means of which he or she can be identified when portrayed in art. Saints are often shown with the instruments of their martyrdom or torture—for example, Catherine with her wheel and Lawrence with his gridiron. Other examples are Jove's thunderbolts, the club of Hercules, the scales of Justice, or the anchor of Hope. While some of these were widely used and in many contexts, other attributes were more variable, and in certain periods the invention of esoteric or enigmatic attributes was rife.

attribution. A term in art history and criticism for the assignment to an artist of a work of uncertain authorship. Attributions are usually made either on the evidence of documents or on that of style alone. A painting of unknown authorship may, for example, be found to accord with a description in an inventory where the artist is named, and depending on how closely particularized the description is, it may be a likely assumption that the two are one and the same. Attribution is, however, most usually made on stylistic evidence, and is based on the notion that an artist, consciously or unconsciously, expresses his individuality through his work to such an extent that, to the expert eye, not even his closest contemporary or most talented imitator will be indistinguishable from him. Given a work that is authenticated beyond reasonable doubt by external evidence such as signatures, contracts, or contemporary accounts, we can therefore proceed to group around it works of a similar character and attribute them to the same master.

In the 19th cent. an attempt was made to put attribution on a scientific footing by closely studying small points of detail such as the way a painter represents finger-nails, but although this kind of system (advocated particularly by Giovanni *Morelli) has its uses, it is now felt that we recognize the work of individual artists more by the general effect than by details, and that the details rather than the general effect are what an imitator will be able to reproduce most closely. Attribution, then, is necessarily a highly subjective business, which explains why experts so often disagree and not infrequently change their minds (Bernard *Berenson, the most famous of all connoisseurs, often changed his attributions in the course of his long career). The uncertainty of attribution can have important financial as well as scholarly consequences, even now that the days are gone when the certificate of authenticity of a man such as Berenson could add several noughts to the price of a painting. Moreover, disputes over the authenticity of prominent (which usually means expensive) works of art are one of the few ways in which art becomes a public issue. George de *La Tour's *Fortune Teller* (which many people consider to be a fake) has been the subject of television documentaries on both sides of the Atlantic, and in 1982 its publicity-conscious owner (the Metropolitan Museum of Art in New York) billed it as 'the world's most controversial painting'.

Various terms are used in connection with attribution, very rarely with any precision. Ascription is sometimes used as a synonym for attribution, but some writers prefer to use it to imply a greater degree of doubt (it is often found in the expression 'tentatively ascribed to') or to indicate an old but not firmly accepted attribution. When a work is described as 'autograph' it is thought to be entirely the work of the artist named. The terms 'studio (or workshop) of', 'school of', and 'circle of' all imply that the work was done in more or less close contact with the artist named, but 'follower of' and 'imitator of' may be much later in date. Auction rooms and dealers have for about 200 years used a system to catalogue works whereby the use of an artist's full name indicates that the work in question is 'in our opinion a work by the artist', the use of his surname and initials indicates that the work is from the artist's period and 'may be in whole or part the work of the artist', and the use of his surname alone may imply no more than that the work is in the style of the artist. Thus a painting catalogued simply 'Rubens' may be no more than a 20th-cent. pastiche. In saleroom and other contexts, the term 'after' indicates a copy of a known work of the artist in question.

Aubusson tapestries. Tapestries produced at Aubusson in central France, an ancient centre of the art that was granted the title of Royal Manufactory in 1665. Its greatest period was the second half of the 18th cent. when great quantities of hangings and furniture coverings were woven there with pastoral and *chinoiserie designs by *Oudry, *Boucher, and *Huet. During the 19th cent. skilful reproductions of earlier tapestries were made, and more recently Aubusson played a leading part in weaving the cartoons of such modern tapestry-designers as Jean *Lurçat. Floral-patterned carpets woven by the tapestry process have also been extensively made at Aubusson.

Audran, Claude III (1658–1734). The best-known member of a family of French painters and decorators. He was one of the most prominent decorators of the time, but most of his work has been destroyed or obscured by later additions. A great number of drawings survive, however (many in the National Museum, Stockholm), and these—with their repertory of *arabesques, *grotesques, and trellises—show him to have been one of the leading figures in the creation of the *Rococo style. In 1704 he was appointed curator at the Luxembourg, where *Rubens's *Marie de Médicis* cycle was then housed, and he introduced his pupil *Watteau to this masterpiece by an artist who was to have an enormous influence on his work.

Audubon, John James (1785–1851). American painter-naturalist. Born at Haiti, the illegitimate son of a chambermaid and a French sea captain, he was educated in France and received instruction in drawing from J.-L. *David. He moved to the USA in 1803 to avoid conscription in Napoleon's army and lived as a naturalist, hunter, and taxidermist, also earning some money as a portraitist and drawing master. His combined interests in art and ornithology grew into a plan to make a complete pictorial record of all the bird species of North America. He was unable to find a publisher in America, so he went to England in 1826. His *The Birds of America, from Original Drawings, with 435 Plates Showing 1,065 Figures* was published in four volumes of hand-tinted aquatints (1827–38) and now ranks among the most famous and prized books of the world. The engraver and publisher was the London firm of Robert Havell and Son. It was followed by *The Viviparous Quadrupeds of America* (1845–8), which was completed by his son **John Woodhouse Audubon** (1812–62) after his sight failed in 1846. His other son, **Victor Gifford Audubon** (1809–60), also assisted his father, mainly with the landscape backgrounds of his works. Audubon's plates reveal a romantic enthusiasm for the splendour and beauty of natural life; many of his original drawings are in the New York Historical Society.

Auerbach, Frank (1931–). German-born painter who came to Britain in 1939 and adopted British nationality in 1947. He studied under David *Bomberg, whom he found an inspiring teacher. His work (characteristically nudes and townscapes) is in the *Expressionist vein of Bomberg's late paintings and is remarkable for its use of extreme *impasto, so thick that the paint at times seems modelled rather than brushed. The Tate Gallery has several examples of his work.

Ault, George. See PRECISIONISM.

Auto-destructive art. Term applied to works of art deliberately intended to self-destruct. Works of art not made to endure are not unique to the 20th cent. (witness the butter sculptures of Tibet and the sand paintings of some North American Indian tribes), but the originator of the modern concept of auto-destructive art is the German-born Gustav Metzger (1926–), who is best known for painting with acid on nylon cloth in front of an audience. The most famous of all auto-destructive works was Jean *Tinguely's *Homage to New York*, which blew itself up at the Museum of Modern Art, New York, in 1960, watched by a distinguished audience.

automatism. Method of producing paintings or drawings in which the artist suppresses conscious control over the movements of the hand, allowing the subconscious mind to take over. The idea is anticipated to some extent in Alexander *Cozens's *blot drawings, but automatism in its fully developed form is associated particularly with the *Surrealists and *Abstract Expressionists. With *Action painters such as Jackson *Pollock, the automatic process in principle permeated the whole process of composition, but with the Surrealists, once an interesting image or form or texture had been achieved by automatic or chance means it was often exploited deliberately with fully conscious purpose.

Automatistes, Les. A radical group of seven Montreal abstract painters active *c.*1946–54. The oldest of them, mainly responsible for the formation of the group, was Paul-Émile *Borduas. The others were: Marcel Barbeau (1925–), Roger Fauteaux (1920–), Pierre Gauvreau (1922–), Fernand Leduc (1916–), Jean-Paul Mousseau (1927–), and Jean-Paul *Riopelle. Members of the group were influenced by the *Surrealists, from whom they took over their techniques of *automatism. Their first Montreal exhibition, in 1946, was the first exhibition by a group of abstract painters to be held in Canada. In 1948 they caused outrage by publishing *Refus Global* ('Total Refusal'), an anarchistic manifesto attacking art and various aspects of Canadian life and culture, including the Church.

Automne, Salon d'. See SALON D'AUTOMNE.

Avanzo. See ALTICHIERO.

Aved, Jacques-André-Joseph-Camelot (1702–66). French portrait painter. He was trained in Amsterdam and the influence of Dutch *naturalism can be seen in his work. He achieved something of the simplicity and directness of *Chardin, whose friend he was, and was the chief exponent of a style of portraiture which depicted middle-class sitters in the avocations of daily life. His best-known portrait is perhaps *Madame Crozat* (Musée Fabre, Montpellier, 1741). A portrait of Aved by Chardin is in the Louvre.

Avercamp, Hendrick (1585–1634). Dutch painter, active in Kampen, the most famous exponent of the winter landscape. He was deaf and dumb and known as 'de Stomme van Kampen' (the mute of Kampen). His paintings are colourful and lively, with carefully observed skaters, tobogganers, golfers, and pedestrians, giving a good idea of sport and leisure in the Netherlands around the beginning of the 17th cent. Avercamp's work enjoyed great popularity and he sold his drawings, many of which are tinted with watercolour, as finished pictures to be pasted into the albums of collectors (an outstanding collection is at Windsor Castle). His nephew and pupil **Barent Avercamp** (1612–79) carried on his style in an accomplished manner.

Averlino, Antonio. See FILARETE.

Avery, Milton (1885–1965). American painter, active mainly in New York. Through the 1930s and 1940s he perpetuated in America *Matisse's post-*Fauvist style of figure painting, with flat areas of interacting colour enclosed in flowing outlines. Eliminating detail in search of formal simplicity he painted in thin veils of restrained but scintillating colour. His favourite subjects included landscapes and beach scenes, but some of his late works are so broadly conceived and ethereally painted that at first glance they can be mistaken for abstracts. Avery was the main and practically the only channel through whom Matisse's sophisticated innovations in the decorative use of colour survived in America until new interest was taken in them by younger artists such as *Rothko (his close friend) and *Gottlieb. Rothko in particular acknowledged the debt that he and other abstract painters owed to the 'sheer loveliness' of Avery's work, in which he had 'invented sonorities never seen nor heard before'.

Avignon, School of. School of painting associated with the city of Avignon originating during the period when the papal court was transferred there from Rome (1309–77). The presence of this great source of patronage drew many artists to the city, mainly Italian masters. *Simone Martini of Siena is the best known of these, but *Giotto himself may have gone there. Avignon thus became one of the channels by which Italian trecento art reached France. The centre of artistic activity was the Palace of the Popes, begun *c.*1340. After the departure of the popes Avignon became the centre of a school of painting which amalgamated Italian with northern Flemish influences (see CHARONTON, FROMENT). The greatest single work which the School produced is the *Pietà* from Villeneuve-lès-Avignon (Louvre, Paris, *c.*1460), now generally attributed to Charonton. Many works which were at one time attributed to the School of Avignon have since been reassigned and the School is no longer a clearly defined stylistic entity.

Axis. English quarterly magazine of *abstract art inaugurated in 1935 and edited by Myfanwy Evans in conjunction with her husband John *Piper. The magazine ran until 1937, eight numbers appearing, and was influential in introducing into England a knowledge of contemporary trends in abstract and *Constructivist art. Later issues of the magazine, however, reverted to a *Neo-Romantic nostalgia for the English landscape tradition.

Aycock, Alice. See LAND ART.

Ayrton, Michael (1921–75). British painter, sculptor, theatre designer, book illustrator, writer on art, and broadcaster. His career was often marked by ill health, but he travelled widely and had a long and varied list of works to his credit. He was an erudite, inventive, and highly individual artist, much of whose work revolved around his obsession with the myth of Daedalus and Icarus, which he treated as analogous to his own artistic endeavours. The most extreme expression of his obsession is the enormous maze of brick and stone he built at Arkville in New York State, imitating the labyrinth Daedalus built for King Minos of Crete at Knossos. He wrote several books, including *The Testament of Daedalus* (1962), and illustrated several others. From 1944 to 1946 he was art critic of *The Spectator* (succeeding John *Piper), and at about the same time he became a regular radio broadcaster on art (he also appeared often in the BBC's *Round Britain Quiz*).

azulejo (Arabic, *al-zulaycha*, the tile). The Spanish and Portuguese word for a glazed polychrome tile, usually about 10–15 cm. square. Such tiles were used in Islamic architecture for facing walls and paving floors. They were introduced into Spain by the Arabs, and a thriving export industry grew up in the Middle Ages. Early designs were predominantly geometrical with each tile forming a complete unit in a repetitive pattern, but at the beginning of the 16th cent. a Pisan artist, who signed himself Niculoso Italiano, settled at Seville and began the method of making the tiles into panels with a single pictorial design.

Up to the mid-17th cent. Portuguese *azulejos* differed little from the Spanish in design and used a wide range of colours. By the end of the 17th cent., however, a characteristic Portuguese style emerged, in which blue-on-white tiles were used for large compositions of religious, historical, mythological, or *genre scenes, framed by such favourite *Baroque motifs as garlands, shells, and urns. Throughout the 18th cent. these exuberant tile pictures were applied with lavish profusion to the walls of buildings in both Portugal and Brazil, going out of fashion only in the early 19th cent. The production of *azulejos* in Spain declined after the 17th cent. but the art was transplanted to Mexico, where in the region of Puebla interiors and exteriors were covered with *azulejo* facings in a wide range of brilliant colours, including vermilion, on a scale unequalled elsewhere. In the 19th cent. mass-produced *azulejos* drove out the hand-painted tile and artistic quality suffered. There have, however, been 20th-cent. revivals of the art, among the most successful being the large pictorial *azulejo* designs of the Brazilian painter *Portinari used on the walls of buildings designed by Oscar Niemeyer (1907–).

Baburen, Dirck van (*c.*1595?–1624). Dutch painter of religious works and *genre scenes. After training in Utrecht with *Moreelse, he went (*c.*1612) to Rome, where his style became strongly influenced by *Caravaggio. He returned to the Netherlands in about 1621 and although he died only a few years after this he played a leading role, with *Honthorst and *Terbrugghen, in establishing Utrecht as a stronghold of the Caravaggesque style. His best-known work is *The Procuress* (Mus. of Fine Arts, Boston, 1622). This picture is seen in the background of two paintings by *Vermeer, whose mother-in-law evidently owned it.

Baciccio, Il (Giovanni Battista Gaulli) (1639–1709). Italian painter, born in Genoa and active mainly in Rome, where he settled in 1657 and became a protégé of *Bernini. He achieved success as a painter of altarpieces and portraits (he painted each of the seven popes from Alexander VII to Clement XI), but is remembered mainly for his decorative work and above all for his *Adoration of the Name of Jesus* (1674–9) on the ceiling of the nave of the Gesù. This is one of the supreme masterpieces of *illusionistic decoration, ranking alongside *Pozzo's slightly later ceiling in S. Ignazio. The *stucco figures that are so brilliantly combined with the painted decoration (from the ground it is not always possible to tell which is which) are the work of Bernini's pupil Antonio Raggi (1624–86).

Backer, Jacob Adriaensz. (1608–51). Dutch portrait and history painter, active mainly in Amsterdam. He is said to have studied with *Rembrandt, and he imitated his style so successfully that attributions have sometimes been disputed between them (see BOL). His best-known painting is the beautiful *Portrait of a Boy in Grey* (Mauritshuis, The Hague, 1634). Backer had a prosperous career and at the time of his early death was probably more generally esteemed than Rembrandt. His nephew, **Adriaen Backer** (1630/2–84), also had a successful career as a portraitist in Amsterdam.

Baço, Jaime (called Master Jacomart) (*c.*1413–61). Spanish painter. He was the leading artist of his day in Valencia and also worked for Alfonso V (1396–1458) of Aragon in Naples, but only one documented work by him survives. This is a *polyptych in the church at Cati near Valencia, commissioned in 1460 and representing the martyrs St Lawrence and St Peter. Many attributions are based on the style of this particular work, but the actual execution of it was largely the work of Baço's frequent collaborator Juan Rexach (active 1431–82). The polyptych is very little influenced by the Italian

*Renaissance, remaining faithful to the Hispano-Flemish style.

Bacon, Francis (1909–92). British painter, born in Dublin of English parents. He left home in 1925 at the age of 16 and moved to London, where he worked for a time as an interior decorator. In the late 1920s he lived in Berlin and then Paris (where he was powerfully affected by an exhibition of *Picasso's work), then returned to London in 1929. He had no formal training as a painter; his first drawings and watercolours were done in 1926–7 and his first oils *c.*1929. Bacon took part in two group exhibitions at the Mayor Gallery in 1933, held a one-man show at the Transition Gallery in 1934, and showed in 1937 in the Agnew Gallery 'Young British Painters' exhibition. However, he destroyed much of his early work and dropped out of sight until 1945, when his *Three Studies for Figures at the Base of a Crucifixion* (Tate, London), painted in the previous year, was exhibited at the Lefevre Gallery and made him overnight the most controversial painter in the country. Visitors to the exhibition were shocked by 'images so unrelievedly awful that the mind shut snap at the thought of them. Their anatomy was half-human, half-animal, and they were confined in a low-ceilinged, windowless and oddly proportioned space. They could bite, probe, and suck, and they had very long eel-like necks, but their functioning in other respects was mysterious. Ears and mouths they had, but two at least were sightless. One was unpleasantly bandaged' (John Russell, *Francis Bacon*, 1971). The emotional impact that Bacon's work evokes depends not only on his imagery—characteristically single figures in isolation and despair—but also on his handling of paint, by means of which he smudged and twisted faces and bodies into ill-defined jumbled protuberances suggestive of formless, slug-like creatures of nightmare fantasy. They are presented, typically, against empty space-frames of unreal, dream perspective. Bacon himself said: 'Art is a method of opening up areas of feeling rather than merely an illustration of an object . . . I would like my pictures to look as if a human being had passed between them, like a snail, leaving a trail of the human presence and memory trace of past events as the snail leaves its slime.'

From the 1950s Bacon's work was shown in numerous exhibitions in Europe and the USA and won him an international reputation as one of the giants of contemporary art. In the catalogue of a major retrospective exhibition of his work held at the Tate Gallery in 1985, the Director of the gallery, Alan Bowness, called Bacon the 'greatest living painter', a judgement in which many critics at the time concurred, although others found the unrelenting pessimism of his vision hard to take. Alongside his reputation as a painter he built up a sulphurous personal legend on account of his promiscuous homosexuality, hard drinking, and heavy gambling.

Bacon, John (1740–99). English sculptor. He started work as a modeller in a porcelain factory and this experience left a permanent mark on his style as a sculptor, as his work, even in marble, is soft, often showing much pretty detail. The favour of George III (1738–1820) gained him the important commission for the monument to the Earl of Chatham (the Elder Pitt) (Westminster Abbey, 1779–83), and he also carried out much sculpture at Somerset House, but his finest work is the monument to Thomas Guy (Guy's Hospital, 1779), showing the founder succouring a sick man.

His practice was carried on by his son **John** (1777–1859), who finished some of his father's work such as the equestrian statue of William III in St James's Square, London.

Bacon, Sir Nathaniel (1585–1627). The first English *amateur painter of note. He was a high-born country gentleman, nephew of the Lord Chancellor Francis Bacon (1561–1626). Only a handful of paintings survive (mainly at Gorhambury House, Herts.); with the exception of a miniature landscape on copper in the manner of *Bril, described as the earliest British landscape (Ashmolean, Oxford), they are either portraits or kitchen still lifes. His ambitious full-length self-portrait at Gorhambury displays a more coherent realization of space, greater subtlety of colouring, and greater power of characterization than had hitherto been achieved in English portraiture and shows he was the equal of *Mytens or Cornelius *Johnson, the finest portraitists in England before the arrival of van *Dyck.

'Bad Painting'. See NEO-EXPRESSIONISM.

Baen, Jan de (1633–1702). Dutch portrait painter. He was a pupil of Jacob *Backer, but van *Dyck's works were his main source of inspiration. In 1660 he settled in The Hague and there became the leading portrait painter of the House of Orange. He also worked for Charles II (1630–85) of England and Frederick William I (1620–88), Elector of Brandenberg, but for patriotic reasons he refused a commission from Louis XIV of France in 1672.

Baerze, Jacques de (active end of 14th cent.). Netherlandish wood-carver from Termonde, near Ghent. His only known works are two altarpieces commissioned by Duke Philip the Bold (1342–1404) of Burgundy for the Chartreuse de Champmol, now in the Musée des Beaux-Arts, Dijon. The more elaborate of the two was decorated with painted wings by Melchior *Broederlam between 1394 and 1399. De Baerze's style was more conservative than Broederlam's and his flat compositions, dominated by the richly gilded architectural framework, make a strong contrast with the solidly three-dimensional style of his great contemporary Claus *Sluter.

Baglione, Giovanni (*c.*1566–1644). Italian painter and writer, born and mainly active in Rome. He had a successful career as a painter but is now remembered mainly as the author of *Le vite de' pittori, scultori, ed architetti* . . ., published in 1642. One of the fundamental sources of information for the period covered (1572–1642), this collection of more than 200 biographies deals mainly with Roman artists but also discusses foreign artists working in the city such as *Bril and *Rubens. In 1603 Baglione sued *Caravaggio and three other painters for circulating coarse satirical poems about him. His hatred of Caravaggio comes out in the biography devoted to him in his *Vite*, but ironically Baglione's best pictures come from his brief Caravaggesque phase, *c.*1600–3.

Baillairgé. The leading family of French-Canadian wood-carvers in the later 18th and early 19th cents., active in Quebec City, where several churches have examples of their work. The founder of the family was **Jean Baillairgé** (1726–1805); others were **François** (1759–1830), the outstanding individual, and his son **Thomas** (1791–1859). Their style, with its combination of elegance and provincialism, reflected the *Classicism of the *Louis XVI style with added touches from the English *Georgian. Their technique was fresh and vigorous.

Baily, Edward Hodges (1788–1867). English sculptor, the son of a carver of ships' figureheads at Bristol. He was *Flaxman's favourite pupil and enjoyed a successful career with public sculpture and portraits. His two most conspicuous works are also his most inaccessible—the statues at the top of Nelson's Column in Trafalgar Square (1843) and the almost equally tall Grey's Monument in Newcastle upon Tyne (1837). *Eve at the Fountain* (City Art Gal., Bristol, 1822) was highly regarded by his contemporaries. His prolific output of portrait *busts included many of fellow artists (examples are in the NPG, London).

Bakhuyzen, Ludolf (1631–1708). Dutch marine painter, active mainly in Amsterdam. After the van de *Veldes moved to England in 1672 Bakhuyzen became the most popular marine painter in Holland. He captures the drama and movement of ships, whether in harbour, or in a regatta, or in a high wind on the open sea, but seldom achieves the poetic effects of either van de Velde the Younger or Jan van de *Cappelle.

Bakst, Léon (originally **Lev**) (1866–1925). Russian painter, graphic artist, and stage designer, active for much of his career in Paris. He was a founder member of the *World of Art group in 1898. Originally he made his reputation as a portraitist, but from about 1902 he turned increasingly to stage design and is now remembered above all for his costumes and sets for *Diaghilev's *Ballets Russes*. The uninhibited splendour of his spectacles revolutionized European stage design with their combination of oriental magnificence and the gaudy colour of Russian peasant art.

baldachin (or **baldacchino**). A canopy over an altar or other hallowed object; it may be portable for use in processions or fixed, and in the latter case it may be supported on columns or suspended from the ceiling. The most famous baldacchino is that designed by *Bernini for St Peter's in Rome. The type, with twisted columns and fringed canopy, became popular in *Baroque churches, and when the high altar of St Paul's Cathedral, London, was restored in 1958 it was given a baldacchino of that type, as had been Sir Christopher Wren's (1632–1723) original intention.

Baldinucci, Filippo (1624–96). Florentine painter, art historian, antiquarian, and philologian, who was prominent in artistic circles at the court of the *Medici Dukes. His *Notizie de' professori del disegno* (1681–1728), lives of artists from *Cimabue to his own time, is valuable for the information it gives about 16th-cent. Florentine artists as well as his own contemporaries. He was an innovator as an art historian in making use of every kind of document. Apart from the *Notizie*, his most important work is his biography of *Bernini (1682), the primary source for the artist's life. He also wrote a history of engraving (1686), which contains one of the earliest accounts of *Rembrandt's life.

Baldovinetti, Alesso (*c.*1426–99). Florentine painter, mosaicist, and worker in stained glass. His training is unknown, but his graceful and refined style shows the influence of *Domenico Veneziano and Fra *Angelico. He had a taste for the current repertoire of flowers, fruit, shrubs, brocades, and so on, which resulted in a curious and engaging blend of naïvety and sophistication. His finest works include a damaged but still enchanting fresco of the *Nativity* (1460–2) in the forecourt of the SS. Annunziata, Florence, a *Madonna and Child* (*c.*1460) in the Louvre, and an *Annunciation* (*c.*1460) in the Uffizi. They show his remarkable sensitivity to light and landscape. Attributed works include the *Portrait of a Lady in Yellow* in the National Gallery, London. In his *History of Italian Renaissance Art* (1970), Frederick Hartt writes that Baldovinetti was 'the finest painter in Florence' in the 1460s, and considers him 'a very gifted master who somehow never quite seemed to fulfil his great initial promise'.

Balduccio, Giovanni di (active 1315–49). Pisan sculptor who brought the style of Nicola and Giovanni *Pisano to north Italy. He is first documented working in the cathedral at Pisa, after which he worked in various places in the 1320s before settling in Milan in 1334. In Milan he made his major work, the Shrine of St Peter Martyr (1335–9) in S. Eustorgio, and the Shrine of St Augustine in S. Pietro in Ciel d'Oro, Pavia, is sometimes attributed to him. His style had considerable influence in north Italy.

Baldung Grien, Hans (1484/5–1545). German painter and graphic artist. He probably trained with *Dürer in Nuremberg, but his brilliant colour, expressive use of distortion, and taste for the gruesome bring him closer in spirit to his other great German contemporary, *Grünewald. His output was varied and extensive, including religious works, allegories and mythologies, portraits, designs for stained glass and tapestries, and a large body of graphic work, particularly book illustrations. He was active mainly in Strasburg, but from 1512 to 1517 he lived in Freiburg-im-Breisgau, where he worked on his masterpiece, the high altar for

Freiburg Cathedral, the centre panel of which is a radiant *Coronation of the Virgin*. His most characteristic paintings, however, are fairly small in scale—erotic allegories such as *Death and the Maiden*, a subject he treated several times. Eroticism is often strongly present in his engravings, the best known of which is *The Bewitched Stable Boy* (1544), which has been interpreted as an allegory of lust.

Balen, Hendrick van (1575?–1632). Flemish *Mannerist painter, active mainly in his native Antwerp. He was in Italy in the 1590s. His speciality was mythological scenes painted in the highly finished manner of Jan *Brueghel, one of the numerous artists with whom he collaborated. Van Balen was a popular teacher, his most important pupil being van *Dyck. He also had three painter sons.

Ball, Thomas (1819–1911). American sculptor. He is best known for his equestrian statue of George Washington (1869) in Boston's Public Gardens and *Lincoln as the Emancipator* (Lincoln Park, Washington, 1875), which was paid for with contributions from freed slaves.

Balla, Giacomo (1871–1958). Italian painter and sculptor, active mainly in Rome, one of the leading *Futurist artists. From a visit to Paris in 1900–1 Balla brought back to Italy a feeling for *divisionism and for colour and light, which he passed on to his pupils *Boccioni and *Severini. His early works included landscapes and portraits, but after the turn of the century he became increasingly interested in portraying aspects of modern industrialized life. In 1910 he joined the Futurists and soon became preoccupied with their characteristic aim of portraying movement. Unlike the other Futurists, however, Balla was not interested in machines and violence and his paintings tend towards the lyrical and the witty. The most famous is the delightful *Dynamism of a Dog on a Leash* (Albright-Knox Art Gal., Buffalo, 1912), in which the multiple impressions of the dog's legs and tail convey movement in a manner that later became a cartoon convention. By 1913 he was painting semi-abstract pictures, such as *Mercury Passing in front of the Sun* (private collection, 1914), originating in observations of reality (*Abstract Speed—the Car Has Passed*, Tate, London, 1913). After the First World War, Balla stayed true to the ideals of Futurism after his colleagues had abandoned them, but he turned to a more conventional style in the 1930s. By the end of his long life he was much admired as the last survivor of a brilliant phase of modern Italian art.

Balthus (Count Balthasar Klossowski de Rola) (1908–). French painter (and occasional stage designer), born in Paris of aristocratic Polish parents, both of whom painted. He had no formal training but had the reputation of being an infant prodigy and was encouraged by the family friends *Bonnard and *Derain (of whom Balthus painted a memorable portrait, MOMA, New York, 1936), while the poet Rainer Maria Rilke (1875–1926) also interested himself in his early productions. Since the late 1930s Balthus has been obsessed with the theme of the adolescent girl awakening to sexual consciousness, usually depicted in languid but powerful interiors such as *The Living Room* (Minneapolis Inst. of Arts, 1941–3). From 1961 to 1977 he was Director of the French Academy in Rome. Otherwise he has spent most of his life living in seclusion in France or Switzerland. He works slowly and his output is small, but his highly distinctive, erotically charged images have made him internationally famous, indeed something of a cult figure. His work has appealed to critics as well as the public and he is widely regarded as one of the 20th cent.'s leading upholders of the great tradition of figure painting.

bambocciate. See LAER, PIETER VAN.

Bandinelli, Baccio (1493–1560). Florentine sculptor, painter, and draughtsman. He was a favourite of the *Medici family, but he is remembered more for his unattractive character and the antipathy of his contemporaries than for the quality of his work. His most famous and conspicuous sculpture is *Hercules and Cacus* (Piazza della Signoria, Florence, finished 1534), a pendant to *Michelangelo's *David*. The commission had originally been intended for Michelangelo himself, and Bandinelli's ponderous figure, which he had boasted would surpass *David*, was ridiculed by *Cellini and others. Bandinelli had a habit of failing to fulfil his commissions and Cellini's accusations of incompetence had much justification. In return, Bandinelli attempted to sabotage Cellini's career, as he also did with another rival, *Ammanati. Bandinelli was a fine draughtsman (several of his drawings were engraved) and he played a part in the development of art *academies. His paintings include a pompous self-portrait in the Isabella Stewart Gardner Museum, Boston.

Banks, Thomas (1735–1805). English sculptor. He studied in Rome 1772–9 and in 1781 was employed by Catherine the Great in Russia. Back in England, he became with *Flaxman the leader of the *Neoclassical movement in sculpture. He had a very high reputation with his contemporaries: *Reynolds called him 'the first British sculptor who has produced works of classic grace' and Queen Charlotte (queen consort of George III) is said to have wept when she saw his most famous work, the monument to Penelope Boothby (Ashbourne Church, Derbyshire, 1793), in which the child is shown sleeping rather than dead. Small monuments such as this show Banks at his best, and his few surviving portrait busts demonstrate a gift for characterization. His larger monuments, however, are somewhat ponderous essays in the heroic style (several examples in St Paul's Cathedral, London).

Barbari, Jacopo de' (active *c.*1497, d. 1516?). Venetian painter and engraver. His early career is obscure and he is first documented in 1500 in connection with his huge (nearly 3 m. wide) woodcut view of Venice, which had taken three years to complete. Most of his known career was spent in northern Europe, where he worked for the emperor Maximilian (1459–1519), Frederick the Wise of Saxony (1463–1525), Count Philip of Burgundy, and Margaret of Austria (1480–1530). His delicate engravings, many of mythological figures, helped to spread the Italian conception of the nude in northern Europe. As a painter he is best known for his *Dead Bird* (Alte Pinakothek, Munich, 1504), an early example of still life. He is described as being dead in a document of 1516.

Barbeau, Marcel. See AUTOMATISTES, LES.

Barberini. Tuscan family who came to prominence with the election of **Maffeo** (1568–1644) as Pope Urban VIII in 1623 and became the chief patrons of *Baroque art in 17th-cent. Rome. Urban's favourite artist was *Bernini, from whom he commissioned two great works in St Peter's—the statue of St Longinus and the *baldacchino over the High Altar. Some of the enormous quantity of bronze required for the baldacchino was taken from ancient metal stipped from the Pantheon, giving rise to the adage *Quod non fecerunt barbari fecerunt Barberini* (What the barbarians failed to do, the Barberini did). Urban used most of the metal to cast cannon for the Castel Sant' Angelo, saying it was more important to defend the pope than to keep rain out of the Pantheon. The Barberini also employed Pietro da *Cortona, *Romanelli, Andrea *Sacchi, and others in the decoration of the family palace and of St Peter's. The Palazzo Barberini, in the design of which Bernini and two other great contemporary architects—Carlo Maderno (1556–1629) and Francesco Borromini (1599–1667)—all played parts, now houses part of the national art collection.

Barberini Faun (*Sleeping Satyr*). *Hellenistic marble statue of a satyr sprawled in drunken sleep (Glyptothek, Munich). It has been several times restored, sometimes with substantial changes to its posture. It is first recorded in the possession of Cardinal Francesco *Barberini in 1628, and in the 17th and particularly the 18th cent. it was generally regarded as one of the greatest works of antiquity. It is still an admired work (unlike most once celebrated antique statues) and is considered by some authorities to be an original work of around 200 BC, although others believe it to be a good copy.

Barbieri, Giovanni Francesco. See GUERCINO.

Barbizon School. Group of French landscape painters who took their name from a small village on the outskirts of the Forest of Fontainebleau, where its leader, Théodore *Rousseau, and several of his followers settled in the latter half of the 1840s. The other main members of the group were Charles-François *Daubigny, Narcisse-Virgile *Diaz, Jules Dupré (1811–89), Charles-Emile Jacque (1813–94), and Constant *Troyon. They were united in their attitude of opposition to Classical conventions and by their interest in landscape painting for its own sake, a fairly new development in French art. The inspiration came partly from England (particularly *Constable and *Bonington), where landscape painting had developed earlier, and partly from the 17th-cent. Dutch painters on whom the English tradition was founded. They were advocates of painting direct from nature, but unlike the *Impressionists, they usually painted only studies in the open air; their finished pictures were almost always done in the studio. Their feeling for nature, amounting almost to a cult, may be regarded as a form of *Romantic revolt from the drabness of urban life and coincided with a longing among the urban population in expanding cities to renew the contact with nature. *Corot is often associated with the group, but his work has a poetic and literary quality which sets him somewhat apart. *Millet is also often linked with the School, as he settled in Barbizon in 1849 and during his last period painted pure landscape.

Barendsz., Dirck (1534–92). Netherlandish painter, born and mainly active in Amsterdam. He went to Italy *c.*1555 and worked in *Titian's studio. On his return to the north (*c.*1562) he became one of the leading representatives of the fashionable Italianate style. *The Adoration of the Shepherds* (Municipal Mus., Gouda, *c.*1565) is usually regarded as his finest work.

Bargello. Museum in Florence housing an unrivalled collection of Italian *Renaissance sculpture. The building was originally the Palazzo del Podestà, begun in 1255 as the residence of the chief magistrate of the city. In 1574 it was converted into a prison and assigned to the head of the Police, the 'Bargello'. In 1857–65 it was restored and remodelled as a museum of arts and crafts, now the National Museum. The collection contains works by virtually all the leading Italian Renaissance sculptors (including the celebrated statues of David by *Donatello and *Verrocchio) and also has an extensive collection of minor arts, including armour, *enamels, ivories, *maiolica, medals, seals, and tapestries.

Barker, Thomas (1769–1847). English landscape and *genre painter, called 'Barker of Bath' from his main place of work, where he enjoyed a successful career. He was entirely self-taught, but was clearly influenced by the landscapes and *fancy pictures of *Gainsborough, and made his name as a painter of rustic scenes, some of which were copied on Worcester china plates. His brother **Benjamin** (1776–1838) and his son **Thomas Jones Barker** (1815–82) were also painters.

Barlach, Ernst (1870–1938). German sculptor, graphic artist, and writer. Until he reached his thirties Barlach was as much involved in ceramics as he was in sculpture, but he found his personal style during a visit to southern Russia in 1906, when the sturdy peasant type led him to an interest in medieval German carving, with which he recognized both a spiritual and a technical affinity—even when they are modelled and cast in bronze his figures have the broad planes and sharp edges typical of wood-carving. In 1910 he settled at Güstrow and passed the rest of his life there. Barlach exemplified the sense of man's alienation which was typical of German *Expressionism, believing that, through the creation of visible artistic forms from the 'unknown darkness' within, man can rediscover himself and his lost God. His work has great vigour and expressive power. He executed a number of war memorials under the Weimar Republic, including one at the cathedral of Güstrow which was dismantled when his art was condemned as *degenerate in 1938 by the Nazi regime; it was subsequently restored and a replica made for the Antoniterkirche in Cologne. After his death his studio at Güstrow was made a permanent museum; there is another museum devoted to his work in Hamburg. Barlach also wrote Expressionist plays, which he illustrated with woodcuts and lithographs, and published an autobiography in 1928.

Barlow, Francis (1626?–1704). English painter and illustrator of animal subjects, known as 'the father of British sporting painting'. He painted numerous large canvases for decorative schemes (examples are at Ham House in London and Clandon Park in Surrey) and was also a prolific book illustrator, his etchings for an edition of Aesop's *Fables* published in 1666 being particularly well known. Many of his drawings are in the British Museum. His work is vividly observed, with an almost naïve charm.

Barna da Siena. Sienese painter of the second half of the 14th cent. Nothing is known for certain of his life and no documented works survive, but he was probably the leading Sienese painter of his period. *Ghiberti's *Commentarii* associated him with frescos in the Collegiata at S. Gimignano, and the series on *The Life of Christ* there (*c.*1350–5) is the core around which his work has been reconstructed. As with *Simone Martini (traditionally Barna's master) and Lippo *Memmi, drawing plays a predominant part in his work; where their line is a graceful arabesque, however, Barna's is direct and thrusting. Many of the compositions in S. Gimignano are influenced by *Duccio's *iconography and some are reminiscent of Simone Martini, but his figures have a more dramatic vigour than any previous Sienese painting. Other works given to Barna include a *Crucifixion* in the Episcopal Palace at Arezzo and *Christ Carrying the Cross* in the Frick Collection, New York. Tradition has it that Barna died in a fall from scaffolding.

Barnard, George Grey. See METROPOLITAN MUSEUM OF ART.

Barnes, Dr Albert C. (1872–1951). American drug manufacturer and art collector. He made a fortune with the antiseptic Argyrol, which he created in 1901 (its success depended largely on its being adapted as the standard anti-venereal treatment of the French Army), and by 1913 was devoting his life to collecting. His greatest interest was in modern French painting (his representation of *Matisse being particularly outstanding), but he also bought Old Masters and primitive art. According to Kenneth *Clark he was 'not at all an attractive character. His stories of how he had extracted Cézannes and Renoirs from penniless widows made one's blood run cold.' In 1922 he established the Barnes Foundation at Merion, Pennsylvania, to house his collection and to provide education in art appreciation. Barnes commissioned Matisse to paint a mural decoration for the Foundation in 1931, and when it turned out to be unusable because of an error in the measurements he had been given, Matisse did a new version. The abortive scheme, *The Dance I* (1931–2), is in the Musée d'Art Moderne de la Ville de Paris, and the second scheme, *The Dance II* (1932–3), is *in situ* in the Barnes Foundation.

The museum Barnes created was closed to the public during his lifetime, but after his death legal moves were made to try to force the trustees to open it. An agreement was reached in 1960 allowing restricted public access, but it retained its reputation as a virtually inaccessible treasure house. In 1993–4, however, following court action, a selection of paintings went on tour to Washington, Paris, and Japan.

Barocci (or **Baroccio**), **Federico** (*c.*1535–1612). Italian painter. Barocci was born in Urbino and apart from two trips to Rome early in his career was based there all his life. He is said to have abandoned his frescos in the Casino of Pius IV in the Vatican Gardens (1561–3) for fear that rivals were trying to poison him, and the hypersensitive temperament this suggests comes out in his work. It consists mainly of religious paintings, which combine the influence of *Correggio and *Raphael (also a native of Urbino) in a highly individual and sensitive manner. His colour harmonies are sharp but subtle, and although his paintings often convey a feeling of intimate tenderness, his handling has great vigour. Despite the fact that he worked away from the main centres of art, his work was much sought after, his patrons including the emperor Rudolf II (1552–1612) in Prague. And although Barocci constantly claimed to be ill, he had a long and productive career; he was prolific as a draughtsman as well as a painter and was one of the first artists to make extensive use of coloured *chalks. He also made a few etchings.

Barocci is generally considered the greatest and most individual painter of his time in central Italy; certain features of his work are thoroughly in the *Mannerist tradition (his rather indefinite treatment of space, for example, and his delight in fluttering draperies), but in his directness and freshness he looked forward to the *Baroque. *Bellori, the pre-eminent biographer of the Baroque age, considered him the finest Italian painter of his period and lamented that he had 'languished in Urbino'.

Baroque. A term used in the literature of the arts with both historical and critical meanings and as both an adjective and a noun. The word has a long, complex, and controversial history (it possibly derives from a Portuguese word for a misshapen pearl, and until the late 19th cent. it was used mainly as a synonym for 'absurd' or 'grotesque'), but in English it is now current with three principal meanings. Primarily, it designates the dominant style of European art between *Mannerism and *Rococo, the characteristics of which are discussed below. Secondly, it is used as a general label for the period when this style flourished, broadly speaking, the 17th cent. Hence such phrases as 'the age of Baroque', 'Baroque politics', 'Baroque science', and so on. This usage is probably more confusing than helpful, particularly as, for example, what literary historians call 'Baroque' more often shows characteristics that the art historian would label 'Mannerist'. Thirdly, the term 'Baroque' (often written without the initial capital) is applied to art of any time or place that shows the qualities of vigorous movement and emotional intensity associated with Baroque art in its primary meaning. Much *Hellenistic sculpture could therefore be described as 'baroque'. The older meaning of the word, as a synonym for 'capricious', 'overwrought', or 'florid', still has some currency, but not in serious criticism. The Baroque originated in Italy and the following characteristics are derived chiefly from the style fully developed there by the mid-17th cent. Modifications become necessary when they are applied to other parts of Europe or points in time, and as they amalgamate with other styles and aspirations.

Despite a frequent lack of reticence, the Baroque is a style which expresses a concern for balance and above all wholeness. *Bellori wrote of the fresco by *Lanfranco in the Roman church of S. Andrea della Valle: '. . . this painting has been rightly likened to a full choir, when all the voices together make up the harmony; for then no particular voice is distinguished, but what is pleasing is the blending, and the overall measure and substance of the singing.' The passage nicely underlines the quality of large Baroque design in whatever form. The emphasis is on balance, through the harmony of parts in subordination to the whole. In this the style differs from the High *Renaissance ideal of a balance of parts each of separate perfection, but has more in common with it (and with 17th-cent. *Classicism) than with either the often wilfully unblended style of Mannerism or the nervous fragmentariness of the Rococo (into which the Baroque evolves). This ambition towards a grand unity (which does not, however, preclude a climax within the work) has as an effect the levelling of painting and sculpture to the needs of the architectural unity they serve. The separate properties conventionally supposed to be inherent in each of the three media may be deliberately run together to the end of more complete harmony. Even in a single medium it is not unusual that architecture, for example, should be robbed of its expected inertia and made to take on the suppleness and plasticity of sculpture, or that both should exploit the expressiveness of *chiaroscuro in a way more immediately proper to painting. Yet harmony, once more, precludes that the technical virtuosity characteristic of the

Baroque artist should lapse into display for its own sake.

The unity of the Baroque is more than a formal and self-sufficient one; like a play (and the style had close connections with spectacle) it is incomplete without its audience. It is a mark of a Baroque work that by various means it engineers the bodily, and hence the emotional, participation of the beholder. For painting and sculpture to succeed in doing this at all they must first be persuasive in creating the illusion of the actuality and truth of their subject. The Baroque representation is, then, concerned with the reality of appearances, or at least with verisimilitude. It insists upon the substance, colours, and texture of things; it creates a space in which the subject and the spectator may be joined in a specific and sometimes dramatic moment of time. The physical urgency of the Baroque is achieved most particularly by the calculation of light against dark, mass against void, the use of strong diagonals or curves, breaking down the expected and separating plane. The expression of substance by the use of colour and contrast of light ('painterliness') distinguishes the style from the linear preoccupations of Mannerism and Rococo: its confidence in the force of its assertions further distinguishes it from the insinuating decorativeness of the latter.

Although the once common identification of Baroque as *'le style jésuite'* has been rightly abandoned, it is certainly true that the style flourished chiefly in countries that were strongly affected by the Counter-Reformation, notably Italy, Spain, Austria, and South Germany. It took root much less firmly in northern Europe (and hardly at all in Protestant England and Holland), although *Rubens was the supreme master of Baroque painting and there is no more spectacular example of the Baroque ideal of unity of the arts than Versailles, where architecture, sculpture, painting, the applied arts, and garden design work together to create an overwhelming whole. The archetypal Baroque artist was *Bernini, a man possessed of boundless energy and the utmost virtuosity and whose work reveals total spiritual conviction.

Barr, Alfred H. jun. (1902–81). American art historian and administrator. He became the first Director of the *Museum of Modern Art in New York in 1929 and remained associated with the museum until his retirement in 1967. He not only built up the museum's collections to become the finest in the world in their field, but also organized more than 100 exhibitions and wrote several books that are still considered standard works. Among them are *Cubism and Abstract Art* (1936), *Picasso: Fifty Years of His Art* (1946), and *Matisse: His Art and His Public* (1951).

Barret, George (1732?–84). Irish landscape painter in oils and watercolour. He came to London on the advice of Edmund *Burke in 1762 and quickly made a reputation as a painter of views and *panoramas and was a foundation member of the *Royal Academy. Though he sometimes imitated the Classical manner of his rival Richard *Wilson (who described Barret's foliage as 'spinach and eggs'), his natural gift was for topographical landscape. His most famous work in his day was a panorama of the Lake District painted in *distemper on the walls of a room at Norbury Park in Surrey (*c.*1780; still *in situ*). He initially had a successful career, but his improvident nature forced Burke to come to his rescue at the end of his life by gaining him the sinecure of Painter to Chelsea Hospital. Barret had three painter sons, **George jun.**, **James**, and **Joseph**, and his daughter **Mary** was a *miniaturist.

Barry, James (1741–1806). Irish history painter. Edmund *Burke brought him to London in 1764 and provided money for him to travel to Italy (1766–71). The work of the great masters of the *Renaissance had an overwhelming effect on Barry, and he was the only British painter who adhered consistently to *Reynolds's precepts for history painting in the *Grand Manner. His scheme of large decorative paintings on *The Progress of Human Culture* (1777–83), for the Great Room of the Society of Arts, London, is the most ambitious achievement of this kind by any artist of the British School, but its weak draughtsmanship and flabby rhetoric show that his ambitions far outstripped his talent. Barry was elected Professor of Painting at the *Royal Academy in 1782, but after a series of quarrels was expelled in 1799. His truculent nature, and uncompromising belief in his own genius, come out in his numerous self-portraits (a good example is in the National Gallery of Ireland, Dublin).

Bartlett, Jennifer. See NEW IMAGE PAINTING.

Bartolo, Sebastiano di. See MAINARDI.

Bartolo di Fredi (active 1353–1410). Sienese painter. He continued the narrative style of the *Lorenzetti but gave it a flatter, more decorative character typical of Sienese art at the end of the century. His career was evidently successful, but his work is much less accomplished than that of his contemporary *Barna, whose influence can be seen in Bartolo's major surviving work, the Old Testament fresco cycle in the Collegiata, S. Gimignano (1367). Bartolo's son was the painter **Andrea di Bartolo** (*c.*1370–1428).

Bartolommeo, Fra (Baccio della Porta) (1472/5–1517). Florentine painter. After training with Cosimo *Rosselli, he was deeply influenced by the preaching of Savonarola (1452–98) and entered the Dominican Order in 1500, giving up painting until 1504. From then until 1508 he developed parallel with *Raphael—though Raphael's was the more imaginative genius—each contributing something to the new High *Renaissance type of Madonna with Saints, in which the figure of the Madonna acts not merely as a centre but as a pivot about which the whole composition turns. The two artists also evolved a new treatment, first adumbrated by *Leonardo, of the theme of the *Madonna and Child with the Infant St John in a Landscape*. There is an example by Fra Bartolommeo in the National Gallery, London. Raphael, *Michelangelo, and Leonardo had all left Florence by 1509 and in the second decade of the century Fra Bartolommeo was rivalled only by *Andrea del Sarto as the leading painter in the city, which he left only briefly for visits to Venice in 1508 and Rome in 1514. His style acquired a solemn restraint and monumentality that made him one of the purest representatives of the High Renaissance (*The Mystical Marriage of St Catherine*, Louvre, Paris, 1511). Bartolommeo was a brilliant draughtsman and the mystical element in his nature found clearer expression in his drawings, which escape the tendency to empty rhetoric occasionally shown in his later paintings. His drawings include not only figure studies, but also landscape and nature studies; good examples are in the Courtauld Institute Galleries, London.

Bartolommeo Veneto (active 1502–46). Italian painter. An obscure figure, not mentioned in *Vasari, he is usually classed as a member of the Venetian School, but he signed himself as '*mezo venizian e mezo cremonexe*' (half Venetian and half Cremonese) and seems to have worked in various parts of north Italy. His early religious works are pastiches of Giovanni *Bellini, but later he painted some striking fashionable portraits. There are two examples in the National Gallery, London, including one of Ludovico Martinengo, dated 1546, which is the last known document of his activity.

Bartolozzi, Francesco (1727–1815). Italian engraver, active mainly in England. He was one of the most accomplished engravers of his period, achieving success in Italy, in England, where he became Engraver to George III in 1764, and in Portugal, where he moved in 1802 to become Director of the National Academy in Lisbon. Bartolozzi became a founder member of the *Royal Academy in 1768 and engraved the works of many of his leading contemporaries, such as *Copley, *Kauffmann, and *Reynolds, but he was also celebrated for engravings after the Old Masters. See also STIPPLE ENGRAVING.

Bartsch, Adam von (1757–1821). Austrian engraver. He was custodian of the imperial library in Vienna and an authority on prints. From 1803 to 1821 he published in twenty volumes *Le peintre graveur*, the pioneering work in the systematic study of Dutch, Flemish, German, and Italian painter-engravers from the 15th to the 17th cent. It is now outdated in many respects, but still remains of fundamental importance in the study of certain minor artists and its numbering system has been referred to by most subsequent works in its field. There have been various continuations and supplements and in 1979 a series of illustrations to the work (known as *The Illustrated Bartsch*) began publication; a similar series (*Le peintre graveur illustré*) was begun in 1971, but only the first volume ever appeared.

Barye, Antoine-Louis (1796–1875). French sculptor, celebrated for his portrayal of animals. He laid the basis of his extensive knowledge of animal forms while employed by a goldsmith making models of animals in the Jardin des Plantes, Paris (1823–31). His work was in the spirit of *Romanticism, particularly his preference for rendering violent movement and tense posture. He also did the pediment *Napoleon dominating History and the Arts* on the Pavillon de l'Horloge of the *Louvre and an equestrian statue of Napoleon at Ajaccio, the Emperor's birthplace in Corsica.

Basaldella, Afro. See AFRO.

basalt. A hard and tough igneous stone, which is very durable and takes a fine polish. It is widely distributed and has many varieties differing in consistency and ranging in colour from black, brown or dark green to pale blue. It is almost as difficult to work as *granite. Black basalt was much used by the ancient Egyptians for portrait sculpture and was also used in the Middle East. In the context of the *Neoclassical revival the potter Josiah *Wedgwood experimented with pottery materials simulating the stones favoured for sculpture in antiquity. Basalt-ware, a hard black stoneware, was the first of these materials to be successfully produced, in 1766.

Baschenis, Evaristo (1617–77). Italian painter, the most prominent of a family of artists recorded from 1400. He was ordained *c.*1647 and painted a few religious subjects, but his fame rests chiefly on his beautifully poised and polished still lifes of musical instruments. His predilection for the subject may have been associated with the contemporary fame of the Amati family of violin-makers of Cremona, which is near to Baschenis's native town of Bergamo. The Accademia Carrara there has the best collection of his paintings.

Baselitz, Georg (1938–). German painter and sculptor. He is regarded as one of the leading exponents of *Neo-Expressionism and his work has often been the subject of controversy, particularly since, in 1969, he began painting the images in his pictures upside down. Since 1980 he has also made sculptures; in these the figures are the normal way up.

Basire, James. See BLAKE, WILLIAM.

Baskerville, John (1706–75). British typographer, the originator of the English tradition in fine printing. His editions of the Latin classics, John Milton's poems, his Books of Common Prayer, and above all his folio Bible printed at Cambridge in 1763 are among the masterpieces of British book production. The beauty of his work depends on clear and careful presswork rather than ornament.

Baskin, Leonard (1922–). American sculptor and graphic artist. Baskin's work has been dominated by the themes of death and spiritual decay and he has created images of 'injured and brutalized man, alone, naked, and defenceless', exploding the delusions of optimism and progress. In his woodcuts he developed a distinctive linear style, giving his figures a superficial likeness to anatomical charts, displaying nerves, muscles, and tendons.

Basquiat, Jean-Michel. See GRAFFITI ART.

Bassa, Ferrer (*c.*1285/90–1348). Spanish painter and *miniaturist who worked for the Aragon court. He is considered the founder of the Catalan School, but the only certain surviving work by his hand is a series of *frescos in the chapel of S. Miguel at the convent of Pedralbes near Barcelona, executed in 1345–6 and strongly Italianate in style.

Bassano, Jacopo (Jacopo da Ponte) (*c.*1510/18–92). Italian painter, the most celebrated member of a family of artists who took their name from the small town of Bassano, about 65 km. from Venice, which was the main centre of their activity. Apart from a period in the 1530s when he trained with *Bonifazio Veronese in Venice, Jacopo worked in Bassano all his life. His father, **Francesco the Elder** (*c.*1475–1539), was a village painter and Jacopo always retained something of the peasant artist, even though the influence of, for example, the fashionable etchings of *Parmigianino is evident in his work. He treated biblical themes in the manner of rustic genre scenes, using genuine country types and portraying animals with real interest. In this way he helped to develop the taste for paintings in which the *genre or still-life element assumes greater importance than the ostensible religious subject. From around 1560 his basically country scenes became vested with a more exaggerated search for novel effects of light, taking on something of the iridescent colouring of *Tintoretto. Bassano had four painter sons who continued his style—**Francesco the Younger** (1549–92), **Gerolamo** (1566–1621), **Giovanni Battista** (1553–1613), and **Leandro** (1557–1622). Francesco (who committed suicide by throwing himself out of a window) and Leandro both acquired some distinction and popularity working in Venice—indeed Leandro was knighted by the Doge in 1595 or 1596 (thereafter he sometimes added *Eques* to his signature). The work of the family is well represented in the Museo Civico at Bassano.

Bassen, Bartholomeus van (d. 1652). Dutch architectural painter and architect, active mainly in The Hague. He is best known as a painter of imaginary church interiors and fanciful views of palaces. None of his buildings has survived.

Bastien-Lepage, Jules (1848–84). French painter, best known for his sentimental scenes of rural life such as *The Haymakers* (Musée d'Orsay, Paris, 1877). Émile Zola (1840–1902) described Bastien-Lepage's work as 'Impressionism corrected, sweetened and adapted to the taste of the crowd', and his work had considerable success and influence in spreading a taste for **plein-air* painting, not only in France, but also in England and Scotland, and even—via Tom *Roberts—in Australia. He was also much admired as a portraitist (*Sarah Bernhardt*, Musée Fabre, Montpellier, 1879).

Bath stone. A general name for several limestones found in England which occur in deposits extending from the coast of Dorset through Somerset, Gloucestershire, Oxfordshire, Northamptonshire, to Yorkshire, particularly well developed near Bath. There is considerable variation both in texture and in colour, and it has been used by several noted modern sculptors, for example Frank *Dobson, Eric *Gill, and Henry *Moore.

batik. An art of textile printing, originating in Indonesia. The design is produced by a negative dyeing method, being marked out in wax before the fabric is dipped so that the waxed portions do not take the dye and stand out in the original colour of the fabric. Batik was brought to Europe by Dutch traders and was adopted by Western craftsmen in the 19th cent.

Batlle-Planas, Juan (1911–66). Argentine painter, considered the father of *Surrealist painting in his country. He was an energetic publicist for modern art and trained and influenced numerous Argentinian artists.

Batoni, Pompeo (1708–87). Italian painter, the last great Italian personality in the history of painting at Rome. He carried out prestigious church commissions and painted numerous fine mythological canvases, many for eminent foreign patrons, but he is famous above all as a portraitist. After *Mengs left Rome for Madrid in 1761 his pre-eminence in this field was unchallenged, and he was particularly favoured by foreign visitors making the *Grand Tour, whom he often portrayed in an *antique setting. His style was a polished and learned distillation from the antique, the works of *Raphael, academic French painting, and the teaching of his master, Sebastiano *Conca. His characterization is not profound, but it is usually vivid, and he presented his sitters with dignity and *élan*. Batoni was also an outstanding draughtsman, his drawings after the antique being particularly memorable. He was curator of the papal collections and his house was a social, intellectual, and artistic centre, *Winckelmann being among his friends. Many galleries in Italy, Britain, and elsewhere have examples of his work, and some of his finest portraits are still in the family collections of the sitter.

Battistello. See CARACCIOLO.

Bauchant, André (1873–1958). French *naïve painter. He was a market gardener before the First World War and did not begin painting until 1919, when he was demobilized after war service (in the army he had shown a talent for drawing and had been trained as a map-maker). In 1921 his work was shown at the *Salon d'Automne, and he became one of the best-known painters of his type, promoted by Wilhelm *Uhde and numbering among his admirers *Diaghilev, *Le Corbusier, *Lipchitz, and *Ozenfant. His favourite subjects were history and mythology, for which he found inspiration in old illustrated books (*Greek Dance in a Landscape*, Tate, London, 1937). His style was meticulously detailed.

Baudelaire, Charles (1821–67). French poet and critic. As well as being a major poet, Baudelaire was one of the foremost art critics of his day. He held that there is no absolute and universal beauty but a different beauty for different peoples and cultures. Beauty arises from the emotions, and therefore every man has his personal beauty. Moreover, the individuality of the artist is essential to the creation of beauty and if it is suppressed or regimented, art becomes banal: 'the beautiful is always bizarre' was a favourite maxim. Baudelaire resisted the claims that art should serve social or moral purposes and was one of the leaders of the 'art for art's sake' school with Gustave Flaubert (1821–80), Théophile Gautier (1811–72), and the brothers de *Goncourt (see AESTHETICISM). He sought to assess the stature of an artist by his ability to portray the 'heroism' of modern life. *Delacroix, to whom he devoted some of his most perceptive essays, he found unsuitable owing to his predilection for *Romantic and exotic subject matter. *Courbet seemed to him too materialistic and he finally chose the relatively minor painter Constantin *Guys as the representative *par excellence* of contemporary society, and wrote a long appreciation of his work entitled *Le Peintre de la vie moderne* (1863). He was a friend and supporter of *Manet and he is one of the persons depicted in Manet's *Music in the Tuileries Gardens* (NG, London, 1863) as well as in Courbet's *The Painter's Studio* (Musée d'Orsay, Paris, 1854–5). His writings later had great influence on the *Symbolists.

Baugin, Lubin (*c.*1610–63). French painter, active in Paris. He painted religious works and has earned the nickname 'Le Petit Guide' (Little Guido) because he was strongly influenced by Guido *Reni. This suggests that he visited Italy, but there is no firm evidence. A small group of strikingly austere still lifes, signed simply 'Baugin' (examples are in the Louvre), has also been attributed to him, although there is little in common between these pictures and the religious works.

Bauhaus. A school of art and design founded by Walter *Gropius in Weimar in 1919 and closed by the Nazis in 1933; although it had such a short life it was the most famous art school of the 20th cent., playing key roles in establishing the relationship between design and industrial techniques and in breaking down the hierarchy that had previously divided 'fine' from 'applied' arts. The Bauhaus was created when Gropius was appointed head of two art schools in Weimar in 1919 and united them in one; they were the Kunstgewerbeschule (Arts and Crafts School) and the Hochschule für Bildende Kunst (Institute of Fine Arts). He gave his new school the name Staatliches Bauhaus in Weimar (Weimar State 'Building House'), coining himself the word 'Bauhaus' (an inversion of 'Hausbau'—

house construction). His prospectus formulated three main aims for the school: first, to unite the arts so that painters, sculptors, and craftsmen could in future embark on co-operative projects, combining all their skills harmoniously; secondly, to raise the status of the crafts to that enjoyed by the fine arts; and thirdly, to establish 'constant contact with the leaders of the crafts and industries of the country' (an important factor if the school were to survive in a country that was in economic chaos after the war).

All students had to take a six-month 'Preliminary Course' (*Vorkurs*) in which they studied the principles of form and colour, were acquainted with various materials, and were encouraged to develop their creativity. After that they moved on to workshop training in the field of their choice. Gropius brought together a remarkable collection of teachers at the Bauhaus. The first head of the preliminary course was Johannes *Itten, and when he left in 1923 he was succeeded by László *Moholy-Nagy, who replaced Itten's rather metaphysical approach with an austerely rational one. The other teachers included some illustrious painters, most notably *Kandinsky and *Klee. Several students went on to become teachers at the school, notably Josef *Albers. In 1924 the Bauhaus moved to Dessau, where it was housed in a group of new buildings designed by Gropius. The school had been involved in architectural commissions from the beginning, but it was only in 1927 that an architectural department was established, with the Swiss architect Hannes Meyer (1889–1954) as its first professor. When Gropius resigned in 1928 to devote himself to his own practice he named Meyer as his successor. It was an unpopular choice with the staff, as Meyer was a Marxist and instituted a sociological approach that changed the whole tone of the school, with politics occupying an important place in the curriculum. In 1930 Meyer was forced to resign and was replaced by Ludwig Mies van der Rohe (1886–1969), one of the greatest architects of the 20th cent. Mies tried to rid the Bauhaus of its political associations and thereby make it a less easy target for its right-wing opponents, but in 1932 the Dessau parliament closed the school. In an attempt to keep it alive Mies rented a disused factory in Berlin and reopened the Bauhaus there as a private enterprise, but it was closed by the Nazis in April 1933, soon after Hitler assumed power. In its last few years the Bauhaus was dominated by architecture, but it produced a great range of goods, with many of them (furniture, textiles, and electric light fittings in particular) being adopted for large-scale manufacture. They were highly varied in appearance, but the style that is generally thought typical of the Bauhaus was severe, geometric, and undecorated. The Bauhaus published a journal (*Bauhaus*, 1926–31) and a series of books, and its ideas were spread also by the emigration of many of its teachers before and during the Second World War. It has had an enormous influence on art education in the Western world and on visual creativity in general: 'The look of the modern environment is unthinkable without it. It left an indelible mark on activities as varied as photography and newspaper design . . . [and] achieved a language of design liberated from the historicism of the previous hundred years' (Frank Whitford, *Bauhaus*, 1984). After the Second World War Dessau became part of East Germany and the Bauhaus buildings were left derelict. In 1976 the school was faithfully restored for the 50th anniversary of its opening in Dessau, and after the reunification of Germany in 1990 it was reopened as a design institution.

Baumeister, Willi (1889–1955). German abstract painter. Unlike most significant German painters of his time, he stood outside the ambit of *Expressionism and is regarded as the most 'European' in spirit of his contemporaries. Between 1911 and 1914 he had several stays in Paris (sometimes in company with his close friend Oskar *Schlemmer) and his early work was influenced by *Cubism. After military service in the First World War, he began to develop a personal style in a series (1919–23) of *Mauerbilder* (wall paintings), so called because he added sand, putty, etc., to his pigments to give a textured effect. In the mid-1920s his work became more figurative in a manner recalling *Léger and the *Purists (he met Léger, *Le Corbusier, and *Ozenfant when he revisited Paris in 1924), and his work met with considerable acclaim in France. In 1928 he was appointed Professor of Typography at the Städel School in Frankfurt, but in 1933 he was dismissed by the Nazis, who declared his work *degenerate. From then until the end of the Second World War he worked in obscurity in Stuttgart. During this time his painting became freer, with suggestions of primitive imagery, creating a kind of abstract *Surrealism. His interest in imagery from the subconscious was described in his book *Das Unbekannte in der Kunst* (The Unknown in Art), written in 1943–4 and published in 1947. After the war Baumeister became a hero to a younger generation of German abstract artists. From 1946 until his death he was a professor at the Stuttgart Academy.

Bawden, Edward (1903–89). English watercolour painter, illustrator, and designer of posters, wallpaper, tapestries, and theatre décor. He was an official war artist in the Second World War and painted numerous murals (for example, at Queen's University, Belfast, 1965), but he was best known for his book illustrations.

Baxter, George (1804–67). English engraver and printer. In 1835 he patented a method of making colour prints using oil colours, and around the middle of the century 'Baxter prints' enjoyed a great vogue, making coloured reproductions of paintings widely available. Baxter licensed other firms to employ the process and did not himself profit greatly from the invention, which was eventually supplanted by cheap coloured *lithographs.

Bayer, Herbert (1900–85). Austrian-American graphic artist, designer, and painter. After studying architecture in Linz, 1919, and painting under *Kandinsky at the *Bauhaus, Weimar, he taught typography in the Bauhaus, Dessau, 1925–8. From then until 1937 he ran an advertising business in Berlin and did exhibition planning and typography while continuing to paint. After his first one-man exhibition, at the Gallerie Povolotzki, Paris, in 1929, he took up *photomontage of a *Surrealist kind. He emigrated to New York in 1938 and worked as a typographer, architect, and designer, notably of the Museum of Modern Art's Bauhaus exhibition in 1938. His work as a designer of book jackets, magazine covers, and posters has been much acclaimed and influential.

Bayes, Gilbert (1872–1953). British sculptor. A leading exponent of the *New Sculpture, he concentrated on romantic themes from such sources as medieval chivalry and Wagner, and his work was much appreciated by conservative critics. His other works include the relief of sporting figures (1934) outside Lord's cricket ground. His brother **Walter** (1869–1956) was a painter, a member of the *Camden Town Group.

Bayeux Tapestry. The most famous of all pieces of needlework (technically an embroidery not a tapestry), depicting the Norman Conquest of England in 1066 and the events that led up to it. For centuries it hung in the cathedral at Bayeux in Normandy and it is now in the Musée de la Tapisserie de la Reine Mathilde there. Worked in two kinds of woollen thread on coarse linen, it is 50 cm. wide and, although incomplete, 70.3 m. long. In seventy-nine consecutive scenes, whose purport is elucidated by Latin inscriptions, the tapestry depicts the story of Harold, Earl of Wessex, during the years 1065–6, culminating in the Battle of Hastings. Neither its date nor its place of origin is beyond doubt, but there is evidence to suggest that it was embroidered within a few years of the Conquest, perhaps to the order of William the Conqueror's half-brother, Bishop Odo of Bayeux. It was rediscovered by the French archaeologists M. Lancelot and Dom Bernard de Montfaucon early in the 18th cent. and engravings were first published by the latter in *Monuments de la monarchie française* (1729–30). Stylistically the Tapestry has much in common with English *illumination of the period. The drawing is clear, vivid, and full of action and the composition leads on skilfully from one scene to the next. The colours, of which there are eight (three shades of blue, a bright and a dark green, buff yellow, red, and grey), are used for decorative rather than descriptive purposes.

Bayeu y Subías, Francisco (1734–95). Spanish painter. He had an eminently successful career, becoming court painter to Charles IV (1748–1819) and director of the Academy of San Fernando, but he is now remembered almost solely because he was the brother-in-law of *Goya, who spent some time in Bayeu's studio in the 1760s and later painted a memorable portrait of him (Prado, Madrid, 1795). Bayeu painted portraits and also did much decorative work, particularly making *cartoons for the royal tapestry factory, where he succeeded *Mengs as director in 1777.

Bazille, Frédéric (1841–70). French painter, one of the early *Impressionist group. As a student in *Gleyre's studio in Paris (1862) he befriended *Monet, *Renoir, and *Sisley, with whom he painted out of doors at Fontainebleau and in Normandy. He was, however, primarily a figure painter rather than a landscapist, his best-known work being the large *Family Reunion* (Musée d'Orsay, Paris, 1867–8). Another group picture, *Mon Atelier* (Musée d'Orsay, Paris), showing *Manet, Monet, Renoir, Edmond Maître, and Émile Zola (1840–1902) in the artist's studio with Bazille himself painted by Manet, was refused by the *Salon in 1870. Bazille was killed in action during the Franco-Prussian War, cutting short a promising career. He came from a wealthy family and had given generous financial support to Monet and Renoir.

Baziotes, William (1912–63). American painter, one of the minor masters of *Abstract Expressionism. From 1936 to 1941 he worked for the *Federal Art Project, and during the war he was attracted to *Surrealism and experimented with various types of *automatism. In 1948–9 he collaborated with *Motherwell, *Newman, and *Rothko in running an art school called 'The Subject of the Artist' in

New York, and from this grew a meeting centre of avant-garde artists known as The Club. In the early 1950s he attained his characteristic style, which was not fully abstract but used strange *biomorphic shapes, akin to those of *Miró, suggesting animal or plant forms in an underwater setting (*Mammoth*, Tate, London, 1957). He said, 'It is the mysterious that I love in painting. It is the stillness and the silence. I want my pictures to take effect very slowly, to obsess and to haunt.'

Beale, Mrs Mary (1633–99). English portrait painter and copyist. Her portraits, often of clergymen, are dull derivations from her friend *Lely, but the diaries of her painting activities kept by her husband Charles, an artist's colourman, afford an interesting picture of contemporary artistic practice. Examples of her prolific output are in the National Portrait Gallery, London. A son, **Charles** (b. 1660), was mainly a *miniaturist.

Beardsley, Aubrey (1872–98). English illustrator, a leading figure in the *fin-de-siècle* *Aestheticism. Beardsley had little formal training, but he read voraciously and studied the art of the past and present, and his highly distinctive style was based on sources including *Burne-Jones (who encouraged him) and Japanese prints (see UKIYO-E). He made a name for himself in 1893 when his illustrations for an edition of Malory's *Morte d'Arthur* were praised in an article by the American illustrator Joseph Pennell (1857–1926) in the first number of *The Studio*. In the following year he became the rage with the publication of his illustrations to the English version of Oscar Wilde's *Salome* and the appearance of the first issue of *The Yellow Book*, a quarterly periodical of which he was art editor. His masterly use of black and white for pattern and *grotesque was highly original and the ornamental quality of his linear rhythms expressed the spirit of *Art Nouveau. Owing perhaps partly to the tuberculosis which carried him off at the age of 25, his work had a morbid suggestion of depravity which made it the most controversial illustration of its day. Some of his work was frankly pornographic, including illustrations for his own *Story of Venus and Tannhauser* (he wrote poetry as well as prose), which was privately published in an unexpurgated edition in 1907. In spite of his ill health and early death, Beardsley's output was prodigious: other well-known works include his illustrations to Pope's *Rape of the Lock* (1896) and for a periodical called *The Savoy*.

Beaumont, Sir George Howland (1753–1827). English collector, connoisseur, and amateur painter, the friend of numerous artists and men of letters. He was a devotee of the *picturesque and *Romantic and his most celebrated work, *Peel Castle in a Storm* (Leicestershire Museum and Art Gallery, Leicester), moved Wordsworth to a sonnet. Beaumont had much to do with the foundation of the *National Gallery, to which he presented the best part of his collection in 1826. His favourite painting was *Claude's *Hagar and the Angel* (NG, London), which he frequently took with him when he travelled and which had a great impact on his friend *Constable. Constable's famous painting *The Cenotaph* (NG, London) shows the memorial to Sir Joshua *Reynolds that Beaumont had erected in the grounds of his house at Coleorton, Leicestershire; it was the last picture that Constable ever exhibited at the *Royal Academy (1836).

Beauneveu, André (active 1360–1403/13). French sculptor and *illuminator from Valenciennes, who worked for the French court, Louis de Mâle, Count of Flanders, and the Duc de Berry. Four of the royal effigies in Saint-Denis came from his workshop (Philip VI; John II, ordered 1364; and Charles V and his queen, both ordered 1367). Of his illuminations the only certain attributions are the Prophets and Apostles of the Duc de Berry's *Psalter* (Bib. nat., Paris, 1380–5). Beauneveu's style in sculpture and painting looked forward to the general northern European trend towards naturalism in the 15th cent.

Beaux, Cecilia (1855–1942). American painter, active mainly in her native Philadelphia and in New York. She was a highly successful and much honoured society portraitist, working in a dashing and fluid style similar to that of *Sargent. Her sitters included celebrities such as Henry James and Theodore Roosevelt. She wrote an autobiography, *Background with Figures*, published in 1930.

Beaux-Arts, École Nationale Supérieure des. The chief of the official art schools of France. The origins of the École des Beaux-Arts go back to 1648, the foundation date of the Académie Royale de Peinture et de Sculpture (see ACADEMY), but it was not established as a separate institution until 1795, during the administrative reforms of the French Revolution. It controlled the path to traditional success with its awards and state commissions, notably the prestigious *Prix de Rome, and teaching remained conservative until after the Second World War. Entry was difficult—among the artists who failed were *Rodin and *Vuillard—and students often preferred the private academies. Many progressive artists, however, obtained a sound technical grounding there—*Degas,

*Manet, *Matisse, *Monet, and *Renoir all attended classes. The École, which is housed in a complex of early 19th-cent. buildings in Paris, has a large and varied collection of works of art. Many of them are primarily of historical interest (including a vast number of copies, portraits of teachers, and paintings that won the Prix de Rome), but the collection of drawings is of high quality. The École also often hosts temporary exhibitions.

Beccafumi, Domenico (*c.*1486–1551). Italian painter, generally considered the outstanding Sienese *Mannerist. He was alive to the developments of Fra *Bartolommeo, *Michelangelo, and *Raphael and combined the new ideas with the bright and decorative colouring of the Sienese tradition. His work is noteworthy for its sense of fantasy and striking effects of light, as in *The Birth of the Virgin* (*c.*1543), one of the several outstanding examples of his work in the Pinacoteca, Siena. Most of his best work is in Siena; he painted decorations for the Town Hall (1529–35) depicting examples of civic virtue, and made designs for the marble pavement of the cathedral.

Becerra, Gaspar (*c.*1520–68/70). Spanish *Mannerist sculptor and painter. He studied in Rome and was one of *Vasari's assistants in his decoration of the Cancelleria Palace there. Soon after 1556 he returned to Spain and in 1558 contracted for the main *reredos of Astorga Cathedral, his most important work as a sculptor. In 1563 he was appointed court painter to Philip II (1527–98) and began mythological ceiling paintings at the palace of El Pardo, near Madrid. His up-to-date knowledge of Italian art gave him a high contemporary reputation.

Becker, Felix. See THIEME.

Beckford, William (1760–1844). English collector, writer, and eccentric. A pampered millionaire from boyhood (at the age of 5 he had piano lessons from the 8-year-old Mozart), he became a legendary figure in his own lifetime. One of his cousins referred to him as 'a second Lucifer' (a reference to both his youthful beauty and his depravity—it was rumoured he dabbled in black magic), and he left England after a homosexual scandal involving a 13-year-old boy in 1784. For the next decade he travelled widely on the Continent and after his return to England he lived in eccentric seclusion at Fonthill in Wiltshire, where the architect James Wyatt (1746–1813) built for him Fonthill Abbey (1796–1807, now destroyed), a huge mansion dressed in ecclesiastical garb and the most extravagant monument of the blossoming *Gothic Revival. He formed an excellent library and a vast collection of objects of every kind, both natural and artificial; it drew from William *Hazlitt the wry comment that 'the only proof of taste he has shown in this collection is his getting rid of it', but it included some outstanding paintings (twenty of them are now in the National Gallery, London). In 1826 when his fortunes had declined Beckford built Lansdowne Tower, Bath, a lesser but still highly impressive *classical folly which now enlivens a cemetery. Beckford's most famous literary work was the fantastic oriental tale *Vathek*, written in French, but published first in English in 1786, a successor to Horace *Walpole's *The Castle of Otranto* in the vogue for the Gothic novel. The nightmarish visions of Beckford's book were inspired partly, as he himself said, by *Piranesi's engravings.

Beckmann, Max (1884–1950). German painter and graphic artist, one of the most powerful and individual of *Expressionist artists. Early in his career Beckmann painted in a conservative, more or less *Impressionist style, with which he made a good living, but his experiences as a medical orderly in the First World War completely changed his outlook and his style. His work became full of horrifying imagery, and his forms expressively distorted in a manner that reflected the influence of German *Gothic art. The combination of brutal realism and social criticism in his work led him to be classified for a time with the artists of the *Neue Sachlichkeit, with whom he exhibited at the Kunsthalle, Mannheim, in 1925, but Beckmann differed from such artists as *Dix and *Grosz in his concern for allegory and symbolism. His paintings were intended as depictions of lust, sadism, cruelty, etc., rather than illustrations of specific instances of those qualities at work, and he ceased to regard painting as a purely aesthetic matter, and thought of it as an ethical necessity.

In 1933 Beckmann was dismissed by the Nazis from his professorship at the Städelsches Kunstinstitut in Frankfurt, and in that year he began *Departure* (MOMA, New York), the first of a series of nine great *triptychs painted between then and his death in which he expressed his philosophy of life and society, and his horror at man's cruelty. He moved to the Netherlands in 1937 (the year in which his work was included in the infamous Nazi exhibition of *Degenerate Art), settling in Amsterdam until 1947. The last three years of his life were spent in the USA, where he taught in Washington and New York. Beckmann's philosophical outlook, which he expressed in *My Theory of Painting* (a lecture delivered in London in 1937 and published in 1941), is somewhat incoherent, but his work has been hailed as the most authentic comment of German culture on the disorientation of the modern world. Apart from his allegorical figure compositions, Beckmann is best known for his portraits, particularly his self-portraits, in which he charted his spiritual experiences.

Beechey, Sir William (1753–1839). English portrait painter. Beechey's style changed little throughout his successful career. He was a careful if sometimes insipid painter and gave great attention to the durability of his pictures. He was appointed Portrait Painter to Queen Charlotte (queen consort of George III) in 1793 and was knighted in 1798 in recognition of his most ambitious painting, *A Review of the Horse Guard with King George III and the Prince of Wales* (Royal Collection).

Beeck, Jan van der. See TORRENTIUS.

Beerbohm, Sir Max (1872–1956). English critic, essayist, and *caricaturist. In both words and drawings he had a brilliantly ironic wit. His drawings first appeared in *The Strand Magazine* in 1892 and he was reproduced in *The Yellow Book* (see BEARDSLEY) of 1894. He became a member of the *New English Art Club in 1909 but after 1910 lived mainly at Rapallo in Italy. He published and exhibited several sets of caricatures, including *Rossetti and His Circle* (1922). His best-known literary work is his only completed novel, *Zuleika Dobson* (1911).

Beerstraten, Jan Abrahamsz. (1622–66). Dutch painter, the best-known member of a family of artists active in Amsterdam. He specialized in topographical scenes, sometimes rearranging parts of the view. Often they are winter scenes (*The Castle of Muiden in Winter*, NG, London, 1658).

Beert, Osias (*c.*1580–1624). Flemish painter of still life and flower pieces. Beert became a master in Antwerp in 1602 and also carried on business as a cork merchant. He is specially noted for his paintings of oysters, where the colours and textures of these and other crustaceans are perfectly captured in simply arranged compositions.

Beggarstaff Brothers. Pseudonym used by the brothers-in-law William *Nicholson and James *Pryde for their poster designs. 'They joined forces in 1894, and for the next five years they produced a series of posters which by their bold simplicity and clarity of design revolutionized certain aspects of poster art throughout Europe . . . they presented the image in its starkest form: the background is stripped bare of unnecessary detail and the fullest use is made of the silhouette . . . Despite the brilliant originality of their work, or perhaps because of it, they received relatively few commissions and several of their designs never reached the hoardings' (Dennis Farr, *English Art: 1870–1940*, 1978).

Beham, Hans Sebald (1500–50) and **Bartel** (1502–40). German engravers, brothers. They were expelled from their native city of Nuremberg in 1525 for their extreme Protestant views. Hans settled in Frankfurt and Bartel worked for Duke William IV (duke 1508–50) of Bavaria. Both brothers produced a great number of illustrations to the Bible, mythology, and history, strongly influenced by *Dürer. Bartel was also a painter, primarily of portraits.

Behnes, William (1795–1864). English sculptor, the son of a German piano manufacturer who had settled in London. After training at the *Royal Academy Schools, he quickly obtained a large practice as a maker of busts, the best of which rival those of Sir Francis *Chantrey, and was appointed Sculptor in Ordinary to Queen Victoria on her accession in 1837. He also made monuments and statues (*Sir Henry Havelock*, Trafalgar Square, London, 1861). His work was uneven in quality and despite his great vogue his extravagance led him to bankruptcy and he died in the Middlesex Hospital after being found lying in the gutter. His brother Henry (d. 1837), who took the name Henry Burlowe, was also a sculptor.

Behrens, Peter (1868–1940). German architect and designer. Beginning as a painter, he came to architecture by way of an interest in the practical arts. In 1907 he joined the firm of AEG (General Electricity Company) and designed for it everything from factories to stationery. His design was in the functionalist tradition and he set himself against the *Art Nouveau principle of applying to machine production forms derived from handicrafts. He has been credited with originating the form of design specialization which is now known as 'Industrial Design' (see INDUSTRIAL ART).

Bell, Clive (1881–1964). British critic and writer on art. With Roger *Fry he was largely instrumental in propagating in Great Britain an appreciation of the *Post-Impressionist painters and particularly *Cézanne. Bell chose the British section of Fry's second Post-Impressionist exhibition (1912), including work by his wife Vanessa *Bell, Fry himself, and Duncan *Grant among *Bloomsbury Group artists, with Spencer *Gore and Wyndham *Lewis representing the more radical wing. His aesthetic ideas were set forth in his book *Art* (1914) and were much concerned with his theory of 'significant form'. He invented this term to denote 'the quality that distinguishes works of art from all other classes of objects'—a quality never found in nature but common to all works of art and existing independently of representational or symbolic content. The book is not now taken seriously as philosophy, and it contains some absurd statements ('The bulk of those who flourished between the high Renaissance and the contemporary movement may be divided into two classes, virtuosi and dunces'); however, it is written with fervour, and there can be no doubt that it was important in spreading an attitude that demanded greater rigour of attention to the sensory and formal qualities of a work of art.

Quentin Bell (1910–), son of Clive and Vanessa Bell, is a painter, sculptor, potter, university professor, and author, probably best known for his writings on art, mainly on the Victorian period and the Bloomsbury Group. Quentin's son **Julian Bell** (1952–) is a painter and writer on art.

Bell, Graham (1910–43). British painter and art critic, born in South Africa. He came to England in 1931 and was a pupil of Duncan *Grant. In the early 1930s he painted abstracts, but from 1934 to 1937 abandoned painting for journalism. When he returned to painting it was in the soberly naturalistic style associated with the *Euston Road School. His work included portraits, landscapes, interiors, and still lifes. He volunteered for war service in 1939 and was killed on a RAF training flight.

Bell, Larry (1939–). American painter and sculptor. Since 1964 he has confined himself to making glass sculptures of refined simplicity, his work representing an extreme application of the tendency among the younger Los Angeles artists to dematerialize the environment and to work with the pure qualities of light. His virtually invisible glass panels were the most dramatically novel exhibit at the exhibition 'Spaces' organized at the Museum of Modern Art, New York, in 1970.

Bell, Vanessa (1879–1961). British painter and designer. She married Clive *Bell in 1907 and like him and her sister, Virginia Woolf (1882–1941), was a central figure of the *Bloomsbury Group. Her early work, up to about 1910, and her paintings produced after the First World War are tasteful and fairly conventional, in the tradition of the *New English Art Club, but in the intervening years she was briefly in the vanguard of progressive ideas in British art. At this time, stimulated by the *Post-Impressionist exhibitions of Roger *Fry (with whom she had an affair), she worked with bright colours and bold designs and by 1914 was painting completely abstract pictures. Her designs for Fry's *Omega Workshops included a folding screen (V & A, London, 1913–14) clearly showing the influence of *Matisse. From 1916—while remaining on good terms with her husband—she lived with Duncan *Grant. Their home, Charleston Farmhouse, at Firle, Sussex, which has much painted decoration by the couple, has been restored as a Bloomsbury memorial and is open to the public.

Bella, Stefano della (1610–64). Italian engraver. His delicate, mannered style was early formed on that of *Callot and remained close to it. Most of his career was spent in his native Florence, where he worked chiefly for the Grand Duke of Tuscany, but he spent a period in Paris (1639–49), where he was patronized by Richelieu, and in 1647 visited the Netherlands, where he admired *Rembrandt's etchings. His output as an engraver was enormous—masque-designs, battle-pieces, animals, landscapes—and he was also a prolific draughtsman. Many of his drawings are in the Royal Library at Windsor.

Bellange, Jacques (active 1600–17). French painter, etcher, and decorator, active in the Duchy of Lorraine. His reputation now rests on his etchings and drawings, as all his decorative work and almost all his paintings have disappeared. His highly individual style represents a last stage of the development of *Mannerist art in Europe. Exaggerating the tradition initiated by *Parmigianino, he expressed a personal religious mysticism through the artificial conventions of aristocratic elegance.

Bellechose, Henri (d. 1440/4). Netherlandish painter from Brabant, who succeeded *Malouel as court painter to the Duke of Burgundy at Dijon in 1415. He is documented in 1416 as completing an altarpiece left unfinished by Malouel, which can be identified as *The Martyrdom of St Denis* in the Louvre. This painting appears to be the work of two hands, and the more naturalistic, less elegant parts are given to Bellechose.

Bellegambe, Jean (*c.*1470/80–*c.*1535). Netherlandish painter of altarpieces and designer of buildings, furniture, frames, and embroidery. He lived in Douai, where he was the only artist of consequence during his lifetime. Douai, today in northern France, was then part of the Spanish Netherlands. It was neither typically Flemish nor typically French, and the same is true of Bellegambe's work. But the rich *Renaissance architectural settings that are such a distinctive feature of it indicate an Antwerp rather than a Paris source. The *Adoration of the Trinity* (Douai Museum) is typical.

Bellini. Family of Venetian painters who played a dominant role in the art of their city for three-quarters of a century.

Jacopo (*c*.1400–70/1) was the father of Gentile and Giovanni and father-in-law of *Mantegna. He was a pupil of *Gentile da Fabriano, with whom he worked in Florence, and he achieved early popularity both in Venice and elsewhere. By the middle of the century he had a flourishing studio with his two sons. His most notable paintings have disappeared, however, and it is not easy to form an assessment from those that survive—mainly fairly simple and traditional representations of the Madonna and Child. Although he has the grace of the late *Gothic, there is a certain dryness and stiffness in his figures. Yet he was obviously keenly alert to contemporary ideas and shared the fashionable interests in archaeology, *perspective, and anatomy. His artistic personality is manifested best in his two surviving sketchbooks (Louvre, Paris, and BM, London), containing more than 230 drawings in all. There are interesting experiments in unusual perspective and developments of open spatial composition in landscape and architectural design.

Gentile (1429?–1507) inherited his father's sketchbooks and took over as head of his studio, so he was presumably the elder son. He carried on the reputation of his father and was greatly admired in his time, but many of his major works have perished. They included erotic scenes painted for the harem of Sultan Mehmet II, when Gentile worked at the court of Constantinople in 1479–81; his portrait of Mehmet, however, survives (albeit much restored) in the National Gallery, London. The most famous of his extant works are probably the *Procession of the Relic of the True Cross* (1496) and the *Miracle at Ponte di Lorenzo* (1500), two huge canvases crowded with anecdotal detail of contemporary Venetian life. Both are in the Accademia, Venice.

Giovanni (called Giambellino) (*c*.1430–1516) was the greatest artist of the family and during his long and prolific career transformed Venice from an artistically provincial city into a *Renaissance centre rivalling Florence and Rome in importance. He was trained by his father Jacopo, but the major influence on his formative years was that of his brother-in-law, Mantegna. This and Bellini's own originality are made clear by a comparison of their pictures of the *Agony in the Garden*, both painted about 1460 and both now in the National Gallery, London. The compositions are clearly related, both deriving from a drawing in one of Jacopo's sketchbooks, but there is great difference in treatment, particularly of the landscape. Mantegna's is sharp, precise, and analytical, Bellini's is lyrical and atmospheric.

To Mantegna's influence was later added that of *Antonello da Messina, who was in Venice in 1475–6. Like him, Bellini became one of the great early masters of the *oil technique, the linear style of his early work mellowing into one of masterly breadth, initiating the characteristically Venetian conception of painting in which colour and light were the primary means of expression. Bellini was remarkably inventive *iconographically and to the end of his long life he continued to learn from new ideas. From the year before his death dates the dreamy *Woman with a Mirror* (Kunsthistorisches Museum, Vienna), which is strongly influenced by his pupil *Giorgione, who predeceased him, and it is easy to credit the remark made by *Dürer on his visit to Venice in 1505–7 that although Bellini was 'very old' he was still 'the best painter'. He painted excellent portraits, of which the *Doge Leonardo Loredan* (NG, London, *c*.1501) is the best-known example, and a few mythologies and allegories, notably the *Feast of the Gods* (NG, Washington, 1514, altered after his death by *Titian), but he was above all a religious painter. His most characteristic subject was the Madonna and Child, and only *Raphael has rivalled his treatment of the theme, which ranges from the wistful melancholy of *The Madonna of the Meadow* (NG, London, *c*.1510), one of the most marvellous examples of his ability to bring together figures and landscape in perfect harmony, to the monumentality of the San Zaccaria Altarpiece (S. Zaccaria, Venice, 1505), perhaps the grandest of all *sacre conversazioni*. His influence on Venetian painting was enormous. He was appointed official painter to the republic in 1483, and from about that time almost all the painters who became eminent in Venice during the next generation (including Giorgione, *Sebastiano del Piombo, and Titian) are either known to or believed to have trained in his workshop.

Bellmer, Hans (1902–75). Polish-French graphic artist, painter, sculptor, and writer, all of whose work is explicitly erotic. In 1922–4 he studied engineering in Berlin, but he gave up the course after becoming friendly with *Dix and *Grosz and began working as a typographer and bookbinder, then as a draughtsman in an advertising agency. In 1933 he constructed an articulated plaster figure of a young girl, inspired partly by an infatuation with his 15-year-old cousin Ursula. He photographed his creation in various attitudes and states of dismemberment (sometimes partly clothed) and published a collection of the photographs as *Die Puppe* (The Doll) in Karlsruhe in 1934; a French edition, *La Poupée*, was published in Paris in 1936. Bellmer sent samples of the photographs to André *Breton in Paris, and the *Surrealists were highly excited by these striking images of 'vice and enchantment'. In

1938, in danger of arrest by the Nazis, Bellmer fled to Paris to join the Surrealists. He was interned at the beginning of the war (with Max *Ernst), then lived in the South of France, 1942–6, before returning to Paris, where he began a long series of drawings and etchings that developed the violent eroticism of his dolls. Bellmer also produced paintings and sculpture in a similar vein. His work includes some of the acknowledged masterpieces of erotic art, but it was not well known until a large retrospective in 1971–2 at the Centre National d'Art Contemporain, Paris.

Bellori, Giovanni Pietro (1615–96). Italian biographer, art theorist, antiquarian, and collector. His most important work—a basic source for the history of the *Baroque period—is *Vite de' pittori, scultori et architetti moderni* (1672), which he dedicated to *Colbert, the founder of the French *Academy, and in the preparation of which he was helped by his friend *Poussin. In contrast to former biographers, his method was to concentrate on artists selected for their importance and only these received comprehensive treatment. The Preface to the work was a lecture given in 1664 to the Academy of St Luke at Rome, which became the seminal statement of the *Classical theory of an art which mirrors the *ideal essence of reality. In the prominence he gave to *Raphael, Annibale *Carracci, and Poussin, his rational Platonism, and his acceptance of the *antique as the model of excellence, his formulation expressed the ideals of the Roman Academy and proved a decisive influence on French academic theory. It later became the theoretical basis of the *Neoclassicism preached by *Winckelmann.

Bellotto, Bernardo (1720–80). Italian painter, nephew, pupil, and assistant of *Canaletto in Venice. Bellotto left Italy for good in 1747, to spend the rest of his life working at various European courts, notably Dresden and Warsaw, where he died. He called himself Canaletto, and this caused confusion (perhaps deliberate) between his work and his uncle's, particularly in views of Venice. Bellotto's style, however, is distinguished from his uncle's by an almost Dutch interest in massed clouds, cast shadows, and rich foliage. His colouring is also generally more sombre, much of his work being characterized by a steely grey. The best collections of his work are in Dresden (Gemäldegalerie) and Warsaw (National Museum). In the re-building of Warsaw after the Second World War his meticulous but picturesque views of the streets and churches were used as guides, even in the reconstruction of architectural ornament.

Bellows, George Wesley (1882–1925). American painter and lithographer. He was a pupil of Robert *Henri and became associated with the *Ash-can School. An outstanding athlete in his youth and noted for his hearty, outgoing character, Bellows is best known for his boxing scenes. The most famous of them is *A Stag at Sharkey's* (Cleveland Museum of Art, 1907), remarkable for its vivid sense of movement and energetic, sketchy brushwork. Such works rapidly won him a reputation, and in 1909—aged 27—he became the youngest person ever elected an associate member of the *National Academy of Design. He took a highly active part in the art life of his day and was one of the organizers of the *Armory Show in 1913. After this, his work tended to become less concerned with movement, placing more emphasis on formal balance. He was a man of strong social conscience, and his work included scenes of the urban poor—of which the most famous is the crowded tenement scene *Cliff Dwellers* (Los Angeles County Museum of Art, 1913)—and a series of paintings and lithographs about First World War atrocities. He did not take up lithography until 1916, but in the nine remaining years of his life he produced almost 200 prints, and he is accorded a high place among modern American printmakers. In the last five years of his life Bellows turned to landscapes and portraits and was considered one of the finest American portraitists of his day. His early death was caused by a ruptured appendix.

Belvedere Torso. A marble torso of a powerful male figure seated on a rock, now in the Vatican Museums and named after the Belvedere court in the Vatican in which it was once displayed. It is signed by a Greek sculptor 'Apollonius, son of Nestor, Athenian', about whom nothing is known, and there is scholarly dispute as to whether it is an original *Hellenistic work or a Roman copy. (It is sometimes alleged that the signature of Apollonius occurs also on the famous and stylistically similar bronze figure of a seated boxer in the Terme Museum in Rome, but most authorities now consider that this is a mirage.) The date of the discovery of the torso is uncertain, but it is first mentioned in the 1430s. It had become well known by 1500 and had a profound influence on *Michelangelo among other *Renaissance artists. From then until the early 19th cent. it was widely regarded as one of the greatest works of art in the world, rivalled in status probably only by the **Apollo Belvedere* and the **Laocoön* among ancient sculptures, although its fame was generally more academic than popular. It was often referred to simply as 'the Torso'. Unusually, the figure has always been left unrestored, but various artists have attempted to reconstruct the statue, notably *Flaxman, who made of it a group as *Hercules and Hebe* (University College, London, on loan to V&A, 1792). Between 1798 and 1815 the *Belvedere Torso* was in Paris, one of the many antique statues taken there by Napoleon.

bench end. The traditional name in Christian church architecture for the upright end part of what are now called 'pews'. Seating for the congregation in the naves of medieval churches became general only towards the end of the 14th and during the 15th cents. This was one of the symptoms of the growing importance of the laity in church life; pews were provided for their convenience and their decoration reflects secular rather than ecclesiastical taste. The subject-matter of the carvings of bench ends derives from popular piety and fables and generally verges on *folk art, although the quality of the carving is often extremely high. In England there are two regions where they are particularly common: East Anglia and the West Country, where the prosperous middle classes were especially strong.

Benedetto da Maiano (1442–97). Florentine sculptor, who carried over into the second half of the 15th cent. many of the motifs and stylistic features characteristic of the first half. His marble tomb designs are variants on patterns established by his master, Antonio *Rossellino; his pictorial *relief style, which found its most eloquent expression in a pulpit executed between 1472 and 1475 in Sta Croce, Florence, belongs to the narrative tradition which is associated with *Ghiberti and *Donatello. Perhaps his most memorable achievement lay not in his figures or reliefs but in the decorative architectural settings in which they were placed. In the design and execution of the exquisite pilasters, *capitals, friezes, niches, and so on which form these settings he was often assisted by his brothers **Giovanni** (1438–78) and **Giuliano** (1432–90), who was primarily an architect. Benedetto's other work includes two outstanding portrait busts—of Pietro Mellini (Bargello, Florence, 1474) and of Filippo Strozzi (Louvre, Paris, *c*.1490).

Benesch, Otto (1896–1964). Austrian art historian, director of the *Albertina in Vienna from 1947 to 1961. An industrious and painstaking scholar, he is remembered chiefly for his complete illustrated catalogue of *Rembrandt's drawings (6 vols., 1954–7, 2nd edn., 1973). His other publications include *The Art of the Renaissance in Northern Europe* (1947).

Bénézit, Emmanuel (1854–1920). French art historian, editor of the *Dictionnaire des peintres, dessinateurs, sculpteurs et graveurs de tous les temps et de tous les pays*, first published in three volumes in 1911–23 and most recently in a ten-volume edition in 1976. Apart from the *Lexikon* of *Thieme-Becker, it is the largest and most comprehensive dictionary of artists' biographies in current use.

Bening (or **Benig**). The name of two Netherlandish book *illuminators, father and son. **Sanders** (sometimes called Alexander) (d. 1519) worked in Ghent and Bruges. No documented works by him are known, but attempts have been made to identify him with the *Master of Mary of Burgundy, which would give him an artistic status appropriate to the success he seems to have enjoyed. **Simon** (1483/4–1561) worked in Bruges and represents one of the final sparks of the tradition of illumination as the manuscript was overtaken by the printed book. He was probably one of the artists who worked on the famous *Grimani *Breviary* (Biblioteca, Marciana, Venice). There is a self-portrait *miniature by Simon in the Victoria and Albert Museum, London. His daughter **Levina Teerlinc** (d. 1576) was also a miniaturist, active at the English court.

Benois, Alexandre (1870–1960). Russian painter, stage designer, art historian, and critic, a leader and spokesman of the *World of Art group. He was a close friend and collaborator of *Diaghilev, both in Russia and later in Paris, and he is best known for his stage designs for the *Ballets Russes*, in which he harmonized the tradition of Russian folk art with French *Rococo elements, one of the most notable examples being Stravinsky's *Petrushka* (1911). Following a difference of opinion with Diaghilev he worked at Stanislavsky's Moscow Arts Theatre, 1912–14. After the Revolution he was appointed curator of paintings at the *Hermitage in Leningrad. He held this post from 1918 to 1925, then settled permanently in Paris. His writings include several volumes of memoirs and books on art, including ones translated into English as *The Russian School of Painting* (1916), *Reminiscences of the Russian Ballet* (1941), and *Memoirs* (1960). Benois's son, **Nikolai** (1901–), was a stage designer at La Scala, Milan; his niece, **Nadia Benois** (1896–1975), settled in London in 1920 and worked as a ballet designer. The actor and writer Peter Ustinov is her son.

Benozzo Gozzoli. See GOZZOLI.

Benson, Ambrosius (d. 1550). Netherlandish painter of religious works and portraits. He was born in Italy, but from 1519 worked in Bruges, where he continued the tradition of Gerard *David. Many of his pictures were done for the export trade to Italy and Spain, and he evidently had a flourishing business. There is a slightly southern flavour to his compositions and for a long time many of them were thought to be by an anonymous Spanish painter working under Flemish influence, known as the Master of Segovia.

Benson, Frank W. See TEN, THE.

Benton, Thomas Hart (1889–1975). American painter, the great-nephew of a famous American statesman of the same name. He studied at the *Académie Julian, 1908–11, and in Paris became a friend of the *Synchromist Stanton *Macdonald-Wright. After his return to the USA (he settled in New York in 1912) he continued painting in the Synchromist manner for some years, but having failed to win success working in an avant-garde style, he abandoned modernism around 1920 and gained fame as one of the leading exponents of *Regionalism. His style became richly coloured and vigorous, with restlessly energetic rhythms and rather flat, sometimes almost cartoonish figures. His work included several murals, notably scenes of American life (1930–1) at the New School for Social Research in New York. In 1935 he left New York to become director of the City Art Institute and School of Design in Kansas City, Missouri, and he lived in that city for the rest of his life. When Regionalism declined in popularity in the 1940s Benton turned more to depicting scenes from American history, and some of his later work introduced American types into representations of Greek myths or biblical stories. Benton wrote two autobiographies, *An Artist in America* (1937) and *An American in Art* (1969). A passage from the second shows how completely he turned his back on the modernism he had espoused in his youth: 'Modern art became, especially in its American derivations, a simple smearing and pouring of material, good for nothing but to release neurotic tensions. Here finally it became like a bowel movement or a vomiting spell.' In view of these words, it is ironic that Benton was influential on Jackson *Pollock, whom he taught at the *Art Students League of New York in the early 1930s.

Bentveughels. See SCHILDERSBENT.

Bérain, Jean (1640–1711). French decorator and designer. He was chief designer at the court of Louis XIV (1638–1715) from 1682 until his death and was perhaps the most important artistic force in designing the elaborate stage settings, costumes, and displays for the celebrations, out-of-door fêtes, ballet-operas, etc., given by the king. His *arabesques and *grotesques inaugurated a new movement which led towards the *Rococo of the following century.

Bérard, Christian (1902–49). French painter and designer born in Paris. He studied painting at the Académie Ranson under *Vuillard. Under the influence of Jean *Cocteau, whose portrait Bérard painted (MOMA, New York, 1928), he later turned to designing for theatre and film and the majority of his work was of this kind. His inventiveness and his gift of fantasy ensured his success in this field.

Berchem, Nicolaes (1620–83). Dutch painter of pastoral landscapes in the Italianate manner, principally active in Haarlem. Berchem was the son of the fine still-life painter Pieter *Claesz., and it is not known why he adopted a different surname. Claesz. was his first teacher, but although Berchem tried his hand at most subjects, no still lifes by him are known. He visited Italy in the 1640s and perhaps again in the 1650s and became, with Jan *Both, the most highly regarded exponent of the Italianate landscape. Successful and well rewarded in his lifetime, he had numerous pupils and his influence on 18th-cent. English and French landscape painters was considerable, *Gainsborough and *Watteau being among the artists who particularly admired his work. He became a great favourite with English collectors and his prolific output is as well represented in British galleries as it is in those of the Netherlands.

Berckheyde, Gerrit Adriaensz. (1638–98). Dutch painter of architectural views, active in Haarlem (his native city), Amsterdam, and The Hague. His representations of those cities have documentary accuracy, but they are never dry, achieving a poetic harmony by a subtle use of light and shade (three of his views of Haarlem are in the NG, London). The work of Gerrit's elder brother, **Job Adriaensz.** (1630–93), is very similar; it is also rarer and more varied. Job painted *genre and biblical scenes as well as vistas of town squares and streets.

Berenson, Bernard (1865–1959). American art historian, critic, and connoisseur. He was born in Lithuania and educated in Boston and at Harvard University (his family emigrated to the USA when he was 10), but he spent most of his long life in Italy, first visiting the country in 1888 and settling permanently near Florence in 1899. He built up a formidable reputation as an authority on Italian *Renaissance painting and was associated with several prominent dealers and collectors, notably Lord *Duveen and Isabella Stewart *Gardner, advising them on purchases. The fortune he earned in the picture trade has caused his impartiality to be questioned, and many of his attributions have been downgraded, but his lists of the work of Renaissance painters formed a basis for further work for many years. His most enduring work of scholarship is *The Drawings of the Florentine Painters* (1903, 2nd edn. 1938, 3rd edn.—in Italian—1961). He amassed a huge library of books and photographs and a fine art collection at his villa I Tatti at Settignano near Florence, which he left to Harvard University. It is now administered as the Harvard Center for Italian Renaissance Studies. Dapper and polylingual, Berenson often played host to visiting art historians and intellectuals at I Tatti and was a renowned conversationalist, diarist, and bon vivant. See also TACTILE VALUES.

Berg, Claus (before 1485–after 1532). German wood-carver, active mainly at Odense in Denmark, where he ran a large workshop. He worked in a vigorous and individual late *Gothic style, excelling in figures of dramatic pathos. The huge altarpiece at Odense church is among his most impressive works.

Berger, John. See KITCHEN SINK SCHOOL.

Berghe, Frits van den (1883–1939). Belgian painter. He began working in an *Impressionist style, but after a stay in the Netherlands (1914–22), where he often collaborated with de *Smet and *Permeke, his work became *Expressionist. Initially, his work in this vein was vaguely derived from *Cubism, with block-like forms and matt surfaces, but from the mid-1920s his paintings took on a more fantastic quality in the tradition of *Bosch and *Ensor. They have been seen by some as an anticipation of *Surrealism.

Bergognone (or **Borgognone**), **Ambrogio** (active 1481–1522). Italian painter, active mainly in and around Milan. He is one of the best examples of the continuance of a native Milanese tradition in painting into the 16th cent., for unlike so many of his Lombard contemporaries he was unaffected by the art of *Leonardo. His style is static and undramatic, aiming at a typically late *quattrocento mood of devotional calm, which is often enhanced by pale landscape backgrounds of great delicacy. There are good examples of his work in the National Gallery, London, the Brera, Milan, and the Pinacoteca, Pavia.

Berlinghieri. A family of Italian painters active at Lucca in the 13th cent. **Berlinghiero Berlinghieri**, the founder of the family, is called 'Milanese' in a document of 1228, which also mentions three sons, **Marco**, **Barone**, and **Bonaventura**. He is not otherwise known, but a painted *Crucifix* (now in the Lucca Pinacoteca) signed 'Berlingeri' without Christian name is attributed to him. It is one of the finest examples of the *Byzantine manner. A *Crucifix* in the Accademia, Florence, is also sometimes assigned to him.

Bonaventura, the most talented of his sons, is known chiefly for his signed and dated altarpiece in the church of S. Francesco at Pescia (1235), which with its combination of solemn images and homely detail has been regarded as one of the most original, as it is one of the earliest, pictorial representations of Franciscan ideas. The *Scenes from the Life of St Francis* in Sta Croce, Florence, and the *St Francis receiving the Stigmata* in the Accademia, Florence, have also been attributed to him.

Berman, Eugene (1899–1972). Russian-born American painter and stage designer. He fled with his family to Paris in 1918 during the Russian Revolution and became friendly with *Tchelitchew (another Russian émigré) and a group of painters who became known as *'Neo-Romantics', painting dreamlike scenes with mournful, drooping figures. In 1935 he emigrated to the USA and became an American citizen in 1937. From this time he was active in designing stage sets for theatre and ballet, working among others for the Metropolitan Opera Company. He also gained some celebrity for his imaginary landscapes, which have a *Surrealist air.

Bermejo, Bartolomé (active 1474–98). Spanish painter and stained-glass designer, born in Cordova, but active in northern Spain—in Aragon and then in Barcelona from 1486. His *Pietà* in Barcelona Cathedral, signed and dated 1490, is one of the earliest Spanish *oil paintings and one of the masterpieces of Spanish art of this period. Representative of the Hispano-Flemish style, his intense naturalism recalls Nuño *Gonçalves.

Bernard, Émile (1868–1941). French painter and writer. In 1884 he entered the studio of Fernand *Cormon, where he was a contemporary of van *Gogh and *Toulouse-Lautrec, and also of Louis *Anquetin, with whom he developed *Cloisonnism, a style in which flat areas of colour and bold, dark outlines emulate the technique of *cloisonné* enamels. He then joined *Gauguin at *Pont-Aven and later claimed that it was he who in 1888 introduced Gauguin to the *Synthetist manner: certainly the two worked closely together between 1888 and 1891 in Pont-Aven and Paris, and Bernard seems to have had a stimulating effect on his great colleague. Thereafter, however, Bernard's work as a painter greatly declined in importance, and he became of interest chiefly for his activities as a writer, playing a significant role as a sponsor of *Post-Impressionism. In 1893 he published van Gogh's letters (three years earlier he had organized the first retrospective exhibition of van Gogh's works), and then left Paris for eight years teaching and travelling in Egypt and the Near East. In 1904 and 1905 he visited *Cézanne and published important interviews with him.

Bernini, Gianlorenzo (1598–1680). Italian sculptor, architect, painter, and designer, the supreme artist of the Italian *Baroque. His father, **Pietro** (1562–1629), was a *Mannerist sculptor of some distinction, active in Naples and then from *c.*1605 in Rome, and Gianlorenzo owed to him not only his early training in the handling of marble but also his introduction to the group of powerful patrons, the *Borghese and the *Barberini, who so promptly fostered and employed his creative genius. For Cardinal Scipione Borghese he executed a remarkable series of life-size marble sculptures; in their dramatic vigour and movement they made a complete break with the Mannerist tradition, and they showed unprecedented virtuosity in making cold stone seem as supple as living flesh. These are *Aeneas, Anchises and Ascanius* (1618–19), the *Rape of Proserpine* (1621–2), *David* (1623), and *Apollo and Daphne* (1622–5), all in the Borghese Gallery in Rome.

After the election of Maffeo Barberini as Pope Urban VIII (1623) Bernini became the principal artist in the papal court and in Rome. According to *Baldinucci, Maffeo had 'scarcely ascended the sacred throne' when he summoned Bernini and told him: 'It is your great fortune to see Cardinal Maffeo Barberini Pope, but our fortune is far greater in that Cavalier Bernini lives during our pontificate.' In 1629 he was appointed architect to St Peter's, for which he made the great *baldacchino over the High Altar (1624–33), and the huge statue of *St Longinus* (1629–38), which stands in a niche in one of the piers of the crossing. He also supervised the lavish marble decoration of the interior and made the tomb of Urban VIII (1628–47) for the same church. However, after Urban's death in 1644 Bernini fell under a cloud. This was partly because of one of his rare failures (a belltower he added to the façade of St Peter's was demolished in 1646 because of structural problems), but as much because of the different artistic tastes of the new pope, Innocent X (1574–1655), who favoured Bernini's rival, *Algardi. During Innocent's papacy Bernini worked mainly for private patrons. The Cornaro Chapel, with the celebrated marble group of the *Ecstasy of St Theresa*, in Sta Maria della Vittoria dates from this period (1645–52). It is a comparatively small work, but an excellent example of Bernini's aims and achievement in the fusion of sculpture, architecture, and painting into a magnificent decorative whole. Bernini never lost his position as architect to St Peter's, however, and he did do some work for Innocent X. He sculpted his portrait (of the several busts of the pope in the Palazzo Doria-Pamphili, Rome, two—in marble—are probably from Bernini's own hand), and designed for him the *Fountain of the Four Rivers* (1648–51) in the Piazza Navona. This is the most celebrated and spectacular of Bernini's fountains, and with these, his buildings and his outdoor statuary he has had a greater effect on the face of Rome than any other artist.

After Innocent's death in 1655 and the accession of Alexander VII (1599–1667) Bernini was restored to full favour and was almost immediately given two major commissions at St Peter's: the decoration of the *Cathedra Petri* (Throne of St Peter), and the building of the vast colonnade round the piazza in front of the church. The *Cathedra Petri* (1657–66), a setting for the wooden chair believed to have been used by St Peter, provides an appropriately spectacular sight to conclude the vista at the east end of the church. Four immense bronze figures of Doctors of the Church support the chair, also encased in bronze, and gilt angels and cherubs float above on stucco clouds; they surround a window bearing an image of the Dove of the Holy Spirit, from which a burst of light seems to emanate. The enclosure of the piazza in front of St Peter's is his greatest achievement as an architect—a design of the utmost dignity and grandeur, expressing the overwhelming authority of the Church. Bernini himself compared the sweeping colonnades to motherly arms that reach out to 'embrace Catholics to reinforce their belief'.

In 1665 Louis XIV invited Bernini to Paris to build the east front of the *Louvre, but his plans were abandoned in favour of a French design, and the trip—which he had made unwillingly—was not a success. He returned to Rome in 1666 and continued to be extremely active into his old age. His late religious works were intensely spiritual, reflecting his own ardent commitment to Catholicism (*The Blessed Lodovica Albertoni*, S. Francesco a Ripa, Rome, 1671–4). As an architect his late work included important secular as well as religious buildings, notably the Palazzo Chigi-Odescalchi (begun 1664), which had great influence on Baroque palace design throughout Europe, and S. Andrea al Quirinale (1658–70), a fairly small church but one of his most sophisticated creations in its use of rich architectural and sculptural decoration to create an appropriate setting for the mysteries of the Catholic faith.

In addition to large works of sculpture and architecture, Bernini executed many portrait busts, among the finest of which are those of his mistress

Costanza Buonarelli (*c*.1645), in the Bargello, Florence, and Louis XIV (1665) at Versailles. He was also a brilliant wit, a *caricaturist, and—for his private pleasure—a painter of such high quality that his rare surviving works (which include several self-portraits) have sometimes been attributed to *Velázquez. Examples of his paintings are in the National Gallery, London, and the Ashmolean Museum, Oxford. Bernini also had a passionate interest in the theatre. There are few material remains of his activity in this field, but the diarist John Evelyn saw a remarkable demonstration of his versatility when visiting Rome in 1644: 'Bernini . . . gave a public opera wherein he painted the scenes, cut the statues, invented the engines [stage machinery], composed the music, writ the comedy, and built the theatre.' Although almost all his major works are in Rome, his drawings are widely distributed among the great collections of the world, particularly fine representations being at Windsor Castle and the Kunstmuseum, Düsseldorf.

Bernini accurately predicted that his reputation would decline after his death. To the *Neoclassical taste of the 18th cent. his approach to sculpture was anathema, to *Ruskin in the 19th cent. it seemed 'impossible for false taste and base feeling to sink lower', and to the devotees of the idea of 'truth to materials' in the 20th cent. he appeared, in the words of his most distinguished apologist, Rudolf *Wittkower, as 'Antichrist personified'. It is only fairly recently that he has come to enjoy a reputation, comparable with his status in his lifetime, as the greatest sculptor since *Michelangelo and one of the giants of Baroque architecture.

Bernward of Hildesheim (*c*.960–1022). German ecclesiastic and art patron. He was abbot of the Benedictine monastery at Hildesheim in Saxony from 993 until his death, and for the church of St Michael there (begun 1001) he commissioned the famous bronze doors (*c*.1008–15) and a great bronze column (*c*.1018–20) probably intended to support the paschal candle. They are important not only for being among the outstanding European works of their period, but also for marking the revival of the **cireperdue* technique of casting, which had virtually disappeared since the time of Charlemagne (?742–814). A contemporary biographer records that Bernward himself practised metalwork and manuscript *illumination, but no work can be attributed to him. He was canonized in 1192 and is the patron saint of goldsmiths.

Berruguete. The name of two Castilian artists, father and son, who are respectively associated with the beginnings of the *Renaissance and *Mannerist styles in Spain.

Pedro (d. 1504) was court painter to Ferdinand (1452–1516) and Isabella (1451–1504). He may have been the 'Pietro Spagnuolo' employed in 1477 with *Melozzo da Forlì and *Joos van Wassenhove on the decoration of the palace library at Urbino. He was working at Toledo from 1483. Ten panels from the Dominican convent at Avila, now in the Prado, demonstrate that his Renaissance style was modified by the Flemish influences then prevailing in Spain.

Alonso (*c*.1488–1561), sculptor and painter, was the son and probably pupil of Pedro. For some years, between 1504 and 1517, he was in Italy, where he completed Filippino *Lippi's *Coronation of the Virgin* (Louvre, Paris). Berruguete was back in Spain by 1518; in that year he was appointed court painter to Charles V (1500–58), but his career flourished mainly as a sculptor. He worked in Valladolid and Toledo, his finest works including the *reredos of the monastery church of S. Benito, Valladolid, dating from 1526 (Valladolid Mus.), and a set of *choir stalls with *alabaster figures above for the choir of Toledo Cathedral (1539–43). The emotional intensity and expressive **contrapposto* characteristic of his style reflected the influence of *Michelangelo (who refers to Berruguete in his letters) and of the **Laocoön*, which he had studied in Rome. He is generally considered the greatest Spanish sculptor of the 16th cent., his work having something of the spirit of El *Greco, who succeeded him as the outstanding artist in Toledo.

Bertoldo di Giovanni (*c*.1420–91). Florentine sculptor. He was a fairly minor talent, but he is remembered for three things. First, he was the pupil and assistant of *Donatello and teacher of *Michelangelo, thus forming the link between the greatest Florentine sculptors of the 15th and 16th cents. Secondly, he was described by *Vasari as the first head of the *academy of art which Lorenzo the Magnificent is said to have founded in the *Medici gardens by the Piazza di S. Marco. Thirdly, he developed a new type of sculpture—the small-scale bronze, intended, like the *cabinet picture, for the private collector. Bertoldo was responsible for the completion of the two pulpits in S. Lorenzo left unfinished by Donatello at his death. His own most noteworthy work is a bronze *relief of a battle scene in the Bargello, Florence, which inspired Michelangelo's *Battle of the Lapiths and Centaurs* (Casa Buonarroti, Florence). Bertoldo was also recognized as one of the leading portrait medallists of his time.

bestiary. A type of medieval manuscript that used illustrations of animals or fabulous beasts to point moral lessons. It was based on the Greek

Physiologus, a pseudo-scientific natural history that was translated into Latin around the 9th cent. In the late 12th and 13th cents. the bestiary was one of the leading picture books, particularly popular in England, and its images exerted a great influence in medieval art, as one can see, for example, in the decoration of initials and later in *misericords and roof *bosses.

Beuys, Joseph (1921–86). German sculptor, draughtsman, teacher, and *Performance artist, regarded as one of the most influential leaders of avant-garde art in Europe in the 1970s and 1980s. Like Yves *Klein, he was one of the leading lights in shifting emphasis from what an artist makes to his personality, actions, and opinions, and he succeeded in creating a kind of personal mythology. (As a Luftwaffe pilot he was shot down in the Crimea in 1943 and was looked after by nomadic Tartars, who kept him warm with fat and felt—materials that came to figure prominently in his work. The hat that he habitually wore hid the head injuries he received in the crash.) After the war he studied at the Düsseldorf Academy, 1946–51, and he became Professor of Sculpture there in 1961. He worked in various media, but is perhaps best known for his performances, of which the most famous was probably *How to Explain Pictures to a Dead Hare* (1965). In this he walked around an exhibition in the Schmela Gallery in Düsseldorf, his face covered in honey and gold leaf, carrying in his arms a dead hare, to which he gave an explanation of various pictures. He described the performance as 'A complex tableau about the problems of language, and about the problems of thought, of human consciousness and of the consciousness of animals.' In 1962 Beuys became a member of *Fluxus, an international group of artists, opposed to tradition and professionalism in the arts, and he was also active in politics, aligning himself with the West German ecology party, the Greens. His 'presumptuous political dilettantism' eventually led to conflict with authority and in 1972 he was dismissed from his professorship. The protests that followed included a strike by his students, and a settlement was eventually reached whereby he kept his title and studio but his teaching contract was ended. He devoted a good deal of his later career to public speaking and debate, and in 1982 he had a meeting with the Dalai Lama in Paris. By the end of his life he was an international celebrity and was regarded by his admirers as a kind of art guru.

Bevan, Robert (1865–1925). English painter. He spent a good deal of time in France and met *Gauguin at Pont-Aven in 1894. Back in London he became a member of *Sickert's circle and he was a founding member of the *Camden Town Group (1911) and the *London Group (1913). His work was much influenced by Gauguin's bold colour and flat patterning, and in his last years his style became increasingly simplified and schematic. He is best known for paintings featuring horses (*The Cab Horse*, Tate, London, *c.*1910). His wife, the Polish-born Stanislava de Karlowska (1880–1952), whom he married in 1897, was also a painter. He made several visits to Poland with her.

Bewick, Thomas (1753–1828). English engraver, active for most of his life in Newcastle upon Tyne. Bewick ran a busy workshop in which most of the day-to-day jobs involved metal-engraving, but for his own projects he preferred *wood engraving, and he was the first artist to show the full potential of this technique. He had a great love of the countryside, and his finest works are natural history illustrations, particularly those to his celebrated books, *A General History of Quadrupeds* (1790) and *A History of British Birds* (2 vols., 1797 and 1804, for which he wrote the texts himself). The animals and birds are characterized with great skill, but Bewick is as much admired for his tailpieces—little (sometimes tiny) vignettes with which he concluded his account of each animal or bird. These miniature scenes give a wonderfully shrewd and sensitive picture of rural life, bringing out its bleakness and cruelty as well as its beauty and humour. Bewick himself punningly called these scenes '*Tale*-pieces', for they were 'seldom without an endeavour to illustrate some truth or point some moral'. The success of his books made wood engraving a popular medium for illustration and his work was carried on by several followers in Newcastle, notably his son **Robert** (1788–1849). Bewick wrote an autobiography, which was published posthumously in 1862 and in a scholarly edition in 1975. A year before his death he was visited by *Audubon, another great artist-naturalist, who left a touching account of his meeting with this 'perfect old Englishman', who was 'kind and attentive' and still—aged 74—'active and prompt in his labours', using 'delicate and beautiful tools . . . all made by himself'.

Beyeren, Abraham van (1620/1–90). Dutch painter, little regarded in his day but now considered one of the greatest of still-life painters. He initially specialized in fish subjects, but around the middle of the 17th cent. he began to devote himself to sumptuous banquet tables laden with silver and gold vessels, Venetian glassware, fine fruit, and expensive table coverings of damask, satin, and velvet. Works of this kind, in which he was rivalled only by *Kalf, gave him even greater opportunity than his fish pieces to demonstrate his ability to show the play of light on surfaces and organize forms and colours into an opulently blended composition. *Still life with a Lobster and Turkey* (Ashmolean, Oxford) shows all his splendour of composition and richness of colour and texture. Van Beyeren worked in various towns before settling in Overschie in 1678.

Bibiena (or **Galli-Bibiena**). Family of Italian architects, *quadraturisti, and stage-designers based in Bologna, members of which practised from the 1680s until the 1780s in practically every country of Europe. The founder of the dynasty was **Giovanni Maria Galli** (1625–65), who adopted the name of his birthplace, Bibbiena, a small town in Tuscany. Other members of the family included **Alessandro** (1687–1769), **Antonio** (1700–74), **Carlo** (1728–87), **Ferdinando** (1657–1743), **Francesco** (1659–1739), and **Giuseppe** (1696–1757). They provided fantastically elaborate stage-settings for operas, balls, state occasions, and religious ceremonies, mainly in the service of the Austrian Imperial family in Vienna and various German princelings. They also built several theatres in Italy and Germany, one of which survives: the Opera House at Bayreuth, decorated by Giuseppe in 1748. The most illustrious and prolific member of the family was Ferdinando, who produced several books on architecture and scenography.

Biblia Pauperum. The first medieval textbook of Christian *typology ('the Poor Man's Bible'), showing in pictures how the principal events from the life of Christ were prefigured in the Old Testament. Unlike the *Speculum Humanae Salvationis*, the *Biblia Pauperum* is strictly scriptural, admitting no historical or secular subjects among its types. It was devised in south Germany in the late 13th cent. and the earliest existing manuscript dates from *c.*1300. Issued as one of the first *block books to be printed, the *Biblia Pauperum* was particularly popular in Germany, but had little vogue in Italy or Spain. It was copied and adapted in late medieval sculpture, tapestries, stained glass, and in easel pictures.

Biederman, Charles Joseph (1906–). American artist and theorist. From the 1930s he produced completely abstract paintings and coloured geometric reliefs in the style of *Mondrian. His promotion of such works in his book *Art as the Evolution of Visual Knowledge* (1948) had an important influence on British *Constructivism, notably on *Pasmore.

Biedermeier. Term applied to a style characteristic of much German and Austrian art and architecture between the Congress of Vienna (1815) and the Revolution of 1848. Gottlieb Biedermeier was a somewhat ludicrous fictional character from the journal *Fliegende Blätter* (Flying Leaves), who personified the solid yet philistine qualities of the bourgeois middle classes. The art to which he has lent his name shares these qualities: the magic poetry of *Romantic landscape painting is replaced by sober realism; heroic themes give place to the illustration of fairy-tales or legends, and portraits concern themselves with the minutiae of appearance and costume rather than the imaginative projection of personality. In architecture, instead of the grandiloquent *Neo-classical forms, we find sound utilitarian structures of good proportions. There were, as is to be expected, no major painters associated with Biedermeier but many excellent practitioners, such as *Waldmüller. The term is sometimes extended to cover the work of artists in other countries, for example *Købke in Denmark.

Biennale (or **Bienale**). An international art exhibition held every two years and adjudicated by an international committee. The first and most famous was the Venice Biennale, instituted in 1895 as the 'International Exhibition of Art of the City of Venice', and claiming to represent 'the most noble activities of the modern spirit without distinction of country'. At this first Biennale artists from 16 different nations were represented, and the committee included such celebrated personalities as *Burne-Jones, *Israëls, *Liebermann, *Moreau, and *Puvis de Chavannes. The exhibition soon acquired world-wide prestige, and after it resumed in 1948 following the Second World War it became the leading show-place for the established international avant-garde. Henry *Moore, for example, set the seal on his reputation when he won the International Sculpture Prize in 1948. Other biennales have been inaugurated on the Venice model, the most prestigious being the São Paulo Bienale, founded in 1951, and the Paris Biennale, founded in 1959.

Bierstadt, Albert (1830–1902). German-born American painter, active mainly in New York. He made several trips to the Far West and was one of the last of those painters (known collectively as the *Hudson River and *Rocky Mountain Schools) who specialized in grandiose pictures of awesome mountain scenery. His paintings—often huge in size—were immensely popular in his lifetime, but his on-the-spot sketches are now generally found much more appealing than the finished studio works.

Bigot, Trophime (1579–1650). French *Caravaggesque painter, active in Rome (*c.*1605–34) and then in the Aix-en-Provence area. His career is obscure, but he has been identified as the author of a number of paintings, mainly intimate candlelit scenes, that were previously grouped under the name of 'the Candlelight Master'. *A Doctor Weighing Urine* (Ashmolean, Oxford) is an example of the unusual subjects he sometimes favoured.

Biguerny, Philippe. See VIGARNY.

Bill, Max (1908–94). Swiss painter, sculptor, architect, designer, and writer. He studied at the Zurich School of Arts and Crafts in 1924–7, then at the *Bauhaus in 1927–9, and in 1932–6 belonged to the *Abstraction-Création group in Paris. In 1936 he took up the term *Concrete Art (*Konkrete Kunst*), proposed by van *Doesburg in 1930, and popularized it in Switzerland in place of 'abstract'. In 1941 he visited Brazil and Argentina, and introduced the concept of Concrete Art there. His work was based upon the use of mathematical formulae to engender the relations between the parts from which a work is constructed, and his ultimate aim was to establish a unity among the individual branches of the visual arts—he once defined art as the 'sum of all functions in harmonious unity'. He had an active career as a publicist of his ideas, but in the exhibitions he organized he covered a broad spectrum of non-figurative art. His sculptures have been considered precursors of the *primary structures of *Minimal art, although they are in fact derived from quite complicated mathematical formulae, and his influence has been repudiated by Minimalists such as *Judd and Robert *Morris. His influence has been acknowledged not only in Switzerland, however, but also in Argentina and Italy, where he was the inspiration of a number of associations of Concrete Art. Bill lived and worked mainly in Zurich, where he designed his own house (1932–3). His other work as an architect includes the much praised Hochschule für Gestaltung in Ulm (1951–5) where, working on a narrow budget, he created an austerely elegant complex of buildings delicately placed in a romantic setting.

Bingham, George Caleb (1811–79). American painter. He worked mainly in Missouri (where he held several political posts), painting the life of the frontier people. Except for a short period studying at the Pennsylvania Academy of Fine Arts, he was self-taught. His finest canvases, particularly the celebrated *Fur Traders Descending the Missouri* (Met. Mus., New York, 1845), distil visual poetry from the commonplace, but after a trip to study in Düsseldorf in 1856–8 his work lost much of its racy freshness and charm, becoming overlaid with sentimentality. In 1877 he became Professor of Art in the University of Missouri.

biomorphic. A term applied to forms in abstract art that derive from organic rather than geometric shapes, as, for example, in the sculpture of Henry *Moore.

Bird, Francis (1667–1731). English sculptor. He trained in Brussels and later visited Rome and could work in a *Baroque idiom more convincingly than most of his English contemporaries, as is seen in his best-known work, *The Conversion of St Paul* (1706), in the west pediment of St Paul's Cathedral, London. His work is uneven, but he ranks as the most significant figure in English sculpture between *Gibbons (for whom he worked for a time) and *Rysbrack. Several of his monuments are in Westminster Abbey.

Birolli, Renato (1906–59). Italian painter who, because of his outspoken political views and his advanced and energetic artistic outlook, exercised an important influence on the Italian avant-garde. In 1938 he took a prominent part in founding the anti-Fascist *Corrente association, and was persecuted and imprisoned for his political activities. In 1947 he joined the *Fronte Nuovo delle Arti association. His work was varied in style, ranging from *Expressionist pictures influenced by *Ensor and van *Gogh to abstracts reflecting his interest in the *Orphists.

Bisschop, Jan (1628–71). Dutch lawyer and dilettante draughtsman (see AMATEUR) who travelled extensively and made accurate *wash drawings of the paintings and sculpture he examined. These are valued today because many of them are records of lost works of art. He also made exquisite small drawings of the scenery which he saw on his travels and a few of these rank among the finest 17th-cent. Dutch landscape drawings.

Bissière, Roger (1888–1964). French painter, born in the province of Lot-et-Garonne. His early paintings were landscapes which imbued his native countryside with a monumental dignity. He moved to Paris in 1910 and earned a living as a journalist while continuing to paint. After a period of experimentation with *Cubism, he was for a time associated with *Ozenfant and *Le Corbusier, contributing to their journal *Esprit Nouveau*. From 1925 to 1938 he taught at the *Académie Ranson, where his influence on many of the younger abstract artists, such as *Manessier, was great, but his own work during the 1920s and 1930s remained almost unknown. In 1938 he retired to Lot and during the war contracted an eye ailment that left him unable to paint. Instead, he produced compositions pieced together from tapestry and other materials. A successful operation in 1948 enabled him to resume painting and during the 1950s he achieved the resounding recognition which had so long escaped him. His large, tapestry-like compositions in rich and glowing colours were exhibited internationally and commanded among the highest prices of any living artist. Abstract in appearance, they resulted from the careful and sensitive reduction of natural scenes to scintillating patterns of interacting colours, and Bissière himself always refused to accept the term 'abstract' for his own work. The recognition accorded to the work of his later years caused a reassessment of the work of his early and middle periods, which Bissière himself was accustomed to say might in course of time be valued even above that which had brought him popularity.

bistre. A transparent brown *pigment prepared by boiling the soot of burned wood. It is often used as a *wash for pen-and-ink drawings, watercolours, and *miniatures. *Rembrandt and *Claude were among the artists who exploited its potentialities. See also SEPIA.

bitumen (asphaltum). A transparent brown *pigment which at the time of use gives a rich glowing quality, but later becomes almost black and increasingly opaque. It never completely hardens and eventually develops a pronounced and often disfiguring **craquelure*. It was most popular in the 18th cent., and its damaging effects can be seen in works by *Reynolds and other British painters of the period.

black-figure vase painting. Technique of vase-painting, originating in Corinth in the 7th cent. BC, in which figures were painted in black silhouette. Details were added by incising through the black pigment to the light red clay background or sometimes by overpainting in red or white. The technique had its finest flowering around the mid-6th cent. BC, notably in the work of *Execias, but then began to give way to *red-figure painting, although black-figure vases continued to be produced for another two centuries.

Black Mountain College. American art educational establishment at Black Mountain, North Carolina, founded by a group of progressive academics in 1933 and closed after long-standing financial problems in 1957. It was run by the teaching staff, with no outside control, and was kept deliberately small (with an average of about fifty students a year) to reduce administration; a variety of arts were taught and interaction between them was encouraged. Mary Emma Harris describes the college as 'a unique combination of liberal arts school, summer camp, farm school, pioneer village, refugee centre and religious retreat' and writes that it was 'a catalyst for the emergence of the American avant-garde after the Second World War' (chapter on Black Mountain College in the catalogue of the exhibition 'American Art in the 20th Century', Royal Academy, London, 1993). In the visual arts, the teacher most associated with Black Mountain College was Josef *Albers, who arrived there with his wife Anni (who taught weaving) soon after it opened and stayed until 1949. Other illustrious figures who taught at Black Mountain include Robert *Motherwell and the composer John Cage (1912–92), whose ideas on chance and indeterminacy in the arts were widely influential. In 1952 he organized there a partly programmed performance (involving paintings and readings) that was later designated the first *happening. Famous former students of the college include John *Chamberlain, Kenneth *Noland, and Robert *Rauschenberg.

Blair, Hugh (1718–1800). Scottish divine and writer, Professor of Rhetoric and Belles Lettres in Edinburgh University from 1762 to 1783. Although not a profound or original thinker, Blair is noteworthy for his *Lectures on Rhetoric and Belles Lettres* (1783), which were of immense importance in his day for popularizing aesthetic and critical speculation. There are more than sixty editions in English and nearly fifty abbreviated editions and translations into German, French, Spanish, Italian, and Russian. He was a friend of Robert Burns (1759–96), David Hume (1711–76), and many other leading Scottish literary and intellectual figures.

Blake, Peter (1932–). British painter, a leading exponent of *Pop art. His use of imagery from comics, pin-up magazines, consumer goods, and advertisements captures the flavour of the times in a manner that now evokes nostalgia for the 'swinging sixties' as was made clear during his enormously successful retrospective exhibition at the Tate Gallery, London, in 1983 (his most famous work is the cover design for the Beatles LP *Sergeant Pepper's Lonely Hearts Club Band*, 1967). In *The Century of Change: British Painting Since 1900* (1977) Richard Shone wrote: 'Blake is an exuberant, highly skilled magpie, gathering images from a wide spectrum—Alice in Wonderland to soft porn—and matching them to, in his own words, "the technical forms that will best recapture the authentic feel of folk art".' Something of the same engaging combination of sophistication and naïvety can be seen also in Blake's work as a member of The Brotherhood of Ruralists, a group of seven painters based in the West Country, of which Blake (then living near Bath) was one of the founders in 1975. The members had several group exhibitions, took working holidays together, and shared a commission to design covers for the New Arden edition of Shakespeare's work, but they had common ideals rather than a common style, taking as their inspiration 'the spirit of the countryside'. A series of winsome fairy paintings are characteristic of this facet of Blake's work, and many critics found the work of the group as a whole insufferably twee—one newspaper review of a 1981 Ruralists exhibition was headed 'Tinkerbell lives'. The Brotherhood last exhibited as a group at Blake's retrospective at the Tate in 1983. The other members were: Ann Arnold (1936–) and her husband Graham Arnold (1932–); the American-born Jann Haworth (1942–), who was married to Blake, 1963–81; David Inshaw (1943–); Annie Ovenden (1945–) and her husband Graham Ovenden (1943–).

Blake, William (1757–1827). English artist, philosopher, and poet, one of the most remarkable figures of the *Romantic period. From childhood he possessed visionary powers, and the engraving of *Joseph of Arimathea* (characteristically based on a figure by *Michelangelo), done at the age of 16, shows him already using a personal symbolism to express his mystical philosophy. His apprenticeship (1772–9) to the engraver James Basire (1730–1802), for whom he made drawings of the monuments in Westminster Abbey and other London churches, led him to a close study of *Gothic art and intensified his love of linear design and formal pattern. In 1779 he entered the *Royal Academy Schools, but his relations with *Reynolds were painful; later he was to find more sympathetic spirits in *Stothard, *Flaxman, *Fuseli, and *Barry.

During the 1780s Blake worked as a commercial engraver, but from about 1787 he became engrossed in a new method of printing his own illustrated poems in colour, which he characteristically claimed to have been revealed to him in a vision by his brother Robert, then recently deceased. The first of these major works of 'illuminated printing', in which handwritten text and illustration were engraved together to form a decorative unit, was *Songs of Innocence* (1789). In 1793 with his wife, Catherine Boutcher, he settled in Lambeth, where he engraved his principal prose work, *The Marriage of Heaven and Hell*. He had little material success and in 1800, at the suggestion of William Hayley (1745–1820), poet and man of letters, he left London to settle for three years at Felpham on the Sussex coast. Here he continued a series of watercolours illustrating biblical subjects for his first and most generous patron, Thomas Butts, and also began to engrave *Jerusalem*, the last and longest of his surviving mystical writings. On his return to London Blake made a series of drawings for Robert Blair's poem *The Grave*, and in 1809 held a small one-man exhibition for which he issued *A Descriptive Catalogue*, eloquently summarizing his aims and convictions about art. This earned him but little recognition, however, and there followed a period of eclipse, during which he appears to have been unproductive, until 1818, when the sympathetic patronage of the painter John *Linnell ensured him a livelihood for the remainder of his life. For Linnell he carried out his engravings for *The Book of Job* and his magnificent designs for *The Divine Comedy*, on which he was working up to the time of his death. Linnell introduced to him a group of younger artists, including *Varley, *Calvert, and Samuel *Palmer, who were inspired and stimulated by Blake's imaginative power. He thus passed his last years surrounded by a group of admiring disciples, who formed themselves into a kind of brotherhood under the name the *Ancients.

In art as in life Blake was an individualist who made a principle of nonconformity. He had a prejudice against painting in oils on canvas and experimented with a variety of techniques in colour printing, illustration, and *tempera. Blake might be called a mystical realist and his work as an artist is almost impossible to divorce from the complex philosophy expressed also through his poetry. He believed that the visible world of the senses is an unreal envelope behind which the spiritual reality is concealed and set himself the impossible task of creating a visual symbolism for the expression of his spiritual visions that would owe nothing to ordinary visual experience. He refused the easy path of vagueness and misty suggestion, remaining content with nothing less than the maximum of clarity and precision. To most of his contemporaries Blake seemed merely an eccentric, and his genius was not generally recognized until the second half of the 19th cent. (*Rossetti—another painter-poet with mystical leanings—was an early champion.) His output was enormous; there are important collections in the British Museum, the Tate Gallery, the Fitzwilliam Museum, Cambridge, and several American museums.

Blanchard, Jacques (1600–38). French painter, active mainly in his native Paris. He studied in Italy (1624–8) and his style reflects both *Baroque developments in Rome and the sensuous richness of the

Venetian school. Back in France he gained a reputation for decorative work, but is now known chiefly as a painter of small religious and mythological subjects in a sensitive but sentimental manner (*Charity*, Courtauld Inst., London, 1637).

Blanche, Jacques-Émile (1861–1942). French painter. The son of Émile Blanche, a noted pathologist, he grew up in a cultured atmosphere and became a well-known figure in artistic and society circles—he was a friend of *Degas, *Renoir, *Whistler, the writers Henry James and Marcel Proust, and many other celebrities. His best-known works are stylish portraits of people from this milieu; the finest collection is in the Musée des Beaux-Arts at Rouen, and there are several examples in the Tate Gallery, London; Blanche lived mainly at Offranville, near the Channel port of Dieppe (the local church has decorative painting by him), and he was a frequent visitor to Britain, painting numerous views of London (the Tate has an example). He wrote several books of criticism and reminiscence.

***Blast*.** See VORTICISM.

Blaue Reiter, Der (The Blue Rider). A loose association of German *Expressionist artists with revolutionary aims formed in Munich in 1911 as a splinter group from the *Neue Künstlervereinigung. The name, deriving from the title of a picture by *Kandinsky, was also used as the title of an 'Almanac' (a collection of essays and illustrations) published by Kandinsky and Franz *Marc in 1912 and of the exhibition they organized at the Gallery Thannhauser in Munich in 1911. Other prominent members of the group were Paul *Klee and August *Macke. The aims of the association were never closely defined and it is obvious that the individual artists associated differed widely in their work and outlook. Their bond was a general desire to embody symbolically in their painting spiritual realities which, it was thought, had been neglected by the *Impressionists. They held two touring exhibitions (1911 and 1912) that brought the work of Der Blaue Reiter to several major cities of Germany, and the association had international affiliations; *Braque, *Derain, *Goncharova, *Larionov, and *Picasso were among those whose works were shown in the exhibitions. With the outbreak of the First World War, however, the group disintegrated. Its short life is considered to mark the high point of German Expressionism.

Blaue Vier, Die (The Blue Four). A group of four painters formed in 1924 by *Kandinsky, *Jawlensky, *Klee, and *Feininger in succession to the *Blaue Reiter, with which they had all been associated. The members were united by a desire to publicize their work and ideas through exhibiting together rather than by stylistic similarity. Exhibitions were held in Germany, the USA, and Mexico between 1925 and 1934. Their patroness, Galka Scheyer, who suggested the formation of the group, formed a large collection of their work, which is now in the Norton Simon Museum in Pasadena.

Blechen, Karl (1798–1840). German painter, primarily of landscapes. For a time he was under the influence of such *Romantics as *Friedrich, but following a visit to Italy in 1828 he developed a more painterly style, producing rather unconventional open-air landscapes which foreshadowed *Impressionism.

bleed. Painters' term for the action of an under layer of paint when it seeps through superimposed layers and comes to the surface, mixing with the upper layers of paint and changing their colour. This happens when the *pigment is soluble in the *medium. It is in order to prevent bleeding that pigments used in *oil painting must be insoluble in oil or 'oil-proof', colours used in *fresco must be 'lime-proof', and so on. The term 'bleed' is also used in book production to refer to an illustration printed without margins on one or more sides so as to run up to the edge of the paper.

Bles, Herri met de (*c.*1500/10–after 1550). Netherlandish painter of landscapes with figures. He is an enigmatic figure presumed to be one and the same as the Herri Patenier who entered the Antwerp Guild in 1535. Herri met de Bles, as van *Mander informs us, is simply a nickname meaning 'Herri with the white forelock', and it is generally assumed that he was a relation of Joachim *Patenier, who certainly had a decisive influence on his work. No signed or documented work by Herri exists, but a *Mountain Landscape with the Holy Family and St John* (Öffentliche Kunstsammlung, Basle) was described as the work of Heinrich Blesii in 1568. A small group of works has been ascribed to him, characterized by panoramic landscapes dominating the figure groups in the manner of Patenier. Yet his style is distinctive. His work was popular with Italian collectors, who called him 'Civetta' (little owl) because he often included owls in his pictures.

Bleyl, Fritz. See BRUCKE.

block book. A book printed from *woodcut blocks on which text and illustrations are combined, rather than by means of movable type. Block books were made in China probably as early as the 6th cent. AD, but in Europe the earliest known examples seem to date from around 1450, that is, at very much the same time that Gutenburg (*c*.1400–68?) introduced printing from movable metal type. (At one time they were thought to predate movable type, but they are now thought to have been introduced very slightly after it.) As the entire text had to be cut letter by letter on wood blocks, the process was extremely laborious and suitable only for short books in continuous popular demand, such as the **Ars Moriendi* or the **Biblia Pauperum*. Block books were at the height of their popularity in the 1470s. Very few were executed after 1480 and their place in the history of printing is as sterile descendants of the woodcut rather than as ancestors of printing from movable type.

Bloemaert, Abraham (1564–1651). Dutch historical and landscape painter and engraver, the son of a sculptor and architect, **Cornelis I Bloemaert** (*c*.1525–*c*.1595). Most of his life was spent in Utrecht, where for many years he was the leading painter and an outstanding teacher. *Both, *Honthorst, *Terbrugghen, and virtually all the Utrecht painters of the period who attained any kind of distinction trained with him. Bloemaert was a good learner as well as a good teacher and rapidly assimilated the new ideas his pupils brought back from Italy. For a time he became a *Caravaggesque painter and late in his career adopted some aspects of the *classicism of the *Carracci. Although his landscape paintings are firmly in the *Mannerist tradition, his landscape drawings are naturalistic and his most original works. All facets of his work can best be studied in the museum in Utrecht. Many of his drawings were etched and published by his son **Frederick** (*c*.1610–69) in a well-known drawing book for the use of art students. Bloemaert had three other painter sons, who like Frederick were his pupils: **Hendrick** (*c*.1601–72), **Cornelis II** (*c*.1603–*c*.1684), and **Adriaen** (1609–66).

Blok. A *Constructivist group formed in Łódź, Poland, in 1922 by Władysław Strzeminski (1893–1952), who had worked with *Malevich during the war. It was also the title of a Constructivist magazine founded at Warsaw in 1924 and fairly close to the outlook of Soviet Constructivism.

Blondeel, Lancelot (1496–1561). Netherlandish artist. He entered the Guild of Painters in Bruges in 1519, and also worked as an architect and designed sculpture, tapestries, and pageant decorations. In 1550 he and Jan van *Scorel were commissioned to restore the van *Eycks' celebrated Ghent Altarpiece. The *triptych of *The Martyrdom of SS. Cosmas and Damian* (S. Jacques, Bruges, dated 1523) is typical of his work as a painter in its profusion of Italianate ornament. Bruges had begun to decline in importance as its port silted up, and Blondeel was one of the last sparks of its great artistic tradition. He devised a plan to link Bruges with the sea to open up trade again, but this was not accomplished until a canal was opened in 1907.

Bloomsbury Group. A loosely knit association of writers, artists, and critics which had an important influence on cultural and intellectual life in Britain during the early decades of the 20th cent. Among the leading members of the group were the writers E. M. Forster (1879–1970) and Virginia Woolf (1882–1941) and the economist John Maynard Keynes (1883–1946); the artists and critics included Clive *Bell, Vanessa *Bell, Dora Carrington (1893–1932), Roger *Fry, Duncan *Grant, and Henry *Lamb. They frequently met at the houses of Clive and Vanessa Bell or of Vanessa's sister, Virginia Woolf, in the Bloomsbury district of London, which had long been a favourite area for artists, musicians, and writers.

The association stemmed from student friendships formed at Cambridge, where many of the group had been 'Apostles'—members of a semi-secret intellectual club. However, the Bloomsbury Group had no formal membership and was unified by no common social or aesthetic ideology; the 'Bloomsberries' were linked rather by attitudes and interests which have caused them to be represented as an intellectual élite in revolt against the artistic, social, and sexual restrictions of Victorian society. A key book for them was *Principia Ethica* (1903) by the Cambridge philosopher G. E. Moore, in which it is argued that 'By far the most valuable things . . . are . . . the pleasures of human intercourse and the enjoyment of beautiful objects . . . it is they that form the rational ultimate end of social progress.'

In the visual arts, it was during the 1920s and early 1930s that the influence of Bloomsbury was most effective. The persistent propaganda of Roger Fry for *Cézanne and the *Post-Impressionists converted ridicule and outraged rejection into interest, if not full understanding. Successive exhibitions of the *London Group were flooded with pastiches in the manner of Duncan Grant. The writings of Fry and Bell tended in the direction of a Formalist aesthetic which played down the importance of content and paved the way for a new method of criticism in the visual arts. If with incomplete consistency, they heralded the reaction from the anecdotal sentimentalism of 19th-cent. criticism and laid the foundation for a more just appreciation of the aims of contemporary art. The group had ceased to exist in its original form by the early 1930s and the suicide of Virginia Woolf in 1941 marked the end of its era. In the 1940s and 1950s the group's aims and achievements fell out of favour and its members were attacked as dilettante and élitist, but since the late 1960s there has been a great revival of interest in all aspects of it.

blot drawing. A technique evolved by Alexander *Cozens and described in *A New Method of Assisting the Invention in Drawing Original Compositions of Landscape* (1785 or 1786). He prescribed the use of an accidental stain or 'blot' on the paper as a basis for an imaginative landscape composition. A similar suggestion had much earlier been made by *Leonardo da Vinci, who proposed that marks on wall surfaces might be used in this way. Somewhat similar techniques were used in the 20th cent. by the *Surrealists for stimulating subconscious imagery.

Blue Rider. See BLAUE REITER.

Blunt, Anthony (1907–83). English art historian. He was Director of the *Courtauld Institute of Art from 1947 to 1974, Surveyor of the King's (later Queen's) Pictures from 1945 to 1972, and one of the leading figures in establishing art history as an academic discipline in Britain. In 1979, however, his career was blighted when—amid clamorous publicity—it was revealed that he had spied for the Soviet Union during his service at the War Office in the Second World War. He wrote on a wide variety of subjects, but is best known for his contributions to the study of French and Italian art and architecture in the 16th cent. and 17th cent., above all for his numerous books and articles on *Poussin. His books include *Artistic Theory in Italy 1450–1600* (1940), *Art and Architecture in France 1500–1700* (1953 and subsequent editions), *The Art of William Blake* (1959), and *Guide to Baroque Rome* (1982). His brother **Wilfrid** (1901–87) was drawing master at Eton 1938–59, Curator of the *Watts Gallery at Compton 1959–85, and the author of numerous books on art and other subjects, notably *The Art of Botanical Illustration* (1950) and *'England's Michelangelo'* (1975), a biography of Watts.

Boccioni, Umberto (1882–1916). Italian *Futurist painter and theorist, and the only sculptor in the movement. In Rome together with *Severini he learnt from *Balla the principles of *Divisionism, which was then the vogue. He joined the Futurists in 1909, helped to draw up their manifestos of painting (1910) and sculpture (1912), and became the most energetic member of the group. Advocating a complete break with the art of the past, Boccioni was centrally concerned with the two main preoccupations of the Futurists—the production of emotionally expressive works and the representation of time and movement. In his book *Futurist Painting and Non-Culture* (1914) he proposed that whereas the *Impressionists painted to perpetuate a single moment of vision, Futurism synthesizes in a picture all possible moments. And in contrast to the objective outlook of *Cubism, he claimed that Futurist painting aspires also to express 'states of the soul'. In common with the other Futurists (following the ideas of the philosopher Henri Bergson, 1859–1941), Boccioni believed that physical objects have a kind of personality and emotional life of their own, revealed by 'force lines' with which the object reacts to its environment. These ideas are perhaps best shown in Boccioni's most famous piece of sculpture, *Unique Forms of Continuity in Space* (casts in Tate, London, MOMA, New York, and elsewhere, 1913), which vividly expresses bodily movement. His theories on sculpture were very forward-looking. He advocated the use of materials such as glass and electric lights and the introduction of electric motors to create movement. However, he died in an accident whilst serving in the Italian army before most of his ideas could be put into practice.

Böcklin, Arnold (1827–1901). Swiss painter. With *Hodler he ranks as the most important Swiss painter of the 19th cent., and in the 1880s and 1890s he was the most influential artist of the German-speaking world, even though from 1850 he had spent most of his time in Italy. He established his reputation with *Pan in the Reeds* (Neue Pinakothek, Munich, 1857), the beginning of his preoccupation with the world of nymphs and satyrs, naiads, and tritons, the results of which are sometimes slightly absurd. Later his style became more sombre and charged with mystical feeling, as in his best-known work, *The Island of the Dead* (Met. Mus., New York, 1880, and four other versions). Such works are considered among the most distinguished *Symbolist paintings produced outside France, and their morbid imagery appealed to the *Surrealists. The best collection of Böcklin's work is in the Kunstmuseum of his native Basle. A curious aspect of his career is that like *Leonardo—whom he disliked—Böcklin spent much of his time experimenting with flying machines.

Bode, Wilhelm von (1845–1929). German art historian. His career was centred on the Berlin Museum; in 1872 he was appointed assistant in the department of sculpture, of which he became director in 1883, in 1903 he became director of the Gemäldegalerie (picture gallery), and in 1905 director general of all the royal museums of Prussia, a post he held until he retired in 1920. Under Bode's administration the Berlin Museum became one of the world's outstanding collections. Apart from being a noted scholar, Bode was one of the pioneers of modern museum organization and display, combining pictures, sculptures, and frames in harmonious arrangements and achieving a balance between creating an up-to-date setting for works of art and reconstructing their historical milieu. 'We had no desire', Bode wrote, 'to model ourselves on some arts-and-crafts museums, but our aim . . . was to give the works of art a modern setting which would heighten their effect . . . Had we merely reproduced old rooms, we would have lessened the monumental effect of the works of art and impaired the character and significance of the museums.' His main publications were in the field of Dutch and Flemish art (particularly *Rembrandt and his contemporaries), but he also wrote on Italian art and assisted *Burckhardt with his *Cicerone*. Bode also wrote a book of memoirs—*Mein Leben* (1930). The Kaiser Friedrich Museum in Berlin, opened in 1904, has been renamed the Bode Museum in his honour.

bodegón. Spanish term, literally meaning 'tavern' or 'chophouse', applied strictly to domestic (particularly kitchen) scenes that have a prominent still-life element. *Velázquez painted several *bodegones* early in his career. More loosely, the term is used as a synonym for still life.

Body art. A type of art in which the artist uses his or her own body as the medium: it is closely related to *Conceptual art and *Performance art, and flourished mainly at the same time that these forms of expression were at their peak—the late 1960s and 1970s. Sometimes works of Body art are executed in private and communicated by means of photographs or films; sometimes the execution of the 'piece' is public. Sometimes the demonstration is pre-choreographed; sometimes it is extemporaneous. Spectator participation is not usually invited. Several leading exponents of Body art have been concerned with self-inflicted pain or ritualistic acts of endurance. For example, in 'Seedbed' (1972) the American Body artist Vito Acconci (1940–) spent several hours daily masturbating under a gallery-wide ramp while the sounds of his activity were relayed via loudspeakers to visitors overhead. The philosophy behind such works is obscure.

body colour. Paint that is opaque rather than transparent, more specifically *watercolour mixed with a white *pigment. *Gouache is an alternative name for body colour used in the more restricted sense.

Boethus. Greek sculptor who worked in the 2nd cent. BC. *Pliny (*Nat. Hist.* xxxiv. 84) mentions a bronze by him representing a *Child Strangling a Goose* by hugging it, though he adds that Boethus was better in silver. The group, known through several copies, appealed to *Renaissance sculptors and echoes of it can be seen in sculpture from *Verrocchio to *Bernini.

Bogart, Martin van den. See DESJARDINS.

Bohemian School. Term applied to art produced in Bohemia in the second half of the 14th cent. during a period of cultural efflorescence associated with Charles IV (1316–78), who was King of Bohemia from 1346 and Holy Roman Emperor from 1355. His favourite residence was Prague, where he founded a university and drew to his court scholars and artists from all over Europe. Manuscripts were imported from France and Italy and inspired a local school of book *illumination, but the main achievements of the Bohemian School were in *panel painting and *fresco.

The masters for the greater part are anonymous and we know nothing about their nationality. Yet they all bear witness to a style in which were fused many influences and which contributed to the formation of *International Gothic. During the 1350s, in the work of the *Master of Vyšší Brod, the flavour was strongly Italian; Sienese models in particular seem to have been used. Later, with the *Master of the Třeboň Altarpiece, who must have worked during the last two decades of the century, French elements became stronger. Under Charles's son Wenceslas painting, and in particular illumination, still flourished, but the importance of Prague as an artistic centre declined early in the 15th cent., when many works perished during the Hussite wars.

Boilly, Louis-Léopold (1761–1845). French painter and engraver. He painted portraits, domestic and *genre scenes, and *scènes galantes*, which brought him into disrepute at the time of the Revolution. Under the **Directoire* he reverted to genre and boudoir scenes and his success lasted under the Empire. In 1823 he took up *lithography and used this technique to popularize his scenes from contemporary life. He was extremely prolific (claiming to have executed 5,000 portraits), and smooth and meticulous in his technique.

Boizot, Simon-Louis (1743–1809). French sculptor. He succeeded *Falconet as Director of Mme de Pompadour's porcelain factory at Sèvres and was at his best working on a small scale (*Cupid*, Louvre, Paris, 1772).

Bol, Ferdinand (1616–80). Dutch painter and etcher. He was a pupil of *Rembrandt in the mid-1630s and in his early work imitated his master's style so well as to create occasional difficulty in distinguishing between them. The portrait of Elizabeth Bas in the Rijksmuseum, Amsterdam, is the best-known instance; it was acknowledged as a Rembrandt until 1911, when it was attributed to Bol by *Bredius, and although this opinion is still generally accepted, there has been renewed support for Rembrandt as the author (as well as some for Jacob *Backer). As Bol's career prospered, both as a portraitist and a painter of historical subjects, his style moved away from that of Rembrandt, becoming blander and more elegant in the manner of van der *Helst. In 1669 he married a wealthy widow and seems to have stopped painting. *Kneller was Bol's most distinguished pupil.

Boldini, Giovanni (1842–1931). Italian painter, one of the most renowned society portraitists of his day. His vivacious brushwork and gift for making his sitters look graceful and poised recalled the work of his even more successful contemporary, John Singer *Sargent, and like Sargent he had an international career, working mainly in Paris, but also in London (1869–72). Apart from portraits, his work includes some excellent street scenes of Paris. There is a Boldini museum in his native Ferrara. See also MACCHIAIOLI.

bole (bolus). A natural clay sometimes used for *grounds, chiefly in *oil painting. Red bole is the commonest, and has served as a ground for gilding since the Middle Ages. It is fairly common, especially in 17th-cent. paintings for the dark ground to 'come through'—particularly in the thinly painted shadows—thus darkening the picture considerably. In general clay grounds are considered unsatisfactory owing to their capacity for retaining moisture.

Bolognese, Il. See GRIMALDI.

Bolotowsky, Ilya (1907–81). Russian-born painter who settled in New York in 1923 and became an American citizen in 1929. In 1933 he was deeply impressed by some paintings by *Mondrian and he became one of America's most committed exponents of geometrical abstraction. He was a founding member of *American Abstract Artists in 1936 and his mural for the Williamsburgh Housing Project, New York, of that year was one of the first abstract murals to be commissioned. Bolotowsky was also a playwright, experimental film maker, and teacher.

Boltraffio, Giovanni Antonio (1466/7–1516). Italian painter, the pupil and assistant of *Leonardo and one of the most talented of his followers in Milan. He painted religious subjects and portraits; the beautiful female portrait in the Louvre known as *La Belle Ferronnière* is attributed to him by some authorities and to Leonardo by others.

Bomberg, David (1890–1957). British painter. He gave up an apprenticeship as a lithographer to devote himself to painting, studying under *Sickert and at the *Slade School. Whilst still a student he showed an advanced understanding of avant-garde Continental painting, particularly *Cubism and *Futurism (he visited Paris in 1913), and he became associated with the *Vorticists, although he disclaimed any formal connection with them. His best-known work of this time is probably *In the Hold* (Tate, London, 1913–14), a dazzlingly coloured abstraction of fragmented geometric forms. After early success, he became dismayed at the failure of a one-man show in 1919, and thereafter worked in isolation. From the 1920s he travelled widely, and at this time began moving away from abstraction to a heavily-worked, somewhat *Expressionist figurative style, painting mainly portraits and landscapes. Much of his later career was devoted to teaching, and he had a strong influence on pupils such as *Auerbach and *Kossoff. At the time of his death he was little appreciated, but his reputation has since soared; a major exhibition was devoted to him at the Tate Gallery in 1988.

Bombois, Camille (1883–1970). French *naïve painter. He passed his childhood on barges along the canals of France (his father was a boatman) and at the age of 12 he became a farmhand. Later he was a labourer and a wrestler in a travelling circus. From 1907 he worked as a porter on the Paris Métro, as a navvy and a docker, and then took a night job in a printing establishment in order to have more time for painting. He served in the First World War and was awarded a Military Medal. In 1922 a pavement exhibition of his pictures attracted the attention of Wilhelm *Uhde and other critics, and thereafter he was able to devote all his time to painting. His pictures have exceptional strength and vitality, particularly his scenes of circus life (*Country Fair Athlete*, Mus. d'Art Moderne, Paris, *c*.1930).

Bonaiuti, Andrea. See ANDREA DA FIRENZE.

Bondol (or **Bondolf** or **Bandol**) **Jean de** (also known as **Jean de Bruges** and **Jan Hennequin**) (active *c*.1368–81). Netherlandish painter and *miniaturist who worked in France and became Court Painter to Charles V (1337–80) of France in 1368. Only two works can certainly be ascribed to him: an illustrated Bible presented to the King (now at the Meermanno-Westreenianum Mus., The Hague); and the design of a celebrated series of tapestries on the Apocalypse (Mus. of Tapestries, Angers, *c*.1375). His style combined French courtly sophistication with Netherlandish realism, looking forward to the *International Gothic style.

Bone, Sir Muirhead (1876–1953). British draughtsman and etcher, mainly of architectural subjects. He studied architecture and painting in his native Glasgow, then settled in London in 1901 and became a member of the *New English Art Club. In 1916 he became the first *Official War Artist to be appointed and he was also an Official War Artist in the Second World War. His son **Stephen** (1904–58) was a painter and art critic.

Bonheur, Rosa (1822–99). French animal painter. Trained by her father, **Raymond Bonheur** (d. 1849), she exhibited regularly at the Paris *Salon from 1841, where her pictures of lions, tigers, wolves, etc., were soon very popular. *The Horse Fair* (1853; Met. Mus., New York, reduced replica in NG, London) gave her an international reputation. She was a colourful and formidable character, outspoken in her feminine independence (she smoked cigarettes and wore trousers), and in 1865 was the first woman to become a member of the Legion of Honour. There is a small museum of her work near Fontainebleau.

Bonifazio Veronese (Bonifazio de' Pitati) (1487–1553). Italian painter. He was born in Verona, but all his recorded activity was in Venice, where he based his style on *Giorgione, *Titian, and *Palma Vecchio. There are few signed, dated, or documented works by him, but he appears to have run a prolific studio. Consequently he has become one of those artists whose names are used as dustbins for dumping difficult attributions. His pupils included Jacopo *Bassano and possibly *Schiavone and *Tintoretto.

Bonington, Richard Parkes (1802–28). English painter, active mainly in France, where his family moved when he was 15. In 1819 he went to Paris, where he entered the École des *Beaux-Arts. He then became a pupil of *Gros, a pioneer of the new *Romantic movement, and formed a friendship with *Delacroix. He was influenced by the medievalism and orientalism of the French Romantics and produced paintings in their manner. However, he established his reputation as a landscapist, particularly with his works exhibited at the *Salon of 1822 and the so-called 'English' Salon of 1824, at which his own paintings (which won him a gold medal) and those of *Constable were the star attractions. In 1825 he accompanied Delacroix to England and sought out pictures by Constable, whose influence is apparent in his subsequent work, and in 1826 he visited Italy, producing some of his finest work in Venice. Bonington was overloaded with work and his delicate health suffered; he died of consumption in London a month before his 26th birthday. Although his career was so brief, Bonington was highly influential, the freshness and spontaneity of his fluid style in both oil and watercolours attracting many imitators. Delacroix wrote of him: 'As a lad he developed an astonishing dexterity in the use of watercolours, which were in 1817 an English novelty. Other artists were perhaps more powerful or more accurate than Bonington, but no one in the modern school, perhaps no earlier artist, possessed the lightness of execution which makes his works, in a certain sense, diamonds, by which the eye is enticed and charmed independently of the subject or of imitative appeal.' These qualities are particularly apparent in the *pochades* (oil sketches done rapidly on the spot as records of transitory effects in nature), a fashion which he together with *Turner and Constable was instrumental in establishing. The best collection of Bonington's work is in the Wallace Collection, London, and he is also well represented in the City Museum and Art Gallery at Nottingham, his home town.

Bonnard, Pierre (1867–1947). French painter and graphic artist. Coming to Paris in 1888 he trained at the *Académie Julian and the École des *Beaux-Arts, his fellow students including *Denis, *Sérusier, and *Vuillard (a lifelong friend), with whom he founded the *Nabis. From 1891 he exhibited fairly regularly at the *Salon des Indépendants and in 1903 he was one of the founder members of the *Salon d'Automne, organized by the *Fauves, exhibiting there regularly from that date. He prospered steadily in his career and his life was quiet and uneventful. Like Vuillard, he is best known for peaceful domestic scenes to which the term *Intimiste is applied. Bonnard generally painted on a larger scale than Vuillard, however, and with greater richness and splendour of colour. His favourite model was his wife, and some of his most characteristic pictures are those in which he depicted her in the bath (she had an obsession with personal cleanliness and spent much of her time in the bathroom). His other subjects included flowers and landscapes. He also did numerous self-portraits. The late ones show his desolation after the death of his wife in 1940, but in general his work radiates a sense of warmth and well-being. This quality and his lively broken brushwork make him one of the most distinguished upholders of the *Impressionist tradition. On the death of his wife Bonnard faked a will in her name to deceive the authorities and this caused grave legal complications as a result of which the substantial part of his output remaining in his possession at his own death was sequestered from public view for many years owing to long-drawn-out lawsuits concerning inheritance. Bonnard was one of the very few foreign painters to be elected a member of the *Royal Academy (1940) and in 1966 the Academy organized the largest and most representative exhibition of his works hitherto seen. The appreciation printed in the catalogue to this exhibition described him as the most important 'pure painter' of his generation.

Bonnat, Léon (1833–1922). French painter and collector. Bonnat's early works were mainly religious paintings in a *tenebrist style influenced by 17th-cent. Spanish painting, but from about 1870 he turned increasingly to portraiture. His portraits are usually as glum as his religious paintings, but their almost photographic realism won them an appreciative audience. Most of the notables of the Third Republic sat for him and the fortune he earned painting them enabled him to form a superb art collection, particularly of Old Master drawings. He donated it to Bayonne, his native city, where it forms the nucleus of the Musée Bonnat. His studio and personal effects can be seen in the nearby Musée Basque. Bonnat was a renowned teacher, his many pupils including *Toulouse-Lautrec and *Braque.

Bontemps, Pierre (*c.*1505/10–68). French sculptor, first documented in 1536 as an assistant to *Primaticcio at *Fontainebleau. By 1550 he was in Paris, working on two important monuments for the royal burial church at S. Denis—the reclining effigies and bas-*reliefs for the tomb of Francis I, designed by the architect Philibert Delorme (*c.*1510–70), and the monument for the heart of Francis I. Only one other work is documented as being by him—the tomb of Charles de Maigny (1557) in the Louvre—but other works of the period are confidently attributed to him on stylistic evidence, and he seems to have been the foremost French tomb sculptor of the mid-16th cent. His style was elegant and decorative.

Book of Hours. A prayer book used by laymen for private devotion, containing prayers or meditations appropriate to certain hours of the day, days of the week, months, or seasons. They originated from prayers gradually added by monks and priests to the Service proper, and became so popular in the 15th cent. that the Book of Hours outnumbers all other categories of *illuminated manuscripts. The most famous Book of Hours and one of the most beautiful of all illuminated manuscripts is the *Très Riches Heures du duc de Berry* (Musée Condé, Chantilly), illuminated by the *Limburg Brothers for Jean de Berry. From the later 15th cent. there are various printed versions illustrated by *woodcuts.

Book of the Dead. Term used with reference to Egyptian funerary literature to describe a miscellaneous collection of formulae and incantations of various dates, selections from which were inscribed inside coffins and later on *papyri buried with the dead. The best examples are finely written, with headings in red, and beautifully illustrated with vignettes, often in colour. They are thus among the earliest examples of *rubrication, *illumination, and the art of book illustration. The name 'Book of the Dead' was invented by the German Egyptologist Richard Lepsius, who was the first to publish a collection of the texts in 1842.

Borch, Gerard the Younger. See TERBORCH.

Bordone, Paris (1500–71). Italian painter. Bordone was from Treviso, but by 1518 he had settled in Venice. *Vasari says he was a pupil of *Titian, but found his teaching disagreeable and soon left (Titian is then said to have stolen his first commission). Whatever the truth of these stories, Bordone's work was certainly strongly influenced by Titian and also by *Giorgione, 'for that master's style pleased him exceedingly' (Vasari). He painted Giorgionesque pastoral scenes and mythologies that now seem rather hard and conventional compared with their inspiration, but they won him great popularity. Commissions came from patrons all over Europe, including the King of Poland, and he also visited France and Germany. His most impressive work is generally agreed to be *The Presentation of the Ring of St Mark to the Doge* (Accademia, Venice, *c.*1535), a large ceremonial composition in Titian's grand manner.

Borduas, Paul-Émile (1905–60). Canadian painter, active mainly in Montreal. He trained as a church decorator under Ozias *Leduc, then studied in Paris. In the early 1940s, under *Surrealist influence, he started to produce 'automatic' paintings (see AUTOMATISM), and with *Riopelle founded the radical *abstract group Les Automatistes. His later paintings have an *all-over surface animation recalling the work of *Pollock, although the only American influence Borduas acknowledged was that of Franz *Kline. In 1953–5 he lived in New York, then spent his final years in Paris. He ranks with Riopelle as one of the most important Canadian abstract painters of the post-war years and like Riopelle he was widely influential in his country.

Borghese Gallery, Rome. Italian state museum housed in the Villa Borghese. The villa was built (1613–15) for Cardinal Scipione Borghese (1576–1633), nephew of Pope Paul V (Camillo Borghese, 1552–1621), and their collections of paintings and sculpture form the nucleus of the museum. Scipione was *Bernini's first important patron, and the Gallery has an unrivalled representation of the sculptor's early work, including two *busts of Scipione. The collection of paintings includes outstanding works by *Caravaggio, *Raphael, and *Titian. The Borghese collection was one of the few Roman patrician collections not dispersed in the 18th cent. Marcantonio Borghese added important antiquities to the collection then, as did Francesco Borghese in the 19th cent., and the villa and its contents were acquired by the Italian government from the Borghese family in 1902.

Borghese Warrior (*Borghese Gladiator*). Marble statue (Louvre, Paris) of a nude warrior in a vigorous attitude of combat (his sword and shield are missing, but he is evidently lunging at an opponent on horseback in a type of pose that has been described as a 'heroic diagonal'). It was discovered in 1611 at Nettuno (near Anzio), had entered the *Borghese collection by 1613, and was bought by Napoleon (brother-in-law of Prince Camillo Borghese) in 1807. The statue is signed by 'Agasias, son of Dositheos, Ephesian' and is generally considered to be a copy of a *Hellenistic work done under the influence of *Lysippus. It became famous soon after its discovery and for two centuries it was one of the most admired and copied of antique statues, praised particularly for its anatomical mastery: *Bernini's *David* is an early instance of a derivation from it and a more curious adaptation is found in *Copley's *Brook Watson and the Shark*, in which the figure of Watson—horizontal in the water—is based, in reverse, on the *Warrior*. It is now much less admired as a work of art, Martin Robertson (*A History of Greek Art*, 1975) describing it as 'harsh and unappealing'.

Borglum, Gutzon (1867–1941). American sculptor of Scandinavian descent. He carried to an extreme the American cult for the colossal (what his wife called 'the emotional value of volume') and gained renown with gigantic reliefs on mountainsides executed with pneumatic drills and dynamite. The first of these, on Stone Mountain in Georgia, commemorating the Confederate Army, was never finished, although he worked on it for 10 years from 1915. He next undertook (1930) the carving of a huge cliff at Mount Rushmore in the Black Hills of South Dakota with colossal portrait busts of Washington, Jefferson, Lincoln, and Theodore Roosevelt. Washington's head was completed in 1930, Jefferson's in 1936, Lincoln's in 1937, and Theodore Roosevelt's in 1939; the final details were added in 1941 after Borglum's death by his son Lincoln. The project was sponsored by the US Government and cost over $1,000,000.

Solon Hannibal Borglum (1868–1922), Gutzon's brother, was also a sculptor, mainly of Wild West subjects.

Borgognone, Ambrogio. See BERGOGNONE.

Borgognone, Il. See COURTOIS, JACQUES AND GUILLAUME.

Borgoña, Juan de (active *c.*1494–d. 1554). Spanish painter who worked mainly at Toledo. Borgoña and Pedro *Berruguete were both representatives of the transition from *Gothic to *Renaissance in Castile and in 1508 he took over work on the main altarpiece of Avila Cathedral which Berruguete had left unfinished at his death. He also executed important works in the Chapter House of Toledo Cathedral. His style recalls late 15th-cent. Florentine painting, more particularly that of *Ghirlandaio. He had several followers, such as Antonio de Comontes (active *c.*1500–19) and Pedro Cisneros (active *c.*1530), and this makes many attributions to Borgoña uncertain.

Borman (or **Borreman**), **Jan I** (active *c.*1479–*c.*1520). Netherlandish sculptor in wood. He was head of the busiest workshop in Brussels, famous for sculptured altars, which were much exported, particularly to Scandinavia. The *Altar of St George* (Musée du Cinquantenaire, Brussels, 1493) was his masterpiece. His two sons, **Jan II** and **Pasquier**, continued his tradition.

Borofsky, Jonathan. See NEW IMAGE PAINTING.

Borrassá, Luis (d. *c.*1425). Spanish painter, a disciple of Pedro *Serra, recorded as active in Barcelona and its neighbourhood from 1388 to 1424. His style shows French and Sienese influences and is representative of the *International Gothic style. Several of his documented works survive, for example, the great composite altarpiece of Sta Clara, executed 1412–15, now in Vich Museum, Barcelona.

Bosboom, Johannes (1817–91). Dutch painter and lithographer, a member of The *Hague School. He specialized in paintings of church interiors and was much inspired by the works of Emanuel de *Witte—figures in many of his church interiors even wear 17th-cent. costumes. He is represented in the Municipal Museums of Amsterdam and The Hague.

Bosch, Hieronymus (*c.*1450–1516). Netherlandish painter, named after his native town of 's Hertogenbosch (Bois-le-Duc) in northern Brabant, where he seems to have lived throughout his life. His real name was Jerome van Aken (perhaps indicating family origins in Aachen, Germany). Bosch married well and was successful in his career (although his town was fairly isolated, it was prosperous and culturally stimulating). He was an orthodox Catholic and a prominent member of a local religious brotherhood, but his most characteristic paintings are so bizarre that in the 17th cent. he was reputed to have been a heretic.

About forty genuine examples of Bosch's work survive, but none is dated and no accurate chronology can be made. It seems likely, however, that the conventional compositions, such as *The Crucifixion* (Musées Royaux, Brussels), are early works. The paintings for which he is famous are completely unconventional and are immediately recognizable by the fantastic half-human half-animal creatures, demons, etc., that are interspersed with human figures in a setting of imaginary architecture and landscape. The basic themes are sometimes quite simple, but heavily embroidered with subsidiary narratives and symbols. Scenes from the life of Christ or a saint show the innocent central figure besieged by horrific representations of evil and temptation—*The Temptation of St Anthony* (Museu Nacional de Arte Antiga, Lisbon) is the most spectacular instance. Other subjects were allegorical representations of biblical texts or proverbs, stressing in morbid vein the greed and folly of human beings and the fearful consequences of their sins. Of these, *The Haywain* and *The Garden of Earthly Delights*, both in the Prado, are perhaps his best-known paintings.

Although his father was a painter, the origins of Bosch's style and technique are far from clear. His manner had little in common with those of Jan van *Eyck or Rogier van der *Weyden, the two painters who most influenced the development of style in the Low Countries until *c.*1500. There is, indeed, something strangely modern about Bosch's turbulent and grotesque fantasy and it is no surprise that his appeal to contemporary taste has been strong. But attempts to discover the psychological key to his motivation or to trace the origin of his imagery or find a coherent interpretation of the symbolism remain inconclusive. In his own time his fame stood high and a generation or so after his death his paintings were avidly collected by Philip II (1556–98) of Spain, whose fine collection of his works is now mainly in the Prado. Through the medium of prints his works reached a wider public and imitators appeared even in his lifetime. But it was not until Pieter *Bruegel the Elder that another Netherlandish artist appeared with a genius strong enough to extend Bosch's vision rather than simply pastiche it. Apart from the riot of fantasy and that element of the grotesque which caused the *Surrealists to claim Bosch as a forerunner, the haunting beauty of his genuine works derives largely from his glowing colour and superb technique, which was much more fluid and painterly than that of most of his contemporaries. Bosch was also an outstanding draughtsman, one of the first to make drawings as independent works.

Boshier, Derek (1937–). British painter, sculptor, designer, and experimental artist. He was one of the *Royal College of Art students who put British *Pop art on the map at the *Young Contemporaries exhibition of 1961. His work at this time was much concerned with the manipulative forces of advertising, treating the human figure in the same way as mass-produced goods and making them coalesce. However, his involvement in Pop art was short lived, and from 1966 he worked in sculpture, photography, film, and *Conceptual art before returning to painting in 1979. In 1980 he became Assistant Professor of Painting at the University of Texas. Although his career has been fairly low key compared with those of some of his Pop art colleagues from the early 1960s, many critics think that Boshier's work has stood the test of time at least as well as theirs.

boss (or **roof boss**). Architectural term for a block of wood or projecting keystone to mask the junction of vaulting ribs. French *Gothic vaults were usually high so their bosses were seldom elaborately decorated, but in England from the 13th to the 16th cent. there is a wealth of sculptural decoration on bosses.

Bosschaert, Ambrosius (1573–1621). Flemish flower and still-life painter, active mainly in the Netherlands, where he is recorded in Middelburg from 1593 to 1613 and later in the Utrecht Guild in 1616. Although he spent the major part of his life in the Netherlands, Bosschaert's style was basically Flemish—similar to that of Jan *Brueghel, with whom he ranks in quality and as one of the pioneers of flower painting as an independent genre. His bouquets have a rich variety of flowers from different seasons arranged in a formal way. He often painted on copper and the surface of his paintings has a mysterious sheen in which individual brush strokes are not apparent. The degree of finish and exactitude, and the subtlety of the colour, are exceptional. His *Vase of Flowers* (Mauritshuis, The Hague, *c.*1620) is one of the most reproduced of all flower pieces. Bosschaert may fairly be said to have initiated the Dutch tradition of flower painting and his style was continued by his three sons, **Ambrosius the Younger** (1609–45), **Abraham** (1613–43), and **Johannes** (*c.*1610–*c.*1650), and his brother-in-law, Balthasar van der *Ast.

Bosse, Abraham (1602–76). French engraver. His large output (more than 1,500 prints) provides a rich source of documentation on 17th-cent. French life and manners. Many of his engravings (such as *Les Métiers*, a series on tradespeople) are *genre scenes, and even his religious works are in modern dress. Bosse taught perspective at the Académie Royale (see ACADEMY) from its foundation in 1648 until 1661, when he was expelled for quarrelling with his colleagues over his opposition to *Lebrun's dogmatic theories. He also wrote treatises on engraving, painting, and architecture and occasionally painted.

Botero, Fernando (1932–). Colombian painter. His early work was influenced by various styles, including *Abstract Expressionism, but in the late 1950s he evolved a highly distinctive style in which figures look like grossly inflated dolls; sometimes his paintings are sardonic comments on modern life, but he has also made something of a speciality of parodies on the work of Old Masters. Since the early 1970s he has lived mainly in New York and he has acquired an international reputation accompanied by extravagant prices for his paintings in the saleroom.

Both, Jan (*c.*1618–52). Dutch painter, with Nicolaes *Berchem the most celebrated of the Italianate landscape painters. He came from Utrecht, where he studied with *Bloemaert before moving to Italy for a period of about four years, *c.*1637–41. Although he died young, his output was large, but none of the more than 300 paintings attributed to him can be convincingly dated to his Italian sojourn. His landscapes are typically peopled by peasants driving cattle or travellers gazing on Roman ruins in the light of the evening sun. Such contemporary scenes were an innovation, for *Claude Lorraine and the earlier Dutch painters of the Italian countryside had populated it with biblical or mythological figures. They express the yearning of northerners for the light and idyllic life of the south, and proved immensely popular with collectors, not least in England, helping to shape ideas about Italy for two centuries. Jan's brother **Andries** (*c.*1612–41) lived with him in Rome from 1639 to 1641 and they are said to have collaborated; but Andries is best known for paintings and drawings of lively peasant scenes which have little in common with Jan's idyllic tone. He was drowned in an accident in Venice.

Botticelli, Sandro (Alessandro di Mariano Filipepi) (1444/5–1510). Florentine painter, long neglected but now probably the best-loved painter of the *quattrocento. His nickname, meaning 'little barrel', was originally given to an older brother, presumably because he was portly, but it became adopted as the family surname. He trained with Filippo *Lippi, who was the most important influence on his style. By temperament he belonged to the current of late 15th-cent. art which reacted against the scientific naturalism of *Masaccio and his followers and revived certain elements of the *Gothic style—a delicate sentiment, sometimes bordering on sentimentality, a feminine grace, and an emphasis on the ornamental and evocative capabilities of line.

Almost all Botticelli's life was spent in Florence, his only significant journey from the city being in 1481–2, when he worked on the decoration of the Sistine Chapel in the Vatican, where he painted side by side with *Perugino, *Rosselli, and *Ghirlandaio. The fact that he was called to Rome for such a prestigious commission shows that he must have had a considerable reputation, and by this time the most characteristic idiosyncrasies of his style had already gained shape in the celebrated poetic allegory known since *Vasari as the *Primavera* (Uffizi, Florence, *c.*1478). There is evidence that the patron who commissioned this and two of his other famous mythological paintings (*The Birth of Venus* and *Pallas and the Centaur*, both in the Uffizi) was Lorenzo di Pierfrancesco de' *Medici (second cousin of Lorenzo the Magnificent), a wealthy Florentine with strong interests in Platonic philosophy. It has been suggested that it was this philosophy that prompted the new idea of large-scale pictures with a secular content; the classical deities represented are not the carefree Olympians of Ovid's tales but the symbolic embodiment of some deep moral or metaphysical truth. Given that the Neo-Platonists regarded Beauty as the visible token of the Divine, there would be no blasphemy in using the same facial type and expression for Venus and for the Holy Virgin.

According to Vasari, Botticelli later fell under the sway of Savonarola's sermons, repented of his 'pagan' pictures, and gave up painting. The final part of this statement is definitely incorrect and the rest is doubtful, but it is certainly true that Botticelli's later paintings are more obviously 'serious'—solemn, intense, sometimes ecstatic—than his early work. The most telling monument of this phase is the *Mystic Nativity* (NG, London, 1500), which bears a cryptic inscription seeming to imply that Botticelli expected the end of the world and the dawn of the millennium.

Botticelli ran a busy studio (his most important pupil was Filippino *Lippi) and his surviving output is large for a painter of his period. Apart from religious and mythological pictures, he produced some memorable portraits and also some marvellously delicate drawings—mainly in pen outline—for a lavish manuscript of Dante's *Divine Comedy* (now divided between the Staatsbibliothek, Berlin, and the Vatican Library). Although little is known of his life, it seems clear that at the peak of his career he was the most popular painter in Florence. His patrons included some of the city's finest churches and most distinguished families, and several of his paintings (particularly those on the theme of the Virgin and Child) exist in several versions or copies, attesting to the vogue they enjoyed. After *Leonardo's return to the city in 1500, however, Botticelli's linear style must have looked archaic and he died in obscurity. His fame was not resurrected until the second half of the 19th cent., when the *Pre-Raphaelites imitated his wan, elongated types, *Ruskin sang his praises, and Walter *Pater dedicated to his art one of the most eloquent essays in his *Studies in the History of the Renaissance* (1873). At the end of the century his work was a major influence on *Art Nouveau.

Botticini, Francesco (Francesco di Giovanni) (*c.*1446–97). Florentine painter. His style consists almost entirely of elements drawn from his more illustrious contemporaries—*Botticelli, Domenico *Ghirlandaio, Filippino *Lippi, *Verrocchio. He painted one remarkable work, however, the *Assumption of the Virgin* (NG, London, *c.*1474), which has the distinction of being the only picture from the *quattrocento known to have been painted to illustrate a heresy. The *donor, Matteo Palmieri, believed that human souls are the angels who stayed neutral when Satan rebelled against God.

boucharde (or **bush hammer**). A sculptor's mallet or hammer studded with V-shaped indentations, used for wearing down the surface of hard stones. It is inclined to stun or bruise the stone, leaving marks that show through the final polish and for this reason is rarely used when the work approaches the final surface.

Bouchardon, Edmé (1698–1762). French sculptor whose work marks the beginning of the *Neoclassical reaction against the *Rococo style. From 1723 to 1732 he worked in Rome, where he made a marble bust of the antiquarian Philippe Stosch (Staatliche Museen, Berlin, 1727) that is very consciously in the *antique manner. Although his style later softened somewhat, notably in the famous *Cupid Making a Bow from Hercules' Club* (Louvre, Paris, *c.*1750), it remained too severe for court taste. Bouchardon had many supporters, however, and his contemporary reputation stood high—indeed he was generally regarded as the greatest French sculptor of his time (subsequent taste has inclined more towards artists with greater warmth, such as *Falconet and *Pigalle). His most important work was an equestrian statue of Louis XV, commissioned by the City of Paris in 1748. It was cast in 1758 but not erected until 1763, a year after Bouchardon's death. It stood in the Place Louis XV (later the Place de la Concorde) and was destroyed during the Revolution. Several small copies exist, as well as engravings, showing that it was based on the famous antique statue of Marcus Aurelius in Rome. Bouchardon's father, **Jean-Baptiste** (1667–1742), and his brother, **Jacques-Philippe** (1711–53), were also sculptors.

Boucher, François (1703–70). French *Rococo painter, engraver, and designer, who best embodies the frivolity and elegant superficiality of French court life at the middle of the 18th cent. He was for a short time a pupil of François *Lemoyne and in his early years was closely connected with *Watteau, 125 of whose pictures he engraved for Jullienne's *Œuvre de Watteau*. In 1727–31 he was in Italy, and on his return was soon busy as a versatile fashionable artist. His career was hugely successful and he received many honours, becoming Director of the *Gobelins factory in 1755 and Director of the Academy and King's Painter in 1765. He was also the favourite artist of Louis XV's most famous mistress, Mme de Pompadour (1721–64), to whom he gave lessons and whose portrait he painted several times (Wallace Coll., London; NG, Edinburgh). Boucher mastered every branch of decorative and illustrative painting, from colossal schemes of decoration for the royal châteaux of Versailles, Fontainebleau, Marly, and Bellevue, to stage settings for the opera and designs for fans and slippers. In his typical paintings he turned the traditional mythological themes into wittily indecorous *scènes galantes*, and he painted female flesh with a delightfully healthy sensuality, notably in the celebrated *Reclining Girl* (Alte Pinakothek, Munich, 1751), which probably represents Louis XV's mistress Louisa O'Murphy.

Towards the end of his career, as French taste changed in the direction of *Neoclassicism, Boucher was attacked, notably by *Diderot, for his stereotyped colouring and artificiality; he relied on his own repertory of motifs instead of painting from the life and objected to nature on the grounds that it was 'too green and badly lit'. Certainly his work often shows the effects of superficiality and overproduction, but at its best it has irresistible charm and great brilliance of execution, qualities he passed on to his most important pupil, *Fragonard.

Boucicaut (or **Boucicault**) **Master** (active early 15th cent.). Franco-Flemish manuscript *illuminator, named after a *Book of Hours done for Jean II le Meingre Boucicaut (1365–1421). (Musée Jacquemart-André, Paris). This manuscript, which was presumably commissioned before 1415, when Boucicaut was captured at the Battle of Agincourt, is a magnificent example of the *International Gothic style, but in its accomplished handling of space and *aerial perspective and its delightful *genre detail it heralds the achievements of the 15th-cent. Netherlandish School. Many other manuscripts are attributed to the workshop of the Boucicaut Master, who is sometimes identified with Jacques Coene, an artist with a high contemporary reputation but by whom no works are known.

Boudin, Eugène (1824–98). French painter. Son of a sailor, he ran a stationery and picture-framing business at Le Havre (1844–9), where his clients included Jean-François *Millet, who encouraged him to paint. *Courbet, *Jongkind, and *Corot were among his friends. He was a strong advocate of direct painting from nature, and had a great influence on the young *Monet, whom he introduced to **plein-air* painting. Boudin's own paintings consist mainly of beach scenes and seascapes from the coast of northern France and are distinguished by the prominence given to luminous skies. He is regarded as a link between the painters of the generation of Corot and the *Impressionists, and he exhibited in the first Impressionist exhibition of 1874. There is a Boudin Museum in Honfleur, his native town, and examples from his prolific output can be found in many other galleries.

Bouguereau, Adolphe-William (1825–1905). French painter. In 1850 he won the *Prix de Rome, and after his return to France in 1854 he became an immensely successful and influential exponent of academic art, upholding traditional values and contriving to exclude avant-garde work from the *Salon—*Cézanne once expressed regret at being excluded from the 'Salon de Monsieur Bouguereau'. He painted portraits of photographic verisimilitude, slick and sentimental religious works, and coyly erotic nudes. For many years damned unequivocally as a 'master in the hierarchy of mediocrity' (J.-K. Huysmans) and an opponent of all progressive ideas, Bouguereau has recently achieved something of a rehabilitation, his work becoming the subject of serious study and fetching huge prices in the saleroom.

Boullongne, Louis de the Elder (1609–74). French painter, chiefly of religious works, the father of two painter sons and two painter daughters. **Bon** (1649–1717) and **Louis the Younger** (1654–1733) had considerable success as decorative painters (both worked at Versailles) and are considered among the pioneers of the *Rococo style. **Geneviève** (1645–1709) and **Madeleine** (1646–1710) were still-life painters.

Bourdelle, Émile-Antoine (1861–1929). French sculptor, born at Montauban. As a boy he obtained practical experience of carving in the workshop of his father, a cabinet maker. In 1876 he began to study at the École des Beaux-Arts, Toulouse, from where he won a scholarship to the École des *Beaux-Arts, Paris, in 1884. However, he shortly left the school and worked for a while with Jules *Dalou before becoming *Rodin's chief assistant from 1893 to 1908. Bourdelle's work has been somewhat overshadowed by his association with Rodin (whom he revered), but he was already an accomplished artist when he started working for him and developed an independent style. His energetic, rippling surfaces owe much to Rodin, but his flat rhythmic simplifications of form, recalling *Romanesque art, are more personal. He was particularly interested in the relationship of sculpture to architecture, and his reliefs for the Théâtre des Champs-Élysées (1912), inspired by the dancing of Isadora Duncan (1878–1928), are among his finest works. Bourdelle had many other prestigious public commissions and also achieved great distinction as a teacher, his studio becoming a school called La Grande Chaumière. From about 1910 he was generally regarded as the outstanding sculptor in France apart from Rodin himself. He was also a talented painter and draughtsman. His house and studio in Paris have been converted into the Musée Bourdelle; the first part opened in 1961 to mark the centenary of his birth.

Bourdichon, Jean (*c*.1457–1521). French painter, the most important pupil of *Fouquet. He was active in Tours, where he worked for several royal patrons, including Charles VIII (1470–98), Louis XII (1462–1515), and Anne of Brittany, who married each of them in turn. For this queen consort Bourdichon produced his most celebrated work—the *Hours of Anne of Brittany* (completed 1508), now in the Bibliothèque Nationale, Paris (see BOOK OF HOURS). It contains numerous exquisite borders of plants and insects, together with fifty-one large scenes—mainly from the New Testament and lives of the saints, but also including a portrait of Anne at prayer. Some of the religious scenes show such strong Italianate influence that it seems almost certain Bourdichon had visited Italy. Bourdichon is recorded as having painted works on a larger scale, but apart from a *triptych of *The Madonna and Child with Saints* in the Museo di Capodimonte, Naples, all his other known works are manuscript *illuminations. He effectively ends the great French tradition of illumination.

Bourdon, Sébastien (1616–71). French painter. In 1634–7 he worked in Rome, where he developed a talent for imitating the work of other painters—*Claude, *Dughet, van *Laer—sometimes probably with intent to deceive. He continued in this vein when he returned to France and his *œuvre* is still ill-defined. From 1652 to 1654 he was court painter to Queen Christina of Sweden, of whom he did two portraits (Prado, Madrid, and Nationalmuseum, Stockholm), and after his return to France he worked mainly as a portraitist, developing a more personal style in which soft tonalities and skilful play with cascading draperies create a languorous, romantic effect (*Self-portrait*, Louvre, Paris). He was one of the founder members of the Académie Royale in 1648 (see ACADEMY).

Bourgeois, Louise (1911–). French-American sculptor, born in Paris, where part of her training was with *Léger. She married the American art historian Robert Goldwater in 1938 and settled in New York. Bourgeois started as a painter and engraver and turned to sculpture only in the late 1940s. She first achieved recognition in the 1950s for her wood constructions painted uniformly black or white, which preceded the similar works of Louise *Nevelson. Since then Bourgeois has worked in various materials, including stone, metal, and latex, and has built up a reputation as one of the leading contemporary American sculptors. Although her work is abstract, it is often suggestive of the human figure, and sexual significance is sometimes ascribed to it.

Bourgeois, Sir Peter Francis, Bt (1756–1811). English painter and collector of Swiss parentage. His work as a painter is now forgotten, and he is remembered for his bequest of 371 paintings to Dulwich College; they formed the nucleus of the Dulwich College Picture Gallery, which opened to the public in 1814. This was the first public art gallery to be opened in England, predating the National Gallery in London by a decade.

Boursse, Esaias (1631–72). Dutch painter, active mainly in his native Amsterdam. He joined the Dutch East India Company in 1661 and made two voyages to the Indies, on the second of which he died at sea. His few surviving paintings include some exquisite interior scenes that invite comparison with *Vermeer because of their tranquil beauty and subtle colour harmonies (*Interior: Woman Cooking*, Wallace Coll., London, 1656).

Bouts, Dirk (or **Dieric**) (d. 1475). Netherlandish painter, born probably in Haarlem and active mainly in Louvain, where he was city painter from 1468. His major commissions there were the *Last Supper* altarpiece for the church of S. Pierre (still *in situ*, 1464–7) and two panels (out of a projected set of four) on the *Justice of Emperor Otto* for the Hôtel de Ville (Musées Royaux, Brussels, 1470–5). Apart from these, there are no documented works, but his style is highly distinctive and a convincing *œuvre* has been built up for him. His static figures are exaggeratedly slender and graceful, and often set in landscapes of exquisite beauty. There is little action, but deep poetry of feeling. Sources for his work have been sought in the mysterious Albert van *Ouwater (who likewise seems to have had Haarlem connections), Rogier van der *Weyden, and Petrus *Christus, but the individuality of Bouts's work transcends any models. His style was highly influential and was continued by his two sons, **Dieric the Younger** (*c*.1448–90/1) and **Aelbrecht** (*c*.1450/60–1549). Particularly popular were small devotional images of the *Mater Dolorosa* and *Christ Crowned with Thorns*; there are three examples in the National Gallery, London, one described as 'style of Aelbrecht Bouts', the other two as 'studio of Dieric Bouts'.

Bowes, John (1811–85). English collector. The illegitimate son of the 10th Earl of Strathmore, he married Josephine Benoîte Coffin Chevallier (1825–1874), a French actress and amateur painter, in 1852, and they devoted much of their wealth and energy to collecting. In 1869 they began to build an enormous museum at Barnard Castle in County Durham (near to the Strathmores' home at

Streatlam) and it was opened to the public in 1892 (by which time the founders were dead). The building was designed by a French architect, Jules Pellechet (1829–1903), and *Pevsner describes it as 'big, bold, and incongruous, looking exactly like the town hall of a major provincial town in France . . . gloriously inappropriate for the town to which it belongs'. It has also been described as a 'Taj Mahal on the Tees'. The Bowes Museum is particularly rich in French paintings and applied art of the 18th cent. (it has been called 'the *Wallace Collection of the North'), but the most remarkable area of the collection is the fine representation of Spanish painting—the best in Britain outside London.

Boyd, Arthur (1920–). Australian painter, potter, etcher, lithographer, and ceramic artist, a member of a family who have made a name in many of the arts. His father was a sculptor and potter, his mother a painter, but he was largely self-taught as an artist. After holding his first one-man show at the age of 17, his artistic career was interrupted by the Second World War. In the 1950s he became well known in Australia, particularly for his large ceramic totem pole at the entrance to the Olympic Pool, Melbourne (for the 1956 Olympic Games), and for his series (twenty pictures) *Love, Marriage and Death of a Half-Caste* (1957–9) concerned with the life and death of an aboriginal stockman and his half-caste bride. They are done in a style combining elements of *Expressionism and *Surrealism. Boyd moved to England in 1959 and soon established a reputation with a one-man show, followed in 1962 by a successful retrospective exhibition at the Whitechapel Gallery. With *Nolan he has become probably the best-known Australian artist of his generation.

Boydell, John (1719–1804). English engraver and print publisher. He made a fortune in the 1740s by publishing views of England and Wales, which he engraved from his own drawings. Later he published the work of other engravers and by developing a large foreign trade spread the fame of English artists and engravers on the Continent. In 1790 he was Lord Mayor of London. His most ambitious undertaking was the celebrated Shakespeare Gallery: from 1786 he commissioned from major artists (including *Fuseli, *Reynolds, and *Romney) 162 oil paintings illustrating Shakespeare's plays, exhibiting them in a purpose-built gallery in Pall Mall, opened in 1789. The engravings after them were published as illustrations to a nine-volume edition of Shakespeare in 1802 and separately in 1803. Boydell hoped by this venture to encourage the rise of a 'great national school of *history painting', and he intended to leave the collection to the nation, but he had heavy losses during the French wars and it was sold by lottery in 1805 shortly after his death (William *Tassie won the main prize). Few of the paintings survive. Boydell's nephew, **Josiah Boydell** (1752–1817), was a painter and engraver, John's partner and successor in his engraving business.

Boys, Thomas Shotter (1803–74). English watercolour painter and lithographer. For some time he lived in France, where he was a friend of *Bonington, but he settled in England in 1837. He specialized in continental urban scenes and in 1839 he published *Picturesque Architecture in Paris, Ghent, Antwerp, Rouen, etc.*, a work which marked the transition from hand-tinted *lithography to chromolithography. In 1842 he published *Original Views of London as it is*, the plates of which, drawn and lithographed by himself, constitute a fine topographical record of Regency London.

bozzetto. Italian term for a sculptor's small-scale model, usually in wax or clay, for a larger work in more durable material. The term is sometimes also applied to a painted sketch.

Bracquemond, Félix (1833–1914). French engraver, painter, and designer. He was an accomplished painter, particularly of *Impressionist landscapes, but his main importance was as an engraver. He helped to found the Société des Aquafortistes (Society of Etchers) in 1862 and played a major role in reviving engraving as a creative rather than merely a reproductive technique. His specialities were portraits and scenes involving birds.

Brailes, W. de (active *c.*1230–*c.*1260). English manuscript *illuminator. His signature on his work gives his name only as 'W. de Braile', but he is almost certainly to be identified with a William de Brailes, who is recorded in various civic records in Oxford, living or working in the illuminators' quarter in Catte Street. In an illustration of the Last Judgement from a Bible or Psalter in the Fitzwilliam Museum, Cambridge, he depicts himself with a tonsure, from which it may be concluded that he was an ecclesiastic, but whether monastic or secular cannot be certain. The character of his work, however, has led to the assumption that he was secular. Several manuscripts and a number of loose leaves have been attributed to him, and differences in quality of execution suggest that he was head of a workshop with various assistants. His style was lively and inventive, and with Matthew *Paris he ranks as one of the few distinctive personalities in the field of medieval English illumination.

Bramante, Donato (Donato di Angelo) (1444–1514). Italian architect and painter. Bramante was the creator and greatest exponent of the High *Renaissance style in architecture, but most of his early career, which is ill-documented, seems to have been devoted to painting. He probably trained in Urbino and is first documented in 1477 working on fresco decorations at the Palazzo del Podestà in Bergamo. In about 1480 he settled in Milan, and in 1481 produced his earliest surviving dated work, the design of an engraving of an elaborate architectural fantasy (the British Museum, London, possesses one of only two known impressions). At about the same time he began his first building, Sta Maria presso S. Satiro, Milan, in which his knowledge of *perspective was used to create an illusion of recession in the choir, which is in reality only a few inches deep. His only certain surviving paintings are poorly preserved frescos of armed men (*c.*1480–5) in the Brera, Milan, which also houses the finest painting attributed to him (on the testimony of *Lomazzo), a sombre and poignant *Christ at the Column*, which shows some influence from his friend *Leonardo. In 1499 Bramante left Milan for Rome, where in 1506 he began the rebuilding of St Peter's. There is no evidence of any activity as a painter in Rome, but *Vasari says that Bramante designed the majestic architectural setting of *Raphael's fresco *The School of Athens* in the Vatican Stanze. Certainly Raphael paid tribute to Bramante by introducing his portrait into this painting as the mathematician Euclid. Bramante had an enormous influence as an architect, and his interest in perspective and **trompe-l'œil* left a mark on Milanese painting, notably in the work of his follower *Bramantino.

Bramantino (Bartolomeo Suardi) (*c.*1460–1530). Milanese painter and architect, a follower of *Bramante, from whom he takes his nickname. He was appointed court painter and architect to Duke Francesco II *Sforza in 1525. His style as a painter is complex and eclectic, drawing on *Piero della Francesca and *Leonardo as well as Bramante; at its best is has a certain stolid dignity. Perhaps his most individual characteristic is his use of sombre classical architectural backgrounds, as in *The Adoration of the Magi* (NG, London). His only surviving work as an architect is a burial chapel for the Trivulzio family in S. Nazaro Maggiore, Milan.

Bramer, Leonaert (1596–1674). Dutch *genre and history painter, active mainly in his native Delft. He travelled widely in Italy and France, 1614–28, and drew on a variety of influences for his most characteristic paintings—small nocturnal scenes with vivid effects of light. Works such as the *Scene of Sorcery* (Musée des Beaux-Arts, Bordeaux) have earned him the reputation of 'an interesting independent who cannot easily be pigeonholed' (J. Rosenberg, S. Slive, and E. H. ter Kuile, *Dutch Art and Architecture, 1600–1800*, 1966). Bramer was also one of the few Dutch artists to paint frescos in Holland, but none of his work in the medium has survived. He evidently knew well the greatest of his Delft contemporaries, *Vermeer, for he came to the latter's defence when his future mother-in-law was trying to prevent him from marrying her daughter. In fact, it is likely that Bramer, rather than Carel *Fabritius, was Vermeer's teacher.

Bramley, Frank. See NEWLYN SCHOOL.

Brancusi, Constantin (1876–1957). Romanian sculptor, active mainly in Paris, one of the most revered and influential of 20th-cent. artists. After studying in Bucharest and afterwards in Vienna and Munich, in 1904 he settled in Paris, where he spent many years of poverty and hardship. In 1906 he was introduced to *Rodin, whose offer to take him on as assistant Brancusi refused, with the famous comment: 'No other trees can grow in the shadow of an oak.' His work of this time was in fact influenced by Rodin's surface animation, but from 1907 Brancusi began creating a distinctive style, based on his feeling that 'what is real is not the external form but the essence of things'. From this time his work (in both stone and bronze) consisted largely of variations on a small number of themes (heads, birds, a couple embracing—*The Kiss*) in which he simplified shapes and smoothed surfaces into immaculately pure forms that sometimes approach complete abstraction. He was particularly fond of ovoid shapes—their egg-like character suggesting generation and birth and symbolizing his own creative gifts. (His woodcarvings, on the other hand, are rougher—closer to the Romanian folk-art tradition and to African sculpture.) His reputation abroad began to grow after five of his sculptures were shown at the *Armory Show, New York, in 1913, and during the 1920s his name became newsworthy when he was involved in two celebrated art scandals. In 1920 his *Princess X* was removed by police from the *Salon des Indépendants because it had been denounced as indecent (there is a clear resemblance to a phallus); and in 1926 he became involved in a dispute with the US Customs authorities. They attempted to tax his *Bird in Space* (one of his most abstract works) as raw metal, rather than treat it as sculpture, which was duty-free. Brancusi was forced to pay up to get the work released for exhibition, but he successfully sued the Customs Office, winning the court decision in 1928. During the 1930s he travelled widely. In 1937 he made sculpture (including the enormous *Endless Column*, nearly 30 m. high) for the public park at Tirgu Jiu near his birthplace and in the same year he visited India to design a Temple of Meditation (never built) for the Maharajah of Indore. In 1955–6 he had retrospective exhibitions at the Guggenheim Museum, New York, and at the Museum of Art, Philadelphia. By the time of his death he was widely regarded as the greatest sculptor of the 20th cent.

Brancusi's originality in reducing natural forms to their ultimate—almost abstract—simplicity had profound effects on the course of 20th-cent. sculpture. He introduced *Modigliani to sculpture, *Archipenko and *Epstein owed much to him, and *Gaudier-Brzeska was his professed admirer. Later, Carl *Andre claimed to have been inspired by *Endless Column*, converting its repeated modules into his horizontal arrangements of identical units. More generally, Henry *Moore wrote of Brancusi: 'Since the Gothic, European sculpture had become overgrown with moss, weeds—all sorts of surface excrescences which completely concealed shape. It has been Brancusi's special mission to get rid of this undergrowth and to make us once more shape-conscious.' On his death Brancusi bequeathed to the French Government his studio and its contents, which included versions of most of his best works (they often exist in multiple replicas in different materials). The studio has now been reconstructed in the *Pompidou Centre in Paris. There is another outstanding Brancusi collection in the Philadelphia Museum of Art.

Brangwyn, Sir Frank (1867–1956). British painter, etcher, and designer, born at Bruges of Welsh parentage. Brangwyn was apprenticed to William *Morris (1882–4), and like his master was active in a variety of fields. He was an *Official War Artist in the First World War and was a skilful etcher and lithographer, but he became best known for his murals. His most famous undertaking in this field was a series of 18 panels on the theme of the British Empire, commissioned by the House of Lords. They were begun in 1926 and rejected—amid great controversy—in 1930. Offers for the panels came from all over the world, and in 1934 they were installed in the Guildhall in Swansea. During his lifetime Brangwyn had a great reputation on the Continent, and there is a museum devoted to him in Bruges and another almost entirely given over to his work in Orange. He was one of the finest draughtsmen of his time, though his painting tended to the decoratively sentimental.

Braque, Georges (1882–1963). French painter, graphic artist, and designer. Initially he followed his father's trade of house painter, but in 1902–4 he took lessons at various art schools in Paris, including briefly the École des *Beaux-Arts. Through his friendship with his fellow students *Dufy and *Friesz, he was drawn into the circle of the *Fauves, and in 1905–7 he painted in their brightly coloured, impulsive manner. In 1907, however, two key events completely changed the direction of his work: first, he was immensely impressed by the *Cézanne memorial exhibition at the *Salon d'Automne; and secondly, he met *Picasso, in whose studio he saw *Les Demoiselles d'Avignon*. Although he was initially disconcerted by it, he soon began experimenting with the dislocation and fragmentation of form it had introduced, and the two men worked in close association until the outbreak of the First World War, jointly creating *Cubism. *The Portuguese* (Kunstmuseum, Basle, 1911) is one of the best-known paintings of this phase of Braque's career, and the first picture to incorporate *stencilled lettering. Braque also took the lead with the **papier collé* technique of introducing pieces of imitation wood engraving, marbled surfaces, etc., stuck on to the canvas.

In 1914 he enlisted in the French Army and was twice decorated for bravery before being seriously wounded in the head in 1915 and demobilized in 1916. After the war his work diverged sharply from that of Picasso. Whereas Picasso went on experimenting restlessly, Braque's painting became a series of sophisticated variations on the heritage of his pre-war years. His style became much less angular, tending towards graceful curves. He used subtle muted colours and sometimes mixing sand with his paint to produce a textured effect. Still life and interiors remained his favourite subject, and many critics regard his *Studio* series, begun in 1947, as the summit of his achievement. Braque also did much book illustration, designed stage sets and costumes, and did some decorative work. By the end of his career he enjoyed immense prestige. He was made a Commander of the Legion of Honour in 1951, and ten years later he had the honour of being the first living artist to have his work exhibited in the Louvre. He was given a state funeral—an occasion that seemed at odds with his life of unassuming dedication to his art.

brass, monumental. A funerary monument consisting of an engraved brass sheet mounted on a stone slab. Brasses, which were cheaper than sculptured tombs, were probably first introduced during the early 13th cent. in western Europe, and had their greatest popularity in England; the earliest surviving English examples date from the late 13th cent., and more than 7,000 are extant from the period before 1600. The material used was not pure brass, but an alloy called at the time 'latten', composed of approximately 60 per cent copper, 30 per cent zinc, 10 per cent lead and tin. Latten was produced in the Low Countries and Germany and imported into England, where it was engraved; large-scale English production of brass began only in the 16th cent., probably made by immigrant German and Flemish craftsmen. The main centre of production seems to have been London, and the majority of medieval brasses are in the home counties. They have been of great importance as a source of detailed information on the evolution of armour and costume in medieval England.

Bratby, John (1928–92). British painter and writer. Bratby was a versatile artist: he painted portraits, still lifes, figure compositions, landscapes, and flower pieces, and also designed film sets. In the years after the Second World War he was one of the group of harsh and austere painters of domestic life who were known as the *Kitchen Sink School. Later his work became lighter and more exuberant. There are numerous examples of his work in the Tate Gallery. Among his publications are the novel *Breakdown* (1960) and a book on Stanley *Spencer (1970).

Bray, Jan de (*c.*1627–97). Dutch painter, principally of portraits. He worked in Haarlem and his vigorously characterized work shows the lasting influence of *Hals in the city, although de Bray's handling is much smoother, in the manner of van der *Helst. Jan's father, **Salomon** (1597–1664), was an architect and painter of biblical and allegorical scenes. He wrote a book, *Architectura Moderna* (1631), describing the buildings of Hendrick de *Keyser.

Bredius, Abraham (1855–1946). Dutch art historian and collector. Bredius was particularly noted for his archival research and published a large amount of new documentation relating to Dutch artists of the 17th cent., his special field of study. His best-known work is his complete illustrated catalogue of *Rembrandt's paintings, originally published in German in 1935 and then in an English edition in 1937; a second English edition, revised by Horst Gerson, Bredius's collaborator on the original edition, appeared in 1969. It is still a standard work (although the illustrations are of poor quality) and the Bredius numbering system is widely used in Rembrandt scholarship. Bredius made a choice collection of Dutch paintings; many were presented or bequeathed by him to the *Rijksmuseum in Amsterdam and the *Mauritshuis in The Hague (of which he was director from 1889 to 1909) and others are in the Bredius Museum in The Hague. His reputation as a connoisseur was somewhat blighted because he was deceived by the work of the Vermeer forger Han van Meegeren (1880–1947); in 1937 Bredius published an article on van Meegeren's now infamous *Christ at Emmaus* (Boymans Museum, Rotterdam), in which he announced it to the world as a newly discovered masterpiece by Vermeer. However, Bredius was very old at the time and almost all his contemporaries were similarly deceived.

Breenbergh, Bartolomeus (1598/1600–1657). Dutch painter, with *Poelenburgh the leading pioneer of the taste for Italianate landscapes in the Netherlands. Breenbergh spent most of the 1620s in Italy and thereafter worked in Amsterdam. His style as a painter is very similar to Poelenburgh's, his biblical and mythological characters set in well-balanced views of the Roman Campagna, often complete with classical ruins. His drawings are much fresher and bolder, and have often passed under the name of *Claude, as is the case with two examples in Christ Church, Oxford. Late in his career Breenbergh turned from landscape to figure painting.

Bregno, Andrea (1418–1506). Italian sculptor, born near Lugano and trained in the Lombard tradition, who became the leading monumental sculptor in Rome during the second half of the 15th cent. With the assistance of a flourishing workshop he produced numerous altars and tombs, showing a command of fashionable antique motifs that no doubt accounted for much of his success. Sir John *Pope-Hennessy described him as 'a sculptor of great taste and technical proficiency, but of limited inventiveness'.

Breitner, George Hendrik (1857–1923). The leading Dutch *Impressionist painter. His most characteristic early works, influenced by the *Hague School, notably by *Mesdag and Willem *Maris, were pictures of horsemen, but a visit to Paris in 1884 brought him under the spell of Impressionism. In 1886 he settled in Amsterdam and became particularly associated with scenes of its busy harbour, its architecture, and its bustling street life. The unposed 'snapshot' compositions of many of these paintings reflect his interest in photography (*Paleisstraat, Amsterdam*, Rijksmuseum, Amsterdam, *c.*1896). After 1910, owing to ill health, he practically ceased to paint.

Breker, Arno. See DESPIAU, CHARLES.

Brera, Milan. Picture gallery, originally the collection of the Accademia di Belle Arti. When Milan became the capital of a French province the Brera was made the centre for paintings from northern Italy displaced by the closing of religious houses after the treaties of Tolentino (1797) and Pressburg (1805). In this way the Brera came into possession of some of the finest works of the *Renaissance, such as *Raphael's *Marriage of the Virgin* (1806) and *Piero della Francesca's *Madonna and Child with Duke Federico of Urbino* (1811), as well as many excellent works of the Venetian School, among them *Tintoretto's *Finding of the Body of S. Mark*. Later some of its treasures were returned to the other Italian states. But the collection has also been continuously enlarged by purchase and bequest. The Brera Palace, in which the collection is housed, is a splendid 17th-cent. building, originally a Jesuit College.

Breton, André (1896–1966). French poet, essayist, and critic, the founder and chief theorist of *Surrealism. He broke with the Paris *Dada movement in 1921 and was instrumental in setting up Surrealism as a separate movement from Dada. He published the first *Manifesto of Surrealism* in 1924 and helped with the first number of the periodical *The Surrealist Revolution*, which he afterwards edited. His *Second Manifesto of Surrealism* was published in the final number of this journal, December 1929, and he wrote numerous other books and articles on Surrealism, including an essay 'Limits not Frontiers of Surrealism' in the English volume *Surrealism* edited by Herbert *Read in 1936. Breton also made *assemblages of surrealistically juxtaposed objects, which he called 'Poem-objects'.

Brett, John (1830–1902). English painter, mainly of coastal scenes and landscapes. He was influenced by the *Pre-Raphaelites and *Ruskin, and a handful of his early paintings such as *The Stonebreaker* (Walker Art Gal., Liverpool, 1857–8) are remarkable *tours de force* of minute and brilliant detail. His later work, however, degenerated into a prosaic catalogue of objects.

Breu, Jörg the Elder (1475/6–1537). German painter and designer of woodcuts, probably a pupil of Ulrich *Apt. Breu was one of the leading painters of his time in his native Augsburg, but his most important works there—a series of frescos in the town hall—are no longer extant. He was patronized by the emperor Maximilian (1459–1519) and by Duke William IV (duke 1508–50) of Bavaria, for whom he painted *The Battle of Zama* (Alte Pinakothek, Munich) in the same series as *Altdorfer's celebrated *Battle of Issus*. His style was complex, sharing something of Altdorfer's passion and love of landscape, and showing strong influence from *Dürer and from a journey he made to Italy in about 1514. His son **Jörg the Younger** (*c.*1510–47) was a prolific book illustrator and worked as court painter at Neuburg.

Breuer, Marcel (1902–81). Hungarian-born American architect and designer. In 1920 he joined the *Bauhaus as a student and by 1924 had become head of the furniture department there. He left the Bauhaus in 1928 to set up in private practice and travelled extensively before settling in the USA in 1937. Breuer worked with Walter *Gropius as an architectural partner and like him taught at Harvard University, where he was instrumental in spreading Bauhaus ideals. He is particularly highly regarded as a furniture designer, his tubular steel and moulded plywood designs being immensely influential.

Breughel. See BRUEGEL.

breviary. A liturgical book containing the hymns, lessons, prayers, etc., to be recited at appointed times in the divine office of the Roman Catholic Church. Originally these various observances were distributed in different books, but from the 11th cent. they began to be collected in one book; *breviarum*, from which the word derives, is Latin for 'abridgement'. Usually neither the small, portable breviary nor the larger choir breviary was much illustrated; sumptuous *illumination was reserved for the breviaries destined for the use of kings, nobles, or Church dignitaries. The most lavish of all is probably the *Grimani Breviary* (Bib. Marciana, Venice), a Flemish work of about 1500, which contains 110 large pictures, including calendar illustrations, scenes from the New Testament, Old Testament, and the lives of saints.

Bril, Paul (1554–1626). Flemish landscape painter, active mainly in Rome, where he settled in about 1575. Bril painted frescos, but his fame rests on his small easel paintings. He lived long enough to assimilate some of the qualities of *Elsheimer's and Annibale *Carracci's landscapes and his work bridges the gap between the fantastic 16th-cent. Flemish *Mannerist style and the more plausible, idealized Italian landscapes of the 17th cent. He also made views of Rome for the tourist trade, and marine pictures. His conception of both of these subjects had considerable influence upon Agostino *Tassi, the teacher of *Claude Lorraine, and upon Claude himself. Paul's brother, **Matthew** or **Mattheus** (1550–83), also worked in Rome, and their work is hard to distinguish.

Brisley, Stuart (1933–). British *Performance artist and (more recently) sculptor. He has won much publicity for his performances concerned with self-inflicted pain and humiliation. In one, he lay in a bath of water for several days in a room in which the floor was scattered with pieces of rotting meat.

Bristol board. Stiff cardboard consisting of sheets of stout drawing-paper pressed together. Its smooth firm surface is attractive to black-and-white illustrators, for it allows a pen line and *hatchings to be drawn with great clarity and clearness, an important factor when a drawing has to be reduced in size for reproduction.

British Museum, London. The national museum of archaeology and antiquities, which also houses the national library of manuscripts and printed books. It was established by Act of Parliament in 1753 when the government purchased the private collection of the physician and naturalist Sir Hans Sloane (1660–1753), consisting of 'books, manuscripts, prints, drawings, pictures, medals, coins, seals, *cameos and natural curiosities'. It was first housed in Montagu House, Bloomsbury, and for nearly 50 years it was necessary to make formal application for admission, only five parties of 15 being admitted on Mondays, Wednesdays, and Fridays. With the acquisition of Sir William Hamilton's collection of classical vases and antiquities (1772), a plethora of Egyptian antiquities (including the Rosetta Stone donated by George III) on the defeat of Napoleon at the turn of the century, the magnificent library of George III (1823) and the Grenville bequest of rare books and manuscripts (1847), the marbles of Charles Townley (1805), the Phrygian marbles (1815), and the *Elgin Marbles (1816), the collections became among the most extensive and valuable in Europe. The present structure (1823–47) was designed by Sir Robert Smirke and the great circular Reading Room, designed by Smirke's brother Sydney, was completed in 1857. Until 1881 the building also housed the collections now in the Natural History Museum, and one of its greatest attractions was a stuffed giraffe in the entrance hall.

The Department of Prints and Drawings led a separate existence from 1808 onwards. It began with over 2,000 drawings from the Sloane collections, which included an album of *Dürer's drawings. Among the most important acquisitions since was the Richard Payne *Knight bequest (1824) of over 1,000 drawings, including works by *Claude (273 drawings), *Rembrandt, and *Rubens. It is now one of the largest and most comprehensive collections in the world containing more than two million items. The British Museum Library (as it was formerly known) was reconstituted by Act of Parliament as part of the British Library in 1973, and will eventually be housed in a new and separate building near St Pancras Station. This was originally scheduled to open in 1989, but the move is now not expected to be complete until 1999.

Briulov, Karl. See BRYULOV.

Brock, Sir Thomas (1847–1922). British sculptor, one of the most successful specialists in monuments and public statues at the turn of the century. His most famous work is the huge Queen Victoria Memorial (unveiled 1911) outside Buckingham Palace, London, and also well known is his equestrian statue of the Black Prince (1903) in Leeds.

Brockhurst, Gerald Leslie (1890–1978). British-born painter and etcher who became an American citizen in 1949. Precociously gifted, an excellent draughtsman, and a fine craftsman, Brockhurst won several prizes at the *Royal Academy Schools and went on to have a highly successful career as a society portraitist, first in Britain and then in the USA, where he settled in 1939, working in New York and New Jersey. He is best known for his portraits of glamorous women, painted in an eye-catching, dramatically lit, formally posed style similar to that later associated with *Annigoni. As an etcher Brockhurst is remembered particularly for *Adolescence* (1932), a powerful study of a naked girl on the verge of womanhood staring broodingly into a mirror—one of the masterpieces of 20th-cent. printmaking.

Broederlam, Melchior (active 1381–1409). Netherlandish painter, court painter to Philip the Bold (1342–1404), Duke of Burgundy, from 1387. In 1391 he was commissioned to paint the backs of the wings for Jacques de *Baerze's *retable at the Chartreuse de Champmol, now in the Musée des Beaux-Arts, Dijon. The two paintings, representing *The Annunciation and Visitation* and *The Presentation and Flight into Egypt*, are Broederlam's only known surviving works, although documents show that he was kept very busy in the Duke's employment, supervising, for example, the construction of pavilions for tournaments and the decoration of ships. In his native Ypres he also designed stained glass and worked as a goldsmith. The Dijon panels are among the first and finest examples of *International Gothic, combining lavish decorative display with realistic touches that look forward to the later development of the Netherlandish School. The figure of St Joseph in *The Flight into Egypt*, for example, is represented as an authentic peasant.

Broeucq, Jacques. See DUBROEUCQ.

bronze. An alloy of copper (usually about 90 per cent) and tin, often also containing small amounts of other metals such as lead or zinc. Since antiquity it has been the metal most commonly used in cast sculpture because of its strength, durability, and the fact that it is easily workable—both hot and cold—by a variety of processes. It is easier to cast than copper because it has a lower melting-point, and its great tensile strength makes possible protrusion of unsupported parts—an advantage over *marble sculpture. Nevertheless, casting a large bronze figure is an extremely complex and time-consuming business, and *Cellini has left a classic account of the difficulties encountered with his figure of *Perseus*. The colour of bronze is affected by the proportion of tin or other metals present, varying from silverish to a rich, coppery red, and its surface beauty can be enhanced when it acquires a *patina. See also CIRE-PERDUE.

Bronzino, Agnolo (Agnolo di Cosimo) (1503–72). Florentine *Mannerist painter, the pupil and adopted son of *Pontormo, who introduced his portrait as a child into his painting *Joseph in Egypt* (NG, London). The origin of his nickname is uncertain, but possibly derived from his having a dark complexion. Bronzino was deeply attached to Pontormo and his style was heavily indebted to his master (in paintings of *c.*1530 it is sometimes difficult to distinguish the hand of one from the other). However, Bronzino lacked the emotional intensity that was such a characteristic of Pontormo's work, his colouring and brushwork were typically harder, and he excelled as a portraitist rather than a religious painter. He was court painter to Duke Cosimo I de *Medici for most of his career, and his work influenced the course of European court portraiture for a century. Cold, cultured, and unemotionally analytical, with superb draughtsmanship and completely controlled effects of complicated pattern, his portraits convey a sense of almost insolent assurance. Bronzino was also a poet, and his most personal portraits are perhaps those of other literary figures (*Laura Battiferri*, Palazzo Vecchio, Florence, *c.*1560). He was less successful as a religious painter, his lack of real feeling leading to empty, elegant posturing, as in *The Martyrdom of S. Lorenzo* (S. Lorenzo, Florence, 1569), in which almost every one of the extraordinarily contorted poses can be traced back to *Raphael or to *Michelangelo, whom Bronzino idolized. It is the type of work that got Mannerism a bad name. Bronzino's skill with the nude was better deployed in the celebrated *Venus, Cupid, Folly, and Time* (NG, London), which breathes the refinement of decadence and with superb technical dexterity conveys suggestions of eroticism under the pretext of a moralizing allegory (one of the subsidiary figures has recently been interpreted as representing the effects of syphilis).

His other major works include the design of a series of tapestries on *The Story of Joseph* for the Palazzo Vecchio. He was a much respected figure who took a prominent part in the activities of the Accademia del Disegno (see ACADEMY), of which he was a founder member in 1563. His pupils included Alessandro *Allori, who—in a curious mirroring of his own early career—was also his adopted son.

Brooking, Charles (1723?–59). English marine painter. Very little is known of his short career, but he was the finest British marine painter of his day, equally adept at calm or rough seas. He is said to have been employed at Deptford dockyard and he had an intimate knowledge of the ships he painted.

Brotherhood of Ruralists. See BLAKE, PETER.

Brouwer, Adriaen (1605/6–38). Flemish painter who spent a great part of his short working life at Haarlem in Holland. He went to Haarlem in about 1623 and was probably a pupil of Frans *Hals. In 1631 he left Holland for Antwerp (where he was for a while detained by the Spaniards as a suspected spy) and apparently spent the rest of his career there. He perhaps died from the plague that swept the city in 1638. Brouwer was an important link between the Dutch and Flemish schools and played a major role in popularizing low-life *genre scenes in both countries in which he worked. Early sources depict him as a colourful bohemian character and his most typical works represent peasants brawling and drinking. Although the subject-matter is humorously coarse, his technique was delicate and sparkling. The virtuosity of brushwork and economy of expression are perhaps surpassed in his landscapes, which are among the greatest of his age. *Rembrandt and *Rubens were among the admirers and collectors of Brouwer's paintings (Rubens at one time owned seventeen), and Adriaen van *Ostade and David *Teniers the Younger were among his many followers.

Brown, Ford Madox (1821–93). English painter. He was born at Calais and trained in Antwerp (under Baron *Wappers), in Paris, and in Rome, where he came into contact with the German *Nazarenes. Settling in England in 1846, he became a friend of the *Pre-Raphaelites and—with his taste for literary subjects and meticulous handling—an influence on their work, though he was never a member of the Brotherhood. *Rossetti studied briefly with him in 1848 and Brown's *Chaucer at the Court of Edward III* (Art Gallery of

New South Wales, Sydney, 1851) contains portraits of several of the Brotherhood. His best-known picture, *The Last of England* (City Art Gallery, Birmingham, 1855), was inspired by the departure of *Woolner, the Pre-Raphaelite sculptor, for Australia. The other famous anthology piece that Brown painted, *Work* (Manchester City Art Gallery, 1852–63), shows his dedicated craftsmanship and brilliant colouring, but is somewhat swamped by its social idealism. In 1861 Brown was a founder member of William *Morris's company, for which he designed *stained glass and furniture. The major work of the later part of his career is a cycle of paintings (1878–93) in Manchester Town Hall on the history of the city. Brown was an individualist and a man of prickly temperament; he opposed the *Royal Academy and was a pioneer of the one-man show.

Brown, Frederick. See SLADE SCHOOL.

Brown, Lancelot (1716–83). The most celebrated of English landscape gardeners, known as 'Capability Brown' from his habit of telling his patrons their estates had 'great capabilities'. A leading figure of the *picturesque movement, Brown replaced formal gardens with parks and lawns broken by serpentine waters and clumps of trees. Temples and *Gothic 'ruins' were often included, the primary intention being to make a garden resemble a landscape painting by *Claude; he succeeded so well that his work is sometimes mistaken for natural landscape. His masterpiece was the creation of the lake at Blenheim Palace, though the park at Chatsworth is perhaps a more complete example of his pastoral style. Other major landscapes created by him are Ashridge Park, Moor Park, Audley End, Bowood, Longleat, and Wardour. He also practised as an architect. He was attacked by Richard Payne *Knight (who favoured a more rugged type of scenery than Brown's artfully informal parks) in *The Landscape*, and defended by Humphry *Repton.

Browne, Hablot Knight ('Phiz') (1815–82). English book illustrator and painter. His name is chiefly remembered for his illustrations for the novels of Charles Dickens (1812–70) under the pseudonym 'Phiz', and he created a visual imagery that is enduringly associated with these. He also illustrated the works of several other novelists and he painted a great number of watercolours and some oils.

Bruce, Edward. See FEDERAL ART PROJECT.

Brücke, Die (The Bridge). Group of German *Expressionist artists founded in 1905 by *Kirchner, *Schmidt-Rottluff, *Heckel, and Fritz Bleyl (1880–1966), who at the time were all architectural students at the Dresden Technical School. The name was chosen by Schmidt-Rottluff and indicated their faith in the art of the future, towards which their own work was to serve as a bridge. Yet they never succeeded in defining this art and their aims remained vague; no clear programme emerged from any of their publications. They were moved by an impulse of revolt and wanted to achieve 'freedom of life and action against established and older forces'. In practice they turned against *Realism and *Impressionism and under the influence of *Munch and *Hodler created an intense and sometimes angst-ridden version of the Expressionism which stemmed from van *Gogh, *Gauguin, the *Nabis, and the *Fauves. Other artists associated with Die Brücke included *Nolde, *Pechstein, and van *Dongen.

Most of the members of the group were without proper training and their handling of paint can have an almost crude vigour. Like the Fauves they were interested in *primitive art, which they saw in the Dresden Ethnological Museum; but the inspiration they derived from it was different. They were interested in figure painting, landscape, and portraiture, but while the subject always remained recognizable, forms were often harshly distorted and colours used symbolically in a violently clashing manner. There is often harsh vigour, too, in their graphic work, and they played an important part in the revival of *wood-cut as a medium for personal expression. Strong contrasts of black and white, bold cutting, and simplified forms were used to great effect. In 1910 Die Brücke shifted its activities to Berlin, where Otto Müller (1874–1930) joined the group. The members had for a time lived together as a community, but the personal rifts that had been present from the beginning became more intense and led to the dissolution of the group in 1913.

Bruegel (or **Brueghel**), **Pieter** (*c.*1525–69). The greatest Netherlandish painter and draughtsman of the 16th cent. There is little documentary evidence concerning his career, but van *Mander's laudatory biography, published in 1604, is a useful source of information, even though it misleadingly projects an image of Bruegel as above all else a comic painter. Far from being the yokel of popular tradition—'Peasant Bruegel'—he seems to have been a man of some culture, as is indicated by his friendship with the great geographer Abraham Ortelius (1527–98). He joined the Antwerp Guild in 1551, having been the pupil of Pieter *Coecke van Aelst, who died in 1550 and whose daughter

Bruegel later married. Between 1551/2 and 1554/5 he made a long journey via France to Italy, where he travelled as far south as Naples and Sicily. In Rome he collaborated with the *miniaturist Giulio *Clovio, who owned a number of Bruegel works that are now lost. On his return journey through the Alps he made accurate and extremely sensitive landscape drawings; the experience of the Alps affected him much more than the example of any art he had seen in Italy. Back in Antwerp he designed a series of landscapes which were engraved and published by Hieronymus *Cock, for whom Bruegel produced many drawings of various subjects, including parables like 'the Big Fish eat Little Fish'. The engraving after Bruegel's drawing of this subject (published in 1557) is inscribed 'Hieronymus Bos Inventor', an attempt by Cock to cash in on the continued popularity of *Bosch, who influenced Bruegel considerably.

In 1563 Bruegel moved to Brussels and married there in that year. From this time until his death six years later he concentrated on painting and produced his best-known works. His patrons included Cardinal Granvella, chief counsellor to Margaret of Parma (1522–86), Philip II's regent in the Netherlands, and the wealthy banker Niclaes Jonghelinck, who in 1565 commissioned the series of *The Months*, of which five survive today. Three of these (including the celebrated *Hunters in the Snow*) are in the remarkable collection of fourteen paintings by Bruegel in the Kunsthistorisches Museum, Vienna, which comprises nearly one-third of his surviving paintings; the other two are in the Metropolitan Museum, New York, and the National Gallery, Prague.

Bruegel's style changed considerably during the last six years of his life in Brussels; he abandoned the crowded panoramas of his earlier years, making his figures bigger and bolder, as is seen most notably in his novel treatment of proverbs, a genre that had previously been of minor account (*The Blind Leading the Blind*, Museo di Capodimonte, Naples, 1568). Bruegel enjoyed a considerable reputation in his lifetime, and his pictorial and spiritual influence, through his original works and the many prints after them, in later Flemish painting, whether landscape or *genre, is incalculable. It is only in the 20th cent., however, that he has come to be recognized also as a profound religious painter and an artist whose human sympathy and understanding has hardly been excelled. His brilliance as a craftsman is also universally acknowledged; his technique was precise yet always fluent, with the paint often thinly applied, giving transparency to a superb variety of colour. Bruegel's two painter sons were infants when he died and so had no training from him (they were reputedly taught by their grandmother—the widow of Pieter Coecke—Mayken Verhulst). Both sons spelled their surname 'Brueghel', retaining the 'h' that their father had dropped in 1559.

Brueghel, Jan (1568–1625). Flemish painter and draughtsman, second son of Pieter *Bruegel the Elder. Early in his career he visited Cologne and Italy, where he was patronized by Cardinal Federico Borromeo, before settling in Antwerp in 1597. He enjoyed a highly successful and honourable career there, becoming Dean of the Guild, working for the Archduke Albert and the Infanta Isabella, and making frequent visits to the Brussels court. His specialities were still lifes, especially flower paintings, and landscapes, but he worked in an entirely different spirit from his father, depicting brilliantly coloured, lush woodland scenes, often with mythological figures, in the manner of *Coninxloo and *Bril. His exquisite flower paintings were rated the finest of the day, and his virtuoso skill at depicting delicate textures earned him the nickname 'Velvet' Brueghel. He often collaborated with other artists, painting backgrounds, animals, or flowers for them. In *The Garden of Eden* (Mauritshuis, The Hague), for example, the background is by Brueghel and the figures are by *Rubens, who was his close friend. He had considerable influence, notably on his pupil Daniel *Seghers, his sons **Jan II** (1601–78) and **Ambrosius** (1617–75), and his grandson Jan van *Kessel. Further descendants and imitators carried his style into the 18th cent.

Brueghel, Pieter the Younger (1564–1638). Flemish painter, the elder son of Pieter *Bruegel. He was born in Brussels but made his career in Antwerp, where he became a guild master in 1585. He is best known for his copies and variants of his father's peasant scenes, which sold well and are often of high quality, in contrast to the work of lesser copyists such as his son **Pieter Brueghel III** (1589–*c.*1640). His other speciality was scenes of fires, which earned him the nickname 'Hell' Brueghel. Frans *Snyders was his most notable pupil.

Brüggemann, Hans (*c.*1485/90–after 1523). North German *Gothic sculptor. He is remembered primarily for one work, the huge *reredos in Schleswig Cathedral (*c.*1514–21), often known as the Bordesholmer Altar because it was originally carved for a church in Bordesholm. Reputedly containing more than 400 figures (many showing influence from *Dürer's engravings), it carries to the furthest extreme the virtuoso German tradition in wood-carving. Brüggemann is said to have died in poverty, possibly because the Reformation killed the market for altar carvings.

Brülloff, Karl. See BRYULOV.

Brunelleschi, Filippo (1377–1446). Florentine architect and sculptor. Brunelleschi was one of the most famous of all architects—a Florentine hero on account of the celebrated dome (1420–36) he built for the city's cathedral—and one of the group of artists, including *Alberti, *Donatello, and *Masaccio, who created the Renaissance style. He trained as a goldsmith and was one of the artists defeated by another goldsmith/sculptor, Lorenzo *Ghiberti, in the competition (1401–2) for the new baptistery doors for Florence Cathedral; their competition panels are in the Bargello. The disappointment of losing is said to have caused Brunelleschi to give up sculpture and turn to architecture, but one important sculptural work of later date is attributed to him—a painted wooden Crucifix in Sta Maria Novella (*c.*1412). Although he was not a painter, Brunelleschi was a pioneer of *perspective; in his treatise on painting Alberti describes how Brunelleschi devised a method for representing objects in depth on a flat surface by means of using a single vanishing point.

Brygos Painter. Greek *red-figure vase-painter, active in Athens (*c.*500–*c.*475 BC), named after a potter called Brygos, several of whose signed works he decorated. He is best known for a cup in the Louvre depicting the *Sack of Troy*. Showing consummate mastery of composition and movement, it is regarded as one of the masterpieces of Greek painting and is sometimes referred to simply as 'the Brygos cup'. More than a hundred other pieces have work attributed to the artist.

Bryulov (or **Brülloff** or **Briulov**), **Karl** (1799–1852). Russian painter. He spent part of his life in Italy (1822–34 and 1849–52), where he painted his chief work, *The Last Day of Pompeii* (Russian Mus., St Petersburg, 1830–3), which was inspired by a performance of the opera of that name by Giovanni Pacini. An enormous (6 m. wide) melodramatic composition, it brought him European fame and inspired Edward Bulwer-Lytton's novel, *The Last Days of Pompeii* (1834).

Bucchero ware. A type of black Etruscan pottery, common in Italy from the 7th to the early 5th cent. BC The term is also applied to the black ware of Pre-Columbian America, characteristic of Chimù.

bucranium (Greek *boukranion*: 'ox-head'). A decorative motif based on the horned head or skull of an ox. It is of immense antiquity, representing the ox that was killed in religious sacrifices, and often appears carved on Classical altars, replacing the actual heads which were hung there in more primitive times. Later it was taken over as a decoration for friezes, etc., on buildings.

Bueckelaer, Joachim (*c.*1535–74). Netherlandish painter of large still lifes—market and kitchen pieces—active in Antwerp. Bueckelaer was the nephew and pupil of Pieter *Aertsen, and he followed his uncle's preference for scenes in which a religious subject is relegated to the background by the still life or *genre content (*Christ in the House of Mary and Martha*, Musées Royaux, Brussels, 1565). He seems to have been the first painter to depict fish stalls.

Buffalmacco (Buonamico Cristofani) (active first half of 14th cent.). Italian painter, a tantalizingly enigmatic figure. Various early sources, not only *Ghiberti and *Vasari, but also the writers Boccaccio (1313–75) and Sacchetti (*c.*1332–*c.*1400), attest to his celebrity as an artist—evidently one of the leading painters of the post-*Giotto generation—and as a burlesque character. Their cumulative testimony is impressive, but as no works can be securely attributed to him, many critics have regarded him as a legendary rather than a historical figure. Recently, however, there have been attempts to give Buffalmacco a stature commensurate with his literary reputation by attributing to him the famous frescos of *The Triumph of Death* in the Campo Santo, Pisa, which are usually considered the work of Francesco *Traini. A rival school of thought has it that Buffalmacco may be identified with another obscure personality, the *Master of St Cecilia.

Buffet, Bernard (1928–). French painter, etcher, lithographer, designer, and occasional sculptor. A precocious artist, he had developed a distinctive style and won considerable critical acclaim by the age of 20. His work, which includes religious scenes, landscapes, still lifes, and portraits, is instantly recognizable, characterized by elongated, spiky forms with dark outlines, sombre colours, and an overall mood of loneliness and despair. It seemed to express the existential alienation and spiritual solitude of the post-war generation, and Buffet enjoyed enormous success in the 1950s. Later, as he found himself overwhelmed by commissions, his work became more stylized and decorative, losing much of its original impact.

Buon, Bartolommeo (*c.*1374–*c.*1465). Venetian sculptor and architect. With his father **Giovanni** (active 1382–d. *c.*1443), he ran the most successful Venetian sculpture workshop of the period. The most important works of the shop include the decoration of the Cà d'Oro (1422–34) and the Porta della Carta (1438–42) of the Doges' Palace. They epitomize the survival of the *Gothic style in Venice into the mid-*quattrocento.

Burchfield, Charles (1893–1967). American painter, mainly in watercolour. In 1921 he settled permanently in Buffalo, where he worked as head designer in a wallpaper factory until he was able to devote himself full-time to art in 1929. Burchfield's work divides into three clear phases. Up to about 1918 he painted scenes of nature that have an obsessive, macabre quality, often based on childhood memories and fantasies. In his second phase—during the 1920s and 1930s—he was one of the leading *American Scene painters, portraying the bleakness of small-town life and the grandeur and power of nature. In the early 1940s he became disenchanted with realism, however, and changed his style again, reviving the subjective spirit of his youthful work but in a more monumental vein, as he turned to a highly personal interpretation of the beauty and mystery of nature (*The Sphinx and the Milky Way*, Munson-Williams-Proctor Institute, Utica, 1946). In the 1950s Burchfield taught at several institutions, including the Buffalo Fine Arts Academy and the University of Buffalo. The Charles E. Burchfield Foundation, Buffalo, possesses his papers and a good collection of his paintings.

Burckhardt, Jacob (1818–97). Swiss historian, professor at the universities of Zurich (1855–8) and Basle (1858–93). He was a pioneer of the cultural approach to history and is best known for his *Die Kultur der Renaissance in Italien* (The Civilization of the Renaissance in Italy), published in 1860. In this survey of the arts, philosophy, politics, etc., of the period he propounds the view that it was at this time that man, previously conscious of himself 'only as a member of a race, people, party, family or corporation', became aware of himself as 'a spiritual individual'. This romanticized view has been highly influential but also much attacked. Burckhardt was pessimistic in outlook, seeing the modern world as undergoing a retrogression in cultural and spiritual values. His other books included *Cicerone* (1855), a guidebook to Italian art that was a popular handbook for German tourists for many years, and *Erinnerungen aus Rubens* (Recollections of Rubens), published posthumously in 1898, but begun many years earlier, in which he championed an artist then much criticized.

Bürger, W. See THORÉ.

Burgkmair, Hans the Elder (1473–1531). German painter and designer of woodcuts. After learning his trade under *Schongauer in Colmar, he had settled in his native Augsburg by 1498. Before then he is presumed to have been to Italy, for his paintings, with their warm glow of colour, their decorative Classical motifs, and their intricate spatial composition, show how decisively he transformed his late *Gothic heritage with *Renaissance influence. Indeed, he occupied a place in Augsburg comparable to that of *Dürer in Nuremberg in introducing the new style. Like Dürer he contributed to the famous series of woodcuts for the Emperor, the *Triumph of Maximilian*. He was also employed to illustrate the Emperor's own writings, the *Teuerdank* and *Weisskunig*, moralizing knightly romances. A certain clarity of characterization, which is typical of all his works, not least his incisive portraits, seems to have influenced Hans *Holbein the Younger. His son, **Hans Burgkmair the Younger** (*c.*1500–59), was a painter and engraver.

burin (or **graver**). The engraver's principal tool. It is a short steel rod, usually lozenge-shaped in section, cut obliquely at the end to provide a point. Its short, rounded handle is pushed by the palm of the hand while the fingers guide the point. The same names are applied to the chipped flints used by palaeolithic man for engraving cave-walls, bones, etc.

Burke, Edmund (1729–97). British statesman and writer. In the history of *aesthetics he is noteworthy for his *A Philosophical Enquiry into the Origin of our Ideas of the Sublime and Beautiful*, first published in 1757. It went through seventeen editions in his lifetime, and after *Addison's essays it was the most influential single work on the course of English aesthetics in the 18th cent. Its influence was also felt in Germany, notably by *Kant. The book marked a move away from the *Classical rationalist ideas of the early 18th cent. in the direction of what would later be called *Romanticism. Burke argued that we are most powerfully affected in art not by what is most clearly stated but by what is suggested: 'It is our ignorance of things that causes all our admiration and chiefly excites our passions.' In particular, he thought that fear was an important ingredient in our enjoyment of the *sublime: 'Whatever is fitted in any sort to excite the ideas of pain, and danger . . . or is conversant about terrible objects, or operates in a manner analogous to terror, is a source of the sublime; that is, it is productive of the strongest emotion which the mind is capable of feeling.'

Burle Marx, Roberto (1909–). Brazilian artist noted particularly as a landscape architect and garden designer. His designs have made an important contribution to modern architectural development in Brazil. They are particularly effective for their pictorial sense and the success with which they integrate tropical colours to emphasize by harmonious contrasts or by picturesque settings the most striking features of the new architecture. Burle Marx has also worked as a painter and designer of fabrics, jewellery, and stage sets.

Burlington, Richard Boyle, third Earl of (1694–1753). English architect and patron. Burlington was the leading figure of the *Palladian movement in English architecture, promoting it through his own buildings, notably his villa at Chiswick, his publication of the drawings of Palladio and Inigo *Jones, and his patronage of artists such as his friend William *Kent. He was generally regarded as an arbiter of taste, thereby provoking the enmity of William *Hogarth, who attacked him in satirical engravings.

Burliuk, David (1882–1967) and **Vladimir** (1886–1916). Russian artists, brothers, leading members of the avant-garde in the period leading up to the First World War. They were closely associated with *Goncharova and *Larionov, adopting a style of deliberate and even exaggerated primitivism akin to theirs, and they were among the first exponents of *Futurism in Russia, *c.*1911. They were also friendly with *Kandinsky, and through him participated in the second *Neue Künstlervereinigung exhibition in Munich in 1910 and in the *Blaue Reiter exhibition there in 1911. Vladimir, who was considered by Kandinsky to be the more talented of the two, was killed in action in the First World War. David settled in New York in 1922 and became an American citizen in 1930. He edited an art magazine, *Color Rhyme*, and ran an art gallery. There was another painter brother, **Nikolai** (1890–1920), and two painter sisters, **Lyudmila** and **Nadezhda**.

Burman, Thomas. See BUSHNELL.

Burne-Jones, Sir Edward Coley (1833–98). English painter, illustrator, and designer. He was destined for the Church, but his interest was turned to art first by William *Morris, his fellow divinity student at Oxford, and then by *Rossetti, to whom Burne-Jones apprenticed himself in 1856 and who remained the decisive influence on him. Like Rossetti, Burne-Jones painted in a consciously aesthetic style (see AESTHETICISM), but his taste was more classical and his elongated forms owed much to the example of *Botticelli. He favoured medieval and mythical subjects and hated such modernists as the *Impressionists, describing their subjects as 'landscape and whores'. His own ideas on painting are summed up as follows: 'I mean by a picture a beautiful romantic dream, of something that never was, never will be—in a light better than any that ever shone—in a land no-one can define or remember, only desire—and the forms divinely beautiful.' He exhibited little before 1877, but then became quickly famous, with a remarkably wide following abroad. His work had considerable influence on the French *Symbolists and the ethereally beautiful women who people his paintings, like the more sensuous types of Rossetti, had a considerable progeny at the end of the century. Some of Burne-Jones's finest work was done in association with William Morris (he was a founder member of Morris and Co. in 1861), notably as a designer of stained glass and tapestries, and as an illustrator of some of the *Kelmscott Press books. The best collection of his work is in the City Art Gallery at Birmingham, his birthplace.

burr. In metal engraving, the rough, upturned edge of the furrow made in the plate by the burin or needle. In *line engraving it is removed to obtain sharpness, but in *drypoint it is allowed to remain because the soft, rich quality it gives to the printed line is considered one of the attractions of the medium. Only a limited number of impressions can be taken before the burr wears down.

Burra, Edward (1905–76). English painter, draughtsman, and stage designer, one of the most delightfully eccentric figures in British art. He suffered chronic ill health continuously from childhood and lived almost all his life in the genteel Sussex seaside town of Rye (he called it an 'overblown gifte shoppe'), but he travelled indomitably and had a tremendous zest for life that comes out in his chaotically misspelt letters as well as his paintings. His life and work, in fact, represent a revolt against his respectable middle-class background, for he was fascinated by low-life and seedy subjects, which he experienced at first hand in places such as the streets of Harlem in New York and the dockside cafés of Marseilles. From 1921 to 1925 Burra trained in various art schools, including the *Royal College in London. He early formed a distinctive style, depicting squalid subjects with a keen sense of the grotesque and a delight in colourful detail. Usually he worked in watercolour, but on a larger scale than is generally associated with this medium and using layer upon layer of pigment so that—in reproduction at any rate—his pictures appear to have the physical substance of oil paintings.

Burra's work has been compared with that of George *Grosz, whom he admired, but whereas the satirical spirit of Grosz is linked with bitter castigation of evil and ugliness, Burra concentrated on the picturesque aspects of his subjects, which he depicted with warmth and humour. Particularly well known are his Harlem scenes of 1933–4, with their flamboyant streetwise dudes and other shady characters (examples are in the Tate, London, and the Cecil Higgins Art Gal., Bedford). Burra's style changed little, but about the mid-1930s his imagery underwent a radical change and he became fascinated with the bizarre and fantastic (*Dancing Skeletons*, Tate, 1934). Many of his recurrent images—such as the bird-man—and his manner of juxtaposing incongruous objects acquired overtones of *Surrealism, and although he generally kept aloof from groups he exhibited with the English Surrealists (he was also a member of *Unit One, organized by his friend Paul *Nash). The sense of tragedy evoked in him by the Spanish Civil War and the Second World War found expression in occasional religious pictures and during the 1950s and 1960s his interest turned from people to landscape. By this time he had achieved critical and financial success, but he reacted with sardonic humour towards his growing fame.

Burrell, Sir William (1861–1958). Scottish art collector. He made an immense fortune from the family shipping business, which he sold in 1917, and devoted most of his life to collecting. His interests were extremely diverse, but his collection became particularly strong in 19th-cent. French painting, some of his greatest treasures being bought from the Glasgow dealer Alexander Reid (1854–1928), who helped to pioneer interest in this field in Scotland. Burrell eventually amassed 8,000 objects, which he presented to the City of Glasgow in 1944, followed by the sum of £450,000 to build a new museum to house them. This was not opened until 1983 in Pollok Country Park, to the south of the city, and in the intervening years the Burrell Collection acquired something of a legendary reputation as a hidden treasure trove; it soon became one of the most popular museums in Britain. Apart from 19th-cent. French painting, it is particularly strong in medieval art, with superb collections of tapestries and stained glass. Burrell also gave paintings to the Museum and Art Gallery at Berwick-upon-Tweed; in his later years he lived nearby at Hutton Castle.

Burri, Alberto (1915–95). Italian painter, *collagist, and designer. Originally a doctor, he was captured while serving with the Italian Army in North Africa and began to paint in 1944 as a prisoner of war in Hereford, Texas. He used whatever materials were to hand, including sacking, and after his return to Italy in 1945 (when he settled in Rome) he frequently used the device of splashing red paint on cloth in a manner that suggested blood-soaked bandages (*Sacking with Red*, Tate, London, 1954). His later works including materials such as charred wood and rusty metal also reflect the direct experience of the carnage of war he had as an army doctor, even though they are elegantly constructed. Burri won international fame for these works, which were among the first to exploit the evocative force of waste and trash, and looked forward to *Junk art in America and *Arte Povera in Italy. He also designed stage décor for La Scala in Milan and other theatres.

Bury, Pol (1922–). Belgian artist, best known as one of the leading exponents of *Kinetic sculpture. From 1947 he exhibited with the *Jeune Peinture Belge group and was also active in the *Cobra group. In 1953, however, he abandoned painting for Kinetic sculpture and took part in the 'Mouvement' exhibition at the Gallerie Denise René, Paris, in 1955. His early works could be rotated at will, inviting spectator participation, but from about 1957 he began to incorporate electric motors. The movement was usually very slow and the impression made was humorous and poetic, in contrast to the violent effects of *Tinguely. Bury has also made films. Since 1961 he has lived mainly in Paris.

Bush, Jack Hamilton (1909–77). One of Canada's leading abstract painters, active mainly in his native Toronto. His early work was in the tradition of the *Group of Seven (*Village Procession*, Art Gal. of Ontario, Toronto, 1946), but in the early 1950s, inspired by Jock *Macdonald and by *Borduas's work, he began to experiment with automatic composition (see AUTOMATISM). In 1952 he made the first of what became regular visits to New York. The influence of these, together with that of the Toronto abstract group *Painters Eleven, with whom he began to exhibit in 1954, brought him, by the mid-1950s, to a type of *Abstract Expressionism. His later works, however, explore the unaffectedly direct use of colour in a more personal way, as in his most famous painting, *Dazzle Red* (Art Gal. of Ontario, 1965), in which the colour is placed in joyous, broadly brushed bands. Bush worked as a commercial designer for most of his career and did not devote his whole time to painting until 1968, but by the end of his career he had an international reputation.

bush hammer. See BOUCHARDE.

Bushnell, John (*c.*1630–1701). English sculptor. He fled to the Continent when he was an apprentice, after his master, Thomas Burman (1618–74), forced him to marry a servant he had himself seduced. In Italy Bushnell assimilated much of the *Baroque style (he probably saw *Bernini's work in Rome) and executed a monument to Alvise Mocenigo (1663) in S. Lazzaro del Mendicanti, Venice. On his return to England, *c.*1670, he received important commissions including a *Sir Thomas Gresham* for the Royal Exchange (now in the Old Bailey), and would have received more but for his difficult and unstable temperament (he died insane). Bushnell's work is extremely uneven, but he is an important figure, for he showed untravelled Englishmen for the first time something of the possibilities of Baroque sculpture.

bust. A representation of the head and upper portion of the body, usually referring to a sculptured portrait, but also by extension to a painting, drawing, or engraving. The word is of uncertain origin, but it is sometimes explained as derived from the Latin *bustum*, 'sepulchral monument'. The forms of busts have varied a good deal; the term covers many types ranging from those which show only the head, the neck, and part of the collar-bone to those which include shoulders, arms, and even hands. Such parts of the body as are shown may serve simply as a base for the portrait head and be so shaped as to give a straight line downward (see HERM), or they may be draped to serve a decorative purpose, or dressed in garments which indicate the sitter's status. The form is particularly associated with the Romans, who from Republican times onwards varied and elaborated the bust form so inventively and so often that archaeologists are able to date a Roman bust from its shape.

Bustelli, Franz Anton (1723–63). Swiss sculptor, the outstanding modeller of porcelain figures at the Nymphenburg porcelain factory housed in the Nymphenburg Palace, outside Munich. After J. J. *Kändler at Meissen he was the most gifted of all the porcelain modellers in the German *Rococo style.

Butler, Elizabeth (Lady Butler, née Thompson) (1846–1933). British painter who concentrated almost exclusively on military scenes. During her heyday in the 1870s she was one of the most popular and talked-about artists in Britain. Her work appealed to popular patriotic sentiment, but she was also praised by critics such as *Ruskin, who in 1875 called her 'this Pallas of Pall Mall' and admitted he had been wrong in believing that 'no woman could paint' and that 'what the public made such a fuss about must be good for nothing'. Lady Butler said 'I never painted for the glory of war, but to portray its pathos and heroism' and, although her pictures often have a glossy, Hollywood quality, they are sincerely felt, and she has been praised for trying to show the experience of the common soldier rather than concentrating—as was then traditional—on the heroic deeds of officers. Her best-known painting is probably *Scotland for Ever!*, showing the charge of the Royal Scots Greys at the Battle of Waterloo (Leeds City Art Gal., 1881). She painted scenes from many of the wars of the 19th cent. and also from the Boer War and the First World War.

Butler, Reg (1913–81). British sculptor. He was an architect by training (his work included the clock-tower of Slough Town Hall, 1936), and architecture remained his main preoccupation until 1950, when he gave up his practice and became the first Gregory Fellow in Sculpture at Leeds University. He stayed at Leeds until 1953, when he suddenly came to prominence on being awarded first prize (£4,500) in the International Competition for a monument to *The Unknown Political Prisoner* (defeating *Calder, *Gabo, and *Hepworth among other established artists). The competition, financed by an anonymous American sponsor and organized by the *Institute of Contemporary Arts, was intended to promote interest in contemporary sculpture and 'to commemorate all those unknown men and women who in our times have been deprived of their lives or their liberty in the cause of human freedom'. His design was characterized by harsh, spindly forms, suggesting in his own words 'an iron cage, a transmuted gallows or guillotine on an outcrop of rock'. The monument was never built (one of the models is in the Tate Gallery), but the competition established Butler's name and he won a high reputation among British sculptors of his generation. He had learned iron-forging when he had worked as a blacksmith during the Second World War (he was a conscientious objector) and his early sculpture is remarkable for the way in which he used his feeling for the material to create sensuous textures. His later work, which was more traditional (and to many critics much less memorable), included some bronze figures of nude girls, realistically painted and with real hair, looking as if they had strayed from the pages of 'girlie' magazines. Butler was an articulate writer and radio broadcaster and he vigorously argued the case for modern sculpture. Five lectures he delivered to students at the *Slade School in 1961 were published in book form the following year as

Creative Development. He was a widely read man, who numbered leading intellectuals among his friends, and his liberal sympathies were shown by his donation of works to such causes as the campaign against capital punishment.

Buys, Cornelis. See MASTER OF ALKMAAR.

Buytewech, Willem (1591/2–1624). Dutch painter and engraver, nicknamed 'Geestige Willem' (Witty Willem). He was active in his native Rotterdam and in Haarlem, where he was closely associated with Frans *Hals, to the extent that they are known to have once worked on the same picture. Although his surviving output as a painter is tiny, he is one of the most interesting artists during the first years of the great period of Dutch painting. His pictures of dandies, fashionable ladies, topers, and lusty wenches are among the most spirited Dutch *genre scenes, and instituted the category known as the 'Merry Company' (*Merry Company*, Boymans Mus., Rotterdam). His engravings are more numerous, and include genre scenes, fashion plates, and etchings of the Dutch countryside that are among the first realistic 17th-cent. landscapes. He had an important influence on painting in Haarlem. His son **Willem the Younger** (1625–70) was also a painter. An example of his very rare work—a landscape—is in the National Gallery, London.

Byrne, William. See HEARNE.

Byzantine art. Art associated with the Eastern Roman Empire, founded in AD 330 by the emperor Constantine and ending in 1453 when his capital Constantinople (formerly named Byzantium) was captured by the Turks and under the name of Istanbul became the capital of the Ottoman Empire. During these eleven centuries the Byzantine territories varied greatly in extent: at one time they embraced almost the whole Mediterranean basin, but from the 7th cent. onwards many provinces were lost, first to the Arabs and later to the Turks.

Byzantine art, however, cannot be defined adequately in political or geographical terms. It did not come suddenly into being, and for a long time it might as properly have been called Roman as Byzantine. Nor did it cease in 1453, for during the second half of the 15th cent. and a good part of the 16th the art of those regions where Greek Orthodoxy still flourished—such as Mount Athos—remained in the Byzantine tradition. And Byzantine art passed far beyond the territorial limits of the empire, to penetrate, for instance, into the Slav countries.

Byzantine art is, above all, a religious art. Not that it treated religious subjects only. There was a fine efflorescence of the 'minor arts' of metal-work, textiles, carved ivories, enamels, jewellery, etc. Secular paintings and *mosaics also adorned the imperial palaces. But these, which have largely disappeared, were few in comparison with the subjects taken from the Old and New Testaments, from the apocryphal books (Gospels of the childhood of Christ or of the Virgin, of Joseph the Carpenter, etc.), and the lives of the saints. It is also a theological art, in the sense that the Byzantine artist did not aspire to freedom of individual interpretation but was the voice of orthodox dogma and subject to the Church which established the dogma. His function was to translate into the language of art, for the instruction and edification of the faithful, the thought of the theologians and the decisions of Councils. Consequently this art was impersonal and traditional. The artist's personality was suppressed, and indeed very few Byzantine masters are known to us by name. The arrangement of mosaics or paintings in a church, the choice of subjects, even the attitudes and expressions of the characters, were all determined according to a traditional scheme charged with theological meaning. If the artist attempted innovation, he risked incurring the guilt of heresy or sacrilege. His role was akin to that of the priesthood and the exercise of his talent a kind of liturgy—liturgy in a sense almost sacramental—rather than a didactic function. It is this which differentiates the essentially theological art of Byzantium from the more didactic art of the West in the medieval period.

Even when it was imperial, Byzantine art hardly diverged from this theocratic and religious character. There was indeed a form of art responsible for glorifying the emperor; but at Byzantium the emperor was an oriental sovereign, an earthly image of the Deity. His court with its carefully contrived hierarchy reflected the hierarchy of heaven. The ceremonies of the Great Palace, meticulously regulated, were a kind of liturgy.

It is not surprising, therefore, that Byzantine art was an art of stylization. It was fundamentally opposed to the spirit of ancient Greek art, whose theme was man and his natural likeness. Byzantium shuns earthly man, the individual, and aspires to the superhuman, the divine, the absolute. By stylization it destroys humanity in art and transfuses forms with the numinous quality of symbols. It is not naturalistic but ritualistic. The conventions of Byzantine art were eventually challenged by the more naturalistic ideals of artists such as *Giotto and *Duccio.

Cabanel, Alexandre (1823–89). French painter. The winner of the *Prix de Rome in 1845, he ranked with *Bouguereau as one of the most successful and influential academic painters of the period and one of the sternest opponents of the *Impressionists. *The Birth of Venus* (Musée d'Orsay, Paris) is his best-known work and typical of the slick and titillating (but supposedly chaste) nudes at which he excelled. It was the hit of the official *Salon of 1863, the year of the *Salon des Refusés, and was bought by the emperor Napoleon III (1808–73), who gave Cabanel several prestigious commissions.

Cabel. See ARENTSZ.

cabinet painting. Term applied to small *easel paintings intended to be viewed at close range. It has no precise limits—Rudolf *Wittkower (*Art and Architecture in Italy 1600–1750*) uses the term of *Caravaggio's 2-m.-wide *Supper at Emmaus* (NG, London)—but is often applied, for example, to 17th-cent. Dutch *genre paintings, which were usually painted to fit into unpretentious bourgeois interiors.

Cadell, F. C. B. See SCOTTISH COLOURISTS.

Cadmus, Paul (1904–). American painter and draughtsman. He paints with an extremely meticulous technique, usually in egg *tempera, and often uses the poses and compositional techniques of the Old Masters. However, his subjects are taken from modern American life, on which he comments pungently and satirically. This has sometimes led to scandal, as with the work that established his reputation, *The Fleet's In!* (Naval Historical Center, Washington, 1934), portraying sailors on shore leave; it was described by the Secretary of the Navy as 'a most disgraceful, sordid, disreputable, drunken brawl, wherein apparently a number of enlisted men are consorting with a party of streetwalkers and denizens of the red-light district'. Because Cadmus works very slowly his output as a painter has been small, but he has been a comparatively prolific draughtsman: 'drawings are more saleable than paintings,' he writes, 'they're less expensive.'

Caffieri. Family of sculptors, bronze-workers, and decorators, of Italian origin, who worked in Paris under Louis XIV, XV, and XVI. The best known are **Philippe I** (1634–1716), who worked on the decoration of the *Louvre under *Lebrun, his son **Jacques** (1678–1735), **Philippe II** (1714–74), son of Jacques, and **Jean-Jacques** (1725–92), another son of Jacques. Jacques and his son Philippe II were considered the outstanding bronze-founders of Louis XV's reign (1715–74) and

in turn held the title of Sculptor, Founder, and Chaser to the King. They worked mainly on small ornamental pieces, whereas Jean-Jacques was best known for lively portrait busts, which in his day were second in popularity only to those of *Houdon.

Cage, John. See BLACK MOUNTAIN COLLEGE.

Cahill, Holger. See FEDERAL ART PROJECT.

Caillebotte, Gustave (1848–94). French painter and collector. He came from a very wealthy family and for many years after his death was remembered primarily for the financial help he gave the *Impressionists, by purchasing their paintings and sometimes by direct gifts of money. Recently, however, his own work as a painter has been reassessed and he is now regarded as an artist of considerable, although uneven, achievement. He exhibited at five of the eight Impressionist exhibitions, concentrating on scenes from everyday life. The most striking feature of his work is his interest in bold perspective effects, as in *Paris, A Rainy Day* (Art Institute of Chicago, 1877), which has become a much reproduced favourite. On his death bequeathed his collection of sixty-five pictures to the State. Against the opposition of various academic artists representing the taste of the École des *Beaux-Arts and the official *Salon (*Gérôme called the works offered 'filth'), thirty-eight of the pictures were accepted after much wrangling and formed the nucleus of the Impressionist collection of the Luxembourg Museum. They were transferred to the *Louvre in 1928 and are now in the Musée d'Orsay.

Calabrese, Il Cavaliere. See PRETI, MATTIA.

Calcar, Jan Joest van. See JOEST.

Caldecott, Randolph (1846–86). English graphic artist and water-colour painter, best known for his illustrations to children's books, beginning with William Cowper's *John Gilpin* (1878). He died in Florida and from 1938 the Caldecott Medal has been awarded to 'the artist of the most distinguished American picture book for children'.

Calder, Alexander (1898–1976). American sculptor and painter, famous as the inventor of the *mobile and thereby as one of the pioneers of *Kinetic art. His grandfather, **Alexander Milne Calder** (1846–1923), and his father, **Alexander Stirling Calder** (1870–1945) were sculptors and his mother was a painter, but he began to take an interest in art only in 1922, after studying mechanical engineering. From 1923 to 1926 he studied at the *Art Students League, New York, where George *Luks and John *Sloan were among the teachers. Calder and his fellow students made a game of rapidly sketching people on the streets and in the subway and Calder was noted for his skill in conveying a sense of movement by a single unbroken line. From this it was but a step to his wire sculptures, the first of which—a sundial in the form of a cockerel—was done in 1925. He also made animated toys in a similar vein. His first exhibition of such works was in New York in 1928. From this point he divided his time between the USA and France and he knew many leading avant-garde artists in France, notably *Miró, who became his lifelong friend.

In 1931 he joined *Abstraction-Création association and in the same year produced his first non-figurative moving construction. The constructions, which were moved by hand or by motor-power, were baptized 'mobiles' in 1932 by Marcel *Duchamp, and *Arp suggested 'stabiles' for the non-moving constructions in the same year. It was in 1934 that Calder began to make the unpowered mobiles for which he is most widely known. Constructed usually from pieces of shaped and painted tin suspended on thin wires or cords, these responded by their own weight to the faintest air currents and were designed to take advantage of effects of changing light created by the movements. They were described by Calder as 'four-dimensional drawings', and in a letter to Duchamp written in 1932 he spoke of his desire to make 'moving Mondrians'. Calder was in fact greatly impressed by a visit to *Mondrian in 1930, and no doubt envisaged himself as bringing movement to Mondrian-type geometrical abstracts. Yet the personality and outlook of the two men were very different. Calder's pawky delight in the comic and fantastic, which obtrudes even in his large works, was at the opposite pole from the messianic seriousness of Mondrian.

After winning first prize for sculpture at the 1952 Venice *Biennale Calder received numerous public commissions. Some of his late works are of very large dimensions: the motorized hanging mobile *Red, Black, and Blue* (1967) at Dallas airport is 14 m. wide. Calder also worked in a variety of other fields, painting *gouaches and designing, for example, rugs and tapestries.

Callcott, Sir Augustus Wall (1779–1844). English painter, a pupil of *Hoppner. He became the most fashionable English landscape painter of his day, and was knighted in 1837. His early work had some freshness, but he settled into a conventional Italianate manner with a carefully modulated reflection of *Turner, for the satisfaction of the larger public who could not yet stomach either Turner himself or *Constable. In 1827 he married **Maria Graham** (née Dundas, 1785–1842), author of *Little Arthur's History of England* (1835) and numerous books on topography and painting, and the two formed a society salon for artistic London.

Callimachus. Greek sculptor of the late 5th cent. BC. His work is not known in the original or in any certain copies, but his fame in ancient writings has led to a number of works being associated with his name. He is said to have pioneered the use of the drill in sculpture and his style was graceful and fastidious, but some critics considered that his labours were excessive and he was known as *catatexitechnus* (the one who spoils his art by overelaboration). According to a well-known story, he invented the Corinthian *capital after seeing some acanthus leaves growing around a basket on a girl's grave; as the earliest known Corinthian capital dates from about 425 BC, at the time Callimachus flourished, there may be some truth in the legend.

Callistratus. Greek writer, the author of a book of *Descriptions* of ancient statues in the manner of the *Imagines* of *Philostratus. Callistratus, who is known only through this book, is believed to have written in the latter part of the 3rd cent. AD. His descriptions are exercises in rhetorical skill rather than sources for the history of art.

Callot, Jacques (1592/3–1635). French engraver and draughtsman. He went to Italy when he was in his teens and, working in Rome and then in Florence at the court of the Grand Duke Cosimo II (1590–1621), he learnt to combine the sophisticated techniques and exaggerations of late *Mannerism with witty and acute observation into a brilliantly expressive idiom. Returning to France in 1621 he became one of the chief exponents of the bizarre and *grotesque which came into vogue in the reign of Louis XIII (1601–43). Most of the remainder of his career was spent in his native Nancy, but he also worked in Paris and the Low Countries. He made a speciality of beggars and deformities, characters from the picaresque novel and the Italian *commedia dell'arte*. In this respect he comes close to *Bellange, also active in Nancy, but Callot's style was more realistic. His last great work, the series of etchings entitled the *Grandes Misères de la Guerre*, followed the invasion of Lorraine by Cardinal Richelieu in 1633, and is a harrowing depiction of the atrocities of war; its themes and imagery were used as a source by *Goya. Callot's output was prodigious; more than a thousand etchings and more than a thousand drawings by him are extant, and some of his plates are large, featuring scores of brilliantly arranged figures. He was one of the greatest of all etchers and one of the first major creative artists to work exclusively in the *graphic arts.

calotype. The first photographic process for producing multiple prints from a single negative image. It was invented by the English scientist William Henry Fox Talbot (1800–77) in 1839 and was sometimes known as Talbotype (see also DAGUERREOTYPE).

Calraet, Abraham. see CUYP.

Calvaert, Denys (called Dionisio Fiammingo) (*c.*1540–1619). Flemish painter from Antwerp who emigrated to Italy in about 1562 and remained there for the rest of his life. Calvaert settled in Bologna, and although his work is in an undistinguished *Mannerist style, he played an important role as a teacher. He established an *academy in 1572 and had more than 100 pupils, among whom were some of the most distinguished artists of the Bolognese School—*Albani, *Domenichino, and *Reni. The more celebrated academy of the *Carracci was probably inspired by Calvaert's.

calvary. A sculptural representation of the scene of Christ's crucifixion on the hill of Calvary, or Golgotha, outside Jerusalem. The term is sometimes applied to any wayside crucifix or to chapels with a series of carvings of the Passion of Christ, but is more appropriate to groups of figures which represent or symbolize the whole scene, such as those found in the open air in Brittany, dating from the late 15th cent. to the early 17th cent. Some are extremely simple, others include great numbers of figures variously arranged, usually on one or more stone bases. The remarkable concentration of calvaries in Brittany is unexplained, but it is clear that only in a remote and isolated region could the creation of works which are essentially medieval have persisted so late. According to one theory the Breton calvaries are translations into stone of the medieval mystery plays in which scenes from the life of Christ were re-enacted in front of the churches.

Calvert, Edward (1799–1883). English painter and engraver. After five years in the navy, he began to study art in 1820 at Plymouth and then in London, coming into contact with *Fuseli, Samuel *Palmer, who became his lifelong friend, and *Linnell, who introduced him to *Blake. He became one of the *Ancients, and under the influence of Blake his imagination was fired, like that of others in the group, to a poetic fervour which was unable to survive the death of the master. The tiny watercolour *The Visionary City* (BM, London) is perhaps the finest of the visionary works he produced during this period. In his later career he painted mostly in oils on paper and after visiting Greece in 1844 developed a sentimental pseudo-Hellenic arcadianism.

camaieu. A painting executed in several shades of a single colour; it is distinct from *grisaille, which is grey or greyish.

Camaino, Tino di. See TINO DI CAMAINO.

Cambiaso, Luca (1527–85). Genoese painter. He was a precocious artist (his highly accomplished frescos in the Doria Palace, now the Prefettura, in Genoa were done in 1544, when he was only 17) and he became the dominating figure in 16th-cent. Genoese painting, running a large and productive workshop. His style derives from *Michelangelo in the massiveness of his figures and *Correggio in the softness of his modelling, but the use of dry paint and the simplification of forms are his own. The latter is particularly noticeable in his drawings, which often utilize geometrical forms that give them a superficially *Cubist look. Another curious instance of antecedence is apparent in his night scenes, which have been claimed as sources for Georges de *La Tour, even though it is not clear by what route they could have become known to him. In 1583 Luca accepted an invitation from Philip II of Spain to decorate the *Escorial. He died there in 1585 and was succeeded by Federico *Zuccaro and then Pellegrino *Tibaldi.

Camden Town Group. Group of British painters formed in 1911 who took their name from the drab working-class area of London (as it was then) made popular as a subject by *Sickert, who lived in the borough for several years. In addition to being the prime inspiration of the group, Sickert suggested the name. The group lasted only two years, but its name is also used in a broader sense to characterize a distinctive strain in British painting from about 1905 to 1920, and as Wendy Baron, the group's leading historian, has written, 'If we define Camden Town painting as the objective record of aspects of urban life in a basically Impressionist-derived handling, and recognize it as a distinct movement in British art, then we must accept that the heyday of Camden Town painting was over by the time the Camden Town Group was born.' Many of Sickert's disciples showed their work at the exhibitions of the *Allied Artists' Association, founded in 1908, and several of them also did so at the *New English Art Club, but for some of them these institutions were not progressive enough, which led to the decision to form the Camden Town Group in 1911. Women were excluded and it was decided to limit the membership to sixteen, who were originally: Walter Bayes (1869–1956), Robert *Bevan, Malcolm Drummond (1880–1945), Harold *Gilman, Charles *Ginner, Spencer *Gore (president), J. D. *Inness, Augustus *John, Henry *Lamb, Wyndham *Lewis, Maxwell Gordon Lightfoot (1886–1911), J. B. Manson (1879–1945), who was secretary, Lucien *Pissarro, William Ratcliffe (1870–1955), Sickert himself, and John Damon Turner (1873–1938), an amateur painter. After Lightfoot's suicide in 1911 he was replaced by Duncan *Grant.

These artists varied considerably in their aims and styles. Their subjects included not only street scenes in Camden Town, but also landscapes, portraits, and still lifes. Several of them painted with a technique that can loosely be described as *Impressionist, with broad, broken touches, but particularly after Roger *Fry's *Post-Impressionist exhibitions of 1910 and 1912, the use of bold, flat areas of colour became characteristic of others, notably Bevan, Gilman, Ginner, and Gore. These four best represent a distinctive Camden Town 'style', one that was much imitated by painters of the urban scene up to the Second World War and beyond. The Camden Town Group held two exhibitions at the Carfax Gallery, London, in 1911 and a third in 1912. They were financially disastrous, and as the gallery then declined to put on more exhibitions, they merged with a number of smaller groups to form the *London Group in November 1913. The new body organized a collective exhibition in Brighton at the end of 1913, but although the exhibition was advertised under the name of the Camden Town Group, it may be regarded rather as the first exhibition of the London Group.

cameo. Engraved work on gemstones, glass, paste, ceramics, or similar materials in which the surrounding ground is cut away so that the design stands up in *relief above the surface. It is the opposite of *intaglio. More specifically, the term is often applied to a small relief carving cut in a banded or multicoloured gemstone, such as an agate, in such a way as to exploit the effect of the different layers of colour.

camera lucida (Latin: 'light chamber'). An apparatus for drawing and copying, patented in 1807 by William Hyde Wollaston (1766–1828), a well-known man of science. It received this misleading name—for it is not a 'chamber' at all—because it performed the same function as the *camera obscura, but in full daylight. It consists essentially of a prism on an adjustable stand. The draughtsman sets the prism between his eye and the paper in such a way that he can see an image of the object apparently lying on the paper and can trace its outline. Various refinements were added to the basic format, including a lens to aid focusing and a system of mirrors to enable it to be used with a microscope.

camera obscura (Latin: 'dark chamber'). An apparatus which projects the image of an object or scene on to a sheet of paper or ground glass so that the outlines can be traced. It consists of a shuttered box or room with a small hole or lens in one side through which light from a brightly lit scene enters and forms an inverted image on a screen placed opposite the opening. The optical principle is essentially that of the photographic camera. For greater convenience a mirror is usually installed, which reflects the image the right way up on to a suitably placed drawing surface. The principle was known as early as Aristotle and medieval astronomers found the device helpful in observing solar eclipses. *Vasari refers to an invention of *Alberti's which sounds like the earliest camera obscura as an instrument for drawing, but the first written account of its use for drawing is ascribed to Giambattista della Porta, a physician of Naples. His description in his work on popular science, *Magiae Naturalis* (1558), did much to make the device widely known. In 1679 the architect and scientist Robert Hooke (1635–1703) built a transportable apparatus for landscape painters and by the 18th cent. the camera obscura had become a craze. Both amateurs and professionals—among them *Canaletto—were using it for topographical painting, and we hear of an apparatus, somewhat like a sedan chair, inside which the artist could sit and draw, at the same time actuating bellows with his feet to improve the ventilation. More modest versions were easily portable and even pocketable.

Campagnola, Giulio (*c*.1482–*c*.1518). Italian artist, active mainly as an engraver. Born in Padua, he trained under *Mantegna in Mantua, and by 1499 was attached to the ducal court at Ferrara. In 1509 he was in Venice and it is with this city that he is chiefly associated, his engravings of idyllic landscape subjects playing a major role in spreading the style of *Giorgione and *Titian. His many copies after *Dürer likewise disseminated knowledge of this artist in Italy.

Giulio's pupil and adopted son, **Domenico** (*c*.1500–64), made some engravings in the manner of his master but became better known as a painter and draughtsman. He sold his drawings (mainly landscapes) as finished compositions, sometimes passing them off as the work of Titian. In the 1520s he moved to Padua and became the city's busiest painter.

Campania, Pedro de (1503–80). Netherlandish painter, active mainly in Seville and known by the Spanish form of his name rather than the Flemish one, Pieter de Kempeneer. Before settling in Seville he had worked in Italy and he exercised a strong influence in Andalusia as a pioneer of *Mannerism and the style of *Raphael. *The Descent from the Cross* (1547) is typical of his style and one of several examples of his work in Seville Cathedral. In 1562 he left Spain to direct a tapestry factory at Brussels, his native city.

Campbell, Steven. See GLASGOW SCHOOL.

Campen, Jacob van (1595–1657). Dutch architect and painter. He was the greatest Dutch architect of the 17th cent. and occupied a role in his country similar to that of his contemporary Inigo *Jones in England by introducing a fully mature classical style; the diplomat Constantin Huygens (1596–1687) described him as the man 'who vanquished Gothic folly with Roman stateliness and drove old heresy forth before an older truth'. His most important building is Amsterdam Town Hall (begun 1648), a triumphant symbol of the city during its greatest period. The building was richly decorated; Artus *Quellin I led a team of sculptors, and *Rembrandt was among those who provided paintings, although his *Conspiracy of Julius Civilis* (Nationalmuseum, Stockholm, 1661–2) was removed soon after installation and replaced with a picture by his pupil Juriaen Ovens (1623–78). Van Campen's other buildings include the beautiful *Mauritshuis in The Hague (1633–5), designed as a royal palace and now a celebrated picture gallery. As a painter he concentrated on historical and decorative work and was one of the team, including *Jordaens and *Lievens, who worked on the decoration of the Huis ten Bosch, the royal villa on the outskirts of The Hague.

Camphuysen, Govert (1623/4–72). Dutch painter of bucolic landscape and *genre scenes, cows, and rustic lovers in a style influenced by Paulus *Potter. In 1652–63 he was in Sweden and in 1655 was made court painter at Stockholm. Most of the rest of his career was spent in Amsterdam. He is represented at Dulwich College Picture Gallery and in the Wallace Collection, London. His brother **Raphael** (1598–1657) painted moonlight and winter scenes in the style of Aert van der *Neer.

Campin, Robert. See MASTER OF FLÉMALLE.

Canadian Group of Painters. A group of 20th-cent. Canadian artists formed in Toronto as a successor to the *Group of Seven. Its policy was 'to encourage and foster the growth of art in Canada which has a national character'. Its first exhibition was held in Atlantic City, New Jersey, in the summer of 1933 and its second at the Art Gallery of Toronto in the following November. The group expanded and many of the best-known Canadian artists exhibited with it from the 1930s to the 1960s. It was disbanded in 1969.

Canaletto (Giovanni Antonio Canal) (1697–1768). Venetian painter, the most famous view-painter of the 18th cent. He began his career as a theatrical scene painter (his father's profession), but he turned to topography (see VEDUTA) during a visit to Rome in 1719–20, when he was influenced by the work of Giovanni Paolo *Panini. By 1723 he was painting picturesque views of Venice, marked by strong contrasts of light and shade and free handling, this phase of his work culminating in the splendid *Stone Mason's Yard* (NG, London, *c*.1730). Meanwhile, partly under the influence of Luca *Carlevaris, and largely in rivalry with him, Canaletto began to turn out views which were more topographically accurate, set in a higher key, and with smoother, more precise handling—characteristics that mark most of his later work. At the same time he became more prolific as a draughtsman, mainly in pen and ink, and began painting the ceremonial and festival subjects which ultimately formed an important part of his work. His patrons were chiefly English collectors, for whom he sometimes produced series of views in uniform size. Conspicuous among them was Joseph Smith (1682–1770), an English merchant in Venice, appointed British Consul there in 1744. It was perhaps at his instance that Canaletto enlarged his repertory in the 1740s to include subjects from the Venetian mainland and from Rome (probably based on drawings made during his visit as a young man), and by producing numerous *capricci. He also gave increased attention to the graphic arts, making a remarkable series of etchings, and many drawings in pen, and pen and wash, as independent works of art and not as preparation for paintings. This led to changes in his style of painting, increasing an already well-established tendency to become stylized and mechanical in handling. He often used the *camera obscura as an aid to composition.

In 1746 he went to England, apparently at the suggestion of Jacopo *Amigoni. For a time he was very successful, painting views of London and of various country houses. Subsequently, with declining demand, his work became increasingly lifeless and mannered, so much so that rumours were put about, probably by rivals, that he was not in fact the famous Canaletto but an impostor. In 1755 he returned to Venice and continued active for the remainder of his life. Legends of his having amassed a fortune in Venice are disproved by the official inventory of his estate on his death. Before this, Joseph Smith had sold the major part of his paintings to George III (1738–1820), thus bringing into the royal collection an unrivalled group of Canaletto's paintings and drawings. Added to examples in English public and private collections this makes England incomparably richer in his work than any other country.

Canaletto was highly influential in Italy and elsewhere. His nephew Bernardo *Bellotto took his style to Central Europe and his followers in England included William *Marlow and Samuel *Scott. Even in its mechanical phases his work is much more than a mere factual record; and by his unobtrusive skill in design, and his power of suffusing his work with light and air, he may justly be regarded as one of the precursors of 19th-cent. landscape painting.

Candlelight Master. See BIGOT.

Cano, Alonso (1601–67). Spanish sculptor, painter, architect, and draughtsman, sometimes called 'the Spanish *Michelangelo' because of the diversity of his talents. He was born and died in Granada, and worked there and in Seville and Madrid. His movements were partly dictated by his tempestuous character, for more than once he fled or was expelled from the city he was working in (once for the suspected murder of his wife). In spite of his violent temperament, his work tends to be serene and often sweet. He studied painting in Seville with *Pacheco (*Velázquez was his fellow-student) and sculpture with *Montañés, and stayed in the city from 1614 to 1638, when he moved to Madrid to become painter to the Count-Duke Olivares (1587–1645) and was employed by Philip IV (1605–65) to restore pictures in the royal collection. Thus he became acquainted with the work of the 16th-cent. Venetian masters, whose influence is apparent in his later paintings; they are much softer in technique than his earlier pictures, which are strongly lit in the manner of *Zurbarán. From 1652 he worked mainly in Granada, where he designed the façade of the cathedral (1667), one of the boldest and most original works of Spanish *Baroque architecture. He was ordained a priest in 1658, as this was necessary for him to further his career at Granada Cathedral. The cathedral has several of his works in painting and sculpture, including a *polychrome wooden statue of the *Immaculate Conception* (1655) that is sometimes considered his masterpiece; he is also well represented in Granada Museum.

Canova, Antonio (1757–1822). Italian sculptor. He was the most successful and the most influential sculptor of the *Neoclassical movement, outdoing even *Thorvaldsen and *Flaxman in international fame and prestige. Born in Possagno, near Treviso, the son of a stonemason, he became assistant to a local sculptor and moved with him to Venice in 1768. His early work is lively and naturalistic (*Daedalus and Icarus*, Museo Correr, Venice, 1779), but after he settled in Rome in 1781 his style became graver and thoroughly imbued with *antique influence. *Theseus and the Minotaur* (V&A, London, 1781–3) was his first major work in Rome, and he soon followed this with the prestigious commission for the tomb of Pope Clement XIV in SS. Apostoli (1783–7). After this Canova never looked back. He ran a large studio and worked for a galaxy of European notables, including Napoleon (who tried unsuccessfully to bring him to Paris), the Duke of Wellington, and Catherine the Great of Russia; his portrait of Napoleon's sister (*Pauline Borghese as Venus*, Borghese Gal., Rome, 1805–7) is one of his most celebrated works, a marble equivalent to *David's *Madame Récamier*. Canova worked much for the papal court and in 1815 he became the Pope's representative in recovering works of art looted by Napoleon; on his visits to Paris and London he was fêted. In 1816 he was created Marchese d'Ischia by the Pope and he retired to Possagno, where he built a studio that is now a museum devoted to him. Canova was immensely influential and was renowned for his generosity to young sculptors. He went out of favour during the *Romantic period, when his work seemed cold and static, but his reputation has greatly revived in the 20th cent. His work, in fact, was always much more individual than that of many of his Neoclassical contemporaries and he placed great importance on the personal handling of his material, whether it was marble or clay.

canvas. A woven cloth used as a *support for painting. The best-quality canvas is made of linen; other materials used are cotton, hemp, and jute. It is now so familiar a material that the word 'canvas' has become almost a synonym for an *oil painting, but it was not until around 1500 that it began to rival the wooden *panel (which was more expensive and took longer to prepare) as the standard support for movable paintings. Canvas is not suitable for painting on until it has been coated with a *ground, which isolates the fabric from the paint; otherwise it will absorb too much paint, only very rough effects will be obtainable, and parts of the fabric may be rotted by the *pigments. It must also be made taut on a *stretcher or by some other means. Nowadays both grounding and stretching are done in the factory, and in that order; but until the 19th cent. the stretching was done first, with the result that on an old canvas the ground ends where the canvas turns over the edge of the stretcher.

canvas board. Sheet of cardboard or pasteboard covered with sized and primed cloth, usually cotton. It was first made commercially in the second half of the 19th cent. and is chiefly used by amateurs as a cheap substitute for canvas or for outdoor sketching. Owing to the doubtful quality of the materials used in some commercially prepared canvas boards, professional artists have usually preferred to prepare their own boards.

Čapek, Josef (1887–1945). Czech painter, graphic artist, designer, and writer. He was a member of an association called the Group of Avant-Garde Artists formed in Prague in 1911 by Otto *Gutfreund and Emil *Filla with the object of combining *Cubism and German *Expressionism into a new national artistic style. The Expressionist current in his work prevailed. Like his brother, the celebrated writer Karel Čapek (several of whose books he illustrated), he was deeply concerned with fundamental moral and social questions. They both fervently opposed the threat from Nazi Germany in the 1930s; Karel died the year before the Second World War began, but Josef lived to see its full horrors and died in a concentration camp.

capital. In architecture, the crowning feature of a column, forming a transition between the shaft of the column and the member it supports. Capitals are often more or less elaborately carved with figurative or decorative elements (and sometimes painted) and in Classical architecture the various types of capitals mark the most obvious distinctions between the different 'Orders'.

Cappelle, Jan van de (*c.*1624–79). Dutch marine and landscape painter. He was a wealthy Amsterdam dyer who taught himself to paint during his spare time, but there is nothing of the Sunday painter in his work. Typically his paintings show handsome vessels on calm rivers or seas; they have a grandeur of composition, a limpid quality of light, and an exquisite sense of tonality that places them among the finest marine paintings of any time or place. Cappelle also painted winter landscapes and beach scenes. His work is rare; the best collection is in the National Gallery in London. He was affluent enough to make a distinguished art collection: he owned works by *Rubens, *Brouwer, van *Dyck, *Jordaens, Hercules *Seghers, Simon de *Vlieger (who influenced his own work), and about 500 of *Rembrandt's drawings. Rembrandt and *Hals are said to have painted his portrait.

capriccio. Italian term meaning 'caprice' that can be applied to any fantasy subject, but is most commonly used of a type of townscape popular in the 18th cent. in which real buildings are combined with imaginary ones or are shown with their locations rearranged. *Canaletto and *Guardi often painted pictures of this type, and there is a painting by William *Marlow in the Tate Gallery showing St Paul's Cathedral above a Venetian canal. *Goya's *Los Caprichos* are etchings of fantastic subjects of a completely different kind.

Caracciolo, Giovanni Battista (called Battistello) (1578–1635). Neapolitan painter. He was one of the greatest of *Caravaggio's followers, and his powerful work was an important factor in making Naples a stronghold of the Caravaggesque style. The decisive impact that Caravaggio made on his style can be seen from his *Liberation of St Peter* (1608–9) painted for the same church (the Chiesa del Monte della Misericordia, still *in situ*) as the master's *Seven Acts of Mercy*. It shows how Caracciolo, unlike so many of the *Caravaggisti, looked beyond the obvious trademarks of Caravaggio's style, emulating it in depth of feeling as well as in mastery of dramatic light and shade. He visited Rome and Florence in the second decade of the century and his later work shows a more classical strain, influenced perhaps by the *Carracci. Unusually for a Caravaggesque artist he was an accomplished fresco painter, and his finest late works are decorations in the Certosa di S. Martino in Naples, finished in 1631.

Caravaggio, Michelangelo Merisi da (1571–1610). The most original and influential Italian painter of the 17th cent., named after his native town near Bergamo. He trained in Milan under the undistinguished Simone Peterzano and by 1592 had moved to Rome, the main centre of his activity, where, apart from the frequent scandals caused by his tempestuous character, he was criticized for his Venetian method of working in oils directly from the natural model on to the canvas, without the careful preparations traditional in central Italy.

Two phases in the Roman phase of Caravaggio's career can be distinguished: an early experimental period (*c.*1592–9) and a mature period (1599–1606) in which he carried out several large commissions. The early works are fairly small, with half-length figures, a preponderance of still life, and a frankly homo-erotic character, e.g. the *Young Bacchus* (Uffizi, Florence, *c.*1595) and *Boy with a Fruit Basket* (Borghese Gal., Rome, *c.*1595). Subsequently his figures gained greater plasticity, and he painted in rich deep colours with strongly accentuated shadows (*The Supper at Emmaus*, NG, London, 1601).

The second Roman period began with a commission (probably gained through his first noteworthy patron, the hedonistic Cardinal Francesco Del Monte) for the Contarelli Chapel in S. Luigi dei Francesi (*Calling of St Matthew* and *Martyrdom of St Matthew*, 1599–1600), in which Caravaggio's extraordinary advance in mastery of construction and ability to handle dramatic action was achieved only after great effort, as X-rays of the paintings make clear. The altarpiece of *St Matthew and the Angel* was rejected because it was thought to lack decorum, but it was bought by the Marchese Vincenzo *Giustiniani, one of the most important patrons in Rome, who also paid for the replacement. (The first altarpiece was formerly in Berlin, but was destroyed in the Second World War; the replacement is still *in situ*. Both were painted in 1602.) Meanwhile Caravaggio had embarked on his second great public commission—two paintings for the Cerasi Chapel in Sta Maria del Popolo (*Crucifixion of St Peter* and *Conversion of St Paul*, 1600–1), which are astounding in the economy of their elements, the force of the pictorial vision, and the new way of seeing old subjects. The Contarelli Chapel and Cerasi Chapel paintings changed the direction of Caravaggio's work, for thenceforth he devoted himself almost exclusively to large-scale religious pictures, among them the *Entombment* (Vatican, 1602–4), *Madonna di Loreto* (S. Agostino, Rome, 1604–5), *Madonna de' Palafrenieri* (Borghese Gal., Rome, 1605–6), and *The Death of the Virgin* (Louvre, Paris, *c.*1605–6). The last two were again refused on grounds of decorum or theological incorrectness. Despite this misunderstanding of his work, Caravaggio was not without powerful supporters and his rejected paintings found ready secular buyers.

Caravaggio fled from Rome in 1606 after killing a man in a brawl over a wager on a tennis match, and spent the last four years of his life wandering from Naples (1606–7) to Malta (1607–8) and Sicily (1609), and back to Naples again (1609–10). He continued to paint large religious compositions in a new style shorn of all inessentials: little colour, thinly applied paint, the crowded drama and movement of the late Roman works replaced by a moving silence and contemplativeness. Remarkable among these is *The Beheading of St John the Baptist* (Valetta Cathedral, Malta, 1608), a work of the utmost tragic power. Caravaggio was not yet 40 when he died from malarial fever while returning to Rome in hope of a pardon, but his last works have all the ineffable qualities of the late works of an aged genius. His short but intense activity is remarkable for its rapid development, and for its impact on painting throughout Europe. He had no pupils, but a legion of followers (the *Caravaggisti), and his work, together with that of the *Carracci, revived Italian painting from the nebulous unreality of late 16th-cent. *Mannerist art.

Caravaggio continued to be a famous name throughout the 17th cent., but he was regarded by many as an 'evil genius' (in the words of Vincenzo *Carducho, writing in 1633), whose influence on other artists was pernicious. *Baglione (1642) wrote that 'Some people consider him to have been the very ruination of painting, because many young artists, following his example, simply copy heads from life without studying the fundamentals of drawing and the profundity of art . . . and are . . . incapable of putting two figures together or of composing a story because they do not understand the high value of the noble art of painting.' In a similar vein, *Bellori (1672) wrote that 'There is no question that Caravaggio advanced the art of painting because he came upon the scene at a time when realism was not much in fashion and when figures were made according to convention and manner and satisfied more the taste for gracefulness than for truth', but he thought that he 'debased the majesty of art', and because of him 'everyone did as he pleased, and soon the value of the beautiful was discounted. The *antique lost all authority, as did *Raphael . . . some artists began to revel in filth and deformity.' Interest in Caravaggio declined in the 18th cent. (he is not mentioned in *Reynolds's *Discourses*), but revived in the mid-19th cent. By this time his rejection of ideal beauty could be seen to have the advantage of truth, although there were still those, like *Ruskin, who saw in him 'perpetual seeking for and feeding upon horror and ugliness, and filthiness of sin' (*Modern Painters*, vol. ii, 1846). Serious historical research on him began in the early years of the 20th cent., since when he has attracted an enormous amount of critical commentary and speculation, so much so that Ellis *Waterhouse has written that 'the innocent reader of art-historical literature could be forgiven for supposing that his place in the history of civilization lies somewhere in importance between Aristotle and Lenin.'

Caravaggisti. Term applied to painters who imitated the style of *Caravaggio in the early 17th cent. Caravaggio's methods, particularly his emphatic use of *chiaroscuro in the interests of dramatic *realism, had extraordinary influence in Rome in the first decade of the 17th cent., on both Italian painters and artists from other countries, who flocked to what was then the artistic capital of Europe. His fame was already widespread by 1604, when Karel van *Mander, in Haarlem, wrote of 'Michelangelo da Caravaggio, who is doing extraordinary things in Rome'. The most prominent of the Italian Caravaggisti included Orazio *Gentileschi, one of the few followers to have close personal contact with the master, and Bartolommeo *Manfredi, who popularized tavern and guard room scenes, subjects that Caravaggio himself had not painted. In Naples, where Caravaggio worked intermittently between 1606 and 1610, *Caracciolo, Artemisia *Gentileschi, and *Ribera, a Spaniard by birth, ensured that the style took firm root. In Rome, Caravaggism went out of favour in the 1620s, but it persisted elsewhere in Italy, and in other parts of Europe, particularly in Sicily (which Caravaggio visited), Utrecht, and Lorraine, lingering into the 1650s in all three places. *Baburen, *Honthorst, and *Terbrugghen were the three most important artists in making Utrecht the Dutch centre of Caravaggism, and in Lorraine George de *La Tour perfected perhaps the most personal and poetic interpretation of the style. Few major painters worked in a Caravaggesque style throughout their careers; some, such as Guido *Reni, had a brief flirtation with it, while others, such as Honthorst (who became a court portraitist), had a complete change of direction. Caravaggism was, however, a phenomenon of great importance, and echoes of it can be found in the work of some of the giants of 17th-cent. art: *Rembrandt, *Rubens, and *Velázquez.

cardboard. A thin but stiff board made from paper pulp or sheets of paper, sometimes used as a *support for paintings. Some of *Etty's finest nudes were painted on the type called millboard. A large number of the best works of *Toulouse-Lautrec are on cardboard as, also, are many paintings by *Bonnard and *Vuillard. Twentieth-century painters have used cardboard which was *sized but not *primed.

Carducho, Vincente (*c.*1576–1638). Spanish painter and writer on art. He was a Florentine by birth (his name was originally Vincenzo Carducci), but he settled in Spain when he was 9, when his brother, **Bartolomé Carducho** (born Bartolommeo Carducci, *c.*1560–1608), went to the *Escorial in 1585 as assistant to Federico *Zuccaro. Vincente was appointed a court painter in Madrid in 1609, and had a successful and prolific career, but he is now remembered mainly for his book *Diálogos de la Pintura* (1633). In this he defended the heroic Italian tradition (championing *Michelangelo in particular), and excoriated the naturalism of *Caravaggio. In a sense this was an attack on *Velázquez, who had completely eclipsed Carducho at court. Bartolomé Carducho also had a successful career in Spain, becoming a court painter in 1598 and working on royal commissions in Madrid, Segovia, and Valladolid. He painted in oils and fresco, and was influential in introducing Italian ideas to Spain.

caricature. A form of art, usually portraiture, in which characteristic features of the subject represented are distorted or exaggerated for comic effect. The term is sometimes used more broadly to denote other forms of pictorial burlesque, *grotesque or ludicrous representation, such as the grotesque heads of *Leonardo. The invention of caricature in the more limited sense is usually credited to Annibale *Carracci, who defended it as a counterpart to idealization (see IDEAL): just as the serious artist penetrates to the idea behind appearances, so the caricaturist also brings out the essence of his victim, the way he should look if Nature wholly had her way. Many other painters of the Bolognese School, such as *Guercino and *Domenichino, were brilliant caricaturists, but the greatest master of economy and expression was *Bernini, who demonstrated his skill before Louis XIV. A generation or so later Pier Leone Ghezzi (1674–1755) became probably the first artist to earn a substantial part of his living as a caricaturist, specializing in portrayals of the many art lovers who congregated in Rome. Political caricature as we know it today emerged in the last three decades of the 18th cent. and was perfected as a distinct genre by the verve and skill of such masters as *Gillray and *Rowlandson, in whose coarse, licentious, and unbridled satires on the personalities of the day the comic likeness of a Pitt, a Fox, or even of George III was distilled into a recognizable stereotype. These masters had imitators but no equals in other

European countries, and English caricature remained in the lead until the gentler atmosphere of early Victorian culture diverted the energies of masters such as George *Cruikshank into humorous illustrations, while the weekly cartoon of *Punch*, sometimes humorous, sometimes rhetorical, institutionalized political caricature for the middle-class home. At the same time Ghezzi's genre of genteel portrait caricature was revived in the pages of the periodical *Vanity Fair*.

It was in France that the political tensions of the 19th cent. produced the greatest master of the genre, *Daumier, who worked for *La Caricature* and *Le Charivari*, a weekly and a daily which remorselessly campaigned against Louis Philippe (1773–1850), transforming his fat face into the famous image of a pear.

Many great artists of the 18th and 19th cents. produced caricatures either as pot-boilers or as a sideline; among the first were *Monet (who began his career as a caricaturist) and *Doré; among the second *Tiepolo, *Puvis de Chavannes, and *Picasso, who made caricatures in his early Barcelona days. Many of the most popular graphic artists of the 20th cent. have combined some gift for caricature with social or political satire.

Carlevaris, Luca (1663–1730). Italian painter, born in Udine and active mainly in Venice. He is regarded as the father of 18th cent. Venetian view-painting (see VEDUTA), for although he was not (as is sometimes asserted) the first to specialize in the genre, he approached it with a new seriousness, his training as a mathematician being reflected in his rigorous perspective settings. His paintings, and his set of over 100 engraved views of the city published in 1703, are the foundation on which *Canaletto and *Guardi built. A number of oil sketches from nature in the Victoria and Albert Museum reveal his powers of lively observation.

Carlstedt, John Birger Jarl (1907–75). Finnish painter and designer, a pioneer of modernism and abstract art in his country. He worked in Paris, where he was influenced by De *Stijl and *Constructivism, and after the Second World War he began painting in a completely abstract style. His best-known work is probably the interior decoration for the restaurant Le Chat Doré in Helsinki (1928).

Carmichael, Franklin. See GROUP OF SEVEN.

carnations. A term, popular in the 18th cent., for the flesh colours in a painting.

Caro, Sir Anthony (1924–). British sculptor, one of the most influential figures in post-war British art. After training as an engineer at Cambridge University and serving in the navy in the Second World War, he studied sculpture in London, then from 1951 to 1953 worked as part-time assistant to Henry *Moore. His early works were figures modelled in clay, but a radical change of direction came after he visited the USA and met David *Smith in 1959. In the following year he began making abstract metal sculpture, using standard industrial parts such as steel plates and lengths of aluminium tubing together with pieces of scrap, which he welded and bolted together and then generally painted a single rich colour. The colour helped to unify the various shapes and textures and often set the mood for the piece, as with the bright and optimistic red of *Early One Morning* (Tate Gallery, London, 1962). This, like many of Caro's sculptures, is large in scale and open and extended in composition; it rests directly on the ground, and Caro has been one of the leading figures in challenging the 'pedestal' tradition. Caro taught part-time at St Martin's School of Art in London 1953–79, and he had a major influence on several of the young sculptors who trained under him, initiating a new school of British abstract sculpture (see NEW GENERATION). In the 1970s his work became much more massive and rougher in texture, sometimes incorporating huge chunks of metal. In the 1980s he returned to more traditional materials and techniques and began making figurative (or semi-abstract) works in bronze, including (in the early 1990s) a series inspired by the Trojan War. His reputation is high in the USA as well as Britain, but he is not without detractors; the critic Peter Fuller described the work with which he became famous as 'nothing if not of its time: it reflected the superficial, synthetic, urban, commercial American values which dominated the 1960s'.

Carolingian art. Term in art history used for the art produced in the reigns of the emperor Charlemagne (800–14) and his successors until *c.*900. The outstanding feature of this period was a revival of interest in Roman antiquity. Inspiration was sought in the art of the emperors Constantine (?288–337) and Theodosius (346–95), and works of the 6th and 7th cents. from *Byzantium served as models.

Carolus-Duran (pseudonym of Charles Durand) (1838–1917). French painter. His early works were influenced by *Courbet, but from about 1870 he concentrated on portraiture, becoming a great fashionable success with his slick style. *Sargent was the most important of his many pupils. In 1905 he was appointed Director of the French School in Rome.

Caron, Antoine (1521–99). French *Mannerist painter. He is one of the few French painters of his time with a distinctive artistic personality, and his work reflects the refined but unstable atmosphere of the Valois court during the Wars of Religion (1560–98). He worked at *Fontainebleau under *Primaticcio in the 1540s and later became court painter to Catherine de Médicis, wife of Henry II of France. His few surviving works include historical and allegorical subjects in the manner of court ceremonies, scenes of magic and prediction, and massacres, as in *Massacres under the Triumvirate* (1566) in the Louvre, his only signed and dated painting. His style is characterized most obviously by extremely elongated, precious-looking figures set in open spaces that seem too large for them. He had a penchant for gaudy colours and bizarre architectural forms. Some of the works attributed to him may be by other hands, however, for French painting of his period is such an obscure area that Caron's name is liable to be attached to anything similar to his known *œuvre*.

Carpaccio, Vittore (*c.*1460–1525/6). Venetian painter. His life is poorly documented, and it is not known with whom he trained, but it is generally agreed that the chief influence on his work was Gentile *Bellini. This is especially apparent in the first of the two great cycles of paintings that are his chief claim to fame—the *Scenes from the Life of St Ursula*, executed in the 1490s and now in the Accademia, Venice. Carpaccio's distinguishing characteristics—his taste for anecdote, and his eye for the crowded detail of the Venetian scene—found their happiest expression in these paintings; indeed one of them, the *Miracle of the Cross*, looks forward to the 18th-cent. compositions of *Canaletto and *Guardi. His other cycle, *Scenes from the Lives of St George and St Jerome*, painted for the Scuola (or 'Society') of S. Giorgio degli Schiavone, Venice, in 1502–7 (still in the Scuola), combines fantasy with detail minutely observed. Carpaccio's altarpieces are generally less memorable, but another deservedly famous work is the delightful *Two Courtesans* (Correr Mus., Venice), probably a fragment of a larger work. It was a favourite work of *Ruskin, who contributed to the great popularity Carpaccio enjoyed in the 19th cent. His reputation has perhaps declined somewhat since, but he is still rated as second only to Giovanni Bellini as the outstanding Venetian painter of his generation.

Carpeaux, Jean-Baptiste (1827–75). French sculptor and painter. The son of a mason, he worked for some months in the studio of *Rude and also studied at the École des *Beaux-Arts, where he won the *Prix de Rome in 1854. His *Ugolino* (Musée d'Orsay, Paris, 1860–2) earned him the acclaim of the French colony in Rome, and upon his return to Paris in 1862 he won favour with the court, receiving many commissions for portrait busts. He also made several large sculpture groups, of which the most famous is *La Danse* (1869) for the façade of the Paris Opéra (the original is now in the Musée d'Orsay). This uninhibitedly dynamic work caused a sensation, was denounced as immoral, and had ink thrown over it. Partially because of such attacks on his work he suffered from a persecution complex in his final years before his early death from cancer. Carpeaux was the outstanding French sculptor of his period and an influential figure. His exuberance of feeling and vivacious modelling made a decisive break with the *Neoclassical tradition and presaged the work of *Rodin. His paintings are well represented in the Petit Palais in Paris, and there are also good examples of his work in the Musée des Beaux-Arts in Valenciennes, his home town.

Carr, Emily (1871–1945). Canadian painter of *Expressionist landscapes. Her artistic development was slow and halting, interrupted by ill health and the need to undertake other work to earn a living. Her training was mainly in San Francisco, 1889–95, England, 1899–1904, and Paris, 1910–11. In France she was impressed by the work of the *Fauves and was probably also influenced by Frances *Hodgkins. After her return to Canada she painted the landscape of her native British Columbia with passionate feeling for the power of nature, executing much of her work out of doors. Discouraged by years of neglect, she had almost ceased to paint when in 1927 she was overwhelmed when she first saw the work of the *Group of Seven in Toronto. Thereafter she worked with renewed energy and deepened spirituality, her ardent spirit given free rein. She was the author of several autobiographical works and overcame her earlier neglect to attain the status of a national heroine. Her work is well represented in the National Gallery of Canada in Ottawa.

Carrà, Carlo (1881–1966). Italian painter and writer on art, a prominent figure in both *Futurism and *Metaphysical painting. He joined the Futurists in 1909, and visits to Paris in 1911 and 1912 introduced a *Cubist influence into his work. In his best-known painting *The Funeral of the Anarchist Galli* (MOMA, New York, 1911), for example, he combined the dynamism typical of Futurism with a sense of Cubist structural severity. In 1915 he met Giorgio de *Chirico and turned to Metaphysical painting, producing about twenty works with de

Chirico's paraphernalia of posturing mannequins, half-open doors, mysteriously significant interiors, etc., though generally without his typically sinister feeling. Carrà broke with de Chirico in 1919 and abandoned Metaphysical painting, adhering in the 1920s and 1930s to the principles of the *Novecento Italiano. He championed a return to traditional values in the journal *Valori Plastici* and also in the Milan newspaper *L'Ambrosiano*, of which he was art critic from 1921 to 1938. From 1941 to 1952 he was professor of painting at the Brera Academy in Milan.

Carracci. Family of Bolognese painters, the brothers **Agostino** (1557–1602) and **Annibale** (1560–1609) and their cousin **Ludovico** (1555–1619), who were prominent figures at the end of the 16th cent. in the movement against the prevailing *Mannerist artificiality of Italian painting. They worked together early in their careers, and it is not easy to distinguish their shares in, for example, the cycle of frescos in the Palazzo Fava in Bologna (*c.*1583–4). In the early 1580s they opened a private teaching *academy, which soon became a centre for progressive art. It was originally called the Accademia dei Desiderosi ('Desiderosi' meaning 'desirous of fame and learning'), but later changed its name to Accademia degli Incamminati (Academy of the Progressives). In their teaching they laid special emphasis on drawing from the life (all three were outstanding graphic artists) and clear draughtsmanship became a quality particularly associated with artists of the Bolognese School, notably *Domenichino and *Reni, two of the leading members of the following generation who trained with the Carracci.

They continued working in close relationship until 1595, when Annibale, who was by far the greatest artist of the family, was called to Rome by Cardinal Odoardo *Farnese to carry out his masterpiece, the decoration of the Farnese Gallery in the cardinal's family palace. He first decorated a small room called the Camerino with stories of Hercules, and in 1597 undertook the ceiling of the larger gallery, where the theme was *The Loves of the Gods*, or, as *Bellori described it, 'human love governed by Celestial love'. Although the ceiling is rich in the interplay of various illusionistic elements, it retains fundamentally the self-contained and unambiguous character of High *Renaissance decoration, drawing inspiration from *Michelangelo's Sistine Ceiling and *Raphael's frescos in the Vatican Loggie and the Farnesina. The full untrammelled stream of *Baroque illusionism was still to come in the work of *Cortona and *Lanfranco, but Annibale's decoration was one of the foundations of their style. Throughout the 17th and 18th cents. the Farnese Ceiling was ranked alongside the Sistine Ceiling and Raphael's frescos in the Vatican Stanze as one of the supreme masterpieces of painting. It was enormously influential, not only as a pattern book of heroic figure design, but also as a model of technical procedure; Annibale made hundreds of drawings for the ceiling, and until the age of *Romanticism such elaborate preparatory work became accepted as a fundamental part of composing any ambitious history painting. In this sense, Annibale exercised a more profound influence than his great contemporary *Caravaggio, for the latter never worked in fresco, which was still regarded as the greatest test of a painter's ability and the most suitable vehicle for painting in the *Grand Manner.

Annibale's other works in Rome also had great significance in the history of painting. Pictures such as *Domine, Quo Vadis?* (NG, London, *c.*1602) reveal a striking economy in figure composition and a force and precision of gesture that had a profound influence on *Poussin and through him on the whole language of gesture in painting. He developed landscape painting along similar lines, and is regarded as the father of *ideal landscape, in which he was followed by Domenichino (his favourite pupil), *Claude, and Poussin. *The Flight into Egypt* (Doria Gal., Rome, *c.*1604) is Annibale's masterpiece in this genre. In his last years Annibale was overcome by melancholia and gave up painting almost entirely after 1606. When he died he was buried according to his wishes near Raphael in the Pantheon. It is a measure of his achievement that artists as great and diverse as *Bernini, Poussin, and *Rubens found so much to admire and praise in his work. Annibale's art also had a less formal side that comes out in his *caricatures (he is generally credited with inventing the form) and in his early *genre paintings, which are remarkable for their lively observation and free handling (*The Butcher's Shop*, Christ Church, Oxford).

Agostino assisted Annibale in the Farnese Gallery from 1597 to 1600, but he was important mainly as a teacher and engraver. His systematic anatomical studies were engraved after his death and were used for nearly two centuries as teaching aids. He spent the last two years in Parma, where he did his own 'Farnese Ceiling', decorating a ceiling in the Palazzo del Giardino with mythological scenes for Duke Ranuccio Farnese. It shows a meticulous but somewhat spiritless version of his brother's lively *classicism.

Ludovico left Bologna only for brief periods and directed the Carracci academy by himself after his cousins had gone to Rome. His work is uneven and highly personal. Painterly and expressive considerations always outweigh those of stability and calm classicism in his work, and at its best there is a passionate and poetic quality indicative of his preference for *Tintoretto and Jacopo *Bassano. His most fruitful period was 1585–95, but near the end of his career he still produced remarkable paintings of an almost *Expressionist force, such as the *Christ Crucified above Figures in Limbo* (Sta Francesca Romana, Ferrara, 1614).

The Carracci fell from grace in the 19th cent. along with all the other Bolognese painters, who were one of *Ruskin's pet hates and whom he considered (1847) had 'no single virtue, no colour, no drawing, no character, no history, no thought'. They were saddled with the label *'eclectic' and thought to be ponderous and lacking in originality. Their full rehabilitation had to wait until the second half of the 20th cent. (the great Carracci exhibition held in Bologna in 1956 was a notable event), but Annibale has now regained his place as one of the giants of Italian painting.

Agostino's illegitimate son **Antonio** (1589?–1618) was the only offspring of the three Carracci. He had a considerable reputation as an artist in his day, but after his early death was virtually forgotten, and it is only recently that his work has been reconsidered.

Carrara marble. See MARBLE.

Carreño de Miranda, Juan (1614–85). Spanish painter, active mainly in Madrid. Until he was appointed one of the royal painters in 1669 he concentrated on religious works, but thereafter he worked mainly as a portraitist. Except for his friend *Velázquez, he was the most important court painter of 17th-cent. Spain; he was of noble birth and his paintings have an aristocratic dignity and something of Velázquez's sensitivity and taste. His religious paintings (which include several frescos, notably in Toledo Cathedral) are, however, more extravagantly *Baroque.

Carriera, Rosalba (1675–1757). Venetian *pastel portraitist and *miniaturist, the sister-in-law of *Pellegrini. She made pastel portraits fashionable and achieved spectacular success throughout the capital cities of Europe, her visits to Paris (1721–2) and to Vienna (1730) being in the nature of royal progresses. She had considerable influence in France and converted Maurice Quentin de *La Tour to the pastel medium. It is now hard to appreciate why her work, which is highly accomplished but generally rather insipid, should have aroused such enthusiasm. After becoming blind in 1745, she had her sight temporarily restored by an operation, but lost it permanently in 1749 and retired into a state of melancholy dejection. Her large output is particularly well represented in the Accademia and the Ca' Rezzonico in Venice, and the Gemäldegalerie in Dresden.

Carrière, Eugène (1849–1906). French painter of portraits and religious pictures. As a young man he was an admirer of *Rubens and *Velázquez; he based his fluid handling on theirs, but often worked in an almost monochromatic manner, close to *grisaille. He painted intimate scenes of middle-class family life and had a special interest in the theme of motherhood, which he treated with an almost mystical reverence. *Rodin called him 'one of the greatest of painters', and a typical example of his work—*Motherhood*—is in the Musée Rodin in Paris. See also ACADÉMIE.

Carrington, Dora. See BLOOMSBURY GROUP.

Carrington, Leonora. See ERNST.

Carstens, Asmus Jacob (1754–98). Danish-born German draughtsman and painter. Apart from some initial training at the Copenhagen Academy he was largely self-educated. He moved to Berlin in 1787 and taught at the Academy. After 1792 he lived in Rome with the help of a grant from the Prussian State. Carstens was totally committed to *Neo-classicism and concentrated on heroic figure compositions. He was uninterested in colour and was much more prolific as a draughtsman than as a painter. His work has a pompous seriousness in tune with his own inflated idea of his genius, but he is a significant figure because of the strictness of his ideals and the influence he had on the next generation of artists, notably *Thorvaldsen and the *Nazarenes.

cartoon. A full-size drawing made for the purpose of transferring a design to a painting or tapestry or other large work. The earlier painters of *fresco simply drew freehand on the wall or copied from a small sketch, but a cartoon was indispensable in the process of making *stained glass, and it was perhaps from this art that the painters borrowed it. Some frescos of the early 15th cent. show clearly that their designs have been traced from cartoons, for their outlines are either indented or punctuated with pin-pricks. The method is described by *Vasari. The drawing was made on stout paper, usually with *charcoal or *chalk, and sometimes heightened with white or coloured with *watercolours. It was then cut into sections. The section which was wanted for the day's painting was laid against the soft fresh plaster of the wall and a *stylus was pressed heavily along the lines, or else pricks were made at intervals and powdered charcoal was rubbed through the holes—a process called *'pouncing'.

Cartoons were used for *easel paintings as well as frescos. A well-known example is *Leonardo's *Virgin and Child with St Anne and the Infant St John* (NG, London). Another, unusually small, is *Raphael's pen-and-ink drawing *The Vision of a Knight*, which hangs in the National Gallery, London, beside the little *panel painting which was made from it. For tapestries, cartoons were made in full colour; famous examples are the series on the Acts of the Apostles lent from the Royal Collection to the Victoria and Albert Museum, London, painted in *distemper by Raphael and his pupils in 1515–16 as designs for tapestries woven for the Sistine Chapel.

In the 19th cent. designs submitted in a competition for frescos in the British Houses of Parliament were parodied in *Punch*. From this the word 'cartoon' acquired its present popular meaning of a humorous drawing or parody.

cartouche. An ornamental panel, typically consisting of a central field with an inscription or coat of arms enclosed by a framework simulating a scroll of cut *parchment. They are found, for example, in architectural and tomb sculpture, and in book illustration. The term is also applied to the oval or oblong figures in Egyptian hieroglyphics, enclosing characters expressing royal or divine names or titles.

caryatid. A carved female figure clad in long robes serving as a column, first used in Greek architecture. The name means 'woman of Caryae', a town in Sparta, and the device may derive from the postures assumed in the local folk dances at the annual festival of Artemis Caryatis. The most famous caryatids are on the Erechtheum at Athens (*c.*421–406 BC). They were not much used by the Romans and were uncommon in *Renaissance architecture. There are some by *Goujon in the *Louvre, and they were often illustrated in editions of *Vitruvius; they returned to favour with the Greek Revival of the 19th cent. St Pancras Church, London (1819–22), by William Inwood and Henry William Inwood, has some famous ones copied from those of the Erechtheum. The male equivalent of the caryatid is the *atlas; the term 'canephorae' is applied to caryatids with baskets on their heads.

casein. A substance with strong adhesive powers made from the curd of milk, used in art as a binding material for paints and *grounds and as a glue for joining parts of a wooden *panel together.

Caslon, William (1692–1766). The most famous of English typefounders. Caslon was the first of the British typefounders who could offer type of his own cut and approximately uniform design to supply the needs of the printing trade. His roman and italic were modelled on the best Dutch work of the 17th cent., and they soon superseded the miscellaneous collection, largely of foreign origin, previously in use in Britain. Large quantities of his types were exported to America and some to European countries. They went out of fashion *c.*1800, but a taste for them revived 50 years later. The Caslon typefoundry continued in business until 1936.

Cassatt, Mary (1844–1926). American painter and printmaker who worked mostly in Paris in the circle of the *Impressionists. Persuaded to exhibit with the Impressionists by *Degas, for whom she had a great admiration, she nevertheless retained the extremely personal character of her art and her affinities with them lay less in technique and theory than in a common attitude towards the rehabilitation of the everyday scene and gesture. Paintings such as *La Loge* (NG, Washington, *c.*1882) and *Lady at the Tea Table* (Met. Mus., New York, 1885) evoke with a delicate beauty the elaborate refinement of the society described by the novelist Henry James (1843–1916). Her draughtsmanship was outstanding and she was as skilful with pastel and the tools of printmaking as she was with oils. Her prints, in which she often combined etching and *drypoint, show the influence of Japanese art (see UKIYO-E). Cassatt's eyesight began to fail when she was in her fifties and she had virtually stopped working by 1914. She came from a wealthy family and exercised an important influence on American taste by urging her rich friends to buy Impressionist works.

cassone. Italian term for a large chest which frequently contained the bride's dowry or was given as a wedding present. Decorated *cassoni* became the fashion in *Renaissance Italy, and *quattrocento Florence saw the development of the painted *cassone* front. These paintings usually represented episodes from the Bible or Classical history or mythology which pointed a lesson or contained a happy augury for the newly-weds. Often the *cassoni* were made as pairs, bearing the coats of arms respectively of the bride and groom, as with a pair, dated 1472, in the Courtauld Institute Galleries, London (this pair is particularly noteworthy in retaining the original backboards—*spallieri*). Though rarely of very high quality, *cassone* paintings reflect the taste of the Florentines for lively narrative and gay display and were therefore eagerly collected (and repainted) in the 19th cent. Some major artists such as *Domenico Veneziano, *Uccello, and

*Botticelli also seem to have decorated *cassoni* once in a while (Uccello's *Night Hunt* in the Ashmolean Museum, Oxford, was very probably once a *cassone* panel). *Vasari attributed much of the production to Dello Delli (*c.*1404–71?), but in this he was certainly mistaken; the chief documented exponent was Apollonio di Giovanni (1415–65). Painted *cassoni* went out of fashion towards the end of the 15th cent. when carved oaken chests came in, and the form was displaced altogether by the chest of drawers in the 17th cent.

Castagno, Andrea del. See ANDREA DEL CASTAGNO.

Castelli, Leo (1907–). Italian-born American art dealer. He settled in New York in the late 1940s and originally sold modern European works, acquired mainly through contacts he had established during the 1930s when he had been a dealer in Paris. However, finding this market dominated by more established dealers, he turned to American art, and in 1958 he first showed work by Jasper *Johns and Robert *Rauschenberg, the two artists with whom he is most closely identified. In the catalogue of the exhibition 'American Art in the 20th Century' (Royal Academy, London, 1993), Castelli is described as 'perhaps the most influential art dealer of the twentieth century . . . The reputations of most of the major artists of the 1960s were made under his guidance. He established the careers of Cy *Twombly, Frank *Stella and Roy *Lichtenstein, before enhancing the standing of Andy *Warhol . . . With his first exhibition of Claes *Oldenburg in 1974, Castelli completed his group of *Pop celebrities . . . In the early years he helped to change the American gallery system by introducing a European-type retainer (monthly wages advanced against royalties from future sales), enabling his artists to concentrate exclusively on their art.'

Castiglione, Giovanni Benedetto (called Il Grechetto) (*c.*1610–65). Genoese painter, etcher, and draughtsman. His style of painting owed something to *Rubens, van *Dyck, and Bernardo *Strozzi, all of whom worked in Genoa, whilst his etchings depend particularly on *Rembrandt. This openness to foreign influence was unusual for an Italian artist of this period. He was extremely versatile as well as eclectic, being equally at home with *Grand Manner history paintings and rustic genre scenes (he was a superb animal painter). Some of his best works have a sense of fantasy recalling Salvator *Rosa, notably the etching *The Genius of Castiglione* (1648). He was highly prolific as a graphic artist as well as a painter and is credited with inventing the *monotype. From 1648 he was court painter at Mantua, a post in which he was succeeded by his son **Francesco** (d. 1716). His work had wide influence, for example, on *Fragonard and Giambattista *Tiepolo.

Castiglione, Giuseppe (Chinese name Lang Shih-Ning) (1688–1766). Italian Jesuit missionary and amateur painter who settled in China *c.*1715. It is said that he studied Chinese painting by imperial command, and his landscapes and animal and *genre paintings, in which Chinese brushwork and Western *realism were combined for the first time, enjoyed great success at the court. He was the first Western painter to be appreciated by the Chinese, and was commissioned by the emperor to paint portraits, scenes of court life, and imperial military expeditions. His work is well represented in the Musée Guimet, Paris.

Catena, Vincenzo (*c.*1480–1531). Venetian painter. Catena was a man of good birth and independent means who moved in humanist circles and may have been the link between these circles and *Giorgione. He is first mentioned in 1506 in an inscription on the back of Giorgione's portrait *Laura* (Kunsthistorisches Mus., Vienna), according to which they had entered into some kind of partnership. Nothing else is known of this arrangement. The main influence on his style was Giovanni *Bellini. His early paintings can be awkward and stiff, but from *c.*1510 his work matured under the influence of the late Bellini, *Cima, and *Titian into a style that was derivative but handsome, with pleasing handling of diffused light and warm colours. He painted religious pictures and portraits, good examples of which are *Holy Family with an Adoring Warrior* (NG, London) and the signed portrait of a man (Kunsthistorisches Mus.).

Catlin, George (1796–1872). American painter and writer, renowned for his portrayal of Red Indian life. He practised law before becoming an artist in the early 1820s (initially as a portraitist in Philadelphia) and was completely self-taught. In 1830 he began a series of visits to various Indian tribes (he had been fascinated with the subject since childhood) and from 1837 to 1845 he exhibited the resulting paintings as the 'Gallery of Indians' in the USA and Europe. He was better received in England and France than in his native country (*Baudelaire wrote about him enthusiastically) and he lived in Europe from 1858 to 1870. In addition to his paintings, he published various illustrated books on Indian life. The major part of his output is now housed in the Smithsonian Institution in Washington.

Cattermole, George (1800–68). English watercolour painter and book illustrator, mainly of architectural and antiquarian subjects. He worked in a detailed, accurate style, but with *picturesque rather than archaeological aims. Some of the best of his many book illustrations were for Dickens.

Caulfield, Patrick (1936–). British painter. Influenced by Roy *Lichtenstein and by popular illustration, he has sometimes been classed as a *Pop artist, but his style is personal and distinct. His typical subjects are banal interiors and still lifes, treated in a deadpan style of flat colours and black outlines.

Cavalcanti, Emiliano di (1897–1976). Brazilian painter, draughtsman, and writer, a pioneer of modern art in his country. In 1922 he helped to organize the *Semana de Arte Moderna* (Modern Art Week), which included a series of dance spectacles, poetry readings, and an art exhibition and is regarded as a turning-point in Brazilian culture. He was in Paris 1923–5 as a newspaper correspondent (he had begun his career as a *caricaturist) and again in 1935–40, and met numerous leading artists, including *Braque, *Matisse, and *Picasso. His work shows the combined influence of Diego *Rivera, the *Fauves and Picasso's monumental nudes of the 1920s, all of which Cavalcanti applied to high-keyed Brazilian themes. Subjects included sensuous mulatto women, carnival and festival scenes, poor fishermen, and prostitutes (a dominant theme) in extravagantly coloured compositions. He published two volumes of memoirs (1955 and 1964).

Cavalcaselle, Giovanni Battista. See CROWE.

Cavaliere d'Arpino. See CESARI, GIUSEPPE.

Cavallini, Pietro (active 1273–*c.*1330). Italian painter and mosaic designer, active mainly in Rome, where he must have been the leading artist of his day. His two major surviving works are mosaics of the *Life of the Virgin* (Sta Maria in Trastevere) and a fragmentary fresco cycle, the most important part of which is a *Last Judgement* (Sta Cecilia in Trastevere, Rome); both works probably date from the 1290s. In 1308 Cavallini was in Naples serving the Angevin kings, and was probably responsible for the design and possibly some of the execution of a fresco cycle in Sta Maria Donnaregina. Although he is such an obscure figure, Cavallini occupies an important place in the history of Italian painting. He was the first artist to make a significant break with the stylizations of *Byzantine art, and his majestic figures have a real sense of weight and three-dimensionality. His work undoubtedly influenced his great contemporary *Giotto, whose *Last Judgement* in the Arena Chapel at Padua features Apostles enthroned exactly as in Cavallini's fresco of the subject.

Cavallino, Bernardo (1616–56?). Neapolitan painter. He was the most individual and sensitive Neapolitan painter of his time, but his career is somewhat obscure. About eighty paintings by him are extant, but only one is dated, *St Cecilia in Ecstasy* (Palazzo Vecchio, Florence, 1645; a **modello* is in the Museo di Capodimonte, Naples). Most of his pictures are small-scale religious works, peopled by exquisitely elegant and refined figures who evoke a feeling of tender melancholy. Their fragile sensitivity is in complete contrast to the earthy vigour of much of Neapolitan painting of his period. Cavallino trained with Massimo *Stanzione, but his style has more in common with that of van *Dyck, whose work was fairly well known in Naples. He is presumed to have died in the plague that devastated Naples in 1656.

Caylus, Anne-Claude-Philippe de Tubières, Comte de (1692–1765). French antiquarian and collector. A member of a prominent military family, he abandoned a promising career in order to indulge a life-long passion for the arts and antiquity. Caylus was an active champion of the younger artists working in a *Classical style, notably *Bouchardon, of whom he published a *Life* in 1762, and he supported the ideals of Classical purity and simplicity in contrast to the decorative artificiality of *Rococo. He is credited with being the first to conceive archaeology as a scientific discipline and in this respect *Winckelmann acknowledged indebtedness to him. His own collection of antiques, which he began in 1729, formed the basis of his *Recueil d'antiquités égyptiennes, étrusques, grecques, romaines et gauloises* (7 vols., 1752–67), the most serious work of antiquarian research in the 18th cent. and one of the most influential in spreading knowledge and enthusiasm for the works of Classical antiquity.

Cellini, Benvenuto (1500–71). Florentine sculptor, goldsmith, and metal-worker. His autobiography, written in a racy vernacular, has been famous since the 18th cent. (it was first published in 1728) for its vivid picture of a *Renaissance craftsman proud of his skill and independence, boastful of his feats in art, love, and war, quarrelsome, superstitious, and devoted to the great tradition embodied in *Michelangelo. It has given him a wider reputation than could have come from his artistic work alone; but to modern eyes he also appears as one of the most important *Mannerist sculptors, and his statue *Perseus* is one of the glories of Florentine art.

His life can be roughly divided into three periods. From the first, spent mainly in Rome, nothing survives but some coins and medals and the impressions of two large seals. During the second (1540–5), which he spent in the service of Francis I of France (see FONTAINEBLEAU, SCHOOL OF), he created the famous salt-cellar of gold enriched with enamel, exquisitely worked with two principal and many subsidiary figures. This (now in the Kunsthistorisches Mus., Vienna) is the most important piece of goldsmith's work that has survived from the Italian Renaissance. He also made for the king a large bronze *relief, the *Nymph of Fontainebleau* (Louvre, Paris). The remainder of Cellini's life was passed in Florence in the service of Cosimo I de' *Medici, and it was only in this period that he took up large-scale sculpture in the round. The bronze *Perseus* (Loggia dei Lanzi, Florence 1545–54) is reckoned his masterpiece. His other sculptures include the *Apollo and Hyacinth* and *Narcissus* (both in the Bargello, Florence) and the *Crucifix* (Escorial, near Madrid), all in marble. His two portrait busts, *Bindo Altoviti* (Gardner Mus., Boston), and *Cosimo I* (Bargello), are in bronze. Their somewhat dry, niggly quality shows that the exquisite precision of handling of his goldsmith's work did not always transfer easily to a larger scale. Because of his fame, many pieces of metalwork have been attributed to him, but rarely on secure grounds.

Cennini, Cennino (*c.*1370–*c.*1440). Florentine painter and writer. None of his paintings has survived, but he is remembered as the author of *Il Libro dell' Arte* (translated by Daniel V. Thompson as *The Craftsman's Handbook*, 1933), the most important source concerning artistic practice in the late Middle Ages. Cennini states in the book that he was a pupil of Agnolo *Gaddi, who learnt from his father Taddeo Gaddi, who in turn was a pupil of *Giotto, so his detailed descriptions of *tempera and *fresco painting no doubt reflect, even if at several removes, the technical procedures of the founder of the great tradition of Florentine painting. The earliest extant manuscript is dated 1437 'in the debtors' prison in Florence', but most authorities put the date of composition at around 1400.

Cephisodotus. Athenian sculptor of the early 4th cent. BC, perhaps the father of *Praxiteles. A group of *Eirene* [Peace] *Holding the Infant Ploutos* [Wealth], known through Roman copies, is attributed to him; the heavy draperies and intimate expressions are features that were to characterize 4th-cent. sculpture in general and that of Praxiteles in particular. Another sculptor called Cephisodotus was the son of Praxiteles. He was active around 300 BC and noted for his portraits; a head of *Menander* in the Museum of Fine Arts in Boston is possibly a copy of a statue he made with his brother Timarchos for the theatre at Athens.

Cerano, Il. See CRESPI, GIOVANNI BATTISTA.

Cercle et Carré (Circle and Square). A discussion and exhibition society for *Constructivist artists formed in Paris in 1929 by the critic Michel *Seuphor and the painter Joaquin *Torres-García. Three numbers of a journal of the same name appeared in 1929–30. *Mondrian contributed an article. In Paris Cercle et Carré was superseded in 1931 by the more important *Abstraction-Création group, but some years later Torres-García formed an Asociación de Arte Constructivo in Montevideo and edited a journal *Circulo y Cuadrado*, of which seven numbers appeared between 1936 and 1938.

cerography. See ENCAUSTIC PAINTING.

Cerquozzi, Michelangelo (1602–60). Italian painter, known as 'Michelangelo of the Battles' because of his predilection for battle scenes. He spent all his career in Rome, but had considerable contact with Northern painters; it was in the studio of the Fleming Jacob de Haase (1575–1634) that he learnt to paint battle scenes, and his friendship with the Dutchman Pieter van *Laer led to his becoming the leading Italian exponent of *bambocciate* (small pictures of low-life and peasant scenes).

César (César Baldaccini) (1921–). French sculptor. His work is highly varied, but he has become best known for his ingenious use of scrap material. In the mid-1950s he began to make sculptures from material that he found in refuse dumps—scrap iron, springs, tin cans, etc.—building these up with wire into strange winged or insect-like creatures. These had closer affinities, however, with the insect-creatures of Germaine *Richier than with the expressionistic industrial forms of the California *Junk school. During the 1960s he became internationally known mainly for sculptures made by crushing car bodies. There are several examples of his work in the Tate Gallery.

Cesare da Sesto (*c.*1477–1523). Milanese painter, active in Rome and Naples as well as his native city. His style was heavily indebted to *Leonardo and his *Leda and the Swan* (Wilton House, Wiltshire) is an important copy of a painting by Leonardo of which the original has not survived.

Cesari, Giuseppe (also called **Cavaliere d'Arpino**) (1568–1640). Italian *Mannerist painter, active mainly in Rome. He had an enormous reputation in the first two decades of the 17th cent., when he gained some of the most prestigious commissions of the day, most notably the designing of the *mosaics for the dome of St Peter's (1603–12). Although some of his early work is vigorous and colourful, his output is generally repetitious and vacuous, untouched by the innovations of *Caravaggio (who was briefly his assistant) or the *Carracci. He was primarily a fresco painter, but he also did numerous *cabinet pictures of religious or mythological scenes in a finicky Flemish manner (*The Expulsion from Paradise*, Louvre, Paris, with smaller versions in Christ Church, Oxford, and Wellington Mus., London).

Cézanne, Paul (1839–1906). French painter, with *Gauguin and van *Gogh the greatest of the *Post-Impressionists and a key figure in the development of 20th-cent. art. He was born at Aix-en-Provence, son of a hat dealer who became a prosperous banker, and his financial security enabled him to survive the indifference to his work that lasted until the final decade of his life. His schoolfellow Émile Zola (1840–1902) introduced him to *Manet and *Courbet, and persuaded him to take up the study of art in Paris. There at the *Académie Suisse in 1861 he met Camille *Pissarro, and the following year he got to know *Monet, *Bazille, *Sisley, and *Renoir. His painting at this time was in a vein of unrestrained and uncouth *Romanticism, with a predilection for themes of violence or eroticism (*The Murder*, Walker Art Gallery, Liverpool). It was completely different from his mature work and gave little hint of greatness to come. In 1869 he met Hortense Fiquet, a model and seamstress, who became his mistress and bore him a son, Paul, in 1872. Cézanne kept them a secret from his family—he was terrified of his domineering father—but eventually married Hortense in 1886, shortly before his father's death. From about 1870 Cézanne started painting directly from nature and began to impose a more disciplined restraint on his natural impetuosity. In 1872 he settled in Auvers-sur-Oise, near Pontoise, the home of Camille Pissarro, and entered upon a long and fruitful association with him (in the last year of his life he even described himself as a 'pupil of Pissarro'). He exhibited with the *Impressionists in 1874 and again in 1877, but never identified himself with the Impressionist group or wholly adopted their aims and techniques (he was a touchy character and hid his insecurities by posing as a provincial boor, once refusing to shake hands with the elegant *Manet because he claimed he had not washed for days and did not wish to dirty the great man). Cézanne was less interested in the realistic representation of casual and fleeting impressions and the fugitive effects of light, devoting himself rather to the structural analysis of nature, looking forward in this respect to the *Neo-Impressionists. His own aims are summarized in two of his sayings: that it was his ambition 'to do *Poussin again, from Nature' and that he wanted to make of Impressionism 'something solid and enduring, like the art of the museums'. He trod a solitary and difficult path towards his goal of an art which would combine the best of the French *classical tradition of structure with the best in contemporary *Realism, an art which appealed not superficially to the eye but to the mind.

After the death of his father in 1886 and his inheritance of the family estate (the Jas de Bouffan, which features in many of his paintings), Cézanne lived mainly in Aix. He devoted himself principally to certain favourite themes—portraits of his wife Hortense, still lifes, and above all the landscape of Provence particularly the Monte Ste-Victoire. His painstaking analysis of nature differed fundamentally from Monet's exercises in painting repeated views of subjects such as *Haystacks* or *Poplars*. Cézanne was interested in underlying structure, and his paintings rarely give any obvious indication of the time of day or even the season represented. His later paintings are generally more sparsely composed and open, permeated with a sense of air and light. The third dimension is created not through perspective or foreshortening but by extraordinarily subtle variations of tonality. He worked in comparative obscurity until he was given a one-man show by the dealer *Vollard in 1895. From that time his painting began to excite the younger artists and he attempted to explain his theories and aims in letters written to Émile *Bernard and others. By the end of the century he was revered as the 'Sage' by many of the avant-garde and in 1904 the *Salon d'Automne gave him a special exhibition. Since his death his reputation has increased among critics and art historians and he has exercised an enormous influence on 20th-cent. art, most notably on the development of *Cubism. His work was introduced to England with the Post-Impressionist exhibitions organized by Roger *Fry in 1910 and 1912, and in 1914 Clive *Bell wrote that: 'He was the Christopher Columbus of a new continent of form.' The belief that the picture surface has an integrity of its own irrespective of what it represents—a characteristic of so much modern painting—stems directly from him.

Although Cézanne was a laboriously slow worker—he is said to have had over 100 sittings for a portrait of Ambroise Vollard (Petit Palais, Paris,

1899) before abandoning it with the comment that he was not displeased with the shirt front—he left a substantial body of work (drawings and watercolours as well as oils). There are works in many major museums, with particularly fine collections in, for example, the Courtauld Institute Galleries, London, the Musée d'Orsay, Paris, and the Barnes Foundation, Merion, Pennsylvania. His studio in Aix is now a Cézanne museum, reconstructed as it was at the time of his death and displaying personal mementoes such as his hat and clay pipe.

Chadwick, Lynn (1914–). One of the leading English sculptors of his generation. He trained as an architectural draughtsman and took up sculpture after serving as a pilot in the Second World War. At first he experimented with *mobiles and these were followed by what he called 'balanced sculptures', ponderous metal structures supported on thin legs, bristling and rough-finished. His work has been shown in a number of international exhibitions and in 1956 he was awarded the International Sculpture Prize at the 28th Venice *Biennale. During the 1960s and 1970s his work became more block-like and monumental, and in his more recent sculpture he has exploited highly polished surfaces and facetings.

Chagall, Marc (1887–1985). Russian-born painter and designer, active mainly in France. In 1910–14 he lived in Paris, where he was a member of an avant-garde circle including *Apollinaire, *Delaunay, *Léger, *Modigliani, and *Soutine. After going to Berlin in 1914 for his first one-man show (at the *Sturm gallery) he visited Russia and had to remain because of the outbreak of war. After the Revolution in 1917 he was appointed Fine Arts Commissar for his home province of Vitebsk, where he founded and directed an art academy. *Malevich was among the other teachers there, and after disagreements with him Chagall moved to Moscow in 1920 and there designed for the newly founded Jewish Theatre. He returned to Paris in 1923 at the invitation of Ambroise *Vollard, who commissioned much work from him, including illustrations for Gogol's *Dead Souls* (1923) and other books. In 1941 he moved from occupied France to the USA, where he lived for the next seven years. He returned to Paris in 1948 and from 1950 lived at St-Paul-de-Vence near Nice, working to the end of his very long life—the last survivor of the generation of artists who had revolutionized painting in the years leading up to the First World War.

Chagall was prolific as a painter and also as a book illustrator and designer of stained glass (in which he did some of his most impressive late work) and of sets and costumes for the theatre and ballet. His work was dominated by two rich sources of imagery: memories of the Jewish life and folklore of his early years in Russia; and the Bible (he was born into a deeply religious family). He derived some of his spatial dislocations and prismatic colour effects from *Cubism and *Orphism, but he created a highly distinctive style, remarkable for its sense of fairy-tale fantasy. This caused André *Breton to claim him as one of the precursors of *Surrealism, but Chagall himself stated in his autobiography *Ma vie* (1931) that however fantastic and imaginative his pictures appeared, he painted only direct reminiscences of his early years. There is a museum devoted to Chagall's religious art in Nice. The work there does not always show him at his best, for he could be sentimental and overblown, but his finest paintings have won him an enduring reputation as one of the greatest masters of the School of *Paris.

chalk. Drawing material made from various soft stones or earths. There are three main types: black chalk (made from stones such as carbonaceous shale); red chalk, also called sanguine (made from red ochre or other red earths); and white chalk (made from various limestones). Chalk drawings are known from prehistoric times, but the medium really came into its own in the late 15th cent., notably in the hands of *Leonardo, who made many drawings in red and black chalk. The terms 'chalk', 'crayon', and *'pastel' are not always distinguished from one another, and there is much ambiguity in the historical literature of the subject. Crayons, as the term is now generally understood, are sticks of colour made with an oily or waxy binding substance, and pastels are sticks of powdered pigment bound with gum. In other words, they are both manufactured products, whereas chalk needs only to be cut to a suitable shape and size to be usable.

Chamberlain, John (1927–). American sculptor. His early sculpture, made largely from metal pipes, was influenced by that of David *Smith, but in 1957 he began introducing scrap metal parts from cars in his work and from 1959 he concentrated on sculpture made entirely from crushed automobile parts welded together. Usually he retains the original colours, and the expressive energy of his work, with its twisted planes and crumpled surfaces, has been compared to that of *Action painting. Many of his compositions are intended for wall hanging rather than to stand on the ground. An example is *Dolores James* (Guggenheim Museum, New York, 1962). Although he has continued with work of this type, which has been widely acclaimed since the early 1960s, Chamberlain has also experimented with other types of sculpture and other media. In 1966, for example, he started using urethane foam, as in *Koko-Nor II* (Tate Gallery, London, 1967). He has also done abstract paintings and made experimental films.

Champaigne, Philippe de (1602–74). Flemish-born painter who came to Paris with his master Jacques Fouquières (*c*.1580/90–1659) in 1621 and took French citizenship in 1629. He became the outstanding French portraitist of the 17th cent. and was patronized by Louis XIII (1601–43), the Queen Mother (Marie de *Médicis), and Cardinal Richelieu (1585–1642). Two of his finest portraits of Richelieu (late 1630s) are in the National Gallery, London; they bring the personality of the cardinal vividly to life and show how Champaigne moderated the *Baroque idiom of *Rubens towards a classical simplicity in line with French artistic trends of the middle of the 17th cent. He was a friend of *Poussin, and Anthony *Blunt has written that 'His portraits and his later religious works are as true a reflection of the rationalism of French thought as the classical compositions of Poussin in the 1640s.' His style became even more severe after he was influenced by the Jansenists—a Catholic sect of great austerity—in the early 1640s. Some of his finest work was done for the Jansenist convent at Port-Royal, where his daughter became a nun: he commemorated her miraculous recovery from paralysis in his most celebrated work, the **Ex-Voto de 1662* (Louvre, Paris). His masterpiece in portraiture might well have been his self-portrait of 1668, which is lost, but survives in a copy by his nephew **Jean-Baptiste de Champaigne** (1631–81) in the Louvre and in a superb engraving (1676) by Gerard *Edelinck.

Champfleury (pseudonym of Jules Husson) (1821–89). French writer, one of the first novelists to call himself a *Realist and the leading spokesman for the painting of *Courbet. Champfleury opposed traditional, religious, and classical themes, declaring that art should depict the social scene as it is without moralizing or idealizing it. He defended the 'ugliness' of figures in Courbet's *Burial at Ornans* (Musée d'Orsay, Paris, 1850) on the ground that they were true to life. Later he turned his hand to art history, publishing studies of *caricature (1865–80) and the *Le Nain brothers (1862). An avid art collector, he was also director of the Sèvres porcelain factory.

Chantrey, Sir Francis (1781–1841). English sculptor. The son of a carpenter, he was apprenticed to a wood carver in Sheffield but left to come to London, *c*.1802, to study at the *Royal Academy Schools. Until about 1804 his work included painted portraits, but after that date he confined himself to sculpture. His portrait bust of the Revd J. Horne-Took, exhibited at the Royal Academy in 1811 (Fitzwilliam, Cambridge), brought him to fame, and he succeeded *Nollekens as the most successful sculptor of portrait busts in England. Once he was well established, Chantrey, like Nollekens, did little of the cutting of the marble himself, for by this date it had become customary for the sculptor to model the bust in clay, leaving the transference to marble to assistants (see POINTING). His enormous practice included statues and church monuments as well as busts; his monument to the Robinson children (1817) in Lichfield Cathedral and his bronze equestrian statue of George IV (1828) in Trafalgar Square are his best-known works in these fields. He became extremely wealthy, and besides being very generous during his life he left the bulk of his fortune of £150,000 to the Royal Academy, the interest to be used for the purchase of 'works of Fine Art of the highest merit executed within the shores of Great Britain'. These are now housed in the *Tate Gallery.

charcoal. Charred twigs or sticks used for drawing. Its use dates back to Roman times and possibly much earlier. An essential characteristic of charcoal is that it is easily rubbed off the drawing surface unless a *fixative is used, so it has been much favoured for preparatory work, either for sketches or *cartoons or for outlining on wall or panel a design that could be gone over with a more permanent medium. The soft-edged effect it produces has been notably exploited by 16th-cent. Venetian painters, *Baroque artists, and the *Impressionists. *Pencils and *chalks have now taken its place to some extent, but it remains well suited to large-scale work and broad, energetic draughtsmanship, and in the 20th cent. has been memorably used by *Barlach and *Kollwitz.

Chardin, Jean-Baptiste-Siméon (1699–1779). French painter of still life and *genre, in which fields he was one of the greatest masters of all time. He was the contemporary of *Boucher and he taught *Fragonard, but his work is a contrast to theirs in every way, representing the naturalistic tendency which persisted through the 18th cent. alongside the more fashionable *Rococo. He was received into the Académie in 1728 on the strength of a still life (*The Rayfish*, Louvre, Paris), which drew forth the extravagant praises of *Diderot for its realism, and he was Treasurer of the Académie for twenty years. His small canvases depicting modest scenes and objects from the everyday life of the middle classes to which he belonged were in the tradition of the Dutch *cabinet pictures, which were having a great commercial success in France at the time. Chardin, however, developed a technique of his own, achieving great richness of tone by successive applications of the loaded brush and a subtle use of *scumbled colour. He was praised for his verisimilitude of detail, but his work goes far beyond matter-of-fact realism and through its simplicity and directness of vision achieves a sense of deep seriousness, in spite of the humble objects he portrayed (*Pipe and Jug*, Louvre). His genre paintings, which usually contain only one or two figures, are likewise completely without sentimentality or affectation (*The Young Governess*, versions in the National Galleries of London and Washington). In his last years, when his sight was failing, he turned his hand to pastel portraits and in the 1775 Salon exhibited two self-portraits and a portrait of his wife (Louvre).

He was well known during his lifetime through engravings of his works, which the historian Mariette remarked were selling better than high-flown allegories in the manner of *Lebrun, noting this as a significant shift in public taste. In the present century admiration for his work has increased on account of the abstract strength of his compositions, and contemporary painters of many schools, from *Cubist to *Abstract Expressionist, have drawn inspiration from him.

Chares of Lindos. See COLOSSUS OF RHODES.

Charonton (or **Quarton**), **Enguerrand** (*c*.1410–66). French painter, active in Provence. His career is unusually well documented for a provincial artist of his date (he worked in Aix, Arles, and Avignon), but there are only two extant works that are certainly by him. These are the *Virgin of Mercy* (1452) in the Musée Condé at Chantilly, painted in collaboration with an obscure artist called Pierre Villatte, and the *Coronation of the Virgin* (1454) in the Musée de l'Hospice at Villeneuve-lès-Avignon. They are both highly impressive works, uniting Flemish and Italian influence and having something of the monumental character of the sculpture of Charonton's region. Indeed, they show Charonton to have been a painter of such commanding stature that there is an increasing tendency to attribute to him the celebrated *Avignon Pietà* (Louvre, Paris), the greatest French painting of the period (see AVIGNON, SCHOOL OF).

Chase, William Merritt (1849–1916). American painter. He settled in New York in 1878 after five years studying in Munich and became the most important American teacher of his generation. He taught at the *Art Students League of New York and then at his own Chase School of Art, founded in 1896. The vigorous handling and fresh colour characteristic of much of the best American painting of the early 20th cent. owes a good deal to his example. His pupils (whom he encouraged to paint in the open air) included *Demuth, *O'Keefe, and *Sheeler. Chase was himself a highly prolific artist (his output of more than 2,000 paintings included still lifes, portraits, interiors, and landscapes), and his work is represented in many American museums. See also TEN, THE.

Chassériau, Théodore (1819–56). French painter. He was the most gifted pupil of *Ingres, whose studio in Rome he entered when he was 11, but in the 1840s he conceived an admiration for *Delacroix and attempted, with considerable success, to combine Ingres's *classical linear grace with Delacroix's *Romantic colour. His chief work was the decoration of the Cour des Comptes in the Palais d'Orsay, Paris, with allegorical scenes of Peace and War (1844–8), but these were almost completely destroyed by fire. There are other examples of his decorative work, however, in various churches in Paris. Chassériau was also an outstanding portraitist and painted nudes and North African scenes (he made a visit there in 1846).

Chavannes, Pierre Puvis de. See PUVIS DE CHAVANNES.

Cheere, Sir Henry (1703–81). English sculptor, possibly of French descent. He went into partnership with Henry *Scheemakers and after Scheemakers left England about 1733 Cheere extended his practice. His output included much work for Oxford University, notably a statue of Christopher Codrington (1732) in the library named after him at All Souls College. Cheere was also prominent in public affairs; he was knighted in 1760 when he presented an address to George III on behalf of the County of Middlesex, and he was created a baronet in 1766. He also had the distinction of being the first sculptor to become a Fellow of the Society of Antiquaries (1750), and devoted efforts to an abortive scheme to found an academy of arts a decade before the *Royal Academy came into being. His art is markedly *Rococo in feeling, with an interest in small rhythms, and in his charming smaller monuments (*Dean Wilcocks*, Westminster Abbey, 1756) he often used coloured marbles. He was well thought of by his fellow artists (in 1748 he accompanied *Hogarth on the continental trip on which the latter was arrested as a spy) and furthered the career of *Roubiliac by gaining him his first important commission in England—the statue of Handel for Vauxhall Gardens (now V&A,

London, 1738). His brother **John** (1709–87) had a yard near Hyde Park Corner which turned out a great number of garden figures (examples are at Stourhead and Longford Castle, Wilts.), some in the *antique manner and some Rococo, many of the latter being originally painted in natural colours.

Chéron, Louis. See VANDERBANK.

Chia, Sandro. See NEO-EXPRESSIONISM.

chiaroscuro (Italian: bright-dark). Term describing the effects of light and shade in a work of art, particularly when they are strongly contrasting. *Leonardo was a pioneer of bold chiaroscuro, but the term is most usually associated with 17th-cent. artists, particularly the *Caravaggisti and *Rembrandt, whose handling of light and shade, not only in his paintings, but also in his etchings and drawings, is unsurpassed.

chiaroscuro woodcut. A type of *woodcut in which tonal effects are created by printing successively on to the same sheet from different blocks of varying tone. Two or more tones of a single colour are used, or of two nearly related colours, one of which is darker than the other. It is usual to make a key block with the design in outline, and to cut this first so that the main lines can be transferred to the other blocks to ensure correct registration.

The method dates from the early 16th cent., when it was chiefly used for the reproduction of drawings in light and shade. It developed more or less simultaneously in Germany and Italy, though there is an interesting difference of approach in the work of the two schools. In Germany great importance was given to the key block, which was to all intents and purposes a complete design in itself, the resulting print being a richly worked woodcut with the addition of background tints. In Italy the medium was handled with much greater breadth, the design being visualized in large areas of tone punctuated by dark accents.

The earliest dated chiaroscuro woodcut is *The Emperor Maximilian on Horseback* of 1508, designed by Hans *Burgkmair: other notable German exponents were *Cranach, *Baldung Grien, and *Altdorfer. In Italy, where the medium was used more extensively, *Ugo da Carpi (who is sometimes credited with inventing the technique) made many prints after designs by *Raphael and *Parmigianino, the latter artist being a prolific designer for the process. Although in some cases an accurate facsimile was intended, in others the cutter interpreted his original with some freedom. The technique was little used after the 17th cent. and its later history tends to overlap with that of colour woodcut and colour *wood engraving, but even today any *relief print cut on several blocks with the intention of rendering light and shade as opposed to colour may be claimed as a descendant of the chiaroscuro woodcut.

Chinese ink. See INK.

Chinese taste. See CHINOISERIE.

Chinnery, George (1774–1852). English painter, active for almost all his career in the Far East. His movements after sailing for India in 1802 were: 1802–7 in Madras, 1807–27 in Calcutta, 1827–*c.*1830 in Canton, *c.*1830–52 in Macao. Chinnery painted a number of portraits while abroad and from time to time sent to Royal Academy exhibitions, but his reputation rests today on the large number of landscapes and decorative studies he made of oriental scenes. He developed a calligraphic style and his rapid and often fragmentary sketches show him to have been among the most visually perceptive of all European artists who travelled and worked in the East.

chinoiserie. Term sometimes used to embrace any aspect of Chinese influence on the arts and crafts of Europe, but more usefully confined to a style reflecting fanciful and poetic notions of China that was particularly popular in the *Rococo period. From the time of Marco Polo (1254–1324) onwards the idealized concepts of 'Cathay' varied from age to age and often bore so little resemblance to the real thing that oriental crafts were sometimes designed specifically for the European taste. Indeed, in the 17th cent. textiles designed by Indian craftsmen for the English 'China fashion' were exported from India to China as a novelty and Chinese weavers then began to produce for the European market basing their designs on these Indian models.

The expansion of trade through the East India Companies in the 16th and 17th cents. produced a lively vogue for Chinese fashions. The Delft blue-and-white pottery was closely imitated from Chinese porcelain, while at Nevers a less closely imitated decorative style may be regarded as the true origin of chinoiserie in European ceramics. The *Trianon de porcelaine* built for Louis XIV in the park at Versailles (1670–1) became the prototype of innumerable Chinese pagodas, kiosks, etc., throughout Europe. Towards the end of the century Jean *Bérain introduced Chinese motifs into his *arabesques and originated that peculiar offshoot of chinoiserie, the **singerie*.

The style was not confined to architecture and the decorative arts. *Watteau painted a series of *Figures chinoises et tartares* (*c.*1709) for the Cabinet du Roi in the Château de la Muette, which are still known from prints, and the spirit of delicate fantasy which enlivened his works was further developed by the chinoiseries and *singeries* of Christophe *Huet. A more original concept of the voluptuous Orient imbued the chinoiserie of *Boucher in his tapestry designs (the *tentures chinoises* woven at Beauvais about the middle of the century), his décor for Noverre's ballet *Les Fêtes chinoises*, drawings which were engraved and published as *Livre des chinois*, and also easel pictures such as *Chinese Fishing Party* (Musée des Beaux-Arts, Besançon, 1742). The taste for chinoiserie faded during the dominance of the *Neoclassical style in the second half of the century, but there was something of a revival in the early 19th cent.

Chirico, Giorgio de (1888–1978). Italian painter, sculptor, designer, and writer, the originator of *Metaphysical painting. He was born in Greece, and trained in Athens, Florence, and Munich, where he was influenced by the *Symbolist work of Böcklin and *Klinger, with their juxtaposition of the commonplace and the fantastic. In 1909 he moved to Italy (dividing his time between Florence, Milan, and Turin) and there painted his first 'enigmatic' pictures, which convey an atmosphere of strangeness and uneasiness through their empty spaces, illogical shadows, and unexpected perspectives. From 1911 to 1915 he lived in Paris, becoming friendly with many members of the avant-garde, including *Apollinaire (who championed his work) and *Picasso. During this period he developed a more deliberate theory of 'metaphysical insight' into a reality behind ordinary things by neutralizing the things themselves of all their usual associations and setting them in new and mysterious relationships. In order to empty the objects of his paintings of their natural emotional significance he painted tailors' dummies as human beings (from 1914).

In 1915 de Chirico was conscripted into the Italian Army and sent to Ferrara. There he suffered a nervous breakdown, and in 1917 met *Carrà in the military hospital and converted him to his views, launching Metaphysical painting as a movement. It was short lived, virtually ending when de Chirico and Carrà quarrelled in 1919, but it was highly influential on *Surrealism, and it was during the later 1920s, when Surrealism was becoming the most talked-about artistic phenomenon of the day, that de Chirico's international reputation was established. However, it was his early work that the Surrealists admired and they attacked him for adopting a more traditional style in the 1920s, when his output included some distinctive pictures featuring horses on unreal seashores with broken classical columns. In the 1920s and 1930s he spent much of his time in Paris (and in 1935–7 he lived in the USA) before settling permanently in Rome in 1944. By this time his paintings had become repetitive and obsessed with technical refinement. His other work included a number of small sculptures and set and costume designs for opera and ballet; his writings include a Surrealistic novel, *Hebdomeros* (1929), and two volumes of autobiography (1945 and 1960). In the catalogue of the exhibition 'Italian Art in the 20th Century' (Royal Academy, London, 1989) Wieland Schmied writes: 'Giorgio de Chirico remains among the most controversial of all twentieth-century artists. There is no other figure of such seminal importance on whom the experts' opinions are so divided or their interpretations so widely divergent.'

Chodowiecki, Daniel Nikolaus (1726–1801). Polish-German painter and illustrator, born in Danzig (now Gdansk), and active mainly in Berlin. He began his career by painting *enamels. His early oil paintings were imitations of the French manner, and his fame rests on the more individual book illustrations and graphic work of all kinds which he produced prolifically from *c*.1770 onwards. Most attractive are the little intimate sketches he made of the bourgeois life around him, not least of his own family. In 1797 he became Director of the Berlin Academy.

choir stalls. Term in Christian church architecture for seats arranged in one or several rows on either side of the choir for the use of clergy, from the Middle Ages a major feature of church furnishing. By the 12th cent. a form of stalls had evolved which remained basically constant throughout the Middle Ages. The back was continuous and the long bench, with ends, was divided into separate seats by low partitions. Hinged seats, which could be raised to a vertical position when the occupant was standing, were in use by this period. In the later Middle Ages they often had gabled canopies rising to great height, and these and the *misericords on the seats became vehicles for virtuoso feats of woodcarving. Among later choir stalls those carved by Grinling *Gibbons in St Paul's Cathedral in London are probably the most celebrated.

Christie's. The popular name for the firm of Christie, Manson & Woods, the oldest fine art auctioneers in the world (*Sotheby's was founded earlier, but originally sold only books). It was founded by James Christie (1730–1803), who gave up a commission in the Navy to become an auctioneer and held his first sale on 5 December 1766 in rooms in Pall Mall, in the same premises in which the exhibitions of the *Royal Academy were held until 1779. He was a friend of *Reynolds and *Gainsborough, and Christie's developed a tradition of holding the studio sales of prominent artists, among them (as well as the two just mentioned) *Romney, *Raeburn, and Augustus *John. The firm acquired its present name in 1859, when James Christie's grandsons took new partners. Its headquarters are still in London and there are branches in New York, Amsterdam and Geneva.

Christmas, Gerard (or **Garrett**) (d. 1633). English sculptor. He made an equestrian figure of James I on the city gate at Aldersgate and carved and perhaps designed an elaborate three-tiered frontispiece on Northumberland House at Charing Cross, but these are both destroyed. About 1614 he was appointed Carver to the Navy, and was also much employed in devising pageantry for the Lord Mayor's Shows. His sons **John** and **Matthias** worked with him, the latter (d. 1654) succeeding him as Master Carver to the shipyard at Chatham. The family were responsible for a number of monuments, the most important being the tomb of Archbishop Abbot (d. 1633) in Holy Trinity, Guildford, which was commissioned from Gerard but finished by his sons after his death. There are several other monuments by the sons dating from the second half of the 1630s. They are competent but uninspired, giving a good idea of the average level of production of the day.

Christo (Christo Javacheff) (1935–). Bulgarian-born sculptor and environmental artist who settled in New York in 1964 and became an American citizen in 1973. After brief periods in Prague, Vienna (where he studied sculpture with *Wotruba), and Geneva, he moved to Paris, where he lived from 1958 to 1964. Initially he earned his living there as a portrait painter, but soon after his arrival he invented 'empaquetage' (packaging), a form of expression he has made his own and for which he has become world-famous. It consists of wrapping objects in materials such as canvas or semi-transparent plastic and dubbing the result art. He began with small objects such as paint tins from his studio (in this he had been anticipated by *Man Ray), but they increased in size and ambitiousness through trees and motor cars to buildings and sections of landscape. He spends a great deal of time and effort negotiating permission to carry out such work and then (if negotiations are successful) in planning the operations, which can involve teams of professional rock-climbers as well as construction workers. He finances such massive enterprises through the sale of his smaller works. The buildings that he has succeeded in wrapping include the Pont Neuf in Paris (1985, after nine years of negotiations) and the Reichstag in Berlin (1995). Among the landscape projects he has carried out is Running Fence, something like a fabric equivalent of the Great Wall of China, undulating through 24 miles of Somona and Martin Counties, California (1976). Christo says of his work: 'You can say it's about displacement. Basically even today I am a displaced person, and that is why I make art that does not last . . . Unlike steel, or stone, or wood, the fabric catches the physicality of the wind, the sun. They are refreshing. And then they are quickly gone.'

Christus, Petrus (d. 1475/6). Netherlandish painter. He is first documented at Bruges in 1444, and he is thought by some authorities to have been the pupil of Jan van *Eyck and to have completed some of the works left unfinished by the master at his death in 1441 (e.g. *St Jerome*, Detroit Inst. of Arts). It is certainly true that he was overwhelmingly influenced by van Eyck, and his copies and variations of his work helped to spread the Eyckian style. Christus's work is more summary than van Eyck's, however, his figures sometimes rather doll-like and without van Eyck's feeling of inner life. The influence of Rogier van der *Weyden is also evident in Christus's work; the *Lamentation* (Musées Royaux, Brussels) is clearly based on van der Weyden's great Prado *Deposition*, but the figures have completely lost their dramatic impact. Christus's most personal works are his portraits, notably *Edward Grimston* (Earl of Verulam Coll., on loan to NG, London, 1446) in which he abandons the dark backgrounds of van Eyck and van der Weyden and places his sitter in a clearly defined interior. His interest in representing space comes out also in his *Virgin and Child with Sts Jerome and Francis* (Städelsches Kunstinstitut, Frankfurt, 1457), the earliest dated example in the north of the use of geometric *perspective with a single vanishing point.

chromolithography. The process of making coloured prints by *lithography, using a separate stone or plate for each colour.

chryselephantine. Term describing statues in which the drapery is made of gold (Greek *chrysos*) and the flesh of ivory (Greek *elephantinos*). The technique was used on a small scale in ancient Egypt, Mesopotamia, and Crete, and in colossal statues by the Greeks from the 6th cent. BC. Most famous of chryselephantine statues were the enormous cult images of Athena and Zeus that *Phidias made respectively for the Parthenon at Athens and the Temple of Zeus at Olympia.

Church, Frederick Edwin (1826–1900). American landscape painter. He was a pupil and close friend of *Cole and continued the preoccupations of the *Hudson River School with the most spectacular aspects of natural scenery. Church looked and travelled beyond his native country, however, painting not only the Niagara Falls, for example, but also the tropical forests of South America, icebergs, and exploding volcanoes, often on a huge scale. He was immensely popular in his day, and after a period of neglect is returning to favour again. His house, Olana, on the Hudson River, is now a museum.

Churrigueresque. Term applied to an extravagant style of architecture and ornament popular in Spain in the late 17th cent. and early 18th cent. and sometimes used more loosely to refer to the *Rococo period as a whole in Spanish architecture. It is named after the Churriguera family of architects and sculptors, who were active mainly in Seville. The most important member of the family was José Benito (1665–1725), whose work is well represented by the *reredos (1693–1700) of the church of San Esteban in Salamanca, an important early example of the style, full of lavishly decorated barley-sugar columns. It still retains a sense of architectural solidity, however, and is restrained compared with later manifestations of the style, in which surface ornament runs riot to such a degree that the underlying structure is hidden. To *Neoclassical taste the Churrigueresque style represented the last word in decadence and it died out completely in the last quarter of the 18th cent., in South America, where it had also flourished, as well as in Spain.

Cibber, Caius Gabriel (1630–1700). English sculptor of Danish birth. Cibber was the son of the cabinet-maker to the King of Denmark, who is said to have sent him to study in Italy. He arrived in England before the Restoration in 1660, probably via Amsterdam, and worked for John Stone, son of Nicholas *Stone. His first important work was the large *relief (1674) on the base of the Monument erected in memory of the Great Fire of London (1666). Other works in London included the dramatic figures of *Raving* and *Melancholy Madness* (*c.*1675) for the gate of old Bedlam Hospital (now in the Bethlem Royal Hospital Museum, Beckenham) and a fountain in Soho Square, which originally showed Charles II and the four major rivers of England (only the rather battered figure of the king remains in the square). Much of his later career was taken up with decorative sculpture, notably for Hampton Court and St Paul's Cathedral, where he was working at the time of his death. With the exception of the figures of *Raving* and *Melancholy Madness*, which are powerful and original pieces, and the dignified and moving tomb of Thomas Sackville at Withyham, Sussex (1677), Cibber's work is usually competent but uninspired; it is of interest, however, in reflecting *Baroque influence from Italy (still unusual in England at this time) and also from the Netherlands. He was the father of Colley Cibber (1671–1757), the actor-manager and dramatist.

ciborium. Term applied to both a liturgical vessel used for holding the consecrated Host and an altar canopy supported on columns, popular particularly in Italy in the *Romanesque and *Gothic periods. In the latter sense the word is not easily distinguished from *baldacchino.

Cignani, Carlo (1628–1719). Bolognese painter. He was a pupil of *Albani, but his style is closer to that of Guido *Reni and he became the main force in upholding the tradition of Bolognese classicism into the 18th cent. In 1711 he became the first president of the Accademia Clementina in Bologna.

Cignaroli, Giambettino (1706–70). Italian historical, religious, and decorative painter, active mainly in and around Verona. He was the leading artist of his period there, working in an elegantly academic style.

Cigoli, Il (Ludovico Cardi) (1559–1613). Florentine painter, architect, and poet. A pupil of Alessandro *Allori, he was the outstanding Florentine painter of his generation and his work represents the complex stylistic crosscurrents in the period of transition from *Mannerism to *Baroque. His sensuous colour and handling shows the influence of *Barocci, *Correggio, and the art of Venice, and his dramatic handling of light and shade that of *Caravaggio, especially after he moved to Rome in 1604 (*Ecce Homo*, Pitti, Florence, 1606). As an architect, his work includes the courtyard of the Palazzo Nonfinito in Florence.

Cimabue (properly Cenni di Peppi) (*c.*1240–1302). Florentine painter. His nickname means 'Ox-head'. He was a contemporary of Dante (1265–1321), who refers to him in *The Divine Comedy* (*Purg.* xi. 94–6) as an artist who was 'believed to hold the field in painting' only to be eclipsed by *Giotto's fame. Ironically enough this passage, meant to illustrate the vanity of short-lived earthly glory, has become the basis for Cimabue's fame; for, embroidering on this reference, Dante commentators and later writers on art from *Ghiberti to *Vasari made him into the discoverer and teacher of Giotto and regarded him as the first in the long line of great

Italian painters. He was said to have worked in the 'Greek' (i.e. *Byzantine) manner, but to have begun the movement towards greater realism which culminated in the *Renaissance.

Documentary evidence is insufficient to confirm or deny this estimate of Cimabue's art. The only work that can be proved to be by his hand is a *St John* forming part of a larger mosaic in Pisa Cathedral (1301), but tradition has tended to attribute to Cimabue many works of outstanding quality from the end of the 13th cent., such as the *Madonna of Sta Trinità* (Uffizi, Florence), a cycle of frescos in the Upper Church of S. Francesco in Assisi, and a majestic *Crucifix* in Sta Croce (badly damaged in the Florence flood of 1966). If these highly plausible attributions are correct, Cimabue was indeed the outstanding master of the generation before Giotto; but the more extreme claims sometimes made for him may still be in need of qualification. The movement towards greater naturalism may owe more to contemporary Roman painters and mosaicists (*Cavallini, *Torriti) than to him; he is documented in Rome in 1272 and could have known their work.

Cima da Conegliano, Giovanni Battista (1459/60–1517/18). Italian painter, named after the town of his birth (Conegliano), and active mainly in nearby Venice, where he was one of the leading artists from about 1490 to 1510. His paintings are mostly quiet devotional scenes, often in landscape settings, in the manner of Giovanni *Bellini. He has been called 'the poor man's Bellini', but because of his calm and weighty figures he was also known in the 18th cent. (rather incongruously) as 'the Venetian Masaccio'. Nine of his works are in the National Gallery, London.

Cimon of Cleonae. Greek painter, a native of Cleonae, a city which lies between Corinth and Argos. According to *Pliny he invented foreshortened or 'three-quarter' views, represented human beings looking backwards or upwards or downwards, and was the first to introduce wrinkles and folds in the drapery. Pliny dates him in the 8th cent. BC, but since foreshortening first appears in Greek art at the very end of the 6th cent. BC, those who consider Cimon its inventor date him about that time.

cinquecento. See QUATTROCENTO.

Cione, Andrea Nardo and **Jacopo di.** See ORCAGNA.

Cipriani, Giovanni Battista (1727–85). Florentine decorative painter and designer, active mainly in England. In 1756 he was brought to London by the architect Sir William Chambers (1723–96) and the sculptor *Wilton whom he had met in Rome. He was employed in the decoration of many public buildings and private houses and in some cases designed such architectural details as plasterwork, woodwork, and stone carving. Good examples of his paintings are at Somerset House (where he worked for Chambers) and in the Philadelphia Museum of Art (a series originally executed for Lansdowne House, London). He was also active as a teacher at the *Royal Academy (he was a foundation member in 1768 and designed its diploma) and his numerous decorative designs (many engraved by *Bartolozzi, his friend since student days) had wide influence. Cipriani's work is accomplished rather than inspired, but he was, in the words of Sir Ellis *Waterhouse, 'one of the great backroom figures of the *Neoclassic style in England'.

Circle. A collective manifesto of *Constructivism edited by the architect Sir Leslie Martin (1908–), the painter Ben *Nicholson, and the sculptor Naum *Gabo, published in London in 1937. Subtitled 'International Survey of Constructive Art', it is nearly 300 pages long with numerous illustrations and was originally intended as an annual. The volume contains an editorial by Gabo entitled 'The Constructive Idea in Art', Piet *Mondrian's seminal essay 'Plastic Art and Pure Plastic Art', and a short statement by *Moholy-Nagy on 'Light Painting'. There are also essays or statements by, among others, *Hepworth (who took much of the responsibility for the layout and production), *Le Corbusier, *Moore, and *Read. The artists illustrated included (in addition to those already mentioned) *Malevich, *Lissitzky, Antoine *Pevsner, and many others, such as *Braque, *Brancusi, *Giacometti, and *Picasso whose work did not conform with the theoretical concept of Constructivism. *Circle* was reprinted in 1971.

cire-perdue (French 'lost wax'). Term used to describe a method of hollow metal casting in which a thin layer of wax corresponding to the shape of the final sculpture is encased within two layers of heat-resistant clay or plaster, melted and drained off and then replaced with molten metal poured into the cavity that the 'lost wax' has created. The technique, found in every continent except Australasia, was used by the Egyptians, Greeks, and Romans and is still the main means of casting used for traditional bronze sculpture. Casting sculptures of any size is an industrial process requiring great expertise, and there is a celebrated account in *Cellini's autobiography of the difficulties he encountered (and heroically overcame) with his figure of *Perseus*.

Cisneros, Pedro. See BORGOÑA.

Civetta. See BLES, HERRI MET DE.

Claesz., Pieter (*c.*1597–1660). Dutch still-life painter, born in Germany and active in Haarlem, where he settled in 1617. He and Willem Claesz. *Heda, who also worked in Haarlem, were the most important exponents of the *ontbijt* or breakfast piece. They both painted with subdued, virtually monochromatic palettes, the subtle handling of light and texture being the prime means of expression. Claesz. generally chose objects of a more homely kind than Heda, although his later work became more colourful and decorative. The two men founded a distinguished tradition of still-life painting in Haarlem, but Claesz.'s son, Nicolaes *Berchem, became famous as a landscape painter.

Clark, Kenneth (Baron Clark) (1903–83). British art historian, administrator, patron, and collector, born into a wealthy family whose fortune had been made in thread-manufacturing: 'My parents belonged to a section of society known as "the idle rich", and although, in that golden age, many people were richer, there can have been few who were idler.' After working with *Berenson in Florence, he was Keeper of Fine Art at the *Ashmolean Museum in Oxford (1931–3), then Director of the *National Gallery, London (1934–45), and at the same time Surveyor of the King's Pictures (1934–44). He also served on numerous boards and committees and he was Chairman of the *Arts Council (1953–60) and of the Independent Television Authority (1954–7). He published more than twenty books, his forte being appreciation and interpretation rather than exact scholarship, although his monographs on *Leonardo da Vinci (1939) and *Piero della Francesca (1951), both of which have been issued in revised editions, still remain standard works many years after they were originally published. Clark himself wrote that his 'only claim to be considered a scholar' was his catalogue of Leonardo drawings at Windsor Castle, which was first published in 1935. His other books include *The Gothic Revival* (1928), *Landscape into Art* (1949), and *The Nude* (1956). He regarded *The Nude* as 'without question my best book, full of ideas and information, simplifying its complex subject without deformation, and in places eloquent'. A polished television performer as well as an elegant and stimulating writer, he did a great deal to popularize art history, most notably with his television series *Civilisation* (1969, also published then as a book), which was shown in over sixty countries. The part he played as a patron and collector (he inherited substantial wealth from his parents) is less well known, but was of considerable importance. He bought the work of *Moore, *Pasmore, *Piper, and *Sutherland in the 1920s and 1930s when they were little known and helped to establish their reputations (he also made a regular allowance—in strict secrecy—to several artists), and during the Second World War he had a major influence as chairman of the War Artists' Advisory Committee (see OFFICIAL WAR ART). His two volumes of autobiography—*Another Part of the Wood* (1974) and *The Other Half* (1977)—are highly entertaining, if not always accurate in detail, but some of the potboilers that appeared in his old age would have been better left unpublished.

Clark, Lygia (1920–88). Brazilian sculptor, painter, and experimental artist. After studying painting under Roberto *Burle Marx in Rio de Janeiro she went to Paris in 1950 and studied under *Léger. Returning to Brazil in 1952, she became a leading member of the Brazilian movement of *Concrete art. In 1959 she turned to sculpture, making hinged pieces that could be manipulated by the spectator. These set the course for much of her later work. In 1964 she initiated 'vestiary' sculpture made of soft materials and designed to be worn by the spectator, and by the mid-1970s she had become one of the most widely known of the experimental artists concerned with creating 'situations' and 'environments' and with stimulating spectator participation.

Clarke, Geoffrey (1924–). British sculptor. He is best known for his church furnishings (for example the pulpit in Chichester Cathedral), successfully showing the potentialities of abstract art in religious settings. Most of his output has been in metal, but he has also worked in enamel, stained glass, and other media. His secular work has included commissions for numerous public buildings.

Clarke, Harry (1889–1931). Irish artist, chiefly famous as one of the 20th cent.'s greatest designers of stained glass, but also a mural painter, textile designer, and book illustrator. He was born in Dublin, the son of a church decorator, and had his main training at the Dublin Metropolitan School of Art, 1910–13. Scholarships then enabled him to study medieval glass in France. He took over the family business on his father's death in 1921 and had a large output in spite of his short life (he died from tuberculosis). The Harry Clarke Stained Glass Studios Ltd continued in business until 1973. Clarke's glass was sumptuous and often rather bizarre in style—in the spirit of French *Symbolist painters. As an illustrator he had a taste for the macabre and is particularly remembered for his black-and-white drawings for an edition of Edgar Allan Poe's *Tales of Mystery and Imagination* (1923).

classicism. Term that, with the related words 'classic' and 'classical', is used in various (and often confusing) ways in the history and criticism of the arts. In its broadest sense, classicism is used as the opposite of *Romanticism, characterizing art in which adherence to recognized aesthetic ideals is accorded greater importance than individuality of expression. In this sense, *Alberti defined beauty in architecture as 'the harmony and concord of all the parts achieved by following well-founded rules and resulting in a unity such that nothing could be added or taken away or altered except for the worse'. The rules that Alberti referred to were those embodied in Greek and Roman architecture, and the use of the word 'classicism' often implies direct inspiration from *antique art, but this is not a necessary part of the concept, and according to context the word might be intended to convey little more than the idea of clarity of expression, or alternatively of conservatism.

In the Western tradition, the term 'classical' usually does suggest a line of descent from the art of Greece and Rome, however indirect or impure, and 'classical beauty' is sometimes used to indicate a facial and bodily type reduced to mathematical symmetry about a median axis and freed from the irregularities which are normally present in living people. 'Classical architecture' is that which uses the repertoire of forms developed in Greece and Rome, as opposed to, say, the pointed arches of the *Gothic period, and the term thus covers most of European architecture from about 1500 to about 1900. In the context of Greek art, the term 'Classical' (with a capital C) has a more precise meaning, referring to the period between the *Archaic and *Hellenistic periods, when Greek culture is thought to have attained its greatest splendour.

The term 'classic' is used to refer to the best or most representative example of its kind in any field or period. This is what *Wölfflin meant when he gave the title *Classic Art* to his book on the Italian High *Renaissance. Thus, in this sense, it would be legitimate, if wilfully confusing, to refer to *Delacroix as the classic Romantic artist. The three terms 'classic', 'classical', and 'classicism' are, then, often not used with discrimination or exactness, the conflation of historical term and value judgement reflecting the idea (dominant for centuries) that the art of the Greeks and Romans set a standard for all future achievement. To clear up (or perhaps add to) the confusion, the rather ungainly word 'classicistic' and 'classicizing' have also entered the lists—they convey the idea of dependence on ancient models but without any sense of qualitative judgement.

Claude Gellée (1604/5?–82). French painter, often called Le Lorrain (in France), or Claude Lorrain(e) (in the English-speaking world), after his place of birth, but usually referred to simply as Claude, a familiarity reflecting his enormous fame as the most celebrated of all exponents of *ideal landscape. When about 12 years of age he moved to Rome, where he is said to have entered the household of Agostino *Tassi as a pastry-cook (a favourite trade of Lorrainers), then became his studio assistant. Around 1620 he made a two-year visit to Naples, where he studied under the German-born landscapist Gottfried or 'Goffredo' Wals. Claude was deeply impressed by the beauty of the Gulf of Naples and used the coastline in his paintings to the end of his life. In 1625 he returned to Lorraine, and worked at Nancy with Claude *Deruet, but in 1627 he was back in Rome, where except for local journeys he remained for the rest of his life.

In Tassi's decorative paintings he came into contact with the conventionalized late *Mannerist style of landscape painting. The influence of the two leading exponents of the style, *Bril and *Elsheimer, comes out in the lively paraphernalia of architectural fragments, figures, and animals that often animates the foreground of his early paintings (*The Mill*, Mus. of Fine Arts, Boston, 1631). But his profound sensitivity to the tonal values of light and atmosphere lent an unpremeditated classical harmony to his pictures which matured with the years. During the 1630s he became well known and successful, working for illustrious patrons, and as early as 1634 *Bourdon had thought it worthwhile to pass off a painting of his own as a work by Claude. To combat such forgeries Claude began to compile his *Liber Veritatis* (Book of Truth) in 1635–6; it contains drawings of virtually all his paintings made from that date, making his *œuvre* exceptionally well documented. It is now in the British Museum, which also has the greatest collection of Claude's drawings from nature, done on his frequent excursions into the countryside around Rome and often showing a freedom of brushwork that has led them to be compared with Chinese art. Claude was also an accomplished etcher, but not nearly as prolific as he was as a painter and draughtsman.

From 1640 to 1660 Claude developed steadily towards the mature style of the poetic landscapes on which his enormous reputation was built. His painting shed the affectations of Mannerism and became an expression of his deep feeling for the beauty of the Roman countryside with its richness of Classical associations of *antique grandeur. He used this landscape not to create a heroic vision of ancient Rome, as did his friend *Poussin, but to evoke a sense of the pastoral serenity of a Golden Age. The ostensible subjects of his pictures, taken frequently from the Bible, Virgil, Ovid, or medieval epics, are subordinate to the real theme, which was the mood of the landscape presented poetically in terms of light and colour. In his earlier paintings Claude, like Elsheimer, used light for the sake of dramatic effects; as his style matured he began to use it for its own sake, letting it play on forms and explore their texture, and opposing to it trees, ruins, or the porticoes of temples so that the light enhanced their outlines. In the landscapes of the last two decades of his life everything was depicted in terms of light: the eye-level was raised and the view kept as open as possible so that the eye can roam at will over a spacious panorama to the distant horizon and beyond it into infinity. Forms melt and lose their material solidity and the figures become unnaturally elongated and insubstantial, as in *Ascanius and the Stag* (Ashmolean, Oxford), painted in the last year of Claude's life.

Claude has nowhere been more admired and more influential than in England. Not only were his works keenly sought after by late 17th- and 18th-cent. collectors, but they had great influence on such artists as *Wilson and *Turner. His name became virtually synonymous with the ideals of the *Picturesque, he inspired a revolution in English landscape gardening about the middle of the 18th cent., and much descriptive verse paid conventional tribute to the ideal of beautiful natural scenery which derived from him.

Claude glass. A small black convex glass used for reflecting landscapes in miniature so as to show their broad tonal values, without distracting detail or colour. It was popular in the 17th and 18th cents., and not only with artists, for the poet Gray carried one with him in his travels round Britain in search of the *Picturesque. *Claude Lorraine was said to have used such a glass, and in the 19th cent. the device was used by *Corot, who regarded tonal unity in painting as supremely important.

Clausen, Sir George (1852–1944). British painter (mainly of landscapes and scenes of rural life), born in London, the son of a decorative painter of Danish descent. His training included a few months at the *Académie Julian, Paris, in 1883 and his work was influenced by French **plein-air* painting. He was preoccupied with effects of light, often showing figures set against the sun, but he always retained a sense of solidity of form. With other like-minded painters he became a member of the *New English Art Club in 1886. From 1904 to 1906 he was Professor of Painting at the *Royal Academy. His lectures were published as *Six Lectures on Painting* (1904) and *Aims and Ideals in Art* (1906); a collected edition appeared as *Royal Academy Lectures on Painting* in 1913. In them he urged the traditional study of the Old Masters.

Clemente, Francesco. See NEO-EXPRESSIONISM.

Clerck, Hendrik de (*c.*1570–1630). Flemish painter, born in Brussels, where he spent the greater part of his successful career. He was the pupil of Martin de *Vos and carried the *Mannerist tradition far into the 17th cent. In 1606 he was appointed Court Painter to Archduke Albert. He was primarily a painter of altarpieces and a characteristic example of his style is the *Family of the Virgin* (Musées Royaux, Brussels, 1590), with its Italianate figures clad in restless draperies and placed in a coldly classical building.

Clichés-verre. See GLASS PRINTS.

Clodion (1738–1814). French sculptor, whose real name was Claude Michel. He was the son-in-law of *Pajou and the nephew of L.-S. *Adam, and had his training with the latter and with *Pigalle. His best work is found in his small statuettes and *terracotta figures and groups. They are often of light-hearted Classical subjects—nymphs and satyrs and so on—and have the wit and verve of the best *Rococo art. After the Revolution he changed his style completely to suit the sterner *Neoclassical taste and worked on the Arc de Triomphe du Carrousel (1806–9) in Paris, which was built to commemorate Napoleon's victories.

Cloisonnism. Style of painting associated with the *Pont-Aven School, characterized by dark outlines enclosing areas of bright, flat colour, in the manner of stained glass or cloisonné enamels (*cloison* is French for 'partition'). *Anquetin and *Bernard first developed the style, and *Gauguin also worked in it. The term was coined by the critic Edouard Dujardin (1861–1949) in 1888 and has remained in use for work done in a similar style during the 20th cent.

Cloisters, The. See METROPOLITAN MUSEUM OF ART.

Close, Chuck. See SUPERREALISM.

Clouet. A family of painters descended from **Jean Clouet** (or **Jan Cloet**) **the Elder**

(b. *c.*1420), a Fleming who came to France *c.*1460. Almost nothing is known for certain of his life and works. The more famous **Jean Clouet** (d. 1540/1) is thought to have been his son. He was celebrated in his lifetime (in 1539 he was praised by the poet Clément Marot (1496–1544) as the equal of *Michelangelo), but no documented works survive. A handful of portraits, however, including the *Dauphin François* (Musées Royaux, Antwerp) and *Man holding Petrarch's Works* (Royal Coll., Windsor), and a number of drawings (mainly in the Musée Condé, Chantilly) are attributed to him on fairly strong circumstantial evidence. The paintings belong to the school of Flemish *naturalism that dominated French portraiture at this time, but the drawings are more personal and often of very high quality. They have often been compared to those of Clouet's contemporary Hans *Holbein the Younger, with which they share a keenness of observation; whereas Holbein's drawings are overwhelmingly linear, however, Clouet's are subtly modelled in light and shade with a delicate system of *hatching that recalls *Leonardo, whose work he could well have known.

Jean's son, **François** (*c.*1510–72), succeeded him as court painter in 1541. His work is somewhat better documented than his father's, but his career is still very obscure (they used the same nickname, 'Janet', which has caused much confusion, and one of the finest works attributed to him, the celebrated portrait of Francis I in the Louvre, showing the king in a lavish gold doublet, has also been given to Jean). François, too, was mainly a portraitist, his signed works including *Pierre Quthe* (Louvre, Paris, 1562), much more Italianate than any of his father's paintings, and *Lady in Her Bath* (NG, Washington, *c.*1570). This mysterious and captivating work has been traditionally identified as representing Diane de Poitiers (1499–1566), but it is more probably a likeness of Marie Touchet, mistress of Charles IX (1550–74). A number of drawings, mostly in the Musée Condé, are also attributed to François.

Clovio, Giulio (1498–1578). Italian painter and *illuminator, born in Croatia. He went to Rome in 1516 and spent most of the rest of his career there, though with frequent breaks when he worked in other cities. In his illuminations he made frequent use of motifs from the work of *Michelangelo and *Raphael, adapting the fashionable *Mannerist style to a miniature scale. Amongst them are the *Towneley Lectionary* (New York, Public Library) and St Paul's *Epistle to the Romans* (Sir John Soane's Mus., London). He also did some work in oils (*Pietà*, Uffizi, Florence, 1553). Clovio enjoyed a very high reputation in his lifetime. *Bruegel worked with him when he visited Rome and El *Greco painted his portrait (Museo di Capodimonte, Naples).

clunch. A generic name for harder grades of chalkstone, or soft *limestone above beds of chalk, more particularly those occurring in the chalk marl of Cambridgeshire. It varies in colour from white to greenish-grey, is easy to work, and takes a good surface. It has occasionally been used as a building stone, but is more suitable for interior carved work and sculpture, for which purposes it was much favoured by the late medieval sculptors of the west of England.

Coade stone. An artificial stone manufactured in London in the late 18th and early 19th cents., used for figure sculpture, monuments, architectural dressings, and decorative work. Essentially a type of clay, fired in a kiln at high temperature, it was named after Eleanor Coade (1733–1821), who set up in business in Lambeth in 1769. She claimed that it resisted frost and therefore retained sharpness of outline better than natural stone, and time has proved her right. It was mixed into a kind of paste and formed into the required shape with moulds, so popular designs could be more or less mass produced. The business was an immediate success; Robert *Adam was one of the notable architects who used the material and several good sculptors, particularly John *Bacon the Elder, worked for the firm. Monuments made of Coade stone exist in many English churches, and some garden sculpture remains. The most remarkable work in the material is perhaps the organ gallery in St George's Chapel, Windsor, designed by Henry Emlyn in 1787, built in 1790–2, and described by Horace *Walpole as 'airy and harmonious'. Mrs Coade's successor in the business, her distant relation William Croggon, died bankrupt in 1835 and Coade stone soon vanished from the market.

Cobra. A group of *Expressionist painters formed in Paris in 1948 by a number of Netherlandish and Scandinavian artists. The name derived from the first letters of the capital cities of the three countries of the artists involved—Copenhagen, Brussels, and Amsterdam. The Dane Asgar *Jorn, the Dutchman Karel *Appel, and the Belgian *Corneille were the leading members, and were joined by Jean *Atlan and Pierre *Alechinsky. Their aims were to exploit free expression of the unconscious, unimpeded and undirected by the intellect. In their emphasis upon unconscious gesture the group had affinities with American *Action painting, but they tended to put more emphasis upon the development of strange and fantastic imagery, related in some cases to Nordic mythology and folklore, in others to various magical or mystical symbols of the unconscious. Their approach was similar to the *Art Informel of *Fautrier and *Wols, but its savage and vigorous expressiveness exceeded theirs. The group published *Cobra Revue*, which ran to eight issues, and some fifteen monographs. It arranged Cobra exhibitions at Copenhagen (1948), Amsterdam (1949), and Liège (1951), but the members soon went their own ways and the group disbanded in 1951.

Cochin, Charles-Nicolas the Younger (1715–90). French engraver, trained by his father, **Charles-Nicolas the Elder** (1688–1754). His huge output forms an invaluable record of contemporary society, particularly life at Louis XV's court. He was also a prolific book illustrator. In 1751 he was appointed keeper of the king's drawings and thereafter exercised considerable influence in artistic circles, even *Diderot consulting him before reviewing the **Salons*. In Cochin's work are seen the beginnings of the *classicism that was to overwhelm French art at the Revolution.

Cock, Jan Wellens de (d. *c.*1526). Netherlandish painter, active in Antwerp, where he was dean of the painters' guild in 1520. Although he is a fairly shadowy figure and the reconstruction of his *œuvre* is controversial, he is noteworthy as one of the earliest followers of *Bosch, his penchant seemingly being small pictures of hermits and saints in weird landscapes. He had two artist sons, **Matthys** (*c.*1509–48) and **Hieronymus** or **Jerome** (*c.*1510–70), both of whom worked in Antwerp. Matthys was renowned in his day as a landscapist and is mentioned by *Vasari as well as van *Mander, but little is known for certain of his work. Jerome was an engraver, printer, and editor who ran a printselling business, 'Aux Quatre Vents' (At the Sign of the Four Winds), that became internationally renowned. Pieter *Bruegel was much employed by Cock in the earlier part of his career, when he excelled at prints in the tradition of Bosch.

Cockerell, Sir Sydney. See FITZWILLIAM MUSEUM.

Cocteau, Jean (1889–1963). French writer, film director, designer, painter, and draughtsman. One of the most dazzling figures of his time in the intellectual avant-garde, he was the friend of leading artists such as *Modigliani and *Picasso, and in his work for the theatre collaborated with, for example, *Diaghilev and the composers Eric Satie (1866–1925) and Igor Stravinsky (1882–1971). His work included poetry, novels, plays, films, and a large amount of paintings, drawings, theatrical designs, and pottery articles. In his painting and drawing he was much influenced by Picasso, and his favourite themes included the figures of Harlequin, embodying the theatre, and Orpheus, the personification of the poet. His most lasting achievement was in the cinema, his reputation resting mainly on his beautiful adaptation of the famous story of Beauty and the Beast (*La Belle et la Bête*, 1946), and on three films dealing with the role of the artist and the nature of his inspiration—Cocteau's recurrent preoccupation—*Le Sang d'un poète* (1930), *Orphée* (1950), and *Testament d'Orphée* (1960).

Codazzi, Viviano. See LAER.

Codde, Pieter (1599–1678). Dutch *genre and portrait painter of the fashionable world and barrack-room life, active in Amsterdam. His best works are usually on a small scale, marked by subtle silvery-grey tonalities, but he achieved one memorable feat on a larger scale. In 1637 he was called upon to finish the group portrait of the Amsterdam Civic guards known as the *Meagre Company* (Rijksmuseum, Amsterdam) that Frans *Hals began in 1633 and refused to finish because he would not come to Amsterdam for sittings, and Codde succeeded so well in capturing Hals's spirit and the touch of his brush that experts still disagree where the work of the one ends and the other begins. Codde also wrote poetry.

Coecke van Aelst, Pieter (1502–50). Netherlandish painter, architect, sculptor, designer of tapestries and stained glass, writer, and publisher. A pupil of Bernard van *Orley, he entered the Antwerp Guild in 1527. Some time before then he had been to Rome and in 1533 he visited Constantinople. His mission to gain business there for the Brussels tapestry works was unsuccessful, but the drawings he made on his journey were later published by his widow Mayken Verhulst as woodcut illustrations in *Les Mœurs et Fachons de Faire des Turcz* (The Manners and Customs of the Turks, 1553). He ran a large workshop and was regarded as one of the leading Antwerp painters of his day, but his work is fairly run-of-the-mill and he is generally more important for his publishing activities. Like his paintings, his books are saturated in Italian influence, and the translation of the architectural treatise of Sebastiano *Serlio that he issued from 1539 played a large part in spreading *Renaissance architecture in the Netherlands (it was from the Dutch edition, too, rather than from the Italian original, that the English translation of 1611 was made). Pieter *Bruegel the Elder was his son-in-law, and, according to van *Mander, his pupil, but there is no trace of Coecke's influence in his work.

Coello, Claudio (1642–93). Spanish painter, the last important master of the Madrid school of the 17th cent. In 1686 he succeeded *Carreño as court painter, and worked in Madrid and at the *Escorial. His masterpiece, *Charles II Adoring the Host* (Escorial, 1685–90), combines a mystical religious subject with realistic portraiture and is an outstanding example of *Baroque *illusionism, mirroring the architecture of the sacristy in which it hangs. He had other noteworthy successes, particularly in work he carried out for Toledo Cathedral, but he died a disappointed man because he was passed over in favour of the Italian Luca *Giordano for the commission to carry out a huge programme of decoration at the Escorial. Coello had travelled to Italy as a young man and also studied the work of *Titian in the royal collection, and his skill as a colourist and painterly brushwork reflect the influence of the great Venetian masters.

Coene, Jacques. See BOUCICAUT MASTER.

Colbert, Jean-Baptiste (1619–83). French statesman who, after the death of Cardinal Mazarin in 1661, became Louis XIV's chief instrument in organizing the machine of state control through which, among other things, he exercised a dictatorship over the arts. It was the object of Colbert and the King to present to the world a visible image of the magnificence and power of the French crown and in carrying out this purpose the grandiose reconstruction of Versailles and its splendidly flamboyant decoration became a symbol of central importance. Colbert's right-hand man in exercising control was Charles *Lebrun, who in 1663 was appointed director of the *Gobelins tapestry factory, which, reorganized as the *Manufacture royale des meubles de la Couronne*, employed the cream of craftsmen in the royal service. In addition to direct patronage, an official body of artistic dogma and official standards of taste were enforced by the re-establishment in 1663 of the Royal *Academy of Painting and Sculpture with Lebrun as President and Colbert as Vice-Protector in 1661 and Protector in 1672. In 1666 a French Academy was founded in Rome under Charles *Errard to provide approved training for young artists from France. Colbert was himself a man of taste and a collector of some importance and it was largely owing to his able support for the King's ambitions that France replaced Italy as the artistic capital of Europe.

Coldstream, Sir William (1908–87). British painter, mainly of portraits, but also of landscape and still life. He studied at the *Slade School, 1926–9, and worked on documentary films for the GPO, 1934–7. In 1937 he was one of the founders of the *Euston Road School, which helped to establish a tradition of sober figurative painting of which he was one of the main representatives. In 1939 he joined the Royal Artillery and in 1943–5 he was an *Official War Artist, working in the Near East and then Italy. After the war he taught at Camberwell School of Art, 1945–9, then was Professor at the Slade School, 1945–75. He exerted an important influence not only through his teaching but also through his appointment as Chairman of the National Advisory Council on Art Education in 1959. The Coldstream Report of 1960 helped to change the structure of art school teaching in Britain, introducing the compulsory study of art history for art students and eventually leading to degree status being awarded to recognized art school courses. Coldstream's own work was typically austerely naturalistic: 'I find I lose interest unless I let myself be ruled by what I see.' Kenneth *Clark described his attitude to art as one of 'dismal rectitude'.

Cole, Sir Henry. See EXHIBITION, GREAT.

Cole, Thomas (1801–48). American *Romantic landscape painter, a founder of the *Hudson River School. His family migrated to America from England in 1819 and he became passionately devoted to the natural scenery of his new country. He spent two years at the Academy of Fine Arts in Pennsylvania and made his living as a portrait painter and engraver there and in New York until some of his landscapes attracted the attention of *Dunlap, *Durand, and *Trumbull, in 1825, assuring his success. In the following year he moved to Catskill on the Hudson River, journeying into the mountains, often on foot, to make sketches of the scenery and working his studies up into finished paintings in the studio. He had two stays in Europe, 1829–32 and 1841–2, living mainly in Florence with *Greenough. These European visits, during which he came under the influence of *Turner and John *Martin, turned him increasingly from the depiction of natural scenery towards grandiose historical and allegorical themes, notably the two great series *The Course of Empire* (New York Hist. Society, 1836) and *The Voyage of Life* (Munson-Williams-Proctor Inst., Utica, 1840).

Colla, Ettore (1896–1968). Italian sculptor and printmaker. He settled in Rome in 1926 after working in Paris with *Laurens and *Brancusi. During the 1950s he was an outstanding figure in Italian sculpture, making his works from machine parts and rusted scrap iron, or broken relics of war objects. From these he constructed symbolic machines which had the evocative effect of new life arising from the ruin and rubble, sometimes incorporating a fetishistic suggestion.

collage. A pictorial technique in which photographs, news cuttings, and other suitable objects are pasted on to a flat surface, often in combination with painted passages (*coller* is French for 'to gum'). Long popular as a leisure-time occupation for children and amateurs, it first became an accredited artistic technique in the 20th cent., when it drew its main material from the proliferation of mass-produced images such as newspapers, advertisements, cheap popular illustrations, etc. The *Cubists were the first to incorporate real objects such as pieces of newspaper into their pictures, often deliberately giving them a dual function both as the real things they were and as contributing to the picture image (see PAPIER COLLÉ). Collage was given a social and ideological direction by the *Futurists and was used by the *Dadaists for their own anarchical purposes. It was adopted by the *Surrealists, who emphasized the juxtaposition of disparate and incongruous imagery. See also MONTAGE.

Collins, William (1788–1847). English painter. He was initially taught by *Morland (of whom his father wrote a biography) and specialized in sentimental rustic landscapes and *genre scenes that won him great popularity. As with Morland, his work became very repetitive. He was a lifelong friend of *Wilkie, after whom he named his elder son, the novelist Wilkie Collins (1824–89), who published a biography of his father, *Memoirs of the Life of William Collins*, in 1848. His second son was **Charles Allston Collins** (1828–73), a friend of *Millais and painter of the well-known *Convent Thoughts* (Ashmolean, Oxford, 1851), which *Ruskin rated highly because of its botanical detail, done in a fastidious *Pre-Raphaelite manner. He married one of the daughters of Charles Dickens and abandoned painting for writing.

Collinson, James (1825–81). English painter. He was one of the original members of the *Pre-Raphaelite Brotherhood in 1848, but left it in 1850 and trained to be a Roman Catholic priest. Another change of heart followed, and in 1854 he returned to painting, specializing in pretty and sentimental *genre scenes, the best known of which is *The Empty Purse* of 1857 (versions in the Tate, London, and Graves Art Gallery, Sheffield). He is probably the least known of the PRB and the most remarkable talent he showed at their meetings was his ability to fall asleep at any time.

Cologne School. Term applied to painting produced in Cologne from the late 14th cent. to the early 16th cent. There was not, as the term misleadingly suggests, a definite local style at this time, but early in the 19th cent., when romantically inclined collectors began to look for old German pictures, it so happened that many of their panels came from the Cologne region, and the town, with its unfinished *Gothic cathedral, became a symbol for the spirit of the Middle Ages.

The alleged founder of the 'School', Master Wilhelm, is a semi-legendary figure and no pictures can be attributed to him with certainty. During the first quarter of the 15th cent. painting in Cologne had all the characteristics of the lyrical 'soft style', that is of *International Gothic. This style may have been perpetuated by guild rules and guild supervision; in any case it was still employed by *Lochner in the middle of the 15th cent. A little later Netherlandish *naturalism was taken up by artists in Cologne. The great Flemish centres are not far away, and one of Rogier van der *Weyden's principal works (the so-called *Columba* Altar, now in the Alte Pinakothek, Munich) was in a Cologne church at the time. Of many anonymous masters the *Master of the Life of the Virgin is perhaps the most attractive.

Colombe, Jean. See LIMBURG BROTHERS.

Colombe, Michel (*c.*1430–after 1512). French sculptor. He had a great name in his day, but only two works that are certainly by him survive, both from very late in his life. They are: the tomb of Francis II of Brittany and Marguerite de Foix (Nantes Cathedral, 1502–7), done in collaboration with Jean *Perréal and Girolamo da Fiesole, an Italian sculptor working in France; and a relief of *St George and the Dragon* (Louvre, Paris, 1508–9), done for an altarpiece for the Château de Chaillon. The latter also was a work of collaboration, for the frame was carved by Jérôme Pacherot, another Italian expatriate. Colombe's fame rests mainly on the *St George*, for it is a captivating work, blending the fantasy of the French *Gothic style with elements of the Italianate *Renaissance taste that was coming into vogue in southern France at this time, yet without copying particular Italian models.

Colonna, Gerolamo Mengozzi. See QUADRATURA and TIEPOLO.

Colossus of Rhodes. Celebrated bronze statue of Helios the Sun-god, regarded as one of the Seven Wonders of the World. Over 30 m. high it stood beside (not astride as it is shown in some reconstructions) the harbour at Rhodes and commemorated the raising of a siege in 304 BC. The sculptor was Chares of Lindos, a pupil of *Lysippus, and the statue was finished in about 280 BC after 12 years' work. In about 225 BC it was overthrown in an earthquake, but the fallen figure remained until AD 653, when it was broken up by Arab raiders and sold as scrap.

Colour Field painting. A type of abstract painting characterized by large expanses of more or less unmodulated colour, with no strong contrasts of tone or obvious focus of attention. This type of painting developed in the USA in the late 1940s and early 1950s, leading exponents including *Newman *Rothko. It is thus an aspect of *Abstract Expressionism, and has also been seen as a type of *Minimal art. From 1952 Helen *Frankenthaler developed colour field painting by soaking or staining very thin paint into raw unprimed canvas, so that the paint is integral with it rather than superimposed. The term **Colour Stain painting** is applied to paintings of this type.

Colquhoun, Robert (1914–62). British painter, graphic artist, and designer. In 1933–8 he studied at Glasgow School of Art, where he became the inseparable companion of his fellow student Robert MacBryde (1913–66). They settled in London in 1941. During the war Colquhoun was an ambulance driver in the Civil Defence Corps by day and painted by night. Within a few years the studio of the Roberts (as they were generally known) at 77 Bedford Gardens, Campden Hill, had become a centre for a group of young artists and writers, among them Keith *Vaughan. Following a successful one-man show in 1943 Colquhoun developed a reputation as one of the outstanding British painters of his generation; his characteristic angular figure compositions owed something to *Cubism, but had an expressive life of their own. After he and MacBryde were evicted from their studio in 1947, however, his fortunes began to decline and he died (of a heart attack) in relative obscurity. Apart from painting, both Colquhoun and MacBryde produced many lithographs and worked together on stage designs.

Colt (or **Coulte**), **Maximilian** (d. after 1645). French-born sculptor and mason, who came to England from Arras *c.*1595 and Anglicized his surname from Poultrain. In 1605 he made the large tomb of Queen Elizabeth I in Westminster Abbey, a splendid example of a conventional type. Much more original is his tomb of Robert Cecil, first Earl of Salisbury, at Hatfield, in which the effigy lies on a black marble bier carried on the shoulders of four solemn and beautifully cut kneeling Virtues. Before this Colt had carried out much work at Hatfield House for Cecil, including two magnificent fireplaces. In 1608 he became Master Sculptor to the Crown, an office he held until his death, though his art, excellent as it was by *Jacobean standards, was too conservative for Charles I (1600–49), from whom he received only minor commissions. Sculptors such as Nicholas *Stone were more to Charles's taste, and in 1641 Colt was imprisoned for debt.

Colville, Alexander (1920–). Canadian painter, regarded as one of the leading exponents of *Magic Realism in his country. Most of his paintings show figures in juxtaposition with inanimate objects, animals, or other human beings, carefully placed at the most concentrated point in that space, drawing substance from the charged atmosphere (e.g. *Nude and Dummy*, New Brunswick Mus., Saint John, 1950). He was an official war artist and his work is well represented in the Canadian War Museum, Ottawa. His other work has included a mural, *History of Mount Allison* (1948), for Mount Allison University, Sackville, where he both studied and taught, and the designs for the special issues of Canadian coinage commemorating the centenary of Confederation (1967).

Combine painting. A term coined by *Rauschenberg for a type of work he originated in the mid-1950s in which a painted surface is 'combined' with various real objects, or sometimes photographic images, attached to it. It may be regarded as a radical development of the technique of *collage used by *Schwitters. See also JUNK ART.

Conca, Sebastiano (1680–1764). Italian painter. A pupil of *Solimena at Naples, he moved to Rome in 1706, and with *Trevisani he was one of the most important decorative painters there during the first half of the 18th cent. Out of the High *Baroque tradition he developed a distinctively Roman *Rococo style, which through its polished elegance gives an impression of superficiality despite the grandeur of its conception. Good examples of his work are the decorations in the churches of S. Clemente (1714) and Sta Cecilia (1725), Rome. *Batoni became the most renowned of his many students.

Conceptual art. Term embracing various forms of art in which the idea for a work is considered more important than the finished product, if any. The notion goes back to *Duchamp, but it was not until the 1960s that Conceptual art became a major international phenomenon (it first made an impact in Britain with the exhibition 'When Attitudes Become Form' at the *Institute of Contemporary Arts, London, in 1969). It often overlaps with other avant-garde forms of expression (*Body art, *Land art, *Minimal art, *Performance art, for example), and its manifestations have been very diverse. Their common characteristic is the claim that the 'true' work of art is not a physical object produced by the artist but consists of 'concepts' or 'ideas'. When a material thing is hung in a gallery or otherwise presented to spectators, this is regarded as no more than a vehicle for the communication of ideas or a means of reference to events or situations removed in space and time from the presentation. Photographs, texts, maps, diagrams, sound cassettes, video, etc., have been used as communication media. Most artists in the field of Conceptual art deliberately render their productions uninteresting, commonplace, or trivial from a visual point of view in order to divert attention to the 'idea' they express. This may be accepted. The notion that aesthetic qualities may belong to *concepts* is not new. Aesthetic delight in the elegance or economy or consistency of mathematical theorems or scientific and philosophical theories has long been recognized. Beauty (or 'brilliancy') prizes are even awarded in chess tournaments. But artists have not usually been notable for either profundity or beauty and elegance in the conceptual field, and the ideas to which works of Conceptual art lead the spectator seem in most cases to be themselves shallow, commonplace, or trivial. A well-known example is Joseph *Kosuth's *One and Three Chairs* (MOMA, New York, 1965), which combines a real chair, a photograph of a chair, and a dictionary definition of 'chair'. This work certainly makes no attempt to be a visually interesting composition in the traditional sense. Yet the idea which it presents—real, tangible chair; photograph; definition of concept—is trivial in the extreme to anyone accustomed to operate in the realm of ideas. There are no readily apparent means for discriminating successful from unsuccessful, professional from amateurish, good from bad, examples of Conceptual art, and to many people the abbreviation 'Con art' that is sometimes used must seem remarkably apposite.

Concrete art. Term applied to abstract art that repudiates all figurative reference. The term was coined by van *Doesburg, who in 1930 issued a manifesto entitled *Art Concret*, disguised as the first number of a review (no other numbers were issued), and it is often particularly associated with Max *Bill. Although Concrete art is often severely geometrical, it is not necessarily so; Bill's sculpture, for example, often uses graceful spiral or helix shapes. Bill gave the following definition: 'Concrete painting eliminates all naturalistic representation; it avails itself exclusively of the fundamental elements of painting, the colour and form of the surface. Its essence is, then, the complete emancipation of every natural model; pure creation.'

Conder, Charles (1868–1909). English painter, a descendant of the sculptor *Roubiliac. After living in Australia for some time, he moved to France and studied with *Cormon. He was a friend of *Toulouse-Lautrec and his work was admired by *Degas and *Pissarro. From 1897 he lived mainly in England. His paintings, well represented in the Tate Gallery, were mostly of arcadian fantasies and landscapes, and included watercolours on silk and designs for painted fans. They are often tinged with a *fin de siècle* decadence reminiscent of that of *Beardsley.

Condivi, Ascanio (d. 1574). Italian painter, sculptor, and writer, a pupil and friend of *Michelangelo. He was an insignificant artist and his only claim to fame is his *Life of Michelangelo*, published in 1553. Three years earlier the first edition of *Vasari's *Lives* had appeared, and Michelangelo seems to have taken exception to some of the statements made there. Condivi's *Life* was meant as a corrective, and he writes not only from intimate personal knowledge, but obviously at times almost at dictation from the master. Unlike *Holanda he has no aesthetic theories of his own and his account is the most trustworthy we have today. Johannes *Wilde (*Michelangelo*, 1978, pp. 9–12) considers that Condivi was a 'simpleton' who could not have composed 'such an eminently readable book' unaided, and thinks that Annibale Caro (1507–66), a humanist man of letters, was probably the ghost writer. One of Condivi's few extant works is a painting of the Holy Family (also called 'Epifania') in the Casa Buonarroti, Florence, done from a cartoon by Michelangelo in the British Museum, London; Wilde says it 'shows an appalling degree of incompetence'.

Coninxloo, Gillis van (1544–1607). The most important member of a large and prolific family of Flemish painters, many of whom are not clearly distinguishable personalities. He was born at Antwerp and in 1587 emigrated to Frankenthal, where he became a leader of a group of landscape painters established there. In 1595 he settled permanently in Amsterdam. Coninxloo's early landscapes are panoramic views of vast valleys and great mountain ranges populated by biblical or mythological personages. In later works, such as the majestic *Forest* (Kunsthistorisches Mus., Vienna), he narrows his field of vision and takes as his subject the mood evoked by luxuriant nature. His younger countrymen, Roelant *Savery and David *Vinckboons, who had also come to Holland at about the same time, were influenced by Gillis's late works. Among his many pupils were two major Dutch artists, Esaias van de *Velde and

Hercules *Seghers. His paintings are rarely signed and lesser artists are often confused with him.

Constable, John (1776–1837). English painter, who is ranked with *Turner as one of the greatest British landscape artists. Although he showed an early talent for art and began painting his native Suffolk scenery before he left school, his great originality matured slowly. He committed himself to a career as an artist only in 1799, when he joined the *Royal Academy Schools and it was not until 1829 that he was grudgingly made a full Academician, elected by a majority of only one vote. In 1816 he became financially secure on the death of his father and married Maria Bicknell after a seven-year courtship and in the face of strong opposition from her family. During the 1820s he began to win recognition: *The Hay Wain* (NG, London, 1821) won a gold medal at the Paris *Salon of 1824 and Constable was admired by *Delacroix and *Bonington among others. From this time also his pictures began modestly to sell. His wife died in 1828, however, and the remaining years of his life were clouded by despondency.

After spending some years working in the *Picturesque tradition of landscape and the manner of *Gainsborough, Constable developed his own original treatment from the attempt to render scenery more directly and realistically, carrying on but modifying in an individual way the tradition inherited from *Ruisdael and the Dutch 17th-cent. landscape painters. Just as his contemporary William Wordsworth rejected what he called the 'poetic diction' of his predecessors, so Constable turned away from the pictorial conventions of 18th-cent. landscape painters, who, he said, were always 'running after pictures and seeking the truth at second hand'. Constable thought that 'No two days are alike, nor even two hours; neither were there ever two leaves of a tree alike since the creation of the world', and in a way that was then new he represented in paint the atmospheric effects of changing light in the open air, the movement of clouds across the sky, and his excited delight at these phenomena stemming from a profound love of the country: 'The sound of water escaping from mill dams, willows, old rotten planks, slimy posts and brickwork, I love such things. These scenes made me a painter.' He never went abroad, and his finest works are of the places he knew and loved best, particularly Suffolk and Hampstead, where he lived from 1821. To render the shifting flicker of light and weather he abandoned fine traditional finish, catching the sunlight in blobs of pure white or yellow, and the drama of storms with a rapid brush that disdained worn-out symbols. Henry *Fuseli was among the contemporaries who applauded the freshness of Constable's approach, for C. R. *Leslie records him as saying: 'I like de landscapes of Constable; he is always picturesque, of a fine colour, and de lights always in de right places; but he makes me call for my great coat and umbrella.'

Constable worked extensively in the open air, drawing and sketching in oils, but his finished pictures were produced in the studio. For his most ambitious works—'six-footers' as he called them—he followed the unusual technical procedure of making a full-size oil sketch, and in the 20th cent. there has been a tendency to praise these even more highly than the finished works because of their freedom and freshness of brushwork. (The full-size sketch for *The Hay Wain* is in the V&A, London, which has the finest collection of Constable's work.) In England Constable had no real successor and the many imitators (who included his son **Lionel**, 1825–87) turned rather to the formal compositions than to the more direct sketches. In France, however, he was a major influence on *Romantics such as Delacroix, on the painters of the *Barbizon School, and ultimately on the *Impressionists.

Constructivism. Geometric abstract art movement, founded in Russia in about 1914 by Vladimir *Tatlin. He was joined by the brothers Antoine *Pevsner and Naum *Gabo, who in 1920 published their *Realist Manifesto*, in which one of the directives was 'to construct' art; it is from this that the name derives. Pevsner and Gabo rejected the idea that art must serve a socially useful purpose and conceived a purely abstract art that reflected modern machinery and technology and used industrial materials such as plastic and glass. Tatlin and Alexander *Rodchenko, on the other hand, were among those who applied Constructivist principles to architecture and design. Gabo and Pevsner left Russia in 1922 after Constructivism had been condemned by the Soviet regime, and they and other exiles helped to spread the ideals of the movement throughout Europe. They were influential, for example, on the *Bauhaus in Germany, De *Stijl in the Netherlands, and the *Abstraction-Création group in France, and Gabo was one of the editors of the English Constructivist manifesto, **Circle*, in 1937.

consular dyptych. See DIPTYCH.

conté crayon. A very hard, grease-free type of crayon, named after Nicolas-Jacques Conté (1755–1805), the French scientist who invented it. Conté, who worked as a portrait painter in his youth, was also the inventor of the modern graphite *pencil.

continuous representation or **continuous narrative.** A pictorial convention whereby two or more incidents from a narrative are combined in the same image. It is most common in medieval art, but occasionally occurs later. In *Pontormo's *Joseph in Egypt* (NG, London, *c.*1515), for example, four episodes from Genesis are shown in the same panel and the figure of Joseph appears separately in each scene.

contrapposto. Term (Italian for 'set against') applied to poses in which one part of a figure twists or turns away from another part. It was originally applied, during the *Renaissance, to a relaxed asymmetrical pose characteristic of much Greek and Roman sculpture in which the body's weight is borne mainly on one leg, so that the hip of that leg rises relative to the other (the *Doryphoros* of *Polyclitus is a classic example). The term is now, however, used in a much broader sense and applied equally to painting as to sculpture. The acknowledged master of *contrapposto* was *Michelangelo, and his *Mannerist followers (for example *Bronzino) often went to absurd lengths to devise poses of wilful complexity to demonstrate their skill in the field.

conversation piece. A portrait group in a domestic or landscape setting in which two or more sitters are engaged in conversation or other polite social activity. Conversation pieces are usually, though not always, fairly small in scale, even when the figures are depicted full-length. They were especially popular in Britain during the 18th cent. (the term is first recorded in print in English in 1706), but the use of the term is not confined to British painting or to this period. Arthur *Devis and the young *Gainsborough were notable practitioners.

Cook, Beryl (1926–). British *naïve painter. She took up painting seriously in early middle age and in 1975 had her first exhibition, at the Plymouth Arts Centre. It was a great success and within a few years she was well known through other exhibitions, television appearances, and the publication of the first of several collections of her work in book form (*The Works*, 1978), with the paintings accompanied by her own amusing captions. Her chubby, usually jovial characters have also been much used on greetings cards. Cook's subjects are drawn from everyday life and frequently involve the kind of saucy humour associated with seaside holidays (she used to run a boarding house in Plymouth) and tabloid Sunday newspapers (often she incorporates newsprint in her work).

Cooper, Douglas (1911–84). British art historian and collector. He lived in France for much of his life and was severely critical of the British for what he regarded as their failure to appreciate or patronize modern art. His main interest was *Cubism, and in 1932 he decided to devote part of his inheritance to forming a collection of its four main protagonists—*Picasso, *Braque, *Gris, and *Léger—in its greatest period, 1907–14. He later added works by other artists, but the Cubists remained the core. In the Second World War he worked in intelligence and helped to identify, protect, and repatriate works of art. Picasso was a neighbour and visitor in southern France, but their friendship turned to hostility. Cooper, indeed, had a notoriously difficult temperament and enjoyed controversy; in the 1950s he became particularly well known for his attacks on the *Tate Gallery and its Director Sir John *Rothenstein. He was a formidable scholar and published substantial books on all four major Cubists.

Cooper, Samuel (1609–72). English *miniaturist, the nephew and probably the pupil of John *Hoskins. Cooper was the greatest English miniaturist of the 17th cent. and he enjoyed a European reputation and the patronage of both Oliver Cromwell (1599–1658) and Charles II (1630–85). His portraits are almost always of the *bust only, but within this limitation his range is remarkable: he presents each sitter with a force and individuality beside which the life-size portraits by contemporaries such as *Lely appear doll-like, and his ambitious composition and *Baroque sense of design mark a complete breach with the tradition of *Hilliard and Hoskins. He is well represented at the Victoria and Albert Museum, the Royal Collection at Windsor Castle, and in the Fitzwilliam Museum, Cambridge. His brother **Alexander** (before 1609–60) was also a miniaturist. He worked mainly on the Continent, in the Netherlands and at the Swedish court.

Coornhert, Dirck. See GOLTZIUS.

Coorte, Adriaen (active 1683–1708). Dutch still-life painter, active around Middelburgh. Nothing is known of his life, and his work was completely forgotten for more than two centuries after his death. Only a handful of paintings by him survive, but they show him to have been one of the most individual still-life painters of his time. They are the complete opposite of the lavish pieces by such celebrated contemporaries as Jan van *Huysum and Rachel *Ruysch, for they are small in scale and depict a few humble objects, characteristically placed on a bare ledge. The intensity of his scrutiny is such, however, that they take on something of the mystical quality of the still lifes of *Sánchez Cotán or *Zurbarán, and the hovering butterfly that Coorte sometimes incorporates in his work may have allegorical significance. One of his favourite subjects was a bundle of asparagus (examples in the Rijksmuseum, Amsterdam, the Fitzwilliam, Cambridge, and the Ashmolean, Oxford).

Copley, John Singleton (1738–1815). The greatest American painter of the 18th cent. He was the stepson of the engraver Peter Pelham (*c.*1695–1751), from whose large collection of engravings he gained a considerable knowledge of European art, but he was virtually self-taught as a painter. While still in his teens he had set up as a painter in his native Boston and by his early twenties he was painting portraits that, in their sense of life and character, completely outstripped anything previously produced by Colonial portraitists (*Epes Sargent*, NG, Washington, *c.*1760). Though he became extremely successful, Copley was diffident and self-doubtful by nature and came to see himself as an artist afflicted with provincialism, cut off from the great European tradition of painting. For a long while he hesitated to leave the security of Boston (where he earned 'as much as if I were a *Raphael or a *Correggio'), even after his portrait of his half-brother Henry Pelham (*The Boy with a Squirrel*, Mus. of Fine Arts, Boston, 1765) had been highly praised by both *Reynolds and *West when it was exhibited in London in 1766. He finally left in 1774, when revolutionary activity was beginning to threaten his practice, and settled in London in 1775 after a study trip to Italy.

In England his style changed radically, sacrificing the forthright vigour of his Colonial work for a more fashionable and ornate manner. He continued to paint fine portraits that were more than a match for the work of most of his contemporaries, but it is generally agreed that those he painted in America have much greater originality and conviction than their English counterparts. In compensation for this decline as a portraitist, he was able to turn his hand to history painting, in which he had long been eager to make a success but for which the opportunities in America were severely limited. The first was *Brook Watson and the Shark* (NG, Washington, 1778; a copy Copley made for himself is in the Mus. of Fine Arts, Boston, and a smaller variant, 1782, is in the Detroit Institute of Arts). In this he followed the innovation of his countryman West in using modern dress, and went beyond him in depicting a subject not because it was of historical importance or moral significance, but merely because it was exciting (see also BORGHESE WARRIOR). It was not until a generation later that the French *Romantics took up this revolutionary idea. His other history paintings took more conventional themes, mainly patriotic and military, such as *The Death of Major Peirson* (Tate, London, 1783). Copley revealed a magnificent gift for depicting heroic action in multi-figure compositions that none of his British contemporaries could approach, and his history paintings won him considerable acclaim. They also brought on him the wrath of the *Royal Academy (of which Copley had become a full member in 1779) when they were shown privately with great success, for this constituted a rival attraction to the Academy's own exhibitions. Copley's success in England, however, was fairly short-lived and his work gradually went out of favour. His final years were marked by a sad decline. *Morse visited him in 1811 and wrote: 'His powers of mind have almost entirely left him; his late paintings are miserable; it is really a lamentable thing that a man should outlive his faculties.' He died leaving debts that had to be paid off by his son and namesake, who became Baron Lyndhurst and was three times Lord Chancellor.

copper point. See METAL POINT.

Coppo di Marcovaldo (b. *c.*1225). Italian painter, one of the earliest about whom there is a body of documented knowledge. He served in the army of Florence and settled in Siena after his capture at the Battle of Montaperti (1260). In 1261 he painted the signed and dated *Madonna and Child Enthroned* (called the *Madonna del Bordone*) for the Servite church at Siena, and in 1274 he and his son **Salerno** painted a Crucifix for Pistoia Cathedral; both paintings still remain in their original locations. On the basis of these documented works two other outstanding paintings are attributed to Coppo: a *Madonna and Child Enthroned* in Sta Maria dei Servi in Orvieto, and a Crucifix in the Pinacoteca at San Gimignano. He introduced new solidity and humanity to the *Byzantine tradition, in the way, for example, that he represents the Virgin with her head inclined towards the Child, and with *Guido da Siena he ranks as the founder of the Sienese School and one of the most important forerunners of *Cimabue and *Duccio.

Coques, Gonzales (1618 or less likely 1614–84). Flemish *genre and portrait painter, known as the 'little van *Dyck', although his style is much closer to *Terborch. He was born in Antwerp and was mainly active there, but he also travelled to Holland and England. His best works, charming and daintily executed, are small-scale fashionable group portraits such as *A Family Group* (NG, London).

corbel. A bracket or other projection on the surface of a wall intended to support a weight. Corbels often provided the opportunity for elaborate carving.

Corinth, Lovis (1858–1925). German painter and graphic artist. Part of his training was with *Bouguereau in Paris (1884–7), but he was more strongly influenced by the painterly work of French artists such as *Courbet and *Manet, as well as by *Hals, *Rembrandt, and *Rubens. On his return to Germany he lived mainly in Munich before settling in Berlin in 1901. With *Liebermann he became recognized as the leading representative of the German *Impressionist school. An apoplectic stroke incapacitated him in 1911, however, and when he began to paint again (with great difficulty) it was in a much looser and more powerful *Expressionist manner, to which he had previously been strongly opposed (he has been described as 'an eleventh-hour convert to modern art'). Corinth was varied and prolific as a painter and graphic artist, his prints being mainly *drypoints or *lithographs. Apart from landscapes, portraits, and still lifes, he had a fondness for voluptuous allegorical and religious subjects. His late works were declared *degenerate by the Nazis.

Cormon, Fernand (pseudonym of Fernand-Anne Piestre) (1845–1924). French painter. He had a successful career both as a painter and a popular teacher (his pupils included *Matisse, *Toulouse-Lautrec, and van *Gogh), but his reputation has not lasted well. His work included decorations at the Museum of Natural History and the Petit Palais in Paris and some excellent portraits, and he also had a penchant for paintings of prehistoric history (*The Age of Stone*, Musée du Prieuré, Saint-Germainen-Laye, 1884).

Corneille (Cornelis van Beverloo) (1922–). Belgian painter, active mainly in Paris. He was a founder member of *Cobra in 1948 and his paintings are marked by brilliant colour and vigorous brushwork.

Corneille de Lyon (active 1533–74). Netherlandish-born painter, active mainly at Lyons in France. He was a native of The Hague (in France he is still often known as 'Corneille de La Haye'), but he settled in Lyons in 1534 and in 1540 he became court painter to the Dauphin, later Henry II (1519–59). Contemporary references to him indicate that he had a considerable reputation as a portrait painter, but only one work survives that is unquestionably from his hand, a portrait of Pierre Aymeric (Louvre, Paris, 1553), authenticated by an inscription in the sitter's handwriting on the back of the picture. Many other works in a similar style go under his name. They are mostly small in scale and sharply naturalistic in manner, with the sitter usually set against a plain green or blue background. The National Gallery, London, has four examples of the type, catalogued as 'attributed to' or 'style of Corneille de Lyon'.

Cornelis van Haarlem (1562–1638). Dutch painter who ranks with Hendrik *Goltzius and Karel van *Mander as one of the leading representatives of *Mannerism in the Netherlands. He is best known for his large biblical and historical pictures packed with athletic, life-size Italianate nudes in wrenched and sharply foreshortened positions. But he also did a few forceful portraits of individuals and groups which show that he was an important forerunner of Frans *Hals. Both facets of his work can best be seen in the Frans Hals Museum in Haarlem.

Cornelisz. van Oostsanen, Jacob (also called Jacob van Amsterdam) (*c*.1470–1533). Netherlandish painter and designer. He was born in Oostsanen and worked mainly in nearby Amsterdam, where he was a leading designer of *woodcuts, liberating the Dutch woodcut from the *miniature tradition and giving it a new power and breadth. Comparatively few of his works have been preserved: among the woodcuts is a series illustrating the *Passion* (1512–17) and among the paintings are a *Self-portrait* (Rijksmuseum, Amsterdam, 1533) and an *Adoration of the Shepherds* (Museo di Capodimonte, Naples, 1512) which contains pudgy angels playing toy-like instruments, singing, and decorating with garlands an improbable *Renaissance manger. Although his work is somewhat provincial, he marks the beginning of the great artistic tradition of Amsterdam, and his keenness of observation was to be one of the trademarks of later Dutch art. Jan van *Scorel was his most important pupil.

Cornelius, Peter von (1783–1867). German painter, best known for the major part he played in the revival of *fresco in the 19th cent. After training at the Düsseldorf Academy, Cornelius moved to Italy in 1811 and joined the *Nazarenes in Rome. In 1819 he was called to Munich by Crown Prince Ludwig of Bavaria (later Ludwig I, 1786–1868), for whom he worked extensively, notably on a series of frescos in the Ludwigskirche (1836–9), including a *Last Judgement* that is larger than *Michelangelo's in the Sistine Chapel. When this work was not well received Cornelius left Munich to work for Frederick William IV (1795–1861) of Prussia in Berlin. His major undertaking there was a commission for frescos in a mausoleum for the royal family modelled on the Campo Santo in Pisa. The project was officially cancelled after the revolution in 1848, but Cornelius continued to work on his drawings for it for the rest of his life. Cornelius's work is undoubtedly impressive, but rather self-conscious in its desire to revive the heroic language of *Raphael and Michelangelo, and combine it with the didactic philosophy of German *Romanticism. He was director of the academies at Düsseldorf and Munich and his influence was considerable; it may well be claimed that his works in Munich sparked off the revival of large-scale fresco decoration in Germany and perhaps elsewhere. His advice was sought when frescos were painted in the Houses of Parliament, London, in the 1840s.

Cornell, Joseph (1903–73). American sculptor, one of the pioneers and most celebrated exponents of *assemblage. He had no formal training in art and his most characteristic works are his highly distinctive 'boxes'. These are simple boxes, usually glass-fronted, in which he arranged surprising collections of photographs or Victorian bric-à-brac in a way that has been said to combine the formal austerity of *Constructivism with the lively fantasy of *Surrealism. Like Kurt *Schwitters he could create poetry from the commonplace. Unlike Schwitters, however, he was fascinated not by refuse, garbage, and the discarded, but by fragments of once beautiful and precious objects, relying on the Surrealist technique of irrational juxtaposition and on the evocation of nostalgia for his appeal. His work was exhibited from 1932 at the Julian Levy Gallery in New York, a major venue for the showing of European Surrealism, and he befriended several members of the movement who settled in the USA during the Second World War. Cornell also painted and made Surrealist films. His work is well represented in the Art Institute of Chicago.

Corot, Jean-Baptiste-Camille (1796–1875). French painter. At the age of 26 he abandoned a commercial career for the practice of art and from the first showed a strong vocation for landscape painting. He lived in Paris, but travelled about France making sketches from nature and from these he composed in his studio. In addition to his journeys in France, he visited England, the Low Countries, Switzerland, and Italy three times (1825–8, 1834, and 1843). Throughout his life Corot found congenial the advice given to him by his teacher Achille-Etna *Michallon 'to reproduce as scrupulously as possible what I saw in front of me'. On the other hand he never felt entirely at home with the ideals of the *Barbizon School, the members of which saw *Romantic idealization of the countryside as a form of escapism from urban banality, and he remained more faithful to the French *classical tradition than to the English or Dutch schools. Yet although he continued to make studied compositions after his sketches done direct from nature, he brought a new and personal poetry into the classical tradition of composed landscape and an unaffected naturalness which had hitherto been foreign to it. Though he represented nature realistically, he did not idealize the peasant or the labours of agriculture in the manner of *Millet and *Courbet, and he was uninvolved in ideological controversy.

From 1827 Corot exhibited regularly at the *Salon, but his greatest success there came with a rather different type of picture—more traditionally Romantic in its evocation of an Arcadian past, and painted in a misty soft-edged style that contrasts sharply with the luminous clarity of his more topographical work. Late in his career Corot also turned to figure painting and it is only fairly recently that this aspect of his work has emerged from neglect—his female nudes are often of high quality. It was, however, his directness of vision that was generally admired by the major landscape painters of the latter half of the century and influenced nearly all of them at some stage in their careers. His popularity was (and is) such that he is said to be the most forged of all painters (this in addition to an already prolific output). In his lifetime he was held in great esteem as a man as well as an artist, for he had a noble and generous nature; he supported Millet's widow, for example, and gave a cottage to the blind and impoverished *Daumier.

Correggio (Antonio Allegri) (*c.*1490–1534). Italian painter, named after the small town in Emilia where he was born. His career is poorly documented and his training has to be conjectured on stylistic grounds. Echoes of *Mantegna's manner in many of his early paintings indicate that he may have studied that master's work in Mantua, and he was influenced in these works also by Lorenzo *Costa and *Leonardo, adopting Costa's pearly Ferrarese colouring and, in the St John of the *St Francis* altarpiece (Gemäldegalerie, Dresden, 1514), his first documented work, Leonardo's characteristic gesture of the pointing finger. Later he developed a style of conscious elegance and allure with soft **sfumato* and gestures of captivating charm. Correggio may well have visited Rome early in his career, although *Vasari maintains that he never went there and the obvious inspiration of the paintings of *Raphael and *Michelangelo could be accounted for by drawings and prints that were known all over Italy. Although he worked mainly in provincial centres, he was one of the most sophisticated artists of his time, blending disparate sources into a potent synthesis.

He was probably in Parma, the scene of his greatest activity, by 1518. His first large-scale commission there was for the decoration of the abbess's room in the convent of S. Paolo. The theme of the decorations is Diana, goddess of chastity and the chase, and the vaulted ceiling uses Mantegna's idea of a leafy trellis framing *putti and symbols of the hunt. The S. Paolo ceiling was followed by two dome paintings in which Correggio developed the *illusionist conception—already used by Mantegna—of depicting a scene as though it were actually taking place in the sky above (see SOTTO IN SU). The first of these domes was commissioned for the church of S. Giovanni Evangelista in 1520. The twelve Apostles sit on clouds round the base, while Christ is shown in steep foreshortening ascending to heaven. In the commission six years later for an *Assumption of the Virgin* in the dome of Parma Cathedral he used the same principle, but on a much larger scale and with still more daring foreshortening. These works reveal Correggio as one of the boldest and most inventive artists of the High *Renaissance and they were highly influential on the development of *Baroque dome painting (one of his most important successors, *Lanfranco, was a native of Parma). Other aspects of Correggio's work were even more forward-looking. His extraordinarily sensuous mythologies, notably the series on the *Loves of Jupiter* (*Io* and *Jupiter and Ganymede* in the Kunsthistorisches Mus., Vienna; *Leda* in the Staatliche Museen, Berlin; *Danae* in the Borghese Gallery, Rome; painted for Federigo *Gonzaga *c.*1530–3), foreshadow *Rococo artists such as *Boucher.

Corrente. An anti-Fascist association of young Italian artists formed by Renato *Birolli in Milan in 1938 with *Guttuso, Mirko Basaldella, and *Afro among its members. The association had no fixed artistic programme beyond a desire to oppose what they regarded as the provincialism of the *Novecento and the official art. They stood for the defence of 'modern' art at a time when the Nazi campaign against *degenerate art (*entartete Kunst*) was spreading to Italy. The association arranged an exhibition in March 1939 in which older Milanese artists were included, and a second exhibition in December of the same year in which only the younger set participated. Its activities, which included the publication of a review of literature, politics, and the arts, were dissipated by the Second World War.

Cortese. See COURTOIS.

Cortona, Pietro da (Pietro Berrettini) (1596–1669). Italian painter, architect, decorator, and designer, second only to *Bernini as the most versatile genius of the full Roman *Baroque style. He was named after his birthplace in Tuscany and probably had some training with his father, a stonemason, before being apprenticed as a painter in Florence. In 1612 or 1613 he moved to Rome. His first major works were frescos in Sta Bibiana, Rome (1624–6), commissioned by Urban VIII (Maffeo *Barberini), and the patronage of the Barberini family played a major part in his career. For their palace he painted his most famous work, the huge ceiling fresco, *Allegory of Divine Providence and Barberini Power*. This was begun in 1633, but he interrupted the work in 1637 to go to Florence and paint two of four frescos commissioned by the Grand Duke of Tuscany for the *Pitti Palace. He returned to finish the Barberini ceiling in 1639. This, one of the key works in the development of Baroque painting, is a triumph of *illusionism, for the centre of the ceiling appears open to the sky and the figures seen from below (*di *sotto in su*) appear to come down into the room as well as soar out of it. It demonstrates Pietro's belief, which came out in a celebrated controversy with Andrea *Sacchi in the Accademia di S. Luca, that a history painting could be compared with an epic and was entitled to use many figures; Sacchi, intent on *classical simplicity and unity, argued for using as few figures as possible. In 1640–7 Pietro was back in Florence to finish his decorations in the Pitti Palace, where he received a new commission for seven ceilings (completed by his pupil Ciro Ferri, 1634–89). These *Allegories of Virtues and Planets* have elaborate *stucco accompaniments uniting the painted ceilings with the framework of the rooms, and this form of decoration was widely influential, not only in Italy, but also in France. (Pietro turned down an invitation to visit Paris from Cardinal Mazarin (1602–61), but his style was taken there by his pupil *Romanelli.) From 1647 until his death Pietro again worked in Rome, his major paintings from this period being an extensive series of frescos in Sta Maria in Vallicella (the Chiesa Nuova, 1647–65) in which, as in his Pitti decorations, paint and stucco are magnificently combined. Throughout his career he also painted easel pictures of religious and mythological subjects.

Pietro once wrote that architecture was merely a pastime for him, but he ranks among the greatest architects of his period. His masterpiece is the church of SS. Martina e Luca in Rome (1635–50), which was the first Baroque church designed and built as a complete unity. Although his architecture has all the vigour of his painting, there is less correspondence between the two fields than might be imagined. He never decorated any of his own churches, and indeed they were not designed with fresco decoration in mind.

Pietro's great contemporary reputation sank in the next century with that of many other Baroque artists. In a famous passage in his *Dizionario delle belle arti* (1797), Francesco Milizia wrote: 'Borromini in architecture, Bernini in sculpture, Pietro da Cortona in painting, and the Cavalier Marini in poetry represent a diseased taste—one that has infected a great number of artists.'

Corvus, Joannes (probably identical with Jehan Raf and John Raven) (active *c.*1512–*c.*1544). A painter, perhaps from Bruges, who worked in England *c.*1520–30, later in France and then perhaps again in England. Two works by him are known: *Bishop Foxe* (Corpus Christi College, Oxford) and *Princess Mary Tudor* (Sudeley Castle, Glos.). They are unexceptional portraits, but the very fact of the survival of the artist's name from this obscure period in English painting has given him a certain interest.

Cosmati work. A type of coloured decorative inlay work of stone and glass that flourished mainly in Rome between *c.*1100 and 1300. It is characterized by the use of small pieces of coloured stone and glass in combination with strips of white marble to produce geometrical designs. The term derives from two craftsmen called Cosmas, whose names are inscribed on several works (e.g. the cloister of Sta Scolastica at Subiaco), but there were several families of 'Cosmati' workers (e.g. those of Vassallettus and Laurentius) and many individual craftsmen. Cosmati work was applied to church furnishings such as tombs and pulpits and was also used for architectural decoration. The style spread as far as England, for example in the tomb of Henry III in Westminster Abbey (*c.*1280), executed by imported Italian craftsmen.

Cossa, Francesco del (*c.*1435–*c.*1477). Italian painter, active mainly in Ferrara, where with Cosimo *Tura and Ercole de' *Roberti he was the leading artist of the period. His style has many affinities with that of Tura and the same background of development from *Mantegna and *Piero della Francesca, but Cossa's work reveals a more genial temperament and relaxed urbanity. This found expression in the delightful frescos of the *Months* in the Palazzo Schifanoia at Ferrara; Cossa, Roberti, and Tura are all thought to have contributed to the scheme, but Cossa seems to have been the leading master. In the early 1470s he moved to Bologna, where he painted an altarpiece for the Griffini Chapel in the church of S. Petronio (1473); the central panel is in the National Gallery, London.

Costa, Lorenzo (*c.*1460–1535). Italian painter. He probably trained in Ferrara and his early work was much influenced by *Tura and Ercole de' *Roberti. By 1483 he had settled in Bologna, where he entered into partnership with *Francia and worked for the ruling Bentivoglio family. In 1504–5 he painted two *Allegories* for Isabella d' *Este (Louvre, Paris) and in 1507 he succeeded *Mantegna as court painter at Mantua. He was the leading artist there until the arrival of *Giulio Romano in 1524, but little of his large-scale work survives. His mature style is often rather sweetly *Peruginesque, with a delicate feeling for landscape, and has been suggested as one of the sources of *Giorgione's work. There are good examples of Costa's work in the National Gallery, London, including *The Concert*, one of the first examples of a type of picture (a close-up of a group of musicians) that was later to have a considerable vogue.

Cosway, Richard (1742–1821). English *miniaturist, a pupil of *Hudson. A friend of the Prince of Wales (later Prince Regent, 1762–1830), Cosway was by far the most fashionable miniature painter of his day, imparting to sitters an air of great elegance. The larger portraits in oils that he occasionally attempted are considered less successful. In 1781 he married **Maria Hadfield** (1759–1838), who was also a miniaturist.

Cotán, Juan Sánchez. See SÁNCHEZ COTÁN.

Cotes, Francis (1726–70). English portrait painter, a pupil of *Knapton. He began as a specialist in *pastel and never altogether gave up the medium, but in the 1760s he returned mainly to oils and became a great fashionable success, the only serious rival to *Gainsborough and *Reynolds. Like them, he was a founder member of the *Royal Academy. His work is charming and vivacious and totally unintellectual; in the words of Sir Ellis *Waterhouse, 'He went all out for health and youth and fine clothes, a strong likeness and no nonsense.' His studio in Cavendish Square in London (and something of his position in the market) was later taken over by *Romney.

Cotman, John Sell (1782–1842). English landscape painter (mainly in watercolour) and etcher.

Son of a well-to-do Norwich merchant, he went to London in 1798 and was employed by Dr *Monro. In 1800, 1801, and 1802 he travelled in Wales and became a member of the circle of artists around the collector Sir George *Beaumont, where he met *Girtin. From 1803 to 1805 he made tours in Yorkshire, where he painted some of his finest work. In 1806 he settled in Norwich, where, together with *Crome, he became the most important representative of the *Norwich School. He made several trips to France, and in 1834 he was appointed Professor of Drawing at King's College, London. Throughout his life Cotman was subject to periods of melancholia and despondency.

Cotman's early watercolours, such as the celebrated *Greta Bridge* (BM, London, *c*.1805; a later version, 1810, is in the Castle Museum, Norwich), include some of the greatest examples of the classic English watercolour technique, showing remarkable boldness and sureness of hand. He used large flat *washes to build up form in clearly defined planes and shapes of almost geometrical simplicity. In his later years, however, he tried to catch the public fancy by large and gaudily melodramatic watercolours, in which he used an *impasto obtained with rice-paste. Cotman also illustrated various books with his etchings, mainly works on architectural antiquities.

Coulte, Maximilian. See COLT.

Counihan, Noel. See ANGRY PENGUINS.

counterproof. See PRINTS.

Courbet, Gustave (1819–77). French painter, born at Ornans, the son of a prosperous farmer. He was a man of independent character and obstinate self-assurance, and claimed to be self-taught. In fact he studied with various minor masters in Ornans, Besançon and in Paris, where he moved in 1839, but he avoided academic instruction and learnt much from copying the work of 17th-cent. *naturalists such as *Caravaggio and *Velázquez. His earliest pictures (including several narcissistic self-portraits) were in the *Romantic tradition, but with three large canvases exhibited at the *Salon of 1850 he established himself as the leader of the *Realist school of painting: these are *The Burial at Ornans* (Musée d'Orsay, Paris), *The Peasants at Flagey* (Musée des Beaux-Arts, Besançon), and *The Stone Breakers* (formerly in Dresden, but destroyed in the Second World War). The huge burial scene in particular made an enormous impact; it was attacked in some quarters for its alleged crudity and deliberate ugliness, but also hailed for its powerful naturalism. Never before had a scene from everyday life been presented in such an epic manner and Courbet was cast in the role of a revolutionary socialist. He gladly accepted this role (although it is unlikely that he painted the picture with political intention) and he became a friend and follower of the anarchist philosopher Pierre-Joseph Proudhon (1809–65) and collaborated on his book *On Art and its Social Significance* (1863). Courbet's boldness and self-confidence are as evident in his technique as in his choice of subjects. He often used a palette knife to apply paint and his work shows an unprecedented relish for the physical substance of his materials.

His unconventionality and hatred of authority was expressed most forcefully in 1855, when, dissatisfied with the representation allotted to him at the Paris Universal Exhibition, he organized a pavilion for his own work, calling it 'Le Réalisme'. Included in the works he showed here was his most celebrated work, *The Painter's Studio* (Musée d'Orsay, Paris, 1854–5). This huge (6-m. wide) canvas was subtitled by Courbet 'a real allegory [a seeming contradiction in terms] summing up seven years in my artistic life'. He wrote a long (but not very clear) account of it, describing it as 'the moral and physical history of my studio' and saying it showed 'all the people who serve my cause, sustain me in my ideal and support my activity'. In it he presents himself as the artist-hero, and in taking as his subject the activity of creating art he sounded a note that reverberated into the 20th cent. Interpretations of the picture have been many and varied, and it has recently been shown that it has covert but carefully thought out political content, attacking Napoleon III (1808–73). After 1855 his work became less doctrinaire. His colours were less sombre and he often chose more obviously attractive subjects—landscapes from the Forest of Fontainebleau, the Jura, or the Mediterranean, seascapes with distant vistas, or comely and sensual nudes.

After the deposition of Napoleon III, Courbet was active in the Paris Commune of 1871 and was appointed head of the arts commission. When the Commune fell he was imprisoned for his role in the destruction of the Vendôme Column, and he fled to Switzerland in 1873, being unable to pay the fine imposed on him (see MEISSONIER). He stayed there for the remaining four years of his life, painting mainly landscapes and portraits.

Courbet did not form a school, but he had an enormous influence on 19th-cent. art because of his resounding rejection of the doctrine of idealization. 'Painting', he said, 'is an art of sight and should therefore concern itself with things seen; it should, therefore, abandon both the historical scenes of the *classical school and poetic subjects from *Goethe and Shakespeare favoured by the Romantic school.' When asked to include angels in a painting for a church he replied: 'I have never seen angels. Show me an angel and I will paint one.' His concentration on the tangible reality of things influenced not only Realist painters, but also, for example, the *Cubists. *Gleizes and *Metzinger began their book *Du Cubisme* (1912) by saying, 'To evaluate the significance of Cubism we must go back to Gustave Courbet', and *Apollinaire stated in *Les peintres cubistes* (1913), '*Cézanne's last paintings and his watercolours appertain to Cubism, but Courbet is the father of the new painters.'

Courtauld, Samuel (1876–1947). British industrialist and art collector. A wealthy director of the family silk firm, he was a pioneer in Britain in the appreciation of French *Impressionist and *Post-Impressionist painters. In 1923 he gave to the *Tate Gallery the sum of £50,000 for the purchase of works by French 19th-cent. painters, who were hardly represented in the gallery. Most of his own superb collection was presented to the University of London in 1932, a year after he had endowed the Courtauld Institute of Art, the first specialist centre in Britain for the study of art history. The co-founders of the Institute were Lord Lee of Fareham (1868–1947), a soldier and politician, who in 1921 had presented his country house—Chequers—to the nation to be the prime minister's country residence, and Sir Robert Witt (1872–1952), a lawyer who formed a library of reproductions of paintings and drawings that is now one of the cornerstones of the Institute's pre-eminence in art-historical studies. Both men left collections to the Courtauld Institute—Lee mainly of paintings, Witt of drawings and watercolours—and there have been several other important bequests, including that of the painter and critic Roger *Fry. The most recent of the major bequests, that of the Anglo-Austrian art historian Count Antoine Seilern (1901–78) in 1978, has raised an already outstanding collection to new heights. Seilern's bequest is varied, reflecting his own scholarly interests, but its chief glory is its superlative group of works by *Rubens. The Courtauld Institute was originally located in Samuel Courtauld's former house at 20 Portman Square, designed by Robert *Adam and possessing one of the best-preserved 18th-cent. domestic interiors in London, while the galleries occupied a building about a mile away next to the *Warburg Institute in Woburn Square. In 1989–90, however, all the Institute's activities and collections were brought together under one roof at Somerset House in the Strand, fulfilling Samuel Courtauld's intention that students should work in intimate contact with original works of art.

Courtois, Jacques (1621–76) and **Guillaume** (1628–79). French painters, brothers, active in Italy and often known by the Italian forms of the names, Giacomo and Guglielmo Cortese. They came from Burgundy and both had the nickname 'Il Borgognone' or 'Le Bourguignon'. Jacques was a prolific painter of battle scenes, fairly close in style to those of Salvator *Rosa, but more colourful. Guillaume was a pupil of Pietro da *Cortona and mainly painted altarpieces. He was also an engraver. Both brothers worked in Rome for much of their careers and they sometimes collaborated. Another brother, a Capuchin priest, Padre **Antonio**, and a sister, **Anna**, were also painters.

Courtois, Marie. See NATTIER.

Cousin, Jean the Elder (*c.*1490–1560/61). French painter, engraver, and designer, active in his native Sens and from about 1538 in Paris. He had a successful career as a painter and a designer of stained glass and tapestries, but very little surviving work can be securely attributed to him. The only certain documented works are three tapestries from a series on the life of St Mammès, which he contracted to design in 1543 (two are in Langres Cathedral, for which they were woven, the other in the Louvre, Paris). The painting *Eva Prima Pandora* (Louvre, Paris), however, can also be confidently given to him as the attribution goes back almost to his lifetime, and two windows in Sens Cathedral are also traditionally attributed to him. In 1560 he published a treatise on *perspective. The career of his son **Jean the Younger** (*c.*1522–*c.*1594), a painter and engraver, follows a similar pattern. He too worked in Sens and Paris and had a great contemporary reputation, but little documented work survives. His most important painting is a *Last Judgement* (Louvre), but he is best known as a book illustrator; his *Livre de Fortune*, a book of *emblem drawings, was published in 1568, and he also illustrated editions of the *Metamorphoses* of Ovid (1570) and the *Fables* of Aesop (1582). The work of both father and son shows strong Italian influence and is remarkable for its independence from the prevailing style of the École de *Fontainebleau.

Coustou, Guillaume I (1677–1746). The best-known member of a dynasty of French sculptors. He was trained by *Coysevox (his mother's brother), and like him worked a good deal for the court. His vigorous style was formed partly on the example of *Bernini, whose work he saw in Rome, where he studied 1695–1703. Guillaume's masterpieces are the celebrated pair of *Horse Tamers* (*The Marly Horses*), originally made for the royal Château at Marly and set up there in 1745, but now in the Place de la Concorde in Paris. **Nicolas** (1658–1733), Guillaume's brother, was also employed in court circles, and his work can be seen at Versailles and in the Tuileries Gardens in Paris. He was probably the teacher of *Roubiliac. **Guillaume II** (1716–77), the son of Guillaume I, inherited his father's technical skill but little of his originality. Nevertheless, he enjoyed a successful career, his most important work being the monument to the Dauphin in Sens Cathedral (*c*.1767). **François** (d. 1690), the father of Guillaume I and Nicolas and the founder of the dynasty, was a minor wood-sculptor working in Lyons.

Couture, Thomas (1815–79). French historical and portrait painter, a pupil of *Gros and *Delaroche. He is chiefly remembered for his vast 'orgy' picture *The Romans of the Decadence* (Musée d'Orsay, Paris), which was the sensation of the *Salon of 1847. As with other 'one-picture painters', his reputation has sunk with that of his big work, which now is often cited as the classic example of the worst type of bombastic academic painting, impeccable in every detail and totally false in overall effect. His more informal works, however, are often much livelier in conception and technique, and as a teacher he encouraged direct study from landscape. *Manet was his best-known pupil, and others included *Puvis de Chavannes and *Fantin-Latour.

Cowie, James (1886–1956). One of the most individual Scottish painters of the 20th cent. Whereas the central tradition of modern Scottish painting has been one of rich colouring and lush, free brushwork (see, for example, SCOTTISH COLOURISTS), Cowie worked in a strong, hard, predominantly linear style—highly disciplined rather than intuitive (he made many preparatory drawings and often worked on a picture for several years). He took his subjects from what he saw around him, but he was also inspired by the Old Masters, often using their compositions as a starting-point, without actually imitating them. Among his contemporaries he was perhaps closest in spirit to John *Nash, an artist he greatly admired. They shared an ability to infuse the ordinary with a sense of the mysterious. Cowie taught at several art schools in Scotland, his pupils including *Colquhoun and *MacBryde and Joan *Eardley.

Cox, David (1783–1859). English landscape painter, mainly in watercolour. In his youth he worked as a scene painter in Birmingham and London, where he received lessons from John *Varley. He lived in Hereford, 1814–27, and in London, 1829–41, before retiring to Harborne, near Birmingham, from where he made annual sketching tours to the Welsh mountains. In spite of a certain anecdotal homeliness, his style was broad and vigorous, and in 1836 he began to paint on a rough Scottish wrapping paper that was particularly suited to it. A similar paper was made commercially and marketed as 'Cox Paper'. He wrote several treatises on landscape painting in watercolour, and in the last two decades of his life also worked quite frequently in oils.

Coypel. Family of French painters of which **Noel** (1628–1707) was the head. He created a successful academic style on the example of *Poussin and *Lebrun, was much employed on the large decorative schemes of Louis XIV, notably at Versailles, and was director of the French Academy in Rome (1672–6) and then director of the Académie Royale in Paris (1695). His son **Antoine** (1661–1722) went to Rome as a child with his father and there is a strong Italian element in his style. This comes out particularly in his most famous work, the ceiling of the Chapel at Versailles (1708), which derived from *Baciccio's ceiling in the Gesù in Rome. This and Coypel's decorations at the Palais Royal in Paris (1702, destroyed) rank as the two most completely *Baroque schemes found in French art of this period. The Versailles ceiling is more successful than much of Coypel's work, which often combines, in the words of Anthony *Blunt, 'the bombast of the Baroque and the pedantry of the *classical style without the virtues of either'. His half-brother **Noel-Nicolas** (1690–1734) painted with much more charm, mainly mythological subjects, but he seems to have had a rather timid personality and did not achieve the worldly success of the other members of the family. Indeed, he was the best painter of the family, but is the least famous. *Chardin was briefly his assistant. Antoine's son **Charles-Antoine** (1694–1752) was a much more forceful character than Noel-Nicolas and had a resoundingly successful career, largely due to his administrative capacity in the various official positions that he held. In 1747 he became director of the Académie Royale and chief painter to the king. He also was an accomplished writer of verse and plays as well as art criticism. As a painter he was versatile and prolific, but the weakest member of the family; his *Supper at Emmaus* (1746) in Saint-Merry, Paris, has been described by Sir Michael Levey as 'pathetically inept'.

Coysevox, Antoine (1640–1720). French sculptor, with *Girardon the most successful of Louis

XIV's reign. His style was more *Baroque than Girardon's and Coysevox overtook his rival in popularity towards the end of the 17th cent. as the king's taste turned away from the *classical. By 1679 Coysevox was working at Versailles, where he made numerous statues for the gardens and did much interior decoration, including a striking *relief of Louis XIV in the Salon de la Guerre. His originality, however, is seen mostly in his portrait busts, which show a naturalism of conception and an animation of expression that look forward to the *Rococo. This is particularly so with his portraits of friends, but even his formal commissions can be remarkably lively. The Wallace Collection, London, has an outstanding example of both his formal and informal portraits: the bronze *Louis XIV* (*c.*1686) and the *terracotta *Charles Lebrun* (1676).

Cozens, Alexander (1717–86). English landscape draughtsman. He was born in Russia, the son of a shipbuilder employed by Peter the Great (1672–1725) (it was rumoured that Peter was his real father) and did not settle in England until 1746. In 1785 or 1786 he published his famous treatise *A New Method of Assisting the Invention in Drawing Original Compositions of Landscape*, in which he explains his method of using accidental blots on the drawing paper to stimulate the imagination by suggesting landscape forms that could be developed into a finished work. Cozens mentions that 'something of the same kind had been mentioned by *Leonardo da Vinci, in his Treatise on Painting' and that reading the passage in question 'tended to confirm my own opinion'. He quotes Leonardo as saying 'If you look upon an old wall covered with dirt, or the odd appearance of some streaked stones, you may discover several things like landscapes, battles, clouds, uncommon attitudes, humorous faces, draperies &c. Out of this confused mass of objects, the mind will be furnished with abundance of designs and subjects perfectly new.' He worked almost exclusively in monochrome, and both his *'blot drawings' and his more formal compositions use intense lights and darks with masterly effect to suggest the power and mystery of nature. Cozens was the first major English artist to devote himself entirely to landscape, and he spent much of his career as a fashionable teacher. He was drawing master at Eton College and to two of George III's sons, and he accompanied his pupil William *Beckford on a visit to Italy in 1762.

His son, **John Robert Cozens** (1752–97), was also a landscape painter. Most of his work derived from two continental journeys in 1776–9 and 1782–3, during which he visited Italy and Switzerland. On the first he was probably draughtsman to Richard Payne *Knight, and on the second he was part of the entourage of William Beckford. In 1793 he became insane and was cared for by Dr *Monro. Although his watercolours were based on sketches made on the spot, he by no means restricted himself to topographical exactitude and he often transposed landscape features in the interests of a more poetical composition. But he does not seem ever to have composed wholly from imagination, as his father did. His narrow but subtly gradated range of subdued colour is intensely evocative of the serene natural effects which appealed so strongly to his poetic melancholy. He was the most talented of the English landscape artists in the *Picturesque tradition and his work was admired and copied by *Turner, *Constable, and *Girtin.

Crabeth, Dirk (d. 1577) and **Wouter** (d. *c.*1590). Netherlandish makers of stained-glass windows, brothers, the most important members of a family of artists active mainly in Gouda. On 1 January 1552 a fire destroyed forty-six of the stained-glass windows in St Jans Church in that city. Dirk was called upon to make nine new windows and Wouter made four. The latter had travelled in France and Italy as well as working in Antwerp and Brussels and his work shows extensive *Renaissance influence. Some of their full-scale drawings for the windows are still extant in Gouda.

Cragg, Tony. See NEW BRITISH SCULPTURE.

Craig, Gordon (Edward Henry Craig) (1872–1966). British theatrical designer and graphic artist, the son of the actress Ellen Terry (1847–1928). Chiefly known as a theatrical designer and producer, Craig was also an artist in watercolour and engraving. He is also remembered as a pioneer of *Luminism and *Kinetic art in view of experiments (*c.*1913–15) with moving scenery and lights made at his theatre school in Florence.

Cranach, Lucas the Elder (1472–1553). German painter. He takes his name from the small town of Kronach in South Germany, where he was born, and very little is known of his life before about 1500–1, when he settled in Vienna and started working in the humanist circles associated with the newly founded university. His stay in Vienna was short (he left in 1504), but in his brief period there he painted some of his finest and most original works. They include portraits, notably those of *Johannes Cuspinian*, a lecturer at the university, and his wife *Anna* (Reinhart Coll., Winterhur), and several religious works in which he shows a remarkable feeling for the beauty of landscape characteristic of the *Danube School. The finest example of this manner is perhaps the *Rest on the Flight into Egypt* (Staatliche Museen, Berlin), which shows the Holy Family resting in the glade of a German pine forest. It was painted in 1504, just before Cranach went to Wittenberg as court painter to Frederick III (the Wise, 1463–1525), Elector of Saxony.

Cranach remained in Wittenberg until 1550, when he followed John Frederick (the Unfortunate, 1503–54) into exile, in Augsburg. During his time in Wittenberg he became extremely wealthy and one of the city's most respected citizens, serving as burgomaster for several years. His paintings were eagerly sought by collectors, and his busy studio often produced numerous replicas of popular designs, particularly those in which he showed his skill at depicting female beauty—more than ten versions are known of his *Reclining Nymph*. He excelled at erotic nudes, which sometimes draw on Italian *Renaissance models but are totally different in spirit, and he also had a penchant for pictures of coquettish women wearing large hats, sometimes shown as Judith or the goddesses in the *Judgement of Paris*. The most innovative works of his Wittenberg period, however, are probably his full-length portraits (*The Duke* and *Duchess of Saxony*, Gemäldegalerie, Dresden, 1514). Cranach continued with his religious work, but his woodcut designs (notably those for the first German edition of the New Testament in 1522) are generally more interesting than his paintings in this sphere. He also painted several portraits of Martin Luther (1483–1546). Despite his allegiance to the Protestant cause, he continued to work for Catholic patrons and was a very astute businessman. During the last years of his life Cranach was assisted by his son, **Lucas the Younger** (1515–86), who carried on the tradition of the workshop and imitated his father's style so successfully that it is often difficult to distinguish between their hands.

Cranach Press. A private printing press set up by Count Harry Kessler at Weimar, Germany, in 1913. He had the help of Edward Johnston and Emery Walker of the *Doves Press in designing the type and the assistance of that Press's compositors. Production was interrupted by the First World War but was resumed in the 1920s. The outstanding works of the Cranach Press were the *Eclogues* of Virgil (German edition in 1926, English edition in 1927), for which *Maillol made the woodcuts; and *Hamlet*, printed in a black letter typeface (i.e. one based on medieval 'Gothic' script), designed by Johnston and with woodcuts by Gordon *Craig. In 1931 the Press produced an edition of *Canticum canticorum* with wood engravings by Eric *Gill.

Crane, Walter (1845–1915). English graphic artist and designer. Best remembered today as an illustrator of children's books, he was also a painter and designer; he was associated with *Burne-Jones and William *Morris in their work for the reform of decorative and *applied art, designing wallpapers, printed fabrics, and stained glass, and also making illustrations for Morris's *Kelmscott Press. His work was exhibited in the USA as well as in Europe and he achieved some prestige as a writer and educationalist. He was influential in the *Arts and Crafts Movement, being a founder-member of the Art Workers' Guild as well as of the Arts and Crafts Exhibition Society, and influential also in the spread of *Art Nouveau. He held several teaching posts and was Principal of the *Royal College of Art for one year in 1898, when he introduced Art Workers' Guild members and methods to the College.

craquelure. The network of small cracks which appears on a painting when in the course of time the *pigment or *varnish has become brittle.

Crawford, Thomas (1813–57). American sculptor. He settled in Rome in 1835, studied with *Thorvaldsen and became the most thoroughgoing *Neoclassicist among American sculptors of his generation. Although he stayed in Rome, he attained an extraordinary reputation in America and received numerous prestigious public commissions, among them the equestrian *George Washington* (1857) in Richmond, Virginia, and the *Armed Liberty* on top of the Capitol dome at Washington, which was set in place in 1863 after his early death.

crayon. See CHALK.

crayon manner. An 18th-cent. engraving technique used in conjunction with *stipple engraving for the reproduction of crayon (or *pastel) drawings. It was a variant of *etching and it is sometimes difficult to distinguish from *soft-ground etching. Invented *c.*1750 in France, where a number of engravers made prints after *Boucher, *Fragonard, and others, the crayon manner was also widely used in England. It was rendered obsolete early in the 19th cent. by the invention of *lithography. See also PASTEL MANNER.

Credi, Lorenzo di (*c.*1458–1537). Florentine painter. He was a fellow pupil of *Leonardo in *Verrocchio's workshop and he seems to have stayed there until Verrocchio's death in 1488, managing the painting side of his master's varied business. He was a very fine craftsman, but his style was lacking in individuality. His early work is in an extremely prosaic version of Leonardo's youthful style. He later absorbed some of the ideas of the High *Renaissance, his *Madonna and Saints* in Sta Maria delle Grazie in Pistoia (1510) recalling Fra *Bartolommeo. He had several pupils and seems to have had a fairly successful career with his solid, unspectacular skills. It is said that influenced by the teachings of Savonarola in 1497 he destroyed all his pictures with profane subjects.

Crespi, Daniele (*c.*1598–1630). Milanese painter. Although he died young of the plague, his output was large and his work is considered to be one of the most typical expressions of the zealous spirit of the Counter-Reformation that affected Milan at this time. *St Charles Borromeo at Supper* in Sta Maria della Passione is his best-known work and in its simple composition and emotional directness reflects the ideals of painting advocated by the Council of Trent. Many other examples of his work are in the church. He was probably a relative of Giovanni Battista *Crespi, whose work influenced him.

Crespi, Giovanni Battista (Il Cerano) (*c.*1575–1632). Italian painter, sculptor, engraver, architect, and writer. His nickname derived from his birthplace near Novaro, but he was active mainly in Milan, where he was one of the leading artists of his time. During the 1590s he was in Rome, where he was befriended by Cardinal Federico Borromeo (nephew of St Charles Borromeo), who became his major patron after they returned to Milan together. Borromeo appointed him head of the painting section of the Accademia Ambrosiana, which he founded in 1620, and in 1629 put him in charge of the sculptural decoration of Milan Cathedral. Crespi's paintings, often mystical in feeling, are complex stylistically; there is a strong *Mannerist current in his colouring and in the elegant posturing of his figures, but his work also shows a solidity and a feeling for realistic details that give it a place in the vanguard of the *Baroque.

Crespi, Giuseppe Maria (called Lo Spagnuolo) (1665–1747), Bolognese painter. He reacted against the academic tradition in which he was trained (*Cignani was one of his teachers) and specialized in *genre subjects, with violent *chiaroscuro effects of brilliant colour against dark backgrounds. They are in the tradition of the everyday-life paintings of the *Carracci, but go far beyond them in their sense of unvarnished reality (*The Hamlet*, Pinacoteca, Bologna). He was an outstanding teacher, numbering *Piazzetta and Pietro *Longhi among his pupils, and he exercised a great influence on Venetian 18th-cent. painting. Rudolf *Wittkower called him 'the only real genius of the late Bolognese school'.

Critius. Greek sculptor, active at Athens in the early 5th cent. BC. With another sculptor called Nesiotes he was commissioned to make the bronze *Tyrannicides*, erected in 477 BC, to replace the group by *Antenor, which had been taken as booty after the Persian sack of Athens in 480. The group by Critius and Nesiotes is lost, but a copy survives in the Archaeological Museum, Naples. Other works have been associated with Critius on the basis of this copy, including a marble **kouros* known as 'The Critius boy' in the Acropolis Museum, Athens.

Critz, John de (before 1552–1642). British painter, the son of an Antwerp goldsmith who settled in London to escape religious persecution. He held the post of Serjeant-Painter from 1603 until his death. No works certainly by him survive, but a number of portraits have been given to him on circumstantial evidence, including one of James I (1610) in the National Maritime Museum at Greenwich. A group of portraits mainly of the Tradescant family in the Ashmolean Museum, Oxford (*c.*1640–50) are associated with **Emmanuel de Critz** (before 1609–55), son of the foregoing. They have a weighty gravity combined with a certain eccentric melancholy that puts them among the most remarkable English paintings of this date. Nothing is known of the work of another painter son, **John de Critz the Younger**.

Crivelli, Carlo (active 1457–93). Italian painter. He was born in Venice and always signed himself as a Venetian, but he spent most of his career working in the Marches, particularly at Ascoli Piceno, and he also lived for some time at Zara in Dalmatia (now in Croatia). In 1490 he was knighted by Ferdinand II (1452–1516) of Naples. His paintings are all of religious subjects, done in an elaborate, old-fashioned style that owes much to the wiry Paduan tradition of *Squarcione and *Mantegna and yet is highly distinctive. Their dense ornamentation is often increased by the use of *gesso decoration combined with the paint. The finest collection of his works is in the National Gallery in London and includes the delightful and much reproduced *Annunciation* (1486). **Vittore Crivelli** (d. 1501/2), probably Carlo's brother, was a faithful but pedestrian follower.

Croce, Benedetto (1866–1952). Italian philosopher, historian, and critic. Croce was regarded as the foremost Italian philosopher of his time, and his views about the nature of art are set out most fully in his *Aesthetic* (1902, English translation, 1909). In this work he regards all art as a form of imaging—a conjuring into being of images of particulars—a process which he calls 'intuition' and identifies with 'expression'. He considers that good art is successful expression of emotion. But Croce uses 'expression' in a special sense, as a synonym for 'intuition', and not in the usual sense which involves some form of external manifestation. His theories have been criticized partly on the ground of inherent confusion of concepts, more generally on the ground that his identification of the work of art with the mental process of intuition / expression does less than justice to the concrete work of art (as understood in ordinary language), the importance of embodiment in a material medium, and the problems, theoretical and practical, which derive from the expression of an idea or intuition in a physical, and therefore communicable, form.

Croce founded and edited the successful journal *La Critica* (1903–44), which in 1944 became *Quaderni della Critica*. He was Minister of Education before Benito Mussolini came to power in 1922 and again after the Second World War, and his staunch opposition to Fascism gave him the status of a moral teacher.

Crome, John (1768–1821). English landscape painter and etcher, with *Cotman the major artist of the *Norwich School. He was born, worked, and died in Norwich and his only journey abroad was to Paris in 1814 to see the exhibition of the pictures which had been seized by Napoleon. Of humble origin, he was first apprenticed to a coach-and-sign painter and taught himself principally by copying works in the collection of Thomas Harvey of Catton, a collector and amateur painter who befriended him. The Dutch masters *Ruisdael and *Hobbema were particularly influential on his work and he also admired *Gainsborough and *Wilson. He exhibited intermittently at the *Royal Academy and the British Institution as well as with the Norwich Society of Artists, of which he was a founder member in 1803, but he earned a major part of his living as a drawing master. Together with Wilson, Crome represents the transition from the 18th-cent. *Picturesque tradition to the *Romantic conception of landscape. His larger compositions lack the architectural quality of *classical construction and have been accused of woolliness in realization; but they are often saved by a unity of mood which anticipates the Romantics. His finest paintings, such as *The Poringland Oak* (Tate, London), are marked by a broad handling of the paint, a bold realization of space, and keen appreciation of local characteristics.

As an etcher Crome's accomplishment was modest, and during his life he never published his plates: they were published by his widow and his eldest son **John Bernay Crome** (1794–1842), 13 years after his death. The elder Crome is sometimes referred to as 'Old Crome' to distinguish him from his son, who painted in his manner but with inferior talent.

Croquis, Alfred. See MACLISE, D.

Crowe, Sir Joseph Archer (1825–96). English journalist, diplomat, and art historian. He had a distinguished career as a commercial attaché in Berlin, Paris, and Vienna and was also a war correspondent in the Crimea and elsewhere, but he is best known for his writings on art history done in collaboration with the Italian painter Giovanni Battista Cavalcaselle (1819–97). They met by chance in 1847 and became firm friends when Cavalcaselle was later a political refugee in London; for a time they lived in the same house. Every detail of their books was discussed between them, but Crowe did all the actual writing because Cavalcaselle's English was inadequate. Their prodigious output included *The Early Flemish Painters* (1857), *A New History of Painting in Italy* (3 vols., 1864–8), *A History of Painting in North Italy* (2 vols., 1871), *Titian: His Life and Times* (2 vols., 1877) and *Raphael: His Life and Works* (2 vols., 1882). These works, all of which have appeared in subsequent editions, either in English or translation, set new standards of methodical research, bringing to light masses of new information, and they are still considered valuable.

Cruikshank, George (1792–1878). English painter, illustrator, and *caricaturist. The son of a caricaturist, Isaac Cruikshank (1756?–1811?), he was highly precocious and quickly established himself in succession to *Gillray as the most eminent political cartoonist of his day. The private life of the Prince Regent (later George IV) was one of his first targets. He began to turn to book illustration in the 1820s and his output was immense. The drawings for Grimm's *German Popular Stories* (1823), and later for the works of Harrison Ainsworth (1805–82) and Charles Dickens (1812–70), are amongst his best-known work in this field. In later life he took up the cause of temperance, producing moral narratives in *woodcut (*The Bottle*, 1847; *The Drunkard's Children*, 1848) and a vast painting *The Worship of Bacchus* (Tate, London, 1860–2). His brother, **Isaac Robert Cruikshank** (1789–1856), was a *miniaturist and caricaturist.

Cruz-Diez, Carlos (1923–). Venezuelan painter and *Kinetic artist, active mainly in Paris since 1960. He became interested in optical phenomena in the 1950s, and in the course of experimenting with the *primary colours in arrangements of thin intersecting bands he found that he could create the illusion of a third or fourth non-existent colour. This led him to further experiments in series entitled *Chromatic Inductions, Chromointerference, Additive* and *Physichromy*. In the last, which he began in 1959, he created shifting geometric images that emerge, intensify, change, and dematerialize as the viewer moves from one side of the work to the other. He achieved this effect by means of narrow vertical strips of metal, or plastic bands appended to the flat surface, and vertical painted colour lines.

Cubism. Movement in painting and sculpture, recognized as one of the great turning points in western art. It was originated by *Picasso and *Braque. They worked so closely during this period—'roped together like mountaineers' in Braque's memorable phrase—that at times it is difficult to differentiate their hands. The movement was broadened by Juan *Gris, and later joined by many other artists, including *Léger, *La Fresnaye, *Delaunay, *Metzinger, *Gleizes, and *Kupka. Its main formative period was from *c.*1907 to 1914, though some of the methods and discoveries of the Cubists have remained a lasting accession to the repertory of many different schools of 20th-cent. art. According to Daniel-Henry *Kahnweiler (the Paris dealer who supported the beginnings of Cubism) in his book *Juan Gris* (Eng. trans. by Douglas Cooper, 1947, p. 69, n. 2), the name originated with the critic Louis Vauxcelles (following a *mot* by *Matisse), who, in a review of the Braque exhibition in the paper *Gil Blas*, 14 November 1908, spoke of '*cubes*' and later of '*bizarreries cubiques*'. As with *Impressionism and *Fauvism the name originated in a jibe.

Cubism made a radical departure from the idea of art as the imitation of nature that had dominated European painting and sculpture since the *Renaissance. Picasso and Braque abandoned traditional notions of perspective, foreshortening, and modelling, and aimed to represent solidity and volume in a two-dimensional plane without converting the two-dimensional canvas illusionistically into a three-dimensional picture-space. In so far as they represented real objects, their aim was to depict them as they are known and not as they partially appear at a particular moment and place. For this purpose many different aspects of the object might be depicted simultaneously; the forms of the object were analysed into geometrical planes and these were recomposed from various simultaneous points of view into a combination of forms. To this extent Cubism was and claimed to be realistic, but it was a conceptual realism rather than an optical and *Impressionistic realism. Cubism is the outcome of intellectualized rather than spontaneous vision.

The two most important positive influences on the emergence of Cubism were African sculpture and the later paintings of *Cézanne, and the harbinger of the new style was Picasso's celebrated picture *Les Demoiselles d'Avignon* (MOMA, New York, 1907), with its angular and fractured forms. It is customary to divide the Cubism of Picasso and Braque into two phases—'Analytical' and 'Synthetic'. In the first and more austere phase, which lasted until 1912, forms were analysed into predominantly geometrical structures and colour was extremely subdued—usually virtually monochromatic—so as not to be a distraction. In the second phase colour became much stronger and shapes more decorative, and elements such as stencilled lettering and pieces of newspaper were introduced into paintings (see COLLAGE; PAPIER COLLÉ). Gris was as important as Braque or Picasso in this phase, and brought a systematic and logical approach in place of their more intuitive methods. Fernand Léger is often reckoned the fourth great painter of the movement. This incorporation of collage further emphasized the flatness of the picture surface. The First World War brought an end to the collaboration of Braque and Picasso, but their work had a rich progeny. Cubism, as well as being one of the principal sources for abstract art, was infinitely adaptable, giving birth to numerous other movements, among them *Futurism, *Orphism, *Purism, and *Vorticism, and to personal reinterpretations such as that of Stuart *Davis. Because it was concerned with depicting ideas rather than observed reality it has been one of the foundations of 20th-cent. aesthetic attitudes. Most of the prominent Cubist painters were also distinguished as graphic artists and book illustrators, and major sculptors who worked in a Cubist idiom included *Archipenko, *Lipchitz, and *Zadkine (as well as Picasso himself).

Cullen, Maurice (1866–1934). Canadian painter, whose work was influential in introducing *Impressionism to his country. From 1889 to 1895 he worked in Paris and elsewhere in France, with trips to Venice and North Africa and he made two shorter trips to Europe before settling for good in Canada in 1902. His subjects included city scenes (*Old Houses, Montreal*, Montreal Mus. of Fine Arts, *c.*1900) and landscapes on the St Lawrence, in the Laurentian hills, at St John's, and in the Rocky Mountains. After about 1920 he lived in virtual seclusion in a cabin he built himself at Lac Tremblant in the Laurentian hills. His friend J. W. *Morrice said of his work: 'he gets at the guts of things.'

Cumberland Market Group. Group of painters formed when the *Camden Town Group merged with the *London Group in 1913; it was named after 49 Cumberland Market, where Robert *Bevan had his studio, and where he, *Gilman, and *Ginner used to meet. In 1915 they were joined by John *Nash and later by McKnight *Kauffer and C. R. W. *Nevinson.

Cure, Cornelius (d. 1607). English stonemason and sculptor, the best-known member of a family from the Low Countries who set up a yard in Southwark. His father **William I** (d. 1579) had settled in England in about 1540. Cornelius was Master Mason to Elizabeth I (1533–1603) and James I (1566–1625) and is best known for his tomb of Mary Queen of Scots in Westminster Abbey (1607–12), although most of the work, including the very fine effigy, was presumably done after his death by his son **William** (d. 1632). He succeeded his father as Master Mason and held the post until his death, but he was described as 'careless and negligent' of his duties and in 1619 was replaced as mason to Inigo *Jones's Banqueting House by Nicholas *Stone, who eventually succeeded him as Master Mason in 1632.

Currier and Ives prints. Popular *lithographs published in New York by Nathaniel Currier (1813–88) and James M. Ives (1824–95), who went into partnership in 1857. These lithographs, advertised by their publishers as 'Coloured Engravings for the People', represented almost every aspect of contemporary America, including sporting, sentimental, patriotic, and political subjects, together with portraits, landscapes, disasters, scenes of city life, of railroads, of Mississippi steamboats, and so forth. A number of artists, most of whom specialized in particular subjects, were retained by the firm to draw the lithographs in black and white; afterwards the prints were coloured by hand on a production-line system (one assistant to each colour) and sold cheaply to the public by agents, print-sellers, and pedlars. The business was carried on until 1907 by the sons of the founders.

Currier and Ives prints vary a good deal in artistic quality, but by and large they have all the virtues of good popular art, being unpretentious, vigorously descriptive, colourful, and romantic in feeling. Like Japanese prints (see UKIYO-E) they were considered of little value in their time, but many of them, after becoming rare, are now collectors' pieces.

Curry, John Steuart (1897–1946). American painter. He was born on a farm in Kansas, and never forgot his Midwestern roots. From 1919 to 1926 he worked as an illustrator for pulp magazines, then spent a year in Europe, before settling in New York, where he was encouraged and supported by Gertrude Vanderbilt *Whitney. He believed that art should grow out of everyday life and be motivated by affection, and his subjects were taken from the Midwest he loved. Two of his most famous works are *Baptism in Kansas* (Whitney Museum, New York, 1928) and *Hogs Killing a Rattlesnake* (Art Institute of Chicago, 1930); they show his anecdotal, rather melodramatic style (he often depicted the violence of nature)—sometimes weak in draughtsmanship, but always vigorous and sincere. In the 1930s Curry was recognized—along with *Benton and *Wood—as one of the leading exponents of *Regionalism, and he was given commissions for several large murals; the best known—generally regarded as his masterpieces—are in the state capitol in Topeka, Kansas (1938–40).

Cuvilliés, François (1695–1768). French architect and engraver of ornaments. He was Flemish by birth and worked mainly in Germany (initially as court dwarf to the Elector Max Emanuel of Bavaria), but he maintained close links with Paris throughout his career. His buildings, notably the Amalienburg hunting lodge at Nymphenburg Park near Munich (1734–9), rank among the finest works of Central European *Rococo, and from 1738 he published numerous engravings of his designs that helped to popularize the *Louis XV style of furniture and decoration in Germany.

Cuyp. The name of a family of Dutch painters of Dordrecht, of which three members gained distinction. **Jacob Gerritsz. Cuyp** (1594–1651/2) was the son of a glass painter and a pupil of Abraham *Bloemaert at Utrecht. He is thought of today mainly as a portrait painter—his portraits of children are particularly fine—but in old biographies is lauded principally for his views of the countryside around Dordrecht. **Benjamin Gerritsz. Cuyp** (1612–52) was the half-brother of Jacob. He is noted principally for paintings of biblical and *genre scenes which use *Rembrandtesque light and shadow effects.

Aelbert Cuyp (1620–91) is the most famous member of the family and now one of the most celebrated of all landscape painters, although he also painted many other subjects. He was the son and probably the pupil of Jacob Gerritsz. Cuyp. His early works also show the influence of Jan van *Goyen. Aelbert was born and died at Dordrecht, but he seems to have travelled along Holland's great rivers to the eastern part of the Netherlands, and he also painted views of Westphalia. A prodigious number of pictures are ascribed to him, but his *œuvre* poses many problems. He often signed his paintings but rarely dated them, and a satisfactory chronology has never been established. Although he had little influence outside Dordrecht, Cuyp had several imitators there, and some of the paintings formerly attributed to him are now given to Abraham Calraet (1642–1722), who signed himself 'AC' (the same initials as Cuyp). In 1658 Cuyp married a rich widow, and in the 1660s he seems to have virtually abandoned painting. He was almost forgotten for two generations after his death. Late 18th-cent. English collectors are credited with rediscovering his merits, and he is still much better represented in English collections, public and private, than in Dutch museums. His finest works—typically river scenes and landscapes with placid, dignified-looking cows—show great serenity and masterly handling of glowing light (usually Cuyp favoured the effects of the early morning or evening sun). He approaches *Claude more closely in spirit than any of his countrymen who travelled to Italy.

Cycladic. Name applied to the Bronze Age art and civilization of the Cyclades (the Greek islands of the central Aegean), flourishing from about 2500 BC to about 1400 BC, when the islands were overrun by invaders from the mainland and became assimilated into *Mycenaean culture. Surviving Cycladic art consists mainly of various types of decorated pottery and of white marble figures. The latter are often female fertility figures of a distinctive type, in which the forms of body and facial features are pared down to a radically elegant simplicity that has greatly appealed to 20th-cent. taste. Because of the extensive maritime activities of the natives of the islands, Cycladic art was widely disseminated throughout the Mediterranean.

Dada. A movement in European art (with manifestations also in New York), *c.*1915–*c.*1922, characterized by a spirit of anarchic revolt against traditional values. It arose from a mood of disillusionment engendered by the First World War, to which some artists reacted with irony, cynicism, and nihilism. According to the most frequently cited of several accounts of how the name (French for 'hobby-horse') originated, it was chosen by inserting a penknife at random in the pages of a dictionary, thus symbolizing the anti-rational stance of the movement. Those involved in it emphasized the illogical and the absurd, and exaggerated the role of chance in artistic creation. They went to extremes in the use of buffoonery and provocative behaviour to shock and disrupt public complacency (for an example see ERNST). Dada did not involve a specific artistic style or aesthetic. The methods and manifestos—particularly the techniques of outrage and provocation—owed much to *Futurism, but the movement lacked the militant optimism of Futurism. In painting, the *Cubist techniques of *collage and *montage were adopted, but the archetypal Dada forms of expression were perhaps the nonsense poem and the *ready-made.

European Dada was founded in 1915 in Zurich in neutral Switzerland by a group of artists and writers including Hans *Arp, the German painter, sculptor, and film-maker Hans Richter (1888–1976), and the Romanian poet Tristan Tzara (1896–1963). By the end of the war Dada was spreading to Germany, and there were significant Dada activities in three German cities: Berlin, Cologne, and Hanover. In Berlin the movement had a strong political dimension, expressed particularly through the brilliant *photomontages of Raoul Hausmann (1886–1971) and John *Heartfeld and through the biting social satire of *Dix and *Grosz; eventually it gave way to *Neue Sachlichkeit. In Cologne a brief Dada movement (1919–20) was centred on Max Ernst, who made witty and provocative use of collage, and on Arp, who moved there from Zurich when the war ended. In Hanover Kurt *Schwitters was the only important Dada exponent but one of the most dedicated of all.

Dada in New York arose independently of the European movement and virtually simultaneously. It was mainly confined to the activities of Marcel *Duchamp, *Man Ray, and Francis *Picabia; their work tends to be more whimsical and less violent than that of their counterparts in Europe. Duchamp was the most influential of all exponents of Dada and Picabia was the most vigorous in promoting its ideas, forming a link between the European and American movements. He founded his Dada periodical *391* in Barcelona and he introduced the movement to Paris in 1919. In Paris the movement was mainly literary in its emphasis and its tendency towards the fanciful and the absurd formed the basis for *Surrealism, which was officially launched there in 1924.

Although it was fairly short lived and confined to a few main centres, Dada was highly influential in its questioning and debunking of traditional concepts and methods, setting the agenda for much subsequent artistic experiment. Its techniques involving accident and chance were of great importance to the Surrealists and were also later exploited by the *Abstract Expressionists. *Conceptual art, too, has its roots in Dada. The spirit of the Dadaists, in fact, has never completely disappeared, and its tradition has been sustained in, for example, *Junk sculpture and *Pop art, which in the USA was sometimes known as *neo-Dada.

Dadd, Richard (1817–86). English painter who murdered his father in 1843 and spent the rest of his life in Bedlam and Broadmoor asylums. Before his mental breakdown he was considered a promising young artist (his friend *Frith called him 'a man of genius that would assuredly have placed him high in the first rank of painters') and he continued painting after his incarceration. Although most of his work before the murder had been fairly conventional, he had begun to paint fairy and fantasy subjects and in the asylums he developed these along highly imaginative lines; *The Fairy Feller's Master-Stroke* (Tate, London, 1855–64) is probably the best known. Dadd was long forgotten, but became popular in the 1970s, when a major exhibition was devoted to him at the Tate (1974) and several books on him appeared.

Daddi, Bernardo (d. 1348). Florentine painter. Daddi was the younger contemporary of *Giotto (who was possibly his teacher) and the outstanding painter in Florence after the latter's death. Daddi ran a busy workshop specializing in small devotional panels and portable altarpieces. His signed and dated works include a *polyptych of *The Crucifixion with Eight Saints* (Courtauld Inst., London, 1348) and the works attributed to him include frescos of the *Martyrdoms of SS. Lawrence and Stephen* in Sta Croce.

Daddi's style is a sweetened version of Giotto's, tempering the latter's gravity with Sienese grace and lightness. He favoured smiling Madonnas, teasing children, and an abundance of flowers and trailing draperies. His lyrical manner was extremely popular and his influence endured into the second half of the century.

daguerreotype. The first practicable photographic process, in which the image was produced on a silvered copper plate sensitized by iodine. Each image was unique, as it was made directly on to the plate without an intervening 'negative'. The process was invented by a French artist, Louis-Jacques-Mandé Daguerre (1789–1851), and made public in 1839 only a few weeks before Fox Talbot announced the invention of the *calotype. Daguerre was also the inventor of the *diorama.

Dahl, Johan Christian (1788–1857). Norwegian painter, often called the discoverer of the Norwegian landscape. From 1824 until his death he was a professor at the Academy of Dresden, where he was a friend of C. D. *Friedrich. The landscapes of *Ruisdael were another influence on his *Romantic outlook. Through his deep feeling for the grandeur of the landscape of his native country he was a pioneer of the new spirit of nationalism that characterized much Norwegian art in the 19th cent.

Dahl, Michael (1659?–1743). Swedish portrait painter, active mainly in England. He first came to England in 1682 and settled permanently in London in 1689, becoming *Kneller's principal rival. His work has not the brilliance and dash of Kneller's, but at his best he surpasses him in sincerity and humanity beneath a somewhat *Rococo artificiality. There are several works by him or from his busy studio in the National Portrait Gallery in London, including a *Self-portrait* (1691), in which the personal colour and carefully constructed head contrast with the affected artificiality of the pose.

Dalí, Salvador (1904–89). Spanish painter, sculptor, graphic artist, and designer. After passing through phases of *Cubism, *Futurism, and *Metaphysical painting, he turned to *Surrealism and in 1929 moved to Paris. His talent for self-publicity rapidly made him the most famous representative of the movement. Throughout his life he cultivated eccentricity and exaggerated his disposition towards megalomaniac exhibitionism (one of his most famous acts was appearing in a diving suit at the opening of the London Surrealist exhibition in 1936), claiming that this was the source of his creative energy. He took over the Surrealist theory of *automatism but transformed it into a more positive method which he named 'critical paranoia'. According to this theory one should cultivate genuine delusion as in clinical paranoia while remaining residually aware at the back of one's mind that the control of the reason and will has been deliberately suspended. He claimed that this method should be used not only in artistic and poetical creation but also in the affairs of daily life. His paintings of the 1930s, which include several of the established classics of Surrealism, employed a meticulous academic technique that was contradicted by the unreal 'dream' space he depicted and by the strangely hallucinatory character of his imagery. He described such pictures as 'hand-painted dream photographs' and had certain favourite and recurring images, such as the human figure with half-open drawers protruding from it, burning giraffes and watches bent and flowing as if made of melting wax (*The Persistence of Memory*, MOMA, New York, 1931).

In the late 1930s Dalí made several visits to Italy and adopted a more traditional style; this together with his political views (he was a supporter of General Franco) led *Breton to expel him from the Surrealist ranks. He moved to the USA in 1940 and remained there until 1948. During this time he devoted himself largely to self-publicity and making money (Breton coined the near anagram for his name 'Avida Dollars'). From 1948 he lived mainly at Port Lligat in Spain, but he also spent much time in Paris and New York. Among his late paintings the best known are probably those on religious themes (*The Crucifixion of St John of the Cross*, St Mungo Mus., Glasgow, 1951), although sexual subjects and pictures centring on his wife Gala were also continuing preoccupations. In old age he became one of the world's most famous recluses, generating rumours and occasional scandals to the end.

Apart from painting, Dalí's output included sculpture, book illustration, jewellery design, and work for the theatre. In collaboration with the director Luis Buñuel (1900–83) he also made the first Surrealist films—*Un chien andalou* (1929) and *L'Age d'or* (1930)—and he contributed a dream sequence to Alfred Hitchcock's *Spellbound* (1945). He also wrote a novel, *Hidden Faces* (1944), and several volumes of flamboyant autobiography. Although he is undoubtedly one of the most famous artists of the 20th cent., his status is controversial; many critics consider that he did little if anything of consequence after his classic Surrealist works of the 1930s. There is a museum devoted to Dalí's work in Figueras, his birthplace in Spain, and two in the USA, at Cleveland, Ohio, and St Petersburg, Florida.

Dalmau, Luis (active 1428–60). Spanish painter, active mainly in Valencia. He was court painter to Alfonso V (1396–1458) of Aragon, under whose patronage he went to Bruges in 1431 to study tapestry weaving. Dalmau's visit to Flanders is the first recorded contact of a Spanish painter with the Netherlandish School, and the only surviving painting certainly by him, the *Virgin of the Councillors* (Barcelona Mus., 1445), is clearly inspired by Jan van *Eyck and is the earliest documented work in the Hispano-Flemish style that eventually came to dominate 16th-cent. Spanish painting.

Dalou, Aimé-Jules (1838–1902). French sculptor, the most important pupil of *Carpeaux. Although his name is particularly associated with the *naturalistic movement in French sculpture, he produced many works of *Baroque inspiration, notably his largest completed monument, the allegorical *Triumph of the Republic* (Place de la Nation, Paris, 1879–99). His most ambitious work, a vast *Monument to Labour*, was left uncompleted at his death. It was to have included figures representing all the labours of the fields and factories. Clay models for many of these figures, reminiscent of *Millet's peasants in their rather sentimental view of human toil, are preserved in the Petit Palais, Paris. Dalou's other work included the memorial to *Delacroix in the Luxembourg Gardens (1890). He spent the years 1871–9 as a political exile in England after taking part in the Paris Commune (1871).

Dalwood, Hubert (1924–76). British sculptor. In the mid-1950s Dalwood was doing massive female figures with striated bodies on tapering legs, figures with an austere but powerful beauty. From these he passed to sensitively modelled abstract images and *reliefs. At the end of the 1960s he was making constructed abstracts which give the impression of architectural elements in a landscape setting. Dalwood was one of the most refined of the younger artists who, in the 1960s, were abandoning the exclusivity implied in the phrase 'the autonomy of art' and endeavouring to bring art back into direct contact with life and experience.

Dalziel Brothers. Firm of English wood engravers founded in London in 1839 by **George Dalziel** (1815–1902) and **Edward Dalziel** (1817–1905). Two younger brothers, **John** and **Thomas**, worked for the firm, which was the most prolific source of book illustrations in Victorian England, producing more than 50,000 plates. George and Edward collaborated on the book *The Brothers Dalziel: A Record of Work, 1840–90* (1901).

dammar. A generic name given to various resins obtained from certain species of trees growing mainly in South-East Asia. It is used for varnishes, lacquers, and as a base for painting media. As a varnish it becomes transparent and does not turn yellow, but it is very soft and friable, and for this reason is sometimes mixed with amber varnish.

Damophon. Greek sculptor from Messene, active in the early 2nd cent. BC. A fair amount is known of his work from literary descriptions (he repaired *Phidias' celebrated statue of *Zeus* at Olympia), but the only known surviving works by him are three colossal marble heads from a monumental group at Lycosura, now in the National Museum, Athens.

Danby, Francis (1793–1861). Irish painter. He worked mainly in Bristol and London, but between 1829 and 1841, owing to financial and marital problems, he settled in Switzerland. He is remembered mainly for his bombastic apocalyptic paintings, such as *The Delivery of Israel out of Egypt* (Harris Mus. and Art Gal., Preston, 1825), which were a direct challenge to John *Martin. However, his best works are now usually considered to be the romantic sunset landscapes of his later years, with their mood of melancholy and solemn serenity (*Temple of Flora*, Tate, London, 1840).

Dance, Nathaniel (1735–1811), English painter, primarily of portraits. He studied under *Hayman and spent the years 1755–64 in Rome, where he was much influenced by the sophisticated portrait style of Pompeo *Batoni. In 1768 he became a foundation member of the *Royal Academy, but on inheriting a fortune in 1776 he retired from professional practice. He later became an MP and was created a baronet with the surname Dance-Holland. One of his best-known portraits is *Captain Cook* (Nat. Maritime Mus., Greenwich, 1766). He was the son and brother of architects, both called **George Dance**. His father designed the Mansion House in London, and his brother's work included Newgate Prison and the façade of the Guildhall.

Dandridge, Bartholomew (1691–*c.*1754). English portrait painter who had a considerable practice in London in the 1730s and 1740s (he took over *Kneller's studio in 1731). His best work is free, stylish, and lively, and in such groups as *The Price Family* (Met. Mus., New York) he contributed to the development of the *conversation piece.

Daniele da Volterra (Daniele Ricciarelli) (*c.*1509–66). Italian *Mannerist painter and sculptor, born in Volterra, where he was a pupil of *Sodoma. In about 1536 he moved to Rome, where he became a friend of *Michelangelo and one of his most gifted and individual followers. Michelangelo helped to gain him commissions and (as with *Sebastiano del Piombo) supplied him with drawings to work from, but Daniele's finest picture owes little to the direct influence of the master. This is his fresco of the *Deposition* (commissioned 1541) in the Cappella Orsini in SS. Trinità dei Monti, a powerful and moving work, based compositionally on *Rosso Fiorentino's famous painting of the same subject in Volterra, but with an eloquent richness of its own. It was one of the most admired works of its generation in Rome and continued to be influential into the next century: *Domenichino (Hatton Gallery, Newcastle upon Tyne) was among the artists who copied it, and *Rubens was clearly inspired by it in his painting of the subject in Antwerp Cathedral. Daniele was present at Michelangelo's deathbed and his most famous work of sculpture is a bronze bust of him based on the death mask (casts are in the Casa Buonarroti, Florence, the Louvre, and elsewhere). Ironically, in view of his devotion to the master, Daniele is perhaps best remembered for painting draperies over the nude figures in Michelangelo's *Last Judgement*, a concession to Counter-Reformation ideals that earned him the nickname 'Il Bragghettone' (the breeches-maker).

Danti (or **Dante**), **Vincenzo** (1530–76). Italian sculptor, architect, theoretician, and poet, born in Perugia and active mainly in Florence. His work bears witness to his admiration for *Michelangelo, for whose funeral ceremonies in 1564 he supplied sculpture and paintings. Danti's style, however, is more elegant and much less powerful than the master's. His best-known works are (in Florence) the bronze group of *The Execution of the Baptist* over the south door of the Baptistry (finished 1571), and (in Perugia) the bronze figure of Pope Julius II outside the Cathedral (1555). From 1573 he resided in Perugia, where he was one of the first professors at the newly founded Accademia del Disegno and city architect. He was the author of a treatise on proportion (1567), dedicated to the Grand Duke Cosimo de' *Medici.

Danube School. Term applied to a number of German painters working in the Danube valley in the early 16th cent. who were among the pioneers in depicting landscape for its own sake. *Altdorfer, *Cranach the Elder (in his earlier work), and *Huber are the most important artists covered by the term. They worked completely independently of one another, so 'Danube School' (German *Donauschule*) is a term of convenience rather than an indication of any group affiliation.

Daret, Jacques (1403/6–68 or later). Netherlandish painter from Tournai. From 1427 to 1432 he was apprenticed along with Rogelet de la Pâture (assumed to be identical with Rogier van der *Weyden) to Robert Campin, and the similarity of Daret's style to that of the *Master of Flémalle is one of the main reasons for thinking that Campin and this master are one and the same. Four panels survive from Daret's main work, an altarpiece for the Abbey of St Vaast in Arras (1433–5), and one of these—the *Nativity* (Thyssen Coll., Madrid)—is obviously based on the Master of Flémalle's painting of the subject in Dijon (Musée des Beaux-Arts). Two of the other three panels from the St Vaast Altarpiece are in Berlin (Staatliche Museen), and the fourth is in Paris (Petit Palais). Daret also designed tapestry *cartoons, was an *illuminator, and contributed to the festival decorations in Bruges for the marriage of Charles the Bold (1433–77) and Margaret of York in 1468.

Daubigny, Charles-François (1817–78). French landscape painter of the *Barbizon School and one of the earliest exponents of **plein air* painting in France. He received his introduction to painting from his father **Edmé-François** (1789–1843), also a landscape painter, and in 1838 joined the class of Paul *Delaroche at the École des *Beaux-Arts. Although closely associated with the Barbizon painters, he did not himself live at Barbizon. His landscapes reflect his love of rivers, beaches, and canals (he often painted from a specially fitted boat), and are notable for the uncrowded quality of the composition and an almost Dutch clarity of atmospheric effect. He seems to belong more to the generation of *Monet and *Boudin, who were in fact admirers of his work.

Daumier, Honoré (1808–79). French *caricaturist, painter, and sculptor. In his lifetime he was known chiefly as a political and social satirist, but since his death recognition of his qualities as a painter has grown. In 1830, after learning the still fairly new process of *lithography, he began to contribute political cartoons to the anti-government weekly *Caricature*. He was an ardent Republican and was sentenced to six months' imprisonment in 1832 for his attacks on Louis-Philippe

(1773–1850), whom he represented as 'Gargantua swallowing bags of gold extorted from the people'. On the suppression of political satire in 1835 he began to work for *Charivari* and turned to satire of social life, but at the time of the 1848 revolution he returned to political subjects. He is said to have made more than 4,000 lithographs, wishing each time that the one he had just made could be his last. In the last years of his life he was almost blind and was saved from destitution by *Corot, who gave him a house at Valmondois-sur-Seine-et-Oise.

Daumier's paintings were probably done for the most part after 1860 when lithographs became temporarily difficult to market. Although he was accepted four times by the *Salon, he never exhibited his paintings otherwise and they remained practically unknown up to the time of a collective exhibition held at *Durand-Ruel's gallery in 1878, the year before his death. The paintings are in the main a documentation of contemporary life and manners with satirical overtones, although he also did a number featuring Don Quixote as a larger-than-life hero. His technique was remarkably broad and free. As a sculptor he specialized in caricature heads and figures, and these too are in a very spontaneous style. In particular he created the memorable figure of 'Ratapoil' (meaning 'skinned rat'), who embodied the sinister agents of the government of Louis-Philippe. (A similar political type in his graphic art was 'Robert Macaire', who personified the unscrupulous profiteer and swindler.)

In the directness of his vision and the lack of sentimentality with which he depicts current social life Daumier belongs to the *Realist school of which *Courbet was the chief representative. His watercolours and drawings are vivid calligraphic impressions, with effective use of light and shadow. As a caricaturist he stands head and shoulders above all others of the 19th cent. He had the gift of expressing the whole character of a man through physiognomy, and the essence of his satire lay in his power to interpret mental folly in terms of physical absurdity, to rise beyond the individual idiosyncrasy and create the image which illuminates a concept. Although he never made a commercial success of his art, he was appreciated by the discriminating and numbered among his friends and admirers *Delacroix and Corot, *Forain, *Baudelaire, the historian Michelet (1798–1874), and Balzac (1799–1850). *Degas was among the artists who collected his works.

Davey, Grenville. See NEW BRITISH SCULPTURE.

David, Gerard (d. 1523). Netherlandish painter. He was born at Oudewater, now in southern Holland, but he worked mainly in Bruges, where he entered the painters' guild in 1484 and became the city's leading painter after the death of *Memlinc in 1494. At this time the economic importance of Bruges was declining, but it still maintained its prestige as a centre of art and David played an important role in the flourishing export trade in paintings that it developed in the first quarter of the 16th cent. His work—extremely accomplished, but conservative and usually rather bland—was very popular and his stately compositions were copied again and again. Among his followers were *Ysenbrandt and *Benson, who carried on his tradition until the middle of the 16th cent. Most of his work was of traditional religious themes, but his best-known paintings are probably the pair representing *The Judgement of Cambyses* (Groeningemuseum, Bruges, 1498), a gory subject to which his reflective style was not ideally suited.

David, Jacques-Louis (1748–1825). French painter, one of the central figures of *Neoclassicism. He had his first training with *Boucher, a distant relative, but Boucher realized that their temperaments were opposed and sent David to *Vien. David went to Italy with the latter in 1776, Vien having been appointed director of the French Academy at Rome, David having won the *Prix de Rome with *Antiochus and Stratonice* (École des Beaux-Arts, Paris, 1774). In Italy he was able to indulge his bent for the *antique and came into contact with the initiators of the new classical revival, including Gavin *Hamilton.

In 1780 David returned to Paris, and in the 1780s his position was firmly established as the embodiment of the social and moral reaction from the frivolity of the *Rococo. His uncompromising subordination of colour to drawing and his economy of statement in the rejection of the irrelevant were in keeping with the new severity of taste. His themes gave expression to the new cult of the sterner civic virtues of stoical self-sacrifice, devotion to duty, honesty, and austerity. Seldom have paintings so completely typified the sentiment of an age as David's *The Oath of the Horatii* (Louvre, Paris, 1784), *Brutus and his Dead Sons* (Louvre, 1789), and *The Death of Socrates* (Met. Mus., New York, 1787). With these pictures he became the recognized symbol of the new France and head of a powerful school whose influence spread far beyond the borders of France. They were received with acclamation by critics and public alike. *Reynolds compared the *Socrates* with *Michelangelo's Sistine Ceiling and *Raphael's Stanze, and after ten visits to the *Salon described it as 'in every sense perfect'.

David was in active sympathy with the Revolution; he served on various committees and voted for the execution of Louis XVI. His position was unchallenged as the painter of the Revolution. In 1789 he was commissioned to eternalize in painting the 'Tennis Court Oath', though the project did not go beyond preliminary sketches and had to be modified as various participants fell from power. His three paintings of 'martyrs of the Revolution', though conceived as portraits, raised portraiture into the domain of universal tragedy. They were: *The Death of Lepeletier* (now known only from an engraving), *The Death of Marat* (Musées Royaux, Brussels, 1793), and *The Death of Bara* (Musée Calvet, Avignon, unfinished).

David was active in founding the new Institut which took the place of the *Academy. After the fall of his friend Robespierre (1794) he was imprisoned, but was released on the plea of his wife, who had previously divorced him because of his Revolutionary sympathies (she was a royalist). They were remarried in 1796, and David's *Intervention of the Sabine Women* (Louvre, 1794–9), begun while he was in prison, is said to have been painted to honour her, its theme being one of love prevailing over conflict. It was also interpreted at the time, however, as a plea for conciliation in the civil strife that France suffered after the Revolution and it was the work that re-established David's fortunes and brought him to the attention of Napoleon (1769–1821), who appointed him his official painter.

David became an ardent supporter of Napoleon and retained under him the dominant social and artistic position which he had previously held. Between 1802 and 1807 he painted a series of pictures glorifying the exploits of the emperor, among them the enormous *Coronation of Napoleon* (Louvre, 1805–7). These works show a change both in technique and in feeling from the earlier Republican works. The cold colours and severe composition of the heroic paintings gave place to a new feeling for pageantry which had something in common with *Romantic painting, although he always remained opposed to the Romantic school. With the fall of Napoleon, David went into exile in Brussels, and his work weakened as the possibility of exerting a moral and social influence receded. (Until recently his late history paintings were generally scorned by critics, but their sensuous qualities are now winning them a more appreciative audience. Indeed, one such painting, *The Farewell of Telemachus and Eucharis*, 1818, was bought by the Getty Museum, Malibu, in 1987 for more than £2,500,000.) He continued to be an outstanding portraitist, but he never surpassed such earlier achievements as the great *Napoleon Crossing the Alps* (Kunsthistorisches Mus., Vienna, 1800, one of four versions) or the coolly erotic *Madame Récamier* (Louvre, 1800). His work had a resounding influence on the development of French—and indeed European—painting, and his many pupils included *Gérard, *Gros, and *Ingres.

David, Pierre-Jean (1788–1856). French sculptor, known after his birthplace as David d'Angers. In 1811 he won the *Prix de Rome, and spent 5 years in Italy, where he met and admired *Ingres and was also influenced by *Canova and *Thorvaldsen. However, *Neoclassical influence was tempered by a strong inclination towards naturalism, and his contemporaries considered him a *Romantic. His most prestigious commission was the high-relief on the pediment of the Pantheon in Paris, which shows an allegorical figure of France distributing wreaths to great Frenchmen (1837), but his best works are to be found among his busts and medallions of famous men. He left a large collection of them to his native city to found the Musée des Beaux-Arts there.

Davie, Alan (1920–). British painter, graphic artist, poet, musician, silversmith, and jeweller. After service in the army and a short period as a professional jazz musician (he plays several instruments), he travelled in Europe, 1948–9. This gave him the chance to see works by Jackson *Pollock and other American painters in Peggy *Guggenheim's gallery in Venice, and he was one of the first British artists to be affected by *Abstract Expressionism. Other influences on his eclectic but extremely personal style are African sculpture and Zen Buddhism. His work is full of images suggestive of magic or mythology (some based on ancient forms, some of his own invention) and he uses these as themes around which—like a jazz musician—he spontaneously develops variations in exuberant colour and brushwork. From the 1960s he developed an international reputation. He is well represented in the Tate Gallery.

Davies, Arthur Bowen (1862–1928), American painter, printmaker, and tapestry designer. Davies was a member of the circle of Robert *Henri, a member of the *Eight group and president of the Association of American Painters and Sculptors that was created to organize the *Armory Show. He was a man of wide and liberal culture, and his enthusiasm for the project of presenting contemporary European art to the narrow provincialism which prevailed in the USA during the first decade of the 20th cent. was largely responsible for the scope of the show and the force of its impact. Although his own work was not radical, it was varied and embraced remarkably diverse influences (unlike the other members of the Eight he did not specialize in modern urban scenes). In his early career he showed an enthusiasm for the *Pre-Raphaelites, *Whistler, and *Puvis de Chavannes, and specialized in idyllic landscapes inhabited by dreamlike, visionary figures of nude women or mythical animals (*Unicorns*, Met. Mus. of Art, New York, 1906). After the Armory Show his work showed superficial *Cubist influence, but in the 1920s he returned to a more traditional style, devoting much of his time to graphic work and also designing for the *Gobelins tapestry factory.

Davis, Stuart (1894–1964). American painter. He grew up in an artistic environment, for his father was art director of a Philadelphia newspaper who had employed *Glackens, *Luks, *Shinn, and *Sloan and his mother was a sculptor. In 1910–13 he studied with Robert *Henri in New York, and in 1913 was one of the youngest exhibitors in the *Armory Show, which made an overwhelming impact on him: 'All my immediately subsequent efforts went toward incorporating Armory Show ideas into my work.' After this he began experimenting with a variety of modern idioms and in the 1920s he achieved a sophisticated grasp of *Cubism, but it was only after spending a year in Paris in 1928–9 that he forged a distinctive style. Using natural forms, particularly forms suggesting the characteristic environment of American life, he rearranged them into flat poster-like patterns with precise outlines and sharply contrasting colours (*House and Street*, Whitney Mus., New York, 1931). In this way he became the only major artist to treat the subject-matter of the *American Scene painters—extraordinarily popular at the time—in avant-garde terms; he was both distinctly American and distinctively modern—a rare combination that won him wide admiration. Later he went over to more purely abstract patterns, into which he often introduced lettering, suggestions of advertisements, posters, etc. (*Owh! in San Pao*, Whitney Mus., 1951). The zest and dynamism of such works reflect his interest in jazz, and in 1960 he said: 'For a number of years jazz had a tremendous influence on my thoughts about art and life.' However abstract his work became he always claimed that every image he used had its source in observed reality: 'I paint what I see in America, in other words I paint the American Scene.' Davis was one of the outstanding American painters of the 20th cent. and an important link between the pioneering avant-garde artists of the Armory Show generation and the triumphant New York art scene of the post-war years. He was an articulate defender of modern art, a major influence on many younger artists, including his friends *Gorky and *de Kooning, and a precursor of *Pop art.

Dayes, Edward (1763–1804). English watercolour painter and *mezzotint engraver. His treatise on landscape painting (*Instructions for Drawing and Colouring Landscapes*) appeared in his posthumously published *Works* in 1805, which also included an attack on the 'wild effusions of the perturbed imaginations' of *Fuseli. The treatise was already beginning to be out of date when it was published, especially in assuming that drawing and colouring were necessarily distinct processes, and that watercolour landscapes were therefore 'tinted drawings'. *Girtin was his pupil.

Deacon, Richard. See NEW BRITISH SCULPTURE.

Dean, Graham. See SUPERREALISM.

De Andrea, John (1941–). American *Superrealist sculptor. As with Duane *Hanson, his figures are made of fibreglass and are realistic to the last detail, but De Andrea specializes in nude figures and his models are usually young and attractive.

decalcomania. A technique for producing pictures by transferring an image from one surface to another. Its invention *c.*1936 is credited to the Spanish *Surrealist artist Oscar Dominguez (1906–58) although a similar idea had earlier been used in ceramic design. Splashes of colour were laid with a broad brush on moderately thin white paper. This was then covered with another sheet of paper and was rubbed gently so that the wet pigment flowed haphazardly, typically producing effects resembling fantastic grottoes or jungles or underwater growths. The point of the process, which had the blessing of *Breton and was used most memorably by *Ernst, who applied it to oil painting, was that the picture was made without any preconceived idea of its subject or form (*sans objet préconçu*).

de Camp, Joseph R. See TEN, THE.

decorative arts. Term embracing *applied art and also including objects that are made purely for decoration.

de Critz, John. See CRITZ.

Degas, Edgar (1834–1917). French painter, graphic artist, and sculptor. He was the son of a wealthy art-loving banker (the family name was originally de Gas, but Degas adopted the less pretentious form) and was initially trained for the law. In 1855, however, he entered the École des *Beaux-Arts and studied under Louis Lamothe (1822–69), a pupil and admirer of *Ingres, who laid the foundation of Degas's superb draughtsmanship. His real artistic education, however, was gained through assiduous study of the Old Masters, and between 1854 and 1859 he spent much of his time in Italy. Most of his early works were portraits or history paintings on Classical themes (*Young Spartans*, NG, London). In 1861 Degas met *Manet while copying a *Velázquez in the Louvre and was introduced by him to the circle of the young *Impressionists. During the next few years he abandoned historical pictures and turned to contemporary subjects, with a special predilection for racing scenes, ballet, theatre, circus, rehearsals, café scenes, and laundresses. It seems that he was influenced towards this change of direction largely by Manet and the writer Edmond *Duranty.

Degas exhibited in seven out of the eight Impressionist exhibitions and is regarded as one of the prominent members of the Impressionist School. He was, however, Impressionist only in certain restricted aspects of his work and like Manet (who also came from an upper-middle-class background) stood somewhat aloof from the rest of the group. He had little interest in landscape and therefore did not share the Impressionist concern for rendering the effects of changing light and atmosphere. He was more interested in draughtsmanship than most of the others and—apart from Manet—he alone had a thoroughly academic background. As with the other Impressionists he liked to give the suggestion of accidental, spontaneous, and unplanned scenes, and Degas's pictures often cut off figures in the manner of a badly executed snapshot or used unfamiliar viewpoints. Like them he was influenced by the new techniques of photography and by Japanese colour prints (see UKIYO-E) and he was interested in conveying the impression of movement. But he did not paint out of doors or directly from nature. The appearance of spontaneity and accidental effects was an appearance only; in reality his pictures were carefully composed. 'Even when working from nature, one has to compose,' he said, and 'No art was ever less spontaneous than mine.'

Degas always worked much in *pastel and when his sight began to fail in the 1880s his preference for this medium increased. His colours grew stronger and his compositions more simplified. He was a restless experimenter, mixing tempera and pastel, for example, and using a technique called *peinture à l'essence*, in which pigment from which the oil has been removed is thinned with turpentine to promote rapid drying. From 1880 Degas also modelled in wax, but he exhibited only one sculpture in his lifetime, the famous *Little Fourteen-year-old Dancer* (1881) dressed in a real tutu (it was cast in bronze after his death; one cast is in the Tate, London). During the 1890s, as his fears of failing sight increased, he devoted more time to modelling, doing mostly horses in action, women at their toilet, or nude dancers in characteristic postures. These were cast after his death.

For the last twenty years of his life Degas was virtually blind and lived a reclusive life. He was a formidable personality and his complete devotion to his art made him seem cold and aloof. His genius compelled universal respect among other artists, however; *Renoir ranked him above *Rodin as a sculptor, and in 1883 Camille *Pissarro wrote that he was 'certainly the greatest artist of our epoch'. He was the first of the Impressionist group to achieve recognition and his reputation as one of the giants of 19th-cent. art has endured undiminished.

degenerate art (*entartete Kunst*). Term applied by the Nazis to all modern art that did not correspond with their ideology. Adolf Hitler made his first speech against 'degenerate art' at Nuremberg in 1934 and the first of a number of exhibitions designed to bring modern art into disrepute and ridicule was held the year before at Karlsruhe. The systematic suppression of modern ideas in art (which included the closing of the *Bauhaus—'a breeding ground of cultural Bolshevism'—in 1933) culminated in 1937 with the notorious exhibition of 'Entartete Kunst' in the arcades of the Hofgarten, Munich, alongside the exhibition of the academic, politically indoctrinated 'German Art' at the House of German Art nearby. The artists whose work was exhibited as 'degenerate' included *Beckmann, *Dix, *Grosz, *Kandinsky, *Mondrian, and *Picasso, and their paintings were mocked by being shown alongside those by inmates of lunatic asylums. As a propaganda exercise the exhibition was a huge success, more than two million people going to see it in Munich before it went on tour round Germany. Twenty-five of the leading German museums were required to surrender works for the exhibition and from 1937 works of modern art, both German and foreign, were expropriated from collections throughout the country. It is estimated that more than 16,500 works were expropriated in all under the guise of 'degenerate art'. Some of the confiscated works were sold at auction, Nazi officials helped themselves to others, and the 'unsaleable stock' is said to have been burnt in Berlin.

The attack was directed not only against innovative art but also against artists and others who were in sympathy with it. Such persons were dismissed from their posts in museums and teaching institutions, and deprived of their honours and degrees.

In 1935 a decree brought all exhibitions, public and private, under the control of the Reichskulturkammer. Artists who refused to conform were forbidden to exhibit and in some cases were forbidden to work. The writings of Alfred Rosenberg, the chief theoretical spokesman of Nazism, as well as the speeches of Hitler, served to link artistic production with political doctrines and racial theories, but it is significant that the artist who had the 'distinction' of having the most works declared degenerate was Emil *Nolde, who was racially 'pure' and had even been a member of the Nazi party. The suppression of 'degenerate art' was not, therefore, simply a matter of political expediency, but also a symptom of the general antipathy to new forms of artistic expression that has been such a feature of the history of 20th-cent. art. In the normal course of events such hostility rarely goes beyond verbal abuse, but in Nazi Germany aesthetic revulsion was armed with political power.

Deineka, Alexander. See SOCIALIST REALISM.

Dekkers, Adrian (1938–74). Dutch sculptor and experimental artist. In the 1960s he worked in a Neo-*Constructivist manner, making reliefs and space constructions of polyester. He made great use of the resources of light and specialized in the technique of presenting different views of the same geometrical design in one work.

de Kooning, Willem (1904–). Dutch-born painter (and latterly sculptor) who became an American citizen in 1961, one of the major figures of *Abstract Expressionism. He went to America as a stowaway in 1926 and the following year settled in New York. His early work was conservative, but in 1929 he met Arshile *Gorky, who became one of his closest friends and introduced him to avant-garde circles. During the 1930s and 1940s he experimented vigorously and by the time of his first one-man show in 1948 (at the Egan Gallery, New York) he was painting in an extremely energetic abstract style (often in black and white) close to that of Jackson *Pollock. The exhibition established his reputation (although prosperity was still some years away) and after it he was generally regarded as sharing with Pollock the unofficial leadership of the Abstract Expressionist group. Unlike Pollock, de Kooning usually retained some suggestion of figuration in his work, and in 1953 he caused a sensation when his *Women* series (*Women nos I–VI*) was exhibited at his third one-man show, at the Sidney *Janis gallery. *Woman I* (MOMA, New York, 1950–2), with its grotesque leer and frenzied brushwork, shocked the public and dismayed those critics who believed in a rigorously abstract art. One of these was Clement *Greenberg, but New York's other most influential critic of avant-garde art—Harold *Rosenberg—supported de Kooning. *Woman I* became one of the most reproduced paintings in the USA and de Kooning was enormously influential on young painters at this time. By the end of the 1950s, however, he was beginning to be regarded as an elder statesman whose best days as a creative force were past. From the 1960s he had honours heaped on him. His paintings continued to mix abstract and semi-figurative work and in 1969 he began making sculpture—figures modelled in clay and later cast in bronze. He continued working well into his eighties, until he was incapacitated by Alzheimer's disease.

His wife, **Elaine de Kooning** (1920–89), was also a painter, notably of *Expressionist portraits, and a writer on art. The couple married in 1943 and separated in the mid-1960s. A collection of her writings, *The Spirit of Abstract Expressionism*, was published in 1994.

Delacroix, Eugène (1798–1863). The greatest French painter of the *Romantic movement. He was the son of a politician, Charles Delacroix, but there is some evidence to indicate that his real father was the diplomat Talleyrand (1754–1838), a friend of the family. His mother, Victoire Oeben, came of a family of notable craftsmen and designers. In 1816 he entered the studio of Pierre *Guérin, who had earlier taught *Géricault. His basic artistic education was obtained, however, by copying Old Masters at the *Louvre, where he delighted in *Rubens and the Venetian School. He met *Bonington in the Louvre and was introduced by him to English *watercolour painting. *Constable's *Hay Wain*, exhibited in the 1824 *Salon, also made a great impression on him and in 1825 he spent some months in England. Here he conceived an enthusiasm for English painting, in particular *Gainsborough, *Lawrence, *Etty, and *Wilkie. Among contemporary French painters he felt affinity with Géricault and Baron *Gros rather than with the school of *David and *Ingres. In the Salon of 1822 he had his first public success with *The Barque of Dante* (Louvre, Paris). It was bought by the State (with Talleyrand perhaps pulling strings in the background), as was *The Massacre at Chios* (Louvre) two years later, ensuring the success of his career. Gros called this painting 'the massacre of painting', but *Baudelaire wrote that it was 'a terrifying hymn in honour of doom and irremediable suffering'.

In 1832 Delacroix visited Morocco in the entourage of the Comte de Mornay and there acquired a fund of rich and exotic visual imagery which he exploited to the full in his later work (*Sultan of Morocco*, Musée, Toulouse, 1845). From the late 1830s his style and technique underwent a change. In place of luminous glazes and contrasted values he began to use a personal technique of vibrating adjacent tones and *divisionist colour effects in a manner of which *Watteau had been a master, making colour enter into the structure of the picture to an extent which had not previously been attempted. In spite of being hailed as the leader of the Romantic movement, his predilection for exotic and emotionally charged subject-matter, and his open enmity with Ingres, Delacroix always claimed allegiance to the *classical tradition, and for his large works followed the traditional course of making numerous preparatory drawings. In his later career he became one of the most distinguished monumental mural painters in the history of French art. His commissions included decorations in several major public buildings in Paris: Palais Bourbon (Salon du roi, 1833–7; Library, 1838–47); the Library of the Luxembourg Palace (1841–6); and three paintings in the Chapelle des Anges of S. Sulpice (1853–61). In the last of these, his *Jacob and the Angel* and *Heliodorus Expelled from the Temple* are among the maturest expressions of his decorative richness of colour and grandiose structural integration. Baudelaire said of him that he was the only artist who 'in our faithless generation conceived religious pictures' and van *Gogh wrote, 'only *Rembrandt and Delacroix could paint the face of Christ'.

Delacroix's output was enormous. After his death his executors found more than 9,000 paintings, pastels, and drawings in his studio and he prided himself on the speed at which he worked, declaring 'If you are not skilful enough to sketch a man falling out of a window during the time it takes him to get from the fifth storey to the ground, then you will never be able to produce monumental work.' Among great painters he was also one of the finest writers on art. He was a voluminous letter writer and kept a journal from 1822 to 1824 and again from 1847 until his death—a marvellously rich source of information and opinion on his life and times. His influence, particularly through his use of colour, was prodigious, inspiring *Renoir, *Seurat, and van Gogh among others. Delacroix's studio in Paris is now a museum devoted to his life and work, but the Louvre has the finest collection of his paintings.

Delaroche, Paul (1797–1856). French painter, one of the leading pupils of *Gros. He achieved great popularity with his melodramatic history scenes, engravings of his work hanging in thousands of homes. Often he chose subjects from English history, as with two of his most famous works, *The Little Princes in the Tower* (Louvre, Paris, 1831) and *The Execution of Lady Jane Grey* (NG, London, 1833). They are *Romantic in flavour, but academically impeccable in their draughtsmanship and detailing. After a period when such pictures were totally out of favour, his work is once again being treated seriously.

Delaunay, Robert (1885–1941). French painter, who from about 1906 devoted most of his career to experiments with the abstract qualities of colour. He began his researches *c.*1906 from the *Neo-Impressionist theories of *Seurat, but instead of using the *pointillist technique he investigated the interaction of large areas of juxtaposed and contrasting colour. He was particularly interested in the interconnections between colour and movement. By 1910 he was making an individual contribution to *Cubism, notably with his series of paintings of the Eiffel Tower, which combine fragmented forms with vibrant colour. *Apollinaire gave the name *Orphism to Delaunay's work, which by 1912 had moved on to become completely abstract, as in his lyrically beautiful *Circular Forms* series (an example is in the Kunsthaus in Zurich). In 1913 Delaunay exhibited at the galleries of Der *Sturm in Berlin, and his work was a major influence on German *Expressionists such as *Klee, *Macke, and *Marc. It also powerfully affected the *Futurists in Italy and the American *Synchromists. Delaunay was notoriously competitive and fully aware of the importance of his work; at about this time he drew up a list of all the artists, however minor, he thought he had influenced. The period of his greatest achievements was, however, fairly short-lived; he lived in Spain and Portugal during the First World War and after his return to Paris in 1920 his work lost its inspirational quality and became rather repetitive. His home became a meeting place for *Dada artists, but he continued with work related to colour theories.

Delaunay-Terk, Sonia (1885–1979). Russian painter and textile designer, the wife of Robert *Delaunay. She came to Paris in 1905, married Delaunay in 1910 (after a short-lived marriage to Wilhelm *Uhde) and became associated with him in the development of *Orphism. During the 1920s she worked mostly on hand-printed fabrics and tapestries; as a designer she made a strong impact on the international world of fashion, designing creations for such famous women as Nancy Cunard and Gloria Swanson. The Depression affected her business, however, and in the 1930s she returned primarily to painting and became a member of the *Abstraction-Création association. After the death of her husband in 1941 she continued to work as painter and designer. In 1964 a gift of 49 works by Robert and 58 by Sonia Delaunay to the Musée National d'Art Moderne, Paris, was exhibited at the *Louvre and she thus became the first woman to be exhibited at the Louvre in her lifetime.

Delli, Dello. See CASSONE.

de Loutherbourg, Philippe-Jacques (1740–1812). French painter who settled in England in 1771. He became a designer of stage sets for David Garrick (1717–79) at Drury Lane (maquettes in the V&A, London) and is best known for his invention of the *Eidophusikon, a theatrical presentation of scenic pictures. Although a foreigner, de Loutherbourg is said to have declared that 'no English landscape painter needed foreign travel to collect grand prototypes for his study' and in his landscapes, which are indeed very varied in character, he exalted the English scenery as material for the *Picturesque and the *Sublime. He also painted battle scenes, and literary and biblical subjects in a lively style.

Delphi Charioteer. Greek bronze statue of a standing charioteer, excavated at Delphi in 1896 and now in the museum there. Fragments of the chariot, horses, and the figure of a groom also survive, and an inscription indicates that the group commemorated a victory of Polyzalus (viceroy of Gela and brother of the tyrant of Syracuse) in the games at Delphi in 478 or 474 BC. The charioteer is one of the finest surviving pieces of Greek sculpture; it has an austere simplicity, but the head is of great beauty and refinement, the eyes inlaid and the eyelashes made separately.

Delvaux, Paul (1897–1994). Belgian painter. After working in *Neo-Impressionist and *Expressionist manners, he discovered *Surrealism in 1934 and became an instant convert, destroying much of his earlier work. He was never formally a member of the movement, and was not in sympathy with its political aims, but he became regarded as one of the foremost upholders of its tradition. Most of his paintings show nude or semi-nude women in incongruous settings. The women are always of the same type—beautiful, statuesque, unattainable dream figures, lost in thought or reverie or even in a state of suspended animation. These dream beauties are often placed in elaborate architectural settings, reflecting both de *Chirico's strange perspectives and Delvaux's interest in the buildings of ancient Rome (he visited Italy in 1938 and 1939). Sometimes he included skeletons in his pictures (influenced by *Ensor) and trains were another recurrent motif. A large retrospective of Delvaux's work was held at the Palais des Beaux-Arts, Brussels, in 1944, and this marked the beginning of his international reputation.

De Maria, Walter (1935–). American sculptor and experimental artist. He was one of the earliest exponents of *Minimal Art, producing examples of the type *c.*1960, before the term was current, and was also a pioneer of *Land art. Some of his work belongs to the category of *Conceptual art, as for example *Mile Long Drawing* of 1968, two parallel chalk lines 12 ft. (3.6 m.) apart in the Mojave Desert. In 1968 he had a one-man show at the Heiner Friedrich Gallery, Munich, at which he exhibited a room filled wall to wall with earth, 1,600 cubic ft. (45 cubic m.) in all. This *Earthroom* was re-created for permanent exhibition in the Lone Star Foundation, New York, in 1980.

Demeter of Cnidus. A marble statue of *c.*330 BC found at Cnidus in Asia Minor and now in the British Museum. It represents Demeter seated; originally, perhaps, a figure of her daughter Persephone stood beside her. It is considered one of the finest Greek sculptures to survive from the fourth century and has been attributed to *Leochares.

De Morgan, Evelyn (1855–1919). British painter, born Evelyn Pickering; in 1887 she married **William De Morgan** (1839–1917), famous as a designer of pottery, tiles, and ceramics, but also a painter himself. She specialized in literary subjects, done in a style owing much to the *Pre-Raphaelites and to such Renaissance artists as Botticelli (she and her husband spent each winter in Italy for the sake of his health). At the end of her career she painted several allegories relating to the First World War, exhibiting them to raise money for the Red Cross.

Demuth, Charles (1883–1935). American painter and illustrator. He made visits to Europe in 1904, 1907–8, and 1912–14, staying mainly in Paris, and during the last of these visits he became seriously interested in avant-garde art, particularly *Cubism. Its influence was felt in his paintings of architectural subjects from about 1916 and he became one of the leading exponents of *Precisionism. His most personal paintings are what he called 'poster portraits' (pictures composed of words and objects associated with the person 'represented'). The most famous example is *I Saw The Figure Five in Gold* (Met. Mus., New York, 1928), a tribute to the poet William Carlos Williams and named after one of his poems. Demuth was lame from childhood and in the last decade of his life was debilitated by diabetes. Often he worked on a small scale in watercolour, rather than in more physically demanding media. The fastidious taste and concentrated energy of his work are suggested by his comment: 'John *Marin [another great American watercolourist] and I drew our inspiration from the same source, French modernism. He brought his up in buckets and spilt much along the way. I dipped mine out with a teaspoon, but I never spilled a drop.'

Denis, Maurice (1870–1943). French painter, designer, and writer on art theory. Early in his career he was a *Symbolist and a member of the *Nabis. In his article *Definition of Neo-Traditionalism* (1890), he made a famous pronouncement on art, which has often been regarded as the key to contemporary aesthetics of painting: 'Remember that a picture—before being a war horse or a nude woman or an anecdote—is essentially a flat surface covered with colours assembled in a certain order.' His early work did indeed put great emphasis on flat patterning, but he was also very much concerned with subject-matter, for he was a devout Catholic and set himself to revive religious painting. In 1917 he did frescos for the church of St Paul in Geneva, and in 1919 together with Georges Desvallières (1861–1950) he founded the Ateliers d'Art sacré. He also designed stained glass. His writings on art are for the most part collected in *Théories* (1912) and *Nouvelles Théories* (1922). In 1939 he published a history of religious art.

Denny, Robyn (1930–). British painter, prominent among the artists who came to the fore at the *Situation exhibition of 1960. One of the works shown there, *Baby is Three* (Tate, London), a large (2 m. × 3.5 m.) tripartite picture composed of interlocking rectangles, shows his characteristically subtle exploration of colour harmonies.

Denon, Dominique-Vivant, Baron (1747–1825). French engraver, draughtsman, archaeologist, diplomat, museum official, and writer. He was a much-travelled and much-liked man who had a highly varied career. In 1798 he accompanied Napoleon on his expedition to Egypt, recording his travels in *Voyage dans la Basse et la Haute Égypte* (1802), illustrated from his own drawings (it was published in English in the same year as *Travels in Upper and Lower Egypt*). From 1804 to 1815 he was director of the national museums, and he had an important role in developing the collections of the *Louvre, advising Napoleon on his choice of works of art to be looted from conquered territories. Denon was one of the first French artists to make *lithographs, his earliest example dating from 1809. At his death he left unfinished a history of ancient and modern art, which was posthumously published in four volumes in 1829.

de Piles, Roger (1635–1709). French art historian and amateur painter. De Piles was employed by Louis XIV (1638–1715) on various diplomatic missions and was thus enabled to study the arts at first hand in many European countries. He was an admirer of *Rubens and in the famous controversy of the 'Rubénsistes' against the 'Poussinistes' (see POUSSIN) that split the French *Academy in the second half of the 17th cent. he took the side of those who held that colour and *chiaroscuro are of prime importance in painting against the upholders of the academic emphasis on drawing. He also recognized the value of genius, imagination, and 'enthusiasm' against the excessive domination of formalized rule. His best-known book is *Cours de peinture par principes avec une balance des peintres* (1708), which has become notorious for the section (the 'balance des peintres') in which he awarded marks to great artists of the past for their skill at composition, drawing, colour, etc., then added up the scores to form a sort of league table of genius. His criticism, however, is usually much less crude than this.

Derain, André (1880–1954). French painter, sculptor, graphic artist, and theatrical designer. In the first two decades of the 20th cent. he was near the centre of avant-garde developments in Paris: he was one of the creators of *Fauvism, an early adherent of *Cubism, and one of the first to 'discover' *primitive art. However, his later works, mainly landscapes, portraits, still lifes, and nudes, were increasingly based on the Old Masters. He also produced numerous book illustrations, and also designed for the stage, notably for *Diaghilev's *Ballets Russes*. Although his historical significance is undeniable, there is disagreement about his standing as an artist. John Canaday (*Mainstreams of Modern Art*, 1959) writes: 'His detractors think of him as a parasite on both the past and the present, but . . . some critics award Derain unique status as the only twentieth-century painter to achieve an individual compound of the great tradition of French culture as a whole with the spirit of his own time . . . This opinion is particularly held in France—where, of course, it is most legitimate.'

der Kinderen, Antonius Johannes (Anton) (1859–1925). Dutch painter, designer of stained glass windows and printmaker. He studied with *Toorop and was one of the leading figures of the *Symbolist movement in the Netherlands. His aim was the revival of monumental wall-paintings which would become an integral part of the design of a building. A series of his large murals is in the City Hall at 's-Hertogensbosch (1889–96).

Deruet, Claude (1588–1660). French painter who, like *Bellange and *Callot, worked mainly for the court of the Duke of Lorraine at Nancy. He was more successful than either of these great engravers (he was also employed by Cardinal Richelieu (1585–1642)), but his few surviving works are in a pedestrian *Mannerist style that was a generation out of date at the time of his death. The best known are the allegorical series *The Four Elements* (Musée des Beaux-Arts, Orléans).

Desiderio da Settignano (1428/31–64). Florentine sculptor. Like most of his contemporaries he formed his style on Donatello's Florentine work of the 1430s. He learnt from Donatello the practice of carving in very low *relief, and the lively, thick-set figures of children on the *Singing Gallery* made by Donatello for Florence Cathedral (1433–9) provided models for Desiderio's own reliefs of the *Madonna and Child*. Desiderio's artistic personality, however, was more delicate than Donatello's, and for refinement of handling he is unsurpassed by any Italian sculptor of his period. His only important public work was the tomb of the Florentine humanist and statesman Gregorio Marsuppini in Sta Croce (after 1453). This is architecturally dependent on the tomb of Leonardo Bruni by Bernardino *Rossellino (probably Desiderio's teacher), executed for the same church about ten years earlier, but is sculpturally richer and more animated. His sensitive modelling is best exemplified in his portrait busts of women, good examples of which are in Florence (Bargello) and Washington (NG).

Desjardins, Martin (1637–94). Flemish sculptor originally called Martin van den Bogaert, who settled in Paris in about 1670. His most important work was a bronze statue of Louis XIV for the Place des Victoires (1686), fragments of which survive in the Louvre. He also did much decorative work in Paris and for Louis XIV (1638–1715) at Versailles.

Despiau, Charles (1874–1946). French sculptor, one of *Rodin's assistants from 1907 to 1914. After this he turned from his master's intense, vigorous style to a more static, generalized manner which had affinities with that of *Maillol. His best-known works are his portrait busts, with their intimate delineation of character (*Head of Madame Derain*, Philips Collection, Washington, 1922). He also made several monuments. In the 1920s and 1930s his reputation stood very high in France, but at the end of his life he was ostracized because of his friendship with the Nazi sculptor Arno Breker (1900–91); they had known each other since before the war and in 1942 Despiau attended an exhibition of the German's work in occupied Paris.

Desportes, Alexandre-François (1661–1743). French painter of dogs, game, and emblems of the chase. In his early career he worked much as a portraitist, notably in 1695–6 at the court of Jan Sobieski (John III, 1624–96) in Poland, but on his return to France he took up hunting subjects and won the patronage of Louis XIV (1638–1715) and Louis XV (1710–74). He achieved considerable celebrity (he was well received on a visit to England in 1712) and in his field was rivalled only by *Oudry. Although he continued the lavish Flemish tradition exemplified by *Snyders, Desportes was among the first artists of the 18th cent. to make landscape studies from nature for his backgrounds, and because of this he was considered eccentric. His work is well represented in the Louvre (which has his *Self-portrait as a Huntsman*, 1699) and in the Wallace Collection, London.

De Stijl. See STIJL, DE.

Desvallières, Georges. See DENIS.

Detroy. See TROY.

Deutsch, Niklaus Manuel (*c*.1484–1530). Swiss painter, designer, and poet, active mainly in his native Berne. Deutsch was one of the outstanding Swiss artists of his period, but much of his energy was expended in other activities. He fought as a mercenary in Italy and took an active part in the political and religious affairs of Berne as a passionate supporter of the Reformation, writing satires against the Pope, whom he equated with Antichrist. His paintings are related to *Baldung Grien and *Grünewald in their love of the grotesque (*The Temptation of St Anthony*, Kunstmuseum, Berne, 1520). Deutsch also designed woodcuts and stained glass and a set of *choir stalls for Berne Cathedral.

Deutscher Werkbund (German Association of Craftsmen). An organization of German manufacturers, architects, and designers for the improvement of design and craftsmanship in machine-made products founded in Munich in October 1907, by the architect Hermann Muthesius (1861–1927). Muthesius had been German cultural attaché in London (1896–1903) and was impressed by the ideas of the William *Morris circle and the domestic architecture of Richard Shaw (1831–1912) and *Voysey. On his return to Germany he became Superintendent of the Prussian School of

Arts and Crafts, and became the proponent of a new style in machine industry which adopted as its principles functional design and the abolition of ornament.

The Werkbund was racked by controversy from the time of its foundation, and soon divided into two factions: those who advocated maximum industrialization and standardization of design (Muthesius and Peter *Behrens were of this persuasion); and those who set a higher value on individuality (Henri van de *Velde was the champion of this cause). Behrens's programme was adopted at the Werkbund's annual meeting at Cologne in 1914, when a special exhibition of members' work was arranged. This display brought to prominence some of Behrens's younger associates, in particular Walter *Gropius, whose model factory building was the most discussed exhibit. The Werkbund had immediate influence, providing the inspiration for similar organizations in Austria (Österreichischer Werkbund, 1912) and Switzerland (Schweizerischer Werkbund, 1913); also modelled along similar lines were Sweden's Slöjdsforening (1913) and the Design and Industries Association in England (1915). The activities of the Deutscher Werkbund were interrupted by the First World War, but it revived afterwards and held an important exhibition of model housing projects in Stuttgart in 1927, organized by the architect *Mies van der Rohe. Gropius and *Le Corbusier were among the exhibitors. The Werkbund disbanded in 1933 after the Nazis came to power, but it was revived after the Second World War.

Deutsche Werkstätten (German Workshops). An organization for the machine production of well-designed furniture and other goods. The workshops were established in Dresden in 1898 by Karl Schmidt (1873–1948), a German follower of William *Morris, and were originally called the Dresdener Werkstätten für Handwerkkunst. They pioneered the production of furniture made from standardized parts.

Deverell, Walter Howell (1827–54). British painter. He was a friend of *Rossetti, and was proposed for membership of the *Pre-Raphaelite Brotherhood (to replace *Collinson) but was never actually elected. In 1849 he 'discovered' Elizabeth Siddal, the archetypal Pre-Raphaelite model and Rossetti's future wife. Deverell was noted for his good looks and charm and Elizabeth was probably in love with him before he died aged 26 from Bright's disease. In his brief career he gave promise of becoming perhaps the most painterly of the Pre-Raphaelite followers (*A Pet*, Tate, London, 1852/3).

Devis, Arthur (1711–87). English painter, the best-known member of a family of artists. He was one of the first specialists in the small *conversation piece and also painted single portraits of similar scale. The sitters are usually in repose, often somewhat artificially posed, and the Devis type of portrait group was animated in the next generation by *Zoffany. Devis was a minor figure in his day and virtually forgotten until the 1930s, but since then his work has attained considerable popularity because of the doll-like charm of his figures and the delicate detail of his settings. It has also become of interest to social historians, as most of his clients were from the newly prosperous middle class—merchants and country squires, who are usually shown singly or with their families, in their own homes or grounds. A small representative collection of his work is in the Art Gallery at Preston, his native town. He worked both in London and in Lancashire. **Anthony** (1729–1816), his half-brother, was a landscape painter. Arthur's son, **Arthur William** (1762–1822), spent the years 1785–95 in India, where he painted portraits and a series of pictures representing the arts, manufactures, and agriculture of Bengal (two examples are in the Ashmolean, Oxford), which were engraved. He lived in London from 1795, working mainly as a portraitist, but also painting *The Death of Nelson* (Nat. Maritime Mus., Greenwich, *c.*1806). Another son, **Thomas Anthony** (1757–1810), painted undistinguished portraits and *fancy pictures. Little of his work survives.

Dewing, T. W. See TEN, THE.

De Wint, Peter (1784–1849). English landscape painter of Dutch extraction. He served his apprenticeship with John Raphael *Smith, then studied at the *Royal Academy Schools and frequented the house of Dr *Monro. Although he was an admirable painter in oils, he is best known as one of the finest exponents of watercolour of his generation. He is particularly associated with views of the countryside around Lincoln (where his wife's parents lived), in which he often uses broad *washes of colour somewhat in the manner of *Cotman. De Wint was a popular figure and enjoyed considerable success as a teacher.

Diaghilev, Sergei (1872–1929). Russian impresario, famous above all as the founder of the *Ballets Russes*, through which he exerted great influence on the visual arts as well as on dancing and music. From 1890 to 1896 he studied law in St Petersburg, where he became part of a circle of musicians, painters, and writers including Leon *Bakst and Alexander *Benois. In 1899 he founded the magazine *World of Art, with the object of interchanging artistic ideas with Western Europe. When it ceased publication in 1904 he concentrated for a while on organizing exhibitions, including one of Russian painting at the 1905 *Salon d'Automne in Paris—the most comprehensive to have been seen in the West up to that time. In 1907 he organized a series of concerts of Russian music in Paris, and in 1909 he brought a ballet company for the first time (this is usually described as the *Ballets Russes*, but the name was first used in 1911). The company was a sensational success, as much for the exotic designs of Bakst as for the music and choreography (the dancers included Nijinsky and Pavlova). For the next two decades, until his death in 1929, Diaghilev toured Europe and America with his ballet (he never returned to Russia after the 1917 Revolution and Paris was the main centre of his operations). He was often on the verge of bankruptcy, but he had a remarkable flair for spotting young talent and for integrating various interests and people, enabling him to bring together as his collaborators some of the foremost artistic personalities of his time; the painters who designed for him included *Braque, de *Chirico, *Derain, *Matisse, and *Picasso.

diaper. An all-over pattern based on small repeated units capable of indefinite extension in any direction. It is found carved in low relief on flat wall surfaces of *Romanesque and *Gothic architecture, in *stained glass, and on the backgrounds of manuscript *illuminations, especially of the late 13th and 14th cents.

Diaz de la Peña, Narcisse-Virgile (1807–76). French painter, born at Bordeaux of Spanish parents. He began his career as a porcelain painter, and then painted *Romantic historical subjects, but after meeting Théodore *Rousseau in 1837 he became a member of the *Barbizon group of landscape painters. His style is distinct from that of his Barbizon colleagues, however, for his work lacks the sense of quiet communion with nature that was a characteristic feature of the school and his brushwork is heavy and restless (his detractors call it turgid). He never lost the Romantic leanings of his youth, and continued to paint mythological pictures (typically featuring nymphs) throughout his career. He was helpful to the *Impressionists; *Renoir stated that his meeting with Diaz led him to lighten his palette and his flickering touch had a great influence on Adolphe *Monticelli, who in turn anticipated the *divisionist technique of the Impressionists.

Dickinson, Edwin (1891–1978). American painter. He often treated enigmatic or disquieting subject-matter and he has been described as 'perhaps the first American artist about whom some knowledge of dream theory is essential for decoding his works' (Matthew Baigell, *A Concise History of American Painting and Sculpture*, 1984). His personal symbolism is seen at its most disturbing and provocative in his self-portraits, in which he sometimes painted himself as dead. He is best known, however, for large compositions such as *The Fossil Hunters* (Whitney Museum, New York, 1926–8). Dickinson often worked on his big pictures for a number of years and said that they were never 'really finished'. He has been called a *Surrealist and also seen as a sophisticated culmination of the 19th-cent. *Romantic tradition.

Dickinson, Preston (1891–1930). American painter. He spent five years in Europe, 1910–15, and in Paris he was influenced particularly by the structural features of *Cézanne's work and the high-keyed colour of the *Fauves. In the 1920s, however, his work became less experimental as he became associated with the *Precisionists. Like others of the school, he favoured subjects which were adapted to representation in terms of semi-geometrical abstract design, in particular the machine (*Industry*, Whitney Mus., New York, *c.*1924).

Diderot, Denis (1713–84). French philosopher and critic, mainly remembered in England as the chief editor of the *Encyclopédie* (1751–72), a work of fundamental importance in shaping the rationalist and humanitarian ideals of the Age of Enlightenment. His views on art appear in articles in the *Encyclopédie* and elsewhere, notably his reviews of the *Salons between 1759 and 1781, which are written in a lively conversational style and formed the model for the later criticism of *Baudelaire. Against the intellectualist bias of *Neoclassicism he maintained that our ideas of beauty arise from practical everyday experience of beautiful things, based on a *sentiment* for the 'conformity of the imagination with the object'. He regarded *taste* as a faculty acquired through repeated experience of apprehending the true or the good through immediate impression which renders it beautiful. He opposed the constraints imposed by a priori rules or the tyranny of the ancients, and defended the right of genius to create beauty by the idealization of nature. His views on the relation between poetry and painting provided a basis for the *Laokoon* of *Lessing.

Diebenkorn, Richard (1922–93). American painter. He taught at the California School of Fine Arts, San Francisco, from 1947 to 1950, and there, under the influence of members of the

*New York School, particularly Mark *Rothko and Clyfford *Still, he abandoned the still lifes and interiors which he was painting and adopted a non-figurative style. Passing rapidly from a more geometrical mode of *Abstract Expressionism he evolved a modified form of *Action painting, with the freer expressive brushwork and vigorous calligraphic line of *de Kooning, which inaugurated a distinctive West Coast movement. After 1955 he revolted from the subjective emotionalism of this way of painting and reverted to a mode in which he tried to apply the techniques of Abstract Expressionist brushwork to studies of figures in an environment, organizing his compositions into large rectangular areas of colour which owed much to the example of *Matisse.

Dietrich, Christian Wilhelm Ernst (1712–74). German painter and etcher, most of whose extremely varied output was done in the styles of 17th-cent. masters. He worked mainly in Dresden, and was Director of the art gallery there and of the art school attached to the famous porcelain factory at nearby Meissen. He is represented in the National Gallery and the Wallace Collection, London.

Dietterlin, Wendel (1550/1–99). German architect, painter, and designer. He worked mainly in Strasburg, where he specialized in painting façade, ceiling, and wall decorations, but none of these has survived. His claim to fame is his book *Architectura* (1593–4, 2nd edn., 1598), an extraordinary collection of engravings in which the Orders of architecture are used as the starting-point for weird and extravagant decorative fantasies, full of bizarre animal and plant forms. It was a popular pattern book.

dilettante. See AMATEUR.

Dilettanti, Society of. Society of young noblemen and gentlemen that was originally a purely social club but came to play a prominent role in matters of taste, and particularly in the study of *antique art. From December 1732 the members met on the first Sunday in the month in a tavern (drinking a toast to 'Grecian taste and Roman spirit'), so provoking the sneer of Horace *Walpole, never one of them, that 'the nominal qualification for membership is having been in Italy and the real one being drunk'. One of them at least, Sir Francis Dashwood (1708–81), was a member of the notorious Hell Fire Club. The serious interests of the group soon prevailed and after unsuccessfully supporting Italian opera, the Dilettanti turned to the architectural and archaeological remains of Greece, the Near East, and Italy, which had stirred their interest and imagination on their travels.

Independently of *Winckelmann, and before his posthumous fame, they financed a succession of expeditions and published the results in various works that laid the foundations of the serious and systematic study of Classical antiquities. They stimulated *Lessing and *Goethe, and contributed powerfully to the Classic revival of the late 18th cent. The Department of Classical Antiquities of the *British Museum owes many of its finest and nearly all its earliest treasures to the enthusiasm, discernment, and public spirit of members of the Society—the collections of Sir William Hamilton (1730–1803), Charles Townley (1737–1805), Richard Payne *Knight, Sir Richard Colt Hoare (1758–1838), the fourth Earl of Aberdeen (1784–1860), W. M. Leake (1777–1860), and others. One serious mistake, however, was the failure of the Dilettanti through the misjudgement of Payne Knight to give support to Lord Elgin when he brought the sculptures of the Parthenon to England (see ELGIN MARBLES).

From their earliest meetings the Society appointed a painter, one of whose duties was to provide a portrait of each member on election at his own expense. The first holder of the title was George *Knapton, and his successors have included Sir Joshua *Reynolds, Sir Thomas *Lawrence, Sir Martin Archer *Shee, Sir Charles *Eastlake, Sir Frederic *Leighton, Sir Edward *Poynter, John Singer *Sargent, and Sir William *Coldstream. Many portraits so commissioned are still among the Society's treasured possessions.

Diller, Burgoyne (1906–65). American painter and sculptor. After passing through phases of *Impressionism and *Cubism, he became interested in *Neo-Plasticism and by the mid-1930s had become one of the earliest and most committed American followers of *Mondrian. He was a member of *American Abstract Artists and from 1935 to 1940 Head of the Mural Division of the *Federal Art Project. His sculpture, restricted to rectangular elements and *primary colours, was like his painting deeply influenced by Mondrian.

diluent. The liquid used to dilute a paint and give it the fluidity that the painter desires; e.g. *turpentine in *oil painting, and water in *watercolour painting.

Dine, Jim (1935–). American painter, printmaker, experimental artist, and poet. In 1959 he was one of the pioneers of *happenings and in the early 1960s he became one of the most prominent figures in American *Pop art (he also made an impact in England, where he lived 1967–71). His Pop canvases were vigorously handled in a manner recalling *Abstract Expressionism, but he often attached real objects to them—generally everyday items such as clothes and household appliances (including a kitchen sink). Characteristically the objects were Dine's personal possessions and his work often has a strong autobiographical flavour. In addition to such *assemblages, he also made free-standing works and *environments, but since the mid-1970s he has concentrated more on traditional two-dimensional work, especially drawings (he has written and illustrated several books of poetry).

diorama. A large, partially translucent scenic painting, which by means of varied illumination simulates such effects as sunrise, changing weather, etc. The term is applied by extension to the building in which the display is housed. The diorama was invented by L.-J.-M. Daguerre (see DAGUERREOTYPE) in 1822 and was exhibited in Regent's Park, London, in the following year, when *Constable wrote to a friend: 'I was at the private view of the "Diorama"; it is in part a transparency; the spectator is in a dark chamber, and it is very pleasing, and has great illusion. It is without the pale of art, because its object is deception. The art pleases by *reminding*, not *deceiving*. The place was filled with foreigners, and I seemed to be in a cage of magpies.'

Nowadays the term diorama is more usually applied to a certain type of museum display. It consists of a miniature scene, viewed through a window in a screen or cabinet, in which the foreground details, modelled in the round, join imperceptibly with the more distant parts which are painted in perspective on a vertical panel.

diorite. Igneous stone of the *granite type but lacking quartz in its composition. It is usually black or grey and is sometimes known as 'black granite'. Like granite it is hard, compact, and resistant. It was favoured for sculpture by the ancient Egyptians and in the ancient Middle East.

dipper. A small metal container that clips on to the oil painter's *palette and holds *medium or *diluent. They are often made in pairs—a 'double dipper'.

diptych. A picture or other work of art consisting of two parts facing one another like the pages of a book and usually hinged together. The consular diptych is a type of ivory carving characteristic of the late Roman empire. On appointment to office, consuls distributed these panels to friends, relatives, and persons of rank. The earliest one to survive is of 428, and the practice ceased in 541. Because they can be precisely dated by the name of the official they bear, they are important tools in scholarship of the period. Many were later reused for Christian purposes. See also POLYPTYCH, TRIPTYCH.

Directoire style. Style of decoration and design prevailing in France between the *Louis XVI and *Empire styles. It takes its name from the *Directoire* (1795–9), the period when France was governed by an executive of five *directeurs* and covered roughly the decade 1793–1804. As a stylistic term *Directoire* is applied to the useful arts of furniture, textiles, costume, etc., rather than the field of the fine arts. No interiors have survived from the *Directoire* period, but there is a model of one based on contemporary designs in the Art Institute of Chicago. Decoration was simpler than in the Louis XVI style and embodied Revolutionary emblems and civic symbols such as the Phrygian bonnet or cap of Liberty, the pike of Freedom, the fasces which symbolized strength through union, clasped hands as an emblem of Fraternity, the oak as a symbol of social virtues, and so on. *Antique ornamental motifs were also in vogue and the 'Grecian' vogue in costume and furniture is well illustrated in *David's famous portrait of Madame Récamier (Louvre, Paris, 1800). As a result of Napoleon's campaigns in 1798–9 *Egyptian taste in ornament also prevailed in this period, its main protagonist being Baron Dominique-Vivant *Denon. The latter part of the *Directoire* period, i.e. 1799–*c.*1804, is sometimes distinguished as the '*Consulat*' style.

Discobolus. Statue of a discus-thrower by the Greek sculptor *Myron. The original bronze of *c.*450 BC is lost, but several Roman copies in marble are known, of which the best is in the Museo delle Terme in Rome.

distemper. Type of paint in which the pigment is mixed with water and glue or *size. Its principal use is in scene-painting, as it is cheap but impermanent. Whitewash is a form of distemper.

divisionism. A method and technique of painting by which colour effects are obtained not by mixing pigments on the *palette but by applying small areas or dots of unmixed pigment on the canvas in such a way that to a spectator standing at an appropriate distance they appear to react together. This method, which produces greater luminosity and brilliance of colour than if the colours are physically mixed, has been employed to some extent by many artists in **alla prima* painting, although it is contrary to the traditional principles of painting by superimposed *glazes and *scumbles. Notable precursors of divisionism were *Watteau and *Delacroix. It was also employed empirically by the *Impressionists and in particular by those who adopted the 'rainbow palette' of *Renoir. It was not developed systematically and scientifically, however, until *Seurat and the *Neo-Impressionists. Seurat (in common with other contemporaries) spoke of an 'optical mixture', but (contrary to what is usually stated) the dots do not really fuse in the viewer's eye to make different colours, for they remain visible as dots. Rather, they seem to vibrate, creating something of the shimmering effect experienced in strong sunlight. The effect is noted in Ogden Rood's *Modern Chromatics* (1879), a treatise on colour theory well known to Seurat. Camille *Pissarro, who was closely associated with Seurat at this time, said that the optimum viewing distance for a picture painted by the divisionist method was three times the diagonal measurement. The terms divisionism and *pointillism are not always clearly differentiated, but whereas divisionism refers mainly to the underlying theory, pointillism describes the actual painting technique associated with Seurat and his followers. 'Divisionism' (usually with a capital 'D') was also the name of an Italian movement, a version of *Neo-Impressionism, that flourished in the last decade of the 19th cent. and the first decade of the 20th cent. It was one of the sources of *Futurism.

Dix, Otto (1891–1969). German painter and printmaker. In the 1920s he was, with George *Grosz, the outstanding artist of the *Neue Sachlichkeit movement, his work conveying his disillusionment and disgust at the horrors of war and the depravities of a decadent society with complete psychological truth and devastating emotional effect. *The Match Seller* (Staatsgalerie, Stuttgart, 1920), for example, is a pitiless depiction of indifference to suffering, showing passers-by ignoring a blind and limbless ex-soldier begging in the street, and Dix's fifty etchings entitled *The War* (1924) have been described by G. H. Hamilton (*Painting and Sculpture in Europe, 1880–1940*, 1967) as 'perhaps the most powerful as well as the most unpleasant anti-war statements in modern art'. Another favourite theme was prostitution and he was a brilliantly incisive portraitist (*Sylvia von Harden*, Musée National d'Art Moderne, Paris, 1926). In 1927 he was appointed a teacher at the Dresden Academy and in 1931 he was elected to the Prussian Academy. His anti-military stance drew the wrath of the Nazi regime and he was dismissed from his academic posts in 1933 and his work declared *degenerate. He went to live quietly in the country near Lake Constance and painted traditional landscapes, yet he still aroused suspicion; in 1939 he was arrested on a charge of complicity in a plot on Hitler's life, but was soon released. He was conscripted into the *Volkssturm* (Home Guard) in 1945 and was a prisoner in France 1945–6; he then returned to Lake Constance. His work after the war lost much of the strength of his great Neue Sachlichkeit period and much of it was inspired by religious mysticism.

Dobell, Sir William (1899–1970). Australian painter. In 1929 he won a travelling scholarship that enabled him to study at the *Slade School under Henry *Tonks and did not return to Australia until 1938. His work reveals the broad artistic education he gained on his travels in Europe, and his rich textures and colours, and exaggerated emphasis on characteristic forms show, in particular, a debt to the *Expressionism of Chaïm *Soutine. In 1943/4 he won the Archibald Prize for portraiture, awarded annually by the Art Gallery of New South Wales, with a portrait of his fellow-artist, Joshua Smith. The award, which was contested in the courts by two of the unsuccessful competitors on the grounds that it was not a portrait but a caricature, created a *cause célèbre* for modernism in Australia and Dobell's name became a household word. During the next decade he was much sought after as a painter of portraits, among the finest being that of the poet *Dame Mary Gilmore* (Art Gal. of New South Wales, 1957). In 1949 and again in 1950 he visited the highlands of New Guinea, and as a result of this experience his work became broader in execution and more decorative. Dobell was unique among Australian artists in combining successfully a mastery of *Renaissance tradition (particularly in portraiture) with a profound insight into the character and values of 20th-cent. Australians.

Dobson, Frank (1886–1963). British sculptor. His early work consisted mainly of paintings, the few surviving examples showing how impressed he was by the *Post-Impressionist exhibitions organized by Roger *Fry. After the First World War (when he was on active service with the Artists' Rifles), he turned increasingly to sculpture, and in the 1920s and 1930s gained an outstanding reputation: in 1925 Roger Fry described his work as 'true sculpture and pure sculpture . . . almost the first time that such a thing has been even attempted in England'. He worked in both bronze and stone (he was one of the earliest to revive direct carving) and his sophisticated stylizations made him one of the pioneers of modern British sculpture. The monumental dignity of his work was in the Classical tradition of *Maillol, and like him Dobson found the female nude the most satisfactory subject for three-dimensional composition, as in *Cornucopia* (University of Hull, 1925–7), described by Clive *Bell as 'the finest piece of sculpture by an Englishman since—I don't know when'. He was also outstanding as a portrait sculptor, as witness his head of Sir Osbert Sitwell in polished brass (Tate, London, 1923), and besides stone carving produced many exquisitely beautiful *terracottas. His craftsmanship was superb and he played an important role as a liberal-minded and kind-hearted teacher at the *Royal College of Art, where he was Professor of Sculpture, 1946–53. With the rise of a younger generation led by Henry *Moore, however, Dobson's prestige as an artist dropped and he was regarded as 'dated'; the memorial exhibition of his work organized by the Arts Council in 1966 was not well received. Since then he has again been recognized as one of the outstanding figures in 20th-cent. British sculpture.

Dobson, William (1611–46). English portrait painter. He is regarded as the most accomplished English painter before *Hogarth, and was described by John Aubrey in his *Brief Lives* (*c.*1690) as 'the most excellent painter that England hath yet bred'. Some sixty paintings by him are known, all from the years 1642–6, when he was painter to the wartime court at Oxford. He is thought to have returned to London after the surrender of that city in 1646. Said to have been 'somewhat loose and irregular in his way of living', he was thrown into prison for debt, and his early death followed shortly after his release.

His style is superficially similar to van *Dyck's, but his colouring is richer and his paint texture rougher, very much in the Venetian tradition. He also had an uncompromisingly direct way of presenting character (as in his most celebrated work—*Endymion Porter*, Tate, London) that is considered quintessentially English. Various paintings by Dobson other than portraits are mentioned by early writers, but only two survive: *The Executioner with the Baptist's Head* (Walker Art Gallery, Liverpool), a copy after the Dutch painter Matthias Stomer (*c.*1600–after 1650); and an allegory *The Civil Wars of France* (Rousham House, Oxfordshire).

Doesburg, Theo van (Christian Emil Maries Küpper) (1883–1931). Dutch painter, architect, and writer on art. His early work was influenced variously by *Impressionism, *Fauvism, and *Expressionism, but in 1915 he met *Mondrian and rapidly underwent a transition to complete abstraction. In 1917 he founded the association of artists called De *Stijl and the periodical of the same name, and for the remainder of his life the propagation of the ideas of the group and its austerely geometrical style was his dominant interest. He went on an extended lecture tour outside the Netherlands in 1921 and his ideas made a considerable impression at the *Bauhaus, where he taught irregularly from 1922 to 1924. In 1930 he moved to Paris and built himself a studio at Meudon that became a new focus of the De Stijl movement. The movement collapsed with his death in 1931, but its influence survived in many fields, notably architecture, where his ideal was the use of geometrically simple elements and *primary colours. His writings include several books and many articles in *De Stijl*. See also CONCRETE ART and ELEMENTARISM.

Dolci, Carlo (1616–86). Florentine painter, active in his native city for virtually his whole career. He was intensely devout and most of his paintings are of religious subjects, done in a cloyingly sweet and meticulously smooth style. They were enormously popular in his lifetime (his reputation spread to England and elsewhere), but have appealed much less to 20th-cent. taste. His portraits, on the other hand, are now much admired for their sober objectivity (*Sir Thomas Baines*, Fitzwilliam, Cambridge, *c.*1665–70). It was as a portraitist that he made his one significant journey from Florence, when in 1675 he went to Innsbruck to paint Claudia Felice de' *Medici when she married the emperor Leopold I (1640–1705).

Domenichino (Domenico Zampieri) (1581–1641). Bolognese painter. He was Annibale *Carracci's favourite pupil and one of the most important upholders of the tradition of Bolognese *classicism. After studying with *Calvaert and Ludovico Carracci he went to Rome (1602) and joined the colony of artists working under Annibale Carracci at the Palazzo *Farnese. His only undisputed work there is the *Maiden with the Unicorn*, a charming, gentle fresco over the entrance of the Gallery. By the second decade of the century he was established as Rome's leading painter and had a succession of major decorative commissions, among them scenes from the life of St Cecilia in S. Luigi dei Francesi (1613–14). The dignified frieze-like composition of the figures reflects his study of *Raphael's tapestries, and in turn influenced *Poussin. The frescos in the *pendentives and apse of S. Andrea della Valle (1624–8), his chief work of the 1620s, show a move away from this strict classicism towards an ampler *Baroque style; but compared with his rival *Lanfranco (who at this time was overtaking him in popularity) Domenichino never abandoned the principles of clear firm drawing for the sake of more painterly effects. In 1631 Domenichino moved to Naples, and in his ceiling frescos of the S. Gennaro chapel in the cathedral he made even greater concessions to the fashionable Baroque. He met with considerable hostility in Naples from jealous local artists and was forced to flee precipitately in 1634. He later returned, but died before completing his work in the cathedral.

Domenichino was important in fields other than monumental fresco decoration, particularly as an exponent of *ideal landscape, in which he formed the link between Annibale Carracci and *Claude (four of his landscapes are in the Louvre). He was one of the finest draughtsmen of his generation (the Royal Library at Windsor Castle has a superb collection of his drawings) and also an excellent portraitist (*Monsignor Agucchi*, City Art Gal., York, c.1610). In the 18th cent. his reputation was enormous—his *Last Communion of St Jerome* (Vatican, 1614) was generally regarded as one of the greatest pictures ever painted—but he fell from grace in the 19th cent. along with other Bolognese painters under the scathing attacks of *Ruskin.

Domenico Veneziano (d. 1461). Italian painter. His name indicates that he came from Venice, but he was active mainly in Florence. *Vasari credits him with introducing *oil painting into Tuscany. Although this is incorrect, it seems to be true that he was responsible for initiating the interest in colour and texture and the style of painting which used colour rather than line as the basis of *perspective and composition. His fresco cycle *Scenes from the Life of the Virgin* (1439–45) in S. Egidio, Florence, on which *Piero della Francesca was one of his assistants, has disappeared, and only two fully authenticated works have survived. These are three much-damaged and repainted fragments from a frescoed street tabernacle (NG, London, c.1440) and the celebrated *St Lucy Altarpiece* of c.1445 (the central panel is in the Uffizi, Florence, and the *predellas dispersed in Cambridge (Fitzwilliam), Washington (NG), and Berlin (Staatliche Mus.)). The delicate beauty of its colouring, mastery of light, and airy lucidity of spatial construction are reflected in the work of his assistant Piero, and also, for example, in that of *Baldovinetti.

Dominguez, Oscar. See DECALCOMANIA.

Donatello (Donato di Niccolo) (1386?–1466). Florentine sculptor. He was the greatest European sculptor of the 15th cent. and one of a remarkable group of artists—including his friends *Alberti, *Brunelleschi, and *Masaccio—who created the *Renaissance style in Florence. Between 1404 and 1407 he was working as an assistant to *Ghiberti, but he developed a style that was radically different to his master's *Gothic elegance. He was unconcerned with the surface polish or linear grace so typical of Ghiberti, and excelled rather in emotional force. *Vasari expressed his admiration in the language of the day by asserting that Donatello had equalled the sculptors of antiquity. Refining upon this we may say that he possessed an imaginative understanding of certain aspects of Classical sculpture, probably acquired from a very few examples and conditioned by a Christian approach. He was thus the counterpart of the early Florentine humanists such as Poggio Bracciolini (1380–1459), with whom he is known to have been in touch. His revolutionary conception of sculpture is exemplified in the great series of standing figures in niches which he made for Or San Michele and Florence Cathedral. The series began with the *St Mark* of 1411–13 (Or San Michele), included the celebrated *St George* (c.1415–17), now in the Bargello, Florence, and culminated in the so-called *Zuccone* ('baldpate') in the Cathedral Museum (probably completed in 1436, although because various figures of prophets, of which this is one, are not identified unambiguously in the documents, some authorities think it was carved in 1423–5). Vasari conveys the brilliance of Donatello's characterization in his description of the *St George*: 'The head exhibits the beauty of youth, its spirit and valour in arms, a proud and terrifying lifelikeness, and a marvellous sense of movement within the stone.' With this acute psychological insight went a technique of daring originality that shows how concerned Donatello was with the optical effects of his works. He carefully took into consideration the position from which they would be viewed, adjusting the proportions of a figure when it would be seen from below, for example, and carving with almost brutal power and boldness when it was positioned to be seen at a distance. On the other hand, his relief of *St George and the Dragon* (Or San Michele, 1417), done for the base of his *St George* statue, is executed with great delicacy in the technique Donatello invented called *rilievo schiacciato* (*relief so low it is like 'drawing in stone'); situated on the north side of the building, the relief is seen in a soft, diffused light, so the subtlety of the carving can be appreciated.

In 1430–2 Donatello visited Rome, probably with Brunelleschi, and the impact of *antique art can be seen most clearly in his famous *Cantoria* (singing gallery) for Florence Cathedral (now Cathedral Mus., 1433–9), which makes a lavish show of freely interpreted classical motifs. The bronze statue of *David* (Bargello), which is credited with being the first free-standing nude statue since antiquity, is also sometimes seen as a response to Donatello's visit to Rome and assigned to the 1430s, but some scholars date it much later. The subject as well as the date is controversial, for it has been proposed that it represents *Mercury with the Head of Argus* rather than *David*.

From 1443 to 1453 Donatello was based in Padua, his reason for moving there presumably being the commission to execute the Gattamelata monument in the Piazza del Santo—the first life-size equestrian statue since antiquity. His other major work in Padua was the High Altar of the church of S. Antonio (the Santo), which features free-standing figures and reliefs that are not now in their original positions.

From 1454 until his death Donatello was based mainly in Florence, although he also worked in Siena on an abortive project for a set of bronze doors for the Cathedral. In his late work he gave more prominence to the emotional intensity that was already such a feature of his style. The most important works from his final years are *Judith and Holofernes* in the Piazza della Signoria, a harrowing and emaciated *Mary Magdalene* in wood (Baptistry), and two pulpits with bronze reliefs in S. Lorenzo, which were unfinished at his death. These sublime late works show how freely Donatello exploited the expressive possibilities of distortion, and in them he created what has been called 'the first style of old age in the history of art'. His work had enormous influence, on painters as well as sculptors, and his true spiritual heir was *Michelangelo.

Dongen, Kees van (1877–1968). Dutch-born painter who settled in Paris in 1897 and took French nationality in 1929. His early work was *Impressionist, but he became a member of the *Fauvist group in 1906 and in 1908 exhibited with the German *Expressionist group Die *Brücke. Nudes and female portraits were his favourite themes. After the First World War he became well known for his paintings of insolently glamorous women, in which he created a type that has been described as 'half drawing-room prostitute, half sidewalk princess'. He kept the brilliant colouring of his Fauve days, but his great facility led to repetition and banality, and it is generally agreed that his best work was done before 1920. From 1959 he lived in Monaco.

Donkey's Tail (in Russian: *Oslinyi Khvost*). Title of an exhibition organized in Moscow in 1912 by *Larionov and *Goncharova after they had dissociated themselves from the *Knave of Diamonds group in 1911. They accused that group of being too much under foreign influence, and advocated a nationalist Russian art. At this time Larionov and Goncharova were painting in their 'primitivist' manner based upon Russian peasant art and *icon painting. *Malevich and *Tatlin also showed at the exhibition, which took its name from Larionov's having heard that a group of artists in Paris had exhibited a picture that a donkey had 'painted' by means of a brush tied to its tail. The exhibition caused an outcry because it was thought to be irreverent to show religious works under such a frivolous title (the police ordered several to be removed). It was followed by the *Target exhibition in 1913.

Donner, Georg Raphael (1693–1741). The outstanding Austrian *Baroque sculptor, active mainly in Salzburg, Bratislava, and Vienna. His masterpieces are acknowledged to be the figures from the fountain in the Mehlmarkt, Vienna (Österreichisches Barockmuseum, Vienna, 1737–9), and the group of *St Martin and the Beggar* (Bratislava Cathedral, *c.*1735), in which the saint is dressed in hussar's uniform rather than the traditional armour. These works are in lead, which Donner preferred to the Austrian speciality of wood, and his smooth surfaces have led some critics to see his work as presaging *Neoclassicism, although his elongated figures seem to place him closer to Italian *Mannerism.

donor. The commissioner of a work of religious art, a term used particularly when a portrait of this person is incorporated in the work. By having themselves included in the picture donors sought to associate themselves in a special way with the sacred figures portrayed there, either in thanks for favours received, or in the hope of future protection and salvation. Famous examples of donor portraits include that of Enrico Scrovegni (d. 1336), who had the Arena Chapel in Padua built in expiation of his father's sin of usury, and is shown kneeling in supplication in *Giotto's fresco of the *Last Judgement* there; and those of Jodocus Vijd and his wife, who are shown in prayer on van *Eyck's *Ghent Altarpiece*.

Doré, Gustave (1832–83). The most popular and successful French book illustrator of the middle 19th cent. Doré became widely known for his illustrations to such books as Dante's *Inferno* (1861), *Don Quixote* (1862), and the Bible (1866), and he helped to give European currency to the illustrated book of large format. He was so prolific that at one time he employed more than forty wood engravers. His work is characterized by a rather naïve but highly spirited love of the *grotesque and represents a commercialization of the *Romantic taste for the bizarre. Drawings of London done in 1869–71 were more sober studies of the poorer quarters of the city and captured the attention of van *Gogh. In the 1870s he also took up painting and sculpture (the monument to the dramatist and novelist

Alexandre Dumas in the Place Malesherbes in Paris, erected in 1883, is his work), but was much less successful in these fields, as his style looked overblown and pretentious when transferred to a large scale.

Doryphorus. See POLYCLITUS OF ARGOS.

Dossi, Dosso (Giovanni Luteri) (*c.*1490?–1542). The outstanding painter of the Ferrarese School in the 16th cent. His early life and training are obscure, but *Vasari's assertion that he was born around 1474 is now thought unlikely. He is first recorded in 1512 at Mantua (the name 'Dosso' probably comes from a place near Mantua—he is not called 'Dosso Dossi' until the 18th cent.). By 1514 he was in Ferrara, where he spent most of the rest of his career, combining with the poet Ariosto (1474–1533) in devising entertainments, triumphs, tapestries, etc., for the *Este court. Dosso painted various kinds of pictures—mythological and religious works, portraits, and decorative frescos—and he is perhaps most important for the part played in his work by landscape, in which he continues the romantic pastoral vein of *Giorgione and *Titian. The influence from these two artists is indeed so strong that it is thought he must have been in Venice early in his career. Dosso's work, however, has a personal quality of fantasy and an opulent sense of colour and texture that gives it an individual stamp (*Melissa*, Borghese Gal., Rome, *c.*1523). In his *Orlando Furioso*, Ariosto described Dosso as one of the nine greatest living painters. His brother **Battista Dossi** (*c.*1497–1548) often collaborated with him, but there is insufficient evidence to know whether he made an individual contribution.

dotted manner. See MANIÈRE CRIBLÉE.

Dou, Gerrit (1613–75). Dutch painter. He was born and active in Leiden and in 1628 became the first pupil of the young *Rembrandt. His early work is closely based on Rembrandt's, and *Anna and the Blind Tobit* (NG, London) is thought by some to have been painted by Dou from Rembrandt's design and with his assistance. After Rembrandt moved to Amsterdam, Dou developed a style of his own, painting usually on a small scale, with a surface of almost enamelled smoothness, and attending to every detail of his carefully arranged subject. He was astonishingly fastidious about his tools and working conditions, with a particular horror of dust. Some of his pictures were painted with the aid of a magnifying glass. He painted numerous subjects, but is best known for domestic interiors. They usually contain only a few figures framed by a window or by the drapery of a curtain, and surrounded by books, musical instruments, or household paraphernalia, all minutely depicted. He is at his best in scenes lit by artificial light.

With Jan *Steen, Dou was among the founders of the Guild of St Luke at Leiden in 1648. Unlike Steen he was prosperous and respected throughout his life, and his pictures continued to fetch big prices (consistently higher than those paid for Rembrandt's work) until the advent of *Impressionism influenced taste against the neatness and precision of his style. Dou had a workshop with many pupils who perpetuated his style (notably *Schalken)—and Leiden continued the *fijnschilder* (fine painter) tradition until the 19th cent.

Doughty, Thomas (1793–1856). American painter and lithographer, one of the first American artists to specialize exclusively in landscapes. He was born in Philadelphia and mainly lived there, but also in Boston and New York. In 1837 and again in 1845 he travelled to Europe, visiting England on both occasions. Originally he had a successful business as a leather dealer, and he was self-taught as a painter, but he took up his hobby as a profession when his works began to sell. As one of the first to recognize the American landscape as a viable subject for painting, he is regarded as a forerunner of the *Hudson River School. His best-known painting is probably *In Nature's Wonderland* (Detroit Inst. of Arts, 1835).

Douris. Athenian *red-figure vase painter, active *c.*500 BC. A large number of pots decorated by him have survived and the themes are very varied (cup representing *Eos and Memnon*, Louvre, Paris; cup showing young women undressing, Met. Mus., New York). His work is noted for its fine draughtsmanship and rhythmic composition.

Dove, Arthur (1880–1946). American painter. For most of his career he earned his living as a commercial illustrator and he was often in great financial difficulty (even though he was supported by *Stieglitz and Duncan *Phillips). He visited Europe in 1907–9, coming into contact with *Fauvism and other avant-garde movements, and in 1910 painted the first abstract pictures in American art (*Abstraction No 1–Abstraction No 6*, private coll.), which are somewhat similar to *Kandinsky's work of the same time. He never exhibited these in his lifetime, but he displayed similar work at his first one-man exhibition at Stieglitz's 291 Gallery in 1912. Typically his abstractions are based on natural forms, suggesting the rhythms of nature with their pulsating shapes (*Sand Barge*, Phillips coll., Washington, 1930), and he wrote: 'I should like to take wind and water and sand as a motif and work with them, but it has to be simplified in most cases to color and force lines and substances just as music has done with sound.' In the 1940s he experimented with a more geometric type of abstraction (*That Red One*, William H. Lane Foundation, Leominster, Mass., 1944). In his later years he took a leading part in the campaign to win artists royalty rights for the reproduction of their work.

Doves Press. A private printing press founded in 1900 at Hammersmith by the bookbinder T. J. Cobden-Sanderson (1840–1922) and the printer Sir Emery Walker (1851–1933), both of whom had previously worked for the *Kelmscott Press. Cobden-Sanderson carried on alone after disagreement with Walker in 1909. Doves Press books had no illustrations and were noted for the austere beauty of their typography. Of the fifty-one titles produced the most important was the Bible (1903–5) in five volumes. The Press was closed down in 1916.

Downman, John (*c.*1750–1824). English portrait painter. Most of his paintings are small society portraits, many of them being in a technique he perfected using pencil or charcoal lightly tinted with watercolour (four examples of the type are in the Wallace Collection, London). He practised in Cambridge, Chester, Exeter, London, Plymouth, and Wrexham and travelled widely about the country, staying in great houses and often painting a series of portraits of members of the family.

Doyle, Richard (1824–83). English humorous draughtsman, the son of the portraitist and *caricaturist **John Doyle** (1797–1868). From 1843 to 1850 he was on the staff of *Punch*, for which among other things he made the cover design used for over 100 years. He left because of *Punch*'s attacks on the papacy (he was a devout Catholic) and devoted himself thereafter to book illustration, mainly children's stories and fairy tales.

drawing frame. A rectangular frame which the artist sets up between himself and his subject, at such a distance that his view through it corresponds to the drawing he intends to make. Sometimes the frame has a grille of wires or threads, and if the paper is correspondingly squared the subject can very easily be transferred to it. *Alberti and *Leonardo describe drawing frames and *Dürer made illustrations of the device. Following Dürer's example, van *Gogh constructed one and in a letter to his brother made a sketch of himself using it. A very small frame or viewfinder, which could be held in the hand, was used in the 18th cent. by topographical painters and travellers in search of the *Picturesque. Later it was made adjustable, so that the ratio of upright to horizontal could correspond to that of the drawing.

Dreier, Katherine S. (1877–1952). American painter, patron, and collector, a wealthy heiress remembered mainly for her missionary zeal in organizing exhibitions of modern art. She became an ardent supporter of avant-garde art as a result of the *Armory Show, where she met Marcel *Duchamp, and in 1920 together with him and *Man Ray she founded the *Société Anonyme. 'A domineering woman of tireless energy and Wagnerian proportions, she was the antithesis of Duchamp in every possible way, and they got along famously' (Calvin Tomkins, *The World of Marcel Duchamp*, 1966). The travelling exhibitions that she organized through the Société Anonyme, most of the catalogue entries for which were written by Duchamp, were a potent factor in bringing a knowledge of European avant-garde art to the USA. Her portrait of Duchamp (1918) is in the Museum of Modern Art, New York.

Droeshout, Martin (b. 1601). English engraver of Flemish origin, one of a family of artists who settled in London in the 16th cent. He was a mediocre and obscure craftsman, but his name lives by his engraved portrait of Shakespeare published in the First Folio of his works in 1623. However, it has been suggested that the portrait is the work of his father, also called Martin Droeshout (d. *c.*1642).

drollery. A comic picture or 'clownish representation', as the diarist John Evelyn (1620–1706) expressed it when he saw 'Landscips, and Drolleries' in the annual fair at Rotterdam (13 Aug. 1641). This term is also used for the *grotesques or comic figures to be found in the borders of medieval manuscripts.

Drost, Willem (active mid-17th cent.). Dutch painter. Almost nothing is known of his life, and only a small number of paintings (together with a few etchings and drawings) are recognized as being by him. All the works that are dated are from the period 1652–63, and it would seem that at this time Drost was one of *Rembrandt's closest and most talented followers. His *Portrait of a Young Woman* in the Wallace Collection, London, bears a false Rembrandt signature.

Drouais, François-Hubert (1727–75). French portrait painter. He trained under *Boucher (among others) and became a rival to *Nattier as a fashionable portraitist. His portraits have a gracious and artificial charm and at their best bear comparison with those of Boucher. He was particularly successful with children, but his best-known painting is probably the very grand portrait of Mme de Pompadour in the National Gallery, London (1763–4), completed after the sitter's death.

His father and his son were painters. **Hubert Drouais** (1699–1767) had a successful career as a *miniaturist and *pastel portraitist. **Germain Drouais** (1763–88) was a favourite pupil of *David's and won the *Prix de Rome, but his promising career was cut short when he died of malaria aged 25.

Drummond, Malcolm. See CAMDEN TOWN GROUP.

Drury, Alfred (1856–1944). British sculptor. As an assistant to *Dalou he worked on the *Triumph of the Republic* (Place de la Nation, Paris, 1879–99), and much of his own sculpture was for public places. Among his works are decorative figures for the façade of the *Victoria and Albert Museum (1908), and the full-length bronze *Sir Joshua Reynolds* in the courtyard of Burlington House outside the *Royal Academy (1931). He also made portrait busts and small bronze figures in a similar traditional style. See also NEW SCULPTURE.

drying oils. Fatty oils of vegetable origin which harden into a solid transparent substance on exposure to air and are much used as *vehicles in paints and varnishes. They do not dry in the sense of losing moisture but by oxidation together with certain molecular changes. It is this chemical character which has made them suitable as *media for *oil painting, so that when *pigments are ground into them to form paints, which are then applied to a *ground, the oil sets hard and binds the particles of pigment firmly in position, fixing them to the ground.

The vegetable oils which have been in commonest use since the Middle Ages are *linseed, *walnut, and *poppy. Almond and olive oil are not suitable as they do not harden. Sunflower oil has been used in Russia but has never become popular. Small amounts of castor oil have sometimes been used to give elasticity to *varnishes. 'Stand oil', or linseed oil polymerized by boiling with air excluded, has been used since medieval times both as a medium and as a varnish or *glaze.

drypoint. A method of engraving on copper in which the design is scratched into the plate with a sharply pointed tool. This is either a thick steel needle or a diamond point, and it is held like a pen or an *etching needle, though considerable force is needed to scratch the metal to any depth. Where the plate is deeply scored a *burr is thrown up alongside the furrow; this burr retains the ink when the plate is wiped, giving to the drypoint line its characteristic rich and furry appearance, as when a pen is drawn across damp paper. Drypoint is therefore a sensitive medium, for the degree of blackness of the burred line changes with the variation in pressure of the artist's hand. It is not, however, quite so fluent as etching, for the point is continuously up against the resistance of the metal, so that the lines take on a slightly angular quality as they change direction.

A drypoint plate is easily damaged and will yield only a handful of first-class impressions because the burr is flattened as the plate passes between the rollers of the press. A larger edition may be obtained by having the plate steel-faced—an operation which involves depositing, by electrolysis, a thin coating of steel over the surface of the copper, thus strengthening the drypoint work.

Drypoint, which seems to have originated in the last quarter of the 15th cent., has often been used in combination with other processes; more often, probably, than by itself. Since it can be worked quickly and directly, it is frequently used for accenting and touching up plates which are already nearly completed in another medium. *Rembrandt, for example, often touched up his etchings in drypoint. He also made a few prints purely in drypoint, notably one of his most celebrated works, *The Three Crosses*.

Drysdale, Sir Russell (1912–81). Australian painter. He was born in England of a family that had long associations with Australia and he spent several years of his childhood there. The family settled in Melbourne in 1923 and in the late 1930s Drysdale gave up farming to study art. After moving to Sydney in 1940, he devoted himself full-time to painting and his work became well known throughout Australia during the 1940s. It revived in a new fashion the tradition of hardship, tragedy, and melancholy associated with the Australian bush that had been obscured by the much more optimistic interpretation developed by the city-based painters of the *Heidelberg School during the 1890s. In 1949 Kenneth *Clark, on a visit to Sydney, encouraged Drysdale to exhibit in London and late in 1950 he held an exhibition at the Leicester Galleries. This exhibition marks the beginning of a new interest in Australian art in London, which reached a peak in the early 1960s.

In 1954 Drysdale, with *Dobell and *Nolan, was chosen to represent Australia at the Venice *Biennale. During the 1950s he travelled widely throughout Australia, drawing and painting the life of the interior. The plight of the Australian Aborigines in contact with white settlement became an important and continuing theme in such paintings as

Mullaloonah Tank (Art Gal. of South Australia, Adelaide, 1953). During the early 1960s he experienced periods of depression accentuated by the death of his son and his wife. From that time he continued to broaden and develop the themes and methods with which he began in the 1940s.

Dubois, Ambroise (1542/3–1614). Flemish-born painter who moved to France when he was a young man and eventually became a naturalized French citizen in 1601. With *Dubreuil and *Froment he ranks as one of the leading artists of the Second School of *Fontainebleau. Much of his decorative work still survives at the Château de Fontainebleau, showing him to have been an accomplished practitioner of an elegant, conventional *Mannerist style.

Dubreuil, Toussaint (1561–1602). French painter, one of the leading artists of the second School of *Fontainebleau. His most important works, such as the frescos of the story of Hercules for the Gallery of Diana at Fontainebleau, and the decorations in the Galerie d'Apollon of the *Louvre, have been destroyed; but a reasonable idea of his style can be formed from copies and works done from his designs as well as the few original surviving paintings. His style was much less obviously *Mannerist than that of his closest contemporaries (his figures, for example, are much less elongated) and looks forward to the academic *classicism of the following century.

Dubroeucq (or **Broeucq**), **Jacques** (1500/10–1584). Netherlandish sculptor and architect. He was in Italy *c.*1530–5 and his work shows how assiduously he studied the great Italian *Renaissance sculptors. A commission to do carvings, particularly a huge *rood-screen, for the cathedral of St Waltrudis at Mons, brought him back to the Netherlands. These were his principal works (1535–48); most of them were destroyed during the French Revolution, but enough remains to show that he ranks with *Mone and Cornelis *Floris as a leading *Renaissance sculptor of the Low Countries. He was the teacher of *Giambologna, *c.*1545–50.

Dubuffet, Jean (1901–85). French painter, sculptor, lithographer, and writer. He studied painting as a young man but was engaged mainly in the wine trade until 1942 when he took up art seriously again, his first exhibition coming in 1945. He made a cult of *Art Brut ('raw art'), the products of psychotics or wholly untrained persons, and of *graffiti, preferring untrained spontaneity to professional skill. His own work was aggressively reminiscent of such 'popular' art, often featuring subjects drawn from the street life of Paris (*Man with a Hod*, Tate Gallery, London, 1956). Frequently he incorporated materials such as sand and plaster into his paintings, and he also produced large sculptural works made from junk materials. His work initially provoked outrage, and he stands out as the pioneer and chief representative of the tendencies in modern art to depreciate traditional artistic materials and methods and, as he himself said in 1957, to 'bring all disparaged values into the limelight'. Opinions about his work have differed widely, but he has undoubtedly been highly influential, foreshadowing many of the trends of the 1960s and beyond by discrediting all conventional artistic standards.

Duca, Giacomo del (*c.*1520–1604). Sicilian architect and sculptor, active mainly in Rome, where he worked for *Michelangelo on the proposed tomb of Pope Julius II (1542) and the Porta Pia (1562). His tomb of Elena Savelli in S. Giovanni in Laterano shows Michelangelo's influence, but he is more distinguished as an architect than a sculptor. After 1588 he worked in Messina, where earthquakes have destroyed most of his buildings, but the dome he added to Sta Maria di Loreto in Rome shows him to have been the boldest of Michelangelo's architectural disciples.

Duccio di Buoninsegna (active 1278–1318/19). The most famous painter of the Sienese School. Little is known of his life: records of several commissions survive and he is known to have been fined on several occasions for various minor offences (one perhaps involving sorcery), but only one fully documented work by him survives. This is the famous **Maestà* commissioned by Siena Cathedral in 1308 and completed in 1311. Today most of this elaborate double-sided altarpiece is in the cathedral museum, but several of the *predella panels are scattered outside Italy—in London (NG), Washington (NG), and elsewhere. It has been described by John White (*Art and Architecture in Italy: 1250–1400*, 1966) as 'probably the most important panel ever painted in Italy. It is certainly among the most beautiful. Compressed within the compass of an altarpiece is the equivalent of an entire programme for the fresco painting of a church.' The whole of the front of the main panel is occupied by a scene of the Virgin and Child in majesty surrounded by angels and saints, and corresponding to this on the back are twenty-six scenes from Christ's Passion. Originally there were subsidiary scenes from Christ's life above and below the main panel. The whole work is of a superb standard of craftsmanship, and the exquisite colouring and supple draughtsmanship create effects of great beauty. Although Duccio drew much on

*Byzantine tradition, he introduced a new warmth of human feeling that gives him a role in Sienese painting comparable to that of *Giotto in Florentine painting. He re-creates the biblical stories with great vividness, and as no one else before him he succeeds in making the setting of a scene—a room or a hillside—a dramatic constituent of the action, so that figures and surroundings are intimately bound up together.

The other main work attributed to Duccio is the large *Rucellai Madonna* (Uffizi, Florence), which is probably the picture documented as having been painted by him for Sta Maria Novella, Florence, in 1285. Several other smaller panels can be attributed to him or his workshop with a fair degree of confidence, but there is no evidence that he ever worked in fresco. His influence in Siena was enormous (*Simone Martini was his greatest disciple) and reached as far as France, notably in the work of *Pucelle. It is possible that he visited France: a 'Duche de Siene' is documented in Paris in 1296 and 1297.

Duchamp, Marcel (1887–1968). French-American artist and theorist, the brother of Raymond *Duchamp-Villon and Jacques *Villon. Although Duchamp produced few works (most of them are now in the Philadelphia Museum of Art), he is regarded as one of the most influential figures in 20th-cent. art because of the originality and fertility of his ideas. He sprang to notoriety with his *Nude Descending a Staircase, No. 2* (Philadelphia), combining the principles of *Cubism and *Futurism, which was the most discussed (and vilified) work at the *Armory Show in 1913. In the same year he invented the *ready-made with a bicycle wheel mounted on a kitchen stool, and from this time he virtually abandoned painting and other conventional media. From 1915 to 1923 he lived mainly in New York, where he was a leader of the *Dada movement. His works of this period include other ready-mades, notably *Bottle Rack* (1914) and *Fountain* (a urinal signed R. Mutt, 1917). Another of his celebrated provocative gestures was adding a moustache and beard and an obscene inscription to a reproduction of the *Mona Lisa* (1919). So far as it is possible to derive a theoretical basis from the incoherences of Dadaism, the concept of the ready-made seems to derive from Duchamp's conviction that life is meaningless absurdity and from his repudiation of all the values of art. He sometimes said that any object becomes a work of art if he selects it from the limbo of unregarded objects and declares it to be so. At other times he said that these ready-mades were not art but anti-art, affirming that all art is junk. Duchamp's major work of this period was a construction on glass entitled *The Bride Stripped Bare by Her Bachelors, Even*, also known as *The Large Glass* (Philadelphia, 1915–23; a facsimile by Richard *Hamilton is in the Tate, London, 1965–6). This is his most esoteric work and to many people is an incomprehensible joke. After leaving *The Large Glass* 'definitively unfinished' Duchamp virtually abandoned art for chess. He was a good enough player to represent France in four chess Olympiads and his obsession for the game intensified as he grew older. Of his marriage in 1927 his friend *Man Ray wrote: 'Duchamp spent most of the one week they lived together studying chess problems, and his bride, in desperate retaliation, got up one night when he was asleep and glued the chess pieces to the board. They were divorced three months later.' He lived mainly in Paris from 1923 to 1942 and then for the rest of his life mainly in New York. Near the end of his life he revealed that he had worked in secret for twenty years on *Etant Donnés* (Given that . . .), a large mixed-media construction; it is now in Philadelphia.

Duchamp became a legend in his own life-time. He was a man of enormous charm, with a keen sense of irony, and by his character as much as his works he did more than anyone else to change the concept of art in the 20th cent. He tried, unsuccessfully as he himself recognized, to destroy the mystique of taste and collapse the concept of aesthetic beauty, and in 1962 said: 'When I discovered ready-mades I thought to discourage aesthetics . . . I threw the bottle rack and the urinal in their faces and now they admire them for their aesthetic beauty.' Nevertheless, he was among the few artists who can claim to have revolutionized notions of art and beauty.

Duchamp-Villon, Raymond (1876–1918). French sculptor, the brother of Jacques *Villon and Marcel *Duchamp. After studying medicine he took up sculpture in 1898. From 1905 to 1913 he exhibited at the *Salon d'Automne with works of expressive naturalism which at first differed little from those of others who were seeking a way to escape from the influence of *Rodin. From about 1910, however, he came under the influence of the *Cubist group and by 1914 he was recognized as pre-eminent among the small number of Cubist sculptors. His most celebrated work is *The Horse* (casts in Tate, London, Mus. National d'Art Moderne, Paris, MOMA, New York, and elsewhere, 1914), which completed his move towards abstraction. This has been called by G. H. Hamilton (*Painting and Sculpture in Europe 1880–1940, 1967*) 'the most powerful piece of sculpture produced by any strictly Cubist artist', and has been compared with the work of the *Futurists, particularly that of *Boccioni, who had met Duchamp-Villon in 1913. In the success with which it expresses the taut energy of muscular movement in static forms it certainly achieves at least one of the things at which the Futurists were aiming in their attempts to represent 'the dynamics of movement'. Duchamp-Villon served with the French army in the First World War and died from blood-poisoning after contracting typhoid fever.

Dufresnoy, Charles-Alphonse (1611–68). French painter and writer. His paintings (few of which survive) are pastiches of *Poussin, and he is chiefly remembered for his Latin poem, *De arte graphica*, which sets out the doctrines of French academic *classicism in epigrammatic form. It was translated into French and published in 1688 by his friend Roger *de Piles, who in his notes emphasized the importance of colour that Dufresnoy had rather tentatively asserted. The poet John Dryden translated it into English (1695) and it was later annotated by *Reynolds, remaining influential for more than a century as an expression of academic theory.

Dufy, Raoul (1877–1953). French painter, graphic artist, and textile designer. His early work was in an *Impressionist manner, but he became a convert to *Fauvism in 1905 under the influence of *Matisse. In 1908 he worked with *Braque at L'Estaque and abandoned Fauvism for a more sober style influenced by *Cézanne, but thereafter he soon developed the highly personal manner for which he became famous. It is characterized by rapid calligraphic drawing on backgrounds of bright colours thinly washed on a white ground, and was well suited to the glittering scenes of luxury and pleasure he favoured. He was a well-established figure by the mid-1920s and the accessibility and *joie de vivre* of his work helped to popularize modern art. In 1910 he made friends with the fashion designer Paul Poiret (1879–1943), who interested him in textile design, and he worked as a designer for both Poiret and Bianchini-Férier, a silk manufacturer of Lyons. Through his designs Dufy exerted a considerable influence on the world of fashion at this time. He also made numerous book illustrations, notably for *Apollinaire's *Bestiaire* in 1910. His popularity continues undiminished, not least in Japan, where he is a favourite with collectors.

dugento. See QUATTROCENTO.

Dughet, Gaspard (called Gaspard Poussin) (1615–75). French landscape painter, draughtsman, and etcher, born and active in Rome. He was a pupil (*c*.1631–5) of Nicolas *Poussin, who married his sister in 1630 and whose surname he adopted. His work combines something of the romanticism of *Claude with the solidity of Poussin, although he preferred a more rugged type of scenery to that favoured by either of his great contemporaries. The combination brought him considerable success in his lifetime, and in the 18th cent. his reputation stood very high, particularly in England. His work was avidly sought by English collectors and he influenced painters such as Richard *Wilson and the supporters of the *Picturesque. Among the few works by Dughet that are securely dated are his frescos (begun 1647) on the *History of the Carmelite Order* in S. Martino ai Monti, Rome (these too are landscapes), and it has proved difficult to establish a chronology for him. His work is exceptionally well represented in British collections.

Dujardin, Karel (*c*.1622–78). Dutch painter and etcher of landscapes, cattle, *genre scenes, portraits, and religious subjects, active mainly in Amsterdam. He is best known for his small paintings of humble bucolic scenes set in an Italianate or a Dutch landscape and diffused with a clear, warm light. These works reveal the impact of *Berchem (who was probably his teacher) and his admiration for Paulus *Potter and Adriaen van de *Velde. He was an excellent portraitist and his large religious pictures show that his long visit to Italy in the 1640s had made him *au fait* with the picture-making techniques of Italian *Baroque. In 1674 he made a second visit to Italy and died in Venice four years later. Like so many of the 17th-cent. Dutch artists who made the journey to Italy, Dujardin was a Catholic. His work is well represented in the National Gallery, London, and the Rijksmuseum, Amsterdam.

Dulac, Edmund (1882–1953). French-born illustrator, designer, and painter who settled in England in 1904 and became a British citizen in 1912. Dulac is best known as a book illustrator, particularly for fairy-tale and legendary subjects, in which his sense of fantasy and gifts as a colourist were put to brilliant effect (he was much influenced by Middle and Far Eastern art). He was also a portrait painter and caricaturist, and a highly versatile designer; his output included much work for the stage, and one of his last commissions was a stamp commemorating the coronation of Queen Elizabeth II in 1953.

Dulwich College Picture Gallery. See BOURGEOIS.

Dumonstier. Family of French portrait painters, who carried on the tradition of the *Clouets into the middle of the 17th cent. About a dozen members of the family are recorded and several of them held court appointments. The earliest of any significance was **Geoffroy** (*c.*1510–60), who was court painter to Francis I (1494–1547) and Henry II (1519–59), and the last and best-known member of the dynasty was **Daniel** (1574–1646).

Dunlap, William (1766–1839). American artist and writer. He had a varied career, much of it being spent as a successful dramatist and theatrical manager, but in the context of art history he is remembered mainly for his *History of the Rise and Progress of the Arts of Design in the United States* (1834). This is the most valuable sourcebook on the subject, rich in information and anecdote, and has earned him the nickname 'the American *Vasari'. Dunlap also wrote a *History of American Theater* (1832). His varied work as a painter and engraver is now virtually forgotten.

Dunoyer de Segonzac, André (1884–1974). French painter, designer, and graphic artist. Although he went through a period of *Cubist influence early in his career, he became an upholder of the *naturalistic tradition in a period dominated by anti-naturalistic tendencies. His oil paintings (landscape, still life, and figures) were often sombre in tone, usually executed in thick paint, emphasizing the weight and earthiness of the forms. His watercolours and etchings, however, were more elegant and spontaneous, and had a wider range of subject-matter, including dancers and boxers. He was a prolific book illustrator, did theatre and ballet designs, and was in charge of a camouflage unit in the First World War. His work is well represented in the Musée de l'Ile de France at Sceaux.

Dupont, Gainsborough (1754–97). English painter and engraver, the nephew and only assistant of Thomas *Gainsborough. He made copies and *mezzotints of his uncle's pictures, completed others left unfinished at his death, and painted some original works in his manner, notably several portraits of actors, examples of which are in the Garrick Club, London.

Dupré, Jules. See BARBIZON SCHOOL.

Duque Cornejo, Pedro (1678–1757). Spanish *Rococo sculptor, the last important sculptor of the School of Seville. He carved statues and altarpieces for cathedrals and churches at Granada, Seville, and El Paular, and his *choir stalls for Cordoba Cathedral (1748–57) are generally regarded as the finest to be produced in Spain in the 18th cent.

Duquesnoy, François (1594–1643). Flemish sculptor, active mainly in Rome, where he settled in 1618. He was a friend of *Poussin, sharing a house with him for a time, and became a leading figure in circles devoted to classical art. Alongside *Algardi he came to be recognized as the outstanding sculptor in Rome after the great *Bernini (who employed him on the decoration of the *Baldacchino in St Peter's in 1627–8), and as with Algardi, his style was much more restrained and less *Baroque than Bernini's. Duquesnoy's two major works are the statues of *Sta Susanna* (Sta Maria di Loreto, 1629–33) and *St Andrew* (St Peter's, 1629–40). He also produced many small bronzes that spread his fame (in J. T. Smith's book on *Nollekens we read how highly the latter valued his works by Duquesnoy). Duquesnoy was particularly renowned for his handling of *putti, and it is curious that someone who so unaffectedly depicted the beauty and charm of children seems to have been mentally unstable; he was a chronic procrastinator and the diarist John Evelyn (1620–1706), visiting Rome in 1644, said that he 'died mad' because his *St Andrew* 'was placed in a bad light'.

Duquesnoy's father and brother were sculptors; **Jerome I** (before 1570–1641) and **Jerome II** (1602–54). His father is remembered mainly for the famous *Manneken-pis* fountain (1619) behind the town hall in Brussels. His brother worked with François in Rome and took a somewhat diluted Baroque style back to Flanders with him. The tomb of Bishop Anton Trest in Ghent Cathedral (*c.*1640–54) is considered his finest work. He is perhaps best remembered, however, for his sticky end; he was executed by strangulation in Ghent for committing sodomy in a church.

Durand, Asher B. (1796–1886). American painter and engraver. His early work was mainly as an engraver and he established his reputation with his print after John *Trumbull's *Signing of the Declaration of Independence* and with portraits of eminent contemporaries. In the 1830s he turned increasingly to painting. At first he worked mainly as a portraitist, but then devoted himself to landscape, becoming a leading figure of the *Hudson River School. Thomas *Cole was a major source of inspiration, and Durand's most famous painting, *Kindred Spirits* (New York Public Library, 1849), was painted as a memorial to him; it shows Cole and the poet William Cullen Bryant admiring spectacular scenery in the Catskill Mountains, New York State. Durand was a founder member of the National Academy of Design in 1826 and its President from 1845 to 1861.

Durand-Ruel, Paul (1831–1922). The best-known member of a family of French picture dealers, renowned as the first dealer to give consistent support to the *Impressionists. He took over the family firm in 1865 and established himself as the main dealer of the *Barbizon School painters. It was one of these—Charles *Daubigny—who introduced him to *Monet and *Pissarro when all four had taken refuge in England from the Franco-Prussian War of 1870–1. For years Durand-Ruel was such a solitary champion of the Impressionists that he often came near to bankruptcy, but in 1886 he achieved a breakthrough with an exhibition of their work in New York, the success of which encouraged him to open a branch of his firm there. This played a major role in building up some of the great American collections of Impressionists. After Durand-Ruel's death, Monet wrote to the dealer's son: 'I shall never forget all that my friends and I owe to your father, in a very special way.' Durand-Ruel was also noteworthy in being one of the first dealers to handle the work of El *Greco.

Duranty, Edmond (1833–80). French novelist and art critic, chiefly known now as an early champion of the *Impressionists and a friend of *Degas. Inspired by *Courbet and *Champfleury, he came within the ambit of 19th-cent. social *Realism and advocated the view that artists should depict modern city life, 'removing the partition which separates the studio from everyday life'. In his novel *Le Peintre Louis Martin* (1881), *Manet, Courbet, Degas, and other painters figured. Duranty was himself painted by Degas (Burrell Coll., Glasgow, 1879), and was one of the figures in *Fantin-Latour's *Hommage à Delacroix* (Musée d'Orsay, Paris, 1864).

Dürer, Albrecht (1471–1528). German graphic artist and painter, the greatest figure of *Renaissance art in northern Europe. Son of a goldsmith, Albrecht Dürer the Elder, and godson of Anthony Koberger (*c.*1445–1513), one of Germany's foremost printers and publishers, he attended a Latin school where he met the humanist and poet Willibald Pirckheimer (1470–1530), who was to become his lifelong friend and correspondent, and then was apprenticed when 15 years old to the leading painter and book illustrator in his native Nuremberg, Michael *Wolgemut. These four men exercised a powerful influence on Dürer's genius and determined to some degree his artistic career. Albrecht the Elder must not only have instilled into him the devotion to exact and meticulous detail that is the mark of a goldsmith's work, but also taught him the rudiments of drawing, as is borne out by the young boy's self-portrait (Albertina, Vienna, 1484), modelled on one by his father (also Albertina). From Wolgemut, whose influence was mainly technical, he learned the painter's trade and the craft of *woodcut. Through Koberger he had access to the world of books and learning, and Pirckheimer directed these interests towards Italy and the new humanism. From the beginning Dürer's world reached well beyond the normal medieval workshop.

After completing his apprenticeship Dürer set out in 1490 on the usual bachelor's journey. He went to the Upper Rhine in search of Germany's leading painter and engraver Martin *Schongauer, who, however, had just died when Dürer reached Colmar in 1492, and he worked for a while as a book illustrator in Basle and Strasburg, making woodcuts for Sebastian Brant's *Ship of Fools*.

After his return to Nuremberg and his marriage (1494) he went on a short visit to north Italy and then set up a workshop in his native town. Though also active as a painter—the self-portrait of 1500 and the *Paumgärtner Altar* of 1504 (both Alte Pinakothek, Munich) being the most important early works—he was for years preoccupied with woodcuts and engravings, among which the large series of the *Apocalypse* (1498), the *Great Passion* (1510), and the *Life of the Virgin* (1510) take first place. In spite of the traditional subject-matter, they are revolutionary in approach, size, and subtlety of technique. They are alive with dramatic tension and a pathos which is not only the result of his close study of *Mantegna's engravings but also an expression of his participation in the spiritual life of his day.

At the same time he began to be occupied by the Renaissance problems of *perspective, of *ideal beauty, of proportion and harmony. He sought instruction from Jacopo de' *Barbari, a travelling Italian artist visiting Nuremberg, and he read *Vitruvius. In 1505–7 he made another visit to Venice in pursuit of his new search. Dürer returned with a system of human proportions which he must have met with in circles close to *Leonardo. His great admiration for Giovanni *Bellini enhanced his sense of colour and the *Feast of the Rose-garlands* (NG, Prague, 1506), done in Venice, was meant to compete with the best Venetian painting.

Most of his landscape watercolours also belong to this period. They are unique in several ways: as personal records, in their choice of medium and subject, but most of all since they seem to have been made for sheer pleasure and not—as was usual with sketches in those days—with larger works in view.

By now Dürer was a well-established painter, engraver, and woodcutter. Commissions for large altar-paintings came not only from his home town of Nuremberg (*Adoration of the Trinity*, Kunsthistorisches Mus., Vienna, 1511), but from further afield, and after *c.*1512 his most important patron was the emperor Maximilian (1459–1519). For him

Dürer designed an enormous triumphal arch laden with history and allegory, and a triumphal procession, all on paper and executed in woodcut by members of Dürer's workshop and others.

At the same time his creative spirit found outlets entirely of his own choosing in such celebrated engravings as *The Knight, Death, and the Devil* (1513), *St Jerome in his Study* (1514), and the brooding and enigmatic allegory of the *Melencolia I* (1514). In these works he achieved a mastery of *line engraving that has never been surpassed, rivalling the richness and textures of painting. During these years he also experimented with a new technique, *etching, and found in it the means to convey, in an *Agony in the Garden* (1515), his troubled religious feelings. There is other evidence to tell us how deeply Dürer felt during the years of the Reformation and we know from one of his own letters that liberation and consolation came to him finally through Luther's writings.

In 1520–1 a journey to the court of the emperor Charles V to ask for a renewal of his pension took him to the Netherlands, where he was fêted as the acknowledged leader of his profession. The day-to-day diary that Dürer kept on this tour, together with his drawings showing the people and places he saw, is the first record of its kind in the history of art. After his return to Nuremberg Dürer was busy with portraits and with the designs for yet another *Passion* series, but his main task was the two panels with the *Four Apostles* (Alte Pinakothek, Munich) which he presented to his native town in 1526, an action without precedent. Here Dürer summed up his life's work: the study of the ideal human figure and the expression of a deeply felt religious message.

It was only natural that a man of Dürer's cast of mind should pursue theoretical studies throughout his life, and that among them—according to his own testimony—proportion should take first place, other things being attempted only 'if God should give me time'. The *Underweyssung der Messung* (Treatise on Measurement) (1525) was published by Dürer himself, but the *Vier Bücher von menschlicher Proportion* (Four Books on Human Proportion) (1528) was published posthumously.

When Dürer died in 1528, though he was widely known as a painter, his real fame rested on his graphic work, which was used in the north and south very much as pattern-books had been. Erasmus (1466–1536) called him 'the *Apelles of black lines', the highest praise that a student of the ancients could give to any artist.

Dusart, Cornelis (1660–1704). Dutch painter of peasant scenes. He was the pupil and assistant of Adriaen van *Ostade and completed several works left unfinished at his master's death. There is thus sometimes confusion between their hands, although Dusart's work is in general coarser, with a tendency towards *caricature.

Dutch mordant. An acid used for biting the plate in *etching. A solution of dilute hydrochloric acid with potassium chlorate, it has a milder action than nitric acid and is preferred to it for *aquatint, *soft-ground etching, and other fine, close work.

Duveen, Joseph (Baron Duveen) (1869–1939). English art dealer, patron, and philanthropist. In 1886 he entered the firm of his father, Sir **Joseph Duveen** (1843–1908), and with his enormous energy and great gift for salesmanship expanded it into the largest firm of art dealers in the world, operating on an unprecedented scale. He employed Bernard *Berenson to give his seal of authenticity to the *Renaissance paintings he sold and he was the main agent in forming the collections of such fabulously wealthy Americans as *Frick, *Kress, and Andrew *Mellon. Duveen's benefactions to the arts were also on a princely scale. In addition to giving many pictures to national collections, he paid for extensions or new galleries at the *National Gallery, the *National Portrait Gallery, the *Tate Gallery, and the *British Museum (to house the *Elgin Marbles). He also bore the cost of decorations at the *Wallace Collection and of Rex *Whistler's murals at the Tate Gallery and endowed a chair for the history of art at London University.

Duvet, Jean (1485–1561/70). French engraver and goldsmith, sometimes called the Master of the Unicorn from his series of engravings (probably from the 1540s) on the medieval theme of the hunting of the unicorn. Little is known of his life (he lived mainly in his native Langres and in Dijon), but the *Renaissance influence in his early work strongly suggests that he spent some time in Italy. His most famous work, a set of twenty-four engravings illustrating the *Apocalypse*, published at Lyons in 1561, is, however, completely different in style. They borrow many features from *Dürer's series on the *Apocalypse*, but are a world apart in spirit, for Duvet treats the subject with a visionary intensity and expressive freedom that seems to anticipate *Blake (who may well have known Duvet's work). His work reflects the disturbed religious conditions which prevailed in Langres and is in complete contrast to the mannered elegance of the School of *Fontainebleau, then the dominant force in French art.

Duyster, Willem Cornelisz. (1599–1635). Dutch painter of *genre scenes and portraits, active mainly in his native Amsterdam. Most of his paintings depict soldiers, sometimes in action, but more usually drinking, gaming, or wooing. His delicate skill at painting textiles, and even more important his ability to characterize individuals and his power to express subtle psychological relationships between them, suggest that if he had not been carried off by the plague in his mid-30s he might well have rivalled *Terborch. There are two examples of his fairly rare work in the National Gallery, London.

Dyce, William (1806–64). Scottish painter, designer, and art educationalist. He made several visits to Italy and was profoundly impressed by Renaissance painting, especially that of *Raphael, and stimulated by the problem of fresco in relation to its architectural setting. He also came into contact with the German *Nazarenes, whose Christian primitivism he tried to acclimatize in Scotland. Dyce was widely talented; he was an accomplished musician and wrote a prize-winning paper on electromagnetism, but for some time he was successful mainly as a rather conventional portraitist and it was not until after 1840 that he was able to devote himself to more ambitious work, producing decorative schemes for the Houses of Parliament, several churches, and the royal residences Buckingham Palace and Osborne on the Isle of Wight (he was a favourite of Prince *Albert). His importance extended beyond his own work, for 'there was no major [artistic] undertaking in mid nineteenth-century Britain in which he did not play either an executive or advisory role' (David and Francina Irwin, *Scottish Painters at Home and Abroad*, 1975). He was commissioned by the Council of the newly founded Government School of Design, which included *Eastlake, *Chantrey, and *Callcott, to investigate state schools in France, Prussia, and Bavaria, and was appointed Director in 1840 on his return. His ideas on design were much respected, and although he had little influence on manufacturers, his training of teachers was important. As a painter he was versatile, for as well as portraits and history pictures he painted works as varied as the delightfully sentimental *Titian's First Essay in Colour* (Aberdeen Art Gallery, 1856–7) and *Pegwell Bay, Kent* (Tate, London, 1859–60), one of the most remarkable of all Victorian landscapes. His bright colours and naturalistic detail formed a bridge between the Nazarenes and the *Pre-Raphaelites, but his design is usually stronger and his high-mindedness more convincing than in the works of either.

Dyck, Sir Anthony van (1599–1641). Apart from *Rubens, the greatest Flemish painter of the 17th cent. In 1609 he began his apprenticeship with Hendrick van *Balen in his native Antwerp and he was exceptionally precocious: on his earliest dated painting (*An Elderly Man*, Musées Royaux, Brussels, 1613) he has proudly inscribed his own age (14) as well as that of the sitter. Although he did not become a master in the painters' guild until 1618, there is evidence that he was working independently for some years before this, even though this was forbidden by guild regulations. Probably soon after graduating he entered Rubens's workshop. Strictly speaking he should not be called Rubens's pupil, as he was an accomplished painter when he went to work for him. Nevertheless the two years or so he spent with Rubens were decisive and Rubens's influence upon his painting is unmistakable, although van Dyck's style was always less energetic and more highly strung.

In 1620 van Dyck went to London, where he spent a few months in the service of James I, then in 1621 to Italy, where he stayed until 1627. A sketchbook of his drawings (BM, London) contains a record of his Italian impressions. In Italy he toned down the Flemish robustness of his early pictures and created the refined and elegant style which remained characteristic of his work through the rest of his life. He travelled a good deal in Italy, but worked mainly in Genoa, where he painted a series of grand portraits of the nobility in which he established a distinctive aristocratic type with proud mien and slender figure enhanced by the famous 'van Dyck' hands. A superb example is the full-length *Marchesa Elena Grimaldi* (NG, Washington, 1623), of which Sir David Piper wrote (*Van Dyck*, 1968): 'this is how one imagines any feminine aristocrat worthy of her rank must feel herself essentially to be, yet did not know it till van Dyck showed her—aloof and formally regal, but endowed with an elegance and grace that are infinitely seductive.' The years 1628–32 were spent mainly at Antwerp. From 1632 until his death he was in England—except for visits to the Continent—as painter to Charles I (1600–49), from whom he received a knighthood. Perhaps the strongest evidence of his power as a portraitist is the fact that today we see Charles and his court through van Dyck's eyes. Many of his portraits of the king and his family are still in the royal collection; among the others the most famous is probably the huge equestrian portrait of Charles (*c*.1637) in the National Gallery, London. The Royal Collection still has one of the finest representations of his work. It is customary to accuse van Dyck of invariably flattering his sitters, but not all his patrons would have agreed. When the Countess of Sussex saw the portrait (now lost) van Dyck painted of her she felt 'very ill-favourede' and 'quite out of love with myself, the face is so bige and so fate that it pleases me not at all. It lokes lyke on of the windes puffinge—but truly I think tis lyke the originale.'

Van Dyck's fame derives chiefly from his portraits and his influence on English portraiture has been profound and lasting: *Gainsborough, in particular, revered him, but he was an inspiration to many others until the early 20th cent., when society portraiture ceased to be a major form of artistic expression. He also painted religious and mythological subjects, however, and a surprising facet of his activity is revealed by his landscapes in watercolour (BM, London). His *Iconography* (1645) is a series of etchings or engravings of his famous contemporaries. Van Dyck etched some of the plates himself, and many more were engraved after his drawings and oil sketches.

Dying Gaul. Celebrated marble statue in the Capitoline Museum, Rome, showing a wounded warrior supporting himself wearily on one arm. It is a Roman copy of a Greek work in the *Pergamene style of the late 3rd cent. BC. The statue is sometimes called the 'Dying Gladiator', but this is a misnomer as the hairstyle and accessories are scrupulously Gallic. It was first recorded in Rome in 1623 and was soon famous. After visiting Italy in 1644–5, the diarist John Evelyn wrote that it was 'so much followed by all the rare Artists, as the many Copies and Statues testifie, now almost dispers'd through all Europ, both in stone & metall'. It was among the works removed from Italy by Napoleon and was in Paris from 1798 to 1815. Unlike many once famous antique statues, the *Dying Gaul* is still admired, particularly for its pathos.

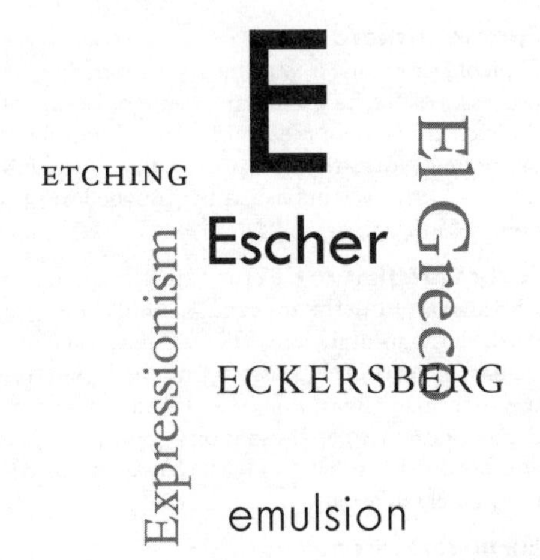

Eakins, Thomas (1844–1916). American painter, active for most of his life in his native city of Philadelphia. Eakins is regarded by most critics as the outstanding American painter of the 19th cent. and by many as the greatest his country has yet produced. In 1866–70 he was in Europe. He studied in Paris with *Gérôme, but learnt most from the Spanish painters *Velázquez and *Ribera, absorbing a precise and uncompromising sense for actuality which he applied to portraiture and *genre pictures of the life of his native city (boating and bathing were favourite themes). He began teaching at the Pennsylvania Academy of Fine Arts in 1876 and was attacked for his radical ideas, particularly his insistence on working from nude models. In 1886 he was forced to resign after allowing a mixed class to draw from a completely nude male model.

Eakins's quest for realism led him to study anatomy and make full use of *Muybridge's photographic researches, but the scientific bent in his work is of less importance than his honesty and depth of characterization. His portraits are often compared with Rembrandt's because of their dramatic play of sombre lighting and sense of inner truth. The most famous of his paintings is *The Gross Clinic* (Jefferson Medical Coll., Philadelphia, 1875), which aroused controversy because of its unsparing depiction of surgery, an experience that was repeated with *The Agnew Clinic* (Univ. of Pennsylvania, 1889). Because of financial support from his father Eakins could continue on his chosen course despite public abuse, but much of his later career was spent working in bitter isolation. It was only near the end of his life that he achieved recognition as a great master, and in the first two decades of the 20th cent. his desire to 'peer deeper into the heart of American life' was reflected in the work of the *Ash-can School and other Realist painters. As well as being a painter and photographer, Eakins also made a few sculptures. Little of his work can be seen outside the USA; the best collection is in the Philadelphia Museum. His wife, **Susan Hannah Macdowell Eakins** (1851–1938), whom he married in 1884, was also a painter and photographer, as well as an accomplished pianist.

Eardley, Joan (1921–63). British painter, born in Sussex but considered Scottish (her mother was Scottish and she lived in Scotland from 1940). One of her teachers was James *Cowie; he perhaps helped to shape her preference for subjects drawn from everyday experience, but her approach was more earthy and sensuous than his. She divided her time between Glasgow (where she painted *kitchen sink subjects) and the fishing village of Catterline, about 20 miles south of Aberdeen on the north-east coast. Her favourite subjects in her later years were the village and the sea, especially in stormy weather (she is said to have set off from her Glasgow home as soon as she heard reports of gales). The freely painted, often bleak and desolate works that resulted are among the most powerful and individual landscapes in 20th-cent. British art. After her early death from breast cancer her ashes were scattered on the beach at Catterline. Her work is well represented in the Scottish National Gallery of Modern Art, Edinburgh.

Earl, Ralph (1751–1801). American painter, active in Connecticut, Massachusetts, New York, and Vermont, and also in England (1778–85), when his loyalty to the British put his life in danger in his homeland. He painted landscapes and battle scenes of the Revolution, but was primarily a portraitist. Although his style became softer and sophisticated after studying with *West in London, his work generally has a sincerity and freshness of vision that makes him one of the finest American artists of the 18th cent. His presentation of character is extremely forthright and his portraits convey the immense pride his New England sitters took in their possessions. For instance in the portrait *Oliver Ellsworth and his Wife* (Wadsworth Atheneum, Hartford, Connecticut, 1792) a window shows a view of the very house in which the proud owners are sitting. Earl's personal life was a disaster. He was imprisoned for debt and died an alcoholic after deserting both of his wives in turn. Other members of his family were artists, notably his brother **James** (1761–96) and his son **Ralph E. W. Earl** (*c.*1785–1838), who is remembered mainly for his portraits of President Andrew Jackson (1767–1845), whose niece he had married.

Earlom, Richard (1743–1822). British engraver, a pupil of *Cipriani. He was the first to combine the processes of *etching and *mezzotint by his use of the etched line on a mezzotint ground. His most celebrated work, the 200 plates after *Claude's *Liber Veritatis*, was published by *Boydell in two volumes in 1777.

Early Christian art. Term generally applied to Christian art from the 3rd cent. AD until about 750, particularly in Italy and the western Mediterranean. The art of the eastern Empire during this time is termed *Byzantine art and the art of the barbarian German tribes *Migration Period art. There are, however, no hard-and-fast demarcations between these divisions.

Earthwork. See LAND ART.

easel. Stand on which a painting is supported while the artist works on it. The oldest representation of an easel is on an Egyptian relief of the Old Kingdom (*c.*2600–2150 BC). *Pliny mentions a *machina* in an anecdote about *Apelles. *Renaissance illustrations of the artist at work show all kinds of contrivances, the commonest being the three-legged easel with pegs such as we still use today. Light folding easels were not made until the 18th and 19th cents., when painters took to working out of doors and sketching became a pastime for the *amateur. The studio easel, a 19th-cent. invention, is a heavy piece of furniture which runs on castors or wheels, and served to impress the clients of portrait painters. The oil painter needs an easel which will support his canvas almost vertically or tip it slightly forward to prevent reflection from the wet paint, whereas the watercolourist must be able to lay his paper nearly flat so that the wet paint will not run down. The term 'easel-painting' is applied to any picture small enough to have been painted on a standard easel.

Easter sepulchre. A representation of the tomb of Christ that figured in Easter celebrations in English churches until 1560. On Good Friday the Host was ceremonially placed in the breast of a figure of Christ; this was placed in the sepulchre and a vigil was kept over it until Easter Sunday, when the Host was carried in triumphant procession. Usually such sepulchres were of wood, temporarily erected, but almost no trace of this type remains (a wooden chest, *c.*1500, at Coity, Mid Glamorgan, carved with the instruments of the Passion, has been plausibly identified as an example). Around 1300 permanent stone structures began to be made for the purpose, usually in the form of a recess in the north wall of the choir. These were often richly sculptured, an outstanding example being at Hawton, Notts. (*c.*1330). Such monuments are found only in England. Elsewhere in Europe a sculptured group of the Entombment served a similar purpose.

Eastlake, Sir Charles Lock (1793–1865). English painter, art historian, and administrator. He studied under *Haydon and early became famous with his *Napoleon on board the Bellerophon* (National Maritime Mus., London, 1815), made from sketches when he witnessed Napoleon on board ship (in Eastlake's native Plymouth) *en route* to exile in St Helena. Using the proceeds from the sale of this work he lived in Rome 1816–30, and there painted picturesque scenes of the Roman Campagna, often peopled by *banditti*, that became very popular in England. After his return to England, however, he turned increasingly to administration and achieved a remarkable record as a public servant. He was Librarian of the *Royal Academy (1842–4), Secretary to the Fine Arts Commission (1842), Keeper of the *National Gallery (1843–7), Commissioner for the 1851 *Great Exhibition, President of the Royal Academy (1850), and Director of the National Gallery (the first holder of this post) from 1855 until his death. Among his writings, *Materials for a History of Oil Painting* (1847) was a pioneer work. His informed purchases of early Italian paintings for the National Gallery were largely responsible for its outstanding representation in this area. His wife, **Elizabeth**, *née* Rigby (1809–93), was in her own right a figure in the literary-artistic world of the day. She wrote several books on art and translated *Waagen's *Treasures of Art in Great Britain*. Her *Memoir* (1870) of her husband was published along with his own *Contributions to the Literature of the Fine Arts*. Eastlake's nephew, **Charles Locke Eastlake** (1836–1906), was Keeper of the National Gallery, 1878–98, and published several works on art and decoration, the best known of which was *Hints on Household Taste* (1868), in which he advocated quality of materials and workmanship. It was highly influential in England and even more so in America, although so-called 'Eastlake furniture' often has little to do with his ideas. In 1878 he stated 'I find American tradesmen continually advertising what they call "Eastlake" furniture, with the production of which I have had nothing whatever to do, and for the taste of which I should be very sorry to be considered responsible.' Eastlake also wrote *A History of the Gothic Revival* (1871), the pioneering work on the subject and the standard work until Kenneth *Clark's *The Gothic Revival* appeared in 1928.

Eckersberg, Christoffer Wilhelm (1783–1853). Danish painter. After being trained in Copenhagen and studying in Paris (1810–13) under J.-L. *David, he continued his studies in Rome (1814), where he executed a masterly portrait of his friend *Thorvaldsen (Royal Academy, Copenhagen, 1815). Returning to Copenhagen in 1816, he occupied himself mainly with portraits, minutely rendering the features of his models with a *Neoclassic feeling for clarity and purity of line. He also painted many landscapes, however (as he had done in Rome), and as an influential teacher at the Copenhagen Academy (from 1818) he introduced painting from nature into the curriculum. His pupils included J. C. *Dahl and Christen *Købke.

eclectic, eclecticism. Terms in criticism for a person or style which conflates features borrowed from various sources. Such a style often arises from the overt or tacit doctrine that the excellences of great masters can be selected and combined in one work of art. After *Vasari had praised *Raphael for his skill in selecting the best from the art of his predecessors, it became commonplace to use the same formula in eulogies of other artists. Thus it was said that *Tintoretto had set himself to combine the drawing of *Michelangelo with the colour of *Titian.

In the 18th cent. 'eclectics' became a label for the *Carracci family and their Bolognese followers, and gradually the term came to be used mainly pejoratively, implying lack of originality. Such usage has been abandoned in serious criticism, and it is clear that the Carracci never adopted eclecticism as the fundamental principle of their school. As Denis *Mahon has said: 'Annibale Carracci, the greatest member of the family and one of the founders of 17th-cent. painting, was . . . contemptuous of art theory, and (far from being a dispenser of learned recipes and synthetic systems) was in practice one of the most insatiable experimentalists known to the history of art.'

École de Paris. See PARIS, SCHOOL OF.

écorché figure (French: 'flayed'). A representation of a figure without the skin, displaying the muscles. Drawings, prints, and statues of such figures, both human and animal (horses were particularly popular), were much used in art teaching from the 16th cent. *Stubbs's *écorché* figures of animals are well known and *Houdon made a celebrated human *écorché* statue (Schlossmuseum, Gotha, 1767).

Edelfelt, Albert (1854–1905). Finnish painter, who with *Gallen-Kallela ranks as his country's leading artist in the 19th cent. He trained in Antwerp and then in Paris, where under the influence of his friend *Bastien-Lepage he took up *plein-air* naturalism. His paintings gave a fresh interpretation of Finnish country life and he sometimes set biblical scenes in the Finnish landscape (*Christ and Mary Magdalene*, Atheneum, Helsinki, 1890). Much of his later work was on themes from Finnish history, a type of patriotic work that was relevant to Russia's growing oppression of his country. Edelfelt was also an outstanding book illustrator and portraitist.

Eddy, Don. See SUPERREALISM.

Edelinck, Gerard (1640–1707). Flemish-born engraver who settled in Paris in 1666 and became a French citizen in 1675. He was the son-in-law of *Nanteuil and like him was celebrated as a portrait engraver. His prints after the Old Masters are also highly distinguished, notably that of *Rubens's copy of *Leonardo's *Battle of Anghiari*.

Edwards, Edward (1738–1806). English landscape and history painter, Professor of Perspective at the *Royal Academy from 1788. His pictures are now forgotten, but he is remembered for his *Anecdotes of Painters*, published in 1808 as a continuation to the *Anecdotes* of Horace *Walpole.

Eeckhout, Gerbrandt van den (1621–74). Dutch painter, born and active in Amsterdam. He was a pupil of *Rembrandt and, according to *Houbraken, his 'great friend'. His religious paintings were deeply influenced by Rembrandt; *St Peter Healing the Lame* (M. H. De Young Memorial Mus., San Francisco, 1667), for example, shows how well he understood the broad touch and warm colours of the master's mature works. Eeckhout's *genre scenes, on the other hand, are close to *Terborch in style (*The Music Lesson*, Statens Museum for Kunst, Copenhagen, 1655). He also practised successfully as a portraitist (*Four Officers of the Amsterdam Coopers' and Wine-Rackers' Guild*, NG, London, 1657).

Egg, Augustus (1816–63). English painter. He painted historical, anecdotal, and literary themes (he was a friend of Dickens and a talented actor), and under the influence of the *Pre-Raphaelites he also turned to overtly moralizing subjects. The most famous of these is the series of three pictures *Past and Present* (Tate, London, 1858), which melodramatically illustrates the dire consequences of adultery.

egg tempera. See TEMPERA.

Egremont, third Earl of. See TURNER.

Egyptian taste. Style of interior decoration and design popular in the early 19th cent. Scholars and connoisseurs had begun to interest themselves in ancient Egypt in the middle of the 18th cent., and in 1769 *Piranesi published a number of designs of Egyptian ornament in his *Diverse Maniere di Adornare i Cammini*. But the fashion did not become widespread until Napoleon's Egyptian campaign in 1798 made it topical and D.-V. *Denon's lavishly illustrated *Voyage dans la Basse et la Haute Égypte* (1802) provided artists and designers with a large repertory of ready-made ornament—Egyptian mummy-cases and idols, cats, hawks, sphinxes, winged globes, *obelisks, hieroglyphs, etc. In England Thomas Chippendale the Younger made furniture in the style for Stourhead in Wiltshire, and Thomas *Hope designed a special Egyptian room in his house, Deepdene, to contain his collection of Egyptian antiquities.

Eidophusikon. An ingenious system of moving pictures within a proscenium arch which, by a clever disposition of lights, coloured gauzes, and the like imitated views in and about London with varying atmospheric effects at different times of day, to the accompaniment of appropriate musical effects. The inventor was P.-J. *de Loutherbourg, who exhibited the Eidophusikon (meaning 'image of nature') in London in 1782 with immediate popular success, appealing to lovers of romantic and *Picturesque scenery and deeply impressing both *Gainsborough and *Reynolds. His masterpiece was the realistic representation of a storm at sea with sound effects which were hailed as a new art: 'the picturesque of sound'. There is an interesting eyewitness account of the Eidophusikon and its effects in *Wine and Walnuts* (1823) by Ephraim Hardcastle.

Eight, The. A group of American painters who exhibited together in 1908, united by opposition to the conservative *National Academy of Design and a determination to bring painting back into direct touch with life. The original group consisted of: Arthur B. *Davies, Maurice *Prendergast, Ernest *Lawson, Robert *Henri, George *Luks, William J. *Glackens, John *Sloan, and Everett *Shinn. It came into being when the National Academy of Design rejected the work of Luks, Sloan, and Glackens, and Henri, the dominant personality in the group, in protest withdrew his own pictures from the exhibition of 1907. Arthur B. Davies was then asked to organize an independent exhibition at the Macbeth Gallery in New York. This exhibition, which took place in February 1908 and was the only occasion on which The Eight exhibited together, was subsequently shown by the Pennsylvania Academy and circulated to eight other museums over a period of a year. The members of the group were not unified stylistically, but they mainly painted contemporary urban life. The exhibition is regarded as an important step in the development of a vigorous native school of American painting in the 20th cent. Glackens, Henri, Luks, Shinn, and Sloan went on to become part of the *Ash-can School.

The Eight was also the name of a group of progressive Czech artists formed in Prague in 1907, and of a group of Hungarian painters, inspired by *Post-Impressionism, founded in Budapest in 1909.

Elementarism. A modified form of *Neo-Plasticism propounded by van *Doesburg in the mid-1920s. While continuing *Mondrian's restriction to the right angle, Elementarism abandoned his insistence on the use of strict horizontals and verticals. By introducing inclined lines and forms van Doesburg sought to achieve a quality of dynamic tension. Mondrian was so offended by this 'heresy' that he left De *Stijl.

Elgin Marbles. A collection of Greek sculpture from the Acropolis in Athens acquired by the British diplomat Thomas Bruce, 7th Earl of Elgin (1766–1841), in 1801–3, when he was Ambassador to the Sultan of Turkey, who at this time ruled Greece. The collection consisted mainly of sculptures from the Parthenon (most of what had survived), but included other pieces, notably a *caryatid from the Erechtheum. They were shipped to Britain over a period of several years and in 1816 they were sold to the nation and installed in the *British Museum. Elgin was paid £35,000, which was only about half the total they had cost him. By their exhibition in London original Greek sculpture of the *Classical age first became generally accessible in modern times; until then people had been familiar only with Roman and late *Hellenistic copies. Their first impact was enormous: *Flaxman was bowled over, declaring that compared with the figure of Theseus (now usually identified as Dionysus or Herakles) from the Parthenon the **Apollo Belvedere* was 'a dancing master', and *Haydon wrote 'I consider it truly the greatest blessing that ever happened to this country their being brought here.' Earlier, when asked to restore them, *Canova said 'it would be a sacrilege in him or any man to presume to touch them with a chisel.' Many other artists voiced similar opinions, although certain connoisseurs associated with the Society of *Dilettanti, led by Richard Payne *Knight, were unenthusiastic or even disparaging. 'You have lost your labour, my Lord Elgin,' said Payne Knight. 'Your marbles are overrated: they are not Greek: they are Roman of the time of Hadrian.'

Although the supporters of the marbles won the day and the sculptures have come to be universally recognized as one of the summits of ancient art, they have continued to be the subject of controversy on another count—that of the morality or legality of their removal when Greece was under the dominion of a foreign power. Byron wrote of them as 'poor plunder from a bleeding land' (*Childe Harold's Pilgrimage*, Canto ii, 1812), and a campaign to have them restored to Greece is strongly active today. See also PHIDIAS.

El Greco. See GRECO, EL.

Eliasz., Nicolaes (called Pickenoy) (1590/1–1654/6). Dutch painter who, with Thomas de *Keyser, was the leading portraitist in Amsterdam until the arrival of *Rembrandt in 1631/2. His work is well represented in the Rijksmuseum, Amsterdam, and in the Louvre.

Elsheimer, Adam (1578–1610). German painter, etcher, and draughtsman, active mainly in Italy. Although he died young and his output was small he played a key role in the development of 17th-cent. landscape painting. He was born in Frankfurt, where he absorbed the *Coninxloo tradition, and moved to Italy in 1598. In Venice he worked with his countryman *Rottenhammer, then settled in Rome in 1600. His early *Mannerist style gave way to a more direct manner in which he showed great sensitivity to effects of light; his nocturnal scenes are particularly original, bringing out the best in his lyrical temperament, and he is credited with being the first artist to represent the constellations of the night sky accurately (*The Flight into Egypt*, Alte Pinakothek, Munich, 1609). He painted a few pictures in which figures predominate, but generally they are fused into a harmonious unity with their landscape settings. They are invariably on a small scale and on copper (the only exception is a self-portrait in the Uffizi, Florence, of doubtful attribution), but although exquisitely executed they have a grandeur out of all proportion to their size.

Elsheimer achieved fame during his lifetime and there are numerous contemporary copies of his works. His paintings were engraved by his pupil and patron, the Dutch amateur artist Count Hendrick Goudt (1573–1648), and Elsheimer himself made a number of etchings. In spite of his popularity he was personally unsuccessful and died in poverty. *Sandrart says he suffered from melancholia and was often unable to work; apparently he was imprisoned for debt. *Rubens was a friend of Elsheimer and after his death lamented his 'sin of sloth, by which he has deprived the world of the most beautiful things'; he also wrote 'I have never seen his equal in the realm of small figures, of landscapes, and of so many other subjects.' Both Rubens (Staatliche Kunstsammlungen, Kassel) and *Rembrandt (NG, Dublin) made paintings of *The Flight into Egypt* inspired by Elsheimer's masterpiece, and his influence is apparent in the work of many other 17th-cent. artists.

emblem. A visual image carrying a symbolic meaning and often accompanied by texts to explain this meaning. Printed collections of emblems ('emblem books') enjoyed a great vogue in the 16th and 17th cents. and were often used by artists as sources of imagery. In its most typical form the emblem consisted of a picture, a motto, and an explanatory verse called an epigram. For example, one of the most famous emblems depicted a dolphin and an anchor with the motto *Festina Lente* ('Make haste slowly'), to symbolize the idea that maturity is achieved by a combination of the speed and energy of the dolphin and the steadiness and gravity of the anchor. The aim of the emblem therefore was to give symbolic expression to a moral adage. Collections of emblems which were borne as personal devices by particular people (e.g. the emperor Charles V's device of the Pillars of Hercules with the motto *Plus Ultra*), known as *imprese*, were also published in great numbers. The sources drawn upon by the emblematists were diverse and included the Bible (e.g. the rainbow as an emblem of peace); medieval lapidaries (treatises on precious stones) and *bestiaries (e.g. the ermine as an emblem of purity); fables (especially Aesop); anecdotes from Classical myth and legend (e.g. Xenophon's story of Hercules' choice between Virtue and Pleasure). The emblem book was part of that vast codification of symbolism which led to Cesare Ripa's *Iconologia* (1593, first illustrated edition, 1603), which became the standard handbook on *iconography for artists.

Empire style. Style of furniture and interior decoration which started in Paris after the French Revolution and spread through Europe. The name is derived from the first Empire (1804–14), but the style continued to be fashionable until a form of *Gothic Revival came into vogue *c.*1830. It corresponds to the *Regency style in England. Its origin was largely due to the architects Charles *Percier, and Pierre-François Fontaine, who decorated the state apartments of Napoleon. Interiors designed by them survive at Fontainebleau, Compiègne, and Malmaison. Basically the style is *Neoclassical, with an attempt to copy what was known of ancient furniture and decorative motifs. There was also an affectation of Egyptian motifs owing no doubt to interest inspired by Napoleon's Egyptian campaigns (see EGYPTIAN TASTE). In women's dress the Empire style coincided with a distinctive high-waisted fashion embellished with dazzling embroidery.

emulsion. A liquid in which water is combined with an oily or resinous substance in such a way that they will not separate out. Oil proverbially will not mix with water, but if an emulsifying agent—such as albumen—is added, it will surround the drops of oil and prevent them from coming together. The *medium of *tempera painting is always an emulsion. The natural emulsions used most commonly in painting are egg-yolk and *casein. Both have the advantage that once they have set they are not soluble in water.

enamel. A smooth, glossy material made by fusing glass to a prepared surface, usually of metal. The term is also applied to any object made with, or decorated by, this material. The study of enamelling belongs mainly to the history of jewellery and the decorative arts, but in the Middle Ages enamel was sometimes used for major works, notably *Nicolas of Verdun's Klosterneuburg Altar. Today the word 'enamel' is loosely used of any glossy protective covering such as durable paint or varnish applied to the surface of objects made from metal, wood, etc., as for example the lining of cooking vessels.

encarnado (Spanish: 'flesh-coloured'). Term applied in Spanish art to the painting of the flesh parts of wooden sculptures in more or less naturalistic colours. The term 'estofado' (literally 'quilted') is applied to the painting of draperies. In the 16th cent. the paint of both flesh and draperies was given a glossy finish, but in the 17th cent. a matt finish was adopted for greater realism. Such work was sometimes done by distinguished painters as well as specialist craftsmen—*encarnadores* and *estofadores*. For example, *Pacheco often painted sculptures by *Montañés. See also POLYCHROME.

encaustic painting. Technique of painting with *pigments mixed with hot wax. Its name derives from a Greek word meaning 'burnt in' and it was one of the principal painting techniques of the ancient world. The technique was said to have been perfected in the 4th cent. BC by *Pausias, who painted with it small figures on ceiling panels. The most remarkable surviving examples are the mummy portraits from *Faiyum, dating from the 1st cent. BC to *c.* 3rd cent. AD. *Pliny describes two methods which were already 'ancient' in his day (one of them on ivory), and a third newer method which had been devised since it became the practice to paint ships, and he records that it stood up to sun, salt, and winds. Plutarch also pays tribute to its durability: 'A beautiful woman leaves in the heart of an indifferent man an image as fleeting as a reflection on water. But, in the heart of one who loves, the image is fixed with fire like an encaustic painting which time can never obliterate.' The older methods described by Pliny were done with a spatula; the newer technique with a brush. Signs of the brush can be seen in some of the Faiyum portraits.

Encaustic painting was the commonest technique in the early centuries of the Christian era but fell into disuse in the 8th or 9th cent. Since then various attempts have been made to revive it (e.g. by Julius *Schnorr von Carolsfeld, who painted several scenes in encaustic in the Residenz at Munich in 1831). Jasper *Johns has used encaustic in his *Flag* and *Target* paintings, but the technique finds few exponents today, probably because it is too troublesome. It may be added that in the past many artists working in oils have added wax to their colours, notably van *Gogh and, it is believed, Sir Joshua *Reynolds.

Endoios. Athenian sculptor of the latter part of the 6th cent. BC. A much damaged marble seated *Athena* in the Acropolis Museum, Athens, has been identified with a statue of his dedicated on the Acropolis in 564 BC. In pose and modelling it is one of the most naturalistic works of its time.

Engelbrechtsz., Cornelis (*c.*1460?–1527). Netherlandish painter active in Leiden, where he was the leading artist of his day. He is thought to have trained in Brussels and seems to have returned home through Antwerp. Although his style shows the influence of the Italianate tendencies prevalent at Antwerp, Engelbrechtsz.'s work has a deeper intensity of feeling that is *Gothic rather than *Mannerist in spirit. Contorted linear rhythms and resonant colouring characterize his highly personal art, which is closer to the *Master of the Virgo inter Virgines than to any Antwerp artist. The altarpieces of *The Crucifixion* and *The Lamentation* in the Municipal Museum, Leiden, are typical of his work. *Lucas van Leyden was his greatest pupil and tends to overshadow his achievements. The other pupils in his large studio included his three sons, **Pieter**, **Cornelis**, and **Lucas**, as well as Lucas van Leyden's brother, Aertgen.

English, Michael. See SUPERREALISM.

engraving. Term applied to various processes of cutting a design into a plate or block of metal or wood, and to the prints taken from these plates or blocks. The different processes of engraving are described under separate headings: *line engraving, *wood engraving, and so on. See also PRINT.

Ensor, James (1860–1949). Belgian painter and printmaker. One of the most original artists of his time, Ensor was one of the formative influences on *Expressionism and was claimed by the *Surrealists as a forerunner, but his work defies classification within any school or group. He was born at Ostend, where his parents (his father was English) kept a souvenir shop, and apart from his training in Brussels rarely left his home town. His early works were mainly bourgeois interiors painted in a thick and vigorous technique. When several were rejected by the Salon in Brussels in 1883, Ensor joined the progressive group Les *Vingt. During the 1880s his subject-matter changed and he began to introduce the fantastic and macabre elements which are chiefly associated with his name. He made much use of carnival masks, grotesque figures, skeletons, and bizarre and monstrous imaginings with a gruesome and ironic humour reminiscent of *Bosch and *Bruegel. The interest in masks probably originated in his parents' shop, but he was also one of the first European artists who appreciated African masks. His paintings, and even more his graphic work, took on a didactic or satirical flavour involving social or religious criticism and his most famous work, the huge *Entry of Christ into Brussels* (Getty Museum, Malibu, 1888), provoked such an outburst of criticism among his associates that he was expelled from Les Vingt. After this Ensor became a recluse and his outlook became even more misanthropic. His work changed little after about 1900, when he was content to repeat his favourite themes. In 1929 he was created a baron when his *Entry of Christ into Brussels* was first exhibited in public. There is an Ensor museum in Ostend.

entartete Kunst. See DEGENERATE ART.

environment art. Type of art in which the artist creates a three-dimensional space pre-programmed or mechanically energized in order to enclose the spectator and involve him in a multiplicity of sensory stimulations—visual, auditory, kinetic, tactile, and sometimes olfactory. Environment art, which began to establish itself in the 1960s, was closely linked with spectator involvement, and environments were deliberately planned with a view to forcing the spectator to participate in the *happenings or the 'game'. Allen *Kaprow, who also pioneered the happening, is sometimes credited with originating environment art, and other leading exponents include *Kienholz and *Oldenburg.

The word has been loosely used and sometimes it has been, wrongly, applied to *Earthworks art or its analogues—that is, to a category of art which consists in manipulating the natural environment on a large scale, instead of to the art which *creates* an environment to enfold and absorb the spectator.

Epstein, Sir Jacob (1880–1959). American-born sculptor (and occasional painter and illustrator) who settled in England in 1905 and became a British citizen in 1911. Before then, in 1902–5, he had studied in Paris and visits to the *Louvre aroused an interest in ancient and *primitive sculpture that lasted all his life and powerfully affected his work. His first important commission was executed in 1907–8: eighteen over life-size figures for the façade of the British Medical Association's headquarters in the Strand. The nude figures aroused a furore of abuse on the grounds of alleged obscenity and were destroyed in 1937 when the building was bought by the government of Southern Rhodesia. Such verbal attacks and acts of vandalism were to become a feature of Epstein's career. The next scandal came with his tomb of Oscar Wilde (Père Lachaise Cemetery, Paris, 1912), a magnificently bold and original piece featuring a hovering angel inspired by Assyrian sculpture. It was banned as indecent until a bronze plaque had been fixed over the angel's sexual organs. (It was removed in a night raid by a band of artists and poets.) While he was in Paris during the controversy Epstein met *Picasso, *Modigliani, and *Brancusi and was influenced by their formal simplifications. Back in England he associated with Wyndham *Lewis and the *Vorticists, and in *The Rock Drill* (Tate, London, 1913–14), a robot-like figure that was originally shown aggressively mounted on an enormous drill, he created his most radical work; he said it symbolized 'the terrible Frankenstein's monster we have made ourselves into'.

Epstein's later work was generally much less audacious than this, but his public sculptures were still attacked with monotonous regularity, their expressive use of distortion being offensive to conservative critics even when they were immune to charges of indecency. *Rima*, a memorial to the naturalist W. H. Hudson in Hyde Park (1922), and the enormous bronze group of *St Michael and the Devil* (1958) at Coventry Cathedral are two of his most famous later works. From the 1920s Epstein devoted himself more and more to bronze portrait busts, and these—unlike his monumental works—have always had an appreciative audience. Many of the great figures of the time sat for him and he portrayed them with psychological insight and great mastery of expressive surfaces, carrying on the tradition of *Rodin. Epstein was knighted in 1954.

Eragny Press. A private printing-press established by Lucien *Pissarro in 1894 at Bedford Park, London, and transferred to Hammersmith in

1903. It was named after the Normandy village where Pissarro lived before coming to England. From 1894 to 1903 the Eragny books were printed in the 'Vale' type designed by Charles *Ricketts. After 1903 the 'Brook' type, designed by Pissarro himself, was used. Characteristic of Eragny books, many of which were in French, were coloured frontispieces, woodcuts, initials, and decoration designed by Pissarro and engraved on wood by himself and his wife Esther. The press closed down in 1914 with the outbreak of the First World War.

Erbslöh, Adolf. See NEUE KÜNSTLERVEREINIGUNG.

Ercole de' Roberti. See ROBERTI.

Ernst, Max (1891–1976). German-born painter, printmaker, collagist, and sculptor who became an American citizen in 1948 and a French citizen in 1958, a major figure of *Dada and *Surrealism. He studied philosophy and psychology at Bonn University, but he became fascinated by the art of psychotics (he visited the insane as part of his studies) and neglected academic work for painting. After serving in the First World War he became with *Arp (his lifelong friend) the leader of the Dada movement in Cologne. In 1920 he organized one of Dada's most famous exhibitions in the conservatory of a restaurant there. Visitors entered through the lavatories and axes were provided so they could smash the exhibits if they felt so inclined. In 1922 he settled in Paris, bringing Dada techniques of *collage and *photomontage with him, and he joined the Surrealist movement on its formation in 1924. Even before then, however, he had painted works such as *The Elephant Celebes* (Tate, London, 1921) that are regarded as Surrealist masterpieces. The irrational and whimsical imagery seen here, in part inspired by childhood memories, occurs also in his highly individual collages. In them he manipulated banal engravings so that the contrast between the archaic appearance of the engraving and the startling novelty he created from it made an impact of strangeness and unreality. In this vein he produced 'collage novels', the best-known of which is *Une Semaine de Bonté* (1934). His work was always imaginative and experimental and he was a pioneering exponent of *frottage and *decalcomania. In the 1930s sculpture also began to occupy a prominent place in his work.

In 1938 he broke with the Surrealist movement, but this did not affect his work stylistically. He was interned for a short while after the German invasion of France and in 1941 went to New York, remaining in America until 1953 (apart from a visit to France in 1949). While in the USA he collaborated with *Breton and *Duchamp in the periodical *VVV*. He settled permanently in France in 1953 and in his late years acquired many honours. His painting of this time became more lyrical and abstract.

Ernst was married four times. His third (very brief) marriage was to Peggy *Guggenheim; his fourth wife (married 1946) was Dorothea Tanning (1910–), one of the outstanding American Surrealist painters. In the late 1930s he lived in Paris with the British-born (later Mexican) Surrealist painter and writer Leonora Carrington (1917–). His son **Jimmy Ernst** (1920–84) was also a painter.

Errard, Charles (*c.*1606–89). French painter, architect, and designer, the son of a painter of the same name (1570–*c.*1629). He spent many of his early years in Rome drawing from the *antique and working in *Poussin's studio, and later returned to be first Director of the French *Academy in Rome (1666). As a painter most of his work was decorative; it has almost all disappeared, but examples remain in the Luxembourg in Paris and the Palais de Justice in Rennes. As an architect he designed the 'clumsy and pedantic' (Anthony *Blunt) church of the Assumption in the rue St Honoré in Paris (1670–6).

Erté (Romain de Tertoff) (1892–1990). Russian-born French designer, painter, and sculptor. He was born in St Petersburg into an aristocratic family and in 1912 moved to Paris, where he studied at the *Académie Julian. Erté was best known for his fashion illustrations (particularly for the American magazine *Harper's Bazaar*) and for his costume and set designs for theatre, cabaret, opera, ballet, and cinema (he designed costumes for Mata Hari among other celebrities). However, he also painted, and in the 1960s he produced lithographs and made sculpture from sheet metal.

Escher, M. C. (Maurits Cornelis) (1898–1972). Dutch graphic artist. His prints (mainly woodcuts and lithographs) make sophisticated use of visual illusion, exploiting ambiguity between figure and ground, and between flat pattern and apparent three-dimensional recession. From *c.*1944 his work took on a *Surrealist flavour as he made brilliant play with optical illusion to represent, for example, staircases that appear to lead both up and down in the same direction. Mathematical concepts played a key role in many of these prints and they have been of considerable interest to mathematicians. An exhibition of them was given at the International Mathematical Congress, Amsterdam, in 1964. He has also had a strong appeal to the general public, particularly since the 1960s, when some young people felt that his work complemented the 'mind-expanding' experiences obtained through taking hallucinogenic drugs.

Escorial. A small village about 50 km. north-west of Madrid that gives its name to the palace and monastery built there by Philip II (1527–98), containing one of the great treasure houses of Spanish art. Begun in 1563 by Juan Bautista de Toledo and finished in 1584 by Juan de Herrera (1530–97), the huge complex of buildings is constructed entirely of granite and is overpoweringly austere. It comprises a royal palace and mausoleum combined with a Hieronymite monastery, college, and church. The plan is a large rectangle, roughly 160 m. × 200 m., and the layout has been compared to a gridiron, the attribute of St Lawrence, to whom the monastery was dedicated in acknowledgement of Philip II's victory over the French at St Quentin on St Lawrence's day, 1557.

Philip II tried unsuccessfully to persuade *Titian to come to Spain to paint altarpieces for the Escorial; and El *Greco was rejected after his *St Maurice* had failed to obtain royal approval (1580). The Italian *Mannerists Pellegrino *Tibaldi and Luca *Cambiaso painted a number of altarpieces and large fresco decorations in the church, cloister, and library. Other altar paintings were executed by Federico *Zuccaro, *Navarrete, and *Sánchez Coello. A school of painting was thus established at the Escorial, in which younger artists such as *Ribalta were trained. The principal sculptors employed by Philip II included Leone and Pompeo *Leoni.

In 1688 Claudio *Coello completed the sacristy altarpiece and a few years later Luca *Giordano painted the vaults of the church and the cloister staircase. Charles III (1716–88) used the Escorial as a hunting seat and furnished the hitherto unoccupied state rooms of the palace. The ceilings were painted by Mariano de Maella (1739–1819) and the walls hung with tapestries, many of which were designed by *Goya. Other outstanding works of art in the Escorial include a crucifix carved by Benvenuto *Cellini and paintings by *Dürer, *Bosch, Titian, El Greco, *Ribera, and *Velázquez.

Esquivel, Antonio María (1806–57). Spanish painter, born in Seville and active in Madrid from 1831. He painted religious, historical, and *genre subjects (his early works imitated the style of *Murillo) and most notably portraits. His best-known work is *Zorilla Reciting his Poems* (Modern Art Mus., Madrid, 1846), showing the poet and dramatist in Esquivel's studio, surrounded by the leading figures of the literary and artistic society of mid-19th-cent. Madrid.

Essex House Press. A private printing-press founded in 1898 by C. R. *Ashbee and the Guild of Handicrafts to continue the traditions of William *Morris's *Kelmscott Press. The ink, paper, and vellum (see PARCHMENT) were the same as those used in the Kelmscott books, though a special watermark was employed and experiments made in printing on grey paper with various colours. Ashbee employed Walter *Crane and other artists of note to illustrate his publications. In 1902 the press was transferred from Essex House, Mile End Road, to Chipping Campden, Gloucestershire. Eighty-four titles were issued before the press closed down in 1909; the first of them (1898) was *Cellini's *Treatises on Goldsmithing and Sculpture*, translated by Ashbee himself.

Este. Italian family, Lords of Ferrara, Modena, and Reggio from the late 13th cent. until 1598 and thereafter of Modena and Reggio until the end of the 18th cent. They were notable patrons of the arts and letters. **Leonello** (1407–50) made Ferrara into an important cultural centre; he was the friend of *Alberti, and a patron of Jacopo *Bellini and *Pisanello; Rogier van der *Weyden painted his illegitimate son **Francesco** (Metropolitan Mus., New York). His brother **Borso** (1413–71) was in touch with the young *Mantegna and *Piero della Francesca, made Cosimo *Tura his chief court painter in 1458, and employed him, Francesco del *Cossa, and Ercole de *Roberti on the frescos in the Palazzo Schifanoia. **Isabella** (1474–1539), daughter of **Ercole I** (1431–1505), the half-brother of Leonello and Borso, was the greatest of the Este patrons and one of the most brilliant women of her time. She secured paintings from Mantegna, *Perugino, *Costa, and later *Correggio to decorate her famous Studiolo in Mantua (she was married to Francesco *Gonzaga) and is said to have implored art dealers not to show her their wares so she would not spend herself even further into debt. A portrait drawing of her by *Leonardo is in the Louvre. Her brother **Alfonso** (1476–1539) commissioned mythological paintings by Giovanni Bellini and *Titian for his Studiolo. Dosso and Battista *Dossi, *Garofalo, *Girolamo da Carpi, and *Scarsellino were the painters principally employed by the Estes during the 16th cent. In the next century they continued their collections in Modena, and **Francesco I** (1610–58) commissioned portraits of himself from *Bernini and *Velázquez. Both portraits are now in the Galleria Estense in Modena, which houses many other works from the Este collections. However, in 1744 **Francesco III** sold 100 of the most splendid pictures to Augustus III (1735–63) of Poland (as Elector of Saxony he had his court in Dresden and these paintings are now among the treasures of the Gemäldegalerie there).

Estes, Richard. See SUPERREALISM.

estofado. See ENCARNADO.

Etchells, Frederick (1886–1973). British painter, architect, and designer. In the second decade of the 20th cent. he was active in some of the most avant-garde developments in British art. He contributed to the 'Second *Post-Impressionist Exhibition' of Roger *Fry in 1912, for example, joined Fry's *Omega Workshops group in 1913, and participated in the London *Vorticist Exhibition in 1915 and in the New York Vorticist Exhibition in 1917. From about 1920, however, he abandoned painting for architecture. He translated *Le Corbusier's *Towards a New Architecture* (1927) and *The City of Tomorrow* (1929), and his most important building—the office block for W. S. Crawford Ltd. in High Holborn (1930)—has been described by Nikolaus *Pevsner as 'a pioneer work in the history of modern architecture in England'. Later Etchells specialized in church architecture and restoration, and with G. W. O. Addleshaw wrote *The Architectural Setting of Anglican Worship* (1948).

etching. Term applied to a method of *engraving in which the design is bitten into the plate with acid, and also to the print so produced. A plate of polished metal—usually copper—is first coated with a substance that will resist the action of acid. This acid resist or 'etching *ground' is usually compounded of beeswax, *bitumen, and *resin and is applied by melting a solid lump on to the heated plate, rolling the mixture flat with a leather roller, and blackening it with the soot of burning tapers. The etcher then draws his design upon the grounded plate with a steel etching needle, which he holds lightly in his hand like a pen, allowing the point to cut through the dark ground and expose the bright metal beneath. After covering the back and edges of the plate with an acid-resisting *varnish called 'stopping-out varnish', he immerses it in a bath of dilute acid, commonly nitric, which bites into the metal wherever the ground has been pierced by the needle. If any parts of the design are to remain lighter than the rest they may be 'stopped out', i.e. painted over with varnish, after which the plate is again immersed in the acid and the remainder of the design bitten to a greater depth. This process of graduated biting by means of 'stopping out' may be repeated any number of times if the etcher wishes to introduce several tones into his design. Finally, when all is bitten as required, the ground is cleaned off and the plate is inked and printed. Etching is frequently combined with other processes, particularly *drypoint, both because by this means additional work may be done on the plate after proofing and without relaying the ground, and because the drypoint lines provide a convenient method of adding strong black accents to the design.

The foregoing account is necessarily simplified, for there are alternative ways of performing some of the actions and different recipes for making up the various substances and chemicals (see e.g. SOFT-GROUND ETCHING). The ground, for example, may be applied in a number of ways or even in liquid form; zinc plates are often used instead of copper; and other acids, such as *Dutch mordant, can take the place of nitric. The principle, however, remains the same in every case.

A characteristic of etching is a spontaneity of line which comes from drawing in the same direct way as with pen or pencil on paper. It thus differs from *line engraving, where the deliberate and indirect nature of the technique tends to produce an air of precision, contrivance, and formality. It is even possible to put a grounded etching plate in one's pocket to be used as the occasion demands like a sketch-book; it seems that *Rembrandt may sometimes have worked in such a way, for the picture dealer Gersaint (*Watteau's friend) recorded that he made his famous etching *Six's Bridge* (1645) 'against time for a wager at the country house of a friend, Jan Six, while the servant was fetching the mustard, that had been forgotten for a meal, from the neighbouring village'. Similarly, a quick portrait sketch can be made direct from the sitter, ready for biting and printing when convenient. This would be unthinkable with line engraving, where the action of pushing the *burin through the metal is clearly incompatible with drawing from life. Close examination of the lines themselves will sometimes reveal further differences. For whereas the engraved line, pointed at its extremities, swells and tapers according to the pressure of the engraver's hand, the etched line remains of constant width because it is produced chemically, by the action of the acid, and not manually. Etched lines, especially if bitten with nitric acid, sometimes have slightly irregular edges; engraved lines are hard and true.

An understanding of these differences often makes it possible to decide whether a print is an etching or a line engraving, but the oldest etchings are not identified so easily. For etching was invented as a labour-saving method of line engraving and consequently in its early days it had to resemble engraving as closely as possible. Moreover, the practice arose, again in order to ease the engraver's labours, of beginning a plate with etching and finishing it by engraving. Thus etching was closely linked to line engraving and was not at the beginning the free and spontaneous art that it became in the hands of Rembrandt. The first etchings date from the early years of the 16th cent., though the basic principle, that of corroding a design into a metal plate, had been utilized earlier for the decoration of armour. *Dürer made a few etchings, of which the best known is the *Cannon* of 1518. He used iron plates, the biting is strong and rather coarse, and there is no stopping out to vary the tone of the lines. Other northern pioneers were Urs *Graf, *Altdorfer, and *Lucas van Leyden. In Italy *Parmigianino was etching soon after 1520. Parmigianino's prints are attractively luminous and free in drawing, indicating the direction etching was to take in later years. The greatest of all etchers was Rembrandt, who made a complete break from the long domination of line engraving, drawing freely on the plate with great vigour and power, correcting and often radically transforming his designs as he went along. His early plates are in the medium of etching alone. Later he added drypoint to the etched lines, and finally he came to rely still more on drypoint in plates that are astonishing for their boldness of handling, breadth, and intensity of feeling. By the early 19th cent. etching was used mainly for commercial illustration, but from the 1860s to the First World War there was a great renewal of interest in it as a medium for original expression, especially in Britain. *Whistler and *Sickert were leading lights of this movement, which is called the Etching Revival. Etching is still a popular technique, *Hockney being a leading contemporary exponent. See also AQUATINT.

Etruscan taste. A fashion in interior decoration and furnishing which formed part of the 18th-cent. Classical revival (see NEOCLASSICISM), stimulated by excavations at Pompeii, Herculaneum, and Paestum. The main feature of the style was the use of contrasting colours of black, white, and terracotta, based on murals and vase paintings found at Pompeii, Herculaneum, and Rome. These were believed to date from around the 9th cent. BC and to belong to the Etruscans, who were thought of as forerunners of the Romans: they are now known to be provincial Greek or Roman.

In France the Etruscan taste was fashionable in the late *Louis XVI and *Directoire periods. In England the style is particularly associated with Robert *Adam, who designed, for example, an Etruscan Room for Osterley Park, Greater London, one of his greatest houses. Later, the Etruscan taste was one of the elements which was taken up into the eclectic fashion of 19th-cent. *Regency.

Etty, William (1787–1849). English painter, one of the few British artists to specialize almost exclusively in the nude. He was born and died in York, but was active mainly in London, where he trained at the *Royal Academy Schools and then with *Lawrence, whose great influence on him was modified by subsequent visits to Italy. His paintings are often of mythological or historical subjects, sometimes on an ambitious scale, but he also made life studies in the RA Schools throughout his career, and these are now probably his most admired works. Etty was poor for much of his life and his pictures were often attacked for their alleged indecency, *The Times* considering them 'entirely too luscious for the public eye'. However, by the time of his death he was a celebrated figure and his works had begun to fetch high prices. He summed up his attitude to his favourite subject thus: 'Finding God's most glorious work to be Woman, that all human beauty had been concentrated in her, I dedicated myself to painting—not the Draper's or Milliner's work—but God's most glorious work, more finely than ever had been done.' His draughtsmanship is often criticized, but it is generally agreed that he attained a glowing voluptuousness in the painting of flesh that few British artists have approached. The best collection of Etty's work is in the City Art Gallery, York.

Euphranor of Corinth. Greek painter and sculptor of the mid-4th cent. BC, active in Athens. Literary accounts show that he was one of the most celebrated artists of his day, but the only surviving work that can be associated with him is a headless and armless marble statue of Apollo (Agora Mus., Athens, *c.*370 BC). His lost works included bronzes of Philip of Macedon (382–336 BC) and his son Alexander the Great (356–323 BC), and he is also said to have written treatises on proportion and colour.

Euphronios. Greek vase painter and potter, active in Athens *c.*520–*c.*500 BC. He signed work as both a potter and a painter and is recognized as one of the outstanding early exponents of the *red-figure style. His best-known work is probably the *krater* (wine bowl) in the Louvre showing Herakles and Antaios. His rival was *Euthymides.

Euston Road School. Group of painters centred round the 'School of Drawing and Painting' that opened at 12 Fitzroy Street, London, in 1937, and soon transferred to nearby 316 Euston Road. Its founding teachers were William *Coldstream, Victor *Pasmore, and Claude *Rogers, and Lawrence *Gowing was also an important member of the circle. These artists were united by a desire to return from abstract and esoteric styles of modernism to a more straightforward *naturalism. The onset of the Second World War caused the closure of the school, but the term 'Euston Road' was used for a decade or so afterwards as a generic description of work in a style similar to that in which the original exponents painted. Coldstream, through his position as Professor at the *Slade School, was the chief upholder of the tradition.

Euthymides. Greek *red-figure vase painter, active *c.*520–*c.*500 BC in Athens. He was the great contemporary of *Euphronios, and was exceptional in his mastery of movement and above all in foreshortening. On an amphora (wine jar) showing revellers (Antikensammlungen, Munich) he inscribed the proud boast 'As Euphronios never could' next to a particularly skilful piece of draughtsmanship.

Evans, Myfanwy. See PIPER.

Evenepoel, Henri (1872–99). Belgian painter. He settled in Paris in 1892 and became a pupil of Gustave *Moreau, with *Matisse and *Rouault among his fellow students. His sombre early work became much more colourful under the influence of *Impressionism, and after a journey to Algeria in 1887–8 he adopted a brilliant palette which brought him to the verge of *Fauvism. Scenes from the street life of Paris formed his main subject-matter and he was a sensitive portraitist with a special gift for depicting children and adolescents. His premature death at the age of 27 deprived Belgium of what might have been one of her greatest talents.

Everdingen, Allart van (1621–75). Dutch landscape and marine painter. He was born in Alkmaar and worked with *Savery in Utrecht and *Molyn in Haarlem. In the 1640s he visited Scandinavia, where he developed a taste for subjects inspired by the scenery there—above all mountain torrents—and helped to popularize such themes in the Netherlands. *Ruisdael, in his pictures of majestic waterfalls, was one of the artists influenced by him. Allart was also a fine etcher. His elder brother **Caesar** (1617–87), who painted portraits and historical pictures, was attracted by the south not the north. Although he never went to Italy, he captured the spirit of Italian art better than many of his countrymen who crossed the Alps: witness his beautiful *Four Muses with Pegasus* (*c.*1650), part of the decoration of the royal villa—the Huis ten Bosch—at The Hague.

Evergood, Philip (1901–73). American painter. He was educated in England, at Eton and Cambridge, and much of his early life was spent travelling and studying in Europe. His early works were mainly of biblical and imaginative subjects, but after settling in New York in 1931 he became a leading figure among the *Social Realists who used their art as an instrument of social protest and propaganda during the Depression years. He was active in several organizations concerned with the civil rights of artists, and under the banner of the *Federal Art Project he produced militant paintings of social criticism, his best-known work in this genre being *American Tragedy* (Whitney Mus., New York, 1937), which commemorates a police attack on striking steel workers in Chicago. Even his allegorical religious painting *The New Lazarus* (Whitney Mus., 1954) has sociological overtones, with its strange Crucifixion and its figures of starving children. His style varied, but his inclination for the bizarre and grotesque sometimes brings his work close to *Surrealism, as is seen particularly in what is perhaps his most famous painting, *Lily and the Sparrows* (Whitney Mus., 1939).

Evesham, Epiphanius (1570–after 1633). English sculptor. He was the first distinctive personality in English sculpture since the Reformation, but details of his career are scanty. From 1601 to *c.*1614 he was working as a master-sculptor and master-painter in Paris, but though he had a studio of some size and several works in both arts are recorded, none has survived. After his return to England he made a number of tombs that stand out for their humanity, freshness of invention, and refinement of handling at a period when most English tomb sculpture was mass-produced. His signed tomb of Lord Teynham (Lynsted, Kent, *c.*1622), for example, has, in addition to the recumbent effigy, a distinguished kneeling figure of the widow and a series of touching *reliefs of mourning children. *Vertue called him 'that most exquisite artist'.

Eworth (or **Ewouts**), **Hans** (*c.*1520–after 1573). Netherlandish painter, active mainly in England, where he settled in the 1540s. About thirty of his paintings survive, almost all portraits (there are also a few allegories), dating from 1549 to 1570. He is known also to have painted for pageants and masques. Although his work is uneven, he was the outstanding figure in the history of English painting in the mid-16th cent.—a sensitive talent responding to changing taste. He painted Queen Mary (1516–58) and Queen Elizabeth I (1533–1603), but his masterpiece is perhaps the striking allegorical portrait of Sir John Luttrell (Courtauld Inst., London, 1550).

Execias. The most famous of Greek *black-figure vase painters, active in the second half of the

6th cent. BC. Among the best-known works with his signature are an amphora (wine jar) showing Achilles and Ajax gaming (versions BM, London, and Vatican Mus.); and a splendid cup showing Dionysus in his boat (Antikensammlungen, Munich). Many of the works signed by Execias contain battle scenes, but he was able to impart an air of dignity and grandeur even to the most ordinary activities, and his greatest gift was perhaps for conveying pathos and psychological insight rather than overt action.

Exhibition, Great. The first international industrial exhibition ever held, open to the public from 1 May to 11 October 1851 in Hyde Park, London. Its full official title was 'The Great Exhibition of the Works of Industry of all Nations, 1851' and the prime movers in its creation were Prince *Albert and the administrator Sir Henry Cole (1808–82). The Exhibition was held at a site on the south side of Hyde Park in a vast building of glass designed by Sir Joseph Paxton (1801–65); it was popularly known as the 'Crystal Palace', a term first applied to it by *Punch* in its issue of 2 November 1850. There were nearly 14,000 exhibitors (7,381 British and 6,556 foreign), with over 100,000 exhibits. The total number of visitors was over six million. During the summer of 1852 the building was taken down and erected in a modified form at Sydenham, where it was opened by Queen Victoria in 1854. It burnt down in 1936. The Royal Commission set up for the organization of the Exhibition was appointed as a permanent body in 1851 to apply the surplus funds, amounting to £186,000, in promoting the knowledge of science and art and their applications in productive industry. This they did by the acquisition of *c.*87 acres (35 ha.) in South Kensington as a site for the group of museums and colleges which included the *Victoria and Albert Museum, the Science Museum, the Natural History Museum, the Imperial College of Science and Technology, the *Royal College of Art, and the Royal College of Music. From the turn of the century the Commission also financed scholarships for the promotion of science and art. In the *Commemorative Album* of the Exhibition issued by the Victoria and Albert Museum in 1950, the Great Exhibition of 1851 is described as 'one of the most outstanding success stories of the nineteenth century'. The intangible benefits were widespread and permanent.

Expressionism. Term in art history and criticism applied to art in which traditional ideas of *naturalism are abandoned in favour of distortions and exaggerations of shape or colour that urgently express the artist's emotion. In its loosest sense, the term can be applied to art of any period or place that elevates intense subjective reactions above the observation of the external world (the work of *Grünewald is an example). When used in this way, the word is usually spelled with a small 'e'. More commonly, the term is applied to a trend in modern European art and more specifically to one aspect of that trend—a movement that was the dominant force in German art from about 1905 until about 1930. In this second sense Expressionism traces its beginnings to the 1880s, but it did not crystallize into a distinct programme until about 1905, and the term itself was not used until 1911, to describe works in an exhibition of *Fauvist and *Cubist paintings in Berlin.

The most important forerunner of Expressionism was van *Gogh, who consciously exaggerated nature 'to express . . . man's terrible passions'. Van Gogh and the Expressionists after him relied on impulse and feeling for their emotional use of colour and line, and in this they are distinguished from *Seurat's attempts to work out a scientific system of formal expression.

*Gauguin broke more consciously and definitely with *Impressionism. Strictly speaking Gauguin was not an Expressionist, but he was the first to accept explicitly the principles of *Symbolism, which in turn became of importance for Expressionism proper as a vehicle of communication. He simplified and flattened all forms, and used colour in a way which gave up all semblance of realism. The fierceness of *Jacob Wrestling with the Angel* (NG, Edinburgh) is made explicit through the red field on which the combat takes place. To the same end Gauguin also abolished the representation of shadow and in this the Expressionists followed his lead. As a counterpart of his new style he sought for simplicity of subject-matter and found it first in the peasant communities of Brittany, and later in the islands of the South Pacific. In turning away from European urban civilization Gauguin discovered *primitive art and *folk art, both of which became of absorbing interest for the later Expressionists.

At the same time the Norwegian Edvard *Munch, who knew the work of van Gogh and Gauguin, began to explore the possibilities of violent colour and linear distortions with which to express the most elemental emotions of anxiety, fear, love, and hatred. His search for pictorial equivalents for his hysterical obsessions led him to realize the potentialities of the simple directness of *graphic techniques such as the woodcut—its revival as an independent art form was one of the distinctive features of Expressionism. Munch had a wide influence, particularly in Germany (he exhibited at the Künstlerverein, Berlin, in 1892), which extended even to sculpture in the work of

Ernst *Barlach, who used Munch's idiom to great effect for religious and social subjects. Among the early Expressionists must also be counted the Belgian painter James *Ensor. He depicted the baseness of human nature by the use of grotesque and horrifying carnival masks, and his weird and uncanny art became widely known, particularly through his etchings.

In 1905 Expressionist groups appeared almost simultaneously in Germany and France. The Fauves combined in their art the theories of van Gogh and Gauguin; in 1908 *Matisse, the leader of the group, summed up their aims: 'What I am after above all is expression. . . . The chief aim of colour should be to serve expression as well as possible. . . . To paint an autumn landscape I will not try to remember what colour suits the season; I will be inspired only by the sensation the season gives me.' Matisse applied these ideals to large figure compositions, *Derain to landscape, and *Rouault to a new religious art of great power and simplicity.

In 1905 Die *Brücke (The Bridge) was founded in Dresden and held its first exhibition in 1906. Whereas the painting of the Fauves, even at their most violent, always retained harmony of design, and colour never lost a certain decorativeness and lyricism, in Germany restraint was thrown to the winds. In spite of undeniable French influence, violence was exploited for its own sake; forms and colours were tortured in the attempt to give psychological and symbolic vent to a vaguely conceived creative urge and a sense of revolt against the established order. In 1913 *Kirchner wrote: 'We accept all the colours which, directly or indirectly, reproduce the pure creative impulse.'

Shortly before the First World War German painters also grafted the schematic forms of *Cubism on to the ideals of earlier Expressionists, and under the influence of theosophy and Indian mysticism attempted to evolve a pictorial system of universal implication. In 1911 Franz *Marc, the Russian *Kandinsky, and others founded the short-lived *Blaue Reiter group, which is regarded as the high point of German Expressionism. After the First World War Expressionism became the fashion in Germany and even artists such as George *Grosz and Otto *Dix, who sought a new and hard realism—*Neue Sachlichkeit—kept much of the distortion and exaggeration which had been one of the chief devices of earlier Expressionism. Expressionism was suppressed by the Nazis in 1933 along with other *degenerate art, but revived after the Second World War, and has well-known exponents today in artists such as Georg *Baselitz. Outside Germany, leading exponents of Expressionism have included *Chagall and *Soutine, and its descendants include *Tachisme and *Abstract Expressionism.

Exter, Aleksandra (1882–1949). Russian painter and theatre designer. From 1908 she was a regular visitor to Paris and other western European cities, forming a link between the Western avant-garde and that in Russia. In 1918 she founded her own studio in Kiev, from which emerged many artists who were to achieve fame in later years, notably Pavel *Tchelitchew. It was here that Exter and her pupils created huge *Suprematist designs for several agit-steamers (propaganda boats) on the Dnieper River. Her stage designs were technically innovative and often highly colourful (examples in the V&A, London).

ex-voto (Latin: 'from a vow'). A painting or other work of art made as an offering to God in gratitude for a personal favour or blessing or in the hope of receiving some miraculous benefit. There is a famous example by Philippe de *Champaigne in the Louvre.

Eyck, Charles Hubert (1897–1962). Dutch painter, sculptor, and designer. He had a preference for working on a large scale and he made impressive decorations for Catholic churches in Limburg. His large bronze sculpture group in Maastricht commemorates the liberation of that city, the first Dutch city freed from the Germans, in 1944.

Eyck, Jan van (d. 1441). The most celebrated painter of the Early Netherlandish School. Within a short time of his death he had a reputation on both sides of the Alps as a painter of great stature and importance, and although he is no longer credited with being the 'inventor' of *oil painting, as was long maintained, his fame has continued undimmed to the present day. Nothing is known of his training and he is first recorded in 1422, entering the service of the Count of Holland, John of Bavaria, in The Hague. In 1425 John died and later in the same year van Eyck moved to Lille when he was appointed court painter and 'varlet de chambre' (equerry) to Philip the Good (1396–1467), Duke of Burgundy, a post he held until his death. Evidently he was highly esteemed by the Duke and he travelled on secret diplomatic missions to Spain and Portugal for him. About 1430 Jan moved from Lille to Bruges, where he lived until his death.

In spite of the documentation on his life, which for a painter of his period is fairly rich, his *œuvre* presents many problems, especially in the reconstruction of his early career, for all his dated paintings come from the last 10 years of his life. The central problem of his career—and one of the most discussed in the history of art—concerns the work that has always been the basis of his resounding fame, the great altarpiece of the *Adoration of the Lamb* (completed 1432) in Ghent Cathedral. An inscription on the frame states that it was begun by 'the painter Hubert van Eyck, than whom none was greater', and completed by 'Jan, second in art'. Jan's brother **Hubert** (died 1426?) is such an obscure figure that some scholars have even denied his existence, and there is certainly no obvious division into the work of two hands in the altarpiece. Thus, Jan's contribution to the central masterpiece of Early Netherlandish painting is uncertain. *Dürer called the Ghent altarpiece 'a stupendous painting' and the comment is appropriate both to the majesty and *iconographical richness of the huge *polyptych, and also to its breathtaking technical mastery. Jan brought the new technique of oil painting to a sudden peak and his ability to depict minute detail with unfaltering sureness of hand and to create glowing effects of colour has never been surpassed.

Apart from the altarpiece, about two dozen other paintings are reasonably attributed to him. They are all either religious works or portraits, although he is known to have painted pictures of other subjects (including a nude woman at her bath), which are now lost. Outstanding among the surviving works are the famous double portrait *Giovanni Arnolfini and his Wife* (NG, London, 1434) and two paintings of the Virgin and Child with *donors—*The Madonna of Chancellor Rolin* (Louvre, Paris, *c.*1435) and *The Madonna with Canon van der Paele* (Groeningemuseum, Bruges, 1436). The Louvre painting, with large figures in the foreground set against a distant panoramic landscape, shows Jan's all-embracing vision of the natural world and his mastery of light and space, as well as detail and texture—in Erwin *Panofsky's words, 'his eye was at one and the same time a microscope and a telescope'. The *Man in a Red Turban* (NG, London, 1433) is generally considered to be a self-portrait. As a portrait painter Jan was preoccupied with the realities and textures of the human face, and in this as in his inanimate interiors he recorded the subtleties of appearances rather than commenting on them as did his great contemporary Rogier van der *Weyden. His portraits do, however, convey a sense of inner life and are not simply coldly objective records.

Jan stands with the *Master of Flémalle as the founder of the Early Netherlandish School and his technique became the accepted model for his successors. His main follower was Petrus *Christus, but his influence was wide (it is seen, for example, in the work of Luis *Dalmau in Spain) and profound. In the Netherlands itself, however, the more emotional work of Rogier van der Weyden came to have even more influence and the very perfection of Jan's work must have made him the most daunting of models.

Fabritius, Carel (1622–54). Dutch painter. He was *Rembrandt's most gifted pupil and a painter of outstanding originality and distinction, but he died tragically young in the explosion of the Delft gunpowder magazine, leaving only a tiny body of work (much may have perished in the disaster). In his youth he worked as a carpenter (the name Fabritius was once thought to have derived from this profession, but it is now known that his father had used it) and he was probably in Rembrandt's studio in the early 1640s. He settled in Delft in about 1650. Although only about a dozen paintings by him are known, they show great variety. His earliest surviving works (*The Raising of Lazarus*, National Mus., Warsaw, *c.*1645) are strongly influenced by Rembrandt, but he broke free from his master and developed a personal style marked by an exquisite feeling for cool colour harmonies and (even though he often worked on a small scale) unerring handling of a loaded brush (*The Goldfinch*, Mauritshuis, The Hague, 1654). These qualities, together with an interest in *perspective, occur in the work of *Vermeer, the greatest of Delft painters, and Fabritius certainly influenced him, although it is not likely (as is sometimes maintained) that he was his master, this distinction perhaps belonging to *Bramer. Carel's brother **Barent** (1624–73) was also a painter, but of much lesser quality. He also may have studied with Rembrandt; he mainly painted portraits and religious works.

Faithorne, William (1616–91). English engraver. He fought as a Royalist and later spent some time in exile in France, where he worked with *Nanteuil; but by 1650 he was established in London and became the most distinguished of English 17th-cent. engravers, especially of portrait heads. He engraved the work of other painters (van *Dyck, *Dobson, *Soest, *Lely) and also made engravings of his own drawings from the life, many published as frontispieces to books. He also drew very sensitive portrait heads as independent works (*John Aubrey*, Ashmolean, Oxford, 1666). In 1662 he published *The Art of Graving and Etching*. His son **William** (1656–1710) was also an engraver.

Faiyum (or **Faiyumic**) portraits. Type of funerary portrait found throughout Egypt but particularly associated with the Faiyum (Al Fayyum) area, about 95 km. south of Cairo. They date from about the 1st to the 4th cent. AD and represent the head and shoulders of the deceased. The portraits are painted in *tempera or *encaustic on canvas or wood and they were enclosed in the wrapping around the corpse's face. Although the quality of the numerous surviving examples varies considerably, the finest are among the most vivid and *naturalistic portraits from the ancient world, suggesting that they were done while the sitter was still alive. The British Museum has excellent examples.

Falca, Pietro. See LONGHI, PIETRO.

Falcone, Aniello (1607–56). Neapolitan painter. He is remembered mainly as the first specialist in battle pieces, a genre that won him an international reputation and in which he inspired his pupil Salvator *Rosa. Falcone also did religious paintings, some in fresco, for churches in Naples.

Falconet, Étienne-Maurice (1716–91). French sculptor and writer on art, a pupil of J. B. *Lemoyne. Falconet was perhaps the most quintessentially *Rococo of all French sculptors, his forté being gently erotic figures such as the celebrated *Bather* (1757) in the Louvre. Like many other of his works, this was reproduced in porcelain by the Sèvres factory, of which he was Director from 1757 to 1766, a post he gained through the influence of his patron Mme de Pompadour (1721–64). Falconet had other sides to his talent, however, and his masterpiece—the equestrian statue of Peter the Great in Leningrad—is in a completely different vein. He went to Russia in 1766, recommended to Catherine II by *Diderot, and left in 1778, the statue being unveiled in 1782. The huge horse is represented with its forelegs raised and unsupported—a daring technical feat—and the heroic vigour of the statue gives it a place among the greatest examples of the type. Falconet suffered a stroke in 1783 and thereafter produced no more sculpture, devoting himself to writing. A six-volume edition of his writings had already appeared in 1781 and in 1761 he had published his best-known literary work, *Réflexions sur la sculpture*. In this he was one of the first to argue that the modern artists were superior to those of the ancient world (it is significant that unlike most of his distinguished contemporaries he never saw the need to visit Italy), and the young *Goethe was among its admirers.

Falk, Robert (1888–1958). Russian painter and designer, a founding member of the *Knave of Diamonds. From 1928 to 1938 he lived in France, and from 1938 to 1944 he worked in Soviet Central Asia. On his return to Moscow in 1944 he continued to work mainly on landscapes and portraits. In 1966 a comprehensive retrospective exhibition in Moscow completely rehabilitated him from the censure of the Stalin years.

Fancelli, Domenico di Alessandro (1469–1519). Florentine sculptor, a pupil of *Mino da Fiesole. He made several visits to Spain and was important in introducing the *Renaissance style to that country. His work there included the tomb of Cardinal Hurtado de Mendoza (Seville Cathedral, 1509); that of Prince John, son of Ferdinand and Isabella (St Thomas, Avila, 1513); and that of Ferdinand and Isabella (Chapel Royal, Granada, 1517).

fancy picture. A term applied in the 18th cent. to a type of sentimental rural *genre picture. Scenes in which idealized peasants behave rather more as if they were in the studio than the countryside are 'fancy pictures', and *Gainsborough's paintings of this type were so called in his own day. The term is elusive and cannot be defined with precision.

Fantastic Realism. A movement in painting that developed in Vienna in the late 1940s and came to be regarded as typical of post-war Austrian painting. The artists involved combined minute realism with a fairy-tale world of fantasy and imagination. Though very different in their ways of painting and in the quality of their work, they had in common an interest in the art of the past, notably that of Pieter *Bruegel (supremely well represented in the Kunsthistorisches Museum in Vienna) and in the literary and anecdotal character of painting. The best-known representative of Fantastic Realism is Ernst Fuchs (1930–), who like several other artists involved in the movement was a pupil of Albert Paris Gütersloh (1887–1973), who was a renowned teacher at the Vienna Academy.

Fantin-Latour, Henri (1836–1904). French painter and lithographer. He is best known for his luxurious flower pieces, but he also painted several group portraits that are important historical documents and show his friendship with leading avant-garde artists. *Homage to Delacroix* (Musée d'Orsay, Paris, 1864) shows Fantin-Latour himself, with *Baudelaire, *Manet, *Whistler, and others grouped round a portrait of *Delacroix; and *A Studio at Batignolles* (sometimes called *Homage to Manet*) (Musée d'Orsay, Paris, 1870) shows *Monet, *Renoir, Émile Zola (1840–1902), and others in Manet's studio. In spite of his associations with such progressive artists, Fantin-Latour was a traditionalist, and his portraits particularly are in a precise, detailed style. Much of his later career was devoted to *lithography; he greatly admired Richard Wagner (1813–83) and did imaginative lithographs illustrating his music and that of other *Romantic composers.

Farington, Joseph (1747–1821). English landscape painter and topographical draughtsman, best known today for his copious diary (1793–1821), which contains valuable information about the London art world of the time. Most of the original manuscript is in the Royal Library at Windsor Castle. Publication of the full text began in 1978; previous editions had been heavily abridged. His views of the English Lakes were engraved and published in book form in 1789 and 1816, but his work as an artist is virtually forgotten.

Farnese. Italian family of humanists and patrons of the arts who rose to importance with the election of Cardinal **Alessandro Farnese** (1468–1549) to become Pope Paul III in 1534. He was the most important patron of *Michelangelo's later years, commissioning from him the *Last Judgement* in the Sistine Chapel and the *Conversion of St Paul* and the *Crucifixion of St Peter* in the Cappella Paolina, and also appointing him architect to St Peter's. Michelangelo also had a hand in the design of the Palazzo Farnese, the finest palace built in Rome in the 16th cent.

Pope Paul's grandson (or 'nephew' as he was known), another **Alessandro** (1520–89), was made a Cardinal in 1534 and was acknowledged to be the greatest patron of his day, surrounding himself with artists and men of letters. He built up the largest collection of antiquities in Rome (now mainly in the Archaeological Museum in Naples; see FARNESE BULL; FARNESE HERCULES), was instrumental in bringing *Titian to Rome (1545–6), encouraged *Vasari to write his *Lives*, engaged Giacoma da Vignola (1507–73) to complete the Palazzo Farnese at Caprarola, and commissioned some of the most important *Mannerist frescos. Alessandro also purchased the villa across the Tiber from the Palace which was henceforth known as the Villa Farnesina. He gave special support to the Jesuits and built for them the church of Il Gesù, Rome (designed by Vignola, begun 1568), one of the most influential buildings in the history of architecture. Alessandro's great-nephew, Cardinal **Odoardo Farnese** (1573–1626), great-great-grandson of Paul III, was responsible for employing Annibale and Agostino *Carracci to decorate the Farnese Gallery in the Palace in Rome.

Elisabetta, second wife of the Bourbon King Philip V (1683–1746), was a Farnese and through this connection their son Charles III of Spain (1716–88) brought the majority of the Farnese collections to Naples, where they still are.

Farnese Bull. Ancient marble sculpture group (probably a Roman copy of a Greek original of *c.*150 BC), once part of the *Farnese collection and now in the Archaeological Museum in Naples. The subject, taken from Greek legend, shows the punishment of Dirce, who for her cruelty to Antiope was tied to the horns of a bull by Antiope's sons (Dirce's stepsons) and trampled to death. The figures are life-size, and the group, which was found in the Baths of Caracalla in Rome in 1545, is one of the most spectacular examples of the virtuosity and love of dramatic movement typical of *Hellenistic art.

Farnese Hercules. Gigantic marble statue of Hercules leaning sideways on his club and resting after his labours, once part of the *Farnese collection and now in the Archaeological Museum in Naples. It is signed by an Athenian sculptor named Glycon and is a copy of an original of the 4th cent. BC, probably by *Lysippus. The figure was discovered in the Baths of Caracalla in Rome in *c.*1546 and became highly influential and much copied. Its powerful musculature and realistic surface treatment were particularly admired by *Baroque artists.

Fattori, Giovanni. See MACCHIAIOLI.

Fauteaux, Roger. See AUTOMATISTES, LES.

Fautrier, Jean (1898–1964). French painter and graphic artist. He came to England as a child in 1909 and studied at the *Royal Academy Schools and at the *Slade School, returning to Paris in 1917. Throughout his life he remained isolated from groups and movements, and his work is difficult to classify, although he is often seen as a forerunner of *Art Informel. His best-known works are the series of *Hostages* (1943), inspired by his horror of war. In these he developed his characteristic technique, building up by layer upon layer of paint, thickened with white, a heavy *impasto into which he worked the representation of his subject. The effect produced is both powerful and mysterious. Fautrier's other works included lithographs illustrating Dante's *Inferno* (1928), and from 1950 he was one of the first to develop the idea of *multiples, printing a basic design on anything up to 300 canvases and then completing the work by hand.

Fauvism. Style of painting based on the use of intensely vivid non-naturalistic colours, the first of the major avant-garde developments in European art between the turn of the century and the First World War. The dominant figure of the Fauvist group was Henri *Matisse, who used vividly contrasting colours as early as 1899, but first realized the potential of colour freed from its traditional descriptive role when he painted with *Signac in the bright light of St Tropez in the summer of 1904 and with *Derain at Collioure in the summer of 1905. The Fauves exhibited together at the *Salon d'Automne of 1905 and their name was given to them by the critic Louis Vauxcelles, who pointed to a *quattrocento-like sculpture in the middle of the same gallery and exclaimed: 'Donatello au milieu des fauves!' (*Donatello among the wild beasts). Among the other major figures of the group were *Marquet, *Rouault, *Vlaminck, *Braque, and *Dufy. The Dutchman van *Dongen was also associated with them. These artists were influenced in varying degrees by van *Gogh, *Gauguin, the *Neo-Impressionists, and *Cézanne, and the outstanding characteristic of their work was the extreme intensity of its colour: pure colours, which they used arbitrarily for emotional and decorative effect, but sometimes also, as Cézanne had done, to mould space. Apart from this, they had little or no programme in common.

The movement reached its peak in the Salon d'Automne of 1905 and the *Salon des Indépendants of 1906. With most of the group Fauvism was a temporary phase through which they passed in the development of widely different styles, and at no later period did their work display again such a degree of similarity. Although short-lived, however, Fauvism was a major influence on the development of German *Expressionism.

Federal Art Project. A project run by the US Government from 1935 to 1943, with the dual purpose of assisting artists who had been hard hit by the Depression, and of deploying the artistic potential of the country in the decoration of public buildings and places. There were also a Federal Writers Project, a Federal Theatre Project, and a Federal Music Project, and collectively they are known as the Federal Arts Projects. They were part of the Works Progress Administration (WPA), a work programme for the unemployed executed as part of President F. D. Roosevelt's New Deal. The Federal Art Project grew out of three previous schemes of a similar nature. In 1933 the Public Works of Art Project was set up to assist artists over the winter of 1933/4, artists being employed on public works at a weekly salary. The following year the Treasury Relief Art Project (July 1935–June 1943) was set up under the painter Edward Bruce (1879–1943) for a similar purpose, while a new section of Painting and Sculpture established in the Treasury Department commissioned artists for specific tasks in connection with the embellishment of Federal buildings. The Federal Art Project was directed by Holger Cahill, a collector of American *folk art, and employed artists on a monthly salary. At its peak the Project employed more than 5,000 people, not only decorating public buildings but also producing prints, posters, various works of craft, and setting up community art centres and galleries in parts of the country where art was virtually unknown. The project also involved an *Index of American Design*, a gigantic documentation of the decorative arts in America. Virtually all the major American artists of the period were involved, either as teachers or practitioners. A huge amount of work was produced, but most of it was unremarkable in quality. This, however, was in line with Cahill's philosophy: 'The organization of the Project has proceeded on the principle that it is not the solitary genius but a sound general movement which maintains art as a vital, functioning part of any cultural scheme. Art is not a matter of rare, occasional masterpieces.' Few of the murals created under the Project survive.

Feininger, Lyonel (1871–1956). American painter who spent most of his career in Europe. He was born in New York into a German-American musical family. In 1887 he went to Germany with the intention of studying music, but he turned instead to art. He had drawings published in Berlin's humorous weeklies and by the turn of the century he was Germany's leading political cartoonist. In 1906–8 he lived in Paris and under the influence of Robert *Delaunay turned seriously to painting. By 1912 he had evolved a personal style (influenced by *Cubism but highly distinctive) in which natural forms were treated in terms of a rhythmic pattern of prismatically coloured interpenetrating planes bounded by straight lines—a manner that he applied particularly to architectural and marine subjects. His work impressed the members of the *Blaue Reiter, who invited Feininger to exhibit with them in 1913. Although he was an alien, he remained in Germany throughout the First World War and afterwards taught at the *Bauhaus from its foundation in 1919 (one of his woodcuts appeared on the cover of its manifesto) until its closure by the Nazis in 1933; he was the only person to be on the staff from start to finish. In 1935 he visited the USA and in 1937 (the year in which the Nazis declared his work *degenerate) he returned there permanently. He settled in New York and adopted the architecture of Manhattan as one of his favourite subjects, working with vigour into his eighties. His son Andreas Feininger (1906–) is a distinguished photographer and writer on photography.

Feke, Robert (active 1740s). American Colonial portrait painter. Nothing is known of his life until 1741 when he executed a large portrait group in Boston, *Isaac Royall and his Family* (Harvard Univ.). He was active from that time in Newport and Philadelphia until 1750, when his life is once more veiled in obscurity. There are about fifteen signed portraits from his hand and about fifty more are reasonably attributed to him. His works are somewhat lacking in characterization, but their strength and clarity of design and delicacy of touch give them a high place among Colonial portraits.

Félibien des Avaux, André (1619–95). French architect and writer, a friend of Nicholas *Poussin in Rome. His *Entretiens sur les vies et sur les ouvrages des plus excellens peintres anciens et modernes*, first published in 1666–88 and often reprinted, is a major source book and contains the best contemporary biography of Poussin. In 1676 he published a textbook on artists' techniques with a dictionary of art terms: *Des principes de l'architecture, de la sculpture, de la peinture, et des autres arts qui en dépendent. Avec un dictionnaire des termes propres à chacun de ces arts*.

Felixmüller, Conrad. See NEUE SACHLICHKEIT.

Feodotov, Pavel (1815–52). Russian painter. He began his brief artistic career as a self-taught amateur, but recognition of his talent enabled him to retire from the army and devote himself to painting. His gently satirical and sentimental scenes from daily life became favourites with the Russian public. He is known as the 'Russian Hogarth' and may have known the work of *Hogarth through engravings, but his works lack Hogarth's moralistic purpose and are smoother in technique. In his last years his work expressed disenchantment with the dreariness of life. He died in a lunatic asylum.

Feofan Grek. See THEOPHANES THE GREEK.

Fergusson, John Duncan (1874–1961). Scottish painter and, to a lesser degree, sculptor, the most renowned of the *Scottish Colourists. He abandoned medicine to study painting and in 1905 settled in Paris. His early work was *Whistlerian, and he then came under the influence of *Manet, but by 1907 he had adopted the bold palette and firm outlines of *Fauvism and became the most uncompromising adherent to the style among British artists (*Blue Beads*, Tate, London, 1910). In 1914 the war brought him back to Britain; he lived in London, 1914–29, in Paris, 1929–40, and finally in Glasgow, 1940–61. Soon after settling in Glasgow he founded the New Art Club to provide better exhibiting facilities for the city's progressive artists, and out of it grew the New Scottish Group (1942), of which Fergusson was first President. In 1943 he published a book entitled *Modern Scottish Painters*.

Fernandez, Alejo (*c.*1470–1543). Spanish painter, probably of German extraction, as he is referred to as 'Maestro Alexos—pintor Aléman'. He married the daughter of a painter called Pedro Fernandez at Cordova and took her name, but he worked mainly at Seville, where he was the leading painter of the first third of the 16th cent. His work, which is represented in Seville Cathedral, was essentially Flemish *Mannerist in style, but it has a personal lyrical quality, and his treatment of architecture and *perspective suggests that he may have visited Italy. He had a busy studio and several followers, among them his son **Sebastián**.

Fernandez (or **Hernandez**), **Gregório** (*c.*1576–1636). Spanish sculptor, active at Valladolid from *c.*1605. Continuing the tradition of painted religious sculpture, he worked in the manner of *Juan de Juni but with greater realism of expressive gesture. He was one of the first and greatest masters of *Baroque naturalism in Spain, abandoning the earlier practice of using gold and brilliant colours and insisting upon realistic colouring from the *polychromists who painted his sculptures. Among the numerous altarpieces emanating from his workshop are those of S. Miguel, Valladolid (1606) and Plasencia Cathedral (1624–34). He is represented in Valladolid Museum by a *Pietà* (1617), a *Baptism of Christ* (1630), and other works.

Ferrari, Gaudenzio (*c.*1471/81–1546). Italian painter, active in his native Piedmont and in Lombardy. His early work was strongly influenced by *Leonardo and his Milanese followers, and throughout his life he remained *eclectic, absorbing into his highly charged, emotional style elements from *Pordenone and *Lotto and also, for example, from the engravings of *Dürer. He was an artist of considerable power and individuality, but his work has remained comparatively little known because much of it is in fairly remote situations. His most remarkable works are *The Stations of the Cross* in a series of chapels at the Sanctuary of Sacro Monte, Varallo, in which life-size foreground figures are carved in the round and the rest of the scene painted behind them. A more sophisticated use of illusion deriving from *Correggio is found in his *Assumption* for the dome of Sta Maria dei Miracoli, Saronno (1534).

Ferri, Ciro. See CORTONA.

fête champêtre (French: 'outdoor feast'). Type of *genre scene in which romantic figures are shown in an idealized outdoor setting, usually eating, dancing, flirting, or listening to music. Since the Gardens of Love represented in medieval manuscripts, the theme has had great popularity in European art, undergoing several transformations. It was particularly favoured in 16th-cent. Venetian painting and the *Concert Champêtre* in the Louvre (traditionally by *Giorgione, but now usually given to *Titian) is the most celebrated of all examples of the type. The term 'fête galante' (courtship party) was invented by the French *Academy in 1717 to describe *Watteau's variants on the theme, in which figures in ball dress or masquerade costume disport themselves amorously in a parkland setting.

Feti (or **Fetti**), **Domenico** (*c.*1589–1623). Italian painter. He was born at Rome, where he studied under Ludovico *Cigoli, was court painter to Vincenzo *Gonzaga at Mantua from 1613 to 1622, and then settled in Venice. His most characteristic works are of religious themes turned into *genre scenes of contemporary life. They are broadly painted, with characteristic 'windswept' brushstrokes, though small in scale. Their great popularity is shown by the fact that they often exist in numerous very similar versions. Feti, who was also an excellent portraitist, was one of a group of non-Venetian artists (including the German *Liss and the Genoan *Strozzi) who revivified painting in the city when there was a scarcity of native talent. Consequently, he is often classed as a member of the Venetian School, even though he spent only the last two years of his life there.

Feuchtmayer, Joseph Anton (1696–1770). German *Rococo *stuccoist and sculptor, the most famous member of a family of artists from Wessobrunn in Bavaria. A virtuoso carver and stuccoist, he did a great deal of decorative work for buildings in the Lake Constance area. The greatest ensemble of his work is in the pilgrimage church of Neu-Birnau (1746–53): it includes his most famous single figure, the *Honey-licking Putto*.

Feuerbach, Anselm (1829–80). German painter. He studied in Düsseldorf, Antwerp, and

Paris (with *Couture), then lived in Italy from 1855 to 1873. His father was a professor of classical archaeology (he had written a book on the **Apollo Belvedere*) and the son grew up in an atmosphere saturated with the high-minded ideals of humanistic philosophy. Violently antagonistic both to purely decorative painting and to an art based on the study of nature, he wished to become the founder of a new school which was to combine noble, didactic, and idealistic subjects with a style derived from the *Grand manner of Venetian 16th-cent. painting. His subjects are usually taken from Greek antiquity—in the case of his most celebrated painting, from one of Plato's Dialogues (*The Banquet*, Karlsruhe, 1869, later version in Berlin). Feuerbach's desire to preach a philosophy through pictorial means was generally a source of weakness rather than strength, and his best works are now generally considered to be his portraits of his model and mistress Nanna Risi, which have a statuesque beauty lacking in his more elaborate paintings; she also posed for subject pictures such as *Iphigenia* (Hessisches Landesmuseum, Darmstadt, 1862), one of several depictions by Feuerbach of this theme. Feuerbach went to Vienna in 1873 to become Professor of History Painting at the Academy, but he returned to Italy in 1876 after criticism of his ceiling of *The Fall of the Titans* for the Academy. Throughout his life Feuerbach complained that he was being misunderstood and not receiving the recognition due to a very great artist. It is this element of self-pity which makes his book *Ein Vermächtnis* (A Testament)—an account of the genesis of some of his works, interspersed with aphorisms about life and art—one of the most pathetic and repellent autobiographies ever written. It was posthumously published in 1882.

Fiammingo (Italian: 'Fleming'). The name given to, or adopted by, a number of Flemish artists working in Italy, especially in the 17th cent., when the Flemish colony in Rome was considerable. Among the important artists who were so known were Denys *Calvaert, who founded an *academy in Bologna, and the sculptor François *Duquesnoy, who frequently signed himself 'Fiammingo'.

Fielding, Anthony Vandyke Copley (1787–1855). English watercolour painter, a pupil of *Varley. He was a popular and prolific artist and much of his work is repetitive. Early in his career he specialized in scenes of Wales and the Lake District, but after 1814 he spent much of his time near the coast because of his wife's health, and turned increasingly to seascapes.

Field painting. A type of painting developed in the USA *c.*1950 in which the picture is no longer regarded as a structure of interrelated elements but as a single indivisible expanse. Field painting has affinities with *Systemic painting and with the *All-over style initiated by Jackson *Pollock. The term *Colour Field painting has been used when emphasis is placed upon brilliance and saturation of colour in monochromatic canvases. Rather than a specific style it may be regarded as an aspect of a very general tendency during the 1950s and 1960s to eschew traditional composition in favour of non-relational presentation of a single 'total' theme.

Figari, Pedro (1861–1938). Uruguayan painter. He had a distinguished career as a lawyer and politician, but from 1920 he devoted himself entirely to painting and spent the years 1925–33 in Paris. Among his favourite subjects were landscapes of the pampas and scenes of Uruguayan life including provincial social events such as Negro and Creole dances. His work, which is notable for its strong colour and simple drawing, is represented in the National Museum at Montevideo.

figurative art. Art in which recognizable figures or objects are portrayed. The term 'representational art' is used synonymously; the opposite is non-figurative or *abstract art.

Filarete (Antonio Averlino) (*c.*1400–*c.*1469). Florentine sculptor, architect, and writer on art. His nickname is derived from the Greek for 'lover of virtue'. He probably trained with *Ghiberti and his most important work in sculpture—the bronze doors of St Peter's in Rome (1433–45)—is heavily indebted to Ghiberti's doors for the Baptistery in Florence, although much less accomplished. After being expelled from Rome for allegedly stealing a relic, Filarete went to Florence and Venice, then in 1450 settled in Milan. There he worked mainly as an architect, his principal work being the Ospedale Maggiore (begun 1457, completed in the 18th cent.), which helped to introduce the *Renaissance style to Lombardy and created new standards of comfort and sanitation in hospital design. His novel ideas came out also in his *Treatise on Architecture*, written in 1461–4. This devotes twenty-one books to architecture and three to painting and drawing. It includes a vision of a new city, Sforzinda (named after his patron, Francesco *Sforza), which is the first symmetrical town-planning scheme of modern times. Among his ingenious proposals for his ideal city was a Tower of Virtue and Vice, a ten-storey structure accommodating a brothel on the ground floor and an astronomical observatory at the top.

Filla, Emil (1882–1953). Czech painter, sculptor, graphic artist, and writer on art. Between 1907 and 1914 he spent much of his time in France, Germany, and Italy, and during this period he turned from his early *Expressionist manner to *Cubism, becoming the pioneer and one of the most distinguished exponents of the style in Czechoslovakia. He passed the First World War in the Netherlands and during the Second World War he was imprisoned in the concentration camp at Buchenwald. In his post-1945 work he moved to a more naturalistic style.

Filonov, Pavel (1883–1941). Russian painter and designer. He was one of the most individual Russian artists of his time, developing a style that has been described as proto-*Surrealist and remaining untouched by the general trend towards *Constructivism. In 1925 he founded his school named 'The Collective of Masters of Analytical Art' in Leningrad which lasted until 1932, when the state disbanded all art groupings. His work, which has been likened to that of Paul *Klee and the German *Expressionists, was of great delicacy and elaborately meticulous finish. Filonov died in the siege of Leningrad during the Second World War.

Finch, Alfred William (1854–1930). Finnish painter, graphic artist, and designer, born in Brussels of Belgian-British extraction. A friend of *Seurat and *Signac, he won recognition as a *Neo-Impressionist painter and was one of the founders of the Belgian Groupe les *Vingt (XX). In 1897 he moved to Finland and was put in charge of the Iris pottery factory at Porvoo, where he had a notable influence on the modernization of Finnish design. As a teacher of graphic art at the Drawing School of the Finnish Arts' Association, 1902–5, and at the Finnish Central School of Applied Art, 1905–30, he was a powerful force for bringing both the fine and the practical arts of Finland into contact with contemporary European trends.

fine arts. Term applied to the 'higher' non-utilitarian arts, as opposed to *applied or *decorative arts. In its most common usage the term is taken to cover painting, sculpture, and architecture (even though architecture is obviously a 'useful' art), but it is often extended to cover poetry and music too. The term did not come into use until the 18th cent., a key work being *Les Beaux Arts réduits à un même principe* (1746), by Charles Batteaux (1713–80). Batteaux divided the arts into the useful arts, the beautiful arts (sculpture, painting, music, poetry), and those which combined beauty and utility (architecture, eloquence). Soon after, in *Diderot's *Encyclopédie*, the philosopher D'Alembert (1717–83) listed the fine arts as: painting, sculpture, architecture, poetry, and music. This list established itself and in England the term 'five arts' was sometimes used in its place with similar meaning. See also LIBERAL ARTS.

Finiguerra, Maso (1426–64). Florentine goldsmith, engraver, designer, and craftsman in **niello* (a type of decorative metal inlay). *Vasari asserts that he was the inventor of copper engraving and although this claim has been discredited, Finiguerra was certainly one of the earliest to use that medium, which was first developed in Italy as an extension of *niello* work, in which he was the most famous practitioner of his day. He was a pupil of *Ghiberti and seems to have collaborated with Antonio del *Pollaiuolo.

Fitzwilliam Museum. The museum and art gallery of the University of Cambridge. It was founded in 1816 and is one of the oldest public museums in Great Britain. Like the *Ashmolean Museum in Oxford it has been built up almost entirely from private benefactions. The founder, the 7th Viscount Fitzwilliam (1745–1816), bequeathed to the University a typical gentleman's collection of the 18th cent., which included Italian High *Renaissance paintings, and the best collection of *Rembrandt etchings then in England. He also left £100,000 for a building, which was begun in 1837 by George Basevi (1794–1845), continued by C. R. Cockerell (1788–1863) after Basevi's death in 1845, and finished by E. M. Barry (1830–80) in 1875. It opened to the public in 1848. Among the bequests to the museum the most noteworthy after the founder's was that of Charles Brinsley Marlay (1831–1912), which enriched all departments. Sir Sydney Cockerell (1867–1962) has been the most remarkable director of the museum (1908–37). According to *The Dictionary of National Biography*, he 'transformed a dreary and ill-hung provincial gallery into one which set a new standard of excellence which was to influence museums all over the world. This he achieved by the skilful and uncrowded display of pictures against suitable backgrounds, and by the introduction of fine pieces of furniture, Persian rugs, and flowers provided and arranged by lady admirers, fired by his enthusiasm.' In Cockerell's own words, 'I found it a pig stye, I turned it into a palace.' The museum's collections are now extremely wide-ranging, embracing archaeology and the *applied arts. Its areas of greatest richness include Italian painting and Greek coins, its holdings in the latter field being second only to the *British Museum.

fixative. A liquid applied to drawings in *chalk, *charcoal, or *pastel (usually by means of spraying) to prevent the *pigments from rubbing off, by binding them together and securing them to the *ground. It is usually a transparent *resin, such as shellac, dissolved in alcohol. It is most needed for pastels, but it tends to reduce their brilliance by making the particles of pigment more transparent.

fixed oils. See DRYING OILS.

Flack, Audrey. See SUPERREALISM.

Flandrin, Hippolyte (1809–64). French painter. He was one of the favourite pupils of *Ingres and won the *Prix de Rome in 1830. In Italy he was influenced by the monumental decorative tradition and after his return to Paris in 1838 he became the leading muralist of his day, painting vast compositions in such churches as St Vincent-de-Paul (1849–53) and St Germain-des-Prés (1856–61) in Paris. He was a zealous but rather frigid upholder of Ingres's theories. Flandrin was an excellent portraitist and also painted historical and mythological works. He came from a family of artists. His brothers **Auguste** (1804–43) and **Paul** (1811–1902) were both pupils of Ingres, and concentrated mainly on portraiture and landscape respectively; his son **Paul-Hippolyte** (1856–1921) painted religious, historical, and *genre scenes.

Flavin, Dan (1933–). American sculptor and experimental artist. He specializes in works involving light fixtures, and in general he eschews complicated effects of pulsating or flashing lights, preferring the bare and simple presentation favoured by *Minimal art.

Flaxman, John (1755–1826). English sculptor, draughtsman, and designer, an outstanding figure of the *Neoclassical movement. He was the son of a moulder of plaster figures, and after studying at the *Royal Academy School (where he met his lifelong friend *Blake) he worked for the potter Josiah *Wedgwood from 1775 to 1787. The designs he produced for Wedgwood not only strengthened his interest in *antique art but also developed the innate sensitivity to line that was his greatest gift. At the same time he gradually built up a practice as a sculptor. In 1787 he went to Rome to direct the Wedgwood studio and stayed for seven years. While there he drew his illustrations, much influenced by Greek vase painting, to the *Iliad* and the *Odyssey*, engraved and published in Rome in 1793, followed by illustrations to Aeschylus (1795) and Dante (1802). These engravings, of exceptional purity of outline, were republished in several editions, and won him international fame almost immediately. His later illustrations to Hesiod (1817) were engraved by Blake. He returned to England in 1794 with a well-established reputation and immediately became a busy sculptor. His monument to the poet William Collins (Chichester Cathedral, 1795) and the more important one to Lord Mansfield (Westminster Abbey, 1795–1801) were commissioned while he was in Rome. His enormous practice as a maker of monuments included large groups with freestanding figures (*Lord Nelson*, St Paul's Cathedral, 1809), but his most characteristic work appears in simpler and smaller monuments, sometimes cut in low *relief, which, though not Christian in imagery, are strongly Christian in sentiment. In these his great gift for linear design was given full play. Flaxman was appointed the first Professor of Sculpture at the Royal Academy in 1810 and his reputation among Neoclassical sculptors was exceeded only by those of *Canova and possibly *Thorvaldsen. He was one of the first English artists to be famous outside his own country, although his reputation and influence were based principally on engravings after his drawings rather than his sculpture. University College, London, has a large collection of Flaxman's drawings and models, and examples of his monuments can be seen in churches throughout England.

Flicke, Gerlach (d. 1558). German portrait painter from Osnabrücke, who worked in England from *c.*1547 till his death. With *Eworth he was the most distinguished painter working in England in the generation after *Holbein, but few of his works survive. The most important are the signed and dated (1547) portrait of an unknown man (sometimes identified with the 13th Lord Grey de Wilton) in the National Gallery of Scotland, Edinburgh, and the signed portrait of Archbishop Cranmer in the National Portrait Gallery, London.

Flinck, Govert (1615–60). Dutch painter, active mainly in Amsterdam. He studied with *Rembrandt in the early 1630s, and his early work was overwhelmingly influenced by his master. From the mid-1640s, however, he adopted the elegant style of van der *Helst, and this new way of painting brought him great success. In 1659 he was awarded the most important commission a Dutch painter of his time could receive: he was asked to paint twelve pictures for van *Campen's new Town Hall of Amsterdam, eight of which (each about 5 m. high) were to represent the story of *The Revolt of the Batavians*. But Flinck died three months after signing the contract and the commission was divided among Rembrandt, *Lievens, and *Jordaens.

Flint, Sir William Russell (1880–1969). British painter and graphic artist. He was trained as a lithographer and was a prolific book illustrator, but is now best remembered for his watercolours (particularly his mildly erotic nudes), painted in a distinctive and rather flashy style. He was President of the Royal Society of Painters in Water-Colour from 1936 to 1956.

flock prints. Prints which imitate patterned velvet. They were apparently made by coating an ordinary carved wood block (see WOODCUT) with glue or paste, impressing it on paper, and then sprinkling the paper with cloth shavings which adhered to the paste. Very few such prints exist, all made probably in south Germany in the third quarter of the 15th cent.; an example is the *Christ on the Cross with the Virgin and St John* in the Ashmolean Museum, Oxford. A similar process was used in the 17th and 18th cents. for wallpapers.

Floris (de Vriendt). Netherlandish family of artists active in Antwerp. The most important members were the brothers **Cornelis** (1514–75) and **Frans** (*c*.1516–70). They both worked in Italy in the early 1540s and returned to Antwerp with a desire to emulate the Italian *Renaissance manner. Both ran flourishing workshops and became principal representatives of 'Romanism' in Flanders.

Cornelis was an architect and sculptor and also published engravings of Italianate motifs, which were used by many northern artists. He is famous principally as the architect of Antwerp Town Hall (1561–5), the finest and most influential building of the 16th cent. in Flanders.

Frans Floris was a painter and studied with Lambert *Lombard before going to Italy, where in 1541 he witnessed the unveiling of *Michelangelo's *Last Judgement* in the Sistine Chapel. This made an indelible impression on him and he concentrated on making large religious and mythological pictures crowded with athletic *Mannerist nudes (*Fall of the Rebel Angels*, Musée Royal, Antwerp, 1554). In his portraits, however, he combined powerful brushwork with forthright characterization in a way that anticipates *Hals (*Portrait of an Old Lady*, Musée des Beaux-Arts, Caen, 1558). According to van *Mander every Flemish youth with artistic leanings studied with him, but in spite of his success he died in debt because of his extravagant lifestyle.

Flötner, Peter (*c*.1495–1546). German sculptor and engraver, active mainly in Nuremberg. He was in Italy in 1530 (perhaps following an earlier visit) and his work was important in spreading the *Renaissance style in Northern Europe. His best-known work is the *Apollo Fountain* (Germanisches Nationalmuseum, Nuremberg, 1532), classical in inspiration, but with a flowing elegance that is Flötner's own.

Fluxus. An international avant-garde art movement founded in Germany in 1962 and flourishing until the early 1970s. Reviving the spirit of *Dada, it was violently opposed to artistic tradition and to everything that savoured of professionalism in the arts. Its activities mainly took the form of *happenings (usually called *Aktions* in Germany), street art, and so on. Fluxus festivals were held in various European cities (including Amsterdam, Copenhagen, Dusseldorf, London, and Paris) and also New York, which became the centre of the movement's activities. The most famous artist involved with Fluxus was Joseph *Beuys; among the others were the Japanese-born American Yoko Ono (1933–) and the German Wolf Vostell (1932–). The movement's chief co-ordinator and editor of its many publications was the Lithuanian-born American George Maciunas (1931–78), who coined its name—Latin for 'flowing', suggesting a state of continuous change.

Focillon, Henri (1881–1943). French art historian. He was a celebrated teacher and held various university appointments, notably at the Sorbonne in Paris, where in 1924 he succeeded Émile *Mâle in the chair of art history. In 1937 he became professor of art history at the Collège de France. He also taught in the USA. Focillon's work ranged from studies of medieval sculpture to 20th-cent. painting; he also wrote much on engraving (his father was an engraver), notably a book on *Piranesi (1918). His best-known work is *Art d'Occident* (The Art of the West, 1938), a study of *Romanesque and *Gothic art in which he placed great emphasis on the technical aspects of artistic creation, stressing how the artist responds to his raw materials, their potentialities, and limitations. This outlook also finds expression in his *Vie des formes* (1934), translated as *The Life of Forms in Art*.

folk art. Term describing objects and decorations made in a traditional fashion by craftsmen without formal training, either for daily use and ornament or for special occasions such as weddings and funerals. Decorative woodcarving, embroidery, lace, basketwork, and earthenware are among the typical products of folk art. The term is not properly extended to include articles which are mass-produced to appeal to popular taste, such as Christmas cards or Coronation mugs.

Folk art is little subject to fashion and changing taste. Its methods are handed down in the home from generation to generation, and traditional patterns and designs persist with little alteration. The perpetuation of a folk art seems to depend upon the continuation of a peasant population or other relatively settled social structure. Attempts to revive or artificially reproduce folk art in the context of *Arts and Crafts movements among the urban intelligentsia are frequent but rarely successful.

Fontaine, Pierre-François. See PERCIER.

Fontainebleau, School of. Term applied to artists working in a style associated with the French court at Fontainebleau in the 16th cent. The palace at Fontainebleau was the most brilliant expression of the ambition harboured by Francis I (1494–1547) to emulate the great humanist princes of Italy and to glorify the prestige of the French crown by bringing about a national revival of the arts under the aegis of lavish court patronage. As France lacked an indigenous tradition of mural painting adequate to his grandiose conceptions, he brought in Italian masters to lead the work, which was carried out between 1528 and 1558. The two most distinguished Italians to work at Fontainebleau were *Rosso, who came to France in 1531, and *Primaticcio, who followed in 1532. Rosso was engaged until his death in 1540 on the decoration of the Great Gallery of the king and Primaticcio's work can be best seen in the Ballroom and the Chamber of the Duchesse d'Etampes. The Italian masters succeeded in adapting their own styles to the courtly ideals of the French taste and were assisted by French and Flemish artists. From the combination was born a distinctive style of *Mannerism, a composite of sensuality and decorative flair, of boudoir voluptuousness and etiolated elegance. Many engravings were made of the work at Fontainebleau and the union of *stucco ornament with mural painting introduced an original feature which had wide influence. Much of the stuccowork was in high *relief, but Rosso also developed a distinctive motif known as *strapwork in which the stucco is formed into shapes resembling leather or parchment that has been rolled and cut into decorative patterns; this became a particularly popular form of ornament in England and the Low Countries. Primaticcio's distinctive figure style—characterized by long limbs, small heads, and sharp, elegant profiles—became virtually canonical in French art until the end of the 16th cent. Other Italian artists who worked at Fontainebleau included Niccolò dell' *Abbate and *Cellini, but much of the work associated with the school is by unknown hands, although often of high quality, such as the celebrated painting of *Diana the Huntress* (*c.*1550) in the Louvre. The mythological subject-matter, elongated elegance, idyllic landscape setting, and air of sophisticated artificiality in this work are wholly typical of the School, the influence of which left few French artists of the time untouched.

After the hiatus caused by the Wars of Religion (1562–98) the decorative painting of royal palaces was revived under the patronage of Henry IV (1553–1610). The name **Second School of Fontainebleau** is usually given to the artists who carried out this work for Henry IV, notably Ambroise *Dubois, Toussaint *Dubreuil, and Martin *Fréminet. Their work was accomplished, but without the inventive brilliance of the best work of the First School.

Fontana, Lucio (1899–1968). Italian painter and sculptor, born in Argentina. His family moved to Milan in 1905 and he passed his youth there, studying sculpture first in the studio of his father, who was a sculptor, and then from 1928 to 1930 at the Accademia di Brera. His exhibition at the Galleria del Milione, Milan, in 1930 was the first appearance of non-figurative sculpture in Italy. In 1934 he joined the *Abstraction-Création group in Paris. During the Second World War he lived in Argentina, where he issued his *White Manifesto* (1946), which introduced a new concept of art called Spatialism (*Spazialismo). This called for co-operation with scientists in synthesizing new ideas and materials. In 1947 he returned to Milan, and in the same year founded the Spatialist movement and issued the *Technical Manifesto of Spatialism* (four more Spatialist Manifestos followed, the last in 1952). His most characteristic works are paintings in which completely plain surfaces are penetrated by gashes in the canvas, but he also made environments, for example using neon lights in blackened rooms.

Fontana, Prospero (1512–97). Italian painter, active mainly in his native Bologna, but also in other cities of Italy, notably Florence, Genoa, and Rome, assisting 16th-cent. decorative masters such as *Pierino del Vaga, *Vasari, and *Zuccaro. He also worked at *Fontainebleau (*c.*1560) under *Primaticcio. Fontana was a favourite artist of Gabriele Paleotti, Archbishop of Bologna, and was the leading painter in the city in the 1570s. Paleotti was one of the churchmen who, in line with the ideals of the Counter-Reformation, called for greater clarity in painting as an aid to devotion, but Fontana's work—elegant but rather spineless—is generally seen as exemplifying everything that the *Carracci opposed in their move towards naturalism. Most of Fontana's work is still in and around Bologna. His daughter and pupil **Lavinia Fontana** (1552–1614) was much esteemed in her day as a portraitist.

Foppa, Vincenzo (*c*.1427–1515). Italian painter, the leading figure in Lombard painting until the arrival of *Leonardo da Vinci in Milan in 1481/2. He was born and died in Brescia, but was active mainly in Milan. According to *Vasari he obtained his training in Padua, and his robust style owed much to *Mantegna, not least in his interest in *perspective. Foppa's importance to the development of art in northern Italy has been compared to that of *Tura in Ferrara. His major works include frescos in S. Eustorgio, Milan, and Sta Maria del Carmine, Brescia.

Forain, Jean-Louis (1852–1931). French painter, lithographer, and *caricaturist. He studied under Jean-Léon *Gérôme at the École des *Beaux-Arts, where he was particularly interested in *Rembrandt and *Goya. In his work, much of it done for Paris journals, he combined the *Realist eye of *Manet with the mordant satire of *Daumier, and he had the gift of expressing disposition in a few lines by a characteristic attitude or gesture. Forain was on good terms with Manet and *Degas and also numbered among his many friends the poets Paul Verlaine (1844–96) and Arthur Rimbaud (1854–91), and *Toulouse-Lautrec. He exhibited in four of the *Impressionist exhibitions between 1879 and 1886. As a painter he was uneven, sometimes influenced by Manet and Degas, sometimes adopting the restricted palette of Daumier, as in the courtroom scenes *Le Tribunal* (*c*.1902–3) and *Counsel and Accused* (1908), both in the Tate Gallery, London.

Forbes, Stanhope. See NEWLYN SCHOOL.

Ford, Edward Onslow. See NEW SCULPTURE.

formalism. In art theory, the belief that aesthetic values are autonomous and self-sufficient and that judgements of art can be detached from other considerations, for example ethical and social ones. It has been particularly influential in the 20th cent., a reflection partly of the dominance of *abstract art. Leading critics who have espoused a basically formalist view of art include Roger *Fry and Clement *Greenberg. In Communist countries, particularly during the early years of the Cold War, 'formalism' was used as a general term of abuse directed at the art of the West.

Forment, Damián (*c*.1475–1540). Spanish sculptor. He probably trained in Florence and is an important figure in the transition between *Gothic and *Renaissance in Spanish art. Early works, such as the altarpiece at the church of the Pilar, Saragossa (completed 1512), incorporate Renaissance-style figure sculpture within Gothic architectural settings, but by 1527, when he contracted for the reredos of the monastery church at Poblet (Tarragona), he had finally abandoned Gothic for Renaissance-*Plateresque. He worked mainly in *alabaster but in 1537–40 he made a large wooden reredos (subsequently gilded and *polychromed) for the church at S. Domingo de la Calzada, in which Alonso *Berruguete's influence may be discerned.

Fortuny y Carbo (or **Fortuny y Marsal**), **Mariano** (1838–74). Spanish painter, son-in-law of Federico de *Madrazo. He worked mainly in Rome, where he was extremely successful with anecdotal costume pieces, often set in the 18th cent. They show great brilliance of colour and brushwork, but their superficiality aroused the condemnation of many artists, who were no doubt jealous of the huge prices they commanded. His son, **Mariano Fortuny y Madrazo** (1871–1949), was a painter, sculptor, designer, photographer, and inventor.

Forum Exhibition (in full, Forum Exhibition of Modern American Painters). An exhibition arranged in New York in 1916 by the critic Willard Huntington Wright (1888–1939) with the support of the magazine *Forum*, to which he was a regular contributor. The purpose of the exhibition was to pinpoint the best of American avant-garde painting in order to convince the public that it could stand up to the European avant-garde, which had captured public interest at the *Armory Show. Both Robert *Henri and Alfred *Stieglitz were on the selection committee. The exhibition consisted of *c*.200 pictures by seventeen artists, including *Benton (ironically, later a vociferous enemy of modern art), *Dove, *Macdonald-Wright (Wright's brother), *Marin, *Sheeler, and *Zorach. Anticipating that some of the work on show might be too advanced for public taste, Wright wrote in the catalogue: 'Not one man represented in this exhibition is either a charlatan or a maniac', and he vigorously defended abstraction.

Foster, Myles Birket (1825–99). English painter and engraver. He was trained as a wood engraver and designed many book illustrations after completing his apprenticeship in 1846. After *c*.1858 he devoted himself primarily to watercolour painting of rustic subjects in a sweetly sentimental style that has made him a favourite artist for manufacturers of greetings cards.

Foucquet, Jean. See FOUQUET.

Foujita, Tsuguharu (or **Léonard**) (1886–1968). Japanese-French painter and graphic artist.

He settled in Paris in 1913 and, except for a world tour in 1929 and residence in Tokyo during the Second World War, he lived in Paris for the rest of his life. In 1959 he became a convert to Roman Catholicism and changed his personal name to Léonard in memory of Leonardo da Vinci. He was a member of the circle of *émigré* *Expressionists in the School of *Paris—*Soutine, *Chagall, *Modigliani—and he developed from *c.*1925 a personal style of delicately mannered Expressionism which combined Western and Japanese traits. He began by doing Parisian landscapes and then became known for his nudes and for his compositions in which still life and figures were combined. A characteristic example of his work is the self-portrait in the Musée National d'Art Moderne, Paris (1928).

Fouquet (or **Foucquet**), **Jean** (*c.*1420–*c.*1481). The outstanding French painter of the 15th cent. He was born at Tours and is known to have been in Rome between 1443 and 1447, when he painted a portrait, now lost, of Pope Eugenius IV. Much has been made of this Italian journey, the influence of which can be detected in the *perspective essays and classical architecture of his subsequent works, but the strongly sculptural character of his painting, which was deeply rooted in his native tradition, did not succumb to Italian influence. On his return from Italy Fouquet entered the service of the French court. His first patron was Étienne Chevalier, the royal secretary and lord treasurer, for whom he produced a *Book of Hours (1450–60), now dismembered but mainly in the Musée Condé at Chantilly, and who appears in the *Diptych of Melun* (*c.*1450), now divided between Antwerp (Musée Royal) and Berlin (Staatliche Museen). The Virgin in this work, at Antwerp, is rumoured to be a portrait of Agnes Sorel (1422–50), Charles VII's mistress, whom Chevalier had also loved. It was not until 1475 that Fouquet became Royal Painter (to Louis XI), but in the previous year he was asked to prepare designs for the king's tomb, and he must have been the leading court artist for many years. The main works from the latter part of his career are illustrations in the *Antiquités Judaïques*, a Josephus manuscript (Bib. nat., Paris, 1470–6), and a *Descent from the Cross* in the parish church at Nouans (Indre-et-Loire). Whether he worked on *miniatures or on a larger scale in panel paintings Fouquet's art had the same clarity and dignity, his figures being modelled in broad planes defined by lines of magnificent purity. He used perspective in order to define his forms more precisely rather than from an interest in space for its own sake. This sculptural sense of form went with a cool and detached temperament, and in his finest works the combination creates a deeply impressive gravity.

Fouquières, Jacques. See CHAMPAIGNE.

Fragonard, Jean-Honoré (1732–1806). French painter whose scenes of frivolity and gallantry are among the most complete embodiments of the *Rococo spirit. He was a pupil of *Chardin for a short while and also of *Boucher, before winning the *Prix de Rome in 1752. From 1756 to 1761 he was in Italy, where he eschewed the work of the approved masters of the High *Renaissance, but formed a particular admiration for *Tiepolo. He travelled and drew landscapes with Hubert *Robert and responded with especial sensitivity to the gardens of the Villa d'Este at Tivoli, memories of which occur in paintings throughout his career. In 1765 he became a member of the *Academy with his historical picture in the *Grand Manner, *Coroesus Sacrificing himself to Save Callirhoe* (Louvre, Paris). He soon abandoned this idiom, however, for the erotic canvases by which he is chiefly known (*The Swing*, Wallace Coll., London, *c.*1766). After his marriage in 1769 he also painted children and family scenes. He stopped exhibiting at the *Salon in 1767 and almost all his work was done for private patrons. Among them was Mme du Barry (1743–93), Louis XV's most beautiful mistress, for whom he painted the works that are often regarded as his masterpieces—the four canvases representing *The Progress of Love* (Frick Coll., New York, 1771–3). These, however, were returned by Mme du Barry and it seems that taste was already turning against Fragonard's lighthearted style. He tried unsuccessfully to adapt himself to the new *Neoclassical vogue, but in spite of the admiration and support of *David he was ruined by the Revolution and died in poverty.

Fragonard was a prolific painter, but he rarely dated his works and it is not easy to chart his stylistic development. Alongside those of Boucher, his paintings seem to sum up an era. His delicate colouring, witty characterization, and spontaneous brushwork ensured that even his most erotic subjects are never vulgar, and his finest work has an irresistible verve and joyfulness.

Frampton, Sir George (1860–1928). British sculptor. Early in his career he was one of the leading avant-garde British sculptors of his day, experimenting with unusual materials and *polychrome and working in a style imbued with elements of Art Nouveau and *Symbolism (*Mysteriarch*, Walker Art Gallery, Liverpool, 1892). Later his work became more traditional and he had a successful career with accomplished but uninspired monuments, the best known of which are *Peter Pan*, erected in Kensington Gardens in 1911, and the Edith Cavell memorial (1920) in St Martin's Place, London. His son **Meredith** (1894–1984) was a painter, primarily of portraits. He gave up painting in 1945 because his sight was failing and was almost entirely forgotten until an exhibition of his work was held at the Tate Gallery in 1982, revealing him as an artist of great distinction. His work is beautifully finished, with a sense of hypnotic clarity (the images seem almost palpable yet at the same time strangely remote), and he excelled in conveying the intellectual qualities of his sitters. He was a slow worker and his output was small.

Francavilla, Pierre. See FRANQUEVILLE.

Franceso da Faenza. See ANDREA DEL CASTAGNO.

Francesco del Cossa. See COSSA.

Francesco di Giorgio (1439–1501/2). Sienese painter, sculptor, architect, military engineer, and writer. He painted mainly during the early part of his career and few works certainly by him survive; the most important are a signed *Nativity* (1475) and a documented *Coronation of the Virgin* (1471), both in the Pinacoteca at Siena. As a sculptor, his major works are four bronze angels (1489–97) on the high altar of Siena Cathedral. Francesco was widely travelled, and the latter part of his career was spent mainly as an architect and engineer, particularly a specialist in fortifications. He is also said to have exploded the first mine. Among his patrons was Federico da *Montefeltro, and Francesco may have had a hand in the designing of his celebrated palace in Urbino. In 1490 he travelled to Pavia with *Leonardo da Vinci to advise on the building of the cathedral there, but his only certain non-military building is Sta Maria del Calcinaio, near Cortona, begun 1484. Francesco wrote a treatise on architecture in the last years of his life (it was not published until 1841) and also translated *Vitruvius.

Francheville, Pierre. See FRANQUEVILLE.

Francia (Francesco Raibolini) (*c.*1450–1517/18). The outstanding Bolognese painter of his period, originally a goldsmith. He entered into a partnership with *Costa after the latter came to Bologna *c.*1483 and was later influenced also by *Perugino. His most characteristic works are sweet, softly rounded Madonnas, which his large workshop produced in some numbers. He was also an accomplished portraitist. There are several examples of his work in the National Gallery, London.

Franciabigio (Francisco di Cristofano) (*c.*1482–1525). Florentine painter, a minor master of the High *Renaissance style. He was a pupil of Mariotto *Albertinelli and collaborated with *Andrea del Sarto, who was the dominant influence on his style. His best works are generally considered to be his portraits, particularly those of young men (*A Knight of Rhodes*, NG, London, *c.*1514).

Francis, Sam (1923–94). American painter, one of the leading second-generation *Abstract Expressionists. While serving in the US Army Air Corps he injured his spine in a plane crash and he took up painting in 1944 when he was recovering in hospital. In 1950 he settled in Paris, where he studied under *Léger and was friendly with *Riopelle and other *Art Informel painters; his style was influenced by these artists as well as by Americans such as Jackson *Pollock. He has visited Japan several times, and the thin texture of his paint, his drip and splash technique, and his asymmetrical balance of colour against powerful voids (he often leaves areas of canvas blank) have led critics to speak of influences from Japanese traditions of contemplative art. In 1961 Francis returned to his native California, settling first at Santa Barbara and then in Santa Monica. From the mid-1960s the feeling of oriental simplicity in his painting increased, bringing his work into closer affinity with *Minimal art. Francis has carried out several mural commissions, but he often works on a small scale in watercolour. He has also made lithographs (from 1960) and sculpture (from 1965).

Francisque. See MILLET.

Francken. Family of Flemish painters active in the 16th and 17th cents., mainly in Antwerp. The individual contributions of the many artists in the family are often difficult to assess, but the two most distinguished members were **Frans I** (1542–1616) and his son **Frans II** (1581–1642). The father mainly painted religious and historical compositions. His early works were frequently life-size; the late ones were small, usually done on copper, and crowded with exotic figures and accessories. Frans II frequently adopted his father's subjects and style, but his range was wider. He painted landscapes and *genre scenes as well as historical pictures, and was also one of the first artists to use the interior of a picture gallery as a subject, giving faithful miniature reproductions of the works in the collection. His paintings were even smaller and more crowded than his father's; they were also more colourful. Frans II was frequently employed by his fellow artists in Antwerp to paint the figures in their landscapes and interiors. There are numerous examples of the work of various members of the family in the Musée Royal in Antwerp.

Frankenthaler, Helen (1928–). American painter, an important figure in the transition from *Abstract Expressionism to *Colour Field painting. In her early work she was influenced by Jackson *Pollock and she developed his drip technique by pouring and running very thin paint on the canvas like washes of watercolour. She first used this method in *Mountains and Sea* (artist's collection, on loan to NG, Washington, 1952), which is regarded as one of the seminal works of post-war American painting. It particularly impressed Morris *Louis and Kenneth *Noland when they saw it in 1953. In 1962 Frankenthaler switched from oil to acrylic paint, which allowed her to achieve more richly saturated colour. Her limpid veils of colour float on the surface of the canvas, but they often evoke suggestions of landscape. Since 1960 Frankenthaler has also made aquatints, lithographs, and woodcuts; in 1964 she began to work in ceramics; and in 1972 she made her first sculpture. From 1958 to 1971 she was married to Robert *Motherwell.

Franqueville (or **Francheville,** or **Francavilla**), **Pierre** (1548–1615). French sculptor who went to Italy in 1574 and became the pupil and then assistant of *Giambologna. Henry IV (1553–1610) recalled him to France (1601), where he executed the slave figures (Louvre, Paris) at the base of Bologna's equestrian statue of Henry IV (now destroyed) set up by Marie de *Médicis on the Pont Neuf, Paris. His style was a somewhat cold and dry version of his master's.

freestone. Any good-quality, fine-grained sandstone or *limestone suitable for carving or masonry.

Fréminet, Martin (1567–1619). French painter, one of the leaders of the Second School of *Fontainebleau. On the death of *Dubreuil in 1602 he was recalled by Henry IV (1553–1610) to Paris from Italy, where he had been working since the late 1580s, mainly in Rome. The ceiling for the chapel of the Trinité at Fontainebleau (begun 1608) is the most important of his few surviving works. Italian *Mannerist influence, particularly that of Giuseppe *Cesari, is apparent in the rather strained poses of the figures, which nevertheless are effectively integrated with the *stucco decoration.

French, Daniel Chester (1850–1931). American sculptor. He made his name with the famous statue of *The Minute Man* (1875) in Concord, Mass., a monument to commemorate the rising of the citizens of the town during the early years of the Revolution (the figure was ready to fight for his country in a minute). After this success, French went on to become the most illustrious sculptor of public monuments of his day, his best-known work being the seated marble figure of Abraham Lincoln (dedicated in 1922) on the Lincoln Memorial in Washington.

fresco (Italian: 'fresh'). A method of wall-painting in which pure powdered *pigments, mixed only in water, are applied to a wet, freshly laid lime-plaster *ground. The colours penetrate into the surface and become an integral part of the wall. This technique is also called *buon fresco* or *fresco buono* (true fresco) to distinguish it from painting on dry plaster, which is called by analogy *fresco *secco* or simply *secco*. *Buon fresco* is exceptionally permanent in dry climates, but if damp penetrates the wall, the plaster may crumble and the paint with it. Consequently the art has been practised chiefly in dry countries, particularly in Italy (though not in Venice), and seldom in northern Europe.

The technique is of great antiquity. *Minoan and Greek wall-paintings were probably in fresco; those at Pompeii certainly are and the Roman writer *Vitruvius describes a method much like that in use during the *Renaissance. Fresco painting is also found outside Europe, for example in China and India. The Italian practice was described in detail by *Cennini in the early 15th cent. The wall was first given a coating of plaster, prepared from lime and sand in water. On this rough surface (the **arricciato*) the design was drawn in charcoal; next the assisting lines and some of the main contours were incised, and the outlines and shading indicated with pigment mixed in water; thirdly the lines of the design were painted in a red ochre called *sinopia*, which was the chief red used in fresco (the other important red, *cinabrese*, was a mixture of *sinopia* and lime white). Until the introduction of the *cartoon *c.*1500 the design was worked out directly on this first plaster ground or copied from a small sketch. A layer of finer plaster, called the **intonaco*, was now applied over one section of the rougher *arricciato*. This was the actual painting ground and was made very smooth. Each day just so much of the design was covered with the *intonaco* as could be painted on that day; no more, because the plaster had to remain wet during the painting. On this small area of fresh plaster the design—perhaps a head or a draped figure—was first drawn with the brush in *verdaccio*, a mixture of black, lime white, and *cinabrese*. Flesh parts received an undercoating of *terra verde*, a green earth pigment. The actual flesh tint was prepared in three tones by mixing *cinabrese* with varying quantities of lime white. For the drapery, or other parts where modelling had to be indicated, similar sets of three tones were prepared, as in the *tempera painting of the time. Since blending was difficult, the final effects were produced by *hatching. Finishing touches were sometimes added after the plaster was dry (*al secco*), but this of course had to be done with egg tempera or *size paint instead of pure pigment and water. *Vasari called it a 'vile practice' and the parts done *al secco* were liable to flake off, but many of the greatest practitioners of the art resorted to it.

The fresco painter thus had to work rapidly, before his plaster could dry; corrections were almost impossible, and he needed a sure hand and purpose. Further he had to work directly because his preliminary design was covered by *intonaco*. The colours available to him were few—in the 15th and 16th cents. painters believed that only natural pigments were suitable for fresco—and apt to become lighter in drying; depth of tone was unattainable. But these difficulties and limitations themselves encouraged him to design his subject broadly and treat it boldly, and did much to foster the purity, strength, and monumentality of Italian Renaissance painting.

*Giotto was the first really great master of fresco, and thereafter many of the leading Italian masters produced works in the medium; *Masaccio, in the Brancacci Chapel, Florence; *Piero della Francesca in S. Francesco, Arezzo; *Raphael in the Stanze at the Vatican; *Michelangelo in the Sistine Chapel; *Correggio in his work at Parma; Annibale *Carracci in the Farnese Gallery. It became less common in the 18th cent. and Giambattista *Tiepolo was the last in the line of great Italian painters who used it. It was revived in the 19th cent., notably by German painters such as the *Nazarenes and *Cornelius, but some notable decorators, such as *Delacroix and *Puvis de Chavannes, preferred to use the method of *marouflage. In the 20th cent. the greatest exponents of fresco have been the Mexican muralists *Orozco, *Rivera, and *Siqueiros.

Freud, Lucian (1922–). German-born British painter. He was born in Berlin, a grandson of Sigmund Freud, came to England with his parents in 1932, and acquired British nationality in 1939. His earliest love was drawing, and he began to work full time as an artist after being invalided out of the Merchant Navy in 1942. In 1951 his *Interior at Paddington* (Walker Art Gal., Liverpool) won a prize at the Festival of Britain, and since then he has built up a formidable reputation as one of the most powerful contemporary figurative painters. Portraits and nudes are his specialities, often observed in arresting close-up. He prefers to paint people he knows well: 'If you don't know them, it can only be like a travel book.' His early work was meticulously painted, so he has sometimes been described as a *'Realist' (or rather absurdly as a *Superrealist), but the subjectivity and intensity of his work has always set him apart from the sober tradition characteristic of most British figurative art since the Second World War. In his later work (from the late 1950s) his handling became much broader, bringing to his flesh painting an extraordinary quality of palpability. In 1993 Peter *Blake wrote that since the death of Francis *Bacon the previous year, Freud was 'certainly the best living British painter'.

Freundlich, Otto (1878–1943). German painter and sculptor. After working as a shop assistant he studied history of art at Munich and Florence and began painting in 1905. From 1909 to 1914 he spent much of his time in Paris, where he became a member of *Picasso's circle. After flirting with *Cubism, he began to do purely abstract painting *c.*1919, composing with interlocking swathes of pure colour. From 1924 to 1939 he lived in Paris, where he was a member of *Cercle et Carré and of the *Abstraction-Création association. In his own country he was classed as a *degenerate artist (before it was destroyed by the Nazis, his sculpture *The New Man* was reproduced on the cover of the catalogue of the infamous exhibition of 'Degenerate Art' held in 1937). After being arrested in the Pyrenees he died in a concentration camp at Lublin.

Frick, Henry Clay (1849–1919). American industrialist, art collector, and philanthropist. He made his fortune in coke and steel operations and assembled a collection of paintings, sculpture, and decorative art under the guidance of Roger *Fry and the dealer Joseph *Duveen. On his death he left his New York mansion and a large fund to form the Frick Collection, which was opened to the public in 1935. It is generally regarded as one of the finest small museums in the world, with a choice collection of works from the Middle Ages to the late 19th cent. *Rembrandt's *Polish Rider* and Giovanni *Bellini's *St Francis* are among the celebrated masterpieces in the collection. Attached to the Frick Collection is the Frick Art Reference library, established in 1920, which has major collections of books and photographs. In 1970 Frick's daughter, Helen Clay Frick, established the Frick Art Museum in Pittsburgh, the city where her father had made his fortune.

Friedlaender, Walter (1873–1966). German-born American art historian. He had a distinguished academic career in Germany and, as professor at the university of Freibourg, numbered Erwin *Panofsky among his pupils. In 1933 he was dismissed by the Nazis and emigrated to the USA, where he became professor at New York University and exercised a great influence on American art historians. His major publications include *Caravaggio Studies* (1955), *David to Delacroix* (1952), *Mannerism and Anti-Mannerism in Italian Painting* (1957), and monographs on *Poussin in German (1914) and English (1966). With Anthony *Blunt he edited a complete catalogue of Poussin's drawings (5 vols., 1939–72).

Friedländer, Max J. (1867–1958). German art historian. The successor to Wilhelm von *Bode as director of the Gemäldegalerie in Berlin, he enriched the collection particularly in his own field of Early Netherlandish painting. In 1934 he retired to Holland. His *magnum opus* is *Die altniederländische Malerei* (14 vols., 1924–37). In a prefatory note to the English edition (*Early Netherlandish Painting*, 1967–76) Erwin *Panofsky described it as 'one of the few uncontested masterpieces produced by our discipline'. Friedländer covered the same ground in a much briefer format in *Die frühen niederländischen Maler von Van Eyck bis Bruegel* (1916), translated as *From Van Eyck to Bruegel* (1956). His other books included *Essays über die Landschaftsmalerei und andere Bildgattungen* (1947), translated as *Landscape, Portrait, Still-Life* (1949), and *On Art and Connoisseurship* (1942).

Friedrich, Caspar David (1774–1840). The greatest German *Romantic painter and one of the most original geniuses in the history of landscape painting. He was born at Greifswald on the Baltic coast, and after studying at the Copenhagen Academy with *Juel and *Abildgaard from 1794 to 1798, he settled permanently in Dresden. There he led a quiet life, interrupted only by occasional excursions to the mountains or the coast of Pomerania. Neither his classical training nor his friendship with *Goethe could induce him to visit Italy. The scientific outlook of his friend J. C. *Dahl was no less antipathetic to him, and he pursued with a rare and instinctive single-mindedness his personal insight into the spiritual significance of landscape. He was intensely introspective and often melancholic (although his marriage at the age of 44 brought him much happiness), and he relied on deep contemplation to summon up mentally the images he was to put on canvas. 'Close your bodily eye, so that you may see your picture first with your spiritual eye', he wrote, 'then bring to the light of day that which you have seen in the darkness so that it may react on others from the outside inwards.'

Friedrich began as a topographical draughtsman in pencil and sepia wash and did not take up oil painting until 1807. One of his first works in the new medium, *The Cross in the Mountains* (Staatliche Kunstsammlungen, Dresden, 1808), caused great controversy because it was painted as an altarpiece, and to use a landscape in this unprecedented way was considered sacrilege by some critics. His choice of subjects often broke new ground and he discovered aspects of nature so far unseen: an infinite stretch of sea or mountains, snow-covered or fog-bound plains seen in the strange light of sunrise, dusk, or moonlight. He seldom uses obvious religious imagery, but his landscapes convey a sense of haunting spirituality. Friedrich had a severe stroke in 1835 and returned to his small sepias. He was virtually forgotten at the time of his death and his immediate influence was confined to members of his circle in Dresden, notably G. F. *Kersting, who sometimes painted the figures in Friedrich's work. It was only at the end of the 19th cent., with the rise of *Symbolism, that his greatness began to be recognized.

Friesz, Othon (1879–1949). French painter. After studying in his native Le Havre alongside *Braque and *Dufy, he moved to Paris in 1898. He met *Matisse and became one of the most enthusiastic and vigorous painters in this *Fauve style. In 1908 he abandoned Fauvism and reverted to a more traditional style, giving more attention to the structural plan of his compositions and using simple, uncomplicated colours. In 1912 he opened his own studio and taught there until 1914. After serving in the First World War he settled in Paris in 1919, but his painting had lost much of its verve and he held aloof from the newer artistic movements.

Frink, Dame Elisabeth (1930–93). British sculptor and graphic artist. Some of her early work—influenced by *Giacometti—was angular and menacing. During the 1960s her figures—typically horses and riders or male nudes—became smoother, but she retained a feeling of the bizarre in the polished goggles that feature particularly in her over-life-size heads. She worked mainly in bronze and had numerous public commissions, for example *Horse and Rider* (1975) in Piccadilly, London, made for Trafalgar House Investments Ltd.

Frith, William Powell (1819–1909). English painter. He began with illustrative paintings of classics such as *The Vicar of Wakefield*, but *c.*1851 he turned to contemporary scenes, with which he had a great commercial success. His crowded, anecdote-packed pictures of Victorian life, among them *Derby Day* (Tate, London, 1858) and *The Railway Station* (Royal Holloway and Bedford New College, Egham, 1862), are among the most familiar images of their age and Frith's pictures were so popular in their day that they had sometimes to be railed off from their masses of admirers at the *Royal Academy. Frith's *Reminiscences* (1887) and *Further Reminiscences* (1888) form a useful record of the conservative academic conception of art and of contemporary gossip.

Froment, Nicolas (active *c.*1450–*c.*1490). French painter, born at Uzès in Languedoc and active in Avignon. Two documented works by him—both *triptychs—survive: *The Raising of Lazarus* (Uffizi, Florence, 1461) and *The Burning Bush* (Aix-en-Provence Cathedral, 1476), which was painted for René of Anjou (1409–80). They show that with *Charonton he introduced Netherlandish naturalism to French art. His figures have strong if sometimes clumsy expressions and gestures, while his draperies have a characteristic angularity reminiscent of some of the works of the Spanish and German followers of Rogier van der *Weyden.

Fromentin, Eugène (1820–76). French painter and writer. As a painter he was a specialist in oriental themes (he visited North Africa in 1846, 1848, and 1852), which were much admired in his day but are now known only to specialists. His reputation now rests on his book *Les Maîtres d'autrefois* (*The Masters of Past Time*, 1876), a study of Dutch and Belgian painting which includes an analysis of the effect of Dutch landscapes on French painters such as *Claude and Théodore *Rousseau. He also wrote a lyrical novel, *Dominique* (1862).

Fronte Nuovo Delle Arti (New Art Front). An association of Italian artists founded in 1946 with the aim of combating the pessimism of the post-war world and advocating a return to an art concerned with human values. *Birolli and *Guttuso were the best-known figures in the group, which combined artists of very different styles and ideologies. The split between Abstractionists and Realists led to the dissolution of the association in 1948.

Frost, Terry (1915–). British painter, one of the leading *St Ives painters. He started painting in 1943 when a prisoner of war encouraged by Adrian *Heath, then studied at St Ives and under *Coldstream and *Pasmore at the Camberwell School of Art, 1947–50. He began painting in the sober, naturalistic tradition of the *Euston Road School, but he soon turned to abstraction, and in the 1950s he created screens of brightly striped colours; more recently he has used circles and ovals in high-pitched and saturated colours, often juxtaposing segments of closely related colour. Characteristically he draws on and transposes direct visual experiences of landscape, boats in harbour, and the human figure. His main period of residence at St Ives was 1959–63. He has taught at various art schools, notably at Reading University, 1965–81.

frottage (French: 'rubbing'). A technique of creating a design by placing a piece of paper over some rough substance such as grained wood or sacking and rubbing with a crayon or pencil until it acquires the surface quality of the substance beneath. The resulting design is usually taken as a stimulus to the imagination, the point of departure for a painting which expresses imagery of the subconscious. Max *Ernst pioneered the technique and it was much used by other *Surrealists.

frottie. The transparent or semi-opaque brushings in of colour with which some artists begin their paintings. English painters sometimes refer to this method of starting a painting as 'rubbing in', since the thin colour is lightly rubbed in over the *ground. *Frotties* can sometimes be seen in the most thinly covered portions of a painting, particularly backgrounds, and are clearly visible, for example, in *David's unfinished portrait of Madame Récamier (Louvre, Paris, 1800).

Fry, Roger (1866–1934). English critic, painter, and designer. He took a first-class degree in natural sciences at Cambridge in 1888, but was already more interested in art, and in the 1890s he built up a reputation as a writer and lecturer (and a much more modest one as a painter). His success as a public speaker depended partly on his mellifluous voice; George Bernard Shaw said it was one of only two he knew that were worth listening to for its own sake—the other was that of the actor Sir Johnston Forbes-Robertson. He was Curator of Paintings at the *Metropolitan Museum of Art, New York, 1906–10, but in the year he took up this appointment he 'discovered' *Cézanne and turned his attention away from the Italian Old Masters, with whom he had established his scholarly reputation, to become the period's most eloquent champion of modern French painting. After returning to London in 1910 he organized two exhibitions of *Post-Impressionist painting at the Grafton Galleries (1910 and 1912) that are regarded as milestones in the history of British taste. They attracted an enormous amount of publicity, most of it unfavourable, and many people thought that Fry was a charlatan or possibly even insane. Certain young artists were immensely impressed by the exhibitions, however, and Fry became an influential figure among them. They included Vanessa *Bell and Duncan *Grant, both of whom worked for the *Omega Workshops, which Fry founded in 1913. He kept up a steady output of writing and lecturing (at the time of his death he was Slade Professor at Cambridge University) and probably did more than anyone else to awaken public interest and understanding of modern art in England. Kenneth *Clark called him 'incomparably the greatest influence on taste since *Ruskin' and said: 'In so far as taste can be changed by one man, it was changed by Roger Fry.' His books include monographs on *Bellini (1899), Cézanne (1927), and *Matisse (1930), an edition of *Reynolds's *Discourses* and several collections of lectures and essays. As a painter Fry was experimental (his work includes a few abstracts), but his best pictures are fairly straightforward naturalistic portraits; his sitters included several of his *Bloomsbury Group friends.

Fuchs, Ernst. See FANTASTIC REALISM.

Führich, Joseph (1800–76). Austrian painter, draughtsman, and engraver. He came at first under the influence of *Dürer, but his style and outlook were really shaped by his contact with the *Nazarenes in Rome (1827–9), where he assisted with frescos in the Villa Massimo. After being appointed Professor of Historical Composition at the Vienna Academy he executed many paintings and frescos in churches and public buildings. He also did various biblical cycles in line engravings which gained him the nickname of 'der Theologe mit dem Stifte' (the theologian with the pencil).

Fuller, Isaac (*c*.1606–72). English decorative and portrait painter. He studied in France with F. *Perrier then worked in Oxford and London. Fuller painted altarpieces for Oxford colleges (that in All Souls was described by the diarist John Evelyn in 1644 as 'too full of nakeds for a chapel') and did decorative painting for taverns in London, including mythological scenes for the Mitre Tavern, Fenchurch Street, but these works have disappeared. His largest surviving works are five canvases, each about 3 m. wide, showing Charles II's escape after the Battle of Worcester in 1651 (NPG, London). Otherwise, Fuller is remembered for his highly idiosyncratic portraits. He was a notorious drunkard and his self-portraits (NPG, and Bodleian Lib., Oxford) are painted with a bravura worthy of a larger-than-life character.

Funk art. Term applied to a type of art that originated in California (specifically the San Francisco area) around 1960 in which tatty or sick subjects—often pornographic or scatological—are treated in a deliberately distasteful way. The word 'funky' originally meant 'smelly' and 'Sick art' is sometimes used as a synonym for 'Funk art'. Although the first funk works were paintings, its most characteristic products are three-dimensional, either sculpture or *assemblages. Edward *Kienholz is the best-known practitioner of the genre.

Fuseli, Henry (Johann Heinrich Füssli) (1741–1825). Swiss-born painter, draughtsman, and writer on art, active mainly in England, where he was one of the outstanding figures of the *Romantic movement. He was the son of a portrait painter, **Johann Caspar Füssli** (1707–82), but he originally trained as a priest; he took holy orders in 1761, but never practised. In 1765 he came to London at the suggestion of the British Ambassador in Berlin, who had been impressed by his drawings. *Reynolds encouraged him to take up painting, and he spent the years 1770–8 in Italy engrossed in the study of *Michelangelo, whose manifestations of the *sublime he sought to emulate for the rest of his life. On his return he exhibited works of a character at once imaginative and grotesque, such as *The Nightmare* (Detroit Institute of Arts, 1781), the picture that secured his reputation when it was exhibited at the Royal Academy in 1782 (there is another version in the Goethe-museum, Frankfurt). Literature provided many of his subjects; he painted several works for *Boydell's Shakespeare Gallery, and in 1799 he followed this example by opening a Milton Gallery in Pall Mall with an exhibition of forty-seven of his own paintings. In the same year he was elected Professor of Painting at the *Royal Academy, and he became Keeper in 1804.

Fuseli's aspirations to the sublime tended towards the horrifying and fantastic. He was a much respected and influential figure in his lifetime, but his work was generally neglected for about a century after his death until the *Expressionists and *Surrealists saw in him a kindred spirit. His work can be clumsy and overblown, but at its best has something of the imaginative intensity of his friend *Blake, who described Fuseli as 'The only man that e'er I knew / who did not make me almost spew'. Fuseli's extensive writings on art include *Lectures on Painting* (1801) and a translation of *Winckelmann's *Reflections on the Painting and Sculpture of the Greeks* (1765).

Futurism. An artistic movement with political implications founded by the poet *Marinetti in Milan in 1909. It sought to free Italy from the oppressive weight of her past, and glorified the modern world—machinery, speed, violence—in a series of exuberant manifestos. The painters Umberto *Boccioni, Carlo *Carrà, Luigi *Russolo,

Giacomo *Balla, and Gino *Severini publicly proclaimed their adherence to the movement in March 1910, following this a month later with a technical manifesto of Futurist painting, and Boccioni published a manifesto of Futurist sculpture in 1912.

A Futurist exhibition at Paris in 1912 was accompanied by a further manifesto which set out the theoretical bases of the movement. At the time the movement was founded the Futurists followed the colour techniques of the *Neo-Impressionists, which Severini and Boccioni had learnt from Balla. By the time of the exhibition, however, the Futurists had adopted *Cubist devices to render motion—one of the primary concerns of the group. As Marinetti said in 1909, 'The splendour of the world has been enriched with a new form of beauty, the beauty of speed.'

As an organized movement Futurism did not last much beyond the death of Boccioni in 1916 or the end of the First World War. It had considerable influence, however, notably in Russia, where there was a Russian Futurist movement which included *Larionov, *Goncharova, and *Malevich. The *Dadaists owed something to it, particularly in their noisy publicity techniques, and in England it had some influence on *Vorticism and C. R. W. *Nevinson. In France Marcel *Duchamp and Robert *Delaunay among others developed in their own ways the Futurist ideas about the representation of movement.

Fyt, Jan (1611–61). Flemish painter and etcher, primarily of still life and hunting pieces. He was born and mainly active in Antwerp, where he was a pupil of *Snyders, but in the course of his successful and prolific career he also travelled in France, the Netherlands, and Italy. Like Snyders, Fyt painted in the elaborate style of large decorative still life associated with the circle of *Rubens. His most characteristic paintings depict trophies of the hunt, dead stags, hares, and birds, all treated with a feeling for texture and detail akin to that often seen in Dutch still life. In his best work his colouring and finish are excellent, and however various the multiplicity of miscellaneous objects he included in a canvas, he could achieve balance and refinement in the whole. The rare flower paintings by Fyt are exceptionally fine and more attuned, perhaps, to modern taste.

Gabo, Naum (Naum Neemia Pevsner) (1890–1977). Russian-born sculptor who became an American citizen in 1952, the most influential exponent of *Constructivism. He was the younger brother of Antoine *Pevsner, and adopted the name Gabo in 1915 to avoid confusion between the two. After studying medicine, natural sciences, and engineering in Munich, he was introduced to avant-garde art when he visited his brother in Paris in 1913 and 1914, and in 1915 he began to make geometrical constructions in Oslo, where they had gone during the First World War. In 1917 the brothers returned to Russia with Antoine and in 1920 they issued their famous *Realistic Manifesto*, which set forth the basic principles of European Constructivism. When it became clear that official policy favoured the regimentation of artistic activity in the direction of industrial design and socially useful work (as exemplified by *Tatlin), rather than the pure abstract art conceived by Gabo, the latter left Russia for Berlin in 1922 and spent the next ten years there in contact with the artists of the *Bauhaus and the De *Stijl group. In 1932 he moved to Paris and was active in the *Abstraction-Création association until 1935, when he went to England, living first in London (where in 1937 he was co-editor of the Constructivist review **Circle*) and then from 1939 in Cornwall (see ST IVES SCHOOL).

In 1946 Gabo moved to the USA, settling at Middlebury, Connecticut, in 1953. In the last three decades of his life he received many prestigious awards and carried out numerous public commissions in Europe and America. He often worked on themes over a long period; his *Torsion Fountain* outside St Thomas's Hospital in London, for example, was erected in 1975, but is a development from models he was making in the 1920s. (Small models are a feature of his work; there are numerous examples in the Tate Gallery, which has an outstanding collection of Gabo material.)

Gabo never trained as an artist but came to art by way of his studies of engineering and physical science, and was one of the first artists to embody in his work modern concepts of the nature of space. He was one of the earliest to experiment with *Kinetic sculpture and to make extensive and serious use of semi-transparent materials for a type of abstract sculpture which, with apparent weightlessness, incorporates space as a positive element rather than displacing or enclosing it. He was throughout his life an advocate of Constructivism not merely as an artistic movement but also as the ideology of a life-style.

Gabriel, Paul Joseph Constantin (1828–1903). Dutch landscape painter of the *Hague School. He specialized in views of meadows, canals, and windmills, but he had an inclination towards more cheerful colours than are normally associated with the Hague School: in 1901 he wrote 'Although I may sometimes seem rather grumpy, I really love it when the sun shines on the water; and, quite apart from that, I think my country is colourful.' *Mondrian admired the almost geometrical quality of Gabriel's compositions and copied his *In the Month of July* (Rijksmuseum, Amsterdam).

Gaddi, Taddeo (*c.*1300–*c.*1366). Florentine painter, the son of a painter and mosaicist **Gaddo Gaddi**, also called Gaddo di Zanobi (*c.*1250–1327/30?). According to Cennino *Cennini, Taddeo was *Giotto's godson and worked with him for twenty-four years. In 1347 he headed a list of the best living painters compiled for the purpose of choosing a master to paint a new high altarpiece for Pistoia Cathedral. His best-known works were painted for Sta Croce, Florence: notably the frescos devoted to the *Life of the Virgin* in the Baroncelli Chapel (finished 1338), and the panels illustrating the *Life of Christ* (*c.*1330), originally meant for the doors of a sacristy cupboard and now scattered among museums in Florence (Accademia), Munich (Alte Pinakothek), and Berlin (Staatliche Museen). Many other panels are attributed to him and he must have had a flourishing workshop. Although transmitting the tradition of Giotto, his style is less heroic and more anecdotal; he strove for vividly picturesque effects and met the popular taste for pictures full of episode and incidental detail.

His son **Agnolo** (active 1369–96) continued the Giotto tradition but modified it still further in the direction of decorative elegance. He is particularly notable for his cool pale colours, which influenced the refined late *Gothic art of artists of the next generation such as *Lorenzo Monaco. Agnolo's works include frescos on *The Story of the Cross* in the chancel of Sta Croce (after 1374) and on *The Story of the Virgin and her Girdle* in the Chapel of the Holy Girdle in Prato Cathedral (1392–5). Many panel paintings also are attributed to him.

gadroon. Type of ornament consisting of a border of repeated slanting curves, like the coils of a rope.

Gainsborough, Thomas (1727–88). English painter of portraits, landscapes, and *fancy pictures, born at Sudbury, Suffolk, one of the most individual geniuses in British art. He went to London in about 1740 and probably studied with the French engraver *Gravelot, returning to Sudbury in 1748. In 1752 he set up as a portrait painter at Ipswich. His work at this time consisted mainly of heads and half-lengths, but he also painted some small portrait groups in landscape settings which are the most lyrical of all English *conversation pieces (*Heneage Lloyd and his Sister*, Fitzwilliam, Cambridge). His patrons were the merchants of the town and the neighbouring squires, but when in 1759 he moved to Bath his new sitters were members of Society, and he developed a free and elegant mode of painting seen at its most characteristic in full-length portraits (*Mary, Countess Howe*, Kenwood House, London, *c.*1763–4). From Bath he sent pictures to the exhibitions of the *Society of Artists in London, and in 1768 he was elected a foundation member of the *Royal Academy. In 1774 he moved permanently to London, where he further developed the personal style he had evolved at Bath, working with light and rapid brush-strokes and delicate and evanescent colours. He became a favourite painter of the Royal Family, even though his rival *Reynolds was appointed King's Principal Painter. Gainsborough sometimes said that while portraiture was his profession landscape painting was his pleasure, and he continued to paint landscapes long after he had left a country neighbourhood, often making imaginative compositions from studio arrangements of glass, twigs, and pebbles. He produced many landscape drawings, some in pencil, some in charcoal and chalk, and he occasionally made drawings which he varnished. He also, in later years, painted fancy pictures of pastoral subjects (*Peasant Girl Gathering Sticks*, Manchester City Art Gal., 1782).

Gainsborough's style had diverse sources. His early works show the influence of French engraving and of Dutch landscape painting; at Bath his change of portrait style owed much to a close study of van *Dyck (his admiration is most clear in *The Blue Boy*, Huntingdon Art Gal., San Marino, 1770); and in his later landscapes (*The Watering Place*, NG, London, 1777) he is sometimes influenced by *Rubens. But he was an independent and original genius, able to assimilate to his own ends what he learnt from others, and he relied always mainly on his own resources. Even when he set out to paint the conventional portrait as a pot-boiler his delight in painting and his individuality won through. With the exception of his nephew Gainsborough *Dupont, he had no assistants and unlike most of his contemporaries he never employed a drapery painter. He was in many ways the antithesis of Reynolds. Whereas Reynolds was sober-minded and the complete professional, Gainsborough (even though his output was prodigious) was much more easy-going and often overdue with his commissions, writing that 'painting and punctuality mix like oil and vinegar'. Although he was an entertaining letter-writer, Gainsborough, unlike Reynolds, had no interest in literary or historical themes, his great passion outside painting being music (his friend William Jackson the composer wrote that he 'avoided the company of literary men, who were his aversion . . . he detested reading'). Gainsborough and Reynolds had great mutual respect, however; Gainsborough asked for Reynolds to visit him on his deathbed, and Reynolds paid posthumous tribute to his rival in his Fourteenth *Discourse*. Recognizing the fluid brilliance of his brushwork, Reynolds praised 'his manner of forming all the parts of a picture together', and wrote of 'all those odd scratches and marks' that 'by a kind of magic, at a certain distance . . . seem to drop into their proper places'.

Gallego, Fernando (*c.*1440–after 1507). Spanish painter. He worked mainly in Salamanca, where *Palomino says he was born, and was the major Castilian painter of his period. Gallego's sober, impassive style has affinities with that of Dirk *Bouts, and it has been suggested that he visited the Netherlands early in his career. His works include a *retable (*c.*1475–80) of *San Idelfonso* in the cathedral of Zamora, a *triptych of *The Virgin, St Andrew and St Christopher* in the new cathedral of Salamanca, and ceiling frescos on astrological subjects (much repainted) in the Old Library in the University of Salamanca. Gallego had considerable influence in Castile. One of his followers, **Francisco Gallego**, was presumably a relative; in 1500 he was paid for a *Martyrdom of St Catherine* in the old cathedral of Salamanca.

Gallen-Kallela, Akseli (1865–1931). Finnish painter, graphic artist, and designer. A major figure in the *Art Nouveau and *Symbolist movements,

Gallen-Kallela travelled widely and was well known outside Finland, particularly in Germany (he had a joint exhibition with *Munch in Berlin in 1895 and exhibited with Die *Brücke in Dresden in 1910). He was deeply patriotic (he volunteered to fight in the War of Independence against Russia in 1918, even though he was in his fifties) and he was inspired mainly by the landscape and folklore of his country, above all by the Finnish national epic *Kalevala* (Land of Heroes). His early work was in the 19th-cent. naturalistic tradition, but in the 1890s he developed a flatter, more stylized manner, well suited to the depiction of heroic myth, with bold simplifications of form, strong outlines, and vivid—sometimes rather garish—colours. Apart from easel paintings, Gallen-Kallela's work included book illustrations (notably for an edition of *Kalevala*, 1922) and he did a number of murals for public buildings (including the Finnish National Museum, Helsinki, 1928). His designs for stained glass, fabrics, and jewellery gave an important stimulus to the development of Finnish crafts. He is regarded not only as his country's greatest painter, but also as the chief figure in the creation of a national art. His former house near Helsinki is now a museum dedicated to his work.

gallery varnish. Preparation used in the early days of the *National Gallery, London, when pictures were exhibited without glass, to protect them from the noxious effects of the atmosphere. It was a mixture of *mastic in turpentine with boiled *linseed oil. The varnish at first imparted a warm golden glow which was thought appropriate to the Old Masters, but it soon darkened disastrously and became almost opaque. Gallery varnish was roundly condemned by the Select Committee which reported on the National Gallery in 1853, but all the pictures which had been treated with it—among them masterpieces such as van *Eyck's *Arnolfini and his Wife*—remained sealed in their dark-brown coating until the 1940s, when means were found of removing it.

Galli-Bibiena. See BIBIENA.

Gambart, Ernest (1816–1902). Belgian picture dealer and print publisher who settled in London in 1840 and became the dominant figure in his profession. In the three decades between his arrival in England and his retirement to the Continent in 1871, 'Gambart, more than any other individual, transformed the London art world. He founded a system for the promotion and sale of pictures on an international scale; contributed increasingly to the status of the artist in society; and brought the London, if not the European art trade from infancy to maturity . . . [He] became a print publisher of international repute, who exploited the possibilities of the print so as to raise it to the level of pure popular art' (Jeremy Maas, *Gambart: Prince of the Victorian Art World*, 1975).

Garamond, Claude (d. 1561). The most celebrated of French letter cutters for typefounding. There is a tradition that he was taught by Geoffroy *Tory and he admired the work of Aldus Manutius (see ALDINE PRESS), reproducing it in a somewhat lighter and more gracious form. His roman and italic types were innovatory in being designed specifically as metal types rather than as imitations of handwriting, and his clear and elegant roman forms were a major factor in establishing this type of lettering as standard in place of black letter or gothic. His beautiful Greek types, on the other hand, commissioned by the French king Francis I and cut between 1541 and 1549, rendered the copying hand of the king's Cretan scribe Angelos Vergetios.

Gardner, Isabella Stewart (1840–1924). American socialite and art collector. She married into a prominent Boston family and spent most of her life in that city, where she patronized numerous artistic causes (including the Boston Symphony Orchestra) and dazzled and occasionally scandalized polite society (she had love affairs and attended boxing matches). Her interest in art was guided by Bernard *Berenson, one of her protégés, who helped her to assemble a superb collection of Italian Renaissance paintings, including *Titian's *The Rape of Europa*, which has often been claimed as the greatest painting in America. Other highlights of her collection include some outstanding 17th-cent. Dutch paintings and her full-length portrait (1888) by *Sargent (to whom she had been introduced by Henry James). The portrait shows off her celebrated figure (she wears her pearls round her waist rather than her neck) and was considered rather shocking—in the spirit of Sargent's earlier *Madame X* (it was rumoured that he and Mrs Gardner were lovers). Her husband, the financier Jack Gardner, died in 1898, and the following year she began building Fenway Court in Boston as both a home and a museum. The building, completed in 1902, is modelled on a Venetian Renaissance palace and incorporates various architectural fragments that she bought on her numerous trips to Europe. She supervised the construction with immense care, acting virtually as site foreman. In her will she left the Isabella Stewart Gardner Museum to Boston as a public institution with the proviso that the collection should be maintained exactly as she had arranged it. The British art historian Sir Philip

Hendy, who published a catalogue of the Gardner Museum in 1931, described it as 'probably the finest collection of its compact size in the world'. In addition to paintings, sculpture, drawings, and prints, it contains objects of many other types, including furniture, textiles, ceramics, glassware, manuscripts, and books.

gargoyle. A spout in the form of a *grotesque figure, animal or human being, projecting from a cornice or parapet and allowing the water from the roof gutters to escape clear of the walls. Many examples on *Gothic cathedrals and churches throughout Europe bear witness to the lively imagination and spirited fantasies of medieval craftsmen. In the 14th and 15th cents. sculptures similar to gargoyles but not serving their function were used to decorate walls, and with the introduction of lead drainpipes in the 16th cent. gargoyles were no longer needed.

Garofalo (Benvenuto Tisi) (1481?–1559). Italian painter, active mainly in Ferrara. *Vasari says that he twice visited Rome, and his work—derivative but beautifully crafted—was heavily influenced by *Raphael. He was the first to paint in such a manner in Ferrara and was influential in spreading the High *Renaissance style. His output was large (frescos, altarpieces, small devotional works, also a few mythologies); there are many examples in Ferrarese churches and, for example, in the National Gallery, London. In 1550 he went blind.

Gaudier-Brzeska, Henri (Henri Gaudier) (1891–1915). French sculptor and draughtsman, active in England for most of his short career and usually considered part of the history of British rather than French art. In 1910 he took up sculpture in Paris without formal training, and in the same year he met Sophie Brzeska, a Polish woman 20 years his senior with whom he lived from that time, both of them adopting the hyphenated name. In 1911 they moved to London, which Gaudier had visited briefly in 1906 and 1908, and lived for a while in extreme poverty. He became a friend of Wyndham *Lewis, Ezra Pound (1885–1972), and other leading literary and artistic figures, and his work was shown in avant-garde exhibitions, such as the *Vorticist exhibition of 1915. In 1914 he enlisted in the French army and was killed in action the following year, aged 23.

Gaudier developed with astonishing rapidity from a modelling style based upon *Rodin towards a highly personal manner of carving in which shapes are radically simplified in a manner recalling *Brancusi (*Red Stone Dancer*, Tate, London, 1913). In England, only *Epstein was producing sculpture as stylistically advanced as Gaudier-Brzeska at this time. In his lifetime his work was appreciated by only a small circle, but since his death he has become widely recognized as one of the outstanding sculptors of his generation and has acquired something of a legendary status as an unfulfilled genius. In addition to his sculptures he left behind some splendid animal drawings.

Gauguin, Paul (1848–1903). French *Post-Impressionist painter, sculptor, and printmaker, born in Paris of a journalist from Orleans and a Peruvian Creole mother. He spent his childhood in Lima, joined the merchant marine in 1865, and from 1872 worked successfully as a stockbroker. In the early 1870s he became a spare-time artist and in 1874 he met *Pissarro and saw the First *Impressionist Exhibition. At about the same time he began to make a collection of Impressionist pictures. He had a landscape accepted by the *Salon in 1876 and his work was shown in the Fifth to Eighth (and last) Impressionist Exhibitions. In 1883 he gave up his employment to become a full-time artist. During the next few years he was unsuccessful in marketing his pictures and sold his collection to support himself and his family. After the last Impressionist Exhibition in 1886 he moved to Brittany, abandoning his family, and until 1890 he spent much of his time at *Pont-Aven, where he became the pivot of a group of artists who were attracted by his picturesque personality and new ideas in aesthetics. The most important work he produced there was *The Vision After the Sermon*, also known as *Jacob Wrestling with the Angel* (NG, Edinburgh, 1888), in which he broke away completely from the Impressionist style, using areas of pure, flat colour for expressive and symbolic purposes. In 1887–8 he went to Panama and Martinique, and in 1888 he spent a short time at Arles with van *Gogh, a visit which ended in a disastrous quarrel as van Gogh suffered one of his first attacks of madness.

Gauguin had had a taste for colourful, exotic places since his childhood in Peru and in 1891 he left France for Tahiti. In the book *Noa Noa*, which he wrote about his life there, he said: 'I have escaped everything that is artificial and conventional. Here I enter into Truth, become one with nature. After the disease of civilization life in this new world is a return to health.' His theory and practice of art reflected these attitudes. He was one of the first to find visual inspiration in the arts of ancient or primitive peoples, and reacted vigorously against the naturalism of the Impressionists and the scientific preoccupations of the *Neo-Impressionists. As well as using colour unnaturalistically for its decorative or emotional effect he reintroduced emphatic outlines forming rhythmic patterns suggestive of Japanese colour prints (see UKIYO-E) or the technique of stained glass. He described his method *c.*1887 as 'Synthetist-Symbolic', using the term 'symbolic' in the sense then current to indicate that the forms and patterns in his pictures were meant to suggest mental images or ideas and not simply to record visual experience. By *'Synthetist' he seems to have meant that the forms of his pictures were constructed from such symbolic patterns of colour and linear rhythms, built up from expressive distortions, and not either empirical or scientific reproductions of what is seen by the eye. Gauguin also did woodcuts in which the black and white areas formed rhythmical, almost abstract, patterns and the tool marks were incorporated as integral parts of the design. Along with those of Edvard *Munch, these prints played an important part in the 20th-cent. revival of the art of woodcut—one of the salient features of modern graphic art. His other work included carving and pottery.

In Tahiti Gauguin endeavoured to 'go native' and despite the constant pressure of poverty he painted his finest pictures there. His colours became more resonant, and his drawing more grandly simplified, and his expression of the mysteries of life more profound. In 1893, however, poverty and ill-health forced him to return to France, but he had a financial windfall when an uncle died and he was back in Tahiti in 1895. At the end of 1897 he painted his largest and most famous picture, the allegory of life, *Where Do We Come From? What Are We? Where Are We Going To?* (Mus. of Fine Arts, Boston) before attempting suicide (although he had deserted his family he had been devastated that year by the news of the death of his favourite daughter). In September 1901 he settled at Dominica in the Marquesas Islands, where he died two years later. Until his death he worked continuously in the face of poverty, illness (he had syphilis), and lack of recognition. During his time in the South Seas he was often unable to obtain proper materials and was forced to spread his colours thinly on coarse sacking, but from these limitations he forged a style of rough vigour wholly appropriate to the boldness of his vision.

At the time of his death few would have agreed with Gauguin's self-assessment: 'I am a great artist and I know it. It is because I am that I have endured such suffering.' His reputation was firmly established, however, when 227 of his works were shown at the *Salon d'Automne in Paris in 1906, and his influence has been enormous. The *Nabis were formed under his inspiration, he was a leading figure of the *Symbolist movement and one of the sources for *Fauvism. Later, he has been one of the major influences on the general non-naturalistic trend of 20th-cent. art. Because of the romantic appeal of his life and personality, particularly his willingness to sacrifice everything for his art, Gauguin has been with van Gogh the most common subject for popular and fictional biography, including the novel *The Moon and Sixpence* (1919) by Somerset Maugham, and the opera (1957) of the same title by John L. Gardner.

Gaulli, Giovanni Battista. See BACICCIO.

Gauvreau, Pierre. See AUTOMASTISTES, LES.

Gavarni (pseudonym of Sulpice-Guillaume Chevalier) (1804–66). French *caricaturist. He is best known for his humorous drawings of social manners, but from the time of his visit to England (1847–51), where he studied the life of the poor in London and graphically contrasted it with that of the rich (*Gavarni in London*, 1849), the benign irony of the earlier works gave way to a more trenchant satire with political implications. He lacked the genius of *Daumier as an artist, but as a satirist of French bourgeois life he has few equals. His work was admired and collected by many writers and artists, among them *Degas.

Geertgen tot Sint Jans (*c.*1460–*c.*1490). Netherlandish painter, born in Leiden but active in Haarlem. Almost nothing is known of his career, but van *Mander says that he was a pupil of *Ouwater and died when he was 28. His name means 'Little Gerard of the Brethren of St John', after the Order in Haarlem of which he was a lay-brother. For the monastery church of the Brethren he painted his only documented work, a *triptych of *The Crucifixion*, of which two large panels (originally two sides of a wing) survive in the Kunsthistorisches Museum in Vienna: the *Lamentation of Christ* and the *Burning of the Bones of St John the Baptist*. Certain features of these paintings—particularly the slender, doll-like figures with smooth, rather egg-like heads—are highly distinctive, and a small *œuvre* of about fifteen paintings has been attributed to Geertgen on stylistic grounds. Unlike the Vienna panels, most of the other pictures given to him are fairly small. They include such remarkably beautiful works as *The Nativity* (NG, London), a radiant nocturnal scene, and *St John the Baptist in the Wilderness* (Staatliche Museen, Berlin), which shows an exquisite feeling for nature. The vein of tender melancholy that pervades Geertgen's work, the beguilingly innocent charm of his figures, and his sensitivity to light are perhaps the salient qualities that make him one of the most irresistibly attractive artists of the Early Netherlandish school.

Gelder, Aert de (1645–1727). Dutch painter, active mainly in his native Dordrecht. After studying there with *Hoogstraten, he became one of *Rembrandt's last pupils in Amsterdam. He was not only one of the most talented of Rembrandt's pupils, but also one of his most devoted followers, for he was the only Dutch artist to continue working in his style into the 18th cent. His religious paintings, in particular, with their imaginative boldness and preference for oriental types, are very much in the master's spirit, although de Gelder often used colours—such as lilac and lemon yellow—that were untypical of Rembrandt, and his palette was in general lighter. One of his best-known works, *Jacob's Dream* (Dulwich College Picture Gal., London), was long attributed to Rembrandt.

Generalić, Ivan (1914–). Yugoslav (Croatian) *naïve painter, the outstanding figure of the school of peasant painters associated with his native village of Hlebine. His repertory is extremely catholic, his favourite themes being scenes from village life, celebrations, festivals, etc., landscapes, still lifes in a landscape, figures, and portraits. Some of his pictures are quiet and almost idyllic depictions of peasant activities, but in others there is an element of grotesque fantasy reminiscent of *Bruegel or Hieronymus *Bosch, while still others have a *Surrealist air of the unexpected. Generalić has all the identifying marks of a genuine naïve painter and despite his occasional contacts with orthodox and historic art, has always remained so. Of peasant stock, he continued to live the life of a peasant even after the 'painting village' of Hlebine was attracting visitors from all over the world, and after he himself had won a world-wide reputation as one of the greatest of naïve painters he continued to paint only in his spare time.

genre. Term in art history and criticism for paintings depicting scenes from daily life. It may be applied to appropriate art of any place or period, but most commonly suggests the type of domestic subject-matter favoured by Dutch 17th-cent. artists. In a broader sense, the term is used to mean a particular branch or category of art; landscape, portraiture, and still life, for example, are all genres of painting, and the essay and the short story are genres of literature.

Gentile da Fabriano (*c*.1370–1427). Italian painter named after his birthplace, Fabriano in the Marches. He carried out important commissions in several major Italian art centres and was recognized as one of the foremost artists of his day, but most of the work on which his great contemporary reputation was based has been destroyed. It included frescos in the Doges' Palace in Venice (1408) and for St John Lateran in Rome (1427). In between he worked in Florence, Siena, and Orvieto. His major surviving work is the celebrated altarpiece of the *Adoration of the Magi* (Uffizi, Florence, 1423), painted for the church of Sta Trinità in Florence, which places him alongside *Ghiberti as the greatest exponent of the *International Gothic style in Italy. It is remarkable not only for its exquisite decorative beauty but also for the naturalistic treatment of light in the *predella, where there is a night scene with three different light sources. Gentile had widespread influence (much more so initially than his great contemporary *Masaccio), notably on *Pisanello, his assistant in Venice, Jacopo *Bellini, who worked with him in Florence, and Fra *Angelico, who was his greatest heir.

Gentileschi, Orazio (Orazio Lomi) (1563–1639). Italian painter. He was born in Pisa, but in about 1576 he settled in Rome. After working in a *Mannerist style he became one of the closest and most gifted of *Caravaggio's followers. He was one of the few *Caravaggisti who was a friend of the master, and in 1603 he and Caravaggio and two other artists were sued for libel by Giovanni *Baglione. Gentileschi's work does not have the power and uncompromising naturalism of Caravaggio, tending rather towards the lyrical and refined. His graceful figures are stately and clearly disposed, with sharp-edged drapery—qualities recalling his Tuscan heritage. In 1621 he moved to Genoa, where he stayed until 1623; while there he painted an *Annunciation* (Galleria Sabauda, Turin) that is often considered his masterpiece. After working for Marie de *Médicis in Paris, he settled in England in 1626 and became court painter to Charles I (1600–49). He was held in great esteem in England and remained until his death. His travels were a factor in spreading the Caravaggesque manner, but by the end of his career he had long abandoned heavy *chiaroscuro in favour of light colours. His major works in England were a series of ceiling paintings commissioned by Charles I for the Queen's House at Greenwich, now in Marlborough House, London (probably after 1635).

His daughter **Artemisia Gentileschi** (1593–1652/3) was one of the greatest of Caravaggesque painters and a formidable personality. She was precociously gifted, built up a European reputation, and lived a life of independence rare for a woman of the time. Born in Rome, she worked mainly there and in Florence until she settled in Naples in 1630 (she also visited her father in England in 1638–40). Artemisia's powerful style—totally different to that of her father—is seen at its most characteristic in paintings of *Judith and Holofernes*, a subject she made her own (one of the finest examples is in the Uffizi, Florence). Her predilection for the bloodthirsty theme has been related to events in her own life. At the age of 19 she was allegedly raped by Agostino *Tassi (who was eventually acquitted of the charge) and was tortured during the legal proceedings; thus the fierce intensity with which she depicted a woman decapitating a man has been seen as pictorial 'revenge' for her sufferings.

Geometric art. Term applied to Greek art in the 9th and 8th cents. BC, named after the decoration associated with the pottery of the period. Geometric vases are characteristically divided into painted horizontal bands filled with various forms of geometric ornament, developing into stylized human and animal forms.

Georgian. A general descriptive term loosely applied to various styles in English architecture, decoration, furniture, silver, etc., from the accession of George I (1714) to the death of George IV (1830). The term 'Early Georgian' is used of the period from *c.*1714 to the 1730s. 'Middle Georgian' (covering the 1740s and 1750s) and 'Late Georgian' (from the 1760s to *c.*1830) are infrequently used, the movements in style and taste during these years being separately designated (*Rococo, *Greek taste, *Neoclassical, *Gothic Revival, *Etruscan taste, etc). The latter part of 'Late Georgian' is known as *Regency. Stylistically the period had little uniformity. So far as overall characterization is possible, its salient features were respect for sound craftsmanship, and a predilection for a Classical system of design and proportion flexible enough to accommodate phases as different as *chinoiserie and Pompeian. Today the term 'Georgian' is also current as a trade description to designate objects manufactured in a vaguely 18th-cent. manner without reference to a date.

Gérard, François (1770–1837). French painter, born in Rome, a favourite pupil of J.-L. *David. In the *Salon of 1796 he won acclaim with his portrait of *Jean-Baptiste Isabey and his Daughter* (Louvre, Paris) and became the most sought after court and society portraitist of his day. He successfully negotiated the various political changes of the day and was made a Baron and a member of the Legion of Honour. In addition to portraits Gérard painted historical and mythological works. His style derived from David, but was much less taut and heroic, tending at times towards a rather mannered gracefulness.

Gere, Charles. See ASHENDENE PRESS.

Gerhaert van Leyden, Nicolaus (active 1462–73). The most powerful and original Netherlandish sculptor of the second half of the 15th cent. He is known to have worked in Strasburg, Trier, and Vienna, and several signed or documented works survive, in both stone and wood, but the details of his life are obscure. His work is extraordinarily vivid and unconventional, capturing an intense feeling of inner life, as in the celebrated bust of a man in Strasburg (Musée de l'Œuvre, Notre-Dame), which is usually considered to be a self-portrait. The voluminous style of his draperies and his boldness of approach suggest that he was trained in a Burgundian workshop where Claus *Sluter's style was still predominant, although his name indicates that he was born in the northern Netherlands. His work had considerable influence, particularly in Germany.

Géricault, Théodore (1791–1824). French painter, one of the prime movers and most original figures of *Romanticism. He studied in Paris with Carle *Vernet and Pierre *Guérin, but was influenced more by making copies of the Old Masters at the *Louvre, developing in particular a passion for *Rubens. In 1816–17 he was in Italy and there became an enthusiastic admirer of *Michelangelo and the *Baroque. On his return to Paris he exhibited the picture for which he is most famous, *The Raft of the Medusa* (Louvre, Paris, 1819), which, although it was awarded a medal at the *Salon, created a furore both on account of its realistic treatment of a horrific event and because of its political implications (it depicts the ordeal of the survivors of the shipwreck of the *Medusa* in 1816, a disaster ascribed by some to government incompetence). The picture, which was remarkably original in treating a contemporary event with epic grandeur, also had a *succès de scandale* in England, where Géricault spent the years 1820–2. He painted jockeys and horse races (*Derby at Epsom*, Louvre, 1821) and was one of the first to introduce English painting to the notice of French artists (he was particularly enthusiastic about *Constable and *Bonington).

Géricault was a passionate horseman and his death at the age of 33 was brought on by a riding accident. In his temperament and life-style as well as his work he ranks (like Byron, for example) as an archetypal Romantic artist. His tempestuous career lasted little more than a decade and in that time he displayed a meteoric and many-sided genius. His fine disregard for the orthodox doctrine of conventional types, his love of stirring action, his sense of swirling movement, his energetic handling of paint, and his taste for the macabre were all to become features of Romanticism. He was, at the same time, forward-looking in his realism: he made studies from corpses and severed limbs for *The Raft of the Medusa* and painted an extraordinary series of portraits of mental patients in the clinic of his friend Dr Georget, one of the pioneers of humane treatment for the insane (*A Kleptomaniac*, Musée des Beaux-Arts, Ghent, *c*.1822–3). His work had enormous influence, most notably on *Delacroix.

Germ, The. The literary organ, in prose and verse, of the *Pre-Raphaelite Brotherhood. Only four issues appeared, the first in January 1850 and the last in April of the same year. The first two were sub-titled *Thoughts towards Nature in Poetry, Literature and Art*, and in the last two the title was changed to *Art and Poetry, being Thoughts towards Nature*. Dante Gabriel *Rossetti was one of the main contributors (his poem 'The Blessed Damozel' appeared in it) and his brother William Michael Rossetti (1839–1919) was the editor. A facsimile edition with an explanatory preface by W. M. Rossetti was published in 1901.

Gérôme, Jean-Léon (1824–1904). French painter and sculptor. He was a pupil of Paul *Delaroche and inherited his highly finished academic style. His best-known works are his oriental scenes, the fruit of several visits to Egypt; two typical examples are in the Wallace Collection. They won Gérôme great popularity and he had considerable influence as an upholder of academic tradition and enemy of progressive trends in art; he opposed, for example, the acceptance by the state of the *Caillebotte bequest of *Impressionist pictures.

Gerstl, Richard (1883–1908). Austrian painter. His early painting was in the style of the Vienna *Sezession, influenced particularly by the decorative linearism of *Klimt. By 1905, however, he had developed a highly personal style of *Expressionism. His finest works are portraits, notably two groups of the family of the composer Arnold Schoenberg (1874–1951), remarkable for their psychological intensity. He was a tormented character, and after running off with Schoenberg's wife he committed suicide. His work, which anticipates that of such painterly Expressionists as *Kokoschka, remained little known until the 1930s.

Gertler, Mark (1891–1939). British painter. He was born in the East End of London to poor Polish-Jewish immigrant parents and he spoke only Yiddish up to the age of 8. In 1908–12 he studied at the *Slade School, where he won several prizes. After the First World War he spent a good deal of time in the South of France for the sake of his delicate health (he had tuberculosis). Gertler was influenced by *Post-Impressionism, but his style was highly individual, with strong elements of Eastern European folk art. His favourite subjects included female portraits, still lifes, and nudes, such as the earthy and voluptuous *The Queen of Sheba* (1922) in the Tate Gallery, London, painted in his characteristic feverishly hot colours. The Tate also has his only piece of sculpture, *Acrobats* (1917), and the painting that is probably his best-known work, *Merry-Go-Round* (1916); in this powerful image—probably a satire on militarism—figures spin on fairground horses in a mad, futile whirl. Gertler had many admirers, including distinguished figures in the literary world; D. H. Lawrence made him the model for the sculptor Loerke in *Women in Love* (1920). The word 'genius' was frequently applied to him, and he was seen by many as the acceptable face of modernism. However, he began to lose popularity in the early 1930s when he adopted a more avant-garde style characterized by a flatter sense of space and a greater emphasis on surface pattern. He had always been subject to fits of depression, and after the failure of an exhibition at the Lefevre Gallery, London, in 1939, he committed suicide.

Gessner, Salomon (1730–88). Swiss painter, etcher, and poet. He illustrated an edition of his own *Idylls* (1772) with charming *Rococo vignettes and painted small landscapes in a similar vein. His *Brief Über die Landschaftsmalerei* (Zurich, 1787; English edition 1798) is an account of his own studies in which he advocates that love of nature and imitation of the great masters, in particular of *Claude Lorraine, are the best training for the landscape painter who wishes to express true feeling.

gesso. Brilliant white preparation of chalky pigment mixed with glue, used during the Middle Ages and the Renaissance as a *ground to prepare a panel or canvas for painting or gilding. In preparing a ground the gesso was applied in several layers. According to *Cennini the underlayers were of relatively coarse and heavy gesso, which he called *gesso grosso*; over this was laid a coat of *gesso sottile*, fine and smooth to paint on and brilliantly white. The gesso could also take the impress of the tools used in the decorative gilding which often adorned a panel painting. When applied to frames and furniture it could be painted and gilded in the same way, and was often modelled (*gesso rilievo*).

In the 20th cent. the term 'gesso' came to be used loosely for any white substance that can be mixed with water to make a ground; in reference to sculpture it often means *plaster of Paris.

Gestel, Leo (Leendert) (1881–1941). Dutch painter. He was in Paris in 1904 and 1910–11 and was one of the first Dutch artists to experiment with *Cubism and *Expressionism. From this basis he elaborated a decorative and lyrical manner for colourful landscapes, nudes, and still lifes. He also made lithographs.

Gestural painting. A term describing the application of paint with expansive gestures so that the sweep of the artist's arm is deliberately emphasized. It carries an implication not only that a picture is the record of the artist's actions in the process of painting it, but that the recorded actions express the artist's emotions and personality, just as in other walks of life gestures express a person's feelings. The term has been applied particularly to *Abstract Expressionism and is sometimes used more or less as a synonym for *Action painting. However, it can also apply to figurative painting, notably *Neo-Expressionism.

Getty, J. Paul (1892–1976). American oil magnate and art collector. Reputedly the richest man in the world, he amassed a large collection of works of art, his main areas of interest being, as he wrote in his book *The Joys of Collecting* (1966), 'Greek and Roman marbles and bronzes; Renaissance paintings; sixteenth-century Persian carpets; Savonnerie carpets and eighteenth-century French furniture and tapestries'. The J. Paul Getty Museum in Malibu, California, was opened in 1954, and in 1974 a new museum, housed in a re-creation of a Roman villa, was opened nearby. One of the archetypes of the eccentric, parsimonious millionaire, Getty lived in England from the 1950s and never saw his museum. On his death it became the most richly endowed museum in the world, and has become famous for its spectacular purchases (see LYSIPPUS), which have aroused fears that it would monopolize the world market for masterpieces.

The J. Paul Getty Trust, founded in 1953 'for the diffusion of artistic and general knowledge', administers various bodies, including the Getty Center for the History of Art and the Humanities (founded 1983) in Santa Monica, California (a centre for advanced research with a large library of photographs and books), and a Grant Program, which assists, for example, with the publication of scholarly works. **J. Paul Getty Jr.** (1932–), one of Getty's five sons by his five wives, lives in England and has been a princely benefactor to British art institutions; most notably he gave £50,000,000 to the National Gallery, London, in 1985.

Gheeraerts, Marcus the Younger (1562–1636). Flemish-born portrait painter, who settled in England in 1568 with his father **Marcus the Elder** (*c.*1530–*c.*1590), an engraver and painter. Marcus the Younger was probably the leading society portraitist in London at the peak of his career (his popularity declined after about 1615), but it is not easy to disentangle his work from that of some of his contemporaries. He was related by marriage to the de *Critz and *Oliver families, and may well have collaborated with his relatives. The best-known work attributed to him is the splendid full-length portrait of Elizabeth I known as the 'Ditchley' portrait (NPG, London, *c.*1592), in which the queen is shown standing on a map of England.

Gheyn, Jacob de II (1565–1629). Dutch draughtsman, engraver, and painter. He was born at Antwerp and was probably a pupil of his father **Jacob de Gheyn I** (*c.*1530–82), a glass painter and *miniaturist. From *c.*1585 to 1590 he studied with Hendrick *Goltzius. He worked for the Court of Orange at The Hague, and designed the grotto (the earliest in the Netherlands) and other ornamentation of Buitenhof, the garden of Prince Maurice (1567–1625). His drawings and engravings are of greater importance than his paintings, for in their spontaneity and informality they are outstanding documents of the period of transition from *Mannerism to *naturalism in Dutch art. His son **Jacob de Gheyn III** (*c.*1596–1641) was also an engraver, specializing in mythological subjects.

Ghezzi, Pier Leone. See CARICATURE.

Ghiberti, Lorenzo (1378–1455). Florentine sculptor, goldsmith, and designer. He came to prominence in 1401 when he competed successfully (defeating *Brunelleschi, Jacopo della *Quercia, and others) for a commission, offered by the merchant guild, to make a pair of gilded bronze doors for the Baptistery of Florence. His competition *relief of *The Sacrifice of Isaac* is in the Bargello, Florence. Work on the doors lasted until 1424 and in 1425 he was asked to make a second pair for the same building, which occupied him until 1452. These two commissions lay at the focus of the civic and religious life of Florence; the fulfilment of them necessitated the formation of a large workshop in which some of the outstanding Florentine artists of the period, including *Donatello, *Masolino, and *Uccello, received at least part of their training. Ghiberti also served on the committee in charge of the architectural works of Florence Cathedral, designed stained-glass windows, goldsmiths' work, and reliquaries, and made several life-size bronze statues. He was also a writer and left a large incomplete manuscript under the title of *Commentarii*. Apart from a survey of ancient art based on *Pliny and notes on the science of optics, this manuscript contains valuable records of Italian painters and sculptors of the trecento, and also Ghiberti's autobiography, the first by an artist that has survived. The same interest in the new humanist ideals that is reflected in Ghiberti's writings also prompted him to collect *classical sculptures and to appreciate any new discovery in this field.

Despite this prominent place which Ghiberti occupies in the classical revival, his style was deeply rooted in the tradition of *Gothic craftsmanship. Not only was his first pair of Baptistery doors closely modelled on the pattern of Andrea *Pisano's earlier doors, but its twenty episodes from the Life of Christ and its eight saints reflect the *International Gothic style with its emphasis on graceful lines, lyrical sentiment, and minute attention to landscape detail. While these traits survive in Ghiberti's second pair of doors, they are here subordinated to the new principles of the *Renaissance. The doors are divided into ten large panels in which episodes from the Old Testament are represented on carefully constructed *perspective stages. As most of these reliefs were planned and laid out by 1437, they must rank among the most 'advanced' works of Florentine art, particularly in the mastery of composition within a spatial framework. The fame of these doors has always stood high. *Michelangelo's dictum, recorded by *Vasari, that they were worthy to form the Gates of Paradise secured their prestige even in times less sympathetic to *quattrocento art.

Ghirlandaio, Domenico (1449–94). Florentine painter. He trained with *Baldovinetti and possibly with *Verrocchio. His style was solid, prosaic, and rather old-fashioned (especially when compared with that of his great contemporary *Botticelli), but he was an excellent craftsman and good businessman and had one of the most prosperous workshops in Florence. In this he was assisted by his two younger brothers, **Benedetto** (1458–97) and **Davide** (1452–1525). His largest undertaking was the fresco cycle in the choir of Sta Maria Novella, Florence, illustrating *Scenes from the Lives of the Virgin and St John the Baptist* (1486–90). This was commissioned by Giovanni Tornabuoni, a partner in the *Medici bank, and Ghirlandaio depicts the sacred story as if it had taken place in the home of a wealthy Florentine burgher. It is this talent for portraying the life and manners of his time (he often included portraits in his religious works) that has made Ghirlandaio popular with many visitors to Florence. But it should not allow us to overlook his considerable skill in the management of complex compositions, and a certain grandeur of conception that sometimes hints at the High *Renaissance.

Ghirlandaio worked on frescos in Pisa, San Gimignano, and Rome (in the Sistine Chapel) as well as in Florence, and his studio produced numerous altarpieces. He also painted portraits, the finest of which is the well-known *Old Man and his Grandson* (Louvre, Paris); this depicts the grandfather's diseased features with ruthless realism, but has a remarkable air of tenderness. Ghirlandaio's son and pupil **Ridolfo** (1483–1561) was a friend of *Raphael and a portrait painter of some distinction. His most famous pupil, however, was *Michelangelo.

Giacometti, Alberto (1901–66). Swiss sculptor and painter, active mainly in Paris. He was the son of **Giovanni Giacometti** (1868–1933), a painter influenced by *Impressionism and *Post-Impressionism. After short periods at the École des Arts et Métiers, Geneva, and in Italy, he went to Paris and there worked under *Bourdelle from 1922 to 1925. He abandoned naturalistic sculpture in 1925, however, and went through a period of restless experimentation. From 1930 to 1935 he participated in the *Surrealist movement, developing a highly individual attenuated manner and open-cage construction exemplified in *The Palace at 4 a.m.* (MOMA, New York, 1933). Giacometti abandoned Surrealism in 1935, however, and began to work again from the model. In 1941–5 he lived in Geneva, but then returned to Paris, and his most characteristic style emerged in 1947, featuring 'transparent constructions' of human figures, sometimes disposed in groups, notable for their emaciated, extremely elongated, and nervous character (*Man Pointing*, Tate, London, 1947). His isolated figures often have a suggestion of existentialist tragedy, and he was indeed a friend of the existentialist philosopher Jean-Paul Sartre (1905–80), who wrote on Giacometti's work, notably the introduction to the catalogue of his exhibition at the Pierre Matisse Gallery, New York, in 1948. It was this exhibition that established Giacometti's post-war reputation, and his work soon had widespread influence, which can be seen, for example, in many of the entries for the *Unknown Political Prisoner* competition of 1953 (see BUTLER, REG).

He impressed many people not only through the quality of his work, but also by his force of personality, integrity, and devotion to his work. Simone de Beauvoir, Sartre's companion, wrote: 'Success, fame, money—Giacometti was indifferent to them all.' He is generally considered one of the outstandingly original sculptors of the 20th cent., and from the late 1950s his reputation as a painter began to increase. Most of his paintings and drawings are portraits of his family and friends; his brother Diego, who was a skilled technician and a life-long assistant, was a favourite model and the subject of dozens of sculptures, paintings, and drawings. Their cousin **Augusto Giacometti** (1877–1947) was a painter, one of the first to produce pure abstracts.

Giacomo del Duca. See DUCA.

Giambellino. See BELLINI, GIOVANNI.

Giambologna (Giovanni Bologna or Jean Boulogne) (1529–1608). Flemish-born Italian sculptor. He was the greatest sculptor of the age of *Mannerism and for about two centuries after his death his reputation was second only to that of *Michelangelo. After training under Jacques *Dubroeucq he went to Italy to study in about 1550 and spent 2 years in Rome. On the way back he stopped in Florence and was based there for the rest of his life. The work that made his name, however, was for Bologna—the *Fountain of Neptune* (1563–6), with its impressive nude figure of Neptune which he had designed for a similar fountain in Florence (*Ammanati defeated him in the competition). Even before working on the fountain in Bologna, however, Giambologna had begun in Florence the first of a series of celebrated marble groups that in their mastery of complex twisting poses mark one of the high-points of Mannerist art: *Samson Slaying a Philistine* (V&A, London, *c.*1561–2); *Florence Triumphant over Pisa* (Bargello, Florence, *c.*1575); *The Rape of a Sabine* (Loggia dei Lanzi, Florence, 1581–2); *Hercules and the Centaur* (Loggia dei Lanzi, 1594–1600). Giambologna worked extensively for the *Medici and his monument to Duke Cosimo I (1587–95) was the first equestrian statue made in Florence and an immensely influential design, becoming the pattern for similar statues all over Europe (for example that of Charles I by Hubert *Le Sueur at Charing Cross in London). Giambologna's similar statue to Henry IV of France, formerly on the Pont Neuf in Paris, has been destroyed. It was for the Medici that he made his largest work—the colossal (about 10 m. high) figure of the mountain god *Appennino* (1577–81) in the gardens of the family's villa at Pratolino. Constructed of brick and stone, the god crouches above a pool and seems to have emerged from the earth, fusing brilliantly with the landscape. Giambologna was as happy working on a small scale as in a monumental vein. His small bronze statuettes were enormously popular (they continued to be reproduced almost continuously until the 20th cent.) and being portable helped to give his style European currency. Many of his preliminary models also survive (uniquely for an Italian sculptor of his period), giving insight into his creative processes. The best collection is in the Victoria and Albert Museum.

Gibbings, Robert (1889–1958). British wood-engraver, book designer, and travel writer. He founded the Society of Wood Engravers in 1919 and ran the Golden Cockerell Press from 1924 to 1933, illustrating many of its books himself and also employing engravers such as Eric *Gill and Eric *Ravilious. He went through a nudist phase at about this time and sometimes typeset in the nude. Gibbings's books typically combine topographical impressions, personal anecdote, and observations of nature, illustrated with his own engravings; they include two on the River Thames—*Sweet Thames Run Softly* (1940) and *Till I End My Song* (1957).

Gibbons, Grinling (1648–1721). Anglo-Dutch wood-carver and sculptor, born in Rotterdam, the son of an Englishman who had business interests there (his mother was probably Dutch). He settled in England *c.*1667 and was 'discovered' by John Evelyn (see *Diary*, 18 Jan., 1671). Evelyn introduced him to King Charles II and to Sir Christopher Wren (1632–1723), who employed him on decorations at Hampton Court and St Paul's Cathedral. In 1714 he was made Master Carver to King George I. Gibbons was unsurpassed in his day for naturalistic decorative carving of fruits, flowers, and shells, strung together in garlands and festoons, with small animals, cherubs' heads, etc.: Horace *Walpole said of him: 'There is no instance of a man before Gibbons who gave to wood the loose and airy lightness of flowers, and chained together the various productions of the elements with the free disorder natural to each species.' Because of his fame an enormous amount of work has been attributed to him, but he lived in a great age of English craftsmanship and much of the carving that is connected with his name was done by artists influenced by his style. Apart from his work for Wren, his documented commissions include outstanding ensembles at Burghley House, Lincolnshire, and Petworth House, Sussex. His virtuosity in wood was not equalled in marble or bronze, and George *Vertue said of him: 'He was a most excellent carver in wood, he was neither well skill'd or practized in Marble or Brass for which works he employd the best artists he coud procure.' About 1684 he took as partner Artus *Quellin III, who is thought to have been responsible for some of the figure sculpture for which Gibbons was officially credited, notably the fine bronze statue of James II (1686) outside the National Gallery in London.

Gibson, Charles Dana (1867–1944). American illustrator and painter. He studied at the *Art Students League of New York, 1884–5, and in the 1890s became a great success with pen-and-ink drawings contributed to such magazines as *Collier's Weekly*, *Harpers*, and *Life*. He specialized in scenes of fashionable social life and achieved immortality with his creation of the 'Gibson Girl', a type (modelled on his wife) representing an ideal of American womanhood—feminine and gracefully attired, but a lover of sports and the outdoor life. His work was immensely popular until about 1914, influencing fashions in women's clothes and hairstyles, and he earned a fortune. He also tried to gain recognition as a portrait painter, but he was much less successful in this field.

Gibson, John (1790–1866). British *Neoclassical sculptor. His early years were spent as a monumental mason in Liverpool, where he became a protégé of the banker and connoisseur William Roscoe. In 1817 he moved to London, where he was taken up by *Flaxman, on whose encouragement he went to Rome the following year with an introduction to *Canova, whose pupil he became. Later he was also taught by *Thorvaldsen. He spent nearly all the rest of his life in Rome apart from occasional visits to England, the longest being from 1844 to 1847. Gibson won recognition internationally as one of the outstanding Neoclassical sculptors, and Lord Lytton (1803–73) wrote of him: 'In you we behold the three great and long undetected principles of Grecian Art, simplicity, calm and concentration.' In his enthusiasm for Greek art, Gibson experimented with the ancient practice of colouring statues (see POLYCHROMY), arousing much controversy. His best-known work of this type is the *Tinted Venus* (Walker Art Gal., Liverpool, 1851). He left the fortune he made from his work to the *Royal Academy.

Giersing, Harald (1881–1927). Danish painter, the most energetic advocate of modern art in Denmark at the beginning of the 20th cent. His broad and roughly painted landscapes, still lifes, and figure subjects, usually constructed within a narrow range of tones and with simplified shapes, are notable for their disciplined compositions.

Gilbert, Sir Alfred (1854–1934). British sculptor and metal worker. After beginning to train as a surgeon he studied art at the *Royal Academy Schools and the École des *Beaux-Arts, Paris, after which he spent six years in Rome. He returned to England in 1884 and worked on several major projects, the best known of which is his Shaftesbury Memorial Fountain in Piccadilly Circus (1887–93). The celebrated figure of *Eros* that surmounts the fountain is cast in aluminium, one of the earliest examples of the use of this metal in sculpture. Its light weight allowed Gilbert to achieve a much more delicately poised pose than if he had been restricted to the traditional medium of bronze. Although Gilbert was hard-working, respected, and sought-after, he was unworldly and a hopeless businessman; his refusal to delegate work or compromise his standards meant that he took on more work than he could handle and sometimes lost money on commissions. In 1901 he became bankrupt, and in 1909 he moved into self-imposed exile in Bruges. However, in 1926 he returned to England at the request of King George V to complete his masterpiece, the tomb of the Duke of Clarence in St George's Chapel, Windsor Castle, which he had begun in 1892. The sinuous and labyrinthine detailing, crafted with consummate skill, reveals Gilbert as one of the major practitioners of *Art Nouveau, although he himself was disparaging about the style. Gilbert's reputation suffered after his death because he was so clearly outside the mainstream of 20th-cent. art, but he is now regarded as the greatest English sculptor of his generation.

Gilbert & George (Gilbert Proesch, 1943– , and George Passmore, 1942–). British artists (Gilbert is Italian born) who met whilst studying at St Martin's School of Art in London in 1967 and since 1968 have lived and worked together as self-styled 'living sculptures': 'Being living sculptures is our life blood, our destiny, our romance, our disaster, our light and life.' They initially attracted attention as *Performance artists, their most famous work in this vein being *Underneath the Arches* (1969), in which—dressed in their characteristic neat suits and with their hands and faces painted gold—they mimed mechanically to the 1930s music-hall song of the title. Although they gave up such 'living sculpture performances' in 1977, they still see themselves as living sculptures, considering their whole lifestyle a work of art. Since the early 1970s their work has consisted mainly of photo-pieces—large and garish arrangements of photographs, usually in black and white and fiery red, and often violent or homo-erotic in content, with scatological titles. The images are often drawn from the street life of the East End of London in which they live. Gilbert & George have become the most famous British avant-garde artists of their generation. Their work has been shown world-wide and has attracted an enormous amount of commentary. In 1986 they won the *Turner Prize. Critical opinion on them is sharply divided, however: to some they are geniuses, to others tedious poseurs.

Gill, Eric (1882–1940). British sculptor, engraver, typographer, and writer. He began to earn his living as a letter cutter in 1903 and carved his first figure piece in 1910. In 1913 he became a convert to Roman Catholicism and was commissioned to make the *Stations of the Cross* at Westminster Cathedral, fourteen *relief carvings which he carried out in 1914–18. These and the *Prospero and Ariel* group on Broadcasting House (1929–31) are his best-known sculptures. Gill was one of the chief protagonists in the movement for the revival of direct carving, and his work usually has an impressive simplicity of conception; he wrote that his 'inability to draw naturalistically was, instead of a drawback, no less than my salvation. It compelled me . . . to concentrate upon something other than the superficial delights of fleshly appearance . . . to consider the significance of things.' He tried to revive a religious attitude towards art and craftsmanship in opposition to the social and economic trends of the time, and in life, as in his work and writing, he was a vigorous advocate of a romanticized medievalism. His unconventional behaviour was well known in his own time, but the most bizarre and unpleasant aspects of his life were not revealed until the publication of Fiona MacCarthy's biography in 1989; he had incestuous relationships with two of his sisters and two of his daughters and sexual congress with a dog. Gill was a major figure in the revival of book design and typography. He illustrated many books, notably for the *Golden Cockerel Press, and his 'Perpetua' and 'Gill Sans-Serif' typefaces, designed for the Monotype Corporation, are among the classics of 20th-cent. typography. His books include *Christianity and Art* (1927), *Art* (1934), and *Autobiography* (1940).

Gillot, Claude (1673–1722). French painter, draughtsman, and etcher. Few of his paintings survive,but his predilection for scenes from the *commedia dell'arte* (*Quarrel of the Cabmen*, Louvre, Paris) was inherited by his pupil *Watteau. His work survives mainly in the form of drawings and etchings, and he excelled at *arabesque designs in the elegant *Rococo manner of *Audran.

Gillray, James (1757–1815). One of the most eminent of English *caricaturists. He began his career as an engraver of letter-heads and although he later studied at the *Royal Academy Schools, he seems to have been largely self-trained. After the publication of his print *A New Way to Pay the National Debt* (1786), a satire on the royal family, he found his bent in caricature and achieved enormous popularity. He enlarged the scope of *Hogarth's satire, making his caricature more personal than Hogarth's general social comment, and he showed great fecundity and vividness of imagination. His fantastic and grotesque inventions, the pointedness and wit behind his bludgeoning attack, and his gift for likeness beneath travesty set the key for English graphic satire during the period of its greatest influence at home and abroad. His career was cut short by insanity in 1811.

Gilman, Harold (1876–1919). English painter. He was a member of *Sickert's circle, a founder of the *Camden Town Group in 1911, and first President of the *London Group in 1913. His early work was rather sombre, but under the influence of Sickert he adopted a higher colour register and a technique of using a mosaic of opaque touches. From Sickert also he derived his taste for working-class subjects. After Roger *Fry's first *Post-Impressionist exhibition (1910) and a visit to Paris (1911) he used very thick paint and bright (sometimes garish) colour. He was one of the most gifted English painters of his generation and one of the most distinctive in his reaction to Post-Impressionism, but his career was cut short by the influenza epidemic of 1919.

Gilpin, Sawrey (1733–1807). English animal painter. He began his career as an apprentice to Samuel *Scott, the marine painter, but turned early to the painting of horses. The Duke of Cumberland employed him to make 'portraits' of celebrated racers, and Gilpin developed this vein, anticipating the work of James *Ward and Benjamin *Marshall in a later generation. In occasional large canvases (*The Election of Darius*, City Art Gal., York) he contrived his own blend of horse and history painting. His son, **William Sawrey Gilpin** (1762–1843), was the first president of the Old Water-Colour Society.

The Revd **William Gilpin** (1724–1804), brother of Sawrey Gilpin, was a writer and amateur draughtsman and one of the most important advocates of the *Picturesque. He was the first to establish the picturesque as an aesthetic category and by his numerous writings, illustrated by his own fine *aquatints, he exerted a profound and lasting influence on both English and European taste in natural and artificial scenery, and landscape painting. In certain respects he prepared the way for the *Romantic outlook on nature and natural beauty. His publications included accounts of his summer tours and an *Essay on Prints* (1768), which was translated into French, German, and Dutch and achieved lasting success as a standard work on print collecting.

Gimson, Ernest (1864–1919). English architect and craftsman in furniture, metalwork, plaster, and embroidery. In 1884 he met William *Morris and came under his influence; he later came into contact with Philip *Webb and leaders of the *Arts and Crafts Movement. Gimson, who designed and did not make furniture, had a masterly understanding of craft processes and respect for materials, and he made a considerable contribution to the emergence of a recognizable early 20th-cent. style.

Ginner, Charles (1879–1952). British painter. He grew up in France (his father, a doctor, practised there) and settled in London in 1910. He was already a friend of *Gilman and *Gore and through them he was drawn into *Sickert's circle, becoming a founder member of the *Camden Town Group in 1911 and the *London Group in 1913. His Continental background made him a respected figure among his associates, who were united by an admiration for French painting. Ginner was primarily a townscape and landscape painter and he is known above all for his views of London (often drab areas, although he also depicted the hustle and bustle of places such as Leicester Square and Victoria Station). He painted with thick, regular brushstrokes and firm outlines, creating a heavily textured surface and a feeling of great solidity. Once he had established his distinctive style (by about 1911) it changed little and he became one of the main upholders of the Camden Town tradition after the First World War (ironically, unlike other members of the group, he never actually lived in Camden Town). He worked for the Canadian War Records Commission in the First World War and was an *Official War Artist in the Second.

Giordano, Luca (1634–1705). Neapolitan painter, the most important Italian decorative artist of the second half of the 17th cent. He was nicknamed '*Luca Fa Presto*' (Luke work quickly) because of his prodigious speed of execution and huge output. His early works were in the *tenebrist manner of *Ribera, but his style became much more colourful under the influence of such great decorative painters as *Veronese, whose works he saw on his extensive travels. Indeed, he absorbed a host of influences and was said to be able to imitate other artists' styles with ease. His work was varied also in subject-matter, although he was primarily a religious and mythological painter. He worked mainly in Naples, but also extensively in Florence and Venice, and his work had great influence in Italy. In 1692 he was called to Spain by Charles II (1661–1700) and stayed there for 10 years, painting in Madrid, Toledo, and the *Escorial. His last work when he returned to Naples was the ceiling of the Treasury Chapel of S. Martino. In his personal self-confidence and courtliness, and in the open, airy compositions and light luminous colours of his work, Giordano presages such great 18th-cent. painters as *Tiepolo.

Giorgione (Giorgio Barbarelli or Giorgio da Castelfranco) (1476/8–1510). Venetian painter. Almost nothing is known of his life and only a handful of paintings can be confidently attributed to him, but he holds a momentous place in the history of art. He had achieved legendary status soon after his early death (probably from plague) and through succeeding centuries he has continued to excite the imagination in a way that few other painters can match. The extraordinary discrepancy between his enormous fame and the tiny scale of his *œuvre* is explained by the fact that he initiated a new conception of painting. He was one of the earliest artists to specialize in *cabinet pictures for private collectors rather than works for public or ecclesiastical patrons, and he was the first painter who subordinated subject-matter to the evocation of mood—it is clear that his contemporaries sometimes did not know what was represented in his pictures. *Vasari, who says that Giorgione earned his nickname—meaning 'Big George'—'because of his physical appearance and his moral and intellectual stature', ranked him alongside *Leonardo as one of the founders of 'modern' painting.

Giorgione was born in Castelfranco, about 30 km. north-west of Venice, and he probably trained with Giovanni *Bellini. He had two important public commissions in Venice: in 1507–8 he worked on a canvas (now lost without trace) for the audience chamber of the Doges' Palace; and in 1508 he collaborated with *Titian on frescos on the exterior of the Fondaco dei Tedeschi (the German warehouse), now known only through engravings and ruinous fragments in the Accademia. Apart from this, the only certain contemporary documentation on any of his surviving paintings is an inscription on the back of a female portrait known as *Laura* (Kunsthistorisches Mus., Vienna), which says it was painted by 'Master Zorzi da Castelfranco' in 1506; it also records that Giorgione was a colleague of Vincenzo *Catena, a partnership about which nothing else is known. (An inscription on *Portrait of a Man* in the San Diego Museum of Art is considered more doubtful.)

The main document for reconstructing Giorgione's *œuvre* is the notebook of the Venetian collector and connoisseur Marcantonio Michiel, written between 1525 and 1543. Michiel, who is a scrupulous and reliable source, mentions a number of paintings by Giorgione, four or five of which can be identified with extant works: *The Tempest*

(Accademia, Venice), *The Three Philosophers* (Kunsthistorisches Mus.), *Sleeping Venus* (Gemäldegalerie, Dresden), *Boy with an Arrow* (a copy?, Kunsthistorisches Mus.), and (an oblique and less explicit reference than the others) *Christ Carrying the Cross* (S. Rocco, Venice). He says *The Three Philosophers* was finished by *Sebastiano del Piombo and the *Sleeping Venus* (the work that founded the tradition of the reclining female nude) by Titian. The problem of attribution was, then, complicated from the start by the fact that some of Giorgione's paintings were completed after his death by other hands, and confusion soon arose; in the first edition of his *Lives* (1550) Vasari attributed the S. Rocco painting to Giorgione, but in the second edition (1568) he gave it in one place to Giorgione and in another to Titian, even though 'many people believed it was by Giorgione'. Distinguishing between the work of Giorgione and the young Titian continues to be one of the knottiest problems in connoisseurship, the celebrated *Concert Champêtre* in the Louvre being the picture most hotly disputed between them.

Among the other paintings given to Giorgione are the *Castelfranco Madonna*, in the cathedral of his home town (first mentioned by *Ridolfi in 1648 and accepted by almost all critics), and several male portraits, including a self-portrait in the Herzog-Anton-Ulrich Museum in Brunswick (perhaps a copy). Giorgione is said to have been handsome and amorous, and he initiated a type of dreamily romantic portrait that became immensely popular in Venice. The powerful influence that his work exerted in the generation after his death (even the venerable Bellini succumbed to it) is one of the main factors in making the construction of a catalogue of his work so difficult, for there are scores of paintings of the period, particularly pastoral landscapes, that can be described as Giorgionesque, and many are of high quality; as S. J. Freedberg has written (*Painting in Italy: 1500–1600*): 'In its treatment of the problem, the criticism of the nineteenth and twentieth centuries has ranged to the very extremes of the inclusive and exclusive positions that may be taken by the practitioners of connoisseurship.'

The problems of *iconography that Giorgione's paintings present are sometimes every bit as difficult as those of attribution. The most famous instance is *The Tempest*. Michiel saw it in 1530 and described it as a 'little landscape with the tempest with the gipsy and soldier', so he evidently did not know what subject, if any, was represented. X-rays have shown that Giorgione radically altered the figures in a way that suggests he was here indulging his imagination rather than illustrating a particular theme, although many ingenious attempts have been made to unravel a subject. This development of the 'landscape of mood' was, indeed, his great contribution to the history of art—an innovation of great originality and influence. Apart from the artists already mentioned, *Palma Vecchio, *Savoldo, and Dosso *Dossi were among the outstanding contemporaries who fell under the Giorgionesque spell, and among later artists *Watteau was his most sensitive heir.

Giottesques. A term applied to the 14th-cent. followers of *Giotto. The best-known of the 'Giotteschi' are the Florentines Taddeo *Gaddi, *Maso di Banco, Bernardo *Daddi, and to a lesser extent the *Master of St Cecilia. They borrowed Giotto's block-like figures and his roomy settings, and like him they studied human action and expression. Giotto's most loyal follower was Maso. He gave only the essential and maintained Giotto's high seriousness and almost stark simplicity.

Giotto di Bondone (*c.*1267–1337). Florentine painter and architect. Giotto is regarded as the founder of the central tradition of Western painting because his work broke free from the stylizations of *Byzantine art, introducing new ideals of naturalism and creating a convincing sense of pictorial space. His momentous achievement was recognized by his contemporaries (Dante praised him in a famous passage of *The Divine Comedy*, where he said he had surpassed his master *Cimabue), and in about 1400 Cennino *Cennini wrote 'Giotto translated the art of painting from Greek to Latin.' In spite of his fame and the demand for his services—he worked for Old St Peter's in Rome and as court painter to Robert of Anjou (king of Naples, 1309–43)—no surviving painting is documented as being by him. His work, indeed, poses some formidable problems of attribution, but it is universally agreed that the fresco cycle in the Arena Chapel at Padua is by Giotto, and it forms the starting-point for any consideration of his work. The Arena Chapel (so-called because it occupies the site of a Roman arena) was built by Enrico Scrovegni in expiation for the sins of his father, a notorious usurer mentioned by Dante. It was begun in 1303 and Giotto's frescos are usually dated *c.*1305–6. They run right round the interior of the building; the west wall is covered with a *Last Judgement*, there is an *Annunciation* over the chancel arch, and the main wall areas have three tiers of scenes representing scenes from the life of the Virgin and her parents, St Anne and St Joachim, and events from the Passion of Christ. Below these scenes are figures personifying Virtues and Vices, painted to simulate stone *reliefs—the first *grisailles. The figures in the main narrative scenes are about half life-size, but in reproduction they usually look bigger because Giotto's conception is so grand and powerful. His figures have a completely new sense of three-dimensionality and physical presence, and in portraying the sacred events he creates a feeling of moral weight rather than divine splendour. He seems to base the representations upon personal experience, and no artist has surpassed his ability to go straight to the heart of a story and express its essence with gestures and expressions of unerring conviction.

The other major fresco cycle associated with Giotto's name is that on the Life of St Francis in the Upper Church of S. Francesco at Assisi. Whether Giotto painted this is not only the central problem facing scholars of his work, but also one of the most controversial issues in the history of art. It is virtually beyond question that Giotto did at some time work at Assisi, and the St Francis frescos are clearly the work of an artist of great stature (their intimate and humane portrayals have done much to determine posterity's mental image of the saint). Nevertheless, the stylistic differences between these works and the Arena Chapel frescos seem to many critics so pronounced that they cannot accept a common authorship. Attempts to attribute other frescos at Assisi to Giotto have met with no less controversy (see also MASTER OF THE ST FRANCIS CYCLE AND MASTER OF ST CECILIA).

There is a fair measure of agreement about the frescos associated with Giotto in Sta Croce in Florence. He probably painted in four chapels there, and work survives in the Bardi and Peruzzi chapels, probably dating from the 1320s. The frescos are in very uneven condition (they were whitewashed in the 18th cent.), but some of those in the Bardi Chapel on the life of St Francis remain deeply impressive. Nothing survives of Giotto's work done for Robert of Anjou in Naples, and the huge mosaic of the Ship of the Church (the *Navicella*) that he designed for Old St Peter's in Rome has been so thoroughly altered that it tells us nothing about his style. In Rome he would have seen the work of Pietro *Cavallini, which was as important an influence on him as that of his master Cimabue.

Several panel paintings bear Giotto's signature or—like the Stefaneschi Altarpiece (Vatican)—are attributed to the master in credible early sources, but it is generally agreed that the signature is a trademark showing that the works came from Giotto's shop rather than an indication of his personal workmanship. On the other hand, the *Ognissanti Madonna* (Uffizi, Florence, *c*.1305–10) is neither signed nor firmly documented, but is a work of such grandeur and humanity that it is universally accepted as Giotto's. Among the other panels attributed to him, the finest is the Crucifix in Sta Maria Novella, Florence.

On account of his great fame as a painter, Giotto was appointed architect to Florence Cathedral in 1334; he began the celebrated campanile, but his design was altered after his death. In the generation after his death he had an overwhelming influence on Florentine painting; it declined with the growth of *International Gothic, but his work was later an inspiration to *Masaccio, and even to *Michelangelo. These two giants were his true spiritual heirs. Boccaccio (1313–75) and Sacchetti (*c*.1330–1400) in their stories make Giotto good-natured, witty, and shrewd—a great man and the greatest painter since antiquity.

Giovanni Bologna. See GIAMBOLOGNA.

Giovanni da Maiano (1438–78). See BENEDETTO DA MAIANO.

Giovanni da Maiano (active first half of 16th cent.). Italian sculptor working in England. His importance lies in his share in the dissemination of Italian *Renaissance art rather than in any outstanding artistic achievement. His *terracotta roundels of Roman emperors for the exterior at Hampton Court (he requested payment for them in 1521), and the relief of *putti holding the Wolsey Arms, confidently attributed to him on the same building, are among the first signs of Renaissance fashion in English architecture. His other work included a black marble *sarcophagus (1531–6) intended for Cardinal Wolsey (1473?–1530) but now used to house the remains of Lord Nelson (1758–1805) in St Paul's Cathedral.

Giovanni da Udine. See UDINE.

Giovanni di Balduccio. See BALDUCCIO.

Giovanni di Paolo (active 1420–82). One of the most attractive and idiosyncratic painters of the Sienese School, sometimes called Giovanni dal Poggio, from the district of the city where he lived. Little is known of his life, but there is a considerable number of surviving works by him—all small-scale religious panels. He may have been taught by *Taddeo di Bartolo and was influenced notably by *Gentile da Fabriano and *Sassetta, but his style is highly personal and engaging, with rather whimsical figures inhabiting strange landscapes. After centuries of neglect his reputation was revived by *Berenson, who called him 'the El *Greco of the *quattrocento'. There are good examples of his work in the National Gallery, London, the Metropolitan Museum, New York, and the Art Institute of Chicago.

Girardon, François (1628–1715). French sculptor. He ranked with *Coysevox as the outstanding sculptor of Louis XIV's reign (1643–1715), but his style was more restrained and *classical, embodying the ideas of the Academy. Much of his work was done for Versailles, where he collaborated with *Lebrun, and his group *Apollo Tended by the Nymphs*, commissioned in 1666, has been considered the most purely classical work of French 17th-cent. sculpture. (The group was originally in a grotto room, but is now in the palace gardens.) His other work includes the monument to Cardinal Richelieu (1675–7) in the church of the Sorbonne. He also made an equestrian statue of Louis XIV made (1683–92) for the Place Vendôme in Paris, but this was destroyed during the French Revolution. Girardon died on the same day as Louis XIV.

Girodet de Roucy, Anne-Louis (1767–1824). French painter and illustrator, usually known as Girodet-Trioson, a name he adopted in honour of a benefactor, Dr Trioson. He studied with J.-L. *David and won the *Prix de Rome in 1789, returning to Paris in 1795. In style and technique he followed David, but for his choice of themes and his emotional treatment he was acclaimed by the young *Romantics. He was particularly interested in unusual colour effects and in the problems of concentrated light and shade, as in *The Sleep of Endymion* (1792) and *The Entombment of Atala* (1808), both in the Louvre. Girodet often favoured literary themes, but he also won renown for his paintings glorifying Napoleon (*The Revolt of Cairo*, Versailles Mus., 1810) and was a fine portraitist. One of his best-known portraits, *Mademoiselle Lange as Diana* (Minneapolis Institute of Arts, 1799), caused a scandal because of its satirical sexual allusions. His book illustrations included work for editions of Jean Racine and Virgil. In 1812 he inherited a fortune and thereafter devoted himself to writing unreadably boring poems on aesthetics.

Girolamo da Carpi (*c*.1501–56). Italian painter and architect, active mainly in Ferrara. There he was the favourite artist of Cardinal Ippolito d'*Este, who also took him to Rome, where he did some architectural work at the Vatican for Pope Julius III (1487–1555). Girolamo was much influenced by the leading masters of the early 16th cent., *Correggio, *Raphael, *Giulio Romano, and others, and his *œuvre* is not well defined (much of his work that *Vasari describes has perished). At his best, however, he was an artist of ability and individuality, as witness the beautiful *Portrait of a Lady in a Green Dress* at Hampton Court.

Girolamo da Fiesole. See COLOMBE.

Giroust, Marie-Suzanne. See ROSLIN.

Girtin, Thomas (1775–1802). English landscape painter in watercolours. He was apprenticed to Edward *Dayes and was later employed together with *Turner by Dr *Monro to copy drawings. His earlier works were tinted drawings in the 18th-cent. topographical tradition, but before the end of his short life he had developed a technique that revolutionized watercolour painting. He used strong colour in broad *washes, influenced to some extent by J. R. *Cozens, but going beyond him in the grandeur with which he created effects of space, the power with which he suggested mood, and the boldness of his compositions. His work stands at the beginning of the classic English tradition of watercolour painting, freed from its dependence on line drawing, and Turner acknowledged his friend's greatness with the words 'If Tom Girtin had lived, I should have starved.' Girtin made tours in various parts of Britain, and spent six months in Paris in 1801–2, making a series of etchings of the city that were published in 1803. In 1802 he exhibited an enormous panorama of London, painted in oils—the *Eidometropolis*—but this is no longer extant. Girtin died of tuberculosis, aged 27.

gisant. French term used from the 15th cent. onwards for a lying or recumbent effigy on a funerary monument. The *gisant* typically represented a person in death (sometimes decomposition) and the *gisant* position was contrasted with the *orant*, which represented the person as if alive in a kneeling or praying position. In Renaissance monuments *gisants* often formed part of the lower register, where the deceased person was represented as a corpse, while on the upper part he was represented *orant* as if alive.

Gislebertus (active first half of 12th cent.). French *Romanesque sculptor. He was one of the great geniuses of medieval art, but his name has survived only because he carved his signature—*Gislebertus hoc fecit*—beneath the feet of the central figure of Christ in the *tympanum of the west doorway of Autun Cathedral in Burgundy. It has been established that the same sculptor was also responsible for most of the carved decoration of the cathedral, including the west and north doorways and the majority of the *capitals of the interior. The unusually prominent position of his signature suggests that his greatness was appreciated in his own time.

It is highly probable that Gislebertus was trained in the workshop that was responsible for the decoration of the abbey of Cluny, the most influential of all Romanesque monasteries, and that he worked at the nearby cathedral at Vézelay before going to Autun. He was already a mature artist when he started at Autun, where he worked *c.*1125–35, and his style changed little while he was there.

Working within the general conventions of the Romanesque style of the school of Cluny, Gislebertus produced some of the most powerful and original sculpture of the period. The tympanum of the west doorway is his greatest work. It represents the Last Judgement and is a masterpiece of expressionistic carving and a superb technical achievement. His carving of Eve, one of the few surviving fragments of the north doorway, and now in the Musée Rolin at Autun, is a large-scale reclining nude without parallel in medieval art. The fecundity of Gislebertus's imagination and range of feeling is vividly displayed in the sixty or so capitals he carved for the interior and the doorways. His influence has been traced in other Burgundian churches at Saulieu, Beaune, and Moûtiers-Saint-Jean, and even at Chartres. Many of his ideas had a long-term effect on the development of French *Gothic sculpture.

Giuliano da Maiano. See BENEDETTO DA MAIANO.

Giulio Romano (prob. 1499–1546). Italian painter, architect, and designer, born in Rome and active mainly in Mantua. He was *Raphael's chief pupil and assistant (although exactly what part he played is controversial) and one of the major figures of *Mannerist art. About 1515 he was working on Raphael's Stanza dell' Incendio in the Vatican and after the master's death in 1520 Giulio became his main artistic executor, completing a number of his unfinished works, including the *Transfiguration* (Vatican), the *Sala di Constantino* frescos in the Vatican, and the decorations of the Villa Madama. His independent works of this time include the *Holy Family* in Sta Maria dell' Anima, Rome, and the *Stoning of St Stephen* in S. Stefano, Genoa. He also designed some pornographic prints that caused such a scandal that their engraver Marcantonio *Raimondi was imprisoned (their notoriety was sustained by the sonnets that the poet Pietro Aretino wrote inspired by them soon after their publication). Giulio had moved to Mantua in 1524 and escaped Raimondi's fate. He remained there for the rest of his life and dominated the artistic affairs of the *Gonzaga court. The great monument to his genius is the Palazzo del Tè, begun in 1526 for Federico Gonzaga. This was one of the first Mannerist buildings, deliberately flouting the canons of classical architecture as exemplified by *Bramante in order to shock and surprise the spectator. The same tendency is continued in Giulio's fresco decoration in the palace, especially in the Sala dei Giganti, where the whole room is painted from floor to ceiling to give an overall illusionistic effect, and the spectator feels himself overwhelmed by the rocks and thunderbolts hurled down on the rebellious Titans who attempted to storm Olympus. Giulio painted several other frescos in the Palazzo del Tè and in the Sala di Troia of the Ducal Palace at Mantua, which testify to his classical learning and exuberant invention. His muscular style owed much to *Michelangelo as well as to Raphael, but was less daunting than that of either and proved widely influential. Indeed, he became one of the most famous painters of his day and has the distinction of being the only artist mentioned by Shakespeare, who called him 'that rare Italian master Julio Romano', but mistakenly imagined him a sculptor (*The Winter's Tale*, v. ii). Among his other architectural works, the most important is his own house in Mantua (1544–6).

Giusti, Giovanni (1485–1549). Italian-born sculptor who settled at Tours in France in about 1504, changing his name to Jean Juste. He was a leading figure in the introduction of the *Renaissance style to France. His masterpiece is the tomb of Louis XII and Anne of Brittany, made at Tours in 1517–18 and set up in the Abbey of Saint-Denis by Francis I in 1531. He was probably assisted in the work by his brother **Antonio** (1479–1519). It features seated figures of the twelve Apostles in purely Italian style, allegorical figures of the Virtues, and *reliefs depicting the King's Italian victories. Perhaps the only concession to the medieval tradition is the presence of *gisants, the corpses on the *sarcophagus; but even these have been subtly idealized.

Giustiniani. Prominent Italian family, branches of which were established in many parts of Italy, especially Genoa and Venice, where they played an important role in politics, literature, and religion. For the arts the most interesting member of the family was the enormously wealthy **Marchese Vincenzo** (1564–1638). He owned the finest collection of *antique sculpture in Rome—published in the *Galleria Giustiniana* (1631), the first ever illustrated catalogue of an art collection—and was an enthusiastic and discriminating patron of painters, especially *Caravaggio and his northern followers.

Glackens, William James (1870–1938). American painter and draughtsman. His early career was spent mainly as a newspaper illustrator, but he was encouraged to take up painting by Robert *Henri, whom he met in 1891. With Henri he became a member of The *Eight and of the *Ash-can school.

Glackens, however, was less concerned with *Social Realism than with representing the life of the people as a colourful spectacle, and he was heavily influenced by the *Impressionists. *Chez Mouquin* (Art Inst. of Chicago, 1905), a high-keyed painting of a restaurant where the New York Realists often resorted, was inspired by *Manet's *Un Bar aux Folies-Bergère*. By the time of the *Armory Show, which he helped to organize and in which he was represented, he was painting in the manner of the early *Renoir. In 1912 Glackens was employed as art consultant to Dr Albert C. *Barnes and toured Europe buying up paintings by Renoir, *Degas, *Cézanne, van *Gogh, and others, which formed the nucleus of the famous Barnes Collection at Merion, Pennsylvania. In 1916 he became the first President of the *Society of Independent Artists.

glair. White of egg when used as the *medium in *illuminating manuscripts, in *tempera painting, and in gilding with gold-dust. It is also used as an adhesive substance to fix gold leaf.

Glasgow School. Term applied to two quite distinct groups of Scottish artists active respectively in the late 19th and early 20th cents. The earlier group was a loose association of painters centred in Glasgow during the second half of the 19th cent. They were in revolt against the conservatism of the Royal Scottish Academy and were advocates of open-air painting (see PLEIN AIR). The school had reached its apogee before 1900 and did not outlast the First World War, but it had some influence on the younger painters at the opening of the 20th cent. The later and more important group created a distinctive version of *Art Nouveau. Its most important member was Charles Rennie *Mackintosh.

Recently the term Glasgow School (or facetiously 'Glasgow pups') has been applied to a number of *New Image painters working in the city from the 1980s. They include Steven Campbell (1953–) and Adrian Wiszniewski (1958–).

glass prints (or **clichés-verre**). Prints made by exposing sensitized photographic paper to the sun beneath a glass plate on which the design has been drawn. The glass is covered with an opaque *ground and the design is drawn on it with a fine point, leaving the glass transparent where the lines are to be printed in black. The resulting print resembles an *etching. The medium was popular in the 1850s, its practitioners including *Corot, *Daubigny, Jean François *Millet, and Théodore *Rousseau.

glaze. A transparent layer of paint applied over another colour or *ground, so that the light passing through is reflected back by the under surface and modified by the glaze. The effect of an undercolour through a glaze is not the same as any effect obtainable by mixing the two *pigments in direct painting, for the glaze imparts a special depth and luminosity. From the 15th to the 19th cent. most oil paintings were built up as an elaborate structure of superimposed layers, glazes, and *scumbles over an *underpainting, but since **alla prima* painting became the norm such a highly deliberate, craftsmanly approach has fallen into disfavour. The word 'glaze' is also applied to the glassy coating used to render ceramics impervious to liquid and smooth to the touch.

Gleizes, Albert (1881–1953). French painter, graphic artist, and writer. His early works were in an *Impressionist style, but in 1909 he became associated with *Cubism, and in 1912 he wrote with *Metzinger the book *Du Cubisme* (an English translation, *Cubism*, appeared in 1913). This was the first book on the movement and is regarded as the most important exposition of the theoretical principles of the Cubist aesthetic. It remains Gleizes's chief claim to fame. In 1912 he was among the founders of the *Section d'Or group and in 1913 he exhibited at the *Armory Show, New York. After serving in the French Army, 1914–15, Gleizes lived from 1915 to 1917 in New York, where he underwent a religious conversion. Much of his later career was devoted to trying to achieve a synthesis of medieval and modern art, expressing Christian ideas through pseudo-Cubist forms. In this he is generally reckoned to have been conspicuously unsuccessful and his modest reputation as a painter rests on his pre-war work.

Gleyre, Charles (1808–74). Swiss painter, active mainly in Paris, where he enjoyed a successful career, particularly with anecdotal scenes, sometimes in an *antique setting, and portraits. He was a renowned teacher and when *Delaroche closed down his teaching studio in 1843, the majority of his students transferred to Gleyre. He taught *Whistler and several of the *Impressionists—*Bazille, *Monet, *Renoir, and *Sisley—and although his own paintings were conventional, he encouraged open-air painting (see PLEIN AIR). Renoir, however, said that his main strength as a teacher was that he left his pupils 'pretty much to their own devices'. Gleyre closed his studio in 1864 because of an eye ailment.

Gobelins. The most famous of all tapestry manufactories. It derives its name from the 15th-cent. scarlet-dyers Jean and Philibert Gobelin, whose works were situated on the Bièvre, a small stream in Paris. In or near these buildings the Flemish tapestry weavers de Comans and de La Planche, summoned by Henri IV (1553–1610), set up their looms at the beginning of the 17th cent. and it was here that Louis XIV reorganized the tapestry weavers of Paris in 1662. Charles *Lebrun was the first Director of the factory, which initially made not only tapestries but also every kind of product (except carpets, which were woven at the Savonnerie factory) required for the furnishing of the royal palaces—its official title was Manufacture royale des meubles de la Couronne. The celebrated tapestry designed by Lebrun showing *Louis XIV Visiting the Gobelins* (Gobelins Mus., Paris, 1663–75) gives a good idea of the range of its activities. In 1694 the factory was closed because of the king's financial difficulties, and although it reopened in 1699, thereafter it made only tapestries.

The grand style and strong simple colouring of Lebrun yielded gradually in the 18th cent. to *Rococo grace of design and soft hues in an enormous variety of delicate shades, a transformation which owed much to the influence of *Oudry and *Boucher, who were successively inspectors of the works (1733–70). The Revolutionary period was unproductive, but numbers of large tapestries, many of them with historical subjects, were woven during the 19th and early 20th cents. The Gobelins continues in production today.

Godefroid de Clair (Godefroid de Huy). *Mosan goldsmith and enamellist, active in the 12th cent. He may have trained in the workshop of *Renier of Huy. Early sources praise his great skill and suggest he was a prolific artist, but his career is obscure and the numerous attributions to him of reliquaries and *enamels are highly speculative. Peter Lasko (*Ars Sacra: 800–1200*, 1972) considers that 'On the whole, the introduction of the personality of Godefroid has hindered rather than helped our understanding of the development of the Mosan style in this vital period.' His name continues to be one to conjure with, however. In 1978 two small enamels attributed to him were bought at *Sotheby's (at the sale of the collection of Robert von Hirsch) for more than £1,000,000 each. They are now in the Germanisches Nationalmuseum, Nuremberg, and the Kunstgewerbemuseum, Berlin.

Godward, J. W. See ALMA-TADEMA.

Godwin, Edward William (1833–86). English architect and designer, mainly of furniture. He designed the White House in Tite Street, Chelsea, for *Whistler in 1877 and like Whistler developed a great enthusiasm for Japanese taste. He produced designs, some in conjunction with Whistler, for light and graceful cabinet work in this style, mainly in ebonized wood, and was partly responsible for the popular fashion for bamboo and wicker furniture. His many notebooks and a selection of his pieces are in the Victoria and Albert Museum.

Goes, Hugo van der (d. 1482). The greatest Netherlandish painter of the second half of the 15th cent. Nothing is known of his life before 1467, when he became a master in the painters' guild at Ghent. He had numerous commissions from the town of Ghent for work of a temporary nature such as processional banners, and in 1475 he became dean of the painters' guild. In the same year he entered a priory near Brussels as a lay-brother, but he continued to paint and also to travel. In 1481 he suffered a mental breakdown (he had a tendency to acute depression) and although he recovered, died the following year. An account of his illness by Gaspar Ofhuys, a monk at the priory, survives; Ofhuys was apparently jealous of Hugo and his description has been called by Erwin *Panofsky 'a masterpiece of clinical accuracy and sanctimonious malice'.

No paintings by Hugo are signed and his only securely documented work is his masterpiece, a large *triptych of the Nativity known as the *Portinari Altarpiece* (Uffizi, Florence, *c.*1475–6). This was commissioned by Tommaso Portinari, the representative of the House of *Medici in Bruges, for the church of the Hospital of Sta Maria Nuova in Florence, and it exercised a strong influence on Italian painters with its masterful handling of the oil technique. There is a great variety of surface ornament and detail, but this is combined with lucid organization of the figure groups and a convincing sense of spatial depth. As remarkable as Hugo's skill in reconciling grandeur of conception with keen observation is his psychological penetration in the depiction of individual figures, notably the awestruck shepherds.

The other works attributed to Hugo include the large and sonorous *Monforte Altarpiece* (Staatliche Museen, Berlin), a *diptych of the *Fall of Adam and Eve* and the *Lamentation* (Kunsthistorisches Museum, Vienna), generally considered an early work, and two large panels probably designed as organ shutters (Royal coll., on loan to NG of Scotland). Hugo's last work is generally considered to be the *Death of the Virgin* (Groeningemuseum, Bruges), a painting of remarkable tension and poignancy that seems a fitting swan-song for such a tormented personality.

Goethe, Johann Wolfgang von (1749–1832). German writer, scientist, and amateur artist, one of the giants of European literature and one of the major figures of the *Romantic movement. Throughout his life he devoted much time to the study of painting and was a prolific draughtsman. Although his talent as an artist was modest, his writings on art were very influential in the upsurge of Romantic ideas in Germany. His first publication, *Von deutscher Baukunst* (1772), was inspired by youthful enthusiasm for Strasburg Cathedral and was one of the earliest appreciations of *Gothic art. Identifying art with 'nature', Goethe held that great art must simulate and carry on the blind creative force in nature. He also advocated the typically Romantic view that art should concentrate on what is individually 'characteristic' rather than the generic type. In 1786–8 he visited Italy, where he came into contact with German artists such as *Tischbein in Rome, and his taste changed to an appreciation of the *classicism of the *Renaissance. Identifying art with 'style', he now maintained that beauty is symbolic expression of the inner laws of nature, and that this expression had been supremely achieved by the art of antiquity. His exaltation of the concept of 'genius' was of central significance for the development of European Romanticism. In contrast to the speculative character of post-*Kantian German Idealism Goethe set his chief emphasis on intuition (*Anschauung*) in regard to the apprehension of beauty. His writings on art included a book on theory of colour (*Zur Farbenlehre*, 1810, English translation 1840), in which he purported to refute the *Optics* of Newton, and a translation into German of *Cellini's *Autobiography* (1798).

Gogh, Vincent van (1853–90). Dutch painter and draughtsman, with *Cézanne and *Gauguin the greatest of *Post-Impressionist artists. His uncle was a partner in the international firm of picture dealers Goupil and Co. and in 1869 van Gogh went to work in the branch at The Hague. In 1873 he was sent to the London branch and fell unsuccessfully in love with the daughter of his landlady. This was the first of several disastrous attempts to find happiness with a woman, and his unrequited passion affected him so badly that he was dismissed from his job. He returned to England in 1876 as an unpaid assistant at a school, and his experience of urban squalor awakened a religious zeal and a longing to serve his fellow men. His father was a Protestant pastor, and van Gogh first trained for the ministry, but he abandoned his studies in 1878 and went to work as a lay preacher among the impoverished miners of the grim Borinage district in Belgium. In his zeal he gave away his own worldly goods to the poor and was dismissed for his literal interpretation of Christ's teaching. He remained in the Borinage, suffering acute poverty and a spiritual crisis, until 1880, when he found that art was his vocation and the means by which he could bring consolation to humanity. From this time he worked at his new 'mission' with single-minded intensity, and although he often suffered from extreme poverty and under-nourishment, his output in the ten remaining years of his life was prodigious. He left about 800 paintings and a similar number of drawings; the finest collection of his work is in the National Museum Vincent van Gogh in Amsterdam. The spontaneous, irrational side of his character has often been stressed, but he was a cultivated and well-read man, who in spite of his speed of work thought deeply about his paintings and planned them carefully.

From 1881 to 1885 van Gogh lived in the Netherlands, sometimes with his parents, sometimes in lodgings, supported by his devoted brother Theo, who regularly sent him money from his own small salary. In keeping with his humanitarian outlook he painted peasants and workers, the most famous picture from this period being *The Potato Eaters* (Van Gogh Museum, 1885). Of this he wrote to Theo: 'I have tried to emphasize that those people, eating their potatoes in the lamp-light have dug the earth with those very hands they put in the dish, and so it speaks of manual labour, and how they have honestly earned their food.'

In 1885 van Gogh moved to Antwerp on the advice of Antoine *Mauve (a cousin by marriage), and studied for some months at the Academy there. Academic instruction had little to offer such an individualist, however, and in February 1886 he moved to Paris, where he met *Pissarro, *Degas, Gauguin, *Seurat, and *Toulouse-Lautrec. At this time his painting underwent a violent metamorphosis under the combined influence of *Impressionism and Japanese woodcuts (see UKIYO-E), losing its moralistic flavour. Van Gogh became obsessed by the symbolic and expressive values of colours and began to use them for this purpose rather than, as did the Impressionists, for the reproduction of visual appearances, atmosphere, and light. 'Instead of trying to reproduce exactly what I have before my eyes,' he wrote, 'I use colour more arbitrarily so as to express myself more forcibly.' Of his *Night Café* (Yale Univ. Art Gal., 1888), he said: 'I have tried to express with red and green the terrible passions of human nature.' For a time he was influenced by Seurat's delicate *pointillist manner, but he abandoned this for broad, vigorous, and swirling brush-strokes.

In February 1888 van Gogh settled at Arles, where he painted more than 200 canvases in 15 months. During this time he lived in poverty, and suffered recurrent nervous crises with hallucinations and depression. He became enthusiastic for the idea of founding an artists' co-operative at Arles and towards the end of the year he was joined by Gauguin. But as a result of a quarrel between them van Gogh suffered the crisis in which occurred the famous incident when he cut off a piece of his left ear, an event commemorated in his *Self-Portrait with Bandaged Ear* (Courtauld Inst., London). In May 1889 he went at his own request into an asylum at St Rémy, near Arles, but continued during the year he spent there a frenzied production of pictures such as *Cornfield with Cypresses* (NG, London) and *Starry Night* (MOMA, New York). He did 150 paintings besides drawings in the course of this year. In 1889 Theo married and in May 1890 van Gogh moved to Auvers-sur-Oise to be near him, lodging with the patron and connoisseur Dr Paul Gachet. There followed another tremendous burst of strenuous activity and during the last 70 days of his life he painted 70 canvases. But his spiritual anguish and depression became more acute and on 29 July 1890 he died from the results of a self-inflicted bullet wound. He sold only one painting during his lifetime (*Red Vineyard at Arles*, Pushkin Mus., Moscow), and was little known to the art world at the time of his death, but his fame grew rapidly thereafter.

Van Gogh's stormy and dramatic life and his unswerving devotion to his ideals have made him one of the great cultural heroes of modern times, providing the most auspicious material for the 20th-cent. vogue in romanticized psychological biography. The voluminous correspondence with his brother Theo (more than 750 of his letters are extant) is an abundant source of information about his aesthetic aims and his mental disturbances. In the history of painting van Gogh occupies a position of the first importance in the movement from the optical realism of the Impressionists to the abstract use of colour and form for symbolical and expressive values. His influence on *Expressionism, *Fauvism, and early abstraction was incalculable, and it can be seen in many other aspects of 20th-cent. art.

Golden Cockerel Press. A private printing-press founded in 1920 at Waltham St Lawrence, Berkshire, by Harold Taylor, taken over by Robert *Gibbings in 1924 and transferred to Staple Inn, London. From 1936 it was directed by Christopher Sandford and Owen Rutter. Under Gibbings's influence illustration received equal emphasis with typography. The *Four Gospels* (1931), the press's outstanding production, which has been compared with the *Kelmscott *Chaucer* and the *Doves *Bible*, contains wood engravings by Eric *Gill, who also designed the special 'Golden Cockerel' type.

Golden Section. A proportion in which a straight line or rectangle is divided into two unequal parts in such a way that the ratio of the smaller to the greater part is the same as that of the greater to the whole. Like the mathematical value pi, it cannot be expressed as a finite number, but an approximation is 8 : 13 or 0.618 : 1. The proportion has been known since antiquity (Euclid and *Vitruvius discuss it) and has been said to possess inherent aesthetic value because of an alleged correspondence with the laws of nature or the universe. The claims have been supported by an immense quantity of data, collected both from nature and from the arts. Statistical experiments are said to have shown that people involuntarily give preference to proportions that approximate to the Golden Section. But this weakens the case for maintaining that when such forms are found in a work of art they have been put there intentionally. In fact, the Golden Section is likely to turn up fairly frequently in any design derived from the square and developed by applying a pair of compasses. The Golden Section was much studied during the *Renaissance, and Luca Pacioli (*c.*1445–*c.*1514), the most famous mathematician of his day and a close friend of *Leonardo and of *Piero della Francesca, wrote a book on it called *Divina Proportione* (1509). In accordance with the tendencies of the time, Pacioli's book, illustrated with drawings by Leonardo, credits this 'divine proportion' with various mystical properties and exceptional beauties both in science and in art. Like many other learned men of the Middle Ages and Renaissance, Pacioli was anxious to harmonize the knowledge of pagan antiquity with the Christian faith, and in the chapter in which he justifies his choice of title he explains that this ratio cannot be expressed by a number and, being beyond definition, is in this respect like God, 'occult and secret'; further, this three-in-one proportion is symbolic of the Holy Trinity.

Goldie, Charles Frederick (1870–1947). New Zealand painter, trained in Paris at the Académie Julian. Brought up at a time when the Maori people were commonly regarded as a 'dying race', he spent his mature years in producing for posterity an immense series of portraits and figure studies of aged Maori models done with a photographic fidelity (Auckland City Art Gal.).

Goltzius, Hendrick (1558–1617). Dutch graphic artist and painter, the outstanding line engraver of his day. His father was a glass painter and he was taught by him and by Dirck Coornhert (1522–90), the Dutch humanist, politician, and theologian who made his living as an engraver. He was the leader of a group of *Mannerist artists who worked in Haarlem, where he founded some kind of 'academy' (a life class?) with *Cornelis van Haarlem and Karel van *Mander. In 1590–1 he visited Rome and on his return to Haarlem he abandoned his Mannerist style for a more classical one. Goltzius's right hand was crippled, but in spite of this handicap he was renowned for his technical virtuosity and his skill in imitating the work of other great engravers such as *Dürer and *Lucas van Leyden. In his early career much of his work was reproductive, but he also produced many original compositions, including a splendid series on *Roman Heroes* (1586). His *miniature portrait drawings were also outstanding, and the landscape drawings he made after 1600 mark him as a forerunner of the great 17th-cent. landscape artists. His paintings are less interesting than his drawings and much less advanced stylistically.

Gombrich, Sir Ernst (1909–). Austrian-born British art historian. He came to England in 1936 and began a long association with the *Warburg Institute in the University of London, where he was Director and Professor of the History of the Classical Tradition from 1959 to 1976. He has also been *Slade Professor at both Oxford and Cambridge. His scholarly work, which shows a remarkable ability to combine great breadth of learning with lucidity and wit, has been devoted largely to the theory of art, the psychology of pictorial representation, and Renaissance symbolism, and has won him a position of the highest esteem in his profession. His writings bear witness to his interest in psychology and scientific method and have helped to break down barriers between art history and other disciplines. Gombrich's best-known book, however, is a popular work, *The Story of Art*, which was first published in 1950 and has ever since held its place as the most congenial introduction to the history of art. It reached its 16th English edition in 1995 and has been translated into 20 languages. Among his other books the best known is probably *Art and Illusion* (1960 and subsequent editions). This highly influential work deals with conventions of representation and examines how styles change and develop, challenging many orthodox views and received opinions about visual perception. The notion of the 'innocent eye'—the idea of the artist simply representing what he sees—is shown to be untenable and the evolution of style ('why different ages and different nations have represented the visible world in . . . different ways') is explained in terms of the modification of schematic images to match the objective reality of the subject. Using the findings of experimental psychology, Gombrich examines the way the viewer looks at works of art and shows that we tend to see what we expect to see. In *Thinkers of the Twentieth Century* (ed. Elizabeth Devine, et al., 1983) J. M. Massing wrote: 'For his scholarly method, his theoretical approach and his defence of cultural values, Gombrich will be remembered as one of the leading art historians of this century. Through his study of the psychology of perception, he is also one of the very few to have widened our understanding of the visible world.'

Gonçalves, Nuño (active 1450–71). Portuguese painter, recorded in 1450 as court painter to Alfonso V (1437–81). No works certainly by his hand survive, but there is strong circumstantial evidence that he was responsible for the *St Vincent* *polyptych (Lisbon Mus., *c.*1460–70), the outstanding Portuguese painting of the 15th cent. The style is rather dry, but powerfully realistic, and the polyptych contains a superb gallery of highly individualized portraits of members of the court, including a presumed self-portrait. There are affinities with contemporary Burgundian and Flemish art, especially the work of *Bouts. Gonçalves is known to have painted another *St Vincent* altarpiece for Lisbon Cathedral, but this was destroyed by earthquake in 1755.

Goncharova, Natalia (1881–1962). Russian painter, graphic artist, and designer, born into an impoverished noble family (she was related to the poet Pushkin). In Moscow in 1900 she met her fellow student Mikhail *Larionov, who became her lifelong companion. In the years leading up to the First World War they were among the most prominent figures in Russian avant-garde art, taking part in and often helping to organize a series of major exhibitions in Moscow. Her early paintings were *Impressionist, but from 1906 she began to develop a *primitivist style combining her interest in peasant art and icon painting with influences from modern French art, particularly *Fauvism and *Cubism, to which was later added *Futurism. By the time of the *Target exhibition of 1913 she was painting in a near-abstract *Rayonist style (*Cats*, Guggenheim Museum, New York, 1913). In 1915 she left Russia with Larionov and after settling in Paris in 1919 she devoted herself mainly to designing settings and costumes for the theatre, particularly *Diaghilev's *Ballets Russes*. Goncharova and Larionov became French citizens in 1938 and were married in 1955. By this time they had been virtually forgotten, but there was a great revival of interest in them in the early 1960s.

Goncourt, Edmond de (1822–96) and **Jules de** (1830–70). French authors, brothers, who wrote in close collaboration. They wrote on various artistic topics, their most important work of criticism being a book made up of a collection of articles, *L'Art du dix-huitième siècle* (1875), which helped to revive the reputation of 18th-cent. French artists such as *Watteau. The brothers inherited a substantial fortune when their mother died in 1848, and their lives were divided between their writing and self-indulgence; the *Journal* that they began in 1851, and which Edmond continued after Jules died until his own death, provides a richly detailed record of Paris in the second half of the 19th cent. Edmond's books on Utamaro (1891) and Hokusai (1896) helped to popularize Japanese art (see UKIYO-E). The brothers also wrote novels, and the Académie Goncourt, founded under Edmond's will, is a body of ten men or women of letters that awards an annual prize (the Prix Goncourt) for imaginative prose.

Gonzaga. Lords of Mantua between 1328 and 1708, who at several different periods attracted to their court some of the greatest Italian and other European artists. Under **Lodovico** (reigned 1445–78) and his immediate successors *Mantegna was employed as court painter and *Alberti began the church of S. Andrea. The presence of Isabella d'*Este, who married **Francesco II** in 1490, helped to make Mantua one of the greatest centres of art collecting and patronage. Under **Federico** (1519–40) *Giulio Romano built and decorated the Gonzaga pleasure house, the Palazzo del Tè, and turned Mantua into one of the main centres of *Mannerist art. Under **Vincenzo I** (1587–1612) *Rubens was made court painter; and the reign of **Ferdinando** (1612–26) saw the employment of van *Dyck, Domenico *Feti, *Albani, and other artists. The spectacular collections built up over the years were sold by **Vincenzo II** in 1628, principally to Charles I (1600–49) of England, and important Gonzaga patronage came to an end after the sack of Mantua in 1630.

González, Julio (1876–1942). Spanish sculptor, metalworker, painter, and draughtsman, the leading pioneer in the use of iron as a sculptural medium. He learnt to work metals under his father, a goldsmith and sculptor, but his early career was spent mainly as a painter. In about 1900 he moved to Paris and formed a lifelong friendship with *Picasso, whom he had earlier met in his native Barcelona. He initially supported himself mainly by making metalwork and jewellery, and it was not until the late 1920s, when he was already 50, that he devoted himself whole-heartedly to sculpture and turned to welded metal as a material. His best-known work, *Montserrat* (Stedelijk Mus., Amsterdam, 1937), is a fairly naturalistic piece, showing a woman with a child in her arms, and commemorates the suffering of the people of Spain in the civil war (Montserrat is Spain's holy mountain). More usually, however, his sculptures are semi-abstract, as in his series of *Cactus People*, formidable pieces with some of Picasso's savage humour. González's work had great influence, notably on Picasso, to whom he taught the techniques of iron sculpture, and on a generation of British and American artists exemplified by Reg *Butler and David *Smith.

Gordon, Douglas. See TURNER PRIZE.

Gore, Spencer (1878–1914). British painter of landscapes, music-hall scenes, interiors, and occasional still lifes. He was the son of Spencer Walter Gore, who won the first Wimbledon tennis championship in 1877, and nephew of Charles Gore, Bishop of Oxford. In 1896–9 he studied at the *Slade School, where he was a particular friend of Harold *Gilman. In 1904 he visited *Sickert in Dieppe; this marked the beginning of his close acquaintance with recent French painting (he returned to France in 1905 and 1906), and his enthusiasm helped to decide Sickert to return to Britain. For the rest of his short career Gore was part of Sickert's circle, becoming a founder member successively of the *Allied Artists' Association in 1908, the *Camden Town Group (of which he was first President) in 1911, and the *London Group in 1913. His early work was *Impressionist in style, but he was strongly influenced by Roger *Fry's *Post-Impressionist exhibitions (Gore's own work was included in the second in 1912) and his later pictures show vivid use of flat, bright colour and boldly simplified forms. He died of pneumonia aged 35 and was much lamented by his many friends in the art world. Sickert said Gore was 'probably the man I love and admire most of any I have known', and his obituary in the *Morning Post* remarked that 'his personal character was so exceptional as to give him a unique influence in the artistic affairs of London in the last dozen years.' His son **Frederick Gore** (1913–) is also a painter.

Gorky, Arshile (Vosdanig Manoog Adoian) (1904–48). American painter, born in Turkish Armenia, who formed a link between European *Surrealism and American *Abstract Expressionism. He emigrated to the USA in 1920 and adopted the pseudonym Arshile Gorky, the first part of the name being derived from the Greek hero Achilles, the second part (Russian for 'the bitter one') from the Russian writer Maxim Gorky, to whom the painter sometimes claimed he was related (evidently not realizing that the writer's name, too, was a pseudonym). In 1925 he settled in New York, where he first studied and then taught at the Grand Central School of Art. Gorky took a romantic view of his vocation and is said to have hired a Hungarian violinist to play during his classes to encourage his students to put emotion into their work. He was among the first to recognize the importance of the abstract work of Stuart *Davis at this time, and owed a debt to *Picasso, seen both in his haunting *The Artist and his Mother* (Whitney Mus., New York, *c*.1926–9) and in the *Cubist abstractions which he did at this period. Gorky was never at home with geometrical abstraction, however, and preferred to adapt Cubist techniques to his own more painterly and expressive purposes. He came into his own when he became a friend of the circle of European immigrant Surrealists who during the early years of the 1940s lived in New York. Under this influence he worked out a style of abstraction using *biomorphic forms akin to those of *Miró, as may be seen in various versions of *Garden in Sochi* (MOMA, New York, 1941) and *Mojave* (Los Angeles County Mus. of Art, 1941–2). At the peak of his powers, however, Gorky suffered a tragic series of misfortunes. In 1946 a fire in his Connecticut studio destroyed a large proportion of his recent work. In the same year he was operated on for cancer. In 1948 he broke his neck in a motor-car accident and, when his wife left him soon after, he hanged himself. Gorky has been called both the last of the great Surrealists and the first of the Abstract Expressionists, and his work in the 1940s was a potent factor underlying the emergence of a specifically American school of abstract art.

Gormley, Antony. See TURNER PRIZE.

Gossaert, Jan (also called Mabuse) (*c*.1478–1532). Netherlandish painter, probably from Maubeuge in Hainault, from which his name 'Mabuse' derives. In 1503 he was registered in the painters' guild at Antwerp and in 1508–9 he visited Rome in the service of Philip Bastard of Burgundy, an ambassador to the Vatican. His work before his Italian journey is in the tradition of Hugo van der *Goes and Gerard *David, whose influences can be seen in the *Adoration of the Magi* (NG, London). On his return to Antwerp, however, his work was transformed by the experience of Italy, although the motifs he learned there were never thoroughly digested and coexisted with Flemish figures and details. *Vasari acclaimed him for being the first 'to bring the true method of representing nude figures and mythologies from Italy to the Netherlands', but in, for example, *Neptune and Amphitrite* (Staatliche Museen, Berlin, 1516), his first dated work, the life-size figures are in fact closer to *Dürer than to any Italian contemporary, and they are set in a curious, totally misunderstood classical temple.

Gossaert was highly thought of by his contemporaries. His commissions took him to various towns in the Netherlands, his patrons included the Royal House of Denmark, and his work was widely influential. However, to modern eyes there seems justice in Dürer's assessment of him as better in execution than in invention ('nit so gut im Haupstreichen als im Gemäl'). Jan van *Scorel was Gossaert's pupil for a short time in Utrecht.

Gotch, Thomas Cooper. See NEWLYN SCHOOL.

Gothic. Style of architecture and art that prevailed in Europe (particularly northern Europe) from the mid 12th cent. to the 16th cent. The word was, like many other stylistic labels, originally a term of abuse; it was coined by Italian artists of the *Renaissance to denote the type of medieval architecture which they condemned as barbaric (implying, quite wrongly, that it was the architecture of the Gothic tribes who had destroyed the classical art of the Roman Empire). In England the word 'Gothick' was often used by 17th- and 18th-cent. writers in the sense of 'tasteless', 'bizarre', or at least contrary to the rules of academic art. But when English antiquarians began to develop an interest in the monuments of the Middle Ages the term gradually lost its derogatory overtones. The *Romantic predilection for the past eventually led to an appreciation of medieval styles of building in their own right, and thus to a study of those stylistic elements which are now regarded as Gothic—elements which were consciously revived in the *Gothic Revival movement. The Gothic style is characterized chiefly in terms of architecture—in particular by the use of pointed arches, flying buttresses, and elaborate *tracery. By extension, however, the term 'Gothic' is applied to the ornament, sculpture, and painting of the period in which Gothic architecture was built, even though it has less precise meaning in these contexts. German critics have even gone so far as to expound the psychology of 'Gothic Man'. See also INTERNATIONAL GOTHIC.

Gothic Revival. Movement in architecture and associated arts in which the *Gothic style of the Middle Ages was revived. Beginning in the mid-18th cent., it was partly of literary origin and partly a breakaway from the rigid *Palladian rules of architectural design then prevailing. Initially, Gothic forms were used in a *Rococo spirit for their picturesque qualities, with no regard for archaeological accuracy. Horace *Walpole's house Strawberry Hill (begun 1748) was the first great monument of the movement, and Walpole was also the author of the first 'Gothic novel', *The Castle of Otranto* (1764), a genre in which the essential ingredients were mystery and horror, and typical settings were graveyards and ruins. In the early 19th cent., however, the romantic interest in medieval forms and fancies gave way and the Gothic style became closely identified with a religious revival; it was advocated by architects such as A. W. N. *Pugin as the only one suitable for churches, as opposed to the 'pagan' *Renaissance style, and churches were built in careful imitation of the constructions of the Middle Ages. Scholarly imitation in turn gave way to more original adaptations and the Gothic style was much used for civic and commercial building as well as churches.

The Gothic Revival had its fullest expression in Britain (where the original Gothic style had never entirely died out—'Gothic survival'), but throughout Europe and America buildings of all kinds were designed in various Gothic styles as part of the vogue for stylistic revivals characteristic of the 19th and early 20th cents. Charles Locke *Eastlake, who himself belonged to the movement, wrote *A History of the Gothic Revival* in 1871. From that time it remained unchronicled until Kenneth *Clark's *The Gothic Revival* (1928), in which he refers to it as 'the most widespread and influential artistic movement which England has ever produced' and as 'perhaps the only purely English movement in the plastic arts'.

Gottlieb, Adolph (1903–74). American painter, one of the leading *Abstract Expressionists. His early work was *Expressionist and in 1935 he was one of the founding members of the Expressionist group The *Ten, with whom he exhibited until 1940. In 1936 he worked for the *Federal Art Project. Some of his landscapes of the late 1930s were influenced by *Surrealism, and from the early 1940s this tendency was enhanced by contact with expatriate European Surrealists and by an interest in Freudian psychology. His personal style began to emerge in 1941 and from then until the end of his life he worked on three main series: *Pictographs* (1941–51), *Imaginary Landscapes* (1951–7, and again in the mid-1960s), and *Bursts* (1957–74). The *Pictographs* use a loose grid- or compartment-like arrangement with schematic shapes or symbols suggesting some mythic face; the *Imaginary Landscapes* feature a zone of astral shapes against a foreground of heavy *gestural strokes; and the *Bursts*, becoming still freer, suggest solar orbs and astral bodies hovering above violently coloured terrestrial explosions (*Orb*, Dallas Mus. of Fine Arts, 1964). Gottlieb also designed stained glass and other works for churches and synagogues, suggesting a religious mood without any specific representation.

gouache. Opaque *watercolour, sometimes also known as *body colour. It differs from transparent watercolour in that the *pigments are bound with glue and the lighter tones are obtained by the admixture of white pigment. Its degree of opacity varies with the amount of white which is added, but in general is sufficient to prevent the reflection of the *ground through the paint and it therefore lacks the luminosity of transparent watercolour painting. The colours sold as poster paints by commercial colourmen are usually a form of gouache. Although gouache lacks the special transparency effected by the white ground of 'true' watercolour, it is easier to handle because earlier trials and errors can be painted over and do not show through to the same extent.

Goudt, Count Hendrik. See ELSHEIMER.

Goujon, Jean (*c.*1510–68). French sculptor. He ranks second only to Germain *Pilon as the greatest French sculptor of the 16th cent. and he created a distinctive *Mannerist style as sophisticated as the finest works of painting and decoration of the contemporary School of *Fontainebleau. Nothing is known of his early life and he is first recorded in 1540 as the carver of the fine columns supporting the organ loft in the church of S. Maclou at Rouen. The pure *classicism of these columns has caused some critics to assume that he had first-hand knowledge of Italian art. The tomb of Louis de Brézé, husband of Diane de Poitiers (1499–1566), in Rouen Cathedral, is generally attributed to him at least in part. He had moved to Paris by 1544, when he was working on the screen of S. Germain-l'Auxerrois, in collaboration with the architect Pierre Lescot. Low-*relief panels (now in the Louvre) from this screen show that Goujon had evolved a style of extreme grace and delicacy, owing something to the influence of Benvenuto *Cellini. The style is seen at its most mature in his decorations (now in the Louvre) for the *Fontaine des Innocents*, Paris (1547–9). The six relief panels of nymphs from the fountain are generally considered Goujon's masterpieces; they are exquisitely carved and the rippling draperies worn by the elegant figures create effects of great decorative beauty. Goujon's most extensive undertaking was on the sculptural decoration of the *Louvre; he worked there from 1549 to 1562 in collaboration with Lescot, mainly on decorative panels forming part of the architectural scheme. Unfortunately all Goujon's work there has been heavily restored, including the famous *caryatids (1550–1) in the Salle des Caryatides. Using caryatids on a monumental scale was a novelty, perhaps inspired by his reading of *Vitruvius (he made illustrations for the first French edition of his treatise in 1547). There is no indication of any work executed after 1562 and it is possible that Goujon left France because of religious persecution and died in Bologna (there is some doubt concerning the documentation).

Gower, George (d. 1596). English portrait painter. He was appointed Serjeant-Painter to Queen Elizabeth in 1581 and seems to have been the leading English portraitist of his day. His *Sir Thomas Kytson and Lady Kytson* (Tate, London, 1573) show his clear and individual, if unsubtle, style. Gower was a gentleman by birth and his *Self-Portrait* (Earl Fitzwilliam Coll., 1579) shows his coat of arms outweighed in a balance by a pair of dividers, a symbol of the painter's craft.

Gowing, Sir Lawrence (1918–91). British painter and writer on art. He had a distinguished academic career, during which he was Deputy Director of the *Tate Gallery (1965–67) and a professor at several universities (notably at the *Slade School, 1975–85), and he wrote books and catalogues, valued for their critical insights, on numerous artists, among them *Cézanne, *Matisse, and *Vermeer. As a painter he began as a pupil of *Coldstream working in the *Euston Road tradition, and much of his subsequent work was in this sombre vein. His work also included abstracts, however, and in 1976 he began producing large pictures in which he traced the outline of his own naked body stretched on the canvas, the paint being applied by an assistant. He was knighted in 1982.

Goya, Francisco de (1746–1828). Spanish painter and graphic artist. He was the most powerful and original European artist of his time, but his genius was slow in maturing and he was well into his thirties before he began producing work that set him apart from his contemporaries. Born at Fuendetodos in Aragon, the son of a gilder, he served his apprenticeship at Saragossa, then appears to have worked at Madrid for the court painter Francisco *Bayeu. In about 1770 he went to Italy but he was back in Saragossa the next year. In 1773 he married Bayeu's sister, and by 1775 had settled at Madrid. Bayeu secured him employment making *cartoons for the royal tapestry factory, and this took up most of his working time from 1775 to 1792. He made sixty-three cartoons (Prado, Madrid), the largest more than 6 m. wide. The subjects range from idyllic scenes to realistic incidents of everyday life, conceived throughout in a gay and romantic spirit and executed with *Rococo decorative charm. During these years Goya also found time for portraits and religious works, and his status grew. He was elected to the Academy of San Fernando in 1780 and became assistant director of painting in 1785. In 1789 he was nominated a court painter to the new king, Charles IV (1748–1819).

A more important turning point in his career than any of these appointments, however, was the mysterious and traumatic illness he experienced in 1792. It left him stone deaf, and while convalescing in 1793 he painted a series of small pictures of 'fantasy and invention' in order, as he said, 'to occupy an imagination mortified by the contemplation of my sufferings'. This marks the beginning of his preoccupation with the morbid, bizarre, and menacing that was to be such a feature of his mature work. It was given vivid expression in the first of his great series of engravings, *Los Caprichos* (Caprices), issued in 1799. The set (executed *c.*1793–8) consists of eighty-two plates in *etching reinforced with *aquatint, and their humour is constantly overshadowed by an element of nightmare. Technically revealing the influence of *Rembrandt, they feature savagely satirical attacks on social customs and abuses of the Church, with elements of the macabre in scenes of witchcraft and diabolism.

In 1795 Goya succeeded Bayeu as director of painting at the Academy of San Fernando and in 1799 he was appointed First Court Painter, producing his most famous portrait group, the *Family of Charles IV* (Prado), in the following year. The weaknesses of the royal family are revealed with unsparing realism, though apparently without deliberate satirical intent. Goya's early portraits had followed the manner of *Mengs, but stimulated by the study of *Velázquez's paintings in the royal collection he had developed a much more natural, lively, and personal style, showing increasing mastery of pose and expression, heightened by dramatic contrasts of light and shade. From about the same date as the royal group portrait are the celebrated pair of paintings the *Clothed Maja* and *Naked Maja* (Prado), whose erotic nature led Goya to be summoned before the Inquisition. Popular legend has it that they represent the Duchess of Alba, the beautiful widow whose relationship with Goya caused scandal in Madrid.

Goya retained his appointment of court painter under Joseph Buonaparte during the French occupation of Spain (1808–14), but his activity as a painter of court and society decreased, and he was torn between his welcome for the regime as a liberal and his abhorrence as a patriot against foreign military rule. After the restoration of Ferdinand VII in 1814 Goya was exonerated from the charge of having 'accepted employment from the usurper' by claiming he had not worn the medal awarded him by the French, and he painted for the king the two famous scenes of the bloody uprising of the citizens of Madrid against the occupying forces—*The Second of May, 1808* and *The Third of May, 1808* (Prado). Equally dramatic, and even more savage and macabre, are the sixty-five etchings *Los Desastres de la Guerra* (The Disasters of War), which he executed in 1810–14. These nightmare scenes, depicting atrocities committed by both French and Spanish, are the most brutally savage protest against cruelty and war which the visual imagination of man has conceived.

Goya virtually retired from public life after 1815, working for himself and friends. He kept the title of court painter but was superseded in royal favour by Vicente *López. Towards the end of 1819 he fell seriously ill for the second time (a remarkable self-portrait in the Minneapolis Institute of Arts shows him with the doctor who nursed him). He had just bought a country house in the outskirts of Madrid, the Quinta del Sordo (House of the Deaf Man); and it was here after his recovery in 1820 that he executed fourteen large murals, sometimes known as the *Black Paintings*, now in the Prado. Painted almost entirely in blacks, greys, and browns, they depict nightmarish scenes, such as *Saturn Devouring One of His Sons*, executed with an almost ferocious intensity and freedom of handling.

In 1824 Goya obtained permission from Ferdinand VII to leave the country for reasons of health and settled at Bordeaux. He made two brief visits to Spain, on the first of which (1826) he officially resigned as court painter. In these last years he took up the new medium of *lithography (in his series the *Bulls of Bordeaux*), while his paintings, such as the portrait of J. B. Muguiro (Prado, 1827), illustrate his progress towards a style which foreshadowed that of the *Impressionists.

Goya completed some 500 oil paintings and murals, nearly 300 etchings and lithographs, and many hundreds of drawings. He was exceptionally versatile and his work expresses a very wide range of emotion. His technical freedom and originality likewise are remarkable—his frescos in San Antonio de la Florida in Madrid (1798), for example, were apparently executed with sponges. In his own day he was celebrated for his portraits, of which he painted more than 200; but his fame now rests equally on his other work.

Goyen, Jan van (1596–1656). Dutch painter, one of the foremost pioneers of realistic landscape painting in the Netherlands. His earliest works are heavily indebted to his master Esaias van de *Velde, but he then created a distinctive type of monochrome landscape in browns and greys with touches of vivid blue or red to catch the eye; gnarled oaks; wide plains, usually seen from a height; low horizons and clouded skies. He was one of the first painters to capture the quality of the light and air in a scene and to suggest the movement of clouds. Most of his paintings seem to be based on drawings made as he travelled about the countryside, and he apparently used the same drawings again and again because the same motifs recur repeatedly in his works. They include river views of Dordrecht and of Nijmegen, sand dunes, and shipping scenes. His finest work has a sense of poetic calm as well as great freshness and luminosity of atmosphere. Van Goyen worked in his native Leiden, Haarlem, and The Hague, where he died. He was hugely prolific and had many pupils and imitators. With Salomon van *Ruysdael, whose paintings are often virtually indistinguishable from his, he was the outstanding master of the 'tonal' phase of Dutch landscape painting, when the depiction of atmosphere was the artist's prime concern.

Gozzoli, Benozzo (Benozzo di Lese) (*c.*1421–97). Florentine painter. Originally trained as a goldsmith, he worked in his early years with *Ghiberti on the doors of the Baptistery in Florence, but he is mentioned as a painter in 1444 and subsequently became Fra *Angelico's assistant in Rome and Orvieto. His reputation rests on only one work—but one of the most enchanting in all Italian *Renaissance art: the decoration of the chapel of the Palazzo *Medici in Florence with frescos of *The Journey of the Magi* (1459–61). This was commissioned by Piero de' Medici, and patron and artist, sharing a taste for pageantry, ceremonial, and Burgundian tapestries, conspired to produce the most glittering fresco paintings of the century, recalling, and perhaps consciously rivalling, *Gentile da Fabriano's *Adoration of the Magi* of 1423. Its secular outlook is far removed in spirit from the work of his master Fra Angelico.

The rest of Benozzo's career was undistinguished. His other major work was a large fresco cycle of Old Testament scenes in the Campo Santo in Pisa; he began work on it in 1467 and spent most of the rest of his life in Pisa. The frescos were badly damaged by bombing in the Second World War. Benozzo also painted altarpieces, one of which (*Madonna and Child with Saints*, 1461–2) is in the National Gallery, London.

Graf, Urs (*c.*1485–1527/8). Swiss graphic artist, designer, and goldsmith, active mainly in Basle. He is best known for his drawings, which survive in considerable number, often signed and dated. They are done in a bold and energetic style, with virtuoso curling strokes of the pen; favourite subjects are soldiers (Graf himself spent some time as a mercenary in Italy), peasants, and flamboyantly dressed ladies of easy virtue. Graf also designed stained glass, made woodcuts, and executed the earliest extant dated etchings (1513). His masterpiece as a goldsmith was a reliquary for the monastery of St Urban (1514), now lost.

Graff, Anton (1736–1813). German portrait painter, active mainly in Dresden. He worked for the courts of Saxony and Prussia, and was at his best in portraits of intellectual, literary, and artistic sitters. These sometimes recall *Reynolds in their direct and penetrating characterization, while his more elegant society pieces are reminiscent of *Gainsborough. A fine example of his work is the self-portrait (1765) in the Gemäldegalerie in Dresden.

Graffiti art. A style of painting based on the type of spray-can vandalism familiar in cities all over the world and specifically in the New York subway system; the term can apply to any work in this vein, but refers particularly to a vogue in New York in the 1980s (several commercial galleries specialized in it at this time and a Museum of American Graffiti opened there in 1989). The best-known figures of Graffiti art are Jean-Michel Basquiat (1960–88) and Keith Haring (1958–90), both of whom enjoyed absurdly inflated reputations (and prices) during their brief careers, which were ended for Basquiat by a drugs overdose and for Haring by AIDS. Basquiat was a genuine street artist who 'crossed over' into the gallery world; Haring had an art school training but adopted a primitivistic style based on graffiti. Robert *Hughes referred to them as 'Keith Boring and Jean-Michel Basketcase'.

graffito (or **sgraffito**) (Italian: 'scratched'). Term now most commonly applied (usually in the plural—'graffiti') to a design or inscription (often obscene) drawn or scratched on a wall; in a broader sense it is applied to any technique of producing a design by scratching through a layer of paint or other material to reveal a *ground of a different colour. In a medieval panel painting, for instance, the ornamental parts would be covered with gold leaf, burnished, and painted, and a design would then be scratched through the paint.

The predilection in certain 20th-cent. aesthetic movements for an appearance of accidental and undesigned decoration combined with a heightened interest in surface texture for its own sake has sometimes directed attention to the effects of random markings and scratchings on walls and other surfaces. These have been photographed and reproduced. Some non-figurative painters have tried to develop this into special kinds of textual effects which formed the predominant motif of their painting.

Granacci, Francesco (1469–1543). Florentine painter, a minor master of the High *Renaissance. He was a pupil of Lorenzo di *Credi and of Domenico *Ghirlandaio and was briefly an assistant of his friend *Michelangelo on the Sistine Ceiling. His work, unadventurous but agreeable, is fairly close in style to Fra *Bartolommeo. It is well represented in the Accademia, Florence.

Grandeville (pseudonym of Jean-Ignace-Isidore Gérard) (1803–47). French *caricaturist and illustrator. He began his career with lithographs of bourgeois life, then achieved notoriety as a political satirist until new regulations against the Press in 1835 made him turn to book illustration. His illustrations to the *Fables* of La Fontaine (1838) achieved lasting popularity for their witty blending of human and animal features (a device which he had already used with telling effect in his political caricatures). Towards the end of his life he showed a predilection for macabre and fantastic subjects, which later attracted the interest of *Surrealists. He died in a lunatic asylum.

Grand Manner. Term applied to the lofty and rhetorical manner of history painting that in academic theory was considered appropriate to the most serious and elevated subjects. The classic exposition of its doctrines is found in *Reynolds's Third and Fourth *Discourses* (1770 and 1771), where he asserts that 'the *gusto grande* of the Italians, the *beau idéal* of the French, and the *great style, genius*, and *taste* among the English, are but different appellations of the same thing'. Cecil Gould (*An Introduction to Italian Renaissance Painting*, 1957) rightly points out that 'the Grand Manner is an attitude rather than a style' and goes on to give a lucid account of some of its characteristics. 'The general aim is to transcend Nature . . . The Subject itself must be on an elevated and elevating plane . . . Similarly, the individual figures in such a scene must be shown purged of the grosser elements of ordinary existence . . . Landscape backgrounds or ornamental detail must be reduced to a minimum and individual peculiarities of human physiognomy absolutely eliminated. Draperies should be simple, but ample, and noble, and fashionable contemporary costume absolutely shunned. Alternatively, the figures should be nude. In the latter case the musculature should be generalized and no single feature stressed unduly . . . the expressive gesture is one of the keynotes of the Grand Manner and on the whole its most constant characteristic.' The idea of the Grand Manner took shape in 17th-cent. Italy, notably in the writings of *Bellori. His friend *Poussin and the great Bolognese painters of the 17th cent. were regarded as outstanding exponents of the Grand Manner, but the greatest of all was held to be *Raphael.

Grand Tour. An extensive journey to the Continent, chiefly to France, the Netherlands, and above all Italy, sometimes in the company of a tutor, that became a conventional feature in the education of the English gentleman in the 18th cent. Such tours often took a year or more. The practice reflects Dr Johnson's pronouncement (1776) that 'A man who has not been in Italy is always conscious of an inferiority.' It had a noticeable effect in bringing a more cosmopolitan spirit to the taste of connoisseurs and virtuosi (see VIRTU) of the arts and laid the basis for many collections among the landed gentry. It also helped the spread of the fashion for *Palladianism and *Neoclassicism and an enthusiasm for Italian painting. Among the native artists who catered for this demand were *Batoni, *Canaletto, *Panini, and *Piranesi, and British artists (such as *Nollekens) were sometimes able to support themselves while in Italy by working for the dealers and restorers who supplied the tourist clientele. There was a flourishing market in guide books, the pioneering work in English being *An Account of Some of the Statues, Bas-Reliefs, Drawings, and Pictures in Italy* (1722) by Jonathan *Richardson and his son.

Granet, François-Marius (1775–1849). French painter. Granet was a pupil of J.-L. *David and subsequently spent the years 1802–19 in Rome. He made a speciality of sombre tonal effects and changing light in dimly lit interiors, his highly individualistic style recalling Dutch interiors rather than the *Neoclassical tradition in which he was trained. His *Choir of the Capuchin Church in Rome* was exhibited at the 1819 *Salon with such success that he made sixteen replicas of it. Granet also painted Italian landscapes, constructed with firm, cubic volumes in which some critics have seen a foreshadowing of *Cézanne. In 1826 he became curator of the *Louvre Museum and was made Keeper of Pictures at Versailles in 1830. During the Revolution of 1848 he retired to his native Aix-en-Provence, where he founded the museum which bears his name. It contains a celebrated portrait of him by *Ingres.

granite. A general term for any crystalline, granular, unstratified igneous stone which is an intimate amalgam of quartz, potash feldspar, and mica. Granite is of world-wide distribution and has many varieties, differing in texture and coarseness. It occurs in a wide range of colours—grey, green, rose, yellow—and the small scales of mica give it a lively sparkle. It takes a brilliant polish on a mirror-smooth surface but is one of the most difficult stones to carve because it is physically very compact; its ingredients are harder than ordinary steel. Nevertheless, its durability and resistance to weather have made it popular for monumental sculpture and at all times when permanence was valued, notably in ancient Egypt. Granite has been used comparatively little for sculpture on a small scale, since its properties preclude delicate carving.

Grant, Duncan (1885–1978). British painter, decorator, and designer. Grant, who studied at the *Slade and in Italy and Paris, was a cousin of the writer Lytton Strachey (1880–1932) and a member of the *Bloomsbury circle of Roger *Fry and Clive and Vanessa *Bell. He exhibited at the *New English Art Club and with the *London Group, and contributed to the second *Post-Impressionist Exhibition in 1912. His work of this period shows that he was among the most advanced of British artists in responding to recent trends in French painting. From about 1913 he was also influenced by African sculpture and he was one of the pioneers of abstract art in Britain; in 1914 he made an *Abstract Kinetic Collage Painting*, which was meant to be unrolled while music by J. S. Bach was playing (this is now in the Tate Gallery, which has had a film made demonstrating the painting being unrolled in the desired fashion). Grant's later painting was generally much more traditional, and many critics consider that he was at his best as a designer. He was the most gifted of the designers who worked for Fry's *Omega Workshops, and he collaborated much with Vanessa Bell (they lived together from about 1916) in interior decoration after the workshops closed in 1919. His work included designs for textiles, pottery, stage scenery, and costumes. The Tate Gallery and the Courtauld Institute Galleries have good representations of his work.

Grant, Sir Francis (1803–78). Scottish painter. He was one of the most fashionable portrait painters of his day and succeeded *Eastlake as President of the *Royal Academy in 1866. His best works are generally considered to be his smaller sporting *conversation pieces. He is well represented in the National Portrait Galleries, Edinburgh and London.

graphic art. Term current with several different meanings in the literature of the visual arts. In the context of the *fine arts, it most usually refers to those arts that rely essentially on line or tone rather than colour—i.e. drawing and the various forms of engraving. Some writers, however, exclude drawing from this definition, so that the term 'graphic art' is used to cover the various processes by which prints are created. In another sense, the term—sometimes shortened to 'graphics'—is used to cover the entire field of commercial printing, including text as well as illustrations.

graphite. Mineral—a form of carbon—used as the 'lead' in *pencils, among other purposes. It is mined in various parts of the world and can also be made synthetically.

Grasser, Erasmus (*c.*1450–1518). German sculptor and wood-carver, active mainly in Munich. He worked in an animated and expressive late *Gothic style and was the leading sculptor of his day in south Bavaria, with a flourishing workshop and numerous pupils. His best-known works are the ten figures (originally sixteen) of morris dancers for the ballroom of the old town hall in Munich (Stadtmuseum, Munich, 1480). Grasser was also an architect and hydraulic engineer.

Gravelot, Hubert-François (1699–1773). French designer and engraver of book illustrations, active for much of his career in England. In London he became a friend of *Hogarth, probably taught *Gainsborough, and caricatured *Walpole and Lord *Burlington. He illustrated Gay's *Fables*, Shakespeare, and Dryden, and was one of the first artists to illustrate the novel, designing engravings for Richardson's *Pamela* (1742) and Fielding's *Tom Jones* (1750). He was at his best depicting scenes of contemporary life rather than tragic subjects or the classical and allegorical themes then fashionable. Although not an artist of outstanding talent, he is important for helping to introduce the French *Rococo style to England.

graver. See BURIN.

Great Exhibition. See EXHIBITION, GREAT.

Greaves, Derrick. See KITCHEN SINK SCHOOL.

Greaves, Walter. See WHISTLER.

Grechetto, Il. See CASTIGLIONE, GIOVANNI BENEDETTO.

Greco, El (1541–1614). Cretan-born painter, sculptor, and architect who settled in Spain and is regarded as the first great genius of the Spanish School. He was known as El Greco (the Greek), but his real name was Domenikos Theotocopoulos; and it was thus that he signed his paintings throughout his life, always in Greek characters, and sometimes followed by *Kres* (Cretan).

Little is known of his youth, and only a few works survive by him in the *Byzantine tradition of *icon painting, notably the recently discovered *Dormition of the Virgin* (Church of the Koimesis tis Theotokou, Syros). In 1566 he is referred to in a Cretan document as a master painter; soon afterwards he went to Venice (Crete was then a Venetian possession), then in 1570 moved to Rome. The *miniaturist Giulio *Clovio, whom he met there, described him as a pupil of *Titian, but of all the Venetian painters *Tintoretto influenced him most, and *Michelangelo's impact on his development was also important.

Among the surviving works of his Italian period are two paintings of the *Purification of the Temple* (Minneapolis Institute of Arts, and NG, Washington), a much-repeated theme, and the portrait of Giulio Clovio (Museo di Capodimonte, Naples). By 1577 he was at Toledo, where he remained until his death, and it was there that he matured his characteristic style in which figures elongated into flame-like forms and usually painted in cold, eerie, bluish colours express intense religious feeling. The commission that took him to Toledo—the high altarpiece of the church of S. Domingo el Antiguo—was gained through Diego de Castilla, Dean of Canons at Toledo Cathedral, whom El Greco had met in Rome. The central part of the altarpiece, a 4-m. high canvas of *The Assumption of the Virgin* (Art Inst. of Chicago, 1577), was easily his biggest work to date, but he carried off the dynamic composition triumphantly. A succession of great altarpieces followed throughout his career, the two most famous being *El Espolio* (Christ Stripped of His Garments) (Toledo Cathedral, 1577–9) and *The Burial of Count Orgaz* (S. Tomé, Toledo, 1586–8). These two mighty works convey the awesomeness of great spiritual events with a sense of mystic rapture, and in his late work El Greco went even further in freeing his figures from earthbound restrictions; *The Adoration of the Shepherds* (Prado, Madrid, 1612–14), painted for his own tomb, is a prime example.

El Greco excelled also as a portraitist, mainly of ecclesiastics (*Felix Paravicino*, Boston Mus., 1609) or gentlemen, although one of his most beautiful works is a portrait of a lady (Pollock House, Glasgow, *c.*1577–80), traditionally identified as a likeness of Jeronima de las Cuevas, his common-law wife. He also painted two views of Toledo (Met. Mus., New York, and Museo del Greco, Toledo), both late works, and a mythological painting,

Laocoön (NG, Washington, *c.*1610), that is unique in his *œuvre*. The unusual choice of subject is perhaps explained by the local tradition that Toledo had been founded by descendants of the Trojans. El Greco also designed complete altar compositions, working as architect and sculptor as well as painter, for instance at the Hospital de la Caridad, Illescas (1603). *Pacheco, who visited El Greco in 1611, refers to him as a writer on painting, sculpture, and architecture. He had a proud temperament, conceiving of himself as an artist-philosopher rather than a craftsman, and had a lavish life-style, although he had little success in securing the royal patronage he desired and seems to have had some financial difficulties near the end of his life. The inventory of his effects compiled after his death records few luxury possessions, but mentions fifty models in plaster, clay, and wax, while his library included a collection of books on architecture.

His workshop turned out a great many replicas of his paintings, but his work was so personal that his influence was slight, his only followers of note being his son **Jorge Manuel Theotocopouli** and Luis *Tristán. Interest in his art revived at the end of the 19th cent. and with the development of *Expressionism in the 20th cent. he came into his own. The strangeness of his art has inspired various theories, for example that he was mad or suffered from astigmatism, but his rapturous paintings make complete sense as an expression of the religious fervour of his adopted country.

Greek Revival. See NEOCLASSICISM.

Greek taste. A fashion for Greek-inspired ornamentation in decoration, furniture, bibelots, costume, etc., popular particularly in France and also in England in the late 18th and early 19th cents. It was often very superficial, with little to do with the serious study of Greek architecture and design that was a feature of the *Neoclassical movement. In his *Correspondance littéraire*, May 1763, Baron Grimm (1723–1807) remarked that 'all Paris is *à la grecque*'. 'Our ladies have their hair done *à la grecque*, our *petits-maîtres* would be ashamed to carry a snuff-box that was not *à la grecque*.' In costume the mode was caricatured in E. A. Petitot's *Mascarade à la grecque* (1771). The absurdities of fashion did not discount the seriousness with which the study of Classical antiquity was pursued, although it is a matter of surprise today to realize how very little from pre-*Hellenistic Greece was actually known even to *Winckelmann. The 'Greek-style' drawings and decorations by *Flaxman, Asmus Jakob *Carstens, and others were based upon vase-paintings found in Italian sites.

Green, Anthony (1939–). British painter. He specializes in scenes from his own middle-class domestic life portrayed with loving attention to detail and an engaging sense of whimsy. Often he uses oddly *shaped canvases that accentuate his strange perspective effects, and his subjects are frequently erotic as well as humorous. His work, which is often on a large scale, is instantly recognizable and is generally highly popular with the public at the *Royal Academy summer exhibition, for he communicates with rare intensity the loving feeling he puts into his paintings.

Green, Valentine (1739–1813). English engraver, chiefly in *mezzotint. His great reputation is largely based on the numerous plates he made from the female portraits and groups by *Reynolds; they ended their association in 1783 after a quarrel. Other painters whose works he engraved included *Romney, *West, and Joseph *Wright, and in 1789 he obtained a patent from the Duke of Bavaria to engrave and publish prints from the Düsseldorf gallery. He ranks with *Earlom as one of the most brilliant, if uneven, of British mezzotint engravers.

Green, William. See ACTION PAINTING.

Greenaway, Kate (1846–1901). English artist famous for her illustrations for children's books. Her delicate skill and fragile sentimentality, often imitated but never rivalled, won her many distinguished admirers, including *Ruskin, and (perhaps for her feeling for flat pattern) Paul *Gauguin. She often illustrated her own texts, issuing for example a series of *Kate Greenaway's Almanacs*, and her work became so popular that the quaint clothes that are such a feature of her illustrations influenced children's costume.

Greenberg, Clement (1909–94). American art critic. With Harold *Rosenberg he was his country's most influential writer on contemporary art in the post-war years when American painting and sculpture first achieved a dominant position in world art. His approach to criticism is sometimes described as *formalist, and the artists to whose works he gave the most powerful advocacy were chiefly uncompromising abstractionists—most famously Jackson *Pollock and David *Smith, and later the *Post-Painterly Abstractionists (Greenberg coined this term) and the British sculptor Anthony *Caro. In painting he laid particular stress on the flatness of the picture surface and the rejection of any kind of illusionistic modelling, and he opposed the mere 'novelty' art of painters such as *Rauschenberg. Although he regarded aesthetic judgements as autonomous, he also believed that history possessed order and purpose (the result of early contacts with Marxism) and this allowed him to endow his 'disinterested aesthetic' verdicts on art with a claim for historical certainty. His influence was at its height in the 1950s and 1960s, but thereafter it waned in the face of such developments as *Conceptual art and *New Figuration. His best-known book is probably *Art and Culture* (1961), an anthology of his writings; his other books include monographs on *Miró (1948), *Matisse (1953), and *Hofmann (1961).

Greene, Balcomb (1904–90). American painter. He began to paint seriously in 1931. During the 1930s he painted geometrical abstracts and was a founder member of the *American Abstract Artists association. In the 1940s his work became more representational, sometimes with figures fragmented against an abstract background, and during the latter 1950s and the 1960s a new note of humanism, almost of existentialism, became noticeable in many of his pictures.

Greenhill, John (*c.*1644–76). English portrait painter, a pupil of *Lely, who evolved a simplified version of his master's style. His work is generally fairly pedestrian, but he made some interesting pastel portraits of actors in costume (*Joseph Harris as Cardinal Wolsey*, Magdalen Coll., Oxford, 1664).

Greenough, Horatio (1805–52). American *Neoclassical sculptor who spent the greater part of his working life in Italy. He is sometimes said to be the first professional American sculptor and his major work, the colossal marble figure of George Washington (1833–41), was the first important state commission given to an American sculptor. It was originally intended for the rotunda of the Capitol in Washington, but is now in the Smithsonian Institution. The seated figure is based on *Phidias' celebrated statue of Zeus at Olympia, but the head follows *Houdon's portrait of Washington—an uneasy mixture of idealism and naturalism. Greenough's work is in general rather stodgy and uninventive, and his writings on art are usually considered more interesting. His views on architecture have been claimed as precursors of modern functionalism and are sometimes thought to have influenced the ideas of the great architect Louis Sullivan (1856–1924). Greenough's brother, **Richard Saltonstall Greenough** (1819–1904), was also a sculptor, best known for his statue of Benjamin Franklin outside Boston City Hall.

Greuze, Jean-Baptiste (1725–1805). French painter. He had a great success at the 1755 *Salon with his *Father Reading the Bible to His Children* (Louvre, Paris) and went on to win enormous popularity with similar sentimental and melodramatic genre scenes. His work was praised by *Diderot as 'morality in paint' and as representing the highest ideal of painting in his day. He also wished to succeed as a history painter, but his *Septimius Severus Reproaching Caracalla* (Louvre, 1769) was rejected by the Salon, causing him acute embarrassment. Much of Greuze's later work consisted of titillating pictures of young girls, which contain thinly veiled sexual allusions under their surface appearance of mawkish innocence; *The Broken Pitcher* (Louvre), for example, alludes to loss of virginity. With the swing of taste towards *Neoclassicism his work went out of fashion and he sank into obscurity at the Revolution in 1789. At the very end of his career he received a commission to paint a portrait of Napoleon (Versailles, 1804–5), but he died in poverty. His huge output is particularly well represented in the Louvre, the Wallace Collection in London, the Musée Fabre in Montpellier, and in the museum dedicated to him in Tournus, his native town.

Grien, Hans Baldung. See BALDUNG.

Grimaldi, Giovanni Francesco (called Il Bolognese) (1606–80). Italian landscape painter. He developed an attractive landscape style in the manner of the mature Annibale *Carracci, and his work, which was popular with collectors and much engraved, helped to spread the tradition of *ideal landscape in Europe. Grimaldi worked mainly in Rome, painting frescos as well as easel paintings, notably at the Villa Doria Pamphili, where he was also employed as an architect. In 1649–51 he worked in Paris.

Grimm, Samuel Hieronymus (1733–94). Swiss-born painter, mainly in watercolour, active in England from 1768. He was a prolific topographical draughtsman, much of his work being done on commission for antiquarian patrons. Grimm also made *caricature drawings and illustrations to Shakespeare, and himself wrote poetry. His work is well represented in the British Museum and Victoria and Albert Museum.

Grimmer. Two Flemish painters, **Jacob** (*c.*1526–90) and his son **Abel** (*c.*1570–*c.*1619), whose styles are so similar that it is often difficult to distinguish between their works. They both worked in Antwerp, painting mainly landscape and *genre subjects in an attractive style, full of lively anecdote, that places them among the best followers of *Breugel. Jacob's work was praised by van *Mander and others. Examples of both may be seen in the Musée Royal in Antwerp.

Grimshaw, Atkinson (1836–93). English painter. He specialized in a distinctive type of nocturnal townscape, usually featuring gas lights and wet streets; *Whistler said of him 'I considered myself the inventor of Nocturnes until I saw Grimmy's moonlit pictures.' Grimshaw's paintings, however, unlike Whistler's, are sharp in focus and rather acidic in colouring, although often remarkably atmospheric. They were very popular (in spite of the fact that he rarely exhibited at the *Royal Academy) and he was much imitated, not least by two of his sons, **Arthur** (1868–1913) and **Louis** (1870–1943?). Their father worked in his native Leeds (there is a good collection of his paintings in the City Art Gallery there) and in other northern towns, as well as in London.

Gris, Juan (1887–1927). Spanish painter, sculptor, graphic artist, and designer, active mainly in Paris, where he settled in 1906. In his early years there he earned his living mainly with humorous drawings for various periodicals and he did not begin painting in earnest until 1910. By this time he was strongly influenced by his fellow Spaniard *Picasso and his serious painting was almost entirely in the *Cubist manner. He made such rapid strides that by 1912 he was becoming recognized as the leading Cubist painter apart from the founders of the movement, Picasso and *Braque. His work stood out at the *Section d'Or exhibition in that year, attracting the attention of collectors and dealers (Gertrude *Stein was among those who bought his paintings and *Kanhweiler gave him a contract). In 1913–14 he developed a personal version of Synthetic Cubism, in which *papier collé played an important part. He said that he conceived of his paintings as 'flat, coloured architecture' and his methods of visual analysis were more systematic than those of Picasso and Braque. His subjects were almost all taken from his immediate surroundings (mainly still lifes, with occasional landscapes and portraits), but he began with the image he had in mind rather than with an object in the external world: 'I try to make concrete that which is abstract . . . *Cézanne turns a bottle into a cylinder, but I make a bottle—a particular bottle—out of a cylinder.' In 1919 Gris had his first major one-man exhibition (at the Galerie l'Effort Moderne in Paris), but in the following year he suffered a serious attack of pleurisy and from then on his health was poor; for this reason he spent much of his time in the South of France. In this last period of his life his style became more painterly (*Violin and Fruit Dish*, Tate Gallery, London, 1924). Apart from paintings, his work included polychrome sculpture, book illustrations, and set and costume designs for *Diaghilev. He wrote a few essays on his aesthetic ideas and a collection of his letters, edited and translated by Douglas *Cooper, was published in 1956.

grisaille. A painting done entirely in shades of grey or another neutral greyish colour. Grisaille is sometimes used for *underpainting or for sketches (notably in the work of *Rubens), and particularly in the *Renaissance it was used for finished works imitating the effects of sculpture. Examples of the latter practice are *Giotto's series of *Virtues* and *Vices* in the Arena Chapel, Padua, and the figures of St John the Baptist and St John the Evangelist on the exterior of the van *Eycks' Ghent Altarpiece.

Gromaire, Marcel. See LURÇAT.

Gropius, Walter (1883–1969). German-born architect, designer, and teacher who became an American citizen in 1944. After working in the office of Peter *Behrens, he set up his own office in Berlin. In 1919 he founded the *Bauhaus, of which he was Principal, first in Weimar and from 1925 in Dessau, until 1928. According to his founding proclamation the purpose of the Bauhaus was to unite all the arts under the primacy of architecture, to ensure that they should be practised as crafts in the sense taught by William *Morris and that they should contribute to the *Gesamtkunstwerk*, the total work of art, which was the building and everything in it.

Gropius left the Bauhaus in 1928 and resumed his architectural practice in Berlin. In 1934, after the National Socialists had come to power, he left Germany for England, where he practised in partnership with the British architect Maxwell Fry (1899–1987). In 1937 he went to the USA, where he had been offered the Chair of Architecture at the Graduate School of Design at Harvard, and he remained President of the Department until 1951. He remained active until the end of his life and had many notable buildings to his credit. His architecture is characterized by an uncompromising use of modern materials, but also by lucidity and gracefulness.

Although Gropius's practical work was in the field of architecture, by his ideas, his teaching, and his personality, through the Bauhaus and afterwards, his influence upon modernist trends in all the visual arts has probably not been exceeded by that of any other man. Nowhere else have so many major artists of outstanding originality been brought into collaboration as those whom Gropius induced to teach at the Bauhaus. Left wing in his political views, he believed that design should be a response to the need of society, utilizing to the full the resources of modern technology and expressing the ideal of a humane community.

Gropper, William (1897–1977). American graphic artist and painter. He established his reputation as a political cartoonist and took up painting in 1921. His work in both media embodied his radical political convictions and was concerned with exposing social injustice: he was involved with the Communist movement and spent a year in Moscow working for the party newspaper *Pravda*. His style was forceful, whether in black-and-white drawings or in his paintings, where he favoured bright primary colours.

Gros, Antoine-Jean (1771–1835). French painter. He trained with his father, a *miniaturist and then with J.-L. *David. Although he revered David and became one of his favourite pupils, Gros had a passionate nature and he was drawn more to the colour and vibrancy of *Rubens and the great Venetian painters than to the *Neoclassical purity of his master. In 1793 Gros went to Italy, where he met Napoleon and was appointed his official battle painter. He followed Napoleon on his campaigns, and his huge paintings such as *The Battle of Eylau* (Louvre, Paris, 1808) are among the most stirring images of the Napoleonic era. Compared to the contemporary war scenes of *Goya, they are glamorous lies, but they are painted with such dramatic skill and panache that they cannot but be admired on their own terms. When David went into exile after the fall of Napoleon, Gros took over his studio, and tried to work in a more consciously Neoclassical style. He never again approached the quality of his Napoleonic pictures, however (although he painted excellent portraits), and haunted by a sense of failure he drowned himself in the Seine. Gros is regarded as one of the leading figures in the development of *Romanticism; the colour and drama of his work influenced *Géricault, *Delacroix, and his pupil *Bonington amongst others.

Grosvenor Gallery, London. A commercial gallery founded in 1877 by the wealthy dilettante painter Sir Coutts Lindsay and Charles E. Hallé, son of the famous musician Sir Charles Hallé. *Whistler showed eight paintings at the opening exhibition, and *Ruskin's notorious outburst against one of them led to the libel trial that caused the painter's financial ruin. The other artists who showed at the gallery included such distinguished academics as *Leighton and *Poynter, but it became particularly associated with the Aesthetic Movement (see AESTHETICISM) and was memorably satirized in Gilbert and Sullivan's *Patience* (1881): 'A greenery-yallery, Grosvenor Gallery, | Foot-in-the-grave young man' (an allusion to the deathly pallor possessed by many of the figures in works by painters such as *Burne-Jones). In 1888 the Grosvenor Gallery was taken over by the New Gallery; by this time it 'had become little more than an overflow from the *Royal Academy' (Dennis Farr, *English Art 1870–1940*, 1978).

Grosz, George (1893–1959). German-born painter and draughtsman who became an American citizen in 1938. He began as a *caricaturist with a strong feeling for social satire and through his drawings he expressed his disgust at the depravity of the Prussian military caste. During the First World War he twice served in the German army and each time was discharged as being unfit for service. In 1917, with *Heartfield, he Anglicized his name (he was born Georg Groß) as a protest against the hatred being whipped up against the enemy. The most famous of the satirical anti-war illustrations he made at this time is the drawing *Fit for Active Service* (MOMA, New York, 1918), in which a fat, complacent doctor pronounces a skeleton fit for duty. From 1917 to 1920 he was prominent among the Berlin *Dada group and during the 1920s, while still working on Dadaist *montages, he became, with *Dix, the leading exponent of the *Neue Sachlichkeit. His collections of drawings *The Face of the Ruling Class* (1921) and *Ecce Homo* (1927) earned him an international reputation as a social satirist and an artist of the Left. In these and in his paintings he ruthlessly denounced a decaying society in which gluttony and depraved sensuality are placed beside poverty and disease; prostitutes and profiteers were frequently among his cast of characters. Grosz was prosecuted several times for obscenity and blasphemy, and in 1933, despairing at the political situation in Germany, he moved to America to take up the offer of a teaching post at the *Art Students League of New York. He had joined the Communist Party in 1918 and after he left Germany he was described by the Nazis as 'Cultural Bolshevist Number One'. In America his satirical manner was largely abandoned for more romantic landscapes and still lifes with from time to time apocalyptic visions of a nightmare future. Although he won several honours in the last decade of his life, he regarded himself as a failure because he was unable to win recognition as a serious painter rather than a brilliant satirist, and he painted several self-portraits showing how isolated and depressed he was in his adopted country (*The Wanderer*, Memorial Art Gallery, the University of Rochester, New York, 1943). He returned to Berlin in 1959, saying 'my American dream turned out to be a soap bubble', and died there shortly after his arrival following a fall down a flight of stairs.

grotesque (Italian *grotteschi*). A type of mural decoration, painted, carved, or moulded in *stucco, which in the early 16th cent. spread from Italy to most countries in Europe. It was characterized by the use of floral motifs, animal and human figures, masks, etc., copied from the ornament found in Roman buildings (called *grotte*) such as the Domus Aurea of Nero, excavated *c.*1500, the whole being imaginatively combined into fanciful and playful schemes. One of the earliest examples of 'grotesque' ornament can be found in the frieze in Carlo *Crivelli's *Annunciation* (NG, London, 1486). The grotesque style was distinguished by its disintegration of natural forms and the redistribution of the parts in accordance with the fantasy of the artist.

From the later 17th cent. this kind of decoration was called *arabesque in France and in elaboration of the earlier arabesques by *Bérain and *Audran it became a characteristic feature of the *Rococo, though it lost much of its initial connection with the Roman motifs. The individual motifs of the *grotteschi* were brought in again by Classicists such as *Piranesi and Robert *Adam in the context of the *Neoclassical movement and became an occasional feature in all the decorative arts. But the distinctive fanciful combinations of the Renaissance grotesques were not revived.

In France the word 'grotesque' was applied to literature and even to people fairly early in the 17th cent. and in 1694 it was defined by the *Dictionnaire de l'Académie française* as 'Bizarre, fantastique, extravagant, capricieux'. This extended sense of the word, which became current also in England after the Restoration, with its connotations of the ridiculous, absurd, and unnatural, carried implications of disapproval for the Age of Reason and by the time Neoclassicism was in vogue both the word and the style had acquired a pejorative sense. It was synonymous with the excessive, the preposterous, and the reprehensible. During the *Gothic Revival and in certain phases of the *Romantic movement the grotesque again came into repute though not in its original connotation. Poe's title *Tales of the Grotesque and Arabesque* (1839) is symptomatic of this change. *Ruskin's treatment of the grotesque had the effect of establishing it as a respectable genre of art both in decoration and more widely, although he was unwilling to allow it a place in the higher branches of art. Of the original grotesque 'which first developed itself among the enervated Romans' he speaks only with disdain. For him the 'true grotesque' was that which revealed an insight into the dreadfulness of nature.

Thus the word 'grotesque', originating as a technical term designating a late Roman type of decoration and a Renaissance decorative style based upon it, came to imply whatever is incongruous with the accepted norm whether in life or in art.

ground. The surface or *support on which a painting or drawing is executed, for example the paper on which a *watercolour is done or the plaster under a *fresco; or, more specifically, the prepared surface on which the colours are laid and which is applied to the panel, canvas, or other support before the picture is begun. The purpose of the ground in the second, more technical, sense is to isolate the paint from the support so as to prevent chemical interaction, to render the support less absorbent, to provide a satisfactory surface for painting or drawing on, and to heighten the brilliance of the colours. The ground should be consistent so that the artist can calculate his effects on any part of it. It should not be too smooth to accept *pigment from brush or pencil nor so rough as to impede handling. It should have an even tone and, unless very opaque pigments are used, a certain luminosity and reflecting power. It must not be too absorbent. And above all it must be durable and not liable to flake or crack. *Gesso is the ground that occurs most often in the literature of art history. In *etching the ground is the acid-resisting mixture which is spread over the plate before work is begun.

Groupe de Recherche d'Art Visuel (GRAV). An association of artists formed in Paris in 1960, the main purpose of which was to research into the aesthetic manipulation of light and movement (see KINETIC ART). The members, who included Julio *Le Parc and *Vasarely's son Yvaral, adopted a scientific approach to the production of art works and investigated the use of modern industrial materials for artistic purposes. In common with other contemporary groups they made it one of their aims to produce works of art which called for closer collaboration on the part of the observer and as well as individual works they collaborated in the production of anonymous group works. The group disbanded in 1968.

Groupe des Vingt (XX). See VINGT.

Group of Seven. Group of 20th-cent. Canadian painters, based in Toronto, who found their main inspiration in the landscape of northern Ontario and created the first major national movement in Canadian art. The group was officially established in 1920, when it held its first exhibition in the Art Gallery of Toronto, the seven painters involved being Franklin Carmichael (1890–1945), Lawren *Harris, A. Y. *Jackson, Frank Johnston (1888–1949), Arthur Lismer (1885–1969), J. E. H. *Mac-

Donald, and Frederick Varley (1881–1969). Some members of the group had, however, been working together since 1913, and Tom *Thomson, who was one of the early leaders, had died in 1917. Other artists joined after the 1920 exhibition. The members made group sketching expeditions and worked in an *Expressionist style characterized by brilliant colour and bold brushwork. After initial critical abuse, they won public favour. The last group exhibition was held in 1931 and two years later the name was changed to the *Canadian Group of Painters; thereafter the members worked more as individuals and developed separately, Harris eventually becoming an abstract artist. In 1966 a gallery dedicated to the Group of Seven—the McMichael Canadian Collection—was opened in Kleinburg, Ontario.

Gruber, Francis (1912–48). French painter. His early work was often of visionary subjects, but from about 1933 he began to paint mainly from the model in the studio; he also did still lifes, views through windows, and from 1937 landscapes painted out of doors. Gruber's mature style was grave and melancholy, featuring long, drooping figures, and he is regarded as the founder of the 'Misérabiliste' strain in French painting, later particularly associated chiefly with *Buffet. A typical work is *Job* (Tate Gallery, London, 1944), painted for the 1944 *Salon d'Automne, which was known as the Salon of the Liberation because it was held soon after Paris was freed from the German Occupation; the picture symbolizes oppressed peoples, who like Job in the Bible had endured a great deal of suffering. In spite of the tuberculosis that caused his early death, Gruber worked with great energy and had a substantial output.

Grüner, Ludwig (1801–82). German engraver and painter. He studied in Dresden and Milan; from 1828 onwards he travelled extensively in France and Spain, and between 1841 and 1856 stayed frequently in England, where he acted as artistic adviser to Prince *Albert and to Sir Charles *Eastlake.

Grünewald, Mathis (*c.*1470/80–1528). German painter, the greatest of *Dürer's contemporaries. His real name was Mathis Gothardt or Neithardt, but this was not discovered until the 1920s; 'Grünewald' is an error of *Sandrart, who published the first biography of the artist in his *Teutsche Akademie* (1675), but it is now hallowed by usage. The obscurity into which he fell reflects the isolation and individuality of his work; he had no known pupils and (unlike most of his German contemporaries) he did not make woodcuts or engravings, which would have spread his name. He was successful for most of his career, working as court painter to two successive archbishops of Mainz, but his reputation did not survive, and in 1597, when the emperor Rudolf II (1552–1612) tried to buy his masterpiece, the Isenheim Altarpiece, the name of the painter had already been forgotten.

The first documentary reference to him (or what appears to be him) is of 1501, when a 'Master Mathis' is recorded working in Seligenstadt, a little town near Frankfurt and within the diocese of the Archbishop of Mainz, Uriel von Gemmingen, for whom Grünewald began working in about 1508. Von Gemmingen died in 1514 and Grünewald was employed by his successor Albrecht von Brandenburg (who was also Archbishop of Magdeburg) from 1516 to 1526. Grünewald was employed as a hydraulic engineer and supervisor of architectural works as well as a painter, and in 1520 he accompanied Albrecht to Aix-la-Chapelle for the coronation of emperor Charles V, an occasion on which he met Dürer. The little that is recorded of his personal life comes from Sandrart, who says he was melancholy and withdrawn and made an unhappy marriage late in life. There is no documentary confirmation of his marriage, but he is known to have had an adopted son called Andreas Neithardt, whose surname the painter sometimes used for himself in documents relating to the boy, thus creating one of the sources of confusion about his identity.

Grünewald's work forms a complete contrast to that of Dürer. Whereas Dürer—an intellectual imbued with *Renaissance ideas—had limitless curiosity about the visual world, Grünewald concentrated exclusively on religious themes, and in particular the Crucifixion, a subject he was to make his own. His most famous treatment of it is the central panel of his masterpiece, the altarpiece for the hospital church of the Anthonite Abbey at Isenheim in Alsace, completed in about 1515 and now in the Musée d'Unterlinden, Colmar. The hospital at Isenheim cared particularly for plague victims, and the concentration on Christ's appalling physical agonies, his body gruesomely mangled and torn, was designed to bolster the faith of the sick by reminding them that he too had suffered horribly before triumphing over death. In the *Resurrection*, Christ displays his nail and lance wounds, but the lacerations that cover his body in the *Crucifixion* have disappeared, affirming that the patients at the hospital could be cleansed of their diseases and sins. The altarpiece is marked by extreme emotional intensity, brought about by expressive distortion and by colouring of an extraordinary incandescent beauty. Grünewald was familiar with Renaissance ideas of *perspective, but spiritually he belongs to the late medieval world. His other work includes Crucifixions in Basle (Öffentliche Kunstsammlung), Karlsruhe (Staatliche Kunsthalle), and Washington (NG), and several drawings survive.

The end of Grünewald's career was marked by a decline in his fortunes. He had Protestant sympathies, and following the Peasants' War in 1525, in which Archbishop Albrecht narrowly escaped death, he was dismissed from his court post. He moved to Frankfurt, where he made a meagre living at a variety of jobs, including selling artist's colours and a curative balm, the latter presumably something he had learnt about at Isenheim. In 1527 he became convinced his life was in danger and fled to Halle, where he died of plague the following year. His effects included 'much Lutheran trash'. Grünewald's influence can be seen in the paintings of contemporaries such as *Baldung Grien and *Ratgeb, but it was not until the advent of *Expressionism in the early 20th cent. that his work started to arouse widespread interest and he began his rise to his present pinnacle of esteem as one of the most awe-inspiring artists of his, or any other, time.

Guardi, Francesco (1712–93). Venetian painter, the best-known member of a family of artists. He is now famous for his views of Venice, indeed next to *Canaletto he is the most celebrated view-painter (see VEDUTA) of the 18th cent., but he produced work on a great variety of subjects and seems to have concentrated on views only after the death of his brother **Gianantonio** (1699–1760). Until then Francesco's personality was largely submerged in the family studio, of which Gianantonio was head and which handled commissions of every kind. Francesco's career was unsuccessful in worldly terms; he was still working for other artists when he was over 40, he never attracted the attention of foreign visitors in the way Canaletto did, and he died in poverty. He often borrowed his compositions from the work of other artists, and but for the quality of his painting would be described as a journeyman. Recognition of his genius came in the wake of *Impressionism, when his vibrant and rapidly painted views were seen as having qualities of spontaneity, bravura, and atmosphere lacking in Canaletto's sharply defined and deliberate works. Francesco was enormously prolific and his work is in many public collections in Italy, Britain, and elsewhere.

The major problem in Guardi studies concerns the authorship of paintings representing *The Story of Tobit* that decorate the organ loft of S. Raffaele in Venice. Critical opinion is sharply divided as to whether these brilliant works, painted with brushwork of breathtaking freedom, are by Francesco or Gianantonio (there is dispute also over the dating), but if they are indeed by the latter, he too must rank as a major figure.

Giambattista *Tiepolo was married to the sister of the Guardi brothers, and it was possibly through his influence that Gianantonio became a founder member of the Venetian Academy in 1756. Francesco was not elected until 1784, during the presidency of his nephew Giandomenico Tiepolo.

Guercino, Il (Giovanni Francesco Barbieri) (1591–1666). One of the outstanding Italian painters of the 17th cent., known as Il Guercino ('Squinting One') on account of an eye defect. He was born at Cento near Ferrara and his early work drew on a variety of north Italian sources, notably Lodovico *Carracci and Venetian painting, to create a highly individual style characterized by dramatic and capricious lighting, strong colour, and broad, vigorous brushwork. In 1621 Guercino was summoned to Rome by the Bolognese pope Gregory XV and among other commissions painted the celebrated ceiling fresco of *Aurora* in the Casino of the Villa Ludovisi for Gregory's nephew. This exuberant work, with its illusionistic architectural framework designed by Agostino *Tassi, is much more *Baroque in style than Guido *Reni's treatment of the subject of a decade earlier. Guercino returned to Cento in 1623 on the death of the pope, but his short stay in Rome introduced a more classical feeling to his work. This trend became more pronounced when he moved to Bologna in 1642 to take over the studio of Reni, who died in that year. For the next quarter of a century, until his own death, he was Bologna's leading painter, and his late works can be remarkably similar to Reni's, calm and light in colouring, with little of the lively movement of his early style (*St Luke Displaying a Painting of the Virgin*, Nelson–Atkins Mus., Kansas City, 1652). Guercino was one of the most brilliant draughtsmen of his age; the finest collection of his drawings is in the Royal Library at Windsor Castle.

Guérin, Pierre-Narcisse (1774–1833). One of the most successful French painters of his period. He won the *Prix de Rome in 1797, and his later successes included becoming director of the French *Academy in Rome in 1822, and being created a baron in 1829. His style was derived mainly from *David, but his scenes from classical history and mythology are less severe and more stagey. As the teacher of *Delacroix and *Géricault amongst others he was an important figure in the transition from *Neoclassicism to *Romanticism. He laid particular emphasis on the painted sketch and as a professor at the Ecole des *Beaux-Arts, was instrumental in establishing a sketch competition as a preliminary to the Prix de Rome.

Guggenheim, Solomon R. (1861–1949). American industrialist, collector, and philanthropist, a member of a famous family of financiers whose fortunes were based on the mining and smelting of metals. Like other members of his family, he devoted much of his vast wealth to philanthropy and in 1937 he founded the Solomon R. Guggenheim Foundation 'for the promotion and encouragement of art and education in art'. In 1943 he commissioned Frank Lloyd *Wright to design a museum in New York City to house his collection, and the Solomon R. Guggenheim Museum was opened in 1959, a decade after the founder's death. It is renowned not only for the outstanding collection of late 19th-cent. and 20th-cent. art it contains, but also for the radical nature of the architecture, which marks a complete departure from traditional museum design; the exhibition space is a continuous spiral ramp, six 'storeys' high, encircling an open central space. It is architecturally exhilarating, but its suitability for displaying paintings and sculptures has been much questioned. Guggenheim's niece, **Peggy Guggenheim** (1898–1979), was a noted patron, collector, and dealer, who played an important role in promoting avant-garde art, in particular by helping to introduce *Surrealism to the USA and by furthering the career of many leading *Abstract Expressionists. She spent much of her life in Europe, but during its brief existence (1942–6) her Art of this Century gallery in New York was the main showcase for Abstract Expressionism in its formative period. In 1941 she married and divorced Max *Ernst. Her own superb collection is open to the public in Venice under the administration of the Solomon R. Guggenheim Foundation.

Guglielmo della Porta (d. 1577). North Italian sculptor, who worked first in Genoa and then (from 1537) in Rome, where he succeeded *Sebastiano del Piombo at the Papal Mint (1547). Guglielmo had a prolific and varied career, his work including several papal busts and tombs in various Roman churches, the most important being that of Paul III in St Peter's (1549–75). He also produced numerous small devotional and pagan statuettes and was known as a restorer and copier of *antique works (both activities typical of his age). The major influence on his style was *Michelangelo and he had a penchant for reclining figures in the manner of the master's *Day* and *Night*, *Dawn* and *Evening* in the Medici Chapel, Florence.

Guido da Siena. Sienese painter active during the 13th cent. Nothing is known of his life, and his only certain work is a *Madonna and Child* in Siena Town Hall. The picture bears the date 1221, but this has been the subject of much controversy as stylistically the painting seems to be about half a century later. It has been suggested that the inscription may have some commemorative purpose, the significance of which is now lost, rather than being a record of the date of execution. Although the painting is majestic in effect and follows *Byzantine conventions of *iconography, the figures are more natural in posture, to some extent relaxing the stiff linear patterns which had been conventional in central Italian painting up to that time. The throne too is set in a deeper picture space, which adds to the realism of the figures.

On the basis of this picture a number of other panels, most of which are in the Siena Pinacoteca, have been assigned to Guido or his school. Despite his great obscurity, he is regarded as sharing with *Coppo di Marcovaldo the honour of founding the Sienese School.

Guillaumin, Armand (1841–1927). French landscape painter, one of the minor figures of the *Impressionist group. Lack of success made him take a post with the department of bridges and causeways until he won a lottery in 1891 and was able to devote all his time to painting. Often his paintings are of industrial subjects, but he also painted seascapes. His style was bold and direct, often brilliantly coloured. He was the last survivor of those who exhibited in the first *Impressionist exhibition in 1874.

guilloche. A mode of ornament in the form of two or more bands interlaced or plaited over each other so as to repeat the same figure in a continued series. Guilloche, also called 'interlacement band', has been very widely used in architecture, textiles, pottery, for the decoration of manuscripts, picture frames, mouldings, etc. It is found in most styles and periods, though more common or varied in some than in others. A very great variety and elaboration of the guilloche were developed at the *Renaissance, based chiefly on elements taken from medieval, Moorish, and *classical models.

Guimard, Hector (1867–1942). French architect and designer, considered the most eminent French exponent of *Art Nouveau. He is most widely known for his entrances to Paris Métro stations, in which he used cast iron in imaginative plant-like forms.

Gully, John (1819–88). British-born painter who migrated to New Zealand in 1852. The scenic beauties there became his exclusive preoccupation throughout a prolific career. His labours earned him in his day the title of the 'New Zealand *Turner', a sobriquet that indicates the source of his inspiration rather than the level of his achievement.

gum. A sticky liquid exuded by certain trees and shrubs, various types of which have been used as painting *media from ancient times. Gum is the normal medium of *watercolour paints and *pastel, and since it readily emulsifies with oil it has long been a medium in *tempera. Gum arabic, obtained from a species of acacia, the best from the Sudan and Senegal, is the variety most favoured.

Gunn, Sir James (1893–1964). British portrait painter. Gunn enjoyed a successful career with portraits of eminent soldiers, academics, judges, and so on, painted in a solid, forthright, traditional style. He was a more interesting painter in less traditional work, notably his portrait of the blind composer Delius (City Art Gallery, Bradford), which was the public's choice as 'Picture of the Year' at the *Royal Academy in 1933. Also well known is Gunn's *Conversation Piece at the Royal Lodge, Windsor, 1950* (NPG, London), showing George VI, Queen Elizabeth (the Queen Mother), and Princesses Elizabeth (later Elizabeth II) and Margaret. Gunn also painted landscapes, but rarely exhibited them.

Günther, Ignaz (1725–75). Bavarian sculptor. After a varied training culminating in some years at the Vienna Academy he settled in Munich in 1754. His short career was productive of a considerable quantity of wood-carving combining a very elegant *Rococo style with a highly emotional religious content. In 1759–62 he produced his chief work, the almost entire furnishing of the church at Rott-am-Inn.

Guston, Philip (1913–80). American painter. After travelling in Mexico in 1934, studying the work of *Orozco and *Rivera in particular, he settled in New York and from 1934 to 1941 worked as a muralist on the *Federal Art Project. In 1941 he moved to Iowa City to teach at the State University there, and from 1945 to 1947 he was artist-in-residence at Washington University, St Louis. After leaving New York he switched from mural to easel painting, and during the 1940s his work changed in another fundamental way, moving from social and political subjects to abstraction; by 1950 (when, after travels in Europe, he settled in New York again) he had eliminated all figurative elements from his work. His most characteristic paintings feature luminous patches of overlapping colours delicately brushed in the central area of a canvas of light background (*Dial*, Whitney Museum, New York, 1956). This manner of his has been described as 'Abstract Impressionism' and he was associated with the more lyrical wing of *Abstract Expressionism—he was the only member of the group who had already had a successful career as a figurative painter. During the 1960s shades of grey encroached on the earlier brilliance of colour and vague naturalistic associations crept in, until in the 1970s he returned to figurative painting in a satirical, garishly coloured, cartoon-like style that has been seen as the source of *New Image Painting. His subjects in this manner included scenes of fantastic social comment, involving, for example, the Ku Klux Klan.

Gutai Group. A group of Japanese artists founded at Osaka in 1954 by Jiro *Yoshihara and fifteen other artists. The members of the group were chiefly known for their *happenings (very early examples of the type) in the spirit of *Dada. They also produced abstract paintings, mainly in an *Abstract Expressionist style. The group broke up following Yoshihara's death in 1972.

Gütersloh, Albert Paris. See FANTASTIC REALISM.

Gutfreund, Otto (1889–1927). Czech sculptor. After training in Prague, he worked with *Bourdelle in Paris, 1909–10, and was attracted by *Cubism. He was among the first to apply the principles of Cubism to sculpture and on his return to Prague in 1911 he formed one of a small group of avant-garde artists attempting a fusion of Cubism with *Expressionism. An example of his work from this time is *Cubist Bust* (Tate Gallery, London, 1912–13). After the First World War he developed a more popular and naturalistic style based upon *folk art. He committed suicide by drowning.

Guttoso, Renato (1912–87). Italian painter. He was a forceful personality and Italy's leading 20th-cent. exponent of *Social Realism; he never subordinated artistic quality to political propaganda, but his art was often the direct expression of his hatred of injustice and the abuse of power. In 1931 he abandoned legal studies for painting, in which he was mainly self-taught. He settled in Rome in 1937 and in the following year became a founder member of the anti-Fascist association *Corrente. Fascism was not his only target, however, for he also pilloried the Mafia and in 1943 published a series of drawings protesting against the massacres that took place under the German Occupation of Italy. After the war (in which he worked with the Resistance) he became a member of the *Fronte Nuovo delle Arti in 1946. His post-war works were often inspired by the struggles of the Sicilian peasantry, and his other subjects included the 1968 student riots in Paris, a city he often visited. Many of his paintings were large, with allegorical overtones, typically painted in a vigorous *Expressionist style.

Guys, Constantin (1805–92). French illustrator. Little is known about his life. He was a soldier as a young man and travelled widely, leading a vagabond life. According to *Baudelaire, who immortalized him as 'The Painter of Modern Life' in his celebrated essay of that name (1863), Guys began to draw without instruction in 1847—but this is probably putting it too late. In 1854 he went to the Crimean War (1853–6) as Special Correspondent of *The Illustrated London News*, for which he had been working in London from 1848. He is most remembered, however, for his pictorial record of Paris life during the Second Empire in witty and lively drawings reinforced by thin washes of tone or colour, depicting all facets from the elegance of the court to the demi-monde. His talent was recognized by *Manet (with whom he had much in common) and *Daumier among others.

Gwathmey, Robert (1903–88). American painter. His favourite subject was rural black workers in the southern states, and his paintings often had strong social implications, castigating poverty and oppression with occasional satire. But he did not paint realistically, converting his subjects instead into strongly patterned designs of flat unmodulated colour with boldly incisive lines.

Haacke, Hans (1936–). West German experimental artist, active mainly in the USA. His work has been much concerned with movement, light, and the reaction of objects with their environment. In the early 1960s, for example, he began making works in which fluids are enclosed in plastic containers in such a way that they can be seen responding to changes in temperature, etc. Wind and water played a large role in his work and he participated in the 'Air Art' exhibition that toured the USA in 1968. From about the mid-1960s he became more interested in spectator participation and the social function of art.

Haase, Jacob de. See CERQUOZZI.

Hackaert, Jan (*c.*1628–99?). Dutch landscape painter. Little is known of his life, but he travelled extensively in Switzerland and Italy in the 1650s and is best known for Italian scenes. The finest is generally regarded as being *Lake Trasimene* (Rijksmuseum, Amsterdam), which shows how well he could capture the golden sunlight of Umbria. The figures in his paintings are often the work of other artists, notably *Berchem and Adriaen van de *Velde.

Hackert, Jakob Philipp (1737–1807). German landscape painter, active in Italy from 1768. In 1786 he became court painter to Ferdinand IV (1751–1825) of Naples. He was a sensitive upholder of the *ideal landscape tradition of *Claude, which he seasoned with touches of *Romanticism. Much of his prolific output was devoted to views of famous sites, which were eagerly sought by foreign visitors to Italy. In 1777 he made a tour of Sicily with the English collectors and connoisseurs Richard Payne *Knight and Charles Gore. He came from a family of artists and often collaborated with his brother **Johann Gottlieb Hackert** (1744–73). *Goethe met Hackert in 1787 and wrote his biography in 1811. His work is exceptionally well represented at Attingham Park in Shropshire.

Hadfield, Maria. See COSWAY.

Haggadah (Hebrew: 'telling'). A Jewish book containing the Exodus narrative ritually recited at Passover, the only Hebrew book with a long and consistent tradition of illustration. Among manuscript Haggadahs the most famous are those of Sarajevo (Spanish, 14th cent.) and Darmstadt (German, 15th cent.). The printed editions of Prague (1526), Mantua (1560, 1568), Venice (1609), and Amsterdam (1695) are also artistically noteworthy.

Hagnower, Niclas (early 16th cent.). German wood-carver, documented in Strasburg from 1493 to 1526. A good deal of sculpture has been associated with his name, but on insecure grounds. The most famous work attributed to him is the carving of *Grünewald's Isenheim Altar.

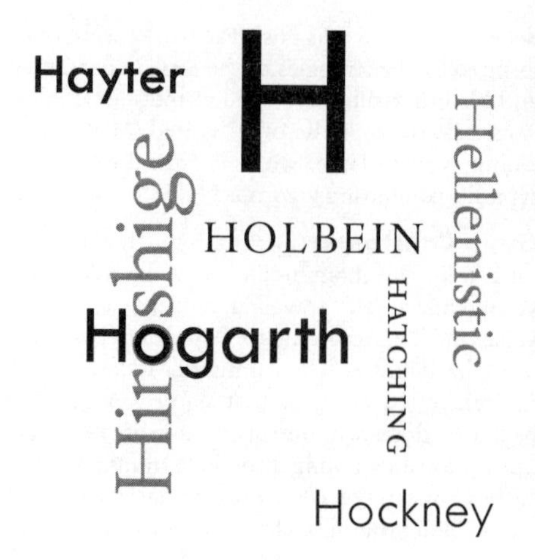

Hague School. Group of Dutch artists who worked in The Hague between about 1860 and 1900. The group is particularly associated with landscapes and beach scenes, but the members also painted street scenes, views of everyday life, and church interiors. In some ways this was a *Romantic revival of the 17th-cent. tradition, and this romantically nostalgic strain—particularly in pictures made during the first years the group worked together—is one of the things which distinguishes them from their French counterparts, the painters of the *Barbizon School and the *Impressionists. They shared with the great Dutch landscapists of the 17th cent. a special sensitivity in recording light and atmospheric effects. Leading members of the group included *Bosboom, *Israëls, the *Maris brothers, *Mauve, *Mesdag, and *Weissebruch. Their works are well represented at the Municipal Museum and the Mesdag Museum at The Hague, and at the Rijksmuseum in Amsterdam.

Hains, Raymond. See AFFICHISTE.

Hall, Peter Adolf (1739–93). Swedish portrait *miniaturist, who went to Paris in 1766 and soon made a reputation, being elected member of the Académie in 1769. He used a spirited technique with comparatively broad brush-strokes—quite rare in the field of miniature painting. Fine specimens of his work may be seen at the National Museum, Stockholm, and in the Wallace Collection, London.

Hals, Frans (1582/3–1666). Dutch painter. He was born in Antwerp of Flemish parents who moved to Holland after the city fell to the

Spaniards in 1585. His parents had settled in Haarlem by 1591 and he spent his long life there. Little is known about his life and character. He was twice married, had at least ten children, and was constantly in financial trouble. *Houbraken says he was 'filled to the gills every evening', but there is no real foundation for the popular image of him as a drunken wife-beater. His second wife, however, was more than once in trouble for brawling. During his last years he was destitute and the municipal authorities awarded him a small annual stipend four years before his death.

Hals was the first great artist of the 17th-cent. Dutch school and is regarded as one of the most brilliant of all portraitists. Almost all his works are portraits and even those that are not (some *genre scenes, and an occasional religious picture) are portrait-like in character. He is said to have been taught in Haarlem by Karel van *Mander, but there is no discernible influence from him in Hals's early works, which are not numerous or well documented. The earliest extant picture is the fragment of a portrait *Jacobus Zaffius* (Hals Mus., Haarlem, 1611), and upon the basis of stylistic evidence one or two paintings can be dated a year or two earlier (*Portrait of a Man holding a Skull*, Barber Institute, Birmingham). Nothing he did before 1616 suggested that he would shatter well-established traditions with his life-size group portrait *The Banquet of the Officers of the St George Militia Company* (Hals Mus.) painted during that year. There is no precedent in either his own work or that of his predecessors for the vigorous composition and characterization of this picture, which has become a symbol of the strength and healthy optimism of the men who established the new Dutch Republic. It demonstrates to the full his remarkable ability—his greatest gift as a portraitist—to capture a sense of fleeting movement and expression and thereby convey a compelling feeling of vivacity.

From 1616 onwards there is no shortage of dated or documented works and his artistic development is clear. He was at the height of his popularity during the 1620s and 1630s. During these decades he made five large group portraits of civic guards; one (finished by Pieter *Codde) is in the Rijksmuseum and the others are in the Frans Hals Museum, Haarlem. The latter gallery, which also contains the 1616 guard piece and Frans's three group portraits of regents, is the only place where one can get a comprehensive view of his full range and power. During the 1630s he painted pictures of greater simplicity, and monochromatic effects took the place of the bright colours of the earlier works (*Lucas de Clercq* and *Feyntje van Steenkiste*, Rijksmuseum, Amsterdam, 1635). The group portrait of the *Regents of the St Elizabeth Hospital* (Hals Mus., 1641) sets the key for the sober restraint of the late period, when his pictures became darker and his brush-strokes more economical. The culmination of this phase—indeed of his entire career—are his group portraits of the *Regents* and the *Regentesses of the Old Men's Alms House* (Hals Mus., *c*.1664). These two pictures rank among the most moving portraits ever painted.

Almost all Frans's genre pictures were done during the first half of his career. It is difficult to establish a precise chronology for them because very few are dated. It is also hard to correlate them with dated portraits because they are always painted more freely than his commissioned portraits—an excellent example of a *Baroque painter following the ancient idea of adjusting his style to the subject he depicted. Only during the last decades of his life did Hals use in his commissioned portraits the bold brushwork and the **alla prima* technique which he had previously reserved for genre pictures. No drawings by him are known and he presumably worked straight on to the canvas.

Hals had two painter brothers and five painter sons, but the only artist of substance among them was his brother **Dirk** (1591–1656), who painted charming small interior scenes. Apart from his sons, Hals taught numerous pupils, including (certainly or probably) Judith *Leyster, Jan Miense *Molenaer, Adriaen van *Ostade, Adriaen *Brouwer, and Philips *Wouwerman. His reputation did not long outlive him, however, and with rare exceptions—*Reynolds was one of them—few critics before 1850 applauded Frans's work. It was only in the second half of the 19th cent. that there was a renewed appreciation of his genius. The spontaneity of his work appealed to the generation of the *Impressionists, and from about 1870 to about 1920 he was one of the most popular of the Old Masters, becoming a model for society portraitists. Lord Hertford's purchase of his most famous work, *The Laughing Cavalier* (Wallace Coll., London, 1624), for the then enormous sum of 51,000 francs in 1865, was a milestone in the revival of his fortunes, and the buoyant confidence of his work later made him a particular favourite with the new generation of fabulously rich American collectors—self-made men—who were beginning to dominate the picture market. This explains why so many works by him are in American collections.

Hamel, Théophile (1817–70). Canadian painter of portraits and religious subjects, active mainly in Quebec. He was the most distinguished pupil of *Plamondon and his earliest portraits combine the latter's *classicism with the simplicity of *folk art

(*Léocadie Bilodeau*, Université Laval, Quebec, 1842). Later he went to Europe (1843–6) and was influenced by the *Romantic painters.

Hamilton, Gavin (1723–98). Scottish painter, archaeologist, and picture-dealer, active mainly in Italy. From 1748 he lived mainly in Rome, where he was a leading member of the *Neoclassical circle of *Mengs and *Winckelmann. His archaeological excavations near Rome resulted in many important additions to contemporary collections, and his interest in antiquity exerted a decisive influence on the young *Canova. Hamilton's history paintings, mostly of Homeric subjects, were influenced by *Poussin as well as by the *antique (*Achilles Lamenting the Death of Patroclus*, NG, Edinburgh, 1763). They were never very numerous and today are generally regarded as rather tepid, but they became well known through engravings, and greatly influenced the development of the Neoclassical style amongst both his contemporaries and the younger generation, including *David. Indeed, Hamilton was much better known on the Continent than in Britain, where his name was more familiar for his activities in selling Old Masters and classical antiquities (he was one of the wealthiest artists in Rome). Together with *Barry, and the Anglo-Americans *West and *Copley, he is one of the few painters to have made a significant contribution to *history painting in Britain.

Gavin Hamilton is not to be confused with **Gawen Hamilton** (1697?–1737), a minor portrait painter remembered for his *Conversation of Virtuosis . . . at the Kings Armes* (NPG, London, 1735), which shows himself and several other artists of the day.

Hamilton, Richard (1922–). British painter, printmaker, teacher, and writer, one of the leading pioneers of *Pop art. As a young man he worked in advertising and commercial art and he is best known for his montages featuring scenes from the fields of advertisement and contemporary life, notably *Just what is it that makes today's homes so different, so appealing?* (Kunsthalle, Tübingen, 1956). This photomontage, which was displayed blown-up to life-size at the entrance to the exhibition 'This is Tomorrow' held at the Whitechapel Art Gallery in London in 1956, is sometimes considered to be the first Pop art work. Hamilton has had an influence on, among others, Peter *Blake and David *Hockney. He has had a distinguished career as a teacher, notably at King's College, Newcastle upon Tyne (which later became Newcastle University), 1953–66, and has organized several exhibitions, including 'The Almost Complete Works of Marcel *Duchamp' at the Tate Gallery, London, in 1966. An anthology of his writings, *Collected Works*, appeared in 1982. See also INDEPENDENT GROUP.

Hammershøi, Vilhelm (1864–1916). Danish painter, active mainly in his native Copenhagen. He painted portraits, architectural subjects (including two murals for Copenhagen Town Hall), and landscapes, but is best known for his quiet interior scenes. They are painted in muted colours, and have a certain affinity with *Vermeer, often featuring a single standing or seated figure. Two of his interiors are in the Tate Gallery, London.

Hanneman, Adriaen (*c.*1601–71). Dutch portrait painter, a pupil of Anthony van *Ravesteyn. From *c.*1623 to 1637 he was in England, where he was influenced particularly by van *Dyck, and on his return to Holland, he achieved great success among the aristocracy of The Hague (his native city) with his elegant and polished portraits. Good examples of his work are *Constantin Huygens and His Five Children* (Mauritshuis, The Hague, 1640) and a self-portrait (1656) in the Rijksmuseum, Amsterdam.

Hanson, Duane (1925–). American sculptor. Hanson is probably the best-known exponent of *Superrealism in sculpture, producing minutely detailed fibreglass resin figures dressed in real clothes and accompanied by real props. He concentrates pungently on the depressing or tasteless aspects of everyday American life—down-and-outs, exhausted shoppers, or, in one of his most famous works, a pair of fat, ageing, and garishly dressed sightseers (*Tourists*, NG of Modern Art, Edinburgh, 1970). 'The subject matter that I like best', he has written, 'deals with the familiar lower and middle class American types of today. To me, the resignation, emptiness and loneliness of their existence captures the true reality of life for these people. Consequently, as a realist I'm not interested in the human form . . . but rather a face or body which has suffered like some weather-worn landscape the erosion of time. In portraying this aspect of life I want to achieve a certain tough realism which speaks of the fascinating idiosyncracies of our time.'

happening. A form of entertainment, often carefully planned but usually including some degree of spontaneity, in which an artist performs or directs an event combining elements of theatre and the visual arts. The term was coined by Allan *Kaprow in 1959 and has been used to cover a diversity of contrived artistic phenomena. The concept of the happening was closely bound up with Kaprow's deliberate rejection of the traditional principles of craftsmanship and permanence in the arts. He thought of the happening as a development mainly from the *assemblage and the *environment. While both the assemblage and the environment were relatively fixed and static—the assemblage something constructed to be contemplated from outside, to be 'handled or walked around', and the environment something to be 'walked into', something by which the observer was enveloped and manipulated—the happening was conceived by contrast as a genuine 'event'. It had close affinities with theatrical and *Performance art, and it was not restricted like the environment to the confines of a gallery or some other site. In conformity with the theories of the composer John Cage (1912–) about the importance of chance in artistic creation, happenings were described as 'spontaneous, plotless theatrical events'.

In America the artists chiefly responsible for the development of the happening in its early stages included, besides Cage and Kaprow, Jim *Dine, Claes *Oldenburg, Robert *Rauschenberg, and Roy *Lichtenstein. The idea of the happening was linked with the principle of spectator participation, which had taken firm hold not only in America but also in Europe and Japan. Outside America the happening was widely exploited, by the *Gutai group in Japan and many others in Europe. Although the very notion of the happening involved the emergence of the artist from the rarefied confines of the galleries and museums into the streets and the market-place, the term was often used to cover staged demonstrations for politico-social propaganda, as for example many of the happenings by Joseph *Beuys, or demonstrations intended to shock established moralities. At the other pole it was held to be characteristic of the happening that it should bring into being situations or events in which the elements of everyday life and everyday technology are invested with the strangeness of the poetic and the fantastic. Bazon Brock, Professor of Non-Normative Aesthetics in Hamburg, described the happening as 'an instrument for the production of contradictions' and the theory of the happening is as diverse as the practice.

Hard Edge painting. Term applied to a type of painting (predominantly abstract) in which forms, although not necessarily geometrical, have sharp contours and are executed in flat colours. The term was coined by the American critic Jules Langsner in 1958, and although it can be retrospectively applied to such styles as *Purism, it is used mainly of the type of painting that emerged as a reaction to the spontaneity and painterly handling of *Abstract Expressionism. Major exponents of Hard Edge painting have included Ellsworth *Kelly and Kenneth *Noland. See also POST-PAINTERLY ABSTRACTION.

Hare, David (1917–91). American sculptor. He began to work as a sculptor in 1942 and before then had been an experimental photographer. His most characteristic works are in welded metal and have been regarded as a sculptural analogue of *Abstract Expressionism in painting. From 1940 Hare edited the *Surrealist magazine *VVV*, and his sculpture was marked by Surrealist and *Dadaist elements, particularly an interest in visual puns.

Haring, Keith. See GRAFFITI ART.

Harnett, William Michael (1848–92). American still-life painter. He was born in Ireland and brought to America as a child, living mainly in Philadelphia until a stay in Europe in 1880–6, and thereafter in New York. He specialized in elaborate **trompe l'œil* compositions, often involving firearms or musical instruments (*After the Hunt*, California Palace of the Legion of Honor, San Francisco, 1885). Although his works were long popular with the public, they were generally dismissed by critics as mere sleight of hand until about 1945, when they began to win favour for their strength of composition.

Harpignies, Henri (1819–1916). French landscape painter and engraver. He did not take up painting until 1846, but he was then very prolific and won considerable success and fame. He is sometimes classed with the *Barbizon School, but his work shows more specifically the influence of *Corot. His work is represented in many public collections in France, Britain, and elsewhere.

Harris, Lawren Stewart (1885–1970). Canadian painter, active mainly in Toronto. From 1904 to 1908 he studied in Berlin, and his work became imbued with bold *Expressionist colour. Until 1920 views of houses and cityscapes were his main interest, but after meeting J. E. H. *Macdonald in 1911 he turned as well to landscape and became one of the founder members of the *Group of Seven. In 1918 Harris had discovered Algoma, in northern Ontario, and the rich lushness of the countryside was suited to the dramatic and colourful style he had developed (*Autumn, Algoma*, Victoria University, Toronto, 1920). Later he sought out more spectacular scenery in the Rockies and even in the Arctic, his work expressing a desire for spiritual fulfilment through immersion in the vital forces of overpowering landscape. This transcendental quality was maintained when Harris turned to abstraction in the 1930s.

Hartley, Marsden (1877–1943). American painter, whose work represented an advanced response to modernist trends in European art. In 1912, help from *Stieglitz (who had given him his first one-man exhibition in 1909) enabled him to travel to Europe but finding *Cubist Paris little to his taste he went to Munich and Berlin, where the *Expressionism of *Kandinsky and *Jawlensky proved more congenial, and he exhibited with the *Blaue Reiter in Berlin in 1913. Later that year he returned to America and exhibited at the *Armory Show. From 1914 to 1916 he was again in Europe, visiting London, Paris, Berlin, and Munich. During these years he painted in an abstract manner with Expressionist overtones, as in the famous *Painting No. 5* (Whitney Mus., New York, 1914–15), a picture painted in rich Expressionist colours and incorporating military emblems and decorations of wartime Germany. The picture represents a remarkable personal synthesis, being more closely structured and objective than German Expressionism yet more freely patterned and highly coloured than French Cubism. Instead of continuing on this path, however, on his return to the USA in 1916, he began to paint near-*Constructivist abstracts in clear pastel hues. In 1917 he abandoned abstraction and from 1918 to 1920 did dramatic pastels of the New Mexico landscape, upon which in the early 1920s he based equally dramatic though more formalized oils. In the 1920s and early 1930s he led a wandering, unsettled life, visiting France, Italy, Germany, and New Mexico. He settled in Maine in 1934 and his late works consisted mainly of rugged mountain and coastal scenes. Hartley was a lonely, reclusive, rather haunted character whose work reveals an intense feeling for the beauty of nature.

Hartung, Hans (1904–89). German-born painter who became a French citizen in 1946. Born in Leipzig, he studied in the Academies of Leipzig, Dresden, and Munich, where he also studied the philosophy and history of art. In Munich he met *Kandinsky and became interested in the work of *Marc. He began painting abstracts in 1922 and had his first exhibition at Dresden in 1931. In 1935 he fled from Germany and settled in Paris. During the war he served in the French Foreign Legion and was seriously wounded. After the war he returned to Paris, and developed a highly original and vibrant style of abstract painting in which thick black lines and blotches predominate in a manner superficially analogous to calligraphic scribbling. They won him a reputation as one of the main precursors of *Art Informel and of *Tachisme.

Hassam, Childe (1859–1935). American painter and printmaker, one of the foremost American exponents of *Impressionism. He discovered Impressionism on his second trip to Europe in 1886–9, when he studied in Paris. On his return to the USA he settled in New York and the life of the city became his favourite subject-matter. His work is fresh and clear but sometimes rather saccharine. See also TEN, THE.

hatching. The use of finely spaced parallel lines to suggest shading. The technique is found mainly in drawing and engraving, but is also used, for example, in *tempera painting. When crossing sets of lines are used, the term crosshatching is applicable.

Hausmann, Raoul. See DADA.

Haworth, Jann. See BLAKE, PETER.

Haydon, Benjamin Robert (1786–1846). English painter, who aspired to bring a new seriousness to British art through historical and religious work in the *Grand Manner preached by *Reynolds. His life was a story of bombastic frustration and intransigent opposition to the establishment (particularly the *Royal Academy), fighting continuously for personal recognition and arguing for the social purpose of art (his lectures in the provinces anticipated those of *Ruskin and *Morris). As a painter, however, his talents fell far short of his ambitions, his multi-figure compositions degenerating into turgid melodrama. His great monument, rather, is the massive collection of autobiographical writings he left behind him (various editions have been published), which gives fascinating insights into the contemporary artistic scene and paints a vividly detailed picture of his disturbed mind and tragi-comical life. He was closely linked with the *Romantic movement in literature, particularly with William Wordsworth (1770–1850), who wrote a sonnet to him, and with John Keats (1795–1821), doing portraits of both of them (NPG, London), and but for his lack of talent he would exemplify all the traits traditionally ascribed to the Romantic concept of genius. In true Romantic fashion his death came by suicide.

Hayez, Francesco (1791–1882). Italian painter, active mainly in Milan. Hayez was the most important figure in the transition from *Neoclassicism to *Romanticism in Italian painting, but his Romantic leanings come out mainly in subject-matter rather than in technique, the clear outlines he favoured revealing his training in Rome in the circle of *Canova and *Ingres. He painted religious, historical, and mythological works in a vein owing something to *Delacroix and *Delaroche, and portraits that are sometimes thought worthy of comparison with those of Ingres. Many of the most eminent Italians of the day sat for him. For many years he taught at the Brera in Milan (he became Director in 1860) and he exercised great influence on his pupils. The Brera has an outstanding collection of his work.

Hayman, Francis (1707/8–76). English painter and book illustrator. He was the most versatile British artist of his period, painting portraits, subjects from literature and the theatre (notably Shakespeare), and scenes of rural folklore; he also had the reputation of being 'unquestionably the best historical painter in the kingdom before the arrival of *Cipriani' (Edward *Edwards, *Anecdotes of Painters . . .*, 1808), but little of his decorative work survives. In addition he was a prolific designer of book illustrations, sometimes collaborating with *Gravelot. His largest undertaking was the painting of decorations for the boxes and pavilions at Vauxhall Gardens, the fashionable London pleasure resort, of which two are now in the Victoria and Albert Museum. His *conversation pieces anticipated those of the early *Gainsborough, who almost certainly worked with him in Gravelot's studio. Hayman was President of the *Society of Artists, 1760–8, and became a foundation member of the *Royal Academy in 1768 and its librarian in 1771. His best work has a certain *Rococo charm, but there is some justification in Horace *Walpole's comment that his paintings are 'easily distinguishable by the large noses and shambling legs of his figures'.

Hayter, Sir George (1792–1871). English historical and portrait painter. The son of a *miniaturist, **Charles Hayter** (1761–1835), he studied at the *Royal Academy Schools and in Rome, and was appointed portrait and history painter to Queen Victoria in 1837. On the death of *Wilkie in 1841 he was made 'principal painter in ordinary to the queen'. He is known chiefly for his royal portraits (several are in the royal collection) and his huge groups (*House of Commons*, NPG, London, 1833), unexciting in their handling, but composed with dexterity and accomplished grandiloquence. In spite of his royal favour he was never a member of the Royal Academy, seemingly because, after an unfortunate early marriage, he lived with a woman who was not his wife.

Hayter, S. W. (Stanley William) (1901–88). British engraver and painter, a descendant of Sir George *Hayter. He spent most of his life in Paris, where in 1927 he founded an experimental workshop for the graphic arts—Atelier 17—that played a central role in the 20th-cent. revival of the print as an independent art form. (The name was adopted in 1933 when Hayter moved his establishment from its original home to 17 rue Campagne-Premiène.) In 1940–50 he lived in New York, taking Atelier 17 with him. Hayter was a chemist by training and had an unrivalled knowledge of the technicalities of printmaking, on which he wrote two major books, *New Ways of Gravure* (1949) and *About Prints* (1962). Although his historical importance has long been acknowledged (probably no modern British artist has been so influential internationally), it is only recently that his own work has won him belated recognition as one of the outstanding graphic artists of his time. His prints are varied in technique and style, but most characteristically are influenced by the abstract vein of *Surrealism and are notable for their experiments with texture and colour.

Hazlitt, William (1778–1830). English essayist. He is known mainly for his literary criticism, but he also wrote much on the fine arts and he ranks as the most important British writer on the subject between *Reynolds and *Ruskin. Although he studied painting and did some portraits (one of Charles Lamb, 1804, is in the National Portrait Gallery, London), he lived mainly by journalism, publishing essays in various radical journals. Hazlitt was a *Romantic critic in placing much more importance on the role of genius in artistic creation than on rules or theories. Thus he admired Reynolds's paintings, but attacked his ideas. A significant characteristic of Hazlitt's writing is that (unlike most previous art criticism) it was written for the general reader rather than for the connoisseur or practising artist.

Heaphy, Charles (*c.*1820–81). British-born painter active mainly in New Zealand. He studied at the *Royal Academy Schools, and in 1839 was appointed artist and draughtsman to the New Zealand Company. For three years he travelled through the country, then a little-known wilderness, and the paintings and sketches he made (Alexander Turnbull Lib., Wellington) form a unique record of early colonial New Zealand. Since a raw colony had no place for professional painters, Heaphy gradually abandoned art for a distinguished career as soldier, administrator, and politician. His father, **Thomas Heaphy sen.** (1775–1835), and his brother, **Thomas jun.** (1813–73), were watercolour painters, and two sisters, **Mary Ann** (Mrs Musgrave) and **Elizabeth** (Mrs Murphy) were *miniaturists.

Hearne, Thomas (1744–1817). One of the outstanding English topographical draughtsmen of the 18th cent. In 1771 he went to the Leeward Islands with the new governor Sir Ralph Payne (?1738–1807) and he sometimes portrayed West Indian subjects after his return. He is best known for his collaboration with the engraver William Byrne (1743–1805) on *The Antiquities of Great Britain* (1786), for which he made fifty-two drawings. Through Dr *Monro, who owned many of his drawings, Hearne influenced *Girtin and *Turner. His work is well represented in the British Museum and Victoria and Albert Museum.

Heartfield, John (Helmut Herzfelde) (1891–1968). German painter, graphic designer, and journalist, a leading light of *Dada in Berlin, best known as one of the pioneers and perhaps the greatest of all exponents of *photomontage. With *Grosz he Anglicized his name during the First World War as a protest against German nationalistic fervour and his finest works are brilliantly satirical attacks—often in the form of book covers and posters—against militarism and Nazism. Harassed by the Nazis he left Germany in 1938 and moved to London, where he lived until 1950. He died in Berlin, his native city.

Heath, Adrian (1920–92). British abstract painter. In 1949 and 1951 he visited St Ives, where he met Ben *Nicholson, and he formed a link between the *St Ives School and London-based *Constructivists such as Victor *Pasmore and Kenneth and Mary *Martin. During the early 1950s he was a significant figure in promoting abstract art—by organizing collective exhibitions at his London studio (at 22 Fitzroy Street) in 1951, 1952, and 1953, and by writing a short popular book on the subject, *Abstract Painting: Its Origin and Meaning* (1953), which begins with the sentence: 'There seems to be little understanding of the values of abstract painting and consequently no general appreciation of its qualities.' The exhibitions helped to inspire Lawrence *Alloway's book *Nine Abstract Artists* (1954). Heath's paintings of this time featured large, block-like slabs of colour, heavily brushed. He also made a few constructions. Later his paintings became freer and more dynamic.

Heckel, Erich (1883–1970). German painter and graphic artist, one of the founders of Die *Brücke. His work was somewhat more lyrical than that of the other members of Die Brücke and he showed a special concern for depicting sickness and inner anguish. His landscapes, too, sometimes displayed a decorative quality which was foreign to most German *Expressionism. In 1911 he settled in Berlin with *Kirchner and other members of Die Brücke. Here, as a result of contacts with *Feininger, *Macke, and Franz *Marc, the formal structure of his painting gained in strength and coherence. But his image of humanity became even more pessimistic, with harshly angular distortions, anguished expressions, and rigid, distracted gestures. The mood was reflected in his colours, which were reduced to harsh contrasts of feverish reds against strident yellows and dull blues. In the First World War Heckel served as a medical orderly in Flanders and came into contact there with *Ensor and *Beckmann, by whom he was again influenced. His landscapes became more sombre in colour, expressing the agony of war through conflict of the elements, and the melancholic and tragic mood of his work was enhanced. After 1920 his style became more conventional, losing its verve and intensity. His work was declared as *degenerate by the Nazis in 1937. In 1944 his Berlin studio was destroyed by fire and he retired to Hemmenhofen on Lake Constance. From 1949 to 1955 he taught at the Karlsruhe Academy.

Heda, Willem Claesz. (1593/4–1680/2). Dutch still-life painter, active in Haarlem. He and Pieter *Claesz. are the most important representatives of *ontbijt* (breakfast piece) painting in the Netherlands. His overall grey-green or brownish tonalities are very similar to those of Claesz., but Heda's work was usually more highly finished and his taste was more aristocratic. He showed a preference for ham, mincemeat pie, and oysters, and after 1629 never included a herring in his pictures. His son **Gerrit** (d. 1702) was his most important pupil.

Heem, Jan Davidsz. de (1606–83/4). Dutch still-life painter, active mainly in Antwerp. He was born in Utrecht and his rare early pictures are in the style of Balthasar van der *Ast, who taught him there. Later he worked in Leiden and showed that he had studied the restrained and simple works of the Haarlem still-life artists *Claesz. and *Heda. In 1636 he moved to Antwerp, became a citizen of that city in 1637, and spent most of his very productive life there. The paintings he did in Flanders are the ones for which he is most renowned and are very different in spirit from his earlier works: splendid flower pieces and large compositions of exquisitely laid tables which breathe all the opulent exuberance of Flemish *Baroque painting. His work formed a link between the Dutch and Flemish still-life traditions and he is claimed by both schools. He came from a large family of painters and his many followers in Flanders and Holland included his son **Cornelis** (1631–95).

Heemskerck, Maerten van (1498–1574). Netherlandish *Mannerist painter, born at Heemskerck and active mainly in nearby Haarlem. His principal training was with Jan van *Scorel in Utrecht, *c.*1527–9. Although Heemskerck was only three years younger than Scorel and was a mature man when he entered his studio (he had already studied with two other teachers), the experience left a distinctive mark on him. In some pictures, particularly the portraits, experts still have difficulty distinguishing their hands. As a rule, however, Heemskerck's paintings are more crowded and nervous than Scorel's balanced and harmonious compositions. Equally important for Heemskerck's development was a visit to Italy (1532–5), where he was impressed—or rather overwhelmed—by *Michelangelo. When he returned to the Netherlands he emulated Michelangelo by painting large works packed with muscle-bound figures in vigorous movement (*Christ Crowned with Thorns*, Hals Museum, Haarlem). During his stay in Rome Heemskerck made drawings of ancient and modern buildings and sculpture. Two of his Italian sketch-books are in Berlin; they are valuable historical documents as well as sensitive impressions of the marvels of Rome. His interest in antiquity also comes out in his self-portrait of 1533 (Fitzwilliam, Cambridge), in which he shows the Colosseum in the background. Heemskerck was one of the leading Netherlandish painters of the 16th cent. and his work was much engraved and highly influential.

Heidelberg School. Group of Australian painters led by Tom *Roberts who met at the painting camp at Eaglemont, Heidelberg, Victoria. The art of the school, based on open-air *Impressionist painting, also featured local subject-matter and was associated with the emergence of a distinctive Australian literature. It flourished, appropriately enough, between 1888 (the centenary of Australia) and 1901 (the foundation year of the Commonwealth). Lack of patronage at home and desire for overseas training and experience had forced most of its members to Europe by 1900. But their vision of Australian life and landscape came to dominate Australian art during the 1920s, and has found a renewed response in the work of some Australian landscape and Social *Realist painters in later decades.

Heizer, Michael. See LAND ART.

Held, Al (1928–). American painter. His early work was in the prevailing *Abstract Expressionist idiom, being particularly influenced by Jackson *Pollock. From about 1960, however, he began to develop a more individual style characterized by clean-edged, bold, brightly coloured geometrical forms. It had affinities with *Hard-Edge painting, but Held's work was distinguished by his use of very heavily textured paint. He often worked on a huge scale, giving his paintings an extremely forceful physical impact. In 1967 he began making black-and-white paintings, using white linear structures on a black ground or black lines on a white ground to create overlapping and interlocking box-like forms that demonstrate his interest in Renaissance perspective. In the 1980s he reintroduced colour with a vengeance, as in his 55-feet-long mural *Mantegna's Edge* (Southland Center, Dallas, 1983), a work of tremendous high-keyed vigour.

Helladic. A term conventionally applied to the culture of the Greek mainland during the Bronze Age from about 2900 BC to about 100 BC. Late Helladic is alternatively called *Mycenaean.

Hellenic. Term applied to the cultures of Greek-speaking societies from the beginning of the Iron Age (late 11th cent. BC) to about 323 BC. It embraces the *Geometric, *Archaic, and *Classical periods. Earlier periods in Greece are 'Prehellenic', or *Helladic, to which *Minoan and *Mycenaean art belong; the subsequent period is called *Hellenistic, although this is sometimes included within the term 'Hellenic'.

Hellenistic. A term conventionally applied to Greek culture in the late 4th to late 1st cent. BC, say from 323 BC, when Alexander the Great died, to 27 BC, when Augustus became the first Roman emperor. During this period Greece itself had lost its political importance and new centres of art and patronage arose in the Greek kingdoms of Asia Minor and Egypt; at its end Rome had extended its power over the whole Mediterranean world. Hellenistic art is more varied in inspiration than that of the *Classical age which preceded it, and the sculpture of the period is often remarkable for its technical bravura and overt display of emotion, as in the celebrated *Laocoön*, the most famous of Hellenistic works of art. After original Greek works of the Classical period became widely known in the course of the 19th cent. much Hellenistic art was generally dismissed as decadent, but it is now recognized as a rich field of study. As J. J. Pollitt has written (*Art in the Hellenistic Age*, 1986) 'Hellenistic art was not tied to a single country or ethnic group: rather, like Hellenistic culture as a whole, it was adopted and produced by diverse peoples in widely separated geographical areas. Further, it throve in a world where many of the familiar figures of the modern "art world"—private patrons, collectors, and even dealers—made their first appearance. The Hellenistic age also seems to have been the first epoch in western art in which an intense sense of "art history" influenced art itself. Systematic histories of art were first written during the period; artists revived the style of earlier centuries; sculptors' workshops began to specialize in the reproduction of "old masters"; different styles came into simultaneous use. The result of these historical conditions was an art which, like much modern art, was heterogeneous, often cosmopolitan, increasingly individualistic, and frequently elitist in its appeal.'

Helst, Bartholomeus van der (1613–70). Dutch portrait painter. He was born in Haarlem, settled in Amsterdam *c.*1636, and in the 1640s took over from *Rembrandt as the most popular portraitist in the city, his detailed, tasteful, and slightly flattering likenesses appealing more to the fashionable burghers than the master's work, which was becoming more individual and introspective. Van der Helst's influence during his lifetime was great. For example, Rembrandt's talented pupils *Bol and *Flinck abandoned the style of their master in order to follow his more popular manner. His reputation endured into the next century and in 1781 Joshua *Reynolds wrote that van der Helst's *Banquet of the Amsterdam Civic Guard in Celebration of the Peace of Munster* (Rijksmuseum, Amsterdam, 1648) 'is, perhaps, the first picture of portraits in the world', adding that it as far exceeded his expectations as Rembrandt's *Night Watch* fell below them.

Hemessen, Jan Sanders van (*c.*1500–*c.*1566). Netherlandish painter of religious and *genre scenes and portraits. The facts of his life are obscure, but in 1524 he was made a master of the Antwerp Guild. He is reputed to have moved to Haarlem *c.*1550 and to have died there. His paintings illustrating popular proverbs and religious parables, and his satirical portraits, link him with Quentin *Massys and *Marinus van Reymerswaele, and Hemessen ranks with them as one of the founders of Flemish genre painting. An example of his work is *The Prodigal Son* (Musées Royaux, Brussels, 1536).

Hendriks, Wybrand (1744–1831). Dutch painter of landscapes, topographical views, still lifes, portraits, and *genre scenes, often done in a manner recalling 17th-cent. Dutch masters. He was custodian of the Teyler Foundation (now the Teylers Museum) in Haarlem from 1786 to 1819, making some important additions to the museum's collections, and his own work is represented in the city's Frans Hals Museum.

Henri, Robert (1865–1929). American painter, teacher, and writer, a major figure in combating conservative attitudes in American art in the early 20th cent. From 1886 to 1888 he trained at the Pennsylvania Academy of the Fine Arts, Philadelphia, under Thomas Anshutz (1851–1912), who passed on the tradition of Thomas *Eakins, an artist Henri came to admire deeply. In 1888–91 he lived in Paris, studying mainly at the *Académie Julian. After returning to Philadelphia he became the leader of a circle of young artists—*Glackens, *Luks, *Shinn, *Sloan—that later became the nucleus of the *Eight and the *Ash-can School. In 1895–7 and 1898–1900 he again lived in Paris, then in 1900 settled in New York. There he became an outstanding teacher, first at the New York School of Art, 1902–9, then at his own school, 1909–12, at the Modern School of the Ferrer Center (a radical educational establishment), 1911–18, and finally at the *Art Students League, 1915–28. The essence of his teaching was that art should grow from life, not from theories. He said that he wanted his own paintings to be 'as clear and as simple and sincere as is humanly possible', and he was a powerful force in turning young American painters away from academism to look at the rich subject-matter provided by modern urban life—'regarded by many of his contemporaries as the most influential single force affecting the development of American art in the generation preceding the *Armory Show of 1913' (William Innes Homer, *Robert Henri and His Circle*, 1969). Henri was open-minded about the new developments seen at the Armory Show but he was not interested in experiment for experiment's sake and his own painting was little affected by it. His early work had been *Impressionist, but in the 1890s he adopted a darker palette, with rapid slashing brushwork geared to creating a sense of vitality and immediacy. From 1909 his work became more colourful again. Apart from scenes of urban life, he painted many portraits, and also landscapes and seascapes (which have been rather neglected). He made frequent visits to Europe and found inspiration there for figure studies of picturesque characters—Irish peasants, gypsies, and so on. His paintings are dashing but rather superficial and they are generally regarded as much less important than his teaching and crusading. Henri wrote numerous articles on art and in 1923 published *The Art Spirit*, a collection of his letters, lectures, and aphorisms, in which art is seen as an expression of love for life.

Hepworth, Dame Barbara (1903–75). English sculptor, one of the most important figures in the development of abstract art in Britain. She trained at Leeds School of Art, where she became a friend of Henry *Moore, and at the *Royal College of Art. Her early sculptures were quasi-naturalistic and had much in common with Moore's work (*Doves*, Manchester City Art Gal., 1927), but she already showed a tendency to submerge detail in simple forms, and by the early 1930s her work was entirely abstract. She worked both in wood and stone, and she described an important aspect of her early career as being 'the excitement of discovering the nature of carving'—this at a time when there was a general antagonism to 'direct carving'. In this, too, she was united with Moore, but her work, unlike his, is not representational in origin but conceived as abstract forms. Yet she consistently professed a *Romantic attitude of emotional affinity with nature, speaking of carving both as a 'biological necessity' and as an 'extension of the telluric forces which mould the landscape'.

From 1925 to 1931 Hepworth was married to the sculptor John Skeaping (1901–80). In 1931 she met Ben *Nicholson, who became her second husband a year later, and through him became aware of contemporary European developments. They joined *Abstraction-Création in 1933, and *Unit One in the same year. During the 1930s Hepworth, Nicholson, and Moore worked in close harmony and became recognized as the nucleus of the abstract movement in England. Hepworth's outlook was already clearly formed in the short introduction she wrote for the book *Unit One* in 1934: 'I do not want to make a stone horse that is trying to and cannot smell the air. How lovely is the horse's sensitive nose, the dog's moving ears and deep eyes; but to me these are not stone forms and the love of them and the emotion can only be expressed in more abstract terms. I do not want to make a machine which cannot fulfil its essential purpose; but to make exactly the right relation of masses, a living thing in stone, to express my awareness and thought of these things. . . . In the contemplation of Nature we are perpetually renewed, our sense of mystery and our imagination is kept alive, and rightly understood, it gives us the power to project into a plastic medium some universal or abstract vision of beauty.'

In 1939 Hepworth moved to St Ives in Cornwall with Nicholson and lived there for the rest of her life (see ST IVES PAINTERS). During the late 1930s and 1940s she began to concentrate on the counterplay between mass and space in sculpture. In 1931 in *Pierced Form* (destroyed in the war) she first introduced into England the use of the 'hole', and she now developed this with great subtlety, making play with the relationship between the outside and inside of a figure, the two surfaces sometimes being linked with threaded string, as in *Pelagos* (Tate, London, 1946). *Pelagos* also shows her sensitive use of painted surface to contrast with the natural grain of the wood. In all her work she displayed a deep understanding of the quality of her materials and superb standards of craftsmanship.

By the 1950s she was one of the most internationally famous of sculptors and she received many honours and prestigious public commissions, among them the memorial to Dag Hammerskjold—*Single Form*—at the United Nations in New York (1963). She now worked more in bronze, especially for large pieces, but she always retained a special feeling for direct carving. Hepworth died tragically in a fire at her studio in St Ives. The studio is now a museum dedicated to her work.

Hering, Loy (*c.*1485–*c.*1554). German sculptor, trained in Augsburg and active mainly in nearby Eichstätt. He excelled in work that was both technically and metaphorically refined and highly polished, using hone-stone as his favourite material and specializing in small figures and *reliefs. His style shows traces of a late *Gothic heritage and at the same time influence from the Italian *Renaissance, knowledge of which he was influential in spreading (he is presumed to have visited Italy early in his career). His subjects were both religious and courtly and he often borrowed motifs from *Dürer's graphic work. Among his few monumental sculptures the figure of *St Willibald* (Eichstätt Cathedral, *c.*1514) may be mentioned.

Herkomer, Sir Hubert von (1849–1914). Bavarian-born English painter. He came to England with his father, a wood-carver, in 1857 and was largely self-taught as a painter. He established his reputation with *The Last Muster—Sunday at the Royal Hospital, Chelsea* (Lady Lever Art Gal., Port Sunlight, 1875), a work appealing to the public taste for patriotic sentiment, and then became a successful and prolific portrait painter. His best-known works today, however, are his scenes of social concern, which were then still something of a novelty in English art (*On Strike*, Royal Academy, London, 1891). Herkomer was a versatile artist and a man of many parts. He founded and directed a school of art at Bushey, Hertfordshire, 1883–1904, was *Slade Professor of Fine Art at Oxford, 1885–94, and published several books. In addition to paintings, his varied artistic output included set designs for the theatre and cinema, and he also acted and composed music.

Herlin, Friedrich (active *c.*1460–1500). German painter, active mainly in Nördlingen in Swabia. His work (which can best be seen in the Städtisches

Museum at Nördlingen) shows how deeply the manner of Rogier van der *Weyden had penetrated into the south of Germany, some of Herlin's figures and compositions being closely modelled on prototypes by the Netherlandish master.

herm. A sculpture in the form of an armless *bust or head of a man surmounting a shaft tapering towards the bottom. It appears in Greek art from the 6th cent. BC (the name derives from the God Hermes) and originally exhibited a phallus on the shaft. Such herms were set up in Athens at street corners and outside the city as milestones. From the 4th cent. BC the herm was increasingly domesticated and used for portrait and other heads, sometimes copied from full-length statues. Since the *Renaissance it has been part of the general vocabulary of decorative art.

Hermitage, St Petersburg. The largest public museum and art gallery in Russia and one of the most important in the world. It takes its name from a pleasure pavilion adjoining the Winter Palace, built to the order of Catherine the Great in 1764–7 (the 'Little Hermitage') for the display of her treasures. In 1787 it was incorporated in a new building (the Old Hermitage). Catherine was one of the most voracious collectors of all time and at her death in 1796 the imperial collections were estimated to total nearly 4,000 pictures. From 1802 pictures by Russian artists began to be added to the imperial collections. In 1837 the Winter Palace was ravaged by fire and the New Hermitage was built by the German architect Leo von Klenze, 1840–9. It was opened to the public by Nicholas I in 1852. In the following year the Czar sold over 1,200 pictures, but the collection continued to grow, doubling the number of its pictures between 1910 and 1932 despite extensive sales by the Soviets (after the Russian Revolution in 1917 the imperial collections came into public ownership).

Western European painting forms only a fraction of the collection, which includes art objects from India, China, ancient Egypt, Mesopotamia, Pre-Columbian America, Greece, and Rome. Special emphasis is laid up on illustrating the continuity of Russia in history and art, from prehistoric art (consisting chiefly of archaeological material from the Soviet Union) to the present day. The representation of Western painting is rich in virtually every period and school, but perhaps most notably in 17th-cent. Dutch painting (the largest collection in the world) and in French painting of the late 19th and early 20th cents. (almost all the great figures of *Impressionism and *Post-Impressionism are well represented). Many of the French paintings come from the collections of two Moscow businessmen who were among the outstanding collectors and patrons of their time: Ivan Morozov (1871–1921) and Sergei Shchukin (1851–1936). They commissioned new works as well as buying through dealers such as Paul *Durand-Ruel and Ambroise *Vollard. Their collections were open to the public at certain times. *Matisse was a particular favourite of both men, and Shchukin's interest also extended to *Cubism. After the 1917 Revolution the collections were nationalized and later distributed between the Hermitage and the Pushkin Museum in Moscow.

Hernandez, Gregório. See FERNANDEZ.

Heron, Patrick (1920–). British painter, writer, and designer. His early paintings were influenced by *Braque and *Matisse, but in 1956 he turned to abstraction; in the same year he settled in Cornwall, becoming a member of the *St Ives School. His abstracts have been varied, including stripe paintings—vertical and horizontal—as well as looser formats with soft-edged shapes, but all his work is notable for its vibrancy of colour. He has written several books, including *The Changing Forms of Art* (1955), *The Shape of Colour* (1973), and studies of *Vlaminck (1947), *Hitchens (1955), and Braque (1958).

Herrera, Francisco the Elder (*c*.1590–1656?). Spanish painter and engraver, a representative of the transition from *Mannerism to *Baroque. With his older contemporary *Roelas, under whose influence he developed, he helped to prepare the way for the new naturalistic style of the School of Seville in the early 17th cent. *St Basil Dictating his Rule* (Louvre, Paris), which is generally considered his masterpiece, shows his work at its most bold and vigorous. About 1638 Herrera moved to Madrid, where he died. According to *Palomino, *Velázquez was Herrera's pupil, but if this was so it could only have been for a short time.

His son, **Francisco Herrera the Younger** (1627–85), painter and architect, spent many years in Italy, where he is said to have fled from his father's notoriously bad temper and may have studied architecture and fresco painting in Rome. He returned to Spain after his father's death, and was appointed *Murillo's deputy of the Academy of Seville when it was founded in 1660. Soon afterwards he moved to Madrid, where he was appointed Painter to the King (Charles II) in 1672 and Master of the Royal Works in 1677. His greatest achievement was the design (subsequently modified) of the church of El Pilar at Saragossa, begun in 1681. His work as a painter, airy and colourful, owed much to the example of Murillo.

Herring, John Frederick sen. (1795–1866). British sporting and animal painter, the best-known member of a family of sporting artists. He had great success as a painter of racehorses, regularly doing portraits of the winners of the Derby and St Leger, and his work enjoyed wide popularity in engravings. His three painter sons included **John Frederick jun.** (died 1907), and he also had a painter brother, **Benjamin Herring sen.** (1806–30). It is often not easy to distinguish between the work of the various members of the family.

Heyden, Jan van der (1637–1712). Dutch painter, active in Amsterdam. He painted some landscapes and still lifes, but is celebrated as one of the greatest of all townscape painters. His views of towns are done with loving attention to detail, but the harmonious colours and sunny light of his elegantly composed pictures prevent the precise way he rendered foliage, bricks, and architectural detail from appearing dull and dry. In spite of the seemingly objective nature of his work, van der Heyden often took liberties with topographical accuracy and he also painted **capricci*. Painting was only a part of his activity, for he was also involved in civic administration in Amsterdam. He organized street lighting and supervised improvements in the Fire Brigade. The fire hose is said to have been his invention and it is included in his *Brandspuiten-boek* ('Fire Engine Book'), a volume about fire-fighting equipment, illustrated with his own engravings, that he published in 1690.

Heysen, Sir Hans (1877–1968). German-born Australian landscape painter, mainly in watercolour. His family emigrated to Australia when he was 6 and he worked mainly in Adelaide. Robert *Hughes (*The Art of Australia*, 1970) writes: 'Heysen's large body of work was immensely popular; it has most of the textbook virtues and, for many years, no Australian business firm was considered quite solid unless it had a Heysen in its boardroom . . . The only deficiency of his art is that it has no imagination . . . He was, in fact, the Alfred *Munnings of the gum-tree.' His work is represented in all Australian state galleries and many provincial galleries.

Hicks, Edward (1780–1849). The best-known American *naïve painter of the 19th cent., active in Bucks County, Pennsylvania. He was a coach- and sign-painter early in life, but for many years he devoted himself to preaching—the pleasure he derived from painting conflicted with his ascetic Quaker outlook and caused him much conscience-searching. Some of his pictures are farm scenes or landscapes, but he is best known for his many versions (he reputedly made more than 100) of *The Peaceable Kingdom*. Exemplifying the pacifism of the Quaker society in which he lived, they depict with a vivid and charming literalness the prophecy in the 11th chapter of Isaiah that all men and beasts will live in peace.

hieratic (Greek *hieros*: sacred). Term applied to a style characteristic of, say, Egyptian or *Byzantine art in which certain fixed types or methods are conventionally adhered to. Hence the term is extended to other religious or even secular painting or sculpture which makes use of rigid or frontal figures.

Highmore, Joseph (1692–1780). English painter, mainly of portraits. He studied at *Kneller's Academy and had a considerable practice as a portraitist by the 1720s. His early work is in the manner of *Richardson, but from the 1730s his portraits became more elegant as he responded to the *Rococo influences that began to pervade English painting at this time. Some of his more informal works, however, have a directness and freshness that recalls *Hogarth (*Mr Oldham and Friends*, Tate, London). Highmore was a friend of the novelist Samuel Richardson (1689–1761) and painted a series of twelve illustrations to *Pamela* (Tate; Fitzwilliam, Cambridge; NG of Victoria, Melbourne), which link him with *Hayman and Hogarth as one of the initiators of a British school of narrative painting. He also painted Richardson's portrait (NPG, London). Highmore, who had originally studied law, was a man of some learning and in 1761 he gave up painting and retired to Canterbury to devote himself to literary pursuits.

Hildebrand, Adolf von (1847–1921). German sculptor and writer on art. He spent much of his career in Italy, where he was closely associated with Hans von *Marées, and is regarded as one of the main upholders in his period of the classical tradition in sculpture. His most characteristic works were nude figures—timeless and rather austere, in the high-minded tradition of Greek art. He is now, however, better known for his treatise *Das Problem der Form in der bildenden Kunst* (1893) than for his highly accomplished but rather bland sculpture. The book went through many editions (an English translation, *The Problem of Form in Painting and Sculpture*, was published in 1907) and its credo of 'pure form' was influential in promoting a move against surface naturalism in sculpture.

Hill, Carl Fredrik (1849–1911). Swedish landscape painter. He went to Paris in 1873 and, inspired by *Corot and the *Barbizon painters, evolved a personal manner of sentimental landscape using intense colour ranges. In 1876 he was struck by an incurable mental illness. This was long considered as a great loss to art, but today Hill's reputation rests no less on the thousands of highly imaginative drawings and pastels which he produced during his years of insanity.

Hilliard, Nicholas (*c*.1547–1619). The most celebrated of English *miniaturists. He was the son of an Exeter goldsmith, and trained as a jeweller. In about 1570 he was appointed Court Miniaturist and Goldsmith by Elizabeth I (1533–1603), and he also worked for James I (1566–1625), but after the turn of the century his position as the leading miniaturist in the country was challenged by his former pupil Isaac *Oliver. The two were head and shoulders above their contemporaries and dominated the *limning of their era. Hilliard's reputation extended to France, which he visited *c*.1577–8.

In his treatise *The Arte of Limning* (written in about 1600 but not published until 1912) Hilliard declared himself as a follower of *Holbein's manner of limning. In particular he avoided the use of shadow for modelling and in his treatise he records that this was in agreement with Queen Elizabeth's taste—'for the lyne without shadows showeth all to good jugment, but the shadowe without lyne showeth nothing'. But whereas for Holbein a miniature was always a painting reduced to a small scale, Hilliard developed in the miniature an intimacy and subtlety peculiar to that art. He combined his unerring use of line with a jeweller's exquisiteness in detail, an engraver's elegance in calligraphy, and a unique realization of the individuality of each sitter. His miniatures are often freighted with enigmatic inscription and intrusive allegory (e.g. a hand reaching from a cloud); yet this literary burden usually manages to heighten the vividness with which the sitter's face is impressed. Apart from the Queen herself, many other of the great Elizabethans sat for him, including Sir Francis Drake, Sir Walter Raleigh, and Sir Philip Sidney. The finest collection of his miniatures is in the Victoria and Albert Museum. He is known also to have worked on a large scale and among the paintings attributed to him are portraits of Elizabeth I in the National Portrait Gallery, London, and the Walker Art Gallery, Liverpool. In spite of his success, Hilliard had considerable financial problems and in 1617 was briefly imprisoned for debt. His son **Lawrence** (1582–after 1640) was also a miniaturist.

Hillier, Tristram (1905–83). British painter. He abandoned an apprenticeship in accountancy to study at the *Slade School under *Tonks, then went to Paris, where he worked under *Lhote. In 1933 he became a member of *Unit One and contributed to Herbert *Read's book of that name. Throughout the 1930s he travelled widely in Europe, conceiving a special affection for Spain and Portugal. During the Second World War he served with the Royal Naval Volunteer Reserve (1940–4) and afterwards settled in Somerset, where he wrote an autobiography, *Leda and the Goose* (1954). Hillier was influenced by *Surrealism, particularly in the achievement of an impression of strangeness and other-worldliness by the juxtaposition of incongruous objects and the use of unreal perspectives. The outstanding qualities of his painting were sharpness of definition, precision, and clarity.

Hilton, Roger (1911–75). British painter of German extraction (Aby *Warburg was his father's cousin). He trained at the *Slade School, and in the 1930s also studied in Paris under *Bissière. In 1942–5 he was a prisoner of war. In Paris Hilton had absorbed a strong feeling for the sensuous qualities of paint, and in the 1950s he became one of the leading British abstract painters, using interlocking coloured shapes. From the mid-1950s he introduced a shallow sense of pictorial space into his work, and no longer eschewed figural elements. In the 1960s he did a series of exuberant, jokey, female nudes that are among his best-known works (*Oi yoi yoi*, Tate, London, 1963). Their sense of good humour is also present in the colourful gouaches he did in the last two-and-a-half years of his life, belying the ill health that by this time kept him bedridden.

Hiltunen, Eila (1922–). Finnish sculptor. She first exhibited in 1944, working in a naturalistic and Classical style, and made her mark by her war memorials, of which the one at Simpele is considered to be the best. At this time she also did portrait busts of Finnish celebrities. About 1958 she discovered the technique of welding and was further inspired to develop it when she met *Archipenko during a visit to the USA. In the early 1960s she continued to work figuratively but in a style deriving from her new welding technique. Her most famous work, however, is the abstract *Sibelius Monument* (Sibelius Park, Helsinki, 1967), consisting of a nest of polished steel tubes which have been likened both to organ pipes and to the pine trunks of the Finnish forests.

Hiroshige, Ando. See UKIYO-E.

Hirshfield, Morris. See JANIS.

Hirschvogel, Augustin (1503–53). German etcher. His numerous landscape etchings, charming but not very original, derive from the so-called

*Danube School, particularly Wolf *Huber, with whom he may have been in personal contact. He was also a painstaking cartographer and made a large map of Austria (1542) for the emperor Ferdinand I (1503–64).

Hirst, Damien. See TURNER PRIZE.

history painting. A term applied not only to scenes representing actual historical events, but also to scenes from legend and literature of a morally edifying kind, treated in a suitably grand and noble way. Thus scenes from the Bible, Greek mythology, Dante, or Shakespeare would usually come under the heading 'history painting', whereas scenes drawn from a domestic novel might be considered as *genre pictures, even if set in a period before the painter's own. In conventional academic theory history painting was considered the highest branch of art, to which the *Grand Manner was appropriate.

Hitchens, Ivon (1893–1979). British painter, mainly of landscapes. He created a highly distinctive style on the borderline between abstraction and figuration in which broad, fluid areas of vibrant colour, typically on a canvas of wide format, evoke but do not represent the forms of the English countryside that were his main inspiration. The style had evolved by 1939 and from 1940, after being bombed in London, he lived in Sussex. His work altered little from then, apart from the fact that his palette changed from naturalistic browns and greens to much more vivid colours such as bright yellows and purples. Contrary to what often happens when an artist remains constant in one style over a period of decades, Hitchens's work did not become stereotyped or banal. Hitchens also painted flowers and figures, and did several large murals, for example at the University of Sussex (1963). His work is represented in the Tate Gallery and many public collections.

Hjorth, Bror (1894–1968). Swedish sculptor and painter. He studied with *Bourdelle in Paris (1921–3), but his earthy, vigorous style is in the tradition of Swedish *folk art. His wooden *reliefs are characteristic—roughly hewn, and gaily painted, usually representing erotic subjects or cult-symbols. He ran a school for sculptors in Stockholm in 1931–4 and in 1949–59 taught at the Academy there. There is a museum dedicated to him in Uppsala.

Hoare, William (*c.*1707–92). English portrait painter. He spent the formative years of his working life in Italy (1728–37), but his style is a continuation of *Richardson's. By 1738 he had settled in Bath, and until the arrival of *Gainsborough in 1759 he was the leading portrait painter of that city. Sir Ellis *Waterhouse has described his style as 'serious, but a little blank'. He worked much in *pastel, and in this was followed by his daughter **Mary** (*c.*1753–1820). There are examples by both of them at Stourhead, Wiltshire.

Hobbema, Meindert (1638–1709). Dutch landscape painter. He worked in his native Amsterdam, where he was the friend and only documented pupil of Jacob van *Ruisdael. Some of his pictures are very like Ruisdael's, but his range was more limited and he lacked the latter's power to capture the majesty of nature. He painted a narrow range of favourite subjects—particularly water-mills and trees around a pool—over and over again. In 1668 he became a wine gauger with the Amsterdam customs and excise, and thereafter seems to have painted only in his spare time. However, his most famous work, *The Avenue at Middelharnis* (NG, London), dates from 1689. Hobbema has long been a popular artist in England (his influence is clear in *Gainsborough's early landscapes) and is outstandingly well represented in English collections, notably the National Gallery.

Hockney, David (1937–). British painter, draughtsman, printmaker, photographer, and designer, active mainly in the USA. After a brilliant prize-winning career as a student at the *Royal College of Art, Hockney had achieved considerable success by the time he was in his mid-twenties, and he has since consolidated his position as by far the best-known and most critically acclaimed British artist of his generation. His phenomenal success has been based not only on the flair and versatility of his work, but also on his colourful personality, which has made him a recognizable figure even to people not particularly interested in art. In 1961 he emerged as one of the leaders of British *Pop art at the *Young Contemporaries exhibition. Hockney himself disliked the label 'Pop', but his work of this time makes many references to popular culture (notably in the use of graffiti-like lettering) and is often jokey in mood. His first retrospective exhibition came as early as 1970, at the Whitechapel Art Gallery, London (it subsequently toured to Hanover, Rotterdam, and Belgrade). By this time he was painting in a weightier, more traditionally representational manner, in which he did a series of large double-portraits of friends, including the well-known *Mr and Mrs Clark and Percy* (Tate Gallery, London, 1970–1). These portraits are notable for their airy feeling of space and light and the subtle flattening and simplification of forms, as well as for the sense of stylish living they capture. Hockney often paints the people and places he knows best (his art is frequently autobiographical) and has memorably celebrated his romance with Los Angeles (he first visited the city in 1963 and settled there permanently in 1976), particularly in his many paintings featuring swimming pools (*A Bigger Splash*, Tate Gallery, London, 1967). In these works he skilfully exploited the qualities of the new *acrylic paint. Hockney has also been outstanding as a graphic artist. His work in this field includes etched illustrations to Cavafy's *Poems* (1967) and *Six Fairy Tales of the Brothers Grimm* (1969), as well as many individual prints, often on homo-erotic themes. In the 1970s he came to the fore as a stage designer, notably with his set and costume designs for Stravinsky's *The Rake's Progress* and Mozart's *The Magic Flute*, produced at Glyndebourne in 1975 and 1978 respectively. The broader style demanded by stage design has been reflected in his subsequent easel paintings. In the 1980s he experimented a good deal with photography, producing, for example, photographic collages and—since 1986—prints created on a photocopier. Hockney is a perceptive commentator on art and has published two substantial books on his own work: *David Hockney by David Hockney* (1976) and *That's the Way I See It* (1993).

Hodges, William (1744–97). English landscape painter and engraver. He was the pupil and assistant of Richard *Wilson *c.*1758–65 and became a skilful imitator of his style. His work took on a more personal character when he travelled as draughtsman with Captain Cook in 1772–5, and his finest paintings are those based on drawings he made of such exotic Pacific islands as Tahiti and Easter Island (examples are in the National Maritime Museum, Greenwich). In 1779–84 he was in India (he was patronized by Warren Hastings (1732–1818) and published books on his travels illustrated with *aquatint illustrations on his return to England) and in 1790 he visited the Continent, going as far as Russia. He did pictures for *Boydell's Shakespeare Gallery and also some allegorical subjects, but in 1795 he gave up painting and opened a bank in Dartmouth. It failed soon before he died.

Hodgkin, Sir Howard (1932–). British painter and printmaker, regarded as one of the outstanding colourists in contemporary art. His paintings, which are usually fairly small, sometimes look completely abstract, but in fact he bases his work on specific events, usually an encounter between people—'one moment of time involving particular people in relation to each other and also to me'. He has travelled widely, making several visits to India, and his preference for flat colours and decorative borders reflects his admiration for Indian *miniatures. A well-known figure in the art world, he has been a Trustee of the Tate Gallery and the National Gallery, and in 1985 he was awarded the Turner Prize.

Hodgkins, Frances (1869–1947). New Zealand painter, active mainly in England, where she settled in 1913 after some time alternating between the two hemispheres. She was the daughter of **William Matthew Hodgkins** (1833–98), an amateur painter who had emigrated from England in 1859 and took a lead in the artistic life of Dunedin. Her father taught her watercolour painting, but she did not begin to paint in oils until 1915. Until that time her work had been conventional, but she gradually developed a more individual style, echoing *Matisse and *Dufy (she spent a good deal of time in France) in its use of vibrant colour. She mainly painted landscapes and still lifes. Her later paintings approach abstraction in a manner akin to *Hitchens's work.

Hodler, Ferdinand (1853–1918). Swiss painter active mainly in Geneva. He ranks alongside *Böcklin as the outstanding Swiss artist of the 19th cent., but his early work (till 1890) was rather unimaginatively naturalistic, and his landscapes were hardly more than rather ambitious colour postcards for tourists. But with his *Night* (Kunstmuseum, Berne) of 1890 began a sudden change of style. From then on Hodler's canvases were filled with monumental and simplified flat figures, composed into a coherent design by a rhythmic and repetitive pattern of lines, forms, and colours—a method which the artist himself called 'Parallelism'. Contacts with the Rosicrucians (he exhibited at the first *Salon de la Rose + Croix in Paris in 1892) and possibly knowledge of the aims of Maurice *Denis gave a markedly *Symbolist flavour to his art (*The Disappointed*, Kunstmuseum, 1892; *Eurythmics*, Kunstmuseum, 1894/5). He applied the same principles to *history painting (*The Return from Marignano*, Kunsthaus, Zurich, 1896–1900) and to his Swiss landscapes. Hodler became immensely popular in the German-speaking world and ranks among the harbingers of *Expressionism.

Hofer, Carl (1878–1955). German painter. In 1903–8 he lived in Rome, basing his style on the idealized *Classicism of painters such as Hans von *Marées, then from 1908 to 1913 was in Paris, where he was influenced by the structural solidity of *Cézanne. Trips to India in 1909 and 1911 introduced a personal *iconography of swooning, lyrical figures. He settled in Berlin in 1913, but was trapped in Paris at the outbreak of war and spent three years in an internment camp. On his return to Berlin in 1918 he was appointed a teacher at the Hochschule für Bildende Künste. Except for a brief experiment with abstract painting in 1930–1, his art was one of disillusionment and pessimism, telling his dark vision of modern life through a small range of obsessively recurrent images. In 1933 he was condemned as *degenerate and removed from his teaching post. His studio and most of his work were destroyed by bombing in 1943. At the end of the war he was reinstated and made director of the Hochschule. Hofer remained throughout his life an individualist, and rejected the label *Expressionist for his work, but he had considerable influence on many German Expressionist artists.

Hofmann, Hans (1880–1966). German-born painter and teacher who became an American citizen in 1941. From 1904 to 1914 he lived in Paris, where he knew many of the leading figures of *Fauvism, *Cubism, and *Orphism. In 1915 he founded his own art school in Munich and taught there successfully until 1932, when he emigrated to the USA (following visits in 1930 and 1931 during which he taught at the University of California, Berkeley). He founded the Hans Hofmann School of Fine Arts in New York in 1934 (followed the next year by a summer school at Provincetown, Massachusetts) and became a teacher of great influence on the minority group of American artists who practised abstract painting during the 1930s. Hofmann continued teaching until 1958, when he closed his schools so that he could concentrate on his own painting. This was to counter opinions that he was merely an academic figure and a symbol of the avant-garde rather than a creative artist. In the course of his career he experimented with many styles, and was a pioneer of the technique of dribbling and pouring paint that was later particularly associated with Jackson *Pollock. His later works, in contrast, feature rectangular blocks of fairly solid colour against a more broken background. He gave a large collection of his pictures to the University of California, Berkeley. As a painter and teacher Hofmann was an important influence on the development of *Abstract Expressionism. The essence of his approach was that the picture surface had an intense life of its own.

Hofstede de Groot, Cornelis (1863–1930). Dutch art historian, the author of the monumental catalogue *Beschreibendes und kritisches Verzeichnis der Werke der hervorragendsten holländischen Maler des XVII Jahrhunderts* (10 vols., 1907–28). An English edition of volumes 1–8 appeared in 1908–27, entitled *A Catalogue Raisonné of the Works of the Most Eminent Dutch Painters of the Seventeenth Century*. Covering the work of forty of the leading painters of the period, the catalogue has been superseded in some areas, but is still regarded as a major source of information and its numbering system is frequently referred to. Hofstede de Groot also published the first complete catalogue of *Rembrandt's drawings (1906).

Hogarth, William (1697–1764). English painter and engraver. He trained as an engraver in the *Rococo tradition, and by 1720 was established in London independently as an engraver on copper of billheads and book illustrations. In his spare time he studied painting, first at the *St Martin's Lane Academy and later under Sir James *Thornhill, whose daughter he married in 1729. By this time he had begun to make a name with small *conversation pieces, achieving a popular success with *A Scene from the Beggar's Opera* (one version is in the Tate Gallery), and about 1730 he set up as a portrait painter. At about the same time he invented and popularized the use of a sequence of anecdotal pictures 'similar to representations on the stage' to point a moral and satirize social abuses. *A Harlot's Progress* (6 scenes, *c.*1731; destroyed by fire) was followed by *A Rake's Progress* (8 scenes, Sir John Soane's Mus., London, *c.*1735), and *Marriage à la Mode* (6 scenes, NG, London, *c.*1743), which each portray the punishment of vice in a somewhat lurid melodrama. Each series was painted with a view to being engraved, and the engravings had a wide sale and were popular with all classes. They were much pirated and Hogarth's campaigning against the profiteers led to the Copyright Act of 1735. 'I have endeavoured', he wrote, 'to treat my subjects as a dramatic writer: my picture is my stage, and men and women my players.' Hogarth, however, was much more than a preacher in paint. His satire was directed as much at pedantry and affectation as at immorality, e.g. *Taste in High Life* (Iveagh Bequest, Kenwood, 1742), and he saw himself to some extent as a defender of native common sense against a fashion for French and Italian mannerisms.

In spite of his rabid xenophobia, Hogarth made some attempts to show he could paint in the Italian *Grand Manner (*Sigismunda*, Tate, London, 1759). These, however, are generally considered his weakest works, and apart from his modern morality subjects he excelled mainly in portraiture. *Captain Coram* (Coram Foundation, London, 1740), which he himself regarded as his highest achievement in portraiture, shows that he could paint a portrait in the *Baroque manner with complete confidence and without artificiality. However, he could not flatter or compromise and had not the disposition for a successful portraitist. In 1753 he published *The Analysis of Beauty*, a treatise on aesthetic theory which he wrote with the conviction that the views of a practising artist should carry greater weight than the theories of the connoisseur or dilettante (see AMATEUR). He there satirized academicism and the 18th-cent. school of taste and proposed a 'precise serpentine line' as a concrete key to pictorial beauty. From 1735 he also ran an academy in St Martin's Lane (independent of the one at which he had studied), and this became the main forerunner of the *Royal Academy.

Hogarth was far and away the most important British artist of his generation. He was equally outstanding as a painter and engraver and by the force of his pugnacious personality as well as by the quality and originality of his work he freed British art from its domination by foreign artists. Because so much of his work has a 'literary' element, his qualities as a painter have often been overlooked, but his more informal pictures in particular show that his brushwork could live up to his inventive genius. The vigour and spontaneity of *The Shrimp Girl* (NG, London, *c.*1740), for example, have made it deservedly one of the most popular British paintings of the 18th cent.

Hokusai, Katsushika. See UKIYO-E.

Holanda (or **Hollanda**), **Francisco de** (1517–84). Portuguese *miniaturist, draughtsman, and writer on art, the son of a Netherlandish miniaturist, **Antonio de Holanda**, who spent part of his career in Lisbon. Francisco grew up in Portugal. In 1538–9 he visited Rome and produced a volume of drawings (now in the Escorial, near Madrid) which contains interesting portraits and drawings of Roman antiquities that are an important source of information about 16th-cent. collecting and archaeology. Although he became Court Painter in Lisbon, his influence in propagating the Italianate style in Portugal was exercised mainly through his writings. In 1548 he completed a manuscript entitled *Da Pintura Antigua (Of Ancient Painting)*, which was not published until 1890–6. It is in two books, the first containing forty-four chapters on art theory and the second four dialogues (Eng. trans., 1928) in which Holanda purportedly discusses theories of art with *Michelangelo, the miniaturist Giulio *Clovio, and others. As an appendage he completed in 1549 ten dialogues entitled *Do tirar polo natural (On Drawing from Nature)*. He had made a study of fortification in Rome and made a number of proposals for improving Lisbon in the Roman manner.

Holbein, Hans (1497/8–1543). German painter and designer, chiefly celebrated as one of the greatest of all portraitists. He trained in his native Augsburg with his father, **Hans Holbein the Elder** (*c.*1465–1524)—another son, **Ambrosius** (1494–*c.*1519), was also a painter—and in about 1514 moved to Basle. There he quickly found employment as a designer for printers, and in 1516 he painted portraits of Burgomaster Meyer and his wife (Öffentliche Kunstsammlung, Basle).

From 1517 to 1519 he was working in Lucerne, assisting his father on the decoration of a house for the von Hertenstein family (now destroyed). It is possible that during this time he crossed the Alps to Lombardy, for on his return to Basle, where he was to remain until 1526, his style was less harsh, his modelling softer, and his composition more monumental. One of the most beautiful of all his works, the portrait of the scholar and collector *Bonifacius Amerbach* (Öffentliche Kunstsammlung, 1519), shows the richness of colour and warmth in the flesh tints which characterize his new style, and the harrowing *Christ in the Tomb* (Öffentliche Kunstsammlung, 1521 or 1522) has a power of expression combined with a mastery of *chiaroscuro that almost rivals *Leonardo.

He was now the leading painter in Basle, and gained an important commission for decorating the Town Hall with scenes of *Justice* taken from classical history. Apart from fragments in the Öffentliche Kunstsammlung, Basle, these are lost, but are known from copies. He also continued to work for printers, producing between 1523 and 1526 his best-known work in this field, the series on the *Dance of Death*. Since these reflected the new critical outlook of the Reformation, they were not published until 1538 in Lyons, when they enjoyed enormous popularity, running into many editions. His most notable portraits in these years are those of Erasmus (Louvre, Paris; Earl of Radnor Coll., Longford Castle, Wiltshire; and Öffentliche Kunstsammlung, Basle; all *c.*1523). In them, perhaps by the sitter's wish, he used for the first time the formula of the scholar in his study, first devised by Quentin *Massys, also for a portrait of Erasmus (Gal. Naz., Rome, 1517).

A visit to France in 1524 gave Holbein further knowledge of *Renaissance painting, especially through the works of *Raphael in the royal collection, and the effect this had may be seen in the *Madonna of Burgomaster Meyer* (Schlossmuseum, Darmstadt, 1526). Mother and Child alike have an ideal beauty which is quite un-German, though the *donor portraits have a splendid naturalism.

The disturbances of the Reformation meant a decline of patronage in Basle, and in 1526, armed with an introduction from Erasmus to Sir Thomas More, Holbein sought work in England. His great group portrait of the More family (lost, but later copies in the NPG, London, and Nostell Priory, West Yorkshire) is a landmark in European art, for no previous artist had produced a group portrait of full-length figures in their own home. A drawing for this group portrait, which was sent by More to Erasmus, is in the Öffentliche Kunstsammlung. A number of single portraits date from this visit, but Holbein obtained no commissions for subject pictures and returned home in 1528.

Basle, however, had changed. Religious pictures were banned and there was much religious strife. Holbein accepted the Reformed religion, continued his work at the Town Hall, and made many designs for stained glass. In 1532, leaving his family, of whom he had painted a penetrating group portrait (Öffentliche Kunstsammlung), he returned to England. More was now out of favour and Holbein found new patrons in the German Steelyard Merchants, for whom he painted several portraits and did decorative work.

Through the Steelyards he probably met Thomas Cromwell (1485?–1540; portrait, Frick Coll., New York, 1532–3), who may have obtained for him the commission for his famous double portrait *The Ambassadors* (NG, London, 1533), and almost certainly helped him to gain royal patronage. By 1536 he was working for Henry VIII (1491–1547), and in the next year undertook his most famous English work, the wall-painting in Whitehall Palace of Henry VIII with his father and mother and his third wife, Jane Seymour. Though the painting perished in 1698, part of the *cartoon survives (NPG, London) and the formidable figure of the King standing four-square and staring at the spectator is well known through copies. The only portrait of the King indisputably from Holbein's hand is a bust-length picture in the Thyssen collection in Madrid, a type of which numerous replicas exist. The King also sent Holbein abroad to paint prospective brides—*Anne of Cleves* (Louvre, 1539), *Christina, Duchess of Milan* (NG, London, 1538).

Numerous other members of the court are portrayed in paintings and in drawings, a marvellous collection of which is in the Royal Library at Windsor Castle. Many designs for decoration survive from Holbein's last years, when he also turned to *miniature painting, to which his exquisitely detailed craftsmanship was well suited. Holbein's portraits were much copied and his decorative work was an important factor in introducing the Renaissance style to England, but none of his followers approached the penetration of his characterization or the virtuosity of his technique. Only in miniature painting did he have a worthy successor in *Hilliard.

Holguín, Melchor Pérez (*c.*1660–*c.*1725). Bolivian painter, active mainly in Potosí. He is regarded as the leading Spanish colonial painter of his period, but his work is crude by European standards. It mixed various European influences, from *Mannerism to *Rococo, with certain native Indian features.

Hollanda, Francisco de. See HOLANDA.

Hollar, Wenceslaus (or **Wenzel**) (1607–77). Bohemian engraver and watercolourist, born in Prague. He trained in the workshop of *Merian in Frankfurt, and became one of the foremost engravers of topographical views in the 17th cent. In 1636, while working in Cologne, he met the English connoisseur, the Earl of *Arundel, who took him on a tour of Europe to make views for his private collections. On account of his English connections Hollar finally settled in London—during the Civil War he fought on the Royalist side—and his views of the city form an invaluable record of its appearance before the Great Fire of 1666. He was very prolific and engraved a wide range of subjects apart from views.

Home, Henry, Lord Kames (1696–1782). Scottish judge, moralist, and writer on *aesthetics. His *Elements of Criticism* (1762) ranks along with Archibald *Alison's *Essay on Taste* as the major production of philosophical aesthetics written in English in the 18th cent. It went through six editions between 1762 and 1785 and was long used as a textbook until the predominance of Germanic philosophy in the 19th cent. brought it into oblivion.

Homer, Winslow (1836–1910). American landscape, marine, and *genre painter. He came to painting from illustration (chiefly for *Harper's Weekly*), and *Prisoners from the Front* (Met. Mus., New York, 1866), one of his first important oils, has a quality of vivid, unromanticized reportage. He aspired to naturalistic recording and expressed his attitude in the words: 'When I have selected the thing carefully, I paint it exactly as it appears.' In 1867 Homer visited Paris; he was influenced by *Manet's broad tonal contrasts, but he explored the rendering of light and colour in a direction other than that of the *Impressionists—instead of dissolving outline into light and atmosphere, he sought luminosity within a firm construction of clear outline and broad planes of light and dark (*Long Branch, New Jersey*, Mus. of Fine Arts, Boston, 1869). The sea was Homer's favourite subject, and after living near Tynemouth on the rugged coast of north-east England in 1881–2 he settled at Prout's Neck on the Maine coast, where he lived in isolation. His pictures of the Maine coast, which represent the power and solitude of the sea and the contest of man with the forces of nature, are his best-known works. Homer was an artist of considerable originality who, through a bold, vivid, and personal naturalism, created an imaginative vision of nature that has come to be accepted as a reflection of American pioneering spirit. He used watercolour with the force and authority of oil (*Inside the Bar, Tynemouth*, Met. Mus., 1883).

Hondecoeter, Melchior d' (1636–95). Dutch painter, the best-known member of a family of artists. He was the Netherlands' most renowned painter of birds, winning an international reputation with his vigorous and brightly coloured canvases. They show both domestic and exotic birds, often in action and sometimes pointing a moral. Hondecoeter also painted still lifes. He was a prolific artist and is represented in many museums, for example in the National Gallery and Wallace Collection in London. His father, **Gysbert** (1604–53), was also a bird painter and his grandfather, **Gillis** (d. 1638), was a landscapist. Melchior trained with his father and with his uncle, Jan Baptist *Weenix; he worked in Utrecht, The Hague, and Amsterdam.

Hondius, Abraham (*c*.1625/30–91). Dutch painter of hunting scenes and animals fighting. He was born in Rotterdam, worked in Amsterdam from 1659 to 1666, then moved to London, where he died. His animal paintings are influenced by Flemish artists such as *Fyt and *Snyders. He also painted some history and religious pictures.

Hone, Nathaniel (1718–84). Irish *miniaturist and portrait painter, who settled permanently in London after studying in Italy, 1750–2, and became a foundation member of the *Royal Academy. He is now remembered mainly for one painting, *The Conjurer* (NG, Dublin, 1775), in which he satirized *Reynolds's practice of borrowing poses from the Old Masters. The picture was accepted at the RA, but was withdrawn after Angelica *Kauffmann (whose name had been linked romantically with Reynolds's) objected that a nude figure in the background was meant to represent her. (Hone painted out the nude figures, but they can be seen in his sketch for the picture in the Tate Gallery.) In protest at the removal of his painting Hone exhibited it in a one-man show in St Martin's Lane, the first of its kind recorded in Britain. Hone's sons, **Horace** (1754–1825) and **Camillus** (1759–1836), were also painters, as was a brother, **Samuel** (born 1726). Camillus was the subject of some of his father's best portraits.

Evie Hone (1894–1955), a descendant of Nathaniel, was one of the outstanding stained-glass designers of the 20th cent. Her masterpiece is the huge east window of Eton College Chapel, commissioned in 1949 to replace glass destroyed by bombing in the Second World War and completed in 1952.

Honnecourt, Villard d'. See VILLARD D'HONNECOURT.

Honthorst, Gerrit van (1592–1656). Dutch painter of biblical, mythological, and *genre scenes, and of portraits. In Utrecht, his birthplace, he was a pupil of *Bloemaert, but his style was formed by a long stay in Italy (*c*.1610–20) and upon his return to the Netherlands he became, along with *Baburen and *Terbrugghen, one of the leading Dutch followers of *Caravaggio. The candlelight effects he favoured in his early pictures (*Christ before the High Priest*, NG, London) earned him the nickname 'Gherardo delle Notti' (Gerard of the Night Scenes). Some of *Rembrandt's early works show that he was impressed by Honthorst's use of *chiaroscuro for dramatic effects. In the late 1620s Honthorst abandoned his Caravaggesque style for a lighter manner in which he achieved international success (rare for a Dutch artist) as a court portraitist. He was employed by the Elector of Brandenburg and by King Christian IV (1577–1648) of Denmark, and in 1628 Charles I called him to England, probably on trial as a court painter (his portrait of Charles is in the NPG, London). From 1637 to 1652 he was court painter at The Hague. A good collection of his portraits is at Ashdown House in Berkshire.

Hooch, Pieter de (1629–84). Dutch *genre painter. He was born in Rotterdam, studied in Haarlem (as a pupil of *Berchem), and spent the last two decades of his life in Amsterdam, but he is particularly associated with Delft. His period of residence there was fairly brief (*c*.1655–*c*.1661), but during this time he painted the pictures on which his reputation depends—a small number of tranquil masterpieces that perfectly evoke the well-being of his peaceful and prosperous country. Typical subjects include a sunny yard (*The Courtyard of a House in Delft*, NG, London, 1658) or light streaming into the interior of a corner of a burgher's house (*The Pantry*, Rijksmuseum, Amsterdam, *c*.1658). There is a kinship of spirit with his great Delft contemporary, *Vermeer, and de Hooch sometimes approaches him in delicate observation of light and lucidity of composition, although not in beauty of brushwork. After de Hooch moved to Amsterdam in the early 1660s, however, the quality of his paintings was less remarkable than their quantity. Instead of the simple brick and plaster backgrounds of his earlier groups he chose sumptuous marble interiors, and towards the end these backgrounds acquired something of the harsh quality of a painted drop-scene. He died in a madhouse.

Hoogstraten, Samuel van (1627–78). Dutch painter and writer on art. He painted *genre scenes in the style of de *Hooch and *Metsu, and portraits, but he is best known as a specialist in *perspective effects. One of his 'perspective boxes', which shows a painted toy world through a peephole, is in the National Gallery, London (see PEEPSHOW BOX). Only in his early works can it be detected that he was a pupil of *Rembrandt. Hoogstraten travelled to London, Vienna, and Rome, worked in Amsterdam and The Hague as well as his native Dordrecht, and was a man of many parts. He was an etcher, poet, director of the mint at Dordrecht, and art theorist. His *Inleyding tot de hooge schoole der schilderkonst* (Introduction to the Art of Painting, 1678) contains one of the rare contemporary appraisals of Rembrandt's work.

Hope, Thomas (1769–1831). British collector, patron, and writer. He was born in Amsterdam to a wealthy banking family, and travelled extensively before and after settling in England in 1795. In 1801 he was described as being reputedly 'the richest, but undoubtedly far from the most agreeable man in Europe', and he used his great wealth to spend lavishly on art for his London mansion in Duchess Street (to which the public could buy admission tickets) and his country seat at Deepdene, Surrey. He was a devotee of *Neoclassicism, and the artists he patronized included *Canova, *Flaxman, and *Thorvaldsen. Hope also had notable collections of paintings and antique statuary. He trained craftsmen to make furniture from his own Greek and Egyptian designs, and his publications included *Household Furniture and Interior Decoration* (1807). Among his other books was a novel *Anastasius*, published anonymously in 1819, which enjoyed considerable popularity in its day.

Hopper, Edward (1882–1967). American painter and etcher. He spent almost all his career in New York, but he travelled extensively in the USA, making long journeys by car. His main training was at the New York School of Art, where Robert *Henri was one of his teachers. Between 1906 and 1910 he made three trips to Europe (mainly Paris), but these had little influence on his style. In 1913 he exhibited (and sold) a picture at the *Armory Show, but from then until 1923 he earned his living entirely by commercial illustration. After turning to painting full-time in 1924, however, he enjoyed a fairly rapid rise to recognition as the outstanding exponent of *American Scene Painting (he was given a retrospective exhibition by the *Museum of Modern Art in 1933 and this set the seal on his reputation). Hopper's distinctive style was formed by the mid-1920s and thereafter changed little. The central theme of his work is the loneliness of city life, generally expressed through one or two figures in a spare setting—his best-known work, *Nighthawks* (Art Institute of Chicago, 1942), has an unusually large 'cast' with four. Typical settings are motel rooms, filling stations, cafeterias, and almost deserted offices at night. He was the first artist to seize on this specifically American visual world and make it definitively his own. However, although his work is rooted in a particular period and place, it also has a peculiarly timeless feel and deals in unchanging realities about the human condition. He never makes feelings explicit or tries to tell a story; rather he suggests weariness, frustration, and troubled isolation with a poignancy that rises above the specific. Hopper himself enjoyed solitude (although he was happily married to another ex-student of Henri) and he disliked talking about his work. When he did, he discussed it mainly in terms of technical problems. Of *Nighthawks* he said: 'I didn't see it as particularly lonely . . . Unconsciously, probably, I was painting the loneliness of a big city.' Deliberately so or not, in his still, reserved, and blandly handled paintings he exerts a powerful psychological impact that makes him one of the great painters of modern life. Hopper worked in watercolour as well as oil and also made etchings, beginning in 1915—in fact his individual vision emerged in this medium before it did in painting. His best-known print is *Evening Wind* (1921), establishing a theme that would later often recur in his paintings—the female nude in a city interior. He virtually abandoned printmaking in 1923, but in spite of his short career in the medium he has been described as 'undoubtedly the greatest American etcher of this century' (Frances Carey and Antony Griffiths, *American Prints 1879–1979*, 1980).

Hoppner, John (1758–1810). British portrait painter. He was trained as a chorister in the Chapel Royal, and later received an allowance from George III (1738–1820) to study at the Royal Academy Schools. This favour and the fact that he received commissions for royal portraits as early as 1785 led to rumours that he was the king's illegitimate son, but these were never proved. In 1789 he was appointed Portrait Painter to the Prince of Wales (later George IV) and from this time was associated with 'the Prince of Wales's Set'. After the death of *Reynolds he and *Lawrence were the leading portraitists in the country. In his attempts to emulate first Reynolds, then Lawrence, Hoppner rarely achieved striking individuality, but his best work, particularly his portraits of women and children, often has great charm. He is represented in many British and American collections.

Hopton Wood stone. A very hard limestone quarried at Middleton, Derbyshire, varying in colour between light grey and light tan and speckled with dark-grey glistening crystalline masses, often geometric in shape. It can be cut to a smooth face and sharp ridge and takes a good polish. Henry *Moore and Barbara *Hepworth are among the sculptors who have used it.

Hornton stone. Limestone named after quarries at Hornton in north-west Oxfordshire. It is usually a rich tawny brown in colour, but green and greyish-blue tints also occur. It was a favourite stone of Henry *Moore, the *Madonna and Child* (1943–4) in St Matthew's, Northampton, being one of his best-known works in this material. The quarries at Hornton are now closed, but similar stone is obtained at nearby Edge Hill in Warwickshire.

Horta, Victor (1861–1947). Belgian architect and designer. Although his chief importance lay in the influence he had on the Modern Movement in architecture, in his early career he created an original vocabulary of ornament which made him a leading figure in European *Art Nouveau. He designed almost exclusively for his own buildings, but for these he worked on every detail down to the light-fittings (which are characteristically in the form of flowers).

Hoskins, John (*c.*1595–1665). The leading English portrait *miniaturist between *c.*1625 and *c.*1640. His early work is a development of *Hilliard's style. He later became a specialist in miniature copies of van *Dyck's portraits in oils—a type which remained much in demand throughout the 17th cent.—but his miniatures often have a charm and originality of their own. He was *'limner' to Charles I in 1640, but thereafter was overshadowed by the work of his nephew and pupil, Samuel *Cooper. Hoskins's work is well represented in the Victoria and Albert Museum.

Houbraken, Arnold (1660–1719). Dutch painter and writer on art. His paintings are now forgotten, but he is important for his large biographical work *De Groote Schouburgh der Nederlantsche Konstschilders en Schilderessen* (The Great Theatre of Netherlandish Painters), 3 vols., 1718–21. This was the first comprehensive study of Netherlandish art since van *Mander published his *Schilderboeck* in 1604 and is the most important source-book on 17th cent. Netherlandish artists. Arnold's son **Jacobus** (1698–1780) was a leading portrait engraver. His work includes engraved plates after his father's designs for the *Groote Schouburgh*.

Houckgeest, Gerrit (*c.*1600–61). Dutch painter of architectural views, active mainly in Delft (1635–52) but also in The Hague, where he was born, and Bergen-op-Zoom, where he died. He painted some fantastic architectural views, but is best known for his luminous and precisely delineated church interiors (*Interior of the New Church at Delft*, Mauritshuis, The Hague, 1651).

Houdon, Jean-Antoine (1741–1828). French sculptor. A pupil of Jean-Baptiste *Lemoyne and of Jean-Baptiste *Pigalle, he won the *Prix de Rome in 1761. He was in Rome 1764–8 and there produced two works of sculpture which assured his renown: an **écorché* figure (Schlossmuseum, Gotha, 1767), casts of which were widely used in art academies, and *St Bruno* (St Maria degli Angeli, 1767), executed in a direct and unpretentious classical style. After returning to Paris in 1769, he was successful in the popular mythological idiom, becoming a member of the Academy in 1777 with his *Morpheus* (Louvre, Paris). His greatest strength, however, was in his portrait busts. He had the highest repute of French 18th-cent. portrait sculptors and his work exemplifies the 18th-cent. interest in the psychological depiction of individuality. By the middle 1780s he was acknowledged as the leading portrait sculptor of Europe and his fame spread also to America. In 1785 the State of Virginia commissioned from him a statue of George Washington, and he visited America in order to study his model face to face before doing his famous marble statue of Washington as the modern Cincinnatus, called from the ploughshare to wield the reins of government (Virginia State Capitol, Richmond, 1788, bronze copy outside the NG, London). In his portraiture Houdon had a knack of catching characteristic tricks of gesture and expression and a brilliant gift of depicting the marks of individuality, rather than a profound penetration into human character. His portraits of Voltaire (e.g. in the Comédie-Française, Paris, and V&A, London) are among his most celebrated works. During the French Revolution Houdon narrowly escaped imprisonment and although he found favour again under Napoleon (a terracotta bust of the emperor is in the Musée des Beaux-Arts, Dijon, 1806), he produced little of importance after the turn of the century. He became senile in 1823.

Howson, Peter. See OFFICIAL WAR ART.

Huber, Wolfgang (or **Wolf**) (*c.*1490–1553). German painter, printmaker, and architect, active mainly in Passau, where he was court painter and architect to the prince-bishop. He was first and foremost a poetic interpreter of his native landscape and is usually counted among the masters of the so-called *Danube School. His landscape drawings are particularly delicate and his religious paintings often have landscape backgrounds.

Hudson, Thomas (1701–79). English portrait painter, a pupil of Jonathan *Richardson, whose daughter he married. From the mid-1740s to the mid-1750s he was the leading fashionable portraitist in London, rivalled only by *Ramsay. His studio produced a great deal of work, with much help from specialist assistants such as the drapery painter Joseph van *Aken, and Hudson has been described by Sir Ellis *Waterhouse as 'the last of the conscienceless artists, of whom *Lely was the first in England, who turned out portraits to standard patterns and executed comparatively little of the work themselves'. Hudson went into semi-retirement in the late 1750s, when his former pupil *Reynolds was rapidly rising in success.

Hudson River School. Term applied retrospectively to a number of American landscape painters, active *c.*1825–*c.*1875, who were inspired by pride in the beauty of their homeland. The early leaders and the three most important figures in the group were Thomas *Cole, Thomas *Doughty, and Asher B. *Durand, who with a reverential spirit painted the Hudson River Valley, the Catskill Mountains, and other remote and untouched areas of natural beauty. These three artists and many of those who followed had studied in Europe and part of their inspiration came from painters of the grandiose and spectacular such as *Turner and John *Martin. The writings of James Fenimore Cooper (1789–1851) and Washington Irving (1783–1859) also played a part in shaping their attitudes. The patriotic spirit of the painters of the Hudson River School won them great popularity in the middle years of the century. Painters of a similar outlook who found their inspiration in the far West are known collectively as the *Rocky Mountain School.

hue. The name of a colour or the attribute by virtue of which it is red, green, blue, etc. The spectrum is conventionally divided into six basic hues—red, yellow, and blue (the primary colours) and green, orange, and violet (the secondary colours, made by mixing the primary colours). In normal parlance the word 'hue' tends to be used so loosely that it is no more than a synonym for colour.

Huerta, Juan de la. Spanish sculptor who in 1443 received the commission left unfinished by Claus de *Werve for the tomb of John the Fearless, Duke of Burgundy (d. 1419), and his wife, at Dijon (now in the Musée des Beaux-Arts). Huerta completed most of the minor figures, which closely follow those of the tomb of Philip the Bold (see SLUTER), before absconding in 1457.

Huet, Christophe (d. 1759). French *Rococo painter, engraver, and designer. He belongs to the tradition of decorative designers stemming from *Bérain and *Audran and is best known for paintings and engravings involving animals, characteristically dressed up and acting like humans (see SINGERIE). Good examples of his work are in the Musée Condé at Chantilly.

Hugh Lane Municipal Gallery of Modern Art. See LANE.

Hughes, Arthur (1832–1915). English painter and illustrator. In the 1850s he was one of the most distinguished of the *Pre-Raphaelite sympathizers, remarkable for his lyrical delicacy of colour and drawing. Two paintings are particularly well known—*April Love* (Tate, London, 1856), which *Ruskin called 'exquisite in every way', and *The Long Engagement* (City Art Gal., Birmingham, 1859). After about 1870, however, his work declined in quality, although he did some good book illustrations, for example for Christina Rossetti's *Sing Song* (1872). He was shy and withdrawn and in later life he lived in suburban obscurity.

Hughes, Robert (1938–). Australian art critic. In 1964 he moved to Europe (first to Italy, then England) and in 1970 settled in New York as art critic of *Time* magazine. His books include *The Art of Australia* (1966, revised 1970), *Heaven and Hell in Western Art* (1969), and *Nothing If Not Critical: Selected Essays on Art and Artists* (1990), He has also made films for television, notably two much-praised series: *The Shock of the New* (1980) and *American Visions* (1996), both with accompanying books. He writes mainly on 20th-cent. art and has a richly deserved reputation as a witty and penetrating observer of the contemporary art scene.

Huguet, Jaime (active *c.*1448–92). Spanish painter, the most prominent figure in the Catalan School during the second part of the 15th cent. Huguet is thought to have settled in Barcelona about 1448; though some historians claim to trace earlier activity in Aragón and Tarragona, where he was born, his fully documented work extends only from 1455. He continued the Catalan tradition of Bernardo *Martorell, but was highly individual in his characterization. His studio produced many sumptuous composite altarpieces of the type that became typical in Spanish art and his work exercised a wide influence on the painting of Catalonia and Aragón. Huguet's work is best represented in Barcelona, particularly in the Museo de Arte de Cataluña.

Humphry, Ozias (1742–1810). English portrait painter. He worked for a time as a *miniaturist in Bath, but settled in London in 1763 on the encouragement of *Reynolds. In 1772 a riding accident which affected his eyes made it necessary for him to abandon miniatures and after a visit to Italy with *Romney in 1773–7 he practised in oils. From 1785 to 1788 he was in India, where he resumed miniature painting, but again he found the work too great a strain. He then took up pastel and in this medium was highly successful (being given the title of Portrait Painter in Crayons to His Majesty in 1792) until he went blind in 1797.

Hundertwasser, Fritz (Friedrich Stowasser) (1928–). Austrian painter and graphic artist. He took the name Hundertwasser in 1949, translating the syllable 'sto' (which means 'hundred' in Czech) by the German 'hundert'. From *c.*1969 he signed his work 'Friedensreich Hundertwasser', symbolizing by 'Friedensreich' (Kingdom of Peace) his boast that by his painting he would introduce the observer into a new life of peace and happiness, a 'parallel world', access to which had been lost owing to the corruption of the instinct for it by the sicknesses of civilization. He often added 'Regenstag' (Rainy Day) to the name—making it in full 'Friedensreich Hundertwasser Regenstag'—on the ground that he felt happy on rainy days because colours then began to sparkle and glow. This exaggerated concern with the name is a symptom of the braggadocio and conceit which are apparent in his work as well as his life. He has a talent for self-advertisement and has built himself up into a picturesque figure given to gnomic utterances about his own significance in the world. Standing outside most contemporary artistic movements, though borrowing from many, he works mainly on a small scale, often in watercolour. His work has sometimes been compared with that of *Klee and *Klimt, but although it is in the same tradition of figurative fantasy it lacks their elegance and wit. In his concern with the dehumanizing aspects of 20th-cent. society, he has been an outspoken critic of modern architecture and his recent work has included the design of an idiosyncratic, multi-coloured, fairytale-like housing unit in Vienna (completed 1986).

Hunt, William Holman (1827–1910). English painter, co-founder of the *Pre-Raphaelite Brotherhood in 1848. He was the only member of the Brotherhood who throughout his entire career remained faithful to Pre-Raphaelite aims, which he summarized as finding serious and genuine ideas to express, direct study from nature in disregard of all arbitrary rules, and envisaging events as they must have happened rather than in accordance with the rules of design. His work was remarkable for its minute precision, its accumulation of incident, and its didactic emphasis on moral or social symbolism, and from 1854 he made several journeys to Egypt and Palestine to paint biblical scenes with accurate local detail. *The Scapegoat* (Lady Lever Art Gal., Port Sunlight, 1854–5), showing the outcast animal on the shore of the Dead Sea, is one of the most famous paintings that resulted from his fanatical devotion to authenticity. His colour tends to be painfully harsh and his sentiment mawkish, but he created some of the most enduring images of the Victorian age, among them *The Hireling Shepherd* (Manchester City Art Gallery, 1851), *The Awakening Conscience* (Tate Gallery, London, 1853), and *The Light of the World* (Keble College, Oxford, 1851–3; a smaller version is in Manchester City Art Gallery, and a larger replica, begun in 1899, is in St Paul's Cathedral, London). Like the other Pre-Raphaelites, Hunt suffered critical attacks early in his career, but the moral earnestness of his work later made it immensely popular with the Victorian public and he made a fortune from the sale of engravings of his paintings. In old age he became a patriarchal figure in the art world and he was awarded the Order of Merit in 1905. In the same year he published his autobiographical *Pre-Raphaelitism and the Pre-Raphaelite Brotherhood*, which is the basic source book of the movement, though somewhat biased.

Hunter, Leslie. See SCOTTISH COLOURISTS.

Hutcheson, Francis (1694–1746). British philosopher. He was the most systematic exponent of the doctrine that beauty is passively perceived by means of an 'inner sense'. In his treatise *An Inquiry into the Original of our Ideas of Beauty and Virtue* (1725) he attempted to work out a law for assessing beauty in terms of the ratio between uniformity and variety.

Huy, Jean de (active first half of 14th cent.). French carver of tomb effigies, active chiefly in Paris. He was a successful example of the prototype of the modern 'monumental mason', working at a time when effigies had become the rule for the ombs of the French nobility and there was a vholesale expansion of the industry and a tenency towards standardization. The tomb of Robert d'Artois in Saint-Denis (*c.*1317) is his work.

Huysum, Jan van (1682–1749). Dutch painter, with Rachel *Ruysch the most distinguished flower painter of his day. He had a European reputation and was much imitated. The light colours he used, the even lighter backgrounds, and the openness of his intricate compositions became distinguishing features of 18th-cent. Dutch flower painting. He occasionally painted subjects other than flowers, including a self-portrait in the Ashmolean Museum, Oxford. His father, **Justus the Elder** (1659–1716), was a flower and landscape painter and he had three painter brothers: **Justus the Younger** (*c.*1684–1707); **Michiel** (d. 1759); and **Jacob** (*c.*1687–1740?), who worked in England and imitated Jan's style.

Hyperrealism. See SUPERREALISM.

Ibbetson, Julius Caesar (1759–1817). English painter. His unusual Christian names were given to him because of his Caesarean birth. He specialized in fairly small landscapes with figures and animals, and his style has been characterized by Sir Ellis *Waterhouse as 'more natural than *de Loutherbourg's, and more civilized than *Morland's'; Benjamin *West called him 'the *Berchem of England'. Ibbetson worked mainly in his native Yorkshire, but also for a time in London and the Lake District, and he visited Java (1789). He worked in watercolour as well as oil and also made etchings. In 1803 he published a treatise on painting. Like his friend Morland, Ibbetson is said to have been given to dissipation, but his work did not obviously suffer because of this as Morland's did.

ICA. See INSTITUTE OF CONTEMPORARY ARTS.

icon. An image of a saint or other holy personage, particularly when the image is regarded by the devotee as sacred in itself and capable of facilitating contact between him or her and the personage portrayed. The term, which derives from the Greek word *eikōn*, meaning 'likeness', has been applied particularly to sacred images of the *Byzantine Church and the Orthodox Churches of Russia and Greece.

iconography. The branch of art history dealing with the identification, description, classification, and interpretation of the subject-matter of the figurative arts. In his book *Studies in Iconology* (1939) Erwin *Panofsky proposed that the term 'iconology' should be used to distinguish a broader approach towards subject-matter in which the scholar attempts to understand the total meaning of the work of art in its historical context. However, in practice an exact distinction between the two terms is rarely made, and 'iconography' is the much more commonly used of the two. The term 'iconography' can also be applied to collections (or the classification) of portraits. Van *Dyck, for example, made a series of etchings of famous contemporaries entitled *Iconography*, and the detailed catalogues of the National Portrait Gallery in London have a section called 'iconography' in the entry for each sitter in which other known portraits of the person represented are listed. Thus it is possible to speak of 'the iconography of Shakespeare' or 'the iconography of Queen Victoria'.

iconology. See ICONOGRAPHY.

iconostasis. In *Byzantine and Russian churches, a screen shutting off the sanctuary from the main body of the church on which *icons were placed.

ideal. A conception of something that is perfect, referring in the visual arts to works that attempt to reproduce the best of nature, but also to improve on it, eliminating the inevitable flaws of particular examples. The notion derives ultimately from Plato's Theory of Ideas, according to which all perceptible objects are imperfect copies approximating to unchanging and imperceptible Ideas or Forms. This idea reappeared with the revival of Platonism in the Italian *Renaissance, and throughout much of subsequent European art the model of ideal beauty was supplied by classical statuary. Its most influential formation was in a lecture by *Bellori delivered before the *Academy of St Luke in Rome in 1664, and published as a Preface to his *Lives* in 1672. Here the true artist is conceived as a seer who gazes upon eternal verities and reveals them to mortal men. It is this gift that separates him from the mere mechanic, the slavish copyist of appearances. The gift can be developed by the study of appropriate models, but the artist must also have a concept of the ideal in his mind. To Bellori, the contemporary artist who best exemplified the doctrine was *Poussin, whose example became binding for the French Academy of the 17th cent. The doctrine provided the philosophical justification for the *Grand Manner, and was the basis of criticism of anti-idealistic artists such as *Caravaggio and *Rembrandt, who were thought to have broken the 'rules' of good art. Although the doctrine has been responsible for much arid art, it has also been an inspiration to such great artists as *Raphael, who said, 'To paint a beautiful woman I must see several, and I have also recourse to a certain ideal in my mind,' and Guido *Reni, who said 'The beautiful and pure *idea* must be in the mind, and then it is no matter what the model is.'

ideal landscape. A type of landscape painting, invented by Annibale *Carracci in the first decade of the 17th cent., in which the elements of the landscape are composed into a grand and highly formalized arrangement suitable as a setting for small figures from serious religious or mythological subjects. It was an extraordinarily influential invention, developed most memorably by *Claude and *Poussin.

illuminated manuscripts. Books written by hand, decorated with paintings and ornaments of different kinds. The word 'illuminated' comes from a usage of the Latin word *illuminare* in connection with oratory or prose style, where it means 'adorn'. The decorations are of three main types: (*a*) *miniatures* or small pictures, not always illustrative, incorporated into the text or occupying the whole page or part of the border; (*b*) *initial letters* either containing scenes (historiated initials) or with elaborate decoration; (*c*) *borders*, which may consist of miniatures, occasionally illustrative, or more often are composed of decorative motifs. They may enclose the whole of the text space or occupy only a small part of the margin of the page.

Manuscripts are for the most part written on skin (*parchment or vellum). From the 14th cent. paper was used for less sumptuous copies. Although a number of books have miniatures and ornaments executed in outline drawing only, the majority are fully coloured. After the preparation of the vellum a sketch was first made. This was often quite rudimentary and omitted many details. The portions destined for gilding were covered with a base in order that the gold would stick more securely, and the gold was burnished. The main blocks of colour were then laid on. Enough of the under-drawing remained visible to guide the later stages of the work. Finally the details of the drapery and complexions were added. The colours were usually mixed with a *tempera medium made with egg and gum dissolved in water.

Though wonderful manuscripts were produced in the 15th cent., illumination by then tended more and more to follow the lead given by painters, and with the invention of printing the illuminated book gradually went out of fashion. During the early years of the 16th cent. fine manuscripts were still being made, but even at this time they were becoming an anachronism. During the 15th and 16th cents. illuminations were added to printed books. These usually consisted of initials and borders; miniatures were less common. some of the finest come from Italy.

illusionism. Term applied in its broadest sense to the basic principle of *naturalistic art whereby verisimilitude in representation causes the spectator in various degrees to seem actually to be seeing the object represented, or the space in which it is represented, even though with part of his mind he knows that he is looking at a pictorial representation and not at the real object or scene. In a somewhat narrower sense 'illusionism' refers to the use of pictorial techniques such as *perspective and foreshortening to deceive the eye (if not the mind) into taking that which is painted for that which is real, or in architecture and stage scenery to make the constructed forms seem visually more extensive than they are. Two specific forms of illusionism in painting are **quadratura*, in which painted architecture appears to extend the real space of a room, and **trompe-l'œil*, in which the spectator is genuinely, if momentarily, tricked into thinking that a painted object (for example, a fly on the picture frame) is a real one.

The term 'illusionism' is also applied to the techniques used for the construction and painting of the 17th-cent. Dutch *peep-show cabinets, which afford entertaining examples of visual deception contrived by means of a rigorous application of the principles of scientific perspective, particularly the principle of the single fixed eye-point. A good example by Samuel van *Hoogstraten is in the National Gallery, London.

impasto. Thickly applied opaque paint retaining the marks of the brush or other instrument of application. Impasto is a feature of *oil painting and certain types of *acrylic, but it is not possible with *watercolour or *tempera. Among artists associated with the use of heavy impasto are *Rembrandt, van *Gogh, and *Auerbach.

Imperial War Museum. See OFFICIAL WAR ART.

Impressionism. Movement in painting originating in the 1860s in France. Impressionism was not a homogeneous school with a unified programme and clearly defined principles, but a loose association of artists linked by some community of outlook and banded together for the purpose of exhibiting. Different artists in the group gave prominence to different ideas within the complex of attitudes which art historians have later regarded as distinctive of the movement. Many artists can strictly speaking be called Impressionists during certain periods of their careers only, and some abandoned Impressionism for a time and later returned to it. Even techniques held to be most characteristic of the Impressionist movement were not uniformly practised by all Impressionists. Yet despite this looseness of structure the movement had a certain coherence.

The first nucleus was formed by *Monet, *Renoir, *Sisley, and *Bazille, who had been fellow students of *Gleyre in the early 1860s and shared a discontent with academic teaching. They established friendships with Camille *Pissarro, *Cézanne, Berthe *Morisot, and Armand *Guillaumin, and the group began to meet regularly in the cafés of Montmartre, and in the studio which Bazille shared with Renoir. The group was joined by *Degas and *Manet. Their meetings included the critics Théodore Duret and Georges Rivière, who in 1877 published a short-lived art review called *L'Impressioniste*, and they were supported by the dealer Paul *Durand-Ruel. In 1873 the *Salon rejected pictures by Pissarro, Monet, Renoir, Cézanne, and Sisley and this served as a spur to their decision to take the then unusual step of organizing independent exhibitions of their works. In the first exhibition, opened in April 1874 at the studios of the photographer Nadar (1820–1910), the artists called themselves 'Société anonyme des artistes peintres, sculpteurs, graveurs'. The title of one of Monet's paintings—*Impression: Sunrise* (Musée Marmottan, Paris)—prompted the journalist Louis Leroy in *Charivari* to dub the whole group 'Impressionists' and the name, coined in derision, was later accepted by the artists themselves as indicative of at least one significant aspect of their aims. There were eight Impressionist exhibitions in all (in 1874, 1876, 1877, 1879, 1880, 1881, 1882, and 1886) and the name 'Impressionist' was used in all except the first, fourth, and last. The exhibitions were not restricted to members of the group proper. The first exhibition included *Boudin, who had encouraged Monet to paint directly from nature. Gustave *Caillebotte joined in the second exhibition and others, such as Mary *Cassatt and *Forain, came in later. Manet never exhibited with the group, preferring to court the favour of the Salon, though with indifferent success. Camille Pissarro was the only one who showed at all eight exhibitions. After the final exhibition the group broke up, and Monet was the only one who continued to pursue Impressionist ideals rigorously.

It is dangerous to lay down rigid criteria for defining so individualistic a group of artists and the Impressionist movement must rather be described in terms of very general attitudes and techniques from which numerous exceptions have to be noted. By and large the group was in opposition to the academic training of the schools, although Manet and Degas at least were well grounded in the principles of *Classical art derived from an extensive study of the older masters. They were in revolt from the basic principle of *Romanticism that the primary purpose of art is to communicate the emotional excitement of the artist and that the recording of nature is secondary. Against this they were generally in sympathy with the *Realist attitude that the emotional condition of the artist is secondary and the primary purpose of art is to record fragments of nature or life in an objective and scientific spirit as impersonally as possible. They repudiated imaginative art, including historical subjects, and were interested rather in the objective recording of contemporary and actual experience. Their outlook was nevertheless distinct from that of Social Realism. Social amelioration was not one of their aims and they saw no merit in the representation of vulgarity or ugliness. Within these limits they varied greatly in the subject-matter of their recording. Renoir gave expression to his delight in pretty women and children. Sisley, Pissarro, and Monet were primarily interested in landscape, but this had little appeal for Degas, who made subjects such as horse races, dances, and laundresses his own.

Their ambition to capture the immediate visual impression rather than the permanent aspects of a subject led the Impressionist landscapists to set great store by painting out of doors (in which they had been anticipated by the *Barbizon School) and on finishing a picture on the spot before the conditions of light should change. Shadows were not painted in grey or black but in a colour complementary to the colour of the object. With the suppression of outline the object tended to lose prominence and Impressionist paintings became paintings of light and atmosphere, a play of direct and reflected colour. Manet, however, was converted to the **plein air* doctrine only after 1870 and Degas never reconciled himself to painting in the open. And even Monet—the most committed exponent of open-air painting—came to rely more and more on working in the studio to revise or retouch his paintings. In doing so, he acknowledged an inherent contradiction in the Impressionist approach, for the more sensitive an artist is to changing atmospheric effects, the less time he has to capture them on canvas before they are gone. In October 1890, when working on his *Haystack* series, he wrote: 'I really am working terribly hard, struggling with a series of different effects, but at this time of year the sun sets so quickly that I can't keep up with it.'

Impressionist painting was at first received with bewilderment or suspicion and frequently abuse. Of the 1876 exhibition *Le Figaro* wrote: 'five or six lunatics, one of them a woman—a collection of unfortunates tainted by the folly of ambition—have met here to exhibit their works. . . . What a terrifying spectacle is this of human vanity stretched to the verge of dementia. Someone should tell M. Pissarro forcibly that trees are never *violet*, that the sky is never the colour of *fresh butter*, that nowhere on earth are things to be seen as he paints them. . . .' Such attitudes died hard amongst conservative artists and critics, and when the academician J.-L. *Gérôme was conducting President Loubet round the exhibitions at the Paris Exposition Universelle of 1900 he stopped him at the door of the Impressionist room with the words: 'Arrêtez, monsieur le Président, c'est ici le déshonneur de la France!' However, Durand-Ruel began to achieve success in the late 1880s, particularly in the USA, and in 1891 an exhibition of Monet's paintings at his gallery in Paris sold out only three days after opening. Acceptance was slow to come in England (during Durand-Ruel's 1905 exhibition at the Grafton Galleries in London—the greatest display of Impressionists ever mounted—not a single picture was sold), but by the time Monet died in 1926 as the grand old man of French painting, sale-room prices for Impressionist paintings had begun climbing towards the astronomical sums commonplace since the 1950s.

The influence of Impressionism was enormous, and much of the history of late 19th-cent. and early 20th-cent. painting is the story of developments from it or reactions against it. The *Neo-Impressionists, for example, tried to put the optical principles of Impressionism on a scientific basis, and the *Post-Impressionists began a long series of movements that attempted to free colour and line from purely representational functions and return to the emotional and symbolic values that the Impressionists had sacrificed in their concentration on the fleeting and the casual.

imprimatura. A thin layer of transparent colour applied to a *ground. It reduced the absorbent quality of the ground and could also be used as a middle tone in the painting.

Indépendants, Salon des. See SALON DES INDÉPENDANTS.

Independent Group. A small and informal discussion group that met intermittently between 1952 and 1955 at the *Institute of Contemporary Arts, London. The members included Lawrence *Alloway, Richard *Hamilton and Eduardo *Paolozzi and the first phase of British *Pop art grew out of the group.

Index of American Design. A project undertaken by the Federal Government of the USA during the administration (1933–45) of President Franklin D. Roosevelt to give relief to unemployed artists. Its aim was to record the *folk arts and crafts of the United States from early Colonial times to the end of the 19th cent. Under a national director individual States organized local groups of artists. With professional guidance they made faithful watercolour renderings of objects in museums and private collections. The Index contains coloured drawings of ceramics, furniture, wood-carving, glassware, metalwork, tools and utensils, textiles, costumes, and other objects not so readily classifiable. Some 15,000 finely executed drawings and about 5,000 photographs may be studied at the National Gallery of Art, Washington, and are also sent out on exhibition.

Indiana, Robert (1928–). American painter, sculptor, and graphic artist. His original name was Robert Clark, but he adopted the name of his native state as his own. He is regarded as one of the leading American *Pop artists, and although he has done some figurative paintings he is best known for pictures involving geometric shapes emblazoned with lettering and signs. His vivid colours often create an effect of optical ambiguity distantly reminiscent of *Op art. In 1964 he collaborated with Andy *Warhol in a film *Eat* and executed a commission for the exterior decoration of the New York State Pavilion at the World's Fair consisting of a 6-m. sign, EAT.

industrial art. Term embracing all design related to objects produced in industry, from bottle tops to aeroplanes. Together with the term 'industrial design' it has advantages over *applied art or *arts decoratifs*, which carry an overtone of artificiality or preciosity and embrace both more and less than 'industrial art'. See also BEHRENS, PETER.

Informalism. See ART INFORMEL.

Inglés, Jorge (active mid-15th cent.). Painter, probably of northern origin (his name means 'George the Englishman'), who was one of the first representatives of the Hispano-Flemish style in Castile. His only documented work is an altarpiece commissioned in 1455 by the Marquess of Santillana (Duke of Infantado Coll.), which suggests knowledge of Rogier van der *Weyden.

Ingres, Jean-Auguste-Dominique (1780–1867). French painter, born at Montauban, the son of a minor painter and sculptor, **Jean-Marie-Joseph Ingres** (1755–1814). After an early academic training in the Toulouse Academy he went to Paris in 1796 and was a fellow student of *Gros in *David's studio. He won the *Prix de Rome in 1801 with a *Neoclassical history painting *The Envoys of Agamemnon* (École des Beaux-Arts, Paris), which was praised by *Flaxman. Owing to the state of

France's economy he was not awarded the usual stay in Rome until 1807. In the interval he produced his first portraits. These fall into two categories: portraits of himself and his friends, conceived in a *Romantic spirit (*Self-portrait*, Musée Condé, Chantilly, 1804), and portraits of well-to-do clients which are characterized by purity of line and enamel-like colouring (*Mlle Rivière*, Louvre, Paris, 1805). He was commissioned to paint *Bonaparte as First Consul* (Musée des Beaux-Arts, Liège, 1804) and *Napoleon as Emperor* (Musée de l'Armée, Invalides, Paris, 1806). These early portraits are notable for their calligraphic line and expressive contour, which had a sensuous beauty of its own beyond its function to contain and delineate form. It was a feature that formed the essential basis of Ingres's painting throughout his life.

During his first years in Rome he continued to execute portraits and began to paint bathers, a theme which was to become one of his favourites (*The Valpinçon Bather*, Louvre, Paris, 1808). He remained in Rome when his four-year scholarship ended, earning his living principally by pencil portraits of members of the French colony. But he also received more substantial commissions, including two decorative paintings for Napoleon's palace in Rome (*Triumph of Romulus over Acron*, École des Beaux-Arts, 1812; and *Ossian's Dream*, Musée Ingres, 1813).

In 1820 he moved from Rome to Florence, where he remained for 4 years, working primarily on his Raphaelesque *Vow of Louis XIII*, commissioned for the cathedral of Montauban. Ingres's work had often been severely criticized in Paris because of its 'Gothic' distortions, and when he accompanied this painting to the *Salon of 1824 he was surprised to find it acclaimed and himself set up as the leader of the academic opposition to the new Romanticism. (*Delacroix's *Massacre of Chios*, dubbed by Gros 'the massacre of painting', was shown at the same Salon.) Ingres stayed in Paris for the next ten years and received the official success and honours he had always craved. During this period he devoted much of his time to executing two large works: *The Apotheosis of Homer*, for a ceiling in the *Louvre (installed 1827), and *The Martyrdom of St Symphorian* (Salon, 1834) for the cathedral of Autun. When the latter painting was badly received, however, he accepted the Directorship of the French School in Rome, a post he retained for 7 years. He was a model administrator and teacher, greatly improving the school's facilities, but he produced few major works in this period. The most important was his *Antiochus and Stratonice* (Musée Condé, Chantilly, 1834–40), minutely executed with sharp colours and described by him as a 'large historic miniature'.

In 1841 he returned to France, once again acclaimed as the champion of traditional values in art. He was heartbroken when his wife died in 1849, but he made a happy second marriage in 1852, and he continued working with great energy into his 80s. One of his greatest works, the extraordinarily sensuous *Turkish Bath* (Louvre, 1863), dates from the last years of his life. At his death he left a huge bequest of his work (several paintings and more than 4,000 drawings) to his home town of Montauban and they are now in the museum bearing his name there.

Ingres is a puzzling artist and his career is full of contradictions. Yet more than most artists he was obsessed by a restricted number of themes and returned to the same subject again and again over a long period of years. He was a bourgeois with the limitations of a bourgeois mentality, but as *Baudelaire remarked, his finest works 'are the product of a deeply sensuous nature'. The central contradiction of his career is that although he was held up as the guardian of classical rules and precepts, it is his personal obsessions and mannerisms that make him such a great artist. His technique as a painter was academically unimpeachable—he said paint should be as smooth 'as the skin of an onion'—but he was often attacked for the expressive distortions of his draughtsmanship; critics said, for example, that the abnormally long back of *La Grande Odalisque* (Louvre, 1814) had three extra vertebrae. It was in the quality of line itself and its use for expressive effect that he excelled.

Unfortunately the influence of Ingres was mainly seen in those shortcomings and weaknesses which have come to be regarded as the hallmark of inferior academic work. He had scores of pupils, but *Chassériau was the only one to attain distinction. As a great calligraphic genius his true successors are *Degas and *Picasso.

ink. Coloured fluid used for writing, drawing, or printing. Inks usually have staining power without body, but printers' inks are pigments mixed with oil and varnish, and are opaque. The use of inks goes back in China and Egypt to at least 2500 BC. They were usually made from lampblack (a pigment made from soot) or a red ochre ground into a solution of glue or gums. These materials were moulded into dry sticks or blocks, which were then mixed with water for use. Ink brought from China or Japan in such dry form came to be known in the West as 'Chinese ink' or 'Indian ink'. The names are also given to a similiar preparation made in Europe, a dispersion of carbon black in water, usually stabilized by some alkaline solution. In drawing ink can be applied in fine lines with the pen or broad washes with the brush.

inlay. A decorative technique in wood- or metal-working, in which a substance of a different kind is embedded in a solid support so that its surface is more or less continuous with that of the matrix. See also NIELLO: INTARSIA.

Innes, J. D. (James Dickson) (1887–1914). British painter. He is best known for his landscapes, especially those painted in his native Wales, where he worked with his friend Augustus *John in 1911 and 1912. He usually worked on a fairly small scale, often on wooden panels. His early work was in an *Impressionist manner, but he later developed a more expressive *Post-Impressionist style combining hot colour and decorative pattern. He died young of tuberculosis.

Inness, George (1825–94). American landscape painter. He was without formal training but developed his style in the course of frequent visits to Europe, being particularly influenced by the *Barbizon School. His work falls into two fairly distinct periods. In the first he attempted to bring greater breadth to American *Romantic realism, dissolving hard outline into a play of atmosphere and colour, but with something of the ordered beauty of *Claude (*Peace and Plenty*, Met. Mus., New York, 1865). From *c.*1859, when he went to live in the village of Medfield outside Boston, his style began to change to a more intimate manner of landscape in which he chose deliberately unpicturesque subjects and relied for pictorial appeal on subtle harmonies of colour and broad massing of light and shade. Inness has often been considered the outstanding American landscape painter of the 19th cent.

His son, **George Inness, jun.** (1854–1926), was also a painter and published an account of his father's career—*Life, Art, and Letters of George Inness*—in 1917.

Inshaw, David. See BLAKE, PETER.

installation. Term that came into vogue during the 1970s for an *assemblage or *environment constructed in the gallery specifically for a particular exhibition.

Institute of Contemporary Arts (ICA), London. Cultural centre founded by Roland *Penrose and Herbert *Read in 1947 to encourage new developments in the arts and cater for some of the functions fulfilled by the *Museum of Modern Art in New York, organizing exhibitions, lectures, films, and so on. Its original home was in Dover Street, but it moved to Nash House, the Mall, in 1968. Many leading artists have been members of the ICA and it has played an important role in certain developments; for example, in the 1950s it was the cradle of British *Pop art (see INDEPENDENT GROUP).

intaglio. Engraved or incised work on gemstones, glass, ceramics, or similar materials in which the design is sunk beneath the surface of the material. It is the opposite of *cameo, in which the design is in relief. An 'intaglio' often means an engraved gem in which figures or other devices are carved into the stone so that when pressed upon a plastic material it produces a likeness in relief. The intaglio is the most ancient form of engraved gem, the earliest known being the Babylonian cylinder seals dating from *c.*4000 BC.

In the graphic arts, 'intaglio printing' refers to any process of printmaking in which the parts of the plate or block that will take the ink are incised into it rather than raised above it ('relief printing'). *Etching is thus a form of intaglio printing (see PRINT). *Lithography, in which prints are made from a flat surface, can be characterized as 'planographic printing'.

intarsia. Method of creating a picture or design on a wooden surface (typically a piece of furniture) by *inlaying it with differently coloured woods and other materials such as mother-of-pearl and ivory.

interlace. A type of pattern in which a number of lines or bands are interlaced or plaited together on the principle that the interlacing elements pass over and under each other alternately. Interlace pattern often takes the form of an 'interlacement band' in which the interlacing elements are symmetrical about a longitudinal axis and can be continued indefinitely. Sometimes, however, the interlace may bend back upon itself as in the 'rosette' or other form of closed pattern.

Interlacement patterns are found in many different styles and periods, though they were more popular in some than in others. They vary considerably and take on many of the characteristics of the style in which they occur. They can be found in pottery and textiles, in manuscript decoration, and in the decoration of craft objects, and they have extensive uses in architecture.

International Gothic. Style in painting, sculpture, and the decorative arts that spread widely over western Europe between *c.*1375 and *c.*1425. The salient characteristics of the style were courtly elegance and delicate naturalistic detail. It was marked also by a new interest in secular themes of aristocratic life, often with an artificially bucolic tone. The antecedents of the style have been traced to the court style which arose in France about the middle of the 13th cent. with its canon of an elongated and supple human form modelled with a new appreciation of sensuous qualities, and to *Simone Martini's merging of French grace with the Italian naturalism of *Giotto and *Duccio. In France Jehan *Pucelle with his linear arabesque of movement and elegance injected new life into the French style, and laid a basis for the Franco-Flemish art which at the court of the Duc de Berry (the brother of Charles V (1337–80) of France) brought into being a fusion of styles which became genuinely international in character. Netherlandish artists working there *c.*1400—Jean de *Bondol, André *Beauneveu, Melchior *Broederlam, the *Limburg brothers—brought the Flemish taste for realism and talent for keen observation into unison with the new Italian naturalism and yet adapted themselves to the aristocratic refinement and spirit of the French court. By the end of the 14th cent. Franco-Flemish Burgundy and Lombardy were established as the main centres of the style.

The Franco-Flemish style flourished particularly at Mehun-sur-Yèvre (where André Beauneveu was in charge of painting and sculpture), Bourges, Dijon (at the Chartreuse de Champmol), and finally in Paris. From France the style spread to other countries of Europe. In northern Italy such artists as *Gentile da Fabriano were attracted by the sinuous line, the polished elegance, the exotic and bizarre features of the style, and with *Stefano da Zevio and *Pisanello International Gothic reached its most ornate and flamboyant. In Bohemia the style is apparent in the *Trebon Altarpiece* (NG, Prague, *c.*1390; see BOHEMIAN SCHOOL). It spread to the Rhineland towns in the early 15th cent. and was modified, for example in the work of *Master Francke, by Germanic middle-class affectation. The style spread to Spain through Avignon and Perpignan, which was at that time Spanish, and was introduced to Catalonia primarily by Luis *Borrassá. In England, the finest work in the style is the celebrated Wilton Diptych (NG, London), which testifies to how genuinely international the International Gothic style was, for although it is a painting of extraordinary beauty and must be from the hand of an artist of the highest rank, authorities disagree as to whether he was English, French, or Italian. It shows Richard II being presented to the Virgin and Child by John the Baptist, Edward the Confessor, and Edmund, king and martyr (his patron saints), but its purpose and significance are uncertain. It may date from late in Richard's reign (heraldic evidence suggests it cannot be earlier than *c.*1395), but some scholars think it is a posthumous memorial.

International Gothic was perpetuated into the 15th cent., long after its decline in France, by the strong commercial links which bound southern Spain with Genoa, Naples, and Venice.

Intimisme. Term applied to a type of intimate domestic *genre painting, more or less *Impressionist in technique, that is particularly associated with *Bonnard and *Vuillard.

intonaco. The final layer of lime plaster on which *fresco painting is done.

Isakson, Karl (1878–1922). Swedish painter, mainly active in Denmark. In Paris (1913–14) Isakson was one of the first Scandinavians to confront the problems of formal structure raised by *Cézanne and the *Cubists and his work had considerable influence on other Scandinavian artists. He painted still lifes, nudes, and landscapes, and also religious subjects, which he treated with the same regard for structure.

Isenbrandt. See YSENBRANDT.

Israëls, Jozef (1824–1911). Dutch painter. He studied in Amsterdam and Paris and began his career as a portrait and historical painter, but in the 1850s he turned to the kind of work for which he is principally known—scenes of peasants and fishermen and the milieu in which they lived. In the 1870s he settled in The Hague and became one of the leading members of the *Hague School. He has been called 'the Dutch *Millet' and during his lifetime he won great popularity because of his piously sentimental approach. His son, **Isaäc** (1865–1934), also worked at The Hague, but in a style almost completely independent of his father's. His pictures of the social life of his time are influenced by *Breitner and characterized by the vivid colours and vigorous brushwork of the *Impressionists. Works by both father and son are in the Municipal Museums of Amsterdam and The Hague.

Itinerants. See WANDERERS.

Itten, Johannes (1888–1967). Swiss painter, designer, writer on art, teacher, and administrator. In 1916 he opened his own school of art in Vienna, then from 1919 to 1923 he taught at the *Bauhaus, where he was in charge of the 'Preliminary Course', obligatory for all students. In 1923 he left the Bauhaus and opened another school of his own in Berlin, then from 1932 to 1938 he taught at the Krefeld School of Textile Design. In 1938 he settled in Zurich, where he held four posts concurrently—as Director of the School of Arts and Crafts, the Museum of Arts and Crafts, the Rietberg Museum, and the School of Textile Design. He held the first three posts until 1953 and retired from the fourth in 1961. Itten wrote several books on art theory and his work as a painter consisted mainly of geometrical abstractions exemplifying his researches into colour. However, he is best remembered as a teacher, especially for his Preliminary Course at the Bauhaus, which had a great influence on instruction in other art schools. He emphasized the importance of knowledge of materials, but also encouraged his pupils to develop their imaginations through, for example, automatic writing (see AUTOMATISM). His mystical ideas were opposed to the technological outlook of *Gropius (their quarrels caused Itten's departure from the Bauhaus) and he had a reputation as a crank (he followed an obscure faith called Mazdazhan, shaved his head, and wore a long robe), but he influenced many of his students. Frank Whitford (*Bauhaus*, 1984) describes him as 'a perplexing mixture of saint and charlatan'.

Ivanov, Alexander (1806–58). Russian painter. He was born in St Petersburg and studied there at the Academy of Fine Arts under his father, the painter **Andrey Ivanov** (1772–1848). Most of his career was spent in Rome, where he settled in 1831. Initially he was preoccupied with subjects from the classical world, but partly under the influence of the *Nazarenes (he was a friend of *Overbeck) he turned to religious painting, and his fame is inseparable from his main work, which occupied him for about 20 years, *Christ's First Appearance to the People* (Tretyakov Gal., Moscow). This enormous painting achieved European celebrity long before its completion, but proved disappointing when it was finally exhibited in St Petersburg in 1858, its *Raphaelesque composition being at odds with the naturalistic setting and details, the result of hundreds of preparatory studies. Ivanov accompanied the painting to St Petersburg and died of cholera a few months afterwards. He had no immediate followers, but the moral sincerity of his work was influential on many Russian painters, notably *Kramskoi.

ivory. A hard, smooth, creamy white substance forming the main part of the tusks of elephants and some other animals, used as a carving material from the earliest times. Elephant and walrus tusks have been the commonest source of ivory, but carvings made from narwhal and rhinoceros horn, stag-horn, and even bone, have been embraced by the term. True ivory is easily worked with saws, chisels, drills, and rasps, but the size and shape of a carving are usually limited by the dimensions of a tusk rarely exceeding 18 cm. in diameter. In statuettes, the curvature of the tusk has sometimes been exploited to give a graceful swing to the figure, notably in *Gothic figures of the Virgin. Though ivory was often painted in the Middle Ages, its natural lustre, translucence, and satin smoothness have always made it the preferred material for small objects such as chessmen that must be handled to be appreciated. In the ancient world it was classed with gold and precious stones as a luxury material, and the Greeks used it for colossal *chryselephantine (gold and ivory) cult-statues. After the 14th cent. ivory-carving declined steadily in Europe, but there were revivals in the 17th and 18th cents. The total eclipse of the art since the beginning of the 19th cent. has been broken only by brief periods of interest.

Jackson, Alexander Young (1882–1974). Canadian landscape painter, active mainly in Toronto, where he settled in 1913 after extensive travels in Europe. He was one of the leading artists in the *Group of Seven and in the latter part of his long career became a venerated senior figure in Canadian painting. Jackson visited virtually every region of Canada, including the Arctic, and responded particularly to the hilly region of rural Quebec along the St Lawrence River. From 1921 he returned there almost every spring, and the canvases he prepared from sketches made there are probably his finest work. *Early Spring, Quebec* and *Laurentian Hills, Early Spring* (Art Gal. of Ontario, Toronto, 1926 and 1931 respectively) are but two that, with their easy, rolling rhythms, and rich and full colouring, had a far-reaching impact on Canadian landscape painting.

Jacobean. Term that in its strictest usage is applied to works produced in England between 1603 and 1625, the reign of James I. It is also applied to slightly later architecture, decoration, and furniture, when this includes lavish ornament derived mainly from Flemish engravings, a somewhat clumsy *strapwork being the dominant form.

Jacomart. See BAÇO.

Jacque, Charles-Emile. See BARBIZON SCHOOL.

Jamesone, George (1589/90–1644). Scottish portraitist, active in Aberdeen and Edinburgh. His name has been indiscriminately applied to a great number of Scottish portraits of the period as he is virtually the only 17th-cent. Scottish painter about whom anything is known. A now discounted tradition has it that he trained with *Rubens and he has been flatteringly called 'the Scottish van *Dyck', but his style was closer to Cornelius *Johnson's. It is difficult to assess, however, as most of the works that are certainly by him are in a bad state of preservation. John Michael *Wright was his pupil.

Janet. See CLOUET.

Janis, Sidney (1896–1989). American art dealer and writer on art. Between the departure of Peggy *Guggenheim from the USA in 1947 and the rise of Leo *Castelli in the 1960s he was the most important figure in promoting the work of avant-garde American artists, particularly the *Abstract Expressionists. Janis wrote *Abstract and Surrealist Art in America* (1944) and (with his wife Harriet) *Picasso: The Recent Years, 1939–1946* (1946). He was also interested in *naïve art and in 1939 'discovered' one of the outstanding American naïve painters, Morris Hirshfield (1872–1946).

Jansens, Cornelis. See JOHNSON, CORNELIUS.

Janssens, Abraham (*c.*1575–1632). Flemish figure and portrait painter, active mainly in Antwerp. He was in Rome in 1598 and back in Antwerp by 1601. A second visit to Italy seems likely, for although in 1601 he was painting in a *Mannerist style (*Diana and Callisto*, Museum of Fine Arts, Budapest), by 1609 (*Scaldis and Antwerpia*, Musée Royal, Antwerp) his work had become much more solid, sober, and classical, suggesting close knowledge of *Caravaggio in particular. For the next decade Janssens was one of the most powerful and individual painters in Flanders, but during the 1620s his work became less remarkable as he fell under the all-pervasive influence of *Rubens. His pupils included Gerard *Seghers and Theodoor *Rombouts.

Japanese prints. See UKIYO-E.

Jawlensky, Alexei von (1864–1941). Russian *Expressionist painter, active mainly in Germany. Originally he was an army officer, but in 1906 he resigned his commission and moved to Munich to devote himself completely to art. Munich was to be his home until the outbreak of the First World War, but he travelled a good deal in this period, making several visits to France, for example (he was the first of his Munich associates to have direct contact with advanced French art). In 1905 he met *Matisse in Paris and was influenced by the strong colours and bold outlines of the *Fauves. He combined them with influences from the Russian traditions of icon painting and peasant art to form a highly personal style that expressed his passionate temperament and mystical conception of art. A mood of melancholy introspection—far removed from the ebullience of Fauvism—is characteristic of much of his work and it has been said that he 'saw Matisse through Russian eyes'. In 1909 he was one of the founders of the *Neue Künstlervereinigung, and apart from *Kandinsky he was the outstanding artist of the group. His most characteristic works of this period are a series of powerful portrait heads, begun in 1910 (*Portrait of Alexander Sacharoff*, Städtisches Museum, Wiesbaden, 1913). On the outbreak of war in 1914 Jawlensky took refuge in Switzerland, where he remained until 1921. His work there included a series of 'variations' on the view from a window—small, semi-abstract landscapes with a meditative, religious aura. Like Kandinsky and others, Jawlensky believed in a correspondence between colours and musical sounds and he named these pictures *Songs without Words*. In 1918 he began a series of nearly abstract heads, in which he reduced the features to a few curves and lines. Unlike Kandinsky, however, he always based his forms on nature. From 1921 he lived in Wiesbaden, and in 1924 he joined with Kandinsky, *Klee, and *Feininger to form the *Blaue Vier. From 1929 he suffered from arthritis and by 1938 this had forced him to abandon painting completely.

Jean de Bruges. See BONDOL, JEAN DE.

Jean de Liège (active 1360–75). Flemish sculptor who worked in Paris on portraits and tomb sculpture for Charles V (1337–80) and other members of the royal family. His works have something of the robust quality of the Netherlandish *illuminator Jean de *Bondol. He is not to be confused with a sculptor and cabinet-maker of the same name active in Dijon *c*.1390.

Jeanneret, Charles-Édouard. See LE CORBUSIER.

Jehan de Paris. See PERRÉAL, JEAN.

Jervas, Charles (*c*.1675–1739). Irish painter, active mainly in London. His surname was pronounced, and often spelled, Jarvis. After studying with *Kneller he spent several years in Italy, particularly Rome, before settling in London in 1709. Jervas had a great reputation and succeeded Kneller as Principal Portrait Painter to King George II in 1723, but his fame depended on his friendship with various literary figures, who trumpeted his praises, rather than on the quality of his work, which does not rise above the level of that of any other of Kneller's pupils or followers.

Jeune Peinture Belge. An avant-garde artists' association founded in Brussels in 1945. Although the members of the group (which included *Alechinsky and *Bury) were strongly individualistic in their aims and had no common programme, they were basically abstract in their outlook and were influenced particularly by *Expressionism and by the expressive abstraction of the post-war School of *Paris. The group dissolved in 1948.

Joest van Calcar, Jan (*c*.1450–1519). Netherlandish painter, recorded in Calcar, from which he takes his name, by 1480. His major work—*The Life of Christ* (1505–8)—is still in the church of St Nicholas there. It shows the influence of *Geertgen, particularly in its delicate handling of landscape, but has a dignity that is personal. From 1509 Joest lived in Haarlem.

John, Augustus (1878–1961). British painter and graphic artist. He studied at the *Slade School, 1894–8. In his early days there 'he appeared a neat, timid, unremarkable personality' (*DNB*), but after injuring his head diving into the sea while on holiday in Pembrokeshire in 1897 he became a dramatically changed figure, described by Wyndham *Lewis as 'a great man of action into whose hands the fairies had placed a paintbrush instead of a sword'. He grew a beard and became the very image of the unpredictable bohemian artist. His work, too, changed dramatically; previously it had been described by *Tonks as 'methodical', but it became vigorous and spontaneous, especially in his brilliant drawings—his draughtsmanship was already legendary by the time he left the Slade. In the first quarter of the 20th cent. John was identified with all that was most independent and rebellious in British art and he became one of the most talked-about figures of the day. In 1911–14 he led a nomadic life, sometimes living in a caravan and camping with gypsies. As well as romanticized pictures of gypsy life he painted deliciously colourful small-scale landscapes, sometimes working alongside his friend J. D. *Innes. During the same period he also painted ambitious figure compositions, with stylized forms that bring him close to French *Symbolist painters (*The Way Down to the Sea*, Lamont Art Gallery, Exeter, New Hampshire, 1909–11). In the First World War he was an *Official War Artist. It is as a portraitist, however, that John is best remembered. He was taken up by society and painted a host of aristocratic beauties as well as many of the leading literary figures of the day. Increasingly, however, the painterly brilliance of his early work degenerated into flashiness and bombast, and the second half of his long career added little to his achievement, although he remained a colourful, newsworthy figure until the end of his life. He was one of the few British artists who have become familiar to the general public, and his image changed from that of rebel to Grand Old Man (he was awarded the Order of Merit in 1942). He wrote two volumes of autobiography,

Chiaroscuro, 1952, and *Finishing Touches*, posthumously published in 1964. A new edition entitled *The Autobiography of Augustus John* appeared in 1975.

John, Gwen (1876–1939). British painter. She was the sister of Augustus *John, but his complete opposite artistically, as she was in personality, living a reclusive life and favouring introspective subjects. After studying at the *Slade School, 1895–8, she took lessons in Paris from *Whistler, and adopted from him the delicate greyish tonality that characterizes her work. In 1899 she returned to London, but in 1904 she settled permanently in France, living first in Paris (earning her living modelling for other artists—including *Rodin, who became her lover), then from 1911 in Meudon, on the outskirts of the city. In 1913 she became a Catholic, and she said 'My religion and my art, these are my life.' Most of her paintings depict single figures (typically girls or nuns) in interiors, painted with great sensitivity and an unobtrusive originality. Good examples are her self-portraits in the Tate Gallery and National Portrait Gallery, London. She had only one exhibition devoted to her work during her lifetime (at the New Chenil Galleries, London, in 1936) and at the time of her death was little known. However, her brother's prophecy that one day she would be considered a better artist than him has been fulfilled, for as his star has fallen hers has risen, and since the 1960s she has been the subject of numerous books and exhibitions.

Johns, Jasper (1930–). American painter, sculptor, and printmaker. His career has been closely associated with that of his friend Robert *Rauschenberg, and they are considered to have been largely responsible for the move away from *Abstract Expressionism to the types of *Pop art and *Minimal art that succeeded it. In the early 1950s he worked as a commercial artist in New York, doing display work for shop windows. He began to emerge on the art scene in 1955 and had his first one-man show at Leo *Castelli's gallery in New York in 1958. This was an enormous success, and since then he has become one of the most famous (and wealthy) living artists. Much of his work has been done in the form of series of paintings presenting commonplace two-dimensional objects—for example, *Flags*, *Targets*, and *Numbers*—and his sculptures have most characteristically been of equally banal subjects such as beer-cans or brushes in a coffee tin. Such works—at one and the same time laboriously realistic and patently artificial—are seen by his admirers as brilliant explorations of the relationship between art and reality; to others, they are as uninteresting as the objects depicted.

Johnson, Cornelius (Cornelis Jonson or Jansens van Ceulen) (1593–1661). Anglo-Dutch portrait painter born in London of Dutch parents. He perhaps trained in Holland, and he settled there in 1643, but he worked mainly in London, where he had an extensive practice. Johnson was at his best when working on a fairly small scale, showing a sensitive gift for characterization. His work is well represented in the National Portrait Gallery and the Tate Gallery, London.

Johnson, Gerard the Elder (originally Garet Janssen) (d. *c*.1611). Dutch-born sculptor and mason who came to England from Amsterdam in about 1567 as a refugee from the Wars of Religion. He settled in Southwark and built up a large practice chiefly as a tomb maker, though chimney-pieces and basins for fountains were also made in the workshop, in which he was assisted by his sons, **Bernard**, **John**, **Nicholas**, and **Gerard the Younger**.

A good example of the accomplished but uninspired work Gerard the Elder produced is the tomb of the 2nd Earl of Southampton at Titchfield, Hampshire (1592). Gerard the Younger's name lives on because he made the monument to Shakespeare (d. 1616) in Holy Trinity church, Stratford-upon-Avon. The most distinguished artist in the family, however, seems to have been Bernard, as he is said to have been the principal mason for Northumberland House in the Strand and Audley End, Essex, which, although partially demolished, is still, in Sir John Summerson's words, 'the most powerful and impressive of Jacobean houses'.

Johnston, Frank. See GROUP OF SEVEN.

Jones, Allen (1937–). British painter, printmaker, sculptor, and designer, one of the most committed exponents of *Pop art. Although he has worked primarily as a painter, printmaker, and designer, he is best known to the public for a distinctive type of sculpture in which figures of women—more or less life-size, dressed in fetishistic clothing, and with what Jones calls 'high definition female parts'—double as pieces of furniture; for example, a woman on all fours supporting a sheet of glass on her back becomes a coffee table, and a standing figure with outstretched hands becomes a hatstand. He began making such sculptures in the late 1960s and was still producing them in the 1990s, although in a manner that he calls 'less aggressive' and 'easier to take' (they have come in for a good deal of criticism for alleged demeaning of women as sex objects; an article in the feminist journal *Spare Rib* in 1973 suggested that they expressed a castration complex). His work as a designer has included sets and costumes for the erotic review *Oh! Calcutta!* (1969).

Jones, David (1895–1974). British painter, engraver, and writer. A convert to Roman Catholicism in 1921, he met Eric *Gill in 1922 and under his influence achieved a sense of purpose (his studies at the Camberwell School of Art, 1909–15, had left him, as he said, 'completely muddle-headed as to the function of art in general'). Gill not only introduced him to the craft of engraving on wood, but also guided him to a conception of art that rejected the current concern with formal properties in favour of an art that aspired to reveal universal and symbolic truths behind the appearance of things. He worked mainly in pencil and watercolour, his subjects including landscape, portraits, still life, animals, and imaginative themes; Arthurian legend was one of his main inspirations. As a writer he is best known for *In Parenthesis* (1937), a long work of mixed poetry and prose on the subject of the First World War (in which he had fought). T. S. Eliot declared this to be a work of genius and it was awarded the Hawthornden Prize. After the Second World War Jones retired to Harrow and devoted himself mainly to calligraphic inscriptions in the Welsh language (he was of Welsh extraction). His work is well represented in the Tate Gallery.

Jones, Inigo (1573–1652). English architect, stage designer, draughtsman, and painter. Jones was one of the greatest of English architects and certainly the most influential, introducing a pure classical style based on the work of the Italian architect Andrea Palladio (see PALLADIANISM) to a country where *Renaissance influence had previously been fairly superficial. It was not until he was in his 40s, however, that he showed his genius as an architect, and the first known mention of him as an artist is as a 'picture maker' in 1603. No paintings certainly by him are known, but his drawings survive in large numbers (the finest collection is at Chatsworth). They are mainly costume and scenery designs for the court masques, on which he worked from 1605 to 1640, and in which he introduced movable scenery and the proscenium arch into England. Inigo's lively and fluent style as a draughtsman reflects two lengthy visits to Italy (*c.*1600 and 1613–14), the second of them accompanying the great collector the Earl of *Arundel. He advised Arundel on the purchase of Italian antiques while developing his own knowledge of Italian and *antique architecture, and his knowledge as well as his skills gave him immense prestige in England at the courts of James I and Charles I. His principal collaborator in the masques was the formidable Ben Jonson (1572–1637), with whom he had a notorious running feud about the rival claims of words and spectacle.

Few of Jones's buildings survive in anything like their original state. The most important are the Queen's House at Greenwich (1616–35) and the Banqueting House in Whitehall (1619–22), with its painted ceiling by *Rubens. Jones's posthumous fame was at its height during the early 18th-cent. Palladian movement, when he was a hero and an inspiration to Lord *Burlington and William *Kent.

Jones, Thomas (1742–1803). Welsh landscape painter, a pupil of Richard *Wilson. He was in Italy 1776–83, and although he painted some ambitious *classical landscapes he is now best known for his remarkably fresh and unaffected oil sketches done in and around Naples and Rome (good examples are in the National Museum of Wales at Cardiff). They are among the earliest British examples of this kind of open-air sketch and have a directness that looks forward to *Corot.

Jongkind, Johan Barthold (1819–91). Dutch landscape painter and etcher who had close affinities with the French *Impressionists. Although he was better appreciated during his lifetime than van *Gogh, in some ways his career is similar to that of his more famous countryman. Both artists made a greater impression abroad than in their own country; both failed to adjust to the society of their time; both were troubled by serious psychological problems; and sensational aspects of their lives—in Jongkind's case it was alcoholism—have interfered with a balanced appraisal of their achievement.

Jongkind studied in The Hague with the *Romantic painter *Schelfhout. In 1846 he moved to Paris, and from then onwards he was in close touch with leading French artists. He worked and exhibited with members of the *Barbizon School, and during the 1860s played an important part in the development of Impressionism. His marine pictures and views of ports, which are beautiful studies of the effects of air and atmosphere, particularly influenced *Monet and *Boudin. Unlike some of the Impressionists he did his oil paintings of open-air scenes in his studio, basing them on the spontaneous drawings and watercolours he made out of doors.

Jonson van Ceulen, Cornelis. See JOHNSON, CORNELIUS.

Joos van Cleve (*c.*1490–1540). Netherlandish painter, born presumably at Cleves in the lower Rhine region and active mainly in Antwerp, where he became a master painter in 1511. He was dean of the painters' guild in 1515 and 1525 and seems to have been one of the most productive Antwerp painters of his time, but his career is ill-defined. He is generally identified with the Master of the Death of the Virgin, so-called after altarpieces of this subject in Cologne (Wallraf-Richartz-Mus., 1515) and Munich (Alte Pinakothek), but his eclectic style and lack of documented works makes attributions difficult. There is a flavour of *Leonardo in some of his works, and he may have visited Italy. Almost certainly he worked in France and there are several portraits of Francis I (1494–1547) and his wife attributed to him, as is a portrait of Henry VIII of England (Royal Coll.). According to van *Mander he collaborated with Joachim *Patenier: a *Rest on the Flight into Egypt* (Musées Royaux, Brussels) is possibly a joint work.

Joos's son, **Cornelis van Cleve** (1520–67), was also a painter. He was known as 'Sotte Cleve' (Mad Cleve) after becoming insane in 1554—a result of failing to win the post of Court Painter to Philip II of Spain.

Joos van Wassenhove (active *c.*1460–80). Netherlandish painter, part of whose career was spent in Italy, where he was known as Giusto da Guanto (Justus of Ghent). He became a member of the Antwerp Guild in 1460, but by 1464 had moved to Ghent, where he was a friend of Hugo van der *Goes. At some time after 1468 he went to Rome, and by 1472 had settled in Urbino, where he worked for Duke Federico da *Montefeltro. Joos's only documented work is *The Communion of the Apostles* (also known as *The Institution of the Eucharist*, 1472–4), which is still at Urbino, in the Galleria Nazionale. Like Hugo's Portinari Altarpiece, it was an important work in spreading knowledge of the Netherlandish oil technique in Italy. Of the other works attributed to Joos, the most important are a series of twenty-eight *Famous Men* (Galleria Nazionale, Urbino, and Louvre, Paris), commissioned for the Ducal Palace. Their authorship is controversial, and they may have been a work of collaboration between Joos and the Spanish painter Pedro *Berruguete, who is identified with the 'Pietro Spagnuolo' recorded at Urbino in 1477.

Jordaens, Jacob (1593–1678). Flemish painter, active in his native Antwerp. He was the pupil and son-in-law of Adam van *Noort. Although Jordaens often assisted *Rubens, he had a flourishing studio of his own by the 1620s, and after Rubens's death in 1640 he was the leading figure painter in Flanders. His style was heavily indebted to Rubens, but was much more earth-bound, using thick *impasto, strong contrasts of light and shade, and colouring that is often rather lurid. His physical types, too, are coarser than Rubens's and his name is particularly associated with large canvases of hearty rollicking peasants. Two of his favourite subjects, which he depicted several times, are *The Satyr and the Peasant*, based on one of Aesop's fables, and *The King Drinks*, which depicts a boisterous group enjoying an abundant Twelfth Night feast. Jordaens's prolific output, however, included many other subjects, including religious works and portraits, and he also etched and made designs for tapestries. He rarely left Antwerp, but commissions came from all over Europe, the most important being *The Triumph of Frederick Hendrik* (1651–2), an enormous composition painted for the Huis ten Bosch, the royal villa near The Hague. In about 1655 Jordaens became a Calvinist; he continued to paint pictures for Catholic churches, but the work of the last two decades of his life is more subdued.

Jorn, Asger (Asger Oluf Jørgensen) (1914–73). Danish painter and graphic artist, active in Paris for much of his career. He was one of the founders of the *Cobra group in 1948, and his mature works are highly coloured abstracts executed with violently expressive brushwork.

Josephson, Ernst (1851–1906). Swedish painter and draughtsman. He travelled widely in Europe early in his career and in 1882–8 he lived in Paris, where he was the leader of a group of anti-academic Swedish artists. At this time he moved from the naturalistic tradition of northern Europe to a much more fantastic style, often inspired by Nordic myth, and his work is particularly distinguished for its intensity and vitality of colouring. In 1888 he became insane and never recovered. None the less, during his insanity he continued to work intensively, and the bizarre works he produced, although little known in his lifetime, were influential on the Swedish *Expressionists. Josephson's work is particularly well represented in the gallery at Göteborg and in the National Museum in Stockholm.

Jouvenet, Jean (1644–1717). French painter, the outstanding member of a family of artists from Rouen. He went to Paris in 1661 and joined the studio of *Lebrun. His early works, including decorations for the Salon de Mars at Versailles, were closely imitative of the style of Lebrun and Eustache *Le Sueur (*St Bruno in Prayer*, Nat. Mus., Stockholm). He was the most distinguished of the group of artists who collaborated with *La Fosse in the decorations at Trianon and Les Invalides, but he is now best remembered as the leading French religious painter of his generation, carrying out numerous major commissions for churches in Paris and elsewhere. His later work was marked both by *Baroque emotionalism and by a realistic treatment of details foreign to the principles encouraged by the Academy. It is recorded, for example, that before painting his *Miraculous Draught of Fishes* (Louvre, Paris, *c.*1706) he studied fishing scenes on the spot at Dieppe.

Juan de Flandes (d. *c.*1519). Flemish painter active in Castile from 1496. He was one of a number of north European artists employed by Queen Isabella (1451–1504), who appointed him court painter in 1498. A miniature altarpiece he painted for her was once much renowned, but is now dismembered and scattered; a characteristic panel from it, *Christ Crowned with Thorns* (Detroit Institute of Arts), shows his delicate *miniaturistic style. After Isabella's death in 1504 he worked for churches in Salamanca and Palencia.

Juan de Juanes. See MACIP.

Juan de Juni (*c.*1507–77). Sculptor, probably of Burgundian origin, active in Spain from *c.*1533. He worked at León and Salamanca before settling at Valladolid in 1540. He was a prolific sculptor of religious subjects, excelling in the dramatic expression of emotion, and is generally ranked with Alonso *Berruguete as the outstanding Spanish sculptor of his period. His most famous works are the two versions of the *Entombment* in Valladolid Museum (1539–44) and Segovia Cathedral (1571).

Juárez, José (*c.*1615–*c.*1665). Mexican painter, son of the painter **Luis Juárez** (*c.*1585–*c.*1645). He was the most distinguished Mexican painter to work in the *tenebrist manner of *Zurbarán, many of whose works were exported to Latin America, and whose style was introduced there also by the Sevillian painter Sebastián de Arteaga (1610–56). A good example of his work, which was often very large, is *Epiphany* (Palacio de Bellas Artes, Mexico City, 1655).

Judd, Donald (1928–94). American sculptor and writer on art, one of the leading exponents of *Minimal art. From 1959 to 1965 he earned his living as an art critic, working mainly for *Arts Magazine*. He began his career as a practising artist as a painter, but in the early 1960s he took up sculpture with heavily textured monochrome reliefs. In 1963 he began making the type of work for which he is best known—arrangements of identical rectangular box-like shapes cantilevered ladder-like from a wall. Initially he worked mainly in wood, but after a successful exhibition at the Green Gallery, New York, in 1963–4 he began having them industrially manufactured in various metals (or sometimes coloured perspex). In 1970 he began making works for the specific space in which they were to be exhibited, and in 1972 he began producing outdoor works. In spite of great financial success, Judd (who was notoriously touchy) disliked the New York 'art crowd' and in 1973 moved to Marfa, Texas, where he converted the buildings of an old army base into studios and installation spaces. *Donald Judd: Complete Writings, 1959–75* was published in 1976.

Juel, Jens (1745–1802). Danish painter. He had a distinguished career both in Denmark and in his travels throughout Europe—he studied in Hamburg and in the 1770s worked in Rome, Paris, Dresden, and Geneva. After settling in Copenhagen he became court painter (1780) and a professor at the Academy, where the German painters *Friedrich and *Runge were among his pupils. Juel painted landscapes, *genre scenes, and still lifes (particularly flowers), but he is most renowned for his sensitive portraits. His work is best represented in Copenhagen.

Jugendstil. See ART NOUVEAU.

Julian, Rodolphe. See ACADÉMIE.

Junk art. Art constructed from worthless materials, refuse, rubbish, and urban waste. In so far as Junk art represented a revolt against the traditional doctrine of fine materials and a desire to show that works of art can be constructed from the humblest and most worthless things, it may be plausibly traced back to Kurt *Schwitters and the *collages of *Cubism. However, it is not possible to speak of a Junk movement until the 1950s, particularly with the work of Robert *Rauschenberg, who in the mid-1950s began to affix to his canvases rags and tatters of cloth, torn reproductions, and other waste materials (see COMBINE PAINTING). The name 'Junk art' was first applied to these by Lawrence *Alloway and was then extended to sculpture made from scrap metal, broken machine parts, used timber, and so on.

The Junk art of the USA had its analogies in the work of *Tàpies and others in Spain, *Burri and *Arte Povera in Italy, and similar movements in most European countries and in Japan, where the litter and refuse left over from the war was sometimes converted to artistic use. In the case of Rauschenberg and others the use of Junk material was objective and unemotional. In other instances, including the Junk sculpture of California and the work of Burri and Tàpies, a nostalgic emotional suggestion was conveyed by the use of discarded machine parts, rotted beams and rusted metal, torn and dirty textile scraps, and the detritus generally of industrialized urban life.

Juste, Jean. See GIUSTI.

Justus of Ghent. See JOOS VAN WASSENHOVE.

Kahlo, Frida (1907–54). Mexican painter. In 1925, at a time when she was preparing to enter medical school, she suffered appalling injuries in a traffic accident, leaving her a permanent semi-invalid, often in severe pain. During her convalescence she began painting portraits of herself and others. She remained her own favourite model and her art was usually directly autobiographical. In 1928 she married Mexico's most famous artist, Diego *Rivera, who was twice her age and twice her size. Their relationship was often strained, but it lasted to her death, through various separations, divorce and remarriage (1939–41), and infidelities on both sides (one of her lovers was Leon Trotsky, who was assassinated while living in her house in Coyoacán, Mexico City, in 1940). Kahlo was mainly self-taught as a painter. She was influenced by Rivera, but more by Mexican folk art, and her work has a colouful, almost *naïve vigour, tinged with *Surrealist fantasy. Her paintings of her own physical and psychic pain are narcissistic and nightmarish, but also—like her personality—fiery and flamboyant. They were widely shown in Mexico and in 1939 she had successful exhibitions in New York and Paris, but during her lifetime she was overshadowed by her husband. Since her death, however, her fame has grown and she has become something of a feminist heroine, admired for her refusal to let great physical suffering crush her spirit or interfere with her art and her left-wing political activities. Her house in Coyoacán was opened as a museum dedicated to her in 1958.

Kahnweiler, Daniel-Henri (1884–1979). German-born art dealer, publisher, and writer, who became a French citizen in 1937. In 1907 he opened a gallery in Paris. His first purchases were of *Fauvist works, but he is best known, however, as the friend and promoter of the *Cubists. In 1912 *Braque and *Picasso signed contracts giving Kahnweiler exclusive rights to buy their entire outputs. He was also a friend and supporter of Juan *Gris, of whom he wrote a standard biography (1947). As a publisher he brought out numerous books illustrated by such artists as *Braque, Derain, and Picasso. In the introduction to Kahnweiler's autobiographical book *My Galleries and Painters* (1961, English translation 1971) John Russell wrote: 'Where the old-style dealers did their artists a favour by inviting them to luncheon, Kahnweiler lived with Picasso, Braque, Gris, Derain and Vlaminck on a day-to-day, hour-to-hour basis. The important thing was not so much that they should sell as that they should be free to get on with their work; and Kahnweiler, by making this possible,

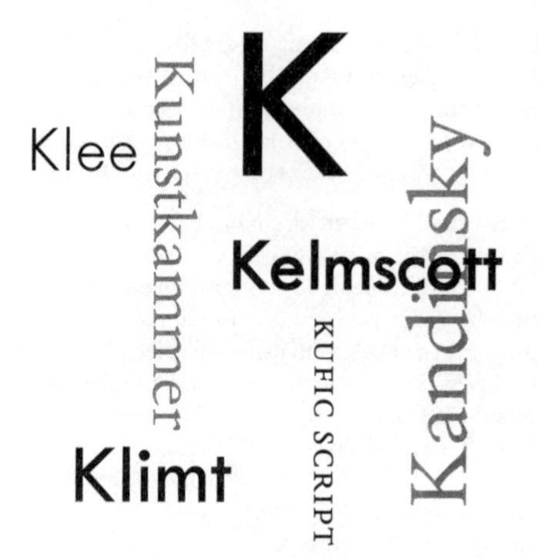

helped to bring into being what now seems to us that last great flowering of French art.'

Kalf, Willem (1619–93). Dutch painter, one of the most celebrated of all still-life painters. He was born in Rotterdam and in 1642–6 worked in Paris. On his return to the Netherlands he lived in Hoorn and then in 1653 settled in Amsterdam. His early works were modest kitchen and courtyard scenes, but he soon became the outstanding exponent of a type of still life in which fruit and precious objects—porcelain, oriental rugs, Venetian glass—are arranged in grand *Baroque displays. His masterly handling of texture and his ability to manipulate warm and cool colours (he frequently contrasts red wine in a glass and the reddish browns in a carpet with the yellow of a peeled lemon and the blue and white of porcelain) have often caused his pictures to be compared with those of *Vermeer.

Kames, Lord. See HOME, HENRY.

Kandinsky, Wassily (1866–1944). Russian-born painter and writer on art, one of the most important pioneers of *abstract art. He abandoned a promising university career teaching law, partly under the impact of an exhibition in Moscow of French *Impressionists, at which one of *Monet's *Haystack* pictures made a particularly lasting impression upon him, and in 1896 went to Munich to study painting. In 1901 he was one of the founders of the avant-garde exhibiting society called the *Phalanx, the main forum for Jugendstil (*Art Nouveau) in Germany, and in 1902 he joined the Berlin *Sezession. Between 1903 and 1908 he travelled widely in western Europe and Africa. His pictures at the turn of the century combined features of Art Nouveau with reminiscences of Russian *folk art, to which he added a *Fauve-like intensity of colour. In 1908 he returned to Munich and in 1909 was one of the founders of the *Neue Künstlervereinigung there. At about this time he began to eliminate the representational element from his work, until, in a series of *Compositions*, *Improvisations*, and *Impressions*, done between 1910 and 1913, he arrived at pure abstraction. (The best collection of works from this period is in the Lenbachhaus in Munich). Kandinsky himself said that his understanding of the power of non-representational art derived from a night when he went into his studio in Munich and failed to recognize one of his own paintings that was lying the wrong way up, seeing in it a picture 'of extraordinary beauty glowing with an inner radiance'. (He is often cited as the first person to paint an abstract picture, but other pioneers, such as the American Arthur *Dove, were making similar experiments at this time and no artist can claim absolute primacy.) From 1911 Kandinsky was one of the most active figures in the *Blaue Reiter, editing with Franz *Marc the *Blaue Reiter* almanac. In 1914 he returned to Russia, where he gained several distinguished academic posts. However, being out of sympathy with the new ideas that subordinated fine art to industrial design, he left Russia in 1921 to take up a teaching post in the *Bauhaus, where he remained until it was closed in 1933. In 1927 he became a German citizen. He left Germany for France in 1933 and settled at Neuilly-sur-Seine, a suburb of Paris, becoming a French citizen in 1939.

Kandinsky was one of the most influential artists of his generation both for his painting and for his writing. He was in the forefront of those who investigated the expressive attributes of artistic materials without reference to natural appearances. His progress towards non-figurative abstraction proceeded alongside his philosophical views about the nature of art, which were influenced by theosophy and mysticism. He did not completely repudiate representation, but he held that the 'pure' artist seeks to express only 'inner and essential' feelings and ignores the superficial and fortuitous. His chief works setting forth his theories of abstract pictorial composition are *Über das Geistige in der Kunst* (Concerning the Spiritual in Art, 1912), *Rückblicke* (Reminiscences, 1913), and *Punkt und Linie zu Fläche* (Point, Line and Surface, published in 1926 as a Bauhaus pamphlet).

Kändler, Johann Joachim (1706–75). German sculptor and porcelain modeller. From 1733 until his death he was chief modeller at the Meissen factory near Dresden, which during this time was the arbiter of European taste in porcelain. Kändler's brilliance and inventiveness was a major factor in its success. He is widely regarded as the greatest of all porcelain modellers, and his work has great vivacity as well as a *Rococo delicacy and charm. His prolific output is found in many museums, notably the Porzellansammlung, Dresden.

Kane, John (1860–1934). American *naïve painter, born in Scotland. He emigrated to the USA in 1879 and moved around a good deal working at various labouring jobs. However, he considered Pittsburgh his home. In 1891 he lost a leg when he was struck by a train, but he became so agile with his artificial limb that few realized he was disabled. He took to drink after his son died soon after birth in 1904 and his wife consequently left him, taking their two daughters with her. Kane then led a wandering life, scraping a living by housepainting and carpentry. His first oil paintings were done *c.*1910; he did portraits—an intense *Self-Portrait* (MOMA, New York, 1929) is his best-known work—landscapes, interiors, and cityscapes of industrial Pittsburgh, combining meticulous observation with naïve stylization and imaginative reconstruction. He achieved sudden fame at the age of 67 when one of his paintings was accepted for the Carnegie International Exhibition in Pittsburgh in 1927. Kane was the first American naïve painter to achieve such recognition; some people thought his picture was a hoax, but in the remaining seven years of his life he achieved considerable acclaim and became something of a celebrity (in consequence of which he was reunited with his wife). His autobiography *Sky Hooks* (named after a housepainter's scaffold supports) was posthumously published in 1938.

Kane, Paul (1810–71). Irish-born Canadian painter of landscape and Indian subjects. He was active mainly in Toronto, but he journeyed as far as the Pacific coast, publishing an account of his travels and adventures in *Wanderings of an Artist* (1859). About half his paintings are portraits of Indians and he took great care over accurate recording of details of costume and ornament, giving his work much historical value. He also had considerable skill, however, in composing figure groups in a way akin to the European *Grand Manner. His work is well represented in the National Gallery, Ottawa, and the Royal Ontario Museum, Toronto.

Kanoldt, Alexander. See NEUE KÜNSTLERVEREINIGUNG.

Kant, Immanuel (1724–1804). German philosopher. Kant's writings on *aesthetics are contained in *Observations on the Feeling of the Beautiful and Sublime* (1764), in the second book of *Anthropology from a Pragmatic Viewpoint* (1798), and principally in *The Critique of Judgement* (1790). According to Hegel

(1770–1831), Kant 'spoke the first rational word on aesthetics', and *The Critique of Judgement* has formed the basis of much subsequent writing on the theory of art and beauty. Kant distinguished judgements about beauty from scientific judgements, moral judgements, judgements of utility, and judgements about pleasure. As against the empirical trend of English 18th-cent. aesthetic writing he maintained that the judgement of beauty claims universal acceptance and is not derivable from or reducible to empirical conformity. As against intellectualist aesthetics he maintained that beauty is not reducible to rule or concept but results from a direct verdict of feeling. He seems to have resolved this antinomy by his view that beauty is 'purposiveness without purpose', consisting in the adaptedness of the object to human faculties of contemplation.

Kapoor, Anish. See NEW BRITISH SCULPTURE.

Kaprow, Allan (1927–). American artist and art theorist, best known as the main creator of *happenings. His studies included a period with the musician John Cage (1912–), from whom he took over the idea of chance and indeterminacy in aesthetic organization. In an article published in 1958 in *Art News* he argued for the abandonment of craftsmanship and permanence in the fine arts and advocated the incorporation of perishable materials. His view was that the *assemblage was to be 'handled and walked around', the *environment was to be 'walked into', but the happening was to be a genuine 'event' involving spectator participation and no longer confined to the museum or gallery. His first major happening, called *18 Happenings in 6 Parts*, was staged at the Reuben Gallery, New York, in 1959. He has been an evangelical promoter of his ideas through teaching at various universities and a voluminous output of writings.

Karlowska, Stanislava de. See BEVAN.

Kauffer, E. McKnight (1890–1954). American-born painter and designer who settled in England in 1914. He was a member of Wyndham *Lewis's Group X and of the *Cumberland Market Group, but he abandoned painting in 1921 and is best known for his brilliant and witty poster designs, notably for the London Transport Board and the Great Western Railway. Writing in the *Evening Standard* in 1928, Arnold Bennett said that he had 'changed the face of London streets' and that his success 'proves that popular taste is on the up-grade'.

Kauffmann, Angelica (1741–1807). Swiss painter. From an early age she travelled with her father, the painter **Joseph Johann Kauffmann** (1707–82), in Switzerland and Italy, and she formed her style in Rome. In 1766 she moved to London, where her work and her person were vastly admired. A foundation member of the *Royal Academy in 1768, she was a close friend of the President, Sir Joshua *Reynolds, their relationship giving rise to gossip and a satirical picture by Nathaniel *Hone. (*Canova, *Goethe, and *Winckelmann were among the other distinguished men charmed by her.)

Kauffmann began in England as a fashionable portraitist, but then devoted herself more to historical scenes and also did decorative work for Robert *Adam and other architects. Although her work owes much to the *Neoclassical tradition, it has a prettiness that can be described as *Rococo. At its best it has great charm, but it can be rather insipid, and she was much more successful with ladylike decorative *vignettes than with scenes from Homer or Shakespeare. After an unhappy first marriage, she married the decorative painter Antonio Zucchi (1726–95) in 1781 and settled in Rome, where she continued her successful career. There is a collection of her work, including a portrait of Reynolds, at Saltram Park in Devon.

Kaulbach, Wilhelm von (1805–74). German painter. In his day he was one of the most celebrated of German painters, but the kind of bombastic and didactic historical scenes in which he specialized have won little favour in the 20th cent. Much more to modern taste are his drawings from nature and his charming illustrations to *Goethe's *Reinecke Fuchs*, which admirably catch the spirit of this animal satire.

Keene, Charles Samuel (1823–91). English *caricaturist, from 1861 until his death one of a group of artists associated with *Punch*. His caricature is delicate and reserved, raising a smile rather than a laugh.

Kelly, Ellsworth (1923–). American painter and sculptor. In the mid-1950s he became recognized as one of the leading exponents of the *Hard-Edge style that was one of the reactions against *Abstract Expressionism. His paintings are characteristically very clear and simple in conception, sometimes consisting of a number of individual panels placed together, identical in size but each painted a different uniform colour (he started using this formula in 1952). He was also one of the first artists to develop the idea of the *shaped canvas. Kelly has also worked in various print-making techniques and in sculpture (using painted cut-out metal forms—often industrially manufactured—related to those in his paintings).

Kelly, Sir Gerald (1879–1972). British painter. One of the leading society portraitists of his day, he had many distinguished friends, among them Somerset Maugham, whose portrait by Graham *Sutherland Kelly wittily attacked. From 1949 to 1954 he was President of the *Royal Academy; in this position he devoted much of his time to organizing loan exhibitions, and became well known for his appearances in related television programmes. His popularity helped to revitalize the Academy's image after the damage done by his predecessor, the arch-conservative *Munnings. Apart from portraits Kelly painted landscapes and also pictures of Asian dancing girls (he had spent some time in Burma) that were once very popular in reproduction.

Kelmscott Press. A private printing-press founded in 1890 by William *Morris at Hammersmith and named after the village near Oxford where he had lived since 1871. Between 1891 and 1898, two years after Morris's death, the press issued more than fifty titles, including editions of several of Morris's own works. Deeply influenced by his study of early printing, Morris himself designed most of the type, borders, ornaments, and title-pages, taking as his basic unit the double-page seen when the book lies open. The press's greatest book, and by common consent one of the world's masterpieces of book production, is the 1896 edition of Geoffrey Chaucer's works, with illustrations by *Burne-Jones. Although short-lived, the Kelmscott Press had enormous influence on the private presses that followed in its wake.

Kemény, Zoltan (1907–65). Hungarian-born sculptor and designer who became a Swiss citizen in 1957. Kemény made *reliefs on the borderline between painting and sculpture, incorporating many different materials in them and creating swirling surface rhythms. In 1951 he began to make translucent reliefs to be set up in front of electric lights and about the mid-1950s he began to do the 'Images en Relief' for which he was chiefly known. These were metal reliefs into which he incorporated curious and unexpected conglomerations of industrial metal products and mass-produced articles.

Kemp-Welch, Lucy. See OFFICIAL WAR ART.

Kennington, Eric (1888–1960). British painter, sculptor, and graphic artist. He was an *Official War Artist in the First and Second World Wars and is best known for his studies of the daily life of ordinary soldiers and the RAF in the Second World War. Between the wars he worked mainly as a portraitist but also did book illustration, including illustrations to T. E. Lawrence's *Seven Pillars of Wisdom* (1926). His sculptures included the *Monument to the 24th Division* in Battersea Park (1924) and figures (in carved brick) for the Shakespeare Memorial Theatre in Stratford-upon-Avon (1930).

Kent, William (1685–1748). English architect, designer, landscape gardener, and painter. He began his career as a painter and studied in Rome, where he made some reputation, painting a ceiling in the church of S. Giuliano dei Fiamminghi in 1717. As a guide and agent for English noblemen on the *Grand Tour he met Lord *Burlington, who in 1719 invited him to return with him to London. From then until Kent's death in 1748 the two were inseparable partners in the conversion of England to *Palladianism. After some rather unsuccessful decorative painting at Burlington House and Kensington Palace in London, Kent began to find himself as an architectural impresario, interior decorator, and garden designer, and he abandoned his first profession. The richness and fantasy of his interior decoration and the imagination displayed in his garden designs are probably his chief claim to fame, and are the perfect complement to the somewhat rigid Classicism of Burlington's architecture. Horace *Walpole said that Lord Burlington, 'the Apollo of arts, found a proper priest in the person of Mr. Kent. . . . He was a painter, an architect, and the father of modern gardening. In the first character, he was below mediocrity; in the second, he was a restorer of the science; in the last, an original, and the inventor of an art that realizes painting, and improves nature. . . . He leaped the fence, and saw that all nature was a garden.' His ideas on informal landscape were developed by Lancelot *Brown. As an architect Kent's most important work is Holkham Hall, Norfolk, begun in 1734 for the first Earl of Leicester, whom he had first met in Rome. Although he was a painter of little talent, Kent has the distinction of painting the earliest medieval history subjects in British art—a series of three pictures for Queen Caroline depicting scenes from the life of Henry V (Royal Coll., *c.*1730). There is no attempt to re-create the scenes accurately, and in fact there is some doubt whether one of the pictures is intended to represent the Battle of Agincourt (1415) or the Battle of Crécy (1346).

Kersting, Georg Friedrich (1785–1847). German painter. After studying at the Copenhagen Academy he settled in 1808 at Dresden, where he specialized in small portraits set in delicately rendered interiors. He was a friend of *Friedrich and painted several versions of a portrait showing him in his studio (Kunsthalle, Hamburg, Staatliche Museen, Berlin, and Städtische Kunsthalle, Mannheim). In 1818 he was made supervisor

(*Malervorsteher*) of the designers in the famous Meissen manufactory. In this capacity he was responsible for the dinner-set presented by King Frederick Augustus IV of Saxony to the first Duke of Wellington (Wellington Mus., London).

Kessel, Jan van (1626–79). Flemish still-life and flower painter active in Antwerp, where he became a Guild member in 1645. He continued the traditions of his grandfather, Jan 'Velvet' *Brueghel, and was also influenced by Daniel *Seghers. Van Kessel painted garlands and bouquets of flowers, but is best known for small, jewel-like pictures, often on copper, of insects or shells against a light background, executed with strong colour and great exactitude. Good examples of his prolific output are in Oxford (Ashmolean), Cambridge (Fitzwilliam), and Madrid (Prado).

Ketel, Cornelis (1548–1616). Dutch portrait and history painter. He worked mainly in Gouda, where he was born, and in Amsterdam, where he died, but also in France and in England, where he lived 1573–81. Van *Mander, who was well informed about him, mentions that he painted a portrait of Queen Elizabeth I for the Earl of Hertford in 1578, but the picture is not known. The portrait of Martin Frobisher (Bodleian Lib., Oxford), however, is a good example of his work from his English period. Ketel's finest works are his group portraits (examples in the Rijksmuseum, Amsterdam), which prefigure those of Frans *Hals.

Kettle, Tilly (1735–86). English portrait painter. After working in the Midlands and London, he became one of the first British painters to risk a long visit to India, where he spent the years 1769–76 and made a fortune painting nabobs and princes. He died on his way out a second time, having found it much harder to achieve success in England. Kettle's style was derivative from that of *Reynolds, *Cotes, and *Romney. His work is well represented in London galleries.

ketubah (Hebrew: 'writing'). A Jewish marriage contract, guaranteeing the bride certain rights. From the *Renaissance period at least *ketubot* were frequently *illuminated, some extant specimens (particularly from Italy) being of considerable artistic merit.

Key. Family of Netherlandish painters best known for their portraiture. **Willem Key** (*c*.1515–68) was a pupil of Lambert *Lombard *c*.1540 in Liège. In 1542 he was made a master of the Antwerp Guild, where he spent the rest of his working life. His nephew, **Adriaen Thomasz. Key** (*c*.1544–after 1589), was probably his pupil. The latter became a master of the Antwerp Guild in 1568. Both artists were highly regarded in their day and did assured and solid portraits of famous people; a portrait of William the Silent (1533–84), which appears in several versions, has been attributed to Adriaen. Their religious works are less well known, and many of Willem's are known to have perished at the hands of iconoclasts.

Keyser, Hendrick de (1565–1621). The outstanding Dutch sculptor and one of the leading Dutch architects of his period. Most of his career was spent in Amsterdam, where he was appointed municipal sculptor and architect in 1594. His most important buildings are the Zuiderkerk (South Church, 1606–14), Holland's first large Protestant church, and the Westerkerk (West Church, 1620–38), which broke free from the *Mannerist tradition, looking forward to the *classicism of Jacob van *Campen. The splendid towers of these two churches are still among Amsterdam's chief landmarks. As a sculptor, de Keyser excelled particularly as a portraitist in a soberly realistic style (*Unknown Man*, Rijksmuseum, Amsterdam, 1608), but his best-known work is the tomb of William the Silent (begun 1614) in the Niewe Kerk at Delft.

Thomas de Keyser (1596 or 1597–1667), Hendrick's son and pupil, was municipal architect to the City of Amsterdam from 1662 until his death (he added the cupola to van Campen's Town Hall), but he is better known as a portrait painter. With *Eliasz., he was Amsterdam's leading portraitist before being overtaken in popularity by *Rembrandt in the 1630s. His life-size portraits look stiff compared with Rembrandt's and he is more attractive and original on a small scale. *Constantin Huygens and His Clerk* (NG, London, 1627) is an excellent example of one of his small portraits of full-length figures in an interior, a forerunner of the *conversation pieces which became so popular in Europe during the 18th cent. His small equestrian portraits were also a new type (*Pieter Schout*, Rijksmuseum, 1660).

Two other sons of Hendrick, **Pieter** (1595–1676) and **Willem** (1603–after 1678), were sculptors. Willem worked for some years in England, probably with Nicholas *Stone, Hendrick's son-in-law and former pupil.

Keyser, Nicaise (or **Nicasius**) **de** (1813–87). Belgian painter of portraits and historical scenes—particularly battles. Popular in his day and much honoured, he was a leader of the Belgian *Romantic artists who reacted against the *classicism of *David and sought a link with the great colouristic tradition of 17th-cent. Flemish painting.

Kiefer, Anselm. See NEO-EXPRESSIONISM.

Kienholz, Edward (1927–94). American sculptor, specialized in life-size three-dimensional tableaux. He belonged to the California school of *Funk art, using the bizarre and shoddy detritus of contemporary life, and creating situations of a horrific and shockingly gruesome character. His brutal images of murder, sex, death, and decay have both attracted and repelled the imagination. A typical example of his work is *The State Hospital* (Moderna Mus., Stockholm, 1964–6), showing a mental patient strapped to his bed, with his own self-image in a thought bubble strapped to the bunk above. Both figures are modelled with revolting realism but have glass bowls for heads. In 1975 Kienholz began to work in Berlin and his work took on more precise social implications. In 1977 he staged exhibitions consisting of imitation radio-receiving apparatus in a poor or decrepit state, which when activated played raucous music from Wagner, supposedly recalling the National Socialist spirit.

Kinderen, Antonius Johannes der. See DER KINDEREN.

Kinetic art. Term describing art incorporating real or apparent movement (from the Greek *kinesis*, 'movement'). In its broadest sense the term can encompass a great deal of phenomena, including cinematic motion pictures, *happenings, and the animated clockwork figures found on clock-towers in many cities of Europe. More usually, however, it is applied to sculptures such as *Calder's mobiles that are moved either by air currents or by some artificial means—usually electronic or magnetic. As well as works employing actual movement, there is another type of Kinetic art that produces illusory movement when the spectator moves relative to it (and *Op art paintings are sometimes included within the field of Kinetic art because they appear to flicker). The idea of moving sculpture had been proposed by the *Futurists as early as 1909, and the term 'kinetic' was first used in connection with the visual arts by *Gabo and *Pevsner in their *Realistic Manifesto* in 1920. Gabo produced an electrically driven oscillating wire construction in this year, and at the same time Marcel *Duchamp was experimenting with *Rotative Plaques* that incorporated movement. Various other works over the next three decades made experiments in the same vein, for example *Moholy-Nagy's *Light-Space-Modulator* (Busch-Reisinger Museum, Harvard University, 1922–30), one of a series of constructions he made using reflecting metals, transparent plastics, and sometimes mechanical devices to produce real movement. However, for many years Calder was the only leading figure who was associated specifically with moving sculpture (and many people regarded him as eccentric), and it was not until the 1950s that the phrase 'Kinetic art' became a recognized part of critical vocabulary; the exhibition 'Le Mouvement' at the Denise René Gallery, Paris, in 1955 was a key event in establishing it as a distinct genre. The artists represented in the exhibition included *Agam, *Bury, Calder, Duchamp, *Tinguely, and *Vasarely.

King, Phillip (1934–). British sculptor. He is probably the most renowned of the generation of avant-garde British sculptors (mainly, like himself, pupils of *Caro) who came to prominence at the *New Generation exhibition in 1965. His work at this time was characteristically in smooth, man-made materials such as plastic or fibreglass, often brightly coloured, with the cone shape being a favourite motif (*And the Birds Began to Sing*, Tate Gallery, London, 1964). At the end of the decade he began using more rugged materials, including steel and wood. With Bridget *Riley he represented Britain at the Venice *Biennale of 1968 and his work has been included in many other international exhibitions. He has taught at various art schools in Britain and elsewhere and in 1980 he was appointed Professor of Sculpture at the *Royal College of Art.

Kip, Johannes (1653–1722). Dutch topographical engraver who migrated to England in about 1690. He is best known for his engravings of country houses in the sumptuous *Britannia Illustrata* (1708 and subsequent volumes). They are of little artistic merit but of great historical interest.

Kiprensky, Orest (1783–1836). Russian painter, active for much of his career in Italy. He is best known for his *Romantic portraits in which he combined the then fashionable attitude of 'Byronic' melancholy with an elegance that earned him the epithet 'the Russian van *Dyck'. Examples of his work are in the Russian Museum, St Petersburg.

Kirchner, Ernst Ludwig (1880–1938). German *Expressionist painter, graphic artist, and sculptor, the dominant figure in the *Brücke group.

He was the first of the group to discover an enthusiasm for Polynesian and other *primitive art, which he admired in the ethnographical department of the Zwinger Museum at Dresden, but this had less apparent effect on his own painting or sculpture than on the work of other members of the Brücke. In painting he was first influenced by the Post-Impressionists, whom he saw exhibited at Munich, and particularly by *Gauguin and van

*Gogh. But under the influence chiefly of *Munch he developed a style similar to that of the *Fauves with simplified drawing and boldly contrasting colours. By 1907–10 he had matured a manner of painting superficially similar to that achieved by *Matisse and his colleagues in 1905, but he was more impetuous and direct in his approach to his subjects, less concerned with pictorial values, than the Fauves. He was more committed to theme than they and attempted to express in paint the emotional atmosphere distilled by the life of the circus and the music-hall, the gaiety and the sadness with its overtones of sexuality in the human detritus of the urban scene. As he himself later put it, it was his concern to distil directly from nature what he referred to as primordial signs and hieroglyphs and to use these harshly simplified and often distorted forms to express contemporary states of mind.

In 1911 he moved to Berlin and during 1912 and 1913 created the series of street scenes which are regarded as the most mature manifestation of German Expressionism. In a style which had become more frenzied, with more ruthless fragmentation of the object, he gave visible expression to the pace, the morbidity, the glare, and the exhibitionist eroticism of megapolitan man (*Street, Berlin*, MOMA, New York, 1913). Kirchner had always been highly strung and he was invalided out of the army in the First World War after suffering a physical and nervous breakdown. In 1917 he was sent to convalesce in Switzerland and stayed there for the rest of his life. Living in solitude among the Swiss mountains near Davos, Kirchner began to paint again, mainly monumental mountain landscapes and scenes from the life of mountain peasants. In the new environment his work gained in serenity what it lost in vigour. From 1928 his style underwent another change. It became more abstract as he painted less directly from nature, using his 'hieroglyphic' forms as a kind of picture-writing in which direct representation played a smaller part. Kirchner's work was declared *degenerate by the Nazis in 1937, adding to his already acute depression at the political events in Germany, and in the following year he killed himself. Throughout his life print-making was as important as painting to him and he ranks as one of the 20th cent.'s greatest masters in this field. He produced a large body of work in woodcut, etching, and lithograph, but each print usually exists in only a few impressions as he liked to print his work himself. In addition to paintings and prints, he made coloured wooden sculpture.

Kisling, Moïse (1891–1953). Polish-born French painter. In 1910 he established himself in Montparnasse, where he became a well-known figure and a friend particularly of *Modigliani and *Chagall, and was influenced by *Cubism. After the First World War, during which he served with the Foreign Legion, he evolved a personal style marked by polished and elegant draughtsmanship and delicately modulated colours, and he achieved considerable success as a portraitist.

Kitaj, Ron B. (1932–). American painter and graphic artist, active mainly in England, where he has been one of the most prominent figures of the *Pop art movement. Before becoming a student in London Kitaj had travelled widely (he was a merchant seaman, then served in the US army) and his wide cultural horizons gave him an influential position among his contemporaries, particularly in holding up his own preference for figuration in opposition to the prevailing abstraction. Kitaj was a contemporary of David *Hockney and Allen *Jones at the *Royal College, influencing them and being influenced by them. He admired *Paolozzi and was a friend of Richard *Hamilton. After a visit to Paris in 1975 he was inspired by *Degas to take up pastel, which he has used for much of his subsequent work. Late 19th-cent. French art has been a major source of inspiration, as has a preoccupation with his Jewish identity, and he has said: 'I took it into my cosmopolitan head that I should attempt to do *Cézanne and Degas and Kafka over again, after Auschwitz.' Unlike the majority of Pop artists, Kitaj has had relatively little interest in the culture of the mass media and has evolved a multi-evocative pictorial language, deriving from a wide range of pictorial and literary sources—indeed he has declared that he is not a Pop artist. Typically his work uses broad areas of flat colour within a strong linear framework, creating an effect somewhat akin to comic strips.

kit-cat. A canvas measuring 36 × 28 in. (92 × 71 cm.). The name derives from *Kneller's portraits of the Kit-Cat Club (NPG, London), all but one of which (there are forty-two) are of this size. Members of the club, founded in the last years of the 17th cent., met originally at a tavern near Temple Bar kept by Christopher Cat (or Kat) that was famous for its mutton pies known as 'Kit-cats'. One of the members, the architect Sir John Vanbrugh (1664–1726), described it as 'the best club that ever met'; it included many of the leading Whigs of the day. Kneller's portraits, painted *c.*1702–21, were commissioned by the publisher Jacob Tomson, the club's secretary and moving spirit, for a room in which its meetings were held in his house at Barn Elms near Putney. In 1735 Tomson published a folio volume of *mezzotint engravings of the portraits;

the paintings remained in the possession of his descendants until 1945. The kit-cat size canvas is particularly suited to life-size portraits showing the sitter's head and shoulders and one or both hands, and Kneller's portraits popularized the format. Previously canvases of 30 × 25 in. (76 × 64 cm., bust length) and 50 × 40 in. (127 × 102 cm., three-quarter length) had been virtually standardized.

Kitchen Sink School. A group of British *Social Realist painters active in the 1950s who specialized in drab working-class subjects, notably interior scenes and still lifes of domestic clutter and debris. The main artists covered by the term were John *Bratby, Derrick Greaves (1927–), Edward Middleditch (1923–), and Jack Smith (1928–), who were supported by the Beaux Arts Gallery in London (they became known as the Beaux Arts Quartet) and by the left-wing critic John Berger (1926–); in 1956 they exhibited together at the Venice *Biennale. By their choice of dour and sordid themes and their harsh aggressive style they expressed the same kind of dissatisfaction with the social and moral values of post-war British society as the 'Angry Young Men' in literature (writers such as John Osborne, whose *Look Back in Anger* was first produced in 1956, were sometimes referred to as 'kitchen sink dramatists'). The mood did not last and from the late 1950s the painters of the Kitchen Sink School developed in different ways, Bratby, for example, emphasizing his *Expressionist handling and Smith eventually turning to abstraction. Berger denounced his former protégés.

kitsch. A German term for 'vulgar trash' which became fashionable in the early 20th cent. Its application ranged from commercial atrocities such as touristic souvenirs to any pretended art which is considered lacking in honesty or vigour. A museum of such products was organized at Stuttgart. Although the battle against kitsch was healthy in its origin, in Germany it frequently led to an unbalanced fear of all obvious beauty or sentiment.

Kitson, Linda. See OFFICIAL WAR ART.

Klee, Paul (1879–1940). German-Swiss painter, graphic artist, and writer on art, one of the most individual figures in 20th-cent. art. He trained at the Academy of Fine Art, Munich, 1898–1901, and after travelling in Italy, 1901–2, and visiting Paris in 1905 with Louis *Moilliet he settled in Munich in 1906, in the same year marrying the pianist Lily Stumpf (he was himself a talented violinist). In 1911 he met *Jawlensky, *Kandinsky, *Macke, and *Marc, and in the following year took part in the second Blaue Reiter exhibition. Also in 1912 he visited Paris for the second time; he met *Delaunay and saw *Cubist pictures. In 1914 he travelled with Moilliet and August Macke to Tunisia, a journey which awakened him to a new sense of colour. Most of his work before this date had been in black and white, but two weeks after arriving he wrote: 'Colour possesses me. I no longer need to pursue it: it possesses me forever, I know. That is the revelation of this blessed moment. Colour and I are one—I am a painter.' During the war he served in the German army, being engaged for part of the time on painting aeroplane wings. After the war he entered into an agreement with the Munich dealer Goltz, who in 1919 staged an exhibition of 362 works which made Klee internationally famous. Invited by *Gropius to teach at the *Bauhaus, he moved to Weimar in 1921 and followed the Bauhaus to Dessau in 1926. In 1931 he left the Bauhaus to take up an appointment at the Düsseldorf Academy, but he was forced to abandon this in 1933 by the Nazi administration and left Germany for Berne. His works were included in the notorious exhibition of *Degenerate Art in 1937.

Although Klee was not politically inclined, there is no doubt that during his last years in Berne his mood was one of profound disappointment, perhaps bordering on acute depression. In 1935 he suffered the first symptoms of the illness that caused his death in 1940, a rare debilitating disease called scleroderma. Although he remained actively productive until the end, the predominance of a darker scale of colour in the paintings of the last seven years, their preoccupation with malign and malevolent forces and themes of corruption, and the appearance of a more bitter form of satire instead of his earlier playfulness all attest to the mental stress under which he lived during these years. Yet his technical and formal mastery remained unaffected and some critics have thought that the work of these last years was among the finest of his whole career. In the catalogue to an exhibition 'Paul Klee. The Last Years' (1974), Douglas Hall wrote: 'The late work of Paul Klee, besides its enormous psychic interest, was of high importance for the future development of modern art. His disjunctive method of composition, his abnegation of the necessity to focus on a point or an episode of a painting, represent one of the very few new inventions in painting since *cubism.'

Klee was one of the most inventive and prolific of the modern masters, his complete output being estimated at some 8,000 works. It is impossible to categorize his work stylistically as he moved freely between figuration and abstraction, absorbing countless influences and transforming these through his unrivalled imaginative gifts. In spite of this variety, his work—in whatever style—is almost always easily recognizable as his, revealing a joyous spirit that is hard to parallel in 20th-cent. art. The finest collection of his work is at the Paul Klee Stiftung in the Kunstmuseum, Berne.

Klee was a brilliant and undogmatic teacher and a stimulating writer on art. Various collections of his writings (including his notebooks and diaries) have been published. The best-known individual work is *Pedagogical Sketchbook* (1953), English translation of *Pädagogisches Skizzenbuch*, published in 1925 as the second of the Bauhaus Books.

Klein, Yves (1928–62). French painter and experimental artist, one of the most influential figures in European avant-garde art in the post-war period. Both his parents were painters, but he had no formal artistic training, and for much of his short life he earned his living as a judo instructor (in 1952–3 he lived in Japan, where he obtained the high rank of black belt, fourth dan). In the mid-1950s he began exhibiting 'monochromes', non-objective paintings in which a canvas was uniformly painted a single colour, usually a distinctive blue that he called 'International Klein Blue'. He used this also for other works, including sculptured figures, and reliefs of sponges on canvas. In a lecture given at the Sorbonne in 1959, Klein explained his theory of monochrome painting as an attempt to depersonalize colour by ridding it of subjective emotion and so to give it a metaphysical quality. Klein also made pictures by a variety of unorthodox methods, including the action of rain on a prepared paper (these he called *Cosmogonies*), the use of a flame-thrower (*Peintures de Feu*), or imprints of the human body (*Anthropométries*). In 1958 he created a sensation (and almost a riot) at the Galerie Iris Clert in Paris by an 'exhibition of emptiness'—an empty gallery painted white. It was called *Le Vide* (The Void). In 1960 he gave his first public exhibition of the *Anthropométries*: girls smeared with blue pigment were dragged over canvas laid on the floor to the accompaniment of his *Symphonie monotone*—a single note sustained for ten minutes and alternating with ten minutes' silence. Critical reception was very mixed. He became a celebrity in Europe, but an exhibition at the Leo *Castelli gallery in New York in 1961 was a dismal failure. Klein died young of a heart attack, but he produced a large amount of work and had wide influence, particularly on the development of *Minimal art. A great showman, he represents the tendency in 20th-cent. art for the personality of the artist to be of more importance than the things he makes.

Klimt, Gustav (1862–1918). Austrian painter and graphic artist, the first President of the Vienna *Sezession. Early in his career he was highly successful as a painter of sumptuous decorative schemes in the grandiose tradition of *Makart, whose staircase decoration in the Kunsthistorisches Museum in Vienna Klimt completed after Makart's death in 1884. In this and other schemes, he worked in collaboration with his brother **Ernst** (1864–92) and Franz Matsch (1861–1942), who had been fellow students at the Kunstgewerbeschule (School of Applied Art) in Vienna. In spite of his official academic successes, Klimt was drawn to avant-garde art, and he came under the influence of *Impressionism, *Symbolism, and *Art Nouveau. Discontent with the conservative attitudes of the Viennese Artists' Association led him and a group of friends to resign in 1897 and set up their own organization, the Sezession, of which he was elected President. His role as a leader of the avant-garde was confirmed when his allegorical mural paintings for Vienna University on the themes of *Jurisprudence, Medicine and Philosophy* aroused great hostility, being called nonsensical and pornographic. (Klimt abandoned the commission in 1905 and the paintings were destroyed by fire in 1945.) Although official commissions dried up after this he continued to be much in demand with private patrons, as a portraitist as well as a painter of mythological and allegorical themes. He was highly responsive to female beauty (he was a great womanizer) and in both his portraits and his subject pictures he stresses the allure and mystery of womanhood. Notable examples are the magnificent full-length portrait of Emilie Flöge (his sister-in-law and mistress) in the Historisches Museum der Stadt, Vienna (1902) and *Judith I* (Österreichische Galerie, Vienna, 1901), one of the archetypal images of the *femme fatale*. Characteristically, the figures in Klimt's paintings are treated more or less naturalistically but embellished—in the background or their clothing—with richly decorative patterns recalling butterfly or peacock wings, creating a highly distinctive style of extraordinarily lush sensuality. He had the opportunity to show his outstanding decorative gifts in a different vein in the major commission of his later years—the mosaic designs (executed 1909–11) for the dining-room of the Palais Stoclet in Brussels, a luxury home built at huge expense for the Belgian millionaire Adolphe Stoclet. Klimt's work was particularly influential on *Kokoschka and *Schiele.

Kline, Franz (1910–62). American painter, generally considered one of the most individual of the *Abstract Expressionists. He began as a representational painter, notably of urban landscapes, but turned to abstraction at the end of the 1940s. This change of direction reflected the influence of *de Kooning and was also stimulated by his seeing some of his drawings enlarged by a projector, an experience that made him realize their potential as abstract compositions. Once he had embarked on this new path he very quickly developed an extremely original style of expressive abstraction, converting the brush-strokes of these drawings into large-scale abstract paintings, using bold black patterns on a white ground reminiscent of oriental calligraphy, but with a highly distinctive rough vigour (he used commercial paints and house-painters' brushes, sometimes up to eight inches wide). Kline had his first one-man show at the Charles Egan Gallery, New York, in 1950 and made a strong impression particularly with his painting *Chief* (MOMA, New York). Towards the end of his life he sometimes incorporated vivid colours but for the most part remained constant in the black-and-white style perfected in the 1950s. He died of heart disease.

Klinger, Max (1857–1920). German painter, sculptor, and graphic artist, born at Leipzig. He studied at Karlsruhe and Berlin, then after brief periods in Brussels, Berlin, and Munich, he spent the years 1883–6 in Paris, 1886–8 in Berlin, and 1888–93 in Rome. After his return to Germany in 1893 he settled in Leipzig, where he was a leading figure of the city's cultural life. His work reveals a powerful imagination and an often morbid interest in themes of love and death. As a sculptor he experimented with *polychromy in the manner of Greek *chryselephantine statues; the culmination was his statue of Beethoven (Mus. der Bildenden Künste, Leipzig, 1899–1902) in white and coloured marbles, bronze, alabaster, and ivory. As a painter he is best known for his enormous *Judgement of Paris* (Kunsthistorisches Museum, Vienna, 1885–7), in which the frame is part of the decorative scheme. It is as a graphic artist, however, that Klinger is now best known and most clearly showed his originality, especially in *Adventures of a Glove* (three series begun 1881), a grotesque exploration of fetishism that antedated the publication of Freud's theories. These etchings concern a hapless young man and his involvement with an elusive lost glove that has clearly sexual connotations. Together with other works of Klinger, they have been claimed as forerunners of *Surrealism, and his influence can be seen in the work of de *Chirico (one of his greatest admirers), *Dalí, and *Ernst, amongst others.

Knapton, George (1698–1778). English portrait painter, a pupil of *Richardson. He was in Italy 1725–32 and in 1736 he became official painter to the Society of *Dilettanti, of which he was a Foundation Member. His twenty-three portraits of his fellow members (still in the possession of the Society) are considered his finest works. From about 1737 Knapton worked much in pastels, but after the 1750s he appears to have virtually given up painting. In 1765 he was appointed Keeper of the King's Pictures. *Cotes was his most important pupil.

Knave of Diamonds or **Jack of Diamonds** (Bubnovyi Valet). An artists' association and exhibition group formed in Moscow in 1909 that for a short time became the most important of the avant-garde associations in Russia. Its first exhibition, in December 1910, featured work by *Larionov, *Goncharova, and *Malevich, the expatriate Russians *Kandinsky and *Jawlensky, and the French *Cubists *Gleizes and *Le Fauconnier.

In 1911 Larionov, Goncharova, and Malevich broke away from the group, accusing it of being too dominated by the 'cheap orientalism of the Paris School' and the 'Munich decadence', and founded their own association, the *Donkey's Tail, to promote an art based upon native Russian inspiration. The Knave of Diamonds held regular exhibitions until 1916, and then broke up.

Kneller, Sir Godfrey (originally Gottfried Kniller) (1646–1723). German-born painter who settled in England and became the leading portraitist there in the late 17th cent. and early 18th cent. He studied in Amsterdam under *Bol, a pupil of *Rembrandt, and later in Italy, before moving to England, probably in the mid-1670s. The opportune death of serious rivals (notably *Lely in 1680) and his own arrogant self-assurance enabled him to establish himself as the dominant court and society painter by the beginning of the reign of James II (1685). He was appointed Principal Painter jointly with *Riley on the accession of William III and Mary II in 1689 (becoming sole bearer of the title when Riley died in 1691), was knighted in 1692, and created a baronet in 1715. In 1711 he was made Governor of the first art academy in England, a post in which he was succeeded by *Thornhill.

Kneller's output was vast and he made extensive use of assistants. Sitters were required to pose only for a drawing of the face and efficient formulas were worked out for the accessories. He is said sometimes to have accommodated as many as fourteen sitters in a day, but the idea of him running a kind of picture-factory is exaggerated. The average portrait turned out from his studio was slick and mechanical (and the heavy wigs then fashionable make for great monotony in male portraits), but Kneller was capable of work of much higher quality when he had a sitter to whom he especially responded; outstanding examples are *The Chinese Convert* (Kensington Palace, London, 1687) and *Matthew Prior* (Trinity Coll., Cambridge, 1700). Many other examples of his work, including the portraits of the *Kit-Cat Club, are in the National Portrait Gallery, London. His style was less elegant and more forthright than Lely's, but the influence of his mass-produced work was stultifying. He was the last foreign-born artist to dominate English painting, but it needed a *Hogarth and a *Reynolds to break through the conventions that he popularized.

Knight, Dame Laura (née Johnson) (1877–1970). English painter. She is best known for her colourful scenes of circus life and the ballet, which achieved great popularity during the height of her fame but which are now generally considered rather banal. In the first half of the century she was one of the most highly regarded of British artists and in 1936 she became the first woman to be elected a *Royal Academician since the original women members Angelica *Kauffmann and Mary *Moser. During the Second World War she was an *Official War Artist, and she was later sent to make portraits at the War Criminals' Trials in Nuremberg. Her husband **Harold Knight** (1874–1961) was also a painter.

Knight, Richard Payne (1751–1824). English collector and connoisseur. He was a prominent member of the Society of *Dilettanti and one of the principals in the *Elgin Marbles controversy. His collections were highly varied, but particularly outstanding were his drawings and antique coins and bronzes, which he bequeathed to the *British Museum. He wrote a didactic poem in the manner of Pope entitled *The Landscape* (1794) and *An Analytical Inquiry into the Principles of Taste* (1805), both of which were important documents in the literature of the *Picturesque, of which he was a leading advocate. His other publications included *An Account of the Remains of the Worship of Priapus* (1786), which was violently attacked as being obscene. See also ELGIN MARBLES.

Købke, Christen (1810–48). Danish painter, a pupil of *Eckersberg. Although he visited Italy, this had little influence on his work, which is narrow in range. He concentrated on everyday scenes in and around Copenhagen and on portraits of family and friends, displaying great sensitivity as a colourist and a warm intimacy of characterization. Købke was little appreciated in his lifetime, but he began to win recognition at the end of the 19th cent. and is now considered the outstanding Danish painter of his period. His work is well represented in the Statens Museum for Kunst, Copenhagen.

Koch, Joseph Anton (1768–1839). Austrian painter, active mainly in Rome, where he settled in 1795. He was influenced by *Carstens and worked with the *Nazarenes on the decorations of the Casino Massimo (1825–9) in Rome, choosing Dante's *Inferno* for his subject, but is now best known for his landscapes. They were directly descended from the heroic and *ideal landscape of *Poussin, but have a distinctive *Romantic flavour, particularly in his paintings of mountains (*The Schmadribach Waterfall*, Neue Pinakothek, Munich, 1822).

Koekkoek, Barend Cornelis (1803–62). The best-known member of a family of Dutch painters. He frequently travelled in Belgium and Germany, where he found inspiration for the *Romantic views of forests and mountains which he painted in a precise and detailed style, often with rosy light effects recalling the work of Jan *Both. Other members of his large and prolific family specialized mainly in seascapes.

Kokoschka, Oskar (1886–1980). Austrian-born *Expressionist painter, graphic artist, and writer who became a British citizen in 1947. His formative years were spent at Vienna amid the intellectual and artistic ferment brought about by the somewhat belated introduction of *Art Nouveau to Austria in the early years of the century, associated with Gustav *Klimt, the architect Adolf Loos (1870–1933), and the designers for the *Wiener Werkstätte. He made a name for himself *c.*1909–10 by his 'psychological portraits' in which the soul of the sitter was thought to be laid bare, and he worked for the avant-garde Berlin periodical *Der Sturm*. At the same time he produced striking and sometimes shocking lithographs and posters. He was seriously wounded in the First World War and after recovery taught at the Dresden Academy 1919–24. After this he embarked on a seven-year period of wide travel, and his interest turned from portraits to landscape, including a distinctive kind of 'portrait' picture of town scenes from a high viewpoint (*Jerusalem*, Detroit Institute of Arts, 1929–30). In 1931 he returned to Vienna, but he was outspokenly opposed to the Nazis. In 1934 he moved to Prague, and then in 1938 to London. From 1953 he lived mainly at Villeneuve in Switzerland. By this time he was internationally famous. In his later years he continued to paint landscapes and portraits, but his most important works of this time are allegorical and mythological paintings, including the *Prometheus* ceiling (1950) for Count Seilern's house at Princes Gate in London, and the *Thermopylae* *triptych (1954) for Hamburg University. Kokoschka remained steadfastly unaffected by modern movements and throughout his long life he pursued his highly personal and imaginative version of pre-1914 Expressionism. His writings include an autobiography (1971) and several plays.

Kollwitz, Käthe (née Schmidt, 1867–1945). German graphic artist and sculptor. She came from a family of strong moral and social convictions, and after marrying a doctor, Karl Kollwitz, in 1891 she lived in the poorer quarters of northern Berlin, where she gained first-hand acquaintance of the wretched conditions in which the urban poor lived. The two series of etchings that established her reputation were inspired by a spirit of social protest against the working conditions of the day: *Weavers' Revolt* (1895–8) and *Peasants' War* (1902–8). After about 1910 she turned from etching to lithography, and after the First World War she turned from representing particular subjects to depicting abstract concepts and great timeless themes such as the Mother and Child. Her work is uncompromisingly serious and often deeply pessimistic in spirit and many of her later drawings and prints were pacifist in intention (her son was killed in the First World War and her grandson in the Second World War, and her best-known sculpture is the war memorial at Dixmuiden, Flanders, completed in 1932). In line with her left-wing views she visited the Soviet Union in 1927, but was subsequently disillusioned with Soviet Communism. In 1919 she had been made the first woman member of the Prussian Academy in Berlin, but in 1933 she was expelled and suffered harassment from the Nazis, although her work was never declared *degenerate and the Nazis sometimes used her images—without her authorization—in their propaganda. Soon after this (1934–5) she made a moving series of eight lithographs on the theme of Death. In its elimination of the accidental instinctive grasp of the tragic essential, and its poignant concern for human suffering her work represents one of the high-points of German *Expressionism and of 20th-cent. graphic art. Only rarely does she lapse into sentimentality. 'I should like', she wrote in 1922, 'to exert influences in these times when human beings are so perplexed and in need of help.'

Konijnenburg, Willem Adriaan van (1868–1943). Dutch painter of murals and portraits and designer of stained-glass windows, who continued the *Symbolist tradition founded by Jan *Toorop, *der Kinderen, and *Thorn Prikker. His designs are remarkable for their mathematical construction. His books on *aesthetics (*Het wezen der schoonheid*, 1908; *De aesthetische idee*, 1916) expound his belief in a connection between mathematical proportion, rhythm, and stylization on the one hand and the struggle between good and evil on the other.

Koninck (or **de Koninck**), **Philips** (1619–88). Dutch painter, the best-known member of a family of artists. He studied with his brother **Jacob** (1614/15–after 1690) in Rotterdam, and *Houbraken says he was also a pupil of *Rembrandt in Amsterdam, where he settled in 1641. Although he painted various subjects (the poet Vondel praised his portraits and history pictures) his fame now rests on his landscapes. He specialized in extensive views, and his work has a majesty and power that rivals the similar scenes of *Ruisdael; the National Gallery in London has four outstanding examples. Like many Dutch painters he had a second occupation; he ran a prosperous shipping firm and apparently painted little in the last decade of his life. His wealth enabled him to collect drawings, including a *Calvary* by *Mantegna now in the British Museum. He was a prolific draughtsman himself and his sketchy penmanship can be deceptively close to Rembrandt's. **Salomon Koninck** (1609–56), the cousin of Philips and Jacob, was also a painter. He was a follower of Rembrandt, imitating him in pictures of hermits, old men, and philosophers in their studies, as well as in religious scenes, and exaggerating the master's early predilection for rich exotic costumes, emphatic gestures, and dramatic contrasts of light and shadow. During the 18th and 19th cents. his work sometimes passed under Rembrandt's name.

Koons, Jeff. See NEO-GEO.

kore. Greek word for 'maiden', applied to the draped standing female statues characteristic of the *Archaic period. The plural is *korai*.

Kossoff, Leon (1924–). British painter, born of Russian-Jewish immigrant parents in the East End of London, an area that has provided the chief subject-matter of his paintings. His training included evening classes under David *Bomberg, 1950–2, and his work has close affinities with that of another Bomberg student, Frank *Auerbach—in choice of subject, emotional treatment of it, and use of extremely heavy *impasto. Kossoff generally retains a firmer sense of structure than Auerbach, however, often using thick black outlines, and unlike him does not approach abstraction. See also SCHOOL OF LONDON.

Kosuth, Joseph (1945–). American experimental artist, a leading exponent of *Conceptual art. He has been much concerned with linguistic analysis of concepts of art, his best-known work being *One and Three Chairs* (MOMA, New York, 1965), which presents an actual chair alongside a full-scale photograph of a chair and an enlarged photograph of a dictionary definition of a chair. 'Actual works of art', he said, 'are little more than historical curiosities.'

kouros. Greek word for 'young man', applied to the nude standing male statues typical of the *Archaic period. The plural is *kouroi*.

Kraft, Adam (*c.*1460–*c.*1508). German sculptor, active in Nuremberg. He was a virtuoso stone carver and his most celebrated work, the tabernacle in St Lawrence, Nuremberg (1493–6), is a gigantic stone imitation (*c.*18 m. high) of a subtle piece of goldsmith's work. The richly decorated structure houses a multitude of human figures, animals, amphibia, etc. One of the supporting figures at the base is supposed to be a self-portrait. Other examples of Kraft's work are in the Germanisches Nationalmuseum in Nuremberg.

Kramskoi, Ivan (1837–87). Russian painter. In 1863 he led a revolt of fourteen students at the St Petersburg Academy; they left together in protest against academic strictures and in 1870 they formed the nucleus of the *Wanderers, of which Kramskoi was a leading light. A sensitive and highly principled man, he believed that 'only a sense of social purpose can give an artist strength and multiply his powers . . . only confidence that the artist's work is needed and appreciated by society can help those exotic plants called pictures to ripen'. He was one of the outstanding Russian portraitists of his time and also painted deeply serious religious works. The most famous is *Christ in the Wilderness* (Tretyakov Gallery, Moscow, 1872), of which *Tolstoy said 'This is the best Christ I know.' His style was clear and sharply focused, perhaps reflecting the fact that he had been a photographic retoucher in his youth. Kramskoi was a hero and intellectual father to a generation of Russian painters, including *Repin, who called him a 'mighty man'.

Krasner, Lee. See POLLOCK.

Kress, Samuel H. (1863–1955). American businessman, art collector, and philanthropist. He became immensely wealthy from his chain of stores and in 1929 he established the Samuel H. Kress Foundation 'to promote the moral, physical and mental welfare of the human race'. His philanthropic work included contributions to medical research and restoring historic buildings in Europe after the Second World War, but he is best known for donations of works of art to American galleries. Above all, his donation of 375 paintings and eighteen sculptures to the newly formed National Gallery of Art in Washington in 1939 (together with subsequent gifts) formed one of the cornerstones of the collection. Kress was a friend of Bernard *Berenson and his main field of interest was Italian *Renaissance painting.

Krieghoff, Cornelius (1815–72). Dutch-born Canadian painter. He studied in Düsseldorf, went to New York *c.*1837 and a few years later moved to Canada, where he worked mainly in Montreal and Quebec. His pictures of the Indians, French-Canadian life, and the landscape, done in a colourful, detailed, and often anecdotal style, proved highly popular, and he has been much imitated and forged. His work is well represented in the National Gallery at Ottawa.

Kritios. See CRITIOS.

Krohg, Christian (1852–1925). Norwegian painter. He trained in Germany and worked in Paris 1881–2, and inspired by the ideas of *Realism he took his subjects mainly from ordinary life—often from its sombre or unsavoury aspects. Particularly well known are his paintings of prostitutes, and he wrote a novel, *Albertine* (1886), on the same subject. From 1902 to 1909 he taught in Paris and from 1909 until his death he was Director of the Academy in Oslo. Krohg's vigorous and straightforward style made him one of the leading figures in the movement from *Romanticism to *naturalism that characterized Norwegian art of his period.

Per Krohg (1889–1965), son of Christian, grew up and was trained in Paris, where he developed in contact with contemporary French artistic trends. *Cubism, in particular, served him as a vehicle for his rich and fantastic imagination. He was best known as a muralist, decorating many public buildings, particularly in Oslo.

Krøyer, Peder Severin (1851–1909). Danish painter, Norwegian by birth. He studied in Paris and was a significant figure in introducing *Impressionism to Denmark. He used a broad technique and was particularly interested in capturing complex effects of light—the fusion of daylight and lamplight, for example. From 1882 he was an influential leader of a colony of Scandinavian artists at the seaside village of Skagen; his work is well represented in the museum there. He ceased painting after he became mentally ill in 1900.

Kruseman, Cornelis (1797–1857). Dutch painter, active for much of his career in Italy. In his day he was renowned for his scenes of life in Italian villages and his *Classicizing religious and historical scenes, but they now seem sugary and his straightforward portraits (as well as the sketches for his more ambitious works) are considered his best works. He had two painter nephews, the cousins **Jan Adam Kruseman** (1804–62), who specialized in portraits and historical scenes, and **Frederik Marinus Kruseman** (1817–*c*.1860), a landscapist.

Kubin, Alfred (1877–1959). Austrian graphic artist, painter, and writer. From 1906 he lived mainly at Zwickledt in Upper Austria, although he travelled a good deal. He was a friend of *Kandinsky and showed his work in the second *Blaue Reiter exhibition in 1912, but his preoccupations were very different to those usually associated with the group. His work shows a taste for the morbid and fantastic, which he combined with pessimistic social satire and allegory. Often he depicted weird creatures in the kind of murky nightmare world associated with Odilon *Redon, whom he met in 1905. Kubin's imagery reflects his disturbed and traumatic life (he had an unhappy childhood, attempted suicide on his mother's grave in 1896, and in 1903 underwent a mental breakdown after the death of his fiancée). He was obsessed with the theme of death (he is said to have liked to watch corpses being recovered from the river) and with the idea of female sexuality as a symbol of death. In 1909 he wrote a Kafkaesque novel *Die andere Seite* (The Other Side) and he illustrated many books, often ones whose subject-matter matched his own macabre interests, such as the stories of Edgar Allan Poe. From the 1920s his reputation was widespread and he was influential on the *Surrealists. His spidery style changed little throughout his career.

kufic script. A form of Arabic script which because of its angular shape is particularly suited to architectural decoration. Sometimes a meaningless series of Arabic letters is used merely for decorative effect, and this use of the script found its way to Europe, where kufic lettering occurs in the decorative arts of the *Romanesque and *Gothic periods.

Kuhn, Walt (1877–1949). American painter, illustrator, and designer, best known for the major role he played in planning and organizing the *Armory Show of 1913. In 1912 he went to Europe with Arthur B. *Davies in order to select and plan the contributions. In spite of his involvement with this milestone in modern art, his own work was fairly conservative, although influenced superficially by, for example, the bright colours of the *Fauves. His best-known paintings are of clowns and circus life. From the 1920s he worked much as a designer for musical revues and also of industrial products. After suffering a nervous breakdown he died in a mental hospital.

Kulmbach, Hans Suess von (*c*.1480–1522). German painter and designer. He was one of *Dürer's most talented pupils, and also came under the influence of Jacopo de' *Barbari when this Italian artist visited Nuremberg between 1500 and 1503. His altarpieces included several for churches in Cracow in Poland, which he visited in 1514–16.

Kunstkammer. German term (literally 'art chamber') used to describe a type of collection of pictures and curios popular with *Renaissance princes. The cabinet pieces in such collections (which were by no means confined to Germany) might include anything from a watch to a fossil. In 16th- and 17th-cent. inventories the term *Kunstkammerstück* means an object of art, a jewel, or a devotional article of particularly remarkable character or quality ordered specially for display in the *Kunstkammer*.

Kupka, František (Frank, François) (1871–1957). Czech painter and graphic artist, active mainly in France, a pioneer of *abstract art. He studied in Prague and Vienna, and settled in Paris in 1895 or 1896, working first mainly as a satirical draughtsman and book illustrator. From an early age he had been interested in spiritualism and the occult (later in theosophy), and from this grew an interest in the spiritual symbolism of colour. He came to the realization that a picture need not have a 'subject' and this laid the roots of his ambition to create paintings whose linear rhythms and colour schemes would produce effects similar to those of music—in his letters he sometimes signed himself 'colour symphonist'. From 1909 (inspired by high-speed photography) he experimented—in a manner similar to that of the *Futurists—with ways of showing motion, and by 1912 this had led him to complete abstraction in *Amorpha: Fugue in Two Colours* (NG, Prague). This created something of a sensation when exhibited at the *Salon d'Automne in 1912. As with *Delaunay and the *Orphists, to whom his work is closely related, Kupka excelled at this stage in his career in the creation of lyrical colour effects.

During the First World War Kupka enlisted in the Czech Legion; he fought on the Somme and he also did a good deal of propaganda work such as designing posters. After the war the Prague Academy appointed him a professor in Paris with the brief of introducing Czech students there to French culture. In 1923 his theoretical work *Tvoreni v Umeni výtvarném* (Creation in Plastic Art) was published in Prague, and in 1931 he was one of the founder members of the *Abstraction-Création group. His later work was in a more geometrical abstract style. Although Kupka gradually established a considerable reputation, his pioneering role in abstraction was not generally realized before the 1960s.

lacería. The Spanish word for geometrical Islamic decoration of straight lines forming intersecting polygons and star shapes. Inherited as a decorative motif from Moorish sources, it was much used by *Mudéjar craftsmen in Spain and Portugal.

Lachaise, Gaston (1882–1935). French-born sculptor who emigrated to the USA in 1906 and became an American citizen in 1916. He was one of the pioneers of modern sculpture. Lachaise settled first in Boston, then in 1912 moved to New York, where he became assistant to Paul *Manship. Lachaise was a consummate craftsman in stone, metal, and wood (his father was a cabinet maker); he helped to reintroduce the method of direct carving in America, but his most characteristic works are in bronze. He did a number of portrait busts remarkable for their psychological insight, and he earned a good deal of his living from decorative animal sculptures, but is best known for his female nudes—monumental and anatomically simplified figures, with voluptuous forms and a sense of fluid rhythmical movement (*Standing Woman*, Whitney Mus., New York, 1912–27). Their smooth modelling links them with the work of *Nadelman, who was also at this time helping to lead American sculpture away from the 19th-cent. academic tradition, but Lachaise's figures are more powerful than those of Nadelman and have an overt sexuality that has caused them to be compared with the nudes of *Renoir. The inspiration for the figures—Lachaise's embodiment of female beauty—was Isabel Dutaud Nagle, a married American woman with whom he fell in love when he was about 20; she was the reason for his move to America and he was eventually able to marry her in 1917.

Laer, Pieter van (1599–1642?). Dutch painter, active for much of his career (*c.*1625–39) in Rome. There he was nicknamed 'Il Bamboccio' (which may be translated as 'Little clumsy one') on account of his deformed body; his self-portrait in the Pallavicini Gallery in Rome suggests that he bore his handicap bravely and without bitterness, and he was one of the leaders of the *Schildersbent (Band of Painters), a fraternal organization set up by the Netherlandish artists in Rome to protect their interests. Van Laer was the first artist to specialize in scenes of street life in Rome. His work proved popular with collectors and he inspired numerous followers who were known as the 'Bamboccianti'. They were mainly other Northerners working in Rome, such as the Flemings Jan Miel (1599–1663) and Michiel *Sweerts, but also included Italians such as Michelangelo *Cerquozzi and Viviano Codazzi (1611–72). Their pictures are called *bambocciate* (the singular is *bambocciata*—Italian for childish-

ness) or in French *bambochades*; an English equivalent—bambocciade—exists but is rarely used. In 1639 van Laer returned to his native Haarlem.

La Fosse, Charles de (1636–1716). French painter, one of the pre-eminent decorative artists of Louis XIV's reign (1643–1715). He was a pupil of *Lebrun and his assistant at Versailles, but his style was more strongly affected by his stay in Italy (1658–63), where he absorbed the *Baroque manner of Pietro da *Cortona in Rome and was influenced by the colour and warmth of such north Italian artists as *Correggio and *Veronese in Parma and Venice. In the 1680s he turned more to *Rubens as a source of inspiration, and his *Presentation of the Virgin* (Musée des Augustins, Toulouse, 1682) was more completely in the mature style of Rubens than anything which had been done in France up to that time. He was a friend of Roger *de Piles and supported him in the controversy over colour versus drawing (or 'Rubenisme' versus 'Poussinisme'). La Fosse was in London working for the first Duke of Montagu on the decoration of Montagu House (formerly on the site of the British Museum) from 1689 to 1692, in which year he returned to Paris to decorate the church of the Invalides. Originally he was commissioned to paint the entire building, but eventually he did only the dome and *pendentives (1702–4), in a style that heralds something of the lightness and elegance of the ensuing *Rococo. La Fosse's work was much more free and colourful than that of most of his French contemporaries, and Anthony *Blunt described him as 'almost the only 17th-cent. French artist whom *Watteau may have studied with profit'.

La Fresnaye, Roger de (1885–1925). French painter. In 1912–14 he was a member of the *Section d'Or group, and his work shows an individual response to *Cubism; his paintings were more naturalistic than those of *Braque and *Picasso, but he adopted something of their method of analysing forms into planes. The effect in La Fresnaye's work, however, is more decorative than structural, and his prismatic colours reflect the influence of *Delaunay, as in his most famous and personal work, *The Conquest of the Air* (MOMA, New York, 1913), in which he portrays himself and his brother in an exhilaratingly airy setting with a balloon ascending in the background. La Fresnaye's health was ruined during his service in the army during the First World War and he never again had the physical energy for sustained work. In his later paintings he abandoned Cubist spatial analysis for a more linear style.

Laguerre, Louis (1663–1721). French decorative painter, active for almost all his career in England. After working for a short time under Charles *Lebrun in Paris he came to England in 1683/4, working initially with *Verrio but soon branching out on his own. He worked mainly in country houses, notably Burghley House, Chatsworth, and Blenheim Palace (where his masterpiece is the *illusionist decoration of the Saloon, with its design of figures representing the Four Continents looking into the room through a Classical colonnade). He was a better painter than Verrio (although still unexceptional judged by European standards) and was also a more attractive personality, but he never achieved the extravagant worldly success of the Italian. From about 1710 *Thornhill began to succeed him in popularity. Late in his career Laguerre turned increasingly to portraits and history paintings.

La Hyre, Laurent de (1606–56). French painter, active in his native Paris. He painted religious and mythological scenes, portraits, and landscapes and also made engravings. His earlier work was influenced by *Primaticcio and the *Fontainebleau schools, but from *c.*1638 his style became more classical, under the influence of *Poussin and then Philippe de *Champaigne. *The Birth of Bacchus* (Hermitage, St Petersburg, 1638) is a good example of his work—obviously indebted to Poussin, but showing individuality in the soft and romantic treatment of the landscape that is the most attractive feature of La Hyre's style. He was a fairly minor figure, but Anthony *Blunt has written of him that 'he embodies in a small way the good sense and the good taste of French seventeenth-century culture'.

Lairesse, Gerard de (1641–1711). Dutch painter, etcher, and writer on art. Born in Liège, he settled in Amsterdam in 1665, and moved to The Hague in 1684. He was the leading decorative painter in Holland in the second half of the 17th cent., working in an academic *classical style that led his over-enthusiastic contemporaries to call him 'the Dutch *Raphael' and 'the Dutch *Poussin'. In about 1690, however, he went blind and thereafter devoted himself to art theory. His lectures were collected in two influential books—*Grondlegginge der Teekenkonst* (Foundation of Drawing, 1701) and *Het Groot Schilderboek* (The Great Painting Book, 1707)—which were translated and much reprinted during the 18th cent. Lairesse's writings reveal the same academic approach as his paintings and he somewhat naïvely confessed that he had a special preference for *Rembrandt until he learned 'the infallible rules of art'. Rembrandt had painted a portrait of the young Lairesse in 1665 (Met. Mus., New York), sympathetically showing his disease-disfigured face.

Lam, Wifredo (1902–82). Cuban painter. His father was Chinese and his mother of mixed African, Indian, and European origin, and Lam's career was appropriately cosmopolitan. After studying in Havana, he went to Madrid in 1924, then in 1938 moved to Paris, where he became a friend of *Picasso. He also met André *Breton, whose book *Fata Morgana* he illustrated in 1940, and joined the *Surrealist association. In 1941 Lam sailed from Marseilles for Martinique on the same ship as *Masson, Breton, and many other intellectuals who were fleeing the Germans. After his return to Cuba in 1942 he came increasingly under the spell of African and Oceanic sculpture, and following visits to Haiti in 1945 and 1946 he also began incorporating images of Voodoo gods and rites in his work. In 1952 he returned to Paris and from the 1960s also spent much of his time at Albisola Mare, near Genoa. In the 1970s he began making bronze sculpture.

Lam is considered to be among the most talented of those who have reconciled the artistic vigour of Latin America with the European avant-garde and with the powerful mystique of African and Oceanic tradition, fusing human, animal, and vegetable elements in menacing semi-abstract images. He won many prestigious prizes and his work is included in numerous leading collections, including the Tate Gallery in London and the Museum of Modern Art in New York.

Lamb, Henry (1883–1960). British painter, mainly of portraits. Under parental pressure he studied medicine, but abandoned it in 1904 to become an artist. (On the outbreak of the First World War, however, Lamb returned to his medical studies, qualifying at Guy's Hospital, London, in 1916 and then serving as a medical officer in France, Macedonia, and Palestine; he was gassed and won the Military Cross. He also worked as an *Official War Artist, as he did again in the Second World War.) Lamb was associated with the *Bloomsbury Group and is best known for his sensitive portraits of fellow members, done in the restrained *Post-Impressionist style that characterized his work throughout his career. Above all he is remembered for his portrait of Lytton Strachey (Tate Gallery, London, 1914), in which he 'has relished emphasizing Strachey's gaunt, ungainly figure, and the air of resigned intellectual superiority with which he surveys the world from that incredible slab-like head' (*DNB*); Sir John *Rothenstein describes it as 'one of the best portraits painted in England in this century'. Apart from portraits, Lamb also painted landscapes and (especially in later life when his health was failing) still lifes.

Lambert, George (1700–65). The leading English landscape painter of his day. He was a pupil of John *Wootton and learned something of the principles of *ideal landscape composition from him but more from studying the work of Gaspard *Dughet. As well as painting handsome works essentially in Dughet's manner, Lambert also did more realistic topographical views; examples of both types are in the Tate Gallery, London. Sir Ellis *Waterhouse has written of him: 'Lambert was the first native painter to apply the rules of art to the English rural scene, and, in this sense, *Wilson followed him.' The figures in Lambert's paintings were done by other artists—sometimes, according to tradition, by *Hogarth. He also collaborated with Samuel *Scott (who painted the shipping) in views of the East India Company's settlements (India Office Library, 1732).

Lami, Eugène-Louis (1800–90). French painter and lithographer. A pupil of Horace *Vernet and Baron *Gros, he made his name as a painter of battle scenes, but is best known for his watercolours and illustrations of contemporary society, in which he took particular delight in recording details of costume. He was in England in 1826, and again in 1848–52, when he followed Louis-Philippe (1773–1850) into exile.

Lamothe, Louis. See DEGAS.

Lancret, Nicolas (1690–1743). French painter. He was a fellow student of *Watteau in *Gillot's studio and had considerable success in imitating the style and the themes which Watteau had made popular, though he lacked Watteau's sensitivity and subtlety. Several of his pictures are in the Wallace Collection, London.

Land art (or **Earth art** or **Earthworks**). A type of art that uses as its raw materials earth, rocks, soil, and so on. The terms are not usually clearly differentiated, although 'Earthworks' generally refers to very large constructions. Earth art emerged as a movement in the late 1960s and has links with several other movements that flourished at that time: *Minimal art in that the shapes created are often extremely simple; *Arte Povera in the use of 'worthless' materials; *happenings and *Performance art because the work created was often impermanent; and *Conceptual art because the more ambitious earthwork schemes frequently exist only as projects. There are affinities also with the passion at this time for the study of prehistoric mounds and ley lines—part of the hippie back-to-nature ethos that expressed a disenchantment with the sophisticated technology of urban culture. The desire to get away from the traditional élitist and money-orientated gallery world was also very much typical of the time, although large earthworks have in fact necessitated very hefty expenditure, and far from being populist and accessible, such works are usually in remote areas. Moreover, in spite of the desire to sidestep the gallery system, dealers have proved capable of exploiting this kind of art, just like any other, and some land artists at least have made handsome livings from it.

The artist associated more than any other with large-scale earthworks *in situ* was Robert *Smithson, whose *Spiral Jetty* (1970) in the Great Salt Lake, Utah, is easily the most reproduced work of this kind. Most of the other leading exponents are—like Smithson—Americans. They include Alice Aycock (1946–), whose work has included underground mazes, and Michael Heizer (1944–), whose best-known work is *Double Negative* (1969–70) in the Nevada desert—two massive cuts 30 feet wide and 50 feet deep in an area where he said he found 'that kind of unraped, peaceful religious space artists have always tried to put in their work'. Some critics, however, consider that earthworks can themselves constitute a type of rape or violation.

*Christo is sometimes grouped with land artists, although his work really defies classification. Outside the USA, the most noted exponent of land art is probably Richard *Long.

Landseer, Sir Edwin (1802/3–73). English painter, sculptor, and engraver of animal subjects. He was the son of an engraver and writer, **John Landseer** (1769–1852), and was an infant prodigy. His life was one of continuous professional and social success; he was the favourite painter of Queen Victoria (who considered him 'very good looking although rather short') and his friends included

Dickens (1812–70) and Thackeray (1811–63). In 1850 he was knighted and in 1865 he refused the Presidency of the *Royal Academy. The qualities in his work that delighted the Victorian public, however, have subsequently caused his reputation to plummet, for although he had great skill in depicting animal anatomy, he tended to humanize his subjects to tell a sentimental story or point a moral. His most familiar works in this vein include *The Old Shepherd's Chief Mourner* (V&A, London, 1837), *Dignity and Impudence* (Tate, London, 1839), and—most famously—*The Monarch of the Glen* (Guinness plc, Edinburgh, 1850). Other paintings by Landseer have been attacked for their cruelty (he made many visits to the Scottish Highlands and frequently painted scenes of deer-hunting), but here his defenders would claim that he observed rather than gloated over the bloodshed. Apart from animal subjects, Landseer also painted portraits and historical scenes. As a sculptor he is best known for the lions at the base of Nelson's Column in Trafalgar Square, London, unveiled in 1867. Landseer's health broke down in the 1860s, and in his last years he suffered from bouts of madness, aggravated by alcohol.

His brother **Thomas** (1798–1880) was an engraver, whose prints played a great part in popularizing Edwin's work. Another brother, **Charles** (1800–79), bequeathed £10,000 to the Royal Academy to found Landseer scholarships.

Lane, Sir Hugh (1875–1915). Irish dealer, patron, collector, and administrator. He made his fortune as a picture dealer in London and had no particular interest in Ireland until about 1900, when through the influence of Sarah *Purser and the playwright Lady Gregory (his aunt) he became caught up in the rising tide of nationalism in the arts. He commissioned Jack *Yeats to paint a series of eminent contemporary Irishmen (it was completed by *Orpen, a distant cousin and close friend of Lane's) and he helped to found Dublin's Municipal Gallery of Modern Art, opened in temporary premises in 1906. In addition to giving and lending numerous works to the gallery, he offered to bequeath his finest late 19th- and early 20th-cent. French paintings to Dublin, on condition that a suitable gallery was built to house them. This caused arguments with the Dublin City Authorities, however, and he moved the pictures to the National Gallery in London. Lane was killed when the *Lusitania* (on which he was returning from business in the USA) was torpedoed by a German submarine. A codicil to his will expressed his intention of returning the French pictures to Dublin, but it was unwitnessed, creating a long-term legal dispute about their ownership. In 1959 an agreement was eventually reached whereby the paintings were divided into groups to be shown alternately in Dublin and London. This agreement was renewed in 1980. The Municipal Gallery of Modern Art was given a permanent home in Dublin in 1933, and in 1979 it was renamed the Hugh Lane Municipal Gallery of Modern Art.

Lanfranco, Giovanni (1582–1647). Italian painter, who with *Guercino and Pietro da *Cortona ranks as one of the founders of the High *Baroque style of painting. He was born in Parma, where he trained under Agostino *Carracci before going to Rome in 1602 to work under Annibale Carracci in the Palazzo Farnese. After Annibale's death in 1609, he returned for a while to Emilia, but by about 1612 was back in Rome, where he gradually overtook his arch-rival *Domenichino as the leading fresco decorator in the city. Their work can be compared in S. Andrea della Valle, where Domenchino painted the apse and the pendentives of the dome, but Lanfranco was awarded the commission for the *Assumption of the Virgin* (1625–7) in the dome itself; this fresco is one of the key works of Baroque art and it ended the dominance of Bolognese classicism in Rome. The figure style derives from the Carracci, but the *illusionistic foreshortening is based on *Correggio's dome paintings in Lanfranco's native Parma, here carried to new extremes. *Bellori compared the handling of the multitude of figures to the harmonious blending of voices in a choir, and the dynamic design became a pattern for illusionistic decorators throughout Europe.

Between 1633 and 1646 Lanfranco was in Naples, where his work was an inspiration to such Neapolitan masters as Mattia *Preti, Luca *Giordano, and *Solimena. In the S. Gennaro Chapel of the cathedral he painted an even more extravagant dome (1641–3) than his masterpiece in S. Andrea della Valle, and this inspired Pietro da Cortona in the Chiesa Nuova in Rome and *Mignard in the church of the Val-de-Grâce in Paris. He returned to Rome in 1646 and his last (unfinished) work in the apse of S. Carlo ai Catinari exemplifies the airy luminosity of his final style.

Lanfranco is much less renowned as an easel painter, but created some outstanding works in this field also. Particularly remarkable are his *Ecstasy of St Margaret of Cortona* (Pitti, Florence), on which *Bernini may have drawn for his *St Theresa*, and *St Mary Magdalen Transported to Heaven* (Museo di Capodimonte, Naples), a bizarre and highly original work in which the rapturous saint is carried by angels above a poetically evoked view of the Roman Campagna.

Lang Shih-ning. See CASTIGLIONE, GIUSEPPE.

Lanyon, Peter. See ST IVES SCHOOL.

Lanzi, Luigi (1732–1810). Italian art historian, archaeologist, and philologist. As a pioneer in the systematic study of the art of antiquity he ranks next to *Winckelmann, but he is best known for his history of Italian painting from the 13th cent. until his own time, *Storia pittorica dell' Italia* . . . (1792; 2nd edn., 1795–6; 3rd edn., 1800). Lanzi classified his material by regional schools and based his work on a thorough knowledge of previous writings on the subject and of the paintings themselves (he was keeper of the galleries of Florence from 1773 and also visited churches and collections throughout central and northern Italy in the course of his work). His methodical arrangement and his synthesis of solid research with sensitive analysis of style make his work a landmark in art historical writing, and Rudolf *Wittkower has described it as 'still unequalled for knowledge of the material and breadth of approach'. There have been several English translations and editions (the first in 1828), as well as numerous other Italian editions published after Lanzi's death. Lanzi published scholarly but controversial works on the Etruscan language and also a book on ancient vases (1806) in which he correctly perceived that vases traditionally called Etruscan were in fact Greek in origin.

Laocoön. An antique marble group (Vatican Mus.) representing the Trojan priest Laocoön and his two sons being crushed to death by snakes as a penalty for warning the Trojans against the wooden horse of the Greeks, an incident related by Virgil in the *Aeneid* ii. 199–231. It is usually dated to the 2nd or 1st cent. BC or the 1st cent. AD, although whether it is an original *Hellenistic piece or a Roman copy has long been a matter of dispute. *Pliny states that in his time it stood in the palace of the emperor Titus in Rome and records that it was made by the sculptors Hagesander, Polydorus, and Athenodorus of Rhodes (Kenneth *Clark, who thought it was a copy of a work originally done in bronze, claimed that the workmanship 'exhibits three different degrees of skill'). Pliny described it as 'a work to be preferred to all that the arts of painting and sculpture have produced'. This praise echoed long after the sculpture had disappeared, and its dramatic rediscovery in a vineyard in Rome in 1506 made an overwhelming impression, particularly on *Michelangelo, who went to see it immediately. Its liberating influence for the expression of the emotions continued to be important for *Baroque sculpture and until the 19th cent. it was ranked (with the **Apollo Belvedere* and the **Belvedere Torso*) as one of the greatest works in the world. (As early as about 1530 *Titian satirized the adulation it received in a woodcut showing the figures changed to monkeys.)

It was given a new aesthetic significance by *Winckelmann, who saw it as a supreme symbol of the moral dignity of the tragic hero and the most complete exemplification of the 'noble simplicity and quiet majesty' which he regarded as the essence of Greek idealistic art and the key to true beauty. In 1766 *Lessing chose *Laokoon* as the title of the book in which he attacked Winckelmann's conception of the tragic hero and refuted the *Neoclassical idea of *antique beauty. The sculpture was one of the prizes taken from Italy by Napoleon and was in Paris 1798–1815.

Although no longer considered one of the world's greatest masterpieces, the *Laocoön* has slipped in esteem much less than some once-revered antique statues; it continues to be a work with a powerful hold over the imagination and still finds a place in almost all general histories of art. It has been restored several times since its discovery, and a complete renovation was made in the 1950s, when Laocoön's original right arm was returned to the figure and replaced in its correct position behind his head.

Study of the *Laocoön* was revolutionized in 1957 by one of the most spectacular archaeological discoveries of the 20th cent., when several groups of marble figures representing events in Homer's *Odyssey* were found at Sperlonga near Naples; the names Hagesander, Polydorus, and Athenodorus are inscribed on one of the groups (now in the museum at Sperlonga), which are close in style to the *Laocoön*. The cave in which these sculptures were found was evidently used as a banqueting hall by the emperor Tiberius (reigned AD 14–37), and there is other evidence linking them with the 1st cent. AD, so this date is now finding favour among classical archaeologists for the *Laocoön* also.

Largillière (or **Largillierre**), **Nicolas de** (1656–1746). French painter, mainly of portraits. He spent his youth in Antwerp and *c.*1674–80 worked as assistant to Sir Peter *Lely in London. Returning to Paris in 1682, he soon established his position as a leading portraitist, rivalled only by Hyacinthe *Rigaud, his almost exact contemporary. The two men were friends and seldom in direct competition, for Largillière specialized in portraits of the rich middle classes and Rigaud painted the aristocracy. Largillière's long and successful career culminated when he was made Director of the *Academy in 1743 at the age of 87. His output of portraits was prodigious (contemporary sources indicate he painted about 1,500), and he also did religious works (once highly regarded), still lifes, and landscapes. At his best, his paintings are vigorous, forthright, and colourful; at his worst, they are pompous and vacuous. His work is well represented in the Louvre.

Larionov, Mikhail (1881–1964). Russian painter and designer, one of the leading figures in the development of modernism in Russia in the period before the First World War. His early work was influenced by *Post-Impressionism, but from 1908, together with Natalia *Goncharova (his lifelong companion and collaborator), he developed a form of cultivated *primitivism—his style more aggressive than hers—based upon an interest in Russian folk art. In a series of 'Soldier' and 'Prostitute' pictures done in 1908–13 this primitivism was exaggerated into crude distortions and a deliberate flouting of conventional good taste. Larionov was involved in avant-garde groups such as the *Knave of Diamonds, and he helped to organize major shows of progressive art, including the *Donkey's Tail exhibition in 1912 and the *Target exhibition in 1913, at which he launched *Rayonism, a near abstract movement that was a counterpart to Italian *Futurism. In May 1914 Larionov and Goncharova accompanied *Diaghilev's *Ballets Russes* to Paris. They returned to Russia in July on the outbreak of the First World War, and Larionov served in the army and was wounded. After being invalided out, he and Goncharova left Russia permanently in 1915, moving first to Switzerland and then settling in Paris in 1919 (they became French citizens in 1938). In Paris he practically abandoned easel painting and concentrated on theatrical designing for the *Ballets Russes*. After Diaghilev's death in 1929 Larionov gradually sank into obscurity, but his reputation was revived shortly before his death with retrospective exhibitions (jointly with Goncharova) in London (Arts Council, 1961) and Paris (Musée d'Art Moderne de la Ville de Paris, 1963).

Larkin, William (d. 1619). English portrait painter. He emerged from total obscurity in 1952 with the publication of his only documented work—a pair of oval portraits at Charlecote Park, Warwickshire. Subsequently several other portraits in a very different vein—full-lengths featuring elaborate Turkey carpets, dazzling metallic curtains, and poses of a starched magnificence—have been attributed to him, including a breathtaking group at Ranger's House, London. If they are indeed all by Larkin he was the genius of Jacobean painting: Sir Ellis *Waterhouse considered the Ranger's House portraits to be the work of 'at least three different hands', but technical examination in the 1980s indicated that they are all from the same studio.

Laroon, Marcellus (1679–1772). English painter, the son of a Franco-Dutch painter of the same name (1653–1701/2) who came to England as a young man and was one of *Kneller's assistants. The younger Laroon was a colourful character who in his long and strenuous life was a musician, singer, professional soldier, and man of pleasure; he drew and painted 'for diversitions', to use the words of *Vertue. He painted portraits, *conversation pieces, and *genre scenes, usually fanciful in character. His nearly monochromatic, feathery style added a touch of French daintiness to the stolid English manner and he anticipated *Gainsborough by his lightness of touch. After a long period of almost total obscurity he was 'rediscovered' in the 20th cent. Examples of his work are in the Tate Gallery.

Larsson, Carl (1853–1919). Swedish painter and graphic artist. Although he painted large-scale murals (the best known are in the National museum, Stockholm, 1896), he is associated particularly with intimate watercolours of the idyllic everyday life of his own home in Dalecarlia. They became popular in colour reproductions and had a lasting influence on the Swedish attitude towards furnishing and interior decoration.

Lascaux. Site of a cave near Montignac in the Dordogne region of France containing some of the finest examples of prehistoric paintings ever discovered. The cave was found in 1940 by some boys searching for a lost dog and rapidly attracted great interest. Although the paintings (probably dating from about 15,000 BC) were remarkably well preserved, they deteriorated so rapidly when they became a tourist attraction that the cave had to be closed in 1963. Various animals are portrayed, some over life-size, and it is generally believed that the cave served as a centre for magical hunting rites. As well as the paintings there are some engravings cut in the rock. See also ALTAMIRA.

Lassaw, Ibram. See ABSTRACT EXPRESSIONISM.

Lastman, Pieter (1583–1633). Dutch painter, highly esteemed in his day but now remembered mainly as the most significant of *Rembrandt's teachers. Most of his career was spent in his native Amsterdam, but in about 1603–7 he was in Italy, where *Caravaggio and *Elsheimer made a strong impact on his style. He specialized in religious, historical, and mythological scenes and often chose unusual subjects that proclaimed his status as a learned artist (*The Volscian Women and Children Beseeching Coriolanus not to Attack Rome*, Trinity College, Dublin, 1622). When Rembrandt went to study with him in about 1624 he was one of the most prestigious painters in Holland, and he inspired Rembrandt to faithful imitation. The glossy colours, the animated gestures and facial expressions, and the dramatic lighting of Rembrandt's early works all owe much to Lastman, and his *Balaam and the Ass* (Musée Cognacq-Jay, Paris, 1626), for example, is clearly based on a prototype by his master (private coll., 1622). Lastman also taught Rembrandt's friend Jan *Lievens.

László, Philip de (1869–1937). Hungarian-born portrait painter (and occasional sculptor) who settled in London in 1907 and became a British citizen in 1914 (although he was interned during the First World War). He trained in Budapest, Munich, and Paris (at the *Académie Julian) and already had an international reputation as a society portraitist when he moved to England. There his career continued in its successful course, his sitters including Edward VII and numerous members of the aristocracy (examples are in the NPG, London). According to the *Dictionary of National Biography*, he had 'a pleasing, courteous, and exuberant manner, and was very popular in society'. He was a fast and fluent worker and his style was elegant but superficial.

La Thangue, H. H. (Herbert Henry) (1859–1929). British painter. He had his main training at the *Royal Academy in London and the École des *Beaux-Arts in Paris. In 1887 he described the Academy as 'the diseased root from which other evils grow', and he was one of the leading figures in founding the *New English Art Club in opposition to it and in introducing the ideals of French **plein-air* painting to Britain. He lived in the countryside (first in Norfolk, then in Sussex), and *Clausen wrote of him: 'Sunlight was the thing that attracted him: this and some simple motive of rural occupation, enhanced by a picturesque surround.' From about 1898 he turned to peasant scenes set in Provence or Italy, places where he often stayed. As the countryside changed, his work became increasingly nostalgic, as he hankered after what *Munnings called a 'quiet old world village where he could live and find real country models'.

La Tour, Georges de (1593–1652). French painter, active at Lunéville in the duchy of Lorraine. He was patronized by the Duke of Lorraine and had a successful career, but his name sank into oblivion after his death and it is only in the 20th cent. that he has been rediscovered (see VOSS) and hailed as the most inspired of *Caravaggesque painters. Little is recorded of his life (although he is known to have been arrogant and unpopular with his neighbours) and it is a matter of dispute whether he gained his knowledge of Caravaggio's style via painters of the Utrecht school such as *Honthorst or by travelling to Italy. Like Honthorst he is particularly associated with nocturnal scenes and the use of a candle as the light source in a painting. La Tour's handling of light is more subtle and sensitive, however, and he is grander in conception and more sombre in mood. In his mature work he smoothed the forms of his figures until they approached geometric simplicity and achieved a feeling of monumental stillness that is considered to represent the spirit of 17th-cent. French *classicism no less than the paintings of Philippe de *Champaigne and *Poussin in their different fields. Only three of La Tour's paintings are dated—*The Payment of Dues* (Museum, Lvov, Ukraine, 1634?); *Penitent St Peter* (Cleveland Mus. of Art, 1645); *The Denial of St Peter* (Musée des Beaux-Arts, Nantes, 1650)—and there is much scholarly debate about his chronology.

The works associated with the beginning of his career are daylit scenes of such subjects as peasants and card-sharpers; they are very different in spirit from the calm and majestic religious images of his maturity and have become controversial as regards *attribution as well as dating. It has been argued (and hotly disputed) that *The Fortune Teller* (Met. Mus., New York) is a modern fake, and although the status of most of the other early works as authentic (and high-quality) 17th-cent. French paintings is not denied, their attribution to La Tour (which rests almost entirely on stylistic evidence) has been questioned. Another problem in La Tour studies is that many of his undeniably authentic compositions exist in more than one version, and the studio replicas (as they appear to be) are sometimes of extremely high quality; the versions of *St Sebastian Tended by St Irene* in the Louvre and the Staatliche Museen, Berlin, for example, are each extraordinarily beautiful. La Tour's son **Étienne** (1621–92) worked in his father's studio and may have been responsible for some of the replicas. No independent works certainly by him are known, but *The Education of the Virgin* (Frick Coll., New York), signed 'de la Tour', has been attributed to him.

La Tour, Maurice-Quentin de (1704–88). With *Perroneau the most celebrated and the most successful French *pastel portraitist of the 18th cent., active mainly in Paris. His portraits have great vivacity and exploit the resources of the technique to the full. Some are lightly sketched impressions, others elaborate and detailed studies. In both, the essence of his art lies in his swift and accurate draughtsmanship; his colour, which is never very deep, and his superb velvet finish are always subordinate. La Tour portrayed many of the most famous men and women of his day and in 1750 he became portraitist to Louis XV, a position he held until he had a nervous breakdown in 1773. His work is best represented in the Louvre and in the museum at Saint-Quentin, the city where he was born and where he died.

Laurana, Francesco (*c*.1430–1502?). Italian sculptor, born near Zara in Dalmatia, at that time subject to Venice. In 1453 he is recorded working on decorations to the Triumphal Arch of Alfonso I (1396–1458) at Castelnuovo, Naples. He was one of several sculptors who did so and his contribution is uncertain. Thereafter he divided his time between France and Italy, mainly the south of the country, including Sicily, although he also visited Urbino, and was possibly related to the architect Luciano Laurano, who designed the celebrated Ducal Palace there. In France his most important work was the chapel of St Lazare (1475–81) in the church of La Major at Marseilles, described by Anthony *Blunt as 'probably the earliest purely Italian work on French soil'. He is best known, however, for his portrait busts of members, mostly female, of the Neapolitan royal house and their relatives. In these remarkably sensitive works the forms of the face are subtly generalized in a search for basic geometric shapes, and their simple naturalism was sometimes enhanced by heightening the marble with colour, as in the bust of Isabella of Aragon in the Kunsthistorisches Museum in Vienna.

Laurens, Henri (1885–1954). French sculptor, designer, and illustrator, born in Paris, where he trained as an ornamental stonemason. His early work shows the influence of *Rodin, but in 1911 he became a friend of *Braque (he later met *Gris, *Leger, and *Picasso) and he was one of the first artists to adapt the *Cubist style to sculpture. He made collages, reliefs, and constructions of wood and metal, mainly still lifes using the familiar Cubist repertory of bottles, glasses, and fruit. Much of his work was coloured, but he retained a genuine sculptor's feeling for mass, and his distrust of intellectual speculation preserved his independence from Cubist theorizing. In the mid-1920s he moved away from his geometrical style to one that featured curved lines and voluptuous forms, notably in female nudes. Many of his fellow artists regarded him as one of the greatest sculptors of his time, but financial success and official recognition were slow in coming. When he failed to win the first prize for sculpture at the 1948 Venice *Biennale, *Matisse was so disgusted that he offered to share his own painting prize with him. In 1953, however, Laurens won the Grand Prix at the São Paulo Bienal. Apart from sculpture, his work included stage design for *Diaghilev and numerous book illustrations.

Lavery, Sir John (1856–1941). British painter, mainly of portraits. He was born in Belfast and studied in Glasgow, in London, and then in the early 1880s in Paris (at the *Académie Julian and elsewhere). Between 1885 and 1896 he lived mainly in Glasgow (see GLASGOW SCHOOL), then settled in London, although he travelled a good deal and often wintered in Morocco, where he bought a house in about 1903. Lavery had an immensely successful career as a fashionable portraitist (particularly of women), painting in a dashing and fluid, if rather facile, style. He also painted interiors, landscapes, and outdoor scenes such as tennis and bathing parties, and he was an *Official War Artist, 1917–18. His reputation did not long survive his death, but there has recently been a revival of interest in his work.

Lawrence, Sir Thomas (1769–1830). The outstanding English portrait painter of his generation. Lawrence was a child prodigy and was almost entirely self-taught. He was also handsome and charming, and after a resounding early triumph with his portrait of Queen Charlotte (NG, London, 1789) he never looked back in terms of professional and social success. On the death of Reynolds in 1792 he succeeded him as Painter in Ordinary to the King, and on the death of *Hoppner in 1810 he was recognized as the leading portrait painter of the time, and also to some extent as head of the profession of painting in Britain. The high point of his career came in 1818, when he was sent to Europe as the envoy of the Prince Regent (later George IV) to paint the heads of state and military leaders who were involved with the allied victory over Napoleon. As a preliminary gesture he was knighted, and on his return in 1820 he succeeded Benjamin *West as President of the *Royal Academy. The portraits painted on this tour are now in the Waterloo Chamber, Windsor Castle.

Lawrence was devoted to the memory and example of Reynolds and in some respects he was the last of the great portrait painters in the 18th-cent. tradition. In others he was a *Romantic, responding to the glamour of the historic years through which he lived. His fluid and lush brushwork won the admiration of French painters when his work was exhibited at the Paris *Salon of 1824 and after *Delacroix visited London in the following year he paid Lawrence the compliment of painting a portrait in his style (*Baron Schwiter*, NG, London). Lawrence's reputation declined after his death, however, and has never revived to its former heights, perhaps because there is something in the British character suspicious of the manifest brilliance his work displays. He was constantly in debt in spite of his success and took on too many commissions, so his work is uneven and sometimes careless (and like Reynolds he was a failure as a history painter), but at his best he has a feeling for paint that few British artists can rival.

Lawrence was a man of great taste and made one of the finest collections of Old Master drawings ever assembled, particularly rich in works by *Michelangelo and *Raphael (these are now among the greatest treasures of the Ashmolean Museum in Oxford). He played a part in founding the National Gallery and in securing the *Elgin Marbles for the nation, and was noted for the unselfish help he gave to young artists.

Lawson, Ernest (1873–1939). American painter, the least distinguished and most orthodox member of the *Eight. Unlike the other members of the group he was primarily a landscapist (although he did also paint urban scenes), and his work was essentially *Impressionist in style.

lay figure. An articulated model of the human figure, jointed so that it can be given all kinds of poses. It may be anything from a few inches in height to life-size. Articulated dolls and marionettes were known in antiquity, but the first description of an artist's lay figure is given by *Filarete in the third book of his *Treatise on Architecture* (1461–4). Although *Vasari mentions a life-size wooden model made by Fra *Bartolommeo, the early lay figures were mostly small and were called manikins. The 18th-cent. portrait painter used a life-size model, completely jointed and covered with knitted fabric. He could arrange the costumes and drapery on it and work on that part of the picture in the absence of the sitter.

Lazzarini, Gregorio. See TIEPOLO.

lead point. See METAL POINT.

Lear, Edward (1812–88). English artist, author, and traveller. Although he is now remembered principally for his nonsense poems and as the popularizer of the limerick, he earned his living mainly through drawing and painting. He began his career as a draughtsman for the Zoological Society, but when the exacting work began to affect his eyesight he turned to topographical painting in the 1830s, initially in watercolour and later in oils. His style is very clear and brightly lit. He travelled widely and published several illustrated accounts of his journeys. In 1871 he settled in San Remo, Italy, where he died. The first of his collections of nonsense poems, with his own illustrations, was published in 1846.

Lebrun, Charles (1619–90). French painter and art theorist, the dominant artist of Louis XIV's reign (1643–1715). After training with *Vouet he went to Rome in 1642 and worked under *Poussin, becoming a convert to the latter's theories of art. On his return to Paris in 1646 he found his true bent in large and flamboyant decorative paintings. From 1661 he became established in the employ of Louis XIV and *Colbert as the chief impulse behind the grandiose decorative schemes at Versailles, and as Colbert's right-hand man in implementing his policy of imposing unified standards of taste and enforcing a strong centralized control over artistic production. In 1662 he was raised to the nobility and named *Premier Peintre du roi*, and in 1663 he was made director of the reorganized *Gobelins factory. Also in 1663 he was made director under Colbert of the reorganized Académie, which he turned into a channel for imposing a codified system of orthodoxy in matters of art (see ACADEMY). His lectures came to be accepted as providing the official standards of artistic correctness and, formulated on the basis of the *classicism of Poussin, gave authority to the view that every aspect of artistic creation can be reduced to teachable rule and precept. In the controversy concerning the relative importance of colour and drawing that divided the Académie at the time he gave his verdict in favour of the latter. In 1698 his small illustrated treatise *Méthode pour apprendre à dessiner les passions* . . . was posthumously published; in this, again following theories of Poussin, he purported to codify the visual expression of the emotions in painting.

Despite the classicism of his theories, Lebrun's own talents lay rather in the direction of flamboyant and grandiose decorative effects. His activities were many and his influence pervasive. Among the most outstanding of his works for the king were the Galerie d'Apollon at the *Louvre (1663), and the famous Galerie des Glaces (1679–84) and the Great Staircase (1671–8, destroyed in 1752) at Versailles. His importance in the history of French art is twofold: his contributions to the magnificence of the *Grand Manner of Louis XIV and his influence in laying the basis of academicism. Many of the leading French artists of the next generation trained in his studio. Lebrun was a fine portraitist and an extremely prolific draughtsman; the Louvre has 3,000 sheets of drawings by him.

Lebrun, Elisabeth Vigée-. See VIGÉE-LEBRUN.

Leck, Bart van der (1876–1958). Dutch painter and designer. After working for eight years in stained-glass studios, he studied painting in Amsterdam, 1900–4. His early work was influenced by *Art Nouveau and *Impressionism, but from about 1910 he developed a more personal style characterized by simplified and stylized forms: his work remained representational, but he eliminated perspective and reduced his figures (which included labourers, soldiers, and women going to market) to sharply delineated geometrical forms in primary colours. In 1916 he met *Mondrian and in 1917 was one of the founders of De *Stijl. At this time his work was purely abstract, featuring geometrically disposed bars and rectangles in a style close to Mondrian and van *Doesburg. However, he found the dogmatism of the movement uncongenial and left it in 1918, reverting to geometrically simplified figural subjects. In the 1920s he became interested in textile design and during the 1930s and 1940s he extended his interests to ceramics and interior decoration, experimenting with the effects of colour on the sense of space. His work can best be seen at the Rijksmuseum Kröller-Müller, Otterlo.

Le Clerc, Jean. See SARACENI.

Le Corbusier (Charles-Édouard Jeanneret) (1887–1965). Swiss-born architect, painter, designer, and writer who became a French citizen in 1930. Although chiefly celebrated as one of the greatest and most influential architects of the 20th cent., Le Corbusier also holds a small niche in the history of modern painting. With *Ozenfant he published a book entitled *Aprés le Cubisme* (1918) setting forth the view that Synthetic *Cubism was degenerating into an art of empty decoration, and the two men founded *Purism, with its doctrine of a machine aesthetic. Le Corbusier wrote, 'A great epoch has begun, animated by a new spirit: a spirit of construction and synthesis, guided by a clear conception.' Up to 1929 he painted only still life, but from that time the human figure was occasionally introduced into his compositions. Although he retained the Purist dislike of ornament for its own sake, his work became more dynamic, imaginative, and lyrical, though still restrained by his horror of any form of excess. He adopted the pseudonym Le Corbusier (derived from the name of one of his grandparents) in 1920, but continued to sign his paintings 'Jeanneret'. Apart from paintings and architecture, his enormous output included drawings, book illustration, lithographs, tapestry designs, furniture, and numerous books, pamphlets, and articles. His influence on architectural and urbanistic thinking throughout the world has been enormous.

Leduc, Fernand. See AUTOMATISTES, LES.

Leduc, Ozias (1864–1955). Canadian painter, active mainly in his native St. Hilaire, Quebec. He painted still lifes and *genre scenes, but is most notable for his church decorations. In 1897 he visited Paris, and his later work was affected by *Symbolist ideas. He lived modestly and unambitiously away from the main centres of art, but he was an inspiration to many of those who knew him, notably Paul *Borduas.

Lee, Arthur Hamilton (Viscount Lee of Fareham). See COURTAULD.

Leech, John (1817–64). English *caricaturist and illustrator, one of the leading artists of *Punch* from 1841 until his death. He made over 3,000 pictures for *Punch* alone (including 600 cartoons) and was also particularly associated with the sporting novels of R. S. Surtees (1805–64). His other book illustrations included *The Ingoldsby Legends* by R. H. Barham (1788–1845) and Charles Dickens's *Christmas Books*, the former of which in particular displays his gift for the light-heartedly *grotesque. His satire was delicate and Dickens said of his pictures that they were 'always the drawings of a gentleman'; he more than any other set the gentlemanly tone for *Punch*.

Le Fauconnier, Henri (1881–1946). French painter, mainly of figure subjects, including nudes and allegories. In 1911 he exhibited with the *Cubists, but in about 1914 he moved to a more *Expressionist style, although he still retained structural features derived from Cubism. He spent the First World War in the Netherlands, where he laid the basis of a European reputation and exercised considerable influence on the development of northern Expressionism (his work is better represented in Dutch collections than it is in French). After his return to Paris in 1920 he gradually abandoned his Expressionist manner for a more restrained and austere style. He is not now generally highly regarded as a painter, but he played an important role in spreading the mannerisms of Cubism. An example of the wide exposure his work had is that his painting *Abundance* (Gemeentemus., The Hague, 1911), was reproduced in the **Blaue Reiter Almanach* and shown in the *Knave of Diamonds exhibition in Moscow.

Léger, Fernand (1881–1955). French painter and designer. From *c.*1909 he participated in the *Cubist movement. He is generally considered one of its major masters but his curvilinear and tubular forms (he was for a time called a 'tubist') contrasted with the fragmented forms preferred by *Picasso and *Braque. The First World War, during which he was gassed whilst serving as a stretcher-bearer, had a profound effect on Léger. His contact with men of different social classes and different walks of life came as a revelation: 'I was abruptly thrust into a reality which was both blinding and new,' he said. Henceforward he made it his ambition to create an art which should be accessible to all ranks of modern society. In 1920 he met *Le Corbusier and *Ozenfant and in the early 1920s he was associated with their *Purist movement. His paintings were static, with the precise and polished facture of machinery, and he had a fondness for including representations of mechanical parts. During the late 1920s and 1930s he also painted single objects isolated in space and sometimes blown up to gigantic size. In the inter-war years he expanded his range beyond easel painting, with murals and designs for the theatre and cinema. He was also busy as a teacher, notably at his own school, the Académie de l'Art Contemporain, and he travelled widely, making three visits to the USA in the 1930s. The connections he had made there stood him in good stead when he lived in America. During the Second World War he lived in the USA, teaching at Yale University, and at Mills College, California. Acrobats and cyclists were favourite subjects in his paintings of this time. From his return to France in 1945 his painting reflected more prominently his political interest in the working classes. But its static, monumental style remained, with flat, unmodulated colours, heavy black contours, and a continuing concern with the contrast between cylindrical and rectilinear forms. In his later career Léger worked much on large decorative commissions, notably the windows and tapestries for the church at Audincourt (1951).

Many honours came to him late in life, and a museum dedicated to him opened at Biot in France in 1957. In the catalogue of the exhibition 'Léger and Purist Paris' (Tate Gallery, London, 1970), John Golding wrote of Léger: 'No other major twentieth-century artist was to react to, and to reflect, such a wide range of artistic currents and movements . . . And yet he was to remain supremely independent as an artistic personality. Never at any moment in his career could he be described as a follower . . . But his originality lay basically in his ability to adapt the ideas and to a certain extent even the visual discoveries of others to his own ends.' He saw the poetic value that lies in the clear delineation of everyday objects, the intrinsic beauty of modern machinery and the things which are mass-produced by machinery, and he favoured proletarian subjects, depicting them with the same clarity and precision as the themes taken from machine culture.

Legros, Alphonse (1837–1911). French-born painter and engraver who settled in England in 1863 (encouraged by *Whistler) and became a British citizen in 1881, although he never acquired fluency in English. His chief importance was as an influential teacher (particularly of etching) at the *Slade School, where he was Professor 1876–92 in succession to *Poynter. He encouraged a respect for the tradition of the Old Masters.

Legros, Pierre, I (1629–90). French sculptor. He was a pupil of Jacques *Sarrazin and worked chiefly with *Tuby at Versailles. His son, **Pierre II** (1666–1719), settled in Rome, where he was one of the leading representatives of the French influence that affected Italian sculpture from the late 17th cent. His best-known work is the highly realistic *polychrome marble memorial to St Stanislas Kostka (1693) in S. Andrea al Quirinale, showing the saint on his deathbed.

Lehmbruck, Wilhelm (1881–1919). German sculptor. His early work was in a fairly conservative academic manner, but when he was living in Paris from 1910 to 1914 he found a much more personal style, influenced by the formal simplifications of *Archipenko, *Brancusi, and *Modigliani, although still essentially in the tradition of *Rodin and *Maillol. It is exemplified in the extremely attenuated forms, angular pose, and melancholic expression of his *Kneeling Woman* (MOMA, New York, 1911), which was greatly praised when it was exhibited at the Cologne Sonderbund exhibition of 1912. On the outbreak of the First World War he returned to Germany and worked in a hospital, the suffering he witnessed being reflected in the poignancy of his last works. The war brought him to a state of acute depression and he committed suicide in 1919.

Lehmbruck often worked in marble, but he was by temperament a modeller rather than a carver, working in clay over a spindly armature, and several of his works were cast in artificial stone to preserve the texture of the clay. With *Barlach he ranks as the outstanding German *Expressionist sculptor. Lehmbruck also did etchings and lithographs, painted, and wrote poetry. There is a museum dedicated to him in his native Duisburg.

Leibl, Wilhelm (1844–1900). German painter, one of the leading exponents of *Realism in his country. His meeting with *Courbet at Munich in 1869 exercised a decisive influence on his art in both style and subject-matter, and he moved to Paris to work with him, although only briefly because of the outbreak of the Franco-Prussian War in 1870. Disgusted with the intrigues of the Munich art world, from 1873 Leibl withdrew to the Bavarian countryside, where he found his favourite models in the simple country folk, as in his best-known work *Three Women in Church* (Kunsthalle, Hamburg, 1878–82). This is in the hard, objective manner of his so-called '*Holbein period'; later his technique became more fluid. Leibl also painted a number of portraits. His work is well represented in the Wallraf-Richartz-Museum in Cologne, his birthplace.

Leighton, Frederic (Baron Leighton) (1830–96). English painter and sculptor, one of the dominant figures of late Victorian art. He travelled widely in Europe as a boy and studied art in Frankfurt, Rome, and Paris. It was not until 1859 that he settled in London, but he had earlier made his name with *Cimabue's Madonna Carried in Procession through the Streets of Florence*, exhibited at the 1855 *Royal Academy exhibition and bought by Queen Victoria (it is now on loan from the royal collection to the National Gallery, London). Thereafter Leighton was matched in worldly success perhaps only by *Millais, his almost exact contemporary; he became President of the Royal Academy in 1878, was made a baronet in 1886, and a few days before he died was raised to the peerage, the first English artist to be so honoured. Intelligent, cultured, and of distinguished appearance, he was one of the chief adornments of London society. Leighton's varied output included portraits and book illustrations, but he is best known for his paintings of classical Greek subjects, the finest of which are distinguished by magnificently opulent colouring as well as splendid draughtsmanship (*The Garden of the Hesperides*, Lady Lever Art Gal., Port Sunlight, 1892). As a sculptor he is best known for the bronze *Athlete Struggling with a Python* (1874–7), which can be seen in Leighton House (on loan from the Tate Gallery), the sumptuously decorated house and studio he built in Holland Park Road, Kensington, now a Leighton museum.

Leinberger, Hans (*c.*1480/5–*c.*1531/5). German sculptor. He worked mainly at Landshut, then the capital of Lower Bavaria, and was one of the leading sculptors of his period in south Germany. The high altar of the collegiate church at Moosburg (1511–14) is typical of his agitated and emotional late *Gothic style.

Lely, Sir Peter (1618–80). Painter of Dutch origin who spent almost all his career in England and was naturalized in 1662. His family name was originally van der Faes, and the name Lely is said to have come from a lily carved on the house in The Hague where his father was born. Lely was born at Soest in Westphalia (where his father, a captain of infantry, was stationed) and trained in Haarlem. He came to England in the early 1640s (early biographers say 1641 or 1643), and although he first painted figure compositions in landscapes (*Sleeping Nymphs*, Dulwich College Picture Gallery), he soon turned to the more profitable field of portraiture. Fortune shone on him, for within a few years of his arrival the best portraitists in England disappeared from the scene; van *Dyck and William *Dobson died in 1641 and 1646 respectively, and Cornelius *Johnson returned to Holland in 1643. In 1654 he was described as 'the best artist in England'. Lely portrayed Charles I and his children, Oliver Cromwell and his son Richard, and other leading figures of the Interregnum (1649–60), but he is associated chiefly with the Restoration court of Charles II. He was made Principal Painter to the King in 1661 and was able to enjoy a lavish lifestyle, described in Samuel Pepys's *Diary*. With the aid of a team of assistants he maintained an enormous output, and his fleshy, sleepy beauties clad in exquisite silks and his bewigged courtiers have created the popular image of Restoration England. Van Dyck was the strongest influence on his style, but Lely was more earthy and less refined. Much of his work is repetitive (it is sometimes hard to tell sitters apart), but he was a fluent and lively colourist and had a gift for impressive composition. His work can be seen in many public and private collections and includes well-known series at Hampton Court (*The Windsor Beauties*) and the National Maritime Museum at Greenwich (*Flagmen*), the latter being character studies of greater strength than he normally achieved. He completely dominated portraiture in his time, and the tradition of the society portrait which he consolidated, developed by *Kneller, *Jervas, and *Hudson, endured for almost a century until it was challenged by *Hogarth. He amassed one of the finest collections of Old Master drawings ever made, which was sold after his death.

Lemoyne. Family of French sculptors. **Jean-Louis Lemoyne** (1665–1755) was a pupil of *Coysevox and is remembered mainly for portrait busts in his master's manner. His brother **Jean-Baptiste the Elder** (1679–1731) was a figure and portrait sculptor of little distinction. Jean-Louis's son, **Jean-Baptiste the Younger** (1704–78), was the outstanding member of the family, becoming official sculptor to Louis XV (1710–74). He did much large-scale work at Versailles and elsewhere, but is renowned particularly for the vivacity of his portraits. Among his pupils were *Falconet, *Houdon, and Jean-Baptiste *Pigalle.

Lemoyne (or **Lemoine**), **François** (1688–1737). French painter. He was one of the leading decorative artists of the day, continuing the grand tradition of *Lebrun but adapting it to the lighter taste of the court of Louis XV (1710–74), to whom he became official painter in 1736. Much of his work can be seen at Versailles, notably in the Salon d'Hercule. He was a man of wide pictorial culture, learning from *Rubens in his use of colour and from Bolognese painters (see CARRACCI) in his clarity and grace of drawing. The easy fluency of his style belies his disturbed personality; he committed suicide a few hours after completing *Time Revealing Truth* (Wallace Coll., London, 1737).

Lempicka, Tamara de (1898–1980). Polish-born painter active in Paris and the USA. She was born Tamara Gorska in Warsaw to wealthy parents and in 1916 she married Tadeusz Lempicki, a Russian lawyer and socialite. In 1918 they fled the Russian Revolution to Paris, where she studied with Maurice *Denis and André *Lhote. She quickly established a reputation as a painter of portraits, mainly of people in the smart social circles in which she moved—writers, entertainers, the deposed nobility of eastern Europe. Her style owes something to the 'tubism' of *Léger, but is very distinctive in its hard, streamlined elegance and sense of chic decadence—better than anyone else she represents the *Art Deco style in painting. Apart from portraits, her main subjects were hefty erotic nudes and still lifes of calla lilies. She received considerable critical acclaim and also became a social celebrity, famed for her aloof Garboesque beauty, her parties, and her love affairs (with women as well as men). In 1939 she moved to the USA with her second husband Baron Raoul Huffner, repeating her artistic and social success in Hollywood and New York. By the 1950s, however, her work was going out of fashion. She tried painting pictures in a different, much looser style, but these were coolly received. Interest in her earlier work began to revive in the 1970s and by the 1990s she had again become something of a stylish icon, with her paintings fetching huge prices in the saleroom and featuring in television advertisements as a symbol of the high life.

Le Nain, Antoine (d. 1648), **Louis** (d. 1648), and **Mathieu** (*c.*1607–77). French painters, brothers, who were born at Laon but had all moved to Paris by 1630. The traditional birth-dates for Antoine and Louis are 1588 and 1593 respectively, but it is now thought likely that they were born shortly before and shortly after 1600, so that all three brothers were of much the same generation. Mathieu was made painter to the city of Paris in 1633, and all three were foundation members of the *Academy in 1648. Apart from this, little is known of their careers and the assigning of works to one or the other of them is fraught with difficulty and controversy, for such paintings as are signed bear only their surname, and of those that are dated none is later than 1648, when all were still alive. The finest and most original works associated with the brothers—powerful and dignified *genre scenes of peasants—are conventionally given to Louis; Antoine is credited with a group of small-scale and richly coloured family scenes, mainly on copper; and in a third group, attributed to Mathieu, are paintings of a more *eclectic style, chiefly portraits and group portraits in a manner suggesting influence from Holland. The brothers are also said to have collaborated on religious works. Examples of all these types are in the Louvre. In 1978–9 a major exhibition in Paris brought together most of the pictures associated with the brothers, but it raised as many problems as it solved. It also, however, confirmed the stature of 'Louis', whose sympathetic and unaffected peasant scenes are the main reason why the Le Nains have attracted so much attention. It has recently been proposed that the traditional description of the figures in some of these paintings as 'peasants' is a misnomer (they are said to be too well dressed for that) and that in fact they represent members of the bourgeoisie.

Lenbach, Franz von (1836–1904). German painter active in Munich and elsewhere. He painted various subjects, but is remembered as the most successful German portraitist of his day. His rich Venetian technique (he had been employed as a copyist of Old Masters in Italy) combined with his solid, respectful characterization appealed greatly to the prosperous ruling classes of Germany. He painted some eighty portraits of Bismarck, whom he first met in 1878 and with whom he had a reserved friendship. Lenbach was a dominant figure in Munich's artistic life in the late 19th cent. His splendid house there, which he designed himself, is now a museum; it houses many examples of his work as well as works that he collected.

Lenoir, Marie-Alexandre (1762–1839). French painter who distinguished himself during the Revolution by saving works of art from suppressed monasteries and sequestrated noble houses, arranging in the cloister and gardens of the Petits Augustins at Paris a 'Musée des Monuments français'. The Museum was suppressed in 1816 and most of the exhibits divided between the *Louvre and the École des *Beaux-Arts.

Le Nôtre, André (1613–1700). French landscape gardener, one of the most celebrated of all practitioners of his art. Both his father and his grandfather were royal gardeners, and his training included studying not only gardening in all its aspects, but also architecture and painting (with *Vouet). In 1637 he succeeded his father in the post of royal gardener, and from the 1650s he designed, altered, or enlarged gardens for some of the greatest châteaux in France. His grandiose style, characterized by formal arrangements and magnificent unbroken vistas, and exemplified above all by the enormous park at Versailles, dominated European garden design until the rise of the more naturalistic English style (associated particularly with Lancelot *Brown) in the 18th cent.

Lentulov, Aristarkh (1882–1943). Russian painter and designer, one of the founders of the *Knave of Diamonds group. In 1911 he travelled to Italy and France, where he was influenced by *Le Fauconnier, *Gleizes, and *Metzinger, and in 1912 he exhibited *Cubist-inspired works. In the late 1920s, however, he changed from his more experimental and innovative style to a more naturalistic and conventional manner, concentrating on landscapes and portraits until his death.

Leochares. Greek sculptor active in the mid-4th cent. BC. He worked for Philip of Macedon (382–336 BC) and his son Alexander the Great and is recorded in several ancient sources, but is an elusive figure. In about 350 BC he worked with *Scopas and two other sculptors on the friezes of the celebrated Mausoleum of Halicarnassus, but it is not possible confidently to assign any of the surviving portions (BM, London) to him. On rather tenuous evidence, the original of the **Apollo Belvedere* is sometimes attributed to Leochares, and the **Demeter of Cnidus* has been proposed as a work from his own hand.

Leonard, Michael. See SUPERREALISM.

Leonardo da Vinci (1452–1519). Florentine artist, scientist, and thinker, the most variously accomplished man of the Italian *Renaissance. Leonardo was born in or near the small town of Vinci in the Tuscan countryside. His father Piero was a Florentine notary and Leonardo was his illegitimate son by a peasant girl, Caterina. Little is known of his youth and upbringing, though *Vasari and others have provided plausible enough speculations, till in 1472 he was enrolled as a painter in the fraternity of St Luke in Florence, after serving an apprenticeship with *Verrocchio. Vasari attributed to him one of the angels in Verrocchio's *Baptism of Christ* (Uffizi, Florence, *c.*1470), and the head of the angel on the left of the picture does indeed far surpass its companion in spirituality and beauty of technique, giving the first demonstration of that combined languor and intensity which is so characteristic of Leonardo's work. Verrocchio is said to have been so impressed that he gave up painting, and it is possible that he was content to entrust the painting side of his business to Leonardo, who was still living in his master's house in 1476. Leonardo stayed in Florence until 1481 or 1482, when he moved to Milan. Several paintings are reasonably attributed to this early Florentine period, notably an exquisite *Annunciation* (Uffizi) and a portrait of Ginevra de' Benci (NG, Washington), probably painted *c.*1476 for the Venetian ambassador Bernardo Bembo, as well as the altarpiece of *The Adoration of the Magi* (Uffizi) that was commissioned in 1481 by the monks of S. Donato a Scopeto and left unfinished (advanced little beyond the underpainting in most places) when Leonardo moved to Milan. This painting and the numerous preparatory drawings for it show the astonishing fecundity of Leonardo's mind. The range of gesture and expression was unprecedented, and features such as the contrasting figures of wise old sage and beautiful youth who stand at either side of the painting, and the rearing horses in the background, became permanent obsessions in his work.

Leonardo lived in Milan until 1499 (when the French invaded), working mainly at the court of Duke Ludovico *Sforza (Il Moro). He is said to have been initially recommended to the duke as a musician, and in a letter to him listing his accomplishments he gives some idea of his versatility, writing of himself first and foremost as a designer of instruments of war and adding his attainments as an artist almost as an afterthought. From then on schemes of applied science such as flying machines, fortifications, and waterways were to occupy much of his energy, while his notebooks were filled with studies of flowers, birds, skeletons, cloud and water effects, and children in the womb. He never properly formulated his researches, and when he put theories into practice, whether methods of painting or diversions of rivers, the results were generally faulty. It was the quest which absorbed him, and there was never the dedication to his art that characterizes many great masters. Although he surpassed all of his contemporaries in the sheer beauty of his technique as a painter, this 'mechanical' aspect of his work was much less appealing to him than solving problems of composition and characterization in his drawings (incomparably the finest collection of which is at Windsor Castle). His dilatoriness infuriated his paymasters and although his *œuvre* as a painter is small, he left a high proportion of his pictures unfinished. This stress on the intellectual aspects of painting was one of the most momentous features of Leonardo's career, for he was largely responsible for establishing the idea of the artist as a creative thinker, not simply a skilled craftsman (see LIBERAL ARTS).

Leonardo's main artistic undertakings in Milan were a project for a huge equestrian statue to Ludovico Sforza's father, now known only in preliminary drawings, and the wall painting of the *Last Supper* (*c.*1495) in the refectory of Sta Maria delle Grazie. The *fresco method of mural painting was not flexible or subtle enough for the slow-working Leonardo, so he adopted an experimental technique that quickly caused the picture to deteriorate disastrously. It has been many times restored, but although it is only a shadow of Leonardo's original creation it still retains some of the immense authority that has made it, for five centuries, the most revered painting in the world. Leonardo's other works in Milan included portraits, notably the marvellous picture of Duke Ludovico's mistress Cecilia Gallerani known as the *Lady with an Ermine* (Czartoryski Gal., Cracow) and an altarpiece of the *Virgin of the Rocks*, which exists in two problematically related versions, the earlier (Louvre, Paris) possibly painted when Leonardo was still in Florence, the later (NG, London) still being worked on in 1508. There may have been some studio assistance in the London version, but the finest passages, notably the heads of the Virgin and the angel, with their exquisitely curled hair and heavy-lidded eyes, can be by no one but Leonardo himself. The larger, bolder forms of the London picture show Leonardo's move towards the more monumental *classical style of the High Renaissance, of which he was the main creator.

Between 1500, when he returned for a time to Florence, and 1516, when he left Italy for France, Leonardo's life was unsettled. In 1502–3 he worked as a military engineer for Cesare Borgia, in 1506–13 he was based again in Milan, and in 1513 he moved to Rome, but the artistic activity of his later years was chiefly centred in Florence in the years 1500–6. From this time dates his portrait of *Mona Lisa* (Louvre) and the wall painting of the *Battle of Anghiari* in the Palazzo Vecchio, Florence, where he worked in rivalry with *Michelangelo. The battle piece is destroyed, but is preserved in copies; fittingly, the most famous copy is a drawing by *Rubens (Louvre), for Leonardo's painting anticipated the dynamism of the *Baroque and influenced battle painters up to the 19th cent. In Florence also Leonardo worked out variations on a theme that fascinated him at this time and presented a great challenge to his skill in composing closely knit groups of figures. This was *The Virgin and Child with St Anne*, known today mainly through a painting in the Louvre and the incomparably beautiful *cartoon (which includes also the infant John the Baptist) in the National Gallery, London.

In 1516 Leonardo accepted an invitation from Francis I to move to France, and he died at Cloux, near Amboise, in 1519. The legend that he died in the arms of the king has no basis in fact, but it indicates the status he attained. Leonardo did little artistic work in the last decade of his life, the last paintings from his hand generally being accepted as two pictures of *St John the Baptist* (one later converted into a *Bacchus*), both in the Louvre (*c.*1515). They show the enigmatic smile, the dense shadow, the pointing finger, and the thick curling hair that rapidly became clichés in the work of his followers. In painting Leonardo had an enormous influence. His heroic figures and beautifully balanced compositions (particularly his use of pyramidical grouping) were the basis of the High Renaissance style, influencing particularly his two greatest contemporaries, Michelangelo and *Raphael, and his subtle modelling through light and shade (see SFUMATO) showed the potentialities of the *oil medium, which he was one of the first Italians to exploit. *Giorgione and *Correggio were among those most deeply influenced by this aspect of Leonardo's work. His writings on painting were influential too; they were first published from his scattered notes as the *Treatise On Painting* (in Italian and French) in 1651, but were well known before then. In sculpture and architecture no work that is indisputably by Leonardo survives, but his expertise and ideas were important in both fields. When the sculptor Giovanni Francesco Rustici (1474–1554), for example, was making his bronze group of *St John the Baptist between a Pharisee and a Levite* for the Baptistery in Florence, he would, as Vasari tells us, 'allow no one near save Leonardo, who never left him while he was moulding and casting until the work was finished'. Leonardo's friend *Bramante, the greatest architect of the High Renaissance, was influenced by his designs for 'ideal' churches. Leonardo is one of the very few artists whose reputation has from his own times onward constantly remained at the highest level, even though his output of completed works was small—a reflection of his extraordinary force of intellect, and his virtually single-handed creation of the idea of the artist as genius.

Leoni, Leone (1509–90). Italian *Mannerist sculptor who worked in many parts of Italy and in the service of the emperor Charles V (1500–58) in

Germany and the Netherlands. He was trained as a goldsmith, but none of his works in that medium survives, though the Metropolitan Museum, New York, possesses a sardonyx *cameo by him (1550). From 1538 to 1540 he was coin engraver to Pope Paul III (Alessandro *Farnese), but he was then condemned to the galleys for conspiring to murder the papal jeweller. He was released in 1541 and for most of the rest of his life held the office of master of the imperial mint in Milan. His sculpture consists mainly of portraits—both medals and busts. Many of his works for the emperor were sent to Spain, where his son **Pompeo** (*c.*1533–1608), who moved there in about 1556, gave them the finishing touches. The most important was a group of twenty-seven bronze statues (finished 1582) for the high altar of the *Escorial. Pompeo executed several tombs in Spain on his own account, and was, like his father, a goldsmith and medallist. Again like his father, he had a dangerous brush with authority, being briefly imprisoned by the Inquisition.

Le Parc, Julio (1928–). Argentinian painter, sculptor, and experimental artist, regarded as one of the leading exponents of *Kinetic art. In 1958 he went to Paris and after working for a while with Victor *Vasarely became a founder member of the *Groupe de Recherche d'Art Visuel (GRAV) in 1959. Le Parc professes to adopt a rational and objective attitude to his work, repudiating the ideas of artistic creativity or symbolic meaning and working according to scientific principles. He often uses the idea of spectator participation, but tries to eliminate subjective response on the part of the spectator, looking for an objective and predictable perceptual response to a planned stimulus. Much of his work consists of devices for disorienting the spectator (distorting glasses and so on), but he has also made some outstanding *mobiles, using perspex or metallic elements to scatter or reflect the light. His *Continual Mobile, Continual Light* of 1963 is in the Tate Gallery, London.

Lepicié, Nicolas-Bernard (1735–84). French painter of portraits, domestic *genre scenes, and historical subjects. His best works, although not entirely free from the sentimentality of the period, have something of the tranquil beauty of *Chardin (*The Reading Lesson*, Wallace Coll., London). He was a teacher of Carle *Vernet and for a while Secretary of the Académie. His father **François-Bernard** (1698–1755) was an engraver and writer on art.

Leslie, Charles Robert (1794–1859). British painter and writer on art, of American parentage. In his day he was well known for his paintings of literary themes, but he is now remembered mainly as a writer, above all for his *Memoirs of the Life of John Constable* (1843), which is regarded as one of the classics of artistic biography (*Constable was a close friend). His other writings include *A Handbook for Young Painters* (1853).

Lessing, Gotthold Ephraim (1729–81). German writer. He was a man of formidable intellect and great versatility who played a leading role in the development of German theatre, but he is known mainly for his treatise on aesthetics, *Laokoon* (1766). This takes as its starting-point a passage in *Winckelmann's writings in which he discussed the celebrated antique statue **Laocoön*. Winckelmann contrasted what he considered the stoical beauty of Laocoön in the sculpture with the loud cries that Virgil causes him to utter in the *Aeneid*, interpreting the alleged difference (to most people he appears to be howling with pain in the sculpture too) as a superior serenity in Greek art. Lessing dissented, and argued that each art achieves its effects by the means appropriate to its medium and that the artist must exploit the potentialities of his medium to the full, whilst respecting its limitations. Poetry, he held, is most adapted to the representation of human action in time but lacks visual vividness. Painting and sculpture are best adapted to the representation of idealized human beauty in repose. Owing to the non-temporal character of the medium they cannot well represent the body in action. Only by selecting the 'critical' or 'fruitful' moment, which simultaneously preserves physical beauty and concentrates within itself the suggestion of past and future action, can the plastic artist even indirectly represent a sequence of events in action. He thought that the *Laocoön* group was a masterly example of this, a work whose beauty and significance made it at once a delight to the eye and a stimulus to the imagination. The important impact of Lessing's *Laokoon* arose from its emphasis on the aesthetic functions of art in contrast with the traditional view of art as the handmaid of religion and philosophy, whose duty was primarily to instruct. Its achievement lay in the general effect it had on habits of thinking about art, emancipating art from religious and social pressures and directing attention to the realities of the artistic process itself.

Lessing, Karl Friedrich (1808–80). German painter, one of the leaders of the Düsseldorf School of history painting. He combined earnestly melodramatic poses with studiously correct historical detail, so that many of his pictures look like scenes from plays or pageants. His best-known works have subjects taken from the Hussite Rebellion of the 15th cent. (*Hussite Prayer*, Staatliche Museen, Berlin, 1836) and were identified with the spirit of rebellion against the political and religious repression of the day. Lessing also painted landscapes. In 1858 he became Director of the gallery at Karlsruhe, and settled there permanently.

Le Sueur, Eustache (1616–55). French painter, active throughout his career in his native Paris. He was a pupil of *Vouet, whose influence is strong on his early works (*The Presentation of the Virgin*, Hermitage, St Petersburg, *c.*1640). In the 1640s he was profoundly affected by the paintings of *Poussin (who visited Paris 1640–2) and his work became more *classical. He lacked Poussin's heroic grandeur, but he added a tenderness of his own to the master's manner, as in his most important works, a series of paintings (begun 1648), illustrating the life of St Bruno, done for the Charterhouse of Paris and now in the Louvre. In the last years of his life his chief model became *Raphael, whom he imitated in an uninspired manner. Le Sueur was a founder member of the French *Academy in 1648. In his own day and throughout the 18th cent. he was almost as well thought of as Poussin, but he now has the status of an attractive but minor master.

Le Sueur, Hubert (*c.*1580–after 1658). French sculptor, active mainly in England, where he is first recorded in 1626. He worked much for Charles I (1600–49), his most famous work being the equestrian statue of the king (1633) at Charing Cross in London. This shows the skill as a bronze caster for which he was renowned, but also his smooth, lifeless surfaces, which give his works, in the words of Margaret Whinney (*Sculpture in Britain 1530 to 1830*, 1964), 'a curious, inflated appearance, as if they were not modelled, but blown up from within'. He was remarkably conceited, on occasions signing himself '*Praxiteles Le Sueur', but Charles recognized him as a second-rate artist and sometimes reduced the prices he asked for his work. By 1643 Le Sueur was back in Paris. His main influence in England was in popularizing the bust portrait.

Lethaby, William Richard (1857–1931). English architect, teacher, and writer on art. His views on art and society were shaped by *Ruskin and William *Morris and he was a leading figure of the *Arts and Crafts Movement. He was Professor of Design at the *Royal College of Art (1900–18) and Principal of the London County Council's Central School of Arts and Crafts (1896–1911). An authority on medieval building, he was Surveyor to Westminster Abbey from 1906 to 1927, successfully advocating the conservation of the fabric rather than its 'restoration'. His books included *Mediaeval Art* (1904) and *Form in Civilization* (1922), a volume of collected papers containing some penetrating analyses of the role of the designer in contemporary life.

Leu, Hans the Younger (*c.*1490–1531). Swiss painter and graphic artist, the son of a painter of the same name. He studied with *Dürer in Nuremberg and *Baldung-Grien in Freiburg, then settled in his native Zurich by 1514. His work is remarkable mainly for the prominence given to landscape, relating him to the masters of the *Danube School (*Orpheus*, Kunstmuseum, Basle, 1519). Leu's career was ruined by the Reformation.

Leutze, Emanuel Gottlieb (1816–68). German-born painter who lived in America from 1825 to 1841 and again from 1859 and is usually considered a member of the American School. He is remembered mainly for *Washington Crossing the Delaware* (Met. Mus., New York, 1851), painted in Düsseldorf, where he spent most of his career, and for another work that similarly appeals more for its patriotic sentiments than for any aesthetic merit—his large mural *Westward the Course of Empire Takes its Way* (1861–2) in the Capitol at Washington. His portraits and rare landscapes are more distinguished, but remain virtually unknown.

Levasseur. Family of Canadian wood-carvers active in Quebec in the mid-18th cent. They made decorations and figures for a number of churches in a style mainly inspired by the French *Régence (e.g. *retable of the Ursuline Chapel, Quebec, 1734–9). Members of the family included **Noël** (1680–1740), **Pierre-Noël** (1684–1747), **Jean-Baptiste Antoine** (1717–75), and **François-Noël** (1703–94).

Lewis, John Frederick (1805–76). English painter, mainly in watercolour, son of the engraver and landscape painter **Frederick Christian Lewis** (1779–1856). He was a great traveller and spent the years 1841–51 in Cairo. His colourful and highly detailed scenes of life in the harem and bazaar were a huge success in London, and after his return there in 1851 he concentrated on them exclusively, playing a major part in creating the vogue for 'Oriental' subjects. His brother, **Frederick Christian Lewis, Jr.** (1813–75), was also a painter. He worked in India for many years and is sometimes known as 'Indian Lewis' to distinguish him from his brother.

Lewis, Wyndham (1882–1957). British painter, novelist, and critic, born of a British mother and a wealthy American father on their yacht off Nova

Scotia. He came to England as a child, studied at the *Slade School, 1898–1901, then lived on the Continent for seven years, mostly in Paris. In 1909 he returned to England and in the years leading up to the First World War emerged as one of the leading figures in British avant-garde art. From 1911 he developed an angular, machine-like, semi-abstract style that had affinities with both *Cubism and *Futurism. He worked for a short time with Roger *Fry at the *Omega Workshops, but after quarrelling with him in 1914 he formed the Rebel Art Centre, from which grew *Vorticism, a movement of which he was the chief figure and whose journal *Blast* he edited. He served with the Royal Artillery, 1915–17, and as an *Official War Artist, 1917–18, carrying his Vorticist style into works such as *A Battery Shelled* (Imperial War Museum, London). In 1919 he founded Group X as an attempt to revive Vorticism, but this failed, and from the late 1920s he devoted himself mainly to writing, in which he often made savage attacks on his contemporaries (particularly the *Bloomsbury Group). His association with the British Fascist Party and his praise of Hitler alienated him from the literary world. The best-known paintings of his later years are his incisive portraits; the rejection of that of T. S. Eliot (Durban Art Gallery) caused Augustus *John to resign in disgust from the *Royal Academy in 1938.

Lewis was the most original and idiosyncratic of the major British artists working in the first decades of the 20th cent., and he was among the first artists in Europe to produce completely abstract paintings and drawings. He built his personal style on features taken from Cubism and Futurism but did not accept either. He accused Cubism of failure to 'synthesize the quality of LIFE with the significance or spiritual weight that is the mark of all the greatest art' and of being mere visual acrobatics. The Futurists, he wrote, had the vivacity that the Cubists lacked, but they themselves lacked the grandness and the 'great plastic qualities' that Cubism achieved. His own work, he declared, was 'electric with a mastered and vivid vitality'. He wrote several books, including novels, notably *Tarr* (1918), and collections of essays and criticism. *Blasting and Bombadiering* (1937), *Wyndham Lewis the Artist* (1939), and *Rude Assignment* (1950) are autobiographical.

Lewitt, Sol (1928–). American sculptor, graphic artist, writer, and *Conceptual artist. His career did not take off until the early 1960s, when he turned to sculpture and became one of the leading exponents of *Minimal art. His 'structures', as he prefers to call them, characteristically involve permutations of simple basic elements, sometimes arranged in box- or table-like constructions. He is also an exponent of Conceptual art; in 1968, for example, he fabricated a metal cube and buried it in the ground at Visser House at Bergeyk in the Netherlands, documenting photographically the object's disappearance (this has also been considered an example of *Land art). Lewitt has also written numerous articles on Conceptual art. His other work includes prints in various techniques.

Leyden, Lucas van. See LUCAS VAN LEYDEN.

Leyster, Judith (1609–60). Dutch painter of *genre scenes, portraits, and still life, probably a pupil of Frans *Hals in Haarlem, where she spent most of her career (she also worked in Amsterdam). In 1636 she married Jan Miense *Molenaer, with whom she shared a studio, using the same models and props. Leyster was one of Hals's best followers and her work has sometimes passed as his, an example being the *Lute Player* in the Rijksmuseum, Amsterdam. Her monogram includes a star, a play on 'Ley / ster' (lode star).

Lhote, André (1885–1962). French painter, sculptor, and writer on art. He worked initially as a commercial wood-carver and was largely self-taught as a painter. His early work was *Fauvist in spirit, but from 1911 he adopted the stylistic mannerisms of *Cubism to his varied range of subjects, including landscapes, still lifes, interiors, mythological scenes, and portraits. All were deliberately composed in complicated systems of interacting planes and semi-geometrical forms precisely articulated and defined by clear, unmodulated colours. His designs and rhythms were usually intellectualized rather than spontaneous, and some of his still lifes had a quiet charm not unlike the work of the *Purists. Lhote, however, was more important as a teacher and critic of modern art than as a practising artist. He exercised an extensive influence on younger artists both French and foreign through his own academy of art, the Académie Montparnasse, which he opened in 1922, and he founded a South American branch on a visit to Rio de Janeiro in 1952. His writings included treatises on landscape painting (1939) and figure painting (1950).

liberal arts. Term applied to those arts that were traditionally considered primarily as exercises of the mind rather than of practical skill and craftsmanship. The concept of a distinction between 'liberal' and 'vulgar' arts goes back to classical antiquity, and survived in one form or another up to the *Renaissance, forming the basis of secular learning in the Middle Ages. The name *quadrivium* was given to the subjects that studied physical reality (arithmetic, astronomy, geometry, and music—that is the mathematical theory of music) and *trivium* to the arts of grammar, rhetoric, and logic. Collectively these were known as the seven liberal arts, and were subservient to philosophy, the supreme art. In these, as in all classifications which preceded the concept of the *fine arts, the word 'art' carries a very different signification from that which it bears in *aesthetic discourse today, and one closer to the meaning which survives in academic terminology such as 'arts degree'.

In the early Renaissance the lowly position accorded to the visual arts in this intellectual framework was increasingly contested, providing a theoretical basis for the social struggle which took place to raise them from the status of manual skill to the dignity of a liberal exercise of the spirit. The most formidable champion of the visual arts was *Leonardo, who more than anyone else was responsible for creating the idea of the painter as a creative thinker. *Vasari records that when Leonardo was painting his *Last Supper* the prior of the community was puzzled by the way in which 'he sometimes spent half a day at a time contemplating what he had done so far; if he had had his way, Leonardo would have toiled like one of the labourers hoeing in the garden and never put down his brush for a moment'. When the prior complained to the Duke of Milan, Leonardo explained 'that men of genius sometimes accomplish most when they work the least, for they are thinking out inventions and forming in their minds the perfect ideas that they subsequently express and reproduce with their hands'. By about 1500 painting and sculpture were generally accepted as liberal arts by Italian humanists (significantly so in Baldassare Castiglione's influential *Book of the Courtier* of 1528, which was translated into English in 1561). However, as Anthony *Blunt points out (*Artistic Theory in Italy 1450–1600*), 'As soon as the visual arts became generally accepted as liberal, the protagonists began to quarrel among themselves about which of them was the noblest and most liberal.' The acceptance came later in other parts of Europe.

The original seven liberal arts (sometimes paired with the seven principal virtues—faith, hope, charity, etc.) are often represented in painting and sculpture, personified as women holding various *attributes and being followed by famous masters of the arts concerned (e.g. Cicero with Rhetoric). The system was formulated by the 5th-cent. scholar Martianus Capella in his treatise *The Marriage of Philology and Mercury*. For the *Baroque age the types of the liberal arts were codified by Cesare Ripa (see EMBLEM) in his handbook of *iconography.

Libre Esthétique, La. See VINGT.

Lichtenstein, Roy (1923–). American painter, sculptor, and graphic artist. In the late 1950s his style was *Abstract Expressionist, but in the early 1960s he changed to *Pop art and his first one-man exhibition in this style, at the Leo *Castelli gallery, New York, in 1962 was a sensational success. In common with other Pop artists, Lichtenstein adopted the images of commercial art, but he did so in a highly distinctive manner. He took his inspiration from comic strips but blew up the images to a large scale, reproducing the primary colours and dots of the cheap printing processes (*Whaam!*, Tate Gallery, London, 1963). The initial stimulus is said to have come from one of his young children, who pointed to a comic book and challenged 'I bet you can't paint as good as that.' Despite their use of such kitsch material, his paintings show an impressive feeling for composition and colour and Lichtenstein has enjoyed continued critical success as well as popular appeal. In the mid-1960s he began making Pop versions of paintings by modern masters such as *Cézanne and *Mondrian, and in the 1970s he expanded his range to include sculpture, mostly in polished brass and imitating the *Art Deco forms of the 1930s. His work in the 1980s included two large murals in New York (at Leo Castelli's gallery in Greene Street and in the Equitable Building).

Liebermann, Max (1847–1935). German painter and graphic artist. His importance in his day lay in his openness to foreign influences. He was one of the first to overcome the parochiality of the German naturalistic tradition and to broaden the outlook of German painters to newer trends. Living mainly in Paris 1873–8, he found himself more in sympathy with *Courbet, *Millet, and the *Barbizon School than with *Manet or *Renoir, but after his return to Germany in 1878 he came to be considered the leading German *Impressionist painter together with *Corinth and *Slevogt. In 1899 he founded the Berlin *Sezession and became its President. He was unable, however, to keep abreast of developments and a decade later stood as the supporter of that old-fashioned traditionalism against which *Nolde, the members of the *Brücke and other German *Expressionists were in revolt. Nevertheless, his work was declared *degenerate by the Nazis.

Liédet, Loyset (active 1460–78). Netherlandish *miniaturist, active mainly in Bruges, who left a larger *œuvre* than any of his contemporaries. His work includes *illuminations in the *Chronicles of Froissart* (Bib. nat., Paris).

Lievens, Jan (1607–74). Dutch painter and graphic artist. He was extremely precocious, and after training in Amsterdam with *Lastman he was practising independently in his native Leiden when he was in his early teens. From *c.*1625 to 1631/2 he worked in close collaboration with his friend *Rembrandt. They shared the same models (and probably a studio) and even worked together on the same pictures—a *Portrait of a Child* in the Rijksmuseum in Amsterdam, for example, is signed 'Lievens retouched by Rembrandt'. The diplomat and connoisseur Constantin Huygens visited them in 1629 and thought they showed equal promise of greatness. He wrote that Rembrandt surpassed Lievens in vivacity of expression, but that Lievens was superior in 'a certain grandeur of invention and boldness of subjects and forms'. That this was not excessive praise is borne out by Lievens's marvellously melodramatic *Raising of Lazarus* (Brighton Museum and Art Gallery, 1631), in which the only parts of Lazarus shown are his arms emerging from the tomb. After the paths of the two young artists separated in 1631/2, however, Lievens did not sustain his early brilliance. He is said to have made a visit to England and from 1635 to 1644 was in Antwerp, where under the influence of van *Dyck he adopted a more elegant and facile style that brought him renown as a portraitist. In 1644 he returned to the Netherlands, where he remained for the rest of his life and during the last three decades was popular in official circles in Amsterdam and The Hague. Lievens was a talented etcher and also made some woodcuts.

Lightfoot, Maxwell Gordon. See CAMDEN TOWN GROUP.

Ligorio, Pirro (1513/14–83). Italian architect, painter, and antiquarian, born at Naples. For his patron Cardinal Ippolito d'*Este he designed the Villa d'Este at Tivoli (1550–72) and its celebrated gardens, and for Pope Pius IV he built the Casino in the Vatican Gardens (1560–5). In 1564 he was appointed *Michelangelo's successor at St Peter's, but in the following year he was imprisoned on suspicion of defrauding the papacy both in his architectural work and in regard to the purchase of antiquities. He was soon released, but in 1569 he left Rome and settled in Ferrara, where he again worked for the Este family, organizing Duke Alfonso II's collection of antiquities. In 1553 he published a treatise on the antiquities of Rome, which is an important source of information on what Roman remains were visible at the time.

Limburg (or **Limbourg**) **Brothers**. Netherlandish manuscript *illuminators, **Herman**, **Jean (Jannequin)**, and **Paul (Pol)**, all three of whom died in 1416, presumably victims of the plague or other epidemic. Pol was probably the head of the workshop, but it is not possible to distinguish his hand from those of his brothers. They were born in Nijmegen, nephews of Jean *Malouel, and Herman and Jean are first documented in the late 1390s apprenticed to a goldsmith in Paris. In 1402 Jean and Pol were working for Philip the Bold (1342–1404), Duke of Burgundy, and after Philip's death all three Limburgs worked for his brother Jean, Duc de Berry, remaining in his service until their deaths and holding privileged positions at his court, which moved with him around France from one magnificent residence to the next. He was, indeed, one of the most extravagant patrons and collectors in the history of art, and the Limburgs illuminated two outstanding manuscripts for his celebrated library: the Belles Heures (Met. Mus., New York, *c.*1408) and the Très Riches Heures (Musée Condé, Chantilly), which was begun *c.*1413 and left unfinished at their deaths (it was completed by the French illuminator Jean Colombe (*c.*1440–93?) about seventy years later). The Très Riches Heures (see BOOK OF HOURS) is by common consent one of the supreme masterpieces of manuscript illumination and the archetype of the *International Gothic style. Its most original and beautiful feature is the series of twelve full-page illustrations of the months (the first time a calendar was so lavishly treated), full of exquisite ornamentation and beautifully observed naturalistic detail. The *miniatures are remarkable, too, for their mastery in rendering space, strongly suggesting that one or more of the brothers had visited Italy, and they occupy an important place in the development of the northern traditions of landscape and *genre painting.

limestone. A general term for sedimentary rocks composed mainly of calcium carbonate. It varies in hardness from easily worked freestone to fine-grained oolites, some of which weather well and can be carved with precision. Some limestones will take a polish and are incorrectly known as marbles (e.g. *Purbeck marble). Generally, however, it is used much more for building than for sculpture.

limner. A word for a painter that has been used in different ways according to time and place. In the Middle Ages it was applied to manuscript *illuminators, and from the 16th cent. it was used of painters of *miniature portraits (Nicholas *Hilliard's treatise is called *The Arte of Limning*). In American usage it denotes the anonymous and often itinerant painters, particularly portraitists, of the 17th and 18th cents. Limners in this last sense are sometimes given invented names in the same manner in which the term 'Master of' is used in European painting—the Schuyler Limner, named after one of the families he portrayed, is an example.

Lindner, Richard (1901–78). German-born painter who became an American citizen in 1948. He fled from the Nazi regime in 1933 and settled in Paris, then moved to the USA in 1941. At first he worked as a magazine and book illustrator (in Germany he had been art director of a publishing firm), and he did not begin to paint seriously until the early 1950s. His most characteristic works take their imagery from the vulgar and sordid aspects of New York life, often with overtly erotic symbolism, and are painted with harsh colours and hard outlines. The effects he created owed something to *Expressionist exaggeration, *Surrealist fantasy, and *Cubist manipulations of form, but his style is vivid and distinctive and anticipates aspects of *Pop art.

Lindsay. Family of Australian artists. The members included five of the children of Dr R. C. Lindsay of Creswick, Victoria: **Percy Lindsay** (1870–1952), painter and graphic artist; **Sir Lionel Lindsay** (1874–1961), art critic, watercolour painter, and graphic artist in pen, etching, and woodcut, who did much to arouse an interest in the collection of original prints in Australia; **Norman Lindsay** (1879–1969), painter, graphic artist, critic, and novelist; **Ruby Lindsay** (1887–1919), graphic artist; and **Sir Daryl Lindsay** (1889–1976), painter and Director of the National Gallery of Victoria, 1942–56. Norman's son **Raymond** (1904–60) and Daryl's wife **Joan** (1896–1984) were also painters. For over half a century this family, through one or other of its members, played a leading role in Australian art. The most interesting character among them was Norman Lindsay, who according to Robert *Hughes (*The Art of Australia*, 1970) 'has some claim to be the most forceful personality the arts in Australia have ever seen'. He believed that the main impulse of art and life was sex, and his work was often denounced as pornographic. However, when he saw some of Lindsay's works at an exhibition of Australian art in London in 1923, Sir William *Orpen commented that they were 'certainly vulgar, but not in the least indecent. They are extremely badly drawn, and show no sense of design and a total lack of imagination.'

line engraving. Term applied to a method of making prints (and the print so made) in which the design is cut directly into the surface of a metal (usually copper) plate. In normal parlance the term *engraving usually refers to line engraving, but strictly the former is a generic term, covering a variety of processes. The line engraver uses a tool called a *burin, holding it in his right hand (presuming he is right-handed) and pushing it slowly through the surface of the copper, cutting a clean V-shaped furrow, while the shred of metal removed from the line is thrown up in a continuous spiral from the moving point. Both hands are in action, for the engraver steadies the plate with his left hand against the pressure exerted by the burin and, when cutting curves, holds the burin still with the right hand while the left rotates the plate on to the point of the tool. The shreds of metal excavated by the tool and the slight *burr thrown up at the sides of the lines are cut off by the scraper, which is also used for making corrections. The essential character of the medium is linear, though shading and tone may be suggested by parallel strokes, cross-*hatching, or textures compounded of various dots and flicks. Characteristically, line engravings have a quality of metallic hardness and austere precision, which comes partly from the nature of the materials and partly from the slow, calculated driving of the lines, compared with the easy, spontaneous drawing of the *etcher or *lithographer.

Line engraving seems to have originated towards the middle of the 15th cent. in the workshops of goldsmiths, arising independently in Italy (see NIELLO) and Germany, though perhaps slightly earlier in the latter country. The early German engravers are mostly anonymous and have to be designated by a system of initials and *noms de plume*, as with the *Master of the Playing Cards and the *Master E. S., who were certainly goldsmiths as well as engravers. Martin *Schongauer, who died in 1491, was the first major artist to work mainly as an engraver, and the medium had its finest flowering in the early 16th cent. in the work of Albrecht *Dürer and *Lucas van Leyden. Active at the same time was Marcantonio *Raimondi, who was the great pioneer in the use of engraving as a means of reproducing the works of other artists. This function soon took over from the role of engraving as a means of original artistic expression, but if the countless engravings produced during the 16th, 17th, and early 18th cents. were nearly all reproductive, their quality from the technical point of view was often exceedingly high. *Rubens realized the value of having his works reproduced by means of engraving and for this purpose a talented school of engravers existed early in the 17th cent. under his patronage. A little later there arose in France a celebrated school of portrait engraving, in which the greatest names were those of Claude *Mellan, Robert *Nanteuil, and the Flemish-born Gerard *Edelinck.

Certain technical developments date from this period. *Etching, invented in Dürer's time as a labour-saving method of engraving, began to be combined with line engraving. The design was begun by etching, in order to ease the task of the engraver, who then took over and completed the plate with his burin. Later, plates worked by other copper-plate processes, particularly the hazier ones such as *mezzotint, were frequently punctuated and finished with the burin at points where the design required precision.

During the 18th cent. line engraving began to decline in importance even as a reproductive process, especially in England, where tonal processes such as mezzotint and *stipple were popular. Steel engraving, in which extremely fine lines and subtle tones are possible, was widely practised in the first half of the 19th cent. but was rendered obsolete by the discovery that engraved copper-plates could be steel-faced by electrolysis and thus enabled to withstand a large printing. Line engraving reached its lowest ebb during the later 19th cent., when it was challenged by *wood engraving in the popular market and then superseded by photomechanical processes.

In the 20th cent., however, line engraving has been revived as a means of original expression, the greatest impetus coming from the experimental workshop for the graphic arts—'Atelier 17'—established in Paris in 1927 by the English artist S. W. *Hayter.

Linnell, John (1792–1882). English painter. He made his reputation and his fortune as a fashionable portraitist and *miniaturist, but was devoted to landscape painting. His wealth enabled him to patronize *Blake, whom he greatly admired, and some of his early landscapes have something of the visionary quality of the master and of Samuel *Palmer, who married Linnell's daughter. In 1852 he settled in Redhill and most of his prolific output thereafter was devoted to idyllic scenes in Surrey, done in a lush and more conventional pastoral idiom than his early work. Such works were highly popular and he became immensely wealthy. In spite of his success he was denied membership of the *Royal Academy, this being a reflection of his unpopularity with his fellow-artists.

linocut. A technique of making a print from a piece of thick linoleum. It was introduced at the beginning of the 20th cent., but is essentially a development of the *woodcut, although linocuts are much simpler to make because the material is soft and grainless and therefore comparatively easy to work, various knives and gouges being used to cut the soft and even surface.

Linocutting has been much used for teaching art in schools, and this circumstance has caused it to be somewhat lightly regarded, but artists of the stature of *Matisse and *Picasso have made memorable use of it. Both of them took up the technique in the late 1930s. For colour prints it has obvious advantages, since a number of large blocks may be used without undue expense, while the fact that the surface can be cut rapidly and spontaneously means that the process is highly suitable for big prints boldly conceived in large masses of flat colour and decorative texture.

linseed oil. Oil from the seeds of flax, the commonest *medium in *oil painting. Most modern painters have used raw linseed oil diluted with *turpentine as a medium, but the Old Masters generally preferred polymerized oil, known as 'stand oil', which was prepared by heating linseed oil or drying it in the sun. Stand oil was thinned to a painting consistency by mixing it with turpentine, or used as an *emulsion mixed with yolk of egg. Boiled oils withstand atmospheric conditions better than raw oils. Linseed oil tends to turn yellow with age, but has less tendency to crack than *walnut or *poppy oil.

Liotard, Jean-Etienne (1702–89). Swiss *pastel painter and engraver. He travelled widely in Europe and also spent 4 years in Constantinople (1738–42), after which he adopted Turkish dress and beard, his eccentric appearance being familiar from his numerous self-portraits. His delicate and polished style brought him fashionable success in Paris, the Netherlands, and England, which he twice visited (1733–5 and 1772–4). The best collection of his work is in the museum at Geneva, the city where he was born and died.

Lipchitz, Jacques (1891–1973). Lithuanian-born sculptor who worked mainly in France and the USA. After studying engineering in Lithuania he moved to Paris in 1909. By about 1912 he was part of a circle of avant-garde artists including *Matisse, *Modigliani, and *Picasso, and from 1914 he became one of the first sculptors to apply the principles of *Cubism in three dimensions (*Man with Guitar*, MOMA, New York, 1916). During the 1920s his style changed, as he became preoccupied with open forms and the interpenetration of solids and voids.

He took French nationality in 1925 but in 1941—by this time an internationally renowned figure—he fled to the USA, settling at Hastings-on-Hudson, New York. In America he returned to greater solidity of form, but with a desire for greater spirituality. At times the tortured, bloated forms of his late work, as in *Prometheus Strangling the Vulture* (Walker Art Center, Minneapolis, 1944–53), look rather like inflated shrubbery.

Lippi, Filippino (*c.*1457–1504). Florentine painter, the son and pupil of Filippo *Lippi, who died when the boy was about 12. He also studied with *Botticelli and learned much from his expressive use of line, but Filippino's style, although sensitive and poetic, is more robust than his master's. His first major commission (1484) was the completion of *Masaccio's fresco cycle in the Brancacci Chapel of Sta Maria del Carmine, a task he carried out with such skill and tact that it is sometimes difficult to tell where his work begins and that of more than half a century earlier ends. Filippino painted several other frescos, the most important of which are cycles on the life of St Thomas Aquinas (1488–93) in the Caraffa Chapel, Sta Maria sopra Minerva, Rome, and the lives of SS. Philip and John (1495–1502) in the Strozzi Chapel, Sta Maria Novella, Florence. In these he strove for picturesque, dramatic and even bizarre effects that reveal him as one of the most inventive of late *quattrocento painters. Filippino also painted many altarpieces, the most famous of which is *The Vision of St Bernard* (Badia, Florence, *c.*1480), an exquisitely tender work, full of beautiful detail. Although he is now somewhat overshadowed by Botticelli, Filippino enjoyed a great reputation in his lifetime, being described by Lorenzo de' *Medici as 'superior to *Apelles'.

Lippi, Fra Filippo (*c.*1406–69). Florentine painter. He was brought up as an unwanted child in the Carmelite friary of the Carmine, where he took his vows in 1421. Unlike the Dominican Fra *Angelico, however, Lippi was a reluctant friar and had a scandalous love affair with a nun, Lucrezia Buti, who bore his son Filippino and a daughter Alessandra. The couple were released from their vows and allowed to marry, but Lippi still signed himself 'Frater Philippus'. His biography (romantically embroidered to include capture by pirates) is one of the most colourful in *Vasari's *Lives* and has given rise to the picture of a worldly *Renaissance artist, rebelling against the discipline of the Church—an image reflected in Robert Browning's poem about Lippi ('Fra Lippo Lippi' in *Men and Women*, 1855). He must certainly have had a more eventful life than most, but there is little documentary evidence of his character and personality.

Vasari writes that Lippi was inspired to become a painter by watching *Masaccio at work in the Carmine church, and his early work, notably the *Tarquinia Madonna* (Galleria Nazionale, Rome, 1437) is certainly overwhelmingly influenced by him. From about 1440, however, his style changed direction, becoming more linear and preoccupied with decorative motifs—thin, fluttering draperies, brocades, etc. Lippi is associated particularly with paintings of the Virgin and Child, which are sometimes in the form of **tondi*, a format he was among the first to use—a beautiful example, showing the wistful delicacy and exquisite pale lighting that characterizes his best work, is in the Pitti, Florence. Another formal innovation with which Lippi is associated is the **sacra conversazione*—his *Barbadori Altarpiece* (Louvre, Paris, begun 1437) is sometimes claimed as the earliest example of the type. As a fresco painter Lippi's finest achievement is his cycle on the lives of SS. Stephen and John the Baptist (1452–66) in Prato Cathedral.

Lippi was highly regarded in his day (he was patronized by the *Medici, who came to his aid when he was imprisoned and tortured for alleged fraud) and his influence is seen in the work of numerous artists, most notably *Botticelli, who was probably his pupil. Four centuries later he was one of the major sources for the second wave of *Pre-Raphaelitism.

Lisboa, António Francisco (1738–1814). Brazilian mulatto sculptor and architect, the illegitimate son of a Portuguese stonemason **Manuel Francisco Lisboa**. He was known as *O Aleijadinho* (little cripple) on account of a disease which from his late thirties deprived him of the use of his limbs so that he is said to have worked with chisel and mallet tied to half paralysed hands. In spite of his handicap, he is considered the greatest sculptor as well as the greatest architect of colonial Brazil. His masterpiece is the open-air life-size group of statues of the *Twelve Prophets* (1800–5) in front of the church of Nosso Senhor Bom Jesus de Matozinhos at Congonhas do Campo.

Lipton, Seymour. See ABSTRACT EXPRESSIONISM.

Lismer, Arthur. See GROUP OF SEVEN.

Liss (or **Lys**), **Johann** (*c.*1597–1631). German painter, active mainly in Italy. He trained in the Netherlands (probably in Amsterdam, possibly with *Goltzius) and visited Paris before moving to Italy *c.*1620. Venice seems to have been the main centre of Liss's activity, but he also worked in Rome, and *Caravaggesque influence is clearly seen in such vivid and strongly lit works as *Judith and Holofernes* (NG, London). His work enjoyed considerable popularity in Venice (where there was a dearth of talented native painters at this time) and his *Vision of St Jerome* in the church of S. Nicolo da Tolentino was much copied. This painting, generally considered his masterpiece, shows the remarkably free brushwork and brilliant use of high-keyed colour that were the salient features of his style and which were influential on Venetian painting when its glory revived in the 18th cent. It was formerly assumed that Liss, who ranks second only to *Elsheimer as the most brilliant German painter of the 17th cent., perished in the Venetian plague of 1629–30, but it is now known that he died in Verona in 1631.

Lissitzky, El (Eliezer Markowich) (1890–1941). Russian painter, designer, graphic artist, and architect. After studying engineering at Darmstadt and architecture in Moscow he worked in an architect's office and collaborated with *Chagall on the illustration of Jewish books (he was an expert lithographer). In 1918 Chagall became head of the art school at Vitebsk and in the following year he appointed Lissitzky as Professor of Architecture and Graphic Art. One of his colleagues at Vitebsk was *Malevich, whose advocacy of the use of pure geometric form greatly influenced Lissitzky, notably in his series of abstract paintings to which he gave the collective name 'Proun' and which he referred to as 'the interchange station between painting and architecture'. They do indeed look like plans for three-dimensional constructions, and at the same time Lissitzky made ambitious architectonic designs that were never realized. In 1921, after a brief period as professor at the state art school in Moscow, he went to Berlin, where he arranged and designed the important exhibition of abstract art at the Van Diemen Gallery that first comprehensively presented the modern movement in Russia to the West (it was later shown also in Amsterdam). While in Berlin he worked on *Constructivist magazines; he also made contact with van *Doesburg and members of De *Stijl and with *Moholy-Nagy, who spread Lissitzky's ideas through his teaching at the *Bauhaus. In 1923 he went with *Gabo to a Bauhaus exhibition at Weimar and there met *Gropius. From 1923 to 1925 he lived in Switzerland, and (after a short visit to Russia) from 1925 to 1928 in Hanover. He returned to Russia in 1928 and settled in Moscow. By this time he had abandoned painting and devoted himself mainly to typography and industrial design. His work included several propaganda and trade exhibitions, notably the Soviet Pavilion of the 1939 World's Fair in New York, and his dynamic techniques of *photomontage, printing, and lighting had wide influence.

For a considerable time Lissitzky was the best known of the Russian abstract artists in the West. In his mature work he achieved a fusion between the *Suprematism of Malevich (often using his diagonal axis), the Constructivism of *Tatlin and *Rodchenko, and features of the *Neo-Plasticism of *Mondrian.

lithography. A method of printing from a design drawn on a surface of stone or other suitable material. The design is neither cut in relief as in a *woodcut nor engraved in *intaglio as in *line engraving, but simply drawn on the flat surface of (most usually) a slab of special limestone known as lithographic stone. The process is based on the antipathy of grease and water. The artist draws his design with a greasy ink or crayon on the stone, which is then treated by the lithographic printer with certain chemical solutions so that the greasy content of the drawing is fixed. Water is then applied. The moisture is repelled by the greasy lines but is readily accepted by the remainder of the porous surface of the stone. The stone is now rolled with greasy ink which adheres only to the drawing, the rest of the surface, being damp, remaining impervious. A sheet of paper is placed on the stone, the whole is passed through the lithographic press, and an exact replica of the drawing is transferred, in reverse as with all prints, to the paper.

This 'planographic' or surface method of printing, the most recent of the principal *graphic techniques, was discovered in 1798 by Aloys *Senefelder, a Bavarian playwright who was experimenting with methods of duplicating his plays; he records that the idea came to him when he made a laundry list with a grease pencil on a piece of stone. Senefelder, who wrote a book on his invention in 1818, appears to have realized at once what its significance was and how it could be used. He called it 'Chemical Printing', insisting that the chemical principles involved were of more importance than the stone on which the designs were made, and in this he was right, for metal and plastic plates, particularly zinc surfaced in a special way, are frequently used today instead of stone. Senefelder was also responsible for the use of transfer paper whereby the design is drawn on paper and transferred subsequently to the stone for printing—a method much used by artists ever since.

As its inventor foresaw, lithography has proved to be an exceedingly flexible medium. Instead of being drawn with pen or crayon, the design may be painted on the stone with a brush; the *washes may be opaque or dilute, they may be scratched or scraped to produce white lines on a background of black, or they may be textured in any way the artist's ingenuity can suggest. Colour prints may be produced in much the same way as in any other graphic method, that is by preparing a separate stone for each of the colours in the design. Lithographs have in consequence taken on a number of appearances, ranging from simple linear designs made with pen or crayon to colour prints with the most varied effects of transparency and texture.

The lithographic principle has also had extensive commercial applications in the printing industry. Offset lithography, in which the ink is printed from the stone or zinc on to a rubber-coated cylinder before being transferred to the paper, allows the design to be made the right way round instead of in reverse and also enables a very thin film of ink to be used, thus permitting the reproduction of the finest lines. Photo litho offset involves the photographic printing of an image, usually by means of a half-tone screen, on to a sensitized zinc plate, which is then, after certain chemical treatments, printed on an offset lithographic machine.

By contrast with these complex commercial procedures, lithography in its simpler forms has always attracted artists as a means of original expression. It is very direct and, since its technical side can be left to the lithographic printer, the artist need do no more than draw upon the stone.

Lithography was invented in time for *Goya, in his old age at Bordeaux, to make a number of striking designs in the new medium. *Géricault, *Delacroix, and *Daumier were among the early masters of the medium, and Daumier was the first artist of stature to execute the largest part of his life's work in lithography. *Toulouse-Lautrec was another great master of the process; he sometimes created tonal effects by spattering ink on the stone with a toothbrush. Among the French painters of the *Impressionist period *Manet and *Degas made memorable contributions, and the final years of the 19th cent. saw a brilliant flowering of the art of colour lithography. Influenced by Japanese woodcut prints (see UKIYO-E) which had recently appeared in Europe, a number of artists, including *Bonnard and *Vuillard, began making lithographs in colour which were quite unlike anything made before in the West by reason of their avoidance of heavy shading, their use of large areas of flat yet transparent colour, and their unusual and provocative sense of pattern. Contemporary with this movement, other aspects of the great range of expressive effects of the medium were explored in the strange, visionary prints of Odilon *Redon and *Whistler's impressions of the London river with their soft and subtle gradations of tone. Meanwhile, in the USA, the firm of *Currier and Ives was producing a series of lithographs which had little in common with the sophisticated European prints of the period but show us a cross-section of the life of the American nation in terms of a genuinely popular art.

In the 20th cent. lithography, like most of the other graphic processes, has been used by most of the celebrated figures of the School of *Paris for free prints and for book illustration. *Picasso is, as usual, outstanding, and produced, mostly since the Second World War, a vast *œuvre* of lithographic prints, chiefly in black and white, of great imaginative and technical variety.

Little Masters. A once popular but now little-used term (a translation of the German *Kleinmeister*) applied to a group of 16th-cent. engravers who delicately worked plates of small dimensions, illustrating biblical, mythological, or *genre scenes. Outstanding practitioners of work of this kind were the Nuremberg masters Hans Sebald *Beham, his brother Bartel *Beham, and Georg *Pencz, all of whom were strongly influenced by *Dürer. Albrecht *Altdorfer and Heinrich *Aldegrever may be joined with the group on account of some of their work.

local colour. The 'true' colour of an object or area seen under normal daylight, without regard for the modifying effect of such factors as distance or reflections from other objects. Thus, the local colour of a typical grass field is green, although at a certain distance it may appear blue because of *atmospheric perspective.

Lochner, Stephan (active 1442–51). German painter, the leading master of his time in Cologne, where he worked from 1442 until his death. His early life is obscure, but stylistic evidence, particularly his eye for naturalistic detail, suggests he trained in the Netherlands. The most important of his surviving works is *The Adoration of the Magi*, painted for Cologne's town hall (where in 1520 *Dürer saw it 'with wonder and astonishment'), but now in the city's cathedral. It shows the exquisite colouring and delicate sentiment that was characteristic of his work.

Loggan, David (1633–92). Polish-born British engraver and draughtsman. He is best known for his books *Oxonia Illustrata* (1675) and *Cantabrigia Illustrata* (1688), topographical studies of the Universities, but he also made sensitive portrait drawings.

loggia. A room or porch open to the air on one or more sides. Popular in Mediterranean architecture, *loggie* were adopted in the north in the first flush of *Renaissance enthusiasm in defiance of the climate, but most have since been glazed, as at Hatfield House, Hertfordshire. The Roman prototypes have also in many cases been enclosed to protect the paintings within them. *Raphael's works at the Farnesina, Villa Madama, and Vatican were originally open to the sky.

Lohse, Richard (1902–88). Swiss painter and graphic artist. In his early works he experimented with various subjects and styles, but in the 1940s he became one of the leading representatives of *Concrete art. His paintings are mathematically based, often featuring chequer-board or grid-like patterns, but they are not cold or analytical in effect; indeed his work is particularly noted for its beauty and refinement of colour, and has a certain resemblance to *Op art of the kind associated with Bridget *Riley. From about 1950 Lohse gained an international reputation.

Lomazzo, Giovanni Paolo (1538–1600). Milanese painter and writer. At the age of 33 he went blind and took to writing on the theory of art, publishing two treatises: *Trattato dell'Arte de la Pittura, Scoltura, et Architettura* (1584) and *Idea del Tempio della Pittura* (1590). The *Trattato* was the largest and most comprehensive treatise on art published in the 16th cent. and has been described as 'the Bible of *Mannerism'. It is divided into seven books, whose themes are Proportion, Motion, Colour, Light, *Perspective, Practice, and History, the last containing a complete prescriptive guide to Christian and classical *iconography. Throughout the book runs the assumption that the arts can be taught by detailed precepts. It was widely influential and was translated into English (as *A Tracte containing the Artes of Curious Paintinge, Carvinge, and Buildinge*) by the Oxford physician Richard Haydocke in 1598. The translation adds details of English artists such as *Hilliard not mentioned in Lomazzo's original. Lomazzo also wrote poetry. An example of his rare surviving paintings is a self-portrait (1568) in the Brera, Milan.

Lombard, Lambert (1505–66). Netherlandish painter, draughtsman, engraver, architect, and antiquarian, active mainly in his native Liège. He was probably a pupil of *Gossaert and was influenced by Jan van *Scorel. A man of scholarly inclinations, Lombard visited Rome in 1537 (he also travelled in France and Germany) and made drawings of the *antique, some of which were engraved in the workshop of Jerome *Cock. He corresponded with *Vasari, providing him with information about Netherlandish artists, and Vasari said of him: 'Of all the Flemish artists I have named none is superior to Lambert Lombard of Liège, a man well versed in letters, a painter of judgement, a learned architect and—by no means his least title to merit—the master of Frans *Floris, and Willem *Key.' This opinion was confirmed by van *Mander, who wrote in 1604: 'Lombard was no less skilled as a teacher than learned in the arts of painting, architecture and perspective. One can confidently rank him among the best Netherlandish painters, past and present.' Unfortunately, very few paintings that can certainly be assigned to him survive to bear witness to his high contemporary reputation, and his work is known from drawings, copies, and engravings. A formidable *Portrait of the Artist* in the Musée de l'Art Wallon, Liège (another version is in the Staatliche Kunstsammlungen, Kassel) is among the best works given to him, but some critics think it is by his pupil Frans Floris. The portraits associated with Lombard—lively and strongly characterized—generally appeal more to modern taste than his somewhat academic religious paintings.

Lombardo. Family of Italian artists, the leading Venetian sculptors of their period: **Pietro** (*c.*1435–1515) and his sons **Tullio** (*c.*1455–1532) and **Antonio** (*c.*1458–1516). Pietro, who came from Lombardy, settled in Venice in about 1467. He was an architect as well as a sculptor, and his church of Sta Maria dei Miracoli (1481–9), on the sculptural decoration of which he was assisted by his sons, has been called the choicest jewel of *Renaissance work in Venice. Of his numerous tombs in Venetian churches, the best known is that of Doge Pietro Mocenigo (SS. Giovanni e Paolo, *c.*1476–81); he also made the tomb of Dante at Ravenna (1482). His style is distinguished by polished mastery of marble cutting and an interest in the *antique, features that recur in the work of Tullio. Tullio's most imposing work is the Vendramin monument (*c.*1493) in SS. Giovanni e Paolo; the figure of *Adam* from this—a sensuously beautiful freestanding nude—is in the Metropolitan Museum, New York. Antonio had less substance as an independent artist. His work included mythological reliefs for Alfonso d'*Este (mainly in the Hermitage, St Petersburg).

Lomi. See GENTILESCHI.

London, School of. See SCHOOL OF LONDON.

London Group. An exhibiting society of English artists formed in 1913 by an amalgamation of the *Camden Town Group with several smaller groups and various individuals. The first President was Harold *Gilman, and among the members were John and Paul *Nash, *Nevinson, *Wadsworth, and the sculptors *Epstein and *Gill. Roger *Fry became a member in 1918 and brought with him a number of his followers. The strong interest of the early members in *Post-Impressionism was deprecated by *Tonks, who said: 'The leaders of the London Group have nearly all come from me. What an unholy brood I have raised.' Thus began an opposition between 'advanced' art and the semi-academicism of the *Slade. Unlike most other associations the London Group survived opposition, was revived after the Second World War, and came to be looked on as something of an institution. But with the dignity of an institution it lost its early sense of mission and by 1950 it would not have been easy to say what were the artistic principles for which it stood. The group still exists.

Long, Edwin. See ALMA-TADEMA.

Long, Richard (1945–). British avant-garde artist whose work brings together sculpture, *Conceptual art, and *Land art. Since 1967 his artistic activity has been based on long solitary walks that he makes through landscapes, initially in Britain, and from 1969 also abroad, often in remote or inhospitable terrain. Sometimes he collects objects such as stones and twigs on these walks and brings them into a gallery, where he arranges them into designs, usually circles or other fairly simple geometrical shapes (*Circle of Sticks*, 1973, *Slate Circle*, 1979, both Tate Gallery, London). He also creates such works in their original settings, and documents his walks with photographs, texts, and maps. Long has an international reputation (as early as 1976 he represented Britain at the Venice *Biennale) and has attracted a great deal of commentary, much of it laudatory, although the critic Peter Fuller described his work as 'the barren arrangement of gathered stones'. In 1989 Long was awarded the *Turner Prize.

Longhi, Pietro (Pietro Falca) (1702–85). Venetian painter. Although he carried out some fresco commissions he is known principally as a painter of small *genre scenes of contemporary patrician and low life. These charming and often gently satirical scenes were very popular, although surprisingly he does not seem to have been patronized by English visitors to Venice. Longhi occasionally painted more than one version of his own compositions, and these again were often duplicated by pupils and followers. His work can best be seen in Venice (in the Accademia, the Ca' Rezzonico, and the Palazzo Querini-Stampalia), but he was prolific and is represented in many other collections.

Alessandro Longhi (1733–1813), the son of Pietro, was a successful portraitist. He was the official portrait painter to the Venetian Academy, so that he was in a good position for compiling his *Compendio delle Vite de' Pittori Veneziani Istorici* (1762) with portraits etched by himself.

Longhi, Roberto (1890–1970). Italian art historian. Longhi was a scholar of great industry, and published much new material, particularly in the area with which he is most associated—*Caravaggio and Caravaggism (he catalogued the great exhibition of the work of Caravaggio and the *Caravaggisti held in Florence in 1951, which is regarded as a landmark in this field). He was involved with various periodicals, and in 1950 founded a new periodical, *Paragone*, in Florence. In the same year he became professor of art history at Florence University. His villa in Florence is now an art-historical foundation, housing his library of books and photographs, and his picture collection.

Lopes, Gregório (*c.*1490–*c.*1550). Portuguese painter, court painter to Manuel I (1469–1521) and John III (1502–57). Among Portuguese painters of his period he was the most directly inspired by *Renaissance influence. His work is represented in the Museu Nacional de Arte Antiga, Lisbon.

López y Portaña, Vicente (1772–1850). Spanish painter. Primarily a portraitist, he was influenced by *Mengs, whose style he projected into the 19th cent. In 1815 he became first court painter to Ferdinand VII (who preferred him to *Goya), in 1817 director of the Academia of San Fernando, and in 1823 director of the *Prado.

Lorenzetti, Pietro (active 1320–48?) and **Ambrogio** (active 1319–48?). Sienese painters, brothers. They were among the outstanding Italian artists of their time, but their lives are poorly documented—both are assumed to have died in the Black Death of 1348. Pietro is usually said to have been the elder brother, but the evidence is not conclusive. His first dated work is of 1320—a *polyptych of *The Virgin and Child with Saints* in the Pieve di Sta Maria at Arezzo; Ambrogio's earliest reliably attributed work is of a year earlier—a *Virgin and Child* in the Museo Diocesano in Florence. Apart from collaborating in a cycle of the *Life of Mary*, now lost, which they painted in fresco on the façade of Siena's public hospital in 1335, the brothers worked independently. They shared a certain affinity of style, however, the weightiness of their figures, which show the influence of *Giotto, clearly setting them apart from the elegance of their greatest Sienese contemporary *Simone Martini and the tradition of *Duccio.

Ambrogio was the more innovative of the brothers, and his greatest work, the fresco series representing *Good and Bad Government* in the Town Hall at Siena (1338–9), is one of the most remarkable achievements in 14th-cent. Italian art. In it he broke new ground in the naturalistic painting of landscape and townscape, and the talkative crowds of figures show he was an acute observer of his fellow men. Ambrogio's other dated work includes altarpieces of *The Presentation in the Temple* (Uffizi, Florence, 1342) and *The Annunciation* (Pinacoteca, Siena, 1344).

Pietro's work is noted for its emotional expressiveness, his fresco of *The Descent from the Cross* in the Lower Church of S. Francesco at Assisi having remarkable pathos and dramatic power. The extent and the date of Pietro's contribution at Assisi are matters of controversy, but his other work includes dated altarpieces of *The Virgin and Child Enthroned* (Uffizi, 1340) and *The Birth of the Virgin* (Cathedral Museum, Siena, 1342).

Lorenzo Monaco (*c.*1370–*c.*1425). Italian painter. He was born in Siena, but seems to have spent all his professional life in Florence. In 1391 he took his vows as a monk of the Camaldolese monastery of Sta Maria degli Angeli. He rose to the rank of deacon, but in 1402 he was enrolled in the painters' guild under his lay name, Piero di Giovanni (Lorenzo Monaco means 'Laurence the Monk'), and was living outside the monastery. The monastery was renowned for its manuscript *illuminations and several *miniatures in books in the Laurentian Library in Florence have been attributed to him, but he was primarily a painter of altarpieces, good examples of which are in the National Gallery in London and the Uffizi in Florence. His main works in fresco are the scenes of the *Life of Mary* in the Bartolini Chapel of Sta Trinità, Florence. His style is distinguished by luminous beauty of colouring and a graceful, rhythmic flow of line. He stands in complete contrast to his great contemporary *Masaccio and represents the highest achievement of the last flowering of *Gothic art in Florence.

Lorrain (or **Lorraine**), **Claude.** See CLAUDE.

Lotto, Lorenzo (*c.*1480–1556/7). Venetian painter. According to *Vasari and *Ridolfi, he trained with *Giorgione and *Titian in the studio of Giovanni *Bellini, but he worked in many places apart from Venice, had an idiosyncratic style, and stands somewhat apart from the central Venetian tradition. In 1508–12 he was in Rome, then lived mainly in Bergamo until 1526, when he returned to Venice. From 1530 he worked mainly in various towns in the Marches, and in 1554, when he was partially blind, he became a lay-brother at the monastery at Loreto, where he died. His rootless existence reflects his anxious, difficult temperament and his work is extremely uneven. It draws on a wide variety of sources, from northern Europe as well as Italy, but at the same time shows acute freshness of observation. He is now perhaps best known for his portraits, in which he often conveys a mood of psychological unrest (*Young Man in his Study*, Kunsthistorisches Museum, Vienna), but he worked mainly as a religious painter. An outstanding example of how original and poetic his altarpieces could be is *The Annunciation* in the church of Sta Maria sopra Mercanti at Recanati—a bizarre and captivating work full of brilliant colours and lighting effects, odd expressions and poses, and unusual and beautifully painted details.

Louis, Morris (Morris Louis Bernstein) (1912–62). American painter, a major pioneer of the movement from *Abstract Expressionism to Colour Stain painting (see COLOUR FIELD PAINTING). Almost all his career was spent first in Baltimore and then from 1947 in nearby Washington. He isolated himself from the New York art world, concentrating on his own experiments and supporting himself by teaching. However, it was a visit to New York in 1953 with Kenneth *Noland that led to the breakthrough in his art. He and Noland went to Helen *Frankenthaler's studio, where they were immensely impressed by her painting *Mountains and Sea*, and Louis immediately began experimenting with her technique of applying liquid paint on to unprimed canvas, allowing it to flow over and soak into the canvas so that it acted as a stain and not as an overlaid surface of pigment. He was secretive about his technical methods and it is uncertain how he achieved his control over the flow of colour, but towards the end of his life he suffered from severe back problems caused by his constant bending and stooping over the canvas. Whatever his technique, the effect was to create suave and radiant flushes of colour, with no sense of brush gesture or hint of figuration. His method was exacting, allowing no possibility for alteration or modification. For this reason, perhaps, Louis destroyed much of his work of this period.

Louis painted various series of pictures in his new technique, the first of which was *Veils* (1954; he did another series in 1957–60). The other major series were *Florals* (1959–60), *Unfurleds* (1960–1), and *Stripes* (1961–2). The *Veils* consist of subtly billowing and overlapping shapes filling almost the entire canvas, but his development after that was towards rivulets of colour arranged in rainbow-like bands, often on a predominantly bare canvas. It was not until 1959 that his career began to take off and he had little time to enjoy his success before dying of lung cancer. However, his reputation now stands very high and he has had enormous influence on the development of Colour Stain painting. In the introduction to the catalogue of the 1974 Arts Council exhibition of his work, John Elderfield wrote: 'Morris Louis is one of the very few artists whose work has really changed the course of painting . . . With Louis, fully autonomous abstract painting came into its own for really the first time, and did so in paintings of a quality that matches the level of their innovation.'

Louis XV style. Term used for the high period of French *Rococo style, primarily in the minor and decorative arts. It coincides only very roughly with Louis's reign (1715–74), emerging around 1700, reaching its peak from about 1720 to about 1750, and being long outmoded by the time of his death.

Louis XVI style. Term applied to a style of interior decoration and design that prevailed in France from about 1760 to the Revolution. Its development thus antedates the reign of Louis XVI (1774–92) after which it is named. The style marked a reaction against the frivolity of the *Louis XV style and the return of a new *classicism based on a revived interest in the *antique.

Loutherbourg, Philippe-Jacques de. See DE LOUTHERBOURG.

Louvre, Paris. The national museum and art gallery of France, an epitome of the nation's history and culture. The first building on the site, begun *c.*1190 by Philip-Augustus (1165–1223) as a fortress and arsenal, held the royal treasures of jewels, armour, illuminated manuscripts, etc. It was enlarged and beautified by Charles V (1337–80), and his successor Charles VI (1368–1422) used it as a residence for visiting royalty. Francis I (1494–1547) began to demolish it in the 1520s and in 1546 commissioned the architect Pierre Lescot (1510/15–78) to build a new palace of four wings around a square court, roughly of the same size as the old castle and on the same site. Only the west and half of the south wings were completed by Lescot, but his work forms the heart of the present vast structure, and his elegant and sophisticated *classical style set the tone for all the future additions, which were made by virtually every French monarch up to Napoleon III.

Under Louis XIV (1638–1715) *Colbert increased the royal collection from some 200 pictures to over 2,000. In 1661 he acquired the greater part of the collections of Cardinal Mazarin, at that time considered the most magnificent in France; in 1671 he obtained the collections of the banker Jabach (the drawings from which form the backbone of the Louvre's department of drawings); and in 1667 he bought for the crown the enormous collection of prints of Michel de Marolles. In support of the policy for state control of the arts and taste some of the king's pictures were opened to public view in the Louvre from 1681 and the exhibitions of the new Académie were held there from 1673. The court had moved into the Louvre in 1652, but it transferred to Versailles in 1678.

Under Louis XVI (1754–93) the conversion of the Grande Galerie into a museum was begun and as a result of the democratic fervour incidental to the Revolution the Louvre was opened as the first national public gallery in 1793 (though as a public gallery it was preceded by others, including the *Ashmolean in Oxford, the *Vatican, and the Charleston Museum). Napoleon renamed the Louvre the Musée Napoléon in 1803 and exhibited there the works of art he had gathered from conquered territories. Most were restored after his fall from power.

Conceived as a comprehensive collection of European art, the Louvre was reopened by Napoleon III (1808–73) in 1851 with the addition of *Rubens's Medici cycle from the Luxembourg. The inauguration in 1818 of the Luxembourg Palace as a museum of contemporary French art made possible the practice of transferring accepted masterpieces to the Louvre in course of time. In addition to one of the world's greatest collections of paintings, the Louvre houses many other treasures, including large holdings of Greek and Roman antiquities. Among the famous ancient statues are the **Borghese Warrior*, the **Venus de Milo*, and the **Victory of Samothrace*. To relieve congestion after the Second World War a special museum for Impressionist art was formed at the Jeu de Paume in the gardens of the Tuileries. The paintings from the Jeu de Paume, together with certain other works from the Louvre, have now been moved to the Musée d'Orsay in Paris (opened 1986), which is devoted to the art of the late 19th cent., *c.*1848–*c.*1905).

Lowry, L. S. (Laurence Stephen) (1887–1976). British painter. He lived all his life in or near Manchester (mainly in Salford) and worked as a rent collector and clerk for a property company until he retired in 1952. His painting was done mainly at night after his day's work, but he was not a *naïve painter, having studied intermittently at art schools from 1905 to 1925. The most important of his teachers—at Manchester School of Art—was Adolphe Valette (1876–1942), a French painter who settled in Manchester in 1905 and whose work includes some memorably atmospheric views of the city. Lowry, too, concentrated on urban subjects, developing a highly distinctive type of picture featuring firmly drawn backgrounds of industrial buildings bathed in a white haze, against which groups or crowds of figures, painted in his characteristic stick-like manner, move about their affairs isolated in an intensely individual, personal life. There is sometimes an element of humour, but his paintings generally embody a disquieting vision, revealing a sense of alienation and man's inconsequence against the juggernaut of industrialism (he was a solitary character and said 'Had I not been lonely I should not have seen what I did'). Although he achieved national recognition with an exhibition at the Reid and Lefevre Gallery in London in 1939, he remained an elusive figure until a large retrospective exhibition arranged by the Arts Council in 1966 (he led a spartan existence and had no great desire to exhibit or sell his pictures). Ten years later, a few months after his death, a comprehensive retrospective exhibition of his work at the Royal Academy brought considerable divergence of opinion among critics. Some thought of him as a great artist with an important original vision. Others represented him as a very minor talent, although interesting as a social commentator.

Luca della Robbia. See ROBBIA.

Lucas van Leyden (1494?–1533). Netherlandish engraver and painter, born and mainly active in Leiden. He was the pupil of his father, from whose hand no works are known, and of Cornelis *Engelbrechtsz., but both of these were painters whereas Lucas was principally an engraver. Where he learnt engraving is unknown, but he was highly skilled in that art at a very early age; his earliest known print—*Mohammed and the Murdered Monk*—dates from 1508, when he was perhaps only 14, yet reveals no trace of immaturity in inspiration or technique. In 1514 he entered the Painters' Guild at Leiden. He seems to have travelled a certain amount, and visits are recorded to Antwerp in 1521, the year of *Dürer's Netherlandish journey, and to Middelburg in 1527, when he met *Gossaert.

An unbroken series of dated engravings makes it possible to follow his career as a print-maker and to date many of his paintings, but no clear pattern of stylistic development emerges. Dürer was the single greatest influence on him, but Lucas was less intellectual in his approach, tending to concentrate on the anecdotal features of the subject and to take delight in *caricatures and *genre motifs. Van *Mander characterizes Lucas as a pleasure-loving dilettante, who sometimes worked in bed, but he left a large *œuvre*, in spite of his fairly early death, and must have been a prodigious worker.

Lucas had a great reputation in his day (*Vasari even rated him above Dürer) and is universally regarded as one of the greatest figures in the history of *graphic art (he made etchings and woodcuts as well as engravings and was a prolific draughtsman). His status as a painter is less elevated, but he was undoubtedly one of the outstanding Netherlandish painters of his period. He was a pioneer of the Netherlandish genre tradition, as witness his *Chess Players* (Staatliche Museen, Berlin)—which actually represents a variant game called 'courier'—and his *Card Players* (Wilton House, Wiltshire), while his celebrated *Last Judgement* *triptych (Municipal Mus., Leiden, 1526–7) shows the heights to which he could rise as a religious painter. It eloquently displays his vivid imaginative powers, his marvellous skill as a colourist and his deft and fluid brushwork. Lucas left no pupils or direct followers, but it is fitting that his work was a stimulus to an even greater Leiden-born artist, *Rembrandt.

Lucchesino, Il. See TESTA.

Luchism. See RAYONISM.

Lucian of Samosata (2nd cent. AD). Greek satirist and rhetorician, who lampooned the empty virtuosity and exhibitionism of his day. His works contain several descriptions of Greek paintings which are now lost, though as was usual in his time these accounts were mainly devoted to subject-matter, the 'story' of the picture, with appreciation in terms of illusionistic verisimilitude. His writings were well known in the *Renaissance (the first printed collected edition of his works was published in Florence in 1496) and his description of *Calumny* by *Apelles, for example, inspired the painting by *Botticelli (Uffizi, Florence).

Luini, Bernardino (*c*.1480–1532). Milanese painter, one of the most prominent of *Leonardo's followers in Lombardy. Little is known of his life, but his prolific output indicates that he must have enjoyed a successful career (he was unusual among Leonardo's followers in that he painted numerous frescos as well as easel pictures). Luini sentimentalized Leonardo's style, and this helped to win him great popularity with the Victorians. His work is well represented in the Brera in Milan and many of his frescos and altarpieces are in Lombard churches. He painted mythological as well as religious subjects, an example being *Cephalus and Procris* in the National Gallery, Washington. Luini's best work is of high quality, but he ran a busy workshop and some of his followers and copyists vulgarized his style into almost a parody of Leonardo, with sickly smirks and exaggerated *chiaroscuro.

Lukasbrüder or **Lukasbund.** See NAZARENES.

Luks, George (1867–1933). American painter and graphic artist. In 1894, after a decade's travel in Europe, he became an illustrator on the Philadelphia *Press* and became friendly with other newspaper artists—*Glackens, *Shinn, and *Sloan—who introduced him to Robert *Henri. In 1896 Luks moved to New York, where he turned more to painting and became a member of The *Eight and the *Ash-can School. A flamboyant character who identified himself with the poorer classes and made a pose of bohemianism, he was much given to tall tales and sometimes posed as 'Lusty Luks', an ex-boxer. His work was uneven and unpredictable. It had vigour and spontaneity but often lapsed into superficial vitality. One of his best-known works is *The Wrestlers* (Museum of Fine Arts, Boston, 1905), which shows his preference for earthy themes and admiration for the bravura painterly technique of artists such as *Manet. Luks taught for several years at the *Art Students League and also ran his own school.

Luminism. Term coined in 1954 by John Baur, director of the Whitney Museum in New York, to describe an aspect of mid-19th-cent. American landscape painting in which the study of light was paramount. He defined Luminism as 'a polished and meticulous realism in which there is no sign of brushwork and no trace of impressionism, the atmospheric effects being achieved by infinitely careful gradations of tone, by the most exact study of the relative clarity of near and far objects, and by a precise rendering of the variations in texture and color produced by direct or reflected rays' ('American Luminism', *Perspectives USA*, Autumn 1954). At their most characteristic, Luminist paintings are concerned chiefly with the depiction of water and sky. Leading Luminists included *Bingham and *Durand, and aspects of it can be seen in the work of the *Hudson River School. By about 1880 Luminism was becoming outmoded by French influences. In the field of 20th-cent. art the term 'luminism' has also been applied to work incorporating electric light.

Lurçat, Jean (1892–1966). French painter and designer. For a time he was influenced by *Cubism, but more important and lasting influences on his painting came from his extensive travels during the 1920s in the Mediterranean countries, North Africa, and the Middle East. His pictures were dominated by impressions of desert landscapes, reminiscences of Spanish and Greek architecture, and a love of fantasy that led him to join the *Surrealist movement for a short period in the 1930s. Lurçat is chiefly remembered, however, for his work in the revival of the art of tapestry in both design and technique. His designs combined exalted themes from human history with fantastic representations of the vegetable and insect worlds, and he succeeded in reconciling the stylizations of medieval religious tapestry with modern modes of abstraction. In 1939 he was appointed designer to the tapestry factory at Aubusson and together with Marcel Gromaire (1892–1971) he brought about a renaissance in its work. He made more than a thousand designs, the most famous probably being the huge *Apocalypse* for the church of Assy (1948). From 1930 onwards he did a number of coloured lithographs, stage designs, and book illustrations, and in the 1960s he renewed his painting activities. He also wrote poetry and books on tapestry.

Luteri, Giovanni. See DOSSI.

lyrical abstraction. A rather vague term, used differently by different writers, applied to a type of expressive but non-violent abstract painting flourishing particularly in the 1950s and 1960s; the term seems to have been coined by the French painter Georges *Mathieu, who spoke of 'abstraction lyrique' in 1947. European critics often use it more or less as a synonym for *Art Informel or *Tachisme; Americans sometimes see it as an emasculated version of *Abstract Expressionism. To some writers it implies particularly a lush and sumptuous use of colour.

Lysippus. Greek sculptor from Sicyon, active in the middle and later 4th cent. BC. He was one of the most famous of Greek sculptors, with a long and prolific career (he worked from perhaps as early as *c.*360 BC to as late as *c.*305 BC and *Pliny said he made 1,500 works—all in bronze), but nothing is known to survive from his own hand. However, there are numerous Roman copies that can be said with a fair measure of certainty to reproduce his works, the best and most reliable being the *Apoxyomenos* (a young athlete scraping himself with a strigil) in the Vatican Museum. The pose is novel. The Apoxyomenos is shifting his weight from one foot to the other and stretching an arm out into the foreground; so the statue occupies more depth than its predecessors and provides a variety of good views. The figure is tall and slender, bearing out the tradition current in antiquity that Lysippus introduced a new scheme of proportions for the human body to supersede that of *Polyclitus. Lysippus was famous also for his portraits of Alexander the Great (356–323 BC), who is said to have let no other sculptor portray him; many copies survive, including examples in the British Museum, London, and the Louvre, Paris. Among his other works was a colossal statue of Hercules at Sicyon, which was probably the original of the celebrated *Farnese Hercules* in the Archaeological Museum in Naples. It shows the realism that was said to be another hallmark of his work. Of the works associated with him on stylistic grounds, the best known is a bronze statue of a victorious athlete found in the Adriatic Sea in 1964. This was bought by the Getty Museum, Malibu, in 1977 for $3,900,000, then the highest price ever paid for a piece of sculpture, and it is now sometimes known as 'the Getty Victor'. Even from the second-hand evidence that survives, it is clear that Lysippus was an outstandingly original sculptor whose stylistic innovations, like those of his great contemporary *Praxiteles, became common currency; J. J. Pollitt (*Art in the Hellenistic Age*, 1986) describes him as 'probably the single most creative and influential artist of the entire *Hellenistic period'.

Lysistratus. Greek sculptor of the later 4th cent. BC, brother of *Lysippus. No works by him are known to survive in originals or copies, but *Pliny's accounts indicate that he pursued even further the naturalism that was a feature of his brother's work. Pliny says that he tried to make portraits lifelike rather than beautiful, and that he was the first to make life masks and to take casts from statues.

Mabuse. See GOSSAERT.

MacBryde, Robert. See COLQUHOUN.

Macchiaioli. Group of Italian painters, active mainly in Florence *c.*1855–65, who were in revolt against academic conventions and emphasized painterly freshness through the use of blots or patches (*macchie*) of colour. They were influenced by the *Barbizon School, but they painted *genre scenes, historical subjects, and portraits as well as landscapes. Leading members included Giovanni *Boldini, Giovanni Fattori (1825–1908), and Telemaco Signorini (1835–1901). The Macchiaioli met with little critical or financial success, but they are now considered the most important phenomenon in 19th-cent. Italian painting and the name tends to be used loosely to cover the whole trend towards *naturalism in Italian landscape painting in their period and the succeeding decades. Sometimes the Macchiaioli are even claimed as proto-*Impressionists, but the differences between the two groups are as striking as the similarities; there is often a strong literary element in the work of the Macchiaioli, for example, and however bright their lighting effects they never lost a sense of solidity of form. See also NITTIS, GIUSEPPE DE.

MacColl, Dugald Sutherland (1859–1948). British painter, critic, and administrator. He was a member of the *New English Art Club, keeper of the *Tate Gallery (1906–11) and of the *Wallace Collection (1911–24), art critic for various journals, and an energetic controversialist. His books include *Nineteenth Century Art* (1902), one of the first true assessments of French *Impressionist painting, *Confessions of a Keeper* (1931), and *Philip Wilson *Steer* (1945).

McCubbin, Frederick (1855–1917). Australian painter, born in Melbourne, where he spent most of his life. He was a member of the *Heidelberg School and painted 'bush' subjects in a sentimental vein. After a visit to Europe late in life in 1906, his work was more directly influenced by *Impressionism. He was a teacher of drawing at the Melbourne National Gallery School from 1886 until his death. His son **Louis** (1890–1952) was also a painter. His major work is a huge mural of battle scenes for the Australian National War Museum in Canberra (1920–9). From 1936 to 1950 he was Director of the National Gallery of South Australia.

MacDonald, James Edward Hervey (1873–1932). Canadian landscape painter. He was a leading member of the *Group of Seven, both as a painter and a spokesman against reactionary criticism.

Macdonald, Jock (James Williamson Galloway) (1897–1960). Canadian painter, born in Scotland. He emigrated to Canada in 1926 to teach at the new Vancouver School of Decorative and Applied Arts. His early work was in the *Group of Seven tradition, but in 1934 he painted his first abstract or *automatic work, *Formative Colour Activity* (NG of Canada, Ottawa). In the late 1930s he became a friend of Emily *Carr, and in 1940 of Lawren *Harris, who encouraged him in his abstract experiments. These included *automatic paintings in a *Surrealist vein, and in the 1950s he was associated with the abstract group *Painters Eleven. During the last five years of his life MacDonald's output was prodigious, as he threw himself into experimenting with various techniques and media. He taught at various art colleges in the course of his career and is considered to have played a leading role in advancing the cause of modern art in Canada.

Macdonald, Margaret. See MACKINTOSH, CHARLES RENNIE.

Macdonald-Wright, Stanton (1890–1973). American painter, one of the first American abstract artists. In 1907 he moved to Paris, where together with Morgan *Russell he evolved a theory of painting based upon the scientific deployment of colour. Their theory, which was parallel to the *Orphism of Robert *Delaunay and which they called *Synchromism, was put forward in statements made at a joint exhibition held first at the Neue Kunstsalon, Munich, and then at the Galerie Bernheim-Jeune, Paris, in 1913. They claimed that they and not Delaunay and *Kupka were the originators of the new style of abstract colour painting. In 1914–16 Macdonald-Wright lived in London, then returned to the USA. In 1919 he moved to

California, where he abandoned Synchromist painting and became involved with experiments with colour film and various other projects. He was affiliated with the *Federal Art Project, for example, and invented a material which he called *Petrachrome* for the decoration of walls. After visiting Japan in 1937 he taught oriental and modern art at the University of California in Los Angeles, 1942–52. After a second visit to Japan, 1952–3, he gave up teaching and reverted to abstract painting. From 1958 he spent part of every year in a Zen monastery in Japan.

McEvoy, Ambrose (1878–1927). English painter. He began as a painter of poetic landscapes and restful interiors, but from about 1915 he gained success as a portraitist. His most characteristic pictures are of beautiful society women, often painted in watercolour in a rapid, sketchy style. They can be merely flashy or cloyingly sweet (during the First World War one critic joked that at a time of suger shortage McEvoy was 'a positive asset to the nation'), but the finest have something of the romantic air of refinement of *Gainsborough, an artist he greatly admired.

Mach, David. See NEW BRITISH SCULPTURE.

Machuca, Pedro (d. 1550). Spanish architect and painter, active mainly in Granada. He worked in Italy in his early career (which is ill documented) and was one of the first Spanish artists to break entirely with medieval tradition and show a full understanding of *Renaissance ideals. His earliest dated work, *The Virgin with the Souls of Purgatory* (Prado, Madrid, 1517), was painted in Italy and is thoroughly *Raphaelesque in style. Machuca was back in Spain by 1520, and although he worked mainly as a painter, he is most important as the architect of the Palace of Charles V in the grounds of the Alhambra in Granada, begun in 1527. Like his paintings, this is completely Italianate in style, and is as impressive as any comparable building in Italy. Most of Machuca's work as a painter has disappeared.

McIntire, Samuel (1757–1811). American architect and wood-carver. He worked all his life in Salem, Massachusetts, and his gracious and refined exercises in the *Adam style made the town the handsomest in New England during the early years of the American Republic. His ideas were borrowed from English pattern books, but he did not copy complete designs, and his decorative carving shows superb craftsmanship. The Gardner–White–Pingree House (Salem, 1804–5) exemplifies his work at its best.

Macip (or **Masip**), **Vicente** (*c.*1475–before 1550). Spanish painter, one of a dynasty of artists working in Valencia. Little is known of his life, but his major work, the main altarpiece of Segorbe Cathedral (completed 1535), shows him to have been a leading representative of the Italianate style. During his later years he collaborated with the outstanding member of the family, his son and follower **Juan Vicente** (better known as Juan de Juanes, *c.*1523–79). Juan's work combines figures in the Italian *Mannerist style with a polished Netherlandish technique. He was the leading painter of his time in Valencia and had many followers. There are several of his works in the Prado, including an altarpiece series of the life of St Stephen.

Maciunas, George. See FLUXUS.

Macke, August (1887–1914). German *Expressionist painter, one of the founders of the *Blaue Reiter. His training included a period studying with *Corinth in Berlin. Between 1907 and 1912 he visited Paris several times and found himself at home with the ways of painting then in vogue there, particularly *Fauvism, *Delaunay's experimental developments in the 'simultaneous contrast' of colours, and the *Cubist analysis of forms. In his own work he evolved a personal synthesis of *Impressionism, Fauvism, and *Orphism with which to display a basically Expressionist attitude, and he came closer in spirit to French art than any other German painter of the time. In 1909–10 he met *Kandinsky and Franz *Marc in Munich and joined them in the formation of the Blaue Reiter, but apart from a few experiments his work moved less towards abstraction than that of other members of the group. Early in 1914 he made a trip with Paul *Klee and Louis *Moilliet to Tunisia, and the watercolours he did on this trip are considered to be his most personal achievement. They represent a highly individual adaptation of Delaunay's use of colour for expressive purposes and were the starting-point for further development by Klee. Macke was killed in action in the First World War, aged 27.

Mackennal, Sir Bertram. See NEW SCULPTURE.

Mackintosh, Charles Rennie (1868–1928). Scottish architect and designer, leader of the Glasgow School of *Art Nouveau, and precursor of several of the more advanced trends in 20th-cent. architecture. His furniture design and interior decoration, often done in association with his wife, **Margaret Macdonald** (1865–1933), showed the characteristic calligraphic quality of Art Nouveau but avoided the exaggerated floral ornament often associated with that style. As an architect he opposed himself vigorously to the current eclectic academicism and 'period revival' fashion, while he became a pioneer in the new conception of the role of function in architectural design. All his work shows a highly personal combination of the rational and the expressive. His fame rests to a large extent on his Glasgow School of Art (1897–9) and its library block and other extensions (1907–9), but his interior decoration was perhaps most spectacularly expressed in the four Glasgow tea-rooms designed with all their furniture and equipment for Miss Catherine Cranston.

Mackintosh's influence on the avant-garde abroad was very great, especially in Germany and Austria, so much so that the advanced style of the early 20th cent. was sometimes known as 'Mackintoshismus'. His work was exhibited in Budapest, Munich, Dresden, Venice, and Moscow, arousing interest and excitement everywhere. From 1914 he lived in London and Port Vendres. Thereafter, apart from a house in Northampton, none of his major architectural projects reached the stage of execution, though he did complete some work as a designer of fabrics, book covers, and furniture, and painted watercolours.

Mackmurdo, Arthur H. (1851–1942). Scottish architect and designer working in England. His style of domestic design in the 1880s was more advanced than anything else in Europe, in its clarity of structure, but in two-dimensional design he was a pioneer of *Art Nouveau. Particularly noteworthy is his design for the cover of his book on *Wren's City Churches* (1883), which makes luxurious use of tendril-like curves. In 1882 he founded the Century Guild, a group of artist-craftsmen inspired by the ideas of *Ruskin and *Morris, and in 1884 started the magazine *The Hobby Horse*, through which he publicized its ideals. Its high standard of printing and design aroused Morris's interest in printing and was one of the influences that led him to found the *Kelmscott Press. Mackmurdo and his followers helped to establish the architect's leadership in contemporary efforts to reunite the arts and to counteract the antiquarian influence of Morris over the designs produced by the *Arts and Crafts Movement. His work was particularly influential on the founders of the *Deutscher Werkbund. He gave up architecture in 1904 and devoted much of his time to writing on economics.

Maclise, Daniel (1806–70). Irish painter and *caricaturist, active in London from 1827. An outstanding draughtsman, Maclise became the leading history painter of his period, his greatest works being two enormous murals in the House of Lords on *The Meeting of Wellington and Blücher at Waterloo* (completed 1861) and *The Death of Nelson at Trafalgar* (completed 1865). They were done in the *water-glass technique and are poorly preserved (a sketch for the *Nelson* in the Walker Art Gallery in Liverpool gives an idea of the original colouring), but they are powerful works and fully coherent in spite of the huge numbers of figures involved, and they remain the most stirring examples of his heroic powers of design. Maclise was handsome, charming, and popular with fellow artists, and *Frith wrote that he was spoken of in academic circles as 'out and away the greatest artist that ever lived'. Grandiose history painting was only one side to his talent, however, for he also excelled as a caricaturist, and is particularly noted for the series of character portraits of literary men and women (Charles Dickens was a close friend) that he contributed to *Fraser's Magazine* under the pseudonym Alfred Croquis (1830–8). There are several examples of his conventional portraits in the National Portrait Gallery, London, including a well-known one of Dickens (1839).

MacMonnies, Frederick (1863–1937). American sculptor. He studied under *Saint-Gaudens and at the École des *Beaux-Arts in Paris. Much of his career was spent in Paris, but he sent work back to the USA and became one of the leading American sculptors of public monuments in his generation. His most notable works include a set of bronze doors and a statue of Shakespeare for the Library of Congress in Washington.

Maderno, Stefano (1576–1636). Italian sculptor. He was one of the leading sculptors in Rome during the papacy of Paul V (1605–21) before *Bernini came into the ascendancy, working on statues and *reliefs in numerous churches, notably the chapel that the pope had built (the Cappella Paolina, begun 1605) in Sta Maria Maggiore. Most of Maderno's work is now known only to specialists, but one of his sculptures has attained lasting fame: the recumbent figure of St Cecilia (1600) in Sta Cecilia in Trastevere, a lyrical and poignant work which is said to show the body of the Early Christian martyr in the exact position in which it was discovered in the church in 1599.

Madrazo, José (1781–1859). Spanish painter, the best-known member of a dynasty of artists. He studied with J.-L. *David in Paris and also worked in Rome, and he became the leading *Neoclassical painter in Spain. His history paintings are generally rather cold, but he was a spirited portraitist. He had three painter sons, of whom the most important was **Federico** (1815–94), who succeeded his father as director of the *Prado.

Maella, Mariano de. See ESCORIAL.

Maes, Nicolaes (1634–93). Dutch painter. He was born in Dordrecht and in about 1648 became a pupil of *Rembrandt in Amsterdam, staying there until 1654 when he returned to his native town. In his early years he concentrated on *genre pictures, rather sentimental in approach, but distinguished by deep glowing colours he had learnt from his master. Old women sleeping, praying, or reading the Bible were subjects he particularly favoured. In the 1660s, however, Maes began to turn more to portraiture, and after a visit to Antwerp around the middle of the decade his style changed dramatically. He abandoned the reddish tone of his earlier manner for a wider, lighter, and cooler range (greys and blacks in the shadows instead of brownish tones), and the fashionable portraits he now specialized in were closer to van *Dyck than to Rembrandt. In 1673 he moved permanently to Amsterdam and had great success with this kind of picture. Maes was a fairly prolific painter and is well represented in, for example, the Rijksmuseum, Amsterdam, the National Gallery, London, and the Museum at Dordrecht.

Maestà. Term (Italian for 'majesty') used to describe a representation of the Virgin and Child in which the Virgin is enthroned as Queen of Heaven, surrounded by a court of saints and angels.

Mafai, Mario. See SCIPIONE.

Maffei, Francesco (*c.*1600–60). Italian painter, born at Vicenza and active mainly in the Veneto. He had a refreshingly individualistic style, carrying on the great painterly tradition of *Tintoretto and *Bassano, reinforced by the example of *Liss, *Feti, and *Strozzi, to which he added his own note of mysterious and sometimes bizarre fantasy. He painted religious and mythological scenes and also allegorical portraits of local officials. His work, which is well represented in the Museo Civico in Vicenza, remained comparatively unknown until an exhibition there in 1956.

Magic(al) Realism. Term coined by the German critic Franz Roh in 1925 to describe the aspect of *Neue Sachlichkeit characterized by sharp-focus detail. In the book in which he originated the term—*Nach-Expressionismus, Magischer Realismus, Probleme der neuesten Europäischen Malerie* (Leipzig, 1925)—Roh also included a rather mixed bag of non-German painters as 'Magic Realists', among them *Miró and *Picasso. Subsequently critics have used the term to cover various types of painting in which objects are depicted with photographic naturalism but which, because of paradoxical elements or strange juxtapositions, convey a feeling of unreality, infusing the ordinary with a sense of mystery. The paintings of *Magritte are a prime example. In the English-speaking world the term gained currency with an exhibition entitled 'American Realists and Magic Realists' at the Museum of Modern Art, New York, in 1943. The director of the museum, Alfred H. *Barr, wrote that the term was 'sometimes applied to the work of painters who by means of an exact realistic technique try to make plausible and convincing their improbable, dreamlike or fantastic visions'.

Magnasco, Alessandro (called Il Lissandrino) (1667–1749). Italian painter. He was born and died in Genoa, but spent most of his working life in Milan. At the beginning of his career he was a portraitist, but virtually nothing is known of this aspect of his career. Later he turned to the type of work for which he is now known—highly individual melodramatic scenes set in storm-tossed landscapes, ruins, convents, and gloomy monasteries, peopled with small, elongated figures of monks, nuns, gypsies, mercenaries, witches, beggars, and inquisitors. His brushwork is nervous and flickering and his lighting effects macabre. He was very prolific and his work is rarely dated or datable. Marco *Ricci and Francesco *Guardi were among the artists influenced by him.

Magnelli, Alberto (1888–1971). Italian painter. He was self-taught, and in his early work he moved from a naturalistic style to one influenced by *Cubism and *Futurism. After the First World War his style became more representational under the influence of *Metaphysical painting, but in the 1930s he turned to pure abstraction, his work characteristically making use of dynamic interplay of shapes. After the Second World War he became recognized as one of Italy's leading abstract artists.

Magritte, René (1898–1967). Belgian painter, one of the leading figures of the *Surrealist movement. Apart from three years living in a suburb of Paris (1927–30), his entire career was spent in Brussels where he lived a life of bourgeois regularity (the bowler-hatted figure who so often features in his work is to some extent a self-portrait). After initially working in a *Cubist–Futurist style, he turned to Surrealism in 1925 under the influence of de *Chirico and by the following year had already emerged as a highly individual artist with *The Menaced Assassin* (MOMA, New York), a picture that displays the startling and disturbing juxtapositions of the ordinary, the strange, and the erotic that were to characterize his work for the rest of his life. Apart from a period in the 1940s when he experimented first with pseudo-*Impressionist brushwork and then with a *Fauve technique, he worked in a precise, scrupulously banal manner (a reminder of the early days when he made his living designing wallpaper and drawing fashion advertisements) and he always remained true to Surrealism. *Iconographically he had a repertory of obsessive images that appeared again and again in ordinary but incongruous surroundings. Enormous rocks that float in the air and fishes with human legs are typical leitmotivs. He repeatedly exploited ambiguities concerning real objects and images of them (many of his works feature paintings within paintings), inside and out-of-doors, day and night. In a number of paintings, for example, he depicted a night scene, or a city street lit only by artificial light, below a clear sunlit sky. He also made Surrealist analogues of a number of famous paintings—for example *David's *Madame Récamier* and *Manet's *The Balcony*, in which he replaced the figures with coffins. For the fertility of his imagery, the unforced spontaneity of his effects, and not least his rare gift of humour, Magritte was one of the very few natural and inspired Surrealist painters. J. T. Soby summed this up felicitously when he wrote: 'In viewing Magritte's paintings ... everything seems proper. And then abruptly the rape of commonsense occurs, usually in broad daylight.' His work was included in many Surrealist exhibitions, but it was not until he was in his fifties that he began to achieve international success and honours. By the time of his death his work had had a profound influence on *Pop art and it has subsequently been widely imitated in advertising.

mahlstick. See MAULSTICK.

Mahon, Sir Denis (1910–). British art historian and collector. A private scholar with private means, he has devoted himself single-mindedly to the study of 17th-cent. Italian (particularly Bolognese) painting, and has not only built up a choice collection in this field, but also played a greater part than anyone else in the rehabilitation of once-scorned artists such as the *Carracci, *Guercino, and *Reni. His publications include *Studies in Seicento Art and Theory* (1947), a brilliant pioneering work, and catalogues of Carracci drawings (1956) and Guercino paintings (1968) and drawings (1969) for major exhibitions of their work held in Bologna. He is also regarded as a leading *Poussin scholar.

Maiano, Benedetto da. See BENEDETTO DA MAIANO.

Maillol, Aristide (1861–1944). French sculptor, painter, graphic artist, and tapestry designer. His early career was spent mainly as a tapestry designer, but he also painted, exhibiting with the *Nabis. Although he first made sculpture in 1895, it was only in 1900 that he decided to devote himself to it after serious eyestrain made him give up tapestry. In 1902 he had his first one-man exhibition, which drew praise from *Rodin; in 1905 came his first conspicuous public success at the *Salon d'Automne; and after about 1910 he was internationally famous and received a constant flow of commissions. With only a few exceptions, he restricted himself to the female nude, expressing his whole philosophy of form through this medium. Commissioned in 1905 to make a monument to the tribune and revolutionary Louis-Auguste Blanqui (1805–81), he was asked by the committee what form he proposed to give it. He replied: 'Eh! une femme nue.' More than any artist before him he brought to conscious realization the concept of sculpture in the round as an independent art form stripped of literary associations and architectural context, and he ranks as the most distinguished figure of the transition from Rodin to the moderns. 'There is something to be learned from Rodin,' Maillol said, 'yet I felt I must return to more stable and self-contained forms. Stripped of all psychological details, forms yield themselves up more readily to the sculptor's intentions.' He rejected Rodin's emotionalism and animated surfaces, however; instead, Maillol's weighty figures, often shown in repose, are solemn and broadly modelled, with simple poses and gestures. His work consciously continued the classical tradition of Greek and Roman sculpture (Maillol visited Greece in 1908), but at the same time has a quality of healthy sensuousness (his peasant wife sometimes modelled for him).

Maillol took up painting again in 1939 when he returned to his birthplace, Banyuls, but apart from his sculpture the most important works of his maturity are his book illustrations, which helped re-establish the art of the book in the 1920s and 1930s. His finest achievements in this field are the woodcut illustrations (which he cut himself) for an edition of Virgil's *Eclogues* (begun 1912 but not published until 1926), which show superb economy of line. He also made lithographic illustrations. Maillol's work is in many important collections of modern art.

Mainardi, Sebastiano (*c*.1460–1513?). Florentine painter, pupil, and collaborator of his brother-in-law Domenico *Ghirlandaio. His name has often been indiscriminately attached to Ghirlandaio's schoolpieces, but very little is known for certain of his work.

Maino, Juan Bautista. See MAYNO.

maiolica (or **majolica**). A type of earthenware decorated with bright colours over a white ground, particularly popular during the Italian *Renaissance. The name derives from the fact that it is said to have originated in Majorca. Maiolica decoration often featured figurative scenes, more or less freely adapted from engravings. The greatest exponent of this *istoriato* manner was Nicola Pellipario (active *c.*1510–40), who worked first at Castel Durante, then at Urbino, where the style was perpetuated by a number of talented painters. There were factories in many other Italian cities, and the fame of Italian maiolica led to the establishment of other manufactories in northern Europe during the 16th cent.

Maitani, Lorenzo (*c.*1270–1330). Sienese sculptor and architect. In 1310 he was put in charge of the works at Orvieto Cathedral and began the façade. The reliefs on the façade at Orvieto are also attributed to Maitani, but there is no conclusive evidence that he worked as a sculptor, and how large a hand he actually had in them is conjectural.

Makart, Hans (1840–84). Austrian painter. He studied under *Piloty in Munich, then the chief centre of history painting, and from 1869 worked in Vienna, where he was enormously successful and a leading society figure. His exuberant, somewhat *Rubenesque style, which he applied to mythological, historical, and allegorical subjects, often on a huge scale, now seems overblown, but it had great influence in Austria and Germany.

Malatesta, Sigismondo (1417–68). Italian nobleman, effective lord of Rimini from 1432 until his death. A brilliant and totally unscrupulous *condottiere*, he is the prototype of the megalomaniac, paganizing tyrant once thought to be characteristic of the Italian *Renaissance. His name is indissolubly linked to one of the most famous creations of 15th-cent. art—the Tempio Malatestiana which *Alberti created for him in Rimini out of the medieval church of San Francesco. On it between 1450 and 1460 were employed *Piero della Francesca, *Agostino di Duccio, and Sigismondo's favourite, Matteo de' Pasti (*c.*1420–90). Under Sigismondo's direct inspiration it became the most self-conscious return to the *antique yet seen. His glorification under the image of Apollo and the prominent place given to his mistress and later wife, Isotta degli Atti, led to the charge by Pope Pius II (1405–64) that 'it seems a temple for the worship not of Christ but of the pagan gods'.

Mâle, Émile (1862–1954). French art historian. A pioneer in the study of French medieval art, on which he published several books, he became Professor of the History of Art at the Sorbonne in 1912. He explored the sources which inspired the medieval carvers and glass painters and was perhaps the first to point out the Eastern origins of medieval *iconography. His work is distinguished not only by great learning but also by literary merit, his style being praised by Marcel Proust amongst others.

Malevich, Kasimir (1878–1935). Russian painter, designer, and writer, with *Mondrian the most important pioneer of geometric *abstract art. He began working in an unexceptional *Post-Impressionist manner, but by 1912 he was painting peasant subjects in a massive 'tubular' style similar to that of *Léger as well as pictures combining the fragmentation of form of *Cubism with the multiplication of the image of *Futurism (*The Knife Grinder*, Yale Univ. Art Gal., 1912). Malevich, however, was fired with the desire 'to free art from the burden of the object' and launched the *Suprematist movement, which brought abstract art to a geometric simplicity more radical than anything previously seen. He claimed that he made a picture 'consisting of nothing more than a black square on a white field' as early as 1913, but Suprematist paintings were first made public in Moscow in 1915 (there is often difficulty in dating his work and also in knowing which way up his paintings should be hung, photographs of early exhibitions sometimes providing conflicting evidence). Malevich moved away from absolute austerity, tilting rectangles from the vertical, adding more colours and introducing a suggestion of the third dimension and even a degree of painterly handling, but around 1918 he returned to his purest ideals with a series of *White on White* paintings. After this he seems to have realized he could go no further along this road and virtually gave up abstract painting, turning more to teaching, writing, and making three-dimensional models that were important in the growth of *Constructivism. In 1919, at the invitation of *Chagall, he started teaching at the art school at Vitebsk, where he exerted a profound influence on *Lissitzky, and in 1922 he moved to Petrograd (Leningrad), where he lived for the rest of his life. He visited Warsaw and Berlin in 1927, accompanying an exhibition of his works and visited the *Bauhaus. In the late 1920s he returned to figurative painting, but he was out of favour with a political system that now demanded *Socialist Realism from its artists and he ran into trouble with the authorities. However, he remained a revered figure among artists. Malevich wrote various theoretical tracts and several collections of his writings have been published. The best collection of his work is in the Stedelijk Museum, Amsterdam.

In *Cubism and Abstract Art* (1936) A. H. *Barr, jun., gives the following estimate of Malevich's importance in the history of 20th-cent. art. 'In the history of abstract art Malevich is a figure of fundamental importance. As a pioneer, a theorist and an artist he influenced not only a large following in Russia but also, through Lissitzky and *Moholy-Nagy, the course of abstract art in Central Europe. He stands at the heart of the movement which swept westward from Russia after the war and, mingling with the eastward moving influence of the Dutch De *Stijl group, transformed the architecture, furniture, typography and commercial art of Germany and much of the rest of Europe.'

Malouel, Jean (d. 1419). Netherlandish painter. He is documented in Paris in 1396 and from 1397 to 1415 he was court painter to the Dukes of Burgundy. No documented works by him survive, but he has been suggested as the painter of several works, including part of the *Martyrdom of St Denis* (completed by *Bellechose), and a *tondo of the *Trinity*, both in the Louvre, Paris. They have all the refinement of French court art combined with a strength of modelling and a realistic naturalism derived from Flanders.

Malton, Thomas. See TURNER.

Malvasia, Count Carlo (1616–93). Italian painter, art historian, and antiquarian. His *Felsina pittrice: vite dei pittori bolognesi* (1678) is the most important source for knowledge of the great period of the Bolognese School that began with the *Carracci (Felsina was the Etruscan name for Bologna). His guide to the paintings of Bologna (1686) was one of the first books of its kind.

Mander, Karel van (1548–1606). Netherlandish painter and writer on art, born in Flanders and active mainly in Haarlem. He is sometimes known as the 'Dutch *Vasari', for his fame rests primarily on his work as a biographer of artists, published in *Het Schilder-Boeck* (The Book of Painters) in 1604. The most important part of the book is made up of about 175 biographies of Netherlandish and German artists from van *Eyck to van Mander's own younger contemporaries. This is the first systematic account of the lives of northern European artists, and our only source of information about certain 15th- and 16th-cent. Netherlandish artists. The book also contains the lives of Italian artists from *Cimabue up to his own time. Most of this material is a condensed translation into Dutch of Vasari, but some of it deals with artists who worked after 1568 (when the second edition of Vasari's work was published) and has valuable information collected by van Mander himself when he was in Italy in 1573–7 or from friends and correspondents: he was sufficiently up-to-date to mention *Caravaggio, 'who is doing extraordinary things in Rome'. Another part of the book is a long poem that gives practical advice to artists and sums up much of the theory and practice of 16th-cent. Netherlandish art, but made little contribution to the great achievements of 17th-cent. Dutch painting.

Van Mander's own pictures, which were mainly religious and allegorical, adopted the elongated forms of the *Mannerists, but his later works showed a tendency towards *naturalism. With *Cornelis van Haarlem and Hendrick *Goltzius, van Mander founded at Haarlem an *academy where artists could draw from nude models. Frans *Hals is said to have been one of his pupils.

mandorla (Italian: almond), also called aureole or *vesica piscis*. An almond-shaped aura or series of lines surrounding the body of a divine personage. It is most commonly used in portrayals of Christ, particularly in post-Resurrection scenes (but also notably in representations of the Transfiguration) when he is seen in his heavenly glory.

Manessier, Alfred (1911–). French lithographer, and designer of tapestries and stained glass, a pupil of *Bissière. During the 1930s his work was influenced by *Cubism and *Surrealism, but after staying at a Trappist monastery in 1943 he became deeply committed to religion and turned to expressing spiritual meaning through abstract art, Characteristically his paintings feature rich colours within a loose linear grid, creating an effect reminiscent of stained glass (a medium in which he has done some of his best work). He is regarded as one of the leading exponents of expressive abstraction in the post-war School of *Paris and has won numerous awards, notably the main painting prize at the 1962 Venice *Biennale.

Manet, Édouard (1832–83). French painter and graphic artist. He was the son of a senior civil servant in the Ministry of Justice and inherited considerable wealth when his father (who disapproved of his choice of career) died in 1862. His upper middle-class background was important, for although he was cast as an artistic rebel, he always sought traditional honours and success and he cut an impeccable figure as a man-about-town. He entered the studio of *Couture in 1850 and remained a pupil there for six years. His own painting style was, however, based rather on a study of the Old Masters at the *Louvre, and particularly Spanish painters such as *Velázquez (his greatest artistic hero) and *Ribera, than on the *classical training of the schools. During the 1850s he visited museums in the Netherlands, Germany, Austria, and Italy and it is one of the ironies of Manet's career that a painter with such reverence for the art of the past should be so much attacked for his modernity. His first taste of official disfavour came when his first submission to the *Salon—*The Absinthe Drinker* (Ny Carlsberg Glyptothek, Copenhagen, 1859)—was rejected. He had two paintings accepted in 1861, but then in 1863 his *Déjeuner sur l'herbe* (Musée d'Orsay, Paris) caused a scandal. It was turned down by the Salon and was shown instead at the *Salon des Refusés, set up specially for such rejected paintings. Its hostile reception was based on moral as well as aesthetic grounds, for nudity was considered acceptable only if it was sufficiently remote in time or place and this showed a naked woman having a picnic with two contemporary, clothed men. Manet caused even greater outrage two years later when his *Olympia* (Musée d'Orsay, 1863) was exhibited at the Salon. The reclining nude figure was based on *Titian's *Venus of Urbino* (which Manet had copied in Florence 10 years earlier), but her blatant sexuality was thought an affront to accepted standards of decorum, and one critic wrote, 'Art sunk so low does not even deserve reproach.' Manet was denounced also for his bold technique, in which he eliminated the fine tonal gradations of academic practice and created vivid contrasts of light and shade: 'The shadows are indicated by more or less large smears of blacking', wrote another critic, 'The least beautiful woman has bones, muscles, skin, and some sort of colour. Here there is nothing.' 'Insults are pouring down on me as thick as hail,' Manet wrote to his friend *Baudelaire (who with Émile Zola (1840–1902) was one of his main defenders) before seeking temporary refuge in Spain. From this time, Manet reluctantly found himself acquiring a reputation as a leader of the avant-garde. He was a respected and admired member of the group of young *Impressionists, including *Monet, *Renoir, *Bazille, *Sisley, and *Cézanne, who met at the Café Guerbois and elsewhere. But despite their admiration for him, Manet stood somewhat aloof from the group (although he enjoyed going to the races with *Degas, who was also from the upper middle class) and did not participate in the Impressionist exhibitions, continuing with indifferent success to seek recognition from the official Salon. Manet did, however, adopt the Impressionist technique of painting out of doors (persuaded by Berthe *Morisot, who became his sister-in-law in 1874), and his work became freer and lighter in the 1870s under their influence.

In the late 1870s Manet became ill with a disease diagnosed as locomotor ataxia (associated with the late stages of syphilis), which caused him bouts of great pain and extreme tiredness. Increasingly he preferred to work in pastels, which were less physically demanding than oils, but in his last great painting, the glorious *A Bar at the Folies-Bergère* (Courtauld Inst., London, 1882), he rose to heights of painterly brilliance that no other 19th-cent. artist surpassed. He died in appalling pain a week after having a gangrenous leg amputated. The official honours he had craved—a second-class medal at the Salon and the Legion of Honour—came too late (1881) to be enjoyed.

Manet was a varied and complex artist. He painted a great variety of subjects (he was also a skilled etcher and lithographer) and rarely repeated himself. His approach was completely undogmatic and he was reluctant to theorize; Zola wrote of him 'In beginning a picture, he could never say how it would come out.' In spite of the fact that his work often has a feeling of complete freshness and spontaneity, he would often repaint and rework pictures or even cut them into fragments. His greatest strength was with modern-life subjects (he sketched constantly in the boulevards and cafés of Paris), but although he is accused by some critics of having no imagination, of being able to paint something only if he had it in front of him, his pictures are anything but straight transcriptions of nature. They are, indeed, sometimes enigmatic and elusive, as with *A Bar at the Folies-Bergère*, and seem to be more concerned with the act of painting than with the ostensible subject. It is partly in this freedom from the traditional literary, anecdotal, or moralistic associations of painting that he is seen as one of the founders of 'modern' art, and it is significant that the official title of the first *Post-Impressionist Exhibition, organized by Roger *Fry in 1910–11, was *Manet and the Post-Impressionists*.

Manfredi, Bartolommeo (1582–1622). Italian painter, born near Mantua and active mainly in Rome, where he was one of the most important of *Caravaggio's followers. He painted various subjects but he is best known for low-life scenes of taverns, soldiers in guardrooms, card-playing, etc., and it was he rather than Caravaggio himself who was mainly responsible for popularizing this kind of work, particularly with painters from France and the Netherlands who came to Italy. In spite of the high contemporary reputation he enjoyed, no works survive that are signed or documented as his, and several of the forty or so paintings now given to him were formerly attributed to Caravaggio.

manière criblée (or **dotted manner**). Term used to describe an early type of metal engraving in which dots were stamped in the plate with a punch to create a textured effect, the dots showing as white against the inked background. The technique was not common after the end of the 15th cent., but there was something of a revival in the late 18th cent.

Mannerism. Term used in the study of the visual arts (and by transference in the study of literature and music) with a confused medley of combined historical and critical connotations. Even more than with most stylistic labels, there is little agreement amongst scholars as to its delimitations, and John Shearman begins his book on the subject (*Mannerism*, 1967) with the frank admission: 'This book will have at least one feature in common with all those already published on Mannerism; it will appear to describe something quite different from what all the rest describe.' Today the term is most commonly used to cover Italian art in the period between the High *Renaissance and the *Baroque, that is from about 1520 to about 1600. It was mainly the writings of *Vasari, himself a leading painter of this period, that gave the term currency, for he often used the word *maniera* (meaning 'style' or 'stylishness') as a term of approbation, signifying qualities of grace, poise, facility, and sophistication. By extension the term 'mannerist' has been applied to art of any place or time showing features analogous to those of Vasari's schema. Usage in this sense, however, has been vitiated by long-standing prejudices of taste towards 16th-cent. Italian art. From the 17th cent. to the first half of the 20th cent. it was the prevailing attitude that most Italian art of the period under discussion (with the notable exception of Venetian painting) was decadent, marking a falling away from the peak of *Classical perfection reached by *Michelangelo in the Sistine Ceiling and *Raphael in the Vatican Stanze. Faced with such unsurpassable models, their unfortunate successors were deemed to have been reduced to artistic inbreeding, feebly plagiarizing and distorting the work of the masters (a scenario that does indeed fit much undistinguished academic work of the period). Thus, as a general, non-technical critical term 'mannered' in English embodies these prejudices of taste, referring to no specific style or period but suggesting exaggerated or obtrusive cultivation of the superficial tricks of any style; 'mannerism' is defined in the *Oxford English Dictionary* as 'excessive or affected addiction to a distinctive manner'.

The term *'maniériste'* was coined by Roland Fréart de Chambray (brother of *Poussin's friend Paul Fréart de Chantelou) in his book *Idée de la perfection de la peinture* (1662) as a term of abuse for a group which included *Cesari and *Lanfranco and it was used by Luigi *Lanzi (1792), who also coined the term *manierismo*. *Bellori in his life of Annibale *Carracci (1672) associated *'maniera'* with a historical movement which he traced back to Raphael, and *Malvasia with a group of artists roughly contemporary with *Salviati. Both disapproved of it as an elegance too artificial and capricious—'totally chimerical and schematic'—and regarded the *naturalism of the Carracci as a healthy reaction from it. As a general stylistic term combining historical and critical connotations 'Mannerism', starting with Lanzi, carried the implication of excessive virtuosity and capriciousness which through *Burkhardt and *Wölfflin was taken over as an uncritical prejudice by 19th-cent. historians. It was in the period between the two world wars that a more positive attitude towards Mannerism generally emerged, with the growing realization that much 16th-cent. art is extremely exciting to contemporary taste: it could be, in the words of Eric Mercer (*English Art 1553–1625*, 1962), 'rigidly disciplined and wilfully licentious, elegant and farouche, pornographic and ecstatically religious, lost in the "other world" of antiquity and yet vividly conscious of the horrors of the sixteenth century', and artists such as *Pontormo and *Parmigianino were again seen as creators of powerful and distinctive styles, not decadents.

With Mannerism no longer receiving blanket condemnation, more subtle issues occupied the minds of historians, for example to what extent the term could be applied to art outside Italy (e.g. El *Greco in Spain, Romanist painters in the Netherlands, and the École de *Fontainebleau in France) or to architecture (where what might be taken in one context as playful or capricious disregard for the rules of classical architecture might in another be regarded as provincial clumsiness). While some critics wish to expand the use of the term, others wish to contract it, and still others seek to distinguish what they regard as the central elements of the style within the general period label by using the term *'maniera'*. The following sentence from S. J. Freedberg's *Painting in Italy: 1500–1600* (1971) in the Pelican History of Art series shows how potentially bewildering the terminology can be: 'The first generation of Mannerism, its inventors, thus could achieve *maniera*, but this requires to be distinguished not only chronologically but in degree and in some respects of kind from the "high Maniera" or Maniera proper.' Thus while the term 'Mannerism' can generally be taken to imply an elegant, refined, artificial, self-conscious, and courtly style, the shade of meaning to be attached to it varies very much according to the context and the outlook of the writer using it.

Man Ray (1890–1977). American painter, photographer, draughtsman, sculptor, and film maker, born Emmanuel Radinsky. He was secretive about his early life and the origin of his pseudonym is unknown. In 1915 he began a lifelong friendship with Marcel *Duchamp, collaborating with him and *Picabia in founding the New York *Dada movement. He also collaborated with Duchamp and Katherine *Dreier in forming the *Société Anonyme in 1920. In 1921 he settled in Paris, where he continued his Dada activities and then became a member of the *Surrealist movement. As well as painting in a Surrealist manner, he made Surrealist and abstract films. For several years he earned his living mainly as a fashion and portrait photographer, but he painted regularly again from the mid-1930s. In 1940 he went back to America to escape the Nazi occupation of Paris and settled in Hollywood, then in 1951 returned to Paris, where he died.

From the 1940s photography took a secondary place in Man Ray's activities, but it is as a photographer that his reputation is now most secure. In the 1920s and 1930s he was one of the most imaginative artists in this field, particularly for his exploitation of the 'Rayograph' (also known as photogram), a photograph produced without a camera by placing objects directly on sensitized paper and exposing them to light, and for his development of the technique of 'solarization' (the complete or partial reversal of the tones of a photographic image). He gained an international reputation as one of the most prominent figures of Dada and Surrealism, and several of his 'objects' have became icons of the movements, but critics have often been dismissive about his paintings.

Manship, Paul (1885–1966). American sculptor. He worked in an elegant, streamlined style, his beautifully crafted figures characterized by clarity of outline and suave generalized forms, and he achieved great success as a sculptor of public monuments. One of his best-known works is the gilded bronze *Prometheus* (1933) in Rockefeller Center Plaza, New York. Manship was also an accomplished portraitist. Because of the stylization of his work, derived partly from his interest in archaic sculpture, he for a time had a reputation as a pioneer of modern sculpture in America, but his modernism was of a very facile kind and by about 1940 he was being labelled an academic artist.

Manson, J. B. See CAMDEN TOWN GROUP.

Mantegna, Andrea (1430/1–1506). Italian painter. He was the pupil and adopted son of Francesco *Squarcione in Padua, growing up in a humanist atmosphere that was to colour his whole approach to art. Squarcione exploited his pupils for his own ends and at the age of 17 Mantegna gave an early indication of his formidable strength of character by taking him to court and forcing him to recognize his independence. He was remarkably precocious, and the distinctive style he created at the beginning of his career changed little over the next half century. It was a style characterized by sharp clarity of drawing, colouring, and lighting, a passion for archaeology that fed on the abundance of *classical remains in northern Italy, and a mastery of *perspective and foreshortening unequalled in the 15th cent. These qualities were apparent in his first major commission (1448), the decoration of the Ovetari Chapel of the Eremitani Church in Padua with frescos on the lives of SS. Christopher and James (almost totally destroyed in the Second World War) and they can be seen in the celebrated *Agony in the Garden* (NG, London, *c*.1460).

In 1460 Mantegna was appointed court painter to Ludovico *Gonzaga at Mantua, and apart from a visit to Rome in 1488–90 he remained there for the rest of his life. Mantegna was held in the highest esteem by Ludovico, by his son and successor Federico, and by Isabella d'*Este, who married Federico's successor, Francesco. At this time Mantua was rising to take its place among the leading centres of humanist culture in Europe, and Mantegna glorified the Gonzaga family and court in his most famous work—the fresco decoration of the Camera degli Sposi (Bridal Chamber) in the Ducal Palace (completed 1474). Group portraits of the Gonzaga family, arranged in various courtly scenes, line the walls and above them are bust medallions of the Caesars, indicating that the reigning house was worthy of the Roman Empire. The most remarkable feature of the room, however, is the *illusionistic painting of the architecture (particularly of the ceiling), which appears to extend the real space of the room. This was the first time since antiquity that such a scheme had been carried out and Mantegna's work became the foundation for much subsequent decorative painting.

Mantegna's other great work for the Gonzaga was his series of nine paintings on the *Triumph of Caesar* (Hampton Court, London, *c*.1480–95)—it is often said that they were done for Francesco, but in fact it is not known which member of the family commissioned them. These large and fragile canvases have suffered dreadfully at the hands of 'restorers' in former centuries, but they were successfully cleaned in the 1960s and 1970s, and although they are battered and faded they still give a marvellous picture of Mantegna's magnificent powers of invention and design and rank alongside *Raphael's tapestry *cartoons as one of the greatest ensembles of *Renaissance art outside Italy.

Mantegna was an engraver as well as a painter and was one of the first artists to use prints to disseminate his compositions. He also designed his own house in Mantua, and it is generally thought that he modelled the bronze bust of himself in his memorial chapel in the church of S. Andrea. At his death he was a venerated figure and his fame has never declined. His influence was profound, not only on Italian artists such as his brother-in-law Giovanni *Bellini, but also, for example, on *Dürer, one of the many northern artists who found his version of the *antique particularly easy to assimilate.

Manuel, Niklaus. See DEUTSCH.

Manueline. A style of architectural decoration peculiar to Portugal and named after Manuel I with whose prosperous reign (1495–1521) it roughly coincided. Merging *Gothic, Moorish, *Renaissance, and possibly Indian influences, it was characterized by richly encrusted carved ornamentation, the most common feature being a twisted rope motif, used for tracery, string courses, vault ribs, columns, pinnacles, etc. The style also spread to the decorative arts.

Manzoni, Piero (1933–63). Italian experimental artist. He worked in a conventional figurative style until 1956, when he turned to avant-garde work. In 1957 he began to produce *Achromes*, textured white paintings influenced by *Burri and *Klein (whom he met in 1957), and from 1959 he devised a series of provocative works and gestures that included signing people's bodies and designating them works of art, a block on which is inscribed upside down 'The base of the world' (Herning Park, Denmark, 1961), and cans of his own excrement. He is regarded as one of the forerunners of *Arte Povera and *Conceptual art. His early death was caused by cirrhosis.

Manzù, Giacomo (1908–91). Italian sculptor. At the age of 11 he was apprenticed to a wood-carver, then to a gilder and stuccoworker, but he was virtually self-taught as an artist. His early work was influenced by Egyptian and Etruscan art, but then he turned to a more *Impressionistic style owing much to the example of *Rodin and Medardo *Rosso. In the 1940s he simplified his style, so that although the surface of his work is often animated, the feeling it produces is one of Classic calm. He worked much on religious subjects (including numerous figures of cardinals), and was celebrated above all for the set of bronze doors he made for St Peter's in Rome after winning an international competition in 1950 (they were not completed until 1964). In 1958 he also completed a set of doors for Salzburg Cathedral, and in 1968 one for the church of St Laurence in Rotterdam. His work shows the possibility of producing sculpture that fits within a traditional religious context and yet is in a modern and personal idiom. Manzù has also worked as an etcher, lithographer, and painter.

maquette. A small preliminary model, often in clay or wax, for a work of sculpture. The word implies something in the nature of a rough sketch, not so fully worked out as a *bozzetto.

Maratta (Maratti), Carlo (1625–1713). Italian painter, the leading painter in Rome in the latter part of the 17th cent. As the pupil of Andrea *Sacchi he continued the tradition of the classical *Grand Manner, based on *Raphael, and he gained an international reputation particularly for his paintings of the Madonna and Child, which are reworkings of types established during the High *Renaissance. The rhetorical splendour of his work is thoroughly in the *Baroque idiom, however, and the numerous altarpieces he painted for Roman churches (many still *in situ*) give wholehearted expression to the dogmas of the Counter-Reformation. Maratta was also an accomplished fresco painter, and the finest portraitist of the day in Rome. He had a large studio and his posthumous reputation suffered when the inferior works of his many pupils and imitators were confused with his own paintings.

marble. (From the Greek *marmaros*: a crystalline rock; root from *marmairein*: to sparkle.) Word loosely applied among masons and in the building trade to any hard *limestone that can be sawn into thin slabs and will take a good polish so that it is suitable for decorative work; more strictly it refers to metamorphosed limestones whose structure has been recrystallized by heat or pressure. Marbles are widely disseminated and occur in a great variety of colours and patterns, but certain types have been particularly prized by sculptors. The most famous of Greek white marbles in the ancient world was the close-grained *Pentelic*, which was quarried at Mt. Pentelicon in Attica. The *Elgin Marbles are carved in Pentelic. Widely used also were the somewhat coarser-grained translucent white marbles from the Aegean islands of Paros and Naxos. *Parian* marble was used for the celebrated Mausoleum at Halicarnassus.

The pure white Carrara marble, quarried at Massa, Carrara, and Pietra Santa in Tuscany from the 3rd cent. BC, is the most famous of all sculptors' stones. It was used for Trajan's Column and the *Apollo Belvedere*, and was much favoured in the *Renaissance, particularly by *Michelangelo, who often visited the quarries to select material for his work. *Neoclassical sculptors, such as *Canova, also favoured it because of its ability to take a smooth, sleek surface.

Marc, Franz (1880–1916). German painter, one of the founders of the *Blaue Reiter group. The son of a Munich painter, he began working in an academic naturalistic style, but visits to Paris in 1903 and 1908 brought him into contact with *Impressionism and *Post-Impressionism, and responding particularly to van *Gogh he advanced towards a more *Expressionist style. In 1910 he entered into a close friendship with *Macke, who introduced him to the Expressionist use of colour. Through Macke he joined the *Neue Künstlervereinigung at Munich and found in *Kandinsky and *Jawlensky men of congenial artistic views. With them he founded the Blaue Reiter in 1911. Marc was of a deeply religious disposition (in 1906 he visited Mt. Athos in Greece with its famous monasteries) and was troubled by a profound spiritual malaise; through painting he sought to uncover mystical inner forces that animate nature. His ideas were expressed most intensely in paintings of animals, for he believed that they were both more beautiful and more spiritual than man: 'The ungodly human beings who surrounded me did not arouse my true emotions, whereas the inherent feel for life in animals made all that was good in me come out.' Using non-naturalistic symbolic colour and simplified, rhythmic shapes, he said he tried to paint animals not as we see them, but as they feel their own existence (*Blue Horses*, Minneapolis Institute of Arts, 1911). In 1912 he met *Delaunay in Paris and was influenced by the *Orphist experiments in the abstract use of colour. The culminating work of this period was *Animal Destinies* (Kunstmuseum, Basle, 1913), which uses panic-stricken animals to symbolize a world on the edge of destruction; on the back of the picture he wrote: 'Und alles sein ist flammend leid' (And all being is flaming suffering). By 1914, under the influence partly of *Cubism and *Futurism, his paintings had become still more abstract, losing almost entirely any representational content, as in *Fighting Forms* (Neue Pinakothek, Munich, 1914), an image of convulsive fury. These last paintings are considered the culmination of that stream of German Expressionism which, free from social and didactic preoccupations, moved in the direction of expressive abstraction. Marc was killed in action in the First World War.

Marcantonio Raimondi. See RAIMONDI.

Marcks, Gerhard (1889–1981). German *Expressionist sculptor. He completed *Barlach's series of statues on the gable of the Catherine church in Lübeck (1947), but his refined and somewhat sentimental style is generally closer to that of *Lehmbruck.

Marcoussis, Louis (1878–1941). Polish-born French painter, originally called Louis Markus (his adopted name, suggested by *Apollinaire, came from the village of Marcoussis near Paris). He went to Paris in 1903 and after painting in an *Impressionist style and then working as a cartoonist for some years he joined the *Cubist movement. In 1912 he exhibited with the *Section d'Or, and again after the war in 1920. One of the most appealing of the minor Cubist artists, he never turned the Cubist techniques into a mannerism, but combined clarity and simplicity of structure with sensitivity of handling. He was an accomplished etcher and illustrated a number of books.

Marées, Hans von (1837–87). German painter, active mainly in Italy, where he settled in 1864. Like his friends *Böcklin, *Feuerbach, and *Hildebrand, he was one of the Germanic artists working in Italy who turned to the tradition of *ideal art at a time when naturalism was becoming dominant in their own countries. He devoted himself mainly to the theme of the human figure in a landscape setting (*Age*, Staatliche Mus., Berlin, 1873–8). In 1873 he received a major commission to paint a series of large frescos in the Zoological Institute in Naples, but he lacked self-confidence, and died disappointed and practically unknown. It was only at the beginning of the 20th cent. that the statuesque dignity of his work became appreciated and taken up as a symbol of a new spirit of modernity in German art.

Margarito (or **Margaritone**) **of Arezzo** (active *c.*1262). Italian painter from Arezzo. He is one of the very few 13th-cent. Italian painters and the only early Aretine by whom we have signed works (examples are in the NG, London, and the NG, Washington). Margarito's paintings are clumsy, but have something of the vividness and lucid brevity of the comic strip. His freedom from formal *Byzantine conventions of drawing and decoration has led some critics to consider him in advance of his time, though others have held that his was a meagre talent beside that of such contemporaries as *Coppo di Marcovaldo and *Guido da Siena. *Vasari, who also came from Arezzo, wrote his biography in his *Lives* (saying he was an architect as well as a painter), and this is virtually the only source of knowledge on him, although a document of 1262 probably refers to him.

Marin, John (1870–1953). American painter. From 1905 to 1910 he lived in Europe, mainly Paris, where he was influenced by *Whistler's watercolours and etchings, but he first came into contact with avant-garde movements after his return to America, when he became a member of *Stieglitz's circle. The *Armory Show also made a great impact on him. Responding especially to German *Expressionism, and the planimetric structure of the late work of *Cézanne, he developed a distinctive semi-abstract style that he used most characteristically in powerful watercolours of city life and the Maine coast (where he often painted in the summer). His oil paintings (which became more important in his work from the 1930s) are often similar in effect to watercolours, leaving parts of the canvas bare. Marin also made etchings, especially early in his career. From the 1920s he enjoyed a high reputation. He was an individualist, belonging to no movement, and one of the finest watercolourists of the 20th cent.

Marinetti, Filippo Tommaso (1876–1944). Italian poet, the founder and leading ideologist of *Futurism.

Marini, Marino (1901–80). One of the outstanding Italian sculptors of the 20th cent., also a painter, lithographer, and etcher. Until *c.*1928 he worked mainly as a graphic artist. During the 1930s he travelled widely in Europe and also visited the USA. Although his travels brought him into contact with many distinguished artists, Marini did not ally himself with any avant-garde movements, remaining essentially isolated in his artistic aims. Working mainly in bronze, he concentrated on a few favourite themes, most notably the Horse and Rider, a subject in which he seemed to express an obscure but poignant tragic symbolism. He also made numerous portrait busts, remarkable for their psychological penetration and sensuous exploitation of the surface qualities of the material. Often he *polychromed his bronzes, sometimes working with corrosive dyes. In the 1940s he took up painting seriously, many of his pictures being near abstracts. There is a Marini museum in Milan, where he spent much of his career, and his work is represented in many major collections of modern art.

Marinus van Reymerswaele (active *c.*1509–d. 1597?). Netherlandish painter, presumably from Roymerswaele (a place once in Zeeland, but now under the sea). He specialized in two types of picture, both with more or less life-size half-length figures: representations of St Jerome in his study (seemingly deriving from *Dürer) and *genre scenes of bankers, usurers, misers, and tax-collectors. The genre scenes show the sin of avarice and the vanity of earthly possessions: according to a Flemish proverb a banker, a usurer, a tax-collector, and a miller were the four evangelists of the devil. These paintings must have been very popular, for they exist in numerous versions and copies, but it is not known what kind of clientele bought pictures of such unpleasant characters, grotesquely presented in a manner deriving (via *Massys) from *Leonardo's *'caricatures'. *Two Tax Gatherers* in the National Gallery, London, gives a good idea of the fanciful costumes in which he dressed his figures, and the scrupulous attention which he paid to every detail of the miscellany of objects which clutter his pictures. A 'Marinus . . . of Romerswael' is mentioned in Middelburg in 1567, when he was condemned to walk in a penitential procession for his part in the looting of a church, but he may not be identical with the painter, who is described as deceased in that year.

Maris. Family of three brothers who played a leading part in Dutch painting during the second half of the 19th cent. **Jacob (Jacobus Hendricus)** (1837–99) was one of the leaders of the *Hague School, painting principally views of the Dutch countryside, with some portraits and figure studies. His work was influenced by the *Barbizon School, but he did not paint directly from nature. **Matthias (Matthijs, Thijs)** (1839–1917) began in similar fashion but came to specialize in figure compositions of visionary subjects. In 1869 he moved to Paris and then in 1877 to London, where he lived for the rest of his life and was influenced by the *Pre-Raphaelites in his choice of poetic subjects, although not in style. **Willem** (1844–1910) was a pupil of his two brothers and was influenced by *Mauve. His subjects are almost entirely confined to meadows and cattle. In his later years he became a leader of Dutch *Impressionism, urging his pupils—among them *Breitner—to paint in the open air and to use vivid colours. Examples of works by all three brothers are in the Gemeentemuseum in The Hague.

Marlow, William (1740–1813). English landscape and marine painter, a pupil of Samuel *Scott and possibly also of Richard *Wilson. His early landscapes were topographical views, including pictures of country houses, but after a visit to the Continent (1765–8) he painted largely from his memories of France and Italy. He retired from professional practice *c.*1785, but continued to paint for his own pleasure. There are examples of his work in the Tate Gallery, including a **capriccio* of *St Paul's and a Venetian Canal*.

Marmion, Simon (d. 1489). Franco-Flemish manuscript *illuminator and painter. In 1449–54 he was at Amiens, where he may have been born, and he was a member of the Tournai guild in 1468, but the greater part of his working life was spent at Valenciennes. He had a great reputation in his day, but no works certainly from his hand survive. The main work attributed to him is the *retable of S. Bertin (1459; most of it in the Staatliche Museen, Berlin, fragments in the NG, London). His style was very tender, with delicate, almost pastel colouring, and does not belong to the mainstream of either French or Netherlandish art.

Marochetti, Carlo (1805–67). Italian sculptor. He had an internationally successful career, being made a baron in his native country, awarded the Legion of Honour in France, and patronized by Queen Victoria and Prince *Albert in England, where he worked from 1848. His dramatic style, exemplified in his equestrian statue of Richard the Lionheart (1851–60) outside the Houses of Parliament, was, however, considered rather flashy in England.

Marot, Daniel (1661–1752). French architect, designer, and engraver. He left France as a Huguenot refugee and moved to Holland, where he worked for the Stadtholder, later William III (1650–1702) of England. As William's court architect he visited England in 1694–8, working at Hampton Court and Kensington Palace, and his engraved designs were widely influential in England (and also became popular in Germany). His intricate style of ornamentation was similar to that of *Bérain. **Jean Marot** (*c.*1619–79), his father, was a minor architect who published several collections of engravings that are invaluable source material for the history of French architecture and design, often recording buildings and decorative schemes that have disappeared.

marouflage. Term for gluing a *canvas down, whether before painting it or afterwards, on to a wall. In French the word *maroufle* means a sticky mixture of the remains of paint left in the artist's pots, and *maroufler* may describe almost any operation which might be carried out with this powerful glue, for example lining one canvas with another to strengthen it.

Marquet, Albert (1875–1947). French painter and draughtsman. He was one of the *Fauves, and for a time his boldness of colour almost matched that of *Matisse (his lifelong friend). However, he soon abandoned Fauvism and turned to a comparatively naturalistic style. He painted some fine portraits and did a number of powerful female nudes between 1910 and 1914, but he was primarily a landscapist. His favourite—eventually almost exclusive—themes were ports and the bridges and quays of Paris, subjects he depicted with unaffected simplicity and great sensitivity to tone (*Winter on the Seine*, NG, Oslo, *c.*1910). Marquet was an outstanding draughtsman and from 1925 worked mainly in watercolour. He travelled widely and built up an international reputation, but he lived very quietly (he was timid in personality) and refused all honours.

Marsh, Reginald (1898–1954). American painter. He was a newspaper illustrator in the 1920s, but during the 1930s became well known for his pictures depicting the shabbiness and tawdriness of city life. His favourite subjects were Coney Island, the amusement arcades of Times Square, the Bowery, and the poorer aspects of the street life of New York (*Tattoo and Haircut*, Art Institute of Chicago, 1932). He was also capable of bitter satire against the smug complacency of the wealthy, but in general his work shows a love of depicting teeming life through ugly but colourful subjects rather than a desire for social protest. His aim was to depict contemporary life in the manner of the Old Masters and he worked mainly in *tempera, also experimenting with other venerable techniques. He came from a wealthy family and to some extent his work expressed a rejection of the affluent and genteel circumstances in which he grew up.

Marshall, Benjamin (1767–1835). English sporting painter, an able follower of *Stubbs. He was a pupil of the portrait painter L. F. *Abbot, but from *c.*1792 he turned to animal painting. In 1812 he settled in Newmarket, where he concentrated on pictures of horses and hunting and racing scenes.

Marshall, Edward (1578–1675). English mason and sculptor. He became Master Mason to Charles II in 1660 and had a considerable business as a maker of monuments, one of the finest being that to Lady Culpeper at Hollingbourne, Kent (1638). In his later work he was assisted by his son **Joshua** (1629–78), who succeeded him as Master Mason to the Crown (1676–7). Besides working on the royal palaces Joshua was much employed on decorative carving in the rebuilding of the City of London after the Great Fire of 1666. He also carved the pedestal for Hubert *Le Sueur's equestrian statue of Charles I at Charing Cross. His monuments are less accomplished in cutting than those of his father.

Martin, Elias (1739–1818). The leading Swedish landscape painter of his period. He lived for several years in England (1768–80 and 1788–91) and his style was influenced by *Gainsborough and *Wilson as well as *Claude. He painted in watercolour as well as oil and also worked in various engraving techniques. His work is well represented in the museums of Göteborg and Stockholm.

Martin, John (1789–1854). English *Romantic painter and *mezzotint engraver, celebrated for his melodramatic scenes of cataclysmic events crowded with tiny figures placed in vast architectural settings. He caught the public imagination with spectacular paintings such as *Joshua Commanding the Sun to Stand Still* (United Grand Lodge of Great Britain, London, 1816), the work that made him famous, and in 1821 *Lawrence referred to him as 'the most popular painter of the day'. His work was indeed truly popular, for at exhibitions his paintings sometimes had to be railed off from crowds of admirers and he made his living mainly through the sale of engravings of his pictures rather than the pictures themselves. He became famous in France as well as Britain, he was knighted by Leopold I of Belgium (1833), and his influence was felt by American artists such as *Cole. However, while he pleased a vast audience and was regarded by some admirers as one of the greatest geniuses who ever lived, Martin was reviled by *Ruskin and other critics, who considered his work vulgar sensationalism. Few artists, indeed, have been subject to such extremes of critical fortune, and his fame sank to such an astonishing degree after his death that very large and once famous paintings by him were sold in the 1930s for as little as £2. In the 1970s his reputation greatly revived. Martin made mezzotints not only as a means of reproducing his paintings but also as original compositions. Particularly noteworthy are his illustrations to the Bible and John Milton's *Paradise Lost*, which show that although he had great weaknesses as an artist, especially in his drawing of the human figure, he also had a vividness and grandiloquence of imagination not unworthy of such elevated subjects. He is sometimes called 'Mad Martin', but the sobriquet is undeserved and applies more to his brother Jonathan, who was insane and set fire to York Minster. John Martin was eminently sane and in the 1830s almost bankrupted himself with extremely ambitious but entirely practical plans for improving the water supply and sewage system of London. They were unsuccessful but reveal a heroic desire to put the architectural visions of his paintings into a concrete form. His work is best represented in the Tate Gallery, London, and the Laing Art Gallery, Newcastle upon Tyne.

Martin, Kenneth (1905–84). British painter and sculptor. In 1930 he married Mary Balmford, whose artistic development as Mary *Martin had very close affinities with his own. In the 1930s he painted in a naturalistic style and was in close touch with artists of the *Euston Road School, but during the 1940s the representational character of his work receded while the element of structure and design became progressively more prominent until in 1948–9 both he and Mary Martin made their first completely abstract paintings. With Victor *Pasmore the Martins were among the leaders of a new *Constructivist movement which burgeoned both in England and in America in the 1950s, making their first constructions at the beginning of the decade. Thereafter he was a pioneer in England of a number of different forms of geometrical abstraction. Of his conversion to geometrical abstraction he said: 'The moment I became a purely abstract artist I began to realise what I'd been missing. . . . That I'd really missed the whole of the modern movement.' His work is well represented in the Tate Gallery.

Martin, Sir Leslie. See CIRCLE.

Martin, Mary (née Balmford) (1907–69). British painter and sculptor, the wife of Kenneth *Martin, who was a fellow student at the Royal College of Art. She painted landscapes and still lifes, but during the 1940s her work, like that of her husband, moved towards geometrical abstraction. With Kenneth Martin and Victor *Pasmore she belonged to the group of post-war *Constructivists which continued the tradition of geometrical abstraction in England, and her work was shown in many exhibitions of international Constructivist art. Her more important commissions included a Screen for Musgrave Park Hospital, Belfast (1957), and a Wall Construction for the University of Stirling (1969).

Martorell, Bernardo (active 1427–52). Spanish painter and *miniaturist. He worked in Barcelona and was the outstanding painter in Catalonia in the second quarter of the 15th cent., the successor to *Borrassá, who probably taught him. Only one surviving work is securely documented—the *Altarpiece of St Peter of Púbol* (Gerona Mus., 1437)—but on stylistic grounds a group of paintings formerly given to 'the Master of St George' has been attributed to him. The group includes (and formerly took its identity from) an altarpiece of St George; the central panel is in the Art Institute of Chicago and the four wings are in the Louvre. Martorell's work, influenced by Franco-Flemish painting and *illumination, was essentially *International Gothic in style, but reveals a highly distinctive personality, particularly in its vigorous sense of drama and delicate handling of light. Among the painters of the next generation *Huguet was most obviously indebted to him.

Marx, Roberto Burle. See BURLE MARX.

Marzal de Sax, Andrés (active 1393–1410). Painter, probably of German origin (Sax indicating Saxony), who worked in Valencia. Only one fragment survives of his documented works—the *Incredulity of St Thomas* in Valencia Cathedral, part of an altarpiece he completed for the cathedral in 1400. Among the works given to him on stylistic grounds the most important is the huge and sumptuous *retable of St George (V&A, London), featuring the varied tortures of the saint in grisly detail. The somewhat rough vigour of Marzal de Sax's style had considerable influence in Valencia; he is last mentioned in 1410, impoverished and ill, receiving free lodging from the city in recognition of the quality of his work and his generosity in training local painters.

Masaccio (Tommaso di Ser Giovanni di Mone) (1401–28). Florentine painter. Although he died in his mid-twenties he brought about a revolution in painting and he ranks alongside his friends *Alberti, *Brunelleschi, and *Donatello as one of the founding fathers of the *Renaissance. His affectionate nickname, which may be translated as 'Sloppy Tom', was given to him, so *Vasari says, because he was so devoted to art that 'he refused to give any time to worldly causes, even to the way he dressed'. He became a guild member in Florence in 1422, but nothing is known of his training, the tradition that he was taught by *Masolino, later his collaborator, now being discounted. The earliest work attributed to him is the San Giovenale *Triptych (Uffizi, Florence, 1422), which is somewhat uncouth but reveals a totally individual spirit in its rejection of all *Gothic elegance and its concentration on the weight and bulk of the figures. Instead of learning from contemporary painters, Masaccio looked back to *Giotto for inspiration, recapturing the gravity and grandeur that characterized his work. But whereas Giotto set his figures in space intuitively, Masaccio grappled with and solved the problem of creating a completely coherent and consistent sense of three dimensions on a two-dimensional surface, his work thus becoming one of the cornerstones of European painting. His enormous achievement was based on his mastery of the new science of *perspective and his use of a single constant light source to define the construction of the body and its draperies. Among contemporary artists he was closest in spirit to Donatello. Both artists were less concerned with surface appearances and isolated detail than with the underlying construction of objects and both excelled at the depiction of emotion with great force and directness.

Masaccio has left three great works to posterity in which he enunciated his new principles: a *polyptych (1426) for the Carmelite church in Pisa (the central panel is in the National Gallery in London, and other parts in the Staatliche Museen, Berlin, the Getty Museum, Malibu, the Museo di Capodimonte, Naples, and the Museo Nazionale, Pisa); a fresco cycle, done in collaboration with Masolino, on the life of St Peter (with additional scenes of *The Temptation* and *The Expulsion from Paradise*) in the Brancacci Chapel of Sta Maria del Carmine, Florence (*c.*1425–8); and a fresco of the *Trinity* in Sta Maria Novella, Florence (probably 1428).

Masaccio went to Rome in 1428, leaving the frescos in the Brancacci Chapel unfinished, and died so suddenly that Vasari said 'there were some who even suspected he had been poisoned'. Vasari adds that 'during his lifetime he had made only a modest name for himself', and certainly many of his Florentine contemporaries and successors were unmoved by his innovations. He was a great inspiration to the progressive masters of the next generation, however (Filippo *Lippi, *Piero della Francesca), and Vasari records a whole roster of great artists, including *Leonardo, *Michelangelo, and *Raphael, who studied his work with profit. 'In short,' he writes, 'all those who have endeavoured to learn the art of painting have always gone for that purpose to the Brancacci Chapel to grasp the precepts and rules demonstrated by Masaccio for the correct representation of figures.'

Masereel, Frans (1889–1972). Belgian *Expressionist painter and print-maker, best known for his *romans in beelden* (novels in pictures), which are series of woodcuts telling a story without a text.

Masip, Vincente. See MACIP.

Maso di Banco (active second quarter of the 14th cent.). Florentine painter. Almost nothing is known of his career (*Vasari does not mention him), but he is regarded as the greatest of *Giotto's followers on the strength of *Ghiberti's testimony that he was the painter of the frescos illustrating the legend of St Sylvester in the Bardi chapel of Sta Croce, Florence. The stately figures here are sometimes even more massive than Giotto's and the lucid and beautifully coloured compositions are of almost geometric clarity (although it has been argued that some of the effect of monumental simplicity may be due to restoration). On stylistic grounds other works have been attributed to Maso, including panels in Budapest (Mus. of Fine Arts), Chantilly (Musée Condé), and New York (Brooklyn Mus. and Met. Mus.).

Masolino da Panicale (*c.*1383–1447?). Italian painter. He is generally considered to be a member of the Florentine School, but he travelled a good deal and even went to Hungary. His career is closely linked to that of *Masaccio, but the exact nature of the association remains ill-defined. The tradition that he was Masaccio's master is now dismissed, for he became a guild member in Florence only in 1423 (a year after Masaccio) and although he was appreciably the older man it was he who was influenced by Masaccio rather than the other way round. They are thought to have collaborated on *The Madonna and Child with St Anne* (Uffizi, Florence, *c.*1425), but the major undertaking on which they worked together was the decoration of the Brancacci Chapel of Sta Maria del Carmine in Florence. Masolino's style was softer than Masaccio's and there is a fair measure of agreement about the division of hands. After Masaccio's death Masolino reverted to the more decorative style he had practised earlier in his career. At his best he was a painter of great distinction, his masterpiece perhaps being the fresco of the *Baptism of Christ* (*c.*1435) in the Baptistery at Castiglione d'Olona, near Como, a graceful and lyrical work that is a world away from Masaccio's *Baptism of the Neophytes* in the Brancacci Chapel.

Masson, André (1896–1987). French painter, engraver, sculptor, stage designer, and writer, one of the leading figures of *Surrealism. He was severely wounded in the First World War and deeply scarred emotionally. His pessimism was accompanied by a profound and troubled curiosity about the nature and destiny of man and an obscure belief in the mysterious unity of the universe, which he devoted the whole of his artistic activity to penetrating and expressing. In the early 1920s he was influenced by *Cubism, but in 1924 he joined the Surrealist movement and remained a member until 1929, when he left in protest against *Breton's authoritarian leadership.

His work belonged to the spontaneous, expressive semi-abstract variety of Surrealism, and included experiments with automatic drawings (see AUTOMATISM), chance effects, and the incorporation of sand in his paintings. Themes of metamorphosis, violence, psychic pain, and eroticism dominated his work. In 1934–6 he lived in Spain until the Civil War drove him back to France and in 1941–5 he took refuge from the Second World War in the USA. There his work formed a link between Surrealism and *Abstract Expressionism. In 1945 he returned to France and two years later settled at Aix-en-Provence, where he concentrated on landscape painting, achieving something of the spiritual rapport with nature seen in some Chinese paintings.

Massys (or **Matsys** or **Metsys**), **Quentin** (1465/6–1530). Netherlandish painter. He was born at Louvain but worked in Antwerp, where he became the leading painter of his day. Although he became a master in the guild there in 1491, his early career is obscure and his first dated works are the altarpiece of *St Anne* (Musées Royaux, Brussels, 1507–9) and the *Lamentation* (Musée Royal, Antwerp, 1508–11). Massys continued the tradition of the great masters of 15th-cent. Netherlandish art, but he was also clearly aware of Italian art (particularly the work of *Leonardo) and may well have crossed the Alps at some point in his career. In his exquisite *Madonna and Child with Angels* (Courtauld Inst., London), for example, the *iconographic type of the standing Virgin goes back to Jan van *Eyck, but the *putti holding a garland reveal *Renaissance influence. The landscape backgrounds of some of his religious works were possibly done by *Patenier. Massys also painted portraits and *genre scenes. The satirical quality in his pictures of bankers, tax-collectors, and avaricious merchants has been linked with the writings of the great humanist Erasmus (1466–1536). Certainly the two met, for Massys painted a pair of portraits of Erasmus (Gal. Naz., Rome) and Petrus Egidius (Earl of Radnor Coll., Longford Castle, Wiltshire) as a gift for Sir Thomas More in 1517. They instituted a new type—the scholar in his study—that influenced *Holbein among others. Massys had two painter sons, **Jan** and **Cornelis**.

Master Bertram (active 1367–d. 1414/15). German painter from Westphalia, active mainly in Hamburg, where he was the leading master of his day. His main work, the Grabow Altar of 1379 (Kunsthalle, Hamburg), shows how the influence of the *Bohemian School, notably *Master Theodoric, extended as far as Hamburg.

Master E. S. (sometimes called the Master of 1466 from the date on one of his engravings). German engraver working in the mid-15th cent., named after the monogram on eighteen of his surviving prints. He was the most prolific and influential of the early German engravers, working on profane and fantastic subjects as well as religious images and producing more than 300 known prints. Earlier engravers had been content with pure outline, but E. S. managed to produce rich tonal effects by the ingenious use of *hatching, and cross-hatching.

Master Francke (early 15th cent.). German painter, active in Hamburg. His major surviving work is an altarpiece for the guild of merchants trading with England (Kunsthalle, Hamburg, 1424), showing Passion scenes and incidents from the life of St Thomas à Becket. Francke was a leader of the northern German version of *International Gothic. He had close links with the west and Burgundy in particular. His influence was widely spread over north Germany and along the shores of the Baltic.

Master Jacomart. See BAÇO, JAIME.

Master of . . . Term used in art history to label the author of anonymous works for convenience in discussing them. This use of invented names is more common in the study of painting and the *graphic arts than of sculpture, and historians of architecture hardly ever resort to it. It began in Germany in the early 19th cent. with the description of Early Netherlandish painting. The choice of names was then often more lyrical than descriptive: for instance the 'Master of the Pearl of Brabant' (who later was identified with *Bouts) got his name from an altarpiece (Alte Pinakothek, Munich) which had become known by that poetical title simply because it caught the imagination. Nowadays invented names are more prosaic and more directly appropriate. Often the anonymous master is named after a particular picture and/or the collection to which it belongs, e.g. 'Master of the Louvre Annunciation'. Less often the name refers to some aspect of the artist's style, as in the 'Master of the Anaemic Figures' (a 15th-cent. Spanish painter), which shows that the designation 'master' is used neutrally and is not a sign of approbation. The practice of creating artistic personalities in this way has been overdone, but is often useful.

Master of 1466. See MASTER E. S.

Master of Alkmaar (active early 16th cent.). Netherlandish painter named after the altarpiece of the *Seven Works of Mercy* painted for the church of St Lawrence in Alkmaar in 1504 and now in the Rijksmuseum, Amsterdam. Attempts have been made to identify the painter with Cornelis Buys, who was the brother of Jacob *Cornelisz. van Oostsanen. The figure style is reminiscent of *Geertgen and the altarpiece is important as an early instance of the characteristically Dutch delight in the representation of everyday life.

Master of Flémalle. Netherlandish painter named after three paintings in the Städelsches Kunstinstitut in Frankfurt that were wrongly supposed to have come from Flémalle, near Liège. There is a strong consensus of scholarly opinion that he is to be identified with Robert Campin (active 1406–44), who was the leading painter of his day in Tournai but none of whose documented pictures survive. The identification depends on the similarity between the Master of Flémalle's paintings and those of Jacques *Daret and Rogier van der *Weyden, for Daret was Campin's pupil and Rogier almost certainly was. The hypothesis that the Master of Flémalle's paintings are early works by Rogier now has few adherents. While there is still doubt about the Master of Flémalle's identity, there is no argument about his achievement, for he made a radical break with the elegant *International Gothic style and ranks with van *Eyck as one of the founders of the Netherlandish school of painting. None of the paintings given to him is dated—with the exception of the wings of the Werl altarpiece of 1438 in the Prado, a doubtful attribution—but it seems likely that his earliest works antedate any surviving picture by van Eyck. The earliest of all is generally thought to be *The Entombment* (Courtauld Inst., London) of about 1410/20. This still has the decorative gold background of medieval tradition, but the influence of Claus *Sluter is clear in the sculptural solidity and dramatic force of the figures. The most famous work associated with the Master of Flémalle is the Mérode Altarpiece (Met. Mus., New York), and he is indeed sometimes referred to as the Master of Mérode. However, the attribution of this painting has also been questioned. Among the other works associated with him are *The Marriage of the Virgin* (Prado, Madrid), *The Nativity* (Musée des Beaux-Arts, Dijon), and *The Virgin and Child before a Firescreen* (NG, London), which shows the homely detail and down-to-earth naturalism characteristic of the artist (the firescreen behind the Virgin's head forms a substitute for a halo). The National Gallery also has three portraits attributed to the Master of Flémalle. In spite of the many problems that still surround him, he emerges as a very powerful and important artistic personality.

Master of Frankfurt (active *c.*1500). Netherlandish painter, named from the *St Anna Altar* painted *c.*1505 for the Dominican Priory in Frankfurt (now in the Städelsches Kunstinstitut there).

Some forty pictures are attributed to him, their style suggesting he was a follower of Quentin *Massys.

Master of Hohenfurth. See MASTER OF VYŠŠÍ BROD.

Master of Liesborn (active *c.*1470). German painter, named after an altarpiece painted for the Benedictine abbey of Liesborn in Westphalia; six fragments of it are in the National Gallery, London. His attractive gentle style was influenced by the *Cologne School.

Master of Mary of Burgundy. Netherlandish manuscript illuminator, active in the last quarter of the 15th cent. He is named after two *Books of Hours painted for Mary of Burgundy, who married the emperor Maximilian I in 1477 and died in 1482 (Österreichische Nationalbibliothek, Vienna, and Kupferstichkabinett, Berlin). They are among the finest illuminated books of the period, with many charming everyday-life and still-life details, and other fine works in a similar style have been attributed to him, notably a Book of Hours in the Bodleian Library, Oxford. It has been suggested that he is to be identified with Alexander *Bening.

Master of Moulins (active *c.*1480–1500). French painter, named after a *triptych in Moulins Cathedral representing the Madonna with Saints and Donors and datable *c.*1498. The style of this work is quite distinctive, and has enabled a considerable *œuvre* to be built up around it. The earliest of these attributions suggest the influence of Hugo van der *Goes, but the Master of Moulins was much closer to *Fouquet in the exquisite sculptural precision of his forms, the poise of his individual figures, and the harmony of his compositions. With this sculptural conception of form goes the brilliant palette of a glass painter and a taste for splendid and meticulous details which make him one of the outstanding painters of his period in northern Europe. Various attempts have been made to identify the Master of Moulins with named artists (for example Jean *Perréal), but none has met with general acceptance.

Master of Naumburg. Name given to the head of the workshop which produced a group of works in the west choir of Naumburg Cathedral around 1240 that are generally regarded as the finest works of *Gothic sculpture in Germany. They include twelve sensitively characterized statues representing the benefactors of the original cathedral (notably the famous figures of Ekkehard and Uta) and a series of *reliefs of the Passion with a Crucifixion group on the choir screen. The Master of Naumburg was probably trained in northern France, but there is little doubt that he was a German by birth; his Germanic temperament emerges in the pathos and vehement gestures of his figures. It is especially these manifestations of emotion and feeling that make his contribution to Gothic art so important. Yet he loses nothing of the monumental dignity and idealism of his predecessors, and in his art the two worlds of feeling seem to meet and enrich one another as in the work of perhaps no other medieval artist.

Master of St Cecilia (active *c.*1304). Italian painter named after the St Cecilia Altarpiece in the Uffizi, which was originally in the church of Sta Cecilia, destroyed by fire in 1304. Presumably he was a Florentine, but nothing is known about him. Other works have been attributed to him because of their resemblance to the Uffizi picture, the most important being the three concluding scenes of the great fresco cycle of the life of St Francis in the Upper Church of S. Francesco at Assisi (see MASTER OF THE ST FRANCIS CYCLE). The painter of these scenes resembles *Giotto in lucidity of presentation and the solid drawing of his figures, but he is more genial in feeling. His figures are more vivacious, his colour warmer and sweeter. The architectural settings, fastidiously moulded and coffered, are his hallmark and they place him in kinship with contemporary Roman painters, especially with the *Cavallini circle, who were also active in Assisi. The completion of the great cycle in the Upper Church would have been entrusted only to an established master and some critics have attempted to identify the painter of these scenes and the St Cecilia Altarpiece with the famous but tantalizingly elusive *Buffalmacco.

Master of St George. See MARTORELL.

Master of St Giles (active *c.*1480–*c.*1500). Netherlandish (?) painter named after two panels representing scenes from the life of St Giles (NG, London). Other paintings in the same style have been grouped round them. Their attention to detail and their meticulous finish have caused some to consider that the artist was trained in the Netherlands, but the inclusion of views of Paris in some of his works indicates that he worked there, whatever his origin. His work is of high quality and he must have been one of the best painters of the day in northern France.

Master of Segovia. See BENSON.

Master of the Amsterdam Cabinet. See MASTER OF THE HOUSEBOOK.

Master of the Brunswick Monogram (active *c.*1520–40). Netherlandish painter, named after a picture in the Herzog-Anton-Ulrich Museum in Brunswick of *The Parable of the Great Supper* (Luke 14). There is no agreement how the monogram should be read. A dozen or so small pictures have been attributed to the same hand; about half depict religious subjects in the open air, and most of the others scenes in brothels. The quality of the pictures is uniformly high. The artist's observation of nature, his fine drawing, and ability to integrate figures into a landscape make him an important forerunner of Pieter *Bruegel the Elder. Attempts have been made to identify him with various painters, including Jan Sanders van *Hemessen.

Master of the Death of the Virgin. See JOOS VAN CLEVE.

Master of the Housebook (late 15th cent.). German (or according to some authorities Netherlandish) engraver, so called from a number of drawings contained in a kind of commonplace book in Castle Wolfegg in Germany. He used to be called the Master of the Amsterdam Cabinet, since the largest collection of his engravings is in the Print Room of the Rijksmuseum. His *line engravings often represent very worldly subjects and his use of *drypoint gives them a curiously sketchy and lively character. *Dürer must have studied them fairly closely as their influence can be traced in several of his early drawings. A few paintings have also been attributed to this master.

Master of the Legend of St Barbara (active *c.*1470–*c.*1500). Netherlandish painter, one of the ablest followers of Rogier van der *Weyden. He probably came from Bruges and worked in Brussels. He is named after the subject of a *triptych (*c.*1475), the centre panel of which is now in the Musées Royaux, Brussels, the left wing in the Confrérie du S. Sang, Bruges, and the right one lost.

Master of the Life of the Virgin (or **Master of the Life of Mary**) (active *c.*1460–90). German painter, named after a series of eight panels illustrating the Life of the Virgin, of which the *Presentation in the Temple* is in the National Gallery, London, and the remainder in the Alte Pinakothek, Munich. He was one of the outstanding Cologne painters of his time, and his affinities with Dirk *Bouts and Rogier van der *Weyden suggest that he trained in the Netherlands. None of the pictures attributed to him is dated, but a *Crucifixion* in the hospital church at Cues on the Moselle, generally accepted as his work, is probably from 1465. The painter of these pictures was at one time known as the Master of the Lyversberg Passion from a *Passion* series at Cologne (Wallraf-Richartz-Museum), but the latter are now generally thought to be by a different hand, although the relationship between the two groups of paintings is admittedly close.

Master of the Playing-Cards (mid-15th cent.). German engraver, named after a set of playing-cards depicting human figures, animals, flowers, etc., now divided between the Kupferstichkabinett in Dresden and the Bibliothèque Nationale in Paris. They have been dated as early as the 1430s and he was one of the first distinct artistic personalities in the history of engraving. More than a hundred prints have been attributed to him, his style being characterized by closely observed naturalistic detail and the use of short, densely packed parallel strokes to create shading.

Master of the St Francis Cycle. A name for the unidentified painter of the famous cycle of frescos on the nave walls of the Upper Church of S. Francesco in Assisi, depicting the life of St Francis and probably dating from the 1290s, although some critics put them as late as the 1330s. This cycle was praised by *Vasari as one of the principal works of *Giotto and figures as such in many histories of art, but many scholars nowadays challenge this assertion, feeling that Giotto's undoubted works in Padua differ so thoroughly from those of Assisi in both sentiment and formal organization that it is hard to imagine that he should have painted both. The last three scenes differ in style from the rest and have been attributed to the *Master of St Cecilia.

Master of the Třeboň Altarpiece (or **of Wittingau**). Painter of the *Bohemian School, active in the late 14th cent., named after his main work, three panels (*c.*1380–90) from an altarpiece originally in the monastery at Třeboň (Wittingau) in the Czech Republic and now in the National Gallery, Prague. With him the art at the short-lived imperial court of Prague reached its climax. His style points west rather than south and shows particular affinities with Burgundian art and Melchior *Broederlam. He combines a feeling for linear rhythm with a strong sense of colour and an arrangement of his figures suggesting space and depth.

Master of the Unicorn. See DUVET.

Master of the View of St Gudule (active *c.*1470–*c.*1500). Netherlandish painter who is named from a panel on the life of St Géry in the Louvre which has in the background a view of the façade of St Gudule in Brussels. Other Brussels buildings figure in pictures attributed to this artist. In the background of his *Portrait of a Man Holding a Heart-shaped Book* (NG, London), for example, there is a view of Notre-Dame-du-Sablon. His style was influenced by Rogier van der *Weyden.

Master of the Virgo Inter Virgines (active *c.*1470–*c.*1500). Netherlandish painter, named after a picture representing the Virgin Mary surrounded by the virgin saints Barbara, Catherine, Cecilia, and Ursula (Rijksmuseum, Amsterdam). He worked in Delft, and his style is reflected in the woodcut illustrations (which he presumably designed) to several books published there between 1482 and 1498. About twenty paintings have been attributed to him, revealing a highly distinctive and distinguished artist who obtained his highly emotional effects through intense colours, desolate landscapes, and gaunt figures. His work is sometimes awkward (especially in draughtsmanship), but always sincere and involving. Two of his finest paintings are *The Crucifixion* (Bowes Museum, Barnard Castle) and *The Entombment* (Walker Art Gal., Liverpool).

Master of Vyšší Brod (or of **Hohenfurth**). Painter of the *Bohemian School, so called after his main work, a large altarpiece with scenes from the life of Christ (NG, Prague, *c.*1350) painted for the monastery of Vyšší Brod (Hohenfurth). These panels show the beginnings of the Bohemian variant of *International Gothic. Another important painting from his workshop, a *Death of the Virgin*, is in the Museum of Fine Arts, Boston.

Master of Wittingau. See MASTER OF THE TŘEBOŇ ALTARPIECE.

masterpiece. A term now loosely applied to the finest work by a particular artist or to any work of art of acknowledged greatness or of pre-eminence in its field. Originally it meant the piece of work by which a craftsman, having finished his training, gained the rank of 'master' in his guild.

Master Theoderic (active mid-14th cent.). One of the most important painters of the *Bohemian School. He was first head of the painters' guild in Prague, founded in 1348, and was painter to the emperor Charles IV between 1359 and 1367. For Charles he decorated the Holy Cross Chapel of Karlstein Castle near Prague with more than a hundred panels of saints, prophets, and angels, most of them still *in situ*. He shows even more clearly than the *Master of Vyšší Brod the evolution of a distinct Bohemian School from Italian and French antecedents.

Master Wilhelm. See COLOGNE SCHOOL.

mastic. A soft *resin from the mastic tree (*Pistacia lentiscus*), native of the shores of the Mediterranean, much used in the manufacture of *varnishes.

Mateo de Compostela (active late 12th cent.). Spanish *Romanesque sculptor and architect responsible for the Portico de la Gloria, signed and dated 1188, at the cathedral of Santiago de Compostela. A huge composition featuring a triple doorway, this is the greatest work of Spanish sculpture of this period.

Mathieu, Georges (1921–). French painter. In the 1950s he gained an international reputation as one of the leading exponents of expressive abstraction. This was partly because of a flair for publicity that has led to him being described as 'the Salvador *Dalí of *Art Informel'. He works rapidly, often on a large scale with sweeping impulsive gestures, sometimes squeezing paint straight from the tube on to the canvas. He has even painted, dressed in armour, in front of an audience. He has written several books expanding the theories behind his paintings.

Matisse, Henri (1869–1954). French painter, sculptor, graphic artist, and designer. From about 1920 he enjoyed an international reputation alongside *Picasso as the foremost painter of his time, and for sensitivity of line and beauty of colouring he was unrivalled among his contemporaries. In 1891 he abandoned a legal career for painting. Initially he produced still lifes and landscapes in a sober range of colour, but in the summer of 1896, painting in Brittany, he began to adopt the lighter palette of the *Impressionists. In 1899 he began to experiment with the *Neo-Impressionist technique, which he still used five years later in one of his first major works—the celebrated *Luxe, calme et volupté* (Musée d'Orsay, Paris, 1904–5), exhibited at the *Salon des Indépendants in 1905 and bought by *Signac. During the same years he had been painting with *Marquet, had met *Derain and through him *Vlaminck, and in 1905, together with these and other friends from student days, he took part in the sensational exhibition at the *Salon d'Automne that give birth to the name *Fauves. In the same year (Matisse's *annus mirabilis*) he acquired his first important patrons—the expatriate Americans Gertrude, Leo, and Michael *Stein—and they were soon followed by others. Previously he had struggled to earn a living, but he was now freed from financial worries and could afford to travel. His growing reputation also attracted many pupils to the art school he ran in Paris from 1907 to 1911.

Matisse had met Picasso as early as 1906, and like him was excited by African sculpture at this time. Although he never allied himself with the

*Cubists, he was influenced by their work in the second decade of the century, when he painted some of his most austere and formal pictures (*Bathers by a River*, Art Institute of Chicago, 1916–17). In the 1920s, however, he returned to the luminous serenity that characterized his work for the rest of his long career. From 1916 he spent most of his winters on the Riviera, mainly at Nice and also at Vence. The luxuriously sensual works he painted there—odalisques, still lifes of tropical fruits and flowers, and glowing interiors—are irradiated with the strong sun and rich colours of the south. Following two major operations for duodenal cancer in 1941, Matisse was confined to bed or a wheelchair, but he worked until the end of his life and one of his greatest and most original works was created in 1949–51, when he was in his eighties. This is the Chapel of the Rosary at Vence, a gift of thanksgiving for a woman who had nursed him after his operations then became a nun at this Dominican convent. Matisse designed every detail, including the priests' vestments. The stained-glass windows show his familiar love of colour, but the walls feature murals of pure white ceramic tiles decorated with black line drawings of inspired simplicity. Matisse was not a believer, but he created here one of the most moving religious buildings of the 20th cent. and expressed 'the nearly religious feeling I have for life'.

In his bedridden final years Matisse also embarked on another kind of highly original work, using brightly coloured cut-out paper shapes (*gouaches découpées*) arranged into purely abstract patterns (*L'Escargot*, Tate Gallery, London, 1953). 'The paper cut-out', he said, 'allows me to draw in the colour. It is a simplification for me. Instead of drawing the outline and putting the colour inside it—the one modifying the other—I draw straight into the colour.' The colours he used in his cut-outs were often so strong that his doctor advised him to wear dark glasses. They must rank among the most joyous works ever created by an artist in old age. Unlike many of his great contemporaries, Matisse did not attempt to express in his work the troubled times through which he lived. 'What I dream of', he wrote, 'is an art of balance, of purity and serenity devoid of troubling or disturbing subject-matter . . . like a comforting influence, a mental balm—something like a good armchair in which one rests from physical fatigue.'

Matisse made sculptures at intervals throughout his career, the best known probably being the four bronzes called *The Black I–IV* (casts in the Tate and elsewhere, 1909–*c.*1929), in which he progressively removed all detail, paring the figure down to massively simple forms. He also designed sets and costumes for *Diaghilev and was a brilliant book illustrator. His work is represented in most important collections of modern art, the finest holdings being at the *Barnes Foundation in Merion, Pennsylvania, the Hermitage, St Petersburg, and the Pushkin Museum, Moscow. There are also Matisse museums in Le Cateau (his birthplace) and Nice.

Matsch, Franz. See KLIMT.

Matsys, Quentin. See MASSYS.

Matta (Roberto Sebastian Antonio Matta Echaurren) (1911–). Chilean painter and sculptor who has worked mainly in Paris, but also in Italy and the USA. He trained as an architect but turned to painting in 1937 and in the same year joined the *Surrealist movement. In 1939 he fled from Europe to New York, where with other emigrés including *Breton, *Ernst, *Masson, and *Tanguy he formed a strong and influential Surrealist presence. He encouraged *Gorky, *Pollock, and others to experiment with *automatic techniques. From about 1944 Matta began to create his most characteristic works—large canvases bordering on abstraction that evoke fantastic subjective landscapes and take as their theme the precariousness of human existence in a world dominated by machines and hidden forces. In 1948 he broke with the Surrealists and returned to Europe, but his work continued in a similar vein. He lived in Rome in the early 1950s, then mainly in Paris, although he travelled widely. In 1957 he began making sculpture.

Matteo de' Pasti. See MALATESTA.

Matteo di Giovanni (active 1452–d. 1495). Sienese painter. He originally came from Borgo San Sepolcro and he painted the wings and *predella (Pinacoteca, Sansepolcro) of the altarpiece of which *Piero della Francesca's *Baptism of Christ* (NG, London) was the centre panel. His style was elegant, linear, and decorative, revealing affinities with *Pollaiuolo, and he seems to have been one of the most popular and prolific Sienese painters of the second half of the 15th cent. His major works include a large *Assumption of the Virgin* in the National Gallery, London.

Maulbertsch (or **Maulpertsch**), **Franz Anton** (1724–96). The outstanding Austrian decorative painter of the 18th cent. He was active (and extremely productive) over a wide area of Central Europe and most of his works (altarpieces as well as *frescos) are still in the churches in Austria, the Czech Republic, Hungary, and Slovakia for which they were painted. Maulbertsch's vivacious, colourful, and emotional style was completely resistant to *Neoclassical influences, representing the last glorious flowering of the *Baroque and *Rococo tradition. His painterly dash is even more apparent in his oil sketches, which are well represented in the Barock-museum, Vienna, and he was also an outstanding etcher.

maulstick (**mahlstick** or **rest-stick**). A stick with a padded knob at one end, used by painters to support and steady the brush hand, particularly when working on detailed passages. It is first recorded in the 16th cent., after the introduction of oil painting, and often appears in artists' self-portraits.

Mauritshuis. The royal picture gallery in The Hague, opened to the public in 1820. The building, built by Jacob van *Campen 1633–5 as a palace for Prince John Maurice of Nassau, is one of the masterpieces of Dutch architecture. It houses one of the world's choicest collections of 17th-cent. Dutch painting (with pictures also from the 15th, 16th, and 18th cents.), including such celebrated works as *Rembrandt's *Anatomy Lesson of Dr Tulp* and *Vermeer's *View of Delft*.

Mauve, Antoine (1838–88). Dutch painter, one of the leading artists of the *Hague School. Influenced by the French painters *Millet and *Corot, he concentrated on small pictures of unpretentious subjects—dunes, meadows, beaches—painted in light, silvery tones. His sincere and modest spirit made a deep impression upon van *Gogh, who was his wife's cousin and spent some time working with him. Mauve was a prolific and popular artist and is represented in many museums in the Netherlands and elsewhere.

May, Phil (1864–1903). English draughtsman who became famous with humorous pen-and-ink drawings, mainly *caricatures and scenes of Cockney life, for *The Stephen's Review*, *The Daily Graphic*, *Punch*, etc. His work was also published in albums such as *Phil May's Gutter-snipes* (1896).

Mayno (or **Maino**), **Juan Bautista** (1578–1641). Spanish painter. There is a tradition that he was a pupil of El *Greco in Toledo, but there is no suggestion of this in Mayno's clear and firm style, which was formed in Italy *c.*1600–10. In about 1620 he moved from Toledo to Madrid, where he worked for Philip III (1578–1621) and Philip IV (1605–65) (whose drawing master he had been) and was a friend of *Velázquez. Mayno was a Dominican priest as well as an artist, but although he painted religious works, he is most highly regarded for his portraits, outstanding among which is the formidably characterized *Dominican Monk* (Ashmolean, Oxford, *c.*1635).

Mazerolles, Philippe de (d. 1479). Netherlandish *miniaturist, active in Paris and in Bruges, where he worked for Charles the Bold (1433–77). There is little secure knowledge of his work, and the attribution to him of one of the most famous French medieval paintings, the *Paris Parlement Altarpiece* (Louvre, Paris, *c.*1452) is highly speculative.

Mazo, Juan Bautista Martínez del (*c.*1612/15–67). Spanish painter. He was a pupil of *Velázquez, married his daughter in 1633, and succeeded him as court painter in 1661. Among his very few signed works is a portrait of Queen Mariana (NG, London, 1666), and many of the works attributed to him were formerly given to Velázquez, whose mature style he imitated with great assurance.

Mazzoni, Sebastiano (*c.*1611–78). Italian painter, poet, and architect, born in Florence and active mainly in Venice, where he settled in 1648. He was one of the most individualistic of Italian *Baroque painters, often choosing unusual subjects and expressing a vivid sense of movement with his brilliantly free brushwork. His work looks forward to 18th-cent. Venetian painting and he may have taught Sebastiano *Ricci.

Meadows, Bernard (1915–). British sculptor, studio assistant to Henry *Moore, 1936–40. He has worked mainly in bronze, and his sculptures, though abstract, often suggest organic forms, particularly fruits; indeed during the 1960s he sometimes used actual fruits in his casting.

Meckenem, Israhel van the Younger (*c.*1450–1503). German engraver, the son of an engraver of the same name, active *c.*1450–65. He was trained by his father and probably by *Master E. S., whose work he copied. His *œuvre* is bigger than that of any other 15th-cent. engraver; he is known to have made more than 600 plates, and in some instances over a hundred prints have been preserved from each plate. Like many early engravers, he also worked as a goldsmith. Although he was a minor figure as a creative artist (much of his work consisted of copies) he is important in showing the growing popularity of engraving. He was the first artist to engrave his own features (in a double portrait together with his wife) and looks a very shrewd individual.

Medici. Italian family of bankers and merchants which was the ruling house of Florence and later Tuscany for most of the period from 1434 to 1737 and was famous for its patronage of learning and the arts throughout the *Renaissance. The name appears in Florentine records as early as the 12th cent., but the founder of the family fortune was **Giovanni di Bicci de' Medici** (1360–1429), who commissioned *Brunelleschi to build the Old Sacristy of San Lorenzo. His son **Cosimo** (1389–1464) spent lavishly on religious foundations and had the family palace designed by *Michelozzo (1444–52; now known as the Palazzo Medici-Riccardi). He also commissioned important works from *Donatello. His son **Piero** (1416–69) commissioned the sumptuous tabernacles of SS. Annunziata and San Miniato (Florence) and Benozzo *Gozzoli's frescos in the family palace. He was in contact with many of the leading painters of the day, including *Domenico Veneziano and Filippo *Lippi, and commissioned many *illuminated manuscripts and tapestries. His son and successor, **Lorenzo the Magnificent** (1449–92), was a humanist poet and the most famous member of the family, but his interest in art lay mainly in the collecting of classical gems and coins. We have no record of any major painting he commissioned, but he was the first patron of *Michelangelo, whom he set to copy *antique sculpture in his garden near San Marco. His second cousin, **Lorenzo di Pierfrancesco** (1463–1503), seems to have been the principal patron of *Botticelli. Lorenzo the Magnificent's son **Giovanni** (1475–1521) and his nephew **Giulio** (1478–1534) became pope as Leo X and Clement VII respectively, the artists they patronized including Michelangelo, *Raphael, *Giulio Romano, and *Sebastiano del Piombo. A member of another line of the family, **Cosimo I** (1519–74), became the first Grand Duke of Tuscany in 1569. While he failed to persuade Michelangelo to re-enter his native city that had lost its freedom, he gathered around him the leading artists of the late Renaissance in Florence, notably *Pontormo, *Bronzino, *Ammanati, *Cellini, *Giambologna, and *Vasari, and laid the basis for the *Uffizi collection. Of later members of the dynasty, **Leopoldo** (1617–75) made perhaps the most distinctive contribution to the arts, founding the great collection of artists' self-portraits in the Uffizi. **Maria de' Medici** (1573–1642, known in France as Marie de Médicis) was daughter of Francesco, Grand Duke of Tuscany, and queen of France (1600–10) by marriage to Henry IV. *Rubens painted a great cycle of paintings on her life (Louvre, Paris).

Medici Venus. Marble statue of naked Venus (Uffizi, Florence), first recorded for certain in 1638 in the Villa Medici in Rome. It is signed by 'Cleomenes son of Apollodorus', but in the 18th cent. its fame as a model of female beauty was so great that the signature's authenticity was doubted and the statue was attributed to such illustrious names as *Phidias and *Praxiteles (to whose **Aphrodite of Cnidus* it bears some resemblance in pose). Several other statues of similar type exist, but in spite of the *Medici Venus*'s quite severe restorations, it far outdid its rivals in fame, and it was one of the greatest prizes that Napoleon caused to be brought to France when Italy was under his dominion (it was in Paris between 1803 and 1815). As late as 1840 it was described by *Ruskin as 'one of the purest and most elevated incarnations of woman conceivable', but its reputation has since crumbled, Martin Robertson (*A History of Greek Art*, 1975) describing it as being 'among the most charmless remnants of antiquity'. It is now considered to be a copy of *c.*100 BC deriving from an original of the time of Praxiteles.

medium. Term used in its broadest sense to describe the various methods and materials of the artist; thus painting, sculpture, and drawing are three different media, and bronze and marble are two of the media of sculpture. In a more restricted sense the term refers to the substance with which *pigment is mixed to make paint; for example, water in *watercolour, egg yolk in *tempera, *linseed oil (most usually) in *oil painting.

Meegeren, Han van. See BREDIUS.

Meer, Jan van der. See VERMEER.

megillah (Hebrew: 'scroll'). A Hebrew scroll containing sacred writings appointed to be read in the synagogue on certain festivals. The term is applied particularly to scrolls of the Book of Esther, one of the few Jewish works which has a consistent tradition of *illumination in modern times, dating back at least to the 16th cent.

megilp. A painting *medium consisting of *mastic varnish mixed with *linseed oil. It makes paint glossy and easy to work, but it is a dangerous aid, in time rendering the paint yellow and brittle.

Meissonier, Ernest (1815–91). French painter, etcher, lithographer, and sculptor. He was immensely successful with his trite and nigglingly detailed historical paintings and historical *genre pieces (particularly scenes from the Napoleonic campaigns) and from the 1840s received the highest official honours, including the Grand Cross of the Legion of Honour—he was the first painter to win this award. Astonishingly conceited as well as mean-spirited, he cultivated a huge white beard and liked to be photographed or painted in attitudes of fiercely profound thought, as in his self-portrait of 1889 in the Musée d'Orsay. He had a personal enmity for *Courbet and may have been instrumental in inducing the government to impose a fine on him after the suppression of the Commune. Meissonier did his best work when he was at his least pretentious. His landscapes are attractive descriptive exercises and his *Rue de la Martellerie* (Musée d'Orsay, Paris, 1848), which shows a corpse-strewn street during the revolutionary events of 1848, has genuine pathos and impressed *Delacroix. There are large collections of Meissonier's work in the Musée d'Orsay and in the Wallace Collection, London.

Meissonnier, Juste-Aurèle (1695–1750). French goldsmith, designer, and architect, appointed Dessinateur de la Chambre du Cabinet du Roi in 1726. He had great fertility of invention and was perhaps the most important originator of the early *Rococo forms which spread through Europe after becoming the vogue in Paris. His charming and fanciful designs were widely known in engravings and were used in all the minor arts.

Meit, Conrad (*c*.1475–1550/1). German sculptor. He was born at Worms and in 1506–10 worked at the court of Frederick the Wise, elector of Saxony, at Wittenberg, collaborating with *Cranach. Soon afterwards he moved to the Netherlands, where he worked as court sculptor to the Habsburg rulers for most of his career. His work included a number of large monuments for the family of Margaret of Austria in Notre-Dame-de-Brou, Bourg, but his most characteristic sculptures are small figures and portraits, most notably sensuous free-standing nudes such as the well-known *Judith* (Bayerisches Nationalmuseum, Munich, *c*.1510–15). In these he created a distinctive type of German *Renaissance sculpture, combining Italian idealism with northern particularity of detail.

Meldolla, Andrea. See SCHIAVONE.

Meléndez, Luis (1716–80). Spanish still-life painter, active mainly in Madrid. He is regarded as the finest Spanish painter in his speciality in the 18th cent. and is sometimes called 'the Spanish *Chardin', but he spent much of his life in poverty. His work is rare outside Spain, but there is an example in the National Gallery in London and the Louvre, Paris, has a striking self-portrait.

Mellan, Claude (1598–1688). French engraver. One of the most renowned and successful engravers of his period, he is best known for his portraits, but also made prints after *Poussin and *Vouet, for example. He was a technical virtuoso and instead of using cross-*hatching he obtained his effects of light and shade by varying the thickness of his lines. The most famous example of his prowess is his *Sudarium of St Veronica* (1642), a portrait of Christ made up of a single continuous spiral line that goes round from the centre to the edge like the groove on a gramophone record. Mellan also painted, but few of his pictures survive.

Mellon, Andrew W. (1855–1937). American businessman, public official, art collector, and philanthropist. A banker and steel, coke, and oil magnate, he became secretary to the US treasury (1921–32), ambassador to England (1932–3), and one of the richest men in the world. In 1937 he donated his collection (particularly rich in Dutch and English painting) to the nation, together with funds to build a gallery to house them, thus creating the National Gallery of Art in Washington. His son, **Paul Mellon** (1907–), is also one of the greatest collectors of his generation, his main field of interest being British art. In 1966 he founded the Yale Center for British Art at New Haven. Opened in 1977, the Center is not only a major gallery, but also a research institution, with important collections of books and photographs. The Paul Mellon Foundation for British Art was established in London in 1962. It is a charitable trust whose purpose is to advance the study of British painting, sculpture, and the graphic arts, mainly by sponsorship of publications and support of research.

Melozzo da Forlì (1438–94). Italian painter from Forlì in the Romagna, active mainly in Loreto, Rome, and Urbino. He was an attractive and idiosyncratic painter who achieved a high reputation in his time, but little of his work survives intact and he has been a neglected figure until fairly recently. His style was indebted to *Piero della Francesca and he was renowned for his skill in *perspective and *illusionism; he was, indeed, credited with being the inventor of the extreme form of foreshortening known as **sotto in su*, of which *Mantegna was another great exponent. Melozzo's skill in this field is seen in his fresco of the *Ascension* (1478–80) for the dome of SS. Apostoli in Rome, fragments of which are in the Quirinal Palace and the Vatican.

Memlinc (or **Memling**), **Hans** (*c*.1430/40–94). Netherlandish painter, active in Bruges from 1465. He was born in Seligenstadt, near Frankfurt, but nothing of his German heritage survives in his paintings, which show close connections with Rogier van der *Weyden, by whom according to tradition he was taught. Memlinc was a conservative artist, but his softened and sweetened version of Rogier's style (there is some influence also from *Bouts) made him the most popular Netherlandish painter of his day. Whereas Rogier excelled in the depiction of intense emotion, Memlinc's impeccably crafted paintings are quiet, restrained, and pious. Tax records indicate Memlinc was one of Bruges's wealthiest citizens and his large output shows he must have had a busy workshop. His style changed very little and it is difficult to place undated paintings in a chronological scheme. He painted numerous portraits and showed rather more originality in this field than in religious painting. Among his patrons were Italians then living in Bruges (*Tommaso Portinari and his wife*, Metropolitan Museum, New York, *c.*1468), and his portraits seem to have influenced artists such as *Bellini in northern Italy. Memlinc's work is best seen in the museum devoted to him at Bruges.

Memmi, Lippo (active 1317–47). Sienese painter, *Simone Martini's brother-in-law and most able follower. They jointly signed the celebrated *Annunciation* (Uffizi, Florence, 1333) and their respective shares in it are uncertain. Several other works are signed by Memmi, including *Madonnas* in the Staatliche Museen, Berlin, and the Church of the Servites at Siena, showing his refined draughtsmanship, delicate palette, and extremely sensitive modelling. He was not an innovator, but an indication of the high quality of his work is that several paintings are disputed between him and Simone. Like Simone, Memmi worked at the papal court at Avignon, where in 1347 he painted a Madonna for the Franciscan church.

Mena, Pedro de (1628–88). Spanish sculptor. He was born in Granada, where he was an assistant of *Cano (1658–62), but lived mainly in Malaga, running an extremely busy workshop that sold sculptures throughout Spain. His most characteristic works are *polychromed wooden figures of saints—pious, dignified, and graceful, if sometimes rather vapid. He also carved forty *relief panels for the choir stalls of Malaga Cathedral in 1658–62.

Mengs, Anton Raffael (1728–79). German painter, the son of a court painter of Dresden, **Ismael Mengs** (d. 1764). His father brought him up with harsh severity to be a great painter on the models particularly of *Correggio and *Raphael (from which artists he gained his Christian names), and of the *antique. In 1741 he was taken to Rome and there established the reputation of a youthful prodigy. He returned to Germany in 1744 and became painter to the Saxon Court in Dresden in 1745. After another visit to Rome in 1748–9, when he married an Italian girl, he settled there in 1752 and became a close friend of *Winckelmann, who provided much of the theoretical inspiration for his work. It was for Winckelmann's patron, Cardinal *Albani, that Mengs painted his most famous work, the ceiling fresco *Parnassus* (1761) in the Villa Albani, Rome. This now seems flimsy and simpering, but it was the basis of Mengs's enormous reputation as the leader of the *Neoclassical reform in painting (he was widely regarded as the greatest living painter). It breaks completely with *Baroque *illusionism, treating the scene exactly as if it were to be seen at normal eye level, and is full of derivations from the most approved masters of the *Renaissance. In 1761–9 and 1773–7 he worked as court painter in Spain, decorating the royal palaces at Madrid and Aranjuez. His frescos there are dull and sterile; but it is a sign of the move in taste towards Neoclassicism that he prevailed over his rival, Giambattista *Tiepolo. Mengs was influential through his writings on art (which appeared in or soon after his lifetime in English, French, German, Italian, and Spanish) as well as his paintings. His most important literary work was *Considerations on Beauty and Taste in Painting* (1762), dedicated to Winckelmann. Today his portraits are considered vastly more successful than his history paintings, and he was *Batoni's main rival as the leading portraitist in Rome.

menorah (Hebrew: 'candlestick'). A seven-branched candelabrum that was part of the furnishings of the Temple in Jerusalem and has through the ages been used as a symbol of Judaism.

Menpes, Mortimer. See WHISTLER.

Menzel, Adolf von (1815–1905). German painter and engraver, active mainly in Berlin, where in 1832 he took over his dead father's lithographic business. He was extremely industrious and soon achieved fame with 400 illustrations (wood engravings from his lively drawings) for Franz Kugler's *History of Frederick the Great* (1840–2). In painting he worked on similar themes and with comparable success, creating the popular image of the founder of the Prussian state. From the 1860s he turned to subjects from modern life and was one of the first German painters to note the picturesque qualities of industry (*The Steel Mill*, Staatliche Museen, Berlin, 1875). Today Menzel is most highly regarded not for the works that brought him contemporary acclaim, but for a series of informal landscapes and interiors dating from the 1840s that remained virtually unknown in his lifetime. They are remarkably free and fresh in technique, unorthodox in composition, and both bold and refined in their treatment of light, presaging the developments of *Impressionism (*The Artist's Sister with a Candle*, Neue Pinakothek,

Munich, 1847). His attitude towards these paintings was strange; he kept them hidden and referred to Impressionism as 'the art of laziness', and when he visited Paris in 1855, 1867, and 1868 the artist he most admired was the tiresomely meticulous *Meissonier. However, even in his historical *genre scenes, which are sometimes fairly close to Meissonnier in subject, Menzel's technique was much more lively and summary.

Merian, Matthäus (1593–1650). German engraver and publisher. He brought out hundreds of topographical prints of European towns which are of greater historical than artistic interest. Much of this vast output came from the hands of assistants. Among his pupils was Wenzel *Hollar. His daughter **Maria Sibylla Merian** (1647–1717) settled in Holland and visited Surinam in South America from 1699 to 1702. She is best known for her coloured drawings of insects and butterflies, which are as remarkable for their scientific accuracy as for their delicate beauty.

Mérida, Carlos (1891–1984). Guatemalan painter, active mainly in Mexico. In 1910–14 he studied in Paris under van *Dongen, meeting *Modigliani, *Picasso, and other members of the avant-garde. He returned to Guatemala in 1914 and in 1919 moved to Mexico, where he worked as *Rivera's chief assistant for several years. In 1927–9 he was again in Europe, where he became friends with *Klee and *Miró, then returned to Mexico. His early work was in a politically conscious figurative style, but in the 1930s he was influenced by *Surrealism (he took part in the international Surrealist exhibition in Mexico City in 1940), and he eventually developed a completely abstract manner, and he often worked in mosaic as well as in fresco.

Mesdag, Hendrik Willem (1831–1915). Dutch painter and collector. He abandoned the family profession of banking in 1866 and became one of the leading artists of the *Hague School, particularly noted for his beach and sea scenes. His best-known work is the vast panorama (1881) of the fishing village of Scheveningen—about 120 m. in circumference—housed in a specially designed building in The Hague. The Mesdag Museum, in the same city, contains his excellent collection of paintings, rich in works by members of the *Barbizon and Hague Schools of painting.

Mesens, E. L. T. (1903–71). Belgian musician, poet, collagist, exhibition organizer, and dealer. His interest in the visual arts developed under the influence of *Duchamp and *Picabia, whom he met in Paris in 1921, and he was influenced towards *Surrealism by the paintings of de *Chirico. He became a friend and champion of *Magritte and a highly active figure in the Surrealist movement, although more as an organizer than an artist. In 1938 he settled in London and became director of the London Gallery in Cork Street, the headquarters of Surrealism in England, organizing exhibitions of the work of many European artists there (including *Ernst, *Schwitters, and *Tanguy); he also edited the gallery's publication, the *London Bulletin*, an important documentary source for the period (it ran for 20 issues, 1938–40). In his own work as an artist, Mesens was best known for his collages, which he created from an assortment of materials—tickets, ribbons, pieces of paper and print, etc. He made extensive use of printed words to create disconcerting or amusing ambiguities and suggested meanings, some of which might almost be regarded as anticipations of *Conceptual art.

Mestizo style. Term sometimes applied to a decorative style of architectural carving which flourished in the central Andes *c.*1650–1750. It implies a combination of traditional indigenous features with Christian elements, but as the word 'mestizo' (Spanish for half-caste) carries disparaging connotations many writers prefer to speak of 'provincial highland' style. The style is characterized by prolix *relief carving on two levels with edges deeply undercut so that in sunlight they are outlined with heavy shadows. The symbolically abstract designs often echo traditional textile patterns and the carving extends in continuous carpet-like areas over façades, round frames, and over columns, vaults, and cupolas.

Meštrović, Ivan (1883–1962). Yugoslavian-born sculptor who became an American citizen in 1954. He studied sculpture at the Academy in Vienna, 1900–4, and in 1907–8 lived in Paris, where he met *Rodin. On returning to Yugoslavia he began to make his name as a monumental sculptor, working in a variety of classicist styles furbished with a superficial air of modernity. He passed the First World War in Rome, Geneva, Paris, and London, where a large exhibition of his works was held at the Victoria and Albert Museum. In 1919 he returned to Yugoslavia, where he received many public commissions, through which he expressed his ardent patriotism, and built up his fame internationally as a monumental sculptor, his works including an enormous mausoleum outside Belgrade in commemoration of the Unknown Soldier (1934). During the Second World War he obtained several commissions from the Vatican and after living in Switzerland from 1943 to 1946 he went to the USA.

There he obtained the post of Professor of Sculpture at Syracuse University and from 1955 was Professor of Sculpture at the University of Notre Dame, Indiana. He executed a number of monuments in the USA. The rhetoric of his large-scale works now seems rather ponderous and his smaller, more lyrical pieces have dated less. There are Meštrović museums at Split (his former house, which he designed himself) and Zagreb, and there are several examples of his work in the Tate Gallery.

metal cut. A print made from a metal plate in which the design is cut in *relief (as in a *woodcut) rather than incised into the plate as in a *line engraving. Prints done in the **manière criblée* are examples of the type.

metal point. Method of drawing using a small metal rod, pointed at one end, on specially prepared paper. The metal may be copper, gold, lead, or (most commonly) silver, which gives an attractive fine grey line that oxidizes to a light brown. The strength of tone can hardly be varied at all, so the technique depends on the quality of the drawn line and is best suited to work on a small scale. Moreover, it demands great certainty of purpose and hand, for the line cannot be removed except by disturbing the *ground, a coating of opaque white, often tinted by the addition of another pigment. Silver point first appeared in medieval Italy and was particularly popular in the 15th cent.; *Dürer and *Leonardo were perhaps the greatest exponents of the medium. Depictions of the instrument sometimes occur in pictures of *St Luke Drawing the Virgin*, notably that by Rogier van der *Weyden (Museum of Fine Arts, Boston). It went out of fashion in the 17th cent., probably because the graphite *pencil was coming in, but was revived in the 18th cent. by *miniature painters, especially in France.

Metaphysical painting. A style of painting invented by de *Chirico in about 1913 and practised by him, *Carrà (from 1917), *Morandi (from 1918), and a few other Italian artists until about 1920. The term (*Pittura Metafisica*) was coined by de Chirico and Carrà in 1917, when both were patients at a military hospital in Ferrara. Metaphysical painting started with no inaugural programme, as had been the case with *Futurism, although attempts were later made to define a 'metaphysical aesthetic' in the periodical *Valori Plastici* (Plastic Values), which ran from 1918 to 1921. The meaning attached to the term 'metaphysical', which occurs in the titles of several pictures by de Chirico particularly, was never precisely formulated, but the style is characterized by images conveying a sense of mystery and hallucination. This was achieved partly by unreal perspectives and lighting, partly by the adoption of a strange *iconography involving, for example, the use of tailor's dummies and statues in place of human figures, and partly by an incongruous juxtaposition of realistically depicted objects in a manner later taken over by some of the *Surrealists. But the dreamlike quality conveyed by Metaphysical painters differed from that of the Surrealists because of their concern with pictorial structure and a strongly architectural sense of repose deriving from Italian *Renaissance art. Metaphysical painting did not long survive the First World War, but had great influence on Surrealism.

Metcalf, Willard L. See TEN, THE

metope. In Classical architecture the square space or block that alternates with the ornamental features called triglyphs in the frieze of the Doric Order. Metopes may be left plain (they were originally left open), but in Greek art are often carved with *relief sculptures. The most famous examples are the metopes from the Parthenon in Athens (see PHIDIAS).

Metropolitan Museum of Art, New York. The most comprehensive collection of art in the USA and one of the greatest in the world. It was founded in 1870 and the present building in Central Park was opened in 1880. The museum is owned by the city, but is supported mainly by private endowment, and the history of its foundation and growth illustrates the rapid rise of New York at the end of the 19th cent. as the financial and cultural capital of North America, and the growing economic supremacy of America over Europe. Between 1880 and 1925, at a time when the major public collections in Europe were engaged in consolidation, relying largely on their purchase grants and other state aid, the Metropolitan Museum was being built up out of the private fortunes of great businessmen, who collected rather for prestige than out of connoisseurship, but collected only first-class works of art. It has also benefited from a number of endowed purchase grants, many of them unconditional, which have enabled it to progress not only as a collection of outstanding works, but as a comprehensive and representative one. The museum is rich in virtually every field of the *fine and *applied arts from all parts of the world and also houses one of the world's largest art libraries. Much of the collection of medieval art is housed in a separate building called the Cloisters in Fort Tryon Park, overlooking the Hudson River.

Opened in 1938, the Cloisters is a medieval-style structure, largely made up of parts of *Romanesque and *Gothic buildings transported from Europe. Many of the works it houses were collected by the American sculptor George Grey Barnard (1863–1938), who lived in France for much of his career.

Metsu, Gabriel (1629–67). Dutch painter, active in his native Leiden, then in Amsterdam, where he had moved by 1657. *Houbraken says he trained with *Dou, but Metsu's early works are very different from his—typically historical and mythological scenes, broadly rather than minutely painted. Metsu also painted portraits and still lifes, but his most characteristic works are *genre scenes, some of which rank among the finest of their period. He concentrated on scenes of genteel middle-class life, fairly close to de *Hooch and *Terborch in style, but with a personal stamp. One of his best-known works, *The Sick Child* (Rijksmuseum, Amsterdam), is often compared with *Vermeer. His work is rarely dated, so his development and relationships with other artists are difficult to trace.

Metsys, Quentin. See MASSYS.

Metzger, Gustav. See AUTO-DESTRUCTIVE ART.

Metzinger, Jean (1883–1956). French painter and writer on art. After passing through periods of interest in the *Neo-Impressionists and the *Fauves he became one of the earliest devotees of *Cubism and a central figure of the *Section d'Or group. He was undistinguished as a painter and is remembered mainly as the co-author with *Gleizes of the book *Du Cubisme* (1912), an important statement of the principles of the movement.

Meulen, Adam Frans van der (1632–90). Flemish painter and tapestry designer. He moved to Paris in 1664, became an assistant to *Lebrun, and was made one of Louis XIV's court painters, specializing in military scenes. His paintings and designs for *Gobelins tapestries are accurate historical documents of the battles which they represent, for he accompanied the king on his campaigns and made careful drawings of the participants, the uniforms, the disposition of troops, and the terrain. His work of this type is well represented at Versailles. He also made much less grandiose pictures of such subjects as hunting parties and landscapes.

Meunier, Constantin (1831–1905). Belgian sculptor and painter, well known for his sincere but rather heavy-handed glorification of the nobility of labour in his treatment of such subjects as miners, factory workers, and stevedores. In the early 20th cent. he had considerable influence on younger sculptors interested in Social *Realist subjects. There is a museum of his work in Brussels and his ambitious Monument of Labour is in the Place de Trooz there.

Meyer, Hannes. See BAUHAUS.

mezzotint. Term applied to a method of engraving in tone and to a print made by that method. A copper plate is first roughened by means of a tool with a serrated edge known as a 'rocker'. The rocker is systematically worked over the surface of the copper in every direction, raising a *burr as it goes. The design is formed by scraping away the burr where the light tones are required and by polishing the metal quite smooth in the highlights. The plate is then filled with ink and wiped with a series of rags. Where the plate is rough the ink is retained and will print an intense black, but where it has been smoothed by the scraper, less ink will be held and a lighter tone will occur. A mezzotint is thus evolved from dark to light and is characterized by soft and hazy gradations of tone and richness in the dark areas. Engraved or *etched lines are sometimes introduced if greater definition is required; this procedure is known as mixed mezzotint.

The process was invented by Ludwig von *Siegen of Utrecht, whose first dated mezzotint was made in 1642. Another pioneer, formerly thought to be the inventor, was Prince *Rupert, nephew of Charles I. The early mezzotinters, however, whose work was largely in the field of portraiture, did not prepare their plates as described above, but worked from light to dark, using *roulettes to roughen limited areas of the copper as required, and combining this procedure with lines and *stipples engraved with the *burin.

In England taste has always tended to favour the tonal processes of engraving, and the high-point of the mezzotint was the 18th cent., when it proved ideally suited to reproducing the work of the great portraitists such as *Reynolds, *Romney, and *Gainsborough. John *Martin is a rare example of an artist who made original creative use of the medium. A drawback that the technique shares with *drypoint is that it will yield only a small number of good impressions, owing to the wearing down of the burr in printing. An attempt was made to overcome this by using steel plates, but in spite of this mezzotint shared the eclipse suffered by all the reproductive copper-plate techniques in the 19th cent. and has become virtually extinct.

Michallon, Achille-Etna (1796–1822). French landscape painter, the first artist to win the *Prix de Rome in the Historical Landscape category that was established in 1817. He was a pupil of *David and *Valenciennes and in his turn taught *Corot, who was influenced by his severe compositions and cool colour harmonies.

Michel, Georges (1763–1843). French landscape painter. Michel was a picture restorer at the *Louvre and was strongly influenced by 17th-cent. Dutch landscape painters. He was one of the earliest to paint in the open air (see PLEIN AIR) and because of this and his intimate, emotional depiction of nature he has sometimes been regarded as a forerunner of the *Barbizon School. His work, which is well represented in the Louvre, went almost unknown until his landscapes were shown posthumously at the Paris International Exhibition of 1889.

Michelangelo Buonarroti (1475–1564). Florentine sculptor, painter, architect, draughtsman, and poet, one of the greatest figures of the *Renaissance and, in his later years, one of the forces that shaped *Mannerism. His father, a member of the gentry, claimed noble lineage and throughout his life Michelangelo was touchy on the subject; pride of birth too had much to do with the family opposition to his apprenticeship as a painter as well as with Michelangelo's own insistence in later life on the status of painting and sculpture among the *liberal arts. Certainly his own career was one of the prime causes of the far-reaching change in public esteem and social rating of the visual arts, for he represented the archetype of the inspired genius—unsociable and totally absorbed in his work. In 1488 he was apprenticed for a term of three years to Domenico *Ghirlandaio and from him he must have learnt the elements of *fresco technique. He could not have learnt very much else, however, since he seems to have transferred very quickly to the school set up in the *Medici gardens and run by *Bertoldo di Giovanni. One of Michelangelo's earliest works, a marble *relief of *The Battle of the Lapiths and Centaurs* (Casa Buonarroti, Florence, *c.*1491–2), was based on a similar work by Bertoldo. More important than either master, however, was what he learned from the drawings he made of figures in the frescos of *Giotto and *Masaccio. After the death of his patron Lorenzo de Medici in 1492 the political situation in Florence deteriorated, and in October 1494 Michelangelo left for Bologna, where he carved three small figures for the tomb of S. Dominic (see NICCOLÒ DELL' ARCA). By June 1496 he was in Rome, where he remained for the next five years and where he carved the two statues that established his fame—the *Bacchus* (Bargello, Florence, *c.*1496–7) and the **Pietà* (St Peter's, Rome, 1498–9). The latter is the masterpiece of his early years—a tragically expressive and yet beautiful and harmonious solution to the problem of representing a full-grown man lying dead in the lap of a woman. There are no marks of suffering—as were common in northern representations of the period—and the carving has a flawless beauty and polish demonstrating his absolute technical mastery. Still in his mid-twenties, Michelangelo returned to Florence in 1501 to consolidate the reputation he had made in Rome. He remained there until the spring of 1505, the major work of the period being the *David* (Accademia, Florence, 1501–4), which has become a symbol of Florence and Florentine art. The nude youth of gigantic size (approx. 4 m. high) expresses in concrete visual form the self-confidence of the new Republic.

Other works of this productive period include the *Madonna* (Notre-Dame, Bruges) and the unfinished *St Matthew* (Accademia), all that was ever done of a commission for twelve statues of Apostles for Florence Cathedral (the contract was signed in April 1503 and annulled in December 1505). He probably also carved the *Taddei *Tondo* (Royal Academy, London) and painted the *Doni Tondo* (Uffizi) about this time. Soon after the *David* was completed in April 1504, Michelangelo received a commission from the Signoria of Florence to paint a huge mural of the *Battle of Cascina* for the new Council Chamber in the Palazzo Vecchio; here he worked in rivalry with *Leonardo, who was engaged on the *Battle of Anghiari* for the same room. Neither painting was completed, but Michelangelo began work on the full-size *cartoon in the winter of 1504, and the fragment known as the *Bathers* was, while it existed, a model for all the young artists in Florence—including *Raphael—and was one of the prime causes of Mannerist preoccupation with the nude figure in violent action. It is now known from a copy (Earl of Leicester Coll.) and an engraving, as well as some preliminary drawings by Michelangelo himself (e.g. in the BM, London).

Michelangelo left the battlepiece unfinished when Pope Julius II summoned him to Rome in 1505 to make him a tomb. Julius died in 1513, but the project dragged on until 1545 and was rightly described by *Condivi as the 'Tragedy of the Tomb'. It was originally conceived on the most grandiose scale, but was whittled down in successive contracts with Julius's heirs, and of the monument finally erected in S. Pietro in Vincoli in 1545 only the celebrated *Moses* (*c.*1515) was from Michelangelo's own hand. (Two figures of *Slaves*, *c.*1513, carved by Michelangelo for the tomb are now in the Louvre.) The other great work commissioned from Michelangelo by Julius—the frescoing of the ceiling of the Sistine Chapel—was equally daunting, but was brought to sublime fruition. The contract was signed on 10 May 1508 and the finished ceiling was unveiled on 31 October 1512. Michelangelo, who always regarded himself as a sculptor first and foremost, was reluctant to undertake the work, but he made of it his most heroic achievement, not only for its quality as a work of art, but also in terms of the endurance and stamina he showed in completing so quickly and virtually unaided such a huge and physically uncomfortable task. There is still much debate about the exact interpretation of the scores of figures that adorn the ceiling, but the main panels represent scenes from Genesis, from the Creation to the Drunkenness of Noah, forming the background to the frescos on the life of Moses and of Christ on the walls below by a number of 15th-cent. artists. Prophets and Sibyls who foretold Christ's birth are at the sides of the ceiling, and at each corner of the central scenes are figures of beautiful nude youths (usually called the *Ignudi*), whose exact significance is uncertain. They have been thought to represent the Neoplatonic ideal of humanity, and as Kenneth *Clark wrote, 'Their physical beauty is an image of divine perfection; their alert and vigorous movements an expression of divine energy.' From the moment of its completion the Ceiling has always been regarded as one of the supreme masterpieces of pictorial art (the recent cleaning has revealed anew the beauty of the colouring), and Michelangelo was, at the age of 37, recognized as the greatest artist of his day, a position he retained unchallenged until his death.

The problem of designing the architectural parts of the Julius Monument led Michelangelo to consider the art of architecture, and in December 1516 he was commissioned by the new pope, Leo X (Giovanni de' Medici), to design a façade for the Medici parish church in Florence, S. Lorenzo, which had been left unfinished by *Brunelleschi. In fact the project came to nothing and wasted a good deal of Michelangelo's time, but it led to two other works for S. Lorenzo—the Medici Chapel, or New Sacristy, planned as a counterpart to Brunelleschi's Old Sacristy, and the Biblioteca Laurenziana. Both were left unfinished, but they nevertheless rank among Michelangelo's finest creations. The Medici Chapel was planned from November 1520 as a mortuary chapel for the family to contain the monuments of four members, but it was abandoned when the Medici were expelled from Florence in 1527, restarted in 1530, and left incomplete in 1534 when Michelangelo finally settled in Rome. It was intended to be a union of architecture and sculpture (like the projected S. Lorenzo façade), with the view from the altar leading to the climax of the whole composition in the figures of the *Madonna and Child* (unfinished) and with the Active and Contemplative Life symbolized by figures on the wall-tombs of Giuliano and Lorenzo de' Medici. The figures of the Medici are set above the reclining figures symbolizing *Day* and *Night* (for *Vita activa*) and *Dawn* and *Evening* (for *Vita contemplativa*). One of the reasons why the Chapel was unfinished is that when the Medici were expelled from Florence in 1527, Michelangelo declared himself a Republican and took an active part in the defence of the city. After the capitulation to the Medici forces in 1530 he was pardoned and set once more to work on the glorification of the Medici until in 1534 he settled in Rome and worked for the papacy for the thirty years that still remained to him. He was at once commissioned to paint the *Last Judgement* in the Sistine Chapel and began the actual painting in 1536. It was unveiled on 31 October 1541, twenty-nine years to the day after the unveiling of the Sistine Ceiling but a whole world away from it in feeling and meaning, with its massive and menacing figures and mood of wrathful desolation. In the interval the world of Michelangelo's youth had collapsed in the horror of the Sack of Rome (1527), and its confident humanism had been found insufficient in the face of the rise of Protestantism and the new, militant spirit of the Counter-Reformation. For Paul III (Alessandro *Farnese), who commissioned the apocalyptic *Last Judgement*, Michelangelo also executed his last works in painting, the *Conversion of St Paul* and the *Crucifixion of St Peter* (1542–50), frescos in the Cappella Paolina in the Vatican. The figures here are even more blunt, heavy, and unconcerned with physical allure, totally repudiating his own early ideals. Something of the same deep and troubled spirituality is seen in his late drawings of the Crucifixion and two sculptures of the Pietà. One (now in Florence Cathedral) was intended for his own tomb and contains a self-portrait as Nicodemus; it was begun *c.*1546 and mutilated and abandoned by Michelangelo in 1555. The other (Castello Sforza, Milan) was his last work, left unfinished at his death.

For the last thirty years of his life, however, Michelangelo devoted most of his attentions to architecture, and in this field his stature is just as great as in sculpture and painting. His most important commission—indeed the most important in Christendom—was the completion of St Peter's, which had been begun under Julius II in 1506. When Michelangelo became architect in 1546, the building had advanced little since *Bramante's death in 1514. As with the Sistine Ceiling, he was initially unwilling to undertake the task, but he then proceeded with formidable energy and by the time of his death work had advanced so far that the drum of the dome was nearly complete. Michelangelo also designed the dome itself, but this was executed after his death and is probably a good deal steeper in outline than he intended. The addition of a long nave in the early 17th cent. altered Michelangelo's plan for a centralized church, but nevertheless the exterior of the building owes more to him than to any other architect and forms a fitting conclusion to his titanic career.

Giacomo del *Duca was the only one of Michelangelo's immediate followers who maintained something of his power and originality of vision in architecture, but his decorative vocabulary soon attained widespread currency. It was only in the 17th cent., however, that his massive and dynamic style was fully appreciated and emulated. It is fitting that *Bernini, the great sculptor-architect of the age, should complete St Peter's with his glorious piazza, and Francesco Borromini, the most brilliant of Baroque architects, was an avowed disciple of Michelangelo. In painting and sculpture Michelangelo's means of expression was limited almost entirely to the heroic male figure, usually nude, but in this field he reigned supreme as no artist has done before or since. The awe that his contemporaries felt for him has not diminished through the centuries and his influence, for good and ill, has been enormous.

Michelozzo di Bartolommeo (sometimes incorrectly called Michelozzo Michelozzi) (1396–1472). Florentine architect and sculptor. As a sculptor he worked for *Ghiberti (on both his sets of doors for the Baptistery of Florence Cathedral) and in partnership with *Donatello (1425–*c.*1433). With Donatello he produced three major tombs—those of anti-pope John XXIII (Baptistery, Florence), Cardinal Brancacci (S. Angelo a Nilo, Naples), and Bartolommeo Aragazzi (Montepulciano Cathedral, but now disassembled; two angels are in the V&A, London). His style was vigorous and forthright. In his later career Michelozzo worked mainly as an architect, and he ranks as one of the leading figures of the generation after *Brunelleschi, whom he succeeded as *capomaestro* at Florence Cathedral (1446). His most famous building is the Palazzo Medici-Riccardi in Florence (begun 1444), often described as the first *Renaissance palace, but his most beautiful work is perhaps the library at San Marco, Florence (*c.*1440)—a light and airy building which became a model for the type throughout Italy. Michelozzo was influential in spreading the Renaissance style; he worked in Milan, Yugoslavia, and the island of Chios.

Middleditch, Edward. See KITCHEN SINK SCHOOL.

Miel, Jan. See LAER.

Miereveld (or **Mierevelt**), **Michiel van** (1567–1641). Dutch portrait painter, active mainly in his native Delft and The Hague. He was portrait painter to the House of Orange, highly successful, and enormously prolific: *Sandrart reports that Miereveld himself estimated that he made about 10,000 portraits. His portraits are dull and repetitive, but they are meticulously crafted and of great value as historical records.

Mieris, Frans van (1635–81). Dutch painter, the most distinguished member of a family of artists who worked in Leiden. He was one of the best pupils of Gerrit *Dou and followed his master in choice of subjects (mainly domestic *genre scenes) and in his highly polished technique. The tradition was continued by his sons **Jan** (1660–90) and **Willem** (1662–1747) and by Willem's son **Frans II** (1689–1763).

Mies van der Rohe, Ludwig (1886–1969). German-born architect and designer who emigrated to the USA in 1938 and became an American citizen in 1944. One of the supreme architects of the 20th cent., Mies was also a highly influential designer. His work is distinguished by classical poise and elegance, masterly refinement of detail, and an insistence on the highest quality of craftsmanship; one of his famous aphorisms is 'I would rather be good than original.' From 1930 to 1933 (when it was forced to close by the Nazis) Mies was director of the *Bauhaus, and in 1938 he became Professor of Architecture at the Armour Institute (now Illinois Institute) of Technology in Chicago. Most of his buildings are in the USA, but one of his last great works was the New National Gallery in Berlin (1962–8). As a designer he was best known for his furniture, notably the 'Barcelona chair' (so-called because it was designed for the German pavilion at the 1929 International Exhibition in Barcelona). This has become acknowledged as a modern classic, and is still produced commercially.

Mignard, Pierre (1612–95). French painter, one of the most successful of *Vouet's pupils. His career culminated in 1690, when, on the death of *Lebrun, he became Director of the Académie and first painter to the king. He was one of the principal supporters on the side of *De Piles and the 'Rubensistes' in their battle against the *classicism of the 'Poussinistes'. His own historical and religious paintings, however, did not exemplify his theories but fitted into the scheme of academic classicism in the tradition of *Domenichino and *Poussin (he was in Italy 1635–57). His best works are his portraits; he painted many of the members of

Louis XIV's court, sometimes fitting out his sitters with allegorical trappings. Pierre's brother, **Nicolas** (1606–68), also trained with Vouet and had a successful career painting portraits and religious subjects.

Migration Period art. Term applied to the art produced by the Teutonic tribes—Visigoths, Ostrogoths, Lombards, Vandals, Franks—who overran the declining Roman Empire from about AD 370 to 800. The art of the migrant tribes was mainly restricted to portable objects, articles of personal use or adornment. They excelled chiefly in goldsmiths' work and jewellery—fibulae, decorated buckles, and crowns, and a particular feature of Migration style was gold with garnet or enamel inlays showing a predilection for isolated units of colour. Little Migration architecture or carving has survived, the most important architectural monument being Theodoric's Tomb in Ravenna, built before his death in AD 526.

In Britain most surviving traces of Roman culture were erased by the Anglo-Saxon invasion after AD 450. The Christian Church retreated to Ireland and the invading Jutes, Angles, and Saxons developed the Migration style in brooches, *cloisonné ware, and other personal ornament.

Manuscript *illumination declined less completely than most other arts during the Migration period and pockets of continuity were provided by the monasteries. In Ireland and later in northern England a style grew up which is known in art history as the *Hiberno-Saxon* or *northern* in constrast to the *Roman* or *southern*, and which had some affinities with Migration ornament style. It is not appropriate, however, to regard this as a true manifestation of Migration Period art.

Millais, Sir John Everett (1829–96). English painter and book illustrator. A child prodigy who was hardworking as well as naturally gifted, he became the youngest ever student at the *Royal Academy Schools when he was 11, and although he suffered some temporary setbacks when he was in his twenties his career was essentially one of the great Victorian success stories. In 1848, with *Rossetti and Holman *Hunt, he founded the *Pre-Raphaelite Brotherhood and had his share of the abuse heaped against the members until *Ruskin stepped in as their champion. (In 1854 Millais married Effie Gray, formerly Ruskin's wife, after her first marriage had been annulled.) In the 1850s Millais's style changed, as he moved away from the brilliantly coloured, minutely detailed Pre-Raphaelite manner to a broader and more fluent way of painting—with a family to support he said he could not afford to spend a whole day working on an area 'no larger than a five shilling piece'. His subjects changed also, from highly serious, morally uplifting themes to scenes that met the public demand for sentiment and a good story (*The Boyhood of Raleigh*, Tate, London, 1870). He became enormously popular, not only with subject pictures such as this, but also as a portraitist and a book illustrator, his drawings for the novels of Anthony Trollope (1815–82) being such a success that Trollope said they influenced the way he developed the characters in sequels. Millais lived in some splendour on his huge income, in 1885 became the first artist to be awarded a baronetcy, and in the year of his death was elected President of the Royal Academy. To some contemporaries it seemed that he wasted his talents pandering to public taste, and many 20th-cent. critics have presented him as a young genius who sacrificed his artistic conscience for money. Millais, an easy-going and much-liked man, certainly enjoyed his success, but he was far from being a cynic. He was always proud of his skills (near the end of his career he wrote, 'I may honestly say that I have never consciously placed an idle touch upon canvas'), and few of his contemporaries could match his late works for sheer beauty of handling (*Bubbles*, A. & F. Pears Ltd., 1886).

millboard. See CARDBOARD.

Milles, Carl (1875–1955). Sweden's greatest sculptor. From 1897 to 1904 he lived in Paris, where he worked for a time as assistant to *Rodin, then moved to Munich (1904–6), where he was influenced by *Hildebrand. In the following two years he lived in Rome, Stockholm, and Austria, then settled at Lidingö, near Stockholm, in 1908. His travels had given him a wide knowledge of ancient, medieval, and Renaissance art, as well as of recent developments, and he forged from these varied influences an eclectic but vigorous style. He is best known for his numerous large-scale fountains, distinguished by rhythmic vitality and inventive figure types (he would fuse classical and Nordic types such as tritons and goblins), and sometimes by a grotesque humour. From 1931 to 1945 he was Professor of Sculpture at the Cranbrook Academy at Bloomfield Hills, Michigan; his work in the USA includes fountains in Chicago, Kansas City, New York, and St Louis. He became an American citizen in 1945 but returned to Sweden in 1951 and died at Lidingö, where his home is now an open-air museum of his work, known as Millesgården.

Millet, Jean-François (often known by his nickname, 'Francisque') (1642–79). French landscape painter of Flemish birth. Active mainly in Paris, he is also said to have visited England and Holland. No signed or documented works are known, but several are authenticated by early engravings. He worked in a style related to *Poussin and *Dughet, sometimes enlivened by romantic touches in the manner of Salvator *Rosa, as in *Mountain Landscape with Lightning* (NG, London), one of the most original works of an artist usually content to be an able follower. His son **Jean** (1666–1723) was a landscape painter, and he too had a painter son, **Joseph** (1697?–1777).

Millet, Jean-François (1814–75). French painter and graphic artist, born of a peasant family at Gruchy, near Cherbourg in Normandy. He studied locally, then in 1837 entered the studio of *Delaroche in Paris. His early pictures consisted of conventional mythological and anecdotal *genre scenes and portraits, but with *The Winnower* (Musée d'Orsay, Paris), exhibited at the *Salon in 1848, he turned to the scenes of rustic life from which his name is now inseparable. He emphasized the serious and even melancholy aspects of country life, emotionalizing the labours of the soil and the sad solemnities of toil. Hostile critics accused him of being a socialist, but Millet's concerns were aesthetic rather than political and he said his desire was 'to make the trivial serve to express the sublime'. In 1849 he settled at Barbizon, where he remained for the rest of his life apart from a stay in Cherbourg during the Franco-Prussian War of 1870–1. Late in his career he turned increasingly to pure landscape, influenced by his close friend Théodore *Rousseau. Millet passed much of his life in poverty, but his work began to bring him success in the 1860s and *The Angelus* (Musée d'Orsay, 1859) became perhaps the most widely reproduced painting of the 19th cent. This has had a harmful effect on his subsequent critical fortunes, for largely on the strength of it he has been pigeon-holed as a purveyor of pious sentimentality. His greatness lies rather in his drawing, which for its elimination of the inessential, investing the ordinary with weight and dignity, has been compared with that of *Seurat. Van *Gogh and Camille *Pissarro were among those who admired Millet's work. He has long been particularly popular in the USA and the Museum of Fine Arts, Boston, has an outstanding collection of his work.

Mills, Clark (1810–83). American sculptor. A jack-of-all-trades, he was self-taught as a sculptor and first made a name for himself in the 1840s with portrait busts based on life masks. In 1848 he won a competition for the monument to President Andrew Jackson (1765–1845) in Lafayette Square, Washington, and worked for five years on this, the first equestrian statue in the USA. He built his own foundry to cast the statue, which daringly has the rearing horse supported only by hind legs, and eventually succeeded at the seventh attempt. The ecstatic response when the statue was unveiled in 1853 brought him great financial rewards and the commission for an equestrian statue to George Washington in Washington Circle (1860). He spent his final years under a cloud, however, suspected of dishonesty in handling the metal assigned to him. His work at its best is crude but vigorous; the George Washington memorial is timid compared with that to Jackson.

Milne, David Brown (1882–1953). Canadian painter, mainly of landscape. A schoolteacher until he turned to painting in 1904, Milne was a retiring character and spent much of his career working in seclusion in the Berkshire Hills of Massachusetts and from 1928 in various Ontario villages. He lived and studied in New York for a period, however (he exhibited at the *Armory Show in 1913), and after enlisting in the Canadian army became an official war artist in Britain, France, and Belgium. His style throughout his career combined delicacy and strength. A very personal and calligraphic quality is seen at all periods. His work after 1937 was entirely in watercolour, a medium which he used in an oriental way as a sensitive means of expressing his emotional response to nature (*Rites of Autumn*, NG, Ottawa, 1943).

miniature. A very small painting, particularly a portrait that can be held in the hand or worn as a piece of jewellery. The word is applied to manuscript *illuminations as well as portraits and derives from the Latin *minium*, the red lead used to emphasize initial letters, decorated by the *miniator*. Since the 17th cent. the term has been applied to all types of manuscript illustration on account of a mistaken etymology: the word was connected with 'minute' (small). What we today call a 'miniature' was called *historia* in the Middle Ages and the portraits painted by *Hilliard and others were named 'limnings' or 'pictures in little' by the Elizabethans. They were painted on vellum (see PARCHMENT), or occasionally on ivory or card, and in the 17th and 18th cents. there was a vogue for miniatures done in an *enamelling technique. The portrait miniature developed from a fusion of the traditions of medieval illumination and the *Renaissance medal and flourished from the early 16th cent. to the mid-19th cent., when photography virtually killed the art. It still survives as a curiosity, however, and there is always a selection of miniatures at the *Royal Academy summer exhibition.

Minimal art. A type of abstract art, particularly sculpture, characterized by extreme simplicity of form and a deliberate lack of expressive content; it emerged as a trend in the 1950s and flourished particularly in the 1960s and 1970s. The roots of Minimal art can be traced to the geometrical abstractions of *Malevich and the *ready-mades of *Duchamp in the second decade of the 20th cent., but as a movement it developed mainly in the USA and its impersonality is seen as a reaction against the emotiveness of *Abstract Expressionism. 'The theory of minimalism is that without the diverting presence of "composition", and by the use of plain, often industrial materials arranged in geometrical or highly simplified configurations we may experience all the more strongly the pure qualities of colour, form, space and materials' (*The Tate Gallery: An Illustrated Companion*, 1979). Minimal art has close links with *Conceptual art (Minimalist sculpture often has a strong element of theoretical demonstration about it, with the artist leaving the fabrication of his designs to industrial specialists), and sometimes there are affinities with other contemporaneous movements, such as *Land art. There is even a kinship with *Pop art in a shared preference for slick, impersonal surfaces (some Minimal artists, however, have used 'natural' materials such as logs rather than machine-finished products). The leading Minimalist sculptors include Carl *Andre, Don *Judd, and Tony *Smith. Minimalist painters (for whom the immediate precedents were *Albers and *Reinhardt) include Frank *Stella (in his early work) and *Hard-Edge abstractionists such as Ellsworth *Kelly and Kenneth *Noland.

Minoan art. A term applied to the art of ancient Crete, particularly that produced from *c.*2500 BC to *c.*1100 BC. The term was first used in 1894 by the British archaeologist Sir Arthur Evans, who, from 1898 to 1935, conducted extensive excavations at Crete, principally at Knossos, where the royal palace he uncovered was named the Palace of Minos (hence Minoan) after the legendary king of Crete. Minoan civilization is in many ways mysterious (Minoan scripts have not been deciphered), but it must have been settled and sophisticated, for the palace at Knossos is immense and entirely without fortification. Apart from architecture, Minoan art survives in sculpture, pottery, and wall painting, often featuring highly spirited depictions of animals, particularly the bull, which had ritual significance (the best collection is in the Archaeological Museum at Heraklion in Crete). Minoan civilization is believed to have been destroyed partly by earthquake and partly by invasion, and by about 1100 BC Minoan art had been absorbed into the tradition of the mainland.

Mino da Fiesole (1429–84). Florentine sculptor. According to *Vasari he was a pupil of *Desiderio de Settignano, but this has been doubted, as Desiderio was about the same age—possibly a year or so younger. Mino is remembered mainly for his portrait busts. Whereas Desiderio's are all of women, Mino's are almost all of men; the earliest—that of Piero de *Medici (Bargello, Florence, 1453) is the first dated portrait bust of the *Renaissance. Mino also worked as a tomb sculptor, but much of his work in this field has been altered or destroyed or is of uncertain attribution because he collaborated with other sculptors. The one that most clearly shows his own workmanship is that of Count Hugo of Tuscany (Badia, Florence, completed 1481), which Vasari describes as 'the most beautiful work that he ever produced'. Mino had three documented stays in Rome (1454, 1463, and 1474–80) and also worked briefly in Naples (1455). His reputation was at its height in the 19th cent., when his delicate carving of marble was much admired.

Minton, John (1917–57). British painter, graphic artist, and designer. In 1938–9 he spent a year in Paris, where he shared a studio with Michael *Ayrton. Among the artists whose work he saw in Paris, he was particularly influenced by the brooding sadness of *Berman and *Tchelitchew. From 1943 to 1946 he had a studio in London at 77 Bedford Gardens (the house in which *Colquhoun and MacBryde lived) and from 1946 to 1952 he lived with Keith *Vaughan. Minton was a leading exponent of *Neo-Romanticism and an influential figure through his teaching at Camberwell School of Art (1943–7), the Central School of Arts and Crafts (1947–8), and the *Royal College of Art (1948–56). He was extremely energetic, travelling widely and producing a large body of work as a painter (of portraits, landscapes, and figure compositions), book illustrator, and designer. After about 1950, however, his work went increasingly out of fashion. He made an effort to keep up with the times with subjects such as *The Death of James Dean* (Tate Gallery, London, 1957), but stylistically he changed little. He committed suicide with an overdose of drugs.

Mir Iskusstva. See WORLD OF ART.

Miró, Joan (1893–1983). Spanish painter, sculptor, graphic artist, and designer. He first visited Paris in 1919 and from then until 1936 (when the Spanish Civil War began) his regular pattern was to spend the winter in Paris and the summer at his family's farm near Barcelona. His early work shows the influence of various modern movements—*Fauvism, *Cubism (he was a friend of *Picasso), and *Dadaism—but he is particularly associated with the *Surrealists, whose first manifesto he signed in 1924. Throughout his life, whether his work was purely abstract or whether it retained figurative suggestions, Miró remained true to the basic Surrealist principle of releasing the creative forces of the subconscious mind from the control of logic and reason. However, even though André *Breton wrote that he was 'probably the most Surrealistic of us all', Miró stood apart from the other members of the movement in the variety, geniality, and lack of attitudinizing in his work, which shows none of the superficial devices beloved of other Surrealists. One of the works in which he first displayed an unmistakable personal vision is *Harlequin's Carnival* (Albright-Knox Art Gallery, Buffalo, 1924–5), featuring a bizarre assembly of insect-like creatures dancing and making music—a scene inspired by 'my hallucinations brought on by hunger'. Much of his work has the delightful quality of playfulness seen in this picture, but he was inspired to much more sombre and even savage imagery by the Spanish Civil War, during which he designed propaganda posters for the Republicans fighting against Franco.

Miró settled in Paris in 1936 because of the Civil War, but in 1940 he returned to Spain to escape the German occupation of France and thereafter lived mainly on the island of Majorca. It was from about this time that he began to achieve international recognition. For the rest of his long life he worked with great energy in a wide variety of fields. In 1944 he began making ceramics and slightly later he took up sculpture, initially small terracottas but eventually large-scale pieces for casting in bronze. He visited the USA for the first time in 1947 and did a large mural for the Terrace Hilton Hotel in Cincinnati. This fulfilled his desire to communicate with a large public, and several of his major works of the 1950s were in a similar vein: a mural for Harvard University in 1950 (now replaced by a ceramic copy; the original is in MOMA, New York) and two vast ceramic wall decorations, *Wall of the Sun* and *Wall of the Moon* (installed 1958), for the Unesco Building in Paris. Another aspect of his desire to make his art widely accessible is his productivity as a printmaker (etchings and lithographs). He continued to explore new techniques into his old age, taking up stained-glass design when he was in his eighties. In spite of the world-wide fame he acquired he was a modest, retiring character, utterly devoted to his work, and in one of his rare public statements he criticized Picasso for what seemed to him like a mania for publicity. The Foundation Joan Miró was opened in 1975 on the heights of Montjuic overlooking Barcelona. It is designed both as a memorial museum housing a collection of Miró's works and as a centre of artistic activity.

misericord. Term in Christian church architecture for a bracket (often enriched with carving) projecting from the underside of hinged tip-up seats, against which participants could rest during the standing parts of long services. The word derives from the Latin *misericordia*, meaning 'compassion'. Choir seats which could be raised in this manner existed in France and Germany in the 12th cent., although no complete stalls from this period survive and the earliest English misericord is perhaps the single example at Hemingborough in Yorkshire (*c.*1200). Since they were seldom seen, misericords did not necessarily conform to the decorative or didactic schemes of the *choir stalls to which they belonged. It is unusual to find a series carved to a single pattern or theme and sacred subjects do not preponderate—indeed, the carvers often seem to be taking great delight in indulging their own fancies. Scenes from domestic life, medieval romances such as Tristan and Iseult (Lincoln Cathedral), jousting (Worcester Cathedral), animals parodying human activities (Bristol Cathedral), grotesque scenes, and fantastic creatures (Norwich Cathedral) are included in a repertory common to European *Gothic. The practice of carving misericords continued into the 17th cent. in England (Lincoln College, Oxford) but generally died out as *Renaissance principles gained dominance.

missal. A liturgical book usually containing everything to be sung or said at, with ceremonial directions for, the celebration of the Mass throughout the year. It developed from the 10th cent., combining the functions of other books such as the sacramentary and gradual and, to a large extent, eventually replacing them. Generally the missal contained only two large *miniatures—the Crucifixion and Christ in Majesty—but more lavishly illustrated examples exist.

mobile. Term coined by Marcel *Duchamp in 1932 to describe the motor- or hand-powered *kinetic sculptures of Alexander *Calder and soon extended to those he produced where the movement is caused by a combination of air currents and their own structural tension. Typically they consisted of flat metal parts suspended on wires. Many other sculptors (notably Lynn *Chadwick) have experimented with the genre, and mobiles have been adopted as articles of interior decoration and (on a miniature scale) as playthings for babies.

modello. A preparatory drawing or painting for a larger work, usually made to be shown to a patron. Since the object was to impress the patron and give him a clear idea of the picture which the artist had in mind, the *modello* was more elaborate and fully worked out than the sketch. *Rubens's work is particularly rich in *modelli*.

Modernista. See ART NOUVEAU.

Modern style. See ART NOUVEAU.

Modersohn-Becker, Paula (1876–1907). German painter and graphic artist, born Paula Becker. In 1898 she joined the artists' colony at *Worpswede and in 1901 she married **Otto Modersohn** (1865–1943), another member of the group. She shared in a high degree the poetic sensibility for nature which was cultivated by the Worpswede School and she was a friend of the poet Rilke. But she was dissatisfied with the sentimentalized idealization of the Worpswede manner, and made it her object to express by means of the utmost simplicity and economy of form 'the unconscious feeling that often murmurs so softly and sweetly within me'. In 1900, and on three subsequent visits before her early death in 1907, she visited Paris, and in the last years of her life, under the influence of *Gauguin, van *Gogh, and *Cézanne, she found the 'great simplicity of form' for which she had been searching and came to regard painting as a kind of 'Runic script' for the expression of innermost emotion. Up to about 1900 she painted landscapes and scenes of peasant life; thereafter she concentrated on single figures, including self-portraits and portraits of peasants, and also still lifes. In her self-portraits she typically shows herself with wide, staring eyes and often in the nude. Although she had a weak physical constitution, she worked with great discipline and perseverance, producing about 650 pictures in a career that lasted only 10 years. She died soon after giving birth to her first child. At this time she was little known, but she has since become recognized as one of the most important precursors of German *Expressionism because of her symbolic use of colour and pattern, her highly subjective vision, and the almost primitive force of some of her work, which place her among the most important precursors of German *Expressionism.

Modigliani, Amedeo (1884–1920). Italian painter, sculptor, and draughtsman, active mainly in Paris. Although virtually his whole career was spent in France (little survives of his early work, which was in the tradition of the *Macchiaioli), he laid the foundations of his style in Italy with his study of the masters of the *Renaissance. In particular, he is often seen as a spiritual heir of *Botticelli because of the linear grace of his work. He moved to Paris in 1906; apart from visits to his family in Italy and a year spent at Nice and Cagnes, 1918–19, this was his home for the rest of his life, and he became a familiar figure in the café and night life of Montmartre. In about 1909 he met *Brancusi and devoted himself mainly to stone carving until 1914, when the war made it impossible for him to get materials. He returned to painting and his finest works were produced in the last five years of his short life. Both as a sculptor and as a painter his range was limited. With few exceptions his sculptures are heads or crouching caryatid figures and the paintings are portraits or female nudes. Common to virtually all his work are extremely elongated, simplified forms and a superb sense of rhythmic vitality, but there is a great difference in mood between, for example, his sculpted heads (*Head*, Tate, London, *c.*1911–12), which have the primitive power of the African masks that inspired them, and his gloriously sensual nudes (*Reclining Nude*, MOMA, New York, *c.*1919), which were censured for their open eroticism. Modigliani's early death from tuberculosis was hastened by his notoriously dissolute lifestyle, and his mistress Jeanne Hébuterne, pregnant with their second child, committed suicide the day after he died. He had exhibited little during his lifetime; his posthumous fame was established by an exhibition at the Galerie Bernheim-Jeune, Paris, in 1922 and a biography by the poet André Salmon in 1926—*Modigliani, sa vie et son œuvre*. His reputation as one of the outstandingly original artists of his time is now secure, but his fame rests even more on his reputation as the bohemian *par excellence*: in the popular imagination he is the archetypal romantic genius, starving in a garret, the victim of drugs and alcohol, an inveterate womanizer, but painting and carving obsessively.

Moholy-Nagy, László (1895–1946). Hungarian born painter, sculptor, experimental artist, and writer who became an American citizen in 1944. After qualifying in law at Budapest University and serving in the First World War, he moved to Vienna in 1919 and then in 1921 to Berlin, where he painted abstract pictures influenced by *Lissitzky (himself newly arrived from Russia). He also experimented with *collage and *photomontage and in 1922 had his first one-man exhibition at the *Sturm Gallery. From 1923 to 1928 he taught at the *Bauhaus, taking over from *Itten the running of the preliminary course. The difference in approach between these two highly distinctive characters is summed up by Frank Whitford (*Bauhaus*, 1984): 'Even Moholy's appearance proclaimed his artistic sympathies. Itten had worn something like a monk's habit and had kept his head immaculately shaved with the intention of creating an aura of spirituality and communion with the transcendental. Moholy sported the kind of overall worn by workers in modern industry. His nickel-rimmed spectacles contributed further to an image of sobriety and calculation belonging to a man mistrustful of the emotions, more at home among machines than human beings.' Following an argument with Hannes Meyer, *Gropius's successor as Director, Moholy-Nagy resigned from the Bauhaus in 1928. He worked for some years in Berlin chiefly on stage design and experimental film, then moved to Amsterdam and in 1935 to London, where he was a member of the *Constructivist group represented by the publication **Circle*. In London he also began the constructions which he named 'Space Modulators' and worked on designs for the film *Things to Come* (1936), produced by his fellow Hungarian Alexander Korda. In 1937 he emigrated to Chicago, where he became Director of the short-lived New Bauhaus (1937–8), then founded his own School of Design (1939; it changed its name to Institute of Design in 1944), directing it until his death.

Moholy-Nagy was one of the most inventive and versatile artists in the Constructivist school, pioneering especially in the artistic uses of light, movement (see KINETIC ART), photography, film, and plastic materials, and he was one of the most influential teachers of the 20th cent. He also was an emphatic advocate of the Constructivist doctrine that so-called *fine art must be integrated with the total environment. His views were most fully expressed in his posthumously published book *Vision in Motion* (1947).

Moilliet, Louis (1880–1962). Swiss painter. From his youth he was a friend of Paul *Klee; Moilliet introduced Klee to the *Blaue Reiter group in 1911, and in 1914 he accompanied him and *Macke to Tunisia on a journey whose repercussions on the development of expressive abstraction have become famous in the history of 20th-cent. art. He passed the war years in Switzerland and then in 1919 began extensive travel after which he lived in seclusion at La Tour de Peiltz on Lake Geneva. Moilliet was at his best as a watercolourist, achieving a brilliant transparency of colour in his landscapes. He also designed windows for the Lukaskirche at Lucerne from 1934 to 1936 and for the Zwinglikirche at Winterthur in 1943–4.

Mola, Pier Francesco (1612–66). Italian *Baroque painter, active mainly in Rome. Although he trained there with *Cesari and in Bologna with *Albani, his style, characterized by warm colouring and soft modelling, was formed mainly on the example of *Guercino and Venetian art (his early career is not well documented, but he probably spent most of the period 1633–47 in north Italy). He painted frescos in Roman churches and palaces, and his best-known painting is the striking *Barbary Pirate* (Louvre, Paris, 1650), but his most characteristic works are fairly small canvases with religious or mythological figures set in landscapes (two examples are in the National Gallery, London). They are somewhat reminiscent of Albani, but much freer, and closer in spirit to *Rosa, with whom Mola was one of the chief representatives of a distinctively romantic strain in Roman painting in the mid-17th cent.

Molenaer, Jan Miense (*c.*1610–68). Dutch painter, active in his native Haarlem and in Amsterdam. In 1636 he married Judith *Leyster; both belonged in their youth to the circle of Frans *Hals. He and his wife probably collaborated and sometimes it is difficult to differentiate their work. Molenaer specialized in *genre scenes, but his range was wide, from pictures of the crude, indecorous activities of peasants to small, exquisitely finished domestic scenes of well-to-do families. He also did portraits and religious scenes. His early works (which are considered his best) have a grey-blond tonality and touches of bright colour, but his later ones are darker, in the manner of *Brouwer or *Ostade. He had two painter brothers, **Bartholomaeus** (d. 1650) and **Claes** (d. 1676), both active in Haarlem.

Molyn, Pieter de (1595–1661). Dutch landscape painter, born in London and active mainly in Haarlem. With Jan van *Goyen, and Salomon van *Ruysdael, also active in Haarlem, he ranks as one of the pioneers of naturalistic landscape painting in Holland. It is not known if these three painters worked together, if they arrived at similar solutions independently, or if one of them began experiments with monochromatic pictures of dunes and cottages and the others followed his lead. However, *The Sand Dune* (Herzog-Anton-Ulrich Museum, Brunswick, 1626) by Molyn is earlier than any comparable dated picture by van Goyen or Ruysdael. Molyn's later career was less distinguished, and he seems then to have worked more as a draughtsman than a painter, producing landscape drawings that were intended as finished works in themselves rather than as preparatory studies for pictures. He also etched.

Momper, Joos (or **Jodocus**) **II de** (1564–1634/5). Flemish landscape painter, the outstanding member of a family of artists. He worked in his native Antwerp, where he became a master in the guild in 1581, but mountains are so much a prevailing theme in his work that it seems likely he crossed the Alps to Italy at some time in his career. His chief inspiration was *Bruegel and his work stands halfway between the constructed landscapes of the 16th cent. and the naturalistic landscapes of the 17th cent. He enriched the traditional colour scheme of 16th-cent. landscape painting, hitherto largely confined to cool blues and greens, until it ranged from brown to grey via green, yellow, and blue. Judging by the great number of extant pictures in his manner, his work must have enjoyed great popularity. They vary greatly in quality, but the best works that are indisputably from his own hand, for example the majestic *Winter Landscape with the Flight into Egypt* (Ashmolean, Oxford), show that he was one of the greatest landscape painters of his period and worthy to be mentioned in the same breath as Bruegel. His style seems to have changed little and his work is difficult to date. Figures in his paintings are often the work of other artists, notably Jan *Brueghel I.

Monamy, Peter (1681–1749). English marine painter, one of the first English imitators of the van de *Veldes. His work, which is meticulous and has little variety, is well represented in the National Maritime Museum, Greenwich.

Mondrian, Piet (1872–1944). Dutch painter, one of the most important figures in the development of *abstract art. His early painting was naturalistic and direct, often delicate in colour, though greys and dark greens predominated (*Landscape with Mill*, MOMA, New York). Between 1907 and 1910 his painting took on a *Symbolist character, partly under the influence of *Toorop and perhaps partly owing to his conversion to Theosophy. In 1911 he moved to Paris, where he came into contact with *Cubism and executed a now famous series of paintings on the theme of a tree, in which the image became progressively more abstract (*Flowering Apple Tree*, Gemeente Mus., The Hague, 1912). He returned to Holland in 1914 and remained there during the war, continuing his study of abstraction and developing theories about the horizontal–vertical axes. With Theo van *Doesburg he founded the group De *Stijl in 1917 and became the main exponent of a new kind of rigorously geometrical abstract painting that he named *Neo-Plasticism, in which he limited himself to rectangular forms and a range of colours consisting of the three primaries plus black, white, and grey (*Composition in Yellow and Blue*, Boymans Mus., Rotterdam, 1929). From 1919 to 1938 Mondrian lived in Paris, where in 1931 he joined the *Abstraction–Création group. For many years he had struggled to earn a living, but in the 1920s he gradually became known to an international circle of admirers, including the American Katherine *Dreier (from 1926). In 1938 he left Paris because of the threat of war, and for the next two years he lived in London, near Naum *Gabo and Ben *Nicholson. In 1940 he settled in New York, where he died. In America he developed a more colourful style, with syncopated rhythms that reflect his interest in jazz and dancing (*Broadway Boogie-Woogie*, MOMA, 1942–3); he was noted for his immaculate tidiness and rather fussy lifestyle, but he had a passion for social dancing and took lessons in fashionable steps.

Mondrian's concept of 'pure plasticity' consisted partly in the simplification of the means of expression to the bare essentials. He not only banished representation and three-dimensional picture-space but also the curved line, sensuous qualities of texture and surface, and the sensuous appeal of colour. This restrictedness he regarded as a sort of mystical pursuit of the Absolute, which he justified in terms of his theosophical beliefs. His extensive influence was not limited to artists whose style had direct affinities with his own. He also had a profound influence on much industrial, decorative, and advertisement art from the 1930s onwards. His influence was spread by his writings as well as his paintings. Besides articles in *De Stijl* Mondrian wrote *Néo-plasticisme* (Paris, 1920), which was published by the *Bauhaus in German translation under the title *Neue Gestaltung* (1924), and the essay 'Plastic Art and Pure Plastic Art' published in **Circle* (1937). Collected editions of his writings appeared in 1945 and 1986.

Mone, Jean (*c.*1480–*c.*1550). French sculptor who trained in Italy and also worked in Spain before settling in the Low Countries (*c.*1520) and becoming court sculptor to the emperor Charles V (1500–58). He worked mainly at Malines. Mone was the first sculptor working in the Netherlands to break completely with the *Gothic tradition, his major works such as the high altar (1533) in the church of Notre-Dame at Hal, near Brussels, being built up from a repertoire of *Renaissance forms. However, he did not really capture the spirit of Renaissance art, for the plethora of ornamented pilasters, decorated friezes, and **tondi* carved in *relief produced an effect of *horror vacui* that is distinctly anti-Classical.

Monet, Claude (1840–1926). French *Impressionist painter. He is regarded as the archetypal Impressionist in that his devotion to the ideals of the movement was unwavering throughout his long career, and it is fitting that one of his pictures—*Impression: Sunrise* (Musée Marmottan, Paris, 1872)—

gave the group its name. His youth was spent in Le Havre, where he first excelled as a *caricaturist but was then converted to landscape painting by his early mentor *Boudin, from whom he derived his firm predilection for painting out of doors: 'By the single example of this painter devoted to his art with such independence, my destiny as a painter opened out to me.' In 1859 he studied in Paris at the Atelier Suisse and formed a friendship with *Pissarro. After two years' military service in Algiers, he returned to Le Havre and met *Jongkind, to whom he said he owed 'the definitive education of my eye'. He then, in 1862, entered the studio of *Gleyre in Paris and there met *Renoir, *Sisley, and *Bazille, with whom he was to form the nucleus of the Impressionist group. Monet's devotion to painting out of doors is illustrated by the famous story concerning one of his most ambitious early works, *Women in the Garden* (Musée d'Orsay, Paris, 1866–7). The picture is about 2.5 m. high and to enable him to paint all of it outside he had a trench dug in the garden so the canvas could be raised or lowered by pulleys to the height he required. *Courbet visited him when he was working on it and said Monet would not paint even the leaves in the background unless the lighting conditions were exactly right.

During the Franco-Prussian War (1870–1) Monet took refuge in England with Pissarro: he studied the work of *Constable and *Turner, painted the Thames and London parks, and met the dealer *Durand-Ruel, who was to become one of the great champions of the Impressionists. From 1871 to 1878 he lived at Argenteuil, a village on the Seine near Paris, and here were painted some of the most joyous and famous works of the Impressionist movement, not only by Monet, but by his visitors *Manet, Renoir, and Sisley. In 1878 he moved to Vétheuil and in 1883 he settled at Giverny, also on the Seine, but about 40 miles from Paris.

After having experienced extreme poverty, Monet began to prosper. By 1890 he was successful enough to buy the house at Giverny he had previously rented and in 1892 he married his mistress, with whom he had begun an affair in 1876, three years before the death of his first wife. From 1890 he concentrated on a series of pictures in which he painted the same subject at different times of the day in different lights—*Haystacks* or *Grainstacks* (1890–1) and *Rouen Cathedral* (1891–5) are the best known. He continued to travel widely, visiting London and Venice several times (and also Norway as a guest of Queen Christiana), but increasingly his attention was focused on the celebrated water-garden he created at Giverny, which served as the theme for the series of paintings on *Water-lilies* that began in 1899 and grew to dominate his work completely (in 1914 he had a special studio built in the grounds of his house so he could work on the huge canvases). In his final years he was troubled by failing eyesight, but he painted until the end, completing a great decorative scheme of water-lily paintings that he donated to the nation in the year of his death. They were installed in the Orangerie, Paris, in 1927. Monet was enormously prolific and many major galleries have examples of his work.

Monnier, Henri (1799–1877). French *caricaturist, writer, and actor. He turned from painting to caricature in the 1820s and is best known for the creation of the character of Joseph Prudhomme, the personification of middle-class pomposity. He was introduced in *Scènes populaires* (1830), which Monnier wrote and illustrated.

Monnoyer, Jean-Baptiste (1634–99). Franco-Flemish painter, who achieved a considerable reputation as a decorator in France and England, specializing in flower pieces. As assistant to *Lebrun he supplied designs of *grotesques for tapestries executed at Beauvais and painted flower pieces for Lebrun's decorations of the Galerie d'Apollon at the *Louvre and elsewhere. He came to England *c.*1685 with the Duke of Montagu and contributed to decorations at Montagu House (the building is destroyed, but several of the paintings are at Boughton House, Northamptonshire). Monnoyer remained in England for the rest of his life. His work was popular and pictures in his style, sometimes designated by his nickname 'Baptiste', often appear in the saleroom. His son **Antoine** (1672–1747) was also a flower painter.

monotype. A method of making a print (and the print so made) in which a design is painted (usually in oil colours) on a flat sheet of metal or glass and is then transferred directly to a sheet of paper. With glass plates it is necessary to apply the pressure to the back of the paper by hand; in other cases monotypes may be printed in a press. Strictly speaking, only one print may be taken by this process (hence the term 'monotype'); in practice the colour on the slab may be reinforced after printing and another one or two impressions taken, although they will differ considerably from the first. Various modifications of the principle are known.

G. B. *Castiglione is generally credited with the invention of the monotype in the 1640s. *Blake and *Degas are two artists who have made memorable use of the technique. The medium is not infrequently used by 20th-cent. artists, primarily for its interesting and original textural qualities.

Monro, Dr Thomas (1759–1833). English collector, patron, and amateur painter, a physician by profession. He played an important role in encouraging some of the outstanding watercolour painters of his period, his protégés including *Turner, *Girtin, *Cotman, *De Wint, and many others. They used his houses in London and Bushey, Hertfordshire, as places to meet and work, and were given the run of his superb collection. *Farington wrote that 'Dr Monro's house is like an *Academy in the evening', and late in life Turner recalled how he and Girtin had often made 'drawings for good Dr Monro at half a crown apiece and a supper'.

montage (French: 'mounting'). Term applied to a pictorial technique in which cut-out illustrations, or fragments of them, are arranged together and mounted, and to the picture so made. Ready-made images alone are used, and they are chosen for their subject and message; in both these respects montage is distinct from *collage and **papier collé*. The technique has affected advertising. *Photomontage is montage using photographs only. In cinematic usage, the term 'montage' refers to the assembling of separate pieces of film into a sequence or a superimposed image.

Montagna, Bartolomeo (d. 1523). Italian painter. He probably trained in Venice, but he worked mainly in Vicenza, where he was the leading painter of his day. His style has been well characterized by S. J. Freedberg (*Painting in Italy: 1500–1600*, 1971) as 'gloomily impressive'. There are examples of his work in the National Gallery, London.

Montañes, Juan Martínez (1568–1649). The greatest Spanish sculptor of the 17th cent., known as 'el dios de la madera' (the god of wood) on account of his mastery as a carver. He worked for most of his long and productive career in Seville (*Pacheco often painted his figures), his most famous work being the *Christ of Clemency* (1603–6) in the cathedral there, which shows the new naturalism he brought to the *polychromed wooden statue. In this he occupied a role comparable to Gregorio *Fernández in Valladolid, but Montañes was more aristocratic in style, tempering *Baroque emotionalism with a classical sense of dignity. In 1635–6 he was in Madrid to undertake his only recorded secular work, a portrait head (now lost) of Philip IV to serve as model for the equestrian statue of the king executed by Pietro *Tacca in Florence, and it was on this occasion that *Velázquez painted his well-known portrait of Montañes (Prado, Madrid). His work influenced painters such as Velázquez and *Zurbarán as well as sculptors such as *Cano (whom he taught), and his style was spread by his flourishing workshop, which exported statues and altarpieces to South America.

Montefeltro. Italian noble family that ruled Urbino from 1234 with short intervals until 1508, when the family became extinct. Under the guidance of **Federico da Montefeltro** (b. 1422, ruled 1444–82) the city became one of the most important centres of *Renaissance culture. He was a brave *condottiere* and the implacable enemy of Sigismondo *Malatesta, but is significant chiefly as an enlightened patron of literature and the arts. His library was the finest in Italy and his palace is one of the most beautiful buildings of the Renaissance. In his own day he was enthusiastically celebrated by the humanists; for us he chiefly survives—broken nose, warts, and all—in the famous portrait by *Piero della Francesca (Uffizi, Florence), who was one of the leading lights of his court. Federico's ideals lived on in his son **Guidobaldo** (1472–1508), whose court is commemorated in Castiglione's famous book *The Courtier* (1528). Guidobaldo was dispossessed by Cesare Borgia in 1502, but recovered Urbino in 1503. When he died childless in 1508 Urbino passed to the family of his nephew Francesco Maria della Rovere.

Monticelli, Adolphe (1824–86). French painter, active in his native Marseilles and in Paris. He was a pupil of Paul *Delaroche, but he learned more from his studies of Old Masters in the Louvre; he was also influenced by his friends *Delacroix and *Diaz de la Peña. His subjects included landscapes, portraits, still lifes, *fêtes galantes* in the spirit of *Watteau, and scenes from the circus, painted in brilliant colours and thick impasto that influenced van *Gogh. He enjoyed great success in the 1860s, but after the Siege of Paris in 1870 he returned to Marseilles and led a retiring life. His work is represented in many French museums, and he has been much forged.

Moore, Albert (1841–93). English painter, son of a portrait painter, **William Moore** (1790–1851). His early works were in a *Pre-Raphaelite vein, but in the mid-1860s, under the influence particularly of the *Elgin Marbles, he turned to classical subjects. He specialized in elaborately, and sometimes diaphanously, draped female figures, singly or in groups—'subjectless' pictures akin to those of his friend *Whistler. The two men met in 1865 and influenced each other. Like Whistler, Moore was a colourist of great sensitivity, although his colours tend to be much higher-keyed, and he ranks with him as one of the leading figures of *Aestheticism.

Moore's brother, **Henry** (1831–95), was a successful marine painter. There are works by both men in the Tate Gallery.

Moore, Henry (1898–1986). English sculptor and graphic artist. He is recognized as one of the greatest sculptors of the 20th cent. and from the late 1940s until his death was unchallenged as the most celebrated British artist of his time. He was born in Yorkshire, the son of a miner, and after service in the British Army in the First World War, he trained at Leeds School of Art and from there obtained a scholarship to the *Royal College of Art in 1919. After completing his training in 1924, he taught there until 1931. From 1932 to 1939 he was the first head of sculpture in a new department at Chelsea School of Art. During the 1930s he lived in Hampstead in the same area as Ben *Nicholson, Barbara *Hepworth, the critic Herbert *Read, and others of the avant-garde. In 1940, after the bombing of his studio, he moved to Much Hadham in Hertfordshire, where he lived for the rest of his life. Most of Moore's early work was carved, rejecting the academic tradition of modelling in favour of the doctrine of truth to material, according to which the nature of the stone or wood—its shape, texture, and so on—was part of the conception of the work. He also rejected the *classical and *Renaissance conception of beauty and put in its place an ideal of vital force and formal vigour which he found exemplified in much ancient sculpture (Mexican, Sumerian, etc.), which he studied in the British Museum, and also in the frescos of *Giotto and *Masaccio, which he saw in Italy in 1925 in the course of a travelling scholarship. In his contribution to **Unit One* he wrote in 1934: 'For me a work must first have a vitality of its own. I do not mean a reflection of the vitality of life, of movement, physical action, frisking, dancing figures, and so on, but that a work can have in it a pent-up energy, an intense life of its own, independent of the object it may represent. When a work has this powerful vitality we do not connect the word Beauty with it. Beauty, in the later Greek or Renaissance sense, is not the aim in my sculpture.' During the 1930s his work was more directly influenced by European avant-garde art particularly the *Surrealism of *Arp. Although he produced some purely abstract pieces, his work was almost always based on forms in the natural world—often the human figure, but also, for example, bones, pebbles, and shells. The reclining female figure and the mother and child were among his perennial themes.

By the late 1930s Moore was well known in informed circles as the leading avant-garde sculptor in England (Kenneth *Clark and Jacob *Epstein were among his early supporters), but his wider fame was established by the poignant drawings he did as an Official War Artist (1940–2) of people sheltering from air-raids in underground stations. Subsequently his reputation grew rapidly (particularly after he won the International Sculpture Prize at the 1948 Venice *Biennale), and from the 1950s he carried out many public commissions in Britain and elsewhere. During this time there were major changes in his way of working. Bronze took over from stone as his preferred medium and he often worked on a very large scale. There was a tendency also for his works to be composed of several elements grouped together rather than made up of a single object. Some critics discerned a falling away of his powers in his later work, marked in particular by a tendency towards inflated rhetoric, but for others he remained a commanding figure to the end. A man of great integrity and unaffected charm, Moore was held in almost universally high regard. The tributes paid after his death made it clear that he was widely regarded not only as one of the greatest artists of the century, but also as one of the greatest Englishmen. He was a lucid and perceptive commentator on his own and other people's sculpture, and his writings have been collected as *Henry Moore on Sculpture* (1966). His work is represented in collections of modern art throughout the world; among those with particularly fine holdings are the Tate Gallery, London, and the Moore Sculpture Gallery at Leeds City Art Gallery.

Mor, Anthonis (*c.*1517/20–1576/7). Netherlandish portrait painter, a pupil of Jan van *Scorel in his native Utrecht. He was the most successful court portraitist of his day, leading an international career that took him to England, Germany, Italy, Portugal, and Spain. In England he painted a portrait of Mary Tudor (Prado, Madrid, 1554, and other versions), for which he is said to have been knighted—he is sometimes known as Sir Anthony More (it is thus he appears in *The Dictionary of National Biography*) and the Spanish version of his name, Antonio Moro, is also commonly used. His work shows little variation throughout his career; sitters are shown life-size or a little larger, half-, three-quarter-, or full-length, turned slightly to the side, with an air of unruffled dignity. His composition is simple and strong and his grasp of character firm but undemonstrative. He owed much to *Titian, but his surfaces are much more detailed and polished in the northern manner. Mor had great influence on the development of royal and aristocratic portraiture, particularly in Spain, where his ceremonious but austere style ideally suited the rigorous etiquette of the court. *Sánchez

Coello was his pupil. From 1568 Mor worked mainly in Antwerp, where he died.

Mora, José de (1642–1724). Spanish sculptor. He was the most distinguished of *Cano's pupils and was court sculptor to Charles II (1661–1700) of Spain. His works are more mannered than his master's, but nevertheless include some deeply felt figures, such as his *Crucifixion* (San José, Granada).

Morales, Luís de (d. *c.*1586). Spanish painter. He worked for most of his life in Badajoz, a town on the Portuguese border, and his style—formed away from the influence of the court or great religious and artistic centres such as Seville—is highly distinctive. His pictures are usually fairly small and he concentrated on devotional images such as the *Mater Dolorosa* or *Ecce Homo* painted with intense spirituality. The piety of his work has earned him the nickname 'El Divino'. His style owes something to Netherlandish art, but his misty modelling seems to derive from *Leonardo. There are examples of his work in the Prado, Madrid.

Morandi, Giorgio (1890–1964). Italian painter and etcher. He was born in Bologna and lived there all his life. Apart from a brief association with *Futurism, and his adhesion to de *Chirico's idea of *Metaphysical painting for a few years from 1918, he stood aloof from the intellectual turmoil and aesthetic experiments of the 20th cent. After early landscapes he painted almost exclusively still lifes, eschewing literary and symbolic content, and using subtle combinations of colour within a narrow range of tones. His style has something a common with *Purism, but is more subtle and intimate, breathing an air of serenity and cultivated sensibility; the greatest influence on his work was *Cézanne, whom he revered. After the Second World War Morandi won an international reputation, and his work won great respect among younger Italian artists for its pure devotion to aesthetic values and its poetic quality.

morbidezza (Italian: 'delicacy' or 'softness'). The rendering of the flesh-tints in painting with softness and delicacy. It was a popular term in criticism of the 18th cent., when it was applied in particular to *Correggio, but it is now antiquated.

Moreau, Gustave (1826–98). French painter, one of the leading *Symbolist artists. He was a pupil of *Chassériau and was influenced by his master's exotic *Romanticism, but Moreau went far beyond him in his feeling for the bizarre and developed a style that is highly distinctive in subject and technique. His preference was for mystically intense images evoking long-dead civilizations and mythologies, treated with an extraordinary sensuousness, his paint encrusted and jewel-like. Although he had some success at the *Salon, he had no need to court this as he had private means, and much of his life was spent in seclusion. In 1892 he became a professor at the École des *Beaux-Arts and proved an inspired teacher, bringing out his pupils' individual talents rather than trying to impose ideas on them. His pupils included *Marquet and *Matisse, but his favourite was *Rouault, who became the first curator of the Moreau Museum in Paris, which Moreau left to the nation on his death. The bulk of his work is preserved there.

Moreau, Louis-Gabriel (1740–1806). French landscape painter and etcher. He did not attain the popularity of contemporaries such as Hubert *Robert, but his work has a freshness, directness, and sincerity rare in the period, looking forward to the naturalism of the 19th cent. He is often called Moreau the Elder, to distinguish him from his brother, **Jean-Michel** (1741–1814), Moreau the Younger, who was a designer and engraver. His work is interesting for its keenly observed details of contemporary life and manners.

Moreelse, Paulus (1571–1638). Dutch painter and architect, active in Utrecht, where he helped found the St Lucas guild in 1611. He is best known for his portraits, which are similar to those made by his teacher *Mierevelt, but less severe. His portraits of shepherds and blonde shepherdesses with a deep *décolletage* were popular during his lifetime. He designed the Catherine Gate (destroyed) and possibly the façade of the Meat Market in Utrecht.

Morelli, Giovanni (1816–91). Italian critic. He trained as a physician, but from 1873 he began to write on Italian art. As a student at Munich he had acquired such a command of German that he preferred to write in that language, and at first he published his work as supposed translations from the Russian of Ivan Lermolieff (an anagram of his surname with a Russian termination). Morelli concentrated mainly on the problems of *attribution and claimed to have reduced this to scientific principles. He maintained that an artist's method of dealing with subordinate details, such as the treatment of fingernails or ears (here his anatomical training was useful), is tantamount to a signature and that by systematic study of such details attribution can be put beyond doubt. This method, still sometimes referred to as 'Morellian criticism', was influential on connoisseurs such as *Berenson but it has proved much less productive of scientific certainty than Morelli hoped; it is now felt that we recognize the work of individual artists more by general effect than by details, and that the details rather than the general effect are what an imitator will be able to reproduce most convincingly. Morelli himself made some brilliant attributions, but also some noteworthy blunders. He was a patriot and a politician as well as a writer (he declined the directorship of the *Uffizi because he did not want to be distracted from his political duties), and he secured the passing of an act forbidding the sale of works of art from religious or public institutions. In 1875 he helped to stop his rival Wilhelm von *Bode acquiring Giorgione's *Tempest* from an Italian private collection for the Berlin Museum (it is now in the Accademia, Venice). His own collection of pictures was left to the Pinacoteca of his adopted city of Bergamo.

Moretto da Brescia (Alessandro Bonvicino) (*c.*1498–1554). Italian painter, active mainly in his native Brescia and the neighbourhood. *Ridolfi says he was a pupil of *Titian and certainly his influence is apparent in Moretto's work. He was the leading Brescian painter of his day and had a large practice as a painter of altarpieces and other religious works, the best of which display an impressive gravity and a poetic feeling for nature (*St Giustina with a Donor*, Kunsthistorisches Museum, Vienna, *c.*1530). However, his portraits, although much less numerous, are considered to be generally of higher quality and of greater importance historically. It seems likely that he introduced the independent full-length portrait to Italy, for although *Vasari credits Titian with this distinction, Moretto's *Portrait of a Gentleman* of 1526 in the National Gallery, London, antedates any known example by Titian by several years. The National Gallery has an outstanding collection of works by Moretto, including two other portraits, which show the thoughtful qualities he passed on to his pupil *Moroni.

Morgan, John Pierpont (1837–1913). American financier, industrialist, and art collector. The son of a financier and head of one of the most powerful banking houses in the world, Morgan used his personal fortune to spend lavishly on works of art. His main collecting activities were in manuscripts and rare books and after his death his son, also **John Pierpont Morgan** (1867–1943), endowed the Pierpont Morgan Library in New York as a research institute and museum in memory of his father. It has superb collections of *illuminated manuscripts and Old Master drawings and also contains stained glass, sculpture, and metalwork. The *Metropolitan Museum in New York also received an important bequest from the elder Morgan, who was chairman of its governing board for many years.

Morisot, Berthe (1841–95). French *Impressionist painter. *Fragonard was her grandfather and she was brought up in a highly cultured atmosphere. She was a pupil of *Corot, but the chief formative influence on her work was *Manet, whom she met in 1868, and whose brother she married in 1874. In her turn she is said to have persuaded Manet to experiment with the Impressionist 'rainbow' palette and with **plein-air* painting. After Manet's death in 1883 she came under the influence of *Renoir. Her pictures were regularly accepted for the *Salon, but she was a strong opponent of conventional academic teaching and a champion of the Impressionist ideals; she exhibited in all the Impressionist exhibitions except the fourth (1879), when she was prevented from participating by ill health. She specialized in gentle domestic scenes, painted in a delicate, feathery technique, and was also an excellent marine painter. Her watercolours are as accomplished as her oils. She was renowned for her beauty and charm and often posed for Manet.

Morland, George (1762/3–1804). English painter, mainly of scenes from rural life. He was the son of **Henry Morland** (1719?–97), a painter of portraits and *fancy pictures who was also a dealer, forger, and restorer, to whom he was apprenticed 1777–84. He was precocious and a fluent worker and produced a huge amount of work, in spite of leading a dissolute life and often being drunk, in hiding from his creditors, or in prison. His name is particularly associated with small scenes of middle- and lower-class rural life, usually drawn more from the tavern and the stable than the cottage. The quality of his work is uneven, but at his best he showed a spirited technique and a sure sense of tone. His paintings became extremely popular and he was much imitated and forged; his brother-in-law William *Ward made *mezzotints of many of his pictures. With *Wheatley and *Ibbetson he established the village scene in the English painter's repertory.

Morley, Malcolm (1931–). British painter. In the 1960s he was one of the pioneers of *Superrealism (a term he coined), but his work later became extremely loose in handling, often depicting animals in lush landscapes. In 1984 he was the first winner of the *Turner Prize, a decision that occasioned much controversy as Morley has lived in New York since 1964.

Moro, Antonio. See MOR.

Morone, Domenico (*c.*1442– after 1517). Italian painter of the School of Verona. His work shows the influence of Paduan painters such as *Mantegna and *Squarcione, but his most important picture, *The Expulsion of the Bonacolsi* (1494), in the Palazzo Ducale in Mantua, is a wide, many-figured townscape recalling Gentile *Bellini and *Carpaccio. It commemorates the beginning of *Gonzaga rule in Mantua, when the rival Bonacolsi family were defeated in battle in 1328. With his son **Francesco** (*c.*1470–1529) he founded a studio in which many minor artists received their training.

Moroni, Giovanni Battista (*c.*1520/5–1578). Italian painter, son of an architect, Andrea Moroni. He trained under *Moretto in Brescia and worked mainly in his home town of Albino and in nearby Bergamo. His style was based closely on that of his master, but whereas his religious and allegorical paintings are generally heavy-handed, his portraits are worthy successors to Moretto's. They are remarkable for their psychological penetration, dignified air, and exquisite silvery tonality. The National Gallery, London, has the best collection of his work, including the celebrated portrait known as '*The Tailor*'.

Morozov, Ivan. See HERMITAGE.

Morrice, James Wilson (1865–1924). Canadian landscape and figure painter, active mainly in Paris, where he settled in 1890 and became friendly with many leading artists. His first inspiration came from *Whistler and *Conder. Later, his style became gently *Fauvist. He made frequent trips to North Africa, Venice, Brittany, and the West Indies, as well as to Canada, where he sketched in Quebec, Montreal, and along the St Lawrence (*The Ferry, Quebec*, NG, Ottawa, *c.*1907). His work was important in introducing modern trends to Canada.

Morris, Robert (1931–). American artist and writer. He is regarded as one of the most prominent exponents and theorists of *Minimal art, and has also experimented with other fields, including *Conceptual art, *Land art, and *Performance art. His most characteristic sculptures consist of large-scale, hard-edged geometric forms, but he has also made 'anti-form' pieces in soft, hanging materials. During the 1960s he was a leader, with Carl *Andre, in breaking down the traditional mentality linking the work of the artist with the studio, and promoted the idea of executing sculptural works *in situ*.

Morris, William (1834–96). English writer, painter, designer, craftsman, and social reformer. As a student at Oxford University he formed a lifelong friendship with *Burne-Jones and began to write poetry and to study medieval architecture, reading *Ruskin and *Pugin. In 1856 he was apprenticed to the architect G. E. Street (1824–81), but soon left to paint under *Rossetti's guidance—his only completed oil painting, *Queen Guenevere* (Tate, London, 1858), is strongly *Pre-Raphaelite. In 1859 Morris married Jane Burden, who appears in numerous paintings by Rossetti as the archetypal *femme fatale*; Morris's architect friend Philip Webb built the famous Red House, Bexley Heath, for the couple. With Webb, Rossetti, Burne-Jones, Ford Madox *Brown, P. P. Marshall (a surveyor), and Charles Faulkner (an accountant), Morris founded the manufacturing and decorating firm of Morris, Marshall, Faulkner & Co. in 1861. After a shaky start, the firm prospered, producing furniture, tapestry, stained glass, furnishing fabrics, carpets, and much more. Morris's wallpaper designs are particularly well known (they are still produced commercially today) and Burne-Jones did some superb work for the firm, particularly in stained glass and tapestry design. Morris repudiated the concept of *fine art and his company was based on the ideal of a medieval guild, in which the craftsman both designed and executed the work. He defined art as 'man's expression of his joy in labour', and saw it as an essential part of human well-being. As a socialist he wished to produce art for the masses, but there was an inherent flaw in his ambition, for only the rich could afford his expensive hand-made products. His ideal of universal craftsmanship and his glorification of manual skill thus proved unrealistic in so far as it ran counter to or failed to come to terms with modern machine production. But his work bore lasting fruit, in England (see ARTS AND CRAFTS MOVEMENT) and abroad (see WIENER WERKSTÄTTEN, DEUTSCHE WERKSTÄTTEN, and DEUTSCHER WERKBUND), in the emphasis which it laid upon the social importance of good design and fine workmanship in every walk of life. He also had an important part in the development of the private printing-press, through the founding of the *Kelmscott Press. Morris's homes at Walthamstow in London and Kelmscott Manor in Oxfordshire contain good examples of work designed by him and his associates.

Morse, Samuel Finley Breese (1791–1872), American painter and inventor. He had ambitions as a history painter and studied with *West in London 1811–15, but had, to his discontent, to work mainly at portraiture for financial reasons: 'I cannot be happy unless I am pursuing the intellectual branch of the art. Portraits have none of it; landscape has some of it; but history has it wholly.' After his return to the USA he worked in Boston and then in New York, where he was a founder member of the National Academy of Design and its first President from 1826 to 1845. Disenchanted by his failure to achieve major commissions as a history painter, however, he turned to invention, and in the 1830s conceived the idea of the telegraph and developed the Morse Code. His first telegraph line was established between Washington and Baltimore in 1844, by which time Morse had abandoned painting as a profession. He eventually became rich as the result of his invention.

Mortimer, John Hamilton (1740–79). English painter, a pupil of *Hudson at the same time as his friend Joseph *Wright. Like Wright, he worked both at portraiture and at subject pieces of a pioneering *Romantic nature. His *conversation pieces bear comparison with those of *Zoffany, but he found his true bent in the 1770s with pictures representing the exploits of soldiers and *banditti* in the 'savage' style of Salvator *Rosa (*Bandit Taking up his Post*, Detroit Institute of Arts). Many of his paintings have disappeared and are now known only through engravings. Mortimer led an eccentric and irregular life, but he became more settled after marrying in 1775 and his early death cut short the career of one of the most individual British painters of his generation.

mosaic. The art of making patterns and pictures by arranging coloured fragments of glass, marble, and other suitable materials and fixing them into a bed of cement or plaster. It was first developed extensively by the Romans in pavements. But it is also well suited to the adornment of walls and vaults, and great use was made of wall mosaic by the Christian churches of Italy and the *Byzantine Empire throughout the Middle Ages. As an exterior decoration it has sometimes appeared on the façades of medieval churches and in modern architecture. More rarely it has been made into portable pictures, or inlaid in furniture and small objects as in the Aztec art of Mexico.

Mosan School. A term applied to a school of manuscript illumination, metalwork, and enamelwork flourishing from the late 11th to the early 13th cents. in the valley of the Meuse (or Maas). The river rises in France and empties into the Rhine estuary in the Netherlands, but in the context of medieval art the term Mosan refers to the stretch of river and its tributaries in present-day Belgium, particularly the area around Liège and the Benedictine monastery of Stavelot. The most important artists of the school are *Godefroid de Claire, *Nicolas of Verdun, and *Renier of Huy. The Mosan style is part of *Romanesque art, but is distinctive because of its more naturalistic, if idealized, attitude towards the human figure. In Renier of Huy's font at Liège, for example, the figures are three-dimensional and well proportioned and their draperies are notably *antique-like. Mosan art is also noteworthy for its sheer sumptuousness, and Mosan metal-workers in particular were famed throughout Europe; Abbot *Suger employed a number at Saint-Denis.

Moser, Lukas (active 1432). German painter, known only from one work, the *Magdalen* Altarpiece in the church at Tiefenbronn (signed and dated 1432—not 1431, as had previously been read). The altarpiece is remarkably advanced stylistically, showing a detailed naturalism in the treatment of figures and landscape and an interest in light that have much in common with the paintings of *Witz, who worked in nearby Switzerland. Also remarkable is the enigmatic inscription the altarpiece bears, running 'Schri kunst schri vnd klag dich ser din begert iecz niemen mer' ('Cry out, art, cry out and wail! No one wants you now'). This may be no more than the lament of an underpaid artist, for there is no reason to suppose that there was then any lack of interest in art generally.

Moser, Mary (1744–1819). English flower painter, the daughter of **George Moser** (1704–83), a Swiss goldsmith, enameller, and medallist who settled in England and became the first Keeper of the *Royal Academy. Like her father she was a foundation member of the Academy and in 1805 her name was put forward as a candidate for the presidency. Her small flower pieces in the Dutch manner were highly popular and she executed a whole room of flower pieces (*c.*1795) at Frogmore for Queen Charlotte.

Moses, Anna Mary Robertson (called Grandma Moses) (1860–1961). The most famous of American *naïve painters. She took up painting in her seventies (initially copying postcards and *Currier & Ives prints) after arthritis made her unable to continue with embroidery, with which she had regularly won prizes at county fairs. Her first exhibition was held in a drugstore at Hoosick Falls, NY, in 1938. She was then 'discovered' by a collector, Louis J. Calder, and had her first 'one-man' show in New York in 1940 at the age of 80. In 1949 she was received at the White House by President Harry Truman and in 1960 Governor Nelson Rockefeller proclaimed her birthday, 7 September, 'Grandma Moses Day' in New York State. She produced more than a thousand pictures (working on a sort of production line system, three or four at a time, painting first the skies and last the figures), her favourite subjects being scenes of what she called the 'old-timey' farm life she had known in her younger days. From 1946 her works were reproduced on Christmas cards and elsewhere and achieved widespread popularity for their brightly coloured freshness and charm. Examples are in many American collections, notably the Bennington Museum, Vermont.

Mostaert, Jan (*c*.1475–1555/6). Netherlandish painter. He was born in Haarlem and the influence of *Geertgen tot Sint Jans can clearly be seen in his rather stiff and gangling figures. As painter to Margaret of Austria, Regent of the Netherlands, he accompanied her on her travels, making portraits of her courtiers. He also painted religious works. His most remarkable painting, however, is a *Landscape of the West Indies* (Frans Hals Mus., Haarlem, *c*.1525–30). Many of his paintings were destroyed in the Great Fire of Haarlem in 1576, and little is known in detail of his career.

Motherwell, Robert (1915–91). American painter, collagist, writer on art, editor, and teacher, one of the pioneers and principal exponents of *Abstract Expressionism. He took up painting seriously in 1941 after studying aesthetics at Stanford and Harvard universities and the erudite approach of his writings played a large part in setting the serious intellectual tone of the Abstract Expressionist movement. His work was deeply influenced by *Surrealism (particularly in his use of *automatism) and he was close to the wartime expatriates of the movement in New York. Motherwell was unusual among Abstract Expressionists in that his painting was essentially abstract from the beginning of his career. However, his work was deeply influenced by *Surrealism (particularly in the use of *automatism) and there is often a suggestion of figuration in the large amorphous shapes and bold austere colours of his paintings. Moreover, the intellectual sensibilities he brought to his work are reflected in the fact that there was often an underlying inspiration from literature, history, or his personal life; for example, he painted a series of works (more than a hundred pictures) entitled *Elegy to the Spanish Republic*. By the late 1960s his style had moved towards *Colour Field painting. He was an extremely prolific artist and his work is represented in many collections of modern art in the USA and elsewhere. His work as a writer, teacher, and lecturer also showed great energy and in 1944 he became director of a major series of publications entitled *The Documents of Modern Art* devoted to the writings of 20th-cent. artists and critics. From 1958 to 1971 he was married to Helen *Frankenthaler.

Mousseau, Jean-Paul. See AUTOMATISTES, LES.

Moynihan, Rodrigo (1910–90). British painter, born at Tenerife of Irish-Spanish descent. He came to England at the age of 8 and travelled widely before studying at the *Slade School, 1928–31. His work fluctuated between figuration and abstraction. In the early 1930s he was one of the most radical of British abstract painters, but he then turned to figuration and became associated with the *Euston Road School. After being invalided out of the army he became an official war artist in 1943. In 1956 he reverted to abstract art, but in the 1970s he took up painting portraits and still lifes, characteristically painting in pale, muted colours.

Mucha, Alphonse (1860–1939). Czech painter and designer. He had a highly varied career, but is best known for his luxuriously flowing poster designs, which rank among the most distinctive products of the *Art Nouveau style. They often feature beautiful women, but have nothing of the morbid sexuality typical of the period. Some of the best known were made in Paris in the 1890s for the celebrated actress Sarah Bernhardt. Mucha also designed sets, costumes, and jewellery for her. He was successful in the USA also, making four journeys there between 1903 and 1922. A Chicago industrialist and Slavophile, Charles Richard Crane, sponsored his series of twenty huge paintings entitled *Slav Epic* (Moravsky Krumlov Castle, 1909–28). Although he is so strongly associated with Paris, Mucha was an ardent patriot, and in 1922 he settled in Prague. Czechoslovakia had become independent only in 1918 and Mucha did a good deal of work for the new nation (giving his services free), including designing its first banknotes and stamps. After a period of neglect a revival of interest in his work culminated in a large exhibition of paintings, posters, drawings, furniture, and jewellery at the Grand Palais, Paris, in 1980.

Mudéjar. Term applied to a style of architecture and decorative art, part Islamic and part *Gothic, which grew up in Spain and Portugal during the gradual reconquest of the Peninsula from the Moors (12th–15th cents.). The Mudéjars were those Muslims allowed to remain in Spain after the reconquest and the term 'Mudéjar' was originally applied to the characteristic style of work executed by Muslim craftsmen for Christian masters. It has since been extended to cover later Spanish work in the Islamic tradition, especially brick, plaster, wood, and tile treated in the Moorish manner. Mudéjar styles also occur in book-binding, textiles, ceramics, ivory, furniture, and *inlays of wood (*taraceas*) and metal (damascene work).

Müller, Otto. See BRÜCKE DIE.

Mulready, William (1786–1863). Irish-born painter, active in England, the pupil and brother-in-law of John *Varley. After undistinguished beginnings with historical *genre and landscape he turned with great success to scenes of contemporary life in the vein made popular by *Wilkie. *The Fight Interrupted* (V&A, London, 1816), which shows a vicar intervening between two boys who have come to blows, made his reputation and set the course for his career. Although at first his meticulous brushwork showed the influence of 17th-cent. Dutch painting, in the 1820s he began to develop a more distinctive technique, using light clear colours over a white ground. This, together with his clear draughtsmanship and the poetic quality of some of his later paintings (*The Sonnet*, V&A, 1839), has led him to be seen as a precursor of the *Pre-Raphaelites, although they themselves rejected his work as trivial. Mulready also designed the first penny postage envelope (1840).

multiples. Term coined in the 1950s for works of art other than prints or cast sculpture that are designed to be produced in a large—potentially limitless—number of copies. Whereas prints and casts of sculptures are copies of an original work hand made by the artist, multiples are different, for the artist often produces only a blueprint or set of specifications for an industrial process of manufacture. The concept of the multiple represents a revolution of aesthetic attitude according to which craftsmanship is disparaged and works of art are no longer regarded as rare items for collectors and connoisseurs but as consumer goods for the masses like any other industrial product. In practice, however, they are too expensive for the mass market, and have been sold through galleries rather than normal retail outlets.

Multscher, Hans (documented 1427–d. before 1467). German sculptor, active in Ulm. The solid *naturalism of his style, reminiscent of *Sluter, suggests that he trained in the Netherlands or northern France. He ran a large workshop, which was influential in spreading this manner in Swabia. Paintings were often integral to his altarpieces, but it is a matter for debate whether he practised painting himself. Among his most important works was the high altar for the church at Sterzing in the Tyrol (1456–8), parts of which are now in the Multscher museum there.

Munch, Edvard (1863–1944). Norwegian painter, lithographer, etcher, and wood-engraver, his country's greatest artist. He began painting in a conventional manner, but by 1884 was part of the world of bohemian artists in Christiania (now Oslo) who had advanced ideas on ethics and sexual morality, Christian *Krohg being his early mentor. In 1885 he made the first of several visits to Paris, where he was influenced by the *Impressionists and *Symbolists and, above all, by *Gauguin's use of simplified forms and non-naturalistic colours. Munch had endured a traumatic childhood (his father was almost dementedly pious and his mother and eldest sister died of consumption when he was young): 'Illness, madness and death were the black angels that kept watch over my cradle', he wrote, and in his paintings he gave expression to the neuroses that haunted him. Certain themes—jealousy, sickness, the awakening of sexual desire—occur again and again, and he painted extreme psychological states with an unprecedented conviction and an intensity that sometimes bordered on the frenzied.

In 1892 he was invited to exhibit at the Kunstlerverein (Artists' Union) in Berlin and his work caused such an uproar in the press that the exhibition was closed. The scandal made him famous overnight in Germany, so Munch moved there and from 1892 to 1908 lived mainly in Berlin, with frequent stays in Norway and visits also to France and Italy. During this period much of his effort went into an ambitious series of pictures that never had a definitive form and which he called the 'Frieze of Life'—'a poem of life, love and death'. The most famous of the paintings from the series, *The Shriek* (NG, Oslo, 1893), and several others were translated by Munch into etching, lithography, or woodcut. The woodcuts (often in colour) are particularly impressive, exploiting the grain of the wood to contribute to their rough, intense vigour. With those of Gauguin they played a major part in the 20th-cent. revival of the technique.

In 1908 Munch suffered what he called 'a complete mental collapse', the legacy of heavy drinking, overwork, and a wretched love-affair, and after recuperating he returned permanently to Norway. He realized that his mental instability was part of his genius ('I would not cast off my illness, for there is much in my art that I owe to it'), but he made a conscious decision to devote himself to recovery and abandoned his familiar imagery. The anguished intensity of his art disappeared and his work became much more extroverted. In landscapes, portraits, and pictures of workmen in the snow his technique grew more and more sketchy and energetic, his palette bright and vigorous. The great achievement of this period is a series of large oil paintings for the University Hall of Oslo (1910–15) exalting the positive forces of nature, science, and history. In 1916 he settled at Ekely, Oslo, thenceforth living a solitary life. In some of his later work, however, he rekindled the passion and profundity of his early years, as in the last of his numerous self-portraits, *Between the Clock and the Bed* (Munch Mus., 1940–2), in which he shows himself old and frail, hovering on the edge of eternity. At his death he left the large body of his work still in his possession to the City of Oslo to found the Munch Museum.

Munch ranks as one of the most powerful and influential of modern artists. His influence was particularly strong in Scandinavia and Germany, where he and van *Gogh are regarded as the two main sources of *Expressionist art. In the mood of disillusionment with contemporary conditions and the sense of man's alienation which his pictures express; in his desire to paint an *iconography of what he called 'modern psychic life' but with emphasis always on conflict, neurosis, and tension, and by the intensity with which he symbolized and communicated mental anguish through an unrestrained and violent distortion of colours and forms, he opened up new paths for art. 'Just as *Leonardo da Vinci studied human anatomy and dissected corpses', he said, 'so I try to dissect souls.'

Munkácsy, Mihály von (1844–1900). Hungarian painter. His real name was Leo Lieb, but he took the name of his birthplace, Munkács (now Mukatchevo in Russia). After training in Budapest, Munich, and Düsseldorf he lived mainly in Paris. He had a resounding early success when he won a gold medal at the 1870 *Salon with *The Last Day of a Condemned Man* (NG, Budapest) and won an international reputation with his *Milton and his Daughters* (Lenox Library, New York, 1877–8). These theatrical costume pieces were enormously popular with rich collectors and Munkácsy became one of the wealthiest artists of his day. His best works are now, however, considered to be his landscapes, in which although he did not paint out of doors, he continued the tradition of the *Barbizon School. He despised the *Impressionists, but some of his own work is remarkably free in handling and he was particularly admired by *Liebermann, one of the first German painters to respond to Impressionism. The National Gallery in Budapest has easily the best collection of Munkácsy's work and he is generally regarded as his country's greatest painter.

Munnings, Sir Alfred (1878–1959). English painter, a specialist in scenes involving horses. He was an artist of considerable natural ability, but he became rather slick and repetitive and his great popularity is more with lovers of horses than lovers of painting. He was President of the *Royal Academy, 1944–9, and was one of the most outspoken opponents of modern art in England. In particular he is notorious for a splenetic speech he delivered at the RA annual dinner in 1949. It was broadcast live on radio and was a national talking point the next day. His successor as PRA, Sir Gerald *Kelly, did much to restore the damage Munnings did to the Academy's prestige. There is a museum dedicated to him in Dedham, Essex, where he lived for much of his life.

Münter, Gabriele. See NEUE KÜNSTLERVEREINIGUNG.

Murillo, Bartolomé Esteban (1617/18–82). Spanish painter, active for almost all his life in his native Seville. His early career is not well documented, but he started working in a naturalistic *tenebrist style, showing the influence of *Zurbarán. After making his reputation with a series of eleven paintings on the lives of Franciscan saints for the Franciscan monastery in Seville (1645–6, the pictures are now dispersed in Spain and elsewhere), he displaced Zurbarán as the city's leading painter and was unrivalled in this position for the rest of his life. Most of his paintings are of religious subjects, appealing strongly to popular piety and illustrating the doctrines of the Counter-Reformation church, above all the Immaculate Conception, which was his favourite theme. His mature style was very different to that seen in his early works; it is characterized by idealized figures, soft, melting forms, delicate colouring, and sweetness of expression and mood. The term *estilo vaporoso* ('vaporous style') is often used of it. Murillo also painted *genre scenes of beggar children that have a similar sentimental appeal, but his fairly rare portraits are strikingly different in feeling—much more sombre and intellectual (an outstanding self-portrait is in the NG, London).

In 1660, with the collaboration of *Valdés Leal and Francisco *Herrera the Younger, Murillo founded an *academy of painting at Seville and became its first president. He died at Seville in 1682, evidently from the after-effects of a fall from scaffolding while painting a *Marriage of St Catherine* for the Capuchins in Cadiz. He had many assistants and followers, and his style continued to influence Sevillian painting into the 19th cent. His fame in the 18th cent. and early 19th cent. was enormous. With *Ribera he was the only Spanish painter who was widely known outside his own country and he was ranked by many critics amongst the greatest artists of all time. Later his reputation plummeted, and he was dismissed as facile and sugary, but now that his own work is being distinguished from that of his numerous imitators his star is rising again. His work is represented in the Prado, Madrid, the Museo de Bellas Artes, Seville, and many major galleries in Europe and America.

Musée du Luxembourg, Paris. See POMPIDOU CENTRE.

Musée National d'Art Moderne, Paris. See POMPIDOU CENTRE.

Musée d'Orsay. See LOUVRE.

Museum of Modern Art, New York. The world's pre-eminent collection of art from the late 19th cent. to the present day, privately founded in 1929 by a group of collectors. It operated first in rented premises, holding loan shows, but the nucleus of the permanent collection was established with the bequests of Lillie P. Bliss (including nine *Cézannes) and other founders. The present building, in 53rd Street, was opened in 1939 and there have been several major extensions (in 1966 it took over the adjacent premises vacated by the *Whitney Museum when it moved to its new home). Apart from painting, sculpture, and the graphic arts, the museum has collections of photographs, films, and architectural documentation, and a large library. Through its permanent collections, exhibitions, and many other activities it exercises a strong influence both on taste and on artistic production. The many publications it has produced include some of the standard works on modern art, several of them written by Alfred H. *Barr, jun., the first director of the museum.

Muthesius, Hermann. See DEUTSCHER WERKBUND.

Muybridge, Eadweard (1830–1904). British-born photographer and pioneer of motion photography who emigrated to the USA as a young man. His original name was Edward Muggeridge, but he adopted the curious form by which he is known in the belief that it was the original Anglo-Saxon form. He became Director of Photographic Surveys to the US Government, and while surveying the Pacific coast in 1872, he was asked by the railroad magnate Leland Stanford, then Governor of California, to photograph a horse in motion, apparently to settle a bet as to whether a horse ever had all four legs off the ground simultaneously. Muybridge experimented with a battery of cameras with high-speed shutters operated by the horse itself passing across trip threads (he worked at Stanford's stud farm, and at his expense), and succeeded in proving that all four legs of a horse are indeed at times in the air simultaneously. He published his photographs in *The Horse in Motion* (1878), and then went on to study the movement of other animals, including humans, publishing his results in volumes such as *Animal Locomotion* (1887). In 1880 he invented the zoopraxiscope to project the pictures and recreate the movements he had photographed, and this he showed to scientific bodies all over Europe and America. This predecessor of the modern cinema caused a sensation. Muybridge's photographs were much used as a source by artists, among them *Eakins and the *Futurists.

Mycenaean art. A term applied to the art of Greece in the Late Bronze Age (Late Helladic Period), that is, from about 1500 to about 1100 BC. Usually the term embraces the art not only of the mainland, but also of the Greek Islands with the exception of Crete (see MINOAN ART). Mycenaean art is named after the fortress-city of Mycenae, the site of the most important remains, which was excavated by the German archaeologist Heinrich Schliemann in the 1870s.

Mylsbek, Josef (1848–1922). The leading Czech sculptor of the late 19th and early 20th cents. He created numerous statues of national heroes in Prague, most notably the famous St Wenceslas Monument in Wenceslas Square, which was conceived in 1887 but not completed until 1923, the year after Mylsbek's death. His work is in the Renaissance tradition but has a suitably Romantic Slavonic ardour; *Rodin admired his richly worked bronze surfaces.

Myron. Greek sculptor active in Athens in the mid-5th cent. BC. He was one of the leading Greek sculptors of the period and is now remembered mainly for his bronze *Discobolus* (Discus Thrower), which survives in Roman marble copies; the best of them is in the Terme Museum in Rome. As an example of compositional equilibrium it achieved a fame comparable to the *Doryphorus* of *Polyclitus. Copies also exist of Myron's group of *Athena and Marsyas*, but no visual record survives of the work for which he was most renowned in his own time—the bronze *Cow* in the market-place at Athens, which was said to display remarkable naturalism and which was celebrated in no less than thirty-six epigrams surviving in the *Greek Anthology*. See also SEVERE STYLE.

Mytens, Daniel (*c*.1590–1647). Anglo-Dutch portrait painter. He was born in Delft and trained in The Hague (probably under *Miereveld), but almost all of his known career was spent in England, where he is first recorded in 1618 working for the

Earl of *Arundel. By 1620 he was working for James I and in 1625 he was appointed 'one of our Picture Drawers' by Charles I. Mytens introduced a new elegance and grandeur into English portraiture, especially in his full-lengths, and he was the dominant painter at court until the arrival of van *Dyck in 1632. Van Dyck completely outclassed him, however, and he returned to The Hague in about 1634. Few paintings are known from his final years, but he continued to work as Arundel's agent. Mytens's work is well represented in the Royal Collection and in the National Portrait Gallery, but his finest picture is acknowledged to be *The First Duke of Hamilton* (Scottish NPG, Edinburgh, 1629). This commanding and beautifully coloured work has been described by Sir Ellis *Waterhouse as 'the great masterpiece of pre-Vandyckian portraiture in England'. Mytens was one of a dynasty of painters active into the 18th cent. Among the other members of the family was his great-nephew, **Daniel Mytens the Younger** (1644–88), also a portraitist.

Nabis. Group of French painters, active in Paris in the 1890s, whose work was inspired by *Gauguin's expressive use of colour and rhythmic pattern. The name 'Nabis' was coined by the poet Henri Cazalis from a Hebrew word meaning 'prophets' because of their half-serious, half-burlesque pose as adepts and their attitude to the new Gauguin style as a kind of religious illumination. *Sérusier, who met Gauguin at *Pont Aven in 1888, provided the initiative for the group and with *Denis was its main theorist. Other members included *Bonnard and *Vuillard (and Vuillard's brother-in-law, Ker-Xavier Roussel (1867–1944). *Maillol exhibited with the Nabis before he turned to sculpture and the musician Debussy (1862–1918) was also associated with the group, the members of which were active in theatre, poster and stained-glass design, and book illustration as well as painting. The first Nabis exhibition was held in 1892 in the gallery of the dealer Le Barc de Bouteville. After a successful exhibition held in 1899 together with certain of the *Symbolists in the gallery of the dealer *Durand-Ruel, the members of the group gradually drifted apart. They had little essential unity of purpose or outlook apart from their common reaction against the naturalism of *Impressionism and a certain interest in the decorative or emotional use of colour and linear distortion.

Nadelman, Elie (1882–1946). Polish-born sculptor who became an American citizen in 1927. After brief studies in his native Warsaw and in Munich he settled in Paris in 1903 or 1904 and lived there until 1914. Initially he was influenced by *Rodin, but he soon became interested in more avant-garde trends and he was later obsessed by the idea that *Picasso's *Cubism had been stolen from his own researches into the diffraction of forms. With the outbreak of the First World War, Nadelman moved to London and then New York. He had a successful one-man show at *Stieglitz's gallery in 1915 and was befriended by Paul *Manship and Gertrude Vanderbilt *Whitney among others. His patrons included Helena Rubinstein (he had known her in Paris), who commissioned him to make sleek marble heads for her beauty salons. He married a wealthy widow in 1919 and his work has a witty sophistication appropriate to the high society world he moved in, as with the delightful bowler-hatted *Man in the Open Air* (MOMA, New York, 1915). With his humour went a bold simplification and distortion of forms that places him alongside *Lachaise as one of the pioneers of modern sculpture in America. The Depression had a disastrous effect on his market and his career virtually ended when a good deal of his work was accidentally destroyed in 1935.

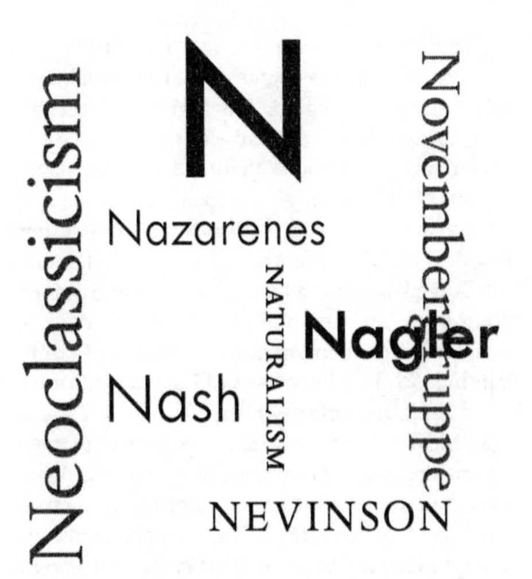

Nagler, Georg Kaspar (1801–66). German art historian. He was the author of two massive works of reference, *Neues allgemeines Künstler-Lexikon* (New General Dictionary of Artists, 22 vols., 1835–52), the largest dictionary of artists before *Thieme-Becker, and *Die Monogrammisten* (5 vols., 1858–79), a dictionary of artists' monograms that was completed by associates after his death. Nagler's other writings included a monograph on *Dürer (1837).

naïve art. Term applied to painting (and to a much lesser degree sculpture) produced in more or less sophisticated societies but lacking conventional expertise in representational skills. Colours are characteristically bright and non-naturalistic, perspective non-scientific, and the vision childlike or literal-minded. The term *'primitive' is sometimes used more or less synonymously with naïve, but this can be confusing, as 'primitive' is also applied loosely to paintings of the pre-Renaissance era as well as to art of 'uncivilized' societies. Other terms that are sometimes used in a similar way are 'folk', 'popular', or 'Sunday painters', but these too have their pitfalls, not least 'Sunday painter', for many amateurs do not paint in a naïve style, and naïve artists (at least the successful ones) often paint as a full-time job. Sophisticated artists may also deliberately effect a naïve style, but this 'false naïvety' (*faux naïf*) is no more to be confused with the spontaneous quality of the true naïve than the deliberately childlike work of say *Klee or *Picasso is to be confused with genuine children's drawing. Naïve art has a quality of its own that is easy to recognize but hard to define. Scottie *Wilson summed it up when he said, 'It's a feeling you cannot explain. You're born with it and it just comes out.'

Henri *Rousseau was the first naïve painter to win serious critical recognition and he remains the only one who is regarded as a great master, but many others have won an honourable place in modern art. The critic Wilhelm *Uhde was mainly responsible for putting naïve painters on the map in the years after the First World War. At first their freshness and directness of vision appealed mainly to fellow artists, but a number of major group exhibitions in the 1920s and 1930s helped to develop public taste for them, notably 'Masters of Popular Painting: Modern Primitives of Europe and America' at the Museum of Modern Art, New York, in 1938. Most of the early naïve painters to make reputations were French (mainly because Uhde was active in discovering and promoting them in France); they included *Bauchant, *Bombois, *Séraphine, and *Vivin. In Britain the best-known figures include Beryl *Cook and Henry *Wallis (two painters who show the huge difference of approach and style that can exist between artists given the same label). L. S. *Lowry is also often claimed as a naïve painter, but some critics regard him as outside this classification because of his many years of study at art school. In the USA the leading figures include John *Kane and Grandma *Moses. The richest crop of naïve painters, however, has been in Croatia, where Ivan *Generalic has been the most famous figure.

Nanni di Banco (d. 1421). Florentine sculptor, one of the major figures of the transition from *Gothic to *Renaissance. Much of his work was designed for one or other of two architectural settings, in Florence, the cathedral and Or San Michele, which together formed the chief source of demand for sculpture in the city during the first quarter of the 15th cent., and he often worked alongside *Donatello. His work has a Gothic elegance, but also shows the influence of the *antique; his masterpiece, the group of the *Quattro Santi Coronati* (Four Crowned Saints) at Or San Michele, for example, features grave and dignified figures modelled on Roman senator figures. Because this and other major works are of controversial dating, however, it is uncertain whether he was in advance of Donatello in his use of *classical exemplars. His early death (he was probably born in the 1380s) cut short an outstanding career.

Nanteuil, Robert (1623–78). French engraver and draughtsman, almost exclusively of portraits. He is considered the greatest European portrait engraver of the 17th cent., and in France stands as the counterpart of Philippe de *Champaigne among painters. He often engraved the work of Champaigne and other painters, but also made original compositions, in which he showed both great technical mastery and penetrating characterization. Louis XIV (1638–1715) appointed him royal draughtsman in 1658 and he made many pastel portraits as well as engravings of the king and royal family. *Colbert and Cardinal Mazarin were also portrayed by him on several occasions, and his work forms an outstanding record of the most prominent figures in French affairs in the mid-17th cent.

Nash, Paul (1889–1946). English painter, book illustrator, writer, photographer, and designer. Nash was one of the most individual British artists of his period, taking a distinguished place in the English tradition of deep attachment to the countryside whilst at the same time responding imaginatively to European modernism. He saw himself as a successor of *Blake and *Turner. After training at the *Slade School he served in the First World War, was wounded, and worked as an *Official War Artist, creating memorable images of the devastation the war wrought on the countryside. In the 1920s and particularly in the 1930s he was influenced by *Surrealism (above all de *Chirico, an exhibition of whose work he saw in London in 1928) and often concentrated on mysterious aspects of the landscape (*Monster Field*, Durban Art Gallery, 1939). For much of this time he lived in rural areas (Kent, Sussex, Dorset), basing his work on scenes he knew well but imaginatively transforming them. In 1933 he was the prime mover in the formation of *Unit One (1933), a group of avant-garde British artists, and he helped to organize and exhibited in the International Surrealist Exhibition in London in 1936. In the Second World War he was again an Official War Artist. He was already very sick with the asthmatic condition that killed him, but he produced one of the best-known works to be inspired by the conflict, *Totes Meer* (*Dead Sea*) (Tate, London, 1940–1), which portrays shot-down aircraft with their wings looking like undulating waves. Nash was regarded as one of the finest book illustrators of his time; he also designed scenery, fabrics, and posters, and was a photographer and writer. *Outline: An Autobiography and Other Writings* appeared posthumously in 1949 and *Fertile Image*, a collection of his photographs taken for use in painting, in 1951. He also wrote the *Shell Guide to Dorset* (1936). His brother **John** (1893–1977) was also a painter and illustrator, excelling in meticulous flower drawings for botanical publications. Like Paul he was an Official War Artist in both world wars.

Nasmyth, Alexander (1758–1840). Scottish painter. He worked mainly in his native Edinburgh, but he was a pupil and assistant of *Ramsay in London 1774–8, and in 1782–5 he visited Italy. There he became interested in landscape painting, which eventually took over from portraiture as his main concern. In his landscapes he blended *classical elements stemming from *Claude with naturalistic observation and became the founder of the Scottish landscape tradition, influencing many younger painters. He was a man of wide culture, interested in science as well as art, and he worked as a stage designer and architectural consultant. One of his friends was the poet Robert Burns (1759–96), whose portrait he painted against a romantic landscape background (NPG, Edinburgh, *c.*1787). Nasmyth had several artist sons, of whom the most important was the eldest, **Patrick** (1787–1831). He worked mainly in London and achieved great popularity with his prolific output of landscapes in the manner of the 17th-cent. Dutch masters, earning the nickname 'the English *Hobbema'.

National Academy of Design, New York. A professional association of artists founded in New York in 1825 in opposition to the conservative American Academy of the Fine Arts (which ran from 1802 to 1841; see TRUMBULL). The National Academy was originally called the Society for the Improvement of Drawing; it adopted its present name in 1828. For most of the 19th cent. it was the leading art institution in America, its annual exhibitions helping to make New York the country's major art centre. *Morse was first President, and most ambitious artists of the time sought membership. However, its views became unprogressive, prompting certain artists to break away (notably to found the *Eight in 1908). Today it exists mainly as a historical institution.

National Gallery, London. The national collection of European paintings from *c.*1300 to *c.*1900 (it also includes a few earlier pictures and has recently started to acquire works from the early 20th cent.). It was founded in 1824 when the government purchased thirty-eight paintings from the collection of John Julius Angerstein, a merchant who died in 1823. They were first displayed in his former house at 100 Pall Mall. Further bequests (including that of Sir George *Beaumont) soon necessitated larger premises, and the present building in Trafalgar Square was opened in 1838; the gallery shared the premises with the *Royal Academy until 1869, by which time it had grown into one of the great collections of the world. Since then there have been various enlargements of the building and in 1991 a major extension was opened—the Sainsbury Wing, the gift of Sir John, Simon, and Timothy Sainsbury. This wing now houses the gallery's early paintings, up to about 1510. The collection as a whole now has about 2,000 pictures. This is a fairly small number compared with some of the great Continental galleries based on former royal collections, but the National Gallery's paintings are distinguished by their very high overall quality. All the paintings are on permanent display, and this is perhaps a unique distinction amongst galleries of comparable stature. In spite of the relatively small size of the collection, it surpasses any other gallery in giving a comprehensive view of the mainstream of European painting from *Giotto to *Cézanne. Best represented of all are the early Italian and Dutch Schools. The representation of the British School is selective because of the existence of the *Tate Gallery as a separate National Gallery of British art; it was founded as an annexe to the National Gallery in 1897 and became independent in 1954. Other well-known national galleries, with their dates of foundation, are: the National Gallery of Scotland in Edinburgh (1859); the National Gallery of Victoria in Melbourne (1859); the National Gallery of Ireland in Dublin (1864); the National Gallery of Canada in Ottawa (1880); the National Gallery of Art in Washington (1937); and the Australian National Gallery in Canberra (1976).

National Museum of Wales, Cardiff. The most important museum in Wales, founded in 1907 by Act of Parliament with the aim of telling 'the world about Wales and the Welsh about their fatherland'. Its chief concern was thus to acquire examples of the work of Welsh painters, pictures illustrating Welsh scenery and customs, portraits of eminent Welsh men and women, and other works of art or objects of historical interest. Works of other types were presented to the museum, however, and after the arrival in 1951 of the first part of the Gwendoline and Margaret Davies Bequest—a choice collection of late-19th-cent. French art—the scope of the museum broadened. It now houses, in addition to its Welsh material (including works by painters such as Gwen and Augustus *John and Thomas *Jones), one of the finest collections of foreign paintings in Britain.

National Portrait Gallery, London. The national collection of portraits of eminent British men and women. It was founded in 1856 at the urging of the historian and politician Philip Henry Stanhope, 5th Earl Stanhope (1805–75), who was the first chairman of the trustees. The collection was originally housed at 29 Great George Street, Westminster, and the present premises, adjoining the National Gallery, were opened in 1896. The gallery includes sculpture, *miniatures, drawings, prints, and photographs as well as paintings, and it also has a vast photographic archive of representations of famous Britons. The criterion for inclusion in the gallery is the celebrity of the sitter rather than the quality of the portrait, so as works of art the pictures in the collection vary enormously, from an acknowledged masterpiece such as *Holbein's cartoon of Henry VIII to the wholly amateurish representation of the three Brontë sisters by their brother Branwell. Nevertheless, because portraiture has played such a great part in the history of British art, many of its most illustrious practitioners are represented by works there. Recently the NPG has begun to commission portraits of living sitters, and since 1980 it has organized a series of annual portrait awards for artists under 40 years old. The NPG has four outstations: Montacute House in Somerset, Beningbrough Hall in Yorkshire, Gawthorpe Hall in Lancashire, and Bodelwyddan Castle in north Wales. The Scottish National Portrait Gallery was established in 1882, and the National Portrait Gallery in Washington in 1962.

Nattier, Jean-Marc (1685–1766). French portrait painter. His father, **Jean** (*c.*1642–1705), was a painter and his mother, **Marie Courtois** (*c.*1655–1703), was a *miniaturist. He was one of the most successful artists at the court of Louis XV (1710–74), excelling in the vogue for painting women in mythological or allegorical fancy dress—or undress—transforming his matrons into goddesses (*Mme de Lambesc as Minerva*, Louvre, Paris, 1737). His portraits are little concerned with individual characterization, but they show fluency, vivacity, and a relaxed charm. He was at his best with women and has been accused of 'painting with make-up', a comment that suggests the pastel-like delicacy of his handling. Taste was turning against him towards the end of his career and some of his later work shows signs of fatigue. His brother **Jean-Baptiste** (1678–1726) was also a painter; he committed suicide after being expelled by the Académie. *Tocqué was Nattier's son-in-law.

naturalism. Term denoting an approach to art in which the artist endeavours to represent objects as they are empirically observed, rather than in a stylized or conceptual manner conditioned by intellectual preconceptions or other factors. In this sense the art of the Greek *Classical period is often claimed to be the first truly 'naturalistic' art and that of the Italian *Renaissance is spoken of as a revival of naturalism. *Bellori (1672) was the first to apply the term to a particular type of painting in discussing the followers of *Caravaggio, with reference to their doctrine of copying nature faithfully whether it seems to us ugly or beautiful. Naturalism, however, is not inconsistent with the *idealization of nature, for Greek sculpture may be naturalistic in its command of anatomy, but idealistic in that it sets up a standard of physical beauty remote from the everyday world. Nor need the term imply minute attention to detail, although this is often part of a naturalistic approach. The shades of meaning to be attached to the word can thus vary greatly according to context; when used in its broadest sense it may suggest little more than that a work is representational rather than *abstract. The terms 'naturalistic' and 'realistic' are often used more or less synonymously, but *Realism with a capital 'r' has a specific meaning in the history of art and should not be used loosely.

Navarrete, Juan Fernández de (*c.*1526–79). Spanish painter from Navarre, called 'El Mudo' (the mute) because he was deaf and dumb. After studying for many years in Italy (traditionally as a pupil of *Titian) he returned to Spain shortly before 1568 and was appointed court painter to Philip II (1527–98). He was commissioned to paint thirty-two altarpieces for chapels in the *Escorial, but at the time of his death had completed only eight (still *in situ*). His eclectic style was instrumental in spreading Italian influence in Spain, and in his use of dramatic lighting effects, notably in his *Burial of St Lawrence* (Escorial, 1579), he anticipated the *tenebrism of the 17th cent.

Nazarenes. Name originally given derisively to a group of early 19th-cent. German painters because of their affectation of biblical dress and hairstyles. The nucleus of the group was established in 1809 when six students at the Vienna Academy, including *Overbeck and *Pforr, formed an association called the Brotherhood of St Luke (*Lukasbrüder*), named after the patron saint of painting. They believed that art should serve a religious or moral purpose and that since about the middle of the 16th cent. this ideal had been betrayed for the sake of artistic virtuosity. In their desire to return to the spirit of the Middle Ages, the Lukasbrüder not only looked to late medieval and early *Renaissance art for stylistic inspiration (they admired in particular *Dürer and *Perugino), but also wished to return to a learning and teaching programme based on the workshop rather than the *academy. To this end they lived and worked together in a quasi-monastic fashion. In 1810 Overbeck, Pforr, and two other members moved to Rome, where they occupied the disused monastery of S. Isidoro. Here they were joined by Peter von *Cornelius and others. One of their aims was the revival of monumental *fresco and they were fortunate in obtaining two important commissions which made their work internationally known (Casa Bartholdy,

1816–17, the paintings are now in the Staatliche Museen, Berlin; Casino Massimo, Rome, 1817–29). These paintings show the clear colour and drawing and the tendency towards overcrowding and insipidness that characterize their work. In general, modern taste has been more sympathetic towards the Nazarenes' simple and sensitive landscape and portrait drawings than to their ambitious, didactic figure paintings.

As a coherent group the Nazarenes had ceased to exist even before the Casino Massimo frescos were completed, but their ideas continued to be influential. Cornelius had moved in 1819 to Munich, where he surrounded himself with a large number of pupils and assistants who in turn carried his style to other German centres. Overbeck's studio in Rome was a meeting-place for artists from many countries (the Russian *Ivanov was his friend, for example); *Ingres admired him and Ford Madox *Brown visited him. William *Dyce introduced some of the Nazarene ideals into English art and there is a kinship of spirit with the *Pre-Raphaelites.

NEAC. See NEW ENGLISH ART CLUB.

Neeffs (or **Neefs**), **Pieter the Elder** (*c.*1578–1656/61). Flemish painter, active in Antwerp. Most of his pictures are interiors of *Gothic churches, some of them night scenes illuminated by artificial light. They are generally small, painted on copper, and executed in a precise, neat way similar in style to those of the *Steenwycks, but more mechanical. His son **Pieter Neeffs the Younger** (1620–after 1675) painted the same subjects, and it is very difficult to distinguish between their hands. Another son, **Lodewijk** (b. 1617), was also a painter, but little is known of his work.

Neer, Aert (or **Aernout**) **van der** (1603/4–77). Dutch landscape painter, active in his native Amsterdam. He had two specialities: moonlit scenes, of which he is the acknowledged master among Dutch painters; and winter landscapes with skaters, as an exponent of which he is in the first rank. In both types he displayed his mastery of light effects and subtle modulations of colour. Paradoxically, although his work was much copied and imitated, he had difficulty earning a living as an artist. In 1658 he opened a wine shop in Amsterdam, but this venture was also a failure and in 1662 he became bankrupt. He was a prolific artist and many galleries have examples of his work, the National Gallery, London, and the Rijksmuseum, Amsterdam, being particularly rich.

Two of his sons were artists. **Eglon** (1634?–1703) is best known for *genre pieces done in the style of *Terborch and *Metsu. He was conspicuously more successful than his father and became court painter at Düsseldorf. The few works which can be attributed to **Jan** (1638–65) show that he was an imitator of his father.

Neoclassicism. The dominant movement in European art and architecture in the late 18th and early 19th cents., characterized by a desire to recreate the heroic spirit as well as the decorative trappings of the art of Greece and Rome. A new and more scientific interest in *classical antiquity, greatly stimulated by the discoveries at Pompeii and Herculaneum, was one of the features of the movement, and it is also seen as mounting a reaction against the light-hearted and frivolous *Rococo style. The order, clarity, and reason of Greek and Roman art appealed greatly in the Age of Enlightenment, and in France the Neoclassical style held strong moral implications, being associated with a change of social outlook and a desire to restore ancient Roman values into civil life. It is, indeed, in the paintings of *David, with their antique grandeur and simplicity of form, and their heroic severity of tone, that Neoclassicism finds its purest expression, but the style was born and had its centre in Rome. *Mengs and *Winckelmann were in the vanguard of the movement there, and other leading figures from all over Europe—Antonio *Canova, John *Flaxman, Gavin *Hamilton, Bertel *Thorvaldsen—spent the main or important parts of their careers in the city. Many American artists worked there too—notably the sculptor Horatio *Greenough, who was a pupil of Thorvaldsen—and took the Neoclassical style back to their country. Winckelmann, the prime propagandist of the movement, saw Classical art as embodying 'noble simplicity and calm grandeur', but there was great stylistic variation within the Neoclassical movement. Angelica *Kauffmann, for example, painted in a delicate and pretty manner that is far removed from David's serenity.

There is, moreover, no firm dividing line between Neoclassicism and Romanticism, even though in some ways they appear to be at opposite spiritual poles. In the revival of interest in antique art, archaeological zeal could easily give way to a nostalgic yearning for a lost Golden Age, and the term 'Romantic Classicism' is sometimes used to characterize an aspect of Neoclassicism in which an interest in antiquity is tinged with Romantic feeling. In fact the antipathy between Classics and Romantics (exemplified by *Ingres and *Delacroix, for example) was unknown before the 19th cent., and it was only in the mid-19th cent., at a time when the antique revival style was out of fashion, that the word 'Neoclassicism' was coined—originally a pejorative term with suggestions of lifelessness and impersonality. These negative connotations have clung tenaciously to the term, and the ardent aspirations of the founders of Neoclassicism have been obscured by the fact that the more decorative aspects of the movement—Wedgwood pottery, for example—have become more closely associated with the word in the public consciousness than have the great masterpieces of David and Canova.

Neoclassicism is related to but can be distinguished from *Greek Taste, which was a fairly superficial fashion for Greek-inspired decoration, and from the Greek Revival, which in architecture was a movement expressing a new interest in the simplicity and gravity of ancient Greek buildings. It began seriously in the 1790s and culminated in the 1820s and 1830s. Greek architecture became widely known in the West only around 1750–60 and in the early days of Neoclassicism it was regarded as primitive and few architects cared to imitate it.

Neo-Dada. A term that has been applied to various styles, trends, or works that are perceived as reviving the methods or spirit of *Dada. It has been used as a synonym for *Pop art, for example, and has been applied to the work of Jasper *Johns.

Neo-Expressionism. Movement in painting (and to a lesser extent sculpture) emerging in the late 1970s, characterized by intense subjectivity of feeling and aggressively raw handling of materials. Neo-Expressionist paintings are typically large and rapidly executed, sometimes with materials such as straw or broken crockery embedded in their surfaces. They are usually figurative, often with violent or doom-laden subjects, but the image is sometimes almost lost in the welter of surface activity. To some extent Neo-Expressionism marked a return to more traditional forms after the 'anything goes' experimentation of the 1970s. Perhaps partly for this reason it was welcomed by art dealers and collectors, but critical reaction to it has been very mixed. Several exponents, above all the American Julian Schnabel (1951–), have become rich and famous, but to many critics their work seems deliberately bad, ignoring all conventional ideas of skill; indeed the term 'Bad Painting' (from the title of an exhibition at the New Museum, New York, in 1978) has been applied to certain works in the vein (Punk Art and Stupid Painting are alternative terms). Distinguishing between good 'Bad' Painting (i.e. that which deliberately cultivates crudeness for its emotional value) and bad 'Bad' Painting (something that is just a mess) is an unenviable critical task. Neo-Expressionism has flourished mainly in Germany (where its exponents are sometimes called Neue Wilden—'New Wild Ones'), Italy, and the USA. Leading exponents include: in Germany, Georg *Baselitz and Anselm Kiefer (1945–); in Italy, Sandro Chia (1946–) and Francesco Clemente (1952–); in the USA, David Salle (1952–) and Julian Schnabel. See also NEW IMAGE PAINTING.

Neo-Geo. Term (short for Neo-Geometric) applied to the work of a group of American artists active in New York in the mid-1980s who employed a variety of styles and media but were linked by the fact that their paintings, sculpture, or other products were predominantly cool and impersonal, in reaction from the emotionalism of *Neo-Expressionism. Jeff Koons (1955–), who exhibited consumer products such as vacuum cleaners in a reworking of *Dada *ready-mades, is the best-known figure of the group. Many critics have seen their work as cynical and empty ('dead on arrival' is one memorable description), but Neo-Geo has been a hit with certain collectors, notably Charles *Saatchi, who has bought it in bulk.

Neo-Impressionism. Term coined by the critic Félix Fénéon (1861–1944) in 1886 for a movement in French painting that was both a development from *Impressionism and a reaction against it. Like the Impressionists, the Neo-Impressionists were fundamentally concerned with the representation of light and colour, but whereas Impressionism was empirical and spontaneous, Neo-Impressionism was based on scientific principles and resulted in highly formalized compositions. *Seurat was the outstanding artist of the movement; *Signac (who was its main theoretician) and (for a while) Camille *Pissarro were the other leading adherents. All three showed Neo-Impressionist pictures at the final Impressionist exhibition in 1886.

The theoretical basis of Neo-Impressionism was *divisionism, with its associated technique of *pointillism—the use of dots of pure colour applied in such a way that when seen from an appropriate distance they achieve a maximum of luminosity. In each painting the dots were of a uniform size, chosen to harmonize with the scale of the work. In Seurat's paintings, this approach combined solidity and clarity of form with a vibrating intensity of light; in the hands of lesser artists, it often produced works that look rigid and contrived. Neo-Impressionism was short-lived, but it had a significant influence on several major artists of the late 19th and early 20th cents., notably *Gauguin, van *Gogh, *Matisse, and *Toulouse-Lautrec.

Neo-Plasticism. Term coined by Piet *Mondrian for his style of austerely geometrical abstract painting and more broadly for the philosophical ideas about art that his work embodied. He claimed that art should be 'denaturalized', by which he meant that it must be purely abstract, with no representational relation to the natural world. To this end he limited the elements of pictorial design to the straight line and the rectangle (the right angles in a strictly horizontal–vertical relation to the frame) and to the primary colours—blue, red, and yellow—together with black, white, and grey. In this way he thought that one might escape the particular and achieve expression of an ideal of universal harmony. Mondrian took the term 'nieuwe beelding' from the writings of Dr Matthieu Schoenmaekers, a Dutch author of popular books on philosophy and religion, whom he admired for a time but later considered to be a charlatan. It was used in Mondrian's first published work, the long essay 'De Nieuwe Beelding in de Schilderkunst' (Neo-Plasticism in Pictorial Art), which appeared in eleven instalments of the periodical *De *Stijl* in 1917–18.

Neo-Romanticism. A term applied to a movement in British painting and other arts *c.*1935–55, in which artists looked back to certain aspects of 19th-cent. *Romanticism, particularly the 'visionary' landscape tradition of *Blake and Samuel *Palmer, and reinterpreted them in a more modern idiom. Painters and graphic artists whose work is embraced by the term include Michael *Ayrton, John *Piper, and Graham *Sutherland, who all worked in a landscape tradition that was regarded as distinctly national, and projected a Romantic image of the countryside at a time when it was under threat from Nazi Germany. Other artists whose work has been dubbed Neo-Romantic include the poet Dylan Thomas, the film director Michael Powell, and photographers such as Bill Brandt and Edwin Smith. The term Neo-Romanticism has also been applied to certain painters working in France in the 1930s, notably *Berman and *Tchelitchew, who typically painted dreamlike imaginary landscapes with rather mournful figures. Their work influenced the British Neo-Romantics. In the 1980s 'Neo-Romanticism' was one of the many terms used as a synonym for *Neo-Expressionism, but it did not catch on in this sense.

Neroccio dei Landi (1447–1500). Sienese painter and sculptor. He was a pupil of *Vecchietta and worked in partnership with *Francesco di Giorgio until 1475. Most of his paintings are representations of the Virgin and Child with Saints, but one of his finest works is a *Portrait of a Girl* in the National Gallery in Washington. He continued the elegant and refined Sienese tradition that stretched back to *Duccio and is particularly noted for his delicate colouring.

Nesiotes. See CRITIUS.

Netscher, Caspar (?1635/9–84). Dutch painter. *Houbraken makes inconsistent statements about his birthplace, mentioning both Heidelberg and Prague, and there is similar doubt about his birth date. Most of his career was spent in The Hague, where he settled in 1661/2, but he trained in Deventer with *Terborch. From his master he took his predilection for depicting costly materials—particularly white satin. He painted *genre scenes and some religious and mythological subjects, but from about 1670 he devoted himself almost exclusively to portraits, often for court circles in The Hague. His reputation was such that Charles II (1630–85) invited him to England (*Vertue says he came, *de Piles and Houbraken that he declined). His work, elegant, Frenchified, small in scale, and exquisitely finished, influenced Dutch portraiture into the 18th cent., his followers including his sons **Constantijn** (1688–1723) and **Theodoor** (1661–1732).

Neue Künstlervereinigung München (NKV) ('New Artists' Association of Munich'). An association of artists founded in Munich in 1909 with *Kandinsky as president. Alexander Kanoldt (1881–1939) was secretary and Adolf Erbslöh (1881–1947) was chairman of the exhibition committee. Other members included *Jawlensky, *Kubin, Kandinsky's lover Gabriele Münter (1877–1962), and Jawlensky's lover Marianne von Werefkin (1870–1938). They were strongly influenced by *Fauvism and the three exhibitions that they held (1909, 1910, and 1911) were far too advanced for the critics and public and were met with torrents of abuse. The second exhibition was European in character, including works by the Russians David and Vladimir *Burliuk, by Le *Fauconnier, *Picasso, and *Rouault, and by members of the Fauves (*Braque, *Derain, van *Dongen, *Vlaminck), some of whom had already moved on to *Cubism. Franz *Marc came to the NKV's defence after this exhibition and it was in this way that he met *Kandinsky. When Erbslöh rejected an abstract painting submitted by Kandinsky for the third exhibition, Kandinsky resigned and with Marc founded the *Blaue Reiter. They moved so quickly that the Blaue Reiter's first exhibition opened on the same day as the NKV's last and stole its thunder.

Neue Sachlichkeit (New Objectivity). Movement in German painting of the 1920s and early 1930s reflecting the resignation and cynicism of the post-war period. It continued the interest in social criticism which characterized much of the prevalent *Expressionism, but repudiated the abstract tendencies of the *Brücke. The name was coined in 1923 by Gustav Hartlaub, director of the Kunsthalle, Mannheim, in connection with an exhibition, held in 1925, of 'artists who have retained or regained their fidelity to positive, tangible reality'. The movement was not characterized by a unified stylistic outlook, but the major trend was towards the use of meticulous detail to portray the face of evil for the purposes of violent social satire. *Dix and *Grosz were the greatest figures of the movement, which was dissipated in the 1930s with the rise of the Nazis. Other artists associated with Neue Sachlichkeit include Conrad Felixmüller (1897–1977), Christian Schad (1894–1982), and Rudolf Schlichter (1890–1955).

Nevelson, Louise (1899–1988). Russian-born American sculptor, painter, and graphic artist. Her family emigrated to the USA in 1905 and settled in New York in 1920. Her serious study of art began at the *Art Students League in 1929–30 and she then studied under Hans *Hofmann in Munich. In 1932–3 she worked with Ben *Shahn as assistant to Diego *Rivera on his frescos in New York. She started to make sculpture in 1932 and in 1944 began experimenting with abstract wooden assemblages. It was towards the end of the 1950s that she began the 'sculptured walls' for which she became internationally famous. These are wall-like *reliefs made up of many boxes and compartments into which abstract shapes are assembled together with commonplace objects such as chair legs and slats, bits of balustrades, finials and other 'found objects' (*An American Tribute to the British People*, Tate, London, 1960–4). These constructions, painted a uniform black, or afterwards white or gold, won her a reputation as a leader in both *assemblage and *environment sculpture. They have great formal elegance, but also a strange ritualistic power. In the late 1960s she began to work in a greater variety of materials—lucite, aluminium, Plexiglas, epoxy, etc.—and sometimes produced small constructions for reproduction as *multiples. From the same time she also began to receive commissions for large open-air and environmental sculptures, which she executed in aluminium or steel.

Nevinson, C. R. W. (Christopher Richard Wynne) (1889–1946). British painter and graphic artist. As a student in Paris in 1912–13 Nevinson met several of the *Futurists and he became the outstanding British exponent of their style. His work included landscapes, urban scenes, figure compositions, and flowers, but he found his ideal subjects during the First World War. He served in France with the Red Cross and the Royal Army Medical Corps, 1914–16, before being invalided out, and his harsh, steely images of life and death in the trenches met with great acclaim when he held a one-man exhibition at the Leicester Gallery, London, in 1916. Stylistically they drew on certain *Cubist as well as Futurist ideas, but they are closer to the work of the *Vorticists (with whom he had exhibited in 1915). In 1917 Nevinson returned to France as an *Official War Artist, and he was the first to make drawings from the air. Some of his work was considered too horrific and was censored, but a second one-man exhibition at the Leicester Gallery in 1918 was another triumph. At the end of the war Nevinson renounced Futurism and his later, more conventional paintings are a sad anticlimax; an example is *Twentieth Century* (Laing Art Gallery, Newcastle upon Tyne), a turgid attempt to portray a world on the brink of catastrophe.

New Contemporaries. See YOUNG CONTEMPORARIES.

New English Art Club (NEAC). Artists' society founded in London in 1886 in reaction against the conservative and complacent attitudes of the *Royal Academy. The founders—largely artists who had worked in France and had been influenced by *plein-air* painting—included *Clausen, *La Thangue, *Sargent, *Steer, and *Tuke. There were about 50 members when the inaugural exhibition was held in April 1886 at the Marlborough Gallery. In 1889 the Club came under the control of a minority group led by *Sickert, who had joined in 1888; he and his associates were interested in the *Impressionists, and in 1889 they held an independent exhibition under the name 'The London Impressionists'. Sickert resigned in 1897 (he returned in 1906) and from then up to about the First World War the NEAC was effectively controlled by Frederick Brown (see SLADE), *Tonks, and Steer. In this period it contained most of the best painters in England. From about 1908, however, it began to lose initiative to progressive groups such as the *Allied Artists' Association and the *Camden Town Group. After the war the Club occupied a position midway between the Academy and the avant-garde groups. With the gradual liberalization of the Academy exhibitions its importance diminished, but the Club still exists.

New British Sculpture. A term sometimes applied to the work of a loosely connected group of British sculptors who emerged in a series of exhibitions at the beginning of the 1980s, notably 'Objects and Sculpture' shown at the *Institute of Contemporary Arts and the Arnolfini Gallery, Bristol, in 1981. There is no single common factor linking these sculptors, but predominantly their work is abstract (although sometimes with human associations), using industrial or junk material, and most of them are represented by the same dealer—the Lisson Gallery, London. Among the leading figures are: Tony Cragg (1949–), Grenville Davey (1961–), Richard Deacon (1949–), Anish Kapoor (1954–) (each of these four has won the *Turner Prize), David Mach (1956–), Julian Opie (1958–), Richard Wentworth (1947–), Alison Wilding (1948–), and Bill Woodrow (1948–). Most of these are well represented in the *Saatchi Collection.

New Figuration. A very broad term for a general revival of figurative painting in the 1960s following a period when abstraction (particularly *Abstract Expressionism) had been the dominant mode of avant-garde art in Europe and the USA. The term is said to have been first used by the French critic Michel Ragon, who in 1961 called the trend 'Nouvelle Figuration'.

New Generation. The title of four exhibitions, sponsored by the Peter Stuyvesant [tobacco company] Foundation, held at the Whitechapel Art Gallery, London, in 1964, 1965, 1966, and 1968, with the aim of introducing young British painters and sculptors to the public. *Op and *Pop art were well represented, but the series is best remembered for the 1965 exhibition, which featured a group of sculptors who were seen as creating a new school of British abstract sculpture, largely under the influence of Anthony *Caro (most of them had been his pupils at St Martin's School of Art). Subsequently their work has been referred to as 'New Generation Sculpture'. The leading figures were Phillip *King, Tim Scott (1937–), William *Tucker, and William *Turnbull (the only one of the group not connected with St Martin's). Their work had in common a liking for simple shapes and strong colours—sometimes close to *Minimal art, sometimes with a Pop flavour.

New Image Painting (or **New Image Art**). A vague term applied since the late 1970s to the work of certain painters who work in a strident figurative style, often with cartoon-like imagery and abrasive handling owing something to *Neo-Expressionism. It was given currency by an exhibition entitled 'New Image Painting' at the Whitney Museum, New York, in 1978. The accompanying catalogue unhelpfully informs us that the New Image painters 'felt free to manipulate the image on canvas so that it can be experienced as a physical object, an abstract configuration, a psychological associative, a receptacle for applied paint, an analytically systemized exercise, an ambiguous quasi-narrative, a specifically non-specific experience, a vehicle for formalist explorations or combinations of any'. Philip *Guston, who in the 1970s abandoned *Abstract Expressionism for a comic-strip style of figuration, is regarded as the progenitor of New Image Painting. Other American artists who have been labelled New Image painters include Jennifer Bartlett (1941–), Jonathan Borofsky (1942–), and Susan Rothenberg (1945–). In Britain the term 'New Image' has been applied particularly to painters of the 1980s *Glasgow School.

Newlyn School. A name applied to the painters who worked in the Cornish fishing village of Newlyn from the 1880s, particularly those directly linked with Stanhope Forbes (1857–1947), who was the founder and leader of the school. One of the attractions of Newlyn was the mild climate, which made it particularly suitable for outdoor work, and Forbes and his associates were among the pioneers of **plein-air* painting in Britain. Apart from Forbes and his wife Elizabeth Armstrong (1859–1912), the artists most closely associated with Newlyn in its heyday include: Frank Bramley (1857–1915); Thomas Cooper Gotch (1854–1931), better known for his later work, particularly his allegorical pictures of children; and Henry Scott *Tuke. Many of the Newlyn artists were associated with the *New English Art Club, but they also showed their work at the *Royal Academy. The golden period of Newlyn was over by the turn of the century; thereafter it was vulgarized by an influx of inferior talent, and *St Ives came to have a greater attraction for 20th-cent. artists. However, distinguished painters continued to be associated with Newlyn: Harold and Laura *Knight lived there, 1908–18, for example, and Dod Procter (1892–1972) and her husband Ernest Procter (1886–1935) spent much of their lives in the village.

Newman, Barnett (1905–70). American painter, one of the leading figures of *Abstract Expressionism and one of the initiators of *Colour Field painting. During the 1930s he had a hard time financially; the Depression almost ruined his father's clothing business, and unlike most American painters of the time Newman did not work for the *Federal Art Project, being unwilling to accept State hand-outs. Part of his living came from teaching art in high schools. He destroyed most of his early work and stopped painting in the early 1940s, but he began again in 1944, and in the second half of the 1940s evolved a distinctive style of mystical abstraction—he considered 'the sublime' to be his ultimate subject-matter. The work with which he announced this style was *Onement I* (private collection, 1948), a monochromatic canvas of dark red with a single stripe of lighter red running down the middle. Such stripes (or 'Zips' as Newman preferred to call them) became a characteristic feature of his work. By the time he painted *Onement I* Newman already had a reputation as a controversialist and a spokesman for avant-garde art (in catalogue essays and in articles in journals). In 1949 he painted his first wall-size pictures (he was one of the pioneers of the very large format) and in 1950 he had his first one-man exhibition, at the Betty *Parsons gallery. This was coolly received by critics and fellow artists, and by the mid-1950s his very spare style had separated him from the predominantly *Gestural idiom of his colleagues. For a time he became a somewhat marginalized figure and he stopped painting in 1956. He had a heart attack in 1957, but the following year a resurgence began with a series of paintings in black and white, and in the last decade of his life his reputation soared and his output was prolific. From 1965 he made steel sculptures (vertical strips recalling his paintings) and in his late years he also experimented with *shaped canvases, painting several triangular pictures.

New Realism. A vague term, of dubious value, that has been used in at least three distinct senses in connection with art of the 1960s and later. First, it has been used in a way similar to the term *New Figuration to describe a revival of figurative art after a dominant period of abstraction; whereas 'New Figuration' has been used very broadly, however, 'New Realism' has often been applied more specifically to works that are objective in spirit, particularly *Superrealist pictures or those of Philip *Pearlstein. In a second sense, 'New Realism' has been used as a straight translation of the French term 'Nouveau Réalisme' and applied to works incorporating three-dimensional objects, usually mass-produced consumer goods, in *assemblages or attached to the surface of a painting. Thirdly—and perplexingly—it has been used as a synonym for *Pop art.

New Sculpture, the. A trend in British sculpture between about 1880 and 1910 characterized chiefly by an emphasis on naturalistic surface detail and a taste for the spiritual or *Symbolist in subject-matter. The name was coined by the critic Edmund Gosse in a series of four articles, 'The New Sculpture, 1879–1894', published in the *Art Journal* in 1894. Leading representatives of the trend include Gilbert Bayes (1872–1953), Alfred Drury (1856–1944), Edward Onslow Ford (1852–1901), Sir George *Frampton, Sir Alfred *Gilbert, the Australian-born Sir Bertram Mackennal (1863–1931), Sir William Reynolds-Stephens (1862–1943), Sir Hamo *Thornycroft, Albert Toft (1862–1949), and Derwent Wood (1871–1926). Their archetypal product was the 'ideal' free-standing figure, often with imagery drawn from mythology or poetry. Most typically these ideal figures were in bronze, but polychromy—using such materials as ivory and coloured stones—was also a feature of the New Sculpture. Although the New Sculpture did not survive the First World War as a major force, some of the practitioners went on working in the idiom long after this.

New Tendency. See NOUVELLE TENDANCE.

New York Realists. Informal name given during the early years of the 20th cent. to Robert *Henri and his disciples. See also the EIGHT.

New York School. Name given to the innovatory painters, particularly the *Abstract Expressionists, who worked in New York during the 1940s and 1950s.

Niccolò dell' Abbate. See ABBATE.

Niccolò dell' Arca (active 1462–94). Italian sculptor. He was of south Italian origin, but his known career was spent in Bologna, where he is first documented in 1462. His name comes from his work on the Arca di S. Domenico (shrine of St Dominic) in S. Domenico Maggiore, Bologna, for which he carved a marble canopy and small free-standing figures (1469 onwards). The work was unfinished at his death and *Michelangelo carved three missing figures. Niccolò's greatest work is a highly emotional group of the *Lamentation over the Body of Christ* in Sta Maria della Vita, Bologna, executed in *terracotta and originally painted. It is variously dated from the 1460s to the 1480s. According to tradition the figure of Nicodemus is a self-portrait.

Nicholson, Ben (1894–1982). British painter and maker of painted *reliefs, one of the most distinguished pioneers of abstract art in Britain. From his father, Sir William *Nicholson, he inherited a feeling for simple and fastidious still lifes, which with landscapes made up the bulk of his early work. They show him responding to the innovations of *Cubism, using the standard Cubist repertoire of objects such as jugs and glasses and arranging them as flat shapes on the picture plane. In 1933, during one of the several long stays in Paris he made at this time, Nicholson made the first of a series of white reliefs using only right angles and circles (*White Relief*, Tate, London, 1935). They show the influence of *Mondrian (whom Nicholson met in Paris in

1934) and were the most uncompromising examples of abstract art that had been made by a British artist up to that date. By this time he was recognized as being at the forefront of the modern movement in England; he was a member of *Unit One and one of the editors of **Circle*. In 1932 he married Barbara *Hepworth (they were divorced in 1951) and in 1939 moved with her to *St Ives, where with John *Piper and others he became the centre of a local art movement. After the war his international reputation grew and he won many prestigious prizes. In 1968 he was awarded the OM. His late work moved freely between abstraction and figuration. From 1958 he lived in Switzerland.

Nicholson's first wife, **Winifred Nicholson** (1893–1981), was a painter of distinction. She is best known for her flower paintings, but she also did other subjects and abstracts, all her work showing her joy in colour and light. Ben and Winifred Nicholson were married from 1920 to 1931, but even after their separation they took a keen interest in each other's work.

Nicholson, Sir William (1872–1949). English painter and graphic artist. He is perhaps best remembered for his brilliant early work as a poster designer, done in collaboration with his brother-in-law James *Pryde under the name 'the *Beggarstaff Brothers'. His finest works as a painter are his still lifes, usually small, unpretentious, and sensitively handled. He also painted portraits and landscapes. His work is well represented in the Tate Gallery.

Nicias. Greek painter, active in Athens during the latter part of the 4th cent. BC, a younger contemporary of the sculptor *Praxiteles, some of whose statues he coloured. None of his work survives, but it is described in some detail by *Pliny. He was famous for his skill in *chiaroscuro, had a reputation for painting female figures in dramatic situations, and held that a great artist should concentrate on noble and heroic themes.

Nicolas of Verdun (active late 12th cent.–early 13th cent.). *Mosan goldsmith, enameller, and metal-worker. He is considered the greatest goldsmith and enameller of his day and a major figure in the transition from *Romanesque to Gothic. Two signed works by him survive: an *enamelled pulpit frontal made for the abbey church at Klosterneuburg, near Vienna (completed in 1181, damaged in 1320, and then remodelled into its present *triptych-altar form); and the Shrine of St Mary for Tournai Cathedral (1205). Among the works attributed to him the most important is the Shrine of the Three Kings, made for Cologne Cathedral in about 1190 and the largest and most sumptuous reliquary of the period. All three works still belong to the churches for which they were commissioned. The Klosterneuburg altar is his masterpiece, featuring forty-five enamel plaques in an elaborate *typological programme, events from the New Testament being paralleled by ones in the Old Testament. His figure style is expressive and dynamic, with individualized faces and richly articulated drapery suggesting influence from the *antique.

niello. Term referring to a black compound (typically sulphur and silver) used as a decorative *inlay on metal surfaces, to the process of making such an inlay, and to a surface or object so decorated. Niello prints are impressions taken from such surfaces and are invariably Italian work of the second half of the 15th cent. They were probably taken as proofs by niellists who wanted to see their work clearly. But it appears that these craftsmen then took to engraving plates with the express purpose of taking impressions from them, and many early examples of Italian *line engraving show the influence of the niello craft. Maso *Finiguerra, whom *Vasari credits with the invention of line engraving, was both niellist and line engraver.

Nike Balustrade. The marble parapet (*c.*410 BC) round the Nike bastion at the south-west corner of the Acropolis at Athens, carved in low *relief with figures of Nike (Victory)—with Athena, sacrificing, tying her sandal, etc. The exquisite remains, graceful women in partly transparent drapery, are in Athens (Acropolis Mus.).

Nittis, Giuseppe de (1846–84). Italian painter, mainly of landscapes and scenes of city life. Early in his career he was associated with the *Macchiaioli. He settled in Paris in 1867, became a friend of *Degas and *Manet, and exhibited with the *Impressionists in 1874 (because de Nittis had had some success at the *Salon, Degas thought the presence of his work among the Impressionists would mean that critics 'won't be able to say that ours is an exhibition of rejected artists'). The best collection of his work is in the Pinacoteca Communale at Barletta, his native town.

NKV. See NEUE KÜNSTLERVEREINIGUNG.

Noguchi, Isamu (1904–88). American sculptor and designer, born in Los Angeles of the Japanese poet Yone Noguchi and an American mother. His training was academic (he was briefly an apprentice of Gutzon *Borglum), but he was inspired by *Brancusi (whose assistant he was in Paris 1927–8) to turn to abstraction. Although he began by making sheet-metal constructions, he was essentially a stone-carver, and his work has a kinship with Brancusi's in its craftsmanship and respect for materials as well as its expressive use of organic shapes. Noguchi made many public sculptures and he also had an extremely successful career as a designer. His theatre and ballet designs included work for Sir John Gielgud (1904–) and Martha Graham (1895–), he designed furniture and interiors, and had a particular interest in sculptural gardens and playgrounds. He made frequent visits to Japan (where he spent his childhood) and his work subtly blends western and eastern tradition.

Nolan, Sir Sidney (1917–92). The most internationally famous of Australian painters. In his youth he was variously employed running a pie shop and working in the office of a gold mining company, and was also a racing cyclist, but he turned from odd jobs to art after attending night classes in his native Melbourne. He became a full-time painter when he was 21. His early work was abstract, but while serving in the Australian army (1941–5) he painted a series of landscapes of the Wimmera District of Victoria that gave the first unmistakable signs of the originality of his vision, capturing the heat and emptiness of the bush. In 1946 he began a series of paintings on the notorious bushranger Ned Kelly, who had become a legendary figure in Australian folk history, and it was with these works that he made his name. Nolan returned to the Kelly theme throughout his career, and he also drew on other events from Australian history. In 1949, for example, he completed a series of paintings about the Eureka Stockade, a gold-miners' uprising in 1854 that possesses legendary significance for Australian labour history. In such works he created a distinctive idiom to express this novel Australian subject-matter and memorably portrayed the hard, dry beauty of the desert landscape. Technically, his work is remarkable for the lush fluidity of his brushwork and sometimes he painted on glass or other smooth materials. Nolan first visited Europe in 1950 (in that year he had his first London exhibition) and from that time he lived mainly in England and travelled widely. Apart from Australian themes, he was inspired by journeys to Antarctica and New Guinea, for example, as well as by literary subjects such as the legend of Leda and the Swan. He was knighted in 1981. His work is represented in many Australian galleries, and the Tate Gallery, London, has numerous examples.

Noland, Kenneth (1924–). American abstract painter and sculptor. In 1949 he settled in Washington, where he became a close friend of Morris *Louis. On a visit to New York in 1953 they were greatly impressed by Helen *Frankenthaler's *Mountains and Sea* and they began experimenting with pouring and staining techniques. They became the leading figures of a group of *Colour Field painters known as the Washington Color Painters, but from the late 1950s Noland tended to use more precisely articulated geometrical forms, and in the 1960s he became one of the chief exponents of *Hard Edge painting. Initially he used concentric circles on a square canvas. This was followed by a chevron motif, sometimes on a diamond or lozenge-shaped canvas, and this again was gradually lengthened into horizontal stripes running across a very long rectangular canvas. In 1966 he began to make sculpture, influenced by his friend Anthony *Caro.

Nolde, Emil (1867–1956). German painter and graphic artist, one of the most powerful representatives of *Expressionism. Born of a peasant family, he was originally trained as a wood-carver and came late to artistic maturity. His studies took him from his native North Germany to Munich and Paris, and from 1906 to 1907 he was a member of the *Brücke in Dresden, but he was essentially an isolated figure, standing apart from his great German contemporaries. He moved around a good deal in Germany and was well travelled elsewhere (in 1913–14 he visited Russia, the Far East, and the South Sea islands as part of an ethnographic expedition), but at times he lived almost like a hermit. His travel broadened his knowledge of the kind of *primitive art that was then beginning to excite avant-garde artists, but Nolde had already established the essential features of his style before his contact with such art, and when the term 'primitive' is applied to his work it refers to its brutal force, not to any exotic trappings. He was a deeply religious man and is most famous for his paintings of Old and New Testament subjects, in which he expresses intense emotion through violent colour, radically simplified drawing, and grotesque distortion. The majority of his pictures, however, were landscapes, and he was also one of the outstanding 20th-cent. exponents of flower painting, working with gloriously vivid colour, often in watercolour. He was also a prolific etcher and lithographer. Although he was a member of the Nazi Party, he was declared a *degenerate artist by the Nazis in 1941 and forbidden to paint. He did, however, execute small watercolours in secret (these are called the 'unpainted pictures') and from these made larger oils after the war. From 1926 he lived at Seebüll, where there is now a Nolde Foundation that has an outstanding collection of his work.

Nollekens, Joseph (1737–1823). English sculptor, the son of an Antwerp painter of the same name (1702–47) who came to London in 1733. He was a pupil of *Scheemakers. From 1760 to 1770 he worked in Rome, making a handsome living copying, restoring, faking, and dealing in *antique sculpture. He also made a few portrait busts including that of Lawrence Sterne (NPG, 1766), a splendid character study in the antique manner, and on his return to England it was chiefly as a portraitist that he built up his great reputation and fortune (he left £200,000 at his death). His finest portraits are vivacious and brilliantly characterized, but there are many inferior studio copies of his more popular works, such as his busts of Fox and Pitt. He also made statues in a slightly erotic antique manner, and had a large practice as a tomb sculptor. In this field, too, he paid lip-service to the antique, but avoided the dedicatory attitude of true *Neoclassical sculptors. He and his wife were well-known figures in the artistic circles of the late 18th and early 19th cents. in London, both of them being notorious skinflints. The life by his pupil and disappointed executor J. T. Smith (1766–1833), *Nollekens and his Times* (1828), gives a remarkable picture of their meanness and has been described by Rupert Gunnis (*Dictionary of British Sculptors 1660–1851*, 1953) as probably 'the most candid, pitiless and uncomplimentary biography in the English language'.

Nonell y Monturiol, Isidro (1873–1911). Spanish painter, one of the pioneers of modern art in Spain. A native of Barcelona, he there belonged to the group of artists, including the young *Picasso, who met at the *Quatre Gats* café—Picasso also shared his studio in Paris for a time. He exhibited with some success at Paris in the 1890s, but did not achieve general recognition until 1910. He is best known for his figure studies of very poor people, idiots, and gypsies, and for his still lifes.

Nonesuch Press. A publishing firm founded in 1923 in London by Sir Francis Meynell (1891–1975), his wife Vera Mendel, and David Garnett (1892–1981) for the production of books characterized by 'significance of subject, beauty of format, and moderation of price'. The books were printed at various presses, as well as a few at the Nonesuch, with the aim of extending the standards of book design already achieved in a more limited field by the *private presses. Although the Nonesuch Press made full use of mechanization and was not part of the individual handicraft school, many of its books were remarkable also for their illustrations, Paul *Nash being among the artists employed. In its original form the Nonesuch Press continued in operation until 1938, publishing over a hundred books.

Non-Objective art. General term for all abstract art that does not depend on the appearance of the visual world as a starting-point, relying solely on the relations between forms and colours. The term seems to have been first used by *Rodchenko (who used it mainly for geometrical abstraction), but it was given currency in the West by *Kandinsky in his book *Über das Geistige in der Kunst* (Concerning the Spiritual in Art) in 1912. Malevich wrote a treatise entitled *Die Gegenstandlöse Welt*, published by the *Bauhaus in 1927 (translated as *The Non-Objective World*, 1959).

Noort, Adam van (1562–1641). Flemish history and portrait painter, remembered chiefly as one of the teachers of *Rubens. Too little is known of the style of either man at the time to know what influence he may have had. Another famous pupil was *Jordaens, who became his son-in-law in 1616.

Northcote, James (1746–1831). English portrait and history painter, *Reynolds's pupil, assistant, and biographer. He spent the years 1777–80 in Rome and settled in London in 1781. Rome had given him ambitions to history painting, but his exercises in this field, notably for *Boydell's Shakespeare Gallery, are ponderous and awkward. As a portraitist he was an uninspired follower of Reynolds, and it is as a writer that he has the main claim to distinction; he was something of a character and a lively commentator on the artistic scene. His *Memoirs of Sir Joshua Reynolds* (1813, supplement 1815) is the most important source of contemporary information on Reynolds, and William *Hazlitt published a book of his *Conversations* in 1830.

Norwich School. English regional school of landscape painting, associated with the Norwich Society of Artists, founded in 1803, with *Crome as president. From 1805 until 1825 exhibitions were held every year at Sir Benjamin Wrench's Court, an old house in Norwich which had become the meeting-place of the Society. This is the first instance of a provincial institution holding regular exhibitions. When Crome died in 1821, his place as president was taken by *Cotman, and the activities of the Society continued until his departure for London in 1834. Whatever the principles professed or debated by the Society, the Norwich artists consisted almost entirely of landscape painters in oil and watercolour, working chiefly under Crome's influence with a bias in favour of Norfolk scenery. They were united by ties of artistic sympathy and often also of family relationship. Their work is best represented in the Norwich Museum.

Nost (or **Van Ost**), **John** (d. 1711/13). Flemish-born sculptor who settled in England in about 1678.

By 1686 he was foreman to Artus III *Quellin, whose widow he later married. He is chiefly notable as a maker of lead garden statues, some based on Italian or *antique models but others of his own creation. Examples remain at Melbourne Hall, Derbyshire, Hampton Court, and many other places. His tombs are less interesting. His son—not, as long thought, his nephew—of the same name (d. 1780) continued the practice.

Notke, Bernt (d. 1509). German sculptor, the leading wood-carver in the Baltic area during his period. He worked mainly in Lübeck, where he is first recorded (as a painter) in 1467, but his masterpiece was executed in Sweden, where he was summoned in 1483 to make a monument commemorating a victory by the Regent, Sten Sture (*c.*1440–1503), over the Danes. The victory was attributed to the assistance of St George, and Notke's stirring group of *St George and the Dragon* (completed 1489) in the Storkyrka (Stockholm's main church) is one of the greatest of all votive images. Its spiky forms represent the most expressionistic strain in late *Gothic art and the vividly naturalistic details include the use of real elk antlers for the dragon's horns.

Nouvelle Tendance (or **New Tendency**). Terms which began to be used in the early years of the 1960s to describe the very diverse *Constructivist tendencies which came to the fore at this time in most countries of western Europe, and in Japan, Argentina, Brazil, and Venezuela. Though very different in their stylistic manifestations, most of these movements had in common the depersonalization of the art work, the appropriation of new materials and new techniques borrowed from industrial science, a cult of group activity and anonymity, the exploitation of direct stimuli such as light, sound, and real movement, and an iconoclastic attitude towards traditional aesthetic standards. Generally the movements were in declared opposition to expressive abstraction.

novecento. See QUATTROCENTO.

Novecento Italiano (Italian 20th Century). Association of Italian artists, founded in 1922 and officially launched in 1923; it aimed to reject European avant-garde movements and revert to a naturalistic type of art based upon classical Italian traditions. It represented a tendency that had already found expression in the interest of both de *Chirico and *Carrà in early Renaissance Italian art, and also coincided with the narrowly chauvinistic outlook of the official regime in artistic matters. The Novecento had no clear artistic programme and numbered within its ranks artists of very different styles and temperaments. Carrà and *Marini were among the most significant artists associated with the group, but it came more and more to be identified with the empty formalism and idealized propaganda encouraged by the Fascist Sindicati delle Belle Arti and during the 1930s it was the main bulwark of reaction. It finally disbanded in 1943.

November Group. An association of Finnish *Expressionist artists founded in Helsinki in November 1917. This was only a month after the declaration of independence from Russia and the members of the November Group were sometimes aggressively nationalistic in outlook, creating a recognizably Finnish form of Expressionism. Tyko *Sallinen was the leading figure of the group.

Novembergruppe. A group of radical left-wing German artists formed in Berlin in December 1918; it was named after the revolution that had broken out in Germany the previous month at the end of the First World War, and the professed aim of the Novembergruppe was to help national renewal by means of a closer relation between progressive artists and the public. Max *Pechstein was among the prime movers. In 1919 the founders of the association created the Arbeitsrat für Kunst (Workers' Council for Art) in an attempt to bring about a dialogue between art and the masses, but this collapsed in 1921 and interest and support came mainly from the middle classes. Artistically the group covered a wide spectrum, from *Expressionism and *Cubism to geometrical abstraction, *Constructivism, and socially critical *Realism. Through its numerous exhibitions the Novembergruppe did much to foster an artistic revival, but it broke up around 1924 with the swing of public opinion towards the right, leaving its aims to be more fully realized at the *Bauhaus. The group disbanded in 1929.

Nuremberg Little Masters. See LITTLE MASTERS.

obelisk. A tall, generally monolithic, stone shaft, square or rectangular in section, slightly tapered, and with a pyramidal apex. The obelisk originated in Egypt as a solar symbol, and many were removed to Rome in imperial times. Their rediscovery during the *Renaissance led to the adaptation of the obelisk form for monuments and architectural ornament. During the 19th cent. others were transferred to Paris, London, and New York (the so-called 'Cleopatra's Needles') and few now remain standing in Egypt.

objet trouvé (French: found object). An object found by an artist and displayed with no, or minimal, alteration as a work of art. It may be a natural object, such as a pebble, a shell, or a curiously contorted branch, or a man-made object such as a piece of pottery or old piece of ironwork or machinery. The essence of the matter is that the finder-artist recognizes such a chance find as an 'aesthetic object' and displays it for appreciation by others as he would a work of art. The practice began with the *Dadaists (especially Marcel *Duchamp) and *Surrealists and was cultivated for a time in England chiefly by Paul *Nash. See also READY-MADE.

Ochtervelt, Jacob (1634–82). Dutch *genre painter. He was born and mainly active in Rotterdam, but he is said to have been a pupil of *Berchem (presumably in Haarlem) and from 1674 he lived in Amsterdam. Apart from a few portraits and some early hunting party and 'merry company' scenes, his paintings are almost all elegant upper-class interiors, in which he showed off a skill in painting silks and satins to rival that of *Terborch. His figures are extremely refined, but there is often a sexual element in his paintings—a couple eating oysters (believed to be an aphrodisiac) was a favourite subject (an example is in the Boymans Museum, Rotterdam).

O'Conor, Roderic (1860–1940). Irish painter and etcher, active for most of his life in France (mainly Paris), where he settled in 1883 after studying in London and Antwerp. He was strongly influenced by *Gauguin and van *Gogh, and by the early 1890s was painting in a full-blooded *Post-Impressionist style with bold colour—often used non-naturalistically—and thick brushwork. He lived a fairly reclusive life (although he was friendly with many British visitors to France, including Clive *Bell and Roger *Fry) and he was virtually unknown in the British and Irish art worlds. It was only after his death that he was recognized as the outstanding pioneer of Post-Impressionism among English-speaking artists. He did, however, notably influence

Matthew *Smith, whom he met in 1919. O'Conor was mainly a landscapist, but he also painted still lifes, portraits, interiors, and figure subjects.

Official War Art. Art sponsored by the British Government during the First and Second World Wars to make a visual record of all aspects of the war effort for information and propaganda purposes. In the First World War the work was directed by the Ministry of Information, which was advised by a committee drawn from distinguished figures in the art world and public life, among them Campbell Dodgson, Keeper of Prints and Drawings at the British Museum, and Eric MacLagan, later Director of the Victoria and Albert Museum. In 1916 the Ministry launched the Official War Artists scheme, under which artists were recruited, with appropriate military ranks, to serve as chroniclers (the Germans already had a similar scheme in operation). The first artist to be commissioned was Honorary 2nd Lieutenant Muirhead *Bone, who left for France on 16 August 1916 and toured the Front in a chauffeur-driven car. Many others soon followed, among them some of the most illustrious British artists of the time. They included men who had already been serving in the armed forces, such as Paul *Nash, C. R. W. *Nevinson, and Stanley *Spencer, and others who were too old for active duty. The works produced varied enormously in style and quality (the committee was admirably broad in its choice of artists) and included imaginative evocations of the war as well as sober factual records. There were many portraits of participants, but two of the most notable portraitists who worked as Official War Artists—*Orpen and *Sargent—showed a different and unexpected side to their talents, powerfully depicting the horrors they saw.

In the autumn of 1939, soon after the outbreak of the Second World War, the Ministry of Information appointed Kenneth *Clark chairman of a small group that became known as the War Artists' Advisory Committee (Muirhead Bone was one of the members). It met weekly at the National Gallery, of which Clark was Director, and its functions were principally 'to draw up a list of artists qualified to record the War at home and abroad . . . [and] . . . to advise on the selection of artists from this list for war purposes and on the arrangement for their employment'. Clark regarded his work on the committee as 'my only worthwhile activity' during the war: 'We employed every artist whom we thought had any merit, not because we supposed that we would get records of the war more truthful or striking than those supplied by photography, but because it seemed a good way of preventing artists being killed' (*Ravilious was one of the rare fatalities). Several painters who had been Official War Artists in the First World War were employed in the same capacity in the Second, among them Nash and Spencer, but the Committee mainly employed men of a younger generation. The terms in which they were employed varied: some were given salaried posts for a specific period, while others were given one-off commissions. The Committee also encouraged artists, whether serving or civilian, to submit pictures for consideration. Generally the commissions in the Second World War were on a smaller scale than those in the First, with many works being executed in watercolour (Spencer's huge canvases of shipbuilding on the Clyde are a conspicuous exception). Henry *Moore's drawings of Londoners sheltering from air raids in Underground stations are perhaps the best known of all the works produced under the auspices of Clark's Committee.

In both wars women were employed as Official War Artists on the home front, notably the animal painter Lucy Kemp-Welch (1869–1958) in the First and Laura *Knight in the Second. A huge number of works was produced. The largest collection (about 10,000 items) is in the Imperial War Museum, London, which was opened in 1920 and moved to its present home (the former Royal Bethlehem Hospital) in 1936. There is another major collection in the Tate Gallery, London, and many provincial museums have good examples.

Since the Second World War the tradition of Official War Art has been maintained on a lesser scale by the Artistic Records Committee of the Imperial War Museum. Linda Kitson (1945–) went on its behalf to the Falkland Islands during the war there against Argentina in 1982, for example, and Peter Howson (1958–) went to Bosnia in 1993. His exhibition 'War in Bosnia' at the Imperial War Museum the following year attracted considerable attention because of its unsparing depiction of atrocities: 'Now that I've actually seen dead bodies, and guts and brains, and starving children, it has made the work authentic.'

offset. A reproduction, also called a counterproof, made from a drawing by dampening it and/or a sheet of paper and pressing them together by machine or hand. The image is therefore reversed left to right. Offsets were often made for working purposes by designers of ornament who needed to have a copy of a design in reverse, or to complete one half of a symmetrical design, and they have also been used to create forgeries. In *engraving, the term 'offset print' is applied to an impression taken from a print in the manner described above for drawings rather than from the block or plate itself. The transfer made in this way is in reverse as regards the print from which it is taken, but in the same sense as the drawing on the block or plate. The term 'offset' is also used as an abbreviation for 'offset *lithography', where the ink is transferred from the lithographic plate or stone on to a rubber roller and thence while still wet to the printing paper.

O'Gorman, Juan (1905–82). Mexican architect and painter. Early in his career he designed a series of houses in Mexico City (notably those for himself and for Diego *Rivera) that were among the first in the Americas to show the functionalist ideas of *Le Corbusier. As head of the Department of Construction in the Ministry of Public Instruction he was in charge of the building of thirty new schools (1932–5) which confirmed his leadership in progressive architecture. In the 1930s, however, he abandoned architecture for painting. His work was strongly nationalistic and his anti-fascist, anti-church frescos at Mexico City airport (1937–8) were destroyed in 1939 during a political swing to the right. In the 1950s he returned to architecture, now advocating a more 'organic' approach inspired in part by Frank Lloyd *Wright. His most celebrated work in this vein is the Library of the National University in Mexico City (1951–3) in which a modern structural design is completely covered externally in mosaics of his own design symbolically representing the history of Mexican culture. In 1953–6 O'Gorman built a second home for himself outside Mexico City. This too was lavishly decorated in mosaics externally and internally and it was designed to harmonize with the lava formation of the landscape. He committed suicide.

oil paint. Paint in which *drying oils are used as the *medium, the dominant material for serious painting in Europe from the 16th cent. to the present day. *Vasari's attribution of the invention of oil painting to Jan van *Eyck was long accorded the status of fact, but it is now known that its origins are older and obscurer. *Theophilus among others specifically advocates grinding *pigments into *linseed oil or *walnut oil, although it is clear from his account that it was the practice to dry each layer in the sun before applying the next layer. The precise nature of van Eyck's innovation is unknown, but there is no doubt that he was responsible for a revolution of the techniques of oil painting in Europe, bringing greater flexibility, richer and denser colour, a wider range from light to dark, and subtle transition and blending of tones. From his time the great potentialities of the method began to be exploited and it gradually increased in popularity over *tempera. In Italy, where *Antonello da Messina was an important pioneer, its use developed more slowly than in the north. Well into the 16th cent. a mixed technique was used and oil painting was still often executed in the meticulous manner traditional to tempera painting. Throughout the *Renaissance pictures were built up in thin transparent *glazes with little or no admixture of white, which allowed light to be reflected back from the luminous ground or *underpainting, and full advantage was taken of the modifications of colour and tone which are produced when an under-layer of paint makes its influence felt through a partially transparent over-layer. Giovanni *Bellini achieved a particular richness of colour and his *Doge Leonardo Loredan* (NG, London, *c.*1501) was one of the earliest attempts to produce texture by the use of *impasto.

Bellini's pupil *Titian was the first artist to give oil paint an expressive life of its own, exploiting its rich textural qualities, and in the final stages of a work, as his pupil *Palma Giovane reported, using 'his fingers more than his brush'. This ability to show an artist's personal 'handwriting' is one aspect of the versatility that has ensured its pre-eminence, for it can attain any variety of surface texture from violent impasto to porcelain smoothness. Its versatility was increased still further in the 19th cent. with the invention of the collapsible tin *tube, which made it convenient to work out of doors, and the introduction of a greater range of bright colours.

O'Keeffe, Georgia (1887–1986). American painter. One of the pioneers of modernism in America, she was a member of the circle of *Stieglitz, whom she met in 1916 and married in 1924. She is best known for her near-abstract paintings based on enlargements of flower and plant forms, works of great elegance and rhythmic vitality, whose sensuous forms are often sexually suggestive (*Black Iris*, Met. Mus., New York, 1926). In the 1920s she also painted townscapes of New York in a manner close to that of the *Precisionists and landscapes done in broad, simple forms. From the 1930s she spent each winter in New Mexico and she settled there after Stieglitz's death in 1946, the desert landscape appearing frequently in her paintings (bleached animal bones were a favourite subject). She began to travel widely in the 1950s and many of her later paintings were inspired by views of the earth, sky, and clouds seen from an aeroplane. She became partially blind in 1971 and did little work thereafter.

Oldenburg, Claes (1929–). Swedish-born sculptor and graphic artist who became an American citizen in 1953. He was educated at Yale University and studied at the Art Institute of Chicago (earning his living with part-time jobs as a reporter and illustrator), then in 1956 settled in New York. There he came into contact with a group of young artists, including Allen *Kaprow, George *Segal, and Jim *Dine, who were in revolt against *Abstract Expressionism and from *c.*1958 onwards he became interested in arranging *happenings, ensembles, *environments, 'situations', etc. His inspiration was drawn largely from New York's street life—shop windows, advertisements, graffiti, and so on—and in 1961 he opened 'The Store', at which he sold painted plaster replicas of foods and other domestic objects. This led to the work with which his name is most closely associated—giant-size sculptures of foodstuffs and 'soft sculptures' of normally hard objects (*Dual Hamburger*, MOMA, New York, 1962). With these he was hailed as one of the leaders of American *Pop art. Oldenburg is also well known for his projects for colossal monuments—for example, *Lipsticks in Piccadilly Circus, London* (Tate, London, 1966), consisting of a magazine cutting of an array of lipsticks pasted on to a picture postcard. The first of these projects to be realized was a giant lipstick erected at Yale University in 1969. Since 1976 he has concentrated almost exclusively on such large-scale projects, for example the 70-foot-high Match Cover erected in Barcelona in 1992.

oleograph. A coloured lithograph impressed with a canvas grain to make it look like an oil painting. Oleographs were popular—but often considered rather vulgar—in the second half of the 19th cent.

Olitski, Jules (1922–). Russian-born American painter and sculptor, one of the leading figures of

*Post-Painterly Abstraction, specifically of *Colour Field painting. His early paintings were influenced by *Fauvism and he then went on to heavily textured abstracts, but in 1960 the direction of his work changed radically when he began experimenting with stain techniques in the manner of *Frankenthaler and *Louis. In 1964 he began using a spray gun and in the second half of the 1960s he developed the type of painting for which he is best known—vast canvases covered with luscious mists of atmospheric colour; he said that ideally he would like 'nothing but some colours sprayed into the air and staying there'. Sometimes there are some heavier touches at the edges of the canvas in a sort of ironic reference to *Abstract Expressionism, and in the 1970s Olitski returned to a more textural handling of paint, often reducing his colour to delicate modulations of greys and brown. He took up sculpture seriously in 1968 and has worked mainly with painted metal.

Oliver, Isaac (before 1568–1617). English *miniaturist of French origin, the son of a refugee Huguenot goldsmith who settled in England in 1568. He trained under *Hilliard (whose main rival he later became) and by 1590 was established in his own practice. Although Hilliard continued to receive royal favour under James I (reigned 1603–1625), Oliver was made *Limner to the Queen, Anne of Denmark, in 1604, and was patronized by Henry, Prince of Wales, and his circle. His style was more naturalistic than that of Hilliard, using light and shade to obtain modelling and generally dispensing with the emblematic trappings so beloved of the Elizabethan age. He was in Venice in 1596, and unlike Hilliard he did history paintings in miniature. Contemporary sources indicate that he probably also painted life-size portraits, and some of the pictures attributed to William *Larkin have been put forward rather as possible works by Oliver. His son, **Peter** (1594?–1647), continued in his style, but specialized also in miniature copies after the Old Masters. There are examples of the work of father and son in the Victoria and Albert Museum.

Olympia Pediments. Greek pedimental sculptures, marble and over life-size, from the temple of Zeus at Olympia and now in the museum there. They date from about 460 BC and form the greatest sculptural ensemble representative of the *severe style of the early *Classical period.

Omega Workshops. Decorative arts company founded by Roger *Fry in 1913. The enterprise was the outcome of new views about the function of art in society developed by Fry perhaps under the influence of Thorstein Veblen's *The Theory of the Leisure Class* (1899), and was an attempt to bring modern art into touch with daily life by the production of decorative art. Among his chief associates in this enterprise were Duncan *Grant (the best of the Omega designers) and Vanessa *Bell, and Fry encouraged amateur as well as professional artists to participate (all work was sold anonymously). In general the work was done by painters and consisted of decorating manufactured objects rather than designing products from scratch. Bright colour predominated in the designs, much of them abstract. The workshops, which operated from 33 Fitzroy Square, London, were a financial failure (Fry had no business aptitude and the First World War had a disastrous effect on sales), and closed in 1919. Examples of their work are in the Courtauld Institute Galleries, London, and in the Victoria and Albert Museum. They include furniture, textiles, pottery, and tiles.

Ono, Yoko. See FLUXUS.

Oostsanen. See CORNELISZ. VAN OOSTSANEN.

Op art (abbreviation of Optical art on the analogy of Pop art). A type of abstract art that exploits certain optical phenomena to cause a work to seem to vibrate, pulsate, or flicker. It flourished mainly in the 1960s; the term was first used in print in the American magazine *Time* in October 1964 and had become a household phrase by the following year, partly through the attention given to the exhibition 'The Responsive Eye' held at the Museum of Modern Art, New York, in 1965. This was the first international exhibition with a predominance of Op paintings. The development of Op art as a recognizable movement had begun a few years earlier than this, in about 1960, the works and theories of Josef *Albers being among the main sources. The devices employed by Op artists (after-images, effects of dazzle and vibration, and so on) are often elaborations on the well-known visual illusions to be found in standard textbooks of perceptual psychology, and maximum precision is sought in the control of surfaces and edges in order to evoke an exactly prescribed retinal response. Many Op paintings employ repeated small-scale patterns arranged so as to suggest underlying secondary shapes or warping or swelling surfaces. This kind of work can retain much of its effect in reproduction, but Op art also embraces constructions that depend for their effects on light and/or movement, so Op and *Kinetic art sometimes overlap. The two most famous exponents of Op art are Bridget *Riley and Victor *Vasarely. Their work illustrates the considerable impact that Op made on fashion and design in the 1960s—its instant popular success (accompanied by a fairly cool critical reception) is hard to parallel in modern art. Op art became something of a craze in women's fashion and in 1965 Riley unsuccessfully tried to sue an American clothing company that used one of her paintings as a fabric design. One of Vasarely's designs was used on the plastic carrier bags of France's chain of COOP stores. Among other exponents of Op art the best known is probably the American Richard Anuszkiewicz (1930–), a former pupil of Albers; his work is typically concerned with radiating expanses of lines and colours.

Opie, John (1761–1807). English painter. He was something of a child prodigy and was discovered by the political satirist John Wolcot (1738–1819; better known by his pen name Peter Pindar), who in 1781 successfully launched him in London as an untaught genius ('The Cornish Wonder'). Opie was then painting strongly modelled portraits and rustic *fancy pictures with *Rembrandtesque lighting. He soon lost the rather rugged freshness of his early work, however, and his later paintings were undistinguished and repetitive. His career continued to flourish, however, and he became Professor of Painting at the *Royal Academy in 1805, his lectures being posthumously published in 1809 (prefaced with a memoir by his wife, the novelist and poet Amelia Opie). Apart from portraits and *genre scenes, he also painted history pictures, notably for *Boydell's Shakespeare Gallery.

Opie, Julian. See NEW BRITISH SCULPTURE.

Oppenheim, Meret (1913–85). German-Swiss painter, sculptor, and maker of objects, born in Berlin. In 1932 she moved to Paris, where she was introduced to the *Surrealist group by *Giacometti and became for a while the model and disciple of *Man Ray. He described her as 'one of the most uninhibited women I have ever known' and she became celebrated among the Surrealists as the 'fairy woman' whom all men desire. She had a long career, but she is remembered mainly for one early work: *Object* (MOMA, New York, 1936), a fur-lined tea cup and saucer. This became famous as a symbol of artistic anarchy after being shown at major Surrealist exhibitions in London and New York in 1936.

opus anglicanum (English work). Term internationally applied in the late 13th and 14th cents. to the English embroidery of that period and all other embroidery in similar style. It was among the most sumptuous and delicate needlework ever made. The surviving ecclesiastical vestments of this kind display small figures or religious scenes, comparable in style with East Anglian manuscript *illumination, framed by foliaged scrolls or in geometrical or architectural compartments; these designs are worked in coloured silks, generally on backgrounds of gold thread. The Vatican inventories of 1295 contain well over one hundred references to such work, while vestments of *opus anglicanum* donated by Pope Nicholas IV to the cathedral of Ascoli Piceno (1295), by Boniface VIII to that of Anagni (*c.*1300), by Clement V to that of Saint-Bertrand-de-Comminges (1309), and by Pius II to that of Pienza (1462) may still be seen in those towns. Among museum collections, that of the Victoria and Albert Museum, which includes the well-known Syon Cope, is the most comprehensive.

opus sectile. The name given by the Romans to *marble *inlay where the pieces are cut into shapes which follow the lines of the pattern or picture, as distinct from *mosaic, where the design is built up from countless small pieces (tesserae) of stone or glass, and where their grouping and not their shape gives the composition its form. It is a rarer and more luxurious art, since the plates of marble are larger, more fragile, and more precious than the mosaic fragments; and it was in Egypt and Asia Minor, lands rich in coloured marble, that it had its origin. The Book of Esther (1: 6) describes 'a pavement of red and blue and white and black marble' in the palace of Ahasuerus. But little survives from before Roman times, except fragmentary remains from the Ptolemaic period which have been found in Alexandria. *Opus sectile* was used in Early Christian and Byzantine churches, and evolved into more specialized crafts such as *cosmati work. Geometrical *opus sectile* continued to be popular in Italy for decorating church floors throughout the Middle Ages and Renaissance.

orant. See GISANT.

Orcagna, Andrea (Andrea di Cione) (d. 1368/9). The leading Florentine artist of the third quarter of the 14th cent., a painter, sculptor, architect, and administrator. His nickname 'Orcagna' was apparently local slang for 'Archangel' (*Arcangelo*). In 1343/4 he was admitted to the guild of the painters and nine years later to that of the masons. His only certain work as a painter is the altarpiece of *The Redeemer with the Madonna and Saints* (1354–7) in the Strozzi Chapel of Sta Maria Novella. This is the most powerful Florentine painting of its period, and in spite of the massiveness of the figures it represents a reversion from *Giotto's naturalism to the *hieratic ideals of *Byzantine art. Colours are resplendent, with lavish use of gold, and the figures are remote and immobile. The major work attributed to Orcagna is a fragmentary fresco trilogy of the *Triumph of Death*, *Last Judgement*, and *Hell* in Sta Croce. As a sculptor and architect he is known through one work, the tabernacle in Or San Michele (finished 1359), a highly elaborate ornamental structure housing a painting of the *Virgin Enthroned* by Bernardo *Daddi. Orcagna was *capomaestro* of Orvieto Cathedral from 1358 to 1362, supervising the mosaic decoration of the façade. He was also an adviser on the construction of Florence Cathedral. During 1368 Orcagna fell mortally ill while painting the *St Matthew* altarpiece (Uffizi, Florence) and this work was subsequently finished by his brother **Jacopo di Cione** (active 1365–98), who worked in his style and continued it to the end of the century. Another brother, **Nardo di Cione** (active 1343/6–65/6), was also a painter. *Ghiberti attributes to him the series of frescos of *The Last Judgement*, *Hell*, and *Paradise* in the Strozzi Chapel, Sta Maria Novella, which houses Andrea's great altarpiece. Orcagna's style was the dominant influence in late 14th-cent. Florentine painting; there are numerous paintings from the period that are called Orcagnesque, but attributions to the hands of his brothers tend to be speculative.

Orchardson, Sir William Quiller (1832–1910). Scottish painter of *genre subjects and portraits, active in London from 1862. He made his name with historical costume pieces, but later did his most memorable work in modern-day scenes of upper-class married life. In these he used large empty spaces to create feelings of psychological tension or despair. The best-known are *The First Cloud* (Tate Gallery, London, 1887) and a pair representing *A Marriage of Convenience* (City Art Gallery, Glasgow, 1883, and Aberdeen Art Gallery, 1886).

Ordóñez, Bartolomé (d. 1520). The first Spanish sculptor to show clearly the impact of the Italian High *Renaissance. On stylistic grounds, he is presumed to have trained in Florence, perhaps with Andrea *Sansovino, and in 1514–15 he was in Naples, working with Diego de *Siloe on the marble *reredos of the Caracciolo Chapel in the church of S. Giovanni a Carbonara. In 1517 he undertook part of the carved decoration for the choir of Barcelona Cathedral, executing in wood and marble a series of *reliefs in a pure Renaissance style. There followed a number of contracts for tombs, including that of Charles V's parents, Philip I and Juana of Castille, commissioned by Charles himself for the chapel royal at Granada. Ordóñez died while carrying out these commissions with Italian assistants at Carrara; but his testament records that he had completed the greater part of the work, including most of the royal tomb. Although he died young (he was perhaps born *c.*1490), his elegant and imaginative style was influential not only in Spain, but also in Naples and in Flanders, where it was taken by Jean *Mone, his assistant at Barcelona.

Orley, Bernard (or **Bernaert** or **Barend**) **van** (*c.*1490–1541). Netherlandish painter of religious subjects and portraits and designer of tapestries and stained glass. He was the leading artist of his day in Brussels, becoming court painter to Margaret of Austria (1480–1530), regent of the Netherlands, in 1518 and to her successor Mary of Hungary in 1532. His work is characterized by the use of ill-digested Italianate motifs. There is no evidence that he visited Italy, and his knowledge presumably came from engravings and from *Raphael's tapestry *cartoons, which were in Brussels *c.*1516–19; he has (very flatteringly) been called 'the Raphael of the Netherlands'. His best-known work is the turbulent *Job* altarpiece (Brussels, 1521). As a portraitist his style was quieter and more thoughtful (*Georg Zelle*, Musées Royaux, Brussels, 1519). None of van Orley's paintings bears a date later than 1530; after that time he was chiefly occupied with designing tapestries and stained-glass windows. Examples of his tapestries are in the Louvre and of his stained glass in St Michel, Brussels.

Orozco, José Clemente (1883–1949). Mexican painter, with his contemporaries *Rivera and *Siqueiros one of the trio of politically and socially committed fresco painters who were the dominant force in modern Mexican art. Following the first outburst of revolutionary activity in Mexico in 1910 (which was to last on and off until 1920), Orozco began working as a political cartoonist. In 1912 he began a series of watercolours called 'House of Tears' dealing with prostitutes (a favourite symbol of human degradation for Orozco). The angry reaction of critics and moralists to these works was one of his reasons for leaving for the USA, where he spent three unhappy and unproductive years, 1917–20. His career as a muralist began after he returned to Mexico in 1920. The country was now relatively stable under the government of Alvaro Obregón, who encouraged nationalistic subjects as a way of creating a positive identity for the country after years of turmoil. Orozco's first murals were in the Escuela Nacional Preparatoria (National Training School), 1923–4. They were controversial because of their caricatural style, and all except *Maternity* and *The Rich Banquet while the Workers Quarrel* were subsequently destroyed or altered. His style matured towards a greater monumentality in frescos in the Casa de los Azulejos (House of

Tiles), Mexico City (1926), and in a second series at the National Training School (1926–7). The work brought him little recognition, however, so in 1927 to 1934 (broken by a brief trip to Europe in 1932) he again worked in the USA. This time he was much more successful, carrying out a number of important mural commissions, most notably a cycle for Dartmouth College, New Hampshire, on *The Coming* and *The Return of Quetzalcoatl* (1932–4). This huge scheme showed his outlook crystallizing into a contrast between a pagan paradise and a capitalist hell. Unlike Rivera and Siqueiros, Orozco did not align himself with a political movement, but his work had an intense humanitarian mission. He returned to Mexico in 1934 with a big reputation after his success in the USA, and he spent most of the rest of his life engaged on mural projects in Mexico City and Guadalajara, the country's second city. In his last years his work became ever more violent in expression, moved by a passionate concern for the suffering and miseries of mankind. His studio in Guadalajara is now a museum dedicated to him.

Orpen, Sir William (1878–1931). British painter, chiefly famous as one of the leading fashionable portraitists of his day. Orpen was a child prodigy and had a brilliant student career at the *Slade School. He worked mainly in London but he kept up links with his native Ireland, teaching part-time at the Metropolitan School in Dublin, 1902–14. His style had much in common with that of his friend Augustus *John, being vigorous and painterly but sometimes rather flashy. He was at his best when he was away from his standard boardroom and drawing-room fare, and his numerous self-portraits are often particularly engaging, as he pokes fun at himself in character roles. Up to the First World War he had a steady rise in worldly success and after the war he earned an average of about £35,000 a year, rising to over £50,000 a year in 1929—a colossal sum then. In 1920 a story appeared in London newspapers that he had refused an offer of £1,000,000 to work for a dealer in the USA, and he was one of the few British artists of his time capable of attracting public attention in such a way. Apart from portraits, Orpen also painted genre subjects, landscapes, interiors, nudes, and allegories, and he did memorable work as an *Official War Artist in France (he also attended the 1919 Peace Conference in Paris and painted *The Signing of the Peace in the Hall of Mirrors, Versailles, 28 June, 1919* (Imperial War Museum, London)). His reputation faded badly after his death but revived greatly in the 1970s.

Orphism (or **Orphic Cubism**). A short-lived movement in French painting that developed out of *Cubism between late 1911 and early 1914. The word 'Orphism', which had previously been used by the *Symbolists, was applied to the movement by Guillaume *Apollinaire at the exhibition of the *Section d'Or in October 1912; the reference to Orpheus, the singer and poet of Greek mythology, reflected the desire of the artists involved to bring a new element of lyricism and colour into the austere intellectual Cubism of *Picasso, *Braque, and *Gris. The painters mentioned by Apollinaire as practitioners of Orphism were Robert *Delaunay, Fernand *Léger, Francis *Picabia, and Marcel *Duchamp; Frank *Kupka was the main member of the movement not named by Apollinaire. They made colour the principal means of artistic expression and Delaunay and Kupka were among the first to paint totally non-representational pictures, seeing an analogy between pure abstraction and music. Although it was short-lived, Orphism had a strong influence on German painting (notably *Klee, who visited Delaunay in 1912, *Macke, and *Marc) and also on *Synchromism.

Orsi, Lelio (*c*.1511–87). Italian *Mannerist painter and architect. He worked mainly in his birthplace, Novellara and in Reggio Emilia, and he is counted a member of the Parmesan School. His large-scale decorative work has almost all perished and he is now known mainly by *cabinet pictures of religious subjects (*The Walk to Emmaus*, NG, London, *c*.1570). He drew on many different influences (*Correggio, *Giulio Romano, *Michelangelo, *Parmigianino, and perhaps also German woodcuts), but his style has a leaning towards the bizarre, with dramatic stage-lighting effects, that gives it a distinctly personal touch.

Os, van. Dutch family of painters active during the late 18th and 19th cents. **Jan** (1744–1808) was the founder. He and his daughter **Maria Margrita** (1780–1862) and his son **Georgius Jacobus Johannes** (1782–1861) specialized in painting flowers and fruits in the lavish detailed manner of Jan van *Huysum. Another of Jan's sons, **Pieter Gerardus** (1776–1839), specialized in pictures of cattle. His son and pupil, **Pieter Frederik** (1802–92), taught *Mauve.

Osona, Rodrigo de (active 1476–84). Spanish painter. He was one of the leading painters of his day in Valencia, introducing both Netherlandish and Italian *Renaissance influence to the area. His most important work is the Crucifixion in St Nicholas, Valencia (1476). On the basis of an altarpiece of *The Adoration of the Magi* (V&A, London) signed 'the son of Master Rodrigo', several works have been attributed to **Rodrigo de Osona the**

Younger, who is documented from 1505 to 1513. He continued his father's style in a weaker and more Italianate manner.

Ostade, Adriaen van (1610–85). Dutch painter, active in his native Haarlem. Although he turned his hand to most types of subject, he was principally a *genre painter. He was a very prolific painter and also made watercolours, etchings, and drawings. According to *Houbraken, both he and *Brouwer were pupils of Frans *Hals. His work, however, is closer to Brouwer's than to that of Hals. His early pictures depict lively scenes of peasants carousing or brawling in crowded taverns or hovels. In these works his *palette is limited to brown tones with a few accents of red, violet, or grey; sharp contrasts of light and shadow heighten the dramatic effect. In his later works (after *c.*1650) his peasants learn better manners and the rooms they live in are tidier. These later pictures are lighter in key and more colourful; thus they follow the general trend of Dutch painting around this time. Ostade was successful, popular, and much imitated (see DUSART). His most talented pupil was his brother **Isaak** (1621–49), also of Haarlem. As well as painting genre scenes in the manner of Adriaen, Isaak was an outstanding practitioner of the winter landscape, and his early death cut short a career of great promise. Good examples of the work of both brothers are in the National Gallery, London.

ottocento. See QUATTROCENTO.

Ottonian art. Term applied to art of the Holy Roman Empire in the 10th cent. and most of the 11th cent. The emperor Otto the Great (936–73), who re-established a strong royal authority, gave his name to the dynasty and the period. Although the Ottonian emperors reigned from 919 to 1024, the style associated with the name continued further into the 11th cent. It represented an amalgam of a revival of *Carolingian art, a renewed interest in Early Christian art and a strong influence from contemporary *Byzantine art with an admixture of certain northern traditions. The Ottonian period saw the revival of large-scale bronze casting (see BERNWARD OF HILDESHEIM) and of life-size sculpture (in the celebrated Gero Crucifix of *c.*970 in Cologne Cathedral), but the most typical sculptural products of the time were in ivory and metalwork, notably for book covers and altar reliefs. Though wall paintings still survive, the character of Ottonian art is better seen in a rich store of *illuminated manuscripts. In spite of differences in local schools, all Ottonian illumination has certain things in common, notably the pre-eminence given to the human figure, which is often imbued with strong expression and marked by exaggerated gestures. Ottonian art was widely influential and was one of the sources out of which *Romanesque grew.

Oudry, Jean-Baptiste (1686–1755). French painter, tapestry designer, and illustrator. He was a pupil of *Largillière and painted some portraits, but he is renowned chiefly as one of the outstanding animal painters of the 18th cent. With *Desportes he was the foremost exponent of hunting scenes and still lifes with dead game. Some of his best work was done as a tapestry designer, and he was head of the Beauvais and *Gobelins factories from 1734 and 1736 respectively. He also did book illustrations, notably for an edition of La Fontaine's *Fables* (1755–9). The best collection of his work is in the State Museum at Schwerin, and he is also well represented in the Louvre and the Wallace Collection. His son **Jacques-Charles** (1720–78), a flower and animal painter, sometimes collaborated with him.

Ouwater, Albert van (active mid-15th cent.). Netherlandish painter. Only one documentary reference to him is known (it refers to the burial of his daughter in 1467) and otherwise knowledge depends on what van *Mander wrote about him. He said Ouwater was the founder of the Haarlem school and the master of *Geertgen tot Sint Jans. He also praised his skill as a landscapist, so it is ironic that the only painting that can be identified as Ouwater's from van Mander's descriptions of his work is an interior scene—*The Raising of Lazarus* (Staatliche Museen, Berlin). The thoughtful, unemotional style has some affinities with the work of Dirk *Bouts, who was a native of Haarlem.

Ovenden, Annie and **Graham.** See BLAKE, PETER.

Ovens, Juriaen. See CAMPEN.

Overbeck, Friedrich (1789–1869). German painter, the leading member of the *Nazarenes. He moved to Rome in 1810 and was based there for the rest of his life, although he made several visits to Germany. In 1813 he was converted to Roman Catholicism and apart from a few portraits (there is a self-portrait in the Uffizi) his work was almost exclusively on religious themes. He painted in a consciously archaic style—clear and sincere but rather pallid—based on the work of *Perugino and the young *Raphael. *The Rose Miracle of St Francis* (Portiuncula Chapel, Assisi, 1829) is perhaps his best-known painting. The high-minded and didactic tone of his work won it a more sympathetic acceptance (particularly in England) than its artistic quality alone merited. *Pugin, *Dyce, and Ford

Madox *Brown were among his supporters and there were affinities between his aspirations and those of the *Pre-Raphaelite Brotherhood.

Ozenfant, Amédée (1886–1966). French painter, writer and teacher. In 1918 he met *Le Corbusier with whom he founded *Purism, but he is more important as a writer and teacher than as a painter. He lived in London, 1935–9, then in New York, 1939–55, founding art schools in both cities. After returning to France he settled in Cannes, where he directed a studio for foreign art students. His *Foundations of Modern Art* (1931, enlarged edition, 1952) is a study of the interrelationship of all forms of human creativity, including science and religion, and is one of the most widely read books by any modern artist. His *Memoires, 1886–1962* was posthumously published in 1968.

Pacheco, Francisco (1564–1644). Spanish painter and writer, active in Seville. He was a man of great culture, a poet and scholar as well as a painter, and his house was the focus of Seville's artistic life. As a painter he was undistinguished, working in a stiff academic style (though his portraits are fresher than his religious works). He was an outstanding teacher, however, for (in spite of his own limitations) he was sympathetic to the more naturalistic style that was then developing. And he was generous enough in spirit to acknowledge openly that his greatest pupil, *Velázquez (who became his son-in-law in 1618), was a much better painter than himself: 'I consider it no disgrace for the pupil to surpass the master.' Alonso *Cano was his other outstanding pupil, and Pacheco often collaborated with the great sculptor *Montañés, painting his wooden figures. In 1649 Pacheco's treatise *El Arte de la Pintura* (The Art of Painting) was posthumously published; part theoretical, part biographical, it is a major source of information for the period (it includes accounts of his meeting with El *Greco in Toledo in 1611 and of Velázquez's early career). Pacheco was an official art censor to the Inquisition and the highly detailed *iconographical prescriptions in his book were often strictly adhered to by contemporary artists. Little of Pacheco's work can be seen outside Seville.

Pacher, Michael (active 1465?–98). Austrian painter and sculptor. He died at Salzburg, but most of his career was spent at Bruneck in the Tyrol, where he is documented 1467–96 and where his house is still preserved. Nothing is known of his training, and the earliest recorded work by him (a signed and dated altarpiece of 1465) is now lost. He worked mainly for local churches, carrying out the carving as well as the painting of his altarpieces, and much of his work is still *in situ*. His most celebrated work is the St Wolfgang altarpiece (1471–81) in the church of S. Wolfgang on the Abersee, a huge and complex *polyptych with some astonishingly intricate painted and gilded wood-carving. Although Pacher's sculpture is thoroughly late *Gothic in spirit, his painting is strongly influenced by Italian art. He is particularly close to *Mantegna, especially in the way dramatic effects are obtained by using a low viewpoint and setting the figures close to the picture plane. There is no documentary evidence that Pacher visited Italy, but because of its proximity to the Tyrol it seems overwhelmingly likely that he did. Pacher's work had wide influence and he was the most important interpreter of *Renaissance ideas for German painting before *Dürer. A **Friedrich** and a **Hans Pacher**, presumably related, were active in the

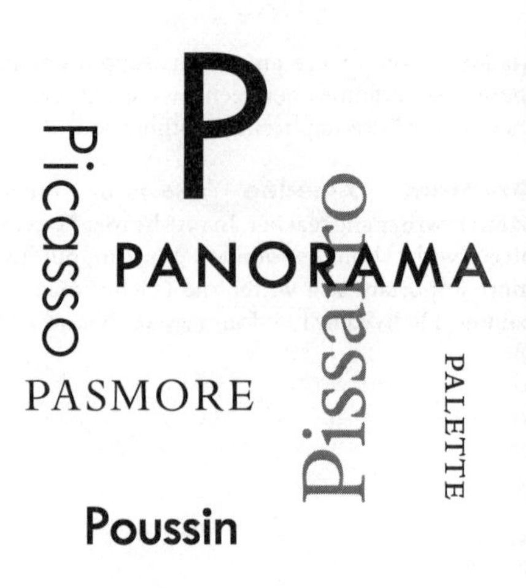

Tyrol at the same time as Michael, and Friedrich collaborated with him.

Pacherot, Jérôme. See COLOMBE.

Paciola, Luca. See GOLDEN SECTION.

Paeonius. Greek sculptor from Mende in Thrace active in the second half of the 5th cent. BC. He was the sculptor of a marble statue of Victory (*c*.420 BC) found at Olympia in 1875 and now in the museum there (an inscription on the pedestal names him). *Pausanias' assertion that he also made the sculpture of the east pediment of the Temple of Zeus, also at Olympia (see OLYMPIA PEDIMENTS), is not accepted as it is appreciably earlier in date and clearly different in style.

Painters Eleven. A group of Toronto abstract painters active from 1953 to 1960. The most important member was William *Ronald. Apart from being abstract rather than figurative artists, the painters had little in common and for this reason the non-committal name of the group was deliberately chosen. They were united mainly by the desire to promote their work in an environment unfavourable to abstract art and in this achieved considerable success. A group of their works is in the Robert McLaughlin Gallery, Oshawa, Ontario.

Pajou, Augustin (1730–1809). French sculptor. He was Keeper of the King's Antiquities at the *Louvre (1777), and director of the scheme of decoration for the Opera at Versailles (1768–70), but like his master J.-B. *Lemoyne worked mainly as a portraitist. His best works are busts of friends (*Hubert *Robert*, Louvre, Paris, 1789).

Palacios, Francisco de. See PEREDA.

pala d'altare. Italian term, usually shortened to '*pala*', for a large *altarpiece, often one consisting of a single picture.

Palamedesz., Anthonie (1601–73). Dutch portrait and *genre painter, active mainly in his native Delft. He was a pupil of *Mierevelt and Dirk *Hals and his paintings resemble those of his masters—his portraits rather wooden, his 'merry company' groups of soldiers, cavaliers, and their ladies livelier and pleasantly coloured. His brother **Palamedes I Palamedesz.** (1607–38) and his nephew **Palamedes II Palamedesz.** (1633–1705) were both painters.

palette. The flat board, usually rectangular, ovoid, or kidney-shaped, and generally with a hole for the thumb, on which artists arrange their paints ready for use: early examples sometimes had a handle, rather like a table-tennis bat, but a thumb-hole is now standard. Palettes first appeared *c.*1400; before then individual containers (sometimes shells) were used for mixing colours. For oil painting, mahogany is traditionally considered the best material for palettes, although other close-grained hardwoods have been used. Materials such as porcelain or ivory have been used by watercolour or *miniature painters and also sometimes by oil painters—*Millais, for example, used a porcelain palette early in his career, when he painted with fastidious detail and wished to avoid muddying his colours.

For many artists choice of their pigments and the order in which they are set out on the palette is a very important and personal matter and much advice was published on how to 'set' a palette in handbooks of the 17th and 18th cents. *Delacroix once said to an assistant 'When I have arranged the whole combination of contrasts of tone on my palette my picture is made', and *Baudelaire describes Delacroix as setting out the pigments on his palette with the fastidious care of a woman arranging a bouquet of flowers. *Courbet is said to have prepared his palette by graduating all his pigments in various strengths with white. By extension the term 'palette' thus refers to the range of colours characteristic of an artist; *Caravaggio has a dark or restricted palette, *Monet a bright or rich palette.

palette knife. A thin, flexible, dull-edged blade, set in a handle, used for mixing paint, scraping it off the *palette or canvas and also as a painting instrument (although more delicate tools—'painting knives'—are often preferred for this purpose). Palette knives became popular in the 18th cent., ordinary knives being used for the purpose before this. The trowel shape commonly used today is said to have been invented by *Courbet.

Palladianism. Movement in architecture and interior design inspired by the buildings and publications of the great Italian architect **Andrea Palladio** (1508–80); in particular the term is applied to British architecture in the period from about 1715 to 1750. During this period Palladio was worshipped by figures such as Lord *Burlington and William *Kent and the exuberant *Baroque style gave way to one characterized by symmetry, regularity, and correctness of detail. Palladio himself did not design furniture, but in early 18th-cent. England it was made in a style thought suitable for Palladian houses with much use of rather heavy architectural features such as pediments, cornices, and swags.

Palma Giovane (properly Jacopo Palma) (*c.*1548–1628). Venetian painter, great-nephew of *Palma Vecchio. He is said to have been a pupil of *Titian, but this tradition has been doubted (it is probably based on the fact that he completed the *Pietà* which Titian left incomplete at his death). In the late 1560s and early 1570s he worked in central Italy, mainly Rome, but thereafter he spent the rest of his life in Venice, and after the death of *Tintoretto in 1594 he was the leading painter in the city. His style was influenced by several of his great Venetian predecessors—*Veronese as well as Titian and Tintoretto—and by central Italian *Mannerism. He was extremely prolific, fulfilling many commissions from abroad as well as for Venetian churches, and his later work is often mechanical. As well as religious pictures, he painted historical and mythological works, and he also made etchings.

Palma Vecchio (properly Jacomo Palma) (*c.*1480–1528). Italian painter, born near Bergamo, but active for all his known career in Venice, where he is first documented in 1510. His original name was Jacomo Negreti, but he was using the name Palma by 1513. He is called Palma Vecchio (Old Palma) to distinguish him from Palma Giovane (Young Palma), his great-nephew. Nothing is known of his training, and there is indeed very little secure knowledge about his life and works, none of his pictures being dated or reliably signed and very few of them being certainly identifiable from early sources. His style is distinctive, however, and in practice the definition of his *œuvre* is much less problematic than with many of his contemporaries. He painted a few altarpieces for Venetian churches, but most of his work was done for private clients, his speciality being half-length portrayals of beautiful, voluptuous, blonde-haired women, sometimes in religious or mythological guise. In opulence of colour and beauty of handling they show the influence of the early work of *Titian, and the finest, such as the celebrated *La Bella* (Thyssen coll., Madrid), are worthy of his name. Palma also painted some *Giorgionesque reclining nudes and some male portraits. His work was influential on painters of the next generation in Venice, notably *Bonifazio Veronese.

Palmer, Erastus Dow (1817–1904). American sculptor. Self-taught, he rarely left Albany in his native New York State, and he was the most successful American sculptor of his period to work in the USA rather than in Europe. He began with *cameo portraits and had a flourishing business with portrait busts and bas-*reliefs on religious subjects, but his most celebrated work, now and in his own day, is *The White Captive* (Met. Mus., New York, 1858). Inspired by *Powers's *Greek Slave*, it shows a naked young girl who has been captured by Red Indians but is sustained by her Christianity—this accompanying storyline undoubtedly contributed to its popularity. The statue is fresher in observation than Powers's *Greek Slave*, for although the marble surfaces are impeccably smooth, the chubby proportions of the figure are unidealized and the strikingly characterized head is a portrait of Palmer's daughter.

Palmer, Samuel (1805–81). English landscape painter and etcher. He showed a precocious talent and exhibited landscape drawings at the *Royal Academy when he was 14. In 1822 he met *Linnell, who introduced him to William *Blake in 1824. Palmer had had visionary experiences from childhood and the effect of Blake upon him was to intensify an inherent mystical bent. In 1826 he moved to Shoreham, near Sevenoaks, Kent, where he was the central figure of the group of artists known as the *Ancients and produced what are now his most famous works—landscapes charged with a sense of pantheistic fecundity and other-worldly beauty. In about 1832 what he called his 'primitive and infantine feeling' for landscape began to fade, and after returning to London in 1835, marrying Linnell's daughter in 1837, and spending a two-year honeymoon in Italy, the break with his visionary manner was complete. His later paintings were in a much more conventional topographical or pastoral mode, highly-wrought and often sentimental in feeling. In his etchings, however, something of his early genius remained; at his death he was working on an edition of Virgil's *Eclogues*, translated and illustrated by himself. His early work was virtually forgotten until the 1920s, but it has subsequently influenced modern romantic landscape artists such as Paul *Nash and Graham *Sutherland. Palmer's work is well represented in London (BM, Tate, V&A), Oxford (Ashmolean), and Cambridge (Fitzwilliam).

Palomino y Velasco, Antonio (1655–1726). Spanish painter and writer on art, active mainly in Madrid, where in 1688 he was appointed a painter to the king. He was famous in his day for the frescos in churches in Madrid and elsewhere, but he is now best known for his book *Museo Pictórico y Escala Optica* (Vol. 1, 1715; Vols. 2 and 3, 1724). The first two volumes are devoted respectively to the theory and the practice of art, and the third volume is a collection of biographies, the most important source for the history of Spanish art from the 16th to the early 18th cents. Abridged translations of the biographies appeared in English (1739) and French (1749) and Palomino has become known as 'the Spanish *Vasari'.

panel. Term in painting for a *support of wood, metal, or other rigid substance, as distinct from *canvas. Wood was used for that purpose in antiquity, which also knew the practice of glueing canvas to wood (*marouflage). In the Middle Ages leather was often stretched on panels before *priming, particularly in Russia. Until the adoption of canvas in the 15th cent. nearly all the movable paintings of Europe were executed on wood, and even up to the beginning of the 17th cent. it is probable that as much painting was done on the one support as on the other. Painters who worked on a small scale often used copper panels (*Elsheimer is a leading example), and in the colonial art of South America copper and tin and even lead and zinc were used. On a larger scale, slate has occasionally been used as a support, notably by *Rubens for his altarpiece for Sta Maria in Vallicella (the Chiesa Nuova) in Rome; the picture he originally painted was said to reflect the light unpleasantly and slate was used for the replacement to produce a more matt finish.

For wood panels the Italian masters of the *Renaissance preferred white poplar, while oak was the most common wood used in northern Europe. Analysis of the contents of art galleries has yielded a long list of other woods, including beech, cedar, chestnut, fir, larch, linden, mahogany, olive, and walnut. In the 20th cent. cedar, teak, and dark walnut are favourites. The panel must be well seasoned to remove resin and gum as otherwise it may warp and split. *Cennini advised that small panels should be boiled to prevent splitting, presumably because this removes some of the resin, and modern experts recommend steaming for the same reason. For a large picture several pieces had to be accurately jointed together and glued with *casein, a difficult operation which is described in medieval treatises. In the 20th cent. painters have used plywood, fibre-board, and other synthetic materials (see also ACADEMY BOARD).

Painting directly on wood is not satisfactory because the wood absorbs too much of the paint and does not reflect enough light, besides reacting chemically with some of the *pigments. Moreover, some woods darken in course of time. Normally, therefore, after the holes had been filled up if necessary, the panel was *sized and coated with several layers of *gesso, or of chalk, so that it presented a smooth, even *ground. The backs of panels also require protection against woodworm and against damp, which can cause warping and rot.

Panini (or **Pannini**), **Giovanni Paola** (or **Gianpaolo**) (1691/2–1765). Italian painter, born in Piacenza and trained in the school of stage designers at Bologna, possibly under one of the *Bibiena family. By 1711 he was in Rome, where he became the pre-eminent painter of real and imaginary views of the city. He was the first painter to make a special feature of ruins—an aspect of his work which links him with Hubert *Robert and *Piranesi—and also did paintings of public festivities and events of historical importance. Panini taught *perspective at the French Academy in Rome and his influence was strong with French as well as Italian artists. He was a prolific painter and many galleries have examples of his work.

Panofsky, Erwin (1892–1968). German-American art historian. He was a professor at Hamburg University 1926–33, then was dismissed by the Nazis. He settled in the USA where he had been a visiting professor at New York since 1931, and was then visiting professor at Princeton University, 1934–5, and from 1935 professor at the Institute for Advanced Study, Princeton. Kenneth *Clark described him as 'unquestionably the greatest art historian of his time', and he is renowned particularly for his immensely learned contributions to the study of *iconography. His many books include *Studies in Iconology* (1939), *Albrecht Dürer* (1943), *Early Netherlandish Painting* (1953), and *Tomb Sculpture* (1964). *Meaning in the Visual Arts* (1955) is a collection of his most important essays and articles representing a cross-section of his work. Panofsky enjoyed teaching and was influential through his work in the classroom and lecture hall as well as through his writings. Many scholars have tried to emulate his way of analysing works of art as part of a broad philosophical, intellectual, and cultural pattern, but few have rivalled his learning or finesse, and some of his followers have been accused of 'over-interpreting' pictures in their desire to uncover 'hidden symbolism'. Panofsky himself warned of this possibility, writing 'there is admittedly some danger that iconology will behave, not like ethnology as opposed to ethnography, but like astrology as opposed to astrography.'

panorama. 'A picture of a landscape or other scene, either arranged on the inside of a cylindrical surface round the spectator as a centre (a *cyclorama*), or unrolled or unfolded and made to pass before him, so as to show the various parts in succession' (*OED*). In 1787 a patent for such a device was granted to Robert Barker (1739–1806), a portrait painter and drawing teacher from Edinburgh, and it soon became a popular form of entertainment: 'Panorama painting seems all the rage', *Constable wrote in 1803. Panoramas were indeed a kind of forerunner of the popular cinema and tended to be remarkable for sheer spectacle rather than artistic merit. Distinguished artists were sometimes associated with them, however; notably *Girtin, who made a panorama of London, now lost, and *Mesdag, whose panorama of Scheveningen can still be seen in The Hague. More typical is the panorama of the Battle of Gettysburg (1883) at Gettysburg National Military Park by the French painter Paul Philippoteaux. In more general parlance, the term 'panorama' is used of any wide, uninterrupted view over a scene, particularly a landscape.

pantograph. An instrument, known since the 17th cent., for copying a drawing on a larger or smaller scale. By a simple system of levers the outline of the original work traced with a point attached to one arm can be repeated on to another surface by a drawing instrument attached to another arm.

Paolozzi, Sir Eduardo (1924–). British sculptor and printmaker of Italian parentage. He had his first one-man exhibition as a sculptor in 1947 and in the same year he began making *collages using cuttings from old American magazines, advertising prospectuses, technological journals, etc. (*I was a Rich Man's Plaything*, Tate, London, 1947). Paolozzi regarded these collages as 'ready-made metaphors' representing the popular dreams of the masses, and they have been seen as forerunners of *Pop art (he eventually amassed a large collection of pulp literature, art, and artefacts, which he presented to the University of St Andrews).

From the 1950s he has worked primarily as an abstract sculptor, often on a large scale. His work of the 1950s was characteristically heavy and bulky, often incorporating industrial components, showing his interest in technology as well as in popular culture. In the 1960s his work became more colourful, including large totem-like figures made up from casts of pieces of machinery and often brightly painted. In the 1970s he made solemn machine-like forms and also box-like low *reliefs, both small and large, in wood or bronze, sometimes made to hang on the wall, compartmented and filled with small carved items. The latter were sharply and precisely cut and had a general resemblance to the work of Louise *Nevelson except that representational likeness was more completely banished from the elements of which they were composed. His more recent work has included several large public commissions, for example mosaic decorations for Tottenham Court Road underground station in London (installed 1983–5). Paolozzi has taught at various art schools and universities in Britain, Europe, and the USA, and he has been awarded many honours. See also INDEPENDENT GROUP.

paper. A tissue of vegetable fibres used for writing, drawing, and printing on. It is an oriental invention carried from the Far to the Middle East by the Turks during the Dark Ages. In Europe it is first traceable in the 12th cent. among the Moors in Spain, where it was made as well as imported. It was known in southern Italy at much the same time. France, southern Germany, and Switzerland had well-developed industries in paper by the end of the 14th cent. White paper was first made in England in 1495, but not on a large scale until the 18th cent.

Until 1800 European paper was made entirely of rags pulped in water by simple water-driven machinery. Drawing-paper of the best quality is still made by hand in the traditional way by dipping a close-meshed wire mould in the pulp. Writing-papers, less expensive drawing-papers, and some book papers are machine-made of a mixture of cotton, hemp, esparto, and wood, with a good deal of china clay added to make them smooth and opaque, *size to make them non-absorbent, and starch to make them stiff. Cheaper papers are machine-made wholly of wood. Oriental papers made of bamboo, rice straw, and mulberry bark are imported for artists' use: very thin sheets of Japanese mulberry paper (*kodsu*), hand-glazed by rubbing with stones, are preferred by *wood-engravers for their proofs.

papier collé (French: 'pasted paper'). A variety of *collage in which pieces of decorative paper are incorporated into a picture or, stuck on a ground such as canvas, themselves constitute the picture. The technique was invented by *Braque in 1913 when he incorporated pieces of wallpaper simulating wood graining in a still life, and was almost immediately adopted by *Picasso. *Matisse was perhaps the greatest exponent of the technique.

papyrus. A writing material prepared from the stem of the marsh plant of the same name, growing in antiquity principally in Egypt and now in the Sudan. It was used in Egypt from the third millennium BC onwards and was the standard writing material in ancient Greece and throughout the Roman empire. From the 4th cent. AD onwards it was increasingly replaced by *parchment though it continued to be used, for example in the Papal Chancery, until the 10th cent. Illustrated papyri survive from the Middle and New Kingdoms (Egypt) and there are a few fragmentary examples of *illuminated Classical texts as well as, for example, weavers' patterns.

parchment. Writing material made from the skins of sheep or calf, less frequently pig, goat, and other animals; it has also been used for painting, and occasionally for printing and bookbinding. *Pliny puts forward the opinion that parchment—meaning, perhaps, a skin prepared by special methods—was discovered by Eumenes II (197–159 BC) of Pergamum, after the Ptolemies had banned the export of *papyrus from Egypt in an attempt to prevent the growth of the Pergamene library; hence the name 'parchment' from the Latin *pergamena*, 'of Pergamum'. Skin had been used as a writing material before this, but the refined methods of cleaning and stretching involved in making parchment enabled both sides of a leaf to be used, leading eventually to the supplanting of the manuscript roll by the bound book. Vellum is a fine kind of parchment made from the delicate skins of young (sometimes stillborn) animals. *Paper began to replace parchment from about the 14th cent., but parchment is still used for certain kinds of documents, and the name is often applied to high-quality writing paper.

Pareja, Juan de. See VELÁZQUEZ.

Paris, Matthew (d. 1259). English chronicler and *miniaturist. He became a monk at St Albans in 1217 and succeeded Roger of Wendover (d. *c.*1236) as monastery chronicler in 1236. In 1248 he visited Norway to reform the monastery of St Benet-Holme, but otherwise seems never to have travelled further than London (the surname Paris is found elsewhere in England in the 13th cent. and does not necessarily imply that he had any French associations). Matthew carried on the abbey's *Chronica Majora* from 1235 until his death (it is now divided between Corpus Christi College, Cambridge, and the BL, London) and also wrote a summary of the chief events between 1200 and 1250 that is known as the *Historia Minor* or *Historia Anglorum* (BL). His historical manuscripts are almost unique in being illustrated with numerous marginal scenes and symbols from his own hand. He also composed several Lives of the Saints (the *Life of St Alban*, Trinity College, Dublin, is autograph) in which the illustrations occupy the upper half of the page and are of equal importance with the text. The frontispiece to his *Historia Anglorum* is his only signed work—a self-portrait showing him on his knees dedicating the manuscript to the Virgin. It is a tinted outline drawing, and apart from a panel painting of St Peter (Nasjonalgalleriet, Oslo) which has been attributed to him, his known artistic production consists entirely of such drawings. Because of his fame there has been a mistaken tendency to assign all mid-13th-cent. English work of this character to St Albans in general and, if at all plausible, to his hand. His lively, anecdotal style was nevertheless certainly appreciated, and 'circle of Matthew Paris' will no doubt remain the standard term for work of the period in a similar manner.

Paris, School of (École de Paris). A term that originally applied to a number of artists of non-French origin, predominantly of Jewish background, who in the years immediately after the First World War lived in Paris and painted in figurative styles that might loosely be called poetic *Expressionism, forming the most distinctive strand in French painting between *Cubism and *Surrealism. *Chagall (Russian), *Foujita (Japanese), *Modigliani (Italian), *Pascin (Bulgarian), and *Soutine (Lithuanian) are among the most famous artists embraced by the term. However, particularly outside France, the meaning of the term was soon broadened to include all foreign artists who had settled in Paris since the beginning of the century (van *Dongen, *Gris, *Picasso, for example), and then it expanded still further to cover virtually all progressive art in the 20th cent. that had its focus in Paris. In this broadest sense, the term reflects the intense concentration of artistic activity, supported by critics, dealers, and connoisseurs, that made Paris the world centre of advanced art during the first 40 years of the 20th cent. In 1951 an exhibition of the works of the École de Paris was held at the New Burlington Galleries, London, covering the period 1900–50. In the Introduction to the Catalogue, written by Frank McEwen, it was said that Paris then had 130 galleries as opposed to 30 in any other capital; the work of some 60,000 artists was shown there, of whom one-third were foreigners; and there were over 20 large salons exhibiting annually an average of 1,000 painters each, mostly semi-professional persons. After the Second World War, New York replaced Paris as the world capital of avant-garde art.

In the context of manuscript *illumination, the term 'School of Paris' is applied to the manuscript illuminators who under St Louis (1226–70) made Paris the leading centre of book illustration in Europe.

Parler, Peter (*c.*1330–99). German architect and sculptor, the most famous member of a dynasty of masons active in the 14th cent. and early 15th cent. In 1353 he was appointed architect of Prague Cathedral and much of the present structure was designed by him. He also built the celebrated Charles Bridge over the River Vltava in Prague. As a sculptor he is best known for a series of portrait busts in the triforium of Prague Cathedral, including a self-portrait and one of the emperor Charles IV. His workshop also produced tombs. Parler had a highly successful career and died a wealthy man.

Parmigianino (Girolamo Francesco Mazzola) (1503–40). Italian *Mannerist painter and etcher, born in Parma, from which he takes his nickname. He was a precocious artist, and in 1522–3 painted accomplished frescos in two chapels in S. Giovanni Evangelista, Parma, showing his admiration for *Correggio, who had worked in the same church a year or two before. The originality and sophistication he displayed from the beginning, particularly his love of unusual spatial effects, is, however, most memorably seen in his celebrated *Self-Portrait in a Convex Mirror* (Kunsthistorisches Museum, Vienna, 1524), in which *Vasari said he looks 'so beautiful that he seemed an angel rather than a man'.

In 1524 Parmigianino moved to Rome, possibly via Florence, and his work became both grander and more graceful under the influence of *Raphael and *Michelangelo. *The Vision of St Jerome* (NG, London, 1526–7) is his most important work of this time, showing the disturbing emotional intensity he created with his elongated forms, disjointed sense of space, chill lighting, and lascivious atmosphere. Parmigianino left Rome after it was sacked by German troops in 1527 and moved to Bologna.

In 1531 he returned to Parma and contracted to paint frescos in Sta Maria della Steccata. He failed to complete the work, however, and was eventually imprisoned for breach of contract. Vasari says he neglected the work because he was infatuated with alchemy—'he allowed his beard to grow long and disordered . . . he neglected himself and grew melancholy and eccentric.' His later paintings show no falling off in his powers, however, and his work reaches its apotheosis in his celebrated *Madonna of the Long Neck* (Uffizi, Florence, *c.*1535). The forms of the figures are extraordinarily elongated and tapering and the painting has a refinement and grace that place it among the archetypal works of Mannerism.

Parmigianino's range extended beyond religious works. He painted a highly erotic *Cupid Carving his Bow* (Kunsthistorisches Museum, Vienna, 1535), and was one of the subtlest portraitists of his age (two superb examples are in the Museo di Capodimonte, Naples). The landscape backgrounds to his religious works have a mysterious and visionary quality that influenced Niccolò dell' *Abbate and through him French art. Parmigianino, whose draughtsmanship was exquisite, also made designs for engravings and for *chiaroscuro woodcuts and appears to have been the first Italian artist to produce original etchings from his own designs. These helped to spread his style, which was influential not only in Italy, but also, for example, on the École de *Fontainebleau.

Parrhasius. Greek painter from Ephesus, active in the later 5th cent. BC. He is said to have been particularly skilful in the use of contour and in depicting character through facial expression, and his mastery of illusionism is recorded in one of *Pliny's most famous anecdotes concerning a contest he had with *Zeuxis. Zeuxis painted some grapes so naturalistically that birds came to peck at them. Victory seeming to be his, he called on Parrhasius to draw back the curtain concealing his picture, but this turned out to be a painted curtain. Zeuxis conceded the contest; he had deceived the birds, but Parrhasius had deceived him.

Parrish, Maxfield (1870–1966). American painter and illustrator. He studied at the Pennsylvania Academy of the Fine Arts with the author-illustrator Howard Pyle (1853–1911), celebrated for his children's books. In 1895 Parrish designed a cover for *Harper's Weekly* and thereafter rapidly made a name for himself with illustrations, posters, and advertisements. His greatest fame and popularity came with colour prints designed for the mass market. Sentimental scenes such as *The Garden of Allah* (copyrighted 1919) and *Dawn* (1920) sold by the million. They are in a lush and romantic style, set in an escapist world combining elements of the Arabian Nights, Hollywood, and classical antiquity, with languorous maidens and idyllic landscape backgrounds. His draughtsmanship and detailing are immaculate and his colouring distinctively high-keyed and luminous. Many of his advertisements were in a similar vein. In the 1930s his style went out of fashion and he retired to paint landscapes, working up to his death at the age of 95. Shortly before this there was a revival of interest in his work, which had long been dismissed as kitsch; in 1964, for example, the Metropolitan Museum, New York, bought his painting *Errant Pan* (*c.*1915).

Parsons, Betty (1900–82). American art dealer, collector, and painter. After Peggy *Guggenheim closed her New York gallery in 1946, Parsons became for a few years the leading dealer of the *Abstract Expressionists, until several of the major figures left her for Sidney *Janis in the early 1950s. She continued to support avant-garde art, and by the time her New York gallery closed in 1977 she had represented many of the most famous names in American art over the previous three decades, especially *Colour Field and *Minimal painters.

Pascin, Jules (Julius Pincas) (1885–1930). Bulgarian-born painter and draughtsman. He led a wandering life, and although he acquired American citizenship when he moved to New York during the First World War, he is chiefly associated with Paris, where he belonged to the circle of artists at Montmartre who gravitated around *Chagall, *Modigliani, and *Soutine. He did portraits of his friends and began a number of large paintings with biblical themes, but most of his work consists of erotically charged studies of the female nude. They have been compared to the work of *Degas and *Toulouse-Lautrec, but Pascin's paintings are less penetrating and more obviously posed. He can be rather repetitive, but his best work has great delicacy of colour and handling and a poignant sense of lost innocence. Pascin's work brought him financial success, but he led a dissolute life and was emotionally unstable. He committed suicide (slashing his wrists and then hanging himself) in spite of his financial success.

Pasiteles. Greek sculptor, active at Rome in the mid-1st cent. BC. He was famous in his time, and *Pliny says that he used preliminary models for his marble sculpture and that he wrote a kind of Greek *Companion to Art*.

Pasmore, Victor (1908–). British painter and maker of constructions who has achieved eminence as both a figurative and an abstract artist. After early experiments with abstraction he reverted to naturalistic painting, and in 1937 he combined with *Coldstream and Claude *Rogers in forming the *Euston Road School. Characteristic of his work at this time and in the early 1940s are some splendid female nudes and lyrically sensitive Thames-side landscapes which have been likened to those of *Whistler (*Chiswick Reach*, NG of Canada, Ottawa, 1943). In the late 1940s he underwent a dramatic conversion to pure abstract art, and by the early 1950s he had matured a personal style of geometrical abstraction. At the same time he made abstract *reliefs, partly under the influence of Ben *Nicholson and partly under that of Charles *Biederman's book *Art as the Evolution of Visual Knowledge*, lent to him in 1951 by Ceri *Richards. His earlier reliefs had a hand-made quality but later, through the introduction of transparent perspex, he gave them the impersonal precision and finish of machine products (examples are in the Tate Gallery). Through work in this style he came to be regarded as one of the leaders of the *Constructivist revival in Britain. Later paintings are less austere and more organic.

Pasmore has been an influential teacher, notably at King's College, Newcastle upon Tyne (University of Durham), where he was head of the painting department, 1954–61. The 'basic design' course he taught there (based on *Bauhaus ideas) spread to many British art schools. He has also been much concerned with bringing abstract art to the general public. In 1955, for example, he was appointed Consulting Director of Architectural Design for Peterlee New Town, County Durham, and designed an urban centre in the form of a Pavilion which integrated architectural design with abstract relief painting. Kenneth *Clark described Pasmore as 'one of the two or three most talented English painters of this century'.

Passarotti, Bartolommeo (1529–92). Italian painter, who except for some years in Rome (*c.*1551–*c.*1565) worked in his native Bologna. There he had a large studio, which became the focal point of the city's artistic life. The religious paintings that were the basis of his success are fairly conventional and undistinguished, and he is now remembered for his pioneering *genre scenes of butchers' shops (one of the few surviving examples is in the Galleria Nazionale, Rome). They reflect the influence of northern painters such as *Aertsen and in their lively observation broke free from the prevailing *Mannerism. Annibale *Carracci (whose brother Agostino studied with Passarotti) was influenced by these genre scenes in his early career, notably in his well-known *Butcher's Shop* (Christ Church, Oxford). In addition to his religious and genre works, Passarotti painted excellent portraits throughout his career. His son **Tiburzio** (d. *c.*1612) imitated his style, and he in turn had two artist sons, **Gaspare** and **Archangelo**.

Passe, van de. Netherlandish family of engravers based in Cologne and then in Utrecht. The founder was **Crispin** (d. 1637). Two of his children, **Simon** (1595?–1647) and **Willem** (1598?–*c.*1637), worked in England. The family produced a large number of engraved portraits, including many of British sitters.

Passeri, Giovanni Battista (1610–79). Italian painter and writer on art. He is of no significance as a painter, but he is important for his collection of biographies of contemporary artists, *Vite de' Pittori, Scultori ed Architetti che anno lavorato in Roma morti dal 1641 fino al 1673*, not published until 1772. His accounts are detailed and on the whole accurate, forming one of the most important sources for the study of Roman *Baroque art.

pastel. A drawing or painting material consisting of a stick of colour made from powdered *pigments mixed with just enough *gum or *resin to bind them. Pastel differs from other methods of painting in that no *medium or *vehicle is used. In other methods the colour as applied is different from the colour when dry; in pastel this is not so, and the artist may know at once what effect his colour will give. The practical disadvantage of pastel is the difficulty of securing adhesion to the *ground and its liability to be disturbed by the slightest touch or even by vibration. As with drawings in *chalk or *charcoal this may be counteracted by spraying with a *fixative, but fixing is apt to impair the characteristic surface quality and reduce the brilliance of the colour. Protection under glass and careful handling are perhaps the best safeguards.

Pastel is opaque, and the colour of the ground does not influence the final effect unless parts of it are left bare. The usual ground is a neutral-toned *paper. Sometimes the design is made up of individual strokes of colour not blended together, and sometimes the pastel is drawn lightly over the rough ground to produce a half-tone—a process known as 'scruffing'; in such cases the ground will show through and its tone is important. In its restricted range of tone pastel resembles *fresco and *gouache, but the brilliant powdery surface is peculiar to pastel.

Pastels originated in Italy in the 16th cent. as a development from the use of chalk for drawing—*Barocci was a noted early exponent. At first colours were generally limited to black, white and red or flesh-colour, and the invention of pastel painting in a full range of colours has been ascribed to the landscape painter and etcher, Johann Alexander Thiele (1685–1752), and also to his contemporaries Mme Vernerin and Mlle Heid of Danzig. Essays in using a full range of colours had been made earlier, but Thiele perhaps used the method more extensively than any of his predecessors, and the first artist to devote herself almost exclusively to it was another contemporary of his, Rosalba *Carriera, who introduced the technique to France. The heyday of pastel was the 18th cent., when it was particularly popular for portraiture (*Chardin, Maurice Quentin de *La Tour, and *Perronneau being famous exponents). During the first half of the 19th cent. the art declined, but in the second half of the century the medium became popular with the French *Impressionists, and there was a general revival. In 1870 the Société des Pastellistes was founded in Paris, and the first exhibition of the Pastel Society in London was held in 1880. The possibilities of the technique began to be more fully exploited. The earlier method had been to blend the colours together by rubbing with the finger or stump. Now, in pastel as in oil painting, a variety of techniques was developed. The new practitioners saw the value of the individual stroke, of the outline enclosing a flat area of colour, and of a more open technique in which the colours were juxtaposed without blending. Their subjects were also far more various. Among the many distinguished artists who have used pastel may be mentioned *Cassatt, *Degas, *Redon, *Renoir, *Toulouse-Lautrec, and *Whistler.

pastel manner. An 18th-cent. engraving technique, a development of the *crayon manner, in which a number of plates were inked with different colours to give the appearance of *pastel drawings.

Pasternak, Leonid (1862–1945). Russian painter and graphic artist. He was a friend of many literary, musical, and political personalities, notably Leo Tolstoy, whose works he illustrated (*War and Peace*; *Resurrection*), and whom he portrayed on many occasions. In 1921 he left the Soviet Union and settled in Berlin, where he worked mainly as a portraitist (his sitters included Max *Liebermann and Albert Einstein). He left Germany because of the rise of Nazism and settled in England in 1938, spending his last years in Oxford. The celebrated writer Boris Pasternak was his son.

pastiche (or **pasticcio**). A picture or other work of art that (often with fraudulent purpose) imitates the style of a particular artist by copying and recombining parts of his authentic works.

Patch, Thomas (1725–82). English painter and engraver who lived in Italy from 1747. In Rome he first made a reputation, especially among the English tourists, as a view painter, but after settling in Florence in 1755 was best known for good-humoured *caricature *conversation pieces by which he is chiefly remembered today. Patch knew *Reynolds, who included a portrait of him in his own well-known caricature of *Raphael's *School of Athens* (NG, Dublin, 1751), and it is uncertain who influenced whom in the genre. In 1770 Patch published a set of engravings of *Masaccio's frescos in the Brancacci Chapel, and they are of considerable historical value in recording the appearance of the paintings before they were damaged by fire in the following year.

Patel, Pierre (*c.*1605–76). French landscape painter. He was a pupil of *Vouet but worked in the manner of *Claude, with whose paintings his own have sometimes been confused. In his day he was well known for his panels set into the decoration of rooms, notably in the Cabinet de l'Amour of the Hôtel Lambert in Paris. His son, **Pierre-Antoine** (1648–1708), painted in his father's manner. Both men often featured classical ruins in their paintings, looking forward to the *Picturesque. Two typical examples by the father are in the Fitzwilliam Museum, Cambridge.

Patenier, Herri. See BLES.

Patenier (or **Patinier,** or **Patinir**), Joachim (d. 1524). Netherlandish painter, a pioneer of landscape as an independent genre. Nothing is known of his early life, but in 1515 he became a member of the Antwerp Guild. In 1521 he met *Dürer, who made a drawing of him and described him as a 'good landscape painter'. There are only a very few signed paintings, but a great many others have been attributed to him with varying degrees of probability. Among the signed works are: *The Baptism of Christ* (Kunsthistorisches Museum, Vienna), *The Flight into Egypt* (Musée Royal, Antwerp), and *St Jerome* (Kunsthalle, Karlsruhe). Patenier also painted landscape backgrounds for other artists and *The Temptation of St Anthony* (Prado, Madrid) was done in collaboration with his friend Quentin *Massys (who after Patenier's death became guardian of his children). Although landscape never constitutes the subject of his pictures, Patenier was the first Netherlandish artist to let it dwarf his figures in religious and mythological scenes. His style combines naturalistic observation of detail with a marvellous sense of fantasy, forming a link between *Bosch and *Bruegel.

Pater, Jean-Baptiste-Joseph (1695–1736). French painter, the only pupil of *Watteau (a fellow native of Valenciennes), with whom he had a somewhat touchy relationship. An unlikely legend has it that Watteau dismissed him from his studio (*c.*1713) because he was disturbed by the threat offered by his progress to his own pre-eminence; whatever the reason for their differences, they were reconciled soon before Watteau's death. Like Watteau's other imitator, *Lancret, Pater repeated the master's type of *fêtes galantes* (see FÊTE CHAMPÊTRE) in a fairly stereotyped fashion. He showed more originality in scenes of military life and groups of bathers (in which he gave freer rein to the suggestiveness often seen in his *fêtes galantes*). Examples of all types of his work are in the Wallace Collection, London.

Pater, Walter (1839–94). English critic and essayist. A fellow of Brasenose College, Oxford, he led an uneventful life and was regarded as an apostle of *Aestheticism, which set a supreme value upon the enjoyment of beauty. His best-known book is *Studies in the History of the Renaissance* (1873), which includes essays on *Winckelmann and the then neglected *Botticelli, and his celebrated evocation of the *Mona Lisa* in his essay on *Leonardo da Vinci: 'She is older than the rocks among which she sits . . .' This volume (which concludes 'To burn always with this hard gem-like flame, to maintain this ecstasy, is success in life'), though attacked by some as unscholarly and morbid, had a profound influence on the undergraduates of the day and was acclaimed by Oscar Wilde (1854–1900) as 'the holy writ of beauty'. His aesthetic creed was also expressed in a philosophical romance *Marius the Epicurean* (1885), and in 1889 he published a collection of critical essays entitled *Appreciations*.

patina. Incrustation, usually green, on the surface of a metal (typically bronze) object, caused by oxidation. Such discoloration occurs naturally with age through exposure to the atmosphere and can be accelerated or modified when an object is buried in the sea or soil, where the particular substances present will cause various chemical reactions. Patination can produce an attractive, mellowing effect, and since the *Renaissance bronze statues have often been artificially patinated, usually by treatment with acid. The French sculptor *Barye made a particular study of the various sorts of bronze patinas, and the Chinese carried the art of patination to great lengths, developing many ornamental, multi-coloured, mottled, and cloudy patinas which have not elsewhere been paralleled. By extension, the term 'patina' can be applied to any form of surface discoloration or mellowing, for example dirty varnish on a painting.

Patinier (or **Patinir**), **Joachim.** See PATENIER.

Paton, Sir Joseph Noël (1821–1901). Scottish painter. A friend of *Millais (a fellow student at the *Royal Academy), he had a kinship with the *Pre-Raphaelites early in his career. He painted mythological and historical scenes and later gained great success with his rather portentous and sentimental religious pictures, which went on tour and were much reproduced. In 1865 he was appointed Her Majesty's *Limner for Scotland.

Pausanias (2nd cent. AD). Greek traveller and geographer, the author of a *Description of Greece* in ten books that is the single most important literary source for the history of Greek art (painting and sculpture as well as architecture). It is a guidebook written for tourists—simple, unpretentious, detailed, and in the main reliable, as is frequently attested by the remains of the monuments he describes. Occasionally he has lapses: he saw several statues said to have been sculpted by the legendary Daedalus and accepted that he really existed. Sir James Frazer, who produced one of the many English translations of the work (6 volumes, 1898), said of Pausanias: 'without him the ruins of Greece would for the most part be a labyrinth without a clue, a riddle without an answer.'

Pausias of Sicyon. Greek painter of the mid-4th cent. BC. He was said to be the first painter who fully mastered the *encaustic technique. According to tradition he also introduced the painting of vaulted ceilings.

Peake, Mervyn (1911–68). British writer and illustrator, born in China, the son of medical missionaries. He is now best known as a novelist, but he studied at the *Royal Academy Schools and spent much of his career teaching drawing in London art schools. His reputation rests mainly on his trilogy of novels *Titus Groan* (1946), *Gormenghast* (1950), and *Titus Alone* (1959), a work of grotesque Gothic fantasy to which his vividly imaginative drawing style was well matched (originally, however, the books were published without his accompanying illustrations). Peake also illustrated numerous other books, including Stevenson's *Treasure Island* (1949) and several by himself (among them an instructional manual, *The Craft of the Lead Pencil*, 1946). In 1946 he was commissioned by the Ministry of Information to make drawings of people liberated from Belsen concentration camp—an experience that left him emotionally scarred. In the last decade of his life he was gradually incapacitated by Parkinson's disease. Peake is described in *The Dictionary of National Biography* as 'Tall, thin, dark, and haggard . . . gentle, gracious, unworldly, and unpractical'.

Peale, Charles Willson (1741–1827). American painter, inventor, naturalist, and patriot, the founder and most distinguished member of a family of artists. A highly versatile craftsman, Peale was a saddler, watchmaker, silversmith, and upholsterer before working briefly in *Copley's studio in Boston. He then spent two years in London (1767–9), where he studied under Benjamin *West. In 1775 he settled in Philadelphia, where he became the most fashionable portraitist in the Colonies, Copley having left for England in 1774. He fought as a colonel of the militia in the War of Independence and became a Democratic member of the Pennsylvania Assembly. In 1782 he opened an exhibition gallery next to his studio, the first art gallery in the United States, and there displayed his own portraits of leading personalities of the Revolutionary War (he painted George Washington several times). This expanded into a natural history museum, which attained a vast size and included as its star exhibit the first mastodon skeleton to be exhumed in America. Two of his most famous paintings celebrate his scientific interests—*The Exhumation of the Mastodon* (1806) in the Peale Museum, Baltimore, and *The Artist in his Museum* (1822) in the Pennsylvania Academy of the Fine Arts, an institution he helped to found in 1805. His inventions included new types of spectacles and false teeth. As a painter, Peale generally worked in a solid, dignified style, but his most celebrated work is a witty piece of **trompe-l'œil*. This is *The Staircase Group* (Philadelphia Museum of Art, 1795), a life-size portrait of two of his sons mounting a staircase, with a real step at the bottom and a real door jamb as a frame; George Washington is said to have been deceived into doffing his hat to the boys' images.

Peale married three times and had seventeen children, of whom several became artists. The most important were **Raphaelle** (1774–1825), one of America's most distinguished still-life painters, **Rembrandt** (1778–1860), a gifted though uneven portraitist, and **Titian Ramsay II** (1799–1885), who continued his father's tradition as an artist-naturalist. Apart from his portraits, Rembrandt won fame with his huge picture *The Court of Death* (3.5 m. × 7 m.), (Detroit Inst. of Arts, 1821), which he called a 'Moral Allegory' and which toured the country with success for over half a century. Charles Willson's brother **James** (1749–1831) was also a painter and his son and four daughters carried on the family tradition. The Peales are undoubtedly the most remarkable family in the history of American art and they were largely responsible for establishing Philadelphia as one of the country's leading cultural centres.

Pearce, Edward. See PIERCE.

Pearlstein, Philip (1924–). American painter, a leading proponent of the return to naturalism and interest in the human figure that was one aspect of the move away from the dominance of *Abstract Expressionism in American art. His early works (mainly landscapes) were painted with a vigorous *Gestural handling, but in the 1960s he developed a cooler, more even type of brushwork. He specializes in starkly unidealized portrayals of the nude figure (singly or in pairs), usually set in domestic surroundings. Because of the clarity of his compositions and the relentlessness of his scrutiny, he is sometimes described as a *Superrealist, but his work has an individuality that puts him outside this classification. He uses harsh lighting, oblique angles, and cropping of the image (heads are often excluded and the body is seen in voyeuristic close-up) in a way that suggests candid photography, but his pictures do not try to counterfeit the effect of photographs; the handling of paint is smooth—but vigorous rather than finicky.

Pechstein, Max (1881–1955). German *Expressionist painter and graphic artist. He studied in Dresden and in 1906 became a member of the *Brücke. In 1908 he moved to Berlin which became his home for the rest of his life. His energy and charm quickly made him a leading figure in the city's artistic life and in 1910 he was elected president of the Neue *Sezession. In 1913–14, showing an interest in the exotic shared with other Expressionists, Pechstein visited the Palau Islands in the Pacific and he painted bright and lively anecdotal pictures depicting the paradisal life of the island fishermen in a near-*Fauvist manner. His other subjects included nudes, landscapes, portraits, and opulent flower pieces. He was the most French in spirit of the German Expressionists (he was particularly influenced by *Matisse) and probably because his work was decorative rather than emotionally intense, he was the first of the Brücke to achieve success and general recognition; the 1920s marked the height of the fashion for his work. Since then his reputation has faded whilst those of his former colleagues have grown. In 1918 he was one of the founders of the *Novembergruppe and he taught in the Berlin Academy from 1923, being dismissed in 1933 and reinstated in 1945. His later work became repetitive and the high quality of his work during his Brücke period is sometimes forgotten.

peep-show box. A cabinet with scenes portrayed on the interior walls which, when viewed through a small opening or eyepiece, give a strong illusion of three-dimensional reality. *Alberti may have been the inventor of the peep-show box, for some such device is ascribed to him in a contemporary, anonymous biography, but they had their greatest popularity in the 17th cent. Samuel van *Hoogstraten, the *perspective virtuoso, was famous for his peep-show cabinets, of which the National Gallery, London, has an example.

Peeters, Bonaventura the Elder (1614–52). Flemish marine painter. He was virtually the only noteworthy practitioner in the genre in his country, seascapes being much more a Dutch than a Flemish speciality. In his early works Peeters was influenced by Dutch masters, such as Simon de *Vlieger, but later he tended to introduce elaborate motifs and bright colour in a decorative vein less impressive than his earlier style. Examples are in many museums, including the National Maritime Museum, Greenwich. Various other members of the family were artists, including his sister **Catharina** (1615–76), a seascape and still-life painter, but **Clara Peeters** (1594–1657?), an outstanding Antwerp still-life painter, was no relation.

peinture à l'essence. See DEGAS.

Pelham, Peter. See COPLEY.

Pellegrini, Giovanni Antonio (1675–1741). Venetian painter, the brother-in-law of Rosalba *Carriera. Pellegrini played a major part in the spread of the Venetian style of large-scale decorative painting in northern Europe, working in Austria, England, France, Germany, and the Netherlands. He was the first Venetian artist to visit England, arriving in 1708 in the train of the fourth Earl of Manchester, for whose country seat, Kimbolton Castle, he painted what are generally regarded as his finest works. His airy, *illusionistic compositions, with their bright, flickering colour and purely decorative intention, set a new standard of *Rococo elegance for English decoration, but he was an extremely prolific painter and by European standards most of his work is routine. *Thornhill defeated him in competition for the commission to decorate the dome of St Paul's Cathedral, London.

Pellipario, Nicola. See MAIOLICA.

pen. Writing and drawing instrument used with ink or a similar coloured fluid. From Early Christian times until the 19th cent. the standard form of pen in Europe was the quill, made from bird feathers, and most of the pen drawings of the Old Masters were done with the quill. Goose, swan, and turkey quills have commonly been used for writing and crow quills provide a very fine point for drawing. The reed pen, made from stems of bamboo-like grasses, was already in use in classical antiquity and is probably older than the quill. The point is much coarser, producing a bold, angular line sometimes slightly blurred at the edges. For drawing it has been used much less than the quill, though *Rembrandt, for example, was a master of the broad energetic technique appropriate to it.

The metal pen dates back at least to Roman times, but steel nibs of the modern type were not made until late in the 18th cent. and began to replace the quill only when they were produced by machine, *c.*1822. Most artists today use a flexible steel nib with a fine point, but they will draw with anything that comes to hand—even a ball-point, whose light, rapid line tones well with *watercolour.

No other drawing tool can produce such a variety of texture or reveal so intimately the personal 'handwriting' of an artist. The pen is the ideal medium for rapidly noting down the first idea and has been used in this way by masters of drawing as different as *Pisanello, *Michelangelo, *Dürer, and Rembrandt. But apart from its use for the hasty or inspired *sketch, in which the fire of execution is reflected in the line, the pen has been used with great effect in a careful, calligraphic manner, as in *Botticelli's illustrations to Dante's *Divine Comedy*.

Pen-and-wash drawing, in which the pen lines are reinforced by brushwork in diluted Indian ink or some similar *pigment, has been practised since the Renaissance (see WASH).

pencil. Writing or drawing instrument consisting of a slender rod of *graphite or similar substance encased in a cylinder of wood (or less usually metal or plastic). Although the material is graphite, the 'lead' pencil took its name from the lead point (see METAL POINT) which it superseded, and is first heard of in the 1560s. Pencils of predetermined hardness or softness were not produced until 1790, however, when Nicolas-Jacques *Conté undertook to solve the problem of making pencils when France was cut off from the English supply of graphite (the mines in Borrowdale, Cumbria, which had opened in 1664, being the main source). He found that the graphite could be eked out with clay and fired in a kiln, and that more clay meant a harder pencil. Conté obtained a patent for his process in 1795. It was only then that the pencil became the universal drawing instrument that it is today. *Ingres, who often used pencil with great delicacy in his portrait drawings, was one of the first to show its potential. Although the *Oxford English Dictionary* records the usage of the phrase 'a pencil of black lead' as early as 1612, until the end of the 18th cent. the word 'pencil' more commonly meant a brush (particularly a small brush) and was often used as a symbol for the painter's art. 'Pencilling' could mean 'colouring' or 'brushwork' as well as 'drawing'. According to a handbook published in 1859 (*Painting Popularly Explained* by Thomas J. Gullick and John Timbs) 'The smaller kinds of brushes are still sometimes termed "pencils"; but the use of the word "pencil" instead of "brush" as distinctive of and peculiar to watercolour is now obsolete.'

Pencz, Georg (Jörg Bencz) (*c.*1500–50). German painter and engraver of religious and mythological subjects and portraits, active in his native Nuremberg, where he was an assistant of *Dürer. He travelled in Italy early in his career and again in 1539, when he is recorded in Florence and Rome, and his work is deeply imbued with Italian influence. The sharp outlines and glossy textures of his portraits show, in particular, a kinship with *Bronzino (*Man Holding a Mirror*, Hessisches Landesmuseum, Darmstadt, 1544). In 1525 Pencz was expelled from Nuremberg with the *Beham brothers, two other 'godless artists', for their radical Protestant views, but the sentence was soon revoked and he returned to the city. In 1550 he was appointed painter to Duke Albrecht of Prussia, but died in the same year. Although as a painter Pencz often worked on a large scale with considerable grandeur, his engravings are often tiny (see LITTLE MASTERS).

pendentive. One of the concave *spandrels by means of which a circular dome can be supported above a square or polygonal compartment; it is often used as a field for painted decoration.

Penni, Giovanni Francesco (*c.*1488–*c.*1528). Italian painter, nicknamed 'il Fattore' (the maker). He was born in Florence and died in Naples, but worked mainly in Rome, where with *Giulio Romano he was *Raphael's main assistant. Various parts of the works of the master have been attributed to him (*Vasari says he worked on the tapestry *cartoons for the Sistine Chapel), but he is a shadowy figure. After Raphael's death he collaborated with Giulio. A rare signed work by Penni is the *Portrait of a Young Man* in Dublin (NG).

Penrose, Sir Roland (1900–84). British writer, exhibition organizer, and artist. As an artist he holds a distinguished place among British *Surrealists (he produced collages and 'objects' as well as paintings), but he is remembered mainly for the missionary zeal with which he promoted Surrealism and contemporary art in general in England. He was one of the organizers of the 1936 Surrealist exhibition in London, and he was closely involved with many leading artists, such as *Ernst, *Man Ray, *Miró, and *Picasso. During the Second World War he was a camouflage instructor. In 1947 he was co-founder and first chairman of the *Institute of Contemporary Arts in London, and he organized several major exhibitions for the *Arts Council. His books include *Picasso. His Life and Work* (1958, 3rd edition 1981), regarded as a standard work on the artist. He also wrote poetry. Penrose's work is represented in the Tate Gallery. His second wife, whom he married in 1947, was the photographer Lee Miller, a pupil and favourite model of Man Ray.

pentimento. Term (Italian for 'repentance') describing a part of a picture that has been overpainted by the artist but which has become visible again (often as a ghostly outline) because the upper layer of pigment has become more transparent through age. The presence of pentimenti is often used as an argument in matters of *attribution, as it is felt that such evidence of second thoughts is much more likely to occur in an original painting than in the work of a copyist.

Peploe, S. J. See SCOTTISH COLOURISTS.

Percellis, Jan. See PORCELLIS.

Perceptual Abstraction. Term sometimes used to cover several schools of abstract art which succeeded *Abstract Expressionism, such as *Colour Field painting, *Hard Edge painting, *Minimal art, and some forms of *Op art. The term indicated the switch from emphasis upon expressive and painterly qualities to perceptual clarity and precision with emotionally neutralized content.

Percier, Charles (1764–1838) and **Fontaine, Pierre-François** (1762–1853). French architects and designers who worked in partnership from the early 1790s until 1814, when Percier retired. They were the favourite architects of Napoleon, carrying out under his orders numerous schemes of restoration, reconstruction, and redecoration, notably in the *Louvre and Tuileries, and the palaces at Malmaison, Fontainebleau, Saint Cloud, Compiègne, and Versailles. Their most famous work is the Arc de Triomphe du Carrousel in Paris (1806–7). In interior decoration and furniture design they were the leading exponents of the so-called *Empire style and their *Recueil de décorations intérieures* (1801, second edition 1812) is the best compendium of early 19th-cent. ornament published in France. After the retirement of Percier and the fall of Napoleon (1815), Fontaine continued to enjoy the patronage of the French crown right up to his death in 1853.

Pereda, Antonio de (1611–78). Spanish painter, active mainly in Madrid. He began as a history painter—his *Relief of Genoa* (Prado, Madrid, 1635) was painted for the Buen Retiro Palace in Madrid as part of the same series as *Velázquez's *Surrender of Breda*—but he is now best known for his still lifes. The most famous painting associated with him is *The Knight's Dream* (also called *The*

Dream of Life or *Life is a Dream*, Academy, Madrid, *c*.1650), a splendidly sensuous composition, full of beautiful still-life details, in which worldly pleasures and treasures are seen to be as insubstantial as a dream. It was a key work in the development of the moralizing still life in Spain, influencing *Valdés Leal in particular. However, the attribution to Pereda has recently been questioned, and Francisco de Palacios (1622/5–52) has been suggested as the author.

Peredvezhniki. See WANDERERS.

Pereira, Manuel (1588–1683). Portuguese sculptor who worked chiefly in Spain. Pereira was one of the outstanding masters of the *polychrome statue in the generation after Gregório *Fernandez and Juan Martínez *Montañés. He worked not only in wood, but also in stone, as in the statue of *St Bruno* (Academy, Madrid, *c*.1635–40). The grand austerity and spiritual conviction he shows here brings to mind the paintings of *Zurbarán. Like *Velázquez, Pereira received from Philip IV the great honour of a knighthood of Santiago.

Pérelle, Gabriel (1603–77). French draughtsman and engraver mainly of topographical views and landscapes. His engravings of buildings are of great importance to the architectural historian. He was assisted by his sons **Adam** (1638–95) and **Nicholas** (1631–95).

Performance art. An art form combining elements of theatre, music, and the visual arts. It is related to *happening (the two terms are sometimes used synonymously), but Performance art is usually more carefully programmed and generally does not involve audience participation. The tradition of Performance art can be traced back to the *Futurists, *Dadaists, and *Surrealists, who often staged humorous or provocative events to promote their work or ideas, then through such activities as Georges *Mathieu painting in front of an audience in the 1950s and Yves *Klein directing nude models smeared with paint in the early 1960s. However, it was only in the later 1960s and particularly in the 1970s that Performance art became recognized as a category of art in itself. 'At that time', RoseLee Goldberg writes, 'Conceptual art was in its heyday and performance was often a demonstration, or an execution, of [its] ideas . . . Art spaces devoted to performance sprang up in the major international art centres, museums sponsored festivals, art colleges introduced performance courses, and specialist magazines appeared' (*Performance Art, From Futurism to the Present*, 1988). The form and tone of Performances have varied enormously. Some practitioners have cultivated sadomasochism and scatology (the abuse of the performer's body is something that often occurs also in *Body art, with which Performance art sometimes overlaps). In Britain, however, the field has more often been characterized by whimsicality (in the 1970s there was a fad for Performance groups with quaint names and for wacky newsworthy stunts). Performance art has also been used as an adjunct to rock music (the American Laurie Anderson (1947–) is the most noted exponent) and as a vehicle for political dissent, as well as for the exploration of private fantasies. Among the artists particularly associated with Performance art are Joseph *Beuys and *Gilbert & George.

Pergamene School. A trend in *Hellenistic sculpture, associated with the city of Pergamum in Asia Minor, whose great period coincided with the attalid dynasty (241–133 BC). Attalus III (d. 133) bequeathed his kingdom to Rome, but before him Attalus I, Eumenes II, and Attalus II aspired to make their capital a magnificent centre of culture. The works associated with the Pergamene School are characterized by an exaggeration of the general tendency of Hellenistic sculpture towards emotional display and virtuoso naturalistic detail. They include a series of Dying Gauls (the most famous is in the Capitoline Museum, Rome) that have been identified as copies of statues dedicated by Attalus I to celebrate a victory over the Gauls, and the Great Altar of Zeus (Pergamum Museum, Berlin, *c*.180–150 BC), which features relief carvings of the fight between gods and giants depicted with an extraordinary sense of movement and dramatic tension. It is perhaps the 'Satan's seat' of Revelation 2: 13. Other Hellenistic works from places other than Pergamum (above all the celebrated **Laocoön*) are clearly similar in their restless energy and some scholars deny that there is adequate reason for supposing there is a separate Pergamene School. Others, as for example T. B. L. Webster in his *Hellenistic Poetry and Art* (1964), take the view that although the artists employed by the Attalid kings came from various places, they 'achieved a unity of style which justifies the name Pergamene'.

Pergamum Altar. See PERGAMENE SCHOOL.

Perino del Vaga (Piero Buonaccorsi) (*c*.1500–47). Florentine painter. He took his name (del Vaga) from a painter with whom he worked after studying with Ridolfo *Ghirlandaio. In about 1518 he moved to Rome, where he became one of *Raphael's assistants working on the Vatican Loggie. After the Sack of Rome in 1527 he moved to Genoa, where he was based until the late 1530s and where his *Mannerist style had great influence. His major work there was a series of mythological frescos in the Palazzo Doria. By 1539 he was back in Rome, where he became the principal decorative artist employed by Pope Paul III (Alessandro *Farnese), his work for him including frescos on the history of Alexander the Great (1545–7) in the Sala del Consiglio of the Castel S. Angelo. Perino's style derives from Raphael and *Giulio Romano, but is ornamental rather than monumental. He was one of the leading decorative artists of his generation, and his work has been aptly described by S. J. Freedberg (*Painting in Italy: 1500–1600*, 1971) as 'intelligent but facile'. He also did a number of devotional pictures in a Raphaelesque vein, notably the unfinished *Holy Family* (Courtauld Inst., London).

Permeke, Constant (1886–1952). Belgian painter and sculptor, one of the leading exponents of *Expressionism in Belgium in the period between the two world wars. He was badly wounded in the First World War in 1914 and was evacuated to England. In 1919 he returned to Belgium, where he lived in Antwerp and Ostend before building his own house and studio at Jabbeke, near Bruges, now a Permeke museum (he called his home De Vier Windstreken, 'The Four Corners of the Earth'). His subjects were taken mainly from the life of the coastal towns of Belgium and he is best known for his strong and solemn portrayals of sailors and fishermen with their women (*The Fiancés*, Musées Royaux, Brussels, 1923). From 1935 he also made sculpture.

Permoser, Balthasar (1651–1732). German sculptor. He was in Italy 1675–89 and his lively *Baroque style was much influenced by *Bernini. In 1689 he became court sculptor at Dresden and spent most of the rest of his life there. His most important work was the sculptural decoration of the Zwinger—Augustus the Strong's pleasure palace in Dresden, designed by the architect M. D. Pöppelmann (begun 1711, badly damaged in the Second World War). Permoser carved in wood and ivory as well as stone and also made ingenious use of coloured marble (*Damned Soul*, Museum der Bildenden Künste, Leipzig, *c.*1715). *Roubiliac is said to have studied with him.

Perov, Vasily. See WANDERERS.

Perréal, Jean (or **Jehan de Paris**) (*c.*1455–1530). French painter, architect, sculptor, and decorator of very considerable reputation in his day. He was in the service of Charles VIII (1470–98), Louis XII (1462–1515), and Francis I (1496–1567) and visited Italy on three occasions, with Charles VIII in 1494 and with Louis XII on his campaigns of 1502 and 1509 (on one of these journeys he met *Leonardo, who mentions Perréal in his notebooks). His activities were manifold. He was employed to design elaborately sculptured tombs such as that of Francis II of Brittany at Nantes; he was a specialist in arranging and designing public ceremonials; he was a designer of medals, and he had a special interest in portraiture. However, little work survives that can confidently be attributed to his own hand—a portrait of Louis XII (1514) at Windsor Castle and a few *miniatures. They show him as a master of a scrupulously observed *naturalism. It has been suggested that he is to be identified with the *Master of Moulins.

Perrier, François (1590?–1650). French history painter and engraver. He visited Rome on two occasions and his style was formed on the example of *Lanfranco (in whose studio he worked), Pietro da *Cortona and the *Carracci. His decorative work helped to introduce the grand *Baroque style to France, but almost all of it has been destroyed or altered. His influence, however, can be seen in the work of *Lebrun, who was briefly his pupil.

Perronneau, Jean-Baptiste (1715?–83). French painter and engraver. He worked mainly as a portraitist and mainly in *pastel. In this medium he was overshadowed in his lifetime by Maurice Quentin de *La Tour, whose style was more vivacious, but posterity has judged the more sober but more penetrating Perronneau to be at least his equal. From about 1755 he led a wandering life, visiting England, the Netherlands several times (he died in Amsterdam), Italy, Poland, and Russia. He was a prolific artist and is represented in the Louvre and in numerous other museums in France and elsewhere.

perspective. Method of representing spatial extension into depth on a flat or shallow surface, utilizing such optical phenomena as the convergence of parallel lines and the apparent diminution in size of objects as they recede from the spectator. Perspective is by no means common to the art of all epochs and all peoples. For example, the pictorial art of the ancient Egyptians, although a richly developed tradition, did not take account of the optical effects of recession. And the drawing of primitive peoples and of young children tends to ignore perspective phenomena. Systematic, mathematically founded perspective, based initially on a fixed central viewpoint, was developed in Italy in the early 15th cent., when it was invented by *Brunelleschi, described by *Alberti in his treatise *De Pictura*, and put into majestic practice by *Masaccio. Various names are given to this type of perspective—geometric, linear, mathematical, optical, *Renaissance or scientific perspective—which remained one of the foundations of European painting until the late 19th cent. In pre-Renaissance Europe and in the East more intuitive systems of representing spatial recession were used. In medieval paintings, for example, lines that would in strict perspective converge are often shown diverging (this 'inverted perspective' can look much more convincing in practice than it sounds in theory); and in Chinese art 'parallel perspective' was a common convention in the depiction of buildings. See also AERIAL PERSPECTIVE.

Perugino, Pietro (Pietro Vannucci) (1445/50–1523). Italian painter, active mainly in Perugia, from which his nickname derives. His early career is obscure, but he seems to have formed his style chiefly in Florence, where *Vasari says he studied with *Verrocchio—this would have been at about the same time that *Leonardo da Vinci was training with him (another tradition has it that Perugino was a pupil of *Piero della Francesca; this could have preceded his training in Florence). In 1472 he was enrolled as a painter in the fraternity of St Luke in Florence (the same year as Leonardo) and in 1475 he was back in Perugia. By 1481 he was sufficiently well known to be commissioned to paint frescos on the walls of the newly built Sistine Chapel, Rome, along with *Botticelli, Domenico *Ghirlandaio, and Cosimo *Rosselli (*Signorelli later completed the work). Vasari says that Botticelli was head of the team, but some modern scholars think that Perugino was more likely to have been leader, partly because of the prominence of his contributions. His main work there is *Christ Delivering the Keys to St Peter*; he also did the frescoed altarpiece, but this was destroyed to make way for *Michelangelo's *Last Judgement*. His reputation firmly established, he travelled extensively in central Italy, and in the 1490s he maintained a workshop in Florence as well as in Perugia. In 1500, when he was decorating the Audience Chamber of the Collegio del Cambio at Perugia, he was called by Agostino Chigi (1465–1520), the wealthy Sienese banker and patron, the best painter in Italy, and he was indeed at his peak at about this time. Perugino was a fine portraitist as well as a fresco painter, but today he is best known for his altarpieces, which are usually gentle, pious, and rather sentimental in manner—even during his lifetime he was criticized in some quarters for excessive 'sweetness'. His style does not seem to have been a reflection of his personality, for Vasari says he 'was not a religious man' and that he 'would have gone to any lengths for money'. In about 1505 he left the competitive atmosphere of Florence, where his work now seemed old-fashioned, and settled permanently in Perugia; his later work is often routine and repetitive. At his best, however, as in the Vatican fresco, the Uffizi *Assumption*, and the *Pietà* in the Accademia, Florence, he has the authority of a great master. The harmony and spatial clarity of his compositions and his idealized physical types had great influence on the young *Raphael, who worked with him early in his career, so Perugino can be seen as one of the harbingers of the High *Renaissance. A second wave of his influence came in the 19th cent., when he was glorified by the *Pre-Raphaelites.

Peruzzi, Baldassare (1481–1536). Sienese architect, painter, and stage designer, active mainly in Rome, where he settled in 1503. He worked under *Bramante on St Peter's, and eventually became architect to the building after *Raphael's death in 1520. Amongst High *Renaissance architects he ranks almost alongside these two great contemporaries, but his style was very different—sophisticated and delicate rather than monumental and grave. In spite of his genius and his open, friendly nature he had little material success, and *Vasari lamented that 'The great abilities and labours of this noble artist benefited him but little, but assisted others, for though he was employed by popes, cardinals and other great wealthy men, not one of them ever rewarded him richly, though this was due more to his own retiring nature than to any want of liberality in his patrons.' His greatest work—indeed the greatest secular building of the High Renaissance—is the Villa Farnesina (1508–11) in Rome, built for the Sienese banker Agostino Chigi (1465–1520). It contains decorations by Raphael, *Sebastiano del Piombo and *Sodoma, as well as Peruzzi's own masterpiece in painting—the Sala delle Prospettive, a brilliant piece of feigned architectural painting that confirms early accounts of his skill in perspective and stage-design.

Pesellino (Francesco di Stefano) (*c.*1422–57). Florentine painter. His only documented work is an altarpiece of the *Trinity with Saints*, begun in 1455, left incomplete at his death and finished in Filippo *Lippi's studio. It was afterwards cut into several pieces, which entered the National Gallery, London, at different dates and have since been reunited (one section is on loan from the Royal Collection). The works attributed to Pesellino are mainly **cassone* panels. Another painter called Francesco di Stefano has sometimes been confused with him.

Peters, Matthew William (1741/2–1814). English painter of portraits, *genre and historical scenes, and *fancy pictures. He studied with *Hudson in the late 1750s and travelled to Italy and

France in 1772–6 and again to Paris in 1783. Foreign influence is clearly seen in the richness of his technique (admirers called him 'the English *Titian') and in the pin-up type of picture of pretty women—distinctly reminiscent of *Greuze—with which he caused mild scandal and attained great popularity in engravings. He took Holy Orders in 1782 and retired from professional practice in 1788, but continued to paint for his own pleasure. His later works include some mawkish religious paintings and some dull pictures for *Boydell's Shakespeare Gallery.

Peto, John Frederick (1854–1907). American still-life painter. With his friend *Harnett he is now considered the outstanding American still-life painter of his period, but he was little known during his lifetime and after his death was completely forgotten until rediscovered in 1947 by the critic Alfred Frankenstein. His work was strongly influenced by Harnett (whose signature has sometimes been fraudulently added to paintings by Peto), but his style was softer and more anecdotal, often depicting discarded objects. Peto worked in his native Philadelphia, then from 1889 in seclusion at Island Heights, New Jersey.

Petrarch Master. See WEIDITZ.

Pettie, John (1839–93). Scottish history and *genre painter, active in London from 1862. He became popular with dramatic and anecdotal historical subjects, particularly military episodes, often from the Civil War period (1642–8) (*Cromwell's Saints*, NG, Edinburgh, 1862). His style was free and dashing, to some extent influenced by van *Dyck.

Pettoruti, Emilio (1892–1971). Argentine painter. He studied and worked in Europe (1913–23), attaching himself to the *Futurist movement in Italy, and experimenting with the *Cubism of Juan *Gris in Paris. After his return to Argentina in 1924 he exercised great influence on the younger generation of painters there through his advocacy of European modernism. He was frowned on by officialdom, however, and his two terms as director of the Museum of Fine Arts in La Plata were short-lived.

Pevsner, Antoine (1886–1962). Russian-born sculptor and painter who became a French citizen in 1930. He was the elder brother of Naum *Gabo and like him one of the pioneers and chief exponents of *Constructivism. Between 1911 and 1914 he made lengthy visits to Paris, where he associated with *Archipenko and *Modigliani. After two years in Norway with Gabo he returned to Russia in 1917, and in that year became professor at the Academy of Fine Arts, Moscow, where *Kandinsky and *Malevich also taught. In 1920 he was co-signatory of Gabo's *Realistic Manifesto*, which set forth the ideals of Constructivism, and in 1922 he helped to organize a major exhibition of Soviet art in Berlin. He moved back to Paris in 1923 because of official disapproval of abstract art and lived there for the rest of his life. Up to this time he had been a painter, but he now turned to sculpture, at first working mainly in plastic, then in welded metal. Initially his sculptures retained vestiges of representation, as in his witty *Portrait of Marcel Duchamp* (Yale University Art Gallery, 1926), but by 1927 he had arrived at pure abstraction. His later work was characterized by bold spiralling forms (*Dynamic Projection in the 30th Degree*, Baltimore Museum of Art, 1950–1). Pevsner was a founder member of *Abstraction-Création and was influential in transmitting Constructivist ideas to other artists in the group.

Pevsner, Sir Nikolaus (1902–83). German-born British art historian. He worked at the Gemäldegalerie in Dresden and taught at Göttingen University (1929–33) before moving to England because of the rise of Nazism. In England he was *Slade Professor at both Oxford and Cambridge universities and also professor at Birkbeck College in the University of London. He is best known for his writings on architecture, above all for the celebrated series of county-by-county guides *The Buildings of England* (46 vols., 1951–74), which he conceived, edited, and largely wrote himself. The series is one of the great achievements of 20th-cent. scholarship, for the books are, in the words of Gerald Randall (*Church Furnishing and Decoration in England and Wales*, 1980), 'such indispensable guides to the traveller and works of reference to the student that it is amazing how people coped without them'. Companion series on the buildings of Ireland, Scotland, and Wales are in the course of publication, the first volumes on each of the three countries appearing in 1978 and 1979. Pevsner's many books included important studies on painting, sculpture, and design as well as architecture. They include *Pioneers of the Modern Movement from William Morris to Walter Gropius* (1936, retitled *Pioneers of Modern Design* in subsequent editions), *Academies of Art, Past and Present* (1940), *An Outline of European Architecture* (1942 and numerous other editions), *The Englishness of English Art* (1956), and *Studies in Art, Architecture and Design* (2 vols., 1968). He also conceived and edited the *Pelican History of Art* (which began publication in 1953), the largest, most comprehensive, and most scholarly history of art ever published in English, many of the individual volumes of which have become classics. Anthony *Blunt, Ellis *Waterhouse, and Rudolf *Wittkower are among the scholars who have written volumes in the series.

Pforr, Franz (1788–1812). German painter. He was one of the founders of the *Nazarenes and went to Rome with other members of the group in 1810. His work evokes a fairy-tale type of medievalism, with bright colours and picturesque details. It is best represented in the Städelsches Kunstinstitut of his native Frankfurt.

Phalanx. An association of artists organized in Munich in 1901 in opposition to the conservative views of the Academy and the *Sezession. *Kandinsky was one of the founders of the association and its leading figure, becoming President in 1902. For the first group exhibition in August 1901 he designed a magnificent poster in *Art Nouveau style showing Greek warriors advancing across a battlefield in phalanx formation. The militaristic name of the association was chosen to suggest its aggressive, progressive spirit. Eleven more exhibitions followed before Kandinsky dissolved Phalanx in 1904 because of lack of public support. The exhibitions featured not only work by members, but also by 'guest' artists, notably French *Post-Impressionists and *Neo-Impressionists at the 10th exhibition in 1904. This was the most important of the exhibitions, confirming Kandinsky's internationalism and having a marked effect on several young artists, notably *Kirchner. In 1902–3 Phalanx ran an art school.

Phidias (d. *c.*430 BC). Greek sculptor, active mainly in Athens, the most famous artist of the ancient world. No work survives that is certainly from his own hand, but through copies, descriptions, and above all the surviving sculpture of the Parthenon at Athens, which Phidias supervised, a fair idea can be gained of his style. In antiquity he was most celebrated for two enormous *chryselephantine cult statues—of Athena, inside the Parthenon, and of Zeus in the temple dedicated to the god at Olympia. The statue of Zeus, a seated figure about 12 m. high and one of the Seven Wonders of the World, is known through reproductions on coins and gems. The statue of Athena (dedicated 438 BC), a standing figure about 9 m. high, is known through several (much smaller) copies. Phidias also made two bronze statues of Athena for the Acropolis in Athens: the huge *Athena Promachos* (champion), in which the goddess was shown holding a spear; and the *Lemnian Athena* (so called because it was dedicated by Athenian colonists going to Lemnos between 451 and 448 BC). The *Athena Promachos* is represented in coins and the *Lemnian Athena* can be partially reconstructed from two fragmentary copies: a remarkably sensitive head in Bologna (Museo Civico Archeologico) and a substantially complete figure in Dresden (Albertinum). Other copies have been credibly associated with works of Phidias mentioned in ancient sources, and recently the two bronze statues of warriors found in the sea near Riace in 1972 ('The Riace bronzes') and now in the archaeological museum at Reggio di Calabria have been linked with his name because of their superlative quality. The greatest testimony to his genius, however, is the sculpture of the Parthenon (447–432 BC), the most ambitious sculptural undertaking of the age, consisting of a low-*relief frieze about 160 m. long, ninety-two *metopes in high relief, and groups of free-standing figures on both pediments. The quality is variable as a team of sculptors was involved and Phidias could not have carved more than a tiny fraction of the work himself, but the finest parts (see ELGIN MARBLES) exemplify the harmony and serene majesty that earned the raptures of ancient commentators and stand as the greatest surviving examples of the *Classical period in Greek art.

The end of Phidias' career is something of a mystery. When his friend and patron, the great statesman Pericles, fell out of favour Phidias was accused of misappropriating gold supplied to him for the statue of Athena. Then, according to Plutarch, having cleared himself of this charge, he was thrown into prison for impiety on the ground that he had introduced portraits of Pericles and himself on the shield of the goddess (a copy of the shield—the 'Strangford Shield'—is in the British Museum). Plutarch says that Phidias died in prison, but according to another ancient source he escaped and went to Olympia to work on his statue of Zeus, the date of which is uncertain. In 1954–8 Phidias' workshop at Olympia was excavated. Moulds, scraps of ivory, and other fragments were discovered, and—remarkably—a cup bearing the inscription 'I belong to Phidias'—the great man's tea-mug, as it were. This poignant relic is in the Olympia Museum. Phidias is said to have been a painter, engraver, and worker in embossed metalwork. His influence is seen in the best of contemporary vase-painting as well as in the sculpture of his successors.

Phillip, John (1817–67). Scottish painter. He began as a specialist in paintings of Scottish life and character in the manner of *Wilkie, but following visits to Spain in 1851, 1856, and 1860 he became celebrated for picturesque Spanish *genre scenes and was known as 'Spanish' Phillip or 'Phillip of Spain'. His style was fluent and colourful, often with a picture postcard flavour. He also painted portraits.

Phillips, Peter (1939–). British painter. He studied at the *Royal College of Art, 1959–62, and with his fellow students Derek *Boshier, David *Hockney, Allen *Jones, and R. B. *Kitaj emerged as one of the leading exponents of British *Pop art at the *Young Contemporaries exhibition in 1961. Typically his imagery is drawn from modern American culture—juke boxes, pinball machines, automobiles, film star pin-ups and so on—painted in the tight, glossy manner of commercial art. However, the images are usually set into bold heraldic frameworks or fragmented into sections and reorganized, so that illusionism and abstraction are combined.

Philostratus. The name of at least three members of a family of Greek writers active in the 3rd and 4th cents. AD, two of whom ('Philostratus the Lemnian' and his grandson 'Philostratus the Younger') wrote series of *Imagines*—descriptions of real or imaginary paintings. They are primarily literary exercises, but they contain some useful information on *Hellenistic painting. *Goethe translated them.

Philoxenos of Eretria. See ALEXANDER MOSAIC.

Phiz. See BROWNE, HABLOT KNIGHT.

Photographic Realism. See SUPERREALISM.

photomontage. Term applied to a technique of making a pictorial composition from parts of different photographs and to the composition so made. Photomontage was popularized by the *Dadaists as a method for political propaganda, social criticism, and generally to assist the shock tactics in which they indulged. *Heartfield was perhaps the most brilliant of all exponents of the technique. Photomontage has also been memorably used by, for example, Max *Ernst and other *Surrealists and by *Pop artists such as Richard *Hamilton, but it is now mainly associated with advertising.

Piazzetta, Giovanni Battista (1683–1754). Venetian painter and graphic artist. After preliminary training in Venice he worked under G. M. *Crespi in Bologna, then settled permanently in his native city by 1711. He was one of the most individual Venetian painters of his period, his sombre and dramatic style looking back to work done a century earlier by *Liss, *Strozzi, and *Feti. Apart from a ceiling fresco of the *Glory of St Dominic* (*c.*1727) in SS. Giovanni e Paolo, he painted in oils and he was a notoriously slow worker, but his pictures seem fresh and spontaneous rather than laboured. He had a large family and although he was not without wealthy patrons (*Algarotti among them) he relied much on drawings and book illustrations to earn money. His illustrations to the edition of Torquato Tasso's *Gerusalemme Liberata* (1581) published by his friend G. B. Albrizzi in 1745 are considered to be among the finest of the 18th cent. As a painter he did religious and historical works and portraits, as well as some hauntingly enigmatic *genre scenes that reflect his training with Crespi. In 1750 Piazzetta became the first Director of the Venice Academy of Fine Arts, but in his last years he was eclipsed by the new generation. The young *Tiepolo was greatly influenced by him, but later the influence was reversed, as Piazzetta's style became softer and lighter—more *Rococo—in feeling. Most of Piazzetta's work is still in Venice—in the Accademia and numerous churches.

Picabia, Francis (1879–1953). French painter, designer, writer, and editor. His talent as an artist was modest, but his restless and energetic personality gave him a significant role successively in the *Cubist, *Dadaist, and *Surrealist movements, and through his publications he helped to disseminate avant-garde ideas. A private income enabled him to carry on his activities without having to worry about earning a living, as well as to indulge his love of fast cars, fast women, and wild living in general. Early in his career he was a successful painter of *Impressionist landscapes. In 1908–9 he experimented with *Neo-Impressionism, and then with *Fauvism and Cubism. In 1911 he met Marcel *Duchamp, who was to be the most important influence on his career, and with him became an exponent of *Orphism. He painted his first purely abstract works in 1912. In 1913 he visited New York as spokesman for the Cubist pictures in the *Armory Show, and he returned in 1915–16, when he, Duchamp, and *Man Ray were involved in the first stirrings of Dada. After moving to Barcelona (where he lived 1916–17), he launched a magazine entitled *391* (1917–24). In 1917 he returned to New York for six months then lived in Zurich (1918–19) before returning to Paris, where he helped to launch the Dada movement. However, in 1921 he denounced Dada for being no longer 'new', and became involved with André *Breton and the nascent Surrealist movement. In 1924 he attacked this, too, but some of his later works are in a Surrealist idiom. From 1925 to 1945 he lived mainly on the Côte d'Azur, experimenting with various styles. In 1945 he settled permanently in Paris and in his final years returned to abstract painting. Apart from his contributions to avant-garde magazines,

Picabia published various pamphlets and wrote poetry. He also conceived the fantasy ballet *Relâche* (1924), with music by Erik Satie, together with the film *Entr'acte* (directed by René Clair), which was used to fill the intermission between the ballet's two acts. Among Picabia's paintings, the most highly regarded today are those in his 'machinist' style, in which mechanistic and *biomorphic forms are combined in dynamic compositions. The most famous is *I See Again in Memory My Dear Udnie* (MOMA, New York, 1914).

Picasso, Pablo (1881–1973). Spanish painter, sculptor, graphic artist, ceramicist, and designer, the most famous, versatile, and prolific artist of the 20th cent. His personality dominated the development of the visual arts during most of the first half of the 20th cent. and he provided the incentive for most of the revolutionary changes during this time. Although it is conventional to divide his work into certain phases, all such divisions are to some extent arbitrary, as his energy and imagination were such that he was at all times working on a wealth of themes and in a variety of styles. He himself said: 'The several manners I have used in my art must not be considered as an evolution, or as steps toward an unknown ideal of painting. When I have found something to express, I have done it without thinking of the past or future. I do not believe I have used radically different elements in the different manners I have used in painting. If the subjects I have wanted to express have suggested different ways of expression, I haven't hesitated to adopt them.'

Picasso was the son of a painter and drawing master and was remarkably precocious, mastering academic draughtsmanship when he was still a child (his first word as a baby is said to have been 'lápiz'—pencil). In 1900 he made his first trip to Paris and by this time had already absorbed a wide range of influences. Between 1900 and 1904 he alternated between Paris and Barcelona, and this time coincides with his Blue Period, when he took his subjects from the poor and social outcasts, and the predominant mood of his paintings was one of slightly sentimentalized melancholy expressed through cold ethereal blue tones (*La Vie*, Cleveland Mus. of Art, 1903). He also did a number of powerful engravings in a similar vein (*The Frugal Repast*, 1904).

In 1904 Picasso settled in Paris and became the centre of an avant-garde circle of artists and writers including *Apollinaire. He had begun to attract the notice of connoisseurs such as the Russian Sergei Shchukin and Leo and Gertrude Stein, and *c.*1907 he was taken up by the dealer *Kahnweiler. A brief phase in 1904–5 is known as his Rose Period. The predominant blue tones of his earlier work gave way to pinks and greys and the mood became less austere. His favourite subjects were acrobats and dancers, particularly the figure of the harlequin. In 1906 he met *Matisse, but although he seems to have admired the work being done by the *Fauves, he did not himself follow their method in the decorative and expressive use of colour. (indeed his work often shows little concern with colour, and it is significant that—unlike most painters—he preferred to work at night by artificial light). The period around 1906–7 is sometimes called Picasso's Negro Period, because of the influence of African sculpture on his work at this point, but *Cézanne was an equally powerful influence at this time, as he concentrated on the analysis and simplification of form. His researches bore fruit in *Les Demoiselles d'Avignon* (MOMA, New York, 1906–7), which in its distortions of form and disregard of any conventional idea of beauty was as violent a revolt against tradition as the paintings of the Fauves in the realm of colour. At the time the picture was incomprehensible to artists, including Matisse and *Derain, and it was not publicly exhibited until 1937. It is now seen not only as a crucial achievement in Picasso's personal development but as the most important single landmark in the development of contemporary painting and as the herald of *Cubism, which he developed in close association with *Braque and then *Gris from 1907 up to the First World War.

In 1917 Picasso went with his friend Jean *Cocteau to Rome to design costumes and scenery for the ballet *Parade* and in the following years he designed for other *Diaghilev productions. The visit to Italy was one factor in introducing the strain of monumental Classicism that was one of the features of Picasso's work in the early 1920s (*Mother and Child*, Art Institute of Chicago, 1921), but at this time he was also involved with *Surrealism—indeed André *Breton hailed him as one of the initiators of the movement. However, his predominant interest in the analysis and synthesis of forms was at bottom opposed to the irrationalist elements of the Surrealists, their exaltation of chance, and equally to the direct realistic reproduction of dream or subconscious material. Following the most serene period in his art, Picasso began to make more violently expressive works, fraught with emotional tension, a mood of foreboding, and an almost clinical preoccupation with anguish and despair. This phase begins with *The Three Dancers* (Tate Gallery, London, 1925), a savage parody of classical ballet, painted at a time when his first marriage was becoming a source of increasing unhappiness and frustration. Following this he took up the mythological image of the Minotaur and images of the Dying Horse and the Weeping Woman. The period which culminated in his most famous work, *Guernica* (Centro Cultural Reina Sofía, Madrid, 1937), produced for the Spanish Pavilion at the Paris Exposition Universelle of 1937 to express horror and revulsion at the destruction by bombing of the Basque capital Guernica during the civil war (1936–9). It was followed by a number of other great paintings, including *The Charnel House* (MOMA, New York, 1945), attacking the cruelty and destructiveness of war. 'Painting is not done to decorate apartments,' he said: 'it is an instrument of war against brutality and darkness.' In treating such themes Picasso universalized the emotional content by an elaboration of the techniques of expression which had been developed through his researches into Cubism.

Picasso remained in Paris during the German Occupation, but from 1946 lived mainly in the South of France, where he added pottery to his many activities. His later output as a painter does not compare in momentousness with his pre-war work, but it remained vigorous, varied, and continuously inventive of new modes for the solution of new problems. It included a number of variations on paintings by other artists, including forty-four on *Las Meninas* of *Velázquez. The theme of the artist and his almost magical powers is one that exercised him greatly throughout his career.

Picasso's status as a painter has perhaps overshadowed his work as a sculptor, but in this field too (although his interest was sporadic) he ranks as one of the outstanding figures in 20th-cent. art. He was one of the first artists to make sculpture that was assembled from varied materials rather than modelled or carved, and he made brilliantly witty use of found objects (see OBJET TROUVÉ). The most celebrated example is *Head of a Bull, Metamorphosis* (Musée Picasso, Paris, 1943), made of the saddle and handlebars of a bicycle. Alan Bowness has written (*Modern Sculpture*, 1965) 'Picasso's sculpture sparkles with bright ideas—enough to have kept many a lesser man occupied for the whole of a working lifetime . . . it is not inconceivable that the time will come when his activities as a sculptor in the second part of his life are regarded as of more consequence than his later paintings.' As a graphic artist (draughtsman, etcher, lithographer, linocutter), too, Picasso ranks with the greatest of the century. He was a prolific book illustrator, and as few other artists had the power to concentrate the impress of his genius even in the smallest and slightest of his works. Picasso's emotional range is as wide as his varied technical mastery. By turns tragic and playful, his work is suffused with a passionate love of life, and no artist has more devastatingly exposed the cruelty and folly of his fellow men or more rapturously celebrated the physical pleasures of love. There are Picasso museums in Paris and Barcelona and other examples of his unrivalled output (which has been estimated at 20,000 works) are in museums throughout the world.

picture plane. In the imaginary space of a picture, the plane occupied by the physical surface of the work. *Perspective appears to recede from it, and objects painted in **trompe l'œil* may appear to project from it.

Picturesque. Term covering a set of attitudes towards landscape, both real and painted, that flourished in the late 18th and early 19th cents. It indicated an aesthetic approach that found pleasure in roughness and irregularity, and an attempt was made to establish it as a critical category between the 'beautiful' and the '*Sublime'. Picturesque scenes were thus neither serene (like the beautiful) nor awe-inspiring (like the Sublime), but full of variety, curious details, and interesting textures—medieval ruins were quintessentially Picturesque. Natural scenery tended to be judged in terms of how closely it approximated to the paintings of favoured artists such as Gaspard *Dughet, and in 1801 the *Supplement* to Samuel Johnson's *Dictionary* by George Mason defined 'Picturesque' as: 'what pleases the eye; remarkable for singularity; striking the imagination with the force of painting; to be expressed in painting; affording a good subject for a landscape; proper to take a landscape from.' The Picturesque generated a great deal of literature (see GILPIN, KNIGHT, PRICE) and the Picturesque Tour in search of suitable subjects was a feature of English landscape painting of the period, exemplified, for example, in the work of *Girtin and (early in his career) of *Turner. Much writing on the Picturesque was pedantic and obsessive and it became a popular subject for satire. It never gave rise to a coherent theory, and a study of the changes and ambiguities and inconsistencies in the concept constitutes a history in itself of developing taste in the 18th cent. and of the momentous transition from *Neoclassical regularity to *Romantic emphasis on the characteristic, the imaginative, and the moving.

Pieneman, Jan Willem (1779–1853). Dutch painter of historical pieces and portraits. His reputation and fortune were established by his painting of *The Battle of Waterloo* (Rijksmuseum, Amsterdam, 1824), which has greater historical interest than artistic merit. Like so many of his Dutch contemporaries he was at his best when painting unpretentious portraits. He was the first director of the Amsterdam Academy and numbered Jozef *Israëls among his pupils.

Pierce (or **Pearce**), **Edward** (*c.*1635–95). English sculptor and mason, son of a painter (d. 1658) of the same name, some of whose decorative work survives at Wilton House. Little is known of his youth, but by 1671 he was well established in London and was much employed by Sir Christopher Wren (1632–1723) on the rebuilding of the City churches, both as a mason and as a stone-carver. He was also a wood-carver, and his work in this field is of such quality that it has sometimes been credited to Grinling *Gibbons. His best work, however, is as a sculptor of portrait busts, the most notable being the splendid marble bust of Sir Christopher Wren (Ashmolean, Oxford, *c.*1673), brilliantly characterized and more convincingly *Baroque than anything else of the date in English art. It is generally considered the best piece of English sculpture of the 17th cent., but it has also been suggested that the workmanship does not live up to the boldness of the conception and that Pierce is here perhaps copying a lost bust by *Coysevox. Also in the Ashmolean is a marble bust of Oliver Cromwell, a bronze version of which (1672) is in the Museum of London. Pierce also worked as an architect, the Bishop's Palace, Lichfield (1686–7), being his chief known work.

Piero della Francesca (Piero dei Franceschi) (1415–92). Italian painter, virtually forgotten for centuries after his death, but regarded since his rediscovery in the early 20th cent. as one of the supreme artists of the *quattrocento. He was born in Borgo San Sepolcro (now Sansepolcro) in Umbria and spent much of his life there. We hear of him also at various times in Ferrara, Rimini, Arezzo, Rome, and Urbino. But he found the origins of his style in Florence, and he probably lived there as a young man for some time during the 1430s, although he is documented there only once, in 1439, when he was assisting *Domenico Veneziano on frescos (now lost) in S. Egidio. His first documented work, the *polyptych of the *Madonna della Misericordia* (Pinacoteca, Sansepolcro), commissioned in 1445 but not completed until much later, shows that he had studied and absorbed the great artistic discoveries of his Florentine predecessors and contemporaries—*Masaccio, *Donatello, Domenico Veneziano, Filippo *Lippi, *Uccello, and even *Masolino, who anticipated something of Piero's use of broad masses of colour. Piero unified, completed, and refined upon the discoveries these artists had made in the previous 20 years and created a style in which monumental, meditative grandeur and almost mathematical lucidity are combined with limpid beauty of colour and light.

Piero's major work is a series of *frescos on the *Legend of the True Cross* in the choir of S. Francesco at Arezzo (*c.*1452–*c.*1465). The subject was a medieval legend of great complexity, but Piero made from its fanciful details some of the most solemn and serene images in Western art—even the two battle scenes have a feeling of grim deliberation rather than violent movement. He was a slow and thoughtful worker and often applied wet cloths to the plaster at night so that—contrary to normal fresco practice—he could work for more than one day on the same section.

Much of Piero's later career was spent working at the humanist court of Federico da *Montefeltro at Urbino. There he painted the portraits of Federico and his wife (Uffizi, Florence, *c.*1465) and the celebrated *Flagellation* (still at Urbino, in the Ducal Palace). The *Flagellation* is his most enigmatic work, and it has called forth varied and sometimes highly involved interpretations; *Gombrich has suggested that the subject is rather *The Repentance of Judas* and *Pope-Hennessy that it is *The Dream of St Jerome*. Piero is last mentioned as a painter in 1478 (in connection with a lost work) and his two final works are probably *The Madonna and Child with Federico da Montefeltro* (Brera, Milan, *c.*1475) and the unfinished *Nativity* (NG, London). Thereafter he seems to have devoted himself to mathematics and *perspective, writing treatises on both subjects. *Vasari said Piero was blind when he died, and failing eyesight may have been his reason for giving up painting, but his will of 1487 declares him to be 'sound in mind, in intellect and in body' and is written in his own clear hand.

After his death, Piero was remembered mainly as a mathematician rather than as a painter. Even Vasari, who as a native of Arezzo must have known the frescos in S. Francesco well, is lukewarm in his enthusiasm for his work. However, he had considerable influence, notably on *Signorelli (in the weighty solemnity of his figures) and *Perugino (in the spatial clarity of his compositions). Both are said to have been Piero's pupils.

Piero di Cosimo (*c.*1462–1521?). Florentine painter, a pupil of Cosimo *Rosselli, whose Christian name he adopted as a patronym. There are no signed, documented, or dated works by him, and reconstruction of his *œuvre* depends on the account given in *Vasari's *Lives*. It is one of Vasari's most entertaining biographies, for he portrays Piero as a highly eccentric character who lived on hard-boiled eggs, 'which he cooked while he was boiling his glue, to save the firing'. The paintings for which he is best known are appropriately idiosyncratic—fanciful mythological inventions, inhabited by fauns, centaurs, and primitive men. There is sometimes a spirit of low comedy about these delightful works, but in the so-called *Death of Procris* (NG, London) he created a poignant scene of the utmost pathos and tenderness. He was a marvellous painter of animals and the dog in this picture, depicted with a mournful dignity, is one of his most memorable creations. Piero also painted portraits, the finest of which is that of Simonetta Vespucci (Musée Condé, Chantilly) in which she is depicted as Cleopatra with the asp around her neck. His religious works are somewhat more conventional, although still distinctive, and Frederick Hartt (*A History of Italian Renaissance Art*, 1970) has written that 'His whimsical Madonnas, Holy Families, and Adorations provide a welcome relief from the wholesale imitation of *Raphael in early Cinquecento Florence'. One of his outstanding religious works is the *Immaculate Conception* (Uffizi, Florence), which seems to have been the compositional model for the *Madonna of the Harpies* by his pupil *Andrea del Sarto.

Pietà. Term, Italian for 'pity', applied to a painting or sculpture showing the Virgin Mary supporting the body of the dead Christ on her lap. Other figures, such as St John the Evangelist or Mary Magdalene, may also be included. The theme, which has no literary source, originated in the early 14th cent. in Germany and was more popular in northern Europe than in Italy. However, the most celebrated of all Pietàs is that by *Michelangelo in St Peter's, Rome; and one of the most sublime is that painted by *Titian for his own tomb (Accademia, Venice). The subject is not always clearly distinguished from the scene known as the Lamentation. However, whereas the Lamentation represents a specific moment from Christ's Passion, between the Descent from the Cross and the Entombment, the Pietà is a timeless devotional image.

pietra dura (Italian 'hard stone'). A term applied to a particular kind of *mosaic work in which coloured stones, such as lapis lazuli, agate, and *porphyry, are used to imitate as far as possible the effect of painting. The main centre of the art was Florence, where in 1588 the Grand Duke Ferdinand established a factory in the *Uffizi which continued to work well into the 19th cent. The production of the workshop consisted mainly of table-tops and panels for cabinets, decorated with birds and flowers or landscapes which were executed with a *naturalism and polychromatic brilliance that made them celebrated all over Europe.

Pietro da Cortona. See CORTONA.

Pietro Spagnuolo. See BERRUGUETE.

Pigalle, Jean-Baptiste (1714–85). French sculptor. He studied under J.-B. *Lemoyne and then in Rome (1736–9). In his early career he endured poverty and sickness (his studies in Rome were made at his own expense and he walked there from Paris), but after he was received into the Académie Royale in 1744 with his rapturously acclaimed *Mercury* (Louvre, Paris; *terracotta model in the Met. Mus., New York), he rapidly went on to become the most successful French sculptor of his period. He was a superb craftsman and highly versatile and inventive, equally adept at small *genre pieces and the most grandiloquent tomb sculpture. As a portraitist he was noted for his warmth and vivacity. His most famous works are the startling nude figure of Voltaire (Institut de France, Paris, 1770–6) and the spectacular and majestic tomb of Maurice of Saxony (designed 1753) in St Thomas, Strasburg.

pigment. Any substance used as a colouring agent, particularly the finely ground particles that when held in suspension in a *medium constitute a paint. Most pigments are now manufactured synthetically, but historically they have been made from a great variety of mineral, plant, and animal sources: the brown colour sepia, for example, comes from the inky secretions of the cuttlefish, and ultramarine blue was originally made from the semi-precious stone lapis lazuli. The history of pigments is a highly specialized field with little practical importance today for artists who use commercial paints, but it is often vitally important to the expert in relation to authentication and *attribution.

Piles, Roger de. See DE PILES.

Pillement, Jean-Baptiste (1728–1808). French painter and designer. As a painter he is best known for his charming landscapes, which are strongly indebted to *Boucher but more atmospheric. His importance, however, lies more in the engravings made after his drawings, which were popular throughout Europe and of influence in spreading the *Rococo style, particularly the taste for *chinoiserie. Pillement himself was well travelled, visiting Austria, England, Poland, Portugal, and Spain.

Pilo, Carl Gustaf (1711–93). Swedish portrait painter. In 1741 he went to Copenhagen, where he became court painter and director of the Academy, but he was expelled from Denmark in 1772 following political intrigues. Returning to Sweden, he settled in Stockholm, and there also worked for the royal family and became director of the Academy. Pilo's best-known work is *The Coronation of Gustavus III* (Nat. Mus., Stockholm, begun 1792), which although it was left unfinished at his death is considered one of the masterpieces of Swedish painting. His style was characterized by nervously sensitive drawing and refined colouring in a personal *Rococo vein.

Pilon, Germain (*c.*1525–90). The most powerful and original French sculptor of the 16th cent. He was born in Paris, the son of a sculptor, **André Pilon**, and spent most of his career there. His first known work is the group of *Three Graces* on the monument for the heart of Henry II (Louvre, Paris, 1561–2), which shows that his early manner was based on the elegant decorative style of the École de *Fontainebleau, in particular *Primaticcio's *stucco-work. Although his work never lost its tendency towards graceful *Mannerist elongation, it developed in the direction of greater naturalism and emotional intensity. These qualities are seen most memorably in his marble **gisants* on the tomb of Henry II and Catherine de Médicis at St Denis (1563–70), poignant works in which the semi-nude figures are shown relaxed in death. The kneeling effigies on the tomb are of bronze and Pilon excelled in this medium as well as marble, both as a portrait sculptor (*Charles IX*, Wallace Coll., London) and a medallist—he was appointed Controller General of the Paris Mint in 1572. Pilon's early work was influential on his successors, but the deeply felt emotion of his later style proved too personal to inspire imitation. Several examples of his sublime late work are in the Louvre, notably *The Virgin* (*c.*1580–5), *The Deposition* (*c.*1580–5), and the tomb of Valentine Balbiani (before 1583).

Piloty, Karl von (1826–86). German painter. After training in Antwerp and Paris Piloty settled in Munich, where in 1856 he became a professor at the Academy and in 1874 its Director. He was highly successful with large, opulent history paintings in which the settings, furnishings, and costumes were reconstructed with great accuracy. They gave the public the same kind of feast for the eye as Hollywood epics and now have a distinctly hollow ring. His influence was felt by many of his pupils.

Pineau, Nicolas (1684–1754). French sculptor and designer, son of **Jean-Baptiste Pineau** (d. 1694) who was employed as a carver at Versailles and elsewhere. Nicolas went to Russia in 1716 with the architect Alexandre Le Blond (1679–1719) and on the latter's death was the leading French decorative artist at the court of the Czar (Peter the Great). After he returned to Paris in about 1730 he became with *Meissonier one of the leaders in the lighter style of *Rococo decoration, his work acquiring wide popularity in engravings. His son **Dominique** (1718–86) was also a designer.

pinprickt pictures. Pictures made by pricking holes into paper to produce a lace-like effect. They may derive from the practice of pricking the outlines of a *cartoon in order to transfer it to the painting *ground. Most examples appear to be English, of the 18th and early 19th cents., when the practice was a fashionable hobby for ladies (the spiked wheel that was commonly used to draw lines was a dressmakers' tool for tracing patterns on cloth).

Pintoricchio (or **Pinturicchio**) (Bernardino di Betto) (*c.*1454–1513). Italian painter, born in Perugia. He was probably a pupil of *Perugino and it is likely that he assisted him with his frescos in the Sistine Chapel, Rome (1481–2). His style was strongly influenced by Perugino, especially in his sweet, elegant figure types, but Pintoricchio lacked Perugino's lucidity of design and was more interested in decorative effects. His chief works are frescos in the Borgia rooms in the Vatican (1492–5) and the colourful *Scenes from the Life of Aeneas Sylvius Piccolomini* (i.e. Pope Pius II) in the Piccolomini Library of Siena Cathedral (1503). In these he showed the brilliant colours, ornamental detail, and fanciful charm that make him at his best, in the words of Frederick Hartt (*A History of Italian Renaissance Art*, 1970), 'one of the most endearing masters of the *Quattrocento . . . a kind of Perugian Benozzo *Gozzoli'. Pintoricchio was a prolific painter of panels as well as frescos—there are several examples of his work in the National Gallery, London.

Piper, John (1903–92). English painter, graphic artist, designer, and writer. He reluctantly became an articled clerk in his father's legal firm, but took up the study of art after his father's death in 1926, first at the Richmond School of Art and then at the *Royal College of Art. From 1928 to 1933 he wrote as an art critic for the *Listener* and the *Nation* and was among the first to recognize such contemporaries as William *Coldstream, Ivon *Hitchens, Victor *Pasmore, and Ceri *Richards. In 1935 he became art editor to the avant-garde quarterly **Axis*, edited by Myfanwy Evans, whom he married in 1937. At this time he was one of the leading British abstract artists, but by the end of the decade he had become disillusioned with non-representational art and reverted to naturalism. He concentrated on landscape and architectural views in a subjective emotionally charged style that continued the English *Romantic tradition. Some of his most memorable works were done as an *Official War Artist when he made pictures of bomb-damaged buildings. A similar stormy atmosphere pervades his famous views of country houses of the same period. Piper's work diversified in the 1950s and he became recognized as one of the most versatile British artists of his generation. He did much work as a stage designer and designer of stained glass (notably at Coventry Cathedral) and was also a prolific printmaker. In addition he made book illustrations and designed pottery and textiles. As a writer he is probably best known for his book *British Romantic Artists* (1942). He also wrote various architectural guidebooks, sometimes collaborating with his friend the poet Sir John Betjeman (1906–84). Piper's work is extensively represented in the Tate Gallery.

Piranesi, Giovanni Battista (1720–78). Italian etcher, archaeologist, and architect, born in Venice but active for almost all his career in Rome, where he settled in 1740. In Venice he had studied perspective and stage design and in Rome he achieved great popularity with his dramatically conceived etchings of the ancient and modern city—the *Vedute*—published from 1745 onwards. He often altered the scale of buildings to make them look even grander than they are in actuality (Horace *Walpole said he 'conceived visions of Rome beyond what it boasted even in the meridian of its splendour') and his work played a major role in shaping the popular mental image of the city. Piranesi produced numerous other plates of Roman antiquities and architectural details, but his most original works are a series of *Carceri d' Invenzione*, fantastic imaginary prisons, begun *c.*1745 and reworked in 1761. These striking and obsessive works were later claimed by the *Surrealists as an anticipation of their principles and their influence can be seen in 20th-cent. horror movies. Only one of Piranesi's architectural designs was built (Sta Maria del Priorato, Rome, 1764–6), but he was important as an architectural polemicist, most notably in his *Della magnificenza ed architettura de' Romani* (1761), in which he championed the superiority of Roman architecture over Greek. Piranesi's influence was felt not only by architects, but also by stage designers and painters of *capricci such as his friend Hubert *Robert, and he had a vivid impact on the literary imagination. William *Beckford, for example, said that in writing his Gothic novel *Vathek* (1786) 'I drew chasms, and subterranean hollows, the domain of fear and torture, with chains, racks, wheels and dreadful engines in the style of Piranesi.' His etchings continued to be published long after his death and his work was continued by his son **Francesco** (1758–1810).

Pisanello (properly Antonio Pisano) (*c.*1395–1455?). Italian painter and medallist. He presumably came from Pisa (hence his nickname), but he spent his early years in Verona, a city with which he kept up his association for most of his life. His successful career also took him to the Vatican and numerous courts of northern Italy. With *Gentile da Fabriano Pisanello is regarded as the foremost exponent of the *International Gothic style in Italian painting, but most of his major works have perished, including frescos in Venice in which he collaborated with Gentile, and in Rome in which he completed work left unfinished by Gentile at his death. His surviving documented frescos are *The Annunciation* (S. Fermo, Verona, 1423–4) and *St George and the Princess of Trebizond* (Sta Anastasia, Verona, 1437–8), and attributed to him are some fragments of murals of jousting knights in the Palazzo Ducale in Mantua, uncovered in 1968 and one of the most spectacular art discoveries of recent years. A very small number of panel paintings is also given to him, two being in the National Gallery, London. On the other hand, a good many of his drawings survive, those of animals being particularly memorable. They show his keen eye for detail and his ability to convey an animal's personality. Pisanello was also the greatest portrait medallist of his period and arguably of the whole *Renaissance, his work setting standards of delicacy, precision, and clarity that have not been surpassed. There are good examples of his drawings in the British Museum, and of his medals in the Victoria and Albert Museum.

Pisano, Andrea (*c.*1290–1348/9?). Italian sculptor and architect, not related to Nicola and Giovanni *Pisano. He probably came from Pontedera near Pisa (he is sometimes called Andrea da Pontedera), but he is first documented in Florence in 1330, when he received the commission to make a pair of bronze doors for the Baptistery. The doors, finished in 1336, are the first of the three great sets for the Baptistery (the other two are by *Ghiberti), and represent twenty scenes from the life of St John the Baptist and eight Virtues. They show a melodious line and a jeweller's refinement of execution. By 1340 Andrea was architect to Florence Cathedral (succeeding *Giotto) and the only other works certainly by him or from his workshop are *reliefs and statues for the cathedral's campanile. In their clear-cut designs the reliefs show the influence of Giotto's painting.

In 1347 Andrea was appointed master of works at Orvieto Cathedral, where he was succeeded by his son **Nino** (d. 1368) in 1349. Nino is known from documents to have been active as goldsmith and architect, but all his surviving works are sculptures in marble. He was much less distinguished as an artist than his father, but noteworthy in being one of the first sculptors to specialize in free-standing life-size statues. There are three signed works by him: *Madonnas* in Sta Maria Novella, Florence, and on the *Cornaro Monument* in SS. Giovanni e Paolo, Venice (*c*.1367); and the statue of a bishop in S. Francesco, Oristona, Sardinia. Others are attributed to him on the basis of these.

Pisano, Nicola (d. 1278/84) and **Giovanni** (d. after 1314). Italian sculptors and architects, father and son. They were the greatest sculptors of their period and stand at the head of the tradition of Italian sculpture in the same way that *Giotto stands at the head of the tradition of Italian painting. They often worked together, but their styles are distinctive. Nicola came from Apulia, where the emperor Frederick II (d. 1250) had encouraged a *classical revival, and his first known work, the pulpit in the Baptistery at Pisa (dated 1260 Pisan style, i.e. 1259) shows his brilliant adaptation of *antique forms to a new context. He transformed a Dionysus into Simeon at the Presentation of Christ, a nude Hercules into a personification of Christian Fortitude, and a Phaedra into the Virgin Mary. But instead of following the *Romanesque convention of separating episodes into compartments arranged in bands, he combined them into the formal unity of single pictures on each side of the pulpit with great power and dramatic effect. Several of the figures were directly inspired by ancient *sarcophagi that Nicola saw in the Campo Santo in Pisa, but they are much more than simple borrowings, for he made them the vehicle for expressing richly varied human feeling.

Nicola followed the Pisa pulpit with a similar but more complex work for Siena Cathedral (1265–8). The carving is deeper, the contrasts between light and shadow sharpened, the *reliefs more densely packed and full of movement. By then Nicola had a large workshop, his assistants including his son Giovanni and *Arnolfo di Cambio. His last great project was the large fountain in the public square of Perugia, which he and Giovanni finished in 1278. The dozens of reliefs are a typical medieval mixture: biblical scenes, heraldic beasts, personifications of seasons and places, and local dignitaries; but the vigour and spontaneity of the carving express a new freedom and naturalness.

By 1284 Nicola was dead. Between the Perugia fountain and this date, Giovanni, alone or in company with his father, had carved the sculpture for the outside of the Pisa Baptistery (now in the Museo Nazionale). Here for the first time in Tuscany a scheme of monumental statuary was incorporated into architecture. Giovanni developed this much further in Siena, where from 1284 onwards he designed the façade of the cathedral and carried out much of the sculptural decoration (some of the figures have been transferred to the cathedral museum and a magnificent fragment is in the Victoria and Albert Museum). It is the most richly decorated of all the great Italian *Gothic cathedral façades, and the statuary has tremendous energy and inner life.

Giovanni's last two great works were pulpits for S. Andrea, Pistoia (1300–1), and Pisa Cathedral (1302–10). They are modelled on those of his father, but more elegant in style (showing French Gothic influence) and also more emotionally charged. The Pisa pulpit was damaged in a fire in 1599, then dismantled and reassembled, some parts being dispersed; several museums, including the Metropolitan Museum, New York, have fragments that are said to come from it. Giovanni also made a number of free-standing statues, the best known of which is the *Madonna and Child* on the altar of the Arena Chapel in Padua (*c*.1305). Its grandeur and humanity suggests a close kinship with Giotto, amid whose frescos it stands.

Pissarro, Camille (1830–1903). French painter and graphic artist. He was born in the West Indies of a Jewish father and a Creole mother, and in 1855 moved to Paris, where he was initially strongly influenced by *Corot. In 1859 he met *Monet, and with him became a central figure of the *Impressionist group. Pissarro in fact was the only artist who exhibited at all eight Impressionist exhibitions and was a much-respected father figure to other members of the group. His great gifts as a teacher made him influential even among artists of greater stature than himself—*Cézanne and *Gauguin, for example, spoke glowingly of him. During the Franco-Prussian War of 1870–1, when his home at Louveciennes was overrun by the invaders and his paintings destroyed, he joined Monet in England and came under the influence of English landscape painters, particularly *Turner and *Constable. In 1872 he settled at Pontoise, where he had lived 1866–9, and for some years painted there in close friendship with Cézanne, who had settled at Auvers-sur-Oise. From 1884 he lived at Eragny, near Gisors, and there met *Signac and came to know *Seurat. For some years during the 1880s he flirted with the *pointillist technique and the methods of the *Neo-Impressionists under Seurat's influence, but later reverted to his earlier style. From about 1895 deterioration of his eyesight caused him to give up painting out of doors and many of his late works are town views painted from windows in Paris. He died blind. Pissarro was extremely prolific not only as a painter but also in various graphic techniques and his work is in many public collections throughout the world.

Lucien Pissarro (1863–1944), eldest son of Camille, was taught by his father and other leading members or associates of the Impressionist group, particularly *Manet and Cézanne. He exhibited in the final Impressionist exhibition (1886) and with Seurat in the Second *Salon des Indépendants, adopting the pointillist technique for a time. In 1890 he settled in England, and he became a British citizen in 1916. From 1905 he was part of *Sickert's circle and he was a member of the *Camden Town Group and afterwards of the *London Group. He was a distinguished book illustrator and from 1894 to 1914 ran the *Eragny Press, one of the best of the private presses that flourished at this period. A modest and unassuming character, he has been overshadowed by his more famous father, but he was an important link between French Impressionism and Neo-Impressionism and English art. His daughter, **Orovida Pissarro** (1893–1968), was a painter and etcher, mainly of animal subjects.

Pitti Palace, Florence. Art gallery, originally built as a palace for Luca Pitti, a wealthy rival of the *Medici. It was probably begun in the 1450s, so the traditional attribution of its design to *Brunelleschi (who died in 1446) is unlikely; it has also been attributed to *Alberti. It remained unfinished until it was acquired in 1549 by Duke Cosimo I, who made it the Medici residence and had it enlarged by *Ammanati. Most of the present vast structure dates from the 16th and 17th cents. The interior includes opulent decoration by Pietro da *Cortona.

The Pitti Gallery on the upper floor in the left wing contains about 500 masterpieces from the Medici collections which rival those of its sister institution, the *Uffizi. The two galleries are connected by a corridor designed by *Vasari, carried over the River Arno on top of the Ponte Vecchio. The State Apartments contain a profusion of art treasures including sculptures and tapestries, and on the ground floor is the Museo degli Argenti, which contains outstanding collections of plate, goldsmiths' work, ivories, *enamels, vestments, etc., most of which were originally from the Medici collections.

Pittoni, Giovanni Battista (1687–1767). Venetian painter of religious, historical, and mythological pictures. He was very popular in his day and ranks as one of the best contemporaries of *Tiepolo, whom he succeeded as President of the Venice Academy of Painting 1658–61. Pittoni never left Italy, but he nevertheless received important foreign commissions, notably from the English playwright and impresario Owen McSwiny (d. 1754) and from the Swedish, Austrian, and German courts. His early work was much indebted to *Piazzetta and Sebastiano *Ricci, but his style later became lighter and more colourful under the influence of Tiepolo.

Place, Francis (1647–1728). English gentleman artist, a member of a distinguished circle of virtuosi (see VIRTU) at York. He was a close friend of *Hollar, and his topographical and architectural drawings are close to his in style. His later drawings rely on a fuller use of *wash and are among the more important anticipations of English 18th-cent. watercolour style. Place also made portraits and was one of the pioneers of the *mezzotint technique. His work is best represented in the City Art Gallery, York.

Plamondon, Antoine (1804–95). Canadian painter of portraits, religious subjects, and figure compositions, active mainly in Quebec. He was the first Canadian to study in France after Quebec was ceded to Britain in 1763, training in Paris, 1826–30. His portraits, which are his best work, are painted in an austerely classical style. An example is the *Ingres-like *Sister Saint-Alphonse* (NG, Ottawa, 1841), one of a sensitive series of portraits of nuns. Plamondon also executed numerous church commissions in Quebec.

plaquette. Small decorative *relief in metal, usually bronze or lead. They were almost always cast by the **cire-perdu* process and new editions could be made from the wax image of an existing plaquette; if this wax were altered a new 'state' would result. A very few were struck like coins. After casting, the best plaquettes were usually chiselled and chased, and finished with a *patina or gilding. Plaquettes were mounted as sword-hilts, ink-wells, or caskets, and applied as decoration to a variety of objects; small ones served as buttons, and a religious subject might be mounted as a pyx for use in the Mass. But they must also have been collected for their own beauty.

During the brief period in which plaquettes flourished (the end of the 14th to the middle of the 16th cent.) one finds that, being so easily transportable, they were copied both in their country of origin and abroad, and not only in other minor arts but in important sculptural works as well. Like the engravings of which they are the plastic counterpart, they helped to disseminate the taste of the Renaissance. *Flötner's plaquettes, for example, helped to make his designs common property among German artists. *Donatello is the greatest name connected with the art, but the greater number of plaquettes, whatever their place of origin, are of unknown authorship.

plaster of Paris. A fine white or pinkish powder, made by the calcination or dehydration of gypsum (see ALABASTER), which when mixed with water forms a quick-setting paste that dries to form a uniform, solid, and inert mass. It is used in sculpture for making moulds and casts.

plaster print. A print produced from an *intaglio plate that has been converted to a *relief print by making a plaster cast of it, thereby causing lines that were originally incised to stand out above the surface.

Plastov, Arkady. See SOCIALIST REALISM.

Plateresque. Architectural style peculiar to Spain in the early 16th cent., characterized by extremely lavish ornament—*Gothic, *Renaissance, and sometimes Moorish in inspiration—unrelated to the structure of the building on which it is used. The term literally means 'silversmith-like'. It is as much a sculptural as an architectural style, and some exponents, most notably Diego de *Siloe, practised both arts.

plein air (French: open air). Term used for a painting done in the open air instead of in the studio, or more generally for a picture that gives a strong feeling of the open air. Although there are earlier instances, painting out of doors did not become common until the 19th cent., when the development of portable equipment—above all the collapsible metal *tube—made it much easier in practical terms. It became a central feature of *Impressionism and is especially associated with *Monet.

Pleydenwurff, Hans (d. 1472). German painter. He was active in Nuremberg and his work is typical of the kind of painting produced there by the generation before *Wolgemut (who married Pleydenwurff's widow). The springs of *International Gothic were drying up and a new naturalism was beginning to come in from the Netherlands. A large *Crucifixion* (Alte Pinakothek, Munich) is characteristic of his work. Pleydenwurff's son **Wilhelm** (d. 1494) was a painter and engraver, also active in Nuremberg.

Pliny the Elder (Gaius Plinius Secundus) (AD 23 or 24–79). Roman encyclopaedist. A writer of extraordinary industry, he produced as a side-line to his career in public office, among many other things, the *Historia Naturalis* (Natural History), a massive compilation in thirty-seven books intended to embrace not only the whole of the natural sciences but also their application to the arts and crafts of civilized life. The Natural History, his only extant work, has been condemned as uncritical, unreliable, and superficial, but it contains much information that would otherwise have been completely lost. The chapters on the history of painting and sculpture in the Natural History are especially interesting because earlier treatises on classical art have perished and Pliny is thought to have preserved (either directly or at second hand) critical and historical views from an earlier date. He also reveals the estimation in which Greek artists were held at Rome in his time. It is typical of Pliny's thirst for knowledge that he perished through asphyxiation when making observations of the eruption of Vesuvius that destroyed Pompeii and Herculaneum.

plumbago. An old-fashioned term for *graphite.

pochade. See BONINGTON.

Poelenburgh, Cornelis van (1594/5–1667). Dutch painter, mainly of landscapes. He studied in his native Utrecht with *Bloemaert and from about 1617 to 1625 was in Rome, becoming one of the leading members of the first generation of Dutch painters of Italianate landscapes. His paintings are typically small scale (he often painted on copper) with biblical or mythological figures set in Arcadian landscapes, sometimes scattered with antique remains. They are strongly influenced by *Elsheimer, but cooler in colour than the German artist's work and without his sense of mystery. After returning to Utrecht Poelenburgh enjoyed a career of great success. He was *Rubens's guide when he visited Utrecht in 1627, was popular in aristocratic and even royal circles (Charles I called him to England in 1637), and was imitated until the early 18th cent. There are examples of Poelenburgh's work (and of the work of imitators) in the Fitzwilliam Museum, Cambridge, including his portrait of Jan *Both, in whose landscapes he sometimes painted the figures (an example of their collaboration is in the National Gallery, London).

pointillism. Technique of using regular small touches of pure colour in such a way that when a picture is viewed from a suitable distance they seem to react together optically, creating more vibrant colour effects than if the same colours were physically mixed together. The term ('peinture au point') was coined in 1886 by the French critic Félix

Fénéon (1861–1944) in reference to *Seurat's *La Grande Jatte*, but Seurat himself, and also *Signac, preferred the word *'divisionism'.

pointing. A method of creating an exact copy of a statue or of enlarging a model into a full-size sculpture by taking a series of measured points on the original and transferring them by means of mechanical aids to the copy or enlargement. An elementary method of pointing using callipers was developed in the 1st cent. BC, when the copying of Greek statues for Roman patrons had become an industry, and various more sophisticated techniques, using for example a frame and plumb-line, have been used since the *Renaissance. It was not until the early 19th cent., however, that the task was rationalized by the perfection of a pointing machine. This device consists of an upright stand carrying movable arms, each arm having attached to it an adjustable measuring rod which shows the depth to which each point must be drilled. Sometimes hundreds and even thousands of points will be taken to ensure a meticulously exact copy. In the late 19th and early 20th cents. most stone sculpture was produced by this indirect and mechanical method, but modern sculptors increasingly reject it in favour of direct carving.

Poliakoff, Serge (1906–69). Russian-born painter who became a French citizen in 1962. He left Russia in 1919, settled in Paris in 1923, and took up painting in 1930 (from 1935 to 1937 he lived in London, and studied at the Chelsea School of Art and then at the *Slade School). However, it was not until 1952 that he was able to devote himself full time to painting and give up his career as a professional musician. By this time he had matured his style of abstract painting, begun in 1937 on his meeting with *Kandinsky, and his position was established as one of the most important painters in the French school of expressive abstraction. He adopted an almost religious attitude towards painting, saying: 'You've got to have the feeling of God in the picture if you want to get the big music in.'

Polidoro da Caravaggio (Polidoro Caldara) (*c.*1500–43). Italian painter, named after his birthplace, Caravaggio in Lombardy. At an early age he moved to Rome, where he assisted *Raphael in the decoration of the Vatican Loggie and then achieved great success painting palace façades with monochrome scenes imitating *classical sculpture. They have almost all perished (only the heavily restored decoration of the Palazzo Ricci remains *in situ*), but they became well known through engravings and drawings and were much imitated. Polidoro's other claim to fame is his decoration of the chapel of Fra Marino Fetti in S. Silvestro al Quirinale (1525) with two murals—oil not fresco—(one each from the life of St Mary Magdalene and St Catherine of Siena) in which he gave an entirely new prominence to the landscape, which dominates the figures. In this he foreshadowed the 'heroic landscape' of *Claude and *Poussin. Polidoro fled Rome after the sack of the city in 1527, moving to Naples and then Messina, where he was murdered by a thief.

Pollaiuolo, Antonio (*c.*1432–98) and **Piero** (*c.*1441–*c.*1496). Florentine artists, brothers, who jointly ran a flourishing workshop. They are both recorded as being painters, sculptors, and goldsmiths, but there are considerable problems in attempting to disentangle their individual contributions. Antonio was apparently primarily a goldsmith and worker in bronze, Piero primarily a painter. Several documented paintings by Piero are known, all of fairly mediocre quality, but none by Antonio, and as certain pictures from the studio of the two brothers are so much better than Piero's independent work, Antonio's collaboration, at least, has usually been assumed. The most important of these pictures is the *Martyrdom of St Sebastian* in the National Gallery, London, probably painted in 1475. The figures of the archers in the foreground reveal a mastery of the nude paralleled in certain bronzes generally accepted as Antonio's (e.g. the *Hercules and Antaeus* in the Bargello, Florence, *c.*1475–80), in his only surviving engraving (*The Battle of the Nude Men*, *c.*1460), and in his numerous pen drawings in which his typically wiry figures are seen in vigorous and expressive movement. His main contribution to Florentine painting lay in his searching analysis of the anatomy of the body in movement or under conditions of strain, but he is also important for his pioneering interest in landscape, seen in the National Gallery *St Sebastian* and other works. He is said to have anticipated *Leonardo in dissecting corpses in order to study the anatomy of the body, and his drawings in particular look forward to Leonardo's both in their style and their exploratory function.

Antonio's two principal public works were the bronze tombs of Pope Sixtus IV (signed and dated 1493) and Pope Innocent VIII (*c.*1492–8), both in St Peter's, Rome. The latter contains the first sepulchral effigy that simulated the living man.

Pollock, Jackson (1912–56). American painter, the commanding figure of the *Abstract Expressionist movement. In 1929–31 he studied at the *Art Students League under Thomas Hart *Benton and was influenced not only by Benton's restlessly energetic style, but also by his image as a virile, hard-drinking macho-man (Pollock began treatment for alcoholism in 1937 and in 1939 he started therapy with Jungian psychoanalysts, using his drawings in sessions with them). During the 1930s he painted in Benton's *Regionalist vein, and he was influenced also by the Mexican muralists (see OROZCO; RIVERA; SIQUEIROS) and by certain aspects of *Surrealism, particularly the use of mythical or totemic figures as archetypes of the unconscious. From 1935 to 1942 he worked for the *Federal Art Project, and in 1943 he was given a contract by Peggy *Guggenheim; his first one-man show was held at her Art of this Century gallery in that year. A characteristic work of this time is *The She-Wolf* (MOMA, New York, 1943), a semi-abstract picture with vehemently handled paint and ominous imagery recalling the monstrous creatures of *Picasso's *Guernica* period. By the mid-1940s Pollock was painting in a completely abstract manner, and the 'drip and splash' style for which he is best known emerged with some abruptness in 1947. Instead of using the traditional easel, he laid his canvas on the floor and poured and dripped his paint from a can (using commercial enamels and metallic paint because their texture was better suited to the technique); instead of using brushes, he manipulated the paint with 'sticks, trowels or knives' (to use his own words), sometimes obtaining textured effects by the admixture of 'sand, broken glass or other foreign matter'. In line with Surrealist theories of *automatism, this method of *Action painting was supposed by artists and critics alike to result in a direct expression or revelation of the subconscious mind of the painter. Pollock's technique was bound up with the creation of the *All-over method of painting, which avoids any points of emphasis and abandons the traditional idea of composition in terms of relation among parts. The design of the painting had no relation to the size or shape of the canvas—indeed in the finished work the canvas was sometimes docked or trimmed to suit the image. These characteristics were important for the development of American abstract painting in the late 1940s and early 1950s, and when the drip paintings were first publicly shown, at Betty *Parsons's gallery in 1948, Willem *de Kooning commented 'Jackson's broken the ice'.

Pollock's drip period lasted only from 1947 to 1951 (in the 1950s he went back to quasi-figurative work), but it is on the paintings of these four years that his enormous reputation rests. Among the most celebrated are *Autumn Rhythm* (MOMA, New York, 1950) and *Lavender Mist* (NG, Washington, 1950), which Robert *Hughes describes as 'his most ravishingly atmospheric painting'. Pollock was supported by advanced critics, particularly Clement *Greenberg and Harold *Rosenberg, and in 1949 the French painter Georges *Mathieu said that he considered him the 'greatest living American painter'. However, he was also subject to much abuse as the leader of a still little comprehended movement (in 1956 *Time* magazine called him 'Jack the Dripper'). By 1960 he was generally recognized as the most important figure in the most important movement in the history of American painting, but a movement from which artists were already in reaction. His unhappy personal life and his premature death in a car crash contributed to his status as one of the legends of modern art; he was the first American painter to become a 'star'.

In 1945 Pollock married Lee Krasner (1911–84), who was an Abstract Expressionist painter of some distinction, although it was only after her husband's death that she received serious critical recognition. She was also an important source of encouragement and support to Pollock, whose attitude to his work fluctuated from supreme confidence to dismal uncertainty.

polychrome. Term meaning 'painted in many colours' applied in art historical writing particularly to sculpture treated in this fashion. Until the *Renaissance the practice was extremely common, whether the colours were conventionalized, as in much ancient art, or more or less naturalistic, as became more common during the Middle Ages, but because of fading of the pigments the colouring is rarely obvious today in the great majority of cases. Spain has a particularly rich tradition of polychrome sculpture (see ENCARNADO).

Polyclitus (or **Polycletus**) **of Argos** (active *c.*450–*c.*420 BC). One of the most celebrated of Greek sculptors. No original works by him survive, but several are known through Roman copies. He is now best known for his *Doryphorus* (Spear Carrier), the best copy of which is in the Archaeological Museum in Naples. In this figure he is said to have embodied the system of mathematical proportions on which he wrote a book, and the statue—long regarded as a standard for ideal male beauty—is sometimes referred to as 'The Canon'. Copies also exist of a *Diadumenus* (a youth wreathing a band round his head) by Polyclitus and of an Amazon with which, according to *Pliny, he defeated *Phidias in a competition. In antiquity his greatest work was held to be the colossal *chryselephantine statue of Hera in her temple—the Heraeum—near Argos. It is now known only through descriptions and representations on coins, but ancient writers compared it favourably with Phidias' statue of Zeus at Olympia. Strabo said the Zeus was more magnificent but the Hera more beautiful in workmanship.

Polygnotus of Thasos. Greek painter, active chiefly in Athens in the mid-5th cent. BC. None of his works survive, but ancient sources credit him with being the first great figure in Greek painting. He painted large compositions (mainly mythological) with many figures and some indication of landscape. His style was said to be serious and dignified, and he showed an advance on earlier art by the life and expression of his faces.

polyptych. A picture or other work of art consisting of four or more leaves or panels. It was a popular form for altarpieces from the 14th cent. to the early 16th cent. Italian polyptychs of this time typically include a *predella beneath the main panel or tier, the whole enclosed in an elaborate frame. In northern Europe polyptychs often had panels hinged together and painted on both sides, so that they could be folded to create different pictorial compositions (sometimes ones that were appropriate for particular liturgical seasons or events). See also DIPTYCH, TRIPTYCH.

Pomarancio, Il. See RONCALLI.

Pompidou Centre (in full, Centre National d'Art et de Culture Georges-Pompidou). Cultural centre in Paris named after Georges Pompidou (1911–74), President of France from 1967 to 1974. In 1969 he expressed a 'passionate wish' for 'a place where the plastic arts, music, cinema, literature, audio-visual research, etc., would find a common ground'. The site chosen for the Centre was the Plateau Beaubourg (hence its colloquial name 'Beaubourg Centre'), a once thriving area near the centre of Paris that had become derelict between the world wars. An international competition for the building produced almost 700 submissions, including bizarre ideas such as a giant egg and an enormous hand extended towards the sky, each finger being intended to house a separate department. The winning design was submitted by the Italian-British team of Renzo Piano and Richard (later Sir Richard) Rogers. Their huge building was constructed in 1971–7 and soon became one of the most famous sights of the city. A leading example of 'high-tech' architecture, it has been described as looking like a 'crazy oil refinery' and has attracted extremes of praise and censure. The large plaza in front of the building is conceived as part of the Centre and is the main forum for the city's street performers. Also outside the building is the ebullient Beaubourg Fountain (1980) by Jean *Tinguely and Niki de *Saint Phalle. The Centre is divided into various departments, including a library, an industrial design centre, and an institute for the development and promotion of avant-garde music. The largest of the departments and the main reason for the Centre's popularity is the national collection of modern art—the Musée National d'Art Moderne—which was formerly housed in the Palais de Tokyo. It was opened there in 1947, but its origins are much older, for it is the heir to the Musée du Luxembourg, opened in 1818 as a showcase for the work of living artists. Its collection of modern art is exceeded in scope and quality probably only by that of the *Museum of Modern Art in New York.

pompier, l'art. A term applied pejoratively to French academic art (more particularly pretentious *history painting) of the late 19th cent. It is said to derive from the habit of posing nude models wearing firemen's helmets to substitute for ancient helmets ('pompier' is French for 'fireman').

Pompon, François (1855–1923). French sculptor. He worked for fifteen years as an assistant of *Rodin and success came to him very late, when he made a name with his *Polar Bear* (Musée National d'Art Moderne, Paris) at the 1922 *Salon d'Automne. This became enormously popular, reproduced in a variety of forms, and Pompon was hailed as the greatest animal sculptor since *Barye. On his death he bequeathed more than 300 pieces to the Musée d'Histoire Naturelle, which founded a museum in his name, later transferred to Dijon, where his studio was turned into a museum.

Pont-Aven, School of. Term applied to painters associated with *Gauguin during his periods of work at Pont-Aven in Brittany, 1886–94, and inspired by his anti-naturalistic style. Emile *Bernard was among them, and he and Gauguin together developed *Synthetism.

Pontormo (Jacopo Carucci) (1494–1557). Italian painter, born in the Tuscan village of Pontormo, near Empoli, and active in and around Florence. According to *Vasari, he studied successively with *Leonardo da Vinci, *Albertinelli, *Piero di Cosimo, and *Andrea del Sarto, whose workshop he is said to have entered in 1512. Andrea was certainly a major influence on his early work. Pontormo was precocious (he was praised by *Michelangelo whilst still a youth) and by the time he painted his *Joseph in Egypt* (NG, London) in about 1515 he had already created a distinctive style—full of restless movement and disconcertingly irrational effects of scale and space—that put him in the vanguard of *Mannerism. The emotional tension apparent in this work reaches its peak in Pontormo's masterpiece, the altarpiece of the *Entombment* (*c*.1526–8) in the Capponi Chapel of

Sta Felicità, Florence. Painted in extraordinarily vivid colours and featuring deeply poignant figures who seem lost in a trance of grief, this is one of the key works of Mannerism.

Pontormo was primarily a religious painter, but he was also an outstanding portraitist (he was a major influence on his pupil and adopted son *Bronzino) and in 1520–1 for the *Medici villa at Poggio a Caiano he painted a memorable mythological work (*Vertumnus and Pomona* according to Vasari, but the identification is disputed) in which an apparently idyllic scene reveals a strong undercurrent of neurosis. In Pontormo's later work his style was enriched by the study of Michelangelo and *Dürer's prints, but this stage of his career is known mainly through his superb drawings (best represented in the Uffizi) as the great fresco scheme in S. Lorenzo, Florence, that occupied him from 1546 until his death, was destroyed in the 18th cent. Pontormo's diary for part of 1554–6 remains, giving a day-to-day account of his progress in his great undertaking. It tells us much of his neurotic character, melancholy and introspective, dismayed by the slightest illness.

Pop art. A movement based on the imagery of consumerism and popular culture, flourishing from the late 1950s to the early 1970s, chiefly in the USA and Britain. The term was coined *c.*1955 by Lawrence *Alloway. Comic books, advertisements, packaging, and images from television and the cinema were all part of the iconography of the movement, and it was a feature of Pop art in both the USA and Britain that it rejected any distinction between good and bad taste. In the USA Pop art was initially regarded as a reaction from *Abstract Expressionism because its exponents brought back figural imagery and made use of *Hard Edge techniques. It was seen as a descendant of *Dada (in fact Pop art is sometimes called *Neo-Dada) because it debunked the seriousness of the art world and embraced the use or reproduction of commonplace subjects (comic strips, soup tins, highway signs) in a manner that had affinities with *Duchamp's *ready-mades. The most immediate inspiration, however, was the work of Jasper *Johns and Robert *Rauschenberg, both of whom began to make an impact on the New York art scene in the mid-1950s. They opened a wide new range of subject-matter with Johns's paintings of flags, targets, and numbers and his sculptures of objects such as beer cans and Rauschenberg's *collages and *combine paintings with Coca-Cola bottles, stuffed birds, and photographs from magazines and newspapers. While often using similar subject-matter, Pop artists generally favoured commercial techniques in preference to the painterly manner of Johns and Rauschenberg. Examples are Andy *Warhol's silkscreens of soup tins, heads of Marilyn Monroe, and so on, Roy *Lichtenstein's paintings in the manner of comic strips, and Mel *Ramos's brash pin-ups. Claes *Oldenberg, whose subjects include ice-cream cones and hamburgers, has been the major Pop art sculptor. John Wilmerding (*American Art*, 1976) writes that Pop art 'cannot be separated from the culmination of affluence and prosperity during the post-World-War-II era. America had become a ravenously consuming society, packaging art as well as other products, indulging in commercial manipulation, and celebrating exhibitionism, self-promotion, and instant success . . . Pop's mass-media orientation may further be related to the acceleration of uniformity in most aspects of national life, whether restaurants or regional dialects. Shared by all Americans were the principal preoccupations of Pop art—sex, the automobile, and food.'

In Britain, too, Pop art revelled in a new glossy prosperity following years of post-war austerity. British Pop was nurtured by the *Independent Group and the work that is often cited as the first fully-fledged Pop art image was produced under its auspices—Richard *Hamilton's collage *Just what is it that makes today's homes so different, so appealing?* (Kunsthalle, Tübingen, 1956). However, British art first made a major impact at the *Young Contemporaries exhibition in 1961 (at about the same time that American art became a force). The artists in this exhibition included Derek *Boshier, David *Hockney, Allen *Jones, R. B. *Kitaj, and Peter *Phillips, who had all been students at the *Royal College of Art. In the same year the BBC screened Ken Russell's *Monitor* film 'Pop goes the easel', in which Peter *Blake was one of the featured artists. Although there are exceptions (notably the erotic sculptures of Allen Jones), British Pop was generally less brash than American, expressing a more romantic view of the subject-matter in a way that can now strike a note of nostalgia. Much of the imagery, however, came directly from the American world of pin-ups and pin-ball machines. Richard Hamilton defined Pop art as 'popular, transient, expendable, low-cost, mass-produced, young, witty, sexy, gimmicky, glamorous, and Big Business', and it was certainly a success on a material level, getting through to the public in a way that few modern movements do and attracting big-money collectors. However, it was scorned by many critics. Harold *Rosenberg, for example, described Pop as being 'Like a joke without humour, told over and over again until it begins to sound like a threat . . . Advertising art which advertises itself as art that hates advertising.'

Pope-Hennessy, Sir John (1913–95). English art historian. He was Director of the *Victoria and Albert Museum 1967–73, Director of the *British Museum 1974–6, and from 1977 to 1987 was Consultative Chairman, Department of European Paintings, *Metropolitan Museum, New York, and Professor of Fine Arts at New York University. His many publications made him perhaps the doyen in the field of Italian Renaissance art among English writers. They include the magisterial *An Introduction to Italian Sculpture* in three parts (all of which have subsequently appeared in revised editions), *Italian Gothic Sculpture* (1955), *Italian Renaissance Sculpture* (1958), and *Italian High Renaissance and Baroque Sculpture* (1963), an edition of *Cellini's *Autobiography* (1949), and monographs on *Giovanni di Paolo (1937), *Sassetta (1939), *Uccello (1950, revised 1972), Fra *Angelico (1952, revised 1974), *Raphael (1970), Luca della *Robbia (1980), and Cellini (1985). The many honours bestowed on him include being made an honorary citizen of Siena in 1982, and receiving in 1986 the Galileo Galilei Prize, awarded each year for meritorious contributions to Italian culture.

poppy oil. Oil extracted from poppy seeds, one of the most popular of the *drying oils used as a *medium for *oil painting. It is less viscous than *linseed and *walnut oil and does not easily turn rancid. It is, however, slow drying. This turned out to be an advantage rather than a disadvantage when the method of **alla prima* painting came into vogue about the middle of the 19th cent. and for a time during the second half of that century poppy oil was much used by commercial colourmen.

Porcellis (or **Percellis**), **Jan** (*c.*1584–1632). Flemish painter and etcher of marine subjects, active in Holland. He was regarded as the greatest marine painter of his day and his work marks the transition from the busy and brightly coloured seascapes of the early 17th cent., with their emphasis upon the representation of ships, to monochromatic paintings which are essentially studies of sea, sky, and atmospheric effects. His favourite theme was a modest fishing-boat making its way through a choppy sea near the shore. Both *Rembrandt and Jan van de *Cappelle collected his works. His son **Julius** (*c.*1609–45) was also a marine painter.

Pordenone (Giovanni Antonio de Sacchis) (*c.*1483?–1539). Italian painter, named after the town of his birth, Pordenone in the Friuli, and active in various parts of northern Italy. After working in a provincial style at the very start of his career (his master is unknown and *Vasari says he was self-taught), by the beginning of the second decade of the 16th cent. he had come close to the contemporary Venetian (specifically *Giorgionesque) manner of painting. In the second half of the decade, however, he was in central Italy, and his style changed under the impact particularly of *Michelangelo, acquiring great weight and solidity. Pordenone was influenced also by *Mantegna's illusionism and by German prints, and the style he forged from these diverse influences was highly distinctive and original. He always retained something of provincial uncouthness—at times vulgarity—but he was, in Vasari's words, 'very rich in invention . . . bold and resolute', and he excelled at dramatic spatial effects. These qualities are seen at their most forceful in his fresco of the *Crucifixion* (1520–1) in Cremona Cathedral; the densely packed, bizarrely expressive figures are seen as if on a stage through a painted proscenium arch and they lunge violently out into the spectator's space. From 1527 Pordenone was based in Venice and for a while he was a serious rival to *Titian. His major works in Venice have been destroyed, however. Pordenone died in Ferrara, where he had gone to design tapestries for Ercole II d'*Este.

porphyry. A hard volcanic stone, difficult to carve and polish, varying in colour from red to green. The porphyry used by ancient sculptors was a very hard, durable stone of a deep purplish red and the only known source was at Mt Porphyrites in Egypt. Purple was the royal colour of the Ptolemies and the quarries were a royal possession. When Cleopatra died (30 BC) Egypt became a Roman province and the Roman emperors took possession of the quarries and control of the stone, purple also being the Imperial colour. Large numbers of Roman works in porphyry have been preserved—gems, vases, urns, sarcophagi, busts, and statues, some of the last having only the clothing in porphyry and the flesh parts in some other stone. The Eastern emperors in Byzantium used porphyry, often despoiling Roman monuments to get it, as when columns were taken from Rome and set up in Sta Sophia. During the Italian *Renaissance this costly stone of the ancients (*pórfido rosso antico*) came again into favour, especially with the *Medici. The name 'porphyry' is now loosely used to cover several other hard igneous stones used for monumental sculpture.

Portinari, Cândido (1903–62). Brazilian painter of Italian descent. He is best known for his portrayals of Brazilian workers and peasants, but he dissociated himself from the revolutionary fervour of his Mexican contemporaries, and painted in a style that shows affinities with *Picasso's 'classical' works of the 1920s (Portinari was in Paris 1928–31). In the 1940s his work took on greater pathos and he also turned to biblical subjects. He acquired an international reputation and his major commissions included murals for the Hispanic section of the Library of Congress in Washington (1942) and for the United Nations Building in New York (two panels representing *War* and *Peace*, 1953–5).

Portland stone. A type of *limestone, quarried mainly at Portland in Dorset. It varies appreciably in texture, but the best-quality, comparatively close-grained type is a favourite with masons and has been much employed for buildings in London, notably St Paul's Cathedral. Some modern sculptors have also used it.

Posada, José Guadalupe (1851–1913). Mexican graphic artist. His enormous output was largely devoted to political and social issues, attacking, for example, the regime of Porfirio Díaz (1876–1911), and revealing the dreadful conditions in which the poor lived. From 1890 he made his studio in Mexico City an open shop fronting the street, and turned out sensational broadsheets and cheap cartoons that spread among the illiterate throughout the country. His work had the vigour and spontaneous strength of genuinely popular art with the inborn Mexican taste for the more gruesome aspects of death—one of his recurring motifs is the *calavera* or animated skeleton. He made a lasting impression on both *Orozco and *Rivera during their student days.

Post, Frans (1612–80). Dutch landscape painter, born in Leiden and active mainly in Haarlem. In 1637–44 he was a member of the Dutch West India Company's voyage of colonization to Brazil and became the first European to paint landscapes in the New World. He observed the unfamiliar flora and fauna with an appropriate freshness, creating scenes of remarkable vividness and charm, and he continued to paint Brazilian landscapes after his return to the Netherlands (indeed he is not known to have painted any other type of picture). Because of his *'naïve' style, he has been called the Douanier *Rousseau of the 17th cent., and he was virtually forgotten or regarded as a curiosity until the 20th cent. Examples of his fairly rare work are in the Louvre, the National Gallery of Ireland, and Ham House, London. His brother **Pieter** (1608–69) was one of the outstanding Dutch architects of the 17th cent. (the Huis ten Bosch near The Hague is his most famous work) and also occasionally painted.

Post-Impressionism. Term applied to various trends in painting, particularly in France, that developed from *Impressionism or in reaction against it in the period *c.*1880–*c.*1905. Roger *Fry coined the term as the title of an exhibition, 'Manet and the Post-Impressionists', which he organized at the Grafton Galleries, London, in 1910–11. The exhibition was dominated by the work of *Cézanne, *Gauguin, and van *Gogh, who are considered the central figures of Post-Impressionism. These three artists varied greatly in their response to Impressionism: Cézanne, who wished 'to make of Impressionism something solid and enduring, like the art of the museums', was preoccupied with pictorial structure; Gauguin renounced 'the abominable error of naturalism' to explore the symbolic use of colour and line; and van Gogh's uninhibited emotional intensity was the fountainhead of *Expressionism. Georges *Seurat, a figure of almost equal importance, concentrated on a more scientific analysis of colour (see NEO-IMPRESSIONISM). The general drift of Post-Impressionism was to lead away from the naturalism of Impressionism towards the series of avant-garde movements (such as *Fauvism and *Cubism) that revolutionized European art in the decade leading up to the First World War. (Some writers extend the notion of Post-Impressionism to cover these developments, making the term embrace the period *c.*1880–*c.*1914, but this makes an already broad concept less rather than more useful.)

Fry organized his first Post-Impressionist exhibition at short notice and in an almost casual atmosphere, but he brought together a highly impressive (if far from balanced) collection of pictures, mainly loaned by leading French dealers. The exhibition created what the *Daily Mail* called 'an altogether unprecedented artistic sensation' or what *Sickert more succinctly described as a 'rumpus'. The reviews were mainly unpleasant, sometimes viciously so. Some visitors were angry (Duncan *Grant recalled people shaking their umbrellas at the pictures) and others mocked. The prevailing opinion was that the pictures on show were childish, crude, and the product of moral degeneracy or mental derangement. Duncan Grant, however, said that he and Vanessa *Bell were 'wildly enthusiastic' about the exhibition, and it powerfully affected the work of several of the painters in Sickert's circle (see CAMDEN TOWN GROUP), in general encouraging the use of strong, flat colours.

Post-Painterly Abstraction. A term coined by the critic Clement *Greenberg to characterize a broad trend in American painting, beginning in the 1950s, in which abstract painters reacted in various ways against the *Gestural 'painterly' qualities of *Abstract Expressionism. Greenberg used the term as the title of an exhibition he organized at the Los Angeles County Museum of Art in 1964. He took the word 'painterly' (in German *malerisch*) from Heinrich *Wölfflin, who had discussed it in his book *Principles of Art History*. By it he understood 'the blurred, broken, loose definition of colour and contour'; Post-Painterly Abstractionists, in contrast, moved towards 'physical openness of design, or toward linear clarity, or toward both'. The characterization was never a very exact one, but essentially it described a rejection of expressive brushwork in favour of broad areas of unmodulated colour. The term thus embraces more precisely defined types of abstract art, including *Colour Field painting and *Hard-Edge painting. Among the leading figures of the trend are Helen *Frankenthaler, Al *Held, Ellsworth *Kelly, Morris *Louis, Kenneth *Noland, Jules *Olitski, and Frank *Stella.

Pot, Hendrik Gerritsz. (*c.*1585–1657). Dutch portrait and *genre painter who was probably a contemporary of Frans *Hals in van *Mander's studio. In 1631 Pot painted Charles I (1600–49) in England; versions of this portrait are in Buckingham Palace and the Louvre.

Potter, Paulus (1625–54). Dutch painter and etcher of animals in landscapes, active in Delft, The Hague, and Amsterdam. His best-known work, the life-size *Young Bull* (Mauritshuis, The Hague, 1647), was in the 19th cent. one of the most famous paintings in Dutch art. Subsequent taste has found its detailed and precise manner a trifle dry and laboured and preferred his smaller, more typical work. His speciality was scenes of cattle and sheep in sunlit meadows.

pouncing. A method of transferring a drawing or design to another surface (typically a *cartoon to a wall for *fresco painting) by dabbing pounce (a fine powder of charcoal or similar substance) through a series of pinpricks in the outlines of the drawing, thus creating a 'join up the dots' replica of it on the surface below.

Pourbus. Family of Netherlandish painters, distinguished mainly as portraitists. **Pieter** (*c.*1523–84) was born in Gouda and settled *c.*1543 in Bruges, where he became the pupil and son-in-law of Lancelot *Blondeel. He was a civil engineer, surveyor, and cartographer as well as painter. Van *Mander wrote 'I have never seen a better equipped studio than his.' As well as portraits he painted religious and allegorical scenes, one of the most splendid of which is the *Love Feast* (Wallace Coll., London). Pieter's son, **Frans the Elder** (1545–81), was active in Antwerp and one of the chief pupils and a close follower of Frans *Floris. The most famous member of the family is his son **Frans the Younger** (1569–1622), who was one of the principal court portraitists of Europe. He first worked for the court at Brussels, and from 1600 to 1609 was employed in Mantua (at the same time as *Rubens) by Vincenzo I *Gonzaga. In 1609 he was called to Paris by Marie de *Médicis and worked as her court painter until his death. His style—more concerned with the meticulous reproduction of rich costumes and jewellery than with interpretation of character—was typical of international court portraiture of the day.

Poussin, Gaspard. See DUGHET.

Poussin, Nicolas (1594–1665). French painter, active mainly in Rome. He is regarded not only as the greatest French painter of the 17th cent., but also as the mainspring of the *classical tradition in French painting. His interest in painting was aroused by the visit to his home town, Les Andelys, in 1611 of Quentin Varin (*c.*1570–1634), a mediocre late *Mannerist painter, and in 1612 he settled in Paris. His early years there are obscure, but he was able to familiarize himself with Classical works in the form of *Renaissance pictures and engravings, particularly of *Raphael and his school, and Roman statuary and *reliefs, in the royal collection. Between commissions of various kinds—the most notable being for work for the Luxembourg palace, with Philippe de *Champaigne (*c.*1621)—he made two unsuccessful attempts to go to Rome. He was also commissioned by the Italian poet Marino to make drawings to illustrate Ovid's *Metamorphoses* (Royal Library, Windsor, *c.*1623), and these are the first surviving works certainly by him. In 1623 he once more set out for Rome; travelling via Venice, he arrived early in 1624. Through Marino he came to the notice of Cardinal Francesco *Barberini and his secretary Cassiano dal Pozzo, who became his patrons. His first Roman works show him to have been still dominated by the Mannerism of the mid-16th cent. He worked for a time in the studio of *Domenichino and under the influence of Cassiano, who had a keen interest in antiquity, he studied Roman sculpture. With the predominance of these interests his style began to shed the earlier Mannerist features, becoming more restrained and classical. He was never at home with the *Baroque style that was coming to the fore in Rome, and the only public picture he painted in Rome, the altarpiece of *The Martyrdom of St Erasmus*, commissioned by Cardinal Barberini for St Peter's (Vatican Pinacoteca, 1628), was coolly received. His most personal work during this period is the *Inspiration of the Poet* (Louvre, Paris, *c.*1628), classical in design but Venetian in colouring.

In 1629–30 Poussin was seriously ill and was nursed back to health by the family of Jacques

Dughet, a French chef working in Rome, whose daughter he married. The illness coincided with a change of direction in his work. Abandoning the competition for public commissions and rivalry with the current Baroque, he gave himself up to his dominating passion for the *antique and during the next ten years brought to maturity the manner which has become recognized as peculiarly his own creation. Instead of religious subjects he painted themes from ancient mythology seen through the eyes of Ovid or Torquato Tasso, which he treated in a pastoral and poetic mood. Until about 1633 the influence of *Titian was paramount. During the latter part of the 1630s he turned to Old Testament and historical subjects which afforded scope for more elaborate pageantry (*The Worship of the Golden Calf*, NG, London, *c*.1635). In the paintings of these years the influence of Titian waned and he moved towards a more austere classicism which echoed the later Raphael and *Giulio Romano. He was preoccupied with the depiction of emotion by the gestures, pose, and facial expression of his figures, and pondered a literary and psychological conception of painting which he elaborated in a letter sent to his friend Paul Fréart de Chantelou, a civil servant, with the picture *The Gathering of the Manna* (Louvre, 1639).

His reputation was very high by the end of the 1630s and in 1640 he reluctantly succumbed to strong pressure and returned to Paris. He was commissioned to superintend the decoration of the Grande Galerie of the *Louvre (work wholly alien to his temperament), to paint altarpieces, and to design frontispieces for the royal press. His visit was ruined by jealousy and intrigue, and in September 1642 he left again for Rome, remaining there for the rest of his life.

The most important outcome of the visit for Poussin was that he had come into contact with the intellectual bourgeoisie of Paris, the public of Pierre Corneille and René Descartes, who patronized him for the remainder of his life. During the next decade he painted for such patrons the works that in his own day were considered to be his finest achievement and are still recognized as the purest exemplification of the classical spirit. The emphasis is on clarity of conception, moral solemnity, and obedience to rule. Poussin also made it his endeavour to achieve a rational unity of mood in each picture and developed a theory of *modes* (later taken up by the Académie) akin to the current theory of musical 'modes' supposed to be derived from antiquity. According to this theory the subject of the picture and the emotional situations depicted dictate the appropriate treatment, which can be worked out rationally and consistently according to principles expressible in language. He formulated what became the central doctrines of the Classicism taught in the *academies. Painting must deal with the most noble and serious human situations and must present them in a typical and orderly manner according to the principles of reason. The typical and general is to be preferred to the particular, and trivial sensuous allures, such as glowing colours, are to be eschewed. Painting should appeal to the mind not to the eye. His working procedure was as methodical as his theoretical approach, for he not only made numerous drawings but also employed wax models on a kind of miniature stage-set so he could study the composition and lighting with great deliberateness. The series of paintings on the *Seven Sacraments* (Earl of Ellesmere Coll., on loan to NG, Edinburgh, 1644–8), painted for Chantelou, show the solemnity and rational economy of his work at this time, and make a fascinating comparison with an earlier, more sensuous series on the same subjects, painted for Cassiano dal Pozzo in 1636–40 (five in collection of Duke of Rutland, Belvoir; one in NG, Washington; one destroyed).

During the second half of the 1640s Poussin displayed a new interest in landscape, applying to animate and inanimate nature the principles of quasi-mathematical lucidity and order he sought elsewhere. He achieved an impression of monumental simplicity and calm, exemplified in two great works of 1648 illustrating the story of Phocion (Earl of Plymouth Coll., on loan to National Mus. of Wales; and Walker Art Gal., Liverpool). Together with the work of his friend *Claude and his brother-in-law *Dughet, Poussin's paintings in the genre were the basis for *ideal landscape for the next two centuries. The *Self-portrait* (Louvre, 1650), showing the artist half-length against a series of overlapping vertical planes created by picture frames and wall, is also typical of this period.

By 1650 Poussin had become something of a hermit, but he had achieved European fame and his position in the world of art was unique. Between 1653 and his death in 1665 his style underwent yet a further development. Psychological expression, even if rationally controlled, was underplayed and his compositions took on a timeless allegorical quality. A motionless solemnity took the place of action and gesture and his pictures became symbols of eternal truths instead of representations of historical events. In some of the works figures attain a superhuman grandeur (*The Holy Family*, Hermitage, St Petersburg, *c*.1655), and in others nature takes on a new wildness and grandeur, as in his last great works, *The Four Seasons* (Louvre, 1660–4), in which he combined the descriptive idea with biblical references; thus Winter is represented by the *Deluge*. The cold rationalism of his earlier works was left behind, and a poetical, imaginative, almost mystical approach took its place.

Poussin's example was the basis of *Lebrun's academic doctrine and has been of enormous influence on the development of French art. In the later 17th cent. Poussin's name was used in the Académie (see DE PILES) to give support to those who believed in the superior importance of design in painting (Poussinistes) in opposition to that of *Rubens, who stood for the importance of colouring, and although the Rubensistes won the day, Poussin continued to be the inspiration of classically minded artists right into the early 19th cent. In the *Romantic era his influence declined, but his spirit was revived again by *Cézanne, who declared that he wanted 'to do Poussin again, from Nature'.

Powers, Hiram (1805–73). American sculptor, active in Italy from 1837. He first achieved success with portrait busts, but his great international fame came with his marble statue *The Greek Slave* (Corcoran Gal., Washington, 1843), which caused a sensation at the Great *Exhibition of 1851 and was for a time one of the most talked about and reproduced works of art of the 19th cent. The naked girl is bound in chains and the astonishing popularity of the statue (which now seems rather insipid) no doubt depended on the way in which its sentimentality licensed its eroticism.

Poynter, Sir Edward (1836–1919). English painter and administrator, son of the architect Ambrose Poynter (1796–1886). He formed his academic style in Italy (1853), where he met *Leighton and admired *Michelangelo above all other artists, and in Paris (1856–9), where he studied with *Gleyre. His reputation was made with the huge *Israel in Egypt* (Guildhall, London, 1867) and he became one of the most popular painters of the day with similar elaborate historical tableaux in which he displayed his great prowess as a draughtsman. In the latter part of his career, however, he confined himself to much smaller works, similar to *Alma-Tadema's classical *genre scenes, as he devoted himself much more to administration. He was first *Slade Professor of Fine Art at University College, London, 1871–5; Director for Art at the South Kensington Museum (now Victoria and Albert Museum) and Principal of the National Art Training School, 1875–81; Director of the National Gallery, 1894–1904; and President of the Royal Academy, 1896–1918.

Pozzo, Andrea (1642–1709). Italian painter and architect, one of the greatest exponents of the *Baroque style of *illusionist ceiling decoration. He became a lay brother in the Jesuit order in 1665 (he is sometimes given the courtesy title 'Padre Pozzo') and worked much for Jesuit churches, both as a painter and architect. His masterpiece is the huge ceiling fresco, *Allegory of the Missionary Work of the Jesuits* (1691–4), in S. Ignazio, Rome, perhaps the most stupendous feat of **quadratura* ever painted. Pozzo worked in several other Italian cities apart from Rome, and from 1702 until his death he was in Vienna, where he decorated the Jesuit Church, the University Church, and the Liechtenstein Garden Palace. His influence was spread not only by his paintings, but also by his treatise *Perspectiva pictorum et architectorum* (2 vols., 1693 and 1700), which was soon translated into several European languages and also (by Jesuit missionaries) into Chinese. As an architect, he designed several churches and numerous altars, but his work in this field was unexciting compared with his paintings and engraved designs.

Prado, Madrid. Spain's national museum of art, founded in 1818 by Ferdinand VII (1784–1833) and opened to the public in 1819. The building, one of the finest examples of Spanish *Neoclassical architecture, had been intended for a Museum of Natural Science but had never served that purpose. The major part of the collection derives from the royal collections made in the course of three centuries by the Habsburg and Bourbon kings of Spain, who were some of the most discriminating and lavish patrons in Europe. The museum is remarkable less for its all-round comprehensiveness than for unequalled representation in certain fields. Above all, it contains what is far and away the world's greatest collection of Spanish painting, the three giants, El *Greco, *Velázquez, and *Goya, being supremely well represented. It is among the richest of all museums in works by Hieronymus *Bosch, *Titian, and *Patenier. It also has superb collections of *Tintoretto, *Veronese, *Rubens, and van *Dyck, and splendid examples of the *Master of Flémalle, Rogier van der *Weyden, *Bruegel, *Raphael, and *Mantegna.

Praxiteles. Greek sculptor, active in the mid-4th cent. BC, perhaps the son of *Cephisodotus. His fame among Greek sculptors, to posterity as in his own time, is second only to that of *Phidias. Various works by him described by ancient authors are known through Roman copies, and a marble statue of *Hermes and the Infant Dionysus*, found at Olympia in 1877 in the position where it was described by *Pausanias (now in the Olympia Museum), is considered by many authorities to be from his own hand. If this is so, it is the only surviving original statue by a Greek sculptor of the first rank. Certainly it has a delicacy in the modelling of forms and a subtlety of finish far removed from the workmanship seen in most Roman copies, and it shows the sensuous charm and gentle grace for which he was renowned. In antiquity his most famous work was the **Aphrodite of Cnidus*, known through several copies (the best is perhaps that in the Vatican Museum). This much imitated work was the first free-standing life-size female nude in Greek art and *Pliny described it as 'the finest statue not only by Praxiteles but in the whole world'. Praxiteles' influence was profound. The tenderness and intimacy of his work marked a move away from the remote idealization of the *Classical period, to an art more concerned with human emotion, and his graceful, sinuous poses, with the figure often shown leaning on a support, became part of the general currency of *Hellenistic sculptors. His preference for working in marble made the material popular again after it had long been eclipsed by bronze.

Precisionism. A movement in American painting, originating *c.*1915 and flourishing in the inter-war period, particularly the 1920s, in which urban and especially industrial subjects were depicted with a very smooth and precise technique, creating clear, sharply defined, sometimes quasi-*Cubist forms. The terms 'Cubist-Realists', 'Immaculates', and 'Sterilists' have also been applied to Precisionist painters. They were not a formal group, but they often exhibited together. *Demuth, *O'Keeffe, and *Sheeler were among the leading figures. In Precisionist painting the light is often brilliantly clear (although George Ault (1891–1948) was best known for his night scenes), and frequently forms are chosen for their geometric interest. Human presence is excluded and there is no social comment. Rather, the American industrial and technological scene is endowed with an air of epic grandeur. The degree of Cubist influence varied greatly. Some of Sheeler's paintings are in an almost photographically realistic style, whereas other works are semi-abstract. Precisionism was influential in both imagery and technique on American *Magic Realism and *Pop art.

Preda (or **Predis**), **Ambrogio da** (*c.*1455–after 1508). Milanese painter. He was appointed court painter to the *Sforza in 1482 and worked mainly as a portraitist, but he is chiefly remembered for his association, together with his elder half-brother **Evangelista da Preda** (d. after 1490), with *Leonardo da Vinci in the 1483 contract for *The Virgin of the Rocks* (NG, London). The wings of this altarpiece, depicting angels with musical instruments, are of much lower quality than the centre panel and are presumed to be by Ambrogio and/or Evangelista. The National Gallery also has a female portrait by Ambrogio, showing his rather wooden imitation of Leonardo's style.

predella. A subsidiary picture forming an appendage to a larger one, especially a small painting or series of paintings beneath the main part of an *altarpiece.

Predis, Ambrogio da. See PREDA.

Prendergast, Maurice (1859–1924). American painter, mainly in watercolour. He was a member of the *Eight, but stood somewhat apart from the rest of the group. He was a Bostonian and spent much of his career travelling and painting abroad, and it was only in the last few years of his life that he lived in New York, the centre of the Eight's activities. The main thing he had in common with the other members was a desire to move American art away from academic stagnation, and his work is notable for its brilliant decorative colour. His paintings were often of people enjoying themselves in innocent pleasures (*Central Park in 1903*, Met. Mus., New York, 1903). He was one of the first American artists to be influenced by *Post-Impressionism, notably in the way in which he emphasized flat pattern rather than illusionistic space. In 1913 he showed seven works at the *Armory Show, and at this time stood out as one of the most stylistically advanced American artists.

Pre-Raphaelite Brotherhood. The name adopted in 1848 by a group of young English artists who shared a dismay at what they considered the moribund state of British painting and hoped to recapture the sincerity and simplicity of early Italian art (i.e. before the time of *Raphael, whom they saw as the fountainhead of academism). The nucleus of the group was formed by three fellow students at the *Royal Academy—John Everett *Millais, William Holman *Hunt, and Dante Gabriel *Rossetti (to whom, son of an Italian ex-revolutionary, the sealing of the group into a secret Brotherhood was due). The other four initial brethren were James *Collinson, the sculptor Thomas *Woolner, and the art critics W. M. Rossetti (1829–1919) and F. G. Stephens (1828–1907). Ford Madox *Brown was closely allied with them, though not at any time a member of the Brotherhood. The movement had a strong literary flavour from the start: the members were roused to lyrical excitement by the poets John Keats and Alfred Lord Tennyson and published a journal called *The *Germ*. Rossetti was distinguished as a poet as well as a painter. His brother defined the aims of the Brotherhood as follows: '(1) To have genuine ideas to express; (2) to study Nature attentively, so as to know how to express them; (3) to sympathise with what is direct and serious and heartfelt in previous art, to the exclusion of what is conventional and self-parading and learned by rote; and (4) and most indispensable of all, to produce thoroughly good pictures and statues.' Their desire for fidelity to nature was expressed through detailed observation of flora, etc., and the use of a clear, bright, sharp-focus technique; and their moral seriousness is seen in their choice of religious or other uplifting themes. The kind of pictures they hated were academic 'machines' and trivial *genre scenes.

The initials PRB were first used on Rossetti's picture *The Girlhood of Mary Virgin* (Tate, London), exhibited in 1849, and were adopted by the other members of the Brotherhood. When their meaning became known in 1850 the group was subjected to violent criticism and abuse. Charles Dickens led the attack in his periodical *Household Words*, calling Millais's *Christ in the House of His Parents* (Tate) 'mean, odious, revolting and repulsive'. He was outraged by the implied rejection of Raphael (still unquestioningly thought of by many critics as the greatest painter who ever lived), and he regarded the claim to go behind Raphael as an antiprogressive reversion to primitivism and ugliness. The Pre-Raphaelites were, however, defended by *Ruskin and attracted numerous followers, including *Deverell and *Hughes.

By 1853, however, the group had virtually dissolved. Apart from their youthful revolutionary spirit (they were very young in 1848) and their romantic if uninformed medievalism, the prime movers had little in common as artists and they went their separate ways. Of the original members only Hunt remained true to PRB doctrines. Millais adopted a much looser style and went on to become the most popular and successful painter of the day. Curiously, however, it was Rossetti, the least committed to PRB ideals (he never cultivated painstaking detail), who continued the name. Although his later work, made up principally of languorous depictions of *femmes fatales*, is entirely different from his Pre-Raphaelite pictures, the name stuck to him and to his followers such as *Morris and *Burne-Jones. Thus in the popular imagination the term 'Pre-Raphaelite' conjures up pictures of medieval romance, and ironically a movement that began as a rebellion against artificiality and sentimentality is now itself identified with a kind of escapism. This second wave of pseudomedieval Pre-Raphaelitism began in the 1860s and survived into the 20th cent., in the work, for example, of Evelyn *De Morgan.

Preston, Margaret (1893–1963). Australian painter, printmaker, and wood engraver. Her early work was naturalistic, but after the First World War she became interested in modern art and devoted herself to a style based on schematic design and strongly contrasted colours. Her advocacy of a distinctive national art is reflected in her bush landscapes and still-life paintings of native flora, such as *Australian Gum Blossoms* (Art Gal. of NSW, Sydney). During the 1920s she became aware of the beauty of Aboriginal art and was probably the first artist of European origin (she trained in Munich and Paris) actively to champion it as an independent art form worthy of emulation. Its influence permeates many of her works of the 1940s. Her adoption of earth colours for prints and paintings developed from this interest.

Preti, Mattia (also called Il Cavaliere Calabrese) (1613–99). Italian *Baroque painter. He came from Taverna in Calabria (hence his nickname) and his prolific career took him to many different parts of Italy (and according to an early biographer to Spain and Flanders). His early work includes groups of musicians and card-players, strongly *Caravaggesque in style, but later he excelled mainly in frescos on religious subjects. In this field his main model was *Lanfranco, whom he succeeded in the fresco decoration of S. Andrea della Valle in Rome (1650–1). After the plague of 1656 carried off virtually a whole generation of artists in Naples, Preti worked with great success there, gaining many important commissions. They included a series of seven frescos commemorating the plague for the city gates; they no longer survive, but two **modelli* for them are in the Museo di Capodimonte in Naples and give some idea of how powerful the huge frescos must have been. In 1661 Preti moved to Malta, where he lived for the rest of his life. Several churches on the island, including the cathedral of Valletta, have decorations by him.

Price, Sir Uvedale (1747–1829). English landed gentleman and writer on the *picturesque. He laid out his Herefordshire estate, Foxley, on picturesque principles, and his *Essay on the Picturesque* (2 vols. 1794–8) is one of the principal monuments of the picturesque doctrine. Price also translated *Pausanias (1780).

Prieur, Barthélemy (d. 1611). French *Mannerist sculptor, best known for his elegant life-size bronze figures of *Virtues* for the tomb of Constable Montmorency (Louvre, Paris), in a style recalling the second school of *Fontainebleau. He was appointed Royal Sculptor in 1594 and worked on the decoration of the *Louvre.

primary colours. In painting, those colours—blue, red, and yellow—that cannot be made from mixtures of other colours. When two of the primaries are mixed together, the result is known as a secondary colour, red and yellow making orange, red and blue making purple, and yellow and blue making green.

Primary Structures. Name given to a type of sculpture that came to prominence in the mid-1960s, characterized by a preference for extremely simple geometrical shapes and frequently a use of industrially fabricated elements. The term was popularized by an exhibition entitled 'Primary Structures' at the Jewish Museum, New York, in 1966. Among the artists who worked in this vein were Carl *Andre, Donald *Judd, Sol *Lewitt, Robert *Morris, and Tony *Smith. Primary Structures comes within the scope of *Minimal art and the two terms have sometimes been used synonymously.

Primaticcio, Francesco (1504/5–70). Italian painter, architect, and decorator, mainly active in France. He was born in Bologna and developed his all-round skills as *Giulio Romano's assistant in Mantua. In 1532 he was called to France by Francis I (1494–1547) and worked with *Rosso at Fontainebleau. Together they provided the main impetus for the distinctive French type of *Mannerism known as the School of *Fontainebleau. Their respective shares in the creation of the new manner—particularly the highly influential combination of paintings with *stucco ornament—are uncertain. Rosso is often accorded primacy, but *Vasari said 'the first works in stucco that were done in France, and the first labours in fresco of any account, had their origin, it is said, in Primaticcio.' His elongated figure style also had wide influence in France. Primaticcio took over the direction of the work at Fontainebleau on Rosso's death in 1540 and in the 1540s he twice visited Rome to buy antiquities or have casts made for Francis. In his later years Primaticcio turned more to architecture and his work helped to introduce *Renaissance elements into France. His major work, the Valois Chapel at St Denis, was, however, completed by others after his death and destroyed in the 18th cent.

priming. A thin layer applied on top of the *ground to make it more suitable to receive paint. For example, if the ground is too absorbent the priming may make it less so, or the priming may supply a tint. The Italian word *'imprimatura' was commonly used in this sense.

primitive. Term used with various meanings in the history and criticism of the arts. In its widest sense it is applied to art of societies outside the great Western, Near Eastern, and Oriental civilizations, even though much of it was produced by highly sophisticated peoples. Pre-Columbian art, North American Indian art, African art south of the Sahara, and Oceanic art are the main areas embraced by the term. By extension it has been applied to other fields of art that appear unsophisticated relative to some particular standard. Thus the term was once widely used of pre-*Renaissance European painting, particularly of the Italian and Netherlandish schools, but this usage (as in the expression 'the Flemish primitives') is now much less common and no longer has derogatory implications: a major series of books on early Netherlandish painting, published in Belgium (1951 onwards), is called *Corpus des Primitifs Flamands*. The term 'primitive' is also used more or less as a synonym for *naïve.

In the context of 20th-cent. art the term 'primitivism' refers to the use by Western modernist artists of forms or imagery derived from the art of so-called primitive peoples (in the first sense mentioned above), especially those who had been colonized by Western countries. For centuries such art was valued mainly for its ethnographic interest or (in the case of articles made from precious materials) for its monetary worth. However, in the 1890s *Gauguin had tried to escape 'the disease of civilization' among the natives of Tahiti, and in the next generation many avant-garde artists followed his example in cultivating primitive art as a source of inspiration, finding in it a vitality and sincerity that they thought had been polished out of Western art. Usually they followed Gauguin in spirit rather than body, although *Nolde and *Pechstein, for example, visited Oceania. Many artists collected African works, notably *Picasso, and their influence is clear in his *Les Demoiselles d'Avignon*. Other artists studied primitive art in museums—Henry *Moore, for example, was impressed by the powerful block-like forms of Mayan sculpture he saw in the British Museum.

print. A picture or design made (usually on paper) from an inked impression of an engraved metal plate, wooden block, etc. Prints are made by a great variety of processes, which are briefly classified here, and described more fully, with something of their history, in separate entries. These processes, or *graphic techniques, fall into four main groups.

(*a*) *Relief Methods*. In these, the parts of the wood block or metal plate that are to print black are left in relief and the remainder is cut away. The block is then rolled over with a stiff printing ink and, as the pressure required for printing is not great, impressions can be taken in an ordinary printing-press or even by hand.

The principal relief methods are *woodcut, *wood engraving, and *linocut. To these we may add certain techniques such as *metal cut, *manière criblée, and *relief etching, in which metal plates are engraved and printed like woodcuts.

(*b*) *Intaglio Methods*. In intaglio printing the principle is the reverse of that which operates in the relief methods, for the surface of the plate does not print, the ink being held only in the engraved furrows. The lines of the design are incised in a plate of copper or zinc; ink is rubbed into these lines and the surface of the metal wiped clean with a series of rags. A sheet of damp paper is laid on the plate, together with two or three thicknesses of felt blanket, and the whole run through a copper-plate press. This press, which consists of a bed-plate passing between two rollers, applies heavy pressure to the plate; the paper, backed by the blankets, is forced into the lines and picks up the ink that is in them, thus receiving an impression of the entire design.

The various ways in which the design may be incised in the metal plate constitute the different intaglio techniques. These techniques are: *line engraving, in which the design is engraved on the metal plate with a *burin; *drypoint, where the lines are drawn by scratching the plate with a strong steel needle; *etching, *soft-ground etching, and *aquatint, where the designs are bitten into the plate by means of acid. In addition there are the defunct, or nearly defunct, reproductive processes of *mezzotint, *stipple, and *crayon manner. Certain of the intaglio methods—mezzotint, aquatint, and stipple—have as their aim the production of a continuous tone; these are called the 'tone processes', though tonal effects may also be obtained by soft-ground etching and crayon manner. The various intaglio processes are frequently used in combination with one another on the same plate.

(*c*) *Surface or Planographic Methods*. These are *lithography and its variants. Lithographs are neither incised nor raised in relief, but are printed from a perfectly flat slab of limestone or from prepared metal plates. The process utilizes the antipathy of grease and water to separate areas which receive and areas which reject the printing ink.

(*d*) *Stencil Methods*. The principle involved here is simply that of cutting a hole in a protecting sheet and brushing colour through the hole on to a surface beneath. *Stencils have long been used for colouring prints in quantity and for fabric printing. The principal modern development is the *silk-screen print or 'serigraph'.

Two processes which cannot be classified with the rest may be added, namely the *monotype, and the *glass print or *cliché verre*, which is related to photography. The term print is also more loosely applied to reproductions of works of art made by photomechanical methods.

private presses. Printing presses usually operated by hand on non-commercial lines by private individuals with freedom to experiment in the production of books distinguished for their typography, illustration, format, and binding. In a general sense private presses are almost as old as printing itself, but the private press movement of more recent times was initiated by William *Morris, whose *Kelmscott Press provided both the impetus and the model for the revival of interest in typography and book production around 1900. Although aims and accomplishment varied considerably in the different presses set up after his death, some emphasizing typography and layout to the exclusion of illustration, the movement as a whole was characterized by (1) an interest in type design shown in the use by individual presses of types produced specially for them; (2) an interest in illustration; (3) the appeal to bibliophiles implicit in the production of limited editions of finely printed and carefully designed books. The English private presses had the credit of stimulating a revival of *wood-engraving for illustration, a development continued after the First World War by the *Golden Cockerel, *Nonesuch, and other presses. The decline of the movement, already apparent in the 1930s, was completed by the Second World War. Improved standards in commercial printing and the high cost of the books, which were usually printed in limited editions for subscribers only, were among the reasons which made the private press an anachronism by the mid-20th cent. See also DOVES, ERAGNY, and VALE PRESS.

Prix de Rome. A scholarship, founded concurrently with the French *Academy in Rome (1666), that enabled prize-winning students at the Académie Royale de Peinture et de Sculpture in Paris to spend a period (usually 4 years) in Rome at the state's expense. Prizes for architecture began to be awarded regularly in 1723, and prizes for engravers and musicians were added in the 19th cent. The prizes were meant to perpetuate the academic tradition and during the 18th and 19th cents. winning the award was the traditional stepping stone to the highest honours for painters and sculptors. Many distinguished artists (as well as many nonentities) were Prix de Rome winners, notably *David, *Fragonard, and *Ingres among painters and *Clodion, *Girardon, and *Houdon among sculptors.

The prizes are still awarded and the system has been adopted by other countries.

Procaccini, Giulio Cesare (1574–1625). Italian painter and sculptor, the most distinguished member of a family of artists. He was born in Bologna and worked mainly in Milan, where the family settled when he was a child, and also in Modena and Genoa. Initially he worked mainly as a sculptor, but after about 1600 he concentrated on painting and became one of the leading painters in Milan. His style was eclectic but often very powerful, combining something of the emotional tension of *Mannerism with the dynamism and sense of physical presence of the *Baroque. His colours tend to be acidic, his handling of light and shade dramatic. Many of his paintings are still in Milan, but two large scenes from Christ's Passion (perhaps part of a series) are in Edinburgh (NG) and Sheffield (Graves Art Gal.). His father, **Ercole** (1515–95), and his brothers, **Camillo** (*c.*1560–1629) and **Carlo Antonio** (1555–1605), were also painters.

Procter, Dod and **Ernest**. See NEWLYN SCHOOL.

Proto-Renaissance. Term that can be applied to any revival of the style or spirit of Classical antiquity before the *Renaissance proper. In particular it is used of the 'proto-Renaissance of the twelfth century', often simply called 'the twelfth-century Renaissance', which Erwin *Panofsky described as a Classicizing movement which began in the latter part of the 11th cent., reached its climax in the 12th, and continued into the 13th cent. It was a 'Mediterranean phenomenon, arising in southern France, Italy and Spain'.

Prout, Samuel (1783–1852). English painter best known for his watercolour views of Picturesque buildings and streets in Normandy, many of which appeared as engravings in illustrated books. He was an important popularizer of picturesque landscape and helped to build up the English *Romantic image of the Continent. His work was greatly admired by *Ruskin.

provenance. The record of the ownership of a work of art. A complete provenance accounts for the whereabouts of a work from leaving the artist's studio to the present day, and the nearer a work's pedigree approaches this ideal, the more secure its *attribution is likely to be.

Provost, Jan (*c.*1465–1529). Netherlandish painter. He was born at Mons and worked in Valenciennes and briefly in Antwerp before settling in Bruges in 1494. His style was fairly close to Gerard *David, then the leading painter in Bruges, but Provost, although clumsier, was more inventive. In 1521 Provost met and entertained *Dürer, who is reputed to have drawn his portrait.

Prud'hon, Pierre-Paul (1758–1823). French portrait and historical painter. He was trained at the Dijon Academy and in 1784 went to Rome, where he was a friend of *Canova and formed his style on the example of the **sfumato* and sensuous charm of *Leonardo and *Correggio. In 1787 he returned to Paris and after working in obscurity for some time he became a favourite of both empresses of the French, Josephine (*Empress Josephine*, Louvre, Paris) and Marie-Louise, designing the decorations for the bridal suite of the latter. His friendship with the statesman Talleyrand enabled him to remain in favour even after the fall of Napoleon in 1815, but he painted little in his final years. He had a neurotic personality and the shock of the suicide of his mistress—his pupil Constance Mayer—in 1821 led to his own death.

Prud'hon belongs both to the 18th and to the 19th cents. In his elegance, his grace, and his exquisite fancy he is akin to the epoch of Louis XVI, and *David referred to him slightingly as 'the *Boucher of his time'. But his deep personal feeling aligns him with the *Romantics. *Gros said of him: 'He will bestride the two centuries with his seven league boots.' Among his best-known pictures are *Justice and Divine Vengeance pursuing Crime* (Louvre, 1808), for which he received the Legion of Honour, and *Venus and Adonis* (Wallace Coll., London, commissioned 1810, but still in Prud'hon's studio at his death), for which the empress Marie-Louise, for whom it was painted and whose drawing master he was, is said to have been the model. Prud'hon was an outstanding draughtsman, but many of his paintings are in poor condition because of his use of *bitumen.

Pryde, James (1866–1941). British painter and designer. In the 1890s he designed posters with his brother-in-law William *Nicholson under the name *Beggarstaff Brothers and these are probably his most famous works. Pryde sometimes supplemented his income at this time by taking small parts on the stage. As a painter he is best known for dramatic and sinister architectural views, with figures dwarfed by their gloomy surroundings. They have something of the spirit of *Piranesi's prison etchings, but they are broadly brushed. Pryde—'tall and handsome', but 'dilatory, extravagant, and unproductive for long periods' (*DNB*)—produced little after 1925. However, in 1930 he designed the sets for Paul Robeson's memorable *Othello* at the Savoy Theatre, London.

psalter. A manuscript (particularly one for liturgical use) or a printed book containing the text of the Psalms. The earliest extant example is a 3rd-cent. AD Greek *papyrus roll from Egypt (BM, London). In the Latin West the distinction must be made between the biblical and the liturgical psalter. The former consisted simply of the Psalms without any additions. The liturgical psalter, which was much more common, was divided into eight sections in accordance with the Benedictine rule that monks must recite the Psalms in full every week (the extra section is for Vespers). The Psalms were generally preceded by a Calendar and followed by various hymns and prayers and by a litany of saints. Both types were illustrated in a number of ways. The great popularity and copious illustration of the psalter make it the most important *illuminated book from the 11th to the 14th cents.

Public Works of Art Project. See FEDERAL ART PROJECT.

Pucelle, Jehan (*c.*1300–*c.*1350). French manuscript *illuminator. Little is known of his career, but his large workshop dominated Parisian painting in the first half of the 14th cent. He enjoyed court patronage and his work commanded high prices. Certain features of his work—particularly his mastery of space—indicate that he probably travelled in Italy early in his career, and he had a genius for enriching his style from a variety of sources. The disposition of the numerous *drolleries (a feature he popularized in France) in the *Hours of Jeanne d'Évreux* (Met. Mus., New York, 1325–8), for example, shows that he must also have been familiar with Flemish developments. It was the synthesis of these two elements, allowing for an increasing penetration of naturalistic representation into traditional *iconography, which formed the basis for Pucelle's individual style.

Puget, Pierre (1620–94). The greatest French sculptor of the 17th cent. He worked mainly in his native Marseilles and in Toulon, for although he sought success at court, his work was much too impassioned to fit into the scheme of *Colbert and *Lebrun's artistic dictatorship. Moreover, he was arrogant and headstrong in temperament and he fell victim to the intrigues of fellow artists. His *Baroque style was formed in Italy, where in 1640–3 he worked with Pietro da *Cortona in Rome and Florence. Subsequently he made several journeys to Genoa, where he established a considerable reputation. His first major work was a pair of *atlas figures for the entrance to Toulon Town Hall (1656) and in these he showed the physical vigour and emotional intensity that were the hallmarks of his style. They occur most memorably in his celebrated *Milo of Crotona* (Louvre, Paris, 1671–82), which was one of his few works accepted for the palace at Versailles. Puget spent his final years embittered by his failures. He worked as a painter, architect, and decorator of ships as well as a sculptor, and was an outstanding draughtsman.

His son **François** (1651–1707) was a painter, working mainly in Toulon and Marseilles. He did a few religious works but was mainly a portraitist; a portrait by him of his father is in the Louvre.

Pugin, Augustus Welby Northmore (1812–52). English architect, designer, writer, and medievalist, one of the key figures of the *Gothic Revival and of Victorian design as a whole. He was the son of **Augustus Charles Pugin** (1769–1832), a refugee from the French Revolution and a gifted draughtsman who published two illustrated works on *Gothic architecture. The young Pugin was a devotee of Gothic art and an established designer before he was 20, his work including furniture for Windsor Castle. After his conversion to Roman Catholicism in 1835 he devoted himself with furious energy to building churches and houses, and designing over a very wide field. He believed that the art of a country reflected its spiritual state and that English art had been in decline since the Reformation. This notion was forcefully promulgated in his book *Contrasts* (1836), with which he became famous; it compares the meanness of contemporary buildings with the glories of the medieval (i.e. Catholic) past. His ideas were influential on, for example, *Ruskin. Pugin was a brilliant and prolific draughtsman but because of financial constraints his architectural work was rarely as rich as he would have liked it to be. In his designs for the Houses of Parliament in London, however, cost was not a restriction and he created one of the great decorative ensembles of the 19th cent. (Sir Charles Barry (1795–1860) was responsible for the basic architectural forms, but Pugin designed the exterior and interior ornament—down to details such as inkstands.)

Purbeck marble. A hard, greyish-brown, shelly *limestone from the royal quarries in the Isle of Purbeck, Dorset. It is the nearest approximation to a true *marble quarried in England, can be cut to moderately fine detail, and polishes to a darkish hue. By about the middle of the 12th cent. Purbeck marble began to come into use for architectural details, columns, etc., gradually replacing the more costly foreign stones previously used. In the 13th cent. Purbeck marble fonts began to supersede the black marble fonts imported from Tournai and column bases, shafts, and *capitals were increasingly made of this material. It seems to have come into use for tombs and tomb effigies before the middle of the 12th cent. and through the 13th cent. was very popular for this purpose. The increasing use of painted *gesso as a finish on the more easily carved *freestone (e.g. for the Eleanor Crosses set up by Edward I, marking the route of Queen Eleanor's funeral procession in 1290) gradually brought the funerary trade in Purbeck marble to a finish, however, and it went out of fashion early in the 14th cent. It has been conjectured that another reason for this may have been that the elaborate decoration, painting, and gilding done by the workers in Purbeck marble obscured the material so that the less expensive freestone seemed as good to clients.

Purism. A movement in French painting linked with the new aesthetic of 'machine art' and flourishing *c.*1918–25. Its founders and protagonists were Amédée *Ozenfant and *Le Corbusier, who met in Paris in 1918. Feeling that *Cubism had missed its path and was degenerating into an art of decoration, they regarded their association as 'a campaign for the reconstitution of a healthy art', their object being to 'inoculate artists with the spirit of the age'. They set great store by 'the lessons inherent in the precision of machinery' and held that emotion and expressiveness should be strictly excluded apart from the 'mathematical lyricism' which is the proper response to a well-composed picture. Their characteristic paintings are still lifes—cool, clear, and impersonally finished.

Despite the anti-emotionalism of this Functionalist outlook, it was advocated by Ozenfant with passionate missionary fervour. The Purist aesthetic was expounded in the books *Après le Cubisme* (1918) and *La Peinture Moderne* (1925), joint works of Ozenfant and Le Corbusier, and in the journal *L'Esprit Nouveau*, which they ran from 1920 to 1925. It won a considerable measure of support from artists and writers of widely different persuasions, but Purism did not establish a continuing school of painting. Both the main protagonists seemed to realize that it represented something of a dead end pictorially and moved in to much looser styles. Its main sequel is to be found in the architectural theories and achievements of Le Corbusier and more generally in the field of design. Purism is significant also in that it gave expression to aesthetic ideas which were becoming dominant elsewhere (e.g. in *Neo-Plasticism, European *Constructivism, and the *Bauhaus) but had otherwise little support in Paris.

Purser, Sarah (1848–1943). Irish painter, designer, patron, collector, and administrator. She was a successful society portraitist, who in her own words 'went through the British aristocracy like the measles', but she is more important for her other roles in Irish art. She knew everyone who mattered (from 1911 she held regular social gatherings for Dublin's intelligentsia at her home, Mespil House) and in 1924 she founded the Friends of the National Collections of Ireland. One of its aims was to campaign for the return of Sir Hugh *Lane's pictures from London to Dublin and she helped to secure Charlemont House as the home for what became the Hugh Lane Municipal Gallery of Modern Art. Perhaps most importantly, she was the founder of the stained-glass workshop An Túr Gloine (The Tower of Glass), which operated in Dublin from 1903 to 1944; the work it produced, which can be seen in so many Irish churches, is her finest memorial.

putto (Italian: 'little boy'). Term applied to a representation of a chubby, naked child, sometimes winged, appearing—usually as a subsidiary figure—in a work of art. Putti have been a frequent motif of decorative art since Classical antiquity and may have a pagan, human, or divine status. They derived from a type of figure used in ancient art to represent Eros, the Greek god of love, and from the *Renaissance onwards a putto has often been used to represent his Roman counterpart, Cupid. More commonly, putti are anonymous figures pictured attending Classical gods, or, for example, the Virgin Mary.

Puvis de Chavannes, Pierre (1824–98). The foremost French mural painter of the second half of the 19th cent. He decorated many public buildings in France (for example, the Panthéon, the Sorbonne, and the Hôtel de Ville, all in Paris) and also Boston Public Library (*Abbey and *Sargent did murals here too). His paintings were done on canvas and then affixed to the walls (see MAROUFLAGE), but their pale colours imitated the effect of *fresco. He had only modest success early in his career (when a private income enabled him to work for little payment), but he went on to achieve an enormous reputation, and he was universally respected even by artists of very different aims and outlook from his own. *Gauguin, *Seurat, and *Toulouse-Lautrec were among his professed admirers. His reputation has since declined, his idealized depictions of antiquity or allegorical representations of abstract themes now often seeming rather anaemic. He remains an important figure, however, for his influence on younger artists. His simplified forms, respect for the flatness of the picture surface, rhythmic line, and use of non-naturalistic colour to evoke the mood of the painting appealed to both the *Post-Impressionists and the *Symbolists.

Pyle, Howard. See PARRISH.

Pynacker, Adam (1622–73). Dutch landscape painter, active chiefly in Delft (he was born in nearby Pijnacker) and in Amsterdam. He was in Italy for three years (before 1649) and he was one of the outstanding Dutch exponents of Italianate landscapes. His style resembles that of Jan *Both and Jan *Asselyn, but his mature work often has a distinctive and attractive silvery tonality. A splendid example of his work, showing his ability to compose boldly on a large scale, is *Landscape with Sportsmen and Game* (Dulwich College Picture Gallery), which features some unnaturalistically (but attractively) blue leaves, caused by yellow pigment fading in the greens.

Pynas, Jan (1583/4–1631) and **Jacob** (1590–1650?). Dutch painters, brothers. They worked in Rome in the first decade of the 17th cent. and brought back a style of small-scale history painting in the manner of *Elsheimer. This in turn may have had some influence on *Rembrandt, who is said to have studied briefly with Jacob.

Pythagoras. Greek sculptor of the 5th cent. BC, who emigrated from Samos to Rhegium (now Reggio di Calabria) in Italy. Ancient writers rank him among the greatest Greek sculptors, but no work survives—either original or copy—that can securely be attributed to him. *Pliny describes him as 'the first to represent sinews and veins and to bestow attention on the treatment of hair' and the Greek biographer Diogenes Laertius writes that he is 'thought to have been the first to aim at rhythm and proportion'. Though doubtless exaggerations, these comments indicate that his work must have seemed remarkably naturalistic and harmonious. He made several statues of athletes and in this field he is said to have excelled *Myron.

pyxis (pl. pyxides). A covered cylindrical box, in *Classical times made of box-wood (Gk. *pyxos*: box-tree, whence the name), but later of metal or ivory, used originally as a toilet box, or a container for jewels or incense, but adopted by the Church as early as the 4th cent. to hold relics or the reserved Host. A section of elephant tusk lends itself to this form when hollowed out and closed at one end with an ivory disc, and most Early Christian pyxides were of this material. As a result of the Crusades, both Early Christian and Muslim pyxides found their way into Western churches in the 12th cent., and were used to hold relics or the Host. But there were few attempts to copy them, since the small enamelled copper pyxides with conical covers produced in Limoges, so popular throughout the 13th and 14th cents., were already beginning to take their place.

quadratura. A type of *illusionistic decoration in which architectural elements are painted on walls and/or ceilings in such a way that they appear to be an extension of the real architecture of a room into an imaginary space. It was common in Roman art, was revived by *Mantegna in the 15th cent., and reached its peaks of elaboration in *Baroque Italy. The greatest of all exponents of *quadratura* was probably *Pozzo, in whose celebrated ceiling in S. Ignazio, Rome, architecture and figures surge towards the heavens with breathtaking bravura. Unlike Pozzo, many artists relied on specialists called *quadraturisti* to paint the architectural settings for their figures; Agostino *Tassi, for example, did this for *Guercino's *Aurora*, and Giambattista *Tiepolo worked much in collaboration with the brilliant *quadraturista* Gerolamo Mengozzi Colonna (*c.*1688–1766).

quadro riportato (Italian: 'carried picture'). Term applied to a ceiling picture that is painted without *illusionistic foreshortening, as if it were to be viewed at normal eye level. *Mengs' *Parnassus* (1761) in the Villa Albani, Rome, is a famous example—a *Neoclassical manifesto against *Baroque illusionism.

Quarton, Enguerrand. See CHARONTON.

Quast, Pieter Jansz. (1605/6–47). Dutch painter of peasant *genre in the manner of *Brouwer and Adriaen van *Ostade, active in his native Amsterdam and in The Hague. Some of his drawings on *parchment of rotund, *caricature-like figures were probably sold as finished pictures to be used as amusing and cheap decorations for taverns and homes. Similar sheets hang in many of the interiors painted by 17th-cent. Dutch artists.

quattrocento. Term (literally 'four hundred') applied to the 15th cent. (the 1400s) in Italian art. It can be used as a noun ('painting of the quattrocento') or as an adjective ('quattrocento sculpture'). Terms used in the same way for other centuries are: dugento (or duecento) for the 13th cent.; trecento for the 14th cent.; cinquecento for the 16th cent.; seicento for the 17th cent.; settecento for the 18th cent.; ottocento for the 19th cent.; and novecento for the 20th cent.

Quellin (or **Quellinus**), **Artus I (Arnoldus)** (1609–68). The most distinguished member of a family of Flemish sculptors. He was born in Antwerp and was a pupil of François *Duquesnoy in Rome. He was back in Antwerp *c.*1640, then moved to Amsterdam, where from *c.*1650 to 1664 he directed the sumptuous sculptural decoration of the Town Hall. His dignified style was singu-

larly appropriate for van *Campen's great building, and the decoration forms the most impressive sculptural ensemble of the time in northern Europe. The high quality of his portrait busts can be seen in the Rijksmuseum. His collaborators at Amsterdam Town Hall included his cousin, **Artus II Quellin** (1625–1700), whose independent work was more *Baroque in style. The commanding figure of *God the Father* (1682) for the *rood screen at Bruges Cathedral is perhaps his finest work. **Artus III Quellin** (1653–86), usually called Arnold, son of Artus II, settled in England about 1678. By 1684 he was working with Grinling *Gibbons, and the drop in quality of Gibbons's large-scale figure work (not his forte) after Quellin's death indicates that the latter was probably the dominant personality in producing such fine statues as the bronze *James II* (1686) outside the National Gallery in Trafalgar Square. Quellin's outstanding independent work is the tomb of Thomas Thynne (Westminster Abbey, 1682), which features a *relief of Thynne's murder in his coach in Pall Mall.

Other artist members of the Quellin family included **Erasmus I** (*c.*1584–1639/40), father of Artus I and likewise a sculptor, and Artus I's two brothers, **Erasmus II** (1607–78), a painter who was a pupil and collaborator of *Rubens, and **Hubert** (1619?–87), an engraver.

Quercia, Jacopo della (Jacopo di Piero di Angelo) (*c.*1374–1438). The greatest sculptor of the Sienese school, the son of an undistinguished goldsmith and wood-carver, **Piero di Angelo** (Quercia, from which he takes his name, is a place near Siena). He was one of the outstanding figures of his generation in Italian sculpture, alongside *Donatello and *Ghiberti, but his career is difficult to follow, as he worked in numerous places and sometimes left one commission unfinished while he took up another elsewhere. Contrary to *Vasari's assertions that he led a 'well-ordered life', he seems to have been inveterately dilatory. He is first firmly documented in 1401, as unsuccessfully competing for the commission (won by Ghiberti) for the Baptistery doors in Florence in 1401. His first surviving work is usually considered to be the tomb of Ilaria del Carretto, wife of the ruler of Lucca, Paolo Guinigi (Cathedral, Lucca, *c.*1406), which was eulogized by *Ruskin. There are Renaissance *putti and swags round the sides of the coffin, but the serene and graceful effigy is in the northern manner and suggests Quercia had knowledge of work done in the circle of Claus *Sluter in Burgundy.

His major work for his native city was a fountain called the *Fonte Gaia* (commissioned in 1409, executed in 1414–19), which is now—much damaged—in the loggia of the Palazzo Pubblico. Its *relief carvings include some beautifully draped female figures and a terribly battered but still awesomely powerful panel of *The Expulsion from Paradise*. Between 1417 and 1431 he worked together with Donatello and Ghiberti on reliefs for the font in the Baptistery at Siena.

In 1425 Quercia received the commission for his last great work (left unfinished at his death), the reliefs on the portal of S. Petronio, Bologna, with scenes from Genesis and the nativity of Christ. The figures—usually only three to a relief, in contrast to the crowded panels of Ghiberti—have a directness and strength which won the admiration of *Michelangelo, who visited Bologna in 1494. Several of the motifs are to be found, reinterpreted, on the Sistine Ceiling.

Quesnel, François (*c.*1545–1619). The best-known member of a dynasty of French painters active in the 16th and 17th cents. He was born in Edinburgh, where his father **Pierre** (d. *c.*1547) was employed at the Scottish court, but he worked in France as a portraitist, carrying on the *Clouet tradition into the 17th cent. His work is known mainly through his polished drawings. Numerous paintings are attributed to him, but only one is securely identified as his by his monogram—*Mary Ann Waltham* (Earl Spencer Coll., Althorp, Northamptonshire, 1572).

RA. See ROYAL ACADEMY.

Rabin, Oskar (1928–). Russian painter, one of the leading representatives of 'unofficial' art in the Soviet Union. In a review of the exhibition of contemporary Soviet art at the Grosvenor Gallery, London, in 1964, the art critic of *Time* magazine wrote: 'The only Russian painter who might be at home in any Western city's art museum is Oskar Rabin, an outcast.' Rabin paints in thick *impasto mainly with a palette knife and generally in subdued ochre and umber tones. His favourite themes include fantastic cityscapes juxtaposed with incongruous objects. For many years he worked as a railway porter and engine driver, painting in his spare time, and he made use of themes taken from the context of the railway. Rabin left Russia in 1978 and settled in Paris.

Rackham, Arthur (1867–1939). British artist, celebrated for his illustrations to children's books. He said he believed in 'the greatest stimulating and educative power of imaginative, fantastic and playful pictures and writings for children in their most impressionable years', and he worked in a striking vein of Nordic fantasy, creating a world populated by goblins, fairies, and weird creatures. At the peak of his career Edmund *Dulac was his only serious rival as an illustrator of fairy stories.

Raeburn, Sir Henry (1756–1823). The leading Scottish portrait painter of his period, active mainly in his native Edinburgh. On leaving school he was apprenticed to a goldsmith, then worked as a *miniaturist, and appears to have been largely self-taught as a painter in oils. Between 1784 and 1787 he had periods in London (where he met *Reynolds) and Italy, but his distinctive style was by this time already formed—one of his finest works, the *Rev. Robert Walker Skating* (NG, Edinburgh), is traditionally said to have been painted in 1784. He painted directly on to the canvas without preliminary drawings, and his vigorous, bold handling—sometimes called his 'square touch'—could be extraordinarily effective in conveying the character of rugged Highland chiefs or bluff legal worthies. He also had a penchant for vivid and original lighting effects (*William Glendonwyn*, Fizwilliam Mus., Cambridge, *c.*1795) and could be remarkably sensitive when painting women (*Isabella McLeod, Mrs James Gregory*, National Trust for Scotland, Fyvie Castle, Aberdeenshire, *c.*1798). At times, however, his technical facility can degenerate into empty virtuosity. In 1822, on the occasion of George IV's visit to Edinburgh, he was knighted and in 1823 he was appointed His Majesty's *Limner for Scotland. Since he had all the sitters he

needed in Scotland, there was no need for him to compete with *Lawrence and *Hoppner in London (although he did consider moving there after Hoppner's death in 1810), and in the history of British portraiture he is an isolated and perhaps underrated figure.

Raedecker, John (1885–1956). Dutch sculptor whose works are represented in all the principal collections of the Netherlands. He and the architect J. J. P. Oud (1890–1963) did the national monument on the Dam, the main square of Amsterdam.

Raf, Jehan. See CORVUS JEHAN.

Raggi, Antonio. See BACICCIA.

Raibolini, Francesco. See FRANCIA.

Raimondi, Marcantonio (*c.*1480–1534). Italian engraver, a pioneering figure in the use of prints to reproduce the work of other artists. He was born near Bologna, studied there with *Francia and learned much from *Dürer's engravings in Venice (Dürer is said to have complained to the Venetian senate about being plagiarized). In about 1510 he settled in Rome, and thereafter worked principally for *Raphael, his engravings helping to spread the master's High *Renaissance style throughout Europe. His works were also often copied on *maiolica dishes. Apart from his association with Raphael, Raimondi is best known for his series of obscene engravings (after designs by *Giulio Romano) that led to his imprisonment in 1524. He left Rome after the Sack of 1527 and died in obscurity in Bologna.

Ramos, Mel (1935–). American painter. Ramos is usually described as a *Pop artist, but his smooth, impersonal handling (in oils and watercolour) brings him also within the orbit of *Superrealism.

He specializes in paintings of nude women of the calendar pin-up or 'playmate' type. Sometimes they are posed with oversized products such as pieces of cheese and sometimes they allude to the work of leading painters of the past (more rarely the present). The jokey quality of his work is reflected in his titles; two typical series are 'You Get More Spaghetti with Giacometti' and 'You Get More Salami with Modigliani'.

Ramsay, Allan (1713–84). Scottish portrait painter, active mainly in London. He was the outstanding portraitist there from about 1740 to the rise of *Reynolds in the mid-1750s. Ramsay studied in London, in Rome, and in Naples (under *Solimena), and when in 1739 he settled in London he brought a cosmopolitan air to British portrait painting. His portraits of women have a decidedly French grace (*The Artist's Wife*, NG, Edinburgh, *c*.1755) and in this field he continued to be a serious rival to Reynolds, who was upset when Ramsay was appointed Painter-in-Ordinary to George III in 1760. Ramsay, however, gradually gave up painting during the 1760s to devote himself to his other interests. He was the son of Allan Ramsay (1686–1758) the poet, and inherited his father's literary bent. Political pamphleteering, classical archaeology (he revisited Rome in 1754–7), and conversation took up much of his later years. He was successful in literary circles (the philosophers David Hume, 1711–76, and Jean-Jacques Rousseau, 1712–78, were among his sitters) and Dr Samuel Johnson (1709–84) said of him: 'You will not find a man in whose conversation there is more instruction, more information, and more elegance.' Ramsay was one of the finest draughtsmen among 18th-cent. British artists; there is a good collection of his drawings in the National Gallery of Scotland.

Rand, John G. See TUBE.

Ranson, Paul. See ACADÉMIE.

Raphael (properly Raffaello Sanzio) (1483–1520). Italian painter, architect, and designer, the artist in whose works the ideals of the High *Renaissance find their most complete expression. He was born in Urbino, where in the court of Federico da *Montefeltro Italian culture had one of its most distinguished settings, and his father, Giovanni *Santi, was a writer as well as a painter, and would have introduced his son to humanist ideas. *Vasari says that 'Raphael came to be of great help to his father in the numerous works that Giovanni executed in the state of Urbino', but Santi died in 1494, when Raphael was only 11, and the overwhelming influence on his early work was *Perugino. It is often said that Raphael was Perugino's pupil but this is probably not strictly true. He was highly precocious and working as an independent artist by 1500; his close contact with Perugino came a little later (*c*.1502–3), when he was probably his colleague rather than assistant. That he soon completely outstripped Perugino is best seen by comparing Raphael's *Marriage of the Virgin* (Brera, Milan, 1504) with Perugino's painting of the same subject (Musée des Beaux-Arts, Caen). The two compositions are closely similar, but Raphael far surpasses Perugino in lucidity and grace.

In his early career Raphael worked in various places in Tuscany and Umbria. From 1504 to 1508 he worked much in Florence, and this is usually referred to as his Florentine period, although he never took up permanent residence in the city. Under the influence particularly of *Leonardo and *Michelangelo his work became grander and more sophisticated. To the Florentine period belong many of his most celebrated depictions of the Virgin and Child. In these and his paintings of the Holy Family he showed his developing mastery of composition and expression. He paints the sacred figures as splendid, healthy human beings, but with a serenity, a sense of some deep inner integrity, that removes any doubt as to the holiness of the subject. This sense of well-being distinguishes the art of Raphael from the more disturbingly intellectual work of Leonardo or the overwhelmingly powerful creations of Michelangelo, and evidently reflects his own balanced nature. Unlike his two great contemporaries, he was not a solitary genius but a sociable and approachable figure, whom Vasari describes as 'so gentle and so charitable that even animals loved him'. Vasari also writes that 'although there were many who considered that he surpassed Leonardo in sweetness and a certain natural facility, none the less, Raphael never achieved the sublimity of Leonardo's basic conceptions or the grandeur of his art'. In 1508, though he was only 25 years old, his reputation was sufficiently established for him to be summoned to Rome by Pope Julius II (1453–1513) and entrusted with the frescos for one of the papal rooms in the Vatican, the Stanza della Segnatura. He continued to work in Rome till his death in 1520 although he is known to have returned briefly to Florence in 1515.

The Stanza della Segnatura is based on a complex theological programme of the relationship between classical learning and the Christian revelation. On one main wall, in the celebrated painting known as *The School of Athens*, are shown the ancient philosophers, led by Plato and Aristotle, enclosed in a majestic architectural setting, a masterpiece of *perspective drawing: opposite, is the painting called the *Disputà* (the Disputation over the Sacrament); the doctors of the church adore the Sacrament, while above the Trinity is surrounded by the saints and martyrs. Here the setting is a wide open space. They are works in which grandeur and gracefulness seem effortlessly combined and they have had a profound and continuing influence on European art. After the completion of the Stanza della Segnatura in 1511 Raphael was entrusted with three other apartments in the Vatican, but by the time he had completed the first of these—the Stanza d'Eliodoro—in 1514, his services were so much in demand that he had to rely increasingly on assistants (of whom *Giulio Romano was the most distinguished) for the execution of his work. The *cartoons for tapestries in the Sistine Chapel (Royal Coll., on loan to V&A, London, 1515–16), for example, rank among his noblest designs, but probably comparatively little of the execution is from his own hand. In addition to the tapestries Raphael supervised various other great decorative schemes in Rome: the Villa Farnesina with its ceiling showing the story of Psyche, and a wall fresco, *Galatea* (1511–12), which is from Raphael's own hand at its most expert; the Chigi Chapel in Sta Maria del Popolo (begun *c.*1512), where he designed the entire scheme, comprising architecture, sculpture, painting, mosaic, stucco-work, and marble inlay; and the ceiling and wall *arabesques of the Vatican Loggie (*c.*1515 onwards), which left a permanent imprint on European interior decoration. The richness of effect in the Chigi Chapel was an important source for similar works in the *Baroque era, and it is fitting that the structure was in fact completed by *Bernini.

Raphael also painted many portraits and it is in these that the quality of his own workmanship in his later years is best seen. They rival Leonardo in subtlety of characterization and *Titian in richness of colouring, show great inventiveness in creating psychological situations, and provide a remarkable record of the intellectual circles in which he moved (*Baldassare Castiglione*, Louvre, Paris, *c.*1515). Other important commissions from his Roman period that are largely from his own hand are the *Sistine Madonna* (Gemäldegalerie, Dresden, *c.*1512–14), his most famous painting of the Virgin and Child, and the great altarpiece of the *Transfiguration* (Vatican Gal.), on which he was working at his death, and which presages the *Mannerist style. From about 1512 he began to work as an architect, and he also embarked on an archaeological survey of ancient Rome, although little evidence of his scheme survives. After the death of *Bramante in 1514 he became architect to St Peter's, and Raphael ranks second only to him among High Renaissance architects. It is difficult to appreciate his status, however, as little of his work survives as he designed it.

Vasari says that Raphael's early death (on his 37th birthday) 'plunged into grief the entire papal court'. He was rich, famous, and honoured (Vasari says the Pope, Leo X, 'who wept bitterly when he died', had intended making him a cardinal), and his influence was widely spread even during his own lifetime through the engravings of Marcantonio *Raimondi. His posthumous reputation was even greater, for until the later 19th cent. he was regarded by almost all critics as the greatest painter who had ever lived—the artist who expressed the basic doctrines of the Christian Church through figures that have a physical beauty worthy of the *antique. *Reynolds said: 'It is from his having taken so many models that he became himself a model for all succeeding painters: always imitating and always original.' He became the ideal of all *academies (it was against his authority that the *Pre-Raphaelites revolted), and today we approach him through a long tradition in which Raphaelesque forms and motifs have been used with a steady diminution of their values. Many lesser artists have imitated him emptily, but he has been a major inspiration to great *classical painters such as Annibale *Carracci, *Poussin, and *Ingres.

Ratcliffe, William. See CAMDEN TOWN GROUP.

Ratgeb, Jörg (*c.*1480–1526). German painter, active in Swabia. His few surviving paintings show him to have been closer in spirit to *Grünewald than practically any of his contemporaries. The most important is the Herrenberg Altarpiece (Staatsgalerie, Stuttgart, 1519), which has some harrowing Passion scenes. Ratgeb himself met a gruesome end—executed at Pforzheim for his part in the Peasants' War.

Rauch, Christian Daniel (1777–1857). German sculptor. In Rome (1804–11) he came under the influence of *Canova and *Thorvaldsen, but he modified their *Neoclassical idealism with a close study of nature, particularly noticeable in his portraits. He is best known for his equestrian monument of Frederick the Great in Berlin (1839–52), and for his *busts of prominent contemporaries such as *Goethe and *Kant.

Rauschenberg, Robert (1925–). American painter, printmaker, designer, and experimental artist. With his friend Jasper *Johns, whom he met in 1954, he is regarded as one of the most influential figures in the move away from the *Abstract Expressionism that had dominated American art in the late 1940s and early 1950s. He studied at various art schools, most notably *Black Mountain

College. In the mid-1950s he began to incorporate three-dimensional objects into what he called *'combine paintings'. The best-known example is probably *Monogram* (Moderna Museet, Stockholm, 1959), which features a stuffed goat with a rubber tyre around its middle, splashed with paint in a manner recalling *Action painting. Other objects he used included Coca-Cola bottles, fragments of clothing, electric fans, and radios, and because of his preoccupation with such consumer products he has been hailed as one of the pioneers of *Pop art. In 1958 he had a one-man show at Leo *Castelli's gallery and from this time his career began to take off. By the early 1960s he was building up an international reputation, and in 1964 he was awarded the Grand Prize at the Venice *Biennale. This caused great controversy, the Vatican newspaper *L'Osservatore Romano* describing the award as 'the total and general defeat of culture'. In the 1960s Rauschenberg returned to working on a flat surface and was particularly active in the medium of *silk screen. He has been interested in combining art with new technological developments and in 1966 he helped to form EAT (Experiments in Art and Technology), an organization to help artists and engineers work together. In 1985 he launched Rauschenberg Overseas Cultural Interchange (ROCI), an exhibition dedicated to world peace that toured the world and included works created specifically for each place visited.

Raven, John. See CORVUS, JOHANNES.

Raverat, Gwen (1885–1957). British graphic artist, theatre designer, painter, and writer, born in Cambridge, the daughter of Sir George Darwin, Professor of Astronomy, and granddaughter of Charles Darwin. She studied at the *Slade School, but was mainly self-taught in engraving, which was her primary activity. Her best-known works are her book illustrations, notably of collections of poems by her cousin Frances Cornford. Her style was bold and vigorous and she was uninterested in technical virtuosity. Although she was art critic for *Time and Tide* from 1928 to 1939, she did not consider herself a writer and was surprised at the success of *Period Piece* (1952), an account of her childhood with her own illustrations, which became a bestseller on both sides of the Atlantic.

Ravesteyn, Jan Anthonisz. van (*c*.1570–1657). Dutch portrait painter active in court circles in The Hague, the best-known member of a family of artists. Like the more famous *Mierevelt (with whose work his own is sometimes confused) he painted a great many portraits that are of more historical than artistic interest.

Ravilious, Eric (1903–42). British painter, graphic artist, and designer. His highly varied output included (apart from paintings) book illustrations and book-jackets, designs for the *Wedgwood pottery factory (including a mug commemorating Edward VIII's coronation in 1937), and designs for furniture, glass, and textiles. He was one of the outstanding wood engravers of his time, his book illustrations in this medium making striking use of bold tonal contrasts and complex patterning. In 1940–2 he was an *Official War Artist, and he produced some memorable watercolours of naval scenes off Norway (*Norway, 1940*, Laing Art Gal., Newcastle upon Tyne). While on a flying patrol near Iceland in September 1942 his plane disappeared, and he was officially presumed dead the following year.

Ray, Man. See MAN RAY.

Rayonism (Rayonnism, Rayism, Luchism). A type of abstract or semi-abstract painting practised by the Russian artists *Goncharova and *Larionov and a few followers in the years 1912 to 1914 and representing their own adaptation of *Futurism. Rayonism was launched at the *Target exhibition held in Moscow in 1913. In the same year Larionov published a manifesto stating: 'Rayonism is a synthesis of *Cubism, Futurism and *Orphism.' The style was bound up with a very unclear theory of invisible rays, in some ways analogous to the 'lines of force' which were postulated by Italian Futurist theory, emitted by objects and intercepted by other objects in the vicinity: the artist, it was said, must manipulate these rays to create form for his own aesthetic purposes. 'The rays which emanate from the objects and cross over one another give rise to rayonist forms. The artist transfigures these forms by bending them and submitting them to his desire for aesthetic expression.' Many Rayonist paintings are very similar in style to those of the Futurists, with particular emphasis on breaking up the subject into bundles of slanting lines. In other works the subject virtually or completely disappears. The movement was short-lived as both Goncharova and Larionov virtually abandoned easel painting after 1914 and they founded no school.

Read, Nicholas. See ROUBILIAC.

Read, Sir Herbert (1893–1968). British poet and critic, regarded as his period's foremost advocate and interpreter of modern art. After serving with distinction in the army in the First World War he worked at the Treasury, then in the ceramics department of the Victoria and Albert Museum, 1922–31, before becoming Watson Gordon Professor of Fine Arts at Edinburgh University, 1931–3. By this time he had published several collections of his verse as well as various art-historical studies (including *English Stained Glass*, 1926, still a standard work), critical works on English literature, and the first of his philosophical works on art, *The Meaning of Art* (1931). In 1933 he returned to London as editor (1933–9) of *The Burlington Magazine*, Britain's foremost scholarly art journal, and his attention turned increasingly to contemporary art; in 1933 he published *Art Now*, the first comprehensive defence in English of modern European art, and in 1934 he edited the modernist manifesto **Unit One*. At this time he lived near Henry *Moore, Barbara *Hepworth, and Ben *Nicholson, and he acted as the public mouthpiece of the group of artists of which they were the centre. He was interested in *Surrealism as well as abstraction and was one of the organizers of the 1936 International Surrealist Exhibition in London. In 1947 he was co-founder of the *Institute of Contemporary Arts in London. Among his many books the most influential was probably *Education Through Art* (1943), which used the insights of psychoanalysis to promote the idea of teaching art as an aid to the development of the personality. His other books include *A Concise History of Modern Painting* (1959) and *A Concise History of Modern Sculpture* (1964), much used as introductory surveys by students. By the time he wrote them he was becoming disenchanted with contemporary artistic developments, but he was known as 'The Pope of Modern Art' and was regarded as 'an international authority and indeed something of a sage. It was not a role to which he ever pretended, for he was a man of conspicuous modesty' (*DNB*).

ready-made. A name given by Marcel *Duchamp to a type of work he invented consisting of a mass-produced article selected at random and displayed as a work of art. His first ready-made (1913) was a bicycle wheel, which he mounted on a kitchen stool, and this was soon followed by others, notably *Bottle Rack* (1914), *In Advance of the Broken Arm* (a snow shovel, 1915), and the celebrated *Fountain* (a urinal signed R. Mutt, 1917); most of the originals have disappeared, but several replicas exist. Duchamp himself distinguished the ready-made from the **objet trouvé* (found object), pointing out that whereas the *objet trouvé* is discovered and chosen because of its interesting aesthetic qualities, its beauty and uniqueness, the ready-made is one—any one—of a large number of indistinguishable mass-produced objects. Therefore the *objet trouvé* implies the exercise of taste in its selection, but the ready-made does not. When *Fountain* was rejected by the hanging committee of the *Society of Independent Artists, Duchamp defended his creation by writing: 'Whether Mr Mutt with his own hands made the fountain or not has no importance. He CHOSE it. He took an ordinary article of life, placed it so that its useful significance disappeared under a new title and point of view . . . [he] created a new thought for that object.' The ready-made was one of *Dada's most enduring legacies to modern art. It was much used in *Pop art, for example, and Robert *Rauschenberg calls *Bicycle Wheel* 'one of the most beautiful pieces of sculpture I've ever seen'.

realism. Term used with various meanings in the history and criticism of the arts. In its broadest sense the word is used as vaguely as *naturalism, implying a desire to depict things accurately and objectively. Often, however, the term carries with it the suggestion of the rejection of conventionally beautiful subjects or of idealization in favour of a more down-to-earth approach, often with a stress on low life or the activities of the common man. In a more specific sense the term (usually spelled with a capital R) is applied to a movement in 19th-cent. (particularly French) art characterized by a rebellion against the traditional historical, mythological, and religious subjects in favour of unidealized scenes of modern life. The leader of the Realist movement was *Courbet, who said 'painting is essentially a concrete art and must be applied to real and existing things'.

Since the 1950s the term has also been used in a contrasting sense, of art which eschews representation and depiction altogether and avoids all forms of illusionism. In this sense art is called 'Realist' when the materials or objects from which the work is constructed are presented for exactly what they are and are known to be. This usage is found particularly in the terms *New Realism and Nouveau Réalisme.

The term *Social Realism has been applied to 19th- and 20th-cent. works that are realistic in the second sense described above and make overt social or political comment. It is to be distinguished from *Socialist Realism, the name given to the officially approved style in the USSR and some other Communist countries; far from implying a critical approach to social questions, it involves toeing the Party line in an academic style.

*Magic Realism and *Superrealism are names given to two 20th-cent. styles in which extreme realism—in the sense of acute attention to detail—produces a markedly unrealistic overall effect. See also VERISM.

Recco, Giuseppe (1634–95). Neapolitan still-life painter, the outstanding member of a family of artists. He specialized in pictures of fish, painted in an impressively grand style, but more austere than those of *Ruoppolo, with whom he ranks as the most distinguished Italian still-life painter of his period. Recco's renown led him to be invited to Spain by Charles II (1661–1700), but he died soon after arriving.

red-figure vase painting. One of the two major divisions of Greek vase painting, the other being *black-figure. In the red-figure technique, the background was painted black, leaving the figures in the unpainted red colour of the pottery. Details of the figure could thus be added with a brush rather than incised through the black paint, allowing much greater flexibility and subtlety of treatment. Because of this advantage the red-figure technique, which developed in Athens from about 530 BC, rapidly superseded the black-figure technique.

Redgrave, Richard (1804–88). English painter, writer, and art administrator. He began as a painter of anecdotal literary subjects, often in 18th-cent. costume, but in the 1840s he became a pioneer of scenes of contemporary social concern (*The Poor Teacher*, Shipley Art Gal., Gateshead, 1845). 'It is one of my most gratifying feelings', he wrote, 'that many of my best efforts in art have aimed at calling attention to the trials and struggles of the poor and the oppressed.' Much of his later career was taken up with administration—the posts he held included surveyor of crown pictures and keeper of pictures at the South Kensington (later Victoria and Albert) Museum—and as a painter he turned mainly to landscapes done in his spare time. His brother, **Samuel Redgrave** (1802–76), was a writer on art, the author of a still-useful *Dictionary of Artists of the English School* (1874). The brothers collaborated on *A Century of Painters of the English School* (1866), which has been several times reprinted and is a valuable source of information on 18th- and 19th-cent. British artists.

Redon, Odilon (1840–1916). French painter and graphic artist, one of the outstanding figures of *Symbolism. He led a retiring life, first in his native Bordeaux, then from 1870 in Paris, and until he was in his fifties he worked almost exclusively in black and white—in charcoal drawings and lithographs. In these he developed a highly distinctive repertoire of weird subjects—strange amoeboid creatures, insects, and plants with human heads and so on, influenced by the writings of Edgar Allan Poe (1809–49). He remained virtually unknown to the public until the publication of J. K. Huysmans's celebrated novel *A Rebours* in 1884; the book's hero, a disenchanted aristocrat who lives in a private world of perverse delights, collects Redon's drawings, and with his mention in this classic expression of decadence, Redon too became associated with the movement. During the 1890s Redon turned to painting and revealed remarkable powers as a colourist that had previously lain dormant. Much of his early life had been unhappy, but after undergoing a religious crisis in the early 1890s and a serious illness in 1894–5, he was transformed into a much more buoyant and cheerful personality, expressing himself in radiant colours in visionary subjects, flower paintings, and mythological scenes (*The Chariot of Apollo* was one of his favourite themes). He showed equal facility in oils and pastel. His flower pieces, in particular, were much admired by *Matisse, and the *Surrealists regarded Redon as one of their precursors. He was a distinguished figure by the end of his life, although still a very private person, and was awarded the Legion of Honour in 1903.

Régence style. Term, named after the Regency of Philip of Orleans (1715–23), sometimes applied to early *Rococo art in France. Its usage is now mainly confined to sale rooms and auctioneers' catalogues. It is not to be confused with the English *Regency style.

Regency style. Term applied to the style of architecture, decoration, costume, etc., in Britain associated with the time when the future George IV was Prince Regent (1811–20) and by extension with the period of his reign (1820–30). Although this was a period of great variety, the visual arts were in general marked by extreme elegance, a refinement, and often an attenuation of *Classical forms.

Regionalism. A movement in American painting, flourishing chiefly in the 1930s, concerned with the depiction of scenes and types from the American Midwest. The term is often used more or less interchangeably with *American Scene painting, but Regionalism can be more precisely thought of as the Midwestern branch of this broader category. Like all American Scene painters, the Regionalists were motivated by a patriotic desire to establish a genuinely American art by using local themes and repudiating avant-garde styles from Europe. Specifically they were moved by a nostalgic desire to glorify, or at least to record, rural and small-town America, and it was on this that their widespread popularity depended. The period when they flourished coincided with the Great Depression, and at this time of profound national doubt, they reasserted America's faith in itself, giving the public pictures with which they could readily identify. Their work was often produced under the auspices of the *Federal Art Project and it was supported by the fanatically patriotic critic Thomas Craven. The three major Regionalists were Thomas Hart *Benton, John Steuart *Curry, and Grant *Wood, who were all Midwesterners but differed greatly in temperament and style. They scarcely knew one another personally, but the idea of a group identity was skilfully promoted by Maynard Walker, a Kansas art dealer. Walker got a Benton self-portrait on to the cover of the Christmas 1934 issue of *Time* magazine and this largely created the image of Regionalism in the public eye; as Benton put it: 'A play was written and a stage erected for us. Grant Wood became the typical Iowa small towner, John Curry the typical Kansas farmer, and I just an Ozark hillbilly. We accepted our roles.' On the fringes of the Regionalist movement were Charles *Burchfield and Ben *Shahn. Burchfield's work has a streak of fantasy absent from that of the others, and Shahn was driven by a spirit of social protest. Specifically local styles did not develop anywhere and Regionalism died out in the 1940s in the more international spirit that prevailed during and after the Second World War.

Rego, Paula (1935–). Portuguese-born British painter of figure, animal, and fantasy subjects. She studied at the *Slade School, where she met her husband, the painter Victor Willing (1928–88). In the 1980s Rego became well known for her colourful expressionistic paintings, which often have a feeling of caricature as well as of fantasy. In 1990 she was appointed the first Associate Artist of the National Gallery, London. She painted murals in the restaurant of the gallery's new Sainsbury Wing in 1991, and in 1992 she became the first living artist to be given an exhibition in the gallery.

Reichenau School. Term used to cover *illuminated manuscripts of the *Ottonian period that have been associated with the abbey of Reichenau situated on an island in Lake Constance. It was once commonly believed that Reichenau was the pre-eminent centre of illumination in this period, but this view was strongly challenged in the 1960s, notably by C. R. Dodwell, who in his Pelican History of Art volume, *Painting in Europe: 800 to 1200* (1971), writes 'my own view is that Reichenau was a small artistic outpost rather than a central headquarters'. He considers Lorsch and Trier to be more important centres.

Reid, Alexander. See BURRELL.

Reid, Robert. See TEN, THE.

Reims School. *Carolingian School of manuscript *illumination and ivories centred on Reims. Its earliest and most important work is the famous *Utrecht *Psalter* (University Library, Utrecht) containing many pen-drawings in a lively and expressive style (produced under Archbishop Ebbo, 816–35). The *Psalter* influenced English drawing of the 11th and 12th cents., and a copy of it made in Canterbury is in the British Library.

Reinhardt, Ad (1913–67). American painter. From the beginning his painting was abstract but it changed radically in style over the years. Through the 1930s he used a crisp, boldly contoured geometrical style which owed something both to *Cubism and to the *Neo-Plasticism of *Mondrian. In the 1940s he passed through a phase of *all-over painting which has been likened to that of Mark *Tobey, and in the late 1940s he was close to certain of the *Abstract Expressionists, particularly *Motherwell, with whom he jointly edited *Modern Artists in America* (1950), a book based on conversations with contemporary artists. During the 1950s he turned to monochromatic paintings. At first they were usually red or blue, but from the late 1950s he devoted himself to all-black paintings with geometrical designs of squares or oblongs barely perceptibly differentiated in value from the background colour. This reduction of his work to 'pure aesthetic essences' reflects his belief in the complete separation between art and life—'Art is Art. Everything else is everything else.' Reinhardt was particularly influential on the development of *Minimalism. His views were extremely uncompromising and he was a noted critic of trends in modern art of which he did not approve, both as a polemical writer and a satirical cartoonist. There are examples of his work in many collections of modern art, including the Tate Gallery, London, and the Museum of Modern Art, New York.

relief (or **relievo**). Term, from the Italian *rilevare* (to raise), applied to sculpture that projects from a background surface rather than standing freely. According to the degree of projection, reliefs are usually classified as high (*alto rilievo*), medium (*mezzo rilievo*), or low (*basso rilievo* or bas-relief). A peculiar type of low relief was sometimes used by the Egyptians in which the surface of the block is allowed to remain as background and the figures are carved within deeply incised outlines, never projecting beyond the surface. It is sometimes called sunken, incised, or coelanaglyphic relief. The name *rilievo stiacciato* (also *schiacciato*) is given to a form of very low relief practised in the 15th cent. especially by *Donatello and *Desiderio da Settignano. *Intaglio is the reverse of relief, in which the design is incised and sunk below the surface of the block.

relief etching. A method of *etching plates for relief printing (ordinary etching being in *intaglio). The design is drawn on the plate in an acid-resisting *varnish. The plate is then immersed in acid, which eats away the unprotected parts so that the design stands out in relief and prints can be taken in the same way as from a *woodcut block. The method dates from the 18th cent., but was little used except by *Blake, who called it 'woodcut on copper'. In the 20th cent. the process has been revived and combined with intaglio printing by S. W. *Hayter and Joan *Miró.

Rembrandt (Rembrandt Harmensz. van Rijn) (1606–69). Dutch painter, etcher, and draughtsman, his country's greatest artist. He was born in Leiden, the son of a prosperous miller. After attending a Latin school he registered at the city's university in 1620, but probably never actually studied there. At about this time he was apprenticed to the mediocre local painter Jacob van *Swanenburgh, with whom he is said to have studied for about three years. More important for Rembrandt's development were six months spent in Amsterdam with Pieter *Lastman, *c.*1624. From Lastman he took not only his predilection for mythological and religious subjects, but also his manner of treating them, with dramatic gestures and expressions, vivid lighting effects, and a meticulous, glossy finish, as in his earliest dated work—*The Stoning of St Stephen* (Musée des Beaux-Arts, Lyons, 1625). *Houbraken says that Rembrandt also studied with Jacob *Pynas and Joris van Schooten (1587–1651), but this can have been only briefly, for by 1625 he was working as an independent master in Leiden. There he had a close association with Jan *Lievens, and both artists were highly praised by the scholar and diplomat Constantin Huygens (1596–1687), who visited them in 1629. Rembrandt's paintings of his Leiden period (which lasted until he settled permanently in Amsterdam in 1631/2) are mainly figure subjects, often involving old men depicted as philosophers or biblical characters. He also did portraits of himself and of members of his family, but it is not until 1631 that he painted his earliest known formal commissioned portrait. The sitter was Nicolaes Ruts, a wealthy Amsterdam merchant (Frick Coll., New York), and Rembrandt no doubt realized that here he had a recipe for success, as this type of work dominated his output in his early years in Amsterdam. It was the busiest period of his life as he quickly established himself as the leading portraitist in the city—about fifty of his paintings are dated 1632 or 1633 and all but a handful are portraits. The most important—the work that most clearly demonstrated his superiority to rivals such as Thomas de *Keyser—is the *Anatomy Lesson of Dr Tulp* (Mauritshuis, The Hague, 1632), which brought a wholly new vitality to the group portrait. Rembrandt's great energy in his early years in Amsterdam comes out also in his religious works. The most important commission he received during the 1630s was from the Stadholder Prince Frederick Henry (1584–1647) of Orange for five pictures depicting scenes of the Passion (Alte Pinakothek, Munich), and the *Baroque tendencies of his work at this time are even more emphatically expressed in his sensational, life-size *Blinding of Samson* (Städelsches Kunstinstitut, Frankfurt, 1636). Rembrandt presented this painting to Huygens, who had probably secured the commission from the Stadholder.

Rembrandt's success in the 1630s was personal as well as professional. In 1634 he married Saskia van Uylenburch, cousin of a picture-dealer associate, and from the evidence of his marvellously tender portraits of her it must have been a blissful union. In 1639 he bought an imposing house (now a Rembrandt museum), and he spent lavishly on works of art and anything else that took his fancy or looked as if it might be useful as a prop—armour, old costumes, etc. His domestic happiness was, however, marred by a succession of infant deaths; of the four children Saskia bore him, only his son Titus (1641–68), who became one of his favourite models, lived longer than two months.

Saskia died in 1642, and in this year Rembrandt painted his most famous picture, *The Night Watch* (Rijksmuseum, Amsterdam), which Jacob Rosenberg, in his standard monograph on the artist, calls 'a thunderbolt of genius'. The erroneous title dates from the late 18th cent. when the painting was so discoloured with dirty varnish that it looked like a night scene. Its correct title is *The Militia Company of Captain Frans Banning Cocq and Lieutenant Willem van Ruytenburch*, and it is the culminating work of the Dutch tradition of civic guard portraits (a genre particularly associated with Frans *Hals). Rembrandt showed remarkable originality in making a pictorial drama out of an insignificant event. To do this he subordinated the individual portraits to the demands of the composition, and according to popular legend the sitters who had paid for the picture were appalled at this and demanded that Rembrandt make radical changes, paint a new picture, or refund their money. Rembrandt's refusal is supposed to have been his downfall and to have led him into penniless obscurity. There is, however, no basis in fact for the story, which is a 19th-cent. invention; indeed all the available evidence suggests that the picture was well received. Samuel van

*Hoogstraten, for example, wrote 'It is so painter-like in thought, so dashing in movement, and so powerful' that the pictures beside which it hung were made to seem 'like playing cards'.

Nevertheless, in the 1640s Rembrandt's worldly success did decline as the direction of his art changed. Formal portraiture took up much less of his time and he concentrated more on religious painting, while his style grew less flamboyant and more introspective. The change has been explained as a response to the death of Saskia (and of his mother in 1640), and religion may well have been a solace to him in this difficult period. At the same time, some of his market in portraiture must have gone to pupils such as *Bol and *Flinck, who imitated his style so well. It seems just as likely, however, that Rembrandt was tired of routine portraiture and wanted to return to his first love—painting subjects from the Bible.

In the 1640s Rembrandt also developed an interest in landscape and it has been suggested that he spent more time in the countryside during this period to escape from the domestic problems he encountered after Saskia's death. A widow called Geertge Dircx was employed as Titus's nurse, and she sued Rembrandt for breach of promise after his affections turned to Hendrickje Stoffels, a servant some 20 years his junior who entered the household in about 1645. After some unpleasant legal proceedings Rembrandt succeeded in having Geertge shut up in a reformatory and the litigation did not end until her death in 1656. Hendrickje remained with Rembrandt for the rest of her life and bore him two children, including a daughter, Cornelia, born in 1654, who was the only one of his children to outlive him. Rembrandt's portrayals of Hendrickje are just as loving as those of Saskia, but he was unable to marry her because of a clause in Saskia's will.

After he turned his back on fashionable portraiture, Rembrandt's extravagance led him into financial difficulties, which became acute by the early 1650s. In 1656 he was declared insolvent; his collections were sold and by 1660 he had to leave his house and move to lodgings in a poorer district of the city. Houbraken says that 'in the autumn of his life he kept company mainly with common people and such as practised art', but the romantic image of him as a pauper and a recluse is grossly exaggerated. He continued to be a respected figure who received important commissions (his patrons included the Sicilian nobleman Don Antonio Ruffo), and Hendrickje and Titus established an art firm with Rembrandt technically their employee, a device that protected him from his creditors. Indeed, with some weight thus lifted from his shoulders, Rembrandt may well have felt renewed energy, and there are more dated paintings from 1661 than from any year since the early 1630s. In 1661–2 he painted two of his greatest works—*The Sampling Officers of the Cloth-Makers' Guild* (sometimes called *The Syndics*, Rijksmuseum) and *The Conspiracy of Claudius Civilis*, painted for Amsterdam Town Hall, but for unknown reasons removed in 1663 and cut down (apparently by Rembrandt himself)—the magnificent fragment is now in the Nationalmuseum, Stockholm.

Rembrandt's final years were clouded by the deaths of Hendrickje in 1663 and Titus in 1668, but his art was in no way impaired. On the contrary, his work seemed to grow in human understanding and compassion to the very end, and his last self-portraits, the culmination of an incomparable series that began 40 years earlier, show him facing his hardships with the utmost dignity—someone who has no illusions about life, but equally no bitterness. Two self-portraits date from the last year of his life (NG, London, and Mauritshuis), but the painting that best stands as his spiritual testimony is perhaps *The Return of the Prodigal Son* (Hermitage, St Petersburg, *c*.1669), a work of the utmost tenderness and poignancy, which Kenneth *Clark described as 'a picture which those who have seen the original in Leningrad may be forgiven for claiming as the greatest picture ever painted'.

The emotional depth and range of Rembrandt's work was matched by the total expressive mastery of his technique. Even as a young man, when surface polish and attention to detail were a necessary part of his skill as a fashionable portraitist, he had experimented boldly in his more private works, sometimes, for example, using the butt end of the brush to scrape through the paint. When he began to paint much more to please himself in the 1640s, his handling grew much broader, and Houbraken wrote that 'in the last years of his life, he worked so fast that his pictures, when examined from close by, looked as if they had been daubed with a bricklayer's trowel'.

It is not only the quality of Rembrandt's work that sets him apart from all his Dutch contemporaries, but also its range. Although portraits and religious works bulk largest in his output, he made highly original contributions to other genres, including still life (*The Slaughtered Ox*, Louvre, Paris, 1655), and he painted some pictures, such as *The Polish Rider* (Frick Coll., New York, *c*.1655), that virtually defy classification. Rembrandt was prodigious, too, as an etcher and draughtsman. He is universally regarded as the greatest of all exponents of etching (and *drypoint), capable of expressing the airy breadth of the Dutch countryside with a few quick strokes, but also prepared to radically rework a complex religious scene such as *The Three Crosses* perhaps a decade after he had begun it in 1653. His drawings were done mainly as independent works rather than as studies for paintings and often with the thick bold strokes of the reed pen, of which he was an unsurpassed master.

Rembrandt was a great teacher; Gerard *Dou became his first pupil in 1628 and Aert de *Gelder, who was with him in the 1660s, continued his master's style into the 18th cent. Between these two, Rembrandt taught such illustrious names as Carel *Fabritius (his greatest pupil), Philips de *Koninck and Nicolaes *Maes, but often his pupils later abandoned his exacting standards for a more facile popularity.

Rembrandt continued to have many admirers after his death, and his work often fetched high prices in the 18th cent. He was generally regarded as incomparable in his mastery of light and shade, but most critics considered him a flawed genius, whose failing was his 'vulgarity' and lack of decorum. The Dutch poet Andries Pels, for example, wrote in 1681: 'If he painted, as sometimes would happen, a nude woman, / He chose no Greek Venus as his model, / But rather a washerwoman . . . / Flabby breasts, ill-shaped hands, nay, the traces of the lacings / Of the corsets on the stomach, of the garters on the legs, / Must be visible, if Nature was to get her due.' It was during the age of *Romanticism, when it was felt that artists should give expression to their innermost feelings and flout conventions, that his reputation began to rise towards its present supremely exalted heights. In 1851 *Delacroix suggested that one day Rembrandt would be rated higher than *Raphael—'a piece of blasphemy that will make every good academician's hair stand on end'; his prophecy came true within 50 years.

Remington, Frederic (1861–1909). American painter, sculptor, and illustrator, the most famous portrayer of the 'Wild West'. He was a burly, athletic man and after attending Yale University he travelled widely in the West, prospecting and cowpunching as he worked to establish himself with his illustrations. Reproductions in *Harper's Weekly* and other popular journals made him a household name, and with his success he was able to turn more to painting and sculpture. His love of horses was so great that he had barn doors built in his studio, enabling them to be brought inside. He continued to travel (to Europe and North Africa as well as in America), and he wrote as well as working as an artist, covering the Indian Wars of 1890–1 and the Spanish-American War (in Cuba) of 1898 as a war correspondent. There is a collection of his work at the Remington Art Memorial, Ogdensburg, New York State.

Renaissance. Term meaning 'rebirth' applied to an intellectual and artistic movement that began in Italy in the 14th cent., culminated there in the 16th, and influenced other parts of Europe in a great variety of ways. The metaphor of a 'rebirth' goes back to the 15th cent. when it was used to describe the revival of Classical learning. It was also applied to a revival of the arts, and it was in the famous *Lives* of *Vasari (1550) that the idea of such a revival was systematically developed. Vasari saw the history of art in Italy from *Giotto to *Michelangelo as a continuous progress, which he compared to the process of organic growth, and his view dominated the conception of art history up to recent times. The extension of the meaning of the term from a movement in art and letters to a period of time began in the 18th cent. and was given final currency when Jules Michelet (1798–1874) called a section of his history of France *La Renaissance* (1855). *Ruskin had used the expression 'The Renaissance period' in his *Stones of Venice* in 1851. A few years later, in 1860, Jakob *Burckhardt published *Die Kultur der Renaissance in Italien* (*The Civilization of the Renaissance in Italy*), an extremely influential book that popularized the rather romantic notion of the period as a great blossoming of the human spirit—'the discovery of the world and of man'.

This wide definition of the term made it particularly difficult to indicate the limits of the period. If the 'discovery of the world' were taken to mean progress towards *'naturalism' in art, such tendencies might be detected in the 13th-cent. French art usually described as *Gothic. If the 'discovery of man' were taken as the touchstone, such thoroughly medieval figures as St Francis of Assisi might be hailed as the harbingers of the new age. These and similar inconsistencies in delimiting the period under discussion have led to much discussion of the whole concept by 20th-cent. historians.

In the visual arts, the term is generally taken to imply a deliberate imitation of Classical patterns or a conscious return to Classical standards of value. It is in architecture that the imitation of Classical models is most easily discerned, for the vocabulary of Roman architecture is quite distinct from that of the Gothic style, and *Brunelleschi was without doubt the first architect to revive the antique manner. (In describing architecture the term 'Renaissance' has sometimes been used to cover all post-medieval building that uses a Classical idiom; thus Sir Reginald Blomfield's *History of Renaissance*

Architecture in England 1500–1800, published in 1897, embraces a period that would now generally be subdivided into other categories such as *Baroque and *Neoclassicism.) In painting and sculpture, it is harder to define the Renaissance in terms of imitation of ancient models.

Vasari saw the beginnings of Renaissance sculpture in Nicola *Pisano because he had borrowed motifs from Classical sarcophagi. But we know that Gothic sculptors also borrowed motifs from Classical statuary, and we now see that there is in fact so much in common between the art of the Pisani and that of their northern contemporaries that we classify their style as *proto-Renaissance rather than Renaissance proper. Moreover, Italian sculpture of the 14th cent. remains Gothic. It is only in the early 15th cent. that we can see a reflection of Renaissance taste in sculpture. *Donatello was not content with the copying of individual Classical motifs but aimed at a revival of the spirit of Classical sculpture whose glory had been extolled by the ancient authors. His *David* (Bargello, Florence) was the first life-size nude statue since Classical antiquity, just as his *Gattamelata* (Padua) was a deliberate revival of the Classical equestrian monument.

The problem of defining what constitutes Renaissance painting, in similar terms, is considerably more complex, since there were virtually no ancient models surviving in painting as there were in sculpture and architecture. The works of such famous masters as *Apelles and *Zeuxis were known only through literary records, but *Pliny's descriptions of them influenced the painters as well as patrons and critics. From Classical authors the Renaissance public learned to expect of painting a high degree of fidelity to nature and a search for the perfect form. *Giotto, who made great advances in naturalism, is sometimes put at the head of the Renaissance tradition, but it is more consistent to give this position to *Masaccio, who brought a new scientific rigour to the problems of representation. Masaccio, like his friends Brunelleschi and Donatello, was a Florentine, and it is thus reasonable to see Florence as the cradle of the Renaissance, and the period around 1425 when they were producing some of their most innovative works as a major turning point in European art.

The term 'High Renaissance' is applied to the brief period (*c*.1500–*c*.1520) in which later centuries saw the fulfilment of all the ideals that painters had pursued since Giotto; *Leonardo, Michelangelo, and *Raphael are the dominant figures of the age. Vasari points out that the perfect mastery of means achieved in this period led to an ease of manner and a graceful harmony that stand in marked constrast to the strained efforts of preceding generations. The harsh sculptural outline gave way to the mellow modelling (**sfumato*) of Leonardo, the rigid symmetry of *Perugino to the balanced variety of Raphael's pyramidal compositions.

The term 'Northern Renaissance' is applied to the transmission of Italian imagery and ideals to the rest of Europe, but here again exact demarcation is far from easy. North of the Alps painting developed independently of the Italian Renaissance movement throughout the 15th cent. There were indeed Italian influences, such as that of *Mantegna on the Austrian Michael *Pacher, but they were exceptions. The acceptance of Renaissance ideals as such was due to the conscious effort of a handful of men, above all to *Dürer, who saw it as his mission to transplant the arts 'reborn' in Italy on to German soil. He visited Italy twice with the aim of acquiring the 'secrets' of the Italian masters, that is the mathematical principles of *perspective and proportion; and he struggled all his life both in his theoretical writings and in his art to fathom the mystery of Classical beauty.

The Netherlands with their strong Gothic tradition were slow to accept the Renaissance and at first they assimilated it only through the medium of Dürer's prints. *Lucas van Leyden best represents this phase. In the work of Quentin *Massys we find the direct influence of Leonardo's art both in his types and in his treatment. But only when the Flemings travelled to Italy did they experience the full impact of the Renaissance: *Gossaert tried somewhat ostentatiously to emulate Classical statuary and the more sensitive Jan van *Scorel absorbed something of southern poise and serenity.

The later imitators of Italian painting, such as *Heemskerck and the *Floris brothers, belong to *Mannerism rather than the Renaissance. The same is true of painting in France. There was a strong influx of Renaissance influence in France around 1500, following the Italian campaigns of Louis XII (1462–1515) and Francis I (1494–1547), but *Primaticcio and *Rosso, who were called to Fontainebleau by Francis I, represent the Mannerist phase no less than does *Cellini, and the School of *Fontainebleau cannot therefore be said to represent pure Renaissance ideals, nor can the courtly portrait style of the *Clouet family.

These distinctions reflect once more the elusiveness of the term 'Renaissance', particularly in its application to the north. While in the field of literature and thought we are used to speaking of the Renaissance in Elizabethan England, few historians of art would now consider such painters as *Hilliard or *Eworth as representatives of Renaissance style. Like their French contemporaries, they embody rigid courtly ideals far removed from the wide sympathies and spirit of intellectual adventure from which the Renaissance had originally sprung in Florence.

Reni, Guido (1575–1642). Bolognese painter. From about 1584 to 1593 he was a pupil of *Calvaert, then he entered the *Carracci academy, where he inherited their tradition of clear, firm draughtsmanship. Reni's style was also strongly influenced by the several visits he made to Rome, the first of them soon after 1600. He flirted briefly with the *Caravaggesque manner (*Crucifixion of St Peter*, Vatican, 1603), but *Raphael and the *antique were the main inspiration for his graceful classical style, as is seen in his most celebrated work, *Aurora* (1613–14), a captivatingly beautiful ceiling fresco in the Casino Rospiglioso in Rome. After Ludovico Carracci's death in 1619, Reni became the most important painter in Bologna, running a large and prosperous studio, whose products (mainly religious works) were sent all over Europe. He cut an impressive, aristocratic figure, always fashionably and expensively dressed and usually attended by servants, but he was noted for his modesty and hated profanities and obscenities. According to his biographer *Malvasia 'It was generally thought that he was a virgin . . . When observing the many lovely young girls who served as his models, he was like marble.' (He was, however, addicted to gambling.)

His fame in his lifetime was great and in the 18th and early 19th cents. many critics went into raptures, ranking him second only to Raphael. *Winckelmann, for example, compared him to *Praxiteles, and *Reynolds wrote that 'his idea of beauty . . . is acknowledged superior to that of any other painter'. He fell from favour under the scornful attacks of *Ruskin (who detested the Bolognese painters in general), and until well into the 20th cent. a just appreciation of his stature was prevented by the failure to distinguish between his own works and those by his countless (often extremely insipid) imitators. The great exhibition devoted to him in Bologna in 1954 was a turning point in his critical fortunes, and his status as one of the greatest Italian painters of the 17th cent. is now firmly re-established. His late works in particular show an ethereal beauty of colouring that sets him apart from any of his contemporaries. The best collection of his work is in the Pinacoteca in Bologna.

Renier of Huy. *Mosan metal-worker active in the early 12th cent. Only one work is documented as being by him, but this is one of the great masterpieces of the period—a bronze font originally made for Notre Dame des Fonts, Liège, and now in St Bartholomew, Liège. This can be fairly accurately dated as it was commissioned by Abbot Hellinus, who was in office 1107–18. It is a large bowl supported on ten (originally twelve) oxen (a reference to the 'sea of cast metal . . . mounted on twelve oxen' made for King Solomon (1 Kings 7: 23–5)) and adorned with scenes appropriate to the sacrament of baptism. The figures are much more naturalistic and classical than in most *Romanesque art. Very little other work can be attributed to Renier or his workshop, but he had great influence on Mosan art.

Renoir, Pierre-Auguste (1841–1919). French *Impressionist painter, born at Limoges. In 1854 he began work as a painter in a porcelain factory in Paris, gaining experience with the light, fresh colours that were to distinguish his Impressionist work and also learning the importance of good craftsmanship. His predilection towards light-hearted themes was also influenced by the great *Rococo masters, whose work he studied in the Louvre. In 1862 he entered the studio of *Gleyre and there formed a lasting friendship with *Monet, *Sisley, and *Bazille. He painted with them in the *Barbizon district and became a leading member of the group of Impressionists who met at the Café Guerbois. His relationship with Monet was particularly close at this time, and their paintings of the beauty spot called La Grenouillère done in 1869 (an example by Renoir is in the Nationalmuseum, Stockholm) are regarded as the classic early statements of the Impressionist style. Like Monet, Renoir endured much hardship early in his career, but he began to achieve success as a portraitist in the late 1870s and was freed from financial worries after the dealer Paul *Durand-Ruel began buying his work regularly in 1881.

By this time Renoir had 'travelled as far as Impressionism could take me', and a visit to Italy in 1881–2 inspired him to seek a greater sense of solidity in his work. The change in attitude is seen in *The Umbrellas* (NG, London), which was evidently begun before the visit to Italy and finished afterwards; the two little girls on the right are painted with the feathery brush-strokes characteristic of his Impressionist manner, but the figures on the left are done in a crisper and drier style, with duller colouring. After a period of experimentation with this 'hard' manner in the mid-1880s he developed a softer and more supple kind of handling. At the same time he turned from contemporary themes to more timeless subjects, particularly nudes, but also pictures of young girls in unspecific settings. As his style became grander and simpler he also took up mythological subjects (*The Judgement of Paris*, Hiroshima Museum of Art, c.1913–14), and the female type he preferred became more mature and ample.

In the 1890s Renoir began to suffer from rheumatism, and from 1903 (by which time he was world-famous) he lived in the warmth of the south of France. The rheumatism eventually crippled him (by 1912 he was confined to a wheelchair), but he continued to paint until the end of his life, and in his last years he also took up sculpture, directing assistants (usually Richard Guino, a pupil of *Maillol) to act as his hands (*Venus Victorious*, Tate, London, 1914).

Renoir is perhaps the best-loved of all the Impressionists, for his subjects—pretty children, flowers, beautiful scenes, above all lovely women—have instant appeal, and he communicated the joy he took in them with great directness. 'Why shouldn't art be pretty?' he said, 'There are enough unpleasant things in the world.' He was one of the great worshippers of the female form, and he said 'I never think I have finished a nude until I think I could pinch it.'

One of Renoir's sons was the celebrated film director **Jean Renoir** (1894–1979), who wrote a noted biography published in both French and English (*Renoir, My Father*) in 1962.

Repin, Ilya (1844–1930). The most celebrated Russian painter of his day. He received his first training from a provincial *icon painter, but later studied at the St Petersburg Academy, and became involved with the *Wanderers. His *Volga Boatmen* (Russian Mus., St Petersburg), exhibited in Vienna in 1873, made him internationally famous, and from then on he was regarded as the leading Russian painter. Repin excelled at scenes from Russian history, such as *Ivan the Terrible with the Body of his Son* (Tretyakov Gal., Moscow, 1885), painted in a colourful, full-blooded (sometimes melodramatic) style. He could also handle modern social themes, however, as in *They Did Not Expect Him* (Tretyakov Gal., 1884), showing the unexpected return of a political exile from Siberia. But it is as a portraitist that he is now most highly regarded. He painted *Tolstoy several times and numerous other distinguished contemporaries. Repin became professor of history painting at the St Petersburg Academy in 1894 and was an influential teacher. After the 1917 Revolution he retired to his country home at Kuokkala (now in Finland), but he continued to be regarded as a figure of massive authority. With the introduction of *Socialist Realism he has been established as the model and inspiration for the Soviet painter.

repoussoir (French: to push back, set off). A figure or object in the foreground of a picture (and usually at the side) used to 'push back', give depth to, and enhance the principal scene or episode.

Repton, Humphry (1752–1818). English landscape gardener, the successor to Lancelot *Brown as the acknowledged leader of his profession. He defended Brown's methods in his *Sketches and Hints on Landscape Gardening* (1795) against the criticisms of Richard Payne *Knight and Uvedale *Price, the intellectual champions of the *picturesque, who demanded a much wilder and more rugged style. Repton's reconstructions were less sweeping than Brown's and latterly he swung back to regular bedding and straight paths close to the house; but his parks were carefully informal, as Brown's were.

Repton thought of himself as an *eclectic, inheriting the best of the modern style in landscape gardening from Brown and combining this with the best of the older style to form a modern renaissance. In his *Observations on the Theory and Practice of Landscape Gardening* (1803)—in which he introduced the phrase to the English language—he wrote: 'I do not profess to follow either *Le Nôtre or Brown, but, selecting beauties from the style of each, to adopt so much of the grandeur of the former as may accord with a palace, and so much of the grace of the latter as may call forth the charms of natural landscape.' Repton also developed an architectural side to his practice, his work consisting mainly of alterations. He was assisted by his sons **John Adey** (1775–1860) and **George** (1786–1858) and about 1795 he joined forces with the architect John Nash (1752–1835), a fruitful partnership for both men which, however, ended in misunderstanding in 1803. Repton was a talented watercolourist and was in the habit of making views with flaps, which the client could lift to make a direct comparison between his park in its 'before' and 'after' state. For his more important commissions he had such illustrations and a report bound in red leather to present to the client, and over seventy of these 'Red Books' are recorded.

reredos. An altarpiece that rises from ground level behind an altar. See also RETABLE.

resin. An adhesive substance, insoluble in water, secreted by many trees and plants and used in art particularly as a constituent of *varnish. The resins used by painters in the past are not always easy to identify but they include both soft resins from living trees, such as *mastic, *dammar, *sandarac, Canada balsam, Venice *turpentine, and hard fossil resins, such as copal and *amber. Synthetic vinyl resins are used mainly for industrial purposes, but have sometimes been used by modern painters.

rest-stick. See MAULSTICK.

retable. An altarpiece that stands on the back of an altar or on a pedestal behind it, rather than rising from ground level. See also REREDOS.

Rethel, Alfred (1816–59). German painter and graphic artist. His biggest work, a cycle of frescos from the life of Charlemagne (Town Hall, Aachen), was once much admired as a great achievement of heroic *history painting but today seems hollow and theatrical. The cycle was begun in 1847 but left unfinished because of the madness that ended his career in 1853. Rethel is now mainly remembered for his series of woodcuts *Another Dance of Death* (1849), much in the spirit of *Holbein's famous depictions of the subject, but satirizing the revolutionary events of 1848, with Death seen as the embodiment of anarchy. Two related woodcuts, *Death as a Strangler* (1847) and *Death as a Friend* (1851), were once extremely popular.

retroussage. Technique of passing a fine muslin cloth over an inked plate, thereby drawing out a little of the ink and spreading it over the edges of the lines. It produces a soft effect in printing.

Rewald, John (1912–). German-born American art historian; a professor at the University of Chicago 1963–71 and thereafter at the City University, New York. He is the doyen of the field of *Impressionist and *Post-Impressionist scholarship, and is famous for two magisterial works, *The History of Impressionism* (1946, 4th edn. 1973) and *Post-Impressionism: From Van Gogh to Gauguin* (1956, 3rd edn. 1978). These show his total command of the voluminous material and remarkable powers of organization and exposition in forming it into a highly readable narrative, and are, by common consent, among the greatest works of art history ever written (even though some critics have accused him of being strong on facts but weak on interpretation); *The History of Impressionism* used to enjoy the 'distinction' of being the most stolen book from the *Courtauld Institute library. Rewald's other writings include studies of many leading 19th-cent. French artists, particularly *Cézanne (he received a gold medal from Cézanne's home town, Aix-en-Provence, in 1984), and two collections of his articles have appeared, *Studies in Impressionism* (1985) and *Studies in Post-Impressionism* (1986).

Rexach, Juan. See BAÇO.

Reymerswaele, Marinus van. See MARINUS VAN REYMERSWAELE.

Reynolds, Sir Joshua (1723–92). English painter and writer on art, the first President of the *Royal Academy, the leading portraitist of his day, and perhaps the most important figure in the history of British painting. He established a tradition of portraiture in the *Grand Manner, expressed most fully the whole range of the English 18th-cent. aesthetic outlook, and did more than any other man to raise the status of the artist to a new level of dignity in England.

Reynolds was born at Plympton in Devonshire, the son of a scholarly clergyman, and he was brought up in an atmosphere of learning. He studied painting in London under *Hudson (another Devonshire man) 1740–3, but set up independently in Devon as a portraitist before his apprenticeship was officially ended. In 1750–2 he was in Italy, where he made an intensive study of the great masters of the 16th and 17th cents. and also of the *antique (it was while copying *Raphael in the cold of the Vatican that he contracted his deafness). He not only absorbed the formal language of his models, but also developed a deliberate cult of learning and classical allusion that coloured his whole approach to art. In tune with established art theory, he thought that *history painting was the highest branch of art, but he believed that portraiture could rise above its traditional status as mere 'face-painting' by improving on the deficiencies of nature and using poses and gestures that allude to the great art of the past. Thus, in the work that established his reputation after he settled in London in 1753—*Commodore Keppel* (National Maritime Mus., Greenwich, 1753–4)—the sitter's heroic pose is based directly (albeit in reverse) on that of the **Apollo Belvedere*, then regarded as the matchless ideal of male beauty.

Reynolds quickly achieved a leading position in his profession. He had 150 sitters a year by 1758, and by 1764 was earning the enormous sum of £6,000 a year. His success was achieved through hard work and careful business management as well as talent; on the day he was knighted (21 April 1769) his visit to St James's Palace was fitted in between two sittings with clients. Moreover, although he always retained traces of his provincial origins (notably his Devonshire accent), he was completely at home with his eminent sitters. His pupil James *Northcote said that 'His general manner, deportment and behaviour were amiable and prepossessing; his disposition was naturally courtly. He . . . contrived to move in a higher sphere of society than any other English artist had done before. Thus he procured for the Professors of the Arts a consequence, dignity and reception which they never possessed in this country.' Reynolds's elevation of the status of the artist depended, then, not only on the intellectual quality of his work, but also on his social acceptability, and it is significant that his friends were mainly men of letters—notably Dr Johnson (1709–84) and Oliver Goldsmith (1730?–74)—rather than other painters. James Boswell dedicated his celebrated *Life of Johnson* to Reynolds. On the foundation of the Royal Academy in 1768, he was the obvious choice for President, and he arranged for Johnson and Goldsmith to be appointed to the honorary positions of Professors of Ancient History and Literature. For the next twenty years, until his blindness stopped him painting in 1790, his authority in the Academy was paramount and his fifteen *Discourses* delivered over that period have become the classic expression of the academic doctrine of the Grand Manner.

As a portraitist Reynolds is remarkable above all for his versatility—his inexhaustible range of response to the individuality of each sitter, man, woman, or child. The celebrated remark of his rival *Gainsborough, 'Damn him! How various he is!', is echoed in the praise of *Ruskin: 'Considered as a painter of individuality in the human form and mind, I think him the prince of portrait painters. *Titian paints nobler pictures and van *Dyck had nobler subjects, but neither of them entered so subtly as Sir Joshua did into the minor varieties of human heart and temper.' In spite of his impressive weight of learning Reynolds could at times be utterly direct, and although his huge output necessitated the employment of assistants and drapery painters, and his experimentation with *bitumen has resulted in some of his pictures being in poor condition, there is much beauty of handling in his work. His finest pictures undoubtedly take their place among the great masterpieces of British portraiture. On the other hand, his history paintings, dating mainly from the end of his career, are generally considered failures. Reynolds's work is in numerous public collections, great and small, in Britain and elsewhere, and many of his finest pictures are still in the possession of the families for which they were painted. His work was so varied that no single collection can be regarded as fully representative.

Frances Reynolds (1729–1807), Joshua's sister and for many years his housekeeper, was an amateur painter, but evidently not a very good one, for he said of her copies of his work: 'They make other people laugh and me cry.' In 1783 she painted a portrait of Dr Johnson, and this has been identified with pictures in Trinity College, Oxford, and the Albright-Knox Art Gallery, Buffalo.

Reynolds-Stephens, Sir William. See NEW SCULPTURE.

Riace bronzes. See PHIDIAS.

Ribalta, Francisco (1565–1628). Spanish painter. He was probably trained at the *Escorial and during most of the 1580s and 1590s he worked in Madrid. His earlier paintings are *Mannerist in character, notably his first known work, *The Nailing to the Cross* (Hermitage, St Petersburg, 1582). By 1599 he was settled in Valencia and there his style became much more sombre and naturalistic. According to *Palomino, Ribalta had studied in Italy and he is known to have made a copy of *Caravaggio's *Martyrdom of St Peter*, but his late *tenebrist style may have been influenced more by *Ribera than by direct knowledge of painting in Rome. In their turn, Ribalta's dramatically lit and powerfully austere mature works (*Christ Embracing St Bernard*, Prado, Madrid) had considerable influence on Spanish painting, notably on *Zurbarán. Ribalta's mature work was more *Baroque in spirit, e.g. *The Vision of Father Simeon* (NG, London, 1612). His son **Juan** (1596/7–1628) was also an able painter in the Caravaggesque manner, but died young.

Ribera, José (or **Jusepe**) **de** (1591–1652). Spanish painter, etcher, and draughtsman, active for all his known career in Italy, where he was called 'Lo Spagnoletto' (the Little Spaniard). Little is known of his life before he settled in Naples (at the time a Spanish possession) in 1616. Naples was then one of the main centres of the Caravaggesque style, and Ribera is often described as one of *Caravaggio's followers. However, although his early work is markedly *tenebrist, it is much more individual than that of most Caravaggesque artists, particularly in his vigorous and scratchy handling of paint. Similarly, his penchant for the typically Caravaggesque theme of bloody martyrdom has been overplayed, enshrined as it is in Byron's lines: 'Spagnoletto tainted/His brush with all the blood of all the sainted' (*Don Juan*, xiii. 71). He undoubtedly painted some powerful pictures of this type, notably the celebrated *Martyrdom of St Bartholomew* (Prado, Madrid, *c.*1630), but he was equally capable of great tenderness, as in *The Adoration of the Shepherds* (Louvre, Paris, 1650), and his work is remarkable for his feeling for individual humanity. Indeed, he laid the foundation of that respect for the dignity of the individual which was so important a feature of Spanish art from *Velázquez to *Goya. This feature of his work is evident also in the secular subjects, such as *The Clubfooted Boy* (Louvre, 1642). He was the first to breach the traditional Spanish dislike for mythological themes (*Apollo and Marsyas*, Musées Royaux, Brussels, 1637), and he broadened the *Baroque repertory by his series of philosophers depicted as beggars or vagabonds

(*Archimedes*, Prado, 1630). Ribera gradually moved away from his early tenebrist style, and his late works are often rich in colour and soft in modelling. He was the leading painter in Naples in his period (Velázquez visited him during his second visit to Italy and probably during his first) and his work was influential in Spain (where much of it was exported) as well as in Italy. His reputation has remained high, and until the Napoleonic Wars (1803–15) he and *Murillo were virtually the only Spanish painters who were widely known outside their native country. Ribera was an outstanding etcher and draughtsman as well as a painter. His work is represented in many major museums, above all in the Prado, Madrid.

Ricci. See RIZI.

Ricci, Sebastiano (1659–1734). Italian decorative painter. He was born at Belluno and is considered a member of the Venetian school, but before he settled in Venice in 1717 he led a peripatetic life, working in numerous Italian cities and also in England, Flanders, France, and Germany. His unsettled existence is a reflection not only of the demand for his talents but also of his penchant for illicit love affairs, which often led to his having to move in haste, and once almost resulted in his execution. In view of this it is not surprising that his work is uneven and sometimes shows signs of carelessness, but he had a gift for vivid, fresh colouring, and his itinerant career was important in spreading knowledge of Italian decorative painting. Little of the decorative work he did in England survives except the *Resurrection* in the apse of the Chelsea Hospital Chapel and some large but damaged canvases on the staircase at Burlington House (now the Royal Academy). He is, however, extremely well represented in the Royal Collection. One very large canvas (over 3 m. × 6 m.) of *The Magdalen Anointing Christ's Feet* hangs at Hampton Court and shows the extent of Sebastiano's debt to *Veronese and to French art.

Marco Ricci (1676–1730), Sebastiano's nephew, was also born in Belluno and travelled extensively. He made two visits to England, and worked there in partnership with his uncle, the collaboration continuing after they returned to Venice in 1717. Examples of their joint works are in the Royal Collection. He was primarily a landscape painter, working in a freely handled style that owed something to *Magnasco.

Riccio, Il (Andrea Briosco) (1470–1532). Italian sculptor, born in Trento, and active in and around Padua, where he was probably trained by Bartolommeo Bellano (*c*.1440–96/7), who in turn is assumed to have been one of *Donatello's assistants. His nickname means 'curly head'. Riccio was a virtuoso bronze-worker and his masterpiece is the great bronze Easter candlestick in the Santo (S. Antonio) at Padua (1507–16), which with its relief scenes of classically draped figures, its satyrs, sphinxes, and decorative conceits, is an endlessly inventive work. He is best known, however, for his small bronze figures, which are done in an *antique manner and greatly appealed to humanist circles in Padua and Venice (he was on intimate terms with leading scholars). They were much imitated, but works from Riccio's own hand are distinguished by a vivacity and delicacy of surface that none of his rivals could match. Examples are in the Victoria and Albert Museum and many other major collections.

Richards, Ceri (1903–71). British painter and maker of *reliefs, born at Dunvant, near Swansea, of a Welsh-speaking family. He was an artist of great versatility, able to absorb many influences without sacrificing his originality. From 1933, under the influence of *Picasso, he worked on a series of relief constructions and *assemblages which were described by John *Rothenstein as 'original creations of a rare order, and unlike anything else done in Britain at the time'. He was influenced by the London *Surrealist Exhibition of 1936, which in his own words 'helped me to be aware of the mystery, even the "unreality", of ordinary things'. Among several examples of his work from this period in the Tate Gallery is *Two Females* (1937–8). After the Second World War his painting drew inspiration from the large exhibition of Picasso and *Matisse at the Victoria and Albert Museum (1945). His love of music showed itself in the many pictures with musical themes done during this time—e.g. *Cold Light. Deep Shadow* (Tate, London, 1950)—culminating in his *Cathédrale Engloutie* series illustrating Debussy's music on this theme. He was also inspired by Dylan Thomas and one of his finest paintings—*'Do not go gentle into that good night'* (Tate, 1956)—is based on his poem of that name. Richards also did work for churches, designed for the stage, and made murals for ships of the Orient Line.

Richardson, Jonathan the Elder (1665–1745). English portrait painter, writer, and collector. John *Riley's most important pupil, he was one of the leading portraitists in the generation after *Kneller's death. This period, however, has been described by Sir Ellis *Waterhouse as 'the most drab in the history of British painting', and Richardson is remembered today more for his writings than his pictures. His most important book is

An Essay on the Theory of Painting (1715), which made claims for the intellectual seriousness of painting and inspired the young *Reynolds. *An Account of Some of the Statues, Bas-Reliefs, Drawings, and Pictures in Italy* (1722), which he wrote in collaboration with his son, **Jonathan the Younger** (1694–1771), also a portraitist, was much used as a guide-book by young Englishmen making the *Grand Tour. Together father and son also wrote *Explanatory Notes and Remarks on Milton's Paradise Lost* (1734). Richardson the Elder made a superb collection of Old Master drawings.

Richier, Germaine (1904–59). French sculptor. She had a traditional training as a carver, working under *Bourdelle from 1925 to 1929, but from about 1940 she began to create a distinctive type of bronze sculpture. Her figures became long and thin, combining human with animal or insect (and sometimes vegetal) forms. The surfaces of these powerful and disquieting works have a tattered and lacerated effect, creating a macabre feeling of decomposition, and she was one of the pioneers of an open form of sculpture in which enclosed space becomes as important and alive as the solid material. Such figures were often extremely difficult to cast and she showed great technical resourcefulness in bringing them to completion. The public sometimes found her work shocking, especially her *Crucified Christ* (church of Nôtre-Dame-de-Toute-Grâce, Assy, 1950), which caused a storm of controversy. Nevertheless, her international prestige grew steadily in the post-war years.

Richier, Ligier (*c.*1500–67). French sculptor, born at St Mihiel in Lorraine and active in the service of the Dukes of Lorraine for most of his life. Richier had been in Italy early in his career, but his best-known work, the tomb of René de Châlons (S. Pierre, Bar-le-Duc, after 1544), shows the *Gothic tradition prevailing over *Renaissance influence. It features a gruesome standing skeleton with shreds of skin attached to its bones. Richier became a convert to Protestantism and died a religious refugee in Geneva.

Richmond, George (1809–96). The best-known member of a family of English painters. He was a pupil of his father, the *miniaturist **Thomas Richmond Sen**. (1771–1837), and also studied at the *Royal Academy, where he became a friend of Samuel *Palmer. With Palmer and others he was one of the group of *Blake's followers known as the *Ancients. His imitation of Blake's mannerisms was heavy-handed and he had nothing of the master's spirit (*The Eve of Separation*, Ashmolean, Oxford, 1830). From about 1830 he turned from poetic and religious themes to portraiture and became a great fashionable success. His brother, **Thomas Richmond Jun.** (1802–74), and his son, **Sir William Blake Richmond** (1842–1921), were also painters. They were principally portraitists, but Sir William was also a sculptor and medallist, and painted ambitious Classical scenes.

Richter, Adrian Ludwig (1803–84). German painter and graphic artist. He worked mainly in his native Dresden (where he was a professor at the Academy), but in 1823–6 he lived in Rome, where he was in close touch with the *Nazarenes. His many paintings and woodcut illustrations—often for children's books—may be called 'romantic' in the popular sense of the word. The pleasures of the countryside, moods of landscape, fairy-tales, and the more idyllic incidents of sacred and legendary art were his favourite subjects. His autobiography, *Lebenserinnerungen eines deutschen Malers* (1885), is not only a charming piece of writing but also an important source for the ideas of German *Romanticism.

Richter, Hans. See DADA.

Ricketts, Charles (1866–1931). British painter, designer, sculptor, collector, and writer on art. In 1882 he began studying wood engraving at Lambeth School of Art, London, and there met fellow student Charles *Shannon, a painter and engraver who became his lifelong companion. Kenneth *Clark writes that 'Ricketts did most of the talking. Shannon was quiet and recessive, but his rare interpolations showed good sense and considerable learning. One could see that Ricketts turned to him as to a reasonable wife.' Ricketts initially made his mark in book production, first as an illustrator, then as the driving force behind the Vale Press (1896–1904), one of the finest private presses of the day, for which he designed founts, initials, borders, and illustrations. After the closure of the Press (following a disastrous fire), Ricketts turned to painting and occasional sculpture, and in 1906 he began to make designs for the theatre. His paintings—typically rather melodramatic, heavy-handed figure subjects—have not worn well, but his colourful stage designs are still much admired. He had a great reputation as an art connoisseur and in 1915 turned down the offer of the directorship of the National Gallery. Later he regretted this decision, but he served on various committees and put much energy into trying to combat modernism in art. Most of the highly varied collection he made with Shannon was bequeathed to the Fitzwilliam Museum in Cambridge, although the gem of the collection—*Piero di Cosimo's *Fight between*

Lapiths and Centaurs—went to the National Gallery. Ricketts's main books were *The Prado and its Masterpieces* (1903), *Titian* (1910), and *Pages on Art* (1913); *Self-Portrait* (taken from his letters and journals) was posthumously published in 1939.

Ridolfi, Carlo (1594–1658). Italian painter, etcher, and art historian. He is insignificant as an artist, and best remembered as the author of *Le Maraviglie dell' arte* (Marvels of the Painter's Art), published in two volumes in 1648. This is a source of great importance for the history of Venetian art, which was somewhat scantily treated by *Vasari. Ridolfi also wrote a life of *Tintoretto (1642).

Riegl, Alois (1858–1905). Austrian art historian, professor of the history of art at Vienna University from 1897 until his death. He is remembered mainly as the originator of the concept of *Kunstwollen* ('will for art' or, as it is more usually translated, 'will to form'). He wished to understand why style changed through the ages, and thought it was inadequate to explain the changes in terms of materials and techniques. Instead he proposed the idea of a dynamic aesthetic impulse, reflecting an innate desire for change, each generation seeing differently from its predecessor. Riegl's main works in which he propounded his ideas are a history of ornament (1893), and books on late Roman industrial arts (1901) and the origins of *Baroque art in Rome (published posthumously in 1907). His *Kunstwollen* concept had considerable influence, which can be seen, for example, in Kenneth *Clark's *The Nude*.

Riemenschneider, Tilman (*c*.1460–1531). German sculptor, active in Würzburg, where he is first recorded in 1483. With *Stoss, he was the outstanding German late *Gothic sculptor, and his workshop was large and productive. He was primarily a wood-carver (he was the first German sculptor to leave the wood unpainted), but he also worked in stone. His style was intricate, but also balanced and harmonious, with none of the extreme emotionalism often seen in German art of the period. He held various offices in city government, and in 1525 he was tortured and briefly imprisoned because he was one of the councilmen who refused to support the use of force against the rebels in the Peasants' War. Much of Riemenschneider's work is still in the churches for which it was carved, but he is also well represented in the Mainfränkisches Museum in Würzburg. Two of his sons, **Jörg** and **Hans**, were sculptors, and two others, **Bartholomäus** (a pupil of *Dürer) and **Tilman**, were painters.

Rietveld, Gerrit Thomas (1888–1964). Dutch designer and architect, a member of the De *Stijl group. An armchair which he made in 1917 was the first piece of furniture constructed according to the group's principles of design, and the Schröder house in Utrecht (1924) was the first realization of their architectural ideas.

Rigaud, Hyacinthe (1659–1743). French portrait painter, the friend and rival of *Largillière. He was born in Perpignan and after working in Montpellier he settled in Paris in 1681. His reputation was established in 1688 with a portrait (now lost) of Monsieur, Louis XIV's brother, and he became the outstanding court painter of the latter part of Louis's reign, retaining his popularity through the Regency (1715–23) and under Louis XV (reigned 1715–74). He was less interested in showing individual character than in depicting the rank and condition of the sitter by nobility of attitude and expressiveness of gesture. These qualities are seen most memorably in his celebrated state portrait of Louis XIV (Louvre, Paris, 1701), one of the classic images of royal majesty. Louis so admired this portrait that, although he had intended it as a present to Philip V (1683–1746) of Spain, he kept it himself. Rigaud's unofficial portraits are much more informal and show a debt to *Rembrandt (*The Artist's Mother*, Louvre, 1695), several of whose works he owned. The output from Rigaud's studio was vast and examples are in many collections.

Rijksmuseum, Amsterdam. The Dutch national art collection. It had its origin in the Royal Museum erected by Louis Bonaparte (Napoleon's brother) as King of Holland in 1808. The idea of the museum was to assemble Netherlandish paintings of national importance and also to stimulate contemporary art. In 1815 the collection was transferred to the Trippenhuis, where it was opened in 1817 as the Rijksmuseum (State Museum). Owing to the inadequacy of the accommodation provided by the Trippenhuis for the growing collection a Commission to Prepare the Establishment of an Art Museum was set up in 1862, but it was not until 1885 that the new Rijksmuseum was opened in a large *Gothic Revival building designed by P. J. H. Cuypers (1827–1921), the leading Dutch architect of the period. In 1922 Schmidt Degener, formerly of the Boymans Museum at Rotterdam, became Director of the Rijksmuseum, modernized the display, and began the acquisition of foreign works of art. The Rijksmuseum has the most comprehensive collection of 17th-cent. Dutch art in the world, and although representation in other areas is uneven, it is strong in certain fields, such as oriental art. Associated with the Rijksmuseum is the Rijksprentenkabinet, which possesses one of the world's finest collections of prints and drawings.

Riley, Bridget (1931–). English painter and designer, the leading British exponent of *Op art. Her interest in optical effects came partly through her study of *Seurat's technique of *pointillism, but when she took up Op art in the early 1960s she worked initially in black and white. She turned to colour in 1966. By this time she had attracted international attention (one of her paintings was used for the cover to the catalogue of the exhibition 'The Responsive Eye' at the Museum of Modern Art, New York, in 1965, the exhibition that gave currency to the term 'Op art'), and the seal was set on her reputation when she won the International Painting Prize at the Venice *Biennale in 1968. Her work shows a complete mastery of the effects characteristic of Op art, particularly subtle variations in size, shape, or placement of serialized units in an all-over pattern. It is often on a large scale and she frequently makes use of assistants for the actual execution. Although her paintings often create effects of vibration and dazzle, her decorative scheme for the interior of the Royal Liverpool Hospital (1983) uses soothing bands of blue, yellow, pink, and white and is reported to have caused a drop in vandalism and graffiti. She has also worked in theatre design, making sets for a ballet called *Colour Moves* (first performed at the Edinburgh Festival in 1983). Unusually, the sets preceded the composition of the music and the choreography. Riley has travelled widely (a visit to Egypt in 1981 was particularly influential on her work, as she was inspired by the colours of ancient Egyptian art) and she has studios in London, Cornwall, and Provence.

Riley, John (1646–91). English portrait painter. His early career is obscure, but he emerged as the most distinguished figure in English portraiture in the interval between the death of *Lely in 1680 and the domination of *Kneller. Although he was appointed Principal Painter to William III and Mary II jointly with Kneller in 1688, his finest works are not court portraits but depictions of sitters from humble callings; the two best known are *The Scullion* (Christ Church, Oxford) and *Bridget Holmes* (Royal Coll., 1686), a full-length portrayal of a nonagenarian royal housemaid who brandishes her broom at a mischievous pageboy. He was generally more successful painting men than women (he was no rival to Lely in depicting fine clothes and soft complexions) and his unassuming sincerity of presentation exemplifies a typically English approach to portraiture that he passed on to his pupil *Richardson.

Rimmer, William (1816–79). English-born American sculptor, painter, teacher, and writer. His family emigrated in the year he was born and eventually settled in Boston, which with New York was his main place of work. Rimmer was an offbeat character and had an eccentric career. He believed he was the heir to the French throne and taught himself medicine, being licensed as a physician in 1855. As an artist too he was self-taught, and although he showed brilliantly precocious talent with his gypsum figure of *Despair* (Mus. of Fine Arts, Boston, *c.*1830), he struggled for recognition, and for years earned his living mainly as a sign and scenery painter and as a cobbler. In his later years, however, he became famous as a teacher, notably for his instructional books, *Elements of Design* (1864) and *Art Anatomy* (1877). Rimmer's output as an artist was very small, but he was the most powerful and original American sculptor of his time; his work excels in dramatic force, and vividly displays his anatomical mastery (*Falling Gladiator*, Mus. of Fine Arts, 1861). As a painter, his best-known work is the nightmarish *Flight and Pursuit* (Mus. of Fine Arts, 1872), which, like his sculptures, shows the freshness and unconventionality of his approach and the richness of his imagination.

Rinehart, William Henry (1825–74). American sculptor. After spending his early career as a stone-cutter in Baltimore, he settled in Rome in 1855, and became one of the leading American expatriate *Neoclassical sculptors of his generation. He did Classical subjects and (in a more naturalistic style) portraits, but his most popular work was perhaps the rather maudlin *Sleeping Children* (1859–60); this was originally made for Greenmount Cemetery, Baltimore, and Rinehart made numerous versions of it. He left his collection of casts and models to trustees, who later transferred them to the Peabody Institute in Baltimore. He also left the bulk of his estate to promote the appreciation of sculpture among Americans and enable American sculptors to study abroad.

Riopelle, Jean-Paul (1923–). Canadian painter, sculptor, and graphic artist. He is considered the leading Canadian abstract painter of his generation, although since 1947 he has lived in Paris. Riopelle's early abstracts were in a lyrical manner, but in the 1950s his work became tauter, denser, and more powerful, often with paint applied with the palette-knife creating a rich mosaic-like effect. International recognition came with pictures such as the huge *triptych *Pavane* (NG of Canada, Ottawa, 1954). Later his handling became more calligraphic. He is prolific, and at home in various media. From the late 1940s he has interchangeably worked in watercolour, ink, oil, crayon, or chalk, made etchings or lithographs or modelled sculpture. He has also produced huge *collage murals.

Ripa, Cesare. See EMBLEM.

Rivera, Diego (1886–1957). Mexican painter, the most celebrated figure in the revival of monumental fresco painting that is Mexico's most distinctive contribution to modern art. He visited Paris in 1909 and after a brief return to Mexico he settled there from 1911 to 1920. During this time he became one of the lions of café society and was friendly with many leading artists. He became familiar with modern movements, but although he made some early experiments with *Cubism, for example, his mature art was firmly rooted in Mexican tradition. At about the time of the Russian Revolution he had become interested in politics and in the role art could play in society. In 1920–1 he visited Italy to study Renaissance frescos (already thinking in terms of a monumental public art), then returned to his homeland, eager to be of service to the Mexican Revolution. In 1920 Alvaro Obregón, an art lover as well as a reformist, had been elected President of Mexico, and Rivera, who was an extremely forceful personality, swiftly emerged as the leading artist in the programme of murals he initiated glorifying the history and people of the country in a spirit of revolutionary fervour. Many examples of his work are in public buildings in Mexico City, and they are often on a huge scale, a tribute to his enormous energy. His most ambitious scheme, in the National Palace, covering the history of Mexico, was begun in 1929; it was still unfinished at his death, but it contains some of his most magnificent work. Rivera's murals were frankly didactic, intended to inspire a sense of nationalist and socialist identity in a still largely illiterate population; their glorification of creative labour or their excoriation of capitalism can be crude, but his best work has astonishing vigour. His skill in choreographing his incident- and figure-packed compositions, in combining traditional and modern subject-matter, and in blending stylized and realistic images is formidable.

In 1927 Rivera visited the Soviet Union and in 1930–4 he worked in the USA, painting several frescos that were influential on the muralists of the *Federal Art Project. His main work in America was a series on *Detroit Industry* (1932–3) in the Detroit Institute of Arts (commissioned by William *Valentiner); another major mural, *Man at the Crossroads* (1933), in the Rockefeller Center, New York, was destroyed before completion because he included a portrait of Lenin. It was replaced by a mural by *Brangwyn. Throughout his career he also painted a wide range of easel pictures, in some of which he experimented with the encaustic (wax) technique. Rivera was an enormous man (standing over 6 feet and weighing over 20 stones), and although he was notoriously ugly he was irresistibly attractive to women. He had numerous love affairs and was married four times, his second wife (and his third, for they divorced and remarried) being the painter Frida *Kahlo.

Rivers, Larry (1923–). American painter, sculptor, graphic artist, and designer, considered one of the leading figures of the movement towards figurative art that succeeded *Abstract Expressionism in the *New York School. He was a professional jazz saxophonist in the early 1940s and began painting in 1945, studying at the Hans *Hofmann School, 1947–8, and at New York University under *Baziotes in 1948. His work of the early and mid-1950s continued the vigorous painterly handling associated with Abstract Expressionism, but was very different in character. Some of his paintings were fairly straightforwardly naturalistic, but others looked forward to *Pop art in their quotations from well-known advertising or artistic sources, their use of lettering, and their deadpan humour. An example is *Washington Crossing the Delaware* (MOMA, New York, 1953), based on the picture by *Leutze. In the late 1950s and 1960s his work came more clearly within the orbit of Pop, sometimes incorporating cut-out cardboard or wooden forms, electric lights, and so on, but his sensuous handling of paint set him apart from other Pop artists. Rivers has also made sculpture, collages, and prints, designed for the stage, acted, and written poetry.

Rizi (or **Ricci**). The name of two Spanish painters, sons of **Antonio Ricci** (d. after 1631), a Bolognese artist who had settled in Spain. The elder, **Juan Andrés** (1600–81), became a Benedictine monk in 1627 and worked mainly for monasteries of his order in Castile in an austere *tenebrist style that at times shows an affinity of spirit with *Zurbarán—indeed he has been called 'the Castilian Zurbarán'. Rizi was also a fine portraitist, with a strong grasp of character. He was a considerable scholar and he produced several treatises on theology, geometry, and art. The one on painting—*De Pintura Sabia* (On learned painting), which he illustrated with his own drawings, was not published until 1930. In 1662 he settled in Italy and was offered a bishopric soon before he died.

Francisco (1614–85) was appointed Painter to the King (Philip IV) in 1656. Renowned for his mastery of *perspective, he was chief designer for the royal theatre at the Buen Retiro Palace, Madrid. His numerous altarpieces are much more colourful than his brother's, at times almost *Rubensian. Paintings by the brothers are rarely seen outside Spain, but there are two attributed to Juan in the Bowes Museum at Barnard Castle.

Robbia, Luca della (1399/1400–82). Florentine sculptor, the most famous member of a family of artists. Nothing is known of his early career, and he was a mature artist by the time of his first documented work—a *Cantoria* (Singing Gallery, 1431–8) for Florence Cathedral, now in the Cathedral Museum. It is a work of considerable originality as well as enormous charm, antedating by a year or two the companion gallery by *Donatello (now also in the Cathedral Museum). Its marble *reliefs of angels and children singing, dancing, and making music reflect *antique prototypes, but conceived in a more cheerful, less heroic spirit than Donatello's figures. In his own time Luca had the reputation of being one of the leaders of the modern (i.e. *Renaissance) style, comparable to Donatello and *Ghiberti in sculpture and *Masaccio in painting, but he is now remembered mainly for his development of coloured, glazed *terracotta as a sculptural medium—in particular for his highly popular invention of the type of the half-length Madonna and Child in white on a blue ground. The family workshop seems to have kept the technical formula a secret and it became the basis of a flourishing business; among the major works by Luca in the medium are the roundels of Apostles (*c*.1444) in *Brunelleschi's Pazzi Chapel in Sta Croce. Luca's business was carried on by his nephew **Andrea** (1435–1525), and later by Andrea's five sons, of whom **Giovanni** (1469–after 1529) was the most important. The famous roundels of infants on the façade of the Foundling Hospital in Florence (1463–6) were probably made by Andrea. His successors tended to sentimentalize Luca's warm humanity, and in course of time the artists' studio became a potters' workshop-industry.

Robert, Hubert (1733–1808). French landscape painter. From 1754 to 1765 he was in Italy (mainly Rome), and in 1761 he travelled to south Italy and Sicily with *Fragonard. He made a vast quantity of drawings in Italy, on which he based his pictures after his return to Paris. His particular interest was in ruins and he was the first to make them the main theme of a picture rather than to use them as *Picturesque accessories. He romanticized the vision of *Panini and *Piranesi, whom he knew in Rome, and often set his ruins in idealized surroundings (although he also painted topographical views). His work was highly successful, satisfying the vogue for rather artificial, idealized landscape that was one aspect of *Rococo taste. Under Louis XVI (1754–93) he became Keeper of the King's Pictures and one of the first curators of the *Louvre, but he was imprisoned during the Revolution. He owed his life to an accident whereby another person of the same name was guillotined in his stead. Robert's work is well represented in the Louvre.

Roberti, Ercole de' (*c*.1450–96). Italian painter, active mainly in Ferrara. He succeeded *Tura as court painter to the *Este in 1486, but little is known of his life. Earlier he appears to have assisted *Cossa for some years, and with Cossa and Tura he ranks as the leading artist of the 15th-cent. Ferrarese school. The only picture reasonably certainly his is the altarpiece with a *Madonna Enthroned with Saints* (1480) painted for Sta Maria in Porto at Ravenna and now in the Brera, Milan. Other works, however, can be confidently given to him because of his distinctive style. He inherited the tradition of Tura and Cossa with their precise line and metallic colours against elaborately fanciful ornamentation, but he developed this manner with great originality, modifying it with a subtlety of handling that seems to derive from Giovanni *Bellini. His work is often remarkable for its almost mystical intensity of feeling, as in his **Pietà* in the Walker Art Gallery, Liverpool.

Roberti has sometimes been confused with Ercole di Giulio Cesare de' Grandi, a painter who is said to have died in 1531 but by whom no pictures are known and whose existence is doubted by some scholars.

Roberts, David (1796–1864). Scottish painter. He was apprenticed to a house painter, then worked as a scene painter for a travelling circus and Glasgow and Edinburgh theatres. In 1822 he settled in London and worked at the Drury Lane Theatre with his friend Clarkson *Stanfield. From 1831 he travelled widely in Europe and the Mediterranean Basin and made a fortune with his topographical views. He worked in oil and watercolour and published lavishly illustrated books, among them the six-volume *The Holy Land, Syria, Idumea, Arabia, Egypt & Nubia* (1842–9). His work can be monotonous when seen *en masse*, but at his best he combines bold design with precise observation.

Roberts, Tom (1856–1931). Australian painter. He was born in England and first went to Australia as a child in 1869. In 1881 he returned to England to study at the *Royal Academy, where he came under the influence of the art of *Bastien-Lepage. In 1883 during a walking tour of Spain he acquired some knowledge of *Impressionism, and when he returned to Melbourne in 1885 he gathered several other painters about him and founded what came to be known as the *Heidelberg School. In 1903 he left for England, and did not return permanently to Australia until 1923. Apart from landscapes, Roberts also painted portraits and genre scenes, particularly of Australian rural life. He was largely responsible for introducing Impressionism to Australia and his work is regarded as beginning the growth of an indigenous school of Australian art. Examples are in many Australian galleries.

Roberts, William (1895–1980). British painter, chiefly of figure compositions and portraits. After travelling in France and Italy, he worked briefly for the *Omega Workshops, then in 1914 joined the *Vorticist movement. His style at this time showed his precocious response to French modernism and was close to that of *Bomberg in the way he depicted stiff, stylized figures through geometrically simplified forms. After the First World War (in which he served in the Royal Artillery and as an *Official War Artist) his forms became rounder and fuller in a manner reminiscent of the 'tubism' of *Léger. Often his paintings showed groups of figures in everyday settings, his most famous work being *The Vorticists at the Restaurant de la Tour Eiffel: Spring* 1915, Tate Gallery, London, 1961–2, an imaginative reconstruction of his former colleagues celebrating at a favourite rendezvous to mark the publication of the first issue of *Blast*. In response to the exhibition 'Wyndham *Lewis and the Vorticists' at the Tate Gallery in 1956, Roberts wrote a series of pamphlets (1956–8) disputing Lewis's claim (in the catalogue introduction) that 'Vorticism, in fact, was what I, personally, did, and said, at a certain period.' However, Roberts later altered his views.

Robusti, Jacopo. See TINTORETTO.

rocaille. Term applied from the mid-16th cent. onwards to fancy rock-work and shell-work for fountains and grottoes, and later to ornament based on such forms. From about 1730 it began to acquire a wider connotation, being applied to the bolder and more extravagant flights of the *Rococo style. Indeed, it preceded the word 'Rococo' itself as an indication of style and the two terms have sometimes been used synonymously by French art historians.

rocker. The tool used to prepare the surface of the plate in *mezzotint.

Rockwell, Norman (1894–1978). American illustrator and painter. He left school at 16 to study at the *Art Students League and by the time he was 18 was a full-time professional illustrator. In 1916 he had a cover accepted by the *Saturday Evening Post*, the biggest-selling weekly publication in the USA (its circulation was then about 3,000,000), and hundreds of others followed for this magazine until it ceased publication in 1969. He also worked for many other publications. Rockwell's subjects were drawn from everyday American life and his style was anecdotal, sentimental, and lovingly detailed; he described his pictorial territory as 'this best-possible-world, Santa-down-the-chimney, lovely-kids-adoring-their-kindly-grandpa sort of thing'. Such work brought him immense popularity, making him something of a national institution. For most of his career critics dismissed his work as corny, but he began to receive serious attention as a painter late in his career. In his later years, too, he sometimes turned to more serious subjects, producing a series on racism for *Look* magazine, for example. From 1953 until his death he lived at Stockbridge, Massachusetts, where there is a museum devoted to him.

Rocky Mountain School. Term applied to 19th-cent. American artists who painted the Rocky Mountains in a spirit similar to that adopted by the *Hudson River School. Alfred *Bierstadt is the best-known artist of the school.

Rococo. Style of art and architecture, characterized by lightness, grace, playfulness, and intimacy, that emerged in France *c.*1700 and spread throughout Europe in the 18th cent. By extension the term is often used simply as a period label—'the age of Rococo'. The word was apparently coined in 1796–7 by one of *David's students, wittily combining **rocaille* and *barocco* (*Baroque), to refer disparagingly to the taste fashionable under Louis XV. Thus, like so many stylistic labels, it began life as a term of abuse, and it long retained its original connotations, implying an art that was, in the words of one of the definitions given to it in the *Oxford English Dictionary*, 'excessively or tastelessly florid or ornate'. The word was used as a formal term of art history from the middle of the 19th cent. in Germany, where those followers of *Burckhardt who tried to establish a rhythmic periodicity for the phases of artistic development applied it to the closing and therefore decadent period of any phase. With more recent changes in aesthetic taste the term has become respectable, and is now used by art historians generally in an objective sense, without implied belittlement, to designate an artistic and decorative style which has a certain coherence and consistency. The concern for colourfully fragile decoration, for trivial instead of significant subjects, for pastoral poetry in art gave it a readily identifiable character. Fiske Kimball, its chief historian (*The Creation of the Rococo*, 1943), has called it 'an art essentially French in its grace, its gaiety and its gentleness'. It was both a development from and a reaction against the weightier Baroque style, and was initially expressed mainly in decoration. It shared with the Baroque a love of complexity of form, but instead of a concern for mass, there was a delicate play on the surface, and sombre colours and heavy gilding were replaced with light pinks, blues, and greens, with white also often being prominent. Elegance and convenience rather than grandeur were the qualities demanded by a society tired of the excesses of Versailles. The early masters of the style were engravers such as *Audran and *Bérain, and *Watteau is generally regarded as the first great Rococo painter. *Boucher and *Fragonard are the painters who most completely represent the spirit of the mature Rococo, and *Falconet is perhaps the quintessential sculptor in the style. It is appropriate that many of his works were reproduced in porcelain, for this art form is much more representative of the age than the heroic statue.

From Paris the Rococo was disseminated by French artists working abroad and by engraved publications of French designs. It spread to Germany, Austria, Russia, Spain, and northern Italy (*Tiepolo, *Longhi, *Guardi). In England it had somewhat less of a vogue, although a substantial exhibition of English Rococo art was held at the Victoria and Albert Museum in 1984 and there are clear reflections of the style even in the work of so xenophobic an artist as *Hogarth. In each country the style took on a national character and in addition many local variants may be distinguished. Outside France, it had its finest flowering in Germany and Austria, where it merged with a still vital Baroque tradition. In churches such as Vierzehnheiligen (1743–72) by Balthasar Neumann, the Baroque qualities of spatial variety and of architecture, sculpture, and painting working together are taken up in a breathtakingly light and exuberant manner. The Rococo flourished in central Europe until the end of the century, but in France and elsewhere the tide of taste had begun to turn from frivolity and lightheartedness towards the sternness of *Neoclassicism by the 1760s. The fate of Fragonard is instructive. His four canvases on *The Pursuit of Love* (Frick Coll., New York), now regarded as his masterpieces, were returned to him in 1773 by Madame du Barry (1743–93), one of Louis XV's mistresses, his style evidently already considered out of date in court circles, and by the end of his career he was virtually forgotten.

Rodchenko, Alexander (1891–1956). Russian painter, sculptor, industrial designer, and photographer, one of the leading *Constructivists. His output was prolific and his artistic evolution was rapid, as he moved from *Impressionistic pictures in 1913 to pure abstracts, made with a ruler and compass, in 1916. He was influenced by *Malevich's *Suprematism, his *Black on Black* (Met. Mus., New York, 1918) being a response to Malevich's *White on White* paintings. Rodchenko, however, was without Malevich's mystical leanings, and he coined the term *'Non-objective' to describe his own more scientific approach. In 1917 he began making three-dimensional constructions under the influence of *Tatlin, and some of these developed into graceful hanging sculptures. Like Tatlin and other Constructivists, however, Rodchenko came to reject pure art as a parasitical activity, and after 1922 he devoted his energies to industrial design, typography, film and stage design, propaganda posters, and photography.

It was perhaps in photography that Rodchenko's originality was most evident. In sharp reaction from his abstract work, his photography was geared towards reportage and a pictorial record of the new Russia. But much of this work was outstanding for its exploitation of unusual angles and viewpoints—'Rodchenko perspective' and 'Rodchenko foreshortening' became current terms in the 1920s—and his innovative use of light and shadow influenced, for example, the great film director Sergei Eisenstein (1898–1948). In the mid-1930s Rodchenko returned to easel painting, and in the early 1940s he produced a series of abstract canvases in an expressionist vein.

Rodin, Auguste (1840–1917). French sculptor and graphic artist, one of the greatest and most influential European artists of his period. He was the first sculptor since the heyday of *Neoclassicism to occupy a central position in public attention and he opened up new possibilities for his art in a manner comparable to that of his great contemporaries in painting—*Cézanne, *Gauguin, and van *Gogh. His beginnings, however, were not auspicious. He came from a poor background, was rejected by the École des *Beaux-Arts three times and for many years worked as an ornamental mason. In 1875 he went to Italy, where (as he later wrote to *Bourdelle) '*Michelangelo freed me from academism'. Michelangelo was the inspiration for his first major work, *The Age of Bronze*, which was exhibited in 1878. (Like many of Rodin's statues, this exists in several casts; the Rodin Museum in Paris has examples of virtually all his work. There is also a Rodin museum in Philadelphia.) It caused a sensation because the *naturalistic treatment of the naked figure was so different from the idealizing conventions then current that he was accused of having cast it from a live model.

Two years later, in 1880, his reputation now established, Rodin was commissioned by the state to make a bronze door for a proposed Musée des Arts Décoratifs. Rodin never finished the huge work—*The Gates of Hell*—in a definitive form (he worked on it intermittently until 1900 and the museum never came into being in its proposed form), but he poured some of his finest creative energy into it, and many of the nearly 200 figures that are part of it formed the basis of famous independent sculptures, most notably *The Thinker*. The several casts of the complete structure that exist were made after Rodin's death. The overall design is a kind of *Romantic reworking of *Ghiberti's *Gates of Paradise* for the Florence Baptistery, the twisted and anguished figures, irregularly arranged, reminiscent of Michelangelo's *Last Judgement* and Gustav *Doré's illustrations for the *Divine Comedy*. His modelling is often rough and 'unfinished' and anatomical forms are exaggerated or simplified in the cause of intensity of expression.

These traits were taken further in some of Rodin's monuments, for example the famous group of *The Burghers of Calais* (1885–95), commissioned by the city of Calais for a site in front of the town hall (another cast is in Victoria Tower Gardens, London). In the figures of the six hostages who face the threat of death Rodin created a profound image of a variety of responses to an extreme emotional crisis. The civic authorities had wanted something in a more traditional heroic-patriotic vein, and the monument was eventually unveiled only after years of wrangling. Even worse hostility was aroused a few years later by his statue of Balzac. This was commissioned by the Société des Gens de Lettres in 1891, but Rodin's design was so radical—an expression of the elemental power of genius rather than a portrait of an individual—that it was rejected. The monument, which ranks as the most original piece of public statuary created in the 19th cent., and which Rodin said was 'the sum of my whole life', was not finally cast and set up—at the intersection of the Boulevards Raspail and Montparnasse—until 1939. In spite of this kind of controversy, which often attended his work, by 1900 Rodin was widely regarded as the greatest living sculptor, and in that year a pavilion was devoted to his work at the Paris World Fair. Apart from his monuments, he did a large number of portraits of eminent personalities and he was a prolific graphic artist, some of his later work especially being notably erotic.

Although the literary and symbolic significance he attached to his work has been out of keeping with the conception of 'pure' sculpture that has predominated in the 20th cent., Rodin's influence on the development of modern art has been immense, for singlehandedly he rescued sculpture from a period of stagnation and made it once again a vehicle for intense personal expression.

Roelas, Juan de Las (1558/60–1625). Spanish painter, the leading painter of his period in Seville. He was a priest and virtually all his work was done for churches and religious houses in and around Seville. Roelas has been called 'the Spanish *Tintoretto' and 'the Spanish *Veronese', and the painterly richness of his large multi-figure compositions suggests he had studied in Italy. His work, however, has a religious fervour that is typically Spanish rather than Italian, and his blending of mysticism with *naturalism was deeply influential in Seville. Most of his work remains there; his masterpiece is perhaps the huge *Martyrdom of St Andrew* (Seville Mus., 1609).

Roerich, Nikolai (1874–1947). Russian painter, designer, archaeologist, anthropologist, and mystical philosopher. He was a prolific painter of landscapes and of imaginary historical scenes that evoke a colourful pagan image of Russia's past. They reveal the same feeling for exotic splendour and bold, sumptuous colour that he displayed in his set and costume designs for *Diaghilev's *Ballets Russes*, notably for Stravinsky's *Rite of Spring* (1913), for which Roerich created the scenario with the composer. A man of immense energy, Roerich combined his career as an artist with one as an archaeologist and anthropologist. In 1925–8 he made a 16,000-mile expedition in Central Asia; his 'investigation of the cultures of the region [is] still the bedrock of anthropological studies of Central Asia' (*The Times Atlas of World Exploration*, 1991). From 1928 until his death he directed a Himalayan research station at Kulu in India, and many of his later paintings feature mountain landscapes. He had a deep interest in esoteric religions and the mysteries of nature, and he developed a philosophy in which art should unite humanity. There are Roerich museums in Moscow and New York (he lived in the USA, 1920–3), but the best collection of his work is in the Russian Museum, St Petersburg.

Roger of Helmarshausen. See THEOPHILUS.

Rogers, Claude (1907–79). British painter. He exhibited with the *London Group in 1931 and became a member in 1938. With *Coldstream and Victor *Pasmore he was a founding member of the *Euston Road School in 1937 and he became one of the main upholders of its sober figurative tradition, although in his later work the underlying abstract quality of the composition became of more importance. He was a distinguished teacher (he lectured at the *Slade School from 1948 to 1963 and was Professor of Fine Art in the University of Reading, 1963–72) and administrator.

Rogers, John (1829–1904). American sculptor. Trained as a machinist and self-taught as an artist, he became highly successful with his small sentimental, plaster *genre groups—characteristic titles are *Checkers up at the Farm* (1875) and *Weighing the Baby* (1876). He ran what amounted to a factory, using a mould of his own design and sometimes turning out thousands of copies, which could be bought in general stores or by mail order. His subjects also covered topical events, so his parlour ornaments were a kind of three-dimensional equivalent of *Currier & Ives prints. Rogers also did a few portrait busts and more ambitious pieces, notably the equestrian statue of *Major-General Reynolds* (1883) in front of the City Hall in Philadelphia.

Rogers, William (active *c.*1589–1605). The first English engraver of note, best known for his allegorical portraits of Queen Elizabeth I. He also engraved title-pages (e.g. Camden's *Britannia*, 1600, and Gerard's *Herbal*, 1597).

Rohlfs, Christian (1849–1938). German painter and graphic artist. Until he was over 50 he worked in a fairly traditional *naturalistic manner, but he then discovered the work of the *Post-Impressionists, in particular van *Gogh, whose brilliant colour and intense feeling were a revelation to him, and he became one of the pioneers of *Expressionism in Germany. His favourite themes were visionary views of old German towns, colourful landscapes, and flower pieces. He received considerable acclaim for work in his new style, but in 1937 he was declared a *degenerate artist by the Nazis and forbidden to paint.

Roldán, Pedro (1624–99). Spanish *Baroque sculptor. He was a fellow student of Pedro de *Mena in Granada under the latter's father, Alonso de Mena. By 1656 he had settled at Seville, where he became the leading sculptor of his period and where from 1664 to 1672 he was director of sculpture at the Academy. His greatest work is the spectacular *reredos for the high altar of the Church of the Hospital de la Caridad (1670–5), which was *polychromed by *Valdés Leal. Roldán's daughter and pupil, **Luisa** (*c.*1656–*c.*1704), was also a sculptor, principally active at Cadiz and Madrid. She was the only woman to hold the position of royal sculptor (to Charles II of Spain). Pedro *Duque Cornejo was Pedro Roldán's grandson.

Rolfsen, Alf (1895–1979). Norwegian painter, notable for his murals, owing something to *Cubism, which decorate many public buildings in Oslo and elsewhere. He was also a book illustrator.

Romanelli, Giovanni Francesco (*c.*1610–62). Italian painter and tapestry designer. He was Pietro da *Cortona's outstanding pupil, and like his master a protégé of the *Barberini family. Romanelli's graceful style was less energetic than Cortona's (he owed much to his first teacher *Domenichino) and his restrained type of Baroque proved particularly popular and influential in France, where he worked 1645–7 and 1655–7. He introduced to Paris Cortona's characteristic manner of decoration, consisting of paintings combined with richly gilded *stuccowork, and this was one of the sources for the great schemes of *Lebrun at Versailles and elsewhere. Examples of Romanelli's work survive in the Bibliothèque Nationale (painted for Cardinal Mazarin) and (much altered) in the Salle des Saisons of the Louvre (painted for Anne of Austria, mother of Louis XIV).

Romanesque. Style of art and architecture prevailing throughout most of Europe in the 11th and 12th cents., the first style to achieve such international currency. The dominant art of the Middle Ages was architecture, and 'Romanesque', like '*Gothic', is primarily an architectural term that has been extended to the other arts of the period. It first came into use in the early 19th cent. as a descriptive term (analogous to 'Romance' in the classification of languages) for all the derivatives of Roman architecture that developed in the West between the collapse of the Roman Empire and the beginning of Gothic (*c.* AD 500–1200). But as knowledge of the period grew and its complexity was appreciated historians found it expedient to distinguish the major style of the 11th and 12th cents., characterized by a new massiveness of scale, from earlier minor styles and to confine the term 'Romanesque' to it. Usage of the term Romanesque in the wider sense is still current, however; thus the phrase 'English Romanesque architecture' can be used to encompass both Anglo-Saxon and Norman styles, as in the titles of Sir Alfred Clapham's standard books *English Romanesque Architecture before the Conquest* (1930) and *English Romanesque Architecture after the Conquest* (1934). The architecture of the renaissance initiated by Charlemagne (742?–814) is often referred to as 'Carolingian Romanesque', but it is usually treated as a distinct stylistic unit—part of *Carolingian art.

Romanesque painting and sculpture are generally strongly stylized, with little of the *naturalism and humanistic warmth of *classical or later

Gothic art. The forms of nature are freely translated into linear and sculptural designs which are sometimes majestically calm and severe and at others are agitated by a visionary excitement that can become almost delirious. Because of its expressionistic distortion of natural form, Romanesque art, as with other great non-naturalistic styles of the past, has had to wait for the revolution in sensibility brought about by the development of modern art in order to be widely appreciated.

Romano, Giulio. See GIULIO.

Romantic Classicism. See NEOCLASSICISM and ROMANTICISM.

Romanticism. Movement in the arts flourishing in the late 18th and early 19th cents. Romanticism is so varied in its manifestations that a single definition is impossible, but its keynote was a belief in the value of individual experience. In this it marked a reaction from the rationalism of the Enlightenment and the order of the *Neoclassical style. The Romantic artist explored the values of intuition and instinct, exchanging the public discourse of Neoclassicism, the forms of which had a common currency, for a more private kind of expression. The German philosopher G. W. F. Hegel in his *Lectures on Aesthetics* (1818) wrote 'in Romantic art the form is determined by the inner idea of the content of substance that this art is called upon to represent'.

Romanticism is commonly seen as the antithesis of *classicism, and the two concepts are sometimes used in a very general sense to designate polarities in attitude that may be seen in the art of any age—thus *Raphael might be described as a 'classical' artist, whereas his contemporary *Giorgione is a 'romantic' one. However, the exponents of both Romanticism and classicism share a concern with the *ideal rather than the real, and that there is sometimes no firm dividing line between the two approaches is shown by the use of the term 'Romantic Classicism' to describe certain works that show a Romantic response to antiquity. Both Romanticism and classicism embrace concepts of nobility, grandeur, virtue, and superiority. But where the classical seems a possible ideal which will adapt man to his society and mould that society into an orderly setting for him, the Romantic envisages the unattainable, beyond the limits of society and human adaptability. The classical hero accepts the fate over which he has no control and triumphs nobly in this acquiescence, otherwise he would not be a hero. The Romantic hero pits himself against a hostile environment and at no time comes to terms with it even if he reaches his goal, otherwise he would not be Romantic.

Romanticism represents an attitude of mind rather than a set of particular stylistic traits and involves the expression of an idea that tends to have a verbal rather than a visual origin. In this context a ruined temple is more significant than a new one because it is more suggestive of the passage of time and human frailty. What is broken or partial can never be archetypally classical because the classical object is whole and coherent, not fragmentary. A view that finds a ruined temple beautiful is a Romantic view, though the temple may once have been a classical masterpiece. Romanticism lends itself more easily to expression through music and literature than through the visual arts, as a sense of the infinite and the transcendental, of forces exceeding the boundaries of reason, must necessarily be vague—suggestive rather than concrete, as it must be in painting and even more so in sculpture. On the other hand, although there is no specific Romantic school in architecture, the *Gothic Revival, especially in its early, non-scholarly phase, is an aspect of Romanticism.

Almost by definition, the leading Romantic artists differ widely from one another—*Blake and *Turner in Britain, *Delacroix and *Géricault in France, *Friedrich and *Runge in Germany. The movement of which they were a part died out in the mid-19th cent., but in a broader sense the Romantic spirit has lived on, representing a revolt against conservatism, moderation, and insincerity and an insistence on the primacy of the imagination in artistic expression.

Rombouts, Theodoor (1597?–1637). Flemish painter, mainly of religious and *genre scenes. He was a pupil of *Janssens in his native Antwerp, then from about 1616 to about 1625 he was in Italy. There his work became strongly *Caravaggesque and he established himself as one of the leading Flemish exponents of the style. Later, he fell under the all-pervasive influence of *Rubens and his work became much lighter in tonality.

Romney, George (1734–1802). English painter, mainly of portraits. Born in Lancashire, the son of a builder and cabinet-maker, he worked in the north of England until 1762, when he settled in London. There he became the most successful portraitist of the day apart from *Reynolds and *Gainsborough. His posthumous reputation was once almost the equal of theirs, but has faded greatly. Much of his work is now considered facile and he was probably at his best with portraits of young people, when his delicate colour sense and graceful line were used to good effect. As with many successful portraitists, his heart lay elsewhere and he had aspirations to be a history painter. His visit to Italy in 1773–5 made a lasting effect on him, but his plans for grandiose literary and historical works rarely advanced beyond his sepia drawings (a large collection is in the Fitzwilliam Museum, Cambridge), although he painted for *Boydell's Shakespeare Gallery. Introspective and nervous by temperament, Romney was attracted to literary circles (the poets William Cowper, 1731–1800, and William Hayley, 1745–1820, were among his friends) and associated little with his fellow artists, never exhibiting at the *Royal Academy. His friends among artists tended to be others of literary temperament, such as *Flaxman. In about 1781 Romney became infatuated with Emma Hart, later Lady Hamilton, Nelson's mistress, and he painted her many times in various guises. He retired to Kendal in 1798 and died insane. Romney was a fast and prolific worker, and his paintings are in many collections in Britain and the USA (he was extremely popular with American collectors in the early 20th cent., when his works fetched huge sums).

Ronald, William (William Ronald Smith) (1926–). Canadian-American painter and radio and television presenter, a leading figure in the development and acceptance of abstract art in Canada. He was one of the founders of the Toronto group of abstract artists *Painters Eleven, but in 1955 he moved to New York, where he enjoyed considerable success among the second generation of *Abstract Expressionists. He lived in the USA until 1965, becoming an American citizen in 1963, but he retained a large following in Toronto and exerted a strong influence on painters there. With the decline of Abstract Expressionism his popularity waned, but after his return to Canada he achieved success as a radio and television personality, presenting chat shows and programmes on art and current affairs. He also developed a kind of road-show performance in which, dressed in immaculate white, he painted on stage before an audience to the accompaniment of rock music.

Roncalli, Cristoforo (1552–1626). Italian painter also called Il Pomarancio (after his birthplace near Volterra). With *Cesari he was one of the leading fresco decorators of his time in Rome. Although he adapted somewhat to the innovations of the *Carracci he remained essentially in the *Mannerist tradition and his work is generally undistinguished. His unbiased, common-sense views on art, however, known from a lecture he gave to the Academy of St Luke in 1594, commended him to Vicenzo *Giustiniani, one of the most enlightened patrons of his time. Roncalli became his artistic adviser and accompanied him on a journey to Germany, Holland, England, and France in 1606.

rood-screen. Term in Christian church architecture for a screen separating the chancel (reserved to the clergy) from the nave (assigned to the laity). It is named after the rood (Old English for 'cross' or 'crucifix') that usually surmounted it, although there is in fact no necessary connection between the rood (which originally stood on a beam) and the screen. A great many screens survive from the Middle Ages, not least in England, where there are some superb examples, particularly of the 15th cent. They represent one of the great flowerings of the art of the medieval carver and in East Anglia they often incorporate painted panels (notably at Ranworth in Norfolk). The screen was often surmounted by a loft that housed the choir and sometimes a small organ. The roods themselves, however, were so thoroughly 'extincted and destroyed' (in the words of an act of 1548) during the Reformation that only a few fragments remain. When the *Renaissance style came to dominate church architecture, an unbroken view from the nave into the chancel became the norm, and the rood-screen became virtually obsolete during the 18th cent.

roof boss. See BOSS.

Rooker, Michael Angelo (1746?–1801). English painter, scene designer, and book illustrator. He studied engraving with his father **Edward** (1724?–1774), who specialized in architectural subjects, and painting with Paul *Sandby. From about 1788 he made regular sketching tours in the southern and midland counties and he was one of the most assiduous watercolour topographers of his time. In 1779 he was appointed scene painter at the Haymarket Theatre in London. Examples of his watercolours are in many public collections.

Rops, Félicien (1833–98). Belgian graphic artist and painter, active mainly in Paris. Rops was primarily a printmaker—one of the most brilliant and technically resourceful of his period—and his work is highly distinctive because of his vividly licentious imagination, which took delight in the morbid, perverse, and erotic. Much of his work was done as illustrations for books or for his own satirical journal *Uylenspiegel*. He was a member of the avant-garde Brussels group Les *Vingt and his work was notably influential on *Ensor.

Rosa, Salvator (1615–73). Neapolitan painter and etcher. He was a flamboyant character—a poet, actor, and musician as well as an artist. Most of his career was spent in Rome, but in the 1640s he lived mainly in Florence (where he worked for the *Medici) after he had been rash enough to satirize the great *Bernini. His colourful personality and unswerving belief in his own genius made him a prototype of the *Romantic artist and his fame was greatest in the 18th and 19th cents. (the story that he was a bandit seems to be a 19th-cent. invention). He was a prolific artist and painted various subjects (including spirited battle-pieces in which he surpassed his teacher, *Falcone), but he is best known for the creation of a new type of wild and savage landscape. His craggy cliffs, jagged, moss-laden trees, and rough bravura handling create a dank and desolate air that contrasts sharply with the serenity of *Claude or the classical grandeur of *Poussin (a situation summed up in the famous lines from James Thomson's *The Castle of Indolence* (1748): 'Whate'er Lorraine light-touched with softening hue,/Or savage Rosa dashed, or learned Poussin drew'). Rosa is also well known for his macabre subjects (notably of witches), but he himself set most store by his large historical and religious compositions, which are now considered his least attractive works. In the 1660s he turned with great success to etching. Rosa was highly influential on the development of the *Picturesque and the *Sublime, and he had a great vogue in England, where *Mortimer was particularly taken with his pictures of bandits. His popularity is reflected in the large number of his paintings still to be seen in British collections. *Ruskin, however, was responsible for the fall of his reputation, condemning his landscapes as artificial.

Rose + Croix, Salon de la. See SALON DE LA ROSE + CROIX.

Rosenberg, Harold (1906–78). American writer, one of the most influential critics in the field of contemporary art from the 1950s until his death. Early in his career he wrote poetry and essays on literary and general cultural issues, and his first important work devoted to the visual arts was an article in *Art News* in 1952 in which he coined the term *Action painting. He was one of the major champions of *Abstract Expressionism, writing monographs on *Gorky (1962), *de Kooning (1974), and *Newman (1978). Unlike his rival Clement *Greenberg, who was concerned only with formal values, Rosenberg had an ethical and political conception of art, believing that the critic should less 'judge it' than 'locate it', subordinating visual analysis to intellectual understanding. He thought that authentic modern art should be perpetually disruptive and he attacked the manipulative fashions created by both the market-place and the museum. *Pop art, for example, he treated with disdain.

Rosenquist, James (1933–). American *Pop artist. During the 1950s he earned his living as an industrial billboard painter and from the late 1950s he began to use techniques derived from commercial painting in his own work. He has often worked on a very large scale, combining fragmented incongruous images blown up in a manner reminiscent of photographic enlargement and given a suave finish.

Roslin, Alexander (1718–93). Swedish portrait painter, active mainly in France. He left his country in 1745, worked at the courts of Bayreuth (1745–7) and Parma (1751–2), and finally settled in Paris in 1752. There he rapidly became one of the leading portraitists of the day, esteemed particularly for his skilful rendering of expensive fabrics and delicate complexions ('Satin, skin? Go to Roslin'). He visited St Petersburg, Vienna, and Warsaw (as well as Stockholm) in the 1770s, but in spite of his international travels his elegant work was entirely French in style. His wife, **Marie-Suzanne Giroust** (1734–72), was a pastellist. One of Roslin's finest works is a portrait of her entitled *The Lady in a Black Veil* (Nationalmuseum, Stockholm, 1768).

Rosselli, Cosimo (1439–1507). Florentine painter. His successful career (the highpoint of which was painting frescos in the Sistine Chapel along with *Botticelli, *Ghirlandaio, and *Perugino) was based on his facility and high standards of craftsmanship rather than on any great distinction or originality as an artist. He ran a busy studio in Florence, his pupils including Fra *Bartolommeo and *Piero di Cosimo.

Rossellino, Bernardo (1409–64) and **Antonio** (1427–79). Florentine sculptors, brothers. Bernardo worked as an architect as well as a sculptor and he combined both arts in his chief work—the tomb of the great humanist and Chancellor of the Florentine Republic, Leonardo Bruni, in Sta Croce, Florence (1444–50). It is based on the monument of the antipope John XXIII (Baldassare Cossa) by *Donatello and *Michelozzo in the Baptistery in Florence, and although less powerful is more graceful and harmonious; the pilasters framing the serene effigy, lying on a bier, have a dignity and elegance almost worthy of *Brunelleschi. It became the model for the niche tomb for the rest of the century.

Antonio was trained by his brother, and his most ambitious work—the tomb of the Cardinal Prince of Portugal in S. Miniato al Monte, Florence (1461–6)—is based on Bernardo's Bruni tomb. It is more elaborate and concerned with movement than Bernardo's masterpiece, but also a less coherent design, and Antonio was a more distinguished artist when working on a smaller scale. He was a fine portraitist (*Giovanni Chellini*, V&A, London, 1456) and also made charming reliefs and statuettes of the Madonna and Child, in which he continued the tradition of Luca della *Robbia of stressing the naturalness and humanity of the Virgin.

Rossetti, Dante Gabriel (1828–82). English painter and poet. He came from a remarkable and talented family: his father was an exiled Italian patriot and Dante scholar, his sister the poet Christina Rossetti (1830–94), and his brother the critic William Michael Rossetti. Growing up in modest circumstances but a strongly literary environment, he at first found it hard to decide whether he should devote himself to poetry or painting. Although painting became his profession (following the advice given to him by the critic Leigh Hunt (1784–1859): 'If you paint as well as you write, you may be a rich man'), he continued to write poetry and make translations from the Italian, and he is accorded a distinguished position as a literary figure.

In 1848 he formed the *Pre-Raphaelite Brotherhood with *Hunt, *Millais, and others. His *Girlhood of Mary Virgin* (Tate, London, 1849), the first picture to be exhibited bearing the Brotherhood's initials, was warmly praised and sold well, but the subsequent abuse that the Pre-Raphaelites received hurt him so much that he rarely again exhibited in public. In the 1850s he virtually gave up oils and concentrated on watercolours of medieval subjects. These found ready buyers (often contacts of *Ruskin, whom Rossetti met in 1854), and Rossetti, who was a skilful businessman, proved Leigh Hunt's prediction true—by the 1860s he was earning the very substantial sum of £3,000 per year.

In 1860 Rossetti married the beautiful but sickly Elizabeth Siddal, the archetypal Pre-Raphaelite 'stunner', after a long and sometimes vexed liaison. Her melancholy face haunted his imagination, and he portrayed 'Guggums' (as he called her) again and again—'It is like a monomania with him', wrote Ford Madox *Brown in 1855. Rossetti immortalized her mainly in drawings, for in spite of the hatred for academic discipline that made him so disdainful of the official art world, he was an outstanding draughtsman. Elizabeth died from an overdose of laudanum, possibly deliberate, in 1862, and Rossetti was devastated; as a gesture of his grief he had the only complete manuscript of his poems placed in her coffin, but he was persuaded to have them exhumed in 1869 and they were published the following year. Rossetti also painted the intensely spiritual *Beata Beatrix* (Tate) as a memorial to Elizabeth, expressing his love for her as a parallel to Dante's for Beatrice (the picture is dated 1864, but was worked on over a period of several years). By the time of Elizabeth's death Rossetti had returned to oil painting, and in the last two decades of his life, his subject-matter was confined almost exclusively to beautiful women, portrayed in a richly sensuous manner and often evoking literary or mythological references. Elizabeth was replaced as his favourite model by William *Morris's wife Janey, who became in Rossetti's pictures one of the archetypal *femmes fatales*—all cascading curls, pouting lips, and smouldering eyes. Rossetti had met Morris and *Burne-Jones in 1856 and entered into partnership with them in 1861 (in the decorative arts firm later known as Morris & Co.), but both business and personal relationships became strained; Rossetti was in love with Janey and he and Morris parted amid rancour in 1875. In his later years Rossetti became an eccentric recluse (he had a menagerie of unusual animals, including a wombat, the death of which occasioned a poem); he fought a losing battle against drugs and alcohol and he died paralysed and prematurely aged.

Rossetti was a commanding personality and his work was highly influential; his romantic medievalism inspired the second wave of Pre-Raphaelitism associated with Burne-Jones and other followers, and his *femmes fatales* appealed to the *Symbolists and had a legion of descendants during the turn-of-the-century taste for 'decadence'.

Rosso, Giovanni Battista (called Rosso Fiorentino) (1495–1540). Florentine painter and decorative artist. *Vasari says that he 'would not bind himself to any master' (a story that fits in with his individuality of temperament), but in his youth he learned most from *Andrea del Sarto, and together with Andrea's pupil *Pontormo (Rosso's friend and exact contemporary) he was one of the leading figures in the early development of *Mannerism. His work was highly sophisticated and varied in mood, ranging from the refined elegance of the *Marriage of the Virgin* (S. Lorenzo, Florence, 1523) to the violent energy of *Moses and the Daughters of Jethro* (Uffizi, Florence, *c.*1523) and to the disquieting intensity of the *Deposition* (Galleria Pittorica, Volterra, 1521). In 1523 Rosso left Florence for Rome, where he worked until the Sack of 1527, and he then worked briefly in several Italian towns until 1530, when he was invited to France by Francis I (1494–1547). With *Primaticcio he was the most important artist to work on the decoration of the royal palace at Fontainebleau and one of the creators of the distinctive style of French Mannerism associated with the School of *Fontainebleau.

Rosso's principal work there is the Gallery of Francis I. Many engravings were made from his designs and his influence on French art was great. Vasari, whose biography of Rosso also includes an entertaining story about his pet baboon, says that he killed himself in remorse after falsely accusing a friend of stealing money from him, but this may well be apocryphal.

Rosso, Medardo (1858–1928). Italian sculptor. In his early career he was a painter, and was virtually self-taught as a sculptor—he was dismissed from the Brera Academy in Milan in 1883 after only a few months' training when he appealed for drawing to be taught from the live model rather than casts of statues. In 1884 he first visited Paris; he lived there from 1889 to 1897, the period of his most intense creative activity. It is indeed with French rather than Italian art that his work has affinity, for in reaction from the Italian *Renaissance tradition of three-dimensional solidity, Rosso applied to sculpture the *Impressionist aesthetic of reproducing the immediate impact and freshness of direct vision in which atmospheric effects and transitory conditions of light break up the permanent identity of the object. He was essentially a modeller rather than a carver and he made subtle use of his preferred medium of wax to express his view that matter was malleable by atmosphere: 'We are mere consequences of the objects which surround us.' He also anticipated later trends by his occasional incorporation of real objects in a sculptural work. His subjects included portraits and single figures and groups in contemporary settings (*The Bookmaker*, MOMA, New York, 1894; *Conversation in a Garden*, Gal. Nazionale d'Arte Contemporanea, Rome, 1893).

In 1897 Rosso returned to Italy and thereafter created little new sculpture, devoting himself instead to organizing exhibitions of his work, through which he gained an international reputation. He was particularly influential on the *Futurists, who took over and developed many of his ideas. *Boccioni called him 'the only great modern sculptor who has attempted to open up a larger field to sculpture, rendering plastically the influences of an ambiance and the atmospheric ties which bind it to the subject'. Today he is regarded as a sculptor of remarkable originality (not even *Rodin challenged so decisively the traditional preoccupations of his art) and one of the precursors of the modern movement because of the emphasis he gave to the direct representation of visual experience and his realization that the ordinary and commonplace could have sculptural expressiveness. His output was fairly small; replicas of several of his works, together with a collection of his drawings, are in the Museo Medardo Rosso at Barzio in Italy.

Roszak, Theodore. See ABSTRACT EXPRESSIONISM.

Rothenberg, Susan. See NEW IMAGE PAINTING.

Rothenstein, Sir William (1872–1945). British painter, graphic artist, writer, and teacher. He studied for a year at the *Slade School (1888–9) under Alphonse *Legros and afterwards at the *Académie Julian, Paris. There he became a close friend of *Whistler and was encouraged by *Degas and *Pissarro. His best works are generally considered to be his early Whistlerian paintings such as *The Doll's House* (Tate, London, 1899), which shows Augustus *John and Rothenstein's wife as characters in a tense scene from Ibsen's play *A Doll's House*. From about 1898, however, he specialized in portraits of the celebrated and those who later became celebrated. In the latter part of his career he was much more renowned as a teacher than a painter. His outlook was conservative (he regarded pure abstraction as 'a cardinal heresy') and as Principal of the *Royal College of Art, 1920–35, he exercised an influence second only to that of *Tonks at the Slade School in earlier decades. His son, **Sir John Rothenstein** (1901–92), had a distinguished career as an art historian. He was Director of the *Tate Gallery, 1938–64, and published numerous books on art, the best known being the three-volume series *Modern English Painters* (1952, 1956, 1973, new edition of the whole work 1985). Another son, **Michael Rothenstein** (1908–93), was a painter, printmaker, and writer on art. His books include *Frontiers of Printmaking* (1966).

Rothko, Mark (1903–70). Russian-born American painter, one of the outstanding figures of *Abstract Expressionism and one of the creators of *Colour Field painting. He emigrated to the USA as a child in 1913. After dropping out of Yale University in 1923 he moved to New York and studied at the *Art Students League under Max *Weber, but he regarded himself as essentially self-taught as a painter. In the 1930s and 1940s he went through phases influenced by *Expressionism and *Surrealism, but from about 1947 he began to develop his mature and distinctive style. Typically his paintings feature large rectangular expanses of colour arranged parallel to each other, usually in a vertical format. The edges of these shapes are softly uneven, giving them a hazy, pulsating quality as if they were floating on the canvas. The paintings are often very large and the effect they produce is one of calmness and contemplation, but in spite of their tranquillity, they cost Rothko enormous emotional effort: 'I'm not an abstract artist... I'm not interested in the relationship of colour or form or anything else. I'm interested only in expressing basic human emotions—tragedy, ecstasy, doom and so on. And the fact that a lot of people break down and cry when confronted with my pictures shows that I can communicate these basic human emotions... The people who weep before my pictures are having the same religious experience as I had when I painted them.'

Rothko was poor for much of his career (from 1929 to 1959 he earned at least part of his living by teaching art), but his reputation grew in the 1950s and in 1961 he was given a major retrospective exhibition at the Museum of Modern Art, New York, that sealed his success. In spite of his soaring fame (and the money it brought), Rothko was plagued by depression. He had a prickly temperament, drank heavily, took barbiturates to excess, was fearful and suspicious of younger artists, had two unhappy marriages, and felt he was misunderstood (he disliked having his paintings discussed in *formalist terms). His early works had often been bright and vivid in colour, but from the 1950s they became increasingly sombre, typically employing blacks, browns, and maroon. He regarded his fourteen paintings for a non-denominational chapel in Houston, Texas (now known as the Rothko Chapel), 1967–9, as his masterpieces. His last paintings were a series of stark black on grey canvases that evoke his painful state of mind leading up to his suicide (he slashed his veins in his studio).

Rottenhammer, Hans (or **Johann**) (1564–1625). German painter. In 1589–96 he worked in Rome and in 1596–1606 in Venice. He specialized in mythological scenes in landscape settings, working on a small scale and often on copper, and his paintings form a link between the styles of Paul *Bril, whom he knew in Rome, and Adam *Elsheimer, whom he knew in Venice.

Rottmann, Carl (1797–1850). German painter. He travelled extensively in Italy and Greece, and is best known for his Greek and Italian landscapes commissioned by King Ludwig I (1786–1868) of Bavaria and painted in fresco on the walls of the *Hofgarten* cloisters in Munich in 1830–3 (now in the Neue Pinakothek, Munich). His panoramic views combine heroic grandeur with accurate observation and an interest in Mediterranean light.

Rouault, Georges (1871–1958). French painter, graphic artist, and designer who created a personal kind of *Expressionism that gives him a highly distinctive place in modern art. In his youth he was apprenticed in a stained-glass workshop, his work including the restoration of medieval glass; the vivid colours and strong outlines characteristic of the medium left a strong imprint on his work. In 1892 he became a fellow pupil of *Matisse and *Marquet under Gustave *Moreau at the École des *Beaux-Arts. He was Moreau's favourite pupil and in 1898 became the first curator of the Musée Moreau in Paris. At about the same time he underwent a psychological crisis and although he continued to associate with the group of artists around Matisse who were later known as *Fauves, he did not adopt their brilliant colours or characteristic subjects; instead he painted clowns, prostitutes, outcasts, and judges in sombre but glowing tones. These subjects expressed his hatred of cruelty, hypocrisy, and vice, depicting the ugliness and degradation of humanity with passionate conviction. Initially they disturbed the public, but during the 1930s he gained an international reputation. From about 1940 he devoted himself almost exclusively to religious art. He was a prolific painter, and his work also included numerous book illustrations, ceramics, designs for tapestry and stained glass and for *Diaghilev's ballet *The Prodigal Son* (1929), the music for which was written by Sergey Prokofiev (1891–1953). By the time of his death he was a much honoured figure and he was given a state funeral.

Roubiliac, Louis-François (1702–62). French-born sculptor, active in England for virtually his entire career. Little is known of his life before he settled in London in the early 1730s, although he is said to have trained under Balthasar *Permoser in Saxony and Nicolas *Coustou in Paris. He made his reputation with a full-length seated statue of the composer Handel (V&A, London, 1738), remarkable for its lively informality, and quickly became recognized as the most brilliant portrait sculptor of the day. His busts have great vivacity, stressing small forms and rippling movement in a manner very different from the broader treatment of his contemporary Michael *Rysbrack. He was especially successful with portraits of old and ugly men (*Dr Martin Folkes*, Wilton House, 1749), and in his series of busts at Trinity College, Cambridge, and the celebrated statue of Newton (1755) there, he showed a remarkable gift for producing lively portraits of men long dead. Roubiliac was also outstanding as a tomb sculptor, several notable examples being in Westminster Abbey, including the marvellously dramatic tomb of Lady Elizabeth Nightingale (1761), who is shown being attacked with a spear by Death (a hideous skeleton emerging from a vault), while her husband vainly tries to keep him at bay (the skeleton was carved by

Roubiliac's assistant Nicholas Read, *c.*1733–87). The Nightingale monument clearly shows the influence of *Bernini, whose work so impressed Roubiliac when he visited Rome in 1752; he said that compared to Bernini's his own sculptures looked 'meagre and starved, as if made of nothing but tobacco pipes'.

Roubiliac is generally regarded as one of the greatest sculptors ever to work in England, certainly the greatest of his period. He had a vivid imagination, he was a superb craftsman, and, as Gerald Randall observes (*Church Furnishing and Decoration in England and Wales*, 1980), 'Unlike even the best of his rivals, Roubiliac seems to have been incapable of indifferent work, and even his most modest commissions are designed and executed with a master's touch.'

roulette. An *engraving tool having at one end a toothed wheel, used for making dotted lines or tonal areas on copper plates. Together with its variants the matting wheel and the chalk roll (which have several rows of teeth) it was evolved in the 18th cent. for making engravings in the *crayon manner and the other tonal processes.

Rousseau, Henri (known as Le Douanier Rousseau) (1844–1910). French painter, the most celebrated of *naïve artists. His nickname refers to the job he held with the Paris Customs Office (1871–93), although he never actually rose to the rank of 'Douanier' (Customs Officer). Before this he had served in the army, and he later claimed to have seen service in Mexico, but this story seems to be a product of his imagination. He began to paint as a hobby, self-taught, when he was about 40, and from 1886 he exhibited regularly at the *Salon des Indépendants. In 1893 he took early retirement so he could devote himself to art. His character was extraordinarily ingenuous and he suffered much ridicule (although he sometimes interpreted sarcastic remarks literally and took them as praise) as well as enduring great poverty. However, his faith in his own abilities never wavered. He tried to paint in the academic manner of such traditionalist artists as *Bouguereau and *Gérôme, but it was the innocence and charm of his work that won him the admiration of the avant-garde. He was discovered by *Vollard and members of his circle in about 1906–7, and in 1908 *Picasso gave a banquet, half serious half burlesque, in his honour.

Rousseau is now best known for his jungle scenes, the first of which was *Tiger in a Tropical Storm (Surprised!)* (NG, London, 1891) and the last *The Dream* (MOMA, New York, 1910). These two paintings are works of great imaginative power, in which he showed his extraordinary ability to retain the utter freshness of his vision even when working on a large scale and with loving attention to detail. He claimed such scenes were inspired by his experiences in Mexico, but in fact his sources were illustrated books and visits to the zoo and botanical gardens in Paris. His other work ranges from the jaunty humour of *The Football Players* (Philadelphia Museum of Art, 1908) to the mesmeric, eerie beauty of *The Sleeping Gypsy* (MOMA, 1897). Rousseau was buried in a pauper's grave, but his greatness began to be widely acknowledged soon after his death.

Rousseau, Théodore (1812–67). French landscape painter, the central figure of the *Barbizon School. He was one of the pioneers of landscape painting in the open air (see PLEIN AIR), and because of the non-academic outlook of his work it was consistently rejected by the *Salon, earning him the nickname 'le grand refusé'. From 1836 he worked regularly in the Forest of Fontainebleau, specializing in wooded scenes, and in 1848 he settled permanently in the village of Barbizon, where he was a close friend of *Millet and *Diaz. Success came to him during the 1850s. His output was prolific and his work is represented in many galleries in France and elsewhere.

Roussel, Ker-Xavier. See NABIS.

Rowlandson, Thomas (1756–1827). English *caricaturist, whose pre-eminence in social satire matched that of *Gillray in political satire. He began as a painter, mainly of portraits, but he turned to caricature to supplement his income (he was a notorious gambler), and finding his sideline highly successful he gave his career over to it completely. His talent for exuberant and flowing line had affinities with the French *Rococo of *Fragonard (Rowlandson had studied in France), but his rollicking humour and delicate tonal effects were distinctively English; the marvel of his art is that there is no inconsistency between the bawdiness or boisterousness of the subject-matter and the beauty of his watercolour technique. He created an instantly recognizable gallery of social types, such as the old maid, the hack writer and the crabbed antiquarian, and his buxom wenches have their descendants in the fat ladies of today's saucy seaside postcards. His repertory of themes was inexhaustible and his *œuvre* has been termed the English equivalent of Balzac's *Comédie humaine*. He was a friend of George *Morland and travelled about England and also in France, Germany, and the Low Countries making rapid and brilliantly illuminating sketches of country life. In addition he produced series of illustrative drawings for publishers, notably *The Comforts of Bath* (1798) and the series on *The Tours of Dr Syntax* (1812–20). Rowlandson's output was huge. Towards the end of his career the quality of his work suffered because of overproduction, but his overall standard is remarkably high.

Royal Academy of Arts, London. The national art *academy of England, founded in 1768. Its Instrument of Foundation was signed on 10 December 1768 by George III, who had been approached by Benjamin *West and the architect William Chambers (1723–96) to lend royal approval to the scheme. Membership was limited to forty Academicians, who had to be artists by profession, and thirty-six were named in the Instrument of Foundation. Their motive in founding the Academy was to raise the status of their profession by establishing a sound system of training and expert judgement in the arts and to arrange for the free exhibition of works attaining an appropriate standard of excellence. Behind this conception was the desire to foster a national school of art and to encourage appreciation and interest in the public based on recognized canons of good taste. The *Society of Artists was a forerunner in those aims. At first twenty, the number of Associates was raised to thirty in 1876.

The Royal Academy was first housed in Pall Mall. Its annual summer exhibition, to which anyone can submit works, has been held every year since 1769 and the RA Schools have also existed from the beginning. The Academy transferred to Somerset House in 1780, shared with the *National Gallery its premises in Trafalgar Square 1837–68, and moved to the present site of Burlington House in 1869. The first President of the Academy was Sir Joshua *Reynolds, who held the office until his death in 1792. His famous *Discourses*, delivered over a period of twenty years, laid down the basic conception of the Academy as a body of professional men which, 'besides furnishing able men to direct the student', was to form 'a repository for the great examples of the Art'. The latter function was important since until the foundation of the National Gallery in 1824 there was no public collection of masterpieces available to students and schools. Although the most celebrated work owned by the Academy is a marble *tondo by *Michelangelo of the Virgin and Child with St John the Baptist (the Taddei Tondo, *c.*1505), the collections are chiefly important for their representation of work by Academicians, the custom being that each new Academician deposits a work (called a Diploma Work) on admission.

Regarding itself as the main depository of national tradition in the arts and the safeguard of sound standards of professional competence in execution, the Royal Academy, in common with other official establishments, has been cautious of innovation and slow to lend the seal of its approval to experimental novelty. During the latter decades of the 19th cent. the reputation of the Academy sank very low and it began to be regarded as the bulwark of orthodox mediocrity in opposition to creative and progressive art. The *Slade School and the *Royal College of Art became more important as teaching institutions, and organizations, such as the *New English Art Club, and later the *London Group, were formed to accommodate progressive trends. Since the presidency (1944–9) of *Munnings, who was notorious for his opposition to modern art, its policy has become more liberal and the conflict between official 'Academy' art and creative art has narrowed. But something of the opinion from the late 19th cent. continued up to the mid-1970s, and the function of the Academy at its inception to provide exhibitions of the best contemporary work from year to year has been challenged by the competitive claims of the commercial galleries and of such bodies as The *Arts Council of Great Britain. The annual Royal Academy summer exhibition still remains a popular event, however, and the Academy regularly organizes historical exhibitions of the highest quality.

Royal College of Art, London. Post-graduate university institution, now Britain's pre-eminent training school for artists and designers. It was founded in 1837 as the School of Design and was originally a school of industrial design, the fine arts being the province of the *Royal Academy. In 1852 it moved from its original home in Somerset House to Marlborough House and was renamed the Central School of Practical Art. It became part of the Government Department of Science and Art in 1853 and in 1857 it moved to join the Museum of Ornamental Art (later the *Victoria and Albert Museum) in South Kensington. In 1863 it moved to new buildings in Exhibition Road, and in 1896 it was renamed the Royal College of Art by Queen Victoria and allowed to grant diplomas. The College moved again to new buildings in Kensington Gore in 1961 and in 1967 was given a Royal Charter and empowered to award degrees. Some of the most illustrious British artists of the 20th cent. have been students at the Royal College of Art, among them *Hepworth, *Hockney, and *Moore.

Rubens, Sir Peter Paul (1577–1640). Flemish painter, designer, and diplomat, the greatest and most influential figure in *Baroque art in northern Europe. He was born at Siegen in Westphalia, the son of a Protestant lawyer from

Antwerp who moved to Germany to escape religious persecution, and he returned to Antwerp in 1587 with his mother soon after his father's death. He had been baptized a Calvinist in Germany, but he became a devout Catholic. His masters were three fairly undistinguished painters of Antwerp, Tobias *Verhaecht, Adam van *Noort, and Otto van *Veen. The first two could teach him no more than the local tradition, but van Veen was a man of some culture, who had spent about five years in Rome, and he no doubt inspired his pupil with a desire to visit Italy. Rubens became a master in the Antwerp painters' guild in 1598, and after working with van Veen for two more years he set out for Italy in 1600. Very little of Rubens's early work survives, and his style was largely formed in Italy, where he was based until 1608. He worked for Vincenzo *Gonzaga, Duke of Mantua, visiting most of the principal art centres of Italy to make copies for the ducal collection and also in 1603–4 travelling to Spain when he accompanied gifts from Vincenzo to Philip III (1578–1621). The most important centres of his activity in Italy, however, were Genoa and Rome. In Genoa he painted some stately aristocratic portraits (*Marchesa Brigida Spinola-Doria*, NG, Washington, 1606) that inspired van *Dyck when he worked in the city, and in Rome he found the basis of his own grandiose style in the *antique, the great *Renaissance masters, and Annibale *Carracci.

On learning that his mother was seriously ill, Rubens returned to Antwerp in 1608, but she died before he arrived. Italy had become Rubens's spiritual home (he usually signed himself 'Pietro Pauolo') and he considered returning for good, but his success in Antwerp was so immediate and great that he remained there, and in spite of his extensive travels later in his career he never saw Italy again. In 1609 Rubens was appointed court painter to the Archduke Albert and his wife the Infanta Isabella, the Spanish Viceroys in the Netherlands, and in the same year he married the 17-year-old Isabella Brant, the daughter of an eminent Antwerp lawyer. The portrait of himself and his wife (Alte Pinakothek, Munich) that he painted presumably to mark the occasion gives a marvellous picture of Rubens on the threshold of his great career—handsome, vigorous, and dashingly self-confident. In the next few years he established his reputation as the pre-eminent painter in northern Europe, his first two resounding successes being the huge *triptychs of the *Raising of the Cross* and the *Descent from the Cross* (Antwerp Cathedral, 1610–11 and 1611–14), which showed his mastery of history painting in the *Grand Manner and the immense vigour of his style.

The demand for Rubens's work was extraordinary, and he was able to meet it only because he ran an extremely efficient studio. It is not known how many pupils or assistants he had because as court painter he was exempt from registering them with the guild. The idea of him running a sort of picture factory has been exaggerated, but even a man of his seemingly inexhaustible physical and intellectual stamina (he habitually rose at 4 a.m. and according to a contemporary account could work whilst dictating a letter and holding a conversation with visitors) could not carry out all the work involved in his massive output with his own hands. Rubens both collaborated with established artists ('Velvet' *Brueghel, van Dyck, *Jordaens, Daniel *Seghers, *Snyders, and others) and retouched pictures by pupils, the degree of his intervention being reflected in the price. Generally his assistants did much of the work between the initial oil sketch and the master's finishing touches. Modern taste has tended to admire these sketches and his drawings (in which his personal touch is evident in every stroke of brush, chalk, or pen) more than the large-scale works, but Rubens himself would surely have found this attitude hard to comprehend, for the sheer scale and grandeur of the finished paintings gives them an extra, symphonic dimension.

Rubens not only painted virtually every type of subject, but also designed tapestries, book illustrations, and decorations for festivals, as well as giving visual directives for sculptors, metalworkers, and architects. 'My talents are such', he wrote in 1621, 'that I have never lacked courage to undertake any design, however vast in size or diversified in subject.' So huge was his output, indeed, that it is difficult to put a figure on it; the *Corpus Rubenianum*, the first attempt in the 20th cent. at a complete scholarly catalogue of his work, began publication in 1968 and is expected to require about thirty volumes. His biggest commission in Flanders was for the decoration of the Jesuit Church in Antwerp (a building he may also have helped to design), but almost all his work there was destroyed by fire in 1718. From outside Flanders, those who sought his services included the royal families of France, England, and Spain. For Marie de' *Médicis (mother of Louis XIII of France) he did a series of twenty-five enormous paintings on her life (Louvre, Paris, 1622–5); for Charles I of England he painted a series of canvases representing the reign of his father James I (completed 1635) for the ceiling of the Banqueting House in London; and for Philip IV of Spain he embarked in 1636 on a series of more than a hundred mythological pictures for his hunting lodge, the Torre de la Parada (the series was incomplete when Rubens died and most of the finished paintings—executed by assistants from his *modelli—were destroyed in 1710 when the building was sacked during the War of the Spanish Succession).

After the death of the Archduke Albert in 1621, Rubens became a trusted adviser to the Infanta Isabella, and he was employed in negotiations to try to gain peace in the Netherlands. In 1628–9 Isabella sent him on a diplomatic mission to Spain (where he met *Velázquez) and this led to him going to England on behalf of the Spanish King, Philip IV, in 1629–30. He played an important part in the arrangement of a peace settlement between England and Spain, and Charles I knighted him. In his diplomatic role his polished manners and his prodigious linguistic skills were put to good advantage—apart from Flemish and Italian, he knew French, German, Latin, and Spanish.

Rubens's wife died in 1626 and in 1630 he remarried; his bride was the 16-year-old Hélène Fourment, daughter of a rich silk merchant and the niece of his first wife. The second marriage was as happy as the first, and Rubens's love of his family shines through many of his late paintings (*Hélène Fourment with Two of her Children*, Louvre, *c.*1637). In 1635 he bought a country house, the Château de Steen, between Brussels and Malines, and in his final years he developed a new passion for painting landscapes—marvellously ripe works that led *Constable to declare 'In no branch of the art is Rubens greater than in landscape.' Superb examples are in the National Gallery and the Wallace Collection, London.

Rubens's influence in 17th-cent. Flanders was overwhelming, and it was spread elsewhere in Europe by his journeys abroad and by pictures exported from his workshop, and also through the numerous engravings he commissioned of his work. In later centuries his influence has also been immense, perhaps most noticeably in France, where *Watteau, *Delacroix, and *Renoir were among his greatest admirers. Because of the unrivalled variety of his work, artists as different in temperament as these three could respond to it with equal enthusiasm.

Rubens's work is found in most major galleries; those with particularly fine collections include: the Prado, Madrid; the Alte Pinakothek, Munich; the Louvre, Paris; and the Kunsthistorisches Museum, Vienna. London is perhaps the richest of all cities in his work, with fine collections at the Courtauld Institute Galleries, Dulwich College Picture Gallery, the National Gallery, and the Wallace Collection, and in the ceiling of the Banqueting House, the only one of Rubens's major decorative schemes still in the position for which it was painted.

Rublev, Andrei (1360?–1430). The most famous of Russian *icon painters. The 600th anniversary of his birth was celebrated by Soviet Russia in 1960, but there is some evidence that he may have been born a decade later and there is little secure knowledge of his life or works. In 1405 he worked as assistant to *Theophanes in the Cathedral of the Annunciation in the Kremlin at Moscow, but it has not been possible to distinguish his share there or in the Cathedral of the Dormition at Vladimir, where he is also said to have painted murals. The work that stands at the centre of his *œuvre* is the celebrated icon of the Old Testament Trinity (that is the three angels who appeared to Abraham) in the Tretyakov Gallery, Moscow (*c.*1411). In its gentle lyrical beauty this marks a move away from the hieratic *Byzantine tradition, and other icons in a similar style have been attributed to Rublev.

rubrication. In calligraphy and typography (particularly *illuminated manuscripts and early printed books) the use of a different colour, usually red, to emphasize initial letters, section headings, etc.

Rude, François (1784–1855). French *Romantic sculptor. He was a fervent admirer of the emperor Napoleon I and his emotionally charged work expresses the martial spirit of the Napoleonic era more fully than that of any other sculptor. In 1812 he won the *Prix de Rome, but he was unable to take it up because of the Napoleonic Wars, and when Napoleon abdicated in 1814 he went into exile in Brussels with *David. On his return to Paris in 1827 he became highly successful with public monuments, most notably his celebrated high *relief on the Arc de Triomphe, *Departure of the Volunteers in 1792*, popularly known as *The Marseillaise* (1833–6). None of Rude's other works matches the fire, dynamism, and heroic bravura of this glorification of the French Revolution, but he created another strikingly original work in his monument *The Awakening of Napoleon* (1845–7) in a park at Fixin, near his native Dijon, which shows the emperor casting off his shroud. In spite of the dramatic movement of his work, it always has a solidity that reveals his Classical training and his lifelong admiration of the *antique.

Ruisdael, Jacob van (1628/9–82). The greatest and most versatile of all Dutch landscape painters. He was born in Haarlem, traditionally called the home of Dutch landscape painting, where he probably received training from his father **Isaac**, who was a painter as well as a frame-maker and picture-dealer (no works by him are known to survive). His uncle Salomon van *Ruysdael (this distinction in spelling occurs consistently in their own signatures) presumably also played a part in his artistic education. Ruisdael was extremely precocious, however; his earliest known paintings date from 1646 and already reveal a mature and distinctive artistic personality. He was also versatile and prolific (about 700 paintings are reasonably attributed to him); he painted forests, grain fields, beaches and seascapes, watermills and windmills, winter landscapes and Scandinavian torrents influenced by Allart van *Everdingen; he could conjure poetry from a virtually featureless patch of duneland as well as from a magnificent panoramic view. Even more than his range, however, it is the emotional force of his work that distinguishes him from his contemporaries. He moved away decisively from the 'tonal' phase of Dutch landscape represented by his uncle; in place of subtle atmospheric effects he favoured strong forms and dense colours and his brush-work is vigorous and *impasted. For him man was insignificant beside the majesty and power of nature, and his emotional, subjective approach found its most memorable expression in *The Jewish Cemetery* (versions in the Detroit Institute of Arts and the Gemäldegalerie in Dresden, *c.*1660), where tombstones and elegiac ruins, symbols of man's transitory and ephemeral existence, are contrasted with nature's power of renewal.

Ruisdael travelled to the Dutch/German border with his friend Nicolaes *Berchem in the early 1650s, and one of the pictures that resulted was the celebrated *Bentheim Castle* (Beit Coll., Blessington, Ireland, 1653), in which the castle heroically crowns the top of a steep, rugged hill, transformed by Ruisdael's imagination from the mild slope it is in actuality. In about 1656 he moved to Amsterdam, where he lived for the rest of his life (although he was buried in St Bavo's Cathedral in Haarlem). He evidently had a reasonably prosperous career, but little is known about his life and it was long thought he had died insane in the workhouse at Haarlem, a fate that is now known to have befallen his cousin and near namesake Jacob van *Ruysdael. There is still uncertainty, however, concerning the story, reported by *Houbraken and supported by other tantalizing evidence, that Ruisdael practised as a surgeon. It seems unlikely that he could have found the time for this (he is said to have taken a medical degree at Caen in Normandy in 1676, when he was in his late 40s), but other prolific Dutch painters, for example *Steen (who ran a tavern), managed to pursue two careers.

Ruisdael's only documented pupil was *Hobbema, but his influence was resounding, both on his Dutch contemporaries and on artists in other countries in the following two centuries—*Gainsborough, *Constable, and the *Barbizon School for example. Examples of his work are in many public collections, the finest representation being in the National Gallery, London.

Runciman, Alexander (1736–85) and **John** (1744–68). Scottish painters, brothers, who painted religious, literary, and historical subjects in a proto-*Romantic manner. John, the more brilliantly gifted, died young during a sojourn by both brothers in Italy. His masterpiece, *King Lear in the Storm* (NG, Edinburgh, 1767), has freshness and originality with nothing of the staginess of most 18th-cent. Shakespearian pictures. Alexander's major work, the decoration of Penicuik House near Edinburgh with romantically treated subjects from Ossian and the history of Scotland, ranked with *Barry's paintings in the Society of Arts as the most ambitious British decorative scheme of the time, but it was destroyed by fire in 1899. Some of its compositions survive in a series of spirited etchings he based on them.

Runge, Philipp Otto (1777–1810). German painter and draughtsman. Although he made a late start to his career and died young, he ranks second only to *Friedrich among German *Romantic artists. He studied under Jens *Juel at the Copenhagen Academy (1799–1801), then moved to Dresden, where he knew Friedrich. In 1803 he settled in Hamburg. Runge was of a mystical, pantheistic turn of mind and in his work he tried to express notions of the harmony of the universe through symbolism of colour, form, and numbers. To this end he planned a series of four paintings called *The Times of the Day*, designed to be seen in a special building and viewed to the accompaniment of music and poetry. He painted two versions of *Morning* (Kunsthalle, Hamburg, 1808 and 1809), but the others did not advance beyond drawings. Runge was also one of the best German portraitists of his period; several examples are in Hamburg. His style was rigid, sharp, and intense, at times almost *naïve. In 1810 he published *Die Farbenkugel* (The Colour Sphere) after doing several years of research on colour, during which he corresponded with *Goethe.

Ruoppolo, Giovanni Battista (1629–93). Neapolitan still-life painter. He specialized in pictures of flowers and food (especially fruit and seafood), depicted in an exuberant and succulent style. With *Recco he was the finest Italian still-life painter of his period.

Rupert, Prince (1619–82). Bohemian-born soldier and amateur artist. Famous as a dashing cavalry commander for his uncle Charles I (1600–49) in the English Civil War, he was also an active dilettante of science and the arts. He was an amateur etcher and introduced to England *mezzotint engraving, which he may have learnt from Ludwig von *Siegen (Rupert was himself long credited as the inventor). He demonstrated the technique to the diarist John Evelyn, who publicized it in his book *Sculptura* (1662) under the auspices of the Royal Society.

Ruralists, Brotherhood of. See BLAKE, PETER.

Rusconi, Camillo (1658–1728). Italian sculptor. He was the outstanding sculptor in Rome during his period, a figure comparable to his friend *Maratta in painting. The vigour and boldness of his style derive from *Bernini, but Rusconi was more restrained and Classical. His most important works are four over life-size statues of Apostles in S. Giovanni in Laterano (1708–18) and the tomb of Pope Gregory XIII (1719–25) in St Peter's.

Rush, William (1756–1833). American sculptor, active in his native Philadelphia. His father was a ship's carpenter and Rush worked mainly in wood, progressing from ships' figure-heads to free-standing figures, such as the *Nymph of the Schuylkill* (1812), a work which almost perished through exposure to the elements in Fairmount Park, Philadelphia, and was preserved only when it was belatedly cast in bronze. His work is vigorous and naturalistic and he marks the transition from the unselfconscious folk carver to the professional artist. He was one of the prime movers in the foundation of the Pennsylvania Academy of the Fine Arts, which has many examples of his work. Thomas *Eakins, another native of Philadelphia, greatly admired Rush's work and called him 'one of the earliest and one of the best American sculptors'.

Rusiñol y Prats, Santiago (1861–1931). Spanish painter, poet, and writer on art. He worked mainly in his native Barcelona, but he studied in Paris and was one of the main channels through which the influence of modern French painting was introduced into Spain.

Ruskin, John (1819–1900). The most influential English art critic of his time, also a talented watercolourist. His output of writing was enormous and he had a remarkable hold over public opinion, as he showed when he successfully defended the *Pre-Raphaelites against the savage attacks to which they were being subjected by Dickens and others. He was the son of a wealthy wine merchant and his artistic education was gained through frequent travel in Britain and on the Continent. Ruskin belonged to the *Romantic School in his conception of the artist as an inspired prophet and teacher. Greatness of art, he thought, is 'the expression of a mind of a God-made man'. Like *Pugin he advocated a revival of the *Gothic style, his rejection of the *classical tradition following from his assumption that art and architecture should mirror man's wonder and delight before the visual creation of God and that this demanded a freely inspired and *naturalistic style to which he felt that Gothic alone was really suited. He saw ornament as an aid to contemplation of the wonders of divinely inspired Nature and he praised the Gothic for its 'noble hold of nature' and for the 'careful distinction of species, and richness of delicate and undisturbed organisation, which characterise the Gothic design'.

Although Ruskin's worship of beauty for its own sake brought him into affinity with the advocates of 'art for art's sake' (see *Aestheticism), his strong interest in social reform and ever-increasing concern with economic and political questions during the second half of his life (he used much of his large inheritance for philanthropic work) kept him from accepting a doctrine of the autonomy of the arts in divorce from questions of social morality. His eloquence in linking art with the daily life of the workman had affinities with the views of William *Morris and his insistence on regarding the state of the arts as a 'visible sign of national virtue' and his constant emphasis on their moral function have sometimes been regarded as a conspicuous instance of the 'moral fallacy' in aesthetics and criticism. He set himself obstinately and shortsightedly against the effects of the Industrial Revolution in supplanting the older craftsmanship and opposed all efforts to raise the standard of design in industry and to institute schools for the application of good principles of design to mass-production.

Ruskin's personal life was deeply unhappy. His marriage was annulled in 1854 on the grounds of non-consummation (his ex-wife married *Millais in the following year) and in middle and old age he made many young girls the objects of his unhealthy affection. He proposed to one of them, the 18-year-old Rose La Touche, in 1866, but was refused; she died mad in 1875. In 1878 he lost a famous libel case against *Whistler, whom he had accused of 'flinging a pot of paint in the public face', and soon after showed the first signs of the mental illness that made his final years wretched. After 1889, living in isolation in the Lake District, where he was cared for by his cousin, John Severn, Ruskin wrote nothing and rarely spoke.

Ruskin's complete works were edited in thirty-nine volumes (1903–12). His most important works of art criticism are: *Modern Painters* (5 vols., 1843–60, epilogue 1888), which began as a defence of *Turner and expanded into a general survey of art; *The Seven Lamps of Architecture* (1849); and *The Stones of Venice* (3 vols., 1851–3). He is accorded a distinguished place amongst English prose writers of the 19th cent., and his finest flights of rhetoric, such as his descriptions of the *Tintorettos in the Scuola di San Rocco, are classics of their kind.

Russell, Morgan (1886–1953). American painter, active mainly in Paris; with Stanton *Macdonald-Wright he was the founder of *Synchromism, one of the earliest abstract art movements. Russell was born in New York, where he studied sculpture at the *Art Students League and painting under Robert *Henri. In 1908 he settled in Paris, where he briefly attended *Matisse's art school. By 1910 he was devoting himself increasingly to painting, and in 1911 he met Macdonald-Wright, with whom he developed theories about the analogies between colours and musical patterns. In 1913 they launched Synchromism, and Russell's *Synchromy in Orange: To Form* (Albright-Knox Art Gallery, Buffalo, 1913–14) won him considerable renown in Paris. His later work, in which he reintroduced figurative elements, was much less memorable than were his pioneering abstract paintings. He lived in Paris until 1946, then returned to the USA.

Russolo, Luigi (1885–1947). Italian painter and musician. He was one of the signatories of the *Futurist painters' manifestos in 1910, but he is remembered mainly as 'the most spectacular innovator among the Futurist musicians (*New Grove Dictionary of Music and Musicians*, 1980). In 1913 he published a manifesto, *L'arte dei Rumari* (*The Art of Noises*, expanded in book form in 1916), and later in the same year he demonstrated the first of a series of *intonarumori* ('noise-makers'), which produced a startling range of sounds. In 1913–14, he gave noise concerts in Milan (causing a riot), Genoa, and London. Others followed after the First World War. Several leading composers, notably Ravel and Stravinsky, thought they opened up interesting possibilities, and Russolo has been regarded as a pioneer of today's electronic music. Unfortunately his compositions and machines have been destroyed. As a painter Russolo made rather crude use of the Futurist device of 'lines of force' in his early work; and after the war his style became more naturalistic.

Rustici, Giovanni Francesco. See LEONARDO DA VINCI.

Rutter, Frank. See ALLIED ARTISTS' ASSOCIATION.

Ruysch, Rachel (1664–1750). Dutch still-life painter, with van *Huysum the most celebrated exponent of flower pieces of her period. The daughter of a botanist and the pupil of Willem van *Aelst, she worked mainly in her native Amsterdam, but also in The Hague (1701–8) and Düsseldorf, where from 1708 to 1716 she was court painter to the Elector Palatine. Her richly devised bouquets were painted in delicate colours with meticulous detail, and their artistry and craftsmanship are worthy of the finest tradition of Dutch flower painting. She continued to use the dark backgrounds characteristic of van Aelst and the older generation long after van Huysum and other contemporaries had gone over to light backgrounds. Examples of her work are in London (NG, V&A), Oxford (Ashmolean), and Cambridge (Fitzwilliam).

Ruysdael, Salomon van (*c.*1600–70). Dutch landscape painter, active in Haarlem, where he became a member of the painters' guild in 1623. Salomon's earliest works show the influence of Esaias van de *Velde and in the 1630s he was so close in style to Jan van *Goyen that it is sometimes difficult to differentiate between the work of the two artists. They both excelled in atmospheric, virtually monochromatic river scenes and are the leading masters of this type of picture. In the 1640s his landscapes became somewhat more solid and colourful, perhaps reflecting influence from his nephew, Jacob van *Ruisdael. Late in his career he occasionally painted still lifes. He was immensely prolific and many galleries have examples of his work. His son **Jacob Salomonsz. van Ruysdael** (1629/30–81) was also a landscape painter. An example of his rare work is *A Waterfall by a Cottage* (NG, London), which shows he worked in a style similar to that of his illustrious cousin and near-namesake, with whom he has sometimes been confused in documentary references.

Ryder, Albert Pinkham (1847–1917). American painter of imaginative subjects. He lived and worked most of his life as a solitary and dreamer in New York, and his methods and approach were largely self-taught. His pictures reflect a rich inner life, with a haunting love of the sea (he was born at the fishing port of New Bedford, Mass.), and a constant search to express the ineffable. He himself is quoted as saying: 'Have you ever seen an inch worm crawl up a leaf or twig, and then clinging to the very end, revolve in the air, feeling for something to reach something? That's like me. I am trying to find something out there beyond the place on which I have a footing.' This imaginative quality and eloquent expression of the mysteriousness of things is expressed typically through boldly simplified forms and eerie lighting (*The Race Track* or *Death on a Pale Horse*, Cleveland Museum of Art). In spite of his self-imposed isolation Ryder's works became well known in his lifetime and he has been much imitated and faked. His own paintings have often deteriorated because of unorthodox technical procedures.

Ryepin. See REPIN.

Ryland, William. See STIPPLE ENGRAVING.

Rysbrack, John Michael (1694–1770). Flemish-born sculptor, a member of an Antwerp family of artists, who settled in England about 1720. He soon achieved success and in the 1730s was the leading sculptor in the country. The highpoint of his career was winning the commission for the monument to William III in Queen Square, Bristol (1735), in preference to *Scheemakers. This is generally regarded as the finest equestrian statue made in England in the 18th cent. However, from about 1740 (the year of Scheemakers's acclaimed Shakespeare monument in Westminster Abbey), Rysbrack began to lose ground to Scheemakers and also to *Roubiliac. He was a versatile and prolific artist, hard-working and popular. His output included tombs, statues, and portraits (often in the form of busts in the *antique manner, a fashion he introduced to England), and he also made architectural elements such as chimney-pieces. His style was vigorous and dignified, less sombre than that of Scheemakers. He did not match the brilliant vivacity that characterizes Roubiliac's work, but he could sometimes rival him in beauty of handling. Rysbrack's best-known work is perhaps the monument to Sir Isaac Newton in Westminster Abbey (1731).

Rysselberghe, Théo van (1862–1926). Belgian painter, graphic artist, and designer. He was a founder of the avant-garde group of Les *Vingt (XX) in 1883. This group encouraged an interest in innovative art largely through contact with France, and Rysselberghe, who met *Seurat in Paris, became the leading Belgian exponent of *Neo-Impressionism. In 1898 he moved to Paris and was associated with the *Symbolist circle of writers and artists; his painting *A Reading* (Museum of Fine Arts, Ghent, 1903) shows several leading literary figures including André Gide and Maurice Maeterlinck. In 1910 he settled in Provence, where he abandoned Neo-Impressionism for a broader style of painting.

Saatchi, Charles (1943–). Iraqi-born British businessman and art collector. In 1970 he was co-founder with his brother Maurice of Saatchi & Saatchi, which became the world's largest advertising agency. He has devoted much of his enormous wealth to buying contemporary art on a huge (almost industrial) scale, and in 1985 the Saatchi Collection was opened to the public in a new gallery (converted from a warehouse) in St John's Wood, north London. While his patronage has been welcomed by many (not least the artists who have benefited from it), others have been critical of the way in which his bulk buying has given him such power in the art market. He helped to create the boom in *Neo-Expressionism and *Neo-Geo.

Sacchi, Andrea (1599–1661). Italian painter, the leading representative of the *classical tradition in Roman painting in the mid-17th cent. He was a pupil of *Albani in Rome and Bologna, but he was inspired chiefly by *Raphael, and with the sculptors *Algardi and *Duquesnoy he became the chief exponent of the style sometimes called 'High *Baroque classicism'. In defence of the classical principles of order and moderation, Sacchi engaged in a controversy in the Academy of St Luke with Pietro da *Cortona on the question of whether history paintings should have few figures (as Sacchi maintained) or many (Cortona). Sacchi's ideas were more immediately influential, but his ponderous ceiling fresco of *Divine Wisdom* (1629–33) in the Palazzo Barberini in Rome is completely outshone by Cortona's exhilarating ceiling of the Gran Salone in the same building. Sacchi, indeed, was at his best on a smaller scale—in altarpieces such as the grave, introspective *Vision of St Romuald* (Vatican, *c.*1631) and in portraits. His most important pupil was *Maratta. Sacchi also worked as an architect, designing the Chapel of St Catherine of Siena (1637–9) in the Sacristy of Sta Maria sopra Minerva, a work of refined classical purity. He was a fine draughtsman. Most of his drawings are in the Royal Collection at Windsor Castle or the Kunstmuseum, Düsseldorf.

sacra conversazione (Italian: 'holy conversation'). A representation of the Virgin and Child with Saints in which all the sacred personages are disposed in a single scene rather than in the separate compartments of a *polyptych and may appear related by attitude or common action. The type originated in Italy in the first half of the 15th cent. Early exponents were Fra *Angelico, *Domenico Veneziano, and Filippo *Lippi, whose Barbadori Altarpiece (Louvre, Paris, begun 1437) is perhaps the first dated example.

sacramentary. See MISSAL.

Saenredam, Pieter Jansz. (1597–1665). Dutch painter of architectural subjects, particularly church interiors, active in Haarlem. Saenredam, the son of an engraver, was a hunchback and a recluse, but he was acquainted with the great architect Jacob van *Campen, who may have played a part in determining his choice of subject. He has been called 'the first portraitist of architecture', as he was the first to concentrate on accurate depictions of real buildings rather than the fanciful inventions of the *Mannerist tradition. Saenredam's pictures were based on painstaking drawings and are scrupulously accurate and highly finished, but they never seem pedantic or niggling and are remarkable for their delicacy of colour and airy grace. The Cathedral of St Bavo (where he is buried) and the Grote Kerk in Haarlem were favourite subjects, but he also travelled to other Dutch towns to make drawings, and Utrecht is represented in several of his paintings. He also made a few views of Rome based on drawings in a sketchbook by Maerten van *Heemskerck that he owned. Saenredam's work, which had great influence on Dutch painting, is best represented in the Rijksmuseum, Amsterdam, and the Boymans Museum, Rotterdam.

Saftleven, Cornelis (1607–81) and **Herman III** (1609–85). Dutch painters, brothers, the best-known members of a family of artists. Cornelis, who was influenced by *Brouwer and *Teniers, painted *genre scenes of peasants and satirical pictures of animals dressed and acting like theologians and jurists. He also did landscapes with sheep and cattle grazing. There are examples in the Fitzwilliam Museum, Cambridge. Herman travelled in Germany and is best known for his highly finished panoramic views of the Moselle and Rhine, done in a distinctive misty blue tonality (*View on the Rhine*, Dulwich College Picture Gallery, 1656).

Sage, Kay (1898–1963). American *Surrealist painter, mainly self-taught as an artist. In 1900–14 and 1919–37 she lived in Italy (she was married to an Italian prince, 1925–35), and Giorgio de *Chirico was an early influence on her work. In 1937 she moved to Paris, where she met *Tanguy in 1939. He followed her to the USA in 1940 and they married later that year. From the time of her return to America architectural motifs became prominent in her work—strange steel structures depicted in sharp detail against vistas of unreal space—and her pictures also included draperies from which faces and figures sometimes mistily emerged (*Tomorrow is Never*, Metropolitan Museum, New York, 1955). Sage also made mixed-media constructions. Tanguy's sudden death in 1955 cast a shadow over her last years and she committed suicide.

Saint-Aubin. Family name of four French *Rococo artists, brothers, who worked primarily as draughtsmen and designers. The eldest, **Charles-Germain** (1721–86), worked on embroidery designs, wrote a treatise on embroidery, and was patronized by the ladies of the court. **Gabriel-Jacques** (1724–80) was the best known of the four and an etcher of some distinction who satisfied the public taste for anecdotal depiction of daily life in its most elegant and appealing aspects. He also painted, and his views of picture sales and the *Salons are valuable documents for the art historian (*An Auction*, Musée Bonnat, Bayonne; *Salon of 1765*, Louvre, Paris). **Louis-Michel** (1731–79), less talented, was a porcelain painter at the Sèvres factory. **Augustin**, the youngest (1736–1807), was a draughtsman of considerable charm (*Young Lady*, Albertina, Vienna), but is best known for his large output as an engraver of the works of *Boucher, *Fragonard, *Greuze, and others.

Saint-Gaudens, Augustus (1848–1907). The leading American sculptor of his period. Born in Dublin of a French father and an Irish mother, Saint-Gaudens was brought to America in infancy. He began his career as a *cameo-cutter in New York, then studied for three years in Paris (1867–70) and three in Rome (1870–3), returning to America in 1874. His first important commission was the *Admiral Farragut Monument* (1878–81) in Madison Square Park, New York; and following its successful reception he quickly achieved a leading reputation among American sculptors and retained this throughout his life. Saint-Gaudens was an energetic artist and he produced a large amount of work in spite of the high standards of craftsmanship he set himself. Some of his decorative work was done in association with the architectural firm of McKim, Mead, and White. His preferred material was bronze and he excelled particularly at memorials. Although his style is generally warmly naturalistic, his most celebrated work, the Adams Memorial (1891) in Rock Creek Cemetery, Washington, is a powerful allegorical figure. It is a monument to a wife of a friend of Saint-Gaudens who had committed suicide, and the mysterious female figure, swathed in magnificent voluminous draperies, has been interpreted as 'Grief', but the sculptor himself saw the elegiac work as embodying 'the Peace of God'. Saint-Gaudens was a highly important figure in the development of American sculpture; he turned the tide against *Neoclassicism and made Paris, rather than Rome, the artistic mecca of his countrymen.

St Ives School. A loosely structured group of artists, flourishing particularly from the late 1940s to the early 1960s, who concentrated their activities in the Cornish fishing village of St Ives. Like *Newlyn, St Ives had been popular with artists long before this, but it did not become of more than local importance in painting and sculpture until Barbara *Hepworth and Ben *Nicholson moved there in 1939, two weeks before the outbreak of the Second World War. They were anxious that their children should be safely outside London, and their friend Adrian *Stokes, who lived at Corbis Bay (virtually a suburb of St Ives), invited the family to stay with him. Hepworth lived in St Ives for the rest of her life (her studio is now a museum of her work) and Nicholson (who had discovered Alfred *Wallis on a day-trip to St Ives in 1928) lived there until 1958. They formed the nucleus of a group of avant-garde artists who made the town an internationally recognized centre of abstract art, and it is to these artists that the term 'St Ives School' is usually applied, even though many of them had little in common stylistically, apart from an interest in portraying the local landscape in abstract terms. The one with the greatest international prestige was Naum *Gabo, who lived in St Ives from 1939 to 1946. After the war a number of abstract painters settled in or near the town or made regular visits. The residents included Terry *Frost and Patrick *Heron; the visitors included Roger *Hilton (who eventually settled in St Ives in 1965), Adrian *Heath, and Victor *Pasmore. Peter Lanyon (1918–64) was the only notable abstract artist to be born in St Ives. The heyday of the St Ives School was over by the mid-1960s, but the town continued to be an artistic centre. In 1993 the Tate Gallery opened a branch museum there (The Tate Gallery St Ives), housing changing displays of the work of 20th-cent. artists associated with the town. The building includes a stained-glass window commissioned from Patrick Heron.

St Martin's Lane Academy. A name of two successive organizations in St Martin's Lane, London, that were important training grounds for English artists in the half century before the *Royal Academy was established in 1768. They were founded in 1720 and 1735 respectively and each was a drawing and painting class rather than an *academy in the sense of a professional institution . The first St Martin's Lane Academy grew out of another academy (London's first) established in 1711 in Great Queen Street, of which *Kneller was the head. *Thornhill replaced Kneller in 1716, and in 1720, when Thornhill himself was deposed (see VANDERBANK), the academy moved to St Martin's Lane, where it was run on democratic principles. It soon became defunct, but *Hogarth reconstituted it in 1735. He described the room in which it met as 'big enough for a naked figure to be drawn after by thirty or forty people'.

Saint Phalle, Niki de (1930–). French sculptor, one of the great entertainers of modern art. She had no formal artistic training. In 1952 she started painting and in 1956 she began making *reliefs and *assemblages. She first made a public name in 1960 with 'rifle-shot' paintings that incorporated containers of paint intended to be burst and spattered when shot with a pistol. After she separated from her husband in 1960 she lived with Jean *Tinguely, with whom she collaborated on numerous projects, notably the enormous sculpture *Hon* (Swedish for 'she') erected at the Moderna Museet, Stockholm, in 1963. It was in the form of a reclining woman (more than 25 m. long) whose interior was a giant *'environment' reminiscent of a fun-fair: visitors entered through the vagina. The attractions inside included a milk bar in the breasts and a cinema showing Greta Garbo movies. Externally the figure was gaudily painted in a manner similar to that of her series of *Nanas*—grotesque fat ladies. Her other works have included *happenings and films, and since 1979, she has worked on a huge sculpture garden at Garavicchio in Italy. Her recent projects include a touching book on AIDS addressed to her son (*AIDS: You Can't Catch it Holding Hands*, 1987) and a huge figure of the Loch Ness Monster, made for an exhibition of her work in Glasgow in 1992.

Salle, David. See NEO-EXPRESSIONISM.

Sallinen, Tyko (1879–1955). The outstanding Finnish *Expressionist painter. He was the son of a tailor belonging to the strict puritan and fundamentalist religious sect of the Hihhulit, and the unhappy background of his early years later formed the basis for some of his paintings. After spending several years as an itinerant jobbing tailor in Sweden, he studied art in Helsinki and in 1909 and 1914 visited Paris, where he was influenced by French modernism, particularly *Fauvism. His favourite subjects were the harsh Karelian landscapes and simple, unsophisticated peasant people of Finland, whom he painted in brilliant but firmly organized colours. He was the leading figure of the *November Group and he came to be regarded as the nationalist Finnish painter *par excellence*. Among his masterpieces are *Washerwomen* (1911) and *The Hihhulit* (1918), both in the Ateneum, Helsinki. His fame was international and he exhibited much both in Finland and abroad.

salomónica. A column with a spirally twisted shaft, like a stick of barley sugar; the word is Spanish (meaning twisted) and the term 'barley sugar column' is often preferred in English. The name derives from a Roman example preserved in St Peter's, Rome, which according to legend came from Solomon's Temple. They were particularly popular in Spain and Portugal and their American colonies in the *Baroque era, when they were a common feature of *Churrigueresque decoration, but the most famous examples are those used by *Bernini for his **baldacchino* in St Peter's (1624–33). A splendid example of their use in England is the pair on the porch of St Mary the Virgin, Oxford (1637), the design of which is usually credited to Nicholas *Stone.

Salon. The official exhibition of members of the French Royal *Academy of Painting and Sculpture (see ACADEMY), first held in 1667. The name derived from the fact that exhibitions were held in the Salon d'Apollon in the *Louvre. Their frequency was originally somewhat irregular (though with stretches when they were held annually or biennially); from 1849 they became annual. The jury system of selection was introduced in 1748. As these were the only public exhibitions in Paris the official academic art obtained through them a stranglehold on publicity, and in the 19th cent. a number of rival Salons were organized by progressive artists. In 1881 the government withdrew official sponsorship and a group of artists organized the Société des Artistes Français to take responsibility for the show with a jury elected from each previous year's exhibitors. It still remained hostile to avant-garde artists, but from this time the Salon began to lose its prestige and influence in the face of the various independent exhibitions. The *Salon des Indépendants, for example, appeared in 1884; the *Salon de la Nationale in 1890; and the *Salon d'Automne in 1903.

Salon d'Automne. Annual exhibition, held in the autumn in Paris, founded in 1903 as an alternative to the official *Salon and the *Salon des Indépendants. *Bonnard, *Rouault, *Matisse, and *Marquet were foundation members and the Salon d'Automne is best remembered for its 1905 exhibition, when Matisse and his associates won notoriety as the *Fauves. Also of importance in the development of modern art were the memorial exhibitions of *Gauguin (1903, the inaugural exhibition) and *Cézanne (1907), which made their work widely known. Cézanne was also strongly represented at the 1905 exhibition.

Salon de la Nationale. Annual exhibition in Paris, founded in 1890 as the Société Nationale des Beaux-Arts (often called the Salon du Champ de Mars) by *Puvis de Chavannes, *Carrière, and *Rodin, together with others who seceded from the official *Salon.

Salon de la Rose + Croix. Art exhibition held annually in Paris from 1892 to 1897. It was organized by the Rosicrucians, an esoteric brotherhood (allegedly founded in the 15th cent. by one Christian Rosenkreuz) which in the late 19th cent. had close connections with the *Symbolist movement. The order's symbol, reflecting the name, was a rose and cross combined. Joséphin Péladan (1859–1918), a man of letters who called himself Sâr (i.e. High Priest) Péladan, founded a lodge of the brotherhood in France and this organized the exhibitions. They became a focal point of Symbolism, and the catalogue to the first exhibition said its objects were 'to destroy *Realism and to bring art closer to Catholic ideas, to mysticism, to legend, myth, allegory, and dreams'.

Salon de Mai. A gallery founded in Paris in 1949 by the art historian and critic Gaston Diehl for the purpose of encouraging and exhibiting the younger abstract artists. It did much to promote interest in abstract art during the 1950s.

Salon des Indépendants. Name given to the annual exhibitions held in Paris by the Société des Artistes Indépendants, an association formed in 1884 by *Seurat, *Signac, and other artists in opposition to the official *Salon. There was no selection committee and any artist could exhibit on payment of a fee. The Salon des Indépendants became the main showcase of the *Post-Impressionists and was a major art event in Paris up to the First World War.

Salon des Refusés. Exhibition held in Paris in 1863 to show work that had been refused by the selection committee of the official *Salon. In that year there were especially strong protests from artists whose work had been rejected, so the emperor Napoleon III (1808–73) ordered this special exhibition. It drew huge crowds, who came mainly to ridicule, and *Manet's *Déjeuner sur l'herbe* was subjected to furious abuse. Other major artists represented included *Cézanne, Camille *Pissarro, and *Whistler. In spite of the unfavourable reaction to the works shown there, the Salon des Refusés was of great significance in undermining the prestige of the official Salon. After this, artists began to organize their own exhibitions (notably the *Impressionists in 1874) and art dealers became of increasing importance. The Salon des Refusés is thus regarded as a turning point in the history of art and 1863 is, in the words of Alan Bowness (*Modern European Art*, 1972), 'the most convenient date from which to begin any history of modern painting'.

Salon du Champ de Mars. See SALON DE LA NATIONALE.

Salviati, Francesco (Francesco de' Rossi) (1510–63). Florentine *Mannerist painter, a pupil of *Andrea del Sarto. He adopted his name from his patron Cardinal Giovanni Salviati, with whom he went to Rome *c.*1530 and for whom he painted the work that established his reputation there—the fresco of the *Visitation* (1538) in S. Giovanni Decollato. In 1539 he travelled to Venice, and from the 1540s led a restless life, working mainly in Florence and Rome but also visiting Fontainebleau in 1556–7. He was one of the leading fresco decorators of his day, specializing in learned and elaborate multi-figure compositions, typically Mannerist in their artificiality and abstruseness, and similar in style to those of his friend *Vasari (although Salviati was an artist of higher calibre). His finest works are perhaps the frescos on the story of the ancient tyrant Furius Camillus (1543–5) in the Sala dell' Udienza of the Palazzo Vecchio, Florence, intended as an allegory of Cosimo de' *Medici's reign. Salviati also made designs for tapestries and was noted for his portraits, which were Florentine in their direct characterization but North Italian in their richness of colour. **Giuseppe Salviati** (*c.*1520–*c.*1575) was his pupil. He was born Giuseppe Porta, but borrowed his master's borrowed name. He worked mainly in Venice, painting numerous altarpieces and also decorations for civic buildings.

Samaras, Lucas (1936–). Greek-born sculptor and experimental artist who settled in the USA in 1948 and became an American citizen in 1955. His work has been extremely varied in scale, medium, and approach. At the end of the 1950s he was producing figures from rags dipped in plaster and pastels combining *Surrealist fantasy and *Pop art iconography. In 1959 he took part in *Kaprow's first *happening. During the 1960s he developed an original style in the use of mixed media, employing thousands of pins, yarns, nails, etc., and he also experimented with *assemblages of diverse objects. He also experimented with light and reflection, notably in his *Mirrored Room* (1966), an environment walled with mirrors in which the spectator was reflected endlessly. Perhaps his best-known works are his 'Autopolaroids'—photographs of the most intimate parts of his own body that he began making in 1970 and which brought him considerable notoriety.

Sambin, Hugues (1515/20–1601/2). French *Mannerist architect, sculptor, and designer, active in Dijon. His work was the finest expression of French provincial architecture during the second half of the 16th cent., notable for its rich and fanciful use of rustication and high-relief carving. Sambin's vivid imagination comes out also in his book of engravings. *Œuvre de la diversité des termes, dont on use en architecture* (1572), which is full of *grotesque and bizarre semi-human figures adorned with flowers, fruits, *putti, and so on. It seems to have been a popular source book for local carvers.

Sánchez Coello, Alonso (1531/2–88). Spanish painter of Portuguese parentage, principally a portraitist. He studied under Anthonis *Mor in Flanders in the early 1550s and by 1555 was working at the court of Philip II (1527–98) of Spain, where he remained for the rest of his life, achieving great success and becoming a personal favourite of the king, who was godfather to his two daughters. His dignified and sober style was strongly influenced by Mor, but was more sensuous, showing his admiration for *Titian, and it is regarded as marking the beginning of the great tradition of Spanish formal portraiture. He also painted religious works, but these are much less distinguished. His work is well represented in the Prado, Madrid.

Sánchez Cotán, Juan (1560–1627). Spanish painter. He was a still-life painter in Toledo until 1603, when he decided to become a monk, and in the following year he entered the Carthusian monastery at Granada as a lay-brother. The religious works he painted after this date are accomplished but unexceptional, but as a still-life painter he ranks with the great names of European painting. Characteristically he depicts a few simple fruits or vegetables, arranged on a ledge or shelf with an almost geometric clarity and standing out against a dark background (*Quince, Cabbage, Melon, Cucumber*, San Diego Museum of Art, 1602). Each form is scrutinized with such intensity that the pictures take on a mystical quality, conveying a feeling of wonder and humility in front of the humblest items in God's creation. Sánchez Cotán's austere style had considerable influence on Spanish painting, notably on *Zurbarán.

sandarac. A soft *resin used in the preparation of *varnishes. It becomes darker and redder with age, so the soft resin *mastic is usually preferred.

Sandby, Paul (1730/1–1809). English topographical watercolourist and graphic artist. He and his brother **Thomas** (1721–98) trained at the Military Drawing Office of the Tower of London and were engaged as draughtsmen on the survey of the Highlands of Scotland after the rebellion of 1745. In 1751 Paul went to live with his brother at Windsor Park, where Thomas held the position of Deputy Ranger. Their work is similar in many respects, but Paul was more versatile as well as a better artist, his work including lively figure subjects as well as an extensive range of landscape subjects. In his later work he often used *body-colour (he also sometimes painted in oils) and he was the first professional artist in England to publish *aquatints (1775). He was singled out by *Gainsborough as the only contemporary English landscape artist who painted 'real views from nature' instead of artificial *Picturesque compositions. The Victoria and Albert Museum has a varied collection of Sandby's aquatints, and also of his etchings, which have a directness of observation anticipating *Bewick. He is also well represented in the British Museum and at Windsor Castle, of which he painted many views. Sandby has rather unjustifiably been called 'the father of watercolour art', but certainly his distinction won prestige for the medium. He was a founder member of the *Royal Academy and his brother was its first Professor of Architecture.

Sandrart, Joachim von (1606–88). German painter, engraver, and writer on art. He travelled widely and was the most highly regarded German artist of his day (he was ennobled in 1653), but he is now remembered almost exclusively for his treatise *Teutsche Academie der Edlen Bau-, Bild- und Mahlerey-Künste* (German Academy of the Noble Arts of Architecture, Sculpture and Painting), published in Nuremberg in 1675–9 (a Latin edition followed in 1683). This treatise, which is divided into three main parts, is a source-book of major importance. The first part is an introduction to the arts of architecture, painting, and sculpture put together largely from material taken from earlier sources such as *Vasari and van *Mander. The second part, consisting of biographies of artists, likewise contains much material borrowed from previous writers but also much that is original, in particular about German artists (it was Sandrart who was the first to use the name *Grünewald) and on contemporary artists that the author knew personally. The third part contains information about art collections and a study of *iconography, and remarkably Sandrart also included in his book a chapter on Far Eastern art. Sandrart was the first director (1662) of the Academy at Nuremberg (the earliest such in Germany). The immense prestige he enjoyed is reflected in his grand *Self-portrait* (Historisches Museum, Frankfurt).

sanguine. Another name for red *chalk.

Sano di Pietro (properly Ansano di Pietro di Mencio) (1406–81). Sienese painter, whose busy workshop was the main source of altarpieces for the village churches around the city. They are in a weakened and unsophisticated version of the style of *Sassetta.

Sansovino, Andrea (*c.*1467/70–1529). Italian sculptor and architect, named after his birthplace near Monte San Savino, his real name being Contucci. He called himself a Florentine, but although Florence contains much of his sculpture his fame rests on what he did elsewhere, notably in Rome, where the companion tombs of Cardinals Sforza and Basso (Sta Maria del Popolo, between 1505 and 1509) and the group of *The Virgin and Child with St Anne* (S. Agostino, 1512) display classical grace combined with human tenderness.

Sansovino spent much of the period 1513–27 in Loreto in charge of the sculpture (and for part of the time the architecture also) of the shrine of the Holy House, originally designed by *Bramante. The *reliefs he supervised and in part carved himself are highly pictorial and rather crowded. *Vasari says that Andrea also worked in Portugal, but his statement is doubted by some scholars. Jacopo *Sansovino was Andrea's main pupil.

Sansovino, Jacopo (Jacopo Tatti) (1486–1570). Florentine sculptor and architect, active mainly in Venice. He trained under Andrea *Sansovino, whose name he adopted as a sign of his admiration. In 1505/6 he followed Andrea to Rome, where he moved in the circle of *Bramante and *Raphael and worked on the restoration of ancient sculpture, gaining a thorough knowledge of Roman art. From 1511 to 1517 he was again in Florence, where he shared a studio with *Andrea del Sarto, and he then returned to Rome until the Sack of 1527, when he moved to Venice. There he was appointed state architect (1529), formed a close friendship with *Titian, and became a dominant figure in the art establishment. Sansovino played a major role in introducing the High *Renaissance style to Venice in both architecture and sculpture, and his sculptures are often important decorative elements of his buildings. His most celebrated work, one of Venice's most familiar sights, is the glorious Library of San Marco (begun 1537), a building that Andrea Palladio (1508–80), his successor as the city's leading architect, described as 'probably the richest ever built from the days of the ancients up to now'. As a sculptor he is best known for the colossal figures of *Mars* and *Neptune* (commissioned 1554) on the staircase of the Doges' Palace. Sansovino's sculptural style was firmly rooted in his study of antiquity, but it was in no way academic and possessed great vitality. He studied assiduously from the life as well as from the *antique, and legend has it that the model for his *Bacchus* (Bargello, Florence, 1511–12) went mad through being made to pose for hours on end with his arm raised and one day was found in this position standing naked on top of a chimney.

Sansovino's son **Francesco** (1521–86) was a scholar of diversified interests. His *Venetia città nobilissima* . . . (Venice, most noble city), published in 1581, is an important sourcebook—the first attempt to give a systematic account of a city's artistic heritage.

Santayana, George (1863–1952). American philosopher, humanist, and poet. His views on art are expressed in *The Sense of Beauty* (1896) and *The Life of Reason* (1905–6) and had an important influence on theory of art and taste in the early 20th cent. Santayana refused to consider theory of art or *aesthetics except as part of an overall theory of values embracing rational living, happiness, and freedom. He included theories of art within theory of morals and on this basis developed them with exceptional sensitivity and wide learning. His style was refreshingly unpedantic and he deliberately avoided technical terms.

Santerre, Jean-Baptiste (1651–1717). French painter, known in his day chiefly for his allegorical portraits. He also painted religious pictures in a nascent *Rococo style that caused some scandal by their use of provocative nudes in a manner which foreshadowed *Boucher and *Fragonard (*Susanna*, Louvre, Paris, 1704).

Santi, Giovanni (d. 1494). Italian painter, the father of *Raphael, active mainly in Urbino, where he worked for the court. He was 'a mediocre painter but an intelligent man' (*Vasari) and no doubt gave his illustrious son his introduction to humanist culture. Santi is now remembered less for his paintings than for his verse chronicle in twenty-three books recounting the exploits of the dukes of Urbino, which he dedicated to the young Duke Guidobaldo da *Montefeltro. This work is interesting to the art historian because it includes incidental comments on the reputations of contemporary artists. Santi's house in Urbino is now a museum—the Casa di Raffaello. It contains a small fresco of the Virgin and Child that some authorities consider to be Santi's portrayal of Raphael and his mother, and others claim as a very early work by Raphael himself. A *Virgin and Child* attributed to Santi is in the National Gallery, London.

Santvoort, Dirck Dircksz. (1610/11–80). Dutch portrait painter, great-grandson of Pieter *Aertsen, active in his native Amsterdam. He worked in an old-fashioned style, but his rather naïve-looking figures often have great charm (particularly his portraits of children) and his career was highly successful. A good example of his work is *A Girl with a Finch* (NG, London, 1631). He had two painter brothers, **Abraham** (*c*.1624–69) and **Peter** (*c*.1603–35).

Saraceni, Carlo (1579–1620). Italian painter. He was born and died in Venice but spent almost all his career in Rome. There he formed his style under the influence of *Caravaggio and *Elsheimer, painting small luminous pictures of figures in landscapes as well as much larger altarpieces, including the replacement for Caravaggio's *Death of the Virgin* (Louvre, Paris), which the church of Sta Maria della Scala had rejected in 1606. Saraceni's replacement is still *in situ*. He painted several other smaller variants or versions of the picture, so the design was evidently popular. His style was sensitive and poetic, showing a delicate feeling for colour and tone. His liking for turbans, tasselled fringes, stringy drapery folds, and his richly *impasted paint may have influenced Dutch artists in Rome such as *Lastman and *Pynas, and through them *Rembrandt. He also possibly had some influence in Lorraine, through his pupil, the French Caravaggesque painter Jean Le Clerc (*c*.1587–1633). Le Clerc was also an engraver; his prints include one of his master's *Death of the Virgin* (after a small version on copper, now in the Alte Pinakothek, Munich, rather than the large altarpiece in Rome).

sarcophagus. The name given in Classical antiquity to a stone or *terracotta coffin or tomb-chest. According to *Pliny the name (Greek for 'flesh-devouring') was derived from the custom of making or lining coffins with a slate stone found in Asia Minor which had the property of destroying the flesh of the corpse. From early *Hellenistic times until the end of the Early Christian period sarcophagi constitute one of the most important surviving sources for our knowledge of funerary sculpture in antiquity.

Sargent, John Singer (1856–1925). American painter, chiefly famous as the outstanding society portraitist of his time: *Rodin called him 'the van *Dyck of our times'. He was born in Florence, the son of wealthy parents, and he had an international upbringing and career—indeed, he has been described as 'an American born in Italy, educated in France, who looks like a German, speaks like an Englishman, and paints like a Spaniard' (William Starkweather, 'The Art of John S. Sargent', *Mentor*, October 1924). His 'Spanishness' refers to his deep admiration for *Velázquez, for although he was encouraged to paint directly by his teacher *Carolus-Duran in Paris (1874–6), the virtuoso handling of paint that characterized his style derived more particularly from Old Masters such as Velázquez and *Hals (he visited Madrid and Haarlem to study their work in 1879–80). In 1884 he became famous when his portrait of Madame Gautreau (Met. Mus., New York) caused a sensation at the Paris *Salon because of what was felt to be its provocatively erotic character. It was exhibited as *Madame X*, but the sitter, a society beauty of strikingly unconventional looks, was unmistakable, and her mother wrote to Sargent imploring him to withdraw the picture, which she said had made her daughter a laughing-stock. The scandal persuaded Sargent to move to London, and he remained based there for the rest of his life; he continued to travel extensively, however, and often visited America. The lavish elegance of his work brought him unrivalled success, and his portraits of the wealthy and privileged convey with brilliant bravura the glamour and opulence of high society life. Even in his lifetime he was deprecated by critics and artists (who no doubt found it easy to be jealous of his success) for superficiality of characterization, but although psychological penetration was certainly not his strength, he was admirably varied in his response to each sitter's individuality.

As with many successful portraitists, Sargent's heart lay elsewhere (he called portraiture 'a pimp's profession'), and after about 1907 he took few commissions. Despite his sophistication and charm and the entrée to high society that his success gave him, he was a very private person, who never married and led a quiet life. He loved painting landscape watercolours, showing a technique as dashing in this medium as in his oil paintings, and from the 1890s he devoted much of his energies to ambitious allegorical murals in the Public Library and Museum of Fine Arts in Boston (as models for the *Danaides* in one of these he used dancing girls from the chorus line of the Ziegfeld Follies). The most unexpected side to his talents is revealed in the enormous *Gassed* (Imperial War Mus., London, 1918), which he painted as an *Official War Artist. It has remarkable tragic power and is one of the greatest pictures inspired by the First World War. Sargent's reputation plummeted after his death but has soared again since the 1970s.

Sarrazin, Jacques (1588–1660). The leading French sculptor of the mid-17th cent. He was in Rome 1610–*c.*1627, during which time he formed his style on the example of the *antique and *classicizing artists such as *Domenichino and *Duquesnoy. On his return to Paris he established himself as head of his profession, running an important workshop through which passed many of the leading sculptors of the following generation (those who worked at Versailles). Much of Sarrazin's best work was done as architectural decoration, the finest example being his *caryatids (1641) on the Pavilion de l'Horloge at the *Louvre. Sarrazin also supervised the decoration (1652–60) of the Château de Maisons, the masterpiece of François Mansart (1598–1667), the greatest French architect of the 17th cent. Although his work does not rank on the same level, Sarrazin's dignified Classicism forms a kind of sculptural parallel to the architecture of Mansart or the paintings of *Poussin.

Sassetta (Stefano di Giovanni) (1392?–1450). With *Giovanni di Paolo the outstanding Sienese painter of the 15th cent. His work continues Sienese tradition in its beautiful colouring and elegant line, but he was also influenced by the *International Gothic style and by contemporary Florentine developments, combining them into a highly personal manner expressive of his mystical imagination. His most important work was the St Francis altarpiece (1437–44) painted for the church of S. Francesco at Borgo San Sepolcro and now dispersed. The central panel, *St Francis in Ecstasy* (Berenson coll., Florence), has a monumental dignity that must have impressed *Piero della Francesca (a native of Borgo), and some of the other panels (seven are in the NG, London) show him at his most lyrical.

Sassoferrato (Giovanni Battista Salvi) (1609–85). Italian painter, known by the name of his town of birth—Sassoferrato—and active in nearby Urbino and other cities of central Italy, notably Rome (where he was a pupil of *Domenichino) and Perugia. He did some portraits, but specialized in religious works painted in an extremely sweet, almost *Peruginesque style. They are very clearly drawn and pure in colouring and totally un-*Baroque in feeling—indeed, they have a deliberately archaic quality that brings the paintings of the *Nazarenes to mind. Little is known of his life (in the 18th cent. it was evidently generally believed he was a contemporary of *Raphael) and few of his pictures are dated or datable; they seem to have been in great demand, however, as his compositions often exist in numerous very similar versions. Most of them were presumably done for private collectors, as few are in churches. Examples of his work are in many galleries, for example, the National Gallery and the Wallace Collection in London. A fine collection of his drawings (virtually his entire surviving output) is in the Royal Library at Windsor Castle.

Saura, Antonio (1930–). Spanish painter. He was self-taught and began painting during a long illness in 1947. During two years in Paris, 1953–5, he made contact with members of the *Surrealist movement, including André *Breton. After his return to Spain, however, he quickly abandoned Surrealism and worked in a semi-abstract *expressionist vein, his powerful, stormy, and thickly textured figures creating a feeling of tortured humanity (*Great Crucifixion*, Boymans Mus., Rotterdam, 1963). His work, at times almost frenzied in its intensity, has been seen as one of the most forceful protests against the Franco regime.

Savery, Roelandt (1576?–1639). Flemish painter and etcher of landscapes, animal subjects, and still life, the best-known member of a family of artists. He was born in Cambrai, grew up in Amsterdam, and in 1619 settled in Utrecht, but he is best known for his association with Prague, where he worked for the emperor Rudolf II from 1603 to 1613. Rudolf's famous menagerie allowed him to study in detail the exotic animals that became the trademark of his work. He painted and drew creatures such as pelicans, ostriches, camels, and the now extinct dodo, and was one of the first artists in the Netherlands to do pictures of animals alone. His favourite subjects, however, were Orpheus and the Garden of Eden, which allowed him to include any number of colourful beasts. A dated example (1628) of *Orpheus with the Animals*, of which there are said to be several dozen extant paintings from his hand, is in the

National Gallery, London. Savery's bright and highly finished style is similar to that of Jan 'Velvet' *Brueghel, but is somewhat more archaic. His rare flower paintings are sometimes of outstanding quality, and with *Bosschaert he was an influential Flemish exponent of this genre in Holland. *Houbraken says that Savery died insane.

Savoldo, Giovanni Girolamo (active 1506–after 1548). Italian painter, perhaps born in Brescia, first documented in Parma in 1506, and active mainly in Venice. His output was small and his career is said to have been unsuccessful, but he is now remembered as a highly attractive minor master whose work stands somewhat apart from the main Venetian tradition. His forté was night scenes, in which he gave his lyrical sensibility and liking for unusual light effects full play. One of the best-known examples is *Mary Magdalen Approaching the Sepulchre*, of which several versions exist, one in the National Gallery, London. The writer Pietro Aretino described Savoldo as 'decrepit' in 1548 and he is not heard of thereafter.

scagliola. Imitation *marble, a compound of marble fragments, *plaster of Paris, colouring matter, and glue. It was known in Classical antiquity, and the secret of its manufacture was rediscovered in the 17th cent. in northern Italy, where it was used for intricate coloured inlay work. In the 18th cent. it was used by British architects such as Robert *Adam.

scarab. The commonest type of Egyptian amulet, made in the form of a sacred beetle (chiefly *scarabaeus sacer*) and associated with ideas of spontaneous generation and renewal of life, as well as with the solar deity Khepri. Scarabs became popular from the end of the 3rd millennium BC, and innumerable examples made of varied materials, including semi-precious stones, have been found in tombs and elsewhere. Some bear the owner's name and were used as seals, but most have decorative designs, messages of good luck, or words of power, including royal *cartouches. Scarabs spread throughout the Near East and the Mediterranean world and were reproduced in many centres during the 1st millennium BC.

Scarsellino, Lo (Ippolito Scarsella) (*c.*1551–1620). The leading Ferrarese painter of his period. He produced a good deal of large-scale work but is now remembered mainly for his small mythological or religious scenes set in landscapes, several of which are in the Borghese Gallery, Rome. Their romantic feeling and spirited technique recall Venetian painting and they were particularly influential on the young *Guercino.

scauper. See SCORPER.

Schad, Christian. See NEUE SACHLICHKEIT.

Schadow, Johann Gottfried (1764–1850). German sculptor, graphic artist, and writer on art. He travelled in Italy, 1785–7, and in 1788 he settled in Berlin, where he became head of the Academy in 1816. His style was *Neoclassical (he knew *Canova in Italy), but retained a degree of *Baroque liveliness. He was active mainly as a portraitist and tomb sculptor, but his best-known work is the *quadriga* (four-horse chariot) (1793) surmounting the Brandenburg Gate in Berlin, which was badly damaged in the Second World War. His finest achievement is perhaps the charming and sensitive group of *The Princesses Luise and Frederika of Prussia* (Staatliche Mus., Berlin, 1795–7). From the 1820s his sight deteriorated and he turned more to graphic work (he was a draughtsman, engraver, and lithographer) and to writing on art theory.

He had three artist sons; **Felix Schadow** (1819–61), **Rudolf Schadow** (1786–1822) and, most importantly, **Wilhelm von Schadow** (1788–1862). Wilhelm was in Italy 1811–19 and a member of the *Nazarenes. He taught at the Berlin Academy, then in 1826 became Director of the Düsseldorf Academy, which he helped to make into a leading centre of history painting.

Schalken, Godfried (1643–1703). Dutch painter of biblical, allegorical, and anecdotal subjects and of portraits, active mainly in Leiden. He was taught by *Hoogstraten and *Dou, and was one of Dou's most distinguished successors in the Leiden *fijnschilder* (fine painter) tradition. His surfaces are as smooth, glossy, and delicately finished as Dou's, but Schalken differed from his master in his lighting effects (he had a strong preference for the red glow of candle-light) and in his penchant for scenes involving coquettish women (*Girl with a Candle*, Pitti, Florence, 1662). As with Dou, Schalken's great contemporary popularity lasted throughout the 18th cent.

Schedoni, Bartolomeo (1578–1615). Italian painter, born near Modena and active mainly in Parma. He was trained in the local *Mannerism of Modena, but in Parma came under the pervading influence of *Correggio and later he was affected by the Roman *Baroque. By the end of his short life he had developed from these diverse influences a distinctive style characterized by bold lighting, poetic feeling, and a striking use of areas of bright, almost metallic colour (*The Three Maries at the Tomb*, Galleria Nazionale, Parma).

Scheemakers, Peter (1691–1781). Flemish sculptor, active mainly in England, where he settled around 1720. Apart from a visit to Rome (1728–c.1730) he was based in London until 1771, when he returned to his native Antwerp. In the 1730s he became the main rival to *Rysbrack, whose prices he undercut. Rysbrack defeated him in competition for the equestrian statue of William III in Bristol, but Scheemakers gained a great success in 1740 with his Shakespeare monument in Westminster Abbey and thereafter received numerous commissions and controlled a large studio with many assistants. He was, indeed, perhaps the most prolific of mid-18th-cent. sculptors, particularly as a maker of monuments, ranging from small tablets to large multi-figure compositions. Rysbrack also made portrait busts, and these, like his monuments, are often severely classical, reflecting his love of the *antique. His Shakespeare monument, which was designed by William *Kent, is much more lively and *Baroque in spirit than most of his work. Scheemakers' brother **Henry** (d. 1748) and his son **Thomas** (1740–1808) were also sculptors working in England.

Scheffer, Ary (1795–1858). Dutch painter, engraver, and book illustrator, active for almost all his career in Paris. His work was immensely popular in his lifetime, but is now generally considered sentimental. Early in his career he favoured literary themes (*Francesca da Rimini*, Wallace Coll., London 1835, and other versions), but later he turned to mawkishly treated religious subjects (*SS. Augustine and Monica*, NG, London, 1854, and other versions). He also painted many portraits. His work is well represented in the museum at Dordrecht, his native town.

Schelfhout, Andreas (1787–1870). Dutch landscape painter, a forerunner of the *Hague School. He is best known for carefully depicted winter scenes reminiscent of those made by 17th-cent. Dutch artists, but he also painted seascapes and was an accomplished water-colourist. His work was influential particularly on *Jongkind, his most famous pupil.

Schiavone, Andrea (Andrea Meldolla) (*c.*1510/15–1563). Italian painter and etcher. His nickname 'Schiavone' means 'Slav', reflecting the fact that he came from Zara, Dalmatia (then under Venetian jurisdiction). He worked mainly in Venice, where he was on friendly terms with *Titian (who along with *Parmigianino was one of the main influences on his style). His most characteristic works were fairly small-scale religious or mythological scenes for private patrons in a vigorous, painterly style.

Schiele, Egon (1890–1918). Austrian painter and draughtsman. He studied at the Vienna Academy, where he met *Klimt, who was a strong influence on his early *Art Nouveau style. By 1909, however, he had begun to develop his own highly distinctive style, which is characterized by an aggressive linear energy expressing acute nervous intensity. He painted portraits, landscapes, and semi-allegorical works, but he is best known for his drawings of nudes (including self-portrait drawings), which have a disturbing and explicit erotic power—in 1912 he was briefly imprisoned on indecency charges, and several of his drawings were burnt. The figures he portrays are typically lonely or anguished, their bodies emaciated and twisted, expressing an aching intensity of feeling. His work was much exhibited, and he was beginning to receive international acclaim when he died (three days after his wife) in the influenza epidemic of 1918. He has since come to be recognized as one of the greatest *Expressionist artists, and his work is admired for its powerful mastery of line as well as for the compassion which infuses his often repulsive images of humanity.

Schildersbent (Dutch: 'band of painters'). A fraternal organization founded in 1623 by a group of Netherlandish artists living in Rome for social intercourse and mutual assistance. Its members called themselves *Bentvueghels* or 'birds of a flock' and they had individual *Bentnames*—for example Pieter van *Laer, one of the early leaders, was called Bamboccio. In 1720 the Schildersbent was dissolved and prohibited by papal decree because of its rowdiness and drunkenness.

Schinkel, Karl Friedrich (1781–1841). German architect, painter, and designer, active mainly in Berlin. Schinkel was the greatest German architect of the 19th cent., a brilliant exponent of both the Greek Revival and *Gothic Revival styles and also of a functional style free from historical associations. Until 1815, however, when he gained a senior appointment in the Public Works Department of Prussia (from which position he effectively redesigned Berlin), he worked mainly as a painter and stage designer. His paintings are highly *Romantic landscapes somewhat in the spirit of *Friedrich, although more anecdotal in detail (*Gothic Cathedral by a River*, Staatliche Museen, Berlin, 1813–14). He continued working as a stage designer until the 1830s, and in this field ranks among the greatest artists of his period. His most famous designs were for Mozart's *The Magic Flute* (1815), in which he combined the clarity and logic of his architectural style with a feeling of mystery and fantasy.

Schjerfbeck, Helène (1862–1946). Finnish painter. She studied in Paris in the 1880s, her teachers including *Bastien-Lepage, and made her mark with **plein air* scenes notable for their fresh colouring. From about 1900 her health began to fail and she lived a solitary life, almost forgotten, developing a much more simplified style. Late in life she was recognized as one of the pioneers of modernism in Finland. She painted landscapes, still lifes, figure compositions, and portraits, including a series of self-portraits that are regarded as being among her finest works.

Schlemmer, Oskar (1888–1943). German painter, sculptor, stage designer, and writer on art. He was an important teacher at the *Bauhaus (1920–9), in the metalwork, sculpture, and stage-design workshops, and later taught in Breslau and Berlin, before he was dismissed by the Nazis, who declared his work *degenerate. Schlemmer had a mystical temperament and his ideas on art were complex. Some of his early work was influenced by *Cubism and he showed a deep concern for pictorial structure; characteristically his paintings represent rather mechanistic human figures, seen in strictly frontal, rear, or profile attitudes set in a mysterious space (*Group of Fourteen Figures in Imaginary Architecture*, Wallraf-Richartz-Museum, Cologne, 1930). He rejected what he considered the soullessness of pure abstraction, but he wished to submit his intuition to rational control: 'Particularly in works of art that spring from the imaginative faculty and the mysticism of our souls, without the help of external subject-matter or content, strict regularity is absolutely indispensable.' His cool, streamlined forms are seen also in his sculpture. Schlemmer did much work for the theatre, notably designs for the *Constructivist *Triadic Ballet*, to music by Paul Hindemith (1895–1963), which was performed at the Bauhaus in 1923.

Schlichter, Rudolf. See NEUE SACHLICHKEIT.

Schlüter, Andreas (*c.*1660–1714). German sculptor and architect, active mainly in Berlin. He was the leading *Baroque sculptor in northern Germany, and his masterpiece is the splendid equestrian monument to the Great Elector Friedrich-Willem (Charlottenburg Castle, Berlin, 1696–1708). He fell under a cloud when some of his architectural work collapsed, and he moved to St Petersburg, where he died soon after arriving.

Schmidt-Rottluff, Karl (né Schmidt) (1884–1976). German *Expressionist painter and graphic artist born at Rottluff near Chemnitz (he added the 'Rottluff' to his name in 1906). He studied architecture at the Technische Hochschule, Dresden, and in 1905 became one of the founders of the *Brücke. His style was harsher than that of the other members of the group, and it was particularly forceful in his superb woodcuts, which are among the finest representatives of Expressionist graphic art. Their abrupt manner was reflected in his paintings with their flat ungraduated planes of contrasting colours. In 1906 he stayed with *Nolde on the island of Alsten, Norway, and in 1907 he painted with *Heckel at Dangast on the coast north-west of Bremen. Apart from landscapes such as the ones he painted in these places, his work included portraits, figure compositions, and still lifes. In 1911, with the other members of Die Brücke, Schmidt-Rottluff moved to Berlin, where he lived for most of the rest of his life. In the 1910s and 1920s he was influenced by the stylized forms of African sculpture (*Dr Rosa Schapire*, Tate, London, 1919). Later, perhaps under the influence of *Kirchner or as a result of a trip to Paris in 1924 and a longer stay in Italy in 1930, his style became somewhat more naturalistic. His work was declared *degenerate by the Nazis and in 1941 he was forbidden to paint. In 1947 he became a professor at the Berlin Hochschule für Bildende Künste. The Brücke-Museum in Berlin was founded on his initiative in 1967 and he gave sixty of his own works to it. There are good collections of his graphic work in the Victoria and Albert Museum, London, and the Leicestershire Museum and Art Gallery, Leicester.

Schnabel, Julian. See NEO-EXPRESSIONISM.

Schnorr von Carolsfeld, Julius (1794–1872). German painter and illustrator. After studying at the Vienna Academy he joined the *Nazarenes in Rome (1817–25) and with them painted frescos in the Casino Massimo. In 1827 he moved to Munich, where he worked for Ludwig I (1786–1868), painting mainly frescos in the royal palace. As head of the Dresden Academy from 1846 onwards, Schnorr exercised a considerable influence in Germany as a representative of Nazarene ideals. As a graphic artist, he was famous in his day for his Bible illustrations, which were also published in England (*Schnorr's Bible Pictures*, 1860).

Schönfeld, Johann Heinrich (1609–82/3). German painter and etcher. His early career is not well documented; he spent about a decade in Italy (mainly Rome and Naples) as a young man, but it is only after he settled in Augsburg in 1652 that his development can be traced through dated works. He was versatile and prolific, painting historical and *genre subjects as well as many altarpieces for churches in southern Germany. His style was lively and *eclectic, drawing on various Italian influences, and in his delicate colouring and lightness of touch he anticipates elements of German *Rococo art.

Schongauer, Martin (d. 1491). German engraver and painter, active in Colmar, Alsace. In his day he was probably the most famous artist in Germany; it was in his workshop that the young *Dürer hoped to study, but when he arrived in Colmar in 1492 the master had recently died. Only one painting certainly by Schongauer survives—*Madonna in the Rose Garden* (St Martin's, Colmar), dated on the back 1473—and he is remembered chiefly as an engraver, the greatest of his period. His work was strongly influenced by Netherlandish art, above all by Rogier van der *Weyden, but Schongauer had a powerful imagination of his own. He concentrated on religious subjects and about 115 plates by him are known. In them he brought a new richness and maturity to engraving, expanding the range of tones and textures, so that an art that had previously been the domain of the goldsmith took on a more painterly quality. The gracefulness of his work, which was widely influential, became legendary, giving rise to the nicknames 'Hübsch ('charming') Martin' and 'Schön ('beautiful') Martin'.

School of London. An expression coined by R. B. *Kitaj in 1976 and given wide currency when it was used as the title of a British Council exhibition that toured Europe in 1987—'A School of London: Six Figurative Painters' (the six were Michael *Andrews, Frank *Auerbach, Francis *Bacon, Lucian *Freud, Kitaj himself, and Leon *Kossoff). In 1989 the term was used as the title of a book by Alistair Hicks—*The School of London: The Resurgence of Contemporary Painting*. The book covers other artists apart from the original six, including Howard *Hodgkin. It is not clear what the label is meant to mean and several critics have denied that any such thing as a School of London exists.

School of Paris. See PARIS, SCHOOL OF.

Schooten, Floris van (active *c*.1610–*c*.1655). Dutch still-life painter active in Haarlem, a specialist in pictures of food laid on a table. His style was originally rather rigid and austere, but became more naturalistic under the influence of his more original contemporaries, *Heda and *Claesz.

Schooten, Joris van. See REMBRANDT.

Schouman, Aert (1710–92). Versatile Dutch artist whose work varied from painting wallpaper to engraving glass, active in Dordrecht and The Hague. His best works are his delicate watercolours of birds and animals.

Schwind, Moritz von (1804–71). Austrian painter, graphic artist, and designer, trained in Vienna and in Munich under *Cornelius. Schwind represents the tail-end of Germanic *Romanticism and his most characteristic works depict an idealized fairy-tale Middle Ages, with knights in armour, damsels in distress, enchanted woods and castles, and much loving depiction of costume, architecture, etc. He was at his best working on a small scale, as in his numerous book illustrations and his woodcuts for *Fliegende Blätter*, a humorous Munich periodical. He visited England in 1857 (for the great exhibition of Old Masters in Manchester) and afterwards designed stained glass for St Michael's, Paddington, and Glasgow Cathedral. In his young days Schwind, who was an accomplished violinist, had been friendly with the composer Franz Schubert (1797–1828) and late in his own life he depicted Schubert's Vienna circle in a number of drawings.

Schwitters, Kurt (1887–1948). German painter, sculptor, maker of constructions, writer, and typographer, a leading figure of the *Dada movement who is best known for his invention of 'Merz'. The word was first applied to collages made from refuse, but Schwitters came to use it of all his activities, including poetry. He used the word as a verb as well as a noun: a fellow artist was once nonplussed when Schwitters asked him to *merz* with him. In his early work Schwitters was influenced by *Expressionism and *Cubism, but after the First World War (in which he served for a time as a draughtsman) he became the chief (indeed, virtually only) representative of Dada in Hanover. In 1918 he began making collages from refuse such as bus tickets, cigarette wrappers, and string, and in 1919 he invented Merz. The name was reached by chance: when fitting the word 'Commerzbank' (from a business letterhead) into a collage, Schwitters cut off some letters and used what was left. He called the collages Merzbilden (Merz pictures) and in about 1923 began to make a sculptural or architectural variant—the Merzbau (Merz building)—in his house in Hanover (it was destroyed by bombing in 1943). From 1923 to 1932 he published a magazine called *Merz* and in this period he was much occupied with typography. In 1937 his work was declared *degenerate by the Nazis and in the same year he fled to Norway, where at Lysaker he began a second Merzbau (destroyed by fire in 1951). When the Germans invaded Norway in 1940, he moved to England, where he lived for the rest of his life—in London (after release from an internment camp) from 1941 to 1945, and then at Ambleside in the Lake District. Here, in an old barn in Langdale, he began work on his third and final Merzbau, with financial aid from the Museum of Modern Art, New York. It was unfinished at his death and is now in the Hatton Gallery, Newcastle upon Tyne.

Scipione (Gino Bonichi) (1904–33). Italian painter. He was the son of a soldier and adopted his pseudonym (in 1927) in homage to Scipio Africanus, the great Roman general who defeated Hannibal. He studied briefly (1924–5) at the Academy in Rome before being expelled with his friend Mario Mafai (1902–65), with whom he introduced a romantic *Expressionist vein into Italian painting in opposition to the prevailing pomposity of officially approved art under Mussolini's Fascist government. His subjects were mainly scenes of modern Rome, painted with violent brushwork and a feeling of visionary intensity. His career was very short, virtually ending in 1931 because of the tuberculosis that killed him, but he was highly influential, becoming a symbol of heroic individuality to Italian artists after the Second World War.

Scopas of Paros. One of the most celebrated Greek sculptors, active in the mid-4th cent. BC. He is recorded as working on the Mausoleum of Halicarnassus, the temple of Artemis at Ephesus, and the temple of Athena Alea at Tegea, of which *Pausanias says he was the architect. Sculptural remains from all three buildings have survived, and although none of them can be certainly associated with Scopas, it is likely that he is the author of certain pieces that show a distinctive style and are of a quality consonant with his elevated reputation. Among them are three slabs from the Mausoleum (BM, London) showing the Battle of Greeks and Amazons, which display the intensity of expression and the characteristically deep-set eyes that are considered typical of his work. Several other pieces have been associated with him on stylistic grounds. In spite of the lack of solid evidence, Scopas is presumed to have ranked with *Praxiteles and *Lysippus as the leading Greek sculptor of the mid to late 4th cent. BC, his work heralding the emotionalism of *Hellenistic sculpture.

Scorel, Jan van (1495–1562). The first Netherlandish artist to bring the ideals of the Italian *Renaissance to the area we today call Holland. He was born in Schoorel near Alkmaar, and after a varied artistic training worked in Utrecht with *Gossaert, who probably encouraged him to visit Italy. Scorel set out in 1519: stopping at Speyer, Strasburg, Basle, and Nuremberg, he eventually reached Venice, where he joined a group of pilgrims and sailed to Cyprus, Rhodes, Crete, and on to Jerusalem. He returned to Venice the following year and finally made his way to Rome, where he found favour with Pope Adrian VI (1459–1523), a native of Utrecht. By 1524 Scorel was back in Utrecht, where he spent most of the rest of his life. His countrymen immediately ranked him as one of their leading artists, and he had contact with the courts of France, Spain, and Sweden.

Most of Scorel's great altarpieces have been lost or destroyed, but enough of his work survives to show how he had studied *Classical sculpture and the works of *Raphael and *Michelangelo while he lived in Rome. His treatment of landscape and portraiture reveals that he was impressed by the Venetian masters *Giorgione and *Palma Vecchio. Unlike the works of many other Netherlandish masters, however, his pictures are no mere jumble of Renaissance motifs. The *Presentation in the Temple* (Kunsthistorisches Museum, Vienna) is set in a *Bramantesque building and the proportion, movement, and drapery of the figures indicate that he understood what his Italian contemporaries were trying to achieve. On the other hand, his interest in atmospheric effects, in the play of light and shadow, and his fine landscape drawings are part of his Netherlandish heritage. The degree to which he was able to synthesize northern and southern elements in his figure compositions varies, but the quality of his portraits is uniformly high. The *Jerusalem Pilgrims* (Utrecht and Haarlem Museums) are the starting-point for the Dutch tradition of group portraiture that culminates in the works of Frans *Hals and *Rembrandt. Scorel's work was highly influential, notably on his pupils *Heemskerck and *Mor.

scorper (or **scauper**). A metal tool used in *wood engraving for clearing away large areas of the block or for engraving broad lines. Its cutting section is usually rounded, occasionally square.

Scott, Kathleen (1878–1947). British portrait sculptor, born Kathleen Bruce. She studied at the *Slade School and in Paris under *Rodin. In 1908 she married Captain Robert Falcon Scott (Scott of the Antarctic), who died on his return from the South Pole in 1912. Her most famous work is the statue commemorating him (unveiled 1915) in Waterloo Place, London. She did portrait busts of many other distinguished contemporaries. After her husband's death she was granted the rank of a widow of a Knight Commander of the Order of the Bath and was known as Lady Scott. In 1922 she married Lieutenant-Commander Edward Hilton-Young, who in 1935 became Baron Kennet; some reference books list her as Lady Kennet. Her son, Sir **Peter Scott** (1909–90), was one of the world's most distinguished ornithologists and also a noted painter and illustrator of wildfowl.

Scott, Samuel (1702?–72). English marine and topographical painter and etcher. He began as a marine painter in the tradition of the van de *Veldes, but he turned to topographical views in the manner of *Canaletto, who was then enjoying great success in England. In particular his views of the Thames and its bridges, which often include small figures, owe much to Canaletto's example. He was not simply a mechanical imitator, however, and had a feeling for the English atmosphere that is lacking in Canaletto, who brought the Venetian light with him to England. Scott also could achieve a distinctive grandeur of design, as in *An Arch of Westminster Bridge* (Tate, London, *c.*1750), which is often considered his masterpiece. He left London in 1765 and settled in Bath for reasons of health, apparently painting little in his later years.

Scott, Tim. See NEW GENERATION.

Scott, William (1913–89). British painter. In 1937–9 he lived in France, and he said:'I picked up from the tradition of painting in France that I felt most kinship with—the still life tradition of *Chardin and *Braque, leading to a certain kind of abstraction which comes directly from that tradition.' His work continued to be mainly based on still life with increasing abstraction, and for a time in the 1950s he painted pure abstracts before returning to his earlier preoccupations. Although forms such as circles and squares feature in his work, they are not geometrically exact but bounded by sensitive, painterly lines. In the late 1960s and 1970s his paintings became more austere. Although his work was restricted in range, Scott was regarded as one of the leading British painters of his generation. His work is represented in the Tate Gallery and numerous other public collections in Britain and elsewhere, and in 1958–61 he painted a large mural for Altnagelvin Hospital, Londonderry.

Scott, William Bell (1811–90). Scottish painter and poet. He was head of the Government School of Design in Newcastle upon Tyne, 1843–64, and his best-known works are near Newcastle, at Wallington Hall, Northumberland: a series (begun 1855) representing Northumbrian history and including the well-known *Iron and Coal*, one of the earliest representations in art of heavy industry. Bell was a close friend of *Rossetti, and his work has affinities with Pre-Raphaelitism in its Romanticism and love of historical detail. His literary output included much poetry and several books on art, among them a memoir (1850) of his brother **David Scott** (1806–49), a *history painter and book illustrator. Their father, **Robert Scott** (1777–1841), was an engraver.

Scottish Colourists. A term applied to four Scottish painters who in the period *c.*1900–14 each spent some time in France and were strongly influenced by the rich colours and bold handling of recent French painting, notably *Fauvism: they are F. C. B. Cadell (1883–1937), J. D. *Fergusson, Leslie Hunter (1879–1931), and S. J. Peploe (1869–1933). The term was popularized by a book by T. J. Honeyman dealing with Cadell, Hunter, and Peploe (*Three Scottish Colourists*, 1950), but it is now usual to add Fergusson to their number, even though he stands apart from the rest in that he returned to live in France after the First World War, whereas the other three remained in Scotland. All four painters knew each other, and they exhibited together as 'Les Peintres de L'Écosse Moderne' at the Galerie Barbazanges, Paris, in 1924, but they did not function as a group. They have been described as the first 'modern' Scottish artists; certainly they were the main channel through which *Post-Impressionism reached their country. None of them was represented in Roger *Fry's Post-Impressionist exhibitions of 1910 and 1912, but this reflects insular English attitudes towards Scottish art rather than the quality of their work.

scraper. A tool, normally in the form of a three-bladed knife, used in engraving and other metal-plate techniques for removing the *burr from engraved lines or for making corrections by smoothing or scraping the surface of the plate.

Scrotes, Guillim (William Stretes) (active 1537–53). Netherlandish portrait painter. He is first heard of in 1537, when he became court painter to Mary of Hungary, Regent of the Netherlands, and by 1546 he was King's Painter to Henry VIII (1491–1547) in England, earning a very high salary. His appointment continued under Edward VI (1537–53). Only a handful of paintings, all done in England, can be confidently attributed to him, among them a full-length portrait of Edward VI in the Royal Collection (it was much imitated—versions are in the Louvre, the Los Angeles County Museum and elsewhere). They show Scrotes to have been a highly accomplished practitioner in the international *Mannerist court style and he was an important figure in introducing the full-length portrait to England. He is last heard of in 1553 and may have left England after Edward's death in that year.

scumbling. Painting technique in which a layer of opaque colour is brushed lightly over a previous layer of another colour in such a way that the lower layer is only partly obliterated and shows through irregularly. With *glazing, scumbling allows a range of textural and colouristic effects that ensured *oil painting's dominance over other media. Similar effects can now be obtained with *acrylic.

Sebastiano del Piombo (Sebastiano Luciani) (*c*.1485–1547). Venetian painter, active mainly in Rome. According to *Vasari, he trained with Giovanni *Bellini, but his early work was most strongly influenced by *Giorgione, whose *Three Philosophers* (Kunsthistorisches Mus., Vienna) Sebastiano is said to have completed after the master's death. Their styles, indeed, can be so close as to cause paintings to be disputed between them, most notably the unfinished *Judgement of Solomon* (National Trust, Kingston Lacy). This large and impressive work was attributed to Giorgione by *Ridolfi, but scholarly opinion now increasingly tends towards giving it to Sebastiano. The half-length figure of *Salome* (or *Judith*?) (NG, London, 1510) shows the magnificent painterly skills of an undoubted work of Sebastiano at this date; it has a sensuous beauty reminiscent of Giorgione, but also a statuesque vigour that is Sebastiano's own.

In 1511 Sebastiano moved to Rome on the invitation of the banker Agostino Chigi (1465–1520), and he remained there for the rest of his life apart from a visit to Venice in 1528–9 after the Sack of Rome. For Chigi he painted mythological frescos at the Villa Farnesina, where *Raphael also worked. It was with *Michelangelo rather than Raphael, however, that Sebastiano formed a friendship and professional relationship. Michelangelo not only recommended him to people of influence, but also made drawings for him to work from, as with *The Raising of Lazarus* (NG, London, 1517–19). This was painted in competition with Raphael's *Transfiguration* (Vatican), both being intended for Norbonne Cathedral, and Vasari suggests that Michelangelo helped Sebastiano in order to discredit the Raphael faction, who had denigrated his powers as a colourist. Under Michelangelo's guidance Sebastiano's work became grander in form whilst losing much of its beauty of handling, the lack of sensuous appeal being accentuated when he began experimenting with painting on slate, as in *The Flagellation* (S. Pietro in Montorio, Rome).

Some of the finest works of Sebastiano's Roman years are his portraits, and after Raphael's death (1520) he had no rival in the city in this field, his work attaining a distinctive sombre grandeur. Clement VII (Giulio de' *Medici), the subject of one of Sebastiano's finest portraits (Museo di Capodimonte, Naples, 1526), appointed him keeper of the papal seals in 1531 and after this he was less active as a painter. The seals were made of lead, 'piombo' in Italian, hence Sebastiano's nickname.

secco (Italian: dry). Term applied to a technique of mural painting in which the colours are applied to dry plaster, rather than wet plaster as in *fresco. The colours were either *tempera, or *pigments ground in lime-water; if lime-water was used, the plaster had to be damped before painting, a method which was described by *Theophilus and which was popular in northern Europe and in Spain. In Italian *Renaissance art the finishing touches to a true fresco would often be painted *a secco*, because it is easier to add details in this way, and because the secco technique is much less permanent, such passages have frequently flaked off with time. Thus in *Giotto's *Betrayal* in the Arena Chapel, Padua, the details of many of the soldiers' weapons are now missing.

secondary colours. See PRIMARY COLOURS.

Section d'Or. Group of French painters who worked in loose association between 1912 (the date of their first exhibition) and 1914, when the war brought an end to their activities. The name, which was also the title of a short-lived magazine published by the group, was suggested by Jacques *Villon in reference to the treatise on the *Golden Section by Luca Pacioli, reflecting the interest of the artists involved in questions of proportion and pictorial discipline. Other members of the group included *Delaunay, *Duchamp, *Duchamp-Villon, *Gleizes, *Gris, *Léger, *Metzinger, and *Picabia. The common stylistic feature of their work was a debt to *Cubism.

sedilia (Latin: seats). Term in Christian church architecture for stone seats on the south side of the chancel, used by officiating priests. They are usually three in number (for the celebrant, deacon, and sub-deacon), but sometimes four or five. Sedilia were in use in England by the 12th cent. and many late medieval examples survive, usually recessed in the thickness of the wall and often richly carved and surmounted by arches and canopies. They are rarely found in other European countries, where movable chairs were favoured.

Segal, George (1924–). American sculptor. His career was slow to mature. He studied at various art colleges and took a degree in art education at New York University in 1950, but until 1958 he worked mainly as a chicken farmer, painting *Expressionist nudes in his spare time. In 1958 he made his first sculpture and in 1960 he began producing the kind of work for which he has become famous—life-size unpainted plaster figures, usually combined with real objects to create strange ghostly tableaux. Examples are: *The Gas Station* (National Gallery of Canada, Ottawa, 1963); *Cinema* (Albright-Knox Art Gallery, Buffalo, 1963); *The Restaurant Window* (Wallraf-Richartz-Museum, Cologne, 1967). The figures are made from casts taken from the human body; he uses his family and friends as models. In the 1970s he began to incorporate sound and lighting effects in his work. Segal has been classified with *Pop art and *Environment art, but his work is highly distinctive and original, his figures and groups dwelling in a lonely limbo in a way that captures a disquieting sense of spiritual isolation.

Segall, Lasar (1891–1957). Lithuanian-born painter and graphic artist who settled in Brazil in 1923 (following an earlier visit in 1912) and became a Brazilian citizen. Before and after the first journey to Brazil he lived in Germany, where he developed an *Expressionist style. His paintings were the first Expressionist pictures to be exhibited in Brazil and had a great influence on younger painters there. Segall's work often expresses his compassion for the persecuted and downtrodden, particularly the Jews who suffered under Nazism. His home and studio in São Paulo were converted into a museum after his death.

Segantini, Giovanni (1858–99). Italian *Divisionist painter. He lived and worked mainly in the Swiss Alps and is best known for his views of mountain scenery. In later life, from *c.*1890, he often painted weirdly symbolic and mystical pictures (*The Bad Mothers*, Walker Art Gal., Liverpool), but most of his work is a more straightforward celebration of nature. There is a Segantini museum at St Moritz.

Seghers, Daniel (1590–1661). The leading Flemish flower painter of the generation after Jan *Brueghel, who was his master. He worked in his native Antwerp, where he joined the Jesuit Order in 1614. The words 'Society of Jesus' usually follow his signature and in 1625–7 he stayed with the Order in Rome. The records of his community in Antwerp, which still survive, give an account of his considerable fame and the distinguished persons who visited him—among them Ferdinand, Governor of the Netherlands, Archduke Leopold-William, and in 1649 the future Charles II of England. His works could not be sold like those of an ordinary artist, but were presented as gifts by the Order, and the princely patrons in return sent lavish treasures including holy relics and a gold palette and brushes.

Seghers's work consists mainly of garlands or 'swags' (festoons) of flowers, often painted around a Madonna and Child or a *Pietà by another artist. He collaborated with his friend *Rubens in this way. Much rarer are his bouquets of flowers in a glass vase, where brilliant colours stand out against a dark background. Their simplicity makes it easier to appreciate his lovely creamy touch (his brushwork was broader than Brueghel's but unerringly sure) and they rank among the most beautiful flower pieces ever painted.

Seghers, Gerard (1591–1651). Flemish painter of religious subjects, active mainly in his native Antwerp, where he is said to have been taught by *Janssens. At some time between 1611 and 1620 he travelled in Italy (and probably also in Spain) and he became one of the very few noteworthy Flemish *Caravaggesque artists. By the time he painted *The Assumption of the Virgin* (Musée de Peinture et de Sculpture, Grenoble, 1629), however, he had fallen under the all-pervasive influence of *Rubens. Seghers had a successful career supplying altarpieces for churches in Antwerp and Ghent and also enjoyed considerable Spanish patronage.

Seghers (or **Segers**), **Hercules Pietersz**. (1589/90–1633/8). Dutch painter and etcher of landscapes, one of the most original but also most enigmatic figures in the history of Dutch art. Few details of his life are known. He was born in Haarlem, studied with Gillis van *Coninxloo in Amsterdam, and worked also in Utrecht and The Hague, where he is last mentioned in 1633. The woman who was evidently his second wife is described as a widow in 1638. Until 1871, when a landscape in the Uffizi, previously attributed to *Rembrandt, was given to him, Seghers was virtually forgotten. Hardly more than a dozen surviving paintings can be securely attributed to him, although contemporary documents show he certainly painted more. None of the paintings are dated and his chronology is difficult to reconstruct. He painted almost exclusively mountainous scenes, fantastic or *Romantic in conception but advanced in the naturalistic treatment of light and atmosphere. They are usually fairly small, but they suggest vast distances, and with their jagged rocks, shattered tree trunks, and menacing skies convey a sense of almost tragic desolation. Only Rembrandt and *Ruisdael among Dutch artists attained a similar degree of emotional intensity in landscape painting, and Seghers certainly influenced Rembrandt, who owned no less than eight of his paintings.

Seghers's etchings, are also rare (the latest catalogue lists 54, with 183 known impressions) and are perhaps even more original than his paintings. He experimented with coloured inks and dyed papers, so that extraordinarily different impressions could be made from the same plate, a dark paper printed with pale ink transforming a daylight scene into a haunting nocturnal view. Sometimes he printed on linen, which emphasized the vigorous and grainy quality of his work. Every print was an individual work rather than an item in a standardized commercial edition: *Hoogstraten said that he 'printed paintings'. They are unique in European art of the time, and it is often pointed out that some of them have a strange spiritual kinship with Chinese art. The best collection of his etchings is in the Print Room at Amsterdam.

In 1678 Hoogstraten published a highly coloured account of Seghers's unhappy career, his desperate experiments with etchings, and his eventual poverty and drunkenness, adding that he was killed by falling downstairs when intoxicated. Although the account may be exaggerated, Seghers certainly had financial problems (he was forced to sell his house in Amsterdam in 1631), and it seems likely that he was little appreciated in his day.

seicento. See QUATTROCENTO.

Seilern, Count Antoine. See COURTAULD.

Seisenegger, Jacob (1504/5–67). Austrian painter, chiefly a portraitist. In 1531 he became court painter to the emperor Ferdinard I (1503–64) at Augsburg and he worked much for the Habsburg family. He was a painter of modest talent, but has some importance in the development of the full-length portrait; his most famous work is the full-length of Charles V (Kunsthistorisches Mus., Vienna, 1532), which served as a model for the much more famous one by *Titian (Prado, Madrid). Seisenegger travelled widely, working in Italy, the Netherlands, and Spain as well as various cities in Central Europe—Innsbruck, Prague, Vienna.

Seligmann, Kurt (1900–62). Swiss-born American painter and graphic artist. He lived in Paris 1929–38 and became a member of the *Surrealist movement in 1934. His paintings of this time had a magical and apocalyptic character with hazy shapes and swirling draperies indistinguishable from the landscape. He also made Surrealist 'objects'. In 1939 he went to the USA and became an American citizen. In America he painted a series of visionary works which he called 'Cyclonic Forms' and which purported to express his reactions to the American landscape. He made a serious study of the occult and wrote a book entitled *The Mirror of Magic* (New York, 1948).

Senefelder, Aloys (1771–1834). German writer who in 1798 invented *lithography as a cheap means of reproducing his plays. He soon realized the artistic possibilities of his invention and wrote of them in his *Vollständiges Lehrbuch der Steindruckerey* (1818), translated into English as *A Complete Course of Lithography* in 1819.

sepia. A brown pigment made from the ink of cuttlefish and other marine creatures. It is used for ink drawings and, because of its semi-transparent quality, in *washes. Its warm, reddish colour distinguishes it from the cooler, more greenish *bistre.

Sequeira, Domingos António de (1768–1837). Portuguese painter, graphic artist, and designer. He formed his *Neoclassical style in Rome 1788–95 and in 1802 he was appointed first court painter in Lisbon. He was a prolific painter of religious and historical compositions as well as portraits (*Dr Neves*, Ashmolean, Oxford, 1825). After 1823, thinking his talent was unappreciated in Portugal, he lived in Paris and Rome. The silver table service presented by the Regent of Portugal to the first Duke of Wellington (Apsley House, London, 1812–16) was designed by Sequeira.

Séraphine (or **Séraphine de Senlis**) (Séraphine Louis) (1864–1934). French *naïve painter. After being orphaned very young, she passed her youth as a farm-hand and later entered domestic service in Senlis. She began painting when she was about 40 and was 'discovered' by Wilhelm *Uhde in 1912. Her pictures are mainly fantastic, minutely detailed compositions of fruit, leaves, and flowers. She was intensely devout and painted in a trance-like state of religious ecstasy, regarding her works as offerings to the Virgin Mary. In the late 1920s her reason began to fail and she became obsessed with visions of the end of the world. She died in a home for the aged at Clermont.

Sergel, Johan Tobias (1740–1814). German-born Swedish sculptor, active mainly in Stockholm. His early works are in a French *Rococo style, but he abandoned this during the period he spent in Rome (1767–78) and became the leading Swedish exponent of *Neoclassicism. He was a much livelier artist than many Neoclassical sculptors, however, and although his mature work has impressive clarity of form, it also possesses warmth and vitality. In Rome he was best known for his spirited sketches in clay and *terracotta, but after his return to Sweden he was mainly a portraitist. He was court sculptor to Gustavus III (1746–92) and his most important work is a bronze statue of the king (1790–1808) in front of the Royal Palace in Stockholm. Other examples of his work are in the National Museum there. Sergel was a prolific draughtsman, many of his drawings being *Romantic in spirit, in a style similar to those of *Fuseli, who was a friend in Rome.

Serial art. A branch of *Minimal art, in which simple, uniform elements, which may be commercially available objects such as bricks, cement blocks, etc., are assembled in accordance with a strict modular principle. Carl *Andre is a noted exponent of Serial art. The term came into common use in the late 1960s.

serigraphy. See SILK-SCREEN PRINTING.

Serlio, Sebastiano (1475–1554). Italian architect and architectural theorist. He was born in Bologna and trained as a painter. In about 1514 he went to Rome, where he worked with *Peruzzi, then after the Sack of 1527 he moved to Venice. In 1540 he was called to France by Francis I (1494–1547) and remained there for the rest of his life, acting as consultant of the building of the palace at Fontainebleau. Only two architectural works by Serlio survive (a doorway at Fontainebleau and the château at Ancy-le-Franc), but he is of great importance as the author of an influential architectural treatise—*L'Architettura*. This was published in six parts (out of order) between 1537 and 1551 (a seventh part appeared posthumously in 1575 and another part—on domestic architecture—was published from his manuscripts in 1967), and there have also been various collected editions, including an English edition (translated from the Dutch version rather than the original Italian) in 1611. It was the first architectural publication designed as a highly illustrated handbook rather than a learned treatise and it was used as a pattern-book by craftsmen all over Europe, playing an important role in spreading the *Renaissance style. *Lomazzo said it created 'more hack architects than he had hairs in his beard'.

Serov, Valentin (1865–1911). Russian painter and graphic artist, son of the composer Alexander Serov. He studied privately under *Repin and then in 1880–5 at the Academy in St Petersburg, where he became a friend of *Vrubel. His work includes landscapes, genre pictures, and historical scenes, as well as book illustrations, but he is best known as a portraitist. In this field he was the greatest Russian painter of his time and a match for any artist in the world. Like *Sargent, he was a cosmopolitan figure, used to moving in high society, and he brought to his work something of the same quality of aristocratic authority and poise associated with the American's portraits. Like Sargent, too, he painted with superb technical freedom and finesse, and he was just as good with informal portraits as with grand showpieces. His two most famous paintings (both in the Tretyakov Gallery, Moscow) are intimate early works, the breathtakingly beautiful *Girl with Peaches* (1887) and the almost equally lovely *Girl in the Sunshine* (1888); later, as his fame grew, he painted many of the leading Russian celebrities of his time, particularly artists, musicians, and writers. Serov was a member of the *World of Art group and some of his later work shows a tendency towards flat *Art Nouveau stylization. The most remarkable example is a nude, almost monochromatic portrait of the dancer Ida Rubinstein (Russian Museum, St Petersburg), painted in Paris in 1910. Serov had a difficult personality (he could be gloomy and rude), but he was greatly admired for the integrity and sincerity of his work.

Serpotta, Giacomo (1656–1732). Sicilian sculptor in *stucco. He was the greatest of all virtuosi in his medium and with the exception of *Antonello da Messina, the most distinguished artist to come from Sicily. Unlike Antonello, he spent almost all his life on the island (although he may have trained in Rome) and his work is mainly found in the churches of his native Palermo. Serpotta's icing-sugar-white figures—elegant, delicate, charming, and joyous in spirit—are amongst the finest expressions of the *Rococo in Italian art. He is particularly well known for his playful *putti, but his finest single figure is generally acknowledged to be the enchantingly coquettish *Fortitude* (1714–17) in the Oratoria del Rosario di San Domenico, Palermo. His brother **Giuseppe** (1653–1719) and his son **Procopio** (1679–1755) were also stuccoists.

Serra, Jaime (d. *c*.1399) and **Pedro** (d. 1404/9). Spanish painters, brothers, who worked for the court of Aragon. A documented altarpiece (1361) by Jaime in the Saragossa Museum is the main basis for further attributions to the brothers. Their work was influenced by Sienese painting, probably via the papal court at Avignon.

Sérusier, Paul (1863–1927). French painter and art theorist. In 1888 he met *Gauguin and Émile *Bernard at Pont-Aven, was converted to their *Symbolist views and founded the *Nabis with *Denis, *Bonnard, *Vuillard, and others. He became the principal theorist of the group, and after visiting the school of religious painting at the Benedictine monastery at Beuron in Germany in 1897 and 1903 his ideas were permeated with concepts of religious symbolism. His theories were set out in his influential book *ABC de la Peinture* (1921), which deals with colour relationships and systems of proportion. Sérusier's paintings, which largely feature Brittany peasants and after about 1900 religious subjects, are generally considered of less interest than his writings.

Servaes, Albert (1883–1966). Belgian *Expressionist painter. He is best known for his religious paintings, which were found offensive by the Roman Catholic Church. His works can be seen at the Museum of Modern Art at Brussels and in the Municipal Museums of Amsterdam and The Hague.

settecento. See QUATTROCENTO.

Seuphor, Michel (1901–). Belgian painter, graphic artist, and writer. He was active in the formation of the associations *Cercle et Carré and *Abstraction-Création, and he edited the periodical *Cercle et Carré*. His artistic work is in a geometrical abstract style, using a variety of graphic techniques, but he is better known for his writings. They include a standard work on abstract art, *L'Art abstrait. Ses origines, ses premiers maîtres* (1949), and dictionaries of abstract painting (1957) and modern sculpture (1959), all of which have been translated into English.

Seurat, Georges (1859–91). French painter, the founder and greatest exponent of *Neo-Impressionism. Seurat was the son of comfortably-off parents and his career took an unusual course; he never had to worry about earning a living and pursued his artistic researches with single-minded dedication. In 1878 he entered the École des *Beaux-Arts in Paris, but his studies were interrupted by military service in 1879. He returned to Paris in 1880 and for the next 2 years devoted himself to drawing (he was one of the subtlest and most original draughtsmen of the 19th cent., typically working with very broad, velvety areas of tone, using a *conté crayon on textured paper). In spite of the mastery of black and white he showed here, as a painter he turned for inspiration to artists in the colourist tradition—notably *Delacroix and the *Impressionists. In addition to studying the work of such painters, Seurat read aesthetic and scientific treatises, and he made it his aim to establish a rational system for achieving the kind of vibrant colour effects that the Impressionists in particular had arrived at instinctively. The method he evolved was to place small touches of unmixed colour side by side on the canvas, producing an effect of greater vibrancy and luminosity by this 'optical mixture' than if the colours had been physically mixed together on the palette. In his first major painting, *Bathers, Asnières* (NG, London, 1883–4, reworked 1887), Seurat was still experimenting with his technique, but his next large work, *Sunday Afternoon on the Island of La Grande Jatte* (Art Institute of Chicago, 1884–6), is a completely mature statement of his ideals. *La Grande Jatte* was shown at the final Impressionist exhibition in 1886 and led to the recognition of Seurat as a leader of the avant-garde. The critic Félix Fénéon (1861–1944) coined the term *pointillism in reference to this painting to describe the technique of using a myriad of tiny dots of colour, but Seurat preferred the term *divisionism.

From 1887 Seurat began to turn his attention to the significance of line in painting, believing that certain directions of lines could express specific emotions—horizontal lines represented calmness, upward- and downward-sloping lines represented happiness and sadness respectively. He embodied his beliefs in paintings such as *Le Chahut* (Rijksmuseum Kröller-Müller, Otterlo, 1890), in which the raised legs of the dancers performing 'Le Chahut' (a sort of can-can) express 'happiness'. Seurat died very suddenly at the age of 31, evidently from meningitis, although his friend *Signac said that he 'killed himself by overwork'. He was so dedicated to his work and kept himself to himself to such an extent that until his death few people knew he had a mistress and son. His mistress, Madeleine Knobloch, is represented in *Woman Powdering Herself* (Courtauld Inst., London, 1890).

Seurat's work was highly influential, but his disciples rarely approached his skill, and even less his level of inspiration, in applying his theories; their paintings often look mechanical and lack his gently satirical humour. In power of composition Seurat stands above not only his followers, but also virtually all other painters of his period. He planned his pictures with extraordinary care, and they have nothing of the sense of the passing moment associated with Impressionism. Rather, they have a highly formalized quality and a conscious grandeur that has caused Seurat to be compared with such revered masters as *Piero della Francesca and *Poussin.

Seven, Group of. See GROUP OF SEVEN.

Severe style. A term applied to the style characteristic of much Greek sculpture in the period *c.*480 BC–*c.*450 BC, transitional between the *Archaic and *Classical periods. Winckelmann had used the term 'severe' of work of the time before *Phidias, but the term 'Severe style' did not come into general usage until Vagn Poulsen's book *Der strenge Stil* of 1937. The characteristics of the Severe style include simplicity and severity of form, grandeur and elevation of spirit, and an increase in characterization compared with the Archaic period. *Myron is the best-known sculptor of the period, but his work is sometimes described as early Classical rather than Severe.

Severini, Gino (1883–1966). Italian painter. In 1906 he settled in Paris, and he played an important role in transmitting the ideas of French avant-garde art to the *Futurists, whose *Technical Manifesto of Painting* he signed in 1910. His work was strongly influenced by *Cubism (he knew *Braque and *Picasso) and he concerned himself with the problem of conveying a sense of movement and action by breaking up his picture space into contrasting and interacting rhythms (*Suburban Train Arriving in Paris*, Tate, London, 1915). He was influenced by theories of mathematical proportion and in 1921 published a book on the subject, *Du Cubisme au Classicisme*. In the 1920s his style became more traditional and he carried out several decorative commissions, including murals for churches in Switzerland. He also worked as a theatrical designer. Severini returned to Italy in 1935, but in 1946 he settled in Paris again. In the late 1940s his style once again became semi-abstract. In 1946 he published an autobiography.

Sezession. Name adopted by several groups of painters in Germany and Austria who in the 1890s broke away ('seceded') from the official academies, which they regarded as tradition-bound, and organized their own, more avant-garde exhibitions. The first of these groups came together in Munich in 1892, its leading members being Franz von Stuck (1863–1928) and Wilhelm Trübner (1851–1917), and there were Sezessionen also in Vienna (1897), with Gustav *Klimt as first President; and Berlin (1899), with Max *Liebermann as first President. When in 1910 a number of young painters were rejected by the Berlin Sezession—among them members of Die *Brücke—they started the Neue Sezession, with Max *Pechstein a prominent figure.

Sezessionstil. See ART NOUVEAU.

Sforza. Italian family, the effective rulers of Milan from 1450, when the *condottiere* **Francesco** (1401–66) seized power, until 1499 when his son **Lodovico** (1452–1508)—the third to succeed him—was expelled by the French. During that time they were lavish patrons of the arts, and the court of Lodovico (who was known as 'Il Moro'—the Moor—because of his dark complexion) was one of the most splendid in Europe. The greatest artists to work for him were *Bramante and *Leonardo da Vinci.

sfumato (from Italian *fumo*: smoke). Term used to describe the blending of tones or colours so subtly that they melt into one another without perceptible transitions—in *Leonardo's words, 'without lines or borders, in the manner of smoke'. Leonardo was a supreme exponent of *sfumato* and *Vasari regarded the capacity to mellow the precise outlines characteristic of the earlier *quattrocento as one of the distinguishing marks of 'modern' painting.

sgraffito. See GRAFFITO.

Shahn, Ben (1898–1969). American painter, illustrator, photographer, designer, teacher, and writer, born at Kovno, Lithuania, then part of Russia. His family emigrated to the USA in 1906 and settled in New York. Shahn's background (his father had been sent to Siberia for revolutionary activities and he grew up in a Brooklyn slum) gave him a hatred of cruelty and social injustice, which he expressed powerfully in his work. He first made a name with a series of pictures (1931–2) on the Sacco and Vanzetti case (these two Italian immigrants had been executed for murder in 1927 on very dubious evidence, and many liberals believed that they had really been condemned for their anarchist political views). Shahn's paintings are in a deliberately awkward, caricature-like style that vividly expresses his anger and compassion. In 1933 he was assistant to Diego *Rivera on the latter's murals for the Rockefeller Center, New York, and subsequently painted a number of murals himself, notably for the Bronx Post Office, New York (1938–9), and the Social Security Building, Washington (1940–1). From 1935 to 1938 he worked as an artist and photographer for the Farm Security Administration, a government agency that documented rural poverty. During the Second World War his work included designing posters for the Office of War Information. After the war he returned to easel painting and was also active as a book and magazine illustrator and as a designer of mosaics and stained glass. His later work tended to be more fanciful and reflective and less concerned with social issues. From the 1950s he gave more time to teaching and lecturing and in 1956–7 he was Charles Eliot Norton Professor of Poetry at Harvard University. His Norton lectures were published as *The Shape of Content* (1957) in which he summarized his humanistic, anti-abstract artistic philosophy.

Shannon, Charles (1863–1937). English lithographer and painter, best known for his lifelong association with Charles *Ricketts. Painting began to take first place with him from 1897, his portraits and figure compositions being indebted to the Old Masters, especially *Titian.

shaped canvas. A term that began to be used in the 1960s for paintings on *supports that departed from the traditional rectangular format. Non-rectangular pictures were of course not new at this time; *Gothic and *Renaissance altarpieces often had irregular pointed tops, the oval was particularly popular in the *Baroque and *Rococo periods, and so on. The phrase 'shaped canvas', however, usually alludes particularly to a type of abstract painting that emphasizes the 'objecthood' of the work, proclaiming it as something that exists entirely in its own right and not as a reference to, or reproduction of, something else. Various artists have been claimed as the 'inventor' of the shaped canvas in this sense, but its most prominent exponent has undoubtedly been Frank *Stella, who has used such shapes as Vs, lozenges, and fragments of circles. The leading British exponent has been Richard Smith (1931–), who has sometimes used a kite-shaped format in which the canvas is stretched on rods that are part of the visual structure of the picture. He was one of the artists represented in an exhibition entitled 'The Shaped Canvas', organized by Lawrence *Alloway at the Guggenheim Museum, New York, in 1974. The shaped canvas has also occasionally been used by modern figurative painters, notably Anthony *Green.

Shchukin, Sergei. See HERMITAGE.

Shee, Sir Martin Archer (1769–1850). Irish portrait painter, active from 1788 in London. There he became second only to *Lawrence as the leading society portraitist, and in 1830 he succeeded him as President of the *Royal Academy, which he guided through a difficult period. Examples of his work—which in style lies between the bravura of Lawrence and the precision of *West—are in the National Portrait Gallery, London. His *Rhymes on Art* (1805), urging the claims of art on national support, enjoyed a considerable popularity. He also published *Elements of Art* (1809).

Sheeler, Charles R. (1883–1965). American painter and photographer, the best-known exponent of *Precisionism. Between 1904 and 1909 he made several trips to Europe, and gradually abandoned the bravura handling of *Chase (his teacher at the Pennsylvania Academy of the Fine Arts) for a manner influenced by European modernism—the paintings he exhibited at the *Armory Show of 1913, for example, were much indebted to *Cézanne. In 1912 Sheeler took up commercial photography for a living while continuing to paint. He worked for a while in fashion photography, but his shy and undemonstrative personality was not suited to this world, and he concentrated more on very mundane subjects such as plumbing fixtures. The clarity needed in such work helped to transform his style of painting to a meticulous smooth-surfaced manner that was the antithesis of his early approach. He began to paint urban subjects in about 1920 and over the next decade shifted from simplified compositions influenced by *Cubism (on 'the borderline of abstraction' in his own words) to highly detailed photographic-like images. In 1927 he was commissioned to take a series of photographs of the Ford Motor Company's plant at River Rouge, Michigan. His powerful photographs, presenting a pristine view of American industry, were widely reproduced and brought him international acclaim. Increasingly, also, he was recognized as the finest painter in the Precisionist style, his work standing out as much for its formal strength as for its technical polish. Sheeler's paintings continued in this realistic vein in the 1930s, but in the mid-1940s his style changed dramatically; he began using multiple viewpoints and bold unnaturalistic colours, although his brushwork remained immaculately smooth. In 1959 he suffered a stroke and had to abandon painting and photography.

Shinn, Everett (1876–1953). American painter and graphic artist. While employed as an illustrator on the *Philadelphia Press* he became a member of the group of friends who were inspired by Robert *Henri, and after he moved to New York in 1896 he became a member of The *Eight and of the *Ashcan School. Shinn differed from his associates in his choice of subject, preferring scenes from the theatre and music-hall to low-life imagery. Typical of his work is *A Winter's Night on Broadway*, showing the Metropolitan Opera House in a snowstorm, which was bought by *Harper's Weekly* in 1908. As a result of the publicity attending this work and of the success of a one-man exhibition in 1902, he obtained a number of commissions for fashionable portraits. However, in 1911 he was commissioned to do murals for Trenton City Hall, New Jersey, and these have been described as the earliest instance of *Social Realist themes in public mural decoration. He also did decorations for the interior of the Belasco Theatre, New York. In addition Shinn illustrated numerous books, wrote plays, and worked as an art director for motion pictures.

Shubin, Fedot (1740–1805). The leading Russian sculptor of the 18th cent. He trained at the Academy in St Petersburg and in 1767 won a scholarship to Paris, where he studied with *Pigalle. Later he travelled in Italy (Turin, Bologna, Rome) and worked for a short while in London under *Nollekens (1773). He is best known for his marble portrait busts.

Siberechts, Jan (1627–1703). Flemish landscape painter who settled in England in the early 1670s.

His landscapes are somewhat *Rubensian, but he is best known for his 'portraits' of English country houses, done in a simple, rather archaic manner; two views of Longleat, Wiltshire (1675 and 1676), are still preserved in the house. He was the first professional exponent of the genre.

siccative. A substance added to oil paint to make it dry more quickly. Siccatives are dangerous aids as they may cause cracking or darken a picture.

Sickert, Walter Richard (1860–1942). British painter, printmaker, teacher, and critic, one of the most important figures of his time in British art. He was born in Munich of a Danish-German father and an Anglo-Irish mother and he remained cosmopolitan. The family settled in London in 1868. Both his father and his grandfather were painters, but Sickert—after a good classical education—initially trained for a career on the stage, 1877–81. He toured with Sir Henry Irving's company but never progressed beyond small parts and in 1881 he abandoned acting and became a student at the *Slade School. In the following year he became a pupil of *Whistler, and in 1883 he worked in Paris with *Degas. Between 1885 and 1905 Sickert spent much of his time in Dieppe, living there 1899–1905, and also visited Venice several times. From 1905, when he returned to England, he became the main channel for influence from avant-garde French painting in British art—the inspiration for the *Camden Town Group and the *London Group. In 1918–22 he again lived in Dieppe, then settled permanently in England, living in London and Brighton before moving to Broadstairs, Kent (1934), and finally to Bathampton, near Bath (1938).

Sickert took the elements of his style from various sources but moulded them into a highly distinctive *œuvre*. From Whistler he derived his subtle modulations of tone, although his effects were richer than the characteristically delicate harmonies of Whistler. To Degas he was indebted particularly for his method of painting from photographs and for the informality of composition this encouraged. His favourite subjects were urban scenes and figure compositions, particularly pictures of the theatre and music-hall and drab domestic interiors. Sickert himself wrote in 1910 that 'The more our art is serious, the more will it tend to avoid the drawing-room and stick to the kitchen.' This attitude permeates his most famous painting, *Ennui*, a compelling image of a stagnant marriage, of which he painted at least four versions, that in the Tate Gallery, London (*c*.1914), being the largest and most highly finished. From the 1920s Sickert received many honours. His later works—often based on press photographs or Victorian illustrations—are very broadly handled, painted in a rough, vigorous technique, with the canvas often showing through the paint in places. The colour is generally much higher keyed than in his earlier work and sometimes almost *Expressionist in its boldness. The prevailing critical opinion for many years was that these late works marked a significant decline, but major claims have recently been made for them, particularly following the 1981 Arts Council exhibition 'Late Sickert: Painting 1927 to 1942'.

As well as painting, Sickert was an outstanding etcher (he learnt the technique from Whistler) and a great teacher (he opened seven private art schools, each of brief duration, and also taught part-time at Westminster School of Art, 1908–12 and 1915–18). He was celebrated for his wit and charm and was a stimulating talker and an articulate writer on art; Osbert Sitwell edited a posthumous collection of his writings entitled *A Free House!* (1947). Sickert was married three times; his third wife (from 1926) was the painter Thérèse Lessore (1884–1945). His brother **Bernard Sickert** (1862/3–1932) was a landscape painter and etcher.

One of the more eccentric theories about the Jack the Ripper murders of 1888 has it that Sickert (against his will) was part of a team that carried out the killings to cover up a scandal about an illegitimate child born to the Duke of Clarence (Queen Victoria's grandson); see Stephen Knight, *Jack the Ripper: The Final Solution* (1976, revised edition, 1984) and Jean Overton Fuller, *Sickert & the Ripper Crimes* (1990).

Siegen, Ludwig von (1609–*c*.1680). German soldier and amateur artist who probably invented the technique of *mezzotint. The earliest of his few surviving prints, a portrait of the Landgravine Amelia Elizabeth of Bohemia, dated 1642, was sent by Siegen to the Landgrave with a letter stating that the invention was his. Prince *Rupert, whom he probably met in Brussels in 1654, learnt the technique from him.

Signac, Paul (1863–1935). French *Neo-Impressionist painter. He began in the *Impressionist manner, but met *Seurat in 1884 and became an ardent disciple of his views and method. After the death of Seurat in 1891 he became the acknowledged leader of the Neo-Impressionist group, and in 1899 he published *D'Eugène Delacroix au néo-impressionnisme*, which was long regarded as the authoritative work on the subject. The book was, however, more in the nature of a manifesto in defence of the movement than an objectively accurate history. It reflected changes which had taken place in Signac's own style since 1890 towards greater brilliance of colour, and his best works are generally considered to be those in which he moved away from the scientific precision advocated by Seurat towards a freer and more spontaneous manner. His work had a great influence on *Matisse.

significant form. See BELL, CLIVE.

Signorelli, Luca (*c.*1440/50–1523). Italian painter from Cortona, active in various cities of central Italy, notably Arezzo, Florence, Orvieto, Perugia, and Rome. According to *Vasari (who was related to him and 'as a child of eight' met 'this good old man'), Signorelli was a pupil of *Piero della Francesca and this seems highly probable on stylistic grounds, for Signorelli's solid figures and sensitive handling of light echo the work of the master. Signorelli differed from Piero, however, in his interest in the representation of action, which put him in line with contemporary Florentine artists such as the *Pollaiuolo brothers. He must have had a considerable reputation by about 1483, when he was called on to complete the cycle of frescos on the walls of the Sistine Chapel in Rome, left unfinished by *Botticelli, *Ghirlandaio, *Perugino, and *Rosselli. (It is not known why these four artists abandoned the work in 1482, but it has been suggested that they simply downed tools because of slow payment.) Signorelli completed the scheme with distinction, but his finest works are in Orvieto Cathedral, where he painted a magnificent series of six frescos illustrating the end of the world and the Last Judgement (1499–1504). In these grand and dramatic scenes he displayed a mastery of the nude in a wide variety of poses surpassed at that time only by *Michelangelo. Vasari says that 'Luca's works were always highly praised by Michelangelo' and several instances of close similarity between the work of the two men can be cited; perhaps most interesting is the enigmatic seated nude youth in Signorelli's *Last Acts and Death of Moses* in the Sistine Chapel, which is remarkably close to some of the *Ignudi* painted by Michelangelo on the ceiling of the chapel a quarter of a century later. By the end of his career, however, Luca had become a conservative artist, working in provincial Cortona, where his large workshop produced numerous altarpieces. Several examples of his work are in the National Gallery, London.

Signorini, Telemaco. See MACCHIAIOLI.

silhouette. An outline image in one, solid, flat colour, giving the appearance of a shadow cast by a solid figure. The term is applied particularly to profile portraits in black against white (or vice versa), either painted or cut from paper, which were extremely popular in the period *c.*1750–*c.*1850. Silhouettists offered the quickest and cheapest method of portraiture (their art has been called 'the poor man's *miniature'), and their popularity was fostered by the *Neoclassical taste and given intellectual prestige by J. C. Lavater's *Essays on Physiognomy* (1775–8, English translation 1789–98), which were held in great respect throughout Europe, notably by *Goethe, and which relied on silhouettes for the exposition of their theme. The usual format was a head profile, but the range extended even to *conversation pieces. After 1800 a greater variety of techniques was developed, but the strength of the silhouette was vitiated by the introduction of colour, gilding, and fancy backgrounds. Its death-blow, like that of the miniature, was dealt by the popularization of photography, although interesting experiments, such as the German film-maker Lotte Reiniger's animated silhouette cartoon films, made in the 1930s, have been attempted. The word 'silhouette' derives from Etienne de Silhouette (1709–67), French finance minister under Louis XV, who was notorious for his parsimony and cut shadow portraits as a hobby; hence the phrase 'à la silhouette' came to mean 'on the cheap'. In Britain silhouette portraits were generally called 'shades' or 'profiles' up to the end of the 18th cent.

silk-screen printing (or **serigraphy**). A modern colour printing process based on stencilling. A cut *stencil is attached to a silk screen of fine mesh which has been stretched on a wooden frame, and the colour is forced through the unmasked areas of the screen on to the paper beneath by means of a squeegee. This method is an improvement on the simple stencil where, for example, a letter O required connecting pieces to prevent the centre from falling out—a problem which does not arise if the stencil is supported by the silk mesh. By a further improvement in the process the cut stencil is dispensed with altogether, its equivalent being painted directly on to the screen with opaque glue or *varnish. The process, which originated in the early 20th cent., has been widely used for commercial textile printing, but in the 1930s it was developed, particularly in the United States, as an artists' medium. Andy *Warhol was a notable exponent.

Siloe, Diego de (*c.*1495–1563). Spanish architect and sculptor, one of the leading figures in the transition from *Gothic to *Renaissance in Spanish art. He was the son of **Gil de Siloe** (d. *c.*1501), who is of uncertain origins (contemporary references suggest both Orleans and Antwerp as his native city) but who settled in Burgos and is regarded as the outstanding Spanish sculptor of the 15th cent. and the last great exponent of the Gothic tradition. Gil's extant work includes two royal tombs (1489–93) in the monastery of Miraflores, Burges: the first of John II of Castile and his wife Isabella of Portugal, the second of their son Prince Alfonso. Diego presumably trained with his father, but he formed his style in Italy, where he collaborated with Bartolomé *Ordóñez. By 1519 he had returned to Burgos, where he carried out several important commissions in the cathedral. They included the tomb of Bishop Luis de Acuña (1519) in the Chapel of St Anne and the altarpiece for the same chapel (1522), but his greatest work there is the *Escalera Dorada* (Golden Stairway) of 1519–23, and it was as an architect rather than as a sculptor that he emerged as one of the great figures of Spanish art. His masterpiece is Granada Cathedral, where he took over as architect in 1528 and constructed a superb Classical building on Gothic foundations. His Classicism is of a highly personal and original kind, for although he used the architectural vocabulary of the Renaissance with grace and assurance, he kept the verticality of the Gothic style. The building was not finished until the 17th cent. (Alonso *Cano designed the façade), but it is the greatest monument to Siloe's genius.

silver point. See METAL POINT.

Silvestre, Israël (1621–91). French etcher, the best-known member of a family of artists active from the 16th to the 18th cent. He worked much for Louis XIII (1601–43) and was drawing-master to Louis XIV (1638–1715). His etchings of architectural subjects and of ceremonies and fêtes are generally considered more valuable as historical records than as works of art.

Simmons, Edward E. See TEN, THE.

Simone Martini (*c*.1285–1344). Next to *Duccio, the most distinguished painter of the Sienese School. He was probably trained in Duccio's circle, and he developed further the decorative use of outline, colour, and patterning that were the marks of Duccio's work. The main features of his style are present in his earliest surviving work, the large fresco of the **Maestà* (1315; reworked 1321) in Siena Town Hall: the sumptuous materials and the aloofness of the Madonna derive from the *Byzantine style of the older generation; the decorative line, gesture, and expression are informed by the gracious *Gothic fashion that was now current in Siena; and the use of foreshortening to create depth shows the awakening desire for more lifelike effects. In 1317 Simone was in Naples, where he executed for Robert of Anjou (king of Naples, 1309–43) a painting (Museo di Capodimonte, Naples) showing his elder brother St Louis of Toulouse, newly canonized, resigning his crown to him. The *predella scenes of this altarpiece contain the boldest compositions in *perspective that had been produced up to that date. The next major work associated with Simone is a fresco, dated 1328, on the wall opposite his *Maestà* in Siena Town Hall, commemorating the *condottiere* Guidoriccio da Fogliano, who in that year had won a great victory for the Sienese and liberated the towns of Montemassi and Sassoforte, shown in the background. This highly original work, which shows the general riding in stately but solitary triumph, is generally regarded as one of the first equestrian portraits since antiquity; however, since the late 1970s it has been the subject of great controversy, some scholars considering that it is not by Simone and considerably later than assumed (this opinion is based on technical evidence allegedly showing that part of the fresco lies on top of another fresco known to date from 1363). Simone's series of frescos on the life of St Martin in the Lower Church of S. Francesco at Assisi are usually dated about the same time as the Guidoriccio painting.

The work that is generally regarded as the epitome of Simone's style is the *Annunciation* (Uffizi, Florence, 1333), although this is jointly signed with his brother-in-law Lippo *Memmi. It is a ravishing blend of etiolated grace and sweet sentiment and for sheer beauty of craftsmanship is unsurpassed in its age. Simone's work is more fully Gothic in spirit than that of any other major Italian painter and it is not surprising that he was appreciated in France. In 1340/1 he went to Avignon to serve the papal court, where he painted the unusual subject *Christ Reproved by his Parents* (Walker Art Gal., Liverpool, 1342) and the frontispiece to a Virgil manuscript owned by the poet Petrarch (1304–74) (Ambrosiana Lib., Milan). He also did a portrait of Petrarch's beloved Laura; it is lost, but mentioned in one of the poet's sonnets. Simone's style and compositions were taken over by *illuminators from France and Flanders and generations of Italian *panel and fresco painters copied him too. He was one of the main sources of the *International Gothic style.

singerie. A depiction of monkeys (French: *singes*) engaged in playful activities, often dressed in human clothes and acting out human roles. The conceit of monkeys involved in human occupations goes back to medieval manuscripts, but the term *singerie* is usually restricted to a type of decorative painting associated with the French

*Rococo. Jean *Bérain first hit on the idea *c.*1695 of replacing the *Classical fauns and statues and *Renaissance grotesques by figures of monkeys, but the origin of the *singerie* as a distinct genre is usually attributed to Claude *Audran, who in 1709 painted an arbour with monkeys seated at table for the Château de Marly. His pupil *Watteau also painted *singeries*.

sinopia. Term applied to a reddish-brown chalk used for the underdrawing of a *fresco and also to the drawing itself. Sinopie are often uncovered during restoration work when the upper layer of a fresco is removed (see, for example, TRAINI).

Siqueiros, David Alfaro (1896–1974). Mexican painter, one of the trio of muralists (with *Orozco and *Rivera) who have dominated 20th-cent. Mexican art. He was a political activist from his youth and in 1914 abandoned his studies at the Academy of San Carlos in Mexico City to join the revolutionary army fighting against President Huerta. His services were appreciated by the victorious General Carranza, who in 1919 sponsored him to continue his studies in Europe, where he was friendly with Rivera (later they became rivals). On returning to Mexico in 1922 he took a leading part in the artistic revival fostered by President Alvaro Obregón. Siqueiros was active in organizing the Syndicate of Technical Workers, Painters and Sculptors and was partly responsible for drafting its manifesto, which set forth the idealistic aims of the revolutionary artists: 'our own aesthetic aim is to socialize artistic expression, to destroy bourgeois individualism . . . We proclaim that this being the moment of social transition from a decrepit to a new order, the makers of beauty must invest their greatest efforts in the aim of materializing an art valuable to the people, and our supreme object in art, which today is an expression for individual pleasure, is to create beauty for all, beauty that enlightens and stirs to struggle.'

Siqueiros's political activities led to his imprisonment or self-imposed exile several times; from 1925 to 1930 he completely abandoned painting for political activity and he later fought in the Spanish Civil War. It was not until 1939 that he eventually completed a mural in Mexico—*Portrait of the Bourgeoisie* for the headquarters of the Union of Electricians in Mexico City (his slow start had prompted Rivera to retort in answer to criticism from him: 'Siqueiros talks: Rivera paints!'). Thereafter, however, Siqueiros's output was prodigious. He painted many easel pictures as well as murals, and though he insisted they were subordinate to his wall paintings, they were important in helping to establish his international reputation. His murals are generally more spectacular even than those of Orozco and Rivera—bold in composition, striking in colour, freely mixing realism with fantasy, and expressing a raw emotional power. In contrast with the sense of disillusionment and foreboding sometimes seen in Orozco's work, Siqueiros always expressed the dynamic urge to struggle; his work can be vulgar and bombastic, but its sheer energy is astonishing. He often experimented technically—working on curved surfaces and using air-brushes and synthetic pigments—and his last major work, the Polyforum Siqueiros in Mexico City (completed 1971), is a huge auditorium integrating architecture, sculpture, and painting. In his late years Siqueiros was showered with honours from his own country and elsewhere; he received the Lenin Peace Prize in 1967, for example, and in the following year became the first President of the newly founded Mexican Academy of Arts.

Sisley, Alfred (1839–99). French *Impressionist painter of English parentage. He was intended by his parents to have a commercial career, but he turned to painting. In 1862 he entered the studio of *Gleyre and there met *Renoir, *Monet, and *Bazille, his friendship with Monet proving particularly important. Like Monet, Sisley devoted himself almost exclusively to landscape, but his work was much less varied. He made several visits to England, but otherwise rarely travelled, and most of his pictures are of scenes in and around Paris. The best are among the most lyrical and gently harmonious works of Impressionism, delicate in touch and with a beautiful feeling for tone. Sisley was given a generous allowance by his wealthy father and his early life was happy: Renoir described him as 'a delightful human being' and said he 'could never resist a petticoat'. However, after the family business failed in 1871 because of the Franco-Prussian War, his father was unable to support him and thereafter he spent much of his life in poverty, never winning the success or renown of his former colleagues. He still remains something of a Cinderella among the Impressionists, claimed as their own by neither the French nor the English. His work is in many public collections, including the Courtauld Institute Galleries and the Tate Gallery in London.

Situation. An exhibition held at the Royal Society of British Artists in London in September 1960 by a group of British painters responding to American *Abstract Expressionism. Paintings included had to be totally abstract and at least 30 sq. ft. in size. Two further Situation exhibitions took place in 1961 and 1962–3. They indicated a move towards a more international context for British painting.

size. Glue made from animal skins, or, more loosely, any fairly weak glue. It is used in art mainly for filling the porous surface of wooden *panels or *canvases to provide a suitable foundation for the *priming (see GROUND). Size is also used as a *medium—paint mixed with it is called *distemper.

Skeaping, John. See HEPWORTH.

sketch. 'A rough drawing or delineation of something, giving the outlines or prominent features without the detail, especially one intended to serve as the basis of a more finished picture, or to be used in its composition . . .' (*OED*). This was the original meaning of the word; not until the latter part of the 18th cent. did it acquire the additional sense of 'a drawing or painting of a slight or unpretentious nature'. In the 18th cent. the cult of the *picturesque led to appreciation of the sketch because of its spontaneous and 'unfinished' character, which stimulated the play of imagination, and in the 20th cent. the oil sketches of *Rubens and *Constable, for example, have often been rated higher than their finished works. A sketch should be distinguished from a 'study', which is a representation of a detail to be used in a large composition and may be highly finished. A **modello* is a particular kind of sketch, done as a proposal for a larger work and often intended to be shown to the patron; it is therefore more elaborate than an ordinary sketch.

Slade, Felix (1790–1868). English art collector and philanthropist. He left a great part of his collection, notable particularly for glass, to the *British Museum and in his will endowed chairs of fine art at the universities of London (University College), Oxford, and Cambridge. The professorships at Oxford and Cambridge involve only the giving of lectures, intended for a general audience, but in London the Slade School of Fine Art, opened in 1871, is an institution giving practical instruction in painting, sculpture, and the graphic arts. The first professor was Sir Edward *Poynter, who founded the Slade tradition of emphasis on drawing from the nude. Rapidly taking over from the *Royal Academy (where the teaching methods were considered arid and academic) as the most important art school in the country, the Slade had its heyday in the period from about 1895 to the First World War. Its students then included some of the most illustrious names in 20th-cent. British art—Augustus and Gwen *John, Wyndham *Lewis, Paul *Nash, Ben *Nicholson, Stanley *Spencer, and so on. After Poynter, the Slade Professors in London have included *Legros, Frederick Brown (1851–1941), who was Professor from 1892 to 1917, presiding over the School's golden age, *Tonks, *Coldstream, and *Gowing. The first Slade Professors at Oxford and Cambridge respectively were *Ruskin and the architect Matthew Digby Wyatt (1820–77). Subsequent professors have included many eminent art historians, including Kenneth *Clark, Ernst *Gombrich, and Nikolaus *Pevsner.

Slevogt, Max (1868–1932). German painter and graphic artist, with *Corinth and *Liebermann one of the leading artists in his country to show the impact of *Impressionism. He was born in Bavaria and studied in Munich, then at the *Académie Julian, Paris, 1889–90. In 1901 he settled in Berlin, where he taught at the Academy from 1917. Although he took over from the Impressionists their fresh, bright palette, he never adopted their fragmentation of colours, and his work always retained something of the Bavarian *Baroque tradition. His loose handling of paint, bold effects of light, and energetic sense of movement give his work great dash, and he was much in demand as a decorative artist. He also made a name for himself as a book illustrator. As a painter his work included landscapes, portraits, and scenes of contemporary life; he loved the theatre and is perhaps best known for portrayals of the singer Francisco d'Andrade in the role of Don Juan. He also painted religious pictures, his last work being a fresco of *Golgotha* in the Friedenskirche at Ludwigshafen (1932); it is considered his masterpiece by some authorities. Many German museums have examples of his work.

Sloan, John (1871–1951). American painter and graphic artist. During the 1890s he worked as a newspaper illustrator in Philadelphia, and he started painting seriously in 1896, influenced by Robert *Henri. In 1904 he settled permanently in New York, where he and Henri were among the members of The *Eight. Sloan was the most political member of the Eight and as well as finding his most characteristic subjects among everyday lower-class New York life, he did illustrations for socialist periodicals including *The Masses*, of which he was editor from 1912 to 1916. However, he was not interested in using his art for what he called 'socialist propaganda' and he resigned from the magazine after a dispute over policy. His paintings are generally marked by a warm humanity, but he could also be sharply satirical and occasionally expressed himself in a totally different vein, as in *Wake of the Ferry* (Phillips Coll., Washington, 1907), which has a romantic, melancholy mood. He taught at the *Art Students League 1914–30 and 1930–8, his students including such distinguished figures as Alexander *Calder, Barnett *Newman, and David *Smith. In 1939 he published an autobiographical–critical book, *The Gist of Art*.

Slodtz. Family of French sculptors, designers, and decorators. They were **Sébastien** (1655–1726), who was Flemish by birth, and his three sons: **Sébastien-Antoine** (*c.*1695–1754), **Paul-Ambroise** (1702–58), and **René-Michel** called Michel-Ange (1705–64). They all worked for the Menus Plaisirs (the office that designed costumes, festivities, etc. for the court), as did another son of Sébastien, the painter **Dominique Slodtz** (1711–64), and Michel-Ange was the only member of the family to attain great distinction as a sculptor. From 1728 to 1747 he was in Rome, where his admiration for *Michelangelo won him his nickname. His best-known work is *St Bruno* (St Peter's, Rome, 1744), which shows the nervous sensitivity of his style. In France he made several major tombs, notably that of Languet de Gergy (S. Sulpice, Paris, completed 1753), which shows the influence of *Bernini in its *Baroque rhetoric and use of different coloured marbles. *Houdon was Slodtz's most important pupil.

Sluter, Claus (d. 1405/6). Netherlandish sculptor, active mainly in Dijon. He was the greatest sculptor of his time in northern Europe and a figure of enormous importance in the transition from *International Gothic to a more weighty and naturalistic style. Sluter is first mentioned in Brussels (*c.*1379) in a document that says he came from Haarlem. In 1385 he entered the service of Philip the Bold (1342–1404), Duke of Burgundy, in his capital, Dijon. All of Sluter's surviving work was done for Philip, and almost all of it remains in Dijon. For the Chartreuse de Champmol, a monastery founded by Philip, Sluter carved figures for the portal of the chapel in the early 1390s, and made a fountain group, the only part of which to survive intact is the base, known as the *Well of Moses* (1395–1403). The monastery was destroyed during the French Revolution, and the portal and the *Well of Moses* are now part of the psychiatric hospital that occupies its site. The Well features six full-length figures of prophets of monumental dignity; they convey an intense sense of physical presence, and as character studies rival the prophets of *Donatello, which they preceded by about 20 years. Originally Sluter's figures were painted (by *Malouel) and the figure of Jeremiah is known to have worn copper spectacles, the record of payment for which still survives. Of the Calvary group that surmounted the Well (symbolizing the 'Fountain of Life') only fragments survive in the Archaeological Museum in Dijon; they include the head and torso of the figure of Christ—one of Sluter's noblest works, in which the expression of suffering stoically endured is deeply moving. Sluter's last work was the tomb of Philip the Bold, begun in 1404 and unfinished at the sculptor's death (Musée des Beaux-Arts, Dijon). Most of the work on it was carried out by Sluter's nephew and assistant Claus de *Werve, but the figures of *pleurants* (weepers or mourners) that form a frieze around the sarcophagus are from the master's own hand, and although they are only about 40 cm. high they possess massive solemnity. They show Sluter's extraordinary ability to use the folds of drapery for expressive effect; indeed some of the mourners are so completely enveloped in their voluminous cowls that they are in effect nothing else but drapery. Erwin *Panofsky has written (*Early Netherlandish Painting*, 1953) that it is 'the concentrated emanation of Claus Sluter's style which we mean when we speak of a "Burgundian school of sculpture of the fifteenth century" ', and his work had great influence also on painters. The emphatic plasticity of the *Master of Flémalle's figures, for example, has Sluter as its source, and in his *Entombment* (Courtauld Inst., London, *c.*1410/20), which stands at the head of the Early Netherlandish tradition of painting, the angel that wipes away a tear with the back of his hand is a quotation from the *Well of Moses*.

Sluyters, Jan (Johannes Carolus Bernardus) (1881–1957). Dutch painter, one of the best-known artists in the Netherlands in the inter-war period, and the one in whom French modernism is most variously reflected. His early works show the influence of van *Gogh and *Breitner, and of *Toulouse-Lautrec and *Matisse. Like many other 20th-cent. painters he also experimented with *Cubism. He finally worked out a lively personal style of colourful *Expressionism which is best seen in his pictures of nudes—he had a predilection for painting nude children. Many Dutch museums have examples of his work.

Smart, John (1742/3–1811). One of the leading British *miniaturists of his period. His style was meticulous, bright, and pretty. He worked mainly in London but was in India 1785–*c.*1796.

Smet, Gustave de (1887–1943). One of the leading Belgian *Expressionist painters. His early work was *Impressionist in style, but he was influenced towards Expressionism by Jan *Sluyters and Henri *Le Fauconnier, whom he met in Holland when he took refuge there during the First World War. Typically de Smet painted scenes of rural and village life in which forms are treated in a schematic way owing something to *Cubism, and there is often an air of unreality reminiscent of that in *Chagall's work (*Village Fair*, Musée des Beaux-Arts, Ghent, *c*.1930). His brother **Léon de Smet** (1881–1966) was also a painter.

Smibert, John (1688–1751). Scottish-born portrait painter who emigrated to America in 1728 and settled in Boston in 1730. Previously he had travelled in Italy and practised successfully in London in an uninspired style derived from *Kneller. After he settled in America his work became somewhat more vigorous. He brought with him a small collection of copies, casts, and engravings and opened a shop where he sold English engravings from the works of well-known artists. These, together with his own paintings, became the corner-stone of the New England Colonial portrait style, and Smibert encouraged the young *Copley.

Smith, David (1906–65). The most original and influential American sculptor of his generation. He began to study art at Ohio University in 1924 but soon dropped out of the course and in the summer of 1925 worked at the Studebaker motor plant at South Bend, Indiana, where he acquired the skills in metalwork that stood him in good stead during his later career. From 1926 to 1930 he studied painting at the *Art Students League, New York, while supporting himself by a variety of jobs. Among his friends were Arshile *Gorky and Willem *de Kooning. He turned to sculpture in the early 1930s, making his first welded iron sculpture in 1933, although he always maintained that there was no essential difference between painting and sculpture and that although he owed his 'technical liberation' to Julio *González, his aesthetic outlook was more influenced by *Kandinsky, *Mondrian, and *Cubism. During the 1930s he was already doing sculpture of considerable originality, constructing compositions from steel and 'found' scrap, parts of agricultural machinery, etc. He had a love of technology, and wrote: 'The equipment I use, my supply of material, comes from factory study, and duplicates as nearly as possible the production equipment used in making a locomotive . . . What associations the metal possesses are those of this century: power, structure, movement, progress, suspension, destruction, brutality.' His first one-man exhibition was at the East River Gallery, New York, in 1938. Smith settled at Bolton Landing, New York, in 1940. In the same year he exhibited a set of fifteen bronze *relief plaques entitled *Medals of Dishonour* stigmatizing the prevalence of violence and greed throughout the world.

After being employed as a welder on defence work, he returned to sculpture *c*.1945. His sculpture during the 1940s and 1950s was open and linear, like three-dimensional metal calligraphy. Perhaps the most noted work in this style is *Hudson River Landscape* (Whitney Mus., New York, 1951). From the end of the 1950s up to his death in a car accident he did the more structural and massive work for which he is best known, often working in series such as *Zig*, *Tank Totem*, *Agricola*, *Cubi*, *Voltri*. Although monumental and intended to be seen in the open, these last works—characteristically consisting of boxes and cylinders of polished metal—were not heavy but had an unstable, dynamic quality which contradicted their sense of density. It was these works which initiated a new era in American sculpture, ushering in the sort of objectivity which characterized *Post-Painterly Abstraction. Smith's work is in many major collections.

Smith, Jack. See KITCHEN SINK SCHOOL.

Smith, John Raphael (1752–1812). English engraver, *miniaturist, and painter. He is best known as one of the leading *mezzotinters of the day, producing (most notably) numerous plates after portraits by *Gainsborough, *Reynolds, and *Romney, and also some after his own designs. Smith also worked as an assistant to various painters, and made pastels and miniatures.

Smith, John 'Warwick' (1749–1831). English landscape watercolourist, nicknamed 'Warwick' Smith because he was patronized by the Earl of Warwick and also lived in Warwick for a time. Lord Warwick paid for Smith to visit Italy in 1776–81, and he later made many sketching tours in Wales and the Lake and Peak Districts. He experimented with directly applied colour without underdrawing and his best work has strength and freshness. He was very prolific, however, and much of his later work is routine. He is represented in most major collections of watercolours.

Smith, Sir Matthew (1879–1959). English painter. He studied at the *Slade School, 1905–7, and for a short time in Paris (1911) under *Matisse. Thereafter he identified strongly with French art and spent much of his time in France. He was delicate in health and of a nervous disposition, but this is hardly apparent from his work, which uses colour in a bold, unnaturalistic manner echoing the *Fauves. His lush brushwork, too, has great vigour, and he was one of the few English painters to excel in painting the nude, his dark saturated colours and opulent fluency of line creating images of great sensuousness. He also painted landscapes (most notably a series done in Cornwall in 1920) and still lifes. Frank *Auerbach and Francis *Bacon are among the many artists who have admired Smith's painterliness. His work is in many public galleries, the best collection being in the Guildhall Art Gallery, London.

Smith, Richard. See SHAPED CANVAS.

Smith, Tony (1912–80). American sculptor, painter, and architect. He served an apprenticeship in architecture as clerk of works to Frank Lloyd *Wright and practised as an architect from 1940 to 1960, during which time he also painted. He began to take an interest in sculpture around 1940, but although he taught at various colleges in the 1940s and 1950s (in addition to his architectural career) and was closely associated with leading avant-garde figures such as Barnett *Newman, Jackson *Pollock, Mark *Rothko, and Clyfford *Still, he did not exhibit sculpture publicly until 1964. From that time he quickly emerged as one of the leading exponents of *Minimal art. His work was sometimes very large in scale, composed of bold geometrical shapes (often repeated modular units) that he had industrially manufactured in steel. Many of his works were placed outdoors, helping to bring to American sculpture a new interest in the environment. A well-known example is *Gracehoper* (Detroit Institute of Arts, 1972), which one can walk through.

Smithson, Robert (1938–73). American sculptor and experimental artist. In the 1960s his work belonged mainly to the category of *Minimal art; he was interested in mathematical impersonality and as well as making block-like steel sculptures he experimented with reflections and mirror images. From the late 1960s he turned to *Conceptual art; he expressed his ideas mainly through *Land art and became the best-known artist working in this field. In 1968 he began a series of 'Sites' and 'Non-Sites'. The latter consisted of photographs and plans of locations he had visited (particularly derelict urban or industrial sites) displayed with specimens of rock or geological refuse he had gathered there, arranged into random heaps or in metal or wood bins: 'Instead of putting a work of art on some land, some land is put into the work of art.' Smithson then moved on to large-scale earthworks, the best known of which is the enormous *Spiral Jetty* (1970), a spiral road running out into Great Salt Lake, Utah. He was killed in a plane crash when he was surveying a work in progress, *Amarillo Ramp* in Texas. Smithson wrote many articles expounding his views on art: *The Writings of Robert Smithson*, edited by Nancy Holt (his widow), was published in 1979.

Smithsonian Institution. A research institution and educational centre founded by the bequest of the English scientist James Smithson (1765–1829) 'for the increase and diffusion of knowledge' and established by congressional act in Washington DC in 1846. Smithson was an illegitimate son of the Duke of Northumberland and it is thought to have been resentment over the circumstances of his birth that caused him to make the bequest to the USA rather than his native country: 'My name shall live . . . when the titles of the Northumberlands . . . are extinct and forgotten.' The Smithsonian administers many prestigious cultural organizations in Washington, including the National Gallery of Art, the National Portrait Gallery, and the Joseph H. Hirshhorn Museum and Sculpture Garden.

Snyders, Frans (1579–1657). Flemish painter of animals, hunting scenes, and still life, active mainly in his native Antwerp. He was a pupil of Pieter *Brueghel the Younger and travelled in Italy 1608–9. Back in Antwerp, he became a close friend of *Rubens, also recently returned from Italy (he was eventually executor of Rubens's will). Snyders often collaborated with Rubens; he painted animals, fruits, and flowers in Rubens's pictures, and Rubens painted figures in his. He later worked out the same kind of reciprocal arrangement with *Jordaens, and collaborated with his brother-in-law Cornelis de *Vos, van *Dyck, and Abraham *Janssens. His independent works show he was the finest animal painter of his time: the best are his scenes of fighting wild animals and hunts, which have a tremendous sense of *Baroque vitality. Examples of his work are in many public collections.

soapstone, also called steatite. A variety of talc, a very soft smooth stone superficially like *marble in appearance but with a soapy texture. It will take a smooth polish, and is so easily worked that it may even be carved with a knife. Its colour is a dull greenish or bluish grey or sometimes brown. It is vulnerable to dampness in the atmosphere and will serve only for indoor sculpture. It has been used, especially for more decorative and less permanent sculpture, in most parts of the world—in Egypt, Europe, China and Japan, in African Tribal art, and by American Indians and Eskimos. Soapstone carvings are sometimes sold to the unwary as jade.

Socialist Realism. The name of the officially approved style of art in Soviet Russia and other Communist countries, involving in theory a faithful and objective reflection of real life and in practice the compulsory and uncritical glorification of the State. Socialist Realism was an aspect of the dictatorship of Stalin, who was leader of the Soviet Union from the death of Lenin in 1924 until his own death in 1953. Alan Bird (*A History of Russian Painting*, 1987) writes that 'He saw all aspects of avant-garde culture, including painting, as subversive infiltrations of the purity of Soviet life' and that his minister Andreii Zhdanov 'made himself responsible for imposing an iron control on artistic expression'. The principles of Socialist Realism began to take shape in the late 1920s and were proclaimed in the 1932 decree 'On the Reconstruction of Literary and Art Organizations' (before this, the term 'Heroic Realism' had often been used, but 'Socialist Realism' now became the official label). In its early days it saw expression in some outstanding works, notably the paintings of Alexander Deineka (1899–1969), but increasingly it became associated with stereotyped images painted in a conventional academic manner. In the 1930s there were four main types of Socialist Realist paintings: domestic scenes, portraits, industrial and urban landscapes, and scenes on collective farms. During the Second World War, patriotic scenes from Russian history were added to the list. After the death of Stalin there was some relaxation of strictures, but the system still remained stifling to creativity, and any form of experiment remained extremely difficult. In the West, Socialist Realism remained synonymous with repression, and its products were generally regarded as morally tragic and aesthetically comic, although the merits of painters such as Arkady Plastov (1893–1972), a specialist in farm scenes, are now being recognized.

Social Realism. A very broad term for painting (or literature or other art) that comments on contemporary social, political, or economic conditions, usually from a left-wing viewpoint, in a realistic manner. Often the term carries with it the suggestion of protest or propaganda in the interest of social reform. However, it does not imply any particular style; Ben *Shahn's caricature-like scenes on social hypocrisy and injustice in the USA, the dour working-class interiors of the *Kitchen Sink School in Britain, and the declamatory political statements of *Guttoso in Italy are all embraced by the term. See also REALISM.

Société Anonyme, Inc. (or **a Museum of Modern Art**). An association founded in 1920 by Katherine *Dreier, Marcel *Duchamp, and *Man Ray for the promotion of contemporary art in America by lectures, publications, travelling exhibitions, and the formation of a permanent collection. In French the term 'société anonyme' means 'limited company', so the name—suggested by Man Ray—was intended as a tautological *Dada jest; as Miss Dreier loved to explain, it meant 'incorporated corporation'. However, the work of the society was serious and trail blazing. Its museum, which opened at 19 East 47th Street, New York, on 30 April 1920, was the first in the USA, and one of the earliest anywhere, to be devoted entirely to modern art. Between 1920 and 1940 the Société organized 84 exhibitions, through which such artists as *Klee, *Malevich, *Miro, and *Schwitters were first exhibited in America. To some extent, therefore, the Société carried on the tradition that had been started by the 291 Gallery of *Stieglitz in the years before the *Armory Show, and to some extent also it prepared the way for the *Museum of Modern Art, which was founded in 1929. The Museum of Modern Art soon eclipsed the Société Anonyme and Miss Dreier's finances were in any case badly hit by the Depression, but she continued officially as President (as Duchamp did as Secretary) until the Société officially closed in 1950. Nine years earlier, in 1941, they had presented the superb permanent collection that the Société had built up (over 600 works) to Yale University Art Gallery.

Society of Artists. An association of artists formed in London in the 1750s and incorporated by Royal Charter in 1765. It emerged in response to the desire for a state-run academy that would provide public exhibitions of contemporary art, and was the main forerunner of the *Royal Academy. Its leaders were *Hayman and George *Lambert. It survived until 1791.

Society of Independent Artists. An American association formed in 1916 to succeed the Association of American Painters and Sculptors, which was dissolved after the *Armory Show. Its object was to hold annual exhibitions in rivalry with the *National Academy of Design and to afford progressive artists an opportunity to show their works. It was organized on the model of the French *Salon des Indépendants without jury or prizes, giving anyone the right to exhibit on payment of a modest fee. The first President was William *Glackens, followed by John *Sloan. The first exhibition, in April 1917, comprised more than 2,000 exhibits and included both American and European artists. It is perhaps best remembered, however, for a work that was not shown, for Marcel *Duchamp, who was one of the Society's officials, resigned after the refusal to exhibit his *ready-made in the form of a urinal signed 'R. Mutt'. Although much recondite aesthetic theory has been read into this gesture, it is likely that the main purpose of it was to demonstrate the artistic incongruity of a Society with the professed purpose of allowing anyone to exhibit anything. Although the Society continued to arrange exhibitions until the mid-1940s, subsequent exhibitions were smaller and inferior in quality and its importance diminished.

Sodoma, Il (Giovanni Antonio Bazzi) (1477–1549). Italian painter, born at Vercelli and active chiefly in and around Siena, where he settled in 1501. *Vasari, who disliked him, explains the origin of his nickname—'the sodomite'—in this fashion: 'His manner of life was licentious and dishonourable, and as he always had boys and beardless youths about him of whom he was inordinately fond, this earned him the nickname of Sodoma; but instead of feeling shame, he gloried in it, writing stanzas and verses on it, and singing them to the accompaniment of the lute.' Sodoma (who was married and had children) himself used the name in his signature, and Vasari's story has been questioned. Vasari also tells us that Sodoma kept a menagerie of strange animals 'so that his home resembled a veritable Noah's ark'. He was a prolific painter of frescos and easel pictures, and he drew on a variety of sources that were not always fully digested; consequently his work often has incongruous juxtapositions and a general air of uncoordination, but it also possesses charm and a flair for decoration. His fresco of the *Marriage of Alexander and Roxane* (1516–17), painted for the Sienese banker Agostino Chigi (1465–1520) in his Villa Farnesina in Rome, is often cited as his finest work. In Rome Sodoma also painted part of the ceiling of the Stanza della Segnatura in the Vatican, where *Raphael later worked. In his time Sodoma was considered the leading artist in Siena, but later critics have come to rank *Beccafumi above him.

Soest, Gerard (*c.*1600–81). Dutch portrait painter who settled in London (probably before 1650). He was much less successful than his countryman *Lely, but he is often a more interesting painter, as he dispensed with the flattery and superficial elegance in vogue at the Restoration court and had a strong grasp of character. There are examples of his work in the National Portrait Gallery and Tate Gallery, London, but his masterpiece is generally held to be *Cecil, Second Lord Baltimore, with a Child and a Negro Page* (Enoch Pratt Library, Baltimore), which shows his considerable powers as a colourist.

Soffici, Ardengo (1879–1964). Italian critic and painter. His formative years, 1900–7, were spent in Paris, where he knew writers and painters such as *Apollinaire, *Picasso, *Braque, and *Modigliani, and wrote in avant-garde periodicals, and in the years before the First World War he was prominent in introducing the discussion of modern art—particularly *Cubism—to Italy. Initially he was hostile to *Futurism, but under the influence of *Boccioni and *Carrà, he became converted to the movement in 1913. After the war, however, Soffici looked increasingly to Italian tradition in painting and in 1918 he associated himself with the attacks on Cubism and Futurism made by de *Chirico and Carrà in their journal *Valori Plastici*. Soffici's paintings are generally regarded as of much less interest and importance than his writings.

Soft art. Term applied to sculpture using non-rigid materials, a vogue of the 1960s and 1970s. The materials employed have been very diverse: rope, cloth, rubber, leather, paper, canvas, vinyl—anything in fact which offers a certain persistence of form but lacks permanent shape or rigidity.

Perhaps the earliest example of 'Soft art' in this century was the typewriter cover which Marcel *Duchamp mounted on a stand and exhibited in 1916. But this belongs rather to the category of *ready-mades, which he introduced. 'Soft art' as a movement is usually traced to Claes *Oldenburg's giant replicas of foodstuffs (ice-cream sundaes, hamburgers, pieces of cake, etc.) made from stuffed vinyl and canvas. Other artists who have experimented with soft materials have been many and diverse, from many different movements—*Pop art, *Surrealism, *Arte Povera, etc.—and an exhibition on the theme, in which some sixty artists were represented, was staged in Munich and Zurich in 1979–80. The movement—if such it can be called—has certain affinities with the *New Realism, since in general, although the work may be representative of something else, the materials are not concealed and make no pretence to be other than what they are.

soft-ground etching. A method of *etching that produces prints characterized by softness of line or a grainy texture. When it was invented late in the 18th cent. it was employed for imitating crayon or *pencil lines. The traditional method was as follows. A copper plate was covered all over with a waxy *resinous mixture, somewhat like the ordinary etching *ground but much softer and sticky enough to adhere to anything pressed into it. Over this ground was laid a sheet of paper on which the artist drew with a pencil. Under the pressure of the pencil strokes the ground stuck to the back of the paper, so that when it was lifted the wax immediately underneath the lines came away with it, while the rest remained in place. The plate was then immersed in acid, which bit into the metal along the lines. Finally it was cleaned and inked and prints were taken from it by the copper-plate press. The printed lines were granular, coarse, or fine according to the texture of the paper used for the drawing: thus if the paper had been smooth, the lines resembled pencil, but if rough, they were more like chalk. Soft-ground etchings bear a strong likeness to prints in the *crayon manner, but are generally a little softer and less regular. Although the technique was used in the late 18th and early 19th cents. chiefly for reproduction, a number of excellent original soft-ground etchings were produced by *Gainsborough, *Cotman, and *Girtin. Gainsborough in particular, in his landscape prints, combined the soft-ground line most effectively with *aquatint in a manner resembling spontaneous drawing in pencil and *wash. Interest in soft-ground etching has revived among modern artists.

Solari (or **Solario**), **Andrea** (active 1495–1524). Milanese painter. His style was strongly influenced by *Leonardo, but his work, which has suggestions also of Flemish and Venetian art, has much more individual character than that of many of the master's followers. The *Madonna with the Green Cushion* (Louvre, Paris, *c.*1507) is his best-known work. In 1507–9 Solario worked on decorations in the Château de Gaillon for Cardinal d'Amboise (1460–1510). He came from a family of artists; the best-known member apart from himself was his sculptor brother, **Cristoforo** (d. 1527), known as Il Gobbo (the hunchback). His best-known work is the tomb of Lodovico *Sforza and Beatrice d'*Este (1497–9) in the Certosa di Pavia.

Solimena, Francesco (1657–1747). The leading Neapolitan painter of the first half of the 18th cent. In a long and extremely productive career he painted frescos in many of the greatest churches in Naples, and he became one of the wealthiest and most famous European artists of his day. His vigorous style, often marked by dramatic lighting, owed much to the example of such *Baroque artists as Luca *Giordano (his great predecessor in Naples), *Lanfranco and *Preti, but it also has a firmness of structure and a clarity of draughtsmanship that shows his allegiance to the *classical tradition of *Raphael and Annibale *Carracci. Solimena's paintings were in demand all over Europe, and his international influence was spread also by his celebrity as a teacher. *Ramsay was among his pupils and *Fragonard copied his work in S. Paolo Maggiore.

Solis, Virgil (1514–62). German engraver and designer working in Nuremberg. He was extremely prolific of patterns of the *applied arts, and also did numerous book illustrations on biblical and Classical themes.

Solomon, Simeon (1840–1905). British painter and graphic artist, a member of a well-known family of artists. He was a member of *Rossetti's circle and a friend of *Burne-Jones, and his work shows strong *Pre-Raphaelite influence. In the 1860s he built up a reputation as a book illustrator, but then sank into a life of idleness and dissipation (in 1873 he was arrested on homosexual charges). His later years were spent in a pathetically bohemian existence, and he died of alcoholism. His sister **Rebecca** (1832–86), who painted portraits and anecdotal historical scenes, also died an alcoholic. Their brother **Abraham** (1824–62) painted scenes from literature and contemporary life scenes, some of which attained great popularity, and were much reproduced as prints.

Somer, Paul van (*c.*1577–1622). Flemish portrait painter and engraver who settled in England in 1616 and became—with Cornelius *Johnson and Daniel *Mytens—the leading portraitist at the court of James I (1566–1625). His finest work is *Queen Anne of Denmark* (Royal Coll., 1617), an imposing full-length, but more archaic in style than similar portraits by Mytens.

Sørensen, Henrik (1882–1962). Swedish-born Norwegian painter. His training included a period studying with *Matisse in Paris, 1909–10, and his colourful and emotional work combines certain modernist features with traditional Norwegian themes. He painted varied subjects, including portraits, landscapes, and religious compositions, and did numerous murals for public buildings, including a huge one in the new Town Hall at Oslo (begun 1950). Sørensen was also a noted book illustrator.

Sorolla y Bastida, Joaquin (1863–1923). Spanish painter and graphic artist, active mainly in Valencia. He was a prolific and popular artist, working on a wide variety of subjects—*genre, portraits, landscapes, historical scenes—and producing many book illustrations. His pleasant and undemanding style was marked by brilliant high-keyed colour and vigorous brushwork, representing a kind of conservative version of *Impressionism. He was well known outside Spain, and in 1910–20 he painted a series of fourteen mural panels for the Hispanic Society of America in New York, representing scenes typical of the various provinces of Spain. His former home in Madrid is now a museum dedicated to his work.

Sotheby's. The oldest and largest firm of auctioneers in the world. It was founded by the London bookseller Samuel Baker, who held his first auction in 1744. On his death in 1778, his estate was divided between his partner George Leigh and his nephew John Sotheby. The last of the Sotheby family to be involved in the firm died in 1861. Although Sotheby's extended its range to take in prints, coins, medals, and antiquities of various kinds, books long remained the primary concern of the company, and it was not until after the First World War that paintings and other works of art became a major part of its business (before this time most important picture sales were held by *Christie's). In 1964 Sotheby's bought Parke Bernet, America's largest fine art auctioneers, and it now has major sale rooms in London, New York, Geneva, and Monaco, with numerous branches throughout the world.

sotto in sù (Italian: from below upwards). Term applied to an extreme form of *illusionistic foreshortening in which figures or objects painted on a ceiling appear to be floating or suspended in space above the viewer. *Mantegna's Camera degli Sposi in the Palazzo Ducale in Mantua is the first major example, but the device is associated particularly with *Baroque decoration.

Soutine, Chaïm (1893–1943). Lithuanian-born painter who settled in France in 1913 and became one of the leading *Expressionists of the School of *Paris. His friends in the circle of expatriate artists there included *Chagall and *Modigliani, who painted a memorable portrait of him (NG, Washington, 1917). Soutine suffered from depression and lack of confidence in his own work (he was reluctant to exhibit and sometimes destroyed his own pictures), and he endured years of desperate poverty until the American collector Dr Albert C. *Barnes bought a number of his paintings in 1923. Thereafter he had a prosperous career. Soutine's work included landscapes, portraits, and figure studies of characters such as choirboys and pageboys. His style is characterized by thick, convulsive brushwork, through which he could express tenderness as well as turbulent psychological states. There is something of an affinity with van *Gogh, although Soutine professed to dislike his work and felt more kinship with the Old Masters whose work he studied in the Louvre; his pictures of animal carcasses, for example, are inspired by *Rembrandt's *Flayed Ox*. However, the gruesome intensity of works such as *Side of Beef* (Albright-Knox Art Gallery, Buffalo, 1925) was not gained simply through study of similar pictures, for Soutine visited abattoirs and even brought a carcass into the studio. His neighbours complained of the smell of the rotting meat and called the police, whom Soutine harangued on the subject of how much more important art was than sanitation. The filthy state in which he lived was notorious: the poet André Salmon recalled that Soutine consulted a specialist about earache and that 'In the canal of the painter's ear the doctor discovered, not an abcess, but a nest of bed bugs.'

Soyer, Moses (1899–1974) and **Raphael** (1899–1987). Russian-born painters, twins. They emigrated to America in 1912 as exiles from tsarist Russia. They are best known for their *Social Realist subjects, particularly those of the Depression years of the 1930s, in which they depicted the lives of working people with sympathy and at times a touching air of melancholy, as in Raphael's well-known *Office Girls* (Whitney Mus., New York, 1936). Both brothers also did many self-portraits and wrote on art. Moses wrote articles defending Social Realism and attacking *Regionalism; Raphael published several autobiographical volumes and a book on Thomas *Eakins (1966). Another brother, **Isaac** (1907–81), who came to America in 1914, was also a painter.

Spagnoletto, Lo. See RIBERA, JOSÉ.

Spagnuolo, Lo. See CRESPI, GIUSEPPE MARIA.

spandrel. Architectural term for the triangular space between the outer curve of an arch and the rectangular frame enclosing the arch or the surface between two adjacent arches; more loosely the term is applied to any triangular area with curved sides, as in the spaces over the windows of the Sistine Chapel. Spandrels are often treated ornamentally, with painting or sculpture, although the shape does not readily lend itself to decorative exploitation. Winged figures, however, have often found favour in this position—victories on triumphal arches and angels in medieval churches.

Spazialismo (or **Spatialism**). A movement founded by Lucio *Fontana in Milan in 1947 in which he grandiosely intended to synthesize colour, sound, space, movement, and time into a new type of art. The main ideas of the movement were anticipated in his *Manifesto Blanco* (White Manifesto) published in Buenos Aires in 1946. In it he spoke of a new 'spatial' art in keeping with the spirit of the post-war age. On the negative side it repudiated the illusory or 'virtual' space of traditional easel painting; on the positive side it was to unite art and science to project colour and form into real space by the use of up-to-date techniques such as neon lighting and television. The new art of Spatialism would 'transcend the area of the canvas, or the volume of the statue, to assume further dimensions and become . . . an integral part of architecture, transmitted into the surrounding space and using new discoveries in science and technology'. Five more manifestos followed; they were more specific in their negative than their positive aspects, and carried the concept of Spazialismo little further than the statement that its essence consisted in 'plastic emotions and emotions of colour projected upon space'. In 1947 Fontana created a 'Black Spatial Environment', a room painted black, which was considered to have foreshadowed *Environment art. His holed and slashed canvases (beginning in 1949 and 1959 respectively) are also considered to embody Spatialism. An example of the slashed type (the slash made with a razer blade) is *Spatial Concept Waiting* (Tate, London, 1960).

Speculum Humanae Salvationis ('The Mirror of Human Salvation'). A medieval textbook of Christian *typology representing how the Incarnation and Passion of Christ (antitype) had been prefigured in the Old Testament, Jewish legend, and secular history (type). Ludolph of Saxony, a Dominican friar, probably wrote the first *Speculum* at Strasburg in 1324 in Latin verse, and versions later appeared in English, French, German, Dutch, and Czech (there was comparatively little interest in it in Italy). Many manuscripts survive, some of them illustrated. The illustrated *Speculum* is normally arranged so that on opening the book there are four pictures at the head of the page—the antitype on the left and the types on the right. In about 1465 the *Speculum* was issued as a *block book, which enjoyed great popularity, and it was also used in painting, sculpture, stained glass, and tapestry throughout Western Europe.

Spencer, Sir Stanley (1891–1959). English painter, one of the most original figures in 20th-cent. British art. He was born in Cookham in Berkshire and lived for most of his life in the village, which played a large part in the imagery of his paintings. His education was fairly elementary, but he grew up in a family in which literature, music, and religion were dominant concerns and his imaginative life was extremely rich. He said he wanted 'to take the inmost of one's wishes, the most varied religious feelings . . . and to make it an ordinary fact of the street', and he is best known for his pictures in which he set biblical events in his own village. For him the Christian religion was a living and present reality and his visionary attitude has been compared to that of William *Blake.

Spencer was a prize-winning student at the *Slade School (1908–12) and served in the army from 1915 to 1918, first at the Beaufort War Hospital in Bristol, then in Macedonia. He was appointed an *Official War Artist in 1918, but his experiences during the war found their most memorable expression a decade later when he painted a series of murals for the Sandham Memorial Chapel at Burghclere in Hampshire (1927–32), built to commemorate a soldier who had died from an illness contracted in Macedonia. The arrangement of the murals consciously recalls *Giotto's Arena Chapel in Padua, but Spencer painted in oil, not fresco, and he concentrated not on great events, but on the life of the common soldier, which he depicted with deep human feeling. There is no violence, and Spencer said that the idea for one of the scenes—*The Dug-Out*—occurred to him 'in thinking how marvellous it would be if one morning, when we came out of our dug-outs, we found that somehow everything was peace and the war was no more'.

By this time Spencer was a celebrated figure, his greatest public success having been *The Resurrection: Cookham* (Tate, London, 1924–6), which when exhibited in 1927 was hailed by the critic of *The Times* as 'the most important picture painted by any English artist in the present century'. He continued: 'What makes it so astonishing is the combination in it of careful detail with modern freedom of form. It is as if a *Pre-Raphaelite had shaken hands with a *Cubist.'

Spencer was again an Official War Artist during the Second World War, when he painted a series of large canvases showing shipbuilding on the Clyde (Imperial War Mus.) that memorably capture the heroic teamwork that went into the war effort. His career culminated in a knighthood in the year of his death, but his life was not a smooth success story, and in the 1930s he somewhat alienated his public with the expressive distortions and erotic content of his work. In 1935 he resigned as an Associate of the *Royal Academy when two of his pictures, considered caricature-like and poorly drawn, were rejected for the annual summer exhibition, but he rejoined the Academy in 1950. In 1937 he divorced his first wife, Hilda Carline, and married Patricia Preece, but Hilda continued to play a large part in his life, and he painted pictures in memory of her and even wrote letters to her after her death. Some of his nude paintings of Patricia vividly express not only the sexual tensions of his life, but also his belief in the sanctity of human love; the best known is the double nude portrait of himself and Patricia known as *The Leg of Mutton Nude* (Tate, 1937). In his later years Spencer acquired a reputation as a landscapist as well as a figure painter. He also occasionally did portraits. There is a gallery devoted to Spencer at Cookham, containing not only paintings, but also memorabilia such as the pram that the eccentric figure used for pushing his painting equipment around the village.

His younger brother **Gilbert Spencer** (1892–1979) was also a painter of imaginative subjects and landscapes, working in a style close to that of Stanley. Examples of his work are in the Tate Gallery, London, and Holywell Manor, Balliol College, Oxford, where he painted murals.

Spinario (Latin: spina, a thorn). Ancient bronze statue of a seated boy extracting a thorn from his left foot (Capitoline Museum, Rome). It is recorded in Rome as early as the 12th cent. and during the *Renaissance it was one of the most influential of ancient sculptures. Among the many copies that were made of it was a statuette by *Antico for Isabella d'*Este. Its fame endured and it was one of the ancient works taken by Napoleon to Paris, where it remained from 1798 to 1815. Various stories grew up from the Renaissance onwards to explain the subject, the most popular being that the statue commemorates a shepherd boy called Martius who delivered an important message to the Roman Senate and only when his task was accomplished stopped to remove a thorn from his foot. It is now generally thought that the *Spinario* is a Roman pastiche of about the 1st cent. BC, combining a *Hellenistic body with a head of earlier date (the way in which the hair falls indicates that the head was meant to be in an upright position rather than looking down).

Spinelli, Parri. See SPINELLO.

Spinello Aretino (active 1373–1410/11). Italian painter. He came from Arezzo (hence the name Aretino) and probably trained in Florence, perhaps under Agnolo *Gaddi. He was the most prolific muralist of his time and undertook large fresco cycles all over Tuscany. He was a quick worker: within 1387 he completed the fresco cycles of St Benedict for the sacristy of S. Miniato, Florence, and of St Catherine of Alexandria for the Alberti family's oratory near Florence. His last series was the cycle devoted to the Sienese pope Alexander III (*c.*1100–81) in Siena Town Hall (1408–10). He also painted altarpieces. Spinello borrowed ideas freely from other painters, notably *Giotto, but his style was sturdy and vigorous. Several fresco fragments by Spinello are in the National Gallery, London. His son **Parri Spinelli** (d. 1452) was his assistant.

Spitzweg, Karl (1808–85). German painter and graphic artist, active in his native Munich. He began his career as a pharmacist, and turned fairly late to art, first as a newspaper *caricaturist, then as a painter. Although he travelled widely (England, France, Italy, and elsewhere), he was provincial in his choice of subjects and is the outstanding representative of the *Biedermeier style. His pictures are generally small, humorous in content, and full of lovingly depicted anecdotal detail (*The Poor Poet*, Neue Pinakothek, Munich, 1839, and other versions). Spitzweg also painted excellent landscapes that show a debt to the *Barbizon School.

Spranger, Bartholomeus (1546–1611). Netherlandish painter and designer who had an international career and played an important part in the spread of *Mannerism in northern Europe. He trained in his native Antwerp, where he came under the influence of Frans *Floris, then travelled via France to Italy, where he spent a decade (1565–75), mainly in Rome. After five years in Vienna, he finally settled in Prague in 1580 at the court of the emperor Rudolf II (1552–1612). His paintings, which are well represented in the Kunsthistorisches Museum, Vienna, are often of mythological or allegorical subjects and are highly polished and sophisticated—close in style to those of his fellow court painter Hans von *Aachen. Spranger had met van *Mander in Rome and through drawings that he gave him his style was carried to Haarlem, where it became a formative influence in the Haarlem Academy. *Goltzius and other engravers made many prints of his designs (Spranger also made a few etchings himself) and his work was widely influential in the years around 1600.

Squarcione, Francesco (1397–*c.*1468). Italian painter, active in Padua. He is an enigmatic figure, who is important in terms of the pupils he trained, rather than for his own work. A Paduan writer of 1560 patriotically described him as a famous and benevolent master, with many pupils and a large collection of *antique sculpture gathered on youthful journeys through Greece and Italy. More recent research, however, gives a picture of a tailor who, turning painter in his middle thirties, was for many years discreditably involved in a series of law-suits with pupils who, like *Mantegna and Marco *Zoppo, resentful of his exploitation of their talents, had broken their apprenticeships with him. No traces of his collection remain, but it is certainly likely that something of the antiquarian erudition of the university town of Padua rubbed off on the young men who spent time in his workshop. It is impossible to assess any stylistic debt to Squarcione himself, however, as so little is known about his work, and his traditional role as the founder of a distinctive 'Paduan style' is highly questionable. Only two paintings are firmly associated with his shop—a *polyptych in the museum in Padua and a half-length *Virgin and Child* in the Staatliche Museen, Berlin—both very dry in style.

Staël, Nicolas de (1914–55). Russian-French painter, born in St Petersburg, son of Baron Vladimir Ivanovich de Staël-Holstein. In 1919 his family was forced to leave Russia (he would later become incensed if anyone suggested they had 'fled') and moved to Poland. Both parents had died by 1922 and Nicolas and his two sisters were adopted by a family of rich Russian expatriates in Brussels, where he studied at the École des Beaux-Arts, 1932–6. In the next two years he travelled widely (France, Italy, Spain, North Africa), then in 1938 settled in Paris, where he studied briefly with *Leger. On the outbreak of war in 1939 he joined the Foreign Legion and was sent to Tunisia. He was demobilized in 1941 and moved to Nice, where he turned from figurative to abstract art, although the forms he used were usually suggested by real objects. In 1943 he returned to Paris and after the war he quickly gained a reputation as one of the leading abstract painters of the School of *Paris, his work showing a sensuous delight in handling paint that was unrivalled at the time. Typically his works feature luscious blocks or patches of thick paint (often applied with a knife), subtly varied in colour and texture. In 1951 he began to reintroduce figurative elements into his work, his subjects including landscapes and still life. From 1952 he spent much of his time working in the bright light of the South of France, and his late works are often very intense in colour. In spite of critical and financial success, de Staël felt that he had failed to reach a satisfactory compromise between abstraction and figuration, and he committed suicide.

staffage. Term applied to small figures and animals in a painting that are not essential to the subject but are used to animate the composition. Landscape painters, notably in 17th-cent. Flanders and Holland, have often employed other artists to paint their staffage (see, for example, POELENBURGH).

stained glass. Glass that has been given translucent colour in any of various ways, used particularly for church windows. As a rule such windows are built up of panels, not too big for convenience in handling, composed of pieces of glass either dyed in its substance or superficially coloured, set together in a framework usually of lead, to form decorative or pictorial designs. The art began in the service of the Christian Church and is of *Byzantine origin, but in its most characteristic development and its highest achievements it is essentially an art of Western Christendom, practised most splendidly in the west and north of Europe as an adjunct to *Gothic architecture. Its early history is traceable only by inference, as there are few survivals before the *Romanesque period (the church of St Paul at Jarrow has some fragments—placed in a window in 1980—that are probably more or less contemporary with the foundation of the monastery there in the late 7th cent.). The earliest known complete windows—in Augsburg Cathedral—are variously dated between 1050 and 1150, and show an art already nearly perfect, with a technique which has endured in principle to the present day, developed or modified only in inessential details.

The broad outlines of ordinary stained-glass technique may be briefly described. A design is first drawn on a small scale; from this a *cartoon is made to the actual size of the window on paper (in earlier times a white-washed board or table was used). In this cartoon the lines to be followed by the 'leads' are heavily drawn in black, whilst the various colours of the glass to be employed are indicated in writing. Pieces of 'pot-metal', as the glass is called by glaziers—whether plain white or stained with colouring oxides in the making—are then cut to the required outlines out of large sheets. This is done by breaking off a piece approximately the requisite size with the help of a heated iron instrument; the piece is then laid over the cartoon and the exact outline required traced with pigment over the underlying lines. To reduce it to this outline the piece of glass is finally nibbled by means of a 'grosing iron'—a tool cut along one edge with notches of various gauges; the invention of a diamond-cutting tool in the 16th cent. made it possible to do this much less laboriously and with greater accuracy.

The pieces of glass, now cut to shape, are ready for painting, an almost invariable accessory for rendering details such as features of a face, folds of drapery, or foliage; the painter follows the lines of the cartoon over which the glass is again laid for this purpose. The pigment is a black or dark brown *enamel which needs firing at a low temperature in a 'muffle' to fix it on the surface of the glass. When this operation has been completed the pieces of glass are once more assembled over the cartoon and fixed in the leads; these are strips of lead, cast—or (from the 16th cent. onwards) moulded in a hand-mill or 'lead-vice'—with flanges forming a groove on each side to clinch the edge of the pieces of glass.

All stained-glass windows of any size are made up of separate panels held in place by iron bars ('saddle bars') to which they are attached by means of lead ribbons or, in modern times, copper wires soldered to the leads. In the very large windows of the 12th and 13th cents. (Canterbury, Chartres, Sens) the space is divided up by a framework of iron ('armature') which serves not only as a support against wind pressure for the leaded panels but also to accentuate the main outlines of the general design of the window.

Such was the technique of glass-painting in practice in the 12th cent. The refinements introduced as time went on had in general the effect of making the art easier for the practitioner, but often at the expense of aesthetic quality. From the 15th cent. stained glass tended towards a greater pictorialism, imitating the effects of oil painting. In the process of change the peculiar character of the art came to be ignored: the artists forgot that light itself, transmitted through a coloured material, is its primary *medium; in their efforts to simulate the mass and solidity of natural forms they cast away the distinctive virtues of an essentially two-dimensional art. The introduction of enamel pigments in the 16th cent. made it possible to paint on glass more or less as one would on canvas, leading to such entirely pictorial windows as those in the chapel of New College, Oxford, designed by *Reynolds and executed in 1778–85. Windows of this kind can have a charm of their own, but they are anathema to many students of the subject: E. Liddell Armitage, in his book *Stained Glass* (1959), describes the use of enamel as 'an artistic poison . . . which killed practically every aesthetic faculty the craftsman of the period might inherently have possessed'.

With the *Gothic Revival in the 19th cent. there came a return to medieval principles, and William *Morris and his associates such as *Burne-Jones were among the artists who designed windows of superb quality. In the 20th cent. many noteworthy artists have designed stained-glass windows, in both figurative and abstract veins—among them *Chagall, *Matisse, and *Piper.

The peculiar beauty of stained glass has been well described by Gerald Randall in *Church Furnishing and Decoration in England and Wales* (1980): 'However intrinsically interesting wall and ceiling paintings may be, there is no doubt that the contribution of glass to our churches is more important. Glass has the advantage of transmuting light instead of merely reflecting it, and at its best has a sparkle and vitality that no opaque surface can match. Its effect changes with the light, from one day to another and from one hour to the next, and there are moments when the whole interior of a church seems to take fire from it.'

Stanfield, Clarkson (1793–1867). English painter, best known for his marine subjects. From 1808 to 1818 he was a sailor, first in the merchant service and then, after being press-ganged in 1812, in the Royal Navy. He was invalided after being disabled in a fall from the rigging, and took up painting as a career on the advice of his captain at the time, the novelist Frederick Marryat (1792–1848). His initial success came as a scene painter (he won renown for his spectacular *dioramas at the Drury Lane Theatre), but from the 1830s he concentrated on easel paintings. With these, too, he gained a considerable reputation; he was praised by *Ruskin and won a gold medal at the Exposition Universelle in Paris in 1855. He has been called 'England's van de *Velde', and apart from *Turner he was indeed the best British marine painter of his period. His work is represented in numerous British collections, including the National Maritime Museum and the art gallery of his native Sunderland.

Stantons of Holborn. A family of English mason sculptors with a large workshop at Holborn in London managed successively by **Thomas** (1610–74), his nephew **William** (1639–1705), and the latter's son **Edward** (1681–1734). They were much patronized by the lesser aristocracy and the professional classes, and their monuments are widespread in England. They range from simple tablets to elaborate tombs. Their designs are usually conservative, but William was one of the most accomplished English sculptors of his generation.

Stanzione, Massimo (1585?–1656?). Neapolitan painter. Like so many artists in his city at this time, he was strongly influenced by *Caravaggism, but his style has a distinctive refinement and grace that has earned him the nickname 'the Neapolitan Guido *Reni'. He was head of the busiest studio in Naples, and many of his works are still in the churches of the city, his masterpiece being the eloquent *Lamentation* (1638) in the Certosa di San Martino. His most important pupil was *Cavallino. Stanzione probably died in the terrible plague that struck Naples in 1656.

Stark, James (1794–1859). English landscape painter in oils and watercolour, a representative of the *Norwich School. A competent, painstaking, and unoriginal follower of his master, *Crome, he worked in his native Norwich and in London. He published *Scenery of the Rivers of Norfolk* (1827–34), engravings from his work by various hands with text by J. W. Robberds. His work may best be studied at the Castle Museum, Norwich.

state. Term applied to any of the stages through which a *print may pass as an artist alters the design. The first state is represented by the first proof taken from the plate. If no alterations are made this can also be described as the 'only state'. Often, however, particularly in *etching, the artist will alter the design several times before reaching the final state; several impressions may be taken from the printing surface each time it is altered, but sometimes only a unique impression may exist. Differences in states may be so subtle that experts can disagree as to whether they are in fact intentional or are simply the result of chance or wearing of the plate. Some of *Rembrandt's etchings, however, exist in radically different states involving a major rethinking of the design.

steatite. See SOAPSTONE.

Steen, Jan (1625/6–79). Dutch painter. He is best known for his humorous *genre scenes, warm-hearted and animated works in which he treats life as a vast comedy of manners. In Holland he ranks next to *Rembrandt, *Vermeer, and *Hals in popularity and a 'Jan Steen household' has become an epithet for an untidy house. But Steen, one of the most prolific Dutch artists, has many other facets. He painted portraits, historical, mythological, and religious subjects (he was a Catholic), and the animals, birds, and still lifes in his pictures rival those by any of his specialist contemporaries. As a painter of children he was unsurpassed. Steen was born in Leiden and is said to have studied with Adriaen van *Ostade in Haarlem and Jan van *Goyen (who became his father-in-law) in The Hague. He worked in various towns—Leiden, The Hague, Delft, Warmond, and Haarlem—and in 1672 he opened a tavern in Leiden. His father had been a brewer, and in the popular imagination Steen was a drunken profligate, but there is nothing in the known facts of his life to justify this reputation. Many of his pictures represent taverns and festive gatherings, but they often feature moralizing allusions, and he also painted scenes of impeccable genteelness. Apart from his versatility, richness of characterization, and inventiveness in composition, Steen is remarkable also for his skill as a colourist, his handling of salmon-red, rose, pale yellow, and blue-green being highly distinctive. His work is in many collections, and he is particularly well represented in London (NG, Wallace Coll., Wellington Mus.). He had no recorded pupils, but his work was widely imitated.

Steenwyck, Hendrick van the Elder (d.1603?) and **Hendrick van the Younger** (d. 1649). Flemish painters, father and son, specialists in architectural views. Little is known of the career of either man, but the father, who was probably a pupil of Vredeman de *Vries, is credited with developing the church interior as a special branch of painting. Both father and son painted small pictures of real and imaginary *Gothic churches, sometimes as eerie nocturnal scenes. There are several examples of the work of Hendrick the Younger in the National Gallery, London; in two of them the figures are credited to Jan *Brueghel the Elder.

Steer, Philip Wilson (1860–1942). English painter, son of a portrait painter **Philip Steer** (1810–71). With *Sickert (his friend and exact contemporary), Steer was the leader in his generation of those progressive British artists who looked to France for inspiration. He trained in Paris 1882–4 (revisiting France four times between 1887 and 1891), and was a founder member of the *New English Art Club in 1886. In 1892 the Anglo-Irish novelist George Moore wrote 'it is admitted that Mr Steer takes a foremost place in what is known as the modern movement' and around this time Steer was indeed at his peak, producing the beach scenes and seascapes that are regarded not only as his finest works but also as the best *Impressionist pictures painted by an Englishman. They are remarkable for their great freshness and their subtle handling of light, and unlike Sickert's paintings they are devoid of any social or literary content. After about 1895 his work became more conventional and more closely linked to the English tradition of *Gainsborough (especially in his portraits), *Turner, and *Constable. In the 1920s he turned increasingly to watercolour. He taught at the *Slade School from 1893 to 1930 and in 1931 was awarded the Order of Merit. His sight began to fail in 1935 and he had stopped painting by 1940. He is well represented at the Tate Gallery.

Stefano da Zevio (or **Stefano di Giovanni da Verona**) (*c.*1375–after 1438). Veronese painter, one of the leading Italian exponents of the *International Gothic style. His work is very much in the northern manner, with rather insubstantial figures and an abundance of pretty details, and has almost no local character (*The Virgin in a Rose Garden*, Verona, *c.*1405).

Stefano della Bella. See BELLA.

Stein, Gertrude (1874–1946). American writer, collector, hostess, eccentric, and self-styled genius. She settled in Paris in 1903 and her home at 27 rue de Fleurus became famous as a literary and artistic salon; many distinguished American visitors to Paris found it their introduction to modern French painting. With her brother, the art critic **Leo Stein** (1872–1947), who lived with her from 1903 to 1912, she was one of the first collectors of the work of *Braque, *Matisse, and *Picasso (who painted a well-known portrait of Gertrude, *c.*1906, Metropolitan Museum, New York); another brother, Michael, and his wife Sarah, were also collectors. Gertrude's writings, which she claimed to be a literary counterpart to *Cubism, are largely unintelligible, concerned with the rhythm and sound of words rather than their meaning. The best-known and most approachable of her many books is *The Autobiography of Alice B. Toklas* (1933), which in fact is her own autobiography, composed as though by

Miss Toklas (1877–1967), her secretary and companion from 1907. Alfred H. *Barr writes that Leo Stein was 'the critic who first felt that Matisse *and* Picasso were the two important artists of his time', but he later turned his back on their work, describing Cubism as 'godalmighty rubbish'. Clive *Bell maintained that 'Neither Gertrude or Leo had a genuine feeling for visual art . . . Pictures were pegs on which to hang hypotheses.'

stele. A Greek word meaning 'standing block' now used principally of an upright slab, usually of stone, that serves as a grave marker or other monument. Grave stelae were sometimes plain, but by the 6th cent. BC were often carved or painted.

Stella, Frank (1936–). American painter, a leading figure of *Post-Painterly Abstraction. In his early work he was influenced by *Abstract Expressionism, but after settling in New York in 1958 he was impressed by the flag and target paintings of Jasper *Johns and the direction of his art changed completely. He began to emphasize the idea that a painting is a physical object rather than a metaphor for something else, saying that he wanted to 'eliminate illusionistic space' and that a picture was 'a flat surface with paint on it—nothing more'. These aims were first given expression in a series of black 'pinstripe' paintings in which regular black stripes were separated by very thin lines. They made a big impact when four of them were shown at the Museum of Modern Art's '16 Americans' exhibition in 1959, inspiring a mixture of praise and revulsion. To identify the patterning more completely with the shape of the picture as a whole he began—from the beginning of the 1960s—to use notched and *shaped canvases, often painting in flat bands of bright colour. In the 1970s he began to experiment with paintings that included cut-out shapes in relief and he abandoned his impersonal handling for a spontaneous, almost graffiti-like manner. He has been an influential figure, not only in painting but also on the development of *Minimal sculpture (his friends have included *Andre and *Judd).

Stella, Jacques (1596–1657). French painter and engraver born at Lyons. From 1619 to 1634 he was in Italy, first in Florence, where he was patronized by Cosimo II de' *Medici and made engravings of his festivities, and then in Rome, where he became a friend of *Poussin. He was one of Poussin's closest followers, and his work has little individual character.

Stella, Joseph (1877–1946). Italian-born American painter. He emigrated to the USA in 1896, but from 1909 to 1912 he lived in Italy and France, where he had his first significant contacts with modern art. He was particularly influenced by *Futurism and he became the leading American exponent of the style. His first and most famous Futurist painting was *Battle of Lights, Coney Island, Mardi Gras* (Yale University Art Gallery, 1913–14), a densely fragmented portrayal of a crowded amusement park at night. In such paintings Stella gave a romanticized image of the industrialized townscape of New York. In particular he was obsessed with Brooklyn Bridge, which he described as 'a shrine containing all the efforts of the new civilization of America' (*Brooklyn Bridge*, Yale University Art Gallery, 1917–18). Stella soon abandoned the Futurist idiom, but industrial and urban themes continued to inspire him. He was active in the administration of two leading avant-garde associations—the *Society of Independent Artists and the *Société Anonyme—and in the early 1920s he experimented with various styles, including *Precisionism. In the 1920s and 1930s he spent much of his time in Italy and France (he lived in Paris 1930–4). From the mid-1920s his work grew more conservative and included mystical and sacred subjects.

stencil. A thin sheet of metal, paper, or other suitable material perforated with a design (or often lettering) that is reproduced on paper or fabric when the sheet is laid on them and colour is brushed through the openings. Until the *silk screen was devised only simple shapes could be printed by the stencil technique; yet from the aesthetic point of view its very simplicity and sharpness of outline may become major virtues, and the process has a long history both for fabric printing and for the colouring of prints, especially *woodcuts. In France, where it is called *pochoir*, stencilling has been much employed in book illustration. A notable English example of this use was in Paul *Nash's illustrations to Sir Thomas Browne's *Urne Buriall* (1932).

Stephens, F. G. See PRE-RAPHAELITE BROTHERHOOD.

stereochromy. The Victorian name for WATERGLASS PAINTING.

Stern, Irma (1894–1966). South African painter. She developed an *Expressionist style in Germany, 1913–20, where she studied and was the most important figure in introducing European modernism to South Africa. She is represented in many collections in South Africa and in Europe.

Stevens, Alfred (1817–75). English sculptor, painter, and designer, born at Blandford in Dorset, the son of a house-painter. With the assistance of the Rector of Blandford, Stevens was sent to study in Italy when he was 15 and remained there until 1842. He worked with *Thorvaldsen in Rome, and laid the foundations of his style in the study of the masters of the *Renaissance, above all *Raphael. After his return to England he taught at the Board of Trade's School of Design, and worked as chief artist to a firm of bronze- and metal-workers in Sheffield, H. E. Hoole & Co. (his designs secured the firm first place in the Great Exhibition of 1851), then settled in London in 1852. In 1856 he entered the competition for the Wellington Monument to be erected in St Paul's Cathedral, and although his design was placed sixth, he was eventually awarded the commission. It occupied him for the rest of his life and was plagued with bureaucratic delays and misunderstandings. It was not finally completed until 1912, when the equestrian group at the top was cast from Stevens's model. Nevertheless, it is not only Stevens's masterpiece (indeed, the only one of his large schemes to come to fruition and survive), but also the greatest piece of sculpture produced in England in the 19th cent. The architectural elements form a splendid, bold composition, fully in keeping with Sir Christopher Wren's building, and the two bronze groups *Valour and Cowardice* and *Truth and Falsehood* have an almost *Michelangelesque grandeur and vigour.

Apart from the Wellington Monument, Stevens's finest work was the decoration for the Dining Room at Dorchester House, London (*c*.1856), a sumptuous residence built for the millionaire R. S. Holford. It was demolished in 1929 (the Dorchester Hotel now occupies the site), but a fireplace is in the Victoria and Albert Museum, London, and many of the fittings are in the Walker Art Gallery, Liverpool, which has an outstanding Stevens collection. He was a masterful craftsman in numerous media—marble, bronze, silver, porcelain—and was also a painter, although he destroyed much of his work because it did not satisfy him. His unexecuted designs included schemes for the decoration of the interior of the dome of St Paul's Cathedral and the Reading Room at the British Museum, recorded in his superb drawings, which are particularly well represented in Cambridge (Fitzwilliam), Liverpool (Walker), London (Tate), Oxford (Ashmolean), and Sheffield (Mappin). These drawings are very much in the High Renaissance tradition, and it is to this era that Stevens belonged in spirit—he was a great artist born out of his time.

Stevens, Alfred-Émile (1823–1906). Belgian painter, active mainly in Paris, where he settled in 1844. From about 1860 he achieved immense success with his pictures of young ladies in elegant interiors dressed in the height of fashion. His skill in rendering fine materials earned him the title 'the *Terborch of France'. He was a friend and supporter of *Manet and influenced *Whistler, with whom he was one of the first enthusiasts for Japanese art. Stevens also painted coastal and marine scenes in a rather freer, more *Impressionistic style, similar to that of *Boudin or *Jongkind. In 1900 he achieved the distinction of becoming the first living artist to be given a one-man exhibition at the École des *Beaux-Arts in Paris. There are examples of his work in the National Gallery, London, but he is much better known and represented in Belgium and France than he is in Britain. His brother **Joseph** (1819–92) was also a painter, mainly of animals, and in his day was almost as famous and successful as Alfred (both brothers received the Legion of Honour). Another brother, **Arthur**, was an art critic and dealer.

Stieglitz, Alfred (1864–1946). American photographer, editor, and art dealer who during the first two decades of the 20th cent., did more than anyone else to bring the works of the European avant-garde before the American public. The son of a German immigrant, he spent most of the 1880s in Berlin and returned to the USA in 1890 with an international reputation as a photographer. His 291 Gallery (at 291 Fifth Avenue, New York), which he opened in 1905, presented the first American exhibitions of *Matisse (1908), *Toulouse-Lautrec (1909), the Douanier *Rousseau (1910), *Picabia (1913), *Severini (1917), and the first one-man exhibition of *Brancusi anywhere (1914). It also gave the first exhibition of children's art and the first major exhibition of African art in America. Stieglitz also championed American artists, among them Georgia *O'Keeffe, whom he married in 1924.

From 1903 to 1917 Stieglitz edited the journal *Camera Work*, which he published from the 291 Gallery. At first devoted to photography, it was later extended to cover all the visual arts, including reviews and criticisms, and opened its pages to avant-garde American writers. The 291 Gallery was closed in 1917 when the building was pulled down, but Stieglitz continued his work with the Intimate Gallery (1925–9) and An American Place (1929–46), both of which promoted American artists of what had come to be known as the Stieglitz group.

Stieglitz was a brilliant innovative artist in his own medium of photography. Learning from the avant-garde paintings which he exhibited, he experimented with various modes of abstraction in photography and his work went a long way to revolutionize the concept of the photographic image and to establish photography as an independent art form. He formed a fine collection of art, much of which was donated to the Art Institute of Chicago.

Stifter, Adalbert (1805–68). Austrian writer and painter. He is now regarded as one of the outstanding Austrian novelists of the 19th cent., but he enjoyed little success in his lifetime, and full recognition of his stature came only after the First World War. For much of his career he worked as a tutor, then an educational administrator. His later years were clouded by the suicide of an adopted daughter in 1859 and he took his own life, cutting his throat with his razor, whilst suffering agonizing pain from what was thought to be cancer. As a painter he was self-taught, and as in his writing went against the prevailing trends of the day in eschewing heroic events in favour of simple, everyday happenings. His sensitive perception of nature comes out in his remarkably fresh landscapes. There is a museum devoted to his work in Vienna.

Stijl, De (Dutch: 'the style'). The name of a group of Dutch artists founded in Leiden in 1917 and of the journal they published to set forth their ideas. Members of the group included painters (most notably *Mondrian and van *Doesburg), sculptors (*Vantongerloo), the architect and designer Gerald Rietveld (1888–1964) and the poet Antony Kok. They sought laws of equilibrium and harmony that would be applicable to life and society as well as art, and their style was one of austere abstract clarity (see NEO-PLASTICISM). The journal was founded by van Doesburg and Mondrian in 1917 and van Doesburg continued to edit it until 1928 (it appeared roughly monthly, but irregularly; the place of publication also varied). A final issue (number 90) was published in 1932 by Mme van Doesburg in memory of her husband, after whose death in 1931 the group disbanded. At first the journal was devoted to the principles of Neo-Plasticism, but in 1924 van Doesburg welcomed *Dadaists such as *Arp and *Schwitters as contributors and with these two and the poet Tristan Tzara (1886–1963) edited a Dadaist supplement under the name of *Mecano*. Mondrian ceased to contribute to *De Stijl* after 1924 and in 1926 van Doesburg published the manifesto of a splinter movement which he called *Elementarism. Despite this lack of cohesion, *De Stijl* was probably the most influential of the many avant-garde publications in Europe between the two wars on both ideas and artistic practice. It was, however, in architecture and the applied arts (including furniture design and typography), rather than painting and sculpture, that it had its greatest influence—notably at the *Bauhaus and in the clean-lined architectural style known as 'International Modern', of which Rietveld's Schröder House in Utrecht (1924) is an early and famous example.

Stile Liberty. See ART NOUVEAU.

Still, Clyfford (1904–80). American painter, one of the major figures of *Abstract Expressionism but the one least associated with the New York art scene. After working in war industries in California, 1941–3, he taught for two years at the Richmond Professional Institute, Richmond, Virginia. He then lived briefly in New York (1945–6), where he had a one-man exhibition at Peggy *Guggenheim's Art of This Century gallery in 1946. Although he stood somewhat apart from the other Abstract Expressionists, he was friendly with Mark *Rothko (they had met in 1943), the two men sharing a sense of almost mystical fervour about their work. In 1946–50 he taught at the California School of Fine Arts in San Francisco, then lived in New York, 1950–61. By the time he returned to New York Still had created his mature style and had a rapidly growing reputation. He was one of the pioneers of the very large, virtually monochromatic painting. But unlike *Newman and Rothko, who used fairly flat, unmodulated pigment, Still used heavily loaded, expressively modulated impasto in jagged forms. His work can have a raw aggressive power, but in the 1960s it became more lyrical. In 1961 Still moved to Maryland to work in tranquillity away from the art world. Scorning galleries, dealers, and critics, and rarely exhibiting, he considered himself something of a visionary who needed solitude to give expression to his high spiritual purpose, and he gained a reputation for cantankerousness and pretentiousness—his comment on his painting *1953* (Tate Gallery, London, 1953) is typical of his high-flown prose: 'there was a conscious intention to emphasize the quiescent depths of the blue by the broken red at its lower edge while expanding its inherent dynamic beyond the geometries of the constricting frame . . . In addition, the yellow wedge at the top is a re-assertion of the human context—a gesture of rejection of any authoritarian rationale or system of politico-dialectical dogma.' Still presented large groups of his paintings to the Albright-Knox Art Gallery, Buffalo, the Metropolitan Museum, New York, and the San Francisco Museum of Art, and his work is represented in many other major collections.

Stimmer, Tobias (1539–84). Swiss painter and graphic artist, active in his native Schaffhausen, in Strasburg (where he died) and in various German towns. He was a versatile artist, whose work included façade decorations (notably that of the

Haus zum Ritter, 1568–70, now in the Museum zu Allerheiligen, Schaffhausen), portraits in a *Holbeinesque style, and numerous book illustrations. He also decorated the astronomical clock in Strasburg Cathedral. There is often strong Italian influence in his work and he may have visited Italy.

stipple engraving. A method of engraving in tone, in which the design is made up of small dots or flecks. It was a popular reproductive technique in the late 18th and early 19th cents., when it was often used in conjunction with the *crayon manner, from which it differs little. Both were rendered obsolete by *lithography.

In principle stipple is merely an element in *line engraving, for dotting the surface of a copper plate with short flicks of the *burin has always been one of the line-engraver's ways of producing a tone. Stipple engraving as practised in the 18th cent. was, however, a combination of the line-engraver's technique with *etching. The plate was first covered with an etching *ground and the design laid in with dotted lines, as if for a print in the crayon manner. After biting, the engraver cleared away the ground and proceeded to build up his tones by means of a multiplicity of dots engraved with a special burin. The work was frequently completed with *roulettes and sometimes with engraved lines.

Stipple engraving, which produces prints characterized by softness and delicately graded tones, was almost entirely confined to England, where Francesco *Bartolozzi made a great reputation with prints after *fancy subjects by Angelica *Kaufmann and G. B. *Cipriani. Other notable stipple engravers were William Ryland (1732?–83), who after serving as engraver to George III was hanged for forging a banknote, and John Raphael *Smith.

Stokes, Adrian (1902–72). English writer on art and painter. An intensely subjective writer with an interest in psychoanalysis, Stokes responded passionately, even ecstatically, to art, believing its task was to show the 'utmost drama of the soul as laid-out things'. Many admirers regard him as the most eloquent and poetic English art critic since *Ruskin, although others find his prose hard going. His best-known books are *The Quattro Cento* (1932) and *The Stones of Rimini* (1934); *The Critical Writings of Adrian Stokes*, edited by Lawrence *Gowing, appeared in three volumes in 1978. Stokes began to paint in 1936 and became a student at the *Euston Road School in 1937. From 1939 to 1946 he lived near St Ives, Cornwall, with its flourishing group of painters. There are examples of his work in the Tate Gallery (see ST IVES SCHOOL).

Stomer, Matthias. See DOBSON, WILLIAM.

Stone, Nicholas (1587–1647). English sculptor, mason, and architect. The son of a Devonshire quarryman, he trained in London, then in 1606 he went as a journeyman to the Amsterdam workshop of Hendrik de *Keyser, whose daughter he married. On his return to London in 1613 he quickly established himself as the outstanding tomb sculptor in the country. He surpassed his contemporaries in technical skill as a marble cutter and outdid them in introducing new ideas: the monument to Francis Holles (d. 1622; Westminster Abbey) has the first English example of a figure in Roman armour, and that to John Donne (made from Donne's own design, St Paul's Cathedral, 1631) shows the great preacher and poet standing erect in his shroud.

In 1619 Stone became Master Mason for Inigo *Jones's Banqueting House at Whitehall, and in 1632 Master Mason to the Crown. His contact with the court gave him a knowledge of the *antique sculpture in Charles I's collection and his work after *c.*1630 shows a change in style, marked by an attempt to imitate antique drapery (Lyttelton Monument, Magdalen Coll., Oxford, 1634). His large workshop produced monuments of many types, and we are unusually well informed about its activities as an office notebook covering the period 1614–41 and an account book for the period 1631–42 still survive (Soane Mus., London). Much less of his work as an architect is extant (and it is sometimes not clear whether he was the designer as well as the mason of the buildings on which he worked), but nevertheless, on the evidence of such buildings as the three gateways in the Botanic Garden, Oxford (1632–3), he is recognized as the creator of 'a vernacular classical architecture of considerable charm and accomplishment' (H. M. Colvin, *A Biographical Dictionary of British Architects 1600–1840*, 1978). Stone may have been the author of *Enchiridion of Fortification, or a Handfull of Knowledge in Martiall Affairs*, published anonymously in London in 1645, but the book has also been attributed to the youngest of his three sons, **John Stone** (1620–67), who ran the family practice after his father's death, even though he does not appear to have been a sculptor himself.

Stone, Reynolds (1909–79). British engraver, letter-cutter, designer, and painter. After working for the Cambridge University Press and a small commercial printer, he set up on his own as a wood-engraver in 1934 and from 1939 also worked (self-taught) as a letter-cutter in stone. His highly varied output included book illustrations (his speciality was quiet rural scenes), memorial tablets (including that to Sir Winston Churchill in Westminster Abbey, 1965), and the design of £5 and £10 notes for the Bank of England (1963–4).

stopping-out varnish. An acid-resisting varnish used in *etching, *aquatint, and similar processes.

Stoskopff, Sébastien (1597–1657). French still-life painter, active mainly in Strasburg. He painted in a spare, stiff, almost archaic style that has appealed greatly to modern taste—most of his paintings have come to light since the 1930s. The best collection is in the Musée de l'Œuvre de Notre Dame, Strasburg.

Stoss, Veit (*c*.1450–1533). German sculptor, with *Riemenschneider the greatest wood-carver of his age. He is first recorded in 1477, when he moved from Nuremberg to Cracow in Poland. There he carved his largest work, the huge altarpiece for St Mary's Church (1477–89), and also made the red marble tomb of King Casimir IV in the cathedral (1492). In 1496 he returned to Nuremberg, where his successful career was blighted when (in an attempt to recoup some money he regarded as having been misappropriated in an investment) he forged a document (1503) and was convicted and branded through both cheeks. He was also confined to the city limits of Nuremberg (he fled but returned), and although he was to some extent rehabilitated, he never regained his former position. He died a wealthy man, but his old age was embittered by disputes with the city authorities.

A good many documented and signed works by Stoss survive and his style is distinctive—bold and powerfully characterized, with exaggerated gestures and expressions and draperies rendered in an ornate, almost calligraphic manner. Indeed, Stoss's work is so individual that the famous figure of *St Roch* in SS. Annunziata, Florence, is almost universally accepted as his, even though it is undocumented and was attributed by *Vasari to 'Janni Francese' (Janni the Frenchman). Vasari wrote eloquently of the virtuosity of the carving, describing the draperies as 'cut almost to the thinness of paper, and with a beautiful flow in the arrangement of the folds, so that nothing more wonderful is to be seen'. Stoss sometimes, as here, left his figures unpainted, but otherwise his work is entirely in the late *Gothic spirit. He is recorded as being a painter and engraver as well as a sculptor and he also declared himself competent as a civil engineer.

Stothard, Thomas (1775–1834). English painter, book illustrator, and designer. He was a prolific and versatile artist, his designs ranging from monuments to jewellery, and although his paintings were mainly small-scale historical pieces in a sentimentalized classical manner, he also did occasional more ambitious works, for example the decoration of the staircase at Burghley House (begun 1794) and the cupola of the upper hall of the Advocates' Library, Edinburgh (1822). His book illustrations were mainly in the field of English novels and poetry. Stothard became librarian of the *Royal Academy in 1812. His work is well represented in the Tate Gallery.

Stradano or **Stradanus, Giovanni.** See STRAET.

Straet, Jan van der (1523–1605). Netherlandish painter and designer, active for almost all his career in Italy, where he was known as Giovanni Stradano (or Stradanus or della Strada). He was born in Bruges, had his main training in Antwerp with *Aertsen, and moved to Italy in the later 1540s. A protégé of *Vasari, he worked in various Italian cities, but mainly in Florence, where he was much employed by the *Medici family. He assisted Vasari with frescos in the Palazzo Vecchio, for example, but his main work for the Medici was as a tapestry designer. His *Mannerist style was influenced by Vasari, but it always retained a Netherlandish accent and he was admired for his skill in *genre painting, a Northern speciality. Several of his designs were published in engravings.

strapwork. A type of ornament consisting of interlaced bands and decorative forms resembling strips of leather or parchment that have been elaborately cut and pierced. First developed by *Rosso and his school at *Fontainebleau in the 1530s, it spread rapidly to Flanders and from there, by means of engraved pattern books and refugee craftsmen, to England. It was profusely used in Elizabethan and *Jacobean decoration in wood, metal, *stucco, stone, and as printer's ornament.

Streeter, Robert (1621–79). English painter, appointed Serjeant-Painter to Charles II in 1660. In the words of Sir Ellis *Waterhouse, 'he left no branch of painting untried and would have been a universal genius had he been endowed with the requisite talent.' The most important of his few surviving works is the allegorical ceiling painting representing *The Triumph of Truth and the Arts* (1668–9) in the Sheldonian Theatre, Oxford. This is a heavy-handed work, but noteworthy as the most ambitious attempt at a piece of *Baroque decoration by any Englishman before *Thornhill (Streeter had travelled in Italy during the Commonwealth). His reputation was evidently high in his day, for Samuel Pepys (1633–1703) in his *Diary* calls him a 'famous history painter' who 'lives very handsomely', and Robert Whitehall (1625–85), whose ability as a poet matched that of Streeter as a painter, eulogized the Sheldonian ceiling in the immortal lines '. . . future ages must confess they owe / To Streeter more than Michael Angelo' (*Urania, or a Description of the Painting of the Top of the Theatre at Oxford*, 1669). Streeter's son **Robert** (d. 1711) succeeded him as Serjeant-Painter.

Streeton, Sir Arthur (1867–1943). Australian painter. He was a prolific landscape painter, working in an *Impressionist style similar to that of his friend Tom *Roberts. Between 1898 and 1924 he spent most of his time abroad (in 1918 he was an *Official War Artist with the Australian forces in France). His work became stereotyped, but he was enormously popular in his own country, regarded as the foremost portrayer of the remote and awesome Australian landscape. By the end of his life he had long enjoyed the status of a national institution.

stretcher. The wooden frame or chassis on which a *canvas is stretched and fixed. Wedges or keys in the inner corners of the stretcher enable the canvas to be tightened if it slackens.

Stretes, William. See SCROTES, GUILLIM.

Strigel, Bernhard (*c*.1460–1528). German painter, active in Memmingen, near Ulm. In the 1490s he worked with *Zeitblom, then the leading painter in Ulm, and he painted religious works in his manner. He is best known, however, as a portraitist, and it was in this field that he worked as court painter to the emperor Maximilian (*The Imperial Family*, Vienna, *c*.1515). His style is transitional between late *Gothic and *Renaissance.

Strozzi, Bernardo (1581–1644). The leading Genoese painter of his period. He entered the Capuchin Order in about 1597, hence his nicknames, Il Prete Genovese (the Genoese priest) and Il Cappuccino (the Capuchin). In about 1610 he was allowed to leave his community to look after his sick and widowed mother, and after she died in 1630 he is said to have been pressurized to return, this accounting for his move in 1631 to Venice (where he spent the rest of his life). Strozzi was successful and prolific in both Genoa and Venice, painting portraits and allegorical and *genre scenes (often of musicians) as well as religious works. The sensuous richness of his style was influenced by *Rubens (who worked in Genoa), but his work is highly distinctive, with an air of refinement and tenderness that recalls van *Dyck (who also worked in Genoa). Strozzi worked in Venice when there was a dearth of native talent, and with two other 'foreigners', *Feti and *Liss, he kept alive the painterly tradition of the 16th cent. His work is represented in many major galleries.

Strudwick, John Melhuish. See PRE-RAPHAELITE BROTHERHOOD.

Strzeminski, Władysław. See BLOK.

Stuart, Gilbert (1755–1828). American portrait painter. With *Copley he was the outstanding American portraitist of his period and he is regarded as the creator of a distinctively American style of portraiture. Much of his early career was spent in Scotland (*c*.1771–2), England (1775–87), and Ireland (1787–92). After he settled permanently in America in 1792, he worked briefly in New York City, then moved to Philadelphia, and finally settled in Boston in 1805. He quickly established himself as the outstanding portraitist in the country and painted many of the notables of the new republic. His portraits of George Washington are his most famous works—he created three types, all of which were endlessly copied: the 'Vaughan' type (NG, Washington, 1795), the 'Lansdowne' type (Pennsylvania Academy of Fine Arts, Philadelphia, 1796), and the 'Athenaeum' type (Mus. of Fine Arts, Boston, 1796), which is one of the most famous images in American art, being used on the country's one-dollar bill. Stuart's style is notable for its strength of characterization (Benjamin *West said he '*nails* the face to the canvas') and its fluent brushwork. His work had great influence on the next generation of American painters.

Stubbs, George (1724–1806). English animal painter and engraver, celebrated as the greatest of all horse painters. He was born in Liverpool, the son of a currier and leather-seller, and his life up to his mid-30s (which is poorly documented) was spent mainly in the north of England. Virtually self-taught as a painter and engraver, he seems to have earned his living mainly as a portraitist early in his career, and in 1751 he made the illustrations (based on his own dissections) for Dr John Burton's treatise on midwifery. In 1754 he visited Rome, then spent 18 months dissecting and drawing horses in preparation for a book on equine anatomy. He moved to London in about 1758 and, unable to find an engraver to do the work, he made the plates himself, and in 1766 published his famous book—*The Anatomy of the Horse*. It was a great success, prized for its beauty as well as its scientific accuracy, and Stubbs was soon in demand as a painter, not only for his 'portraits' of horses with their owners or grooms, but also for his conversation pieces in which the sitters were grouped in and around a carriage.

His command of anatomy was matched by his ability to depict the beauty and grace of his equine subjects without sentimentalizing them and his range of feeling was wide, extending from the lyrical calm of *Mares and Foals in a River Landscape* (Tate, London, *c.*1763–8) to the full-blooded *Romanticism of his series of pictures on the theme of a horse attacked by a lion (the largest—*c.*1762—is in the Yale Center for British Art). Stubbs is said to have derived his fascination for the subject from having witnessed a lion attacking a horse in Morocco on his way back from Italy, but he may also have been familiar with a much copied *antique statue on the theme.

Stubbs painted many other animals apart from horses—among them a moose, a rhinoceros, and a zebra. At his death he was working on *A Comparative Anatomical Exposition of the Structure of the Human Body with that of a Tiger and a Common Fowl*, the drawings for which are in the Yale Center for British Art. His scientific curiosity extended to the materials he used and he experimented with painting in *enamel on earthenware panels manufactured for him by the great potter Josiah *Wedgwood. Stubbs's work became less popular during the 1780s and by the end of his life he was in financial difficulties. He kept his great powers until the end, however, and one of his finest works, painted when he was 75, is the enormous *Hambletonian, Rubbing Down* (National Trust, Mount Stewart, 1799), showing the champion horse looking strained and exhausted after winning a race in which it was 'much cut with the whip' and 'shockingly goaded' with the spur. It is a magnificent, heroic, almost tragic image. Stubbs was for long classified as a superior sporting painter, but he is now placed alongside *Gainsborough and *Reynolds in the front rank of English painters of his age.

stucco. A type of light, malleable plaster made from dehydrated lime (calcium carbonate) mixed with powdered *marble and glue and sometimes reinforced with hair. It is used for sculpture and architectural decoration, external and internal. In a looser sense, the term is applied to a plaster coating applied to the exterior of buildings, but stucco is a different substance from plaster (which is calcium sulphate). Stucco in the more restricted sense has been known to virtually every civilization, the exact ingredients varying greatly from place to place and from time to time. In Europe it was exploited most fully from the 16th cent. to the 18th cent., notable exponents being the artists of the School of *Fontainebleau and Giacomo *Serpotta, an unsurpassed virtuoso in the medium. By adding large quantities of glue and colour to the stucco mixture *stuccatori* were able to produce a material that could take a high polish and assume the appearance of marble. Indeed, sometimes it is difficult to distinguish from real marble without touching it (stucco feels warmer).

Stuck, Franz von. See SEZESSION.

stump (also called tortillon). A short tapered stick usually of cork or tightly rolled leather or paper, used to soften the edges of a drawing or spread the *chalk, crayon, or *pencil in shading, thereby providing very delicate tonal transitions. It was commonly used in the 18th and 19th cents. but has gone out of favour in the 20th cent.

Sturm, Der (The Assault). Name of a magazine and an art gallery in Berlin, both of which were founded and owned by Herwarth Walden (1878–1941?), a writer and composer whose aim was to promote avant-garde art in Germany. The magazine ran from 1910 to 1932 and the gallery from 1912 to 1932. They became the focus of modern art in Berlin, introducing the work of the *Futurists to Germany, for example, and publicizing the *Expressionism of the *Blaue Reiter group. Walden's activities extended also to publishing art books and portfolios of prints, organizing lectures and discussions and experiments with Expressionist theatre. Walden left Germany in 1932 because of the economic depression and the rise of Nazism and moved to the Soviet Union, where he is said to have died as a political prisoner in 1941.

stylus (or **style**). A pointed metal instrument, usually of iron, used to make indented marks on a surface. Originally used in Classical times for writing on wax tablets, it later served many other purposes: to incise the ornament on gold grounds, rule the lines for a manuscript, or trace the outlines of a composition from the *cartoon on to the plaster in *fresco painting or engraving. It was much used for the *perspective lines, and other lines of measurement and guidance, in architectural drawings, and for squaring up compositions.

Suardi, Bartolomeo. See BRAMANTINO.

Subleyras, Pierre (1699–1749). French painter who settled permanently in Rome after winning the *Prix de Rome in 1727. He painted various subjects, including portraits and still lifes, but he is most highly regarded for his religious paintings, which are much more serious in spirit than most French works of the *Rococo period. His most famous work is the *Mass of St Basil*, painted for St Peter's, but now in Sta Maria degli Angeli. This huge picture was highly acclaimed when it was unveiled in 1748, but Subleyras died in the following year before he could follow up his success. He has subsequently been something of an underrated figure, but a major exhibition of his work in Paris and Rome in 1987 did much to establish his reputation as one of the outstanding French painters of his period.

Sublime. Term that came into general use in the 18th cent. to denote a new *aesthetic concept that was held to be distinct from the beautiful and the *Picturesque and was associated with ideas of awe and vastness. It was first used as a term in rhetoric and poetry and gained wide currency after the French translation (1674) of a Greek treatise entitled *On the Sublime* attributed to Longinus (*c.*1st cent. AD), although the word itself had occurred earlier in English—quite frequently in Milton for example. Longinus described the immensity of objects in the natural world—stars, mountains, the ocean—in terms of the sublime, and this idea was of profound importance to the growing feeling for the grandeur and violence of nature. From the literary sphere the term was extended to a wider range of aesthetic reactions, and in particular the new sensibility for the wild, awe-inspiring, and stupendous aspects of natural scenery.

The outstanding work on the concept of the Sublime in English was Edmund Burke's *A Philosophical Enquiry into the Origin of our Ideas of the Sublime and Beautiful* (1757). In keeping with the current practice in his time Burke understood the sublime as a general mode of aesthetic experience found in literature but also far more widely. He did, however (and for this he was ridiculed by Richard Payne *Knight), restrict its nature more systematically than any of his predecessors to the emotion of terror: 'terror', he says, 'is in all cases whatsoever, either more openly or latently, the ruling principle of the sublime.' His work is important for being one of the first to realize (in contrast with the emphasis on clarity and precision during the Age of Enlightenment) the power of suggestiveness to stimulate imagination. Speaking of painting he says that 'a judicious obscurity in some things contributes to the effect of the picture', because in art as in nature 'dark, confused, uncertain images have a greater power on the fancy to form the grander passions than those which are more clear and determinate.'

*Reynolds discussed the sublime in his last *Discourse*, delivered in Burke's presence in 1790. He claimed that the figures of God and the sibyls on *Michelangelo's Sistine Chapel excited the same sensations as 'the most sublime passages of Homer'. 'The Sublime in painting, as in poetry, so overpowers, and takes possession of the whole mind, that no room is left for attention to minute criticism. The little elegancies of art, in the presence of these great ideas thus greatly expressed, lose all their value, and are, for the instant at least, felt to be unworthy of our notice. The correct judgement, the purity of taste which characterize *Raphael, the exquisite grace of *Correggio and *Parmigianino, all disappear before them.' The cult of the Sublime had varied expressions in the visual arts, notably the taste for the 'savage' landscapes of Salvator *Rosa and the popularity among painters of subjects from Homer, John Milton, and Ossian (the legendary Gaelic warrior and bard, whose verses—actually fabrications—were published in the 1760s to great acclaim). In literature, the 'Gothic novel', in which mystery and horror were the essential ingredients, appealed to the same sentiments. The first Gothic novel was Horace *Walpole's *The Castle of Otranto* (1764), and when crossing the Alps in 1739 Walpole expressed the essence of the imaginative appeal of the Sublime in his memorable exclamation: 'Precipices, mountains, torrents, wolves, rumblings—Salvator Rosa.' The vogue for the sublime, with that for the Picturesque, helped shape the attitudes that led to *Romanticism.

sudarium. See VERNICLE.

sugar aquatint. See AQUATINT.

Suger, Abbot (*c.*1081–1151). French churchman and statesman, one of the greatest patrons of the Middle Ages. He was adviser to Louis VI (1081–1137) and Louis VII (1120–80), acting as regent in 1147–9 during the latter's absence on the Second Crusade, but he is remembered mainly for the rebuilding of the abbey of Saint-Denis, near Paris, of which he was abbot from 1122 until his death. Suger's reconstruction and redecoration of the abbey was the most important landmark in the emergence of the *Gothic style in architecture and sculpture. He wrote an autobiographical account of the work between 1144 and 1147—a unique document of the time. It was edited and translated (1946) by Erwin *Panofsky, who wrote that Suger was 'frankly in love with splendour and beauty in every conceivable form: it might be said that his reponse to ecclesiastical ceremonial was largely aesthetic'. He thought that through the splendour and brilliance of material treasure, the soaring architecture, and the effulgence of material light men were led away from the material to an appreciation of the spiritual glories of the divine Being. Suger commissioned numerous items of church furnishing and sacred vessels for the abbey, several of which survive. The finest is an antique *porphyry vase that he had converted into the shape of an eagle (Louvre, Paris).

Suisse, Académie. See ACADÉMIE.

Sully, Thomas (1783–1872). The pre-eminent American portrait painter of his period, active mainly in Philadelphia. He was born in England, but went to America as a child. Subsequently he made two visits to England, in 1809, when he studied with *West, and in 1837–8, to paint Queen Victoria for the St George Society of Philadelphia (a preliminary version is in the Wallace Coll., London). His style, distinguished by fluid, glossy brushwork and romantic warmth and dash, betrays his great admiration for *Lawrence. In his later work, however, he tended towards a genteel sentimentality. Sully was highly successful and extremely productive. He is said to have painted some 2,000 portraits (he also made *miniatures), but his best-known work is probably one of his (also very numerous) history paintings—*Washington Crossing the Delaware* (Mus. of Fine Arts, Boston, 1819).

Superrealism. Style of painting (and to a lesser extent sculpture), popular from the late 1960s, particularly in the USA and Britain, in which subjects are depicted with a minute and impersonal exactitude of detail. Hyperrealism and Photographic Realism (or Photorealism) are alternative names, and some artists who practise the style do indeed work from photographs (sometimes using colour slides projected on the canvas); sharpness of detail is evenly distributed over the whole picture (except where out-of-focus effects in the photograph are faithfully recorded), but the scale is sometimes greatly enlarged. (Some critics prefer to use the terms 'Photographic Realism' or 'Photorealism' only when a picture has been painted direct from a photograph, but most are not so restrictive.) The immediate progenitor of Superrealism was *Pop art; banal subject-matter from the consumer society was common to both, and certain artists, such as Malcolm *Morley (who coined the word Superrealism) and Mel *Ramos, overlap both fields. The kind of humour found in Pop is very rare in Superrealism, however, which tends to be cool and impersonal, with subjects often chosen because they are technically challenging (involving multiple reflections, for example). Like Pop, Superrealism was a hit commercially, but it was less well received critically. Some critics, indeed, regard it as involving a great deal of painstaking work but very little else; others think that its exponents can achieve a strange kind of intensity, the effect of the indiscriminate attention to detail being—somewhat paradoxically—to create a strong feeling of unreality.

The leading American Superrealist painters include: Chuck Close (1940–), whose speciality is giant portrait heads; Don Eddy (1944–), notable for scenes involving reflections in shop windows; Richard Estes (1936–), most of whose work is devoted to the urban landscape; and Audrey Flack (1931–), who is unusual in that she aims for emotional effect in her still lifes of religious symbols and images of vanity and death. Philip *Pearlstein is sometimes labelled a Superrealist but stands somewhat apart. British Superrealist painters include Graham Dean (1951–), Michael English (1943–), and Michael Leonard (1933–), whose work includes highly detailed portrait drawings in a style mimicking the Old Masters. The leading Superrealist sculptors are the Americans John *De Andrea and Duane *Hanson, who often use real clothes or props and attend to minutiae such as body hair.

support. The material—*canvas, wooden *panel, *paper, or other substance—on which a painting is executed; it is usually distinguishable from the *ground.

Suprematism. Russian abstract art movement, launched by *Malevich in 1915. His Suprematist paintings were the most radically pure abstract works created up to that date, for he limited himself to basic geometric shapes—the square, rectangle, circle, cross, and triangle—and a narrow range of colours. Although he somewhat softened his approach in allowing pastel colours and introducing elliptical shapes, Malevich then returned to complete austerity and reached the ultimate distillation of his ideas in a series of paintings of a white square on a white ground (*c.*1918) after which he announced the end of Suprematism. In contrast to *Constructivism, where the stress was on the utilitarian function of art, Suprematism was a vehicle for Malevich's spiritual ideas; he wrote that 'The Suprematists have deliberately given up the objective representation of their surroundings in order to reach the summit of the 'unmasked' art and from this vantage point to view life through the prism of pure artistic feeling.' Although Malevich's direct followers in Russia were of minor account, Suprematism had great influence on the development of art and design in the West.

Surikov, Vasily. See WANDERERS.

Surrealism. Movement in art and literature flourishing in the 1920s and 1930s, characterized by a fascination with the bizarre, the incongruous, and the irrational. It was closely related to *Dada, its principal source; several artists figured successively in both movements, each of which was conceived as a revolutionary mode of thought and action— a way of life rather than a set of stylistic attitudes. Both were strongly anti-rationalist and much concerned with creating effects that were disturbing or shocking, but whereas Dada was essentially nihilist, Surrealism was positive in spirit. Surrealism originated in France. Its founder and chief spokesman was the writer André *Breton, who officially launched the movement with his first *Manifeste du surréalisme*, published in 1924. The central idea of the movement was to release the creative powers of the subconscious mind, or as Breton put it, 'to resolve the previously contradictory conditions of dream and reality into an absolute reality, a super-reality'. Within this general aim Surrealism embraced a large number of different and not altogether coherent doctrines and techniques, characteristically aimed at breaching the dominance of reason and conscious control by methods designed to release primitive urges and imagery. Breton and other members of the movement drew liberally on Freud's theories concerning the subconscious and its relation to dreams. The way in which Surrealist artists set about exploration of submerged impulses and imagery varied greatly (in spite of Breton's demands there was little doctrinal unity, and defections, expulsions, and personal attacks are a feature of the history of the movement). Some artists, for example *Ernst and *Masson, cultivated various spontaneous techniques such as *frottage in an effort to eliminate conscious control. At the other extreme, *Dali, *Magritte, and others painted in a scrupulously detailed manner to give an hallucinatory sense of reality to scenes that make no rational sense.

Paris remained the centre of Surrealism until the Second World War, when the emigration of many European artists to the USA made New York the new hub of its activity. However, it became the most widely disseminated and controversial aesthetic movement of the 1920s and 1930s, spread partly by a series of major international exhibitions. Two of the most important took place in 1936: the International Surrealist Exhibition at the New Burlington Galleries, London, and 'Fantastic Art, Dada, Surrealism' at the Museum of Modern Art, New York. Surrealism did not take root in Germany (Ernst, the major German Surrealist, lived mostly in France and the USA), but it flourished vigorously in Belgium—in the work particularly of Magritte, the most inspired of all Surrealist painters, and *Delvaux, the most long-lived upholder of the tradition. Many artists who were not in sympathy with the political aims of Surrealism (for a time it was associated with the French Communist Party), and who were never formal members of the movement, nevertheless found its ideas stimulating and were influenced by its imagery. In Britain, Henry *Moore and Paul *Nash were among the major artists who went through a Surrealist phase. The English Surrealist Group was founded in 1936, but it was social rather than revolutionary in its aims.

Although it broke up as an organized movement during the war and by this time had spent its main force, the spirit of Surrealism lived on. With its stress on the marvellous and the poetic, Surrealism offered an alternative approach to the formalism of *Cubism and various types of abstract art, and its methods and techniques continued to influence artists in many countries. It was, for example, a fundamental source for *Abstract Expressionism.

Sutherland, Graham (1903–80). English painter, graphic artist, and designer. He abandoned an apprenticeship as a railway engineer to study engraving and etching, 1921–6, and up to 1930 worked exclusively as a graphic artist. His etchings of this period are in the *Romantic and visionary tradition of Samuel *Palmer. In the early 1930s he began experimenting with oils (following a decline in the market for prints), and by 1935 he had turned mainly to painting. His paintings of the 1930s show a highly subjective response to nature, inspired mainly by visits to Pembrokeshire. He had a vivid gift of visual metaphor and his landscapes are not scenic, but semi-abstract patterns of haunting and monstrous shapes rendered in his distinctively acidic colouring (*Entrance to a Lane*, Tate, London, 1939). During the war years he was employed as an *Official War Artist to record the effects of bombing, and his work matured as he wrestled with the problems of finding a visual surrogate for the devastation and the destruction of man-made things. Writing of Sutherland's pictures of ruined and shattered buildings, the critic Eric Newton said: 'they have a bold, crucified poignancy that gives the war a new meaning.' Soon after the war he took up religious painting, with a *Crucifixion* (1946) for St Matthew's, Northampton (he received the commission at the dedication of Henry *Moore's *Madonna and Child* in this church), and also portraiture, with *Somerset Maugham* (Tate, 1949). It was in these two fields that he chiefly made his mark in his later career. The Maugham portrait has an almost *caricature quality and his most famous portrait, that of Winston Churchill (1954), was so hated by the sitter that Lady Churchill (1885–1977) destroyed it. Sutherland's most celebrated work, however, has become widely popular—it is the immense tapestry of *Christ in Glory* (completed 1962) in Coventry Cathedral. Sutherland continued to paint landscapes—his first love—often inspired by the

French Riviera, where he lived for part of every year from 1947. Apart from paintings and graphic art, his work included ceramics and designing posters and stage costumes and decor. He was one of the most famous British artists of the 20th cent. and received many honours.

Swanenburgh, Jacob Isaacsz. van (*c.*1571–1638). Dutch painter of Leiden, a very minor figure now remembered exclusively because he has the distinction of being the first recorded teacher of *Rembrandt. His work had no discernible influence on his great pupil, for although Rembrandt painted a great variety of subjects, he never—so far as is known—tried his hand at either of Swanenburgh's specialities: architectural views and scenes of hell.

Swart van Groningen, Jan (*c.*1500–after 1553). Netherlandish painter, book illustrator (he made 73 of the 97 *woodcuts for the Dutch Bible published by W. Vostermann, Antwerp, 1528), and designer of stained-glass windows. He came from Groningen, worked in Antwerp and Gouda, and also travelled to Italy early in his career. His works show that he was familiar with those of *Dürer, *Holbein, *Scorel, and other northern artists who were impressed by Italian *Renaissance art, but in spite of this his style has a certain archaic charm. He had a predilection for showing people wearing high hats, turbans, and other odd headgear.

Sweerts, Michiel (1618–64). Flemish painter, an enigmatic and exceedingly attractive artist. Nothing is known of his training or early career. From about 1646 to about 1656 he was in Rome, where he came into contact with the *Bamboccianti* (see LAER, PIETER VAN). He painted *genre scenes in their manner, but his work is in a class apart because of the quiet, melancholy dignity of his figures and his exquisite silvery tonality. His other pictures in Rome included views of artists' studios, which are valuable historical documents as well as beautiful works of art (an example dated 1652 is in the Detroit Institute of Arts). By 1656 Sweerts had returned to his native Brussels, where in 1659 he became a member of the painters' guild. In 1661 he was in Amsterdam, where he joined a missionary group, and he sailed from Marseilles to the Orient in the following year. Sweerts was found quarrelsome and unsuitable, however, and was dismissed; he died at Goa in India. Towards the end of his career, Sweerts seemed to have worked mainly as a portraitist. Like his genre scenes, his portraits are distinguished by delicate and subdued colour harmonies and great sensitivity of expression and handling. They have often been compared with the work of *Vermeer, to whom Sweerts's *Portrait of a Girl* (Leicestershire Museum and Art Gallery, Leicester) was formerly attributed.

Symbolism. A loosely organized movement in literature and the visual arts, flourishing *c.*1885–*c.*1910, characterized by a rejection of direct, literal representation in favour of evocation and suggestion. It was part of a broad anti-materialist and anti-rationalist trend in ideas and art towards the end of the 19th cent. and specifically marked a reaction against the naturalistic aims of *Impressionism. Symbolist painters tried to give visual expression to emotional experiences, or as the poet Jean Moréas put it in a Symbolist Manifesto published in *Le Figaro* on 18 September 1886, 'to clothe the idea in sensuous form'. Just as Symbolist poets thought there was a close correspondence between the sound and rhythm of words and their meaning, so Symbolist painters thought that colour and line in themselves could express ideas. Symbolist critics were much given to drawing parallels between the arts, and *Redon's paintings, for example, were compared with the poetry of Baudelaire and Edgar Allan Poe and with the music of Claude Debussy. Many painters were inspired by the same kind of imagery as Symbolist writers (the *femme fatale* is a common theme), but *Gauguin and his followers (see SYNTHESISM) chose much less flamboyant subjects, often peasant scenes. Religious feeling of an intense, mystical kind was a feature of the movement, but so was an interest in the erotic and the perverse—death, disease, and sin were favourite subjects. Stylistically, Symbolist artists varied greatly, from a love of exotic detail to an almost primitive simplicity in the conception of the subject, and from firm outlines to misty softness in the delineation of form. A general tendency, however, was towards flattened forms and broad areas of colour—in tune with *Post-Impressionism in general. By freeing painting from what Gauguin called 'the shackles of probability' the movement helped to create the aesthetic premisses of much 20th-cent. art. Although chiefly associated with France, Symbolism had international currency, and such diverse artists as *Hodler and *Munch are regarded as part of the movement in its broadest sense. Symbolist sculptors include the Norwegian Gustav *Vigeland.

Synchromism. Movement in painting founded in 1912 by Stanton *Macdonald-Wright and Morgan *Russell, two American artists living in Paris. The term 'synchromism' means literally 'colours together' and Russell and Macdonald-Wright were concerned with the purely abstract use of colour; in 1912 Russell said that he wished to do 'a piece of expression solely by means of colour and the way it is put down, in showers and broad patches, distinctly separated from each other, or blended . . . but with force and clearness and large geometric patterns'. Synchromism was very close to *Orphism and the two Americans protested in manifestos that they had primacy. Although the movement petered out with the First World War, Synchromism influenced several American artists, and its founders hold distinguished places in the vanguard of abstract art.

Synthetism. Term applied to a manner of painting associated with *Bernard, *Gauguin, and their associates at Pont-Aven in Brittany. It involved the simplification of forms into large-scale patterns and the expressive purification of colours. Bernard believed that form and colour must be simplified for the sake of more forceful expression, and Gauguin spoke much of 'synthesis', by which he meant a blending of abstract ideas of rhythm and colour with visual impressions of nature. He advised his disciples to 'paint by heart' because in memory coloured by emotion natural forms become more integrated and meaningful. Bernard and Gauguin each claimed credit for developing Synthetism and they probably acted as mutual catalysts. Synthetism was influential on the *Nabis and has affinities with the more literary *Symbolism. See also CLOISONNISM.

Systemic art. Term coined by Lawrence *Alloway in 1966 to refer to a type of abstract art characterized by the use of very simple standardized forms, usually geometric in character, either in a single concentrated image or repeated in a system arranged according to a clearly visible principle of organization. The chevron paintings of *Noland are examples of Systemic art. It has been described as a branch of *Minimal art, but Alloway extended the term to cover *Colour Field painting.

Tacca, Pietro (1577–1640). Florentine sculptor in bronze, the chief pupil and follower of *Giambologna. After the latter's death Tacca completed a number of his works and succeeded him as sculptor to the *Medici Grand Dukes of Tuscany. Tacca's works for them include his masterpieces, the four *Slaves* (1615–24) at the foot of *Bandinelli's statue of Ferdinand I de' Medici at Leghorn. His last work was the equestrian statue of Philip IV of Spain (1634–40) for the Plaza de Oriente, Madrid, in which the King is shown on a rearing horse. This *Baroque pose was imposed on Tacca, having been already used in pictures of the King by *Rubens and *Velázquez (a copy of a painting by one or the other of these artists was sent to Florence to act as a model). The smooth, generalized treatment of the work shows, however, that Tacca remained essentially a *Mannerist sculptor. Tacca's son **Ferdinando** (1619–86) was also a sculptor; his best works are his graceful bronze statuettes.

Tachisme. A style of abstract painting popular in the late 1940s and 1950s characterized by irregular dabs or splotches of colour (*tache* is French for spot or blotch). The term was first used in this sense in about 1951 and was given wide currency by the French critic Michel Tapié in his book *Un Art autre* (1952). Tachisme had affinities with *Abstract Expressionism (although it initially developed independently of it) in that it strove to be completely spontaneous and instinctive, excluding deliberation and formal planning, and the term is often used as a generic label for any European painting of the time that parallels the American movement. It did not try to represent or to express, but to *be* a dramatic or emotional state of mind. Michel Tapié called Tachisme the 'other art' (*Art Autre) to emphasize its complete break with all aesthetic values and techniques prior to 1950, and in 1953 Jean *Atlan gave a vivid idea of the philosophy behind it: 'I believe that there is a common source for the painter and the dancer, this common source is a certain manner of living rhythms . . . For me a picture cannot be the result of a preconceived idea; the role of chance (*aventure*) is too important in it and indeed it is this role of chance which is ultimately decisive in creation. At the beginning there is a rhythm which tends to unfold itself: it is the perception of this rhythm that is fundamental and it is in its development that the vital quality of the work depends.' Tachisme was primarily a French movement (Jean *Fautrier, Georges *Mathieu, and the German-born but Paris-based *Wols were among the leading exponents), and Tachiste paintings are characteristically more elegant and less aggressive than the work of the American Abstract Expressionists. The terms '*abstraction lyrique*' (*lyrical abstraction) and *'*Art Informel*' (art without form) are sometimes used synonymously with Tachisme, although certain critics use them to convey different nuances, sometimes corresponding with niceties of theory rather than with observable differences in practice. To add to the confusion of terminology, the word 'tachisme' was used differently in the 19th cent., being applied pejoratively to the *Impressionists and the *Macchiaioli.

tactile values. Term introduced into art criticism by Bernard *Berenson in his *Florentine Painters of the Renaissance* (1896) to describe those qualities in a painting that stimulate the sense of touch. He thought that *Giotto was the first master since *classical antiquity whose painting demonstrated these qualities, which he considered to be a distinctive feature of Florentine painting. His beliefs prompted some fine passages of criticism, but the psychological theories on which they were based have been largely superseded, and his elevation of tactile values, which he considered 'life-enhancing', into a general requisite of all painting must be rejected as an example of the fallacy into which many critics have fallen of wishing to impose their own interests and likings universally.

Taddeo di Bartolo (*c.*1362/3–1422). Sienese painter, active in Pisa, Perugia, San Gimignano, and Volterra, his native city. He was a conservative artist, but is noteworthy for his series of frescos on Roman Republican heroes and civic Virtues (1406–14) in the Palazzo Publico, Siena, which are early examples of a type that became popular in the Renaissance.

Taeuber-Arp, Sophie (née Taeuber) (1889–1943). Swiss artist, the wife of Jean *Arp. Her prolific output included *collages, embroideries, paintings, puppets, sculpture, and stage designs, much of her work being in an abstract style distinguished by its rhythmic vitality. She met Arp in Zurich in 1915 and married him in 1922. From 1916 they worked in close collaboration with the Zurich *Dada group and also collaborated in joint paintings and compositions of a distinctive *Constructivist type. From 1927 to 1940 they lived at Meudon, near Paris, and they joined *Cercle et Carré in 1930 and *Abstraction-Création in 1932. From Meudon she founded and edited a periodical of abstract art, *Plastique*, of which five numbers appeared between 1937 and 1939, in French, German, and English. The first number was devoted to *Malevich and the third was an American number with an article on the development of abstract art in America. From 1941 to 1942 she lived with Arp at Grasse, forming the centre of a small group of artists which included Sonia *Delaunay and Alberto *Magnelli. In 1942 she fled to Switzerland, where she died from an accident with a leaking stove.

Taft, Lorado (1860–1931). American sculptor, writer, and teacher. He studied at the École des *Beaux-Arts in Paris, 1880–6, and taught sculpture at the Art Institute of Chicago, 1886–1929. In his day Taft was well known for portraits and allegorical public sculpture, particularly fountains such as The Fountain of Time (Washington Park, Chicago, 1922), but he is now remembered mainly for his books *The History of American Sculpture* (1903), the first comprehensive work on the subject, and *Modern Tendencies in Sculpture* (1921), in which he defended the academic tradition and attacked abstraction. In addition to writing and teaching, he spread his ideas as a public lecturer, touring clubs and schools in Illinois. Taft's studio in Chicago has been preserved as a national monument.

Taine, Hippolyte (1828–93). French historian and critic, the best-known exponent of positivism in 19th-cent. aesthetics. In *Philosophie de l'art* (1865) he declared: 'My sole duty is to offer you facts and show how these facts are produced', and he described art history as a sort of applied botany: 'Just as there is a physical temperature which by its variations determines the appearance of this or that species of plants, so there is a moral temperature which by its variations determines the appearance of this or that kind of art.' He defended the view that a work of art is the product of its environment and nothing else and that it can be fully explained by environmental factors, his anti-*Romantic views leading him to declare that 'vice and virtue are products, like vitriol and sugar'. Taine was professor of aesthetics and art history at the École des *Beaux-Arts in Paris and his work, which was translated into various languages, had wide influence.

Takis (Panayotis Vassilakis) (1925–). Greek experimental artist, best known for his highly original work in *kinetic sculpture. His creations often employ magnetic fields in which various metal objects produce changing patterns—the magnet's 'live force and vibration gives life to what has seemed to be dead material'. In 1960 he suspended the poet Sinclair Beiles in a magnetic field at the Galerie Iris Clert in Paris. Sometimes he combines light effects and music with movement. Since 1954 Takis has lived mainly in Paris, and in 1972 a major retrospective exhibition at the Centre National d'Art Contemporain there showed the great variety of his work over the previous 20 years.

Tamayo, Rufino (1899–1991). Mexican painter and graphic artist. Although he painted many murals as well as easel paintings, his work was opposed to the creed of the revolutionary mural painters, *Orozco, *Rivera, and *Siqueiros, who set greater store by the message and theme of a picture than its pictorial qualities, and for many years his reputation was higher abroad than in Mexico. From 1938 to 1957 he lived mainly in New York and thereafter mainly in Paris. Tamayo's fervent and highly distinctive style blended influences from modern European movements such as *Surrealism with the *folk-art traditions of the indigenous Mexican Indians (his parents were Zapotec Indians and in the 1920s he was head of the Department of Ethnographic Drawing at the National Museum of Archaeology in Mexico City). His subjects included animals, still lifes, and portraits, many of his pianist wife, Olga. In 1974 he donated his fine collection of pre-Hispanic art to his native city of Oaxaca when the Tamayo Museum of Pre-Spanish Mexican Art was inaugurated.

Tanagra statuettes. A class of carefully moulded and painted *terracotta figurines mostly of the 3rd cent. BC, named after Tanagra in Boeotia, where many of them were found in 1874. The favourite subjects were elegant draped women. They became enormously popular and have been much forged.

Tanguy, Yves (1900–55). French-born American painter. He decided to become a painter after seeing pictures by de *Chirico in 1923. In 1925 he met *Breton and joined the *Surrealist movement. His art developed quickly and by 1927 his mature style had emerged. In 1939 he met the American Surrealist painter Kay *Sage in Paris. He emigrated to the USA that year, married Sage in 1940, and became an

American citizen in 1948. The couple lived first in New York and then from 1942 at Woodbury, Connecticut. Tanguy's most characteristic works are painted in a scrupulous technique reminiscent of that of *Dalí, but his imagery is highly distinctive, featuring half marine and half lunar landscapes in which amorphous nameless objects proliferate in a spectral dream-space (*The Invisibles*, Tate, London, 1951).

Tàpies, Antoni (1923–). The most important Spanish painter to emerge in the period since the Second World War. He studied law at the university in his native Barcelona from 1943 to 1946 and was largely self-taught as an artist. His early works were in a *Surrealist vein influenced by *Klee and *Miró, but in about 1953, after turning to abstraction, he began working in mixed media, in which he has made his most original contribution to art. He incorporated clay and marble dust in his paint and used discarded materials such as paper, string, and rags, and then (from about 1970) more substantial objects such as parts of furniture. He explained his belief in the validity of commonplace materials in his essay *Nothing is Mean* (1970). Tàpies has travelled extensively and his ideas have had worldwide influence. Apart from paintings he has also made etchings and lithographs. There are examples of his work in the Tate Gallery and many other major collections.

Tarbell, Edmund C. See TEN, THE.

Target exhibition. Exhibition organized in Moscow by *Larionov, *Goncharova, and *Malevich in March 1913, at which *Rayonism was launched.

Tassi, Agostino (Agostino Buonamici) (*c*.1580–1644). Italian painter, active in Rome for most of his career. He was one of the outstanding **quadratura* specialists of the period, his most famous work being the illusionistic architectural setting for *Guercino's celebrated *Aurora* fresco (1621–3) in the Casino Ludovisi, Rome. Tassi also painted small landscapes in the manner of *Bril and *Elsheimer, and for centuries he was remembered because he taught *Claude (who entered his service as a pastry-cook), his significance as a decorative painter being forgotten. His other claim to fame is that he was accused of raping Artemisia *Gentileschi; he was eventually acquitted after spending some time in prison.

Tassie, James (1735–99). Scottish maker of medallion portraits and reproductions of antique gems and cameos. He began his career as a stonemason, and in 1763 moved to Dublin, intending to set up as a sculptor. In Dublin he met Dr Henry Quin, a physician who made casts from antique gems as a hobby, and together they developed a 'white enamel composition' suitable for reproducing gems (and for creating miniature portrait heads) in imitation of marble. Tassie successfully kept the secret of his vitreous paste and learned to use it so skilfully that (as well as varying the colour) he could make his reproductions opaque or transparent and imitate the varied layers of a cameo. In 1766 he settled in London, where he attained a considerable reputation; from 1769 to 1791 he exhibited regularly at the *Royal Academy and he made casts for *Wedgwood. The German-born antiquary Rudolph Eric Raspe (1737–94), who is perhaps best remembered as the creator of the humorous character 'Baron Munchausen', issued a two-volume catalogue of Tassie's 'ancient and modern engraved gems' in 1791. In all, Tassie reproduced more than fifteen thousand gems, cameos, and medallions, and he made more than five hundred medallion busts (his work is best represented in the Scottish National Portrait Gallery in Edinburgh).

Tassie's nephew, **William Tassie** (1777–1860), succeeded to his business. He was equally industrious, but less skilful. Like his uncle, he was a kindly, popular man, and his studio in Leicester Square was much frequented by artists and literary men. In 1805 he won the main prize in the lottery when *Boydell's Shakespeare Gallery pictures and estate were disposed of. Tassie sold the pictures in the same year.

Tate Gallery, London. The national collection of British art and of modern art from *c*.1870 to the present day. It is named after Sir Henry Tate (1819–99), the millionaire inventor of the sugar cube, who in 1890 offered his collection, consisting mainly of the work of Victorian contemporaries, to the nation on condition that the Government found a suitable site for a gallery.

After many difficulties the Tate Gallery was opened at Millbank in 1897 in an undistinguished neo-Baroque building designed by Sidney R. J. Smith (1858–1913). It had eight galleries and housed Tate's gift of sixty-five paintings and two sculptures and certain other works, including eighteen paintings presented by G. F. *Watts. The Tate Gallery on formation was not, however, an independent gallery of British art as had been envisaged by Henry Tate. It was subordinate to the *National Gallery and it was intended only for modern British art. It began to be established as a historical collection of British art in 1910 when a collection of works by Alfred *Stevens was added and a wing was opened to accommodate most of the paintings left in *Turner's studio at his death, which had previously been housed (but in the main unexhibited) at the National Gallery. In 1916 the Tate was given the additional responsibility of forming the national collection of modern art. This coincided with Sir Hugh *Lane's bequest of thirty-seven modern pictures, which formed a nucleus for the collection (although they were later transferred to the National Gallery). In 1923 Samuel *Courtauld created a fund for the purchase of French Impressionist and Post-Impressionist paintings as a result of which the Tate soon had an outstanding collection in this area. It was not until 1954, however, that the Tate Gallery became completely independent of the National Gallery (although transfers are still made between the two institutions). There have been several extensions to the building (some of them paid for by Sir Joseph and Lord *Duveen), and in 1987 a new gallery—the Clore Gallery (named after Sir Charles Clore, 1904–79, a businessman and philanthropist who was one of the Tate's greatest benefactors)—was opened to house works by Turner, including not only the oil paintings already at the Tate, but also watercolours and drawings previously in the British Museum. The Turner collection is the greatest glory of the Tate, but it also has the most complete representation of British art in general, from the mid-16th cent. to the present day, and one of the world's finest collections of modern art. In 1988 a new branch of the Tate Gallery was opened in Liverpool and in 1993 another one in *St Ives. The creation of these two outstations reflected not only a desire to share the Tate's collections with regional audiences, but also the fact that the gallery had outgrown its site in London. In 1994 the Tate Trustees announced a decision to create a new museum—the Tate Gallery of Modern Art—in the decommissioned Bankside Power Station, a huge and impressive building designed by Sir Giles Gilbert Scott, occupying a prime site on the Thames opposite St Paul's Cathedral. It is expected to open to the public in the year 2000. The original Tate Gallery at Millbank will then become the Tate Gallery of British Art.

Tatlin, Vladimir (1885–1953). Russian painter, designer, and maker of abstract constructions, the founder of *Constructivism. He ran away to sea at the age of 18 and until 1914 combined painting with the life of a merchant seaman: many of his earlier pictures are of maritime subjects, notably *The Sailor* (Russian Mus., St Petersburg, 1911–12), a self-portrait. From 1910 he exhibited at several avant-garde exhibitions in Russia and worked in close association with *Goncharova and *Larionov, who had known him as a boy. In 1914 Tatlin visited Berlin and Paris, and met *Picasso, who inspired his revolutionary *Painted Reliefs, Relief Constructions*, and *Corner Reliefs* of 1914 onwards. Only one or two of these survive and most are known only from photographs preserved in Soviet archives. It appears that he used a variety of materials—tin, glass, wood, plaster, etc.—with the object of doing away with pictorial illusion, and used three-dimensional *relief instead of illusory picture-space.

After the October Revolution of 1917 Tatlin's constructions from real materials in real space were felt to be in accordance with the new doctrines and he threw himself whole-heartedly into support of the demand for socially oriented art, declaring: 'The events of 1917 in the social field were already brought about in our art in 1914, when material, volume and construction were laid as its basis.' In 1919 he was commissioned by the Revolutionary Department of Fine Arts to design the *Monument to the Third International* (i.e. the Third International Communist Congress, to be held in Moscow in 1921). The huge monument was intended for a position in the centre of Moscow; it was to be in glass and iron, and the central glass cylinder was to revolve. It was described by Tatlin as: 'A union of purely plastic forms (painting, sculpture and architecture) for a utilitarian purpose.' A model was exhibited in December 1920 at the exhibition of the VIIIth Congress of the Soviets. The design was condemned by *Gabo as impracticable and it was never executed (it was intended to be much bigger than the Eiffel Tower), but it is recognized as the outstanding symbol of Soviet Constructivism.

The *Monument* was the culmination of Tatlin's artistic career, and the rest of it is something of an anticlimax. He taught and was active in the Soviet programme for the organization of museums, schools, etc. for the propagation of modern artistic culture. His own work was mainly in the field of *applied art, designing furniture, workers' clothes, etc. In the late 1920s and early 1930s he devoted his energies to designing a glider, which he called *Letatlin* (a compound of his name and the Russian word for 'to fly'). He seems to have been more concerned with aesthetic qualities than with practicality, for the machine did not in fact fly. From the 1930s his main activity was theatre design, and his later years were spent in lonely obscurity. There has even been some suspicion that he lived for several years after his official death date.

Tchelitchew, Pavel (1898–1957). Russian-born American painter. In 1918 he fled his native Moscow because of the Revolution and in 1923 settled in Paris, where he did theatrical designs, circus pictures, landscapes, and portraits, and was regarded as a leading exponent of *Neo-Romanticism. He went to the USA in 1934 and became an American citizen in 1952. His interest in theatrical and ballet design was reflected in a certain decorative and mannered quality of his painting with traits of a superficial *Surrealism. His best-known work is probably *Hide and Seek* (MOMA, New York, 1942), in which strangely coloured children's heads weirdly metamorphose into vegetable forms. From 1949 he lived mainly in Italy and he died in Rome.

Teerlinc, Levina. See BENING.

tempera. Term originally applied to any paint in which the *pigment is dissolved in water and mixed (tempered) with an organic gum or glue, but now generally confined to the most common form of the medium—egg tempera. *Cennini used the word almost as an equivalent of *medium and spoke of both *size and oil tempera. *Vasari also applied it to all mixtures of pigments, including those bound with oil or *varnish. The name was, therefore, applicable to almost all methods except true *fresco. Egg tempera was the most important technique for *panel painting in Europe from the beginning of the 13th cent. to the end of the 15th cent., when it began to give way to *oil painting. The classic description of painting with egg-yolk was given by Cennini in the early 15th cent. It was used chiefly for painting executed on wood panels. The panel was first prepared with a *ground and *priming. The whole design was then drawn in with *charcoal, the excess charcoal was dusted away, and the outlines and parts of the shading were painted with a *watercolour. If the background was to be in gold, its outlines were incised in the *gesso ground, and some of the contours of the design were incised also. Any gilding was completed first; next came the painting of the drapery, architecture, and other parts. The flesh parts were painted last and often had an *under-painting of the green earth pigment called *terre verte*. Other parts too might be under-painted in order to enhance or subdue the colour. A few local colours only were used—one colour for flesh, and two or three for the different draperies, etc.—and, as in fresco, each was prepared in three different tones by adding varying quantities of white. The darker parts of the flesh might be shaded in *verdaccio*, a mixture of ochre and black, and a pure white might be used for the highlights.

The grinding of the pigments in egg-yolk was of course done beforehand. The yellowness of the egg had little effect on the colours, though Cennini says that town hens produce the palest and best yolks, and adds that darker yolks will do for the flesh colour of 'old people, or such, who are darker in colouring'. Pale yolks are still preferred today.

The method demanded not only knowledge of the materials, but also great certainty of purpose. The paint was not easy to handle, and the colours were few and difficult to blend. So the variety and subtlety that were so often attained depended on a slow building-up process, in which each stage—ground, underpainting, and various layers of semi-transparent paint—would have a calculated effect upon the next. Tempera has more luminosity and depth than fresco, but its range of colour and tone is limited and it cannot achieve the close imitation of natural effects that is attainable in oil painting. In the 15th and 16th cents. it was very common for pictures to be painted in a mixture of tempera and oil, with oil usually supplying the final touches. After being neglected for about 400 years, tempera painting has had a limited revival in the present century, the Americans *Cadmus, *Tooker, and *Wyeth being noted exponents.

Tempesta, Antonio (1555–1630). Italian painter and engraver, born in Florence and active mainly in Rome. He worked on frescos in the Vatican and other Roman palaces, but is remembered mainly for his hunting scenes and battle pieces.

Ten, The. Group of American painters from New York and Boston who exhibited together for about 20 years from 1898. Most of them had studied in Paris in the 1880s and the common factor in their work was an interest in *Impressionism. They were Frank W. Benson (1862–1951); Joseph R. De Camp (1858–1923); T. W. Dewing (1851–1938); Childe *Hassam; Willard L. Metcalf (1858–1925); Robert Reid (1862–1929); Edward E. Simmons (1852–1931); Edmund C. Tarbell (1862–1938); J. H. Twachtman (1853–1902); J. Alden Weir (1852–1919). On the death of Twachtman his place was taken by W. M. *Chase. The Ten was also the name of a group of American *Expressionist painters who exhibited together from 1935 to 1940. *Gottlieb and *Rothko were the best-known members.

tenebrism. Term describing predominantly dark tonality in a painting. It derives from the Italian 'tenebroso' (obscure) and is applied mainly to the 17th-cent. followers of *Caravaggio in Italy and elsewhere.

Teniers, David the Younger (1610–90). Flemish painter, the most important of a family of Antwerp artists. His output was huge and varied (about 2,000 pictures have been attributed to him), but he is best known for his peasant scenes—similar to those of *Brouwer, but less hearty. In 1651 he was appointed court painter to Archduke Leopold Wilhelm of Austria, Governor of the Austrian Netherlands in Brussels, and was also made custodian of the Archduke's art collection. Teniers was an unusual curator. He not only compiled a first-rate catalogue of the Archduke's pictures but he also made paintings of the galleries in which the works of art were installed, and painted small copies of some of the pictures (examples of the latter are in the Courtauld Inst., and Wallace Coll., London).

Teniers' father, **David the Elder** (1582–1649), was primarily a painter of religious scenes. Few pictures are known that are certainly by him, and many formerly attributed to him are now given to his son, with whom he may have collaborated. David the Younger's son, **David III** (1638–85), was one of the many artists who imitated his father's work. Other members of the family were **Julien I** (1572–1615), the brother of David I, **Julien II** (1616–79), the brother of David II, and **Abraham** (1629–70), the brother of David III.

Tenniel, Sir John (1820–1914). English illustrator. He is remembered chiefly for his brilliant illustrations to Lewis Carroll's *Alice's Adventures in Wonderland* (1865) and its sequel *Through the Looking Glass* (1872), which now seem inseparable from the text, and for his long association with *Punch*. He worked for *Punch* from 1850 to 1901, succeeding *Leech as chief cartoonist in 1864, and drew more than 2,000 cartoons for it.

Terborch (or **Ter Borch**), **Gerard the Younger** (1617–81). Dutch painter and draughtsman of interiors and small portraits. A highly precocious artist—his earliest dated drawing (Rijksmuseum, Amsterdam) is from 1625—Terborch studied with his father, **Gerard the Elder** (1584–1662), in his native Zwolle, and with Pieter de *Molyn in Haarlem. Unlike most of the Dutch artists of his time he travelled extensively. In 1635 he was in London, and from *c.*1636 to 1640 he was in Italy. From Italy he probably went to Spain and then to Flanders. In 1648 he was in Germany, where he painted *The Swearing of the Oath of Ratification of the Treaty of Münster* (NG, London, 1648), a group portrait of the signatories to the treaty that gave the Dutch independence from the Spanish. In 1654 he finally settled in Deventer, where he won both professional and social success. After beginning his career with guardroom scenes, he turned to pictures of elegant society, to which his gifts for delicate characterization and exquisite depiction of fine materials were ideally suited. His best-known work, the subject of a charming passage by *Goethe, is the so-called *Parental Admonition* (versions in the Rijksmuseum, Amsterdam, and the Staatliche Museen, Berlin). It is symptomatic of Terborch's unvaryingly tasteful decorum that the true theme of this picture is a man making a proposition to a courtesan (the coin that he proffers to his 'daughter' has been partially erased in the Berlin version and it is omitted in the engraving Goethe knew). Terborch's main pupil was Caspar *Netscher.

Terbrugghen, Hendrick (1588–1629). Dutch painter, one of the earliest and finest exponents of *Caravaggism in northern Europe. Born into a Catholic family, he grew up in Utrecht, studied there with *Bloemaert, then spent about a decade in Rome (*c.*1604–14). On his return to the Netherlands he became with *Honthorst the leader of the Caravaggism associated with the Utrecht School. A second journey to Italy, *c.*1620, has been postulated, as his later works are generally more thoroughly Caravaggesque than his earlier ones. Terbrugghen was chiefly a religious painter, but he also produced some remarkable *genre works, notably a pair of paintings of *Flute Players* (Staatliche Kunstsammlungen, Kassel, 1621), which in their subtle tonality—with dark figures placed against a light background—anticipated by a generation the achievement of painters of the Delft School such as *Fabritius and *Vermeer. Although he was praised by *Rubens, who visited Utrecht in 1627, Terbrugghen was neglected by 18th- and 19th-cent. collectors and historians. The rediscovery of his sensitive and poetic paintings has been part of the general reappraisal of Caravaggesque art in the 20th cent.

terracotta (Italian: baked earth). Clay baked to become hard and compact. Figures and architectural ornaments have been made of it since very early times (see, for example, TANAGRA STATUETTES) and it is to these, rather than pottery vessels, that the word 'terracotta' usually refers. Clay is, of course, found all over the world in many different colours and qualities. Coloured clay is commoner than white. The presence of certain chemicals, such as iron oxide, affects the colour of the baked product, so terracotta works are not necessarily of the reddish-brown colour that is normally associated with the word. Firing may produce a wide range of colour from light buff to deep red or black. The hardness and strength of the baked clay vary according to the temperature at which it has been fired. During the firing the clay shrinks by about one-tenth of its volume, sometimes more, sometimes less, according to its quality and the amount of moisture.

Testa, Pietro (called Il Lucchesino) (1611–50). Italian engraver and painter active in Rome. He trained with *Domenichino and was employed by Nicolas *Poussin's patron Cassiano dal Pozzo to make antiquarian drawings, but his bizarre imagination brings him closer in spirit to his more Romantic contemporaries such as *Castiglione and *Rosa. His paintings are rare and he is better known as an etcher. Testa died by drowning and it was rumoured that he killed himself.

Theed, William (1804–91). English sculptor. He was the son of a painter and sculptor of the same name (1764–1817), best known for his pediment group *Hercules Taming the Thracian Horses* (Royal Mews, London, *c.*1816), one of the earliest works to show the influence of the *Elgin Marbles. Theed the Younger trained under E. H. *Baily and at the *Royal Academy Schools. In 1826 he went to Rome and worked under *Thorvaldsen and *Gibson. He returned to London in 1848 and became one of the most distinguished and prolific of Victorian sculptors. His work includes busts, statues, and the colossal group of *Africa* for the Albert Memorial.

Theophanes the Greek (in Russian Feofan Grek, *c.*1330–*c.*1405). Painter from Constantinople, active mainly in Russia. His first extant work is a cycle of *frescos (1378) in the church of the Transfiguration at Novgorod, one of the outstanding monuments of Russian medieval art. Theophanes brought the style of late *Byzantine painting to Russia but adapted himself also to the native Russian School and evolved a highly personal manner of modelling his figures in light and shade with masterly, almost *Impressionistic, brush-strokes and an individualistic *palette. The wall-paintings in the church of the Dormition at Volotovo, near Novgorod, and in the church of St Theodore Stratilates in Novgorod itself, are in a similar style. They have often been attributed to Theophanes, but are now regarded as the work of anonymous artists of the Novgorod School painting under his influence (*c.*1380). Theophanes executed paintings at Moscow between 1395 and 1405, but none of these survives except some *icons from the Cathedral of the Annunciation on the Kremlin, which were painted with the help of Russian assistants and are poorly preserved. According to a contemporary source Theophanes was also famous as a book-*illuminator. No signed or documented work in this field is known by him, but several examples have been attributed to him. Theophanes made a great impression on painting in Novgorod and Moscow in the 15th cent. See also RUBLEV, ANDREI.

Theophilus. The pseudonym adopted by the author of a treatise on medieval Christian arts and crafts entitled *De Diversis Artibus* (The Various Arts). The work has been variously dated from the 9th to the 13th cent. C. R. Dodwell, who produced the standard edition and English translation (1961), assigns it to a date between 1110 and 1140. It is the most important source of information on medieval arts and techniques and is unusual for its time in its references to the artist's attitude to his work. Various medieval manuscripts are extant but the treatise did not become generally known until the late 18th cent. *Lessing produced the first printed edition, posthumously published in 1781, and in the same year Rudolph Eric Raspe (see TASSIE) published an edition in England from another manuscript. Raspe was mainly interested in the remarks on oil painting and the title-page tells us that the treatise proves 'that the art of painting in oil was known before the pretended discovery of John and Hubert van *Eyck'.

Little is known with certainty about Theophilus, who adopted a pseudonym because he wanted to dedicate his skills to God rather than win fame for himself. However, internal evidence suggests that he was a German Benedictine monk and priest, an educated person conversant with scholastic philosophy and also a practising craftsman whose primary interest lay in metalwork. He has been plausibly identified with Roger of Helmarshausen, a goldsmith and monk at the abbey of Helmarshausen, who is documented as the maker of a portable altar (Paderborn Cathedral Treasury, 1100) for Bishop Henry of Werl.

Thiele, Johann Alexander. See PASTEL.

Thieme, Ulrich (1872–1922), and **Becker, Felix** (1864–1928). German art historians, editors of the *Allgemeines Lexikon der bildenden Künstler von der Antike bis zur Gegenwart* (General Dictionary of Artists from Antiquity to the Present, 37 volumes, 1907–50), the largest and most comprehensive dictionary of artists' biographies ever published. Entries, written by leading scholars, are long and detailed, making it an indispensable reference tool. Architects and decorative artists are covered, as well as painters, sculptors, and engravers. Work began in 1898; Becker retired for reasons of health in 1910 and after Thieme's death Hans Vollmer (1878–1969) became editor. Vollmer also edited the supplement, *Allgemeines Lexikon der bildenden Künstler des XX. Jahrhunderts* (6 vols., 1953–62). The first volume of what is in effect a new edition of the complete dictionary appeared in 1983; it is entitled *Allgemeines Künstler-Lexikon* and is edited by Günter Meissner.

Thoma, Hans (1839–1924). German painter. His early landscapes—usually scenes from his native

Black Forest—are his finest and freshest works, showing the influence of *Courbet and the *Barbizon School. Later, under the impact of *Böcklin, he turned to more ponderous symbolic subjects. Thoma was immensely popular during his lifetime, but has fallen greatly in critical esteem. The largest collection of his works is in the Karlsruhe Kunsthalle, of which he was Director from 1899.

Thomson, Tom (1877–1917). Canadian landscape painter, one of the main creators of an indigenous Canadian school of painting. Most of his career was spent as a commercial artist in Toronto, and it was only in 1914 that he was able to take up painting full time. Much of his painting was done out of doors, notably the fluently spontaneous oil sketches he produced in Algonquin Park. Among his more finished paintings, the most famous is the bold and brilliantly coloured *Jack Pine* (NG, Ottawa, 1917), which has become virtually a national symbol of Canada. Thomson's career ended tragically when he was mysteriously drowned in Algonquin Park, but his ideals were continued by the artists who formed the *Group of Seven, to whom he was an inspiration.

Thoré, Théophile (1807–69). French writer on art. He ranks alongside *Baudelaire as the most perceptive art critic of his time (he was among the first to acclaim *Courbet, *Daumier, *Manet, *Monet, and *Renoir and to see the weakness of *Meissonier), and he also made memorable contributions to the study of earlier art. Above all, he is remembered for virtually rediscovering *Vermeer, a brilliant feat of connoisseurship and historical detective work. He was originally a political journalist and he published much of his work under the pseudonym W. Bürger when he was living in exile.

Thornhill, Sir James (1675/6–1734). English decorative painter. He was the only British painter of his day to understand and successfully to emulate the European formulas for wall and ceiling painting and was the only native English painter who could challenge on their own ground the many foreign decorative painters then at work in England. His two finest works are the *grisaille paintings on the Life of St Paul (1716–19) in the dome of St Paul's Cathedral, London (he beat *Pellegrini and Sebastiano *Ricci in the competition for the commission), and the Painted Hall at Greenwich Hospital, on which he worked intermittently from 1708 to 1727. Thornhill was appointed History Painter to George I in 1718 and Serjeant Painter in 1720, was knighted in 1722, and elected Member of Parliament in the same year. *Hogarth was his son-in-law.

Thorn-Prikker, Johan (1868–1932). Dutch painter and designer. Early in his career he passed through phases of *Impressionism and *Post-Impressionism, then changed to an elaborate linear style with which he became a leading exponent of *Symbolism and *Art Nouveau, as in his most famous painting—the mystical, erotic *The Bride* (Kröller-Müller Museum, Otterlo, 1893). From 1893 he concentrated on the design of mosaics, murals, and stained glass, mainly for churches, continuing in this vein after he settled in Germany in 1904. He taught at several art schools in Germany and was a major figure in the development of modern religious art. His masterpiece is perhaps the cycle of windows in the Romanesque church of St George in Cologne, completed in 1930.

Thornycroft, Sir Hamo (1850–1925). British sculptor, one of the leading exponents of the *New Sculpture. Early in his career he concentrated on idealized figures in which he expressed a 'poetic mood' praised by the critic Edmund Gosse; the best-known example is probably *The Mower* (Walker Art Gallery, Liverpool, 1884; there are also several small-scale versions). From the 1890s, however, he was increasingly preoccupied with portrait sculpture and above all public monuments, becoming perhaps the most distinguished British practitioner in this field in the early 20th cent. His statues, dignified and thoughtful in tone, include those of Oliver Cromwell outside the Houses of Parliament, London (unveiled 1899), Alfred the Great in Winchester (1901), Gladstone in the Strand, London (1905), Lord Armstrong in Newcastle upon Tyne (1906), Lord Curzon in Calcutta (1912), and Edward VII in Karachi (1915). These are all in bronze, but he also worked in stone.

Both his father, **Thomas Thornycroft** (1816–85), and his mother, **Mary Thornycroft** (1809–95), were sculptors. They were primarily portraitists, but Thomas is now chiefly remembered for his dramatic Boadicea monument at Westminster Bridge, London, showing the fearsome warrior queen in her chariot. He began work on the group in 1856 and was encouraged by Prince Albert, who lent him horses as models. The plaster model was complete at Thornycroft's death, but it was not cast in bronze until 1897, after his son had presented it to the nation, and it was finally erected in 1902.

Thorvaldsen (or **Thorwaldsen**), **Bertel** (1768 or 1770–1844). Danish sculptor, next to *Canova the most celebrated sculptor of the *Neo-classical movement. After five years at the Academy in his native Copenhagen, he reached Rome in 1797 on 3 March, a day which he henceforth considered as his birthday. He made his name with the statue *Jason* (Thorvaldsens Mus., Copenhagen, 1802–3), which was based on the *Doryphoros* of *Polyclitus, and his growing reputation resulted in so many commissions that by 1820 he had forty assistants in his Roman workshop. In that year, when visiting Copenhagen, he began planning the decoration of the newly built church of Our Lady with marble statues and *reliefs, a scheme which was to be his principal task for several years. The *Christ* and the *St Paul* were carved entirely by his hand; the other figures were made by his assistants from his models. His other major works include the tomb of Pius VII at St Peter's in Rome (1824–31) and a monument to Lord Byron (Trinity College, Cambridge, 1829). In 1838 he returned finally to Denmark, a celebrity whose authority in the arts was sovereign. He was respected not only as an artist and interpreter of the classical past, but also as an authority on the antique world. In Copenhagen a museum was built for him (1839–48), itself a remarkable piece of neo-antique architecture. In addition to his own sculptures, the museum houses the outstanding art collection he made; he bought works by contemporary painters (notably the *Nazarenes) as well as ancient works.

Thorvaldsen aimed at reviving the sublimity of Greek sculpture, but he never went to Greece and (in common with other artists of his time) bestowed his admiration mainly on late *Hellenistic or Roman copies. He did, however, gain close knowledge of Greek sculpture from the restorations he made to the recently excavated sculptures from the Temple of Aphaia in Aegina, which in 1816 passed through Rome on their way to Munich (they are now in the Glyptothek there); Thorvaldsen's restorations have only recently been removed. Compared with Canova he is cool and calculating; his sculptures are more logically worked out and have great precision and clarity, but they lack Canova's sensitive surfaces.

Thulden, Theodor van (1606–69). Flemish painter, engraver and designer of tapestries, a pupil and collaborator of *Rubens. Although like most contemporary painters of historical and religious themes he was strongly influenced by Rubens, he did succeed in working out a personal idiom. His appealingly sweet style won him numerous commissions both inside and outside Flanders, and he worked in The Hague and Paris, as well as Antwerp, where he was mainly based.

Tibaldi, Pellegrino (1527–96). Italian *Mannerist painter, sculptor, and architect. He was in Rome in the late 1540s and early 1550s, and his style in painting, distinguished by grand, if sometimes rather ponderous, figures, was based mainly on the work of *Michelangelo, for whom he had a lifelong admiration. His finest paintings are frescos illustrating the story of Ulysses (*c.*1555) in the Palazzo Poggi (now University), Bologna. From the mid-1560s Tibaldi worked mainly as an architect, chiefly in and around Milan, where he was much employed by the Archbishop, Charles Borromeo (1538–84), and was appointed chief architect to the cathedral in 1567. His major works in Milan include the church of S. Fedele and the court of the Archbishop's Palace, and his massive style was influential in Lombardy. In 1587 he went to Madrid to superintend building operations at the *Escorial and did sculpture and a vast amount of paintings for its decoration. His work there was influential in the development of Spanish Mannerism. He returned to Milan, ennobled by Philip II (1527–98), in the year of his death.

Tidemand, Adolph (1814–76). Norwegian painter, mainly active in Germany. He became a leading figure among the painters of Düsseldorf, producing idyllic and often rather sentimental peasant scenes in the romantic setting of Norway, based on sketches made there.

Tiepolo, Giovanni Battista (Giambattista) (1696–1770). The greatest Italian (and arguably the greatest European) painter of the 18th cent. His work sums up the splendours of Italian decorative painting, and with him the monumental fresco tradition which had begun with *Giotto is brought to an end. He revived the glories of the Venetian School, especially of *Veronese, and enriched them with the experience of the Roman and Neapolitan *Baroque and the new *perspective techniques of theatre decoration. His deep pictorial culture was drawn from a wide variety of sources, including *Rubens, *Rembrandt, and *Dürer.

Tiepolo trained with the history painter Gregorio Lazzarini (1655–1730), but according to his earliest biographer, Vincenzo da Canal, he soon departed from his master's 'diligent manner, and, being all fire and spirit, adopted one which was rapid and free'. After making his name in Venice with some works in the 'dark' manner of *Piazzetta, he carried out his first important fresco cycle in the Archbishop's Palace in Udine (*c.*1727–8) and with the change from oil to fresco he broke away from the dark tonality of his earliest work and established the clear, sunny palette which became one of his foremost characteristics. In the 1730s much of his work was done outside his native city, and by 1736 his fame was such that he was invited to Stockholm to decorate the Royal Palace—an invitation he declined because the fee offered was too small. Between 1741 and 1750 Tiepolo was active mainly in Venice, where his chief work is the decoration of the Palazzo Labia (1745). Assisted by his expert in **quadratura*, Gerolamo Mengozzi Colonna (*c.*1688–1766), Tiepolo created a sumptuous scheme which embraces the entire space of the Gran Salone. On the walls there are two scenes from the life of Cleopatra, a favourite theme of his, and in the *Embarkation* it seems as if Antony and Cleopatra were actually stepping down into the room.

The next decade represents the peak of Tiepolo's career. In 1750–3 he was in Würzburg to decorate the Kaisersaal and the grand staircase of the Prince Archbishop's palace, and this work is the masterpiece of his maturity. Like Rubens, Tiepolo could make even the most ponderous allegory come alive, and here the unpromising task of paying tribute to the lack-lustre Prince Archbishop brought forth his most glorious work—full of light and colour and perfectly attuned to the superb architectural setting. In 1757 Tiepolo decorated a series of rooms in the Villa Valmarana near Vicenza, with scenes from Homer, Virgil, Ariosto, and Tasso, and this work gives perhaps the most immediate experience of his qualities: grandeur, rich and glowing colour, and fancifulness warmed by humanity in the narrative.

His last large-scale work in Italy was the ceiling of the ballroom in the Villa Pisani at Strà. While there he was called by Charles III (1716–88) to Madrid, where he spent the last eight years of his life. The *Apotheosis of Spain* (1764) on the ceiling of the throne room in the royal palace was his principal commission. Intrigue and jealousy embittered these last years in Madrid, and the final blow was dealt posthumously when his seven altarpieces for the church of S. Pascal Baylon at Aranjuez were displaced by seven canvases done by *Mengs, the champion of *Neoclassicism (Mengs is said to have fallen from a tree whilst lying in wait to attack the aged Tiepolo, but Tiepolo—with characteristic nobility—saw to it that his rival was treated in hospital). Tiepolo's altarpieces are now divided between the Prado and the Royal Palace in Madrid, and the superb oil-sketches for them are mainly in the Courtauld Institute Galleries, London. They display a pathos and psychological intensity that was new to his art. He had painted easel pictures as well as frescos throughout his career, and a work such as the ravishing *A Young Woman with a Macaw* (Ashmolean Mus., Oxford, *c.*1760) shows the masterly fluency of his brushwork. Tiepolo was as prolific a draughtsman as he was a painter (an outstanding collection is in the Victoria and Albert Museum, London) and was one of the finest etchers of his period; *Goya was especially indebted to his graphic work.

Tiepolo was married to Cecilia Guardi, sister of the *Guardi brothers. They had two painter sons. **Giandomenico Tiepolo** (1727–1804) began to assist his father *c.*1745 and accompanied him to Madrid. He was so faithful to his father's work that it is sometimes impossible to say exactly where he has collaborated. In his independent work he has a clearly defined style with a marked bias towards *genre and *caricature, e.g. at the Villa Valmarana. He is famous for his engravings and etchings, especially the twenty-two variations on the theme of the *Flight into Egypt* (1753). **Lorenzo Tiepolo** (1736–before 1776) also assisted his father, but unlike his brother Giandomenico he has no individual substance as an artist.

Tiffany, Louis Comfort (1848–1933). American designer, interior decorator, and architect, his country's most famous exponent of the *Art Nouveau style. He was born in New York, the son of a prosperous jeweller, and spent virtually all his career there. His initial training was as a painter, but in the 1870s he turned more to the decorative arts and founded an interior decorating business in 1879. Tiffany became famous above all for his highly distinctive glass vases and lamps, but until about 1900 his firm was better known for stained glass and mosaic work (it did interiors for many socially prominent New Yorkers as well as for clubs). Most of his architectural work, including his own mansion on Long Island, has been destroyed, but the loggia of the main entrance of his masterpiece, Laurelton Hall (1903–5), has been installed in the Metropolitan Museum, New York. Tiffany was also an art patron and established a foundation that provided study and travel grants for students.

Tijou, Jean (active 1688–1712). French ironworker, all of whose known work was executed in England. His life before he is first recorded (working at Chatsworth) in 1688 and after he left England in 1712 is obscure. Tijou's best-known work—gates, railings, screens, balustrades—is at Chatsworth, Derbyshire, and Hampton Court and St Paul's Cathedral, London, and is universally regarded as being among the finest metalwork of its type ever created. His style was luxurious but dignified—a more Classical version of the *Baroque style of Louis XIV in France. In 1693 he published a book of designs entitled *A New Booke of Drawings Invented and Designed by Jean Tijou . . . for the use of them that will work iron in perfection and with art*, which was the first book of designs for ironwork to appear in England. It had considerable importance in disseminating his style among English ironworkers.

Tinguely, Jean (1925–91). Swiss sculptor and experimental artist. Tingueley's work was concerned primarily with movement and the machine, satirizing technological civilization. His boisterous humour was most fully demonstrated in his auto-destructive works, which turned *Kinetic art into *Performance art. The most famous was *Homage to New York* which was presented at The Museum of Modern Art, New York, on 17 March 1960, but failed to destroy itself as programmed and caused a fire. The object was constructed of an old piano and other junk. Tinguely was an innovator not only in his combination of Kineticism with the *Expressionism of *Junk sculpture, but also in the impetus he gave to the principle of spectator participation, as in his *Cyclograveur*, in which the spectator mounts the saddle and pedals the bicycle, causing a steel nail to strike against a vertically mounted flat surface, and the *Rotozazas*, in which the spectator plays ball with the machine. His most famous work (done in collaboration with Niki de *Saint Phalle) is the exuberant Beaubourg Fountain (1980) outside the *Pompidou Centre in Paris, featuring fantastic mechanical birds and beasts that spurt water in all directions. Such works might be interpreted as an ironic ridicule of the practical functions of machines.

Tino di Camaino (*c.*1285–1337). Sienese sculptor, chiefly of tombs, active in Pisa, Florence, and Naples, as well as his native city. He probably trained with Giovanni *Pisano, but his style was more calm and reserved, with an imposing block-like massiveness. His early career was spent in Pisa and Siena, but his chief works are in Florence (where he worked 1321–4) and Naples (where he worked from 1324 until his death). In Florence his work included the tomb of Bishop Orso in the cathedral—possibly the earliest example of the seated effigy. In Naples, where he is known to have been in touch with *Giotto, who was court painter there at the time, and with the Sienese painter Pietro *Lorenzetti, his work included tabernacled tombs for the Angevin court (e.g. for Queen Mary of Hungary in Sta Maria Donnaregina). He somewhat modified his own rigorous style in the direction of the more decorative grace of the *Gothic style, but none the less his influence was significant as one of the Tuscan artists who carried the new northern developments to the southern parts of Italy.

Tintoretto (Jacopo Robusti) (1518–94). Venetian painter. His nickname derives from his father's profession of dyer (*tintore*). Although he was prolific and with *Veronese the most successful Venetian painter in the generation after *Titian's death, little is known of his life. He is said to have trained very briefly with Titian, but the style of his immature works suggests that he may also have studied with *Bonifazio Veronese, Paris *Bordone, or *Schiavone. Almost all of his life was spent in Venice and most of his work is still in the churches or other buildings for which it was painted. He appears to have been unpopular because he was unscrupulous in procuring commissions and ready to undercut his competitors.

By 1539 he was working independently, but the little that is known of his early work suggests that he was not precocious. The first work in which he announced a distinctive voice is *The Miracle of the Slave* (Accademia, Venice, 1548), in which many of the qualities of his maturity, particularly his love of foreshortening, begin to appear. To help him with the complex poses he favoured, Tintoretto used to make small wax models which he arranged on a stage and experimented on with spotlights for effects of light and shade and composition. This method of composing explains the frequent repetition in his works of the same figures seen from different angles. He was a formidable draughtsman and, according to *Ridolfi, he had inscribed on his studio wall the motto 'The drawing of *Michelangelo and the colour of Titian'. However, he was very different in spirit from either of his avowed models—more emotive, using vivid exaggerations of light and movement. His drawings, unlike Michelangelo's detailed life studies, are brilliant, rapid notations, bristling with energy, and his colour is more sombre and mystical than Titian's.

Tintoretto's greatest works are the vast series of paintings he did for the Scuola di San Rocco in Venice from 1565 to 1587—scenes from the life of Christ in the upper hall and scenes from the life of the Virgin in the lower hall. The complicated scheme was probably not conceived by Tintoretto himself, but he interpreted it with a vividness and economy of colour and detail that give a wonderful cohesion to the whole scheme. Its personal conception of the sacred story overwhelmed *Ruskin, who devoted eloquent pages to it, and Henry James (1843–1916) wrote of the stupendous *Crucifixion* (1565) 'Surely no single picture in the world contains more of human life; there is everything in it, including the most exquisite beauty.' The unorthodox rough brushwork of such paintings incurred the censure of *Vasari, but later generations recognized it as a means of heightening the drama and tension. In his treatment of light and shade and even in his introduction of northern elements of costume Tintoretto never shrank from the bizarre, though in all his greatest works he kept it duly subordinate.

As well as religious works, Tintoretto painted mythological scenes and he was also a fine portraitist, particularly of old men (a self-portrait in old age is in the Louvre). Some of the weaker portraits that go under his name may be the product of his large workshop. His son, **Domenico** (*c.*1560–1635), became his foreman and is said to have painted many portraits, although none can be attributed to him with certainty. Another son, **Marco** (1561–1637), and a daughter, **Maria** (*c.*1556–90), were among his other assistants. The later paintings can thus be divided into those which are largely studio productions on the one hand and the visionary inspirations from Tintoretto's own hand on the other. A prime example of the latter is *The Last Supper* (S. Giorgio Maggiore, Venice, 1592–4), the culmination of a lifetime's development of this subject, from the traditional frontal arrangement of his youth to this startling diagonally viewed composition lit by the flickering glow of a rush lamp and the radiance of Christ. Tintoretto had great influence on Venetian painting, but the artist who most fruitfully absorbed the visionary energy and intensity of his work was El *Greco.

tint tool. A type of *burin used in *wood engraving for cutting fine lines of even thickness. These lines, set close together and parallel to one another, form the grey tones of 'tints' so characteristic of 19th-cent. reproductive wood engraving.

Tischbein. Family of German 18th-cent. painters. The best-known of them, **Johann Heinrich Wilhelm Tischbein** (1751–1829), was most important for his engravings of classical antiquities, but is now remembered almost solely for his famous portrait *Goethe in the Roman Campagna* (Städelsches Kunstinstitut, Frankfurt, 1786–7). Tischbein was *Goethe's friend and he is usually called 'Goethe' Tischbein. The two other leading members of the family, his uncle and cousin, were nicknamed after their main place of work: **Johann Heinrich the Elder**, 'Kassel' Tischbein (1722–89) and **Johann Friedrich** 'Leipzig' Tischbein (1750–1812). They were principally portraitists, and like 'Goethe' Tischbein they had considerable reputations in their day.

Tissot, James (1836–1902). French painter and graphic artist. Early in his career he painted historical costume pieces, but in about 1864 he turned with great success to scenes of contemporary life, usually involving fashionable women. Following his involvement in the Siege of Paris and the Commune (1871) he was obliged to take refuge in London, where he lived from 1871 to 1882. He was just as successful there as he had been in Paris and lived in some style in St John's Wood; in 1874 Edmond de *Goncourt wrote sarcastically that he had 'a studio with a waiting room where, at all times, there is iced champagne at the disposal of visitors, and around the studio, a garden where, all day long, one can see a footman in silk stockings brushing and shining the shrubbery leaves'. His paintings are distinguished most obviously by his love of painting women's costumes: indeed, his work—which has a fashion-plate elegance and a chocolate-box charm—has probably been more often reproduced in works on the history of costume than on the history of painting. He also, however, had a gift for wittily observing nuances of social behaviour. In 1882, following the death of his mistress Kathleen Newton (the archetypal Tissot model—beautiful but rather vacant), he returned to France. In 1885 he underwent a religious conversion when he went into a church to 'catch the atmosphere for a picture', and thereafter he devoted himself to religious subjects. He visited the Holy Land in 1886–7 and 1889, and his illustrations to the events of the Bible were enormously popular, both in book form and when the original drawings were exhibited. For many years after his death Tissot was considered a grossly vulgar artist, but since the 1970s there has been an upsurge of interest in him, expressed in sale-room prices for his work as well as in numerous books and exhibitions devoted to him.

Titian (Tiziano Vecellio) (*c.*1485–1576). The greatest painter of the Venetian School. The evidence for his birthdate is contradictory, but he was certainly very old when he died. He was probably a pupil of Giovanni *Bellini, and in his early work he came under the spell of *Giorgione, with whom he had a close relationship. In 1508 they collaborated on the external fresco decoration of the Fondaco dei Tedeschi, Venice, and after Giorgione's early death in 1510 it fell to Titian to complete a number of his unfinished paintings. The authorship of certain works is still disputed between them.

Titian's first major independent commission was for three frescos in the life of St Antony of Padua in the Scuola del Santo, Padua (1511), noble and dignified paintings suggesting an almost central Italian firmness and monumentality. When he returned to Venice, Giorgione having died and *Sebastiano having gone to Rome, the aged Bellini alone stood between him and supremacy, and that only until 1516 when Bellini died and Titian became official painter to the Republic. He maintained his position as the leading painter in the city until his death sixty years later. In the second decade of the century Titian broke free from the stylistic domination of Giorgione and developed a manner of his own. Something of a fusion between Titian's worldliness and Giorgione's poetry is seen in the enigmatic allegory known as *Sacred and Profane Love* (Borghese Gallery, Rome *c.*1515), but his style soon became much more dynamic. The work that more than any other stamped his new authority is the huge altarpiece of *The Assumption of the Virgin* (Sta Maria dei Frari, Venice, 1516–18). It is the largest picture he ever painted and one of the greatest, matching the achievements of his most illustrious contemporaries in Rome in grandeur of form and surpassing them in splendour of colour. The soaring movement of the Virgin, rising from the tempestuous group of Apostles towards the hovering figure of God the Father looks forward to the *Baroque. Similar qualities are seen in his two most famous altarpieces of the 1520s: the Pesaro altarpiece (Sta Maria dei Frari, Venice, 1519–26), a bold diagonal composition of great magnificence, and *The Death of St Peter Martyr* (completed 1530), which he painted for the church of SS. Giovanni e Paolo, Venice, having defeated *Palma Vecchio and *Pordenone in competition for the commission. The painting was destroyed by fire in 1867, but it is known through copies and engravings; trees and figures together form a violent centrifugal composition suited to the action, and *Vasari described it as 'the most celebrated, the greatest work ... that Titian has ever done'. Titian had important secular as well as ecclesiastical patrons in this energetic period of his career, one of his most important commissions being three mythological pictures (1518–23) for Alfonso d'*Este—the *Worship of Venus*, the *Bacchanal* (both in the Prado, Madrid), and the *Bacchus and Ariadne* (NG, London). Outstanding among his portraits of these years is the exquisitely sensitive *Man with a Glove* (Louvre, Paris, *c.*1520).

About 1530, the year in which his wife died, a change in Titian's manner becomes apparent. The vivacity of former years gave way to a more restrained and meditative art. He now began to use related rather than contrasting colours in juxtaposition, yellows and pale shades rather than the strong blues and reds of his previous work. In composition too he became less adventurous and used schemes which, compared with some of his earlier works, appear almost archaic. Thus his large *Presentation of the Virgin* (Accademia, Venice, 1534–8) makes use of the *relief-like frieze composition dear to the quattrocento.

During the 1530s Titian's fame spread throughout Europe. In 1530 he first met the emperor Charles V (in Bologna, where he was crowned in that year) and in 1533 he painted a famous portrait of him (Prado) based on a portrait by the Austrian *Seisenegger. Charles was so pleased with Titian's work that he appointed him court painter and elevated him to the rank of Count Palatine and Knight of the Golden Spur—an unprecedented honour for a painter. At the same time his works were increasingly sought after by Italian princes, as with the celebrated *Venus of Urbino* (Uffizi, Florence, *c.*1538), named after its owner, Guidobaldo, Duke of Camerino, who later became Duke of Urbino. The pose is based on Giorgione's *Sleeping Venus* (Gemäldegalerie, Dresden), but Titian substitutes a direct sensual appeal for Giorgione's idyllic remoteness.

Early in the 1540s Titian came under the influence of central and north Italian *Mannerism. In the *Ecce Homo* (Kunsthistorisches Museum, Vienna, 1543), for example, his earlier vigour was once more in evidence, but it was now a vigour expressed in crowds and uncomfortable movement and shimmering colours. Titian was reluctant to travel and in 1545–6 he made his first and only journey to Rome. There he was deeply impressed not only by modern works such as *Michelangelo's *Last Judgement*, but also by the remains of antiquity. His own paintings during this visit aroused much interest, his *Danaë* (Museo di Capodimonte, Naples) being praised for its handling and colour and (according to Vasari) criticized for its inexact drawing by Michelangelo. Titian also painted in Rome the famous portrait of *Pope Paul III and his Nephews* (Museo di Capodimonte). The decade closed with further imperial commissions. In 1548 the emperor summoned Titian to Augsburg, where he painted both a formal equestrian portrait (*Charles V at the Battle of Mühlberg*, Prado) and a more intimate one showing him seated in an armchair (Alte Pinakothek, Munich). He travelled to Augsburg again in 1550 and this time painted a full-length portrait (Prado) of Charles's son, the future Philip II of Spain, who was to be the greatest patron of his later career. Titian's work for Philip included a great series of seven erotic mythological subjects (*c.*1550–*c.*1562): *Danaë* and *Venus and Adonis* (Prado), *Perseus and Andromeda* (Wallace Coll., London), *The Rape of Europa* (Gardner Mus., Boston), *Diana and Actaeon* and *Diana and Callisto* (Ellesmere Coll., on loan to NG of Scotland), and *The Death of Actaeon* (NG, London). Titian referred to these pictures as *poesie*, and they are indeed highly poetic visions of distant worlds, quite different from the sensual realities of his earlier mythological paintings.

Titian ran a busy studio, his assistants including his brother **Francesco Vecellio** (*c.*1490–1559/60), his son Orazio, and his cousin Cesare. Of these only Francesco seems to have had any individual substance as a painter, but his *œuvre* is not well defined. During the last twenty years of his life Titian's personal works, as opposed to those produced under his supervision and with his intervention, showed an increasing looseness in the handling and a sensitive merging of subdued colours, so that outlines disappear and the forms become more immaterial. With this went a growing emphasis on intimate pathos rather than external drama. About the same time his interest in new pictorial conceptions waned, but his powers remained undimmed until the end, his career closing with the awe-inspiring *Pietà (Accademia, Venice), which was intended for his own tomb and finished after his death by *Palma Giovane.

Titian was recognized as a towering genius in his own time (*Lomazzo described him as 'the sun amidst small stars not only among the Italians but all the painters of the world') and his reputation as one of the giants of art has never been seriously questioned. He was supreme in every branch of painting and his achievements were so varied—ranging from the joyous evocation of pagan antiquity in his early mythologies to the depths of tragedy in his late religious paintings—that he has been an inspiration to artists of very different character. *Poussin, *Rubens, and *Velázquez are among the painters who have particularly revered him. In many subjects, above all in portraiture, he set patterns that were followed by generations of artists. His free and expressive brushwork revolutionized the oil technique: Vasari wrote that his late works 'are executed with bold, sweeping strokes, and in patches of colour, with the result that they cannot be viewed from near by, but appear perfect at a distance . . . The method he used is judicious, beautiful, and astonishing, for it makes pictures appear alive and painted with great art, but it conceals the labour that has gone into them.' His greatness as an artist, it appears, was not matched by his character, for he was notoriously avaricious. In spite of his wealth and status, he claimed he was impoverished, and his exaggerations about his age (by which he hoped to pull at the heartstrings of patrons) are one of the sources of confusion about his birthdate. Jacopo *Bassano caricatured him as a moneylender in his *Purification of the Temple* (NG, London). Titian, however, was lavish in his hospitality towards his friends, who included the poet Pietro Aretino and the sculptor and architect Jacopo *Sansovino. These three were so close that they were together known in Venice as the triumvirate, and they used their influence with their respective patrons to further each other's careers.

Tobey, Mark (1890–1976). American painter. In 1918 he became a convert to the Baha'i faith and much of his subsequent work was inspired by an interest in Oriental art and thought. He lived mainly in Seattle, but travelled widely and from 1931 to 1938 was artist-in-residence at Dartington Hall, a progressive school in Devon, England. Following a visit to the Far East in 1934–5, he developed a distinctive style of painting that he called 'white writing', characterized by calligraphic white patterns overlying dimly discerned suggestions of colour beneath. Although he painted representational pictures in this style, he turned increasingly to abstractions. Their *All-over manner anticipated and perhaps influenced Jackson *Pollock, but unlike *Action painters, Tobey believed that 'painting should come through the avenues of meditation rather than the canals of action'. Unusually for an American painter, he was more highly esteemed abroad than in his own country, and he was influential on French *Tachisme in the 1950s. In 1960 he moved to Switzerland and settled in Basle.

Tocqué, Louis (1696–1772). French portrait painter, the pupil and son-in-law of *Nattier, active at the courts of Russia, Sweden, and Denmark as well as in Paris. He was a practitioner of the fashionable mode of allegorical society portraits, but his non-official portraits have a more forthright vigour (*Mme Danger Embroidering*, Louvre, Paris). In 1750 he published a treatise on painting that played an important role in the development of a more naturalistic style of portrait painting.

Toft, Albert. See NEW SCULPTURE.

Tolsá, Manuel (1757–1816). Spanish sculptor and architect, trained in Valencia and active from 1791 in Mexico, where he was director of sculpture at the Academy of S. Carlos (founded 1783) and the central figure in the *Neoclassical reaction against the extravagances of Mexican *Churrigueresque. His best-known works are both in Mexico City: the bronze equestrian statue of Charles IV (1748–1819), the last Spanish king to rule Mexico (Plaza de la Reforma, 1803), and the School of Mines (1797–1813).

Tolstoy, Count Leo (1828–1910). Russian writer. He is celebrated as one of the greatest of all novelists, but much of his later career was devoted to writing on a broad range of moral issues, expressing his radical ethical views. His writings in this vein include the book *What is Art?* (1896), in which he stigmatizes as decadent all art of restricted appeal and argues that art is justifiable only if it communicates universal human emotions of a high moral and religious character. In his theory of art, beauty is incidental and the criterion is effectiveness to humanize and implant broadly religious sentiments. Tolstoy's enormous prestige ensured a wide readership for his book, but his extraordinary value judgements—dismissing as worthless *Michelangelo's *Last Judgement* and Shakespeare's *King Lear*, for example—meant that his ideas were generally considered eccentric.

Tomé, Narciso (*c.*1690–1742). Spanish architect, sculptor, and painter, a leading representative of the *Churrigueresque style. He is first recorded in 1715 working with others of his family on the façade of Valladolid University. In 1721 he was appointed architect to Toledo Cathedral, where he executed the Transparente (completed 1732), a sacramental chapel without walls which he made into an astonishing exercise in *Baroque *illusionism and dramatic lighting, combining the arts of architecture, sculpture, and painting. The Transparente was hailed as a wonder when completed, but to *Neoclassical taste it represented the ultimate in artistic decadence.

Tomlin, Bradley Walker (1899–1953). American painter. He studied in Paris, and after painting *Cubist still lifes with occasional *Surrealist figurative elements, he turned towards *Abstract Expressionism in the 1940s, being introduced to *automatism by *Motherwell. By the end of the 1940s he had evolved an original style of composition utilizing ideograms based upon brush-strokes which were regularized with a certain Classical restraint into quasi-geometrical forms.

Tommaso da Modena (*c.*1325–79). Italian painter, one of the leading artists of his day in northern Italy. His earthy, humane, and naturalistic style is well seen in his series of frescos of famous Dominicans (signed and dated 1352) in the Chapter House of S. Niccolò in Treviso. The saintly figures are shown meditating, writing, and reading (the first dated example of spectacles being worn appear here) and Tommaso shows a remarkable ability to depict intellectual activity. His reputation was such that work was commissioned from him by the emperor Charles IV (1316–78) in Bohemia, and two panels by Tommaso are still in Karlstein Castle, near Prague. It is unlikely that he visited Bohemia but there is some kinship between his work and that of his leading Bohemian contemporary, *Master Theoderic.

tondo. A painting or *relief carving of circular shape: the word is Italian for 'round'.

Tonks, Henry (1862–1937). British painter, draughtsman, and teacher. He was interested in art from childhood, and in 1893 he abandoned his successful medical career to become a teacher at the *Slade School. (However, he worked as a plastic surgeon during the First World War.) Tonks remained at the Slade until 1930 (as Professor from 1918), and became the most renowned and formidable teacher of his generation—'in appearance tall, gaunt, and severe' (*DNB*). Under him the Slade maintained its position as the dominant art school in Britain (although it was now challenged by the *Royal College of Art), and he was a major influence as an upholder of traditional values and an opponent of modern ideas: 'I don't believe I really like any modern development.' He set high standards for his pupils, particularly in draughtsmanship (his own forte) and he got on well with them, in spite of being notorious for his sarcasm and abruptness. Because of his refusal to move with the times he was increasingly looked on as a back number by more progressive artists and students, but he remained a dominant presence in the art world. Tonks's own paintings are mainly figure subjects, often consciously (or self-consciously) poetic in spirit: Sir John *Rothenstein refers to 'the sheer prettiness of much of his art', but his pictures often look rather laborious, partly because of his technique of 'Tonking', which involved using an absorbent material to soak excess oil from the canvas after each day's work and so produce a dry surface for the next session. He was at his best in his slighter satirical works and in conversation pieces such as *Saturday Night in The Vale* (Tate Gallery, London, 1928–9), which shows the novelist and critic George Moore reading aloud to a gathering at Tonks's studio in The Vale, Chelsea. Moore complained that he had been made to look like a 'flabby old cook', whereas Tonks had depicted himself as a young and elegant 'demi-god'.

Tooker, George (1920–). American painter. He studied at the *Art Students League, 1943–4. His teachers there included Reginald *Marsh, and it was from him and Paul *Cadmus (with whom he studied privately) that he acquired his preference for painting in egg *tempera. His technique is scrupulously detailed in the manner of the Old Masters, but his subjects express the spiritual desolation and debilitating uniformity of modern life. The figures in his paintings all look more or less like one another and go through life as if on a conveyer belt, tense and drained of energy. They are physically close to one another, but emotionally distant. *Subway* (Whitney Museum, New York, 1950) is perhaps his most famous work—a terrifying vision of Kafkaesque isolation.

Toorop, Jan (1858–1928). Dutch painter, graphic artist, and designer. He was born in Java (at this time a Dutch colony) and moved to the Netherlands with his family when he was 14. Toorop's work reflected many of the main stylistic currents of his time and he was a leading figure in the *Symbolist and *Art Nouveau movements. His most characteristic paintings are literary subjects depicted with flowing lines, as in his masterpiece, *The Three Brides* (Rijksmuseum Kröller-Müller, Otterlo, 1893), a ghostly scene with an exotic feeling that recalls his origins in Java. In 1905 he was converted to Catholicism and thereafter concentrated mainly on religious works. Toorop's prolific output included book illustrations, designs for stained glass, and posters. His daughter, **Charley Toorop** (1891–1955), trained as a musician but began to paint in about 1914. Her early work was influenced by her father's Symbolism, but later her style became more solid and naturalistic.

Topolski, Feliks (1907–89). Polish-born painter and draughtsman who settled in England in 1935 and became a British subject in 1947. A versatile and prolific artist, he is perhaps best known for his large murals, notably *The Coronation of Elizabeth II* (30 m. × 1.5 m.) in Buckingham Palace. He was an *Official War Artist 1940–5, and his other work includes portraits and book illustrations. His style is marked by the use of vigorous swirling line.

Torel, William (active late 13th cent.). English goldsmith and sculptor. In 1291 he began three life-size gilt-bronze effigies for Edward I (1239–1307): one of Edward's father Henry III (1207–72) and two of Edward's wife, Eleanor of Castile. The one of Queen Eleanor at Lincoln has been lost, but the others survive in Westminster Abbey. Although they were the first large-scale figures to be cast in England, Torel produced works of superb quality.

Torrentius, Johannes (1589–1644). Dutch painter whose name before he translated it into Latin was Jan van der Beeck. He specialized in still lifes and bawdy *genre scenes and was a very different character from the usual workmanlike Dutch painter of the 17th cent. In 1627 he was tried in Haarlem for heresy (the authorities charged him with being the leader of the Rosicrucian sect) and immorality. He was tortured and then sentenced to be burned alive, but the sentence was commuted to 20 years' imprisonment. Thanks to the intervention of Charles I of England (1600–49) he was released in 1630 and permitted to go to England. He returned to his native Amsterdam *c.*1641. Ironically, the only known work by him is an *Allegory of Temperance* (Rijksmuseum, Amsterdam, 1614).

Torres-García, Joaquin (1874–1949). Uruguayan painter and art theorist. In 1891 his family moved to Spain and he spent most of his life there. He lived mainly in Barcelona, where he moved in avant-garde circles that included the young *Picasso. After a visit to New York (1920–2) he settled in Paris (1924–32), where he developed a symbolic, severely geometrical, two-dimensional *Constructivist style, and founded the review **Cercle et Carré* in conjunction with the artist and critic Michel *Seuphor. In 1934 he returned to Uruguay, where he opened an art school. He wrote various works on art theory and by his example and teaching did much to promote Constructivist and *Kinetic art in South America. There is a Torres-García Museum in his native Montevideo.

Torrigiano, Pietro (1472–1528). Florentine sculptor. He was trained under *Bertoldo in the *Medici 'academy' and there broke the nose of his fellow student *Michelangelo and permanently disfigured him. Quitting Florence to escape the wrath of Lorenzo de' Medici, he wandered round Italy, producing no work of importance, and serving as a mercenary for Cesare Borgia. He visited the Netherlands, where he worked for Margaret of Austria in 1509–10, then in about 1511 moved to England, where he executed his most important works, chief among them the tomb of Henry VII and his wife Elizabeth of York (1512–18) in Westminster Abbey. The two sensitive effigies, made with the aid of death masks, have a *Gothic elegance, but the figures of angels at the corners of the tomb and the exquisite decorative work introduced a pure *Renaissance style into England. It had little immediate influence, however. In about 1522 Torrigiano left England for Spain, abandoning his projected tomb for Henry VIII. He was arrested by the Inquisition for heresy and died in prison.

Torriti, Jacopo (active 1290s). Italian mosaicist and painter. Nothing is known of his life, but he signed two mosaics commissioned by Pope Nicholas IV for the apses of S. Giovanni in Laterano (1291), and Sta Maria Maggiore (1295?), pupil of *Bonnat and in 1883 of *Cormon. At the age of 21, in 1885, he was given an allowance and set up in a studio of his own at Montmartre. He also began to draw for illustrated journals. He met van *Gogh at Cormon's school in 1886 and came into contact with *Impressionist and *Post-Impressionist painters. From the age of 15 until he was 24 he painted mainly sporting subjects. From *c.*1888 he began to illustrate the theatres, music-halls (particularly the Moulin Rouge), cafés, and low life of Paris; circuses and brothels were among the subjects he returned to again and again. He collected the etchings of *Goya and his painting came much under the influence of *Degas.

In 1888 he had come into contact with *Gauguin and began a continuing enthusiasm for Japanese colour prints (see UKIYO-E). The influence of Gauguin's flat rhythmic patterning and calligraphic use of strong outline is observable in the painting *Au Cirque Fernando* (Art Inst., Chicago) done in that year. But it became dominant chiefly in Lautrec's *lithographs, including his famous posters. In spite of his notoriously dissipated lifestyle, he was a dedicated craftsman, and would arrive early at the workshop—even after a hard night's drinking—to supervise the printing of his designs. His work, with its masterfully bold and arresting forms, was one of the most important influences in gaining acceptance for both lithography and the poster as major art forms. Lautrec's alcoholism and dissolute life (he had syphilis) led to a breakdown in 1899 and the paintings he produced in the brief period between his recovery and his premature death, aged 36, are more sombre in style than his earlier work. In 1922 the Lautrec family presented some 600 of his works to his native town of Albi, where the Musée Toulouse-Lautrec was created to house them. The museum houses not only works by Lautrec but also memorabilia, notably his walking stick, which ingeniously opens up to reveal a tiny glass and flask of brandy.

Tournier, Nicolas (1590–1638/9). French painter. Little is known of his life and most of the attributions to him are speculative, but he seems to have been—with the exception of Georges de *La Tour—the most individual and sensitive of French *Caravaggesque painters. He was in Rome *c.*1619–26, then is recorded in Carcassonne in 1627 and in Toulouse from 1632. The works attributed to him in his Roman period are *genre scenes of music making, dice-playing, etc. (*A Musical Party*, City Art Mus., St Louis), but after he returned to France he concentrated on religious pictures. There are examples in the Louvre and the Musée des Augustins, Toulouse, including a remarkable *Battle of Constantine* influenced by *Piero della Francesca's battle scenes at Arezzo—this at a time when Piero was virtually forgotten. More typical of Tournier's paintings in the museum at Toulouse are *The Lamentation* and *The Entombment*, which show the grace and refinement of his style.

Tours School. Term applied to products of the *Carolingian scriptorium of the Abbey of S. Martin at Tours. It had its beginnings before the emperor Charlemagne's time; Alcuin (abbot, 796–804) reformed it, and it reached its prime under Lothar (d. 855) and Charles the Bald (d. 877). Two of the most important books from the scriptorium are the *Alcuin Bible* (BL, London) and the Bible made for Charles the Bald (Bib. nat., Paris).

Towne, Francis (1739/40–1816). English watercolour painter, active in London and Exeter. He earned his living mainly as a drawing master and his work was little known to his contemporaries, but he is now regarded as being one of the most individual watercolourists of his period. His method of painting in flat *washes over a brown pen-and-ink outline was employed with a severe economy of means and represents the culmination of the 18th-cent. 'tinted drawing' style of watercolour. He brought new life to the stereotyped *picturesque tradition of landscape and he had an eye for geometrical structure that gives his pictures an almost abstract feel and an affinity with such 20th-cent. work as that of John *Nash. His best works are considered to be those done when he passed through Switzerland in 1781 when returning from a visit to Italy. He also visited Wales and the Lake District. Towne's work is represented in many major collections of watercolours.

tracery. Term used first and foremost to describe decorative stonework set in window openings and by extension to describe similar ornament applied to other architectural features or, for example, to furniture. Tracery is one of the most characteristic features of medieval architecture after *c.*1200 and its variations or absence have been used to classify different stages of *Gothic architecture.

Two basic kinds of tracery are distinguished. The earlier form was 'plate' tracery, in which solid areas of flat stone were perforated with decorative openings; the later (and much longer-lived) form was 'bar' tracery, in which thin strips of stone (usually developing the lines of the mullions) were arranged into a meshwork or other pattern. The intricate designs of the latter provided Gothic masons with rich opportunities for displaying their skill as craftsmen, and the tendency for tracery to become more and more intricate, and to be applied more and more extensively not only to windows but to walls and voids as well, persisted down to the end of the Middle Ages in almost every part of Europe except Italy. The earliest known use of the word was made by Sir Christopher Wren (1632–1723); the medieval term was 'forms' or 'form pieces'.

Traini, Francesco (documented 1321–63?). Pisan painter. Only one work is certainly known to be by him—the signed *polyptych of *St Dominic and Scenes from his Life* (Mus. Naz., Pisa, 1345)—but a panel of *St Thomas Aquinas in Glory* (Sta Caterina, Pisa), showing a similarly strong degree of Sienese influence, is also very likely by him. The most important works attributed to him are frescos in the Campo Santo in Pisa—the celebrated *Triumph of*

Death, with accompanying scenes of the *Last Judgement*, *Hell*, and *Legends of the Hermits*. These, among the outstanding Italian paintings of the 14th cent., were badly damaged by bombs in the Second World War, but this brought to light, by way of partial compensation, the beautiful *sinopie, which are now shown in the Museo delle Sinopie. The frescos include many telling details of death's victims and are usually seen as a reflection of the horrors of the Black Death of 1348, but some authorities consider them earlier, and recently there have been attempts to attribute them to the mysterious *Buffalmacco.

Transitional style. Term that is (somewhat confusingly) applied with two distinct meanings to certain aspects of European art in the late 12th cent. and early 13th cent. In architectural writing it is applied to buildings that manifest distinct features of both the *Romanesque and *Gothic styles—in its crudest terms, those that use both round and pointed arches. In the study of the figurative arts, the term is applied to certain works that are distinguished from both Romanesque and Gothic by their strongly *antique flavour and fluent forms: *Nicholas of Verdun's Klosterneuberg altar is a prime example.

Travertine. A *limestone of almost pure calcium carbonate found at Tivoli (Latin *Tibur*, hence *lapis Tiburtinus*—Tiburtine stone) and elsewhere in Italy. Varying in colour from pale buff to orange pink, it is sometimes porous. It has been much used in the buildings of Rome (notable examples are the Colosseum and the colonnade of St Peter's) and has also been employed for sculpture in outdoor works that do not require a smooth finish, as for example by *Bernini in his *Triton* Fountain in Rome (1642–3).

trecento. See QUATTROCENTO.

Tretyakov, Pavel (1832–98). Russian businessman and art collector. From an early age he conceived the idea of a national gallery of Russian art, and he not only bought great numbers of paintings, but also commissioned contemporary artists to produce work for him (he gave much encouragement and assistance to the *Wanderers). In 1892 he and his brother **Sergei**, also a collector, presented their collections to the city of Moscow, and the building in which they are housed, the façade of which was designed by the painter Victor Vasnetsov (1848–1926), was completed in 1902. The Tretyakov Gallery became the property of the State in 1918. It originally displayed foreign as well as Russian paintings, but since 1925 it has been devoted exclusively to Russian art, of which it has the finest collection in the world. In the 1980s and 1990s a major rebuilding and expansion of the gallery was carried out, greatly enlarging the exhibition space. As well as being a museum, it is an important scientific study centre for works of art.

Trevisani, Francesco (1656–1746). Neapolitan painter, trained in Venice and working in Rome from 1678. He specialized in altarpieces and *cabinet pictures and was one of the most popular and prolific Italian painters of his day, his sweet and colourful *Rococo style challenging the monumental classicism of *Maratta. His work was exported all over Europe and his studio was much frequented by French painters.

triptych. 'A picture or carving (or set of three such) in three compartments side by side, the lateral ones being usually subordinate, and hinged so as to fold over the central one' (*OED*). Like the *diptych, this form lent itself to the making of portable religious images, whose delicate inner surfaces were protected by the outer leaves, which might, however, also be painted or carved on their outer as well as their own inner surfaces. See also POLYPTYCH.

Tristán, Luis (*c*.1586–1624). Spanish painter, active mainly in Toledo, where he was a pupil of El *Greco from 1603 to 1606. He is also known to have visited Italy, probably between 1606 and 1613. His style (notably his characteristically elongated proportions) owes much to El Greco, but Tristán is more sober, marking the transition from *Mannerism to a more naturalistic approach (*The Adoration of the Shepherds*, Fitzwilliam, Cambridge, 1620).

trompe-l'œil. Term (French for 'deceives the eye') applied to a painting (or detail of one) that is intended to deceive the spectator (if only briefly) into thinking that it is a real object rather than a two-dimensional representation of it. Such virtuoso displays of skill often have a humorous intent, and anecdotes of almost miraculous feats of *trompe-l'œil* by which some of the most famous painters of the past are said to have tricked and astonished their contemporaries are typical of periods in which *naturalism has been cultivated, such as the *Classical age in Greece (see PARRHASIUS) and the Italian *Renaissance. *Vasari, for example, records in evidence of *Giotto's precocity that as a boy he painted on the nose of a figure on which his master *Cimabue was engaged 'a fly so lifelike that when Cimabue returned to carry on with his work he tried several times to brush it off with his hand, under the impression that it was real, before he realized his mistake'. Petrus *Christus's *Portrait of a Carthusian* (Met. Mus.,

New York, *c*.1446) is an example of a painting with such a *trompe-l'œil* fly, here on a painted ledge at the bottom of the picture, rather than on the sitter's nose. Sometimes the term *trompe-l'œil* is used loosely to refer to any type of pictorial illusionism (for example *quadratura), but such usage deprives a useful term of its precision.

Troost, Cornelis (1697–1750). The outstanding Dutch painter of the 18th cent. He made his name early with a lively group portrait of the *Amsterdam Inspectors of the Collegium Medicum* (Rijksmuseum, Amsterdam, 1724) and continued to enjoy success as a formal portraitist and painter of *conversation pieces. He is best known for his pictures of actors in famous roles and for his witty *genre scenes. His most famous work—made in his favourite technique of pastel and watercolour—is a series of five pictures entitled *NELRI* (Mauritshuis, The Hague, 1740); the name is derived from the first letters of the Latin inscriptions which accompany the five views of the activities of a group of men during the night of a reunion. In style and subject-matter Troost has much in common with his contemporary *Hogarth, but unlike Hogarth, Troost is a humorist without an ulterior motive; he never attempted to teach or preach.

Troostwijck, Wouter Johannes van (1782–1810). Dutch painter, mainly of landscapes. He is regarded as the best Dutch landscapist of his period, but his early death cut short his promising career. His freshness of vision is seen in *The Rampoortje* (Rijksmuseum, Amsterdam, 1809), a view of an Amsterdam gate in winter.

trophy. A carved, painted, or engraved representation of a group of arms and armour (and usually banners), a common decorative motif on, for example, tomb sculpture and buildings. Trophies derive from the ancient practice of displaying the actual weapons of defeated enemies as spoils of war and symbols of conquest.

Troy, Jean-François de (1679–1752). French painter and tapestry designer. His successful career was based initially on large historical and allegorical compositions (*Time Unveiling Truth*, NG, London, 1733), but he is now most highly regarded for his smaller and more spirited scenes of elegant social life. They are among the best of those that rode on the wave of *Watteau's success—indeed *The Alarm* (V&A, London, 1723) was attributed to Watteau in the 19th cent. In 1738 he was appointed Director of the French Academy in Rome, and spent the rest of his life there. He was one of a family of painters, his father and teacher, **François de Troy** (1645–1730), being a successful painter of fashionable portraits and Director of the Academy in Paris.

Troyon, Constant (1810–65). French painter. He was born in Sèvres and began his career as a porcelain decorator (the family trade) at the famous factory there. In the 1830s, however, he took up landscape painting, and became one of the leading figures of the *Barbizon School. He visited Holland in 1847, and the influence of *Cuyp and *Potter is seen in his predilection for including cows in his pictures. Late in his career he turned to seascapes, the freshness and freedom of his work influencing *Boudin and *Monet in this genre. Troyon was a prolific and successful painter, and his work is in many public collections in France and elsewhere.

Trubetskoy, Prince Paulo or **Pavel** (1866–1938). Russian-Italian sculptor. Trubetskoy grew up in Italy, lived in Russia from 1897 to 1906, and afterwards in Paris, the United States, and Italy. He was self-taught and formed his style under the influence of *Rodin. His best works are *Impressionistic portraits and statuettes of animals from the early years of the century. Such works won him a great reputation at the Paris International Exhibition of 1900, and after a period of total neglect are again becoming fashionable.

Trübner, Wilhelm. See SEZESSION.

Trumbull, John (1756–1843). American painter. Trumbull fought in the American War of Independence (for a time he was aide-de-camp to George Washington) and his career was devoted mainly to depicting the outstanding events and personalities of the Revolution. He was strongly influenced by Benjamin *West, with whom he studied in London (he made several visits there, and during the first of them, 1780–3, was imprisoned in reprisal for the hanging of a British agent in America). In 1817 he became President of the American Academy of Fine Arts in New York, but his tyrannical attitude, especially towards young painters, led to many members leaving to set up the *National Academy of Design in 1825. His pictures did not sell well, so in 1831 he assigned those in his studio to Yale University Art Gallery in exchange for an annuity. The most famous of his works there, and one of the most reproduced images in American art, is *The Declaration of Independence* (1786–97), in which most of the portraits were painted from life. His larger works are usually fairly stodgy, but his smaller pictures and sketches can be much livelier. Trumbull, who died an embittered old man, published an autobiography in 1841.

tube. Collapsible container for paints. It was devised in 1841 by John G. Rand (1801–73), an American portrait painter working in London, and was commercially manufactured very soon afterwards. Before this time, artists either mixed their own paints or bought them in rather cumbersome bladders or in metal cylinders that were emptied by means of a piston and could be refilled at the colourman's shop. With the success of the tube, the preparation of paints now passed from the studio to the factory and the painter lost part of his character as craftsman. Moreover, a less fluid consistency had to be given to *oil paints to make them suitable for packing in tubes, and other ingredients had to be added during manufacture in order to ensure that the *pigments would stay suspended in the oil. Consequently manufactured paints do not flow from the brush as those of the Old Masters used to do and a stiffer, shorter brush is needed for handling them or a *diluent must be used. The tube, therefore, has brought about a quiet revolution in painting technique. Furthermore the availability of ready-made paints in tubes made the practice of painting out of doors in oils much easier (see PLEIN AIR) and also encouraged artists to work with a much wider range of colours instead of mixing the smaller number with whose tempering they were familiar.

Tuby, Jean-Baptiste (1635–1700). French sculptor of Italian origin. He worked chiefly at Versailles, often in collaboration with *Girardon. Much of his work there was garden sculpture, including a copy of the *Laocoön* group.

Tucker, William (1935–). British abstract sculptor. He was among the pupils of Anthony *Caro who created a new British sculptural avant-garde in the latter half of the 1960s (see NEW GENERATION). His work is in various materials and often makes striking use of colour. In 1974 he published *The Language of Sculpture* and in 1975 he organized a major Arts Council exhibition at the Wayward Gallery, London, entitled 'The Condition of Sculpture'.

tufa. Soft porous rock formed from the deposits of springs rich in lime. It is easily sawn and worked, but durable, hardening on exposure to air, and has been used since ancient times (notably by the Romans) for buildings and sometimes for sculpture.

Tuke, Henry Scott (1858–1929). British painter. He studied at the *Slade School, 1875–80, then in Italy and Paris, where he was strongly influenced by contemporary French **plein-air* painting. Tuke had known and loved Cornwall since childhood and after he returned to England in 1883 he settled there, living first at Newlyn (see NEWLYN SCHOOL) and then from 1885 in a cottage near Falmouth. His favourite subject—which he made his own—was nude boys in a sunlit atmosphere against a background of sea or shore. At first the freshness of these works—so different to the frigid studio nudes to which the public was accustomed—caused prudish objections (Tuke was a founder member of the *New English Art Club in 1886 and the sight of one of his paintings caused the dealer Martin Colhaghi to withdraw his financial backing for the group's first exhibition). However, they soon became favourites with the public and are now regarded as being among the finest and most individual works of English *Impressionism. Tuke also painted portraits throughout his life.

Tura, Cosmè (or **Cosimo**) (*c.*1430–95). Italian painter, the first major artist of the School of Ferrara, where he was appointed court painter to the *Estes in 1452. His sculptural figure style was derived in the first place from *Mantegna, though its tortuous, metallic quality was a product of Tura's own feverish imagination. He also acquired a feeling for monumentality from *Piero della Francesca, who was painting in Ferrara *c.*1449. Tura was mainly a religious painter, his work including two huge shutters (1469) for the organ of Ferrara Cathedral, now in the Museo del Duomo; they represent *The Annunciation* and *St George and the Princess*. Good examples of his work on a smaller scale are in the National Gallery, London. Tura was an important influence on the other two major painters of the 15th cent. Ferrarese School—*Cossa and *Roberti. The latter replaced him as court painter in 1486 and Tura died poor.

Turnbull, William (1922–). British sculptor and painter. After serving as an RAF pilot in the Second World War he studied at the *Slade School, 1946–8, then in 1948–50 lived in Paris, where he saw a good deal of his fellow Scot and Slade student *Paolozzi and met such illustrious figures as *Brancusi and *Giacometti. As a sculptor Turnbull moved from *Surrealistic and primitivist works to painted steel geometrical constructions in the manner of *Caro. As a painter he was one of the first British artists to work in the manner of the American *Colour Field painters.

Turner, J. M. W. (Joseph Mallord William) (1775–1851). English painter, one of the greatest and most original of all landscape painters. His family called him Bill or William, but he is now invariably known by his initials. Precociously gifted, he became a student at the *Royal Academy Schools in 1789 and first exhibited a watercolour at the Academy in 1790, when he was only 15. He studied at the

Academy for four years, and during this time also had lessons from Thomas Malton (1748–1804), a topographical watercolourist who specialized in neat and detailed town views. From 1792 Turner began making regular sketching tours, producing drawings of *Picturesque views and architectural subjects that he later sold to engravers or worked up into watercolours. At this time his work was more polished but less inventive than that of his friend *Girtin (with whom he worked for Dr *Monro). Initially he painted only in watercolour, but in 1796 he first exhibited an oil at the Academy, *Fishermen at Sea* (Tate, London), a work showing his admiration for 17th-cent. Dutch marine painting. Only three years later, in 1799, he was elected an Associate of the Royal Academy at the youngest permitted age (24), and in 1802 he became the youngest ever full Academician. His career flourished in terms of money as well as prestige, for he was hardworking, a good businessman, and frugal by nature (he lived rather squalidly, but he was not miserly or ungenerous, as is sometimes maintained).

The Dutch influence in Turner's work soon gave way to that of *Claude and *Wilson, but already in the early 1800s it was recognized that he was introducing a new and revolutionary approach to landscape, his painting becoming increasingly *Romantic in its dramatic subject-matter and sense of movement, as in the powerful *Shipwreck* (Tate, 1805). During these years, however, he continued exhibiting pictures in a more conventional manner and was still working for engravers (his most ambitious engraving project was his *Liber Studiorum*, conceived in emulation of Claude's *Liber Veritatis* and intended to show the range of his own work; between 1807 and 1819 he issued 71 of a projected 100 plates).

Turner made his first journey to the Continent in 1802, during a temporary peace in the war with France, visiting Paris like so many other artists to see pictures looted by Napoleon, which were then on exhibition. From Paris he travelled on to Switzerland. The resumption of war made Continental travel impossible for more than a decade, and Turner did not go abroad again until 1817, when he visited Belgium, Holland, and the Rhine. He first visited Italy two years later, and from then until 1845 made fairly regular journeys abroad (including three more to Italy, the last in 1840). Unlike his contemporary *Constable, who concentrated on painting the places he knew best, Turner was inspired to a great extent by what he saw on his travels (he lived in London all his life, but the city appears fairly infrequently in his paintings). The mountains and lakes of Switzerland and the haunting beauty of Venice, in particular, provided him with an enduring fund of subjects. On his journeys he was in the habit of making rapid pencil jottings, which he used later as reminders for imaginative compositions. He was inspired by history (especially ancient history) and literature as well as nature. Many of the paintings he exhibited at the Royal Academy were accompanied by verses printed in the catalogue, and from 1800 he added lines he had composed himself.

From the 1830s Turner's painting became increasingly free, with detail subordinated to general effects of colour and light. His work was often attacked by critics, one of his most celebrated pictures—*Snow Storm: Steam-Boat off a Harbour's Mouth* (Tate, 1842)—being dismissed as 'soapsuds and whitewash'. However, he also had many admirers, including some who regarded him as the outstanding genius of the day. His most important patron was the third Earl of Egremont (1751–1837), who was unusual among collectors of the time in buying contemporary British art (sculpture as well as painting) on a large scale. Turner had a studio at Petworth, Egremont's country house in Sussex, and several of his paintings are still to be seen there (although their ownership was transferred to the Tate Gallery in 1984). Turner's other great champion was the young *Ruskin, who discovered his work in the 1830s and wrote eloquently of him in the first volume of *Modern Painters*, published in 1843. By this time Turner's brushwork had become breathtakingly free and some of his compositions were almost abstract, the forms dissolved in a haze of light and colour: 'He seems to paint with tinted steam, so evanescent and so airy', wrote Constable. Turner's originality lay not only in such handling of colour and light—in which he anticipated *Impressionism—but also in his use of the power, beauty, and mystery of nature to express deep human concerns. For example, *The Fighting Temeraire* (NG, London, 1839), showing a ship that had fought at Trafalgar being towed to the breaker's yard, is a poignant elegy for a passing era.

Turner always led a fairly solitary existence and late in life he became more and more of a recluse, sometimes calling himself Mr Booth (assuming the name of his mistress Sophia Booth). After his death, Ruskin destroyed many erotic drawings that he found among his works, thinking that they tainted his hero's memory. In his will Turner left plans for disposing of his considerable fortune (he wanted to found an almshouse for 'poor and decayed male artists') and for the creation of a special gallery at the National Gallery to display certain of his paintings (he had a huge stock of his work, including not only pictures that had never been sold, but also favourite paintings that he had bought back at auction). Long-forgotten relatives contested the will (which was ambiguously worded) and won the money in 1856; at the same time the Court of Chancery awarded all the works remaining in his possession at his death to the National Gallery—about 300 oils and 19,000 drawings and watercolours. Most of these are now in Clore Gallery at the Tate Gallery, but a few of Turner's most famous oil paintings remain in the National Gallery.

Turner Prize. An annual prize of £20,000 for British achievement in the visual arts, named after J. M. W. *Turner. It was established in 1984 by the Patrons of New Art, a body founded two years earlier (as part of the Friends of the Tate Gallery) to encourage the collection of contemporary art. The regulations have changed somewhat since the prize was inaugurated. Originally it was awarded for 'the greatest contribution to art in Britain in the previous twelve months' and was open to critics and administrators (who were shortlisted but never won) as well as artists; since 1991 those eligible are British artists under the age of 50 who have had 'an outstanding exhibition or other presentation of their work' in the previous twelve months. The original sponsors, Drexel Burnham Lambert, suffered a financial collapse in 1990 and the prize was suspended that year; since then it has been sponsored by Channel 4 Television, which broadcasts the award ceremony live from the Tate Gallery. The Director of the Tate is on the jury that makes the award. Like the Booker Prize in literature, the Turner Prize attracts a great deal of publicity, but much of this attention has been expressed as damning criticism, as it is regarded by many as showcasing all that is most pretentious and self-regarding in contemporary art. The winners of the Prize have been: 1984, Malcolm *Morley; 1985, Howard *Hodgkin; 1986, *Gilbert & George; 1987, Richard Deacon (1949–); 1988, Tony Cragg (1949–); 1989, Richard *Long; 1990, prize suspended; 1991, Anish Kapoor (1954–); 1992, Grenville Davey (1961–); 1993, Rachel Whiteread (1963–); 1994, Antony Gormley (1950–); 1995, Damien Hirst (1965–); 1996, Douglas Gordon (1967–).

turpentine. A *resinous liquid exuded from various species of pine, which in its distilled form ('spirits of turpentine') is the usual *diluent, or thinner, for *oil paint. There is no definite evidence that it was used before the 16th cent. Some painters in the 20th cent. use petrol instead.

Twachtman, J. H. See TEN, THE.

Twombly, Cy (1929–). American abstract painter and draughtsman. His distinctive style is characterized by apparently random scrawls and scribbles on white or black grounds. In his rejection of traditional ideas of composition he shows an affinity with the *All-over style initiated by Jackson *Pollock, though Twombly's work is looser and more disorganized than Pollock's. In 1957 he settled in Rome.

Two nine one (291) Gallery. See STIEGLITZ, ALFRED.

Tworkov, Jack (1900–82). Polish-born American painter. He went to the USA in 1913 and during the Depression he worked for the *Federal Art Project. In 1934 he met Willem *de Kooning and in 1948–53 they worked in adjacent studios in New York. Under de Kooning's influence Tworkov abandoned his figurative style and turned to *Abstract Expressionism, but around 1960 his style changed again and he moved to more geometrical designs. His paintings of the 1970s, with their closely *hatched brush-strokes within a geometrical framework, represent an individual departure from the full style of *Colour Field painting of the 1960s. In 1974 he wrote that he aimed at 'a painting style in which planning does not exclude intuitive and sometimes random play'.

tympanum (Greek: 'drum'). Architectural term for the space enclosed between the lintel of a doorway and an arch over it. Tympana are found over windows or within decorative arcades on walls, and the word can also be applied to the triangle enclosed by a Classical pediment. The tympanum has no structural function and is often richly carved; some of the chief glories of *Romanesque sculpture, for example, are tympanum decorations (see, for example, GISELBERTUS).

typology. In Christian *iconography a system whereby figures and scenes from the Old Testament were thought of as prefiguring those of the New Testament. It became the custom in biblical illustration for Old Testament 'types' to be juxtaposed with or subordinated to New Testament 'antitypes', to demonstrate visually that the promise of the Old Testament was fulfilled in the New. Thus Abraham's willingness to sacrifice his son Isaac was seen as foreshadowing God's sacrifice of Christ, and David was seen as a 'type' of Christ, his fight with Goliath prefiguring Christ's struggle with Satan. Occasionally classical myths or other secular sources were also accepted as types. Typological illustration is found not only in book illustration, but in virtually every field of art throughout the Middle Ages (the earliest examples date from the 3rd cent.), and occasionally afterwards. It reached a wide popular audience in *block books such as the **Biblia Pauperum* and the **Speculum Humanae Salvationis*.

Tytgat, Edgar (Edgard) (1879–1957). Belgian painter and graphic artist whose works were an *Expressionist version of Flemish popular art. His *woodcuts especially show a sophisticated and self-conscious naïvety. He worked in London during the First World War as a book illustrator and was also a sculptor, lithographer, etcher, and designer.

Tzara, Tristan. See DADA.

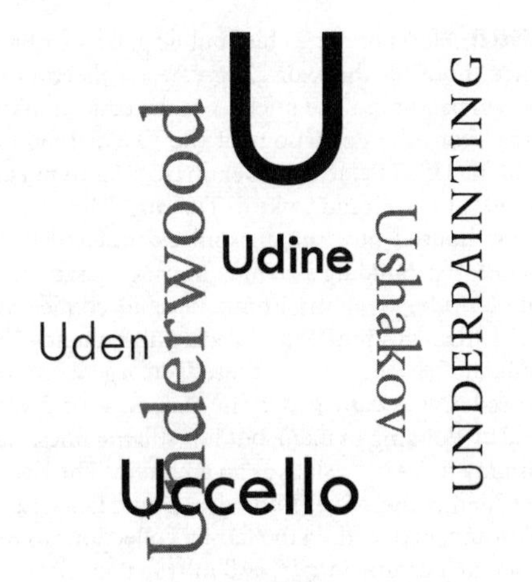

Uccello, Paolo (Paolo di Dono) (*c*.1397–1475). Florentine painter, one of the most distinctive artists of the early *Renaissance. *Vasari says he got his nickname (*uccello* means 'bird') because he loved animals, and birds in particular, and to his contemporaries, as well as to many later critics, he appeared an eccentric figure. He is first documented *c*.1412 in the workshop of *Ghiberti, but he is not known to have worked as a sculptor. In 1425 he moved to Venice, where he worked as a mosaicist, but nothing survives there that can be certainly associated with him. By 1432 he was back in Florence, and in 1436 he painted his first dated surviving work—a huge fresco in Florence Cathedral depicting an equestrian statue, a monument to the English *condottiere* Sir John Hawkwood (d. 1394). It demonstrated the fascination with *perspective that was central to his style. His two other large-scale works are a series of frescos on Old Testament themes (probably 1430s and 1440s) in the 'Green Cloister' of S. Maria Novella and a series of three panels (*c*.1455) on the *Battle of San Romano*, a minor Florentine victory against the Sienese in 1432. The pictures were painted for the Palazzo *Medici and are now separated, with one panel each in the National Gallery, London, the Louvre, Paris, and the Uffizi, Florence. Uccello's other works include the decoration of the clock-face and designs for stained-glass windows in Florence Cathedral, and two enchanting paintings that are generally considered to date from late in his career—*St George and the Dragon* (NG, London), one of the earliest known Italian paintings on canvas, and *The Hunt in the Forest* (Ashmolean, Oxford). Vasari says that 'he came to live a hermit's life, hardly knowing anyone and shut away out of sight in his house for weeks and months at a time', and in his tax return of 1469 Uccello described himself as 'old, without means of livelihood . . . and unable to work'.

Uccello's work presents a striking—and often captivating—combination of two seemingly opposing stylistic currents: the decorative tradition of *International Gothic and the scientific involvement with perspective of the early Renaissance. Vasari maintained that Uccello wasted his time 'on the finer points of perspective' and presents him as an amiable fanatic who worked into the night and when told to come to bed by his wife would reply: 'What a sweet mistress is this perspective!' He undoubtedly took his enthusiasm to extraordinary lengths (in the *Battle of San Romano* the broken weapons and even the corpses recede neatly in accordance with the perspective scheme), but his effects were appropriate to his subjects and to the decorative charm of his pictures rather than mere technical exercises. In *The Hunt in the Forest*, for example, he creates not only an atmosphere of fairy-tale romance, but also, through the way in which the horses and dogs move swiftly back into space, an exhilarating sense of darting energy. Uccello's name became so identified with the subject of perspective that he was often said to have invented it: *Ruskin, for example, wrote in a letter to Kate *Greenaway 'I believe the perfection of perspective is only recent. It was first applied in Italian art by Paul Uccello. He went off his head with love of perspective.'

Uden, Lucas van (1595–1672). Flemish landscape painter and engraver, active mainly in his native Antwerp. Although there is no firm evidence for the tradition that he worked in *Rubens's studio and painted landscape backgrounds for him, he was certainly strongly influenced by the master. His pictures are often large and have something of Rubens's sweep and richness. The figures were often added by other artists—David *Teniers II, for example, in *Peasants Merrymaking before a Country House* (NG, London, *c*.1650).

Udine, Giovanni da (1487–1561/4). Italian painter, decorative artist, and architect, born at Udine near Venice. He was one of *Raphael's leading assistants and is chiefly important for reviving antique techniques of *stucco and the ancient taste for *grotesques, inspired by archaeological discoveries in Rome. His light and graceful style, seen best in the Vatican Loggie (1517–19), was imitated all over Europe, particularly by *Neoclassical designers. From the 1530s he was based in Udine, but also worked in Florence, Venice, and Rome again. In 1552 he was made responsible for all public architectural undertakings in Udine.

Uffizi, Florence. The chief public gallery of Florence, housing the world's greatest collection of Italian paintings. The nucleus of the collection derives from the collections of the *Medici family, and the Uffizi Palace was begun by *Vasari in 1560 for Cosimo I, Grand Duke of Tuscany. The ground floor housed government offices (Italian *uffizi*), hence the building's name. In 1565 Vasari built the corridor over the Ponte Vecchio connecting the Uffizi with the *Pitti Palace. Subsequently the building has been much altered, enlarged, and restored (it was damaged in the Second World War and by flooding in 1966), but it is still the finest testimony to Vasari's skills as an architect. The last of the Medici line, Maria Ludovica, Grand Duchess of Tuscany, bequeathed the family collections to the State of Tuscany in 1737, and in 1789 they were reorganized to allow regular public visiting. In the 19th cent. the Uffizi was subjected to radical reorganization. Much archaeological material was placed in the Archaeological Museum in the Palazzo della Crocetta, while the medieval and *Renaissance sculpture and the rich collection of *applied art was transferred to the *Bargello. The Uffizi collection on the other hand was enriched by early Italian paintings resulting from suppressions of churches and monasteries and confiscations of religious property. In 1919 the Gallery of the Academy of Fine Arts became a subsidiary of the Uffizi. Although it is primarily famous for its incomparable representation of Florentine Renaissance painting, the Uffizi also has outstanding works from other Italian and non-Italian schools (for example, Hugo van der *Goes's Portinari Altarpiece) and important examples of *antique sculpture. The collection of prints and drawings in the Gabinetto dei Disegni e Stampe is one of the finest in the world, and the gallery of artists' self-portraits, begun by Cardinal Leopoldo de' Medici in the 17th cent., is unrivalled.

Uglow, Euan (1932–). British painter. He has become well known for paintings of carefully composed human figures in which the naturalistic tradition stemming from the *Euston Road School is combined with geometrical precision of composition. His work also includes portraits and still lifes.

Ugo da Carpi (d. 1532). Italian painter and wood engraver, a pioneer of the *chiaroscuro woodcut. In 1516 he requested from the Venetian senate a patent for his method 'of making from woodcuts prints that seem as though painted', and although there are German examples earlier than any known by Ugo, he may have discovered the technique independently. Certainly his prints achieve their pictorial effect better than those of the Germans. They are often based on designs by *Raphael and *Parmigianino.

Ugolino di Nerio (active 1317–27). Sienese painter, a close follower of *Duccio. His only certain work is the high altarpiece for Sta Croce, Florence, which formerly bore the signature 'Ugolino da Siena' and is now widely dispersed. Several panels are in the National Gallery, London. *Vasari says he painted 'many pictures and chapels in all parts of Italy' and several paintings are attributed to him on stylistic grounds. His style was charming, but must have been something of an anachronism in the Florence of *Giotto's day.

Uhde, Fritz von (1848–1911). German painter. Trained in academic history painting, he later became a convert to *Impressionism (he visited Paris in 1879–80) and was encouraged by *Liebermann. He painted *genre scenes and portraits in an attractively sentimental style, but he is best known for his New Testament scenes set in a contemporary environment (*Sermon on the Mount*, Museum of Fine Arts, Budapest, 1887)—works that caused great controversy. He was a founder member of the Munich *Sezession (1892).

Uhde, Wilhelm (1874–1947). German collector and writer on art. After studying at Munich and Florence, he settled in Paris in 1904. In 1905 he was buying pictures by *Picasso and *Braque at a time when these artists were practically unknown and in 1906 he was one of the first to discover the Douanier *Rousseau. He published the first monograph on Rousseau in 1910 and in 1912 organized the first retrospective exhibition of his works. Subsequently he was best known for 'discovering' and encouraging other *naïve artists, such as *Bombois and *Séraphine. His collection was seized by the French in 1914 as he was considered an enemy alien, but he was honoured after his death when a room in the Musée d'Art Moderne in Paris was devoted to the works of the '20th-cent. primitives' and named after him.

Ukiyo-e. Japanese term, meaning 'pictures of the floating world', applied to the dominant movement in Japanese art of the 17th to 19th cents. It refers to the subjects from transient everyday life, with its ever-shifting fashions, favoured by printmakers at this time, including such celebrated artists as Ando Hiroshige (1797–1858), Katsushika Hokusai (1760–1849), and Kitagawa Utamaro (1753–1806). Favourite subjects were theatre scenes, actors in well-known roles, and prostitutes and bath-house girls. Japanese prints first came to Europe in the 19th cent., and they had a great influence on avant-garde French artists, to whom their flat decorative colour and expressive pattern came as a revelation.

underpainting. The preliminary blocking out of a composition on the painting surface in which the main shapes and tones of the picture are established before being built up by layers of *glazes, *scumbles, or solid paint. Until the late 19th cent., when *alla prima painting became general, most paintings were built up in a series of layers and the underpainting was often in *grisaille. In *tempera painting, after the design had been drawn on the *ground, the system of light and shade was indicated by a *wash of *terre verte* (green earth colour), over which the other colours were laid, light at first, to be strengthened and enriched by subsequent layers if required. This green underpainting is often to be seen in 14th-cent. paintings, especially in the shadows of the flesh.

In *oil painting different neutral colours were used for the monochrome underpainting, which might be painted over a coloured *imprimatura. In the 15th and 16th cents. it was a common practice to use tempera for the underpainting of works in oil. In the 19th cent. the main design of the painting was usually put in a colour scheme similar to that planned for the finished work, though each colour might be more subdued and extreme darks and lights were avoided.

Underwood, Leon (1890–1975). British sculptor, painter, graphic artist, and writer. A versatile and original figure, Underwood was out of sympathy with the main trends of modernism, describing abstraction as 'artfully making emptiness less conspicuous'. Nevertheless, from the 1960s critics began to speak of him as the 'father of modern sculpture in Britain', in view of the streamlined stylized forms of his stone carvings and bronzes in the 1920s and 1930s and the influence of his teaching. He taught at the *Royal College of Art (where Henry *Moore was among his pupils) from 1920 to 1923, resigning after an argument with William *Rothenstein, and at his own Brook Green School in Hammersmith, which he opened in 1921. Underwood travelled widely and wrote several books, notably on African art.

Unit One. A group of avant-garde British artists formed in 1933, its members including Barbara *Hepworth, Henry *Moore, Paul *Nash, and Ben *Nicholson. The group held one exhibition, in 1933, and published a book, *Unit One: The Modern Movement in English Architecture, Painting and Sculpture*, edited by Herbert *Read, in 1934. In the introduction to this volume (which was originally intended as the first of a series), Read explained that the name was chosen because 'though as persons, each artist is a *unit*, in the social structure they must, to the extent of their common interests be *one*'. As stated by the editor in his Introduction, the group 'arose almost spontaneously among a few artists well-known to each other, out of a consciousness of their mutual sympathies and common necessities'. They had no common doctrine or programme and the group was breaking up by 1935. Despite its short life, however, it made a considerable impact on British art of the 1930s, helping to promote abstraction and *Surrealism.

Ushakov, Semen (1626–86). Russian *icon painter active at Moscow. His talent was modest, but he is noteworthy as the first Russian artist to show the influence of Western painting, his *naturalism being decried by traditionalists. He also designed engravings for book illustration. There are examples of his work in the Tretyakov Gallery, Moscow.

Utamaro, Kitagawa. See UKIYO-E.

Utrillo, Maurice (1883–1955). French painter. The illegitimate son of Suzanne *Valadon, he was given the name by which he is known by the Spanish art critic Miguel Utrillo (1863–1934), who recognized him as his son in order to help him (*Puvis de Chavannes was said by some to have been his real father). He began to paint in 1902 under pressure from his mother, who hoped that it would remedy the alcoholism to which he had been a victim since his boyhood (in 1934 the Tate Gallery wrongly said that he had died of drink in that year, thus bringing on the Trustees' heads a libel suit, settled out of court). Valadon gave him his first lessons, but he was largely self-taught. He first exhibited at the *Salon d'Automne in 1909. In the 1920s he became prosperous and critically acclaimed, and in his later years he devoted as much time to religious devotions as to painting.

Utrillo was highly prolific, painting mainly street scenes in Montmartre. The period from about 1910 to 1916 is known as his 'white period' because of the predominance of milky tones in his pictures, and it is generally agreed that he did his best paintings during this time. They subtly convey solitude and emotional emptiness and have a delicate feeling for tone and atmosphere, even though he often worked from postcards. His later work is livelier but less touching. Utrillo's work is in many museums and (no doubt because of the deceptive simplicity of his paintings) he is among the most forged of modern artists.

Uytewael, Joachim. See WTEWAEL.

Vaenius. See VEEN, OTTO VAN.

Vaillant, Wallerant (1623–77). Franco-Netherlandish painter and engraver. He was born in Lille (then part of Flanders) and worked in Amsterdam, Paris, and various other places (*Vertue says he visited England). Most of his works were portraits, often in chalk, but he also did still lifes and *genre scenes (*A Boy Seated Drawing*, versions in the Louvre, Paris, and NG, London). He was among the first to make *mezzotints, of which more than 200 are attributed to him, for the most part reproductions of the work of other artists. He may have learned the secret of the new art from Prince *Rupert, whom he met at Frankfurt in 1658. Vaillant had several painter brothers and half-brothers: **Jacques** (*c*.1625–91), **Jean** (1627–after 1688?), **Bernard** (1632–98), and **Andries** (1655–93). They were active mainly as portraitists.

Valadon, Suzanne (1865–1938). French painter. As a girl she worked as a circus acrobat, but had to abandon this after a fall and then became a model and the reigning beauty of Montmartre. The artists she posed for included *Renoir, *Puvis de Chavannes, and *Toulouse-Lautrec (each of whom numbered among her lovers). Toulouse-Lautrec brought her drawings to the attention of *Degas, who encouraged her to develop her artistic talent and she became a full-time artist in 1896. She owed little to formal training or to the influence of the artists with whom she associated and her painting has a fresh and personal vision. Her subjects included portraits and still lifes, but she was at her best in figure paintings, which often have a splendid earthy vigour and a striking use of bold contour and flat colour. A child of the people, she has been compared with the writer Colette for her sharpness of eye and avidity for life. Maurice *Utrillo was her son.

Valdés Leal, Juan de (1622–90). Spanish painter and engraver, born in Seville, where he worked from 1656 after some early years in Cordova. With *Murillo he helped to found an *academy of painting there in 1660, and after Murillo's death in 1682 he was the leading artist in the city. Like Murillo, he was primarily a religious painter, but he was very different in style and approach. He had a penchant for macabre or grotesque subject-matter, and his style is characterized by feverish excitability, with a vivid sense of movement, brilliant colouring, and dramatic lighting. His most remarkable works are two large *Allegories of Death* (commissioned 1672) in the Hospital de la Caridad, Seville. He also *polychromed *Roldán's great altarpiece in the Caridad.

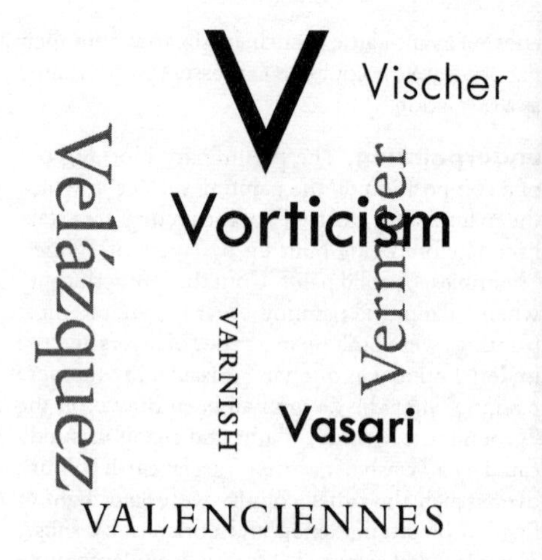

Valenciennes, Pierre-Henri de (1750–1819). French landscape painter and writer on art. From 1769 to 1777 he lived in Italy and his grand, formal style was influenced most notably by Nicolas *Poussin. He became a leading upholder of the classical tradition in landscape painting, but he also encouraged studying direct from nature.

Valentin, Moïse (also called **Le Valentin** and **Valentin de Boulogne**) (*c*.1591–1632). French *Caravaggesque painter, active in Rome from about 1612. His life is obscure; the name 'Moïse' (the French form of Moses) by which he was called was not his Christian name (which is unknown) but a corruption of the Italian form of 'monsieur'. He did, however, have one important public commission—*The Martyrdom of SS. Processus and Martinian* (Vatican, 1629–30), painted for St Peter's as a pendant to *Poussin's *Martyrdom of St Erasmus*. About fifty works are attributed to him. They vary in subject—religious, mythological, and *genre scenes and portraits—but the same models often seem to reappear in them, and all his work is marked by an impressively solemn, at times melancholic, dignity. He was one of the finest of Caravaggio's followers and one of the most dedicated, still painting in his style when it had gone out of fashion. *Baglione says that he died after taking a cold bath in a fountain following a drinking bout; his death was much lamented in the artistic community.

Valentiner, William (1880–1958). German-American art historian. He was an assistant of *Hofstede de Groot on his great catalogue of the work of Dutch painters and likewise gained a reputation as one of the foremost connoisseurs of Dutch painting, on which he published numerous books. From 1906 to 1908 he worked at the Berlin Museum under *Bode, then moved to the *Metropolitan Museum in New York, where one of his innovations was to ban the attendants from smoking cigars while they were on duty. He stayed there until 1914, then served in the German army in the First World War. At this time he knew various German *Expressionist artists, notably *Schmidt-Rottluff, and helped to make their work known in the USA. In 1921 Valentiner began working at the Detroit Institute of Arts, of which he was director from 1924 to 1945 and where he commissioned murals from Diego *Rivera. He was then director of the County Museum of Art in Los Angeles (1946–54) and the first director of the North Carolina Museum at Raleigh (1955–58). Valentiner was also the founder of the periodical *The Art Quarterly* (1938).

Vale Press. A private publishing concern founded in 1896 in London by Charles *Ricketts, who designed the types used and was also responsible for most of the *woodcut illustration and decoration. The books were printed at the commercial firm Ballantyne Press and were published by John Lane. More than forty titles were issued, the most notable production being *The Works of Shakespeare*, published 1900–3 in thirty-nine volumes. The Vale Press closed after a disastrous fire in 1903.

Valette, Adolphe. See LOWRY.

Valkenborch (or **Valkenborgh**). Family of Netherlandish landscape and *genre painters, the most important members of which were **Lucas** (*c.*1530–97) and his younger brother **Marten** (1535–1612). Lucas joined the Malines Guild in 1560 and in 1565 was in Antwerp, whence he fled to Aix-la-Chapelle to avoid persecution (he was a Protestant). He worked for Archduke Matthias and accompanied him to Linz before settling in Frankfurt in 1593. Marten fled with his brother and returned to Antwerp before going to Frankfurt, Venice, and Rome. Both worked in the *Bruegel tradition, and both favoured the subject of the *Tower of Babel* (an example by Lucas is in the Alte Pinakothek in Munich, and an example by Marten is in the Gemäldegalerie in Dresden). They also painted series of *The Seasons*, and their winter scenes are especially notable.

Vallotton, Félix (1865–1925). Swiss-born painter, graphic artist, sculptor, and writer who became a French citizen in 1900. In 1882 he settled in Paris and became a friend of *Bonnard and *Vuillard, exhibiting several times with the *Nabis. He painted portraits, nudes, interiors, and landscapes in a style that is essentially naturalistic but shows an almost abstract feeling for simplified form, influenced by his enthusiasm for Japanese prints (see UKIYO-E). In about 1890 he began to make woodcuts, and with artists such as *Gauguin and *Munch he played a part in the revival of interest in the technique that took place in the early years of the 20th cent.

Valori Plastici. See METAPHYSICAL PAINTING.

Vanderbank, John (1694–1739). English portrait painter. He had a considerable practice in London in the years immediately succeeding the death of *Kneller, and *Vertue says he started the fashion for depicting ladies in Rubens costume. In 1720, with the French-born painter Louis Chéron (1660–1725), Vanderbank deposed *Thornhill as head of what became *St Martin's Lane Academy.

Vanderlyn, John (1775–1852). American painter. He spent much of his career in Paris (he also visited Rome), and his style was nearer to the mainstream of European *Neoclassicism and *Romanticism than that of almost any other American painter of his generation. His best-known painting is *Ariadne Asleep on Naxos* (Pennsylvania Academy of Fine Arts, Philadelphia, 1814), a reclining figure in the tradition of the Venuses of *Giorgione and *Titian that is regarded as the finest American nude before *Eakins. Although such works won him considerable renown in France, he was much less successful in America, where he worked mainly as a portraitist, and he became embittered in his final years.

Vantongerloo, Georges (1886–1965). Belgian sculptor, painter, architect, and writer on art. After being wounded in the First World War he was interned from 1914 to 1918 in the Netherlands. There he joined the De *Stijl group in 1917 and turned from the conventionally naturalistic style he had previously practised to abstract sculptures in which he applied the principles of *Neo-Plasticism to three dimensions. In 1919–27 he lived at Menton and then for the rest of his life in Paris, where he was involved in the *Cercle et Carré and *Abstraction-Création groups. From 1928 he designed architectural projects and in the 1940s he began using wire and perspex in his sculpture, exploring effects of light. His paintings were based on horizontal and vertical lines until 1937, when he introduced rhythmic curving lines. Vantongerloo was one of the pioneers of the mathematical approach to abstract art. He rejected his friend *Mondrian's idea that only constructions based on the right angle reflect the harmony of the universe, believing that this is only one way among many of achieving the formal relations that embody spiritual values. His work is represented in many major collections of modern art.

Varin, Quentin. See POUSSIN, NICOLAS.

Varley, Frederick. See GROUP OF SEVEN.

Varley, John (1778–1842). English landscape painter in watercolour. A protégé of Dr *Monro and later a member of *Blake's circle, he represents the transition between tinted topographical drawing and the bolder, more fully developed manner of watercolour painting characteristic of the 19th cent. He was perhaps the most influential teacher of his day; his pupils included *Cox, *Linnell, and *Palmer, and he published several handbooks on technique. His brothers **Cornelius** (1781–1873) and **William** (1785?–1856) were also watercolourists and several descendants carried on the family tradition.

varnish. Solution, usually of *resin in oil or a synthetic equivalent, used in art either as a protective coat over the paint or as a *medium. Oil varnishes are made by dissolving resins—usually *amber, copal, or *sandarac—in *drying oils such as *linseed. They have been used since the Middle Ages and the method is described by *Cennini. Ethereal varnishes are usually made from *mastic or *dammar dissolved in essential oil of *turpentine or petroleum. They are very quick drying and cannot be safely applied until the paint is thoroughly dry. Albumen varnishes, usually made from white of egg, have been used as a temporary varnish intended to be removed before the final varnish was applied about a year after completion of the painting. Spirit varnishes, in which alcohol was commonly the solvent, originated in the Far East and were not introduced into Europe until the 16th or 17th cent., becoming popular in the 18th cent. They are chiefly used as *fixatives for *watercolour or *tempera paintings but have a solvent action on oils. Waxes, chiefly bleached beeswax or paraffin wax, have been used to give a matt protective coat in place of resin varnishes and so-called 'wax varnishes' are sold for this purpose.

Ideally varnish used as a protective coating should be colourless, transparent, without effect on the colours of the paint, and easily removed if it deteriorates. Varnishes have seldom possessed these qualities. Most have deteriorated and darkened with age, very many have modified the colours of the paint beneath, and in consequence it is not often possible to recover with any high degree of certainty the original appearance of Old Master paintings (see e.g. GALLERY VARNISH). The removal of old and impacted varnishes is a difficult and sometimes impossible task without damage to the underlying paint layers, and revarnishing to restore the appearance intended by the artist is never an easy or straightforward matter. When varnish has been used in the medium or when a coloured varnish has been applied as a final glaze the difficulties of restoration are multiplied.

In the 18th and 19th cents. mastic varnish was mixed with linseed oil to form a jelly-like substance called *megilp. This was used as a painting medium and imparted a rich 'buttery' quality to the colour with which it was mixed. It was discovered, however, that pictures painted with this medium are liable to crack, blister, and turn brown and its use has been discontinued. Pictures painted with megilp are difficult to clean as the solvent which removes the varnish also dissolves the mastic in the paint.

Vasarely, Victor (1908–). Hungarian-born painter, active in France, the main originator and one of the leading practitioners of *Op art. He settled in Paris in 1930, and for the next decade he worked mainly as a commercial artist, particularly on the designing of posters, showing a keen interest in visual tricks such as **trompe-l'œil* and space illusions. From 1943 he turned to painting and *c.*1947 he adopted the method of geometrical abstraction for which he is best known.

From *c.*1955 Vasarely wrote a number of manifestos, which, together with his painting, were a major influence on younger artists in this field. In his own work he explored the means and methods of creating a hallucinatory impression of movement through visual ambiguity, using for this purpose alternating positive–negative shapes interrupted in such a way as to suggest underlying secondary shapes. His fascination with the idea of movement led him to experiment with *Kinetic art and he also collaborated with architects in such works as his relief in aluminium for Caracas University (1954), and the French Pavilion at 'Expo '67' in Montreal, hoping to create a kind of urban *folk art.

Vasarely has lived mainly in the south of France since 1961 and there are two museums dedicated to him in Provence—the Fondation Vasarely at Aix-en-Provence, which he designed himself, and the Château et Musée Vasarely at Gordes. There is also a Vasarely museum at Pécs, his home town in Hungary. Vasarely's son, **Jean-Pierre** (1934–), who works under the name Yvaral, is also an Op and Kinetic artist.

Vasari, Giorgio (1511–74). Italian painter, architect, and biographer, born at Arezzo and active mainly in Florence and Rome. In his day he was a leading painter, architect, and artistic impresario, but his activities in these fields have been completely overshadowed by his role as the most important of all artistic biographers. His great book, generally referred to as *Lives of the Artists*, is not only the fundamental source of information on Italian *Renaissance art, but also a key document in shaping attitudes about the period for centuries afterwards. (The book was first published in 1550 as *Le Vite de' più eccellenti architetti, pittori, et scultori italiani*—The Lives of the Most Eminent Italian Architects, Painters, and Sculptors; in 1568 he published a second, much enlarged edition, in which the painters are mentioned first in the title.) Vasari wrote with a definite philosophy of art and art history. He believed that art is in the first instance imitation of nature and that progress in painting consists of the perfecting of the means of representation. He accepted the belief of Italian humanism that these had been taken to a high level of perfection in *classical antiquity, that art had passed through a period of decline in the Middle Ages, and that it was revived and set once more on its true path by *Giotto. The main theme of the *Lives* was to set forth the revival of art in Tuscany by Giotto and *Cimabue, its steady progress at the hands of such artists as *Ghiberti, *Brunelleschi, and *Donatello, and its culmination with *Leonardo, *Raphael, and above all *Michelangelo, whom Vasari idolized and whose biography was the only one of a living artist to appear in the first edition of his book (the second edition includes accounts of several artists then living, as well as Vasari's own autobiography). The idea of artistic 'progress' he promulgated subsequently coloured most writing on the period. Although Vasari's testimony has often been impugned on particular points (see, for example, *Andrea del Castagno and *Andrea del Sarto), he gathered a vast amount of invaluable information and presented it in a lively style, full of memorable anecdotes. Moreover, his qualitative judgements have generally stood the test of time well. His book became the model for artistic biographies in other countries, such as van *Mander in the Netherlands, *Palomino in Spain, and *Sandrart in Germany.

As a painter, Vasari was one of the most prolific decorators of his period, but is not now highly regarded, his work representing the most in-bred and affected kind of *Mannerism. His best-known paintings are probably the frescos (1546) in the grand salon of the Palazzo della Cancelleria in Rome, illustrating the life of Pope Paul III (Alessandro *Farnese). Pressed to complete the work quickly, Vasari used numerous assistants and accomplished the cycle in under 100 days—hence the nickname Sala dei Cento Giorni (room of a hundred days). As an architect his reputation stands higher, his most important work being the *Uffizi in Florence. He designed and decorated his own house in Arezzo, now a Vasari museum. Vasari was also the first important collector of drawings, using them partly as research material for his biographies, for the insight they gave into the creative process.

Vasnetsov, Viktor. See WANDERERS.

Vatican Museums. Institutions housing the enormous collections of antiquities and works of art accumulated by the papacy since the beginning of the 15th cent. As the leaders of the Christian Church the popes were continually showered with gifts; as political rulers they were, paradoxically, chief guardians of the remains of pagan Rome until Italy was unified in 1870. The Vatican collections are now among the largest and most important in the world, housed in a complex of buildings in the papal palace and elsewhere in the Vatican. There are several separate museums and the visitor to them is also admitted to the exhibition rooms of the Vatican Library and to various suites of *Renaissance painting, of which the most important are the Sistine Chapel, decorated by *Michelangelo and others, and the Stanze, decorated by *Raphael and others. The museums had their origin with Julius II (pontificate 1503–13), who placed some of the most famous works of *classical sculpture in the Cortile del Belvedere, where they were accessible to artists, connoisseurs, and scholars. The Counter-Reformation, however, inaugurated a long period of indifference and hostility to the pagan works of art in the Vatican, and it was not until 1734 that a museum proper was set up by Clement XII (1652–1740). Now, as then, the Vatican Museums are perhaps most famous for their classical statues, including the **Apollo Belvedere*, the **Belvedere Torso*, and the **Laocoön*, but they also contain great riches in, for example, Egyptian and *Early Christian art, jewellery, and vestments. The Pinacoteca (picture gallery) has an impressive if somewhat haphazard collection, devoted mainly to Italian painting of the 13th cent. to the 17th cent. Inevitably it is overshadowed by the frescos of Michelangelo and Raphael, but it contains outstanding works by, for example, *Leonardo da Vinci, *Caravaggio, and *Domenichino. There is also a collection of modern religious art, most of it merely of curiosity value.

Vaughan, Keith (1912–77). British painter. In the 1940s, with his friend John *Minton, he was one of the leading exponents of *Neo-Romanticism. His later work, in which he concentrated on his favourite theme of the male nude in a landscape setting, became grander and more simplified, moving towards abstraction (*Leaping Figure*, Tate, London, 1951). Vaughan also designed textiles and book-jackets. In 1954 he designed an abstract ceramic mural for Corby New Town, Northamptonshire, and in 1963 he was commissioned by the London County Council to do a mural for the Aboyne Road Estate. In 1966 he published *Journal and Drawings*, extracts from a diary he had begun in 1939 (a new edition appeared in 1989). It gives a remarkably frank (and often amusing) account of his homosexual and masturbatory activities and of the struggle with cancer that led to his suicide.

Vecchietta (Lorenzo di Pietro) (*c.*1412–80). One of the outstanding Sienese artists of the 15th cent., a painter, sculptor, architect, and military engineer. He was probably trained by *Sassetta, but he also came under the influence of Florentine art and his large-scale paintings have a monumentality rare in Siena in the *quattrocento. As a sculptor he worked in wood and marble and late in his career in bronze, this change in medium reflecting the influence of *Donatello, who was in Siena 1457–9. *The Risen Christ* (Sta Maria della Scala, Siena, 1476) has something of Donatello's sinewy expressiveness. Donatello's influence may also account for the strength and plasticity of Vecchietta's later paintings, such as the *St Catherine* in the Town Hall, Siena, and the *Assumption* in Pienza Cathedral, both dating from 1461/2. Another side to Vecchietta's talent is seen in his delightful *illuminations in a manuscript of Dante's *Divine Comedy* (BL, London, *c.*1440).

Vecellio, Francesco. See TITIAN.

veduta. Term (Italian: 'view') applied to a representation of a town or landscape that is essentially topographical in conception, specifically one that is faithful enough to allow the location to be identified (an imaginary but realistic-looking view can be called a *veduta ideata*). Painters of *vedute* (for example *Canaletto) are called *vedutisti*.

Veen, Otto van (1556–1629). Flemish painter born in Leiden. From *c.*1575 to *c.*1580 he was in Italy, where he was a pupil of Federico *Zuccaro, and after working in various places in Germany and Flanders he settled in Antwerp in 1592. He was an uninspired *Mannerist painter, but he had a successful career by modelling his work on Italian masters such as *Correggio and *Parmigianino (*The Mystic Marriage of St Catherine*, Musées Royaux, Brussels, 1589). His love of Italian art and his scholarly inclinations (he often Latinized his name to Vaenius) must have been appreciated by *Rubens, who had his final training in van Veen's studio. Rubens continued to work with him for two years after he became a master in 1598 before leaving for Italy himself.

vehicle. The liquid in which *pigment is suspended—the 'carrier'. The word can be applied to either the *medium or the *diluent, or to a mixture of both; thus the vehicle of *oil painting might be either *linseed oil, or *turpentine, or both together.

Velázquez (or **Velasquez**), **Diego** (1599–1660). The greatest painter of the Spanish School. He was born in Seville, where in 1610/11 he was apprenticed to *Pacheco (possibly following a brief period of study with *Herrera the Elder). In 1617 he became a master painter and in the following year he married Pacheco's daughter. Velázquez was exceptionally precocious and while he was still in his teens he painted pictures that display commanding presence and complete technical mastery. Pacheco's style in religious paintings was Italianate, dry, and academic; Velázquez revitalized it by following his master's advice to 'go to nature for everything', and in works such as *The Immaculate Conception* (NG, London, *c.*1618) and *The Adoration of the Magi* (Prado, Madrid, 1619) he developed a more lifelike approach to religious art in which the figures are portraits rather than *ideal types (his wife may be the model for the Virgin in both these pictures). The light, too, is realistically observed, even though it has a mysterious, spiritual quality. These pictures are painted on a warm *ground in the Venetian manner and in their strong *chiaroscuro as well as their naturalism show an affinity with the work of *Caravaggio and his followers. The clotted but supple brush-work is, however, already entirely Velázquez's own. Contemporary with these religious works were a series of **bodegones*, a type of *genre scene to which he brought a new seriousness and dignity, as in *An Old Woman Cooking Eggs* (NG, Edinburgh, 1618) and *The Waterseller of Seville* (Wellington Mus., London, *c.*1620).

In 1622 Velázquez paid a short visit to Madrid, during which he painted a portrait of the poet Luis de Góngora (Museum of Fine Arts, Boston). In the following year he was recalled to the capital by Philip IV's chief minister, the Count-Duke Olivares (1587–1645), and painted a portrait of the king (now lost) that pleased Philip so much that he appointed him one of his court painters and declared that now only Velázquez should paint his portrait. Thus, at the age of 24, he had suddenly become the country's most prestigious painter, and he kept his position as the king's favourite unchallenged for the rest of his life.

With his appointment as court painter, the direction of Velázquez's work changed. He entirely abandoned *bodegones*, and although he painted historical, mythological, and religious pictures intermittently throughout his career, he was from now on primarily a portraitist. Technically, too, his work changed as a result of his move to Madrid, his brushwork becoming broader and more fluid under the influence particularly of the *Titians in the royal collection. Although his portraits of the king and his courtiers are grand and dignified, he humanized the formal tradition of Spanish court portraiture derived from *Mor and *Coello setting his models in more natural poses, giving them greater life and character, and eliminating unnecessary accessories.

The king (who was six years younger than Velázquez) had an extremely high opinion of the artist's personal qualities as well as his artistic skills, and the warmth with which he treated him was considered astonishing, given the stiff etiquette for which the Spanish court was renowned. In 1627 Philip made Velázquez 'Usher of the Chamber', the first of a series of appointments that brought him great prestige but took up much of his time in trivial bureaucratic matters, thus partially accounting for his fairly small output as a painter. He was conscientious in his duties, however, and apparently well suited to them temperamentally.

In 1628–9 *Rubens visited Spain on a diplomatic mission and he and Velázquez became friends. *Palomino records that the contact with Rubens 'revived the desire Velázquez had always had to go to Italy', and the king duly gave him permission to travel there. Velázquez was in Italy from 1629 to 1631, visiting Genoa, Venice, and Naples, but spending most of his time in Rome. Two major paintings date from this period—*Joseph's Coat* (Escorial, Madrid) and *The Forge of Vulcan* (Prado), works that show how his brushwork loosened still further under the influence of the great Venetian masters and how his mastery of figure composition matured.

The 1630s and 1640s (before he again left for Italy) were the most productive period of Velázquez's career. His series of royal and court portraits continued and he expanded his range in a series of glorious equestrian portraits (Prado). In these he showed an unprecedented ability to attain complete atmospheric unity between foreground and background in the landscape. Their rhetorical poses are in the *Baroque tradition, but they are without bombast or allegorical embellishments and as portraits are characteristically direct. The same ability to look beyond external trappings to the human mystery beneath is seen in his incomparable series of portraits of the pitiful court fools (Prado)—dwarfs and idiots whom Philip, like other monarchs, kept for his amusement. Velázquez presents them without any suggestion of caricature, but with pathos and human understanding, as if they too are worthy of his respect.

During the 1630s and 1640s Velázquez painted occasional religious and mythological works, but they are all eclipsed by his great masterpiece of contemporary history painting, *The Surrender of Breda* (Prado, 1634–5), one of a series of twelve paintings by various court artists glorifying the military triumphs of Philip's reign that were executed for the new Buen Retiro Palace in Madrid. The composition is highly organized, but Velázquez creates a remarkable sense of actuality and no earlier picture of a contemporary historical event had seemed so convincing. Characteristically, he concentrates on the human drama of the situation, as Ambrogio Spinola (1569–1630), the chivalrous Spanish commander, receives the key of the town from Justin of Nassau, his Dutch counterpart, with a superb gesture of magnanimity.

Between 1648 and 1651 Velázquez paid another visit to Italy in order to purchase paintings and antiques for the royal collection (he may have been there briefly in 1636 but the evidence is inconclusive). Again, he spent most of the time in Rome, where he painted several portraits, including two of his most celebrated works—*Juan de Pareja* (Met. Mus., New York, 1650) and *Pope Innocent X* (Doria Gal., Rome, 1650). Juan de Pareja (*c.*1610–*c.*1670), who was himself a painter, was Velázquez's mulatto slave (he granted him his freedom while they were in Rome), and Velázquez painted this portrait because he felt he needed some practice before tackling that of the pope. The *Innocent X* is by common consent one of the world's supreme masterpieces of portraiture, unsurpassed in its breathtaking handling of paint and so incisive in characterization that the pope himself said the picture was 'troppo vero' (too truthful). While in Rome Velázquez fathered an illegitimate son, Antonio, by a widow named Martha, but nothing is known of what became of mother or child. They may have been on Velázquez's mind when he applied for (and was refused) permission to return to Italy in 1657, but his life and work continued to unfold with the same serious dignity and the skeleton in his cupboard remained hidden until 1983 when the newly discovered documentation was published.

In his final years in Madrid, Velázquez continued to acquire new honours (the greatest was being made a knight of the Order of Santiago in 1659) and to reach new heights as a painter. His last portraits of the royal family are mainly of the new young Queen, Mariana of Austria, and of the royal children. In these works his brushwork has become increasingly sparkling and free, and the gorgeous clothes the sitters wore (such a change from the sombre costumes of the king and male courtiers) allowed him to show his prowess as a colourist (several examples are in the Kunsthistorisches Museum, Vienna). Velázquez never ceased to base his work on scrutiny of nature, but his means grew increasingly subtle, so that detail is subordinated to overall effect. Thus in his late works space and atmosphere are depicted with unprecedented vividness, but when the pictures are looked at closely the individual forms dissolve into what Kenneth *Clark called 'a fricassee of beautiful brushstrokes'. As Palomino put it, 'one cannot understand it if standing too close, but from a distance it is a miracle'.

The culmination of his career is *Las Meninas* (The Maids of Honour) (Prado, *c.*1656). It shows Velázquez at his easel, with various members of the royal family and their attendants in his studio, but it is not clear whether he has shown himself at work on a portrait of the king and queen (who are reflected in a mirror) when interrupted by the Infanta Margarita and her maids of honour or vice versa. Velázquez's prominence in the picture seems to assert his own importance and his pride in his art, but in the background he has included two pictures by Rubens showing the downfall of mortals who challenge the gods in the arts. Apparently spontaneous but in the highest degree worked out, it is both Velázquez's most complex essay in portraiture and an expression of the high claims he made for the dignity of his art. Luca *Giordano called it 'the Theology of Painting' because 'just as theology is superior to all other branches of knowledge, so is this the greatest example of painting'. Posterity has endorsed his verdict, for in a poll of artists and critics in *The Illustrated London News* in August 1985, *Las Meninas* was voted—by some margin—'the world's greatest painting'.

The number of good contemporary copies of Velázquez's work indicates that he ran a busy studio, but of his pupils only his son-in-law *Mazo achieved any kind of distinction. As with most Spanish painters, Velázquez remained little known outside his own country until the Napoleonic Wars, but from the mid 19th cent. the technical freedom of his work made him an inspiration to progressive artists, above all *Manet, who regarded him as the greatest of all painters. Most of Velázquez's work is still in Spain, and his genius can be fully appreciated only in the Prado, which has most of his key masterpieces. Outside Spain, he is best represented in London—in the National Gallery, which has his only surviving female nude, the *Rokeby Venus* (*c.*1648), in the Wellington Museum, and in the Wallace Collection.

Velde, Esaias van de (1587–1630). Dutch painter, one of the most important figures in the development of the tradition of naturalistic landscape painting in his country. He was born in Amsterdam, where he was perhaps the pupil of Gillis van *Coninxloo, and worked in Haarlem and The Hague. His earliest works are in the *Mannerist tradition, but by 1615 he had already moved away from the panoramic effect and high point of view of his predecessors. The *Winter Scene* (1623) in the National Gallery, London, exemplifies his fresh brushwork and directness of vision, and heralds the subsequent accomplishment of his pupil Jan van *Goyen and of Salomon van *Ruysdael. Esaias was also an excellent etcher and draughtsman. There were other artists in his family, but he was not related to Willem van de *Velde.

Velde, Henri van de (1863–1957). Belgian architect, designer, painter, and writer, one of the leading figures in the development of the *Art Nouveau style. Initially he was a painter—a member of Les *Vingt influenced by *Neo-Impressionism. In about 1890, following a nervous collapse brought on by the death of his mother and under the impact of the ideas of *Ruskin and *Morris, he abandoned painting and turned to design. He received acclaim for the furnished interiors he exhibited at Dresden in 1897, and from 1901 to 1914 he worked in Weimar. In 1908 he became Director of the School of Arts and Crafts there, one of the first institutions of its kind. His immediate successor there was Walter *Gropius, in whose hands it developed into the celebrated *Bauhaus.

Van de Velde, who was also one of the founders of the *Deutscher Werkbund, was a prolific writer and lecturer on architecture and *aesthetics as well as the designer of many buildings in Belgium and Germany. In Belgium he founded the École Nationale Supérieure d'Architecture et des Arts Décoratifs (1926) and was its principal until 1935. In the Netherlands he designed the Kröller-Müller Museum at Otterlo (1937–54). He remained an active designer up to the age of 80, when he retired to live in Switzerland, but most of his significant work was done before the First World War.

Velde, Willem van de the Elder (1611–93). Dutch marine painter. He was the son of a naval captain, his brother was a skipper of merchant vessels, and he himself spent part of his youth as a sailor before devoting himself to the drawing and painting of ships. His pictures, which are frequently *grisailles, contain faithful and detailed portraits of ships. Historians use them as a source of exact knowledge on build and rig and for information about naval battles and manœuvres—he always gives a better account of ships than of the sea and was for a time an official artist for the Dutch fleet. In 1672, when the Netherlands were at war with England, he went to London and entered the service of Charles II; why he left his country at a critical moment in its fortunes remains a mystery.

Willem the Younger (1633–1707), his son, is one of the most illustrious of all marine painters. He was the pupil of his father and Simon de *Vlieger. Like his father, he gave very accurate portrayals of ships, but is distinguished from him by his feeling for atmosphere and majestic sense of composition. He left Amsterdam for England with his father in 1672 and in 1674 Charles II gave them a yearly retaining fee of £100 each; the father received his 'for taking and making draughts of seafights' and the son 'for putting the said draughts into colours for our own particular use'. They did not switch their allegiance to England completely; both subsequently painted pictures of naval battles for the Dutch as well as the English market. Willem the Younger's influence, however, was particularly great in England, where the whole tradition of marine painting stemmed from him. Both artists are well represented in the National Maritime Museum, Greenwich, in the Royal Collection, and in the Rijksmuseum in Amsterdam.

Adriaen (1636–72), Willem II's younger brother, was a versatile and prolific artist in spite of his short life. His father and Jan *Wijnants were his teachers. He painted various types of landscapes (most notably some fresh and atmospheric beach scenes) and also religious and mythological works, portraits, and animal pictures. He also did exceptionally fine etchings of landscapes with cattle and often painted the figures into the landscapes of other artists, notably *Hobbema and *Ruisdael.

Vellert, Dirk (active 1511–44). Netherlandish artist, best known as a designer of stained glass. He was active in Antwerp, where he became a master of the guild of St Luke in 1511. Much of his work, richly ornamented with garlands, masks, vases, and other *Classical motifs, was exported, and some of the windows in King's College Chapel, Cambridge, were done from his designs. Vellert was also a painter and engraver, his prints owing a good deal to *Dürer and *Lucas van Leyden.

vellum. See PARCHMENT.

Venne, Adriaen van de (1589–1662). Dutch painter, born in Delft and active in Antwerp, Middelburg, and The Hague. Early in his career he painted scenes of crowded markets and village fairs—colourful and entertaining works showing a debt to Jan *Brueghel's circle. Later he restricted his palette to greys and browns and turned to more sombre subjects, such as beggars, cripples, and thieves, or did illustrations for moralistic proverbs. He also painted portraits, including many of members of the House of Orange, and wrote poetry.

Venturi, Adolfo (1856–1941). Italian art historian. He was professor at the university of Rome (1890–1931), editor of the periodical *L'Arte*, and the most distinguished Italian art historian of his generation. His primary work is the monumental *Storia dell' arte italiana* (11 vols. in 25 parts, 1901–40; a separate index volume was published in 1975). This covers Italian art from Early Christian times up to the end of the 16th cent. in great detail and with a vast number of illustrations, and it is still a much-used work. Venturi wrote many other works. His son, **Lionello Venturi** (1885–1961), was professor of the history of art at the university of Turin from 1915 to 1931, but resigned because of his opposition to Fascism and moved to Paris and then the USA. Returning to Italy in 1945, he taught at the university of Rome. Like his father, he was a prolific writer, his books including studies of *Caravaggio (1911 and 1952), *Cézanne (2 vols., 1936: the standard catalogue of his works for many years), and *Giorgione (1913).

Venus de Milo. A marble statue of Aphrodite, the best known of all ancient statues, found on the small Greek island of Melos (or Milos) in 1820 and now in the Louvre. A plinth found with the statue is signed '. . . andros [Alexandros or Agasandros] of Antioch on the Maeander', but nothing is known of the sculptor. Originally the statue was thought to date from the *Classical age of Greek sculpture, but it is now put appreciably later—*c.*100 BC—and is thought to be a sophisticated combination of older styles—the goddess's head derives from the later 5th cent. BC, her nudity from the 4th cent., and her spiral, omnifacial posture from the *Hellenistic age. The *Venus de Milo* arrived in the Louvre (1821) soon after the **Medici Venus* had been returned to Italy (1815), and its enormous fame stemmed from French determination to persuade the world that they had gained a greater treasure than they had lost; Martin Robertson (*A History of Greek Art*, 1975) writes that its 'extraordinary reputation, which started by propaganda, has become perpetuated by habit'. The statue's arms are missing and many conjectures have been made as to what the goddess might have been holding: it has been suggested for example that she is intended as Venus Victrix, and so would have been shown with the golden apple presented to her by Paris when he adjudged her more beautiful than her rivals Juno and Minerva.

Verhaecht, Tobias (1561–1631). Flemish painter, now remembered almost solely because he was *Rubens's first teacher. Apart from a sojourn in Italy as a young man, Verhaecht spent his career in his native Antwerp. He was a landscape specialist, working in a style similar to Paul *Bril's, and had no detectable influence on his great pupil.

Verhulst, Mayken. See BRUEGEL.

Verhulst, Rombout (1624–98). Flemish sculptor active mainly in Holland. In the 1650s he was the most important assistant to Artus *Quellin in the sculptural decoration of the Town Hall in Amsterdam. In 1663 he settled in The Hague, where he became a leading sculptor of busts and tombs. The noble and sensitively cut monument to Admiral Michiel de Ruyter (1681) in the Niewe Kerk in Amsterdam is perhaps his masterpiece.

verism. An extreme form of *realism, in which the artist makes it his aim to reproduce with rigid truthfulness the exact appearance of his subject and repudiates idealization and all imaginative interpretation. The term has been applied, for example, to the most realistic Roman portrait sculpture. It has also been applied to the form of *Surrealism that claims to reproduce hallucination in exact and unselective detail.

Vermeer, Jan (1632–75). Dutch painter. Among the great Dutch artists of the 17th cent., he is now second in renown only to *Rembrandt, but he made little mark during his lifetime and then long languished in obscurity. Almost all the contemporary references to him are in colourless official documents and his career is in many ways enigmatic. Apart from a visit to The Hague in 1672 (to act as an expert witness in a dispute about a group of Italian paintings of disputed authenticity), he is never known to have left his native Delft. He entered the painters' guild there in 1653 and was twice elected 'hooftman' (headman), but his teacher is not known. His name is often linked with that of Carel *Fabritius, but it is doubtful if he can have formally taught Vermeer, and this distinction may belong to Leonaert *Bramer, although there is no similarity between their work. Only about thirty-five to forty paintings by Vermeer are known, and although some early works may have been destroyed in the disastrous Delft magazine explosion of 1654, it is unlikely that the figure was ever much larger; this is because most of the Vermeers mentioned in early sources can be identified with surviving pictures, whilst only a few pictures now attributed to him are not mentioned in these sources—thus there are few loose ends. This small output may be at least partially explained by the fact that he almost certainly earned most of his living by means other than painting. His father kept an inn and was a picture-dealer and Vermeer very likely inherited both businesses. In spite of this he had grave financial troubles (he had a large family to support—his wife bore him fifteen children, and she was declared insolvent in the year after his death).

Only three of Vermeer's paintings are dated—*The Procuress* (Gemäldegalerie, Dresden, 1656), *The Astronomer* (Louvre, Paris, 1668), and its companion piece *The Geographer* (Städelsches Kunstinstitut, Frankfurt, 1669). (Another signed and dated work, *St Praxedis mopping up the Blood of the Martyrs* of 1655, appeared in the 1970s, but is of doubtful authenticity. It is in a private collection.) It is difficult to fit his other paintings into a convincing chronology, but his work nevertheless divides into three fairly clear phases. The first is represented by only two works—*Christ in the House of Mary and Martha* (NG of Scotland, Edinburgh) and *Diana and her Companions* (Mauritshuis, The Hague)—both probably dating from a year or two before *The Procuress*. They are so different from Vermeer's other works—in their comparatively large scale, their subject-matter, and their handling—that *Diana and her Companions* was long attributed to the obscure Jan *Vermeer of Utrecht, in spite of a genuine signature. *The Procuress* marks the transition to the middle phase of Vermeer's career, for although it is fairly large and warm in tonality—like the two history paintings—it is a contemporary life scene, as were virtually all Vermeer's pictures from now on.

In the central part of his career (into which most of his work falls) Vermeer painted those serene and harmonious images of domestic life that for their beauty of composition, handling, and treatment of light raise him into a different class from any other Dutch genre painter. The majority show one or two figures in a room lit from the onlooker's left, engaged in domestic or recreational tasks. The predominant colours are yellow, blue, and grey, and the compositions have an abstract simplicity which confers on them an impact out of relation to their small size. In reproduction they can look quite smooth and detailed, but Vermeer often applies the paint broadly, with variations in texture suggesting the play of light with exquisite vibrancy—Jan *Veth aptly described his paint surface as looking like 'crushed pearls melted together'. From this period of Vermeer's greatest achievement also date his only landscape—the incomparable *View of Delft* (Mauritshuis), in which he surpassed even the greatest of his specialist contemporaries in lucidity and truth of atmosphere—and his much-loved *Little Street* (Rijksmuseum, Amsterdam). Another painting of this period is somewhat larger in scale and unusual in subject for him—*The Artist's Studio* (Kunsthistorisches Mus., Vienna), in which Vermeer shows a back view of a painter, perhaps a suitably enigmatic self-portrait.

In the third and final phase of his career Vermeer's work lost part of its magic as it became somewhat harder. There are still marvellous passages of paint in all his late works, but the utter naturalness of his finest works is gone. The only one of his paintings that might be considered a failure is the *Allegory of Faith* (Met. Mus., New York). His wife was a Catholic and he may well have been converted to her religion, but this rather lumbering figure shows he was not at ease with the trappings of *Baroque allegory. There are symbolic references in other of his paintings, but they all—except for this one—make sense on a straightforward naturalistic level.

No drawings by Vermeer are known and little is known of his working method. It is virtually certain, however, that he made use of a *camera obscura; as the exaggerated perspective in some of his pictures (in which foreground figures or objects loom unexpectedly large) and the way in which sparkling highlights sometimes look slightly out of focus are effects duplicated by unsophisticated lenses. The scientist Antony van Leeuwenhoek (1632–1723), celebrated for his work with microscopes, became the executor of Vermeer's estate and it may well have been an interest in optics that brought them together.

Vermeer (or **van der Meer**) **van Haarlem.** Family of Dutch painters. Four painters of the family, members of successive generations, were called Jan Vermeer van Haarlem. The titles of Elder and Younger have become attached to the second and third, very little being known about the other two.

Jan the Elder (1628–91) made panoramic landscapes of the dunes and woods around Haarlem. During the 19th cent. he was confused with the great Jan *Vermeer van Delft, but the name is the only thing they have in common. Jacob van *Ruisdael and Philips de *Koninck have more affinity with his work, but he was a much less powerful artist than either of these.

His son **Jan**, called **the Younger** (1656–1705), was a pupil of his father and of Nicolaes *Berchem. He also was a landscape painter. He spent some time in Italy and most of his pictures are idyllic views of the south. There was also a **Jan Vermeer** of Utrecht (*c.*1630–after 1692), a *genre and portrait painter. The works of all three artists are fairly rare.

Vermeyen, Jan Cornelisz. (*c.*1500–59). Netherlandish painter, engraver, and tapestry designer, born near Haarlem. He travelled much as court painter to the Habsburgs, going to Spain and Tunis with the emperor Charles V in 1534–5. A series of tapestry *cartoons by him depicting Charles's campaign in Tunis is in the Kunsthistorisches Museum, Vienna. Vermeyen may have studied under *Gossaert, but his lively, forthright style shows closer connections with that of his friend Jan van *Scorel. For many years the *Portrait of Erard de la Marck* (Rijksmuseum, Amsterdam) was attributed to Scorel, but now it is generally accepted as Vermeyen's. Many of his portraits are of prominent members of Charles V's court. Vermeyen also painted religious pictures and made engravings.

Vernet. Family of French painters, three members of which attained distinction. **Claude-Joseph Vernet** (1714–89) was one of the leading French landscape painters of his period. From 1733 to 1753 he worked in Rome, where he was influenced by the light and atmosphere of *Claude and also by the more wild and dramatic art of Salvator *Rosa. With Hubert *Robert, he became a leading exponent of a type of idealized and somewhat sentimental landscape that had a great vogue at this time. Vernet was particularly celebrated for his paintings of the sea-shore and ports, and on returning to Paris in 1753 he was commissioned by Louis XV to paint a series of the sea-ports of France. The sixteen which he did are in the Louvre.

Antoine-Charles-Horace, known as 'Carle' (1758–1836), son of the foregoing, painted large battle pictures for Napoleon (*The Battle of Marengo*; *Morning of Austerlitz*; both at Versailles), and after the restoration of the monarchy he became official painter to Louis XVIII (1755–1824), for whom he did racing and hunting scenes.

Émile-Jean-Horace Vernet (1789–1863), known as Horace Vernet, son of the foregoing, was one of the most prolific of French military painters, specializing in scenes of the Napoleonic era. A portrait of Napoleon and four battlepieces by him are in the National Gallery, London. He also did animal and oriental subjects. From 1828 to 1835 he was Director of the French Academy in Rome.

vernicle (or **sudarium**). 'The cloth or kerchief, alleged to have belonged to St Veronica, with which, according to legend, she wiped the face of

Christ on his way to Calvary, and upon which his features were miraculously impressed' (*OED*), or, by extension, any similar picture of Christ's face. The name Veronica was said in the Middle Ages to derive from *vera icon* ('true image'). A cloth that is purported to be the original has been in St Peter's, Rome, since the late 13th cent.; other places have also claimed possession of the original. In art, the vernicle is most commonly shown as being carried by St Veronica, although *Zurbarán, for example, painted several pictures showing the cloth by itself. It was a very popular subject in art from the 15th to the 17th cent. Outstanding examples are *Memling's painting in the National Gallery, Washington, and *Dürer's woodcut. Generally the face conforms to the type of Christ current in the art of the time.

Veronese, Paolo (Paolo Caliari) (1528–88). Italian painter, born at Verona (from which his nickname derives), but active in Venice from about 1553 and considered a member of the Venetian school. With *Tintoretto he became the dominant figure in Venetian painting in the generation after *Titian and he had many major commissions, both religious and secular. He soon established a distinctive style and thereafter developed relatively little. Few of his paintings are dated or can be reliably dated, so his chronology is difficult to construct. Similarly, because he had such a highly organized studio and his output was so large (in terms both of the size of his pictures and of their numbers), there can be problems in distinguishing the work of his own hand. Nevertheless, in spite of difficulties of dating and connoisseurship his status and achievement are clear. He was one of the greatest of all decorative artists, delighting in painting enormous pageant-like scenes that bear witness to the material splendour of Venice in its Golden Age. Marble columns and costumes of velvet and satin abound in his work, and he used a sumptuous but delicate palette in which pale blue, orange, silvery white, and lemon yellow predominate. In his religious works his penchant was for feast scenes from the Bible rather than incidents from Christ's Passion. His love of richness and ornament got him into trouble with the Inquisition in a famous incident when he was taken to task for crowding a painting of the *Last Supper* with such irrelevant and irreverent figures as 'a buffoon with a parrot on his wrist . . . a servant whose nose was bleeding . . . dwarfs and similar vulgarities'. Veronese staunchly defended his right to artistic licence: 'I received the commission to decorate the picture as I saw fit. It is large and, it seemed to me, it could hold many figures.' He was instructed to make changes, but the matter was resolved by changing the title of the picture to *The Feast in the House of Levi* (Accademia, Venice, 1573).

Veronese's secular works include the delightfully light-hearted frescos (including *illusionistic architecture and enchanting landscapes) decorating the Villa Barbaro at Maser, near Treviso (*c.*1561), a series of four canvases, *Allegory of Love* (NG, London, *c.*1575), for the emperor Rudolf II in Prague, and the resplendent *Triumph of Venice* (*c.*1585) on the ceiling of the Hall of the Great Council in the Doge's Palace, Venice. He also painted portraits. His studio was carried on after his death by his brother Benedetto and sons Carletto and Gabriele. He had no significant pupils, but his influence on Venetian painting was important, particularly in the 18th cent., when he was an inspiration to the masters of the second great flowering of decorative painting in the city, above all *Tiepolo.

Verrio, Antonio (*c.*1639–1707). Italian decorative painter who settled in England in about 1671 after working in France and enjoyed an enormously successful and well-remunerated career. He was much employed by the Crown—at Whitehall Palace, Windsor Castle, and Hampton Court—and also worked at great houses such as Burghley and Chatsworth. His success was based on his self-assertiveness and the lack of native talent in his field rather than on his skills as an artist, for his work is at best mediocre (and often dismal) judged by European standards. *Laguerre, his one-time assistant, was a better painter but had less worldly success.

Verrocchio, Andrea del (Andrea di Cioni) (*c.*1435–88). Florentine sculptor, painter, and metalworker, one of the outstanding Italian artists of his period. His nickname—Verrocchio means 'true eye'—refers not to his sharpness of vision, but to the fact that as a youth he had been the protégé of an ecclesiastic of that name. He is said to have studied in *Donatello's workshop, but his main training was as a goldsmith, and delicacy of craftsmanship is one of the salient features of his work. Only one work in precious metal by him survives, however—a silver *relief of the *Beheading of John the Baptist* (1477–80), done for the Baptistery in Florence and now in the Cathedral Museum. His major activity was as a sculptor, principally in bronze, but also in marble and *terracotta, and his two most famous works rank with the statues of Donatello that inspired them among the great masterpieces of Italian sculpture, whilst also showing the great differences in approach between the two artists. Verrocchio's *David* (Bargello, Florence, *c.*1475) is more refined, but less broodingly intense than Donatello's *David* in the same museum, and Verrocchio's masterpiece, the equestrian statue of the *condottiere* Bartolomeo Colleoni in Venice (begun 1481, completed after Verrocchio's death), has a magnificent sense of movement and swagger, but less of the heroic dignity of Donatello's *Gattamelata* statue in Padua. It is much harder to assess Verrocchio's stature as a painter as very few works exist that can be convincingly assigned to his own hand. Nevertheless numerous paintings came from his workshop, which was the largest in Florence at this time, and he trained distinguished painters, most notably *Leonardo da Vinci , who assisted his master with the *Baptism of Christ* (Uffizi, Florence, *c.*1470), one of the few paintings indisputably by Verrocchio. Leonardo took his superb craftsmanship from Verrocchio and also shared his fascination with two contrasting types—the tough old warrior (as in the *Colleoni* monument) and the epicene youth (as in the *David*). Leonardo's enormous fame has tended to cast a shadow over Verrocchio, but he is generally regarded as the greatest Italian sculptor between Donatello and *Michelangelo.

Vertue, George (1684–1756). English engraver and antiquarian. He was a prolific engraver of portraits and antiquarian subjects (he was official engraver to the Society of Antiquaries, 1717–56), but is important chiefly for the voluminous notes he collected for a history of art in England (now in the BL, London). A marvellously rich storehouse of information, they were used by Horace *Walpole as a basis for his *Anecdotes of Painting in England* (1762–71) and have been published separately by the Walpole Society (6 vols., 1930–55).

vesica piscis. See MANDORLA.

Vesperbild. See PIETÀ.

Veth, Jan (1864–1925). Dutch painter and writer on art. He was one of the leading figures in the Dutch art world of his time and had great influence as professor of art history at the University of Amsterdam and as a critic and writer on art. His monograph on *Rembrandt, originally published to mark the 300th anniversary of the painter's birth in 1606, is his best-known book. As a painter he excelled in portraits.

Victoria and Albert Museum, London. One of the world's greatest and most varied collections of *fine and *applied art. It was the brainchild of Prince *Albert and developed out of the Great *Exhibition (1851), the profits from which were used to buy a site in South Kensington for a cultural centre of museums and colleges. The Museum of Ornamental Art (as it was originally called) was opened by Queen Victoria on the present site in 1857. The nucleus of the collection consisted of objects of applied art bought from the Great Exhibition, which had been temporarily displayed in a Museum of Manufacturers in Marlborough House. Albert's ideal was to improve the standard of design in Britain by making the finest models available to study. A new building on the same site was begun in 1899, designed by Sir Aston Webb (1849–1930), and the name of the Museum was changed to 'The Victoria and Albert Museum' at the ceremony of laying the foundation-stone by Queen Victoria. The new building was opened by King Edward VII on 26 June 1909. At this time the scientific collections were assigned to a separate Science Museum.

The collections of the V&A are exceedingly rich, incorporating the national collections of post-*classical sculpture (excluding modern), of British *miniatures, of *watercolours, and English silversmiths' work. The National Art Library is also part of the museum. There are great collections of, for example, ceramics, furniture, and musical instruments, and Oriental art is strongly represented. The highlights of the collection include the tapestry *cartoons by *Raphael (on loan from the Royal Collection), which form the most important ensemble of High *Renaissance art outside Italy; the representation of Italian Renaissance sculpture (again the finest outside Italy); and the collection of *Constable's work—the largest anywhere. The V&A has five branch museums in London: the Bethnal Green Museum of Childhood; Ham House; Osterley Park House; the Theatre Museum; and the Wellington Museum.

Victory of Samothrace. Celebrated larger-than-life Greek marble statue (Louvre, Paris) representing winged Victory (Nike) alighting on the bows of a galley. The figure, discovered on the Greek island of Samothrace in 1863, is lithely outstretched and draped with magnificent swirls. Erected around 200 BC above a rocky pool, it showed its best view obliquely and from below, and now is appropriately and dramatically placed at the top of the main staircase in the Louvre.

Vien, Joseph-Marie (1716–1809). French painter. A winner of the *Prix de Rome, he was in Rome at a time (1743–50) coinciding with excavations at Herculaneum and Pompeii, and in his lifetime he gained a great reputation (partly self-promoted) as a pioneer of the *Neoclassical style. He was enthusiastic for the ideas of *Winckelmann and had a close association with the antiquarian *Caylus, but his *classicism was of a very superficial kind, consisting of prim and sentimentalized anecdote or allegory with pseudo-antique trappings (*The Cupid Seller*, Château de Fontainebleau, 1763). Nevertheless, he gauged the taste of the time well and had a career of exemplary success, becoming director of the French Academy in Rome (1776) and First Painter to the King (1789). He was made a senator by Napoleon after the Revolution, a count in 1808, and was buried in the Panthéon. *David was his most important pupil. His son, **Joseph-Marie the Younger** (1762–1848), was also a painter, mainly of portraits.

Vigarny, Felipe (Philippe Biguerny) (d. 1543). French-born sculptor who worked in Spain from 1498. In that year he contracted to carve three stone *reliefs for the enclosure round the high altar of Burgos Cathedral, executed in a style which combined Flemish-Burgundian with Italian *Renaissance features. Vigarny was a prolific if uneven sculptor. In 1519 he signed a partnership agreement with Alonso *Berruguete: and his colleague's *Mannerist influence is apparent in the *reredos which he carved for the Chapel Royal at Granada (1520–1). In 1539 he was engaged at Toledo Cathedral, together with Berruguete, upon the carving of the upper part of the *choir stalls. He also collaborated with Diego de *Siloe in the cathedrals of Granada and Burgos.

Vigée-Lebrun, Élisabeth (1755–1842). French portrait painter, daughter and pupil of the pastellist **Louis Vigée** (1715–67). She was also taught by Claude-Joseph *Vernet and *Greuze and in 1776 she married the picture dealer Jean-Baptiste Lebrun. Renowned for her beauty, wit, and charm as well as for her talent, she had a highly successful career, becoming a friend of Queen Marie-Antoinette (1755–93), whom she portrayed many times. On the outbreak of the Revolution she left France (1789), travelling to Italy (1789–93), Vienna (1793–4), and St Petersburg (1795–1802), before returning to Paris (1802). Disliking the Napoleonic regime she left almost immediately and stayed in England (1802–5), the Netherlands, and Switzerland (1805–9) before returning once again to Paris. She received distinguished patronage wherever she went and was admitted to several *academies. Her work is graceful, charming, pleasingly sentimental, and delicately executed. There are good examples in the Louvre. Her memoirs give a lively picture of the Europe of her day as well as an account of her own works, and show what a redoubtable woman she was: 'on the day that my daughter was born I never left my studio and I went on working . . . in the intervals between labour pains' (*Souvenirs de Mme Vigée-Lebrun*, 1835–7).

Vigeland, Gustav (1869–1943). The most famous of Norwegian sculptors. He studied in Oslo and Copenhagen and then (1892–5) in Paris and Italy, spending a few months with *Rodin. The painstakingly naturalistic style he developed at this time developed in the direction of expressive stylization when he devoted himself in 1900 to the study of medieval sculpture in preparation for restoration work on Trondheim Cathedral. In the same year he made his first sketches for the massive project that occupied him for the rest of his life and which he left unfinished at his death—a series of allegorical groups at Frogner Park, Oslo. Originally only a fountain was planned, but with the help of assistants he went on to create numerous other groups, including a 17-metre-high column composed of intertwining bodies. The symbolism of the scheme is not clear, but essentially it represents 'a statement of the doubt, disillusion, and physical decline that beset humanity in its passage through this world' (G. H. Hamilton, *Painting and Sculpture in Europe: 1880–1940*, 1967). Reactions to the whole megalomaniac conception, involving scores of bronze and granite figures, have been mixed since the figures were finally installed in 1944, some critics finding it stupendous, others tasteless and monotonous. G. H. Hamilton aptly comments: 'Were its quality only better, the fountain in the Frogner Park might, for its size, allegorical intricacy, and for the artist's life-long dedication to his ideal, be described as sculptural Wagnerism.'

vignette. Term now most commonly applied to an illustration or design (especially a photograph) that fades into the space around it without a definite border. It is also applied to any small illustration placed at the beginning or end of a chapter or book, to fill up a space, and to foliage ornament (from the French *vigne*: 'vine') in a manuscript, book, or decorative carving.

Vignon, Claude (1593–*c.*1670). French painter and engraver born in Tours and active mainly in Paris. His richly eclectic style was formed mainly in Italy, where he worked *c.*1616–*c.*1622, and his openness to very diverse influences was later fuelled by his activities as a picture dealer. Paradoxically, in view of his varied sources of inspiration, his style is the most distinctive of any French painter of his generation—highly coloured and often bizarrely expressive. *Elsheimer and the *Caravaggisti were strong influences on his handling of light, and his richly encrusted brushwork has striking affinities with *Rembrandt, whose work he is known to have sold. There are examples of his work in the Louvre. He is said to have had more than twenty children by his two wives, and his sons **Claude**

the Younger (1633–1703) and **Philippe** (1638–1701) were also painters.

Villalpando, Francisco de (d. 1561). Spanish architect and metalworker, a member of a family of Palencian artists noted for their decorative work in plaster. His best-known works are the gilded and silver-plated chancel screen of Toledo Cathedral (dated 1548), and the great staircase of the Alcazar at Toledo, built after he was appointed royal architect in 1553. He made a Spanish translation of the third and fourth Books of *Serlio's *Architettura*, published at Toledo in 1552.

Villard d'Honnecourt (13th cent.). French architect from Picardy who owes his fame to a manuscript volume now in the Bibliothèque Nationale, Paris. Half sketch-book, half pattern-book or treatise, this volume, compiled *c.*1235 and the only one of its kind prior to the 15th cent. to have survived, offers a unique insight into the working practices of French architects during the period of the great *Gothic cathedrals. It contains sections on technical procedures and mechanical devices, hints for the construction of human and animal figures, notes on the buildings and monuments that the author saw on his travels (which took him as far as Hungary), and some indication of his own work (which includes the rose window of Lausanne Cathedral). The existence of books of this kind helps to explain the rapid spread of Gothic forms throughout Europe.

Villatte, Pierre. See CHARONTON.

Villeglé, Jacques de la. See AFFICHISTE.

Villon, Jacques (Gaston Duchamp) (1875–1963). French painter, the elder brother of Raymond *Duchamp-Villon and Marcel *Duchamp. He changed his name because of his admiration for the 15th-cent. poet François Villon. As a young man he studied law and worked as a newspaper illustrator. From 1904 he exhibited at the *Salon d'Automne of which he was a founding member. In 1911 he began experimenting with *Cubism and he became a leader of the *Section d'Or group. During the 1920s he did graphic work and reproductions for a living. In the late 1920s he evolved a technique of abstraction in which he claimed to represent the essence of objects by 'signs' rather than by reproducing their properties. His work became more naturalistic from *c.*1934, but in the 1950s he developed a style of abstraction without relation to natural appearances. Villon lived much of his life in comparative obscurity, but after the Second World War he achieved recognition and honours.

Vinckboons, David (1576–1630/3). Flemish-born painter active in Amsterdam. He painted landscapes and *genre scenes and is a transitional figure between the decorative and imaginative *Mannerist tradition and the more naturalistic style associated with 17th-cent. Dutch painting. His early genre pictures depicting village festivals reveal the influence of Pieter *Bruegel, and Vinckboons is credited with introducing some of his motifs into Holland. It has been asserted that Vinckboons's works can always be identified by the presence of a finch (*vinck*) in a tree (*boom*), but the painstaking student usually finds that the bird has flown.

Vingt, Les (Les XX). Group of twenty progressive Belgian painters and sculptors who exhibited together from 1884 to 1893. The members included James *Ensor, Jan *Toorop, and Henry van de *Velde. They showed not only their own work, but also paintings by non-Belgian artists such as *Cézanne, van *Gogh, and *Seurat. The group was influential in spreading the ideas of *Neo-Impressionism and *Post-Impressionism and became the main Belgian forum for *Symbolism and *Art Nouveau. Although the group dissolved in 1893 its work was carried on by an association called La Libre Esthétique, which ran from 1894 to 1914. Most of the leading Belgian avant-garde artists of the period were members.

virtu. 'A love of or taste for works of art or curios; a knowledge of or interest in the fine arts; the fine arts as a subject of study or interest' (*OED*). The word, deriving from the Latin *virtus* ('excellence') via the Italian *virtù*, became common in England in the 18th cent. (Horace *Walpole in a letter of 1746 refers to 'my books, my virtu and my other follies and amusements'). It now hardly survives apart from the term 'object of virtu', meaning a curio, which is still current in sale-room language.

The related term 'virtuoso' was originally used of someone who was learned in the arts or sciences, and according to Henry Peacham in *The Compleat Gentleman* (1634) it was applied particularly to those skilled in distinguishing copies of antiquities from the originals. It came to be applied to professional artists, and the 'Society of Virtuosi', founded in 1689, was composed of 'gentleman painters, sculptors and architects'. In modern usage it denotes a person who shows consummate mastery of technique and is applied primarily in music ('a virtuoso violinist') rather than in the visual arts.

virtuoso. See VIRTU.

Vischer. Family of German sculptors active in Nuremberg. **Hermann the Elder** (d. 1488) established the family bronze-foundry, and the business was inherited by his son **Peter the Elder** (*c.*1460–1529), the best-known of the Vischers. He was assisted by five sons: **Hermann the Younger** (*c.*1486–1517), **Peter the Younger** (1487–1528), **Hans** (*c.*1489–1550), **Jakob** and **Paulus**.

The masterpiece of the Vischer workshop is the spectacular bronze shrine over the *sarcophagus of St Sebald in the church dedicated to him in Nuremberg. The first design—a drawing (by Peter the Elder?) in the Academy, Vienna—dates from 1488, but actual work began only in 1508 and was carried on till 1519. Through journeys of the younger generation the workshop had by that time gained a good deal of knowledge of Italian bronzes and in particular the work of north Italian sculptors. Thus the Sebaldus tomb became a fascinating mixture of *Gothic and *Renaissance styles. The canopy remained Gothic, as also the main figures of the Apostles standing before the supports; but both the base and the **baldacchino* were inhabited by biblical, mythological, and decorative figures conceived in a genuine Renaissance spirit. Peter the Elder included a celebrated self-portrait (complete with tools and leather apron) among these figures. Peter the Elder also made two splendid free-standing figures (1512–13) of Theodoric and King Arthur for the tomb of the emperor Maximilian I in the Hofkirche, Innsbruck.

Georg (*c.*1522–92), a son of Hans, was the last artist member of the family.

Visigothic. Term applied to the style of architecture and ornament current in the Iberian peninsula during the rule of the Visigoths (5th–8th cents.), characteristically combining local Roman tradition with *Byzantine influence. Apart from architecture, the remains of Visigothic art consist mainly of goldsmiths' work and jewellery.

Vitruvius Pollio (active second half of 1st cent. BC). Roman architect, the author of a treatise *De Architectura* which is the only work of its kind to have survived from antiquity. It was well known in manuscript throughout the Middle Ages and the first printed edition appeared in Rome in 1486. Since then it has been much edited and translated and for centuries was regarded as the authoritative voice on *Classical architecture. It is still a major source for the history of ancient art, for it contains much incidental information about Greek and Roman painting and sculpture as well as about architecture.

Vittoria, Alessandro (1525–1608). Italian sculptor, born at Trent and active from 1543 in Venice. There he was the pupil of Jacopo *Sansovino, whom he succeeded as the leading monumental sculptor in the city, continuing his master's style but giving it a more *Mannerist character. He worked extensively in the Doges' Palace both before and after the fire of 1577, executing, for example, the *stuccoed ceiling of the Scala d'Oro (1555–9) and three statues in the Sala delle Quattro Porte (1587). Other examples of his work are in many Venetian churches. He also produced small elongated bronze figures of great elegance and realistic portrait busts of Venetian personalities of which there are good examples in the Ca' d'Oro, Venice.

Vivarini. Family of Venetian painters. **Antonio** (*c.*1415–76/84) seldom worked independently. He collaborated first with his brother-in-law, **Giovanni d'Alemagna** (active 1441–49/50), and secondly with his own younger brother, **Bartolomeo** (*c.*1432–*c.*1499). Their pictures usually took the form of large-scale *polyptychs with stiff, archaic figures and very elaborate carved and gilded frames in the *Gothic tradition. Bartolomeo's independent works date from the 1460s onwards. He continued to paint polyptychs, but modernized his style to some extent by imitating *Mantegna. **Alvise** (*c.*1445–1503/5), son of Antonio, was probably trained by his uncle, but later adopted the manner of Giovanni *Bellini, with whom he worked (1488) on paintings (now lost) for the Doges' Palace in Venice. None of the three had much originality. There are examples of the work of all three in the National Gallery, London.

Vivin, Louis (1861–1936). French *naïve painter. He had a passion for painting from childhood, but could not devote himself to it regularly until he retired from the Post Office in 1922. In 1925 he was 'discovered' by Wilhelm *Uhde and won wide recognition. He painted *genre scenes, hunting scenes, flower pieces, and latterly views of Paris with charmingly wobbly perspective effects.

Vlaminck, Maurice de (1876–1958). French painter, graphic artist, and writer. As a young man he earned his living mainly as a racing cyclist and orchestral violinist, painting in his spare time virtually without instruction. Indeed, he delighted to inveigh against all forms of academic training and boasted that he had never set foot inside the *Louvre: 'I try to paint with my heart and my loins, not bothering with style.' In 1901 the exhibition of van *Gogh's work in Paris overwhelmed him, intensifying his love of pure colour, and with *Matisse and *Derain he became a leading exponent of *Fauvism, often using paint straight from the tube in vigorous and exuberant compositions—mainly landscapes. From 1908 his palette darkened and he sought to achieve volume and a more solid basis of composition from a study of *Cézanne (in 1910–14 he was also mildly influenced by *Cubist stylization). After the First World War he moved from Paris and worked in isolation in the countryside, settling at La Tourillière (Eure-et-Loir) in 1925. His later work became rather slick and mannered. Vlaminck was a colourful many-sided character. He wrote novels and volumes of memoirs and was a pioneer collector of African art, although it had no influence on his style. There are examples of his work in many collections of modern art.

Vlieger, Simon de (*c*.1600–53). Dutch painter, mainly of marine subjects, active in his native Rotterdam, Delft, and Amsterdam. One of the outstanding marine painters of his period, he moved from stormy subjects in the manner of *Porcellis to serene and majestic images that influenced van de *Cappelle and Willem van de *Velde the Younger. De Vlieger also painted a few landscapes and *genre pictures.

Vollard, Ambroise (1865–1939). French dealer, connoisseur, art publisher, and writer on art, famous as a champion of avant-garde art. In 1893 he opened a gallery in Paris and in 1895 he gave the first important exhibition of *Cézanne. From that time the gallery became the centre of innovative art in the city, other landmark events including the first one-man exhibitions of *Picasso (1901) and *Matisse (1904). Vollard was the first to commission prominent artists to illustrate literary classics and contemporary works, publishing volumes which were of interest primarily for the illustrations. He thus encouraged many leading artists of the time to engage in graphic work. His portrait was painted many times, among others by *Bonnard, Cézanne, Picasso, *Renoir, and *Rouault. Vollard's writings included books on Cézanne and *Degas and the autobiographical *Recollections of a Picture Dealer* (1936).

Volterra, Daniele da. See DANIELE.

Volto Santo (Italian: 'Sacred Face'). A large wooden crucifix in Lucca Cathedral, on which Christ is shown fully robed. According to an early medieval tradition this crucifix was an actual portrait of Christ made by Nicodemus, who had helped to give him burial. It was said to have been in Lucca from the 8th cent., but there is a suggestion that the present *Volto Santo* is a 13th-cent. copy of an 8th-cent. original. The commercial importance of Lucca in the Middle Ages helps to explain the appearance of a considerable number of 12th–15th-cent. copies of the *Volto Santo* in nearly every country in Europe.

Vorticism. British avant-garde art movement originating just before the First World War. It developed from a rift in October 1913 within the group of artists associated with Roger *Fry in the *Omega Workshops scheme, triggered by a personal quarrel between Fry and Wyndham *Lewis. Lewis and several other artists withdrew from the Workshops amid acrimonious publicity and accusations and in March 1914 formed the Rebel Art Centre. Vorticism grew out of this as a riposte to *Marinetti's *Futurist manifesto *Vital English Art* (June 1914), to which he had appended the names of the 'rebel' artists without their permission. Lewis quickly replied by publishing the first number of *Blast: Review of the Great English Vortex* (there were only two issues, in 1914 and 1915), in which he made a bitter attack on *Vital English Art* in the form of his own *Vorticist Manifesto*. The name Vorticism had been coined by the poet Ezra Pound (1885–1972) and may have had reference to a statement by *Boccioni that all artistic creation must originate in a state of emotional vortex. Apart from Lewis and Pound, the main artists associated with the group included the painters *Bomberg, *Nevinson, *Roberts, and *Wadsworth, and the sculptors *Epstein and *Gaudier-Brzeska. T. S. Eliot (1888–1965) was among the writers who contributed to *Blast*. One Vorticist exhibition was held, in 1915. In spite of Lewis's dissociation from Marinetti, the Vorticists were quite clearly influenced by Futurism in their use of a harsh, angular, diagonally oriented style. Lewis wrote in 1956 (in the introduction to a catalogue of an exhibition 'Wyndham Lewis and Vorticism' at the Tate Gallery) 'Vorticism . . . was what I, personally, did, and said, at a certain period', but this was vehemently rebutted by Bomberg and Roberts. The movement did not survive the First World War and an attempt to revive it in 1920 as Group X proved abortive, but although short-lived it was of great significance as the first organized movement towards abstraction in English art and subsequently had considerable influence on the development of British modernism.

Vos, Cornelis de (1584?–1651). Flemish painter, active in Antwerp. He painted historical, allegorical, mythological, and religious works, but excelled chiefly as a portraitist, in a style derived from *Rubens and van *Dyck. His finest paintings are his portraits of children, which have great sensitivity and charm without lapsing into sentimentality. A splendid self-portrait of *The Artist with his Family* (Musées Royaux, Brussels, 1621) shows him looking happy and proud with his own children. His brother **Paul de Vos** (*c.*1596–1678) painted hunting scenes and still lifes in the style of Frans *Snyders, who was the brother-in-law of the Vos brothers. In 1637 all three helped Rubens execute pictures for the Torre de la Parada, Philip IV of Spain's hunting-lodge near Madrid.

Vos, Maerten de (1532–1603). Netherlandish painter, active mainly in his native Antwerp. In about 1552 he went to Italy (perhaps in company with *Bruegel) and studied in Rome, in Florence, and with *Tintoretto in Venice. By 1558 he was back in Antwerp, and after the death of Frans *Floris in 1570 he became the leading Italianate artist in that city. The altarpieces that make up the bulk of his output are typically *Mannerist in their strained, slender elegance. His much rarer portraits, on the other hand, are notably direct and more in the Netherlandish tradition (*Antoine Anselme and his Family*, Musées Royaux, Brussels, 1577).

Voss, Hermann (1884–1969). German art historian. He worked in museums in Berlin, Dresden, and Leipzig and also taught in Germany, England, and the USA. A major figure in the revival of interest in Italian *Baroque painting, Voss is best known for his great book *Die Malerei des Barock in Rom* (1924), which has lost none of its importance and is described by Rudolf *Wittkower as 'the basic study without which no work in the field can be undertaken'. He also published numerous important articles on—in particular—*Caravaggesque artists, including one on Georges de *La Tour in 1915 that marks the 20th-cent. rediscovery of this artist.

Vostell, Wolf. See FLUXUS.

Vouet, Simon (1590–1649). The leading French painter in the first half of the 17th cent. He spent the years 1613–27 in Italy (mainly in Rome) and achieved a considerable reputation, being made President of the Roman Academy of St Luke in 1624. His early work in Italy was much influenced by *Caravaggio, but he later developed an *eclectic style in which *Baroque tendencies were tempered by the *classicism of Guido *Reni and *Domenichino. In 1627 Louis XIII (1601–43) recalled Vouet to Paris and made him his court painter, launching him on an extremely busy and prosperous career. His compromise style proved exactly to the taste of a French public brought up in the *Mannerist tradition and seeking something new, but unprepared for the dramatic naturalism of the Caravaggesque or the full emotionalism of the Baroque. Only when *Poussin returned from Rome to Paris in 1640–2 was Vouet's dominance threatened. He painted religious and allegorical works and portraits and was employed on many important decorative schemes, sometimes in conjunction with Jacques *Sarrazin. Little of his best decorative work is extant, however. Vouet was a versatile and hardworking artist rather than a great one and his success and influence depended on his having hit upon a style which accorded with the taste of the day at a time when French painting was at a low ebb. He introduced new life and a tradition of solid competence and most of the leading members of the next generation of artists passed through his studio, including *Lebrun, *Le Sueur, and *Mignard.

Voysey, Charles Francis Annesley (1857–1941). English domestic architect and designer of furniture, textiles, wallpapers, etc. The simple, fairly small houses for which he is famous were a revelation after the fussiness and fancifulness of late Victorian period-revival architecture. He was linked to the *Arts and Crafts movement and influenced *Art Nouveau. Though he was the most widely imitated of English designers abroad, Voysey had a decided insularity of outlook, unimpressed by Continental movements. He pioneered simplicity in interior decoration, and he was one of the first designers to appreciate the significance of industrial design.

Vrel, Jacobus (active 1654–62). Dutch *genre painter, an enigmatic figure who has only recently gained recognition as one of the most charmingly idiosyncratic masters of his time. Nothing at all is known of his life, and his paintings—either sparse interiors or quiet street scenes—are very rare. They show a remarkably fresh, almost *naïve vision and have a sense of tranquil poetry that has led them to be compared to the work of *Vermeer. Several of his paintings have, indeed, formerly gone under Vermeer's name, for example *Street Scene* in the Getty Museum, Malibu, which was once owned by *Thoré, who considered it a Vermeer. Because of the similarity of spirit, it is surmised that Vrel, too, worked in Delft.

Vriendt, de. See FLORIS.

Vries, Adriaen de (*c.*1546–1626). Netherlandish sculptor. He was born in The Hague, trained in Italy under *Giambologna and worked mainly in Central Europe, notably for the emperor Rudolf II (1562–1612) in Prague, where he settled in 1601. He was a fine craftsman in bronze and his sleek and elegant figures imitated his master's style with great accomplishment; their popularity showed the international currency of the courtly *Mannerist style. His major works included fountains for

Augsburg (1598, 1602) and Copenhagen (1617). The figures made for the fountain in Copenhagen were taken by the Swedes as war booty in 1660 and are now in the Palace of Drottningholm near Stockholm. None of de Vries's commissions came from the Low Countries.

Vries, Hans Vredeman de (1526/7–1606). Netherlandish painter, architect, engineer, and designer, active in Germany and Prague, as well as Amsterdam, Antwerp, and The Hague. He was famous in his lifetime for his skill in *illusionistic architectural decoration, but much of his work was of a temporary nature (triumphal arches for festivities and so on) and few paintings are known by him (one of them is *Christ in the House of Mary and Martha* in the Royal Collection, which is set in an extremely elaborate interior). He is now remembered primarily for his many books and prints containing *perspective studies of fanciful palaces, courts, gardens, furniture, and decorative work. They had wide circulation in northern Europe and had great influence on architecture and decoration.

Vrubel, Mikhail (1856–1910). Russian painter and designer, the outstanding exponent of *Symbolism in his country. His first important work was the restoration of murals in the Church of St Cyril, Kiev, and in his subsequent career he showed an affinity with the spirituality of medieval religious art. In 1889 he moved from Kiev to Moscow and there was taken up by the wealthy art patron Savva Mamontov (a portrait of him by Vrubel, 1897, is in the Tretyakov Gallery, Moscow). In 1890 he began to do interpretations of Mikhail Lermontov's (1814–41) poem *The Demon* and the theme became central to his work. In treating it he passed from fairly naturalistic depictions to highly idiosyncratic anguish-ridden images rendered in brilliant fragmented brushwork that recalls the effects of medieval mosaics. The obsessive treatment of the theme reflected his own emotional instability; in 1902 the first symptoms of approaching insanity became apparent, in 1906 he went blind, and he died in a lunatic asylum. Although he was little appreciated in his lifetime, Vrubel stands out as the great precursor of much that was best in Russian 20th-cent. painting.

Vuillard, Édouard (1868–1940). French painter and occasional lithographer. In the 1890s he was a member of the *Nabis and at this time painted intimate interiors and scenes from Montmartre, his sensitive patterning of flattish colours owing something to *Gauguin and something to *Puvis de Chavannes, but creating a distinctive manner of his own. From about 1900 he turned to a more naturalistic style and with *Bonnard became the main practitioner of *Intimisme, making use of the camera to capture fleeting, informal groupings of his friends and relatives in the intimate settings of their homes and gardens. He lived quietly, sharing an apartment with his mother until her death in 1928; she often features in his paintings. There are examples of his work in many collections of modern art.

Waagen, Gustav Friedrich (1794–1868). German art historian. The son of a painter, in 1822 he published a book on the van *Eycks that made his reputation and led to his appointment as director of the newly founded Berlin Museum in 1832. Waagen had studied art history at the universities of Breslau and Heidelberg and was one of the first trained art historians in the museum profession. In 1844 he became the first holder of a university chair in art history when he was appointed professor at Berlin University. A frequent traveller, with a great reputation as a connoisseur, Waagen is best remembered for his notes on works of art in public and private collections in various countries, which are a mine of information. Outstanding among them is *Treasures of Art in Great Britain* (3 vols., 1854); a supplementary volume entitled *Galleries and Cabinets of Art in Great Britain* was published in 1857. Waagen was a much-respected figure in England and in 1853 he gave evidence before the royal commission on the condition and future of the National Gallery.

Wadsworth, Edward (1889–1949). English painter and graphic artist. He worked for a short time with Roger *Fry at the *Omega Workshops, then was associated with Wyndham *Lewis in the *Vorticist group. At this time he produced completely abstract works such as the stridently geometrical *Abstract Composition* (Tate, London, 1915). In the First World War he served with the Royal Naval Volunteer Reserve as an intelligence officer in the Mediterranean, 1914–17, then worked on designing dazzle camouflage for ships, turning his harsh Vorticist style to practical use. He documented this war work in the large *Dazzle-Ships in Drydock at Liverpool* (NG, Ottawa, 1919). The lucidity, precision, and clarity of his work were enhanced when *c.*1922 he changed from oil painting to *tempera, and during the rest of the 1920s, abandoning *Cubist leanings, he painted in a more naturalistic manner, showing a penchant for maritime subjects and developing a distinctive type of highly composed marine still life that often has a *Surrealistic flavour, brought about by oddities of scale and juxtaposition and the hypnotic clarity of the lighting (*Satellitium*, Castle Mus., Nottingham, 1932). In 1933 he became a member of *Unit One and again painted abstracts (influenced by *Arp), but he reverted to his naturalistic style. During the 1930s he had several commissions for mural decorations, notably for the liner *Queen Mary* (1936). His graphic work included collections of drawings and engravings published in books as *The Black Country* (1920) and *Sailing-Ships and Barges of the Western Mediterranean and Adriatic Seas* (1926).

Walden, Herwarth. See STURM, DER.

Waldmüller, Ferdinand Georg (1793–1865). One of the leading Austrian painters of the *Biedermeir period. He painted portraits, *genre subjects, and still life, but is perhaps best known for his landscapes, which in their loving attention to detail illustrate his belief that the close study of nature should be the basis of painting. His views were in opposition to the official doctrines of *ideal art promulgated by the Vienna Academy, where he became Professor of Painting in 1829, and he was dismissed from his post in 1857. There are examples of his work in the Österreichische Galerie and other museums in Vienna.

Walker, Dame Ethel (1861–1951). One of the most distinguished British women painters of her period. She did not take up art seriously until she was in her late twenties. In 1900 she became a member of the *New English Art Club and it was there that she exhibited most of her work. She painted portraits, flowerpieces, interiors, and seascapes in an attractive *Impressionist style, but her most individual works are decorative compositions inspired by her vision of a Golden Age. They show the influence of *Puvis de Chavannes as well as of her interest in philosophy and religion. Examples of her work are in the Tate Gallery and many other public collections in Britain.

Walker, Frederick (1840–75). English illustrator and painter. A wood-engraver by training, he drew, 1859–66, for *Good Words*, *Once a Week*, and *The Cornhill Magazine*, then gave up illustration for painting in oil and watercolour. His scenes of peasants in a landscape, treated with grace and pathos, had considerable influence in England.

Walker, Robert (*c.*1605–*c.*1659). English portrait painter. He has a certain niche in history as he was the portraitist most favoured by Oliver Cromwell (1599–1658) and the Parliamentarians during the Interregnum (1649–60), but his work is generally dull and derivative (mainly of van *Dyck). Versions of his portrait of Cromwell are in the National Portrait Gallery, London, and elsewhere, and there is a self-portrait in the Ashmolean Museum, Oxford.

Wallace Collection, London. National museum consisting of the collection built up in the 18th and 19th cents. by the Seymour-Conway family, Earls and later Marquesses of Hertford; it was bequeathed to the nation in 1897 by Lady Wallace, widow of Sir Richard Wallace, the illegitimate son of the 4th Marquess of Hertford, and opened to the public in 1900. It is housed in Hertford House, the former London residence of the family. The collection, the largest of its date to be preserved intact, reflects the tastes of various members of the family, but particularly of the 4th Marquess (Richard Seymour-Conway, 1800–70) and his son Sir Richard Wallace (1819–90), who each spent much of their lives in Paris. The Marquess bought most of the superb representation of 18th-cent. French art (furniture as well as paintings) that is the collection's chief glory; Sir Richard added much *Renaissance decorative art and the magnificent collection of armour, rivalled in Britain only by that in the Royal Armouries. Other areas in which the Wallace Collection is particularly rich are 17th-cent. Dutch and Flemish painting (*Rembrandt, *Rubens, and van *Dyck are all present in strength and Frans *Hals's *Laughing Cavalier* is the most famous picture there); 18th-cent. English portraits; and the works of Richard Parkes *Bonington (the best representation anywhere).

Wallis, Alfred (1855–1942). British *naïve painter of sailing ships and landscapes. He went to sea as a cabin boy and cook at the age of 9, and from about 1880 worked as a fisherman in Cornwall. In 1890 he opened a rag and bone store in St Ives and after retiring from this did odd jobs including selling ice-cream. He began to paint in 1925 to ease the loneliness he felt after his wife's death and was discovered by Ben *Nicholson and Christopher *Wood in 1928, the unselfconscious vigour of his work making a powerful impression on them. Wallis painted from memory and imagination, usually working with ship's paint on odd scraps of cardboard or wood. Although he rapidly became the best known of British naïve artists, he died in a workhouse.

walnut oil. *Drying oil obtained from the common walnut. It was one of the earliest oils used in painting, and perhaps the commonest *medium in the early days of *oil painting, but it is little used today. It dries more slowly than *linseed oil but has less tendency to turn yellow.

Walpole, Horace (1717–97). English collector, connoisseur, man of letters, and amateur architect. He was the fourth son of Sir Robert Walpole, Britain's first prime minister, and in 1791 he became fourth Earl of Orford. In 1739–41 he made the *Grand Tour, travelling in France and Italy with the poet Gray. His own literary fame rests on his voluminous correspondence and on *The Castle of Otranto* (1764), the first 'Gothic novel'. In the history of taste he is primarily important for his house at Twickenham, Strawberry Hill, which he bought in 1747 and extended into a showpiece of the *Gothic Revival, employing professional architects to work from his sketches. Although Walpole was interested in archaeological research, he adapted his medieval models in accordance with the *Rococo taste of the time. His main influence on the Gothic Revival was probably to give social and aristocratic *cachet* to a fashion in taste which might otherwise have remained the foible of a few eccentrics. He filled the building with his collections and it became such a tourist attraction that he had to issue tickets. In 1757 Walpole established his own printing press at Strawberry Hill, from which he issued *Anecdotes of Painting in England* (4 vols., 1762–71), based on *Vertue's notebooks. His other publications include *A Description of the Villa of Horace Walpole at Strawberry Hill* (1774) and *Aedes Walpolianae* (1747), a catalogue of his father's paintings.

Walton, Henry (1746–1813). English painter, a pupil of *Zoffany. He was a gentleman painter and his work is rare. His portraits are unexceptional, but he painted a small number of *genre scenes that are amongst the finest of their time in English art, notable for their unaffected charm and sureness of tone (*Girl Plucking a Turkey*, Tate, London, 1776). They have something of the quiet dignity of *Chardin, whose works he probably saw during visits to Paris.

Wanderers (or **Itinerants**). Group of Russian painters associated with the Society of Wandering (Travelling) Exhibitions, a body founded in 1870 with the aim of bringing art to the people. The nucleus of the group was formed by a number of artists—led by *Kramskoi—who had left the St Petersburg Academy in 1863 because of its rigid traditionalism. The rebels had refused to accept 'The Feast of the Gods in Valhalla' as a competition subject because it was so irrelevant to contemporary social needs. They were united by their belief that art should express humanitarian ideals and encourage social reforms. Thus they painted realistic pictures of peasant and middle-class life, often scenes arousing pity or inspiring the oppressed to better themselves. Their exhibitions travelled about the country, bringing art to a new public. The Wanderers (in Russian *Peredvizhniki*) eventually included most of the leading Russian painters of the last quarter of the 19th cent., notably Vasily Perov (1834–82), Ilya *Repin, Vasily Surikov (1848–1916), and Viktor Vasnetsov (1848–1926). The society lasted until 1923, but by then the once radical group had become conservative.

Wappers, Gustav (1803–74). Belgian painter. He was an enormous success in his day with his elaborate historical costume pieces, catching the public imagination with patriotic themes that appealed to the Belgian people, who had just won their independence (*Episode from the Belgian Revolution of 1830*, Musées Royaux, Brussels, 1834). Such works have dated badly, however. In 1845 he was created a baron, and in 1853 he settled in Paris.

War Artists' Advisory Committee. See OFFICIAL WAR ART.

Warburg, Aby (1866–1929). German art historian. His main field of study was the art of the Florentine *quattrocento, and opposing the *Aestheticism of the late 19th cent. he tried to understand the art of the period not in terms of formal values but as part of the intellectual history of the time. He searched the archives and memoirs of the period for evidence of the taste and preoccupations of the artists and their patrons, and the picture that emerged was far removed from the idyllic springtime envisaged by Victorian travellers. In particular Warburg was impressed by the hold that religious loyalties and astrological superstition retained on the minds of Renaissance merchants and princes, indicating that the subject-matter of the paintings they commissioned was no mere pretext for the display of artistic fancies. Warburg published little, but his aim of understanding every work of art in terms of a tradition modified by the psychological needs of the moment had great influence on his followers. The superb library that he built up in his home in Hamburg developed into a research institute, which was transferred to London in 1933 to escape the Nazi regime and in 1944 was incorporated in the University of London as the Warburg Institute. Its field of study is now officially defined as 'the history of the *classical tradition' (das Nachleben der Antike), but a better idea of the intellectual range of its activities can be gauged from a comment by *Panofsky: 'It stands to reason that an institute like the Warburg . . ., which was founded for the explicit purpose of eliminating the borderlines between the history of art, the history of religion and superstition, the history of science, the history of cultic practices (including pageantry) and the history of literature, could not help being important for the practitioners of all these disciplines.' Ernst *Gombrich, who wrote a biography of Warburg (1970, 2nd edn. 1986), is one of the many distinguished scholars associated with the Institute.

Ward, James (1769–1859). English painter and engraver. Until about the end of the century he painted mainly anecdotal *genre scenes in the manner of his brother-in-law George *Morland, but he then turned to the paintings of animals in landscape settings for which he is remembered. They are often dramatic and *Romantic in character and their rich colouring was influenced by *Rubens (*Bulls Fighting*, V&A, London, *c.*1804). His taste for natural grandeur and the *Sublime often led him to work on a large scale, as in the enormous *Gordale Scar* (Tate, London, 1811–15). Ward had many admirers, including *Delacroix and *Géricault, but he lived in retirement in the country from the 1830s and ended his life in poverty. His brother **William** (1766–1826) was an engraver.

Ward, John Quincy Adams (1830–1910). American sculptor. He was one of the most prolific and successful sculptors of public monuments of his period. His portrait statues, equestrian figures, allegorical groups, war memorials, and architectural decorations have solid dignity, but none of the genius of his contemporary *Saint-Gaudens. He was elected President of the National Academy of Design in 1873 and was the first President of the National Sculpture Society in 1893.

Warhol, Andy (1928–87). American painter, graphic artist, film-maker, and writer, born to Czech immigrant parents. In the 1950s he was enormously successful as a commercial artist in New York (specializing in shoe advertisements): by 1956 he was earning $100,000 a year. In 1960 he began making paintings based on mass-produced images such as newspaper advertisements and comic strips, then in 1962 of Campbell's soup cans. They were exhibited in that year with sensational success and Warhol soon became the most famous and controversial figure in American *Pop art. In the same vein as his soup cans he did pictures of Coca-Cola bottles and made equally banal sculptures of Brillo soap pad boxes and similar cartons. He also embarked on a seemingly endless series of pictures of Marilyn Monroe, Elvis Presley, Elizabeth Taylor, and other celebrities. Similar in method but different in effect were his pictures of disasters such as car crashes and views of the electric chair. Whatever the subject in his pictures, he often made use of rows of repeated images. The silkscreen process allowed infinite replication, and he was opposed to the idea of a work of art as a piece of craftsmanship, hand-made and expressing the personality of the artist: 'I want everybody to think alike. I think everybody should be a machine.' In keeping with this outlook he used clippings of 'dehumanized' illustrations from the mass media as his sources, turned out his works like a manufacturer, and called his studio 'The Factory'. There he was surrounded by a crowd of helpers and hangers-on, described by Robert *Hughes as 'cultural space-debris, drifting fragments from a variety of sixties subcultures'. Warhol liked to give the idea that he took a paternal interest in his followers, but Eric Shanes (*Warhol*, 1991) writes that 'Just how cynical he could be in his dealings with his entourage is demonstrated by an incident that occurred in October 1964 when one of his hangers-on, Freddie Herko, committed suicide by jumping from a fifth-floor window in Greenwich Village while high on LSD: Warhol was heard to complain repeatedly that Herko should have forewarned him so that he could have filmed his death.'

In 1965 Warhol announced his retirement as an artist to devote himself to films and to managing the rock group The Velvet Underground, but in fact he never gave up painting and in the 1970s he made an enormous amount of money churning out commissioned portraits of wealthy patrons. In the 1980s he sometimes collaborated with other painters, including the *Graffiti artist Jean-Michel Basquiat and LeRoi Neiman, who is best known for his illustrations in *Playboy* magazine. As a film-maker, Warhol became perhaps the only 'underground' director to be well known to the general public. His first films were silent and virtually completely static: *Sleep* (1963)—a man sleeping for six hours; and *Empire* (1964)—the Empire State Building seen from one viewpoint for eight hours: 'I like boring things.' Later films, such as the two-screen *Chelsea Girls* (1966), gained widespread attention because of their voyeuristic concentration on sex. In 1968 Warhol was shot and severely wounded by a bit-part player in one of his films, a member of SCUM (The Society for Cutting Up Men), an incident that added to his legendary status. By this time he was perhaps already more famous for his celebrity-courting lifestyle and deliberately bland persona than for his art; indeed, it could be argued that his advertising skills were nowhere more brilliantly deployed than in promoting himself. In purely financial terms his success in promoting himself was prodigious. At his death (following a routine gall bladder operation) he left a fortune estimated at $100,000,000, most of which went to create an arts charity, the Andy Warhol Foundation. His status as an artist, however, is controversial. Even his most fervent admirers tend to admit that he added little to his achievement as a painter after the mid-1960s, but large claims are sometimes made for his earlier works. Warhol published a celebrity magazine called *Interview*, and several books appeared under his name, some genuinely written by him, others put together from tapes. *The Diaries of Andy Warhol* appeared posthumously in 1989. In 1994 a museum dedicated to his work opened in his home town of Pittsburgh.

wash. An application of diluted *ink or transparent *watercolour to paper. The term usually refers to a uniform broad area done with a full brush.

Washington Color Painters. See NOLAND.

watercolour. Term applied in its most general sense to paint in which the pigment is bound with a *medium (generally *gum arabic) which is soluble in water. Its use has been widespread and extends over many centuries. It includes the papyrus rolls of ancient Egypt, the paintings on silk and rice-*paper of the Orient, and (combined with an egg medium) the decorations of European *illuminated manuscripts of the Middle Ages. In normal parlance, however, the term 'watercolour' generally refers specifically to a type of painting in which the lighter tones are not obtained by adding white pigment but by thinning with water so that the light is given by the paper or other support showing more strongly through the thinner layers of paint. It can thus be distinguished from other kinds of painting, such as *gouache, that use water as a medium but are opaque. Although there are isolated earlier examples of leading artists making memorable use of watercolour (*Dürer and van *Dyck, for example), its chief development took place in 18th-cent. and early 19th-cent. England, particularly in landscape. At first it was used mainly for topographical scenes, and the technique consisted essentially of tinting an underlying drawing. Around 1800 a transition was made to a bolder approach in which the colour was used freely and directly. *Girtin and *Turner (both born in 1775) brought watercolour to its greatest heights, Girtin being the consummate master of the classic broad technique and Turner achieving unequalled variety of effect and intensity of expression.

In the wake of *Impressionism the capacity of watercolour to achieve spontaneous effects was more widely appreciated and it ceased to be so much of an English speciality. Among the modern artists who have been great exponents of the technique (in their very different ways) are *Cézanne, *Dufy, *Grosz, *Klee, *Nolde, and *Sargent.

water-glass painting. A method of mural painting. The *pigments are mixed with plain water and painted on the plaster, which is then coated with a solution of water-glass (potassium or sodium silicate). Potassium silicate was first made commercially as a painting *medium in 1825. When the water-glass dries it leaves a thin film which seals the painting. As water-glass is strongly alkaline it can be used only with certain pigments. Some of the mural paintings in the House of Lords (see MACLISE) were executed in it, because it was thought that they would be proof against the damp and dirty atmosphere of London, but they deteriorated within ten years. Afterwards, in the 1880s, Adolf Keim of Munich improved the process, which he called *Mineral-malerei*, 'mineral painting'. The Victorian name was 'stereochromy'.

Waterhouse, Sir Ellis (1905–85). English art historian. In a highly distinguished career he was director of the National Galleries of Scotland (1949–52), director of the Barber Institute of Fine Arts, Birmingham (1952–70), and the holder of many other prestigious posts in Britain and the USA. His publications were centred on two main areas, British painting of the 18th cent. and Italian *Baroque painting, his chief books being *Baroque Painting in Rome* (1937, revised edn. 1976), *Painting in Britain 1530–1790* (1953 and several revised edns.), *Italian Baroque Painting* (1962), *The Dictionary of British 18th-Century Painters* (1981), a monograph on *Gainsborough (1958), and two on *Reynolds (1941 and 1973), and *The Dictionary of British 16th- and 17th-Century Painters* (published posthumously 1988). He wore his great erudition lightly and was one of the most entertaining writers among art historians of his generation.

Waterhouse, John William (1849–1917). English painter. Early in his career he painted Greek and Roman subjects, but in the 1880s he turned to literary themes, painted in a distinctive, dreamily romantic style. In approach he was influenced by the *Pre-Raphaelites, but his handling of paint is quite different from theirs—rich and sensuous. His work includes such classic Victorian anthology pieces as *The Lady of Shalott* (Tate, London, 1888) and *Hylas and the Nymphs* (City Art Gal., Manchester, 1896).

Watson, Homer (1855–1936). Canadian landscape painter, born and active for most of his life at Doon, near Kitchener, Ontario. His early style was straightforward in its rendering of the farmsteads of the peaceful Grand River valley in Ontario. Later he emulated the *Barbizon painters, *Constable, and the *Impressionists. He was internationally famous around the turn of the century (Oscar Wilde called him 'the Canadian Constable'), but his reputation is now much diminished.

Watteau, Jean-Antoine (1684–1721). The greatest French painter of his period and one of the key figures of *Rococo art. He was born at Valenciennes, which had passed to France from the Spanish Netherlands only six years before his birth, and he was regarded by contemporaries as a Flemish painter. There are indeed strong links with Flanders in his art, but it also has a sophistication that is quintessentially French.

He moved to Paris in about 1702 and *c.*1703–7 he worked with *Gillot, who stimulated his interest in theatrical costume and scenes from daily life. Soon afterwards he joined Claude *Audran, Keeper of the Luxembourg Palace, and thus had access to the *Rubens's *Marie de Médicis* paintings, which were of enormous influence on him, even though Rubens's robustness was far removed from the fragile delicacy that characterized Watteau's art. Rubens was one of the prime inspirations for the type of picture with which Watteau is most associated—the *fête galante* (see FÊTE CHAMPÊTRE), in which exquisitely dressed young people idle away their time in a dreamy, romantic, pastoral setting. The tradition of lovers in a parkland setting goes back via *Giorgione to the medieval type known as the Garden of Love, but Watteau was the first painter to make the theme his own, and his individuality was recognized by his contemporaries. In 1717 he submitted a characteristic work, *The Pilgrimage to the Island of Cythera* (Louvre, Paris; a slightly later variant is in Schloss Charlottenburg, Berlin), as his reception piece to the Academy, and owing to the difficulty of fitting him into recognized categories was received as a 'peintre de fêtes galantes', a title created expressly for him. He was, indeed, a highly independent artist, who did not readily submit to the will of patrons or officialdom, and the novelty and freshness of his work delivered French painting from the yoke of Italianate academicism, creating a truly 'Parisian' outlook that endured until the *Neoclassicism of *David. Watteau's world is a highly artificial one (apart from scenes of love he took his themes mainly from the theatre), but underlying the frivolity is a feeling of melancholy, reflecting the certain knowledge that all the pleasures of the flesh are transient. This poetic gravity distinguishes him from his imitators and parallels are often drawn between Watteau's own life and character and the content of his paintings. He was notorious for his irritable and restless temperament and died early of tuberculosis, and it is felt that the constant reminder of his own mortality that his illness entailed 'infected' his pictures with a melancholic mood. In 1719 he travelled to London, almost certainly to consult the celebrated physician Dr Richard Mead (1673–1754), but the hard English winter worsened his condition. His early death came when he may have been making a new departure in his art, for his last important work combines something of the straightforward naturalism of his early pictures in the Flemish tradition with the exquisite sensitivity of his *fêtes galantes*. This last great work, a shop sign painted for the picture dealer Edmé Gersaint and known as *L'Enseigne de Gersaint* (Staatliche Mus., Berlin, 1721), brilliantly synthesizes sharpness of observation with delicacy of handling.

Watteau was careless in matters of material technique and many of his paintings are in consequence in a poor state of preservation. A complete picture of his genius depends all the more, then, on his numerous superb drawings, many of them scintillating studies from the life. He collected his drawings into large bound volumes and used these books as a reference source for his paintings (the same figure often appears in more than one picture). In spite of his difficult temperament, Watteau had many loyal friends and supporters who recognized his genius, and although his reputation suffered with the Revolution and the growth of Neoclassicism, he always had distinguished admirers. It is perhaps as a colourist that he has had the most profound influence. His method of juxtaposing flecks of colour on the canvas was carried further by *Delacroix and reduced to a science by *Seurat and the *Neo-Impressionists. Watteau's principal, but much inferior, followers were *Lancret and *Pater. He also had a nephew and a great-nephew (father and son) who worked more-or-less in his manner. They are both known as 'Watteau de Lille' after their main place of work—**Louis-Joseph Watteau** (1731–98) and **François-Louis-Joseph Watteau** (1758–1823).

Watts, George Frederic (1817–1904). English painter and sculptor. He trained with W. *Behnes and in the Royal Academy Schools. In 1843 he won a prize in the competition for the decoration of the Houses of Parliament and used the money to visit Italy, where the great *Renaissance masters helped shape his high-minded attitudes towards art. After returning to England in 1847, he established a solid reputation in intellectual circles, but popular fame did not come until about 1880. Thereafter he was the most revered figure in British art, and in 1902 became one of the original holders of the newly instituted Order of Merit. His style was early influenced by *Etty, but the *Elgin Marbles, the great Venetian painters (notably *Titian), and *Michelangelo were his avowed exemplars in his desire 'to affect the mind seriously by nobility of line and colour'. He tried to invest his work with moral purpose and his most characteristic paintings are abstruse allegories that were once enormously popular, but now seem vague and ponderous (*Hope*, Tate, London, 1886; and other versions). His portraits of great contemporaries (*Gladstone, Tennyson, J. S. Mill*, etc., NPG, London) have generally worn much better. As a sculptor, he is remembered chiefly for his equestrian piece *Physical Energy* (1904). A cast of it forms the central feature of the Cecil Rhodes Memorial, Cape Town, and another is in Kensington Gardens, London. Watts was twice married, his first wife being the celebrated actress Ellen Terry (1847–1928). She was thirty years his junior and the marriage ended in divorce. His former house at Compton, near Guildford, is now the Watts Gallery, devoted to his work.

wax painting. See ENCAUSTIC PAINTING.

Webb, Philip (1831–1915). English architect and designer. He met William *Morris in 1856 and after the formation of Morris & Co. in 1861 became the firm's chief designer in furniture, glass, metalwork, jewellery, and embroideries. He resigned his partnership in the firm after its reorganization in 1875, but he continued to make designs for it. His work, particularly his furniture, mainly massive oak pieces with the structural elements strongly emphasized, was influential on the *Arts and Crafts Movement.

Weber, Max (1881–1961). Russian-born American painter, sculptor, printmaker, and writer, whose work more than that of any other American artist synthesized the latest European developments at the beginning of the 20th cent. He emigrated to New York with his parents when he was 10. From 1905 to 1908 he travelled in Europe, studying in Paris at the *Académie Julian and with *Matisse, admiring early *Cubism and becoming a friend of Henri *Rousseau. After his return to New York in 1909 he rapidly became a controversial figure—no other American avant-garde artist of the time exhibited his work more widely or was more harshly attacked. His work was influenced by

*Fauvism and *primitive art (he was one of the first American artists to show interest in it), but most importantly by Cubism (in sculpture as well as painting). After about 1917, however, Weber's work became more naturalistic. During the 1930s his subjects often expressed his social concern and in the 1940s his work included scenes with rabbis and Jewish scholars—mystical recollections of his Russian childhood. He published several books, including *Cubist Poems* (1914), *Essays in Art* (1916), and the autobiographical *Max Weber* (1945).

Wedgwood, Josiah (1730–95). The most famous of English potters. He combined organizing ability and flair for business with scientific knowledge and artistic taste and was largely responsible for the great expansion of the Staffordshire pottery industry in his period. His products established the taste for *Neoclassical designs in English ceramics and were influential as far as the USA and Russia. He employed excellent designers, the most distinguished of whom was *Flaxman, and also collaborated with *Stubbs in producing paintings on porcelain plaques.

Weenix, Jan Baptist (1621–*c*.1660). Dutch painter. In 1642–6 he was in Italy, and on his return to his native Amsterdam he painted Italianate landscapes close in style to those of *Berchem, who is said to have been his cousin. Later he turned mainly to pictures of still life with dead game; he also painted portraits. His son, **Jan Weenix** (1642?–1719), specialized in hunting trophy subjects similar to those of his father; indeed, it is often difficult to tell their work apart. Most of his career was spent in Amsterdam, but from 1702 to 1712 he worked at Düsseldorf for the Elector Palatine. Both artists were prolific (Jan told *Houbraken that his father could paint three half-length life-size portraits in a day) and are represented in many public collections.

Weiditz, Hans (*c*.1500?–after 1536). German designer of woodcuts, active in Augsburg (1518–22) and then in Strasburg. A prolific book illustrator, he illustrated many sacred, Classical, humanist, and scientific texts and designed a large number of devotional prints. His illustrations are remarkable for their acute observation, and at a time when *Dürer's influence was all-pervasive Weiditz maintained his own very personal style. Much of the work attributed to him on stylistic grounds was previously given to an artist called 'the Petrarch Master', named after an edition of Petrarch illustrated with his work published in 1532. A few paintings by Weiditz are also known.

Weight, Carel (1908–). British painter. In the Second World War he served with the Royal Engineers and Army Education Corps and was appointed an *Official War Artist in 1945, working in Austria, Greece, and Italy. He began teaching at the *Royal College of Art in 1947 and was Professor of Painting there from 1967 until his retirement in 1973. Weight is something of a maverick figure ('I don't like the art world very much. I don't like the dealers and I don't like the critics') and his work is highly individual. His best-known paintings are imaginative figure compositions, set in suburban surroundings. They are superficially realistic, but feature idiosyncratic perspective effects and strange human dramas, producing an effect that is sometimes humorous and sometimes menacing: 'My art is concerned with such things as anger, love, fear, hate and loneliness, emphasized by the ordinary landscape in which the dramatic scene is set.' Weight has also painted portraits and landscapes.

Weir, J. Alden. See TEN, THE.

Weissenbruch, Hendrik Johannes (1824–1903). Dutch landscape and marine painter, a pupil of *Schelfhout and one of the outstanding artists of the *Hague School. His work is distinguished by its subtle handling of tone and feeling for atmosphere. His cousin **Johannes (Jan) Weissenbruch** (1822–80) painted town views in the detailed manner of 17th-cent. artists such as *Saenredam.

Wentworth, Richard. See NEW BRITISH SCULPTURE.

Werefkin, Marianne von. See NEUE KÜNSTLERVEREINIGUNG.

Werenskiold, Erik (1855–1938). Norwegian painter and graphic artist. He was one of the leading personalities in Norwegian art, the friend of numerous writers and intellectuals and a symbol of national culture. His work included landscapes, in which he showed an affectionate yet unsentimental approach to his native land, portraits of many of the leading Norwegians of his day (*Henrik Ibsen*, NG, Oslo, 1895), and book illustrations. After the turn of the century he was influenced by *Cézanne.

Werff, Adriaen van der (1659–1722). Dutch painter of religious and mythological scenes and portraits, active mainly in Rotterdam. He combined the precise finish of the Leiden tradition (learned from his master Eglon van der *Neer) with the classical standards of the French Academy and became the most famous Dutch painter of his day, winning international success and earning an enormous fortune. *Houbraken, writing in 1721, considered him the greatest of all Dutch painters and this was the general critical opinion for about another century. He is now considered an extremely accomplished, rather sentimental and repetitive minor master. Van der Werff also worked as an architect in Rotterdam, designing elegant house façades. His brother, **Pieter van der Werff** (1655?–1722), was his collaborator and close imitator as a painter.

Werve, Claus de (d. 1439). Netherlandish sculptor, the nephew and pupil of *Sluter, with whom he worked from 1396 and whom he succeeded in 1404 as chief sculptor to Philip the Bold (1342–1404), Duke of Burgundy. It was Werve who carried out most of Sluter's design for the tomb of Duke Philip the Bold at Dijon, finished in 1410. He was then commissioned to produce a similar tomb for Duke John the Fearless (1371–1419), but died before it was begun. His style was modelled very closely on that of Sluter.

Wesselmann, Tom (1931–). American painter, one of the best-known exponents of *Pop art. He frequently incorporates in his work elements of *collage and *assemblage (using household objects such as clocks and television sets), bringing representation and reality together to create a tension or ambiguity between the real world and the world of art. His creations, which sometimes incorporate sound effects, have the unreal reality of shop display windows. Like other Pop artists he favours banal objects and situations and he often favours aggressively sexual subjects and is best known for his continuing series *Great American Nude* (begun in 1961), in which the nude becomes a depersonalized sex symbol set in a realistically depicted commonplace environment. He emphasizes the woman's nipples, mouth, and genitals, with the rest of the body depicted in flat, unmodulated colour. In other works he isolates parts of the body still further, as in his *Smoker* series, in which the mouth—often depicted on a huge scale—becomes a provocatively erotic symbol.

West, Benjamin (1738–1820). American history and portrait painter who spent almost all his career in England. After early success as a portraitist in New York, he sailed for Italy in 1760. He spent three years studying there, chiefly in Rome, and in 1763 he settled in London. Here he soon repeated the professional and social success he had enjoyed in Italy, in part due to the novelty value of his being an American (a blind cardinal enquired whether he was a Red Indian). He became a founder member of the *Royal Academy in 1768 (in 1792 he succeeded *Reynolds as President) and in 1772 he was appointed historical painter to George III, with whom he had a long and lucrative association. Initially West had set up as a portraitist in London, but it was as a history painter that he made his mark. In Rome he had been in contact with the circle of Gavin *Hamilton and *Mengs, and his early work is in a determined but rather flimsy *Neoclassical style (*Agrippina Landing at Brundisium with the Ashes of Germanicus*, Yale Univ., 1768). With his famous *Death of Wolfe* (NG, Ottawa, 1770), however, he broke new ground; this was not (as is sometimes maintained) the first *history painting to feature contemporary costume, but it was the first such to score a triumph at the Royal Academy; his idea was soon adopted by other artists, most notably his countryman *Copley, and it marks an important turning-point in taste. As his style grew away from Neoclassicism, West was in the vanguard of the *Romantic movement with paintings such as the melodramatic *Saul and the Witch of Endor* (Wadsworth Atheneum, Hartford, 1777), and his *Death on a Pale Horse* (Philadelphia Museum of Art, 1817) has been hailed as a forerunner of *Delacroix. Following *Kent, he was also one of the pioneers of medieval subjects; in 1787–9 he did a series of paintings for Windsor Castle showing events from the life of Edward III (1312–77). West's historical importance far outweighs the quality of his work, which (in spite of its modernity in ideas) is generally pedestrian. He was the first American painter to achieve an international reputation and he became the prototype for the American expatriate artist. His studio was a centre for American artists visiting London, his collection was open to students as a gallery, and his influence on his successors, such as *Trumbull, was great.

Westall, Richard (1765–1836). English painter and graphic artist. His history paintings (he contributed to *Boydell's Shakespeare Gallery) are typical of the work of a second-class artist of the generation which was trained in the 18th-cent. tradition and lived to adapt itself to the *Romanticism of the new century. He was more successful in pastoral scenes and particularly in book illustration, in which field he was one of the most prolific artists of the day. As a watercolourist he was noted for his unusually rich colour effects.

Westmacott, Sir Richard (1775–1856). English *Neoclassical sculptor. The son of a sculptor also called **Richard** (1747–1808), he trained first under his father and then in Rome under *Canova (1793–7). After his return to London, he soon had a very large practice, second only to *Chantrey.

His best-known work is the huge *Achilles* statue (unveiled 1822) in Hyde Park; it honours the Duke of Wellington and is made of bronze—33 tons—from captured French cannon. At the time the figure's conspicuous nudity was considered shocking or amusing, especially considering it had been paid for by a group of lady subscribers. Westmacott also did the pediment sculpture on the British Museum (finished 1847). His work is dignified but often rather pedestrian and dead in handling. He was Professor of Sculpture at the *Royal Academy from 1827 to 1854.

Two of his brothers, **George** (active 1799–1827) and **Henry** (1784–1861), were sculptors, as was his son, another **Richard Westmacott** (1799–1872). Richard junior carried on his father's practice, but was much less distinguished. He was Professor of Sculpture at the Royal Academy from 1857 to 1867, though he virtually ceased to practise as a sculptor in 1855, and he was a well-known writer and lecturer on art.

Weyden, Rogier van der (1399/1400–64). The leading Netherlandish painter of the mid-15th cent. In spite of his contemporary celebrity (his work was appreciated in Italy as well as north of the Alps), his reputation later faded, and there is little secure knowledge about his career. There are, in fact, no paintings that can be given to him indisputably on the basis of signatures or contemporary documentation, but several are mentioned in early sources, and the style these show is so distinctive that a coherent *œuvre* has been built up around them. His early career is still somewhat problematic, however. In 1427 a certain Rogelet de la Pâture entered the workshop of Robert Campin at Tournai and left as Maistre Rogier in 1432. It is generally accepted that this is Rogier van der Weyden (the French and Flemish forms of the name both meaning 'Rogier of the Meadow'), although it is uncertain why he should have started his apprenticeship so late. There are no documented pictures surviving from Campin's hand, but he is generally agreed to be identical with the *Master of Flémalle, so the whole question of Rogier's relationship with his master is based on stylistic analysis. Some scholars have assumed that the Master of Flémalle should be identified with the young Rogier rather than with Campin, but the prevailing opinion is now that Rogier's work shows a development from the powerfully naturalistic and expressive style of his master towards greater refinement and spirituality. Rogier's celebrated *Deposition* (Prado, Madrid), for example, is close to the Master of Flémalle's *Crucified Thief* fragment (Städelsches Kunstinstitut, Frankfurt) in its dramatic power and use of a plain gold background, but it has a new poignancy and exaltedness. *The Deposition*, like all of Rogier's works, is undated, but it must be earlier than 1443, when a copy was made.

By 1436 Rogier had moved to Brussels and been appointed official painter to the city. Apart from making a pilgrimage to Rome in 1450, he is never known to have left Brussels again. His work for the city included secular work—four large panels (destroyed in 1695) on the theme of justice for the court room of the town hall, for example—but all his surviving paintings are either religious pictures or portraits. He was extremely inventive *iconographically and compositionally, and was a master of depicting human emotion. Unlike Jan van *Eyck he seems to have had a large workshop with numerous assistants and pupils, and many of his compositions are known in several versions. His influence was strong and widespread; in his own lifetime his paintings were sent all over Europe, to Spain, France, Italy, and Germany, and his emotional and dramatic style found more followers than the quiet perfection of van Eyck. Rogier's portraits, usually serene and aristocratic, were also much imitated, influencing Netherlandish portraiture until the end of the 15th cent.

Wheatley, Francis (1747–1801). English painter. He first practised in London as a painter of small full-length portraits and *conversation pieces in the manner of *Zoffany. After working in Dublin in 1779–83 he returned to London and began to specialize in scenes of rural and domestic life, imparting a certain 18th-cent. elegance to a genre in which *Hogarth had excelled, and exploiting a facility for popular moral sentiment. Engravings of his *Cries of London* (1795), showing street-vendors, milkmaids, and so on had a great sale and it is by these that he is generally remembered. The originals are now dispersed; examples are at Upton House (National Trust) and in the Geffrye Museum, London.

Whistler, James Abbott McNeill (1834–1903). American-born painter and graphic artist, active mainly in England. He spent several of his childhood years in Russia (where his father had gone to work as a civil engineer) and was an inveterate traveller. His training as an artist began indirectly when, after his discharge from West Point Military Academy for 'deficiency in chemistry', he learnt etching as a US navy cartographer. In 1855 he went to Paris, where he studied intermittently under *Gleyre, made copies in the *Louvre, acquired a lasting admiration for *Velázquez, and became a devotee of the cult of the Japanese print (see UKIYO-E) and, in general, oriental art and decoration. Through his friend *Fantin-Latour he met

*Courbet, whose *Realism inspired much of his early work. The circles in which he moved can be gauged from Fantin-Latour's *Homage to Delacroix*, in which Whistler is portrayed alongside *Baudelaire, *Manet, and others. He settled in London in 1859, but often returned to France. His *At the Piano* (Taft Mus., Cincinnati, 1859) was well received at the *Royal Academy exhibition in 1860 and he soon made a name for himself, not just because of his talent, but also on account of his flamboyant personality. He was famous for his wit and dandyism, and loved controversy. His life-style was lavish and he was often in debt. Dante Gabriel *Rossetti and Oscar Wilde (1854–1900) were among his famous friends.

Whistler's art is in many respects the opposite to his ostentatious personality, being discreet and subtle, but the creed that lay behind it was radical. He believed that painting should exist for its own sake, not to convey literary or moral ideas, and he often gave his pictures musical titles to suggest an analogy with the abstract art of music: 'Art should be independent of all claptrap—should stand alone, and appeal to the artistic sense of eye or ear, without confounding this with emotions entirely foreign to it, as devotion, pity, love, patriotism, and the like. All these have no kind of concern with it, and that is why I insist on calling my works "arrangements" and "harmonies".' He was a laborious and self-critical worker, but this is belied by the flawless harmonies of tone and colour he created in his paintings, which are mainly portraits and landscapes, particularly scenes of the Thames. His exquisite taste allowed him to combine disparate sources in a novel, almost abstract synthesis. No less original was his work as a decorative artist, notably in the Peacock Room (1876–7) for the London home of the Liverpool shipping magnate Frederick Leyland (now reconstructed in the Freer Gal., Washington), where attenuated decorative patterning anticipated much in the *Art Nouveau style of the 1890s. In 1877 *Ruskin denounced Whistler's *Nocturne in Black and Gold: The Falling Rocket* (Detroit Institute of Arts), accusing him of 'flinging a pot of paint in the public's face', and Whistler sued him for libel. He won the action, but the awarding of only a farthing's damages with no costs was in effect a justification for Ruskin, and the expense of the trial led to Whistler's bankruptcy in 1879. His house was sold and he spent a year in Venice (1879–80), concentrating on the etchings—among the masterpieces of 19th-cent. graphic art—that helped to restore his fortunes when he returned to London. He made a happy marriage in 1888 to Beatrix Godwin, widow of the architect E. W. *Godwin, with whom Whistler had collaborated, but she died only eight years later. In his fifties Whistler began to achieve honours and substantial success. His portrait of Thomas Carlyle (1795–1881) was bought by the Corporation of Glasgow in 1891 for 1,000 guineas and soon afterwards his most famous work, *Arrangement in Grey and Black: Portrait of the Painter's Mother* (Musée d'Orsay, Paris, 1871), was bought by the French state and he was made a member of the Legion of Honour.

Whistler's work is related to *Impressionism (although he was more interested in evoking a mood than in accurately depicting the effects of light), to *Symbolism, and to *Aestheticism, and he played a major role in introducing modern ideas of British art. Those who were most immediately influenced by him included his pupils Walter Greaves (1849–1930), Gwen *John, the Australian-born Mortimer Menpes (1860–1938), and W. R. *Sickert. His aesthetic creed was explained in his *Ten O'Clock Lecture* (1885) and this, and much else on art and society, was republished in *The Gentle Art of Making Enemies* (1890).

Whistler, Rex (1905–44). English painter, graphic artist, and stage designer. He is best known for his decorations in a light and fanciful style evocative of the 18th cent., notably the series of murals *In Pursuit of Rare Meats* (1926–7) in the restaurant of the Tate Gallery. He also did numerous book illustrations and much work for the stage, including ballet and opera. He was killed in action in the Second World War. His brother **Laurence Whistler** (1912–), a writer and glass engraver, has published several books on him.

Whiteread, Rachel. See TURNER PRIZE.

Whitney, Gertrude Vanderbilt (1875–1942). American sculptor, patron, and collector, the founder of the Whitney Museum of American Art, New York. She was the daughter of Cornelius II Vanderbilt, an immensely wealthy railroad magnate, and turned seriously to art after her marriage to Harry Payne Whitney, a financier and world-class polo player, in 1896. Her training as a sculptor included periods at the *Art Students League and in Paris, where she knew *Rodin. She won several major commissions, notably for monuments commemorating the First World War, including the Washington Heights War Memorial, New York (1921). Her style was traditional, but she was sympathetic towards progressive art and is much more important as a patron than as an artist. In 1907 she opened her New York studio as an exhibition space for young artists, and in 1914 she put her patronage on a more formal basis when she bought the house adjoining her studio, converted it into galleries, and opened it as the Whitney Studio; later she founded a series of organizations in New York with the same aim of helping young artists—the Friends of Young Artists (1915), the Whitney Studio Club (1918), and the Whitney Studio Galleries (1928). In 1929 she offered to donate her own collection of about 500 American paintings, sculptures, and drawings to the *Metropolitan Museum, New York, but the gift was turned down. Consequently in 1930 she announced the founding of the Whitney Museum of American Art and it opened the following year at 10 West 8th Street in a group of converted brownstone buildings. In 1954 the museum moved to a new building at 22 West 54th Street on land provided by the *Museum of Modern Art, and in 1966 to its present home—a spectacular building designed for it by Marcel Breuer at 945 Madison Avenue. The museum now has the largest and finest collection of 20th-cent. American art in the world, as well as a good representation from earlier periods. Every other year it holds the Whitney Biennial, a major showcase for work by living artists. Mrs Whitney donated funds to many other good causes, artistic and otherwise, but she was 'a woman of modest disposition who carried out her public activities quietly' (*Dictionary of American Biography*).

Wiener Werkstätte (German: Viennese workshops). Arts and crafts studios established at Vienna in 1903 by members of the *Sezession. The designers and craftsmen working here aimed, like William *Morris, to combine usefulness with aesthetic quality. Again like Morris, they hoped to reach a wide public but their prices precluded this. They made everything from jewellery to complete room decorations (including *mosaics) and they often worked in an *Art Nouveau style. The workshops closed in 1932.

Wiertz, Antoine (1806–65). Belgian painter, one of the great eccentrics in the history of art. He painted enormous religious, historical, and allegorical canvases in a staggeringly bombastic and almost dementedly melodramatic style and thought that he had surpassed the masterpieces of his models—*Michelangelo and *Rubens. The Belgian government built him a special studio in Brussels (now the Wiertz Museum) to produce these bizarre (and often macabre and erotic) works, and he said they were painted for honour and his portraits for bread.

Wijnants, Jan (d. 1684). Dutch landscape painter, active first in his native Haarlem, then from about 1660 in Amsterdam, where he also ran an inn. He specialized in landscapes with dunes and sandy roads, inspired by the countryside around Haarlem—unpretentious, naturalistic views that were favourites with collectors in the 17th and 18th cents. The figures in his paintings were apparently always painted by other artists, among them *Wouwerman. He, too, was an excellent painter of dunescapes and it is uncertain if one influenced the other. Wijnants was a prolific painter and his work is in many public collections. Adriaen van de *Velde was his pupil and *Gainsborough was among the artists he influenced.

Wilde, Johannes (1891–1970). Hungarian-born art historian who became an Austrian citizen in 1928 and a British citizen in 1947. From 1923 to 1938 he was on the staff of the Kunsthistorisches Museum in Vienna and he gained an international reputation with his work on the Italian (particularly the Venetian) paintings there. He made many contributions to the attribution and dating of pictures, and one of his most important achievements was the systematic use of X-rays not only as a tool for discovering the physical condition of a painting but also as a guide to the individual artist's creative process. Following the annexation of Austria by Nazi Germany in 1938, Wilde resigned from the Kunsthistorisches Museum and moved to England. From 1948 to 1958 he taught at the *Courtauld Institute, where he was an inspirational figure—Kenneth *Clark described him as 'the most beloved and influential teacher of art history of his time'. Wilde published comparatively little, and as Anthony *Blunt wrote 'his wisdom was mainly dispensed in lectures, supervisions and private conversation'. During his lifetime only two of his major contributions appeared in book form in English, both on *Michelangelo's drawings—in the catalogue of the 15th- and 16th-cent. drawings at Windsor Castle (1949, with A. E. Popham) and in his catalogue of Michelangelo's drawings in the *British Museum (1953). In these two works, which demonstrate his keen sensibility as well as his great learning, Wilde effectively reversed the 'revisionist' tendency whereby many genuine drawings by Michelangelo had been rejected. After his death, two collections of lectures were published: *Venetian Art from Bellini to Titian* (1974) and *Michelangelo: Six Lectures* (1978).

Wildens, Jan (*c*.1586–1653). Flemish landscape painter born in Antwerp, where he became a master in 1604. He is best known for painting landscape backgrounds for *Rubens and for many artists in his circle, but his finest independent work—*Winter Landscape with a Hunter* (Gemäldegalerie, Dresden, 1624)—shows he was an accomplished master in his own right.

Wilding, Alison. See NEW BRITISH SCULPTURE.

Wiligelmo (active c.1100). Italian *Romanesque sculptor. His name is known from an inscription on the façade of Modena Cathedral—'Among sculptors, your work shines forth, Wiligelmo. How greatly you are worthy of honours.' He must have been the main sculptor of the *reliefs on the façade (scenes from Genesis together with figures of prophets), which date from soon after 1099, when the cathedral was begun. Nothing else is known of him, but he created a distinctive style that was the fountainhead of Romanesque sculpture in northern Italy. His figures are squat and full of earthy vigour; within the context of Romanesque art they are markedly *Classical and owe much to antique *sarcophagi.

Wilkie, Sir David (1785–1841). Scottish painter. He was trained in Edinburgh and then in 1805 moved to London, where he studied at the *Royal Academy Schools. His *Village Politicians* (private coll.) was the hit of the RA exhibition of 1806 and he established himself as the most popular *genre painter of the day. He was strongly influenced in technique and subject-matter by 17th-cent. Netherlandish artists such as *Ostade and *Teniers, and the public loved the wealth of lively and often humorous incident in his paintings. In 1825–8 he travelled abroad for reasons of health and his style changed radically under the influence particularly of Spanish painting, becoming grander in subject-matter and broader in touch. The change was regretted by many of his contemporaries. In 1840 Wilkie went to the Holy Land to research material for his biblical paintings and on the return journey died at sea; *Turner commemorated him in *Peace: Burial at Sea* (Tate, London).

Wilkie's success did much to establish the popularity of anecdotal painting in England and many Victorian artists were influenced by him. The esteem in which he was held was possible only in an age which looked first to the 'story' of a painting and the moral lesson it contained. In his *Last Judgement* (Tate, 1853), painted twelve years after Wilkie's death, John *Martin portrayed Wilkie in the company of some of the world's greatest artists—*Leonardo, *Michelangelo, *Raphael, and so on—and many contemporaries would have concurred with this verdict.

Williams, Frederick (1927–82). Australian painter and graphic artist. He studied at the National Gallery School in his native Melbourne, 1944–9, and then in London, 1951–6. His earliest etchings, often with music-hall subjects, were produced about this time. He returned to Australia in 1957, and from the late 1950s began to produce paintings revealing a distinctly personal vision of the Australian landscape such as *Charcoal Burner* (NG of Victoria, 1959) and *You Yang's Pond* (Art Gal. of South Australia, 1963). By increasing reductivism, his paintings in the 1970s became uniquely evocative of the primeval mystery and remoteness of Australian landscape, and he was regarded as the most original of Australian landscape painters. His work is represented in many leading public collections including all Australian state galleries.

Willing, Victor. See REGO.

Willumsen, Jens Ferdinand (1863–1958). Danish painter, sculptor, architect, engraver, and potter. He was in Paris in 1888 and again in 1890–4, abandoning his early naturalistic manner under the impact of *Gauguin (whom he met) and *Symbolism. His work became highly individual, notable for its obscure and disturbing subject-matter and glaringly bright colours (*After the Tempest*, NG, Oslo, 1905). His sculpture is often *polychromatic, using mixed media, showing the influence of *Klinger. Willumsen was also influenced by El *Greco, on whom he wrote a long book (2 vols., 1927). There is a museum devoted to Willumsen at Frederikssund in Denmark.

Wilson, Richard (1713/14–82). British painter, born in rural North Wales, the son of a well-connected clergyman who encouraged his interest in art as well as giving him a good education. Wilson became the first major British artist to specialize in landscape, but after moving to London in 1729 he seems to have initially worked mainly as a portraitist. The decisive change in his career did not come until his visit to Italy in 1750–6, when he decided to devote himself exclusively to landscape. He is said to have done this at the urging of Francesco *Zuccarelli, whom he met in Venice and whose portrait (Tate, London, 1751) he painted, but he was more obviously influenced by the painting of *Claude and by the natural surroundings of Rome where Claude had worked—the Campagna, the Alban Hills, Tivoli, and the gardens of some of the villas overlooking the city. Here he found subject-matter suited both to his own taste and to that of his patrons, who were for the most part English noblemen, for whom he made chalk and charcoal drawings as well as paintings.

Back in England Wilson became successful with his Italian landscapes and applied the same classical compositional principles to English and Welsh views, as in his celebrated *Snowdon from Llyn Nantlle* (versions in the Walker Art Gallery, Liverpool, and the Castle Museum, Nottingham, c.1765).

He also painted large historical landscapes more or less in the manner of *Dughet or Salvator *Rosa, trying to establish himself as a master of the *Grand Manner (*The Destruction of the Children of Niobe*, Yale Center for British Art, 1760). Wilson, however, had a prickly nature and a problem with drink, and in the early 1770s his career went into a sharp decline. The *Royal Academy (of which he had been a founder member in 1768) helped him out by appointing him librarian, but in 1781 his family took him (now a pitiable figure) back to Wales. His work is of great importance in the history of British art, for he transformed landscape from an art that was essentially topographical to one that could be a vehicle for ideas and emotions. The only one of his pupils to attain distinction was Thomas *Jones, but he had many imitators and was admired by such later artists as *Cotman, *Crome, *Constable, and *Turner.

Wilson, Scottie (Robert) (1889–1972). British self-taught painter of imaginative subjects, born in Glasgow of working-class parents. A colourful character, he ran away from home at the age of 16, did military service in India and South Africa, and lived in Canada. He started to draw in the 1930s and his work was first exhibited shortly after the end of the Second World War. After the war he settled in London and became something of a character in the art world (he was barely literate and fond of the bottle). He made a good living selling his work to dealers but lived in very modest circumstances. Unlike the work of most *naïve artists (with whom he is sometimes grouped), his pictures (which are often in coloured inks) were not 'realistic' renderings of scenes from the daily life with which he was familiar but pure decorative fantasies incorporating stylized birds, fishes, butterflies, swans, flowers, self-portraits, and totem heads. The last of these he saw at Vancouver, and they were rather fancifully thought by some commentators to represent the powers of evil in contrast to the powers of good symbolized in the images taken from nature. All his work, however, was decorative rather than symbolic or profound. He is represented in a number of major collections, including the Tate Gallery, and the Museums of Modern Art in Paris and New York.

Wilton, Joseph (1722–1803). English sculptor. He trained in Flanders and with *Pigalle in Paris, then was in Italy from 1747 till 1755. On his return to London he rapidly became successful, carved the state coach (still in use) for the coronation of George III (1760), and was then appointed sculptor to the King (1764). He was an intimate friend of the architect Sir William Chambers (1723–96), with whom he often collaborated, and he was one of the original members of the *Royal Academy and Keeper there from 1790 till his death. His portraits are generally regarded as his best works, his monuments showing him hesitating between various styles. Much of his work was executed by assistants. His talents were considerable, but he was more interested in social life than his work; he inherited a fortune from his father (a manufacturer of ornamental plaster) but dissipated it and became bankrupt in 1793. Considering the quality of his training and his friendship with key figures in the art world, his career is a story of wasted opportunity.

Wilton Diptych. See INTERNATIONAL GOTHIC.

Winckelmann, Johann Joachim (1717–68). German art historian and archaeologist, a key figure in the *Neoclassical movement and in the development of art history as an intellectual discipline. He impressed contemporaries and later generations as much through his romantic lifestory as by his writing. The son of a poor cobbler, he early developed a fervent love for *classical antiquity and ancient art. After studying theology and medicine he held lowly positions as a schoolmaster and tutor whilst he taught himself Greek and absorbed himself in ancient culture. Eventually, in 1755, he managed to reach Rome (his conversion to Catholicism in 1754 was allegedly to facilitate this), where he became librarian to the famous collector Cardinal *Albani and soon established himself as a scholar and antiquarian of European fame. In 1768 he was murdered in Trieste, perhaps for the sake of some gold coins he had shown to a fellow guest at his inn.

Winckelmann's two most important books are *Gedanken über die Nachahmung der griechischen Werke in der Malerei und Bildhauerkunst*, published in 1755, shortly before he left for Rome (*Fuseli published an English translation in 1765 under the title *Reflections on the Painting and Sculpture of the Greeks*), and *Geschichte der Kunst des Altertums* (History of Ancient Art), published in 1764 (this is the first occurrence of the phrase 'history of art' in the title of a book). In these immensely influential works he proclaimed the superiority of Greek art and culture, combining rapturous descriptions of individual works (above all the **Apollo Belvedere*) with historical analysis. He never went to Greece and unwittingly based most of his observations on Roman copies, but his account of the stylistic development of Greek sculpture was a milestone in archaeological writing, and he is regarded as having laid the foundations of modern methods of art history. His analysis of ancient Greek culture as a unity, and his interpretation of art as an index of the spirit of the time were novel (he thought that when social conditions in general were good, then art was good, and when one declined the other did also); these ideas were subsequently developed into a proper philosophy of culture by 19th-cent. German writers. He refined the notions of how a work may be dated or its place of origin located and explained the character of works of art by reference to such factors as climate, religious customs and social conditions. His influence on contemporary artists—above all *Mengs—was enormous, and his interpretation of classical antiquity determined aspects of German education right into the 20th cent. *Goethe wrote of him: 'It was Winckelmann who first urged on us the need of distinguishing between various epochs and tracing the history of styles in their gradual growth and decadence. Any true art lover will recognize the justice and importance of this.'

Wint, Peter De. See DE WINT.

Winterhalter, Franz Xaver (1805–73). German painter, the most successful court portraitist of his period. He was based in Paris for most of his career, but he painted most of Europe's royalty and was a particular favourite of Queen Victoria (1819–1901), who called him 'excellent, delightful Winterhalter' (the royal collection has more than a hundred of his paintings). His style was romantic, glossy, and superficial and his portraits have until recently generally been valued more as historical records than as works of art. However, a major exhibition of his work at the National Portrait Gallery, London, and the Petit Palais, Paris, in 1987 brought him into the limelight again. Winterhalter was also an accomplished lithographer. His brother **Hermann** (1808–91) was a painter. A watercolour by him, *A Girl of Frascati* (signed but until recently given to his brother), is in the Wallace Collection.

Wiszniewski, Adrian. See GLASGOW SCHOOL.

Wissing, Willem (1655–87). Dutch portrait painter who moved to London in 1676, became an assistant of *Lely, and took over much of his fashionable practice after Lely died in 1680. His style was coarser and more Frenchified than that of Lely. Wissing was particularly favoured by James II (1633–1701), but his early death opened the way for *Kneller's dominance.

Wit, Jacob de (1695–1754). The outstanding Dutch decorative painter of the 18th cent., active mainly in his native Amsterdam. He had his principal training in Antwerp and learned much from *Rubens's ceiling paintings in the Jesuit church there (his drawings became valuable records after the paintings were destroyed by fire in 1718). De Wit's style, however, was much lighter than Rubens's, with a *Rococo delicacy and charm. He was a Catholic and was the first Dutch artist since the 16th cent. to carry out a good deal of decorative work for Catholic churches, but he was at his best in domestic ceiling decorations (*Bacchus and Ceres in the Clouds*, Huis Boschbeek, Heemstede, 1751). His name has entered the Dutch language to describe a kind of *trompe-l'œil* imitation of marble *reliefs for which he was renowned; such pictures, usually set over a chimney-piece or door, are called 'witjes' (*wit* is Dutch for 'white'). De Wit was also an engraver and a noted collector of Old Master drawings.

Witt, Sir Robert. See COURTAULD.

Witte, Emanuel de (1615/17–91/2). Dutch painter, born at Alkmaar and active there, then in Rotterdam (by 1639), Delft (by 1641), and Amsterdam (by 1652). His range was wide, including history paintings, *genre scenes (notably of markets) and portraits, but after he settled in Amsterdam he concentrated on architectural paintings (primarily church interiors—both real and imaginary). *Houbraken wrote that 'in the painting of churches, no one was his equal with regard to orderly architecture, innovative use of light and well-formed figures', and this verdict has been endorsed by posterity, for his paintings are very different in spirit from the sober views of most Dutch architectural specialists, making powerful use of the dramatic play of light and shadow in the lofty interiors. His life was unhappy (he was constantly in debt) and he evidently committed suicide (his body was found in an Amsterdam canal). There are good examples of his work in the National Gallery, London, the Rijksmuseum, Amsterdam (which has a beautiful sunset seascape), and numerous other museums.

Wittkower, Rudolf (1901–71). German-born art historian who became a British citizen in 1934. He worked at the Biblioteca Hertziana in Rome from 1923 to 1933. From 1934 to 1956 he was on the staff of the *Warburg Institute in London and from 1949 to 1956 was also professor of history of art at University College, London. In 1956 he moved to Columbia University, New York, as head of the Department of Fine Arts and Archaeology, which under his direction became one of the leading centres of art historical scholarship in the USA. After he retired in 1969 he was *Kress Professor at the National Gallery in Washington and *Slade Professor at Cambridge. Wittkower's many books and articles were devoted mainly to Italian art and architecture of the 16th and 17th cents., and his writings form one of the cornerstones in the study of Italian *Baroque art. His major books, several of which have appeared in revised editions, include: *Architectural Principles in the Age of Humanism* (1949), *Gian Lorenzo Bernini* (1955), *Art and Architecture in Italy 1600–1750* (1958), *Sculpture* (1977), and (with his wife **Margot Wittkower**) *Born under Saturn: The Character and Conduct of Artists* (1963).

Witz, Konrad (*c*.1400–44/6). German-born painter from Rottweil in Swabia, active in Switzerland and generally considered a member of the Swiss school. He entered the painters' guild in Basle in 1434 and apparently spent the rest of his career there and in Geneva. Little else is known of him and few paintings by him survive. These few, however, show that he was remarkably advanced in his *naturalism, suggesting a knowledge of the work of his contemporaries Jan van *Eyck and the *Master of Flémalle. In place of the soft lines and lyrical qualities of *International Gothic we find in Witz's work heavy, almost stumpy, figures, whose ample draperies emphasize their solidity. Witz's most famous works are the four surviving panels (forming two wings) from the altarpiece of St Peter he painted for the cathedral in Geneva. These are now in the Musée d'Art et d'Histoire there; the central panel is lost. One of the panels, the *Miraculous Draught of Fishes*, is Witz's masterpiece and his only signed and dated work (1444). The landscape setting depicts part of Lake Geneva (one of the earliest recognizable landscapes in art) and Witz's naturalism is even more remarkable in his observation of reflection and refraction in the water.

Wölfflin, Heinrich (1864–1945). Swiss art historian, professor at the universities of Basle (1893–1901), Berlin (1901–12), Munich (1912–24), and Zurich (1924–34). He was one of the most influential art historians of his period, and several of his books are still widely read. They include *Die klassische Kunst* (Classic Art, 1899), on the art of the High *Renaissance, *Kunstgeschichtliche Grundbegriffe* (Principles of Art History, 1915), and a monograph on *Dürer (1905). Most of his work was devoted to stylistic analysis, and he attempted to show that style—in painting, sculpture, and architecture—follows evolutionary principles. *Principles of Art History* presents his ideas in the most fully developed form, discussing the transformation from *Renaissance to *Baroque in terms of contrasting visual schemes—for example the development from linear to painterly ('malerisch'). Wölfflin's view that style was a force in its own right rather than an intellectual abstraction and his lack of interest in *iconography are out of tune with much modern thinking on art history, and his approach is often over-rigid; however, he was a figure of great importance in establishing his subject as an intellectually demanding discipline. Herbert *Read wrote that 'it could be said of him that he had found art criticism a subjective chaos and left it a science'.

Wolgemut, Michael (1434–1519). German painter and woodcut designer, active in his native Nuremberg. In 1472 he married the widow of Hans *Pleydenwurff and took over his workshop, the most prosperous in the city. It produced numerous large altarpieces in which there is little sign of a distinctive individual personality, and Wolgemut is more important for his book illustrations. Amongst many other books he illustrated was Hartman Schedel's *Weltchronik* (1493), the most enterprising attempt of its time at combining letterpress with woodcut illustration. Hitherto woodcuts had often been embellished by *illumination, but Wolgemut tried to refine the technique of woodcut so that it could achieve its own proper effects without hand painting. His pupils included *Dürer.

Wols (pseudonym of Alfred Otto Wolfgang Schulze) (1913–51). German-born painter active mainly in France. Although he had shown a talent for drawing from early childhood, he was converted to serious interest in art by *Moholy-Nagy and studied for a while at the *Bauhaus. He went to Paris in 1932, where he earned a living by photography under the name Wols. He was interned at the outbreak of war but liberated in 1940 and lived in poverty in the south of France. On the termination of war he returned to Paris and was befriended by Jean-Paul Sartre (1905–80) and Simone de Beauvoir (1908–86), for whom he did illustrations. In the late 1940s he began to make a name for himself as a painter, but his irregular life, poverty, and excessive drinking undermined his health and he died in 1951. His posthumous fame far outstripped his reputation during his lifetime and he came to be regarded as the 'primitive' of *Art Informel and one of the great original masters of expressive abstraction. His work had an important influence on the *Tachiste art of the late 1940s and 1950s. His *œuvre* consisted of a few large paintings, many finely executed watercolours, and a large number of drawings.

Wood, Christopher (1901–30). British painter, mainly of landscapes, harbour scenes, and figure compositions. In 1921 he studied at the *Académie Julian in Paris and subsequently travelled widely on the Continent. To influences from modern French art (*Picasso and *Diaghilev were among his friends), he added an entirely personal lyrical freshness and intensity of vision, touched with what Gwen *Raverat felicitously described as 'fashionable clumsiness'. In a remarkably short time he achieved a position of high regard in the art worlds of London and Paris but he was emotionally unstable and his early death was probably suicide (he was killed by a train). After this he became something of a legend as a youthful genius cut off before his prime. Much of Wood's best work was done in Cornwall, where he and his friend Ben *Nicholson discovered the *naïve painter Alfred *Wallis in 1928. There are examples of Wood's work in the Tate Gallery.

Wood, Derwent. See NEW SCULPTURE.

Wood, Grant (1892–1942). American painter, active mainly in Iowa, his native state. Early in his career he was an artistic jack-of-all-trades. The turning-point in his life came when he obtained a commission to make stained-glass windows for the Cedar Rapids Veteran Memorial Building in 1927 and went to Munich to supervise their manufacture the following year. Influenced by the early Netherlandish paintings he saw in museums there, he abandoned his earlier *Impressionist style and began to paint in the meticulous, sharply detailed manner for which his work is chiefly known. Adapting this style to the depiction of the ordinary people and everyday life of Iowa, he became one of the leading exponents of *Regionalism. He first came to national attention in 1930 when his painting *American Gothic* won a bronze medal at an exhibition of the Art Institute of Chicago, which now owns the painting. Although at the time it aroused violent controversy and was deplored as an insulting caricature of plain country people, the painting gradually gained great popularity. In 1931 Wood introduced an element of humorous fantasy in *The Midnight Ride of Paul Revere* (Met. Mus., New York) and in 1932 he painted his famous satirical *Daughters of the Revolution* (Cincinnati Art Museum), described as 'three sour-visaged and repulsive-looking females, represented as disgustingly smug and smirking because of their ancestral claim to be heroes of the American Revolution'. This may have been Wood's quiet revenge for the opposition to the dedication of his stained-glass windows put up by the ladies of Cedar Rapids because they had been made in Germany. His other work includes some vigorous stylized landscapes, and during the 1930s he supervised several Iowa projects of the *Federal Art Project. In 1934 he became assistant professor of fine arts at the University of Iowa.

woodcut. Term applied to the technique of making a *print from a block of wood sawn along the grain and to the print so made. It is the oldest technique for making prints, and its principles are very simple. The artist draws his design on the smooth, flat surface of a block of wood (almost any wood of medium softness will serve—beech, pear, sycamore for example) and then cuts away, with knife and gouges, the parts which are to be white in the print, leaving the design standing up in *relief. After *inking the surface of the block, he places upon it a sheet of *paper. Finally by applying pressure to the back of the paper, either by hand or in a press, he transfers to the sheet an impression in reverse of his original design. The colour woodcut is usually made by cutting a separate block for each colour and printing the blocks successively on the same sheet of paper.

Woodcuts come into the category of relief prints and differ basically from *etchings and *line engravings, which are printed from the grooves incised in the plate (see INTAGLIO). The characteristics of woodcut are boldness, simplicity, and ruggedness of line, all of which derive from the nature of the wood itself and the simple tools with which it is cut. Subtlety, however, is not unattainable, and though the virtues of the early woodcuts are chiefly those of bold drawing and forceful cutting, modern designers have achieved surprising results by exploiting the surface qualities of the wood. Led by Paul *Gauguin, some have scored and scratched the surface of the block with sandpaper and other unconventional tools, thus imparting to the print a variety of mysterious textures, while others, following the example of Edvard *Munch, have used the wood-grain itself, lightly inked and carefully printed, to add interest to the solid areas of their designs.

The history and practice of woodcutting have always been closely connected with the art of the book. Since the impression is taken from the relief surface of the block, which can be made the same thickness as the standard height of type, the medium is inherently suitable for book illustration. The engraved blocks may be locked with the movable type in the press so that text and illustration are inked and printed together. This procedure is much cheaper and simpler than illustrating with copperplate engravings, which have to be printed in a different press and on separate sheets, and bound into the book afterwards.

The origins of woodcut are obscure, for the medium is very ancient. Stamps cut on wood for fabric printing dating from the 5th cent. AD have been found in the Middle East, and there is evidence of their use in still older times. The earliest known dated woodcut appeared in a book known as the *Diamond Sutra*, a Buddhist scripture printed in China in AD 868. The knowledge of paper-making, on which the *graphic arts as we know them depend, did not reach Europe from the East until much later, and we may assume that the art of woodcutting in the West, for the purpose of taking paper prints, dates approximately from the beginning of the 15th cent. It developed steadily during the next 100 years, in Germany, in Italy, and finally in France, taking three main forms: religious pictures, playing-cards, and book illustrations.

The first extant European cuts are from the first quarter of the 15th cent. They are religious pictures, probably distributed at pilgrimage centres, depicting the Virgin and the saints in simple linear design, and apparently printed by hand pressure. Woodcut playing-cards also appear in the first half of the 15th cent., sometimes hand coloured. The practice of cutting a simple inscription as part of the woodcut, which is found on many of the religious prints of that time, led directly to the *block books of the 15th cent., where each page was printed from a single block on which was cut both text and illustration. However, printed book illustration as we know it was made possible when Johann Gutenberg invented printing with movable type, between 1440 and 1460. By 1470 woodcut illustrations, already common in Germany, were beginning to appear in Italian books.

This early period appears in retrospect to have been the Golden Age of the woodcut. In Italy comparatively few single-sheet prints were issued, but the illustrated books, of which the *Hypnerotomachia Poliphili*, published by Aldus Manutius in Venice in 1499 (see ALDINE PRESS), is the most celebrated, were one of the great achievements of the *Renaissance. The Italian style is calmer and more noble than the northern; there is purity of drawing and design, an interesting use of decorative borders and solid black areas, and above all a feeling of complete unity with the accompanying type.

Although *Dürer was a great master of the woodcut, in the 16th cent. the technique lost ground to line engraving, especially as a reproductive medium. The cross-*hatching, for example, which the taste of the times demanded, could be rendered easily enough by the line engraver, who had only to make intersecting strokes with his *burin, but the woodcutter was condemned to the laborious and inartistic task of cutting out row upon row of lozenges, working with relatively clumsy tools upon a somewhat recalcitrant surface. Not until the end of the 18th cent., when the technique of burin engraving on wood was developed (see WOOD ENGRAVING), did the relief print again challenge the intaglio process for public favour.

During the 17th and 18th cents. woodcut was used solely for popular jobbing work—broadsheets, printers' decorations, and so forth. The revival of the art at the end of the 19th cent. was due to painters such as Gauguin, Munch, and Félix *Vallotton, who saw once again its possibilities as a means of original expression; for now that reproduction was taken over by the process camera, all kinds of hand engraving could make a fresh start. A return was made to the simplicity and directness of the 15th cent., with the difference that the modern artists learned the craft themselves and cut their own designs on wood, both for book illustration and for free prints. Rich and forceful illustrations were cut by *Derain and *Dufy before the First World War, and the sculptor *Maillol, taking as his models the Italian books of the Renaissance, succeeded with designs of a truly classical spirit in achieving again a wonderful balance between type and illustration. The German *Expressionists found the woodcut particularly suited to their needs, and *Kirchner, Franz *Marc, and others made prints in the medium that are among the masterpieces of 20th-cent. graphic art.

wood engraving. Term applied to the technique of making a print from a block of hardwood (usually boxwood) sawn across the grain and to the print so made. It derives from the *woodcut, but because of the harder and smoother surface and the use of the *burin and other tools associated with the copper-plate engraver, the effect is generally finer and more detailed. As in woodcut the printing is done from the surface of the block, and the parts that are not to be printed are cut away. The wood used is a hardwood, generally box, sawn across the grain and very highly polished. The medium is often referred to as 'end-grain' engraving (French *bois de bout*), as distinct from 'side-grain' woodcutting (*bois de fil*). In practice, however, it is not always easy to tell which method produced a given print (there is a middle ground where the results are very similar), and the French, though they have different names for the two techniques, tend to minimize the distinction by referring to both simply as *gravure sur bois*.

Wood engraving developed in the 18th cent. and its first great exponent was Thomas *Bewick (his contemporary William *Blake also made some superb wood engravings, but these were little known at the time). Bewick ushered in the technique's heyday, which coincided with a great expansion in journalism and book publishing from about 1830. From then until about 1890, when it was superseded by photomechanical processes, it was the most popular medium for illustration. The *Dalziel brothers and Gustave *Doré were among the most prolific exponents during this period.

Woollett, William (1735–85). English draughtsman and *line engraver. In 1760 he engraved *Claude's *Temple of Apollo* for *Boydell, and he established his reputation in 1761 with *Wilson's *Niobe*. His greatest success was *The Death of Wolfe* after *West (1776), for which he was given the title Historical Engraver to His Majesty. Woollett was the first English engraver whose work had a considerable market on the Continent.

Woolner, Thomas (1825–92). English sculptor, the only sculptor member of the *Pre-Raphaelite Brotherhood. His early career was unsuccessful, so he decided to try his hand at gold-prospecting in Australia. His departure in 1852 inspired Ford Madox *Brown's picture *The Last of England* (City Art Gallery, Birmingham, 1852–5). Woolner found little gold, but he began to prosper as a portrait sculptor, and after returning to England made a name for himself with his marble bust of Alfred Tennyson of 1857 (Trinity College, Cambridge, and replicas in Westminster Abbey and the NPG). His work was praised for its lifelikeness and he did portraits of many other distinguished sitters. Woolner also did a few figure subjects, occasionally painted, and wrote poetry.

Wootton, John (*c*.1682–1764). English landscape and sporting painter. He specialized very successfully in horse subjects, but his main contribution to British painting was the introduction of the *classical landscape—Horace *Walpole said his works in this vein 'approached towards Gaspar Poussin [i.e. *Dughet], and sometimes imitated happily the glow of *Claude Lorrain'. His landscape manner was continued by his pupil George *Lambert. There are examples of Wootton's work in the Tate Gallery and other public collections, but many of his finest paintings are still in private hands—notably at Althorp, Northamptonshire, and Longleat, Wiltshire.

Works Progress Administration. See FEDERAL ART PROJECT.

World of Art (Mir Iskusstva). The name of an informal association of Russian artists and of the journal they published from 1899 to 1905; the association was formed in St Petersburg in 1898 and lasted in its original form until 1906 (it was revived as an exhibiting society, 1910–24). *Diaghilev was the journal's editor, and his contributors and collaborators included Leon *Bakst and Alexander *Benois. The group encouraged interchange with Western art (many articles published in the journal had previously appeared in European magazines) and became the focus for avant-garde developments in Russia. In particular it promoted the *Art Nouveau style. Some of the artists involved in the group (notably Nikolai *Roerich) were also interested in evoking the spirit of ancient Russia, and this synthesis of old and new was best expressed in their decor for Diaghilev's ballet company, which revolutionized European stage design when he brought it to Paris.

Worpswede. A north German village near Bremen that in the last decade of the 19th cent. became the centre of a group of artists who settled there, following the example of the *Barbizon School in France. The most famous artist to work there was Paula *Modersohn-Becker, and the 'Worpswede School' is sometimes regarded as one of the roots from which German *Expressionism sprang.

Worringer, Wilhelm (1881–1965). German art historian and aesthetician, professor at the universities of Bonn (1920–8), Königsberg (1928–46), and Halle (from 1946). His best-known books are *Abstraktion und Einfühlung* (1907) (translated as *Abstraction and Empathy*, 1953) and *Formprobleme der Gotik* (1912), which has been translated as *Form Problems of the Gothic* (1919) and *Form in Gothic* (1927). His writings, which are somewhat metaphysical in tone, encouraged a more sympathetic response to non-realist styles and expressionist distortion in art.

Wotruba, Fritz (1907–75). The leading Austrian sculptor of the 20th cent. His masterly craftsmanship soon won him acclaim, as when his work was shown in an exhibition of Austrian art in Paris in 1929; Aristide *Maillol is said to have refused to believe that such work could have been done by a 22 year old. In his most characteristic works he carved directly in stone, preferring a hard stone with a coarse texture. He began his career working in a naturalistic style reminiscent of Maillol, but moved towards abstraction by reducing the figure to bare essentials. It was similar to the method used by *Brancusi, but in contrast to Brancusi's smooth, subtle abstractions, Wotruba's figures are solid, block-like structures. They were left in a rough state, creating a feeling of primitive power. Wotruba also worked in bronze, and near the end of his life he branched out into architecture, designing the church of the Holy Trinity on the outskirts of Vienna (constructed 1974–6). After the Second World War he taught at the Vienna Academy and his work was admired and imitated by many younger Austrian artists, bringing about a revival of sculpture in his country.

Wouters, Rik (1882–1916). Belgian painter and sculptor whose great talents reached incomplete development owing to his early death following operations for cancer of the eye. He is regarded as the leading Belgian exponent of *Fauvism, but his work is less violent in colour than that of the French adherents of the style and often (as in his portraits of his wife) has a quality of serene intimacy.

Wouwerman (or **Wouwermans**), **Philips** (1619–68). The most celebrated member of a family of Dutch painters from Haarlem, where he worked virtually all his life. He became a member of the painters' guild in 1640 and is said by a contemporary source to have been a pupil of Frans *Hals. The only thing he has in common with Hals, however, is his nimble brushwork, for he specialized in landscapes of hilly country with horses (usually including a grey)—cavalry skirmishes, camps, hunts, travellers halting ouside an inn, and so on. In this genre he was both immensely prolific and immensely successful—*Houbraken says he left his daughter a dowry of 20,000 guilders. He had many imitators, including his brother **Peter** (1623–82), and his great popularity continued throughout the 18th cent., when he was a favourite with princely collectors and engravings after his work had wide circulation. Subsequently he has perhaps been underrated, for although his work generally follows a successful formula, he maintained a high quality; his draughtsmanship is elegant, his composition sure, his colouring delicate, and his touch lively. Examples of his work are in many galleries; the biggest collections are in Dresden (Gemäldegalerie) and St Petersburg (Hermitage), but he is also very well represented in London (Dulwich College Picture Gal., NG, Wallace Coll., Wellington Mus.).

WPA. See FEDERAL ART PROJECT.

Wright, Frank Lloyd (1867–1959). The outstanding American architect of the 20th cent., also an important designer and theorist. His views, particularly as expressed in his lecture *The Art and Craft of the Machine* (1901), had considerable influence on the *Arts and Crafts Movement.

Wright, John Michael (1617–94). English portrait painter. He was apprenticed to George *Jamesone in Edinburgh in 1636, then spent a long period abroad, chiefly in Rome, where he became a member of the Accademia di S. Luca (see ACADEMY), the only British painter of the 17th cent. to have this distinction. On his return to England in 1656 he won many patrons among his fellow Catholics and became *Lely's chief rival, although he never had his great worldly success and died in modest circumstances. His style was less glossy than Lely's but more penetrating and individual in characterization, his sitters tending to look thoughtful rather than merely glamorous (*Magdalen Aston*, Castle Museum, Nottingham). Wright's most unusual work is an allegorical ceiling painting (Castle Museum) done for Charles II's bedroom at Whitehall Palace; it cannot be considered a success, but it is interesting as an attempt at *Baroque decoration by someone who was familiar with the works of Pietro da *Cortona (whom he described as 'the greatest master of his time'). Wright was a collector, antiquarian, and scholar, a man of considerable culture; in 1685–7 he accompanied an embassy from James II (1633–1701) to Pope Innocent XI (1611–89) and wrote an account of it in Italian (1687, English translation, 1688).

Wright, Joseph (1734–97). English painter, born and principally active in Derby and generally known as 'Wright of Derby'. He was one of the most original, versatile, and accomplished British artists of the 18th cent. and the first major English painter whose career was based outside London. In 1751–3 and again in 1756–7 he trained under *Hudson in London, and returning to Derby he made a name as a portraitist in the Midlands, his works displaying a firm grasp of character. In the 1760s he began to paint candlelit scenes of various types, showing the fascination with unusual lighting effects that was to run throughout his career. He was influenced in some of these by Dutch painting, but in his depictions of the contemporary scientific world he broke new ground—his finest work, *An Experiment on a Bird in the Air Pump* (NG, London, 1768), has been described by Sir Ellis *Waterhouse as 'one of the wholly original masterpieces of British art'. Such works won him a considerable reputation, and in 1772 James *Northcote called him 'the most famous painter now living for candle-lights'. In 1768–71 Wright worked in Liverpool, and in 1774–5 was in Italy, where he assiduously studied the *antique but was even more impressed by the eruption of Vesuvius he witnessed and by a kind of man-made equivalent—the great fireworks display held annually at the Castel Sant' Angelo in Rome (he painted several pictures of both subjects). On his return to England in 1775 he moved to Bath, hoping to fill the gap left by *Gainsborough's departure for London, but his more forthright style did not please sophisticated society there and in 1777 he returned to Derby. There he remained for the rest of his life apart from short journeys when he made tours of the Lake District in 1793 and 1794 (landscape became increasingly important towards the end of his career). Appropriately, Derby Art Gallery has far and away the finest collection of his work.

Wtewael (or **Uytewael**), **Joachim** (1566–1638). Dutch figure painter. After travelling in Italy and France *c.*1588–92, he settled in his native Utrecht, where he became one of the leading Dutch exponents of *Mannerism. His highly distinctive, charmingly artificial style, which remained untouched by the naturalistic developments happening around him, was characterized by acidic colours and elegant figures in wilfully distorted poses. The best collection of his work, including a self-portrait (1601), is in the Centraal Museum, Utrecht.

Wyeth, Andrew (1917–). American painter, son and pupil of a well-known muralist and illustrator of children's books, **Newell Convers Wyeth** (1882–1944). Wyeth's work consists almost entirely of depictions of the people and places of the two areas he knows best—the Brandywine Valley around his native Chadds Ford, Pennsylvania, and the area near Cushing, Maine, where he has his summer home. He usually paints in watercolour or *tempera with a precise and detailed technique, and often he conveys a sense of loneliness or nostalgia (trappings of the modern world, such as motor cars, rarely appear in his work). He became famous with *Christina's World* (1948), which was bought by the Museum of Modern Art, New York, in 1949 and has become one of the best-known images in American art. It depicts a friend of the artist, Christina Olson, who had been so badly crippled by polio that she moved by dragging herself with her arms. She is shown in a field on her farm in Maine, 'pulling herself slowly back towards the house'. Building on the picture's fame, Wyeth has gone on to have an enormously successful career. There is great disparity of critical opinion about him, however: J. Carter Brown, Director of the National Gallery in Washington, has called him 'a great master', whereas Professor Sam Hunter, one of the leading authorities on 20th-cent. American art, has written 'What most appeals to the public, one must conclude, apart from Wyeth's conspicuous virtuosity, is the artist's very banality of imagination.'

xylography. A term (which is now little used) for any kind of printing from a wooden block, thus encompassing both *woodcut and *wood engraving.

Yáñez (or **Yáñez de la Almedina**), **Fernando** (active 1506–26). Spanish painter. He is first documented in 1506 in Valencia, where in 1507–10 he collaborated with Fernando de los Llanos (active 1506–25) in painting twelve panels of the *Life of the Virgin* for the main altarpiece of the cathedral. From the style of these paintings it is inferred that the 'Fernando spagnuolo' who was working in 1505 with *Leonardo da Vinci on the *Battle of Anghiari* was either Yáñez or Llanos. Yáñez is credited with being the more gifted artist and one of the most important influences in introducing the High *Renaissance style to Spain. Examples of his work are in the museum at Valencia and the Prado in Madrid.

Yeames, William Frederick (1835–1918). British painter. He specialized in scenes from British history, particularly of the Tudor and Stuart periods, treated in an anecdotal and often melodramatic way that looks forward to Hollywood 'costume' movies. Most of his work has been forgotten, but one of his pictures has achieved enduring fame as an archetypal image of Victorian sentimentality, the Cromwellian tear-jerker *'And When Did You Last See Your Father?'* (Walker Art Gallery, Liverpool, 1878).

Yeats, Jack Butler (1871–1957). The best-known Irish painter of the 20th cent., son of **John Butler Yeats** (1839–1922), a barrister who became a successful portrait painter, and brother of the poet William Butler Yeats (1865–1939). He initially worked mainly as an illustrator and did not work regularly in oils until about 1905. His early work as a painter was influenced by French *Impressionism, but he then developed a more personal *Expressionistic style characterized by vivid colour and extremely loose brushwork (there is some similarity to the work of *Kokoshka, who became a great friend in the last decade of Yeats's life). He painted scenes of Celtic myth and everyday Irish life, contributing to the upsurge of nationalist feeling in the arts that accompanied the movement for Irish independence. His work is represented in numerous galleries in Ireland, notably the National Gallery and the Hugh Lane Gallery of Modern Art in Dublin and the Sligo County Museum and Art Gallery, which has material relating also to his father and brother (Sligo being the ancestral home). Yeats was a writer as well as a painter—the author of several plays, novels, and volumes of

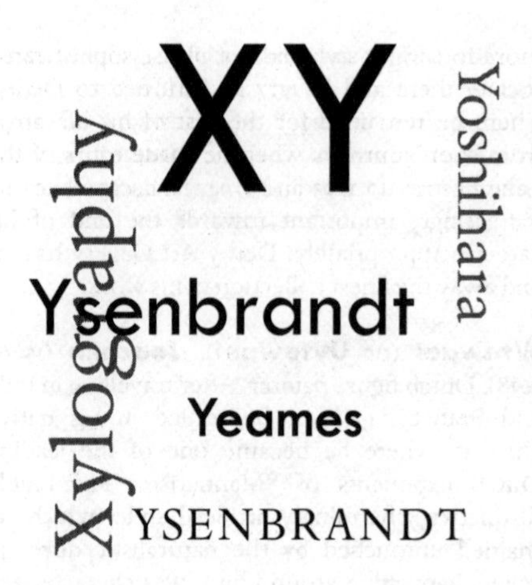

poetry, as well as *Life in the West of Ireland* (1912) and *Sligo* (1930).

Yoshihara, Jiro (1905–72). Japanese painter. During the 1930s he was a pioneer of abstract art in Japan and in the 1950s he took the lead in the formation of the *Gutai group at Osaka. His best-known works—the *Circle* series—date from this time, and feature irregular circular forms left blank against a painted background. They were done under the influence of Zen Buddhism.

Young Contemporaries. Exhibition of works by British art students held in London since 1949 on a roughly annual basis (lack of funds or organization—they are generally arranged by the students themselves—has sometimes prevented the shows taking place). The first Young Contemporaries show was held at the Royal Society of British Artists (RBA Galleries). Other venues have included the *Institute of Contemporary Arts, and some of the exhibitions have toured the provinces. The most famous Young Contemporaries exhibition was that of 1961, when British *Pop art first appeared in force in the work of Derek *Boshier, David *Hockney, Allen *Jones, R. B. *Kitaj, and Peter *Phillips, all of them students or former students at the *Royal College of Art. In 1974 the name was changed to 'New Contemporaries'.

Ysenbrandt (or **Isenbrandt**), **Adriaen** (d. 1551). Netherlandish painter. He became a master in Bruges in 1510 and is said by an early source to have been a pupil of Gerard *David. Otherwise, virtually nothing is known of him and there are no signed or documented works. However, in 1902 the Belgian art historian Georges Hulin de Loo (in the catalogue of a major exhibition of Early Netherlandish painting held at Bruges) proposed

Ysenbrandt as the author of a large group of paintings deriving from David, and the identification has generally been accepted. Previously the paintings had been attributed to Jan *Mostaert and the anonymous Master of the Seven Sorrows of the Virgin, named after a *diptych of that subject divided between the church of Notre-Dame in Bruges and the Musées Royaux in Brussels. Some authorities prefer to use the name Ysenbrandt in inverted commas.

Yvaral. See VASARELY.

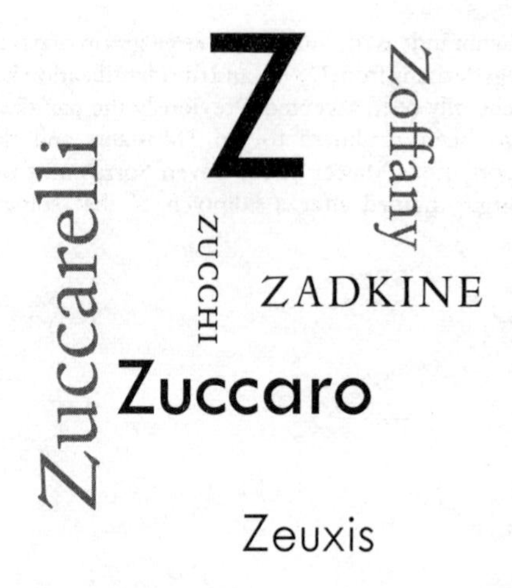

Zadkine, Ossip (1890–1967). Russian-born sculptor who worked mainly in Paris and became a French citizen in 1921. He moved to Paris in 1909 after spending four years in Britain (sent there by his father—a professor of classics—to learn 'English and good manners'). Although he deeply admired *Rodin, *Cubism had a greater impact on his work. His experiments with Cubism, however, had none of the quality of intellectual rigour and restraint associated with *Picasso and *Braque, for although Zadkine was a learned artist, his primary concern was with dramatically expressive forms. The individual style he evolved made great use of hollows and concave inflections, his figures often having openings pierced through them.

In 1915 he joined the French army but was invalided out after being gassed. He worked in Paris through the 1920s and 1930s, and spent the Second World War in New York (where he taught at the *Art Students League), returning to Paris in 1944. Often Zadkine's work can seem merely melodramatic, but for his greatest commission—the huge bronze *To a Destroyed City* (completed 1953) standing at the entry to the port of Rotterdam—he created an extremely powerful figure that is widely regarded as one of the masterpieces of 20th-cent. sculpture. With its jagged, torn shapes forming an impassioned gesture mixing defence and supplication, it vividly proclaims anger and frustration at the city's destruction and the courage which made possible its rebuilding. This work gave Zadkine an internationed reputation and many other major commissions followed it. The house in which he lived in Paris is now a museum dedicated to his work.

Zeitblom, Bartholomaus (Bartel) (*c.*1460–*c.*1520). German painter of religious works and portraits. He was the leading painter of the day in Ulm, and his prosperous workshop supplied large altarpieces for a number of Swabian towns. His late-*Gothic lyricism endeared him to the *Romantics, and in the 19th cent. he was called a 'German *Perugino' or even more incongruously a 'German *Leonardo'.

Zeuxis (active latter 5th cent. BC). Greek painter from Heraclea (probably meaning the town of that name in southern Italy, rather than the one on the Black Sea). None of his works survives, but ancient writers describe him as one of the greatest of Greek painters and there are many anecdotes about his powers of verisimilitude (for an example see PARRHASIUS). Another story tells how when called upon to paint a picture of Helen of Troy for a temple at Croton he assembled the five most beautiful maidens of the city and combined the best features of each into one figure of *ideal beauty—an early example of an idea that later became commonplace in aesthetic theory. He is said to have specialized in *panels rather than murals.

According to legend, Zeuxis died laughing while painting a picture of a funny-looking old woman, and the story has occasionally formed the basis for later artists' self-portraits. Aert de *Gelder painted himself as Zeuxis (Städelsches Kunstinstitut, Frankfurt, 1685) and *Rembrandt's 'Laughing' Self-Portrait in the Wallraf-Richartz-Museum, Cologne (*c.*1665), has been interpreted in this way.

Zoffany, Johann (1733–1810). German-born painter who settled in England in about 1760 after working in Rome. He was patronized by the actor David Garrick (1717–79) and made his name with paintings representing scenes from plays, usually showing Garrick in one of his favourite parts. They show how quickly he adapted to English taste, and he also painted *conversation pieces of much the same small scale and in the same relaxed vein. No doubt because of his German background, he was taken up by George III (1738–1820) and Queen Charlotte and he painted numerous works for the royal family. The two most important (still in the Royal Collection) are *The Academicians of the Royal Academy* (1772) and *The Tribuna of the Uffizi* (1772–8). For the latter he made a long visit to Florence (1772–9) and when he returned to England he found that the demand for conversation pieces had slumped. He went to India in 1783 and made a fortune painting Indian princes and expatriate Britons before returning to England in 1789. The remainder of his career was undistinguished. For a long time Zoffany's paintings were valued chiefly as historical records (they are sharp and clear in detail and contain a wealth of information about costume, etc.), but he is now also appreciated for his charm and recognized as an artist who brought new life to the conversation piece.

Zoppo, Marco (*c.*1432–78). Italian painter of religious subjects born in Cento, trained in *Squarcione's workshop in Padua (*c.*1453–5), and active mainly in Bologna (he called himself Bolognese) and Venice, where he died. His wiry style reflects the influence of *Mantegna, softened by that of Giovanni *Bellini. Two of his works are in the National Gallery, London.

Zorach, William (1887–1966). Lithuanian-born American sculptor. Initially Zorach worked as a painter in a *Fauvist style, but he took up sculpture in 1917 and abandoned painting (apart from watercolours) about five years later. His sculpture is figurative and its salient characteristics are firm contours, block-like bulk, and suppression of details: 'I owe most', he said, 'to the great periods of primitive carving in the past—not to the modern or the classical Greeks, but to the Africans, the Persians, the Mesopotamians, the archaic Greeks and of course to the Egyptians.' He was a pioneer in America of the revival of direct carving in stone and wood and in this as well as in his formal austerity he exercised a powerful influence on modern American sculpture. He had numerous major commissions, including relief carvings for the Municipal Court Building, New York (1958). His most famous work is not a carving, however, but the aluminium *Spirit of the Dance* (1932) for Radio City Music Hall, New York—a heroic female figure that was banished for a time because of its nudity but reinstated by public pressure. Zorach taught at the *Art Students League from 1929 to 1966. His wife, **Marguerite Thompson Zorach** (1887–1968), was one of America's leading modernist painters in the years immediately before and immediately after the *Armory Show (1913), in which both she and her husband exhibited. At this time she painted in a style influenced by Fauvism and *Expressionism. In her later career, however, much of her time was spent selflessly helping her husband with his sculptural commissions, producing many of the preliminary drawings for his work.

Zorn, Anders (1860–1920). Swedish painter and etcher. After leaving the Stockholm Academy in 1881 because of its restrictive and out-dated ideas, he travelled widely, becoming the most cosmopolitan of Scandinavian artists and an international success. He was based in London (1882–5), then Paris (1888–96), and visited Spain, Italy, the Balkans, North Africa, and (on several occasions) the USA, where he painted three Presidents. Originally he worked almost exclusively in watercolour, but in 1887–90 he abandoned the medium for oils. In 1896 he settled permanently at Mora in Sweden, building his own house, which is now a museum dedicated to him. He painted three main types of pictures: portraits, *genre scenes (often depicting the life and customs of the area in which he lived), and female nudes. It is for his nudes—unashamedly healthy and voluptuous works—that he is now best known. He often painted them in landscape settings and delighted in vibrant effects of light on the human body, depicted through lush brushwork that recalls the handling of his friend Max *Liebermann. Zorn also gained a great reputation for his etchings and he occasionally made sculpture. His work is well represented in the Zornmuséet at Mora and the Nationalmuseum in Stockholm.

Zuccarelli, Francesco (1702–88). Italian painter of landscapes, often featuring mythological subjects. He was born near Florence and studied in Rome, but is associated chiefly with Venice, where he settled in 1732 and worked in the vein of pastoral landscape popularized by Marco *Ricci. Like *Canaletto, Zuccarelli was patronized by the English entrepreneur Joseph Smith (1682–1770) and his work found a ready market in England. When Richard *Wilson (still primarily a portrait painter) visited Venice in 1751 he painted Zuccarelli's portrait and it was partly Zuccarelli's encouragement which induced him to turn his attention to landscape. In 1752–62 and again in 1765–71 Zuccarelli worked in England, where his delicate *Rococo style met with great success. He was made a foundation member of the *Royal Academy in 1768, and on his return to Venice he became President of the Venetian Academy. There are many examples of his work in British collections, notably Glasgow Art Gallery and the Royal Collection (George III bought many of his paintings from Joseph Smith and also commissioned work from him).

Zuccaro, Taddeo (1529–66) and **Federico** (*c.*1540–1609). Italian *Mannerist painters, brothers, from the neighbourhood of Urbino. Taddeo worked mainly in Rome and although he was only 37 when he died he had made a great name for himself as a fresco decorator, working notably for the *Farnese family in their palace at Caprarola. His style was based on *Michelangelo and *Raphael and tended to be rather dry and wooden. Federico took over his brother's flourishing studio, continuing the work at Caprarola and also the decoration of the Sala Regia in the Vatican (begun by Taddeo in 1561). His talent was no more exceptional than Taddeo's, but he became even more successful and acquired a European reputation—indeed for a time he was probably the most famous living painter. In 1573/4 he travelled via Lorraine and the Netherlands to England, where he is said to have painted portraits of the Queen and many courtiers, although only two drawings in the British Museum portraying Elizabeth I and Robert Dudley, the Earl of Leicester, can safely be attributed to him. (Many anonymous portraits of the period—usually hanging in English country houses—are improbably attributed to him.) After working in Florence, Rome, and Venice, he was invited to the *Escorial by Philip II (1527–98) of Spain, where he painted a number of altarpieces (1585–8). Back in Rome he was elected the first President of the new Accademia di S. Luca, founded in 1593 (see ACADEMY), to which he later gave his house as headquarters. Like many of his contemporaries he believed that correct theory would produce good works of art and himself wrote *L'Idea de' Pittori, Scultori, et Architetti* (1607). Zuccaro also worked as an architect, designing a doorway in the form of a grotesque face (one enters through the open mouth) for his own house in Rome (the Palazzo Zuccaro, now the Biblioteca Hertziana).

Zucchi, Antonio. See ADAM, ROBERT, and KAUFFMANN, ANGELICA.

Zuloaga y Zabaleta, Ignacio (1870–1945). Spanish painter, the son of **Placido Zuloaga**, a well-known metalworker and ceramicist. He spent much of his career in Paris, where he was on friendly terms with *Rodin, *Gauguin, and *Degas, but his art is strongly national in style and subject-matter. Bull-fighters, gypsies, and brigands were among his subjects, and he also painted religious scenes and portraits. His inspiration came from the great Spanish masters of the past, notably *Velázquez and *Goya, and he is credited with being one of the first to 'rediscover' El *Greco. He was highly popular during his lifetime (he had a resoundingly successful one-man show in New York in 1925) and won the main painting prize at the Venice *Biennale in 1938, but his work now often looks rather stagy.

Zurbarán, Francisco de (1598–1664). Spanish painter, born at Fuente de Cantos and active mainly in Seville. He trained there 1614–17 and after a period at Llerena near his birthplace returned in 1629 as town painter. In 1634–5 he was in Madrid working for Philip IV (1605–65) on a series of ten pictures on *The Labours of Hercules* and a large historical scene, *The Defence of Cadiz* (all now in the Prado, Madrid), but apart from these pictures, a few portraits, and some masterly still lifes, he devoted himself almost entirely to religious works. He worked for churches and monasteries over a wide area of southern Spain and his paintings were also exported to South America. His compositionally simple and emotionally direct altarpieces, combining austere *naturalism with mystical intensity, made him an ideal Counter-Reformation painter.

His most characteristic works are single figures of monks and saints in meditation or prayer, most of which seem to have been executed in the 1630s. The figures are usually depicted against a plain background, standing out with massive physical presence. Many of these monumentally solemn figures are conceived in great series, such as *The Members of the Mercedarian Order* (Academy, Madrid), or *The Carthusian Saints* (Cadiz Mus.). But there are single pictures of the same kind. He painted numerous pictures of *St Francis*, for example (two in the NG, London), and a number of virgin saints.

Towards the end of his career Zurbarán's work lost something of its power and simplicity as he tried to come to terms with the less ascetic style of *Murillo, who in the 1640s overtook him as the most popular painter in Seville. In 1658 he moved to Madrid, where he spent his final years.

His son **Juan** (1620–49) is known from a few accomplished still-life paintings.

Zyl, Gerard Pietersz. van (*c.*1607–65). Dutch painter of portraits and *genre scenes, the best-known member of a dynasty of painters of the 16th and 17th cents. He was a pupil of Jan *Pynas in Amsterdam and from 1639 to 1641 worked in London, where he was a friend of van *Dyck.

Chronology

This chronology shows how certain key examples of Western art fit into a wider historical context. In the left-hand column it lists famous two- and three-dimensional works (excluding whole buildings), almost all of which are discussed in the appropriate entry in the text. The right-hand column gives a selective list of events of cultural and general historical importance.

	Key works	Other events
c.530 BC		Invention of red-figure vase-painting
490 BC		Athenians defeat the Persians at the battle of Marathon
c.470 BC	*Delphi Charioteer*	Temple of Zeus at Olympia begun; it houses Pheidias' statue of the god, one of the Seven Wonders of the World
c.460–429 BC		Pericles presides over a golden age in Athenian culture
c.450 BC	Myron's *Discus Thrower*	
447 BC		Parthenon, Athens, begun
432 BC	Parthenon sculptures completed	
399 BC		Socrates d.
c.350 BC		Mausoleum, Halicarnassus (Caria, Turkey), with sculptural decoration by Scopas and others
c.330 BC	Praxiteles' *Hermes*	
323 BC		Alexander the Great d.
c.300 BC	*Alexander Sarcophagus*	Euclid's *Elements* represent a break-through in mathematics
c.280 BC	*Colossus of Rhodes* completed	
c.240 BC		Chinese start making astronomical recordings
c.200 BC	*Barberini Faun* *Victory of Samothrace*	Romans use concrete for building
180–150 BC	Pergamum altar	
c.100 BC	*Venus de Milo*	
70 BC		Virgil b.
c.50 BC	Wall paintings, Villa of the Mysteries, Pompeii	
44 BC		Julius Caesar assassinated
13–9 BC	*Ara Pacis*, Rome	
AD 17		Ovid d.
c. AD 30	*Laocoön* group	Jesus Christ crucified
79		Pompeii destroyed
c.100		Waterwheels introduced

	Key works	Other events
106–13	Trajan's Column, Rome	
c.118–c.128		Pantheon built in Rome
c.130–8	*Antinous* bas-relief	
161–80	*Marcus Aurelius* equestrian statue, Rome	
c.200	Wall paintings, catacomb of Priscilla, Rome (one of the earliest examples of Christian art)	
c.240	Wall paintings, Dura Europos synagogue, Syria (the earliest continuous cycle of biblical images)	
306–37		Constantine rules Roman empire
c.312–15	Arch of Constantine, Rome	
330		Constantine establishes Byzantium as his capital and renames it Constantinople
402		Ravenna becomes capital of western Roman empire
410		Visigoths sack Rome
432		Traditional date for the beginning of St Patrick's mission in Ireland
476		Western Roman empire falls
c.529		St Benedict founds Benedictine order
532–7		Church of Hagia Sophia, Constantinople
c.540–7	Mosaics, Church of San Vitale, Ravenna (consecrated 547), one of the greatest ensembles of Byzantine art	
590–604		Pontificate of Gregory the Great, whose reforms lay foundation of medieval papacy
597		St Augustine leads Christian mission to England
c.600		Woodblock printing invented in China
early 7th century	First known icon of the *Virgin and Child*, St Catherine, Mt Sinai	
622		Muhammad founds Islamic religion
c.625	Sutton Hoo ship burial treasure, England	
c.698	*Lindisfarne Gospels*	
early 8th century	Ruthwell Cross, Scotland	*Beowulf*, Anglo-Saxon epic poem
711		Moors occupy most of Spain
726		Byzantine emperor Leo III bans images in churches
732		Battle of Poitiers halts Arab advance in Europe
786		Great Mosque, Córdoba, begun (decorations completed *c.*976)
792–805		Charlemagne's palace chapel, Aachen

	Key works	Other events
800		Charlemagne crowned Holy Roman Emperor
c.800	*Book of Kells*	
c.825	Oseberg Viking ship burial treasure	
868		First dated printed book (China)
962–73		Otto the Great rules as Holy Roman Emperor
969–76	*Gero Crucifix*, one of the masterpieces of Ottonian art	
c.1000		Magyars invade Transylvania and Hungary
c.1008–15	Bernward of Hildesheim commissions bronze doors for Hildesheim Cathedral	
1063		St Mark's, Venice, begun
1066		Normans invade England
c.1080	*The Bayeux Tapestry*	
1093		Durham Cathedral begun
1095–9		First crusade
1098		Cistercian order founded
c.1110	Renier of Huy: bronze font, Liège	
c.1120		Cathedral of Santiago de Compostela begun
c.1130	Giselbertus: Romanesque tympanum, Autun Cathedral	
1140–4		Rebuilding of the abbey church of Saint-Denis, near Paris, for Abbot Suger, marking the birth of the Gothic style
1163		Notre-Dame Cathedral, Paris, begun
c.1181		St Francis of Assisi b. (d. 1226)
1181	Nicholas of Verdun: *Klosterneuberg Altar*	
1216		Foundation of Dominican order
1225		St Thomas Aquinas b. (d. 1274)
1259	Nicola Pisano: Pisa pulpit	
1265		Dante b. (d. 1321)
c.1280	Cimabue: *Madonna of Sta Trinità*	
c.1305–6	Giotto: Arena Chapel frescos, Padua	
1308–11	Duccio: *Maestà*	
1328	Simone Martini: *Maestà*	
1337		Start of Hundred Years War between England and France
1338–9	Ambrogio Lorenzetti: *Good and Bad Government*	
1347–50		Black Death devastates Europe
1353		Peter Panter appointed architect of Prague Cathedral

	Key works	**Other events**
1354–7	Orcagna: *The Redeemer with the Madonna and Saints*	
c.1395	*The Wilton Diptych*	
1395–1403	Sluter: *Well of Moses*	
1401		Ghiberti wins competition for bronze doors of Florence baptistery
c.1413–16	Limbourg Brothers: *Très Riches Heures du Duc de Berry*	
1415		Bohemian religious reformer Jan Hus burnt at stake as heretic
c.1415–17	Donatello: *St George*	
1420–36		Dome of Florence Cathedral built (designed by Brunelleschi)
1423	Gentile da Fabriano: *Adoration of the Magi*	
c.1428	Masaccio: *Holy Trinity*	
1431		Joan of Arc burnt at stake
1432	Jan van Eyck: *The Ghent Altarpiece* (completed)	
1436	Uccello: *Sir John Hawkwood*	Alberti writes *Della Pittura*
c.1438–45	Fra Angelico: S. Marco frescos	
1447–53	Donatello: *Gattamelata* equestrian statue	
1448		King's College, Cambridge begun
c.1450	Jean Fouquet: *Diptych of Melun*	
1452		Leonardo b.
1452–65	Piero della Francesca: *Legend of the True Cross*	
1453	Mino da Fiesole: *Piero de' Medici* (first dated portrait bust of the Renaissance)	Turks capture Constantinople, destroying Byzantine empire
1465–74	Mantegna: *Bridal Chamber*, Ducal Palace, Mantua	
1469–92		Lorenzo de' Medici is head of the family and virtual ruler of Florence
1470–5	Dirk Bouts: *Justice of Emperor Otto*	
1475		Michelangelo b.
c.1475	Pollaiuolo: *The Martyrdom of St Sebastian*	
1475–6	Antonello da Messina: *S. Cassiano altarpiece*	
c.1478	Botticelli: *The Birth of Venus*	
1481–2		Botticelli, Ghirlandaio, Perugino, and Rosselli paint frescos in Sistine Chapel
1481–96	Verrocchio: *Colleoni* equestrian statue	
1483		Raphael b.
1492		Collumbus lands in W. Indies Moors driven from Granada, completing the Christian reconquest of Spain

	Key works	Other events
c.1495–7	Leonardo: *The Last Supper*	
1498		Vasco da Gama reaches India
1498–9	Michelangelo: *Pietà*	
c.1503–6	Leonardo: *Mona Lisa*	Julius II becomes pope (1503)
1504	Cranach: *Rest on the Flight into Egypt*	
1505	Giovanni Bellini: *S. Zaccaria Altarpiece*	
1506		Bramante begins rebuilding of St Peter's, Rome
1508–12	Michelangelo: *Sistine Ceiling*	
1508–c.1514	Raphael: *Stanze*, Vatican	
1510		Giorgione d.
1513	Dürer: *The Knight, Death, and the Devil*	
c.1515	Grünewald: *Isenheim Altarpiece* (completed)	
1516		Bosch d.
1516–18	Titian: *Assumption of the Virgin*	
1519		Leonardo d.; Charles V becomes Holy Roman Emperor
1520		Raphael d.; Luther excommunicated
1522		First circumnavigation of the globe completed by Sebastian de Cano (the original leader, Ferdinand Magellan, having died the previous year)
1526		Turks defeat Lajos II at Battle of Mohács and take control of Hungary for the next 150 years
1527		Sack of Rome
1528		Castiglione: *The Book of the Courtier*
c.1530	Correggio: *The Loves of Jupiter*	
1533	Holbein: *The Ambassadors*	Ivan the Terrible becomes Grand Duke of Moscow (he is crowned Czar of Russia in 1547)
c.1535	Parmigianino: *Madonna with the Long Neck*	
1536–41	Michelangelo: *The Last Judgement*	
1540–3	Cellini: *Salt-cellar of Francis I*	
1543		Copernicus' *De revolutionibus orbium coelestium* lays the foundations of modern astronomy
1545		Council of Trent begins
1546		Francis I begins rebuilding Louvre, Paris
1548	Titian: *Charles V at the Battle of Mühlberg*	
1550		Vasari's *Lives of the Artists* published
1563		Accademia del Disegno founded in Florence

	Key works	Other events
1564		Michelangelo d. Shakespeare b.
1565	Pieter Bruegel starts *The Months* series	
1565–87	Tintoretto: Scuola di San Rocco paintings, Venice	
1567		Palladio's Villa Rotunda begun
1573	Veronese: *The Feast in the House of Levi*	
1577		Rubens b.
1577–9	El Greco: *El Espolio*	
1584		William the Silent, chief architect of Dutch independence, assassinated in Delft
1586–8	El Greco: *The Burial of Count Orgaz*	
1588		England defeats Spanish Armada
1597–1600	Annibale Carracci: Farnese ceiling	
1598		Edict of Nantes defines rights of Protestants at end of Wars of Religion in France
1600–1	Caravaggio: *Crucifixion of St Peter* and *Conversion of St Paul*	
1606		Rembrandt b.
1609	Elsheimer: *The Flight into Egypt*	Dutch Republic effectively wins freedom from Spain
1610–11	Rubens: *Raising of the Cross*	
1611		Authorized Version of the Bible
1616		Shakespeare d.
1618–48		Thirty Years War
1619–22		Inigo Jones, Whitehall Banqueting House
1624–33	Bernini: *Baldacchino*	
1633–9	Pietro da Cortona: Barberini ceiling	
1637		Descartes, *Discourse on Method*
1642	Rembrandt: *The Night Watch*	English Civil War begins
1643–1715		Reign of Louis XIV of France
1645–52	Bernini: *Ecstacy of St Teresa*	
1648	Poussin: *The Ashes of Phocion*	Académie Royale de Peinture et de Sculpture founded in Paris
1649		Charles I executed. England becomes a republic under Cromwell
1650	Georges de la Tour: *The Denial of St Peter*	
c.1656	Velázquez: *Las Meninas*	
1660		Restoration of the English monarchy
1661		Louis XIV begins enlarging château of Versailles
1662	Philippe de Champagne: *Ex Voto*	

	Key works	Other events
1666		Fire of London
c.1669	Rembrandt: *The Return of the Prodigal Son*	
1672		Bellori, *Lives of the Modern Painters*
1675–1710		St Paul's Cathedral, London
1682	Claude: *Ascanius and the Stag*	
1685		Johann Sebastian Bach b. Handel b. Louis XIV revokes Edict of Nantes
1687		Newton, *Principia mathematica*
1689	Hobbema: *The Avenue*	
1691–4	Pozzo: ceiling of S. Ignazio, Rome	
1701	Rigaud: *Louis XIV*	
1703		Peter the Great founds St Petersburg
1707		Act of Union unites England and Scotland
1717	Watteau: *The Pilgrimage to the Isle of Cythera*	Winckelmann b.
1728	Chardin: *The Rayfish*	
c.1730	Canaletto: *The Stonemason's Yard*	
c.1735	Hogarth: *A Rake's Progress*	Linnaeus, *System of Nature* (1735)
1740		Frederick II (the Great) becomes ruler of Prussia
c.1745	Piranesi: *Carceri*	
1749		Goethe b.
1750		Bach d.
1750–3	Tiepolo: decorations for the Prince Archbishop's palace, Würzburg	
1751	Boucher: *Reclining Girl*	
1753–4	Reynolds: *Commodore Keppel*	
1756		Mozart b.
1757	Tiepolo: decorations for the Villa Valmarana, Vicenza	
1761	Mengs: *Parnassus*	
c.1766	Fragonard: *The Swing*	
1768	Wright: *An Experiment on a Bird in the Air Pump*	Foundation of Royal Academy, London Winckelmann d.
1770	Gainsborough: *The Blue Boy* West: *Death of Wolfe*	Beethoven b.
1776		American Declaration of Independence
1778	Copley: *Brook Watson and the Shark*	
1784	David: *Oath of the Horatii*	
1787		Mozart, *Don Giovanni*
1789	Blake: *Songs of Innocence*	French Revolution begins
1790	Bewick: *A General History of Quadrupeds*	
1791		Mozart d.

	Key works	Other events
1793	David: *Death of Marat*	
c.1793–8	Goya: *Los Caprichos*	
1797		Schubert b.
1798		lithography invented
1802–3	Thorvaldsen: *Jason*	
1804		Napoleon becomes Emperor of France Beethoven, *Eroica* symphony
1805	Cotman: *Greta Bridge*	Battle of Trafalgar
1805–7	Canova: *Pauline Borghese as Venus*	
1808	Friedrich: *The Cross in the Mountains* Ingres: *The Valpinçon Bather*	
1812		Napoleon retreats from Moscow
1814		George Stephenson constructs the first successful steam locomotive
1815		Battle of Waterloo
1819	Géricault: *The Raft of the Medusa*	First Atlantic steamship crossing
1820–4	Goya: *The Black Paintings*	
1821	Constable: *The Hay Wain*	
1824	Delacroix: *The Massacre at Chios*	National Gallery, London founded
1827		Beethoven d.
1828		Schubert d. Tolstoy b.
1830		July Revolution in France
1832		Daumier imprisoned for political satire
1833		Brahams b.
1837		Queen Victoria accedes to throne in England
1839	Turner: *The Fighting Temeraire*	Daguerre publicizes 'daguerrotype'
1840		Monet b. Rodin b.
1841		Eakins b.
1843–60		Ruskin, *Modern Painters*
1848		The Year of Revolutions Marx and Engels, *Communist Manifesto* Gaugin b. Pre-Raphelite Brotherhood formed
1850	Courbet exhibits *The Burial at Ornans*, *The Peasants at Flagey*, and *The Stone Breakers*	
1851		The Great Exhibition, London
1853	Holman Hunt: *The Awakening Conscience*	Van Gogh b.
1853–6		Crimean War
1858	Frith: *Derby Day*	
1859	Millet: *The Angelus*	Darwin, *The Origin of Species*
1861–5		American Civil War

	Key works	Other events
1863	Ingres: *The Turkish Bath* Manet: *Le Déjeuner sur l'Herbe*	Salon des Réfuses
1866–7	Monet: *Women in the Garden*	
1870		Metropolitan Museum, New York, founded
1870–1		Franco-Prussian War
1871	Whistler: *Arrangement in Grey and Black: Portrait of the Painter's Mother*	
1872	Monet: *Impression: Sunrise*	
1873	Repin exhibits *Volga Boatmen*	
1874		First Impressionist exhibition, Paris
1875	Eakins: *The Grass Clinic*	
1876		Battle of Little Bighorn
1877		Ruskin sues Whistler, leading to famous libel trial
1880	Böcklin: *The Island of the Dead* Rodin begins *Gates of Hell*	Apollinaire b. Epstein b. Kirchner b. Flaubert d. Dostoevsky, *The Brothers Karamazov*
1881	Degas: *Little Fourteen-year-old Dancer* (bronze)	Léger b. Picasso b. Samuel Palmer d. Béla Bartók b. Dostoevsky d. Ibsen, *Ghosts* Tsar Alexander II assassinated
1882	Manet: *A Bar at the Folies-Bergère*	Braque b. D. G. Rosetti d. James Joyce b. Charles Darwin d.
1883		Manet d. Orozco b. Marx d. Wagner d. Nietzsche, *Also sprach Zarathustra*
1884	Sargent: *Madame X*	Beckmann b. Modigliani b. First exhibition of Salon des Indépendents, Paris Huysmans, *A Rebours*
1844–6	Seurat: *Sunday Afternoon on the Island of La Grande Jatte*	
1884–93		Les Vingt group exhibitions
1885	van Gogh: *The Potato Eaters*	Victor Hugo d. Zola, *Germinal* Khartoum falls
1886	Millais: *Bubbles*	New English Art Club founded in London

	Key works	Other events
1886 (*cont.*)		Mies van der Rohe b. Rivera b. Liszt d.
1886–94		Gaugin intermittently active at Pont-Aven
1887		Archipenko b. Chagall b. Duchamp b. Le Corbusier b. Verdi, *Otello*
1888	Ensor: *Entry of Christ into Brussels* Gauguin: *The Vision after the Sermon* van Gogh: *The Night Café*	van Gogh settles is Arles Albers b. de Chirico b. T. S. Eliot b. Strindberg, *Miss Julie* Kaiser Wilhelm II accedes to the German throne
1889		Crown Prince Rudolf of Austria and his mistress commit suicide at Mayerling, near Vienna
1890	Edelfeldt: *Christ and Mary Magdalene*	Gabo b. van Gogh d. Ibsen, *Hedda Gabler*
1891	Monet begins *Rouen Cathedral* series	Gauguin sails to Tahiti Dix b. Max Ernst b. Gaudier-Brzeska b. Stanley Spencer b. Seurat d. Prokofiev b. Edison invents cine camera Trans-Siberian railway started
1892	Leighton: *The Garden of the Hesperides*	Grant Wood b. The Nabis' first exhibition Munch exhibition in Berlin closed amid scandal
1893	Munch: *The Scream* Toorup: *The Three Brides*	Grosz b. Miró b.
1894		Tsar Alexander III forms Franco-Russian alliance President Carnot of France assassinated Sino-Japanese War
1894–7	Beardsley *et al.*: *The Yellow Book*	
1894–1906		The Dreyfus affair
1895		Cézanne has one-man show in Paris Marconi invents wireless telegraph Röntgen discovers X-rays Lumière invents cinematograph
1896		Breton b. J. E. Millais d. William Morris d. Puccini, *La Bohème* Alfred Jarry, *Ubu Roi*

	Key works	Other events
1897	Gauguin: *Where Do We Come From? What Are We? Where are We Going To?*	Tate Gallery opens Brahams d.
1898		Aalto b. Beardsley d. Burne-Jones d. Calder b. Boudin d. Magritte b. Henry Moore b. Spanish-American War
1899–1902		Boer War
1900		John Ruskin d. Oscar Wilde d. Freud, *The Interpretation of Dreams* Max Planck propounds quantum theory Boxer Rising in China
1901	Klimt: *Judith I*	Böcklin d. Alberto Giacometti b. Toulouse-Lautrec d. Queen Victoria d. President McKinley assassinated
1902		Zola d.
1903	Picasso: *La Vie*	Gauguin d. Hepworth b. Camille Pissarro d. Whistler d. Salon d'Automne founded in Paris First powered flight, USA
1904		Dali b. de Kooning b. Dvořák d.
1904–5		Russo-Japanese War
c.1904–14		Ash-can School active in New York
1905		Fauvism launched at Salon d'Automne Die Brücke formed in Dresden Bouguereau d. Burra b. Einstein publishes his special theory of relativity
1906		Cézanne d.
1907	Bellows: *A Stag at Sharkey's* Picasso: *Les Demoiselles d'Avignon*	Salon d'Automne holds Cézanne memorial exhibition Kahnweiler opens gallery in Paris
1907–8	Epstein: figures for façade of British Medical Association, London	
1907–14		Braque and Picasso evolve Cubism
1909		Bacon b. Marinetti launches Futurism First production car Diaghilev brings his Russian ballet company to Paris

	Key works	Other events
1910	Rousseau: *The Dream*	Roger Fry's first Post-Impressionist exhibition, London Tolstoy d.
c.1910		Birth of abstract art
1911	Braque: *The Portuguese* Carrà: *The Funeral of the Anarchist Galli*	Der Blaue Reiter's first show Camden Town Group formed Čapek, Filla, and Gutfreund create avant-garde movement in Prague
1912	Balla: *Dynamism of a Dog on a Leash* Robert Delaunay: *Circular Forms* Kupka: *Amorpha: Fugue in Two Colours*	Jackson Pollock b. Gleizes and Metzinger, *Du Cubisme* 'Donkey's Tail' exhibition, Moscow
1913	Boccioni: *Unique forms of Continuity in Space* Duchamp: *Bicycle Wheel* (first ready-made)	Armoury Show, New York Proust, *Swann's Way* Camus b. Reg Butler b. 'Target' exhibition, Moscow Fry founds Omega Workshops Stravinsky, *Rite of Spring*
1914	Lamb: *Lytton Strachey* Marc: *Fighting Forms*	Birth of Vorticism
1914–18	Gill: *Stations of the Cross*	First World War
1915	Duchamp begins *The Bride Stripped Bare by her Bachelors, Even*	Dada founded
1916	Gertler: *Merry-go Round*	Eakins d.
1917	Thomson: *Jack Pine*	Rodin d. Russian Revolution De Stijl founded Carrà and de Chirico meet
1918	Grosz: *Fit for Active Service*	Apollinaire d. Poland and Czechoslovakia become republics
c.1918	Malevich: *White on White*	Romanovs assassinated
1918–19	Sargent: *Gassed*	
1919		The Bauhaus opens in Weimar
1920	Dix: *The Match Seller*	Group of Seven founded Modigliani d.
1921	Ernst: *The Elephant Celebes*	
1922		Mussolini's Fascists march on Rome James Joyce, *Ulysses*
1923	Frank Dobson: *Osbert Sitwell*	
1924		Lenin d. First Surrealist manifesto
1925	Miró: *The Harlequin's Carnival* Soutine: *Side of Beef*	Eisenstein, *Battleship Potemkin* Kafka, *The Trial*
1926	Magritte: *The Menaced Assassin*	Monet d. First television transmission
1927	Hepworth: *Doves*	*The Jazz Singer* (first sound film)
1928	Curry: *Baptism in Kansas*	Fleming discovers penicillin

	Key works	Other events
1928 (cont.)	Foujita: *Self-portrait*	
1929		Wall Street crash Museum of Modern Art, New York, founded
1930	Grant Wood: *American Gothic*	Buñuel: *L'Age d'Or* (film)
1931	Dali: *The Persistance of Memory*	Abstraction-Création group founded in Paris Courtauld Institute of Art London, founded
1933	Beckmann: *Departure* Gunn: *Delius* Manship: *Prometheus*	Hitler becomes Chancellor of Germany
1934	Burra: *Dancing Skeletons*	
1935	Ben Nicholson: *White Relief*	Italy invades Ethiopia
1936	Oppenheim: *Object*	International Surrealist Exhibition, London
1936–9		Spanish Civil War
1937	Brancusi: *Endless Column* González: *Montserrat* Picasso: *Guernica* Spencer: *The Leg of Mutton Nude*	Germans bomb Guernica, Spain Nazi exhibition of 'Degenerate Art', Munich
1938	Lewis: *T. S. Eliot*	Kirchner d. Germany annexes Austria
1939	Nash: *Monster Field*	Franco becomes Spanish dictator Germany invades Poland
1939–45		Second World War
1940	Klee: *Death and Fire*	
1940–2	Moore: Air-raid shelter drawings	
1941		Japanese attack Pearl Harbor James Joyce d.
1942	Hopper: *Nighthawks*	Grant Wood d.
1943	Dobell: *Joshua Smith* Mondrian: *Broadway Boogie Woogie*	
1944	Bacon: *Three Studies for Figures at the Base of a Crucifixion* Gruber: *Job*	Allies liberate Paris
1945		US drops atomic bombs on Hiroshima and Nagasaki Bartók d. Orwell, *Animal Farm*
1946	Burchfield: *The Sphinx and the Milky Way* Nolan's first *Ned Kelly* paintings	Moholy-Nagy d.
1947	Paolozzi: *I was a Rich Man's Plaything*	India gains independence from UK Institute of Contemporary Arts, London, founded
1948	Newman: *Onement I* Wyeth: *Christina's World*	South Africa adopts apartheid Communists seize Czechoslovakia

	Key works	Other events
1949	Sutherland: *Somerset Maugham*	Orozco d. Carol Reed, *The Third Man*
1950	Pollock: *Lavender Mist*	Beckmann d. Gombrich: *The Story of Art*
1951	Dali: *Crucifixion of St John of the Cross* Davis: *Owh! in San Pao* Freud: *Interior at Paddington*	J. D. Salinger, *The Catcher in the Rye* Festival of Britain Harold Rosenberg coins term 'Action painting'
1952	Frankenthaler: *Mountains and Sea* de Kooning: *Woman I*	Samuel Beckett, *Waiting for Godot*
1953	Matisse: *L'Escargot*	Reg Butler wins international competition for monument to *The Unknown Political Prisoner*
1954	Burri: *Sacking with Red*	Matisse d.
1955	John's first *Flag* paintings	Léger d.
1956	Hamilton: *Just What Is It That Makes Today's Homes so Different, so Appealing?* Richards: *'Do not go gentle into that good night'*	Pollock d. Hungarian uprising crushed by USSR Suez crisis
1957		Rivera d. USSR sends world's first satellite into space Ingmar Bergman, *The Seventh Seal*
1958		Klein's *Le Vide* exhibition scandal, Galerie Iris Clert, Paris
1959	Kaprow's first happenings Rauschenberg: *Monogram*	Epstein d. Grosz d. Spencer d. Fidel Castro seizes power in Cuba
1960	Tinguely: *Homage to New York*	*Lady Chatterley's Lover* obscenity trial Sharpeville massacre, South Africa
1961		Greenberg, *Art and Culture* Joseph Heller, *Catch 22* British Pop art is put on the map at Young Contemporaries exhibition Berlin Wall erected
1962	Caro: *Early One Morning* Oldenburg: *Dual Hamburger* Warhol's first *Soup can* paintings	Birth of Fluxus, Germany Edward Albee, *Who's Afraid of Virginia Woolf?*
1963	Hilton: *Oi yoi yoi* Lichtenstein: *Whaam!*	Braque d. US President Kennedy assassinated
1964	Gottlieb: *Orb* King: *And the Birds Began to Sing*	Archipenko d.
1965	Beuys: *How to Explain Pictures to a Dead Hare*, performance Kosuth: *One and Three Chairs*	Le Corbusier d.
1966	Andre: *Equivalent VIII*	Breton d. Giacometti d. Floods destroy Florentine art treasures England wins football World Cup

	Key works	Other events
1967	Hockney: *A Bigger Splash*	Magritte d. The Beatles, *Sergeant Pepper's Lonely Hearts Club Band* (LP cover design by Peter Blake)
1968		Duchamp d. Students and workers riot in Paris USSR invades Czechoslovakia
1969	Gilbert & George: *Underneath the Arches*	Baselitz paints his first upside-down image Kenneth Clark, *Civilisation* (television series) Dix d. Mies van der Rohe d. First lunar landing
1970	Hanson: *Tourists* Smithson: *Spiral Jetty*	
1971		Intel develops microchip, USA
1971–7		Pompidou Centre, Paris, built
1972	Tony Smith: *Gracehoper*	Palestinian terrorist murder Israeli athletes at Munich
1973		Picasso d.
1974		US President Nixon resigns following Watergate scandal
1975	Frink: *Horse and Rider*	Hepworth d.
1976	Christo: *Running Fence*	Aalto d. Albers d. Burra d. Calder d. Ernst d.
1977		Gabo d.
1978	Hockney: *The Magic Flute* (stage designs)	de Chirico d.
1979		Iranian Revolution deposes Shah Margaret Thatcher becomes Britain's first woman Prime Minister Soviet Army invades Afghanistan
1980		Iraq invades Iran Solidarity union formed, Poland
1981		First known AIDS case in USA Butler d.
1982		Argentinian–British war in Falkland Islands
1983		Miró d. US occupies Grenada
1984		Malcolm Morley is first winner of Turner Prize
1985	Christo: *The Pont Neuf Wrapped*	Chagall d. Mikhail Gorbachev becomes USSR premier, and introduces liberal policies Saatchi Collection, London, opens

	Key works	Other events
1986		Henry Moore d. Chernobyl explosion
1989		Berlin Wall torn down Tiananmen Square massacre, China
1990		Nelson Mandela released from prison Iraq invades Kuwait Hubble space telescope launched
1991	Rego: *Crivelli's Garden* (mural, National Gallery, London)	Gulf War Civil war breaks out in former Yugoslavia
1992	Saint Phalle: *Loch Ness Monster*	Bacon d. USSR dissolved
1993		Frink d. Tate Gallery St Ives opens
1994		Max Bill d. Mandela becomes South African president; apartheid abolished
1995	Christo: *Wrapped Reichstag*	Burri d. Damien Hirst wins Turner Prize

Index of Galleries and Museums

It should be noted that, because the technology is still relatively new, websites are particularly liable to change their addresses and the information that they present.

Australia

National Gallery of Victoria
180 St Kilda Road
Melbourne
3004 Victoria
Tel (0061) (3) 627 411
An outstanding collection which comprises North European schools, Italian, French and British paintings, including 16th-, 17th-, and 18th-century portraiture and works by van Dyck, Constable, Turner, Blake (36 watercolours for Dante's *Divine Comedy*), Bacon, Moore, and Hockney. One of the gallery's most impressive paintings is Rubens's *Hercules and Antaeus*.

Austria

Kunsthistorisches Museum
Burgring 5
1010 Wien
Tel (0043) (1) (222) 52177
Fax 932 7701 / 523 7750
Website (The gallery was temporarily on the web in 1996—unbeknown to most of its curators!—but has no site at present.)
The gallery represents western European art as a whole, and is especially strong in works by Bruegel and Velázquez. The collection focuses on the Habsburg taste for Netherlandish, 16th-century Venetian and 17th-century Flemish art. Early Netherlandish examples include works by van Eyck and van der Weyden, van der Goes and Geertgen tot Sint Jans. It also contains a whole room of Bruegels with pictures from his early, middle and late periods such as *Children's Games* and *Battle between Carnival and Lent*. Its Dutch holdings include paintings by Hals, Aertsen, Rembrandt, and Vermeer; the Flemish section is rich in van Dycks and has two Rubens altarpieces for the Jesuit Church in Antwerp. The German Danube School comprises scenes by Altdorfer, Cranach, and Holbein while the Venetians are here in full force with Bellini, Giorgione (his *Three Philosophers*, possibly finished by Sebastiano del' Piombo), Titian, and Tintoretto, Veronese with an *Adoration of the Magi* while the rest of Italy is represented by, among others, Raphael, Parmigianino, Caravaggio, and Annibale Carracci. The highlight of the Spanish collection is Velázquez's series of portraits of the Infanta Margarita Teresa (*c.*1653–9).

Graphische Sammlung Albertina
A-1010 Wien
Augustinerstrasse 1
Tel (0043) (222) 153483–0
Fax 533 7697
Website http: / / www.telecom.at / albertina / welcomee.html
Houses a superb collection of Old Master drawings, engravings, and other types of print, although only a few are shown at a time.

Belgium

Koninklink Museum voor Schone Kunsten (Musée Royal des Beaux-Arts)
Plaatsnijdersstrasse 2
2000 Antwerp
Tel (0032) (3) 238 7809
Fax 248 0810
Website http: / / kmsk.mrbab-kmsk.be
Represents Flemish painting over five centuries, including paintings from the churches of St Augustinus and St Jacobus. From the former is Rubens's large altarpiece, the *Madonna and Child with Saints* and van Dyck's *The Vision of St Augustine*. From the latter Rubens's *Madonna with Saints* of 1639 is thought to be based on portraits of the artist and his family. Jan van Eyck's *Madonna of the Fountain* is another high point of the collection. The Dutch School is represented by works by Hals and Rembrandt, and there are also some Italian paintings.

Musées Royaux des Beaux-Arts (Musée d'Art Ancien)
Rue de la Régence 3
1000 Bruxelles
Tel (0032) (2) 508 3211
Fax 508 3232
Website same as above
Contains mainly Flemish paintings from the 14th to the 19th century but has some fine examples of Dutch, German, and French art as well. The Netherlandish section has Dirk Bouts's two

paintings of the *Justice of Emperor Otto*, a *Pietà* by Petrus Christus, Bruegel's *Fall of Icarus*. Other attractions include the Cranachs in the German section, landscapes, portraits, and altarpieces by Rubens, and some notable French paintings of the 17th and 18th centuries, including David's *Death of Marat*. The collection also includes 19th-century Belgian and French paintings, including works by Gauguin and Seurat.

Brazil

Museu de Arte de São Paulo 'Assis Chateaubriand'
Avenida Paulista 1578
01310 São Paulo
Tel (0055) (11) 251 5644
Fax 284 0574
A wide ranging collection of European Old Masters and modern works, 20th-century South American paintings among them. It contains Mantegna's *St Jerome* as well as paintings by Memling, Bosch, and Rembrandt. The British School is represented by Holbein, Constable, Gainsborough, and Turner. The Spanish rooms have Velázquez's portrait of Olivares and paintings by Goya and El Greco. Cézanne and Renoir are among the leading 19th-century artists represented.

Canada

Art Gallery of Ontario
317 Dundas Street W
Toronto
Ontario M5T 1G4
Tel (001) (416) 977 0414
Fax 979 6646
Website http://www.ago.on.ca/
The collection contains over 16,000 works but its particular strengths lie in Canadian art and the New York School. Its European holdings include Hals's *Isaak Massa* of 1626, a Poussin, a Claude, and Chardin's *Jar of Apricots*. Includes the Henry Moore Sculpture Center with over 200 examples of his work, comprising sketches and plaster casts as well as finished pieces.

Czech Republic

Národní Galerie v Praze (National Gallery in Prague)
Hrad anské nám sti 15
Praha
Tel (0042) (2) 2451 0594
Fax 536 6469
This is a very interesting collection, which is rich in 17th-century Dutch and 19th-century French paintings. The holdings include Dürer's *Feast of the Rose-garlands*, Bruegel's *Haymaking*, and five Picassos. Prague also has the Galerie hlavniho m sta Prahy (Gallery of the City of Prague on Mickiewiczowa 1, Bílkova vila, 160 00 Praha, tel 02 231 0272), which contains fascinating examples of Bohemian medieval and Renaissance art.

Denmark

Ny Carlsberg Glyptothek
Dantes Plads 7
1556 København
Tel (0045) (33) 418141
Fax 912058
Website (this is not an official website for the gallery but contains rudimentary information in Danish) http://www.knlturnet.dk/museer/haand-bog/mht286.html
No e-mail
Focuses mainly on sculpture, with a large collection of Danish carvings after Thorvaldsen, and a broad collection of 19th-century Danish painting. Rodin, Carpeaux, and Degas are represented in the 19th-century French section and the museum has paintings by the Impressionists as well.

Louisiana Museum of Modern Art
Gl Strandvej 13
3056 Humlebaek
Tel (0045) (49) 190719
Fax 193505
Website http://www.louisiana.dk/museer/haand-bog/mht708.html (rudimentary information in Danish)
E-mail Curatorial@louisiana.dk
Houses an excellent collection of 20th-century American and European works, especially paintings by the CoBrA group. The sculpture park has pieces by Arp, Giacometti, and Calder.

France

Musée du Louvre
34–36 Quai du Louvre
75058 Paris
Tel (0033) (1) 402 05 009
Fax 402 05442
Website http://www.louvre.fr
E-mail naber@louvre.fr (cultural activities); culturel@louvre.fr (collections); devitry@louvre.fr (informatics service)
This is a vast collection of European art, much too large in fact to take in on one visit. It is divided into sections devoted to French, Dutch, and Flemish, Italian, Spanish, English, and German art as well as sculpture. The *Mona Lisa* is its most famous individual painting, but it also has the controversial Giorgione/Titian *Concert Champêtre*, works by van Eyck and Michelangelo's *Slaves*. Many of its former holdings from the late 19th century have been in-

stalled in the Musée d'Orsay along with works from the Jeu de Paume.

Musée d'Orsay
62 rue de Lille (entrance in the rue de Bellechasse)
75007 Paris
Tel (0033) (1) 404 94814
Websites http://www.paris.org.:80/musees/orsay; http://meteora.uscd.edu.:80/norman/paris/musees/orsay
A superb collection of mid- to late 19th-century paintings and sculptures, housed in a converted railway station on the Quai d'Orsay. It features not only celebrated Impressionist and Post-Impressionist pictures, but also more academic works. The museum opened in 1986.

Centre National d'Art et de Culture Georges Pompidou (incorporating the Musée National d'Art Moderne)
19 rue de Renard
75191 Paris
Tel (0033) (1) 447 81233
Fax 447 81300
Website http://www.cnacgp.fr
No e-mail
One of the world's major collections of modern art, housed in one of the most famous buildings of the 1970s. Highlights include a reconstruction of Brancusi's studio, and Yaacov Agam's Op art antechamber for the Elysée Palace. The collection as a whole gives a good overview of the development of 20th-century art.

Musée Granet-Palais de Malte
Pl St-Jean de Malte
13100 Aix en Provence
Tel (0033) 42 38 14 70
A fine collection from the 17th, 18th and 19th centuries from France and the Netherlands. It contains many excellent minor works as well as some major ones, including Ingres's *Jupiter* and *Thetis*.

Musée Toulouse-Lautrec et Galerie d'Art Moderne
Palais de la Berbie
BP 100
81003 Albi
Tel (0033) 63 54 14 09
The most important single collection of Toulouse-Lautrecs, which also houses paintings by some of his well-known contemporaries.

Musée des Beaux-Arts
Palais des Etats de Bourgogne
Place de la St-Chapelle
21000 Dijon
Tel (0033) 80 74 52 70
16th- to 19th-century paintings from Flanders, France, the Rhine, the Netherlands, and Switzerland as well as some Italian works. One of its key exhibits is an altarpiece by Jacques de Baerze and Melchior Broederlam.

Musée National du Château de Fontainebleau
77300 Fontainebleau
Tel (0033) (1) 64 22 49 80
The palace for which François I commissioned a programme of sumptuous murals. The main artists concerned were the 16th-century Italians Rosso, Primaticcio, and Niccolò dell'Abbate.

Musée de Grenoble
Place Lavalett
38000 Grenoble
Tel (0033) 76 63 44 11
Fax 34410
One of France's best collections of painting and sculpture outside Paris, housed in a custom-built museum which opened in 1992. It comprises Old Masters and modern art, focusing on individuals from the 17th century (with masterpieces that include George de La Tour's *St Jerome* and Strozzi's *Disciples at Emmaus*) through to the present-day. Its pre-20th-century holdings concentrate mainly on French artists while the 20th-century section represents European and American schools, such as Pop Art, Minimalism, and Conceptual Art. Its modern collection contains Matisse's *Interior with Aubergines*.

Musée des Beaux-Arts
Place de la République
59000 Lille
Tel (0033) 20 57 01 84
A wide-ranging collection of Dutch and Flemish paintings with groups of Spanish, German and Italian and 19th-century French art. Rubens's *Descent from the Cross* is one of its most famous holdings.

Chapelle du Rosaire, Henri Matisse
06140 Vence
Tel (0033) 93 58 03 26
Matisse's last important decorative scheme; he designed the chapel's whole interior, including murals and stained glass.

Germany

Many German museums do not yet have an individual website, although more are coming on-line all the time. For museums and galleries without a site, internet users can obtain some basic information about their collections, addresses, hours of opening, etc., via a national service called the Webmuseum. Its website is http://Web-Museen.de/

Ludwig Forum für Internationale Kunst
Jülicher Strasse 97–109
52020 Aachen
Tel (0049) (24) 18070
A vast, new branch of the Ludwig Museum in Cologne. It opened in 1991 and concentrates on modern and contemporary movements including Pop Art, Superrealism, international art from the 1980s, Conceptual Art, and Neo-Expressionism. Part of the Ludwig collection is permanently on view in Budapest.

Staatsgalerie, Schäzler-Palais
Maximilianstrasse 46
86150 Augsburg
Tel (0049) (821) 510350
Website http://server.StMUKWK.bayern.de/kunst/zwmuseen/augsbg1.html (rudimentary information)
Houses a collection of early German painting focusing particularly on artists from this area and Swabia, as well as examples of German Baroque art. One of the gallery's key works is Holbein the Elder's *S Paolo fuori le Mura*, one of a series representing major Roman churches.

Schloss Charlottenburg
19 Luisenplatz
10585 Berlin
(Postfach 601462
14414 Potsdam)
Tel (0049) (30) 320 911
The gallery in the castle has a superb collection of 18th-century French painting chosen by Frederick the Great (see especially Watteau's *L'Enseigne de Gersaint*). Other major works include paintings by Cranach and the German Romantics.

Museum Dahlem Gemäldegalerie, Staatliche Museen Preussischer Kulturbesitz
Arnimallee 23–27
14195 Berlin
Tel (0049) (30) 830 1217
Fax 831 6384
No official website as yet but rudimentary information on http://www.informatik.hu-berlin.de/BIW/A-Z/64.html
Holds Dutch, Early Netherlandish, Flemish, French, German, and Italian paintings as well as sculpture. The contentious *Man with the Golden Helmet*, traditionally ascribed to Rembrandt, is part of the collection, which also contains Hugo van der Goes's *Adoration of the Magi*.

Neue Nationalgalerie
Potsdamerstrasse 50
10785 Berlin
Tel (0049) (30) 266 2651
Fax 262 4715
Focuses on 19th- and 20th-century European paintings and sculpture. Its main strength lies in the section devoted to French Impressionism, but it also shows the development of German Romantics through to Expressionists such as Nolde and the Brücke painters.

Kunstsammlung Nordrhein-Westfalen
Grabbeplatz 5
40213 Düsseldorf
Tel (0049) (211) 83810
Fax 838 1201
Website http://www.rp-online.de/Duesseldorf/KunstsammlungNRW.html
An impressive collection of modern art from Fauvism onwards. There is a particularly good representation of works by Paul Klee.

Städelsches Kunstinstitut und Städtische Galerie
Schaumainkai 63
60596 Frankfurt am Main
Tel (0049) (69) 605098–0
Fax 610163
Website http://fortress.wiwi.uni-frankfurt.de/Frankfurt/museum.html
Specializes in Early Netherlandish, French, German, Italian, and Flemish work. Its most famous paintings include Rembrandt's *Blinding of Samson* and Tischbein's portrait of Goethe.

Staatliche Museen Kassel
Schloss Wilhelmshöhe
93777 Kassel
Tel (0049) (561) 93777
Fax 315873
Its main strength lies in its 17th-century Dutch and Flemish paintings, notably a portrait of *Saskia* by Rembrandt and Rubens's *The Flight into Egypt*. It also holds good examples of German and Italian art.

Alte Pinakothek
Barer Strasse 27
80333 München
Tel (0049) (89) 238 05216
Website http://server.StMUKWK.bayern.de/Kunst/museen/pinalt.html (rudimentary information)
One of the world's great collections of Old Masters. It is rich in German art, and Rembrandt, Rubens, and Titian are amongst others represented. The gallery is closed for repairs until the end of 1997.

Städtische Galerie im Lenbachhaus
Luisenstrasse 33
80333 München
Tel (0049) (89) 211 27137

The former house of Franz Lenbach, containing works he painted and collected. Its main attraction, however, is a superb collection of works by Kandinsky, Klee, Macke, Marc, Jawlensky, and Gabriele Münter.

Stiftung Seebull Nolde
D-25927 Neukirchen
Tel (0049) (4664) 364
Fax (0049) (4664) 1475
A collection of paintings by Nolde shown in the house he built for himself at Seebüll in 1927. It includes his polyptych *The Life of Christ*.

Germanisches Nationalmuseum
Kartänsergasse 1
90402 Nürnburg
Tel (0049) (911) 13310
Fax 1331200
Website http://laokoon.inberlin.de/Museum/Nuernburg/germanischesnationalmuseumnuernburg.html (the address is almost as long as the information it contains!)
One of the best collections of early German painting. It focuses particularly on work produced in Nuremburg, which was an important centre of art during the Renaissance. Highlights include six paintings by Dürer who spent most of his life in the city.

Staatsgalerie in der Residenz Würzburg
Schloss und Gartenverwaltung
Residenzplatz 2
97070 Würzburg
Tel (0049) (931) 355170
Fax 51925
Website http://server.StMUKWK.bayern.de/kunst/zwmuseen/wuerz.html (contains only rudimentary information)
One of the world's most spectacular Rococo interiors. The architecture is by Balthasar Neumann and the frescos by Giambattista Tiepolo, two of the greatest artists of the age.

Holland

Rijksmuseum
Stadhouderskade 42
Postbus 74888
1070 DN
Amsterdam
Tel (0031) (20) 673 2121
Fax 679 8146
E-mail Presentatie@Rijksmuseum.hl
Famous for its collection of Rembrandts, it also houses fine examples by most of the other leading figures in 17th-century Dutch art.

Stedelijk Museum of Modern Art
Paulus Potterstraat 13
Postbus 75082
1070 AB
Amsterdam
Tel (0031) (20) 573 2911/573 2737
Fax 675 2716
A superb collection of European painting and sculpture dating from 1850. It is best-known for its works by Malevich and Mondrian. Other artists represented include Chagall, the CoBrA group, de Kooning, and Matisse, and there are examples of Pop, Minimal, and Conceptual Art.

Rijksmuseum Vincent van Gogh
Paulus Potterstraat 7
Postbus 75366
1070 AJ
Amsterdam
Tel (0031) (20) 570 5200
Fax 673 5053
A building by the Dutch architect and designer Gerrit Rietveld houses this collection of around 200 paintings and 500 drawings by van Gogh and his contemporaries. The van Gogh collection includes *The Potato Eaters* and *Crows in the Wheatfields*.

Hungary

Magyar Nemzeti Galéria
(National Gallery of Hungary)
Szent György tér 2
Budavári Pálota Pf 31 (1250)
Tel (0036) (1) 757 533
Website http://origo.hum.hu/ottlap2.html
E-mail (for comments on website) webmaster @hum.hu
An important collection of 19th- and 20th-century Hungarian painting and sculpture. It contains some of Central Europe's most virile history paintings, many of them depicting the Magyars' resistance to the Turks.

Ireland

National Gallery of Ireland
Merrion Square West
Dublin 2
Tel (00353) (1) 661 5133
Fax 661 5372
(No official website or e-mail yet, but the gallery will have them soon)
A good collection of paintings from the major European schools, as well as an outstanding representation of Irish works. The Irish artists on display include James Barry, Francis Danby, Daniel Maclise, William Orpen, Sarah Purser, and Jack Yeats.

Italy

Museo-Tesoro Basilica di San Francesco
Piazza San Francesco 2
06082 Assisi
Tel (0039) (75) 812238
Fax 816187
(No website or e-mail as yet, July 1996)
Famous for its fresco decoration from the late 13th and early 14th centuries, including a series on the life of St Francis attributed by some to Giotto.

Pinacoteca Nazionale
Via delle Belle Arti 56
40126 Bologna
Tel (0039) (51) 243222
A range of Bolognese paintings from the 14th to the 18th century. Holds works by the Carracci family, Guercino, and Guido Reni.

Galleria dell'Accademia
Via Ricasoli 60
50122 Firenze
Tel (0039) (55) 238 8609
Fax 238 8699
Small but select: it contains Michelangelo's famous *David* and four of his unfinished *Slaves*.

Museo di San Marco
Piazza San Marco 1
50121 Firenze
Tel (0039) (55) 238 8608
Fax 238 8699
Superb devotional frescos by Fra Angelico and his assistants.

Museo Mediceo e Palazzo Medici Riccardi
Via Cavour 1
50129 Firenze
Tel (0039) (55) 27601
Its most famous holdings include Benozzo Gozzoli's *Journey of the Magi* (1459), paintings by Luca Giordano, and a series of Bronzino portraits (in the Museo Mediceo).

Museo Nazionale del Bargello
Palazzo del Bargello
Via del Proconsolo 4
50122 Firenze
Tel (0039) (55) 23885
Website http://www.thais.it/scultura/fmndbl.htm (for exhibitions only)
The world's best collection of Italian (particularly Florentine) Renaissance sculpture. Donatello, Michelangelo, Verrochio, and other major figures are represented by key works. Among later works, Bernini's bust of his mistress Constanza Buonarelli stands out.

Galleria degli Uffizi
Piazza degli Uffizi
50122 Firenze
Tel (0039) (55) 2388651/2
Fax 238 8699
Website http://musa.uffizi.frienze.it.
E-mail www @ musa.uffizi.frienze.it.
One of the most famous galleries in the world, with an unrivalled collection of Italian Renaissance painting. Most of the great names are represented by key masterpieces, among them Botticelli's *Primavera* and the *Birth of Venus*. It is less well known for its holdings of Early Netherlandish art (see especially van der Goes's Portinari altarpiece) and for works by Dutch, Flemish, French, and German painters such as Dürer and Claude. There is also an interesting collection of artists' self-portraits, begun in the 17th century. The top floor contains examples of 19th-century Italian art.

Palazzo Pitti (Galleria Palatina)
Piazza Pitti
50125 Firenze
Tel (0039) (55) 218741
This is the main home of the Medici collection. The palace was begun in the 15th century and was opulently decorated by Pietro da Cortona in the 17th century. It still retains the feeling of a great princely collection. Italian artists are best represented, but there are also good examples of the works of Rubens and Van Dyck.

Pinacoteca di Brera
Via Brera 28
20121 Milano
Tel (0039) (2) 800985/808387/862634
One of Italy's best known collections of Old Masters. It includes masterpieces by Raphael (*Marriage of the Virgin*), Piero della Francesca (*Madonna and Child with Duke Frederico of Urbino*), and Tintoretto (*Finding of the Body of St Mark*).

Galleria Borghese
Piazza Scipione Borghese 5
00197 Roma
Tel (0039) (6) 85 85 77
A collection based on the treasures amassed by the pleasure-loving Cardinal Scipione Borghese in the early 17th century. It is housed in a villa built for him in 1613–15; the decoration includes frescos by Lanfranco. Cardinal Borghese was an early collector of Caravaggio and a major patron of Bernini. Both these artists are superbly represented in the collection. Its other attractions include works by Giovanni Bellini, Raphael, and Titian.

Galleria Doria Pamphili
Piazza del Collegio Romano 1/A

00186 Roma
Tel (0039) (6) 679 4365
A superb collection of 17th-century Italian and Northern European landscapists set amongst antique sculpture, tapestries, and 18th-century furniture. The most famous work in the collection is Velázquez's incomparable portrait of Pope Innocent X (Giambattista Pamphili).

Poland
Muzeum Narodowe w Warszawie
(National Museum in Warsaw)
Al. Jerozolimskie 3
00–495 Warszawa
Tel (0048) (22) 621 1031
Fax 622 8559
Website will be available at the end of July 1996
E-mail muznw@plearn.edu.pl
This museum has a marvellous collection of minor Dutch paintings. It is also remarkable for a group of over 20 views of Warsaw in the 18th century by Bernardo Bellutto and a magnificent equestrian portrait of Stanislaw Kostka Potocki by Jacques-Louis David. It also contains Italian, Polish, and Russian works (including paintings by the Wanderers).

Romania
Muzeul Naţional de Artă al Romaniei
Calea Victoriei nr. 49–53
70101 Bucureşti
Tel (0040) (1) 615 5193/312 4327
Website http://www.itc.ro/museum/museum/html
The museum occupies part of the old royal palace opposite the old communist party headquarters in Piaţa Revolutiei. It was in the direct firing line during the December 1989 uprising and more than 100 paintings were destroyed. Lack of funds has delayed the reopening of the building, which is now targeted to the year 2000. However, the museum has a virtual gallery on its website which gives a flavour of what it has to offer: several sculptures by Brancusi, paintings by El Greco, Rembrandt, and Rubens, and fine French Impressionist and Post-Impressionist works. At one end of its historical spectrum, the museum houses a magnificent set of early 16th-century 'royal doors' from an Orthodox church in Moldavia while at the other it claims the pick of 19th- and 20th-century Romanian art.

Russia
Hermitage
Gosudarstvennij Ermitaj
(The State Hermitage Museum)
Dworcowaja Nabereshnaja 34–36
St Petersburg 191065
Tel (007) (812) 212 9545
One of the greatest museums in the world, with collections that are breathtaking in terms of quality and quantity. There are more than 1,000 rooms displaying a vast range of treasures based on the imperial collections built up by Peter the Great, Catherine the Great, and other rulers. The museum is strong in many areas of painting including 17th-century Dutch art (the largest collection in the world), 17th-century Flemish, 17th-, 18th-, late 19th-, and early 20th-century French, Early and High Renaissance Italian, and Venetian. There are also examples of British, German, Early Netherlandish, and Spanish painting. The wonderful representation of French painting from Impressionism to Cubism stems from the collections of two Moscow merchants, Ivan Morozov and Sergei Shchukin.

Gosud Arstvennaja Tretjakovskaja Galerija
(State Tretyakov Gallery)
Lavrushinskij Per 10
117049 Moskva
Tel (007) (095) 230 7788
The largest collection of Russian art in the world, ranging from the 10th century to the present. The highlights include icons by Andrei Rublev and numerous portraits and historical scenes by Repin, the most famous Russian painter of the 19th century.

Spain
Fundació Joan Miró
Barcelona
Tel (0034) (93) 329 1908
Website http://www.bcn.fjmiro.es/
E-mail fjmiro@bcn.fjmiro.es
The world's largest collection of art by this leading Surrealist opened in 1975 in a building designed specially for it by Josep Lluís Sert. The gallery contains over 10,000 paintings, sculptures, prints, drawings, and works in other media created by Miró, as well as a library. Miró himself donated most of the collection which spans the whole of his career.

Museo Nacional del Prado
Paseo del Prado
28014 Madrid
Tel (0034) (91) 486 0950/420 2836
This is a remarkably individual national collection which came into being thanks mainly to the Spanish royal family who started acquiring paintings in the 15th century. It has far and away the greatest collection of Spanish art in the world, with Velázquez and Goya supremely well represented, and virtually all the other major artists present in

force. It is also strong in Early Netherlandish, Flemish, and Italian painting, with particularly fine works by Bosch, Rubens, and Titian.

Centro de Arte Reina Sofia
Calle Santa Isabel 52
28012 Madrid
Tel (0034) (91) 467 5161
A major international collection of contemporary art housed in an 18th-century building. Picasso's *Guernica* is the highlight.

Museo de Bellas Artes
Plaza del Museo 9
41001 Sevila 1
Tel (0034) (954) 221 829
A major collection of Spanish art, particularly rich in the work of painters active in Seville. Highlights include El Greco's portrait of his son, Murillo's *The Charity of St Thomas Villanueva*, and Zurbarán's *St Hugo and the Carthusian Monks*.

Museo Nacional de Escultura
Cadenas de S Gregorio 1
47011 Valladolid
Tel (0034) (983) 250375
Superb collection of sculpture, rich in the painted wooden statues of religious figures that were a Spanish speciality in the 16th and 17th centuries.

Museo Thyssen-Bornemisza
Paseo del Prado 8
28014 Madrid
Tel 00341 369 0151
Fax 4202780
website http://www.offcampus.es/museo.thyssen-bornemisza
E-mail museo.thyssen-bornemisza@offcampus.es
The museum opened in 1992 to give a new, permanent home to the superb Thyssen-Bornemisza collection of European and American paintings. The works range from the 13th to the 20th century and highlights include Italian quattrocento portraits, Dutch and Flemish paintings, 18th-century Venetian scenes, and a wonderful selection of Impressionist, Post-Impressionist, and German Expressionist canvases, as well as modern and contemporary art.

Sweden

Göteborgs Konstmuseum
Götaplatsen
41256 Göteborg
Tel (0046) (31) 612980
Fax 184119
Website http://www.westnet.se/tourist/sights/museums/art.html (This is an unofficial site)
No e-mail address
While it has a wide-ranging collection of European art including works by Rembrandt, Monet, van Gogh, and Picasso, the most unusual section of the gallery is devoted to Scandinavian art from the 17th century. It contains excellent paintings by Carl Larsson and Anders Zorn, for example.

Moderna Museet
Birger Jarlsgatan 57
10327 Stockholm
Box 16382
Tel (0046) (8) 666 4250
Fax 661 8311
Website http://sunsite.kth.se/museums/moderna (rudimentary information)
An excellent collection of Swedish and international 20th-century art. Holds works by Oldenburg and Rosenquist as well as classic examples of the early modern movement by Braque, Léger, and Picasso.

Nationalmuseum
Södra Blasieholmshamnen
10324 Stockholm
Box 16176
Tel (0046) (8) 666 4250
Fax 611 3719
Website http://sunsite.kth.se/museums/national (this is an unofficial site containing rudimentary information)
E-mail nat-se@nordm.se
A wide-ranging collection of European art from the Middle Ages to the 20th century. Recommended especially for its Netherlandish and Swedish paintings from the 17th, 18th, and 19th centuries, it also has one of the best 18th-century French collections outside France. Boucher's sublime *Triumph of Venus* is one of its most spectacular holdings.

United Kingdom

Fitzwilliam Museum
University of Cambridge
Trumpington Street
Cambridge CB2 1RB
Tel (0044) (1223) 332900
Fax 332923
Websites http://www.cam.ac.uk/CambUniv/rep/museums/Fitz.html;
http://www.cam.ac.uk/CambArea/Fitz.html
An archaeological museum and art gallery, it has a fine collection of Old Masters, particularly of Italian works. Domenico Veneziano, Salvator Rosa, Titian, and Veronese are among the artists represented by outstanding works. These are arranged in rooms containing antique furniture and ceramics. The museum also holds a wide-ranging collec-

tion of miniatures, prints, drawings (notably by Holbein) and British watercolours as well as 20th-century art.

National Gallery of Scotland
The Mound
Edinburgh EH2 2EL
Tel (0044) (131) 556 8921
Fax 220 0917
While it is relatively small, this is one of the most attractive national collections to be found anywhere in the world. It includes masterpieces from most of the major European schools. Highlights include Gauguin's *Vision after the Sermon*, Poussin's series on *Seven Sacraments*, and Velázquez's *Old Woman Cooking Eggs*.

Courtauld Institute Galleries
Somerset House
Strand
London WC2R 0RN
Tel (0044) (181) 873 2526
The collection is principally famous for its wonderful holdings of French Impressionist and Post-Impressionist paintings, among them Manet's *A Bar at the Folies-Bergères*, but it also includes a magnificent group of works by Rubens and many other treasures.

Dulwich College Picture Gallery
College Road
London SE21 7BG
Tel (0044) (181) 693 5254
Fax 693 0923
A small but excellent collection of British, Dutch, Flemish, French, and Italian paintings. Artists who are best represented include Poussin, Rubens, and Van Dyck.

National Gallery
Trafalgar Square
London WC2N 5DN
Tel (0044) (171) 839 3321
Fax 930 4764
Website http://ukguide.cs.ucl.ac.uk/local/museums/National Gallery.html
One of the finest of all the national collections of Western art. It contains superb examples of all the major European schools, offering a more balanced view of the mainstream of Western painting from *c.*1300 to *c.*1900 than can be obtained in any other gallery. The Sainsbury Wing is a new extension to the gallery, housing paintings up to *c.*1510, as well as temporary exhibitions.

Tate Gallery
Millbank
London SW1P 4RG
Tel (0044) (171) 887 8000
Fax 887 8008
(No website or e-mail until the gallery opens at Bankside)
Contains the national collection of British art from the 16th century to *c.*1900 and the national collection of modern art, including British, European, and American paintings and sculpture. The Clore Gallery extension, which opened in 1987, is devoted specifically to Turner's paintings, drawings, and sketches. The British section contains work by many of the country's foremost artists and includes a superb collection of Blake's drawings. The 20th-century collection is one of the largest in the world. It has two regional branches (in Liverpool and St Ives) but to solve the problem of overcrowding it has campaigned successfully for a new site at the Bankside power station in Southwark, London. The new gallery will house virtually all of its modern holdings and is due to open in 2000.

Victoria & Albert Museum
South Kensington
London SW7 5RL
Tel (0044) (171) 589 6371
Fax 938 8379
Websites http://www.vam.ac.uk;
http://ukguide.cs.ucl.ac.uk/local/museums/VictoriaAndAlbert.html
National Art Library website http://www.nal.vam.ac.uk
E-mail (for comments on website) 100140.3103@compuserve.com
The V&A was conceived primarily as a museum for the applied and decorative arts but it holds many outstanding examples of fine art as well. These include superb collections of sculpture, miniatures, and watercolours, a celebrated series of tapestry cartoons by Raphael (on loan from the royal collection), and the largest representation anywhere of Constable's work.

Wallace Collection
Hertford House
Manchester Square
London W1M 6BN
Tel (0044) (171) 935 0687
Fax 224 2155
Website http://www.demon.co.uk/heritage/wallace
E-mail administration@wallcoll.demon.co.uk
A collection that is particularly famous for its 18th-century French paintings and also includes many memorable works from other countries. Notable works include Hals's *The Laughing Cavalier*, Fragonard's *The Swing*, and Titian's *Perseus and Andromeda*.

Ashmolean Museum
Beaumont Street
Oxford OX1 2PH
Tel (0044) (1865) 278000
Fax 278018
Website http://www.ashmol.ox.ac.uk/
E-mail jonathan.moffett@ashmus.ox.ac.uk (for comments on website)
One of the greatest museums in Britain, especially rich in ancient sculpture, Italian Renaissance painting, and Old Master drawings. There are also good representations of British art, especially the Pre-Raphaelites and the Camden Town group.

USA

Museum of Fine Arts
465 Huntingdon Avenue
Boston
MA 02115
Tel (001) (617) 267 9300
Fax 267 0280
Website http://www.mfa.org/
E-mail webmaster@mfa.org (for comments on website)
One of the outstanding museums in the USA. The collections are wide ranging, but it is particularly renowned for its holdings in 19th-century French painting from 1825. These include Gauguin's *Where Do We Come From? What Are We? Where Are We Going To?* and a superb array of Monets.

Albright-Knox Art Gallery
1285 Elmwood Avenue
Buffalo
NY 14222
Tel (001) (716) 882 8700
Fax 882 1958
Website http://www.akag.org (forthcoming)
The gallery has built its reputation around contemporary American art (see especially works by Klein, Nevelson, and Samaras). But it also has a fine collection of European paintings, representing the British, Dutch, and French schools. It is particularly strong in early modernist art, of which Gauguin's *Yellow Christ* is an outstanding example.

The Art Institute of Chicago
111 Michigan Avenue
Chicago
IL 60603
Tel (001) (312) 443 3600
Fax 443 0849
Website http://www.artic.edu/
It has a wide-ranging collection which includes Dutch, Flemish, and Italian Old Masters, 19th-century French paintings, and examples of early 20th-century European art. The most famous works in the collection include El Greco's *Assumption of the Virgin*, Seurat's *Sunday Afternoon on the Island of the Grande Jatte*, and Grant Wood's *American Gothic*.

The Cleveland Museum of Art
University Circle
11150 E Boulevard
Cleveland
OH 44106–1797
Tel (001) (216) 421 7340/7350
Fax 421 0411
Website http://www.clemusart.com/
A major collection of Oriental and Western art. The American and European sections include Dutch and Flemish paintings, 18th-, 19th-, and 20th-century French art, Spanish art and important works by Motherwell, Guston, and Rothko.

Detroit Institute of Arts
5200 Woodward Avenue
Detroit
Michigan
MI 48202
Tel (001) (313) 833 7900/7895
Fax 833 2357
Website http://www.dia.org/
E-mail web@www.dia.org (for comments on website)
A world-famous collection of Oriental and Western art from antiquity to the modern age which contains over 30,000 items. Early Netherlandish, Dutch, Italian, and American art from the 16th to the 20th centuries feature prominently in its Western section. Highlights from these holdings include Poussin's *Diana and Endymion*and Whistler's *Nocturne in Black and Gold: The Falling Rocket*, the picture that sparked off his libel trial with Ruskin.

J. Paul Getty Museum
17985 Pacific Coast Highway
Malibu
CA 90265
Postal address: PO Box 2112, Santa Monica, CA 90406
Tel (001) (310) 459 7611 (Admin)/458 2003 (Information)
Fax 454 6633
This gallery and museum is housed in a copy of an ancient Roman villa. Appropriately, it has a fine collection of classical antiquities among its rich and varied holdings of fine and applied art. Italian Renaissance paintings and Old Master drawings are other strongly represented areas.

Yale Center for British Art
1080 Chapel Street
New Haven

Connecticut CT 06520
Postal address: PO Box 208280, CT 06520 8280
Tel (001) (203) 432 2800
A superb collection of British painting from the 16th to the 19th centuries. Contains works by Constable, Stubbs, Turner, and many others.

Frick Collection
1 E 70th Street
NY 10021
New York
Tel (001) (212) 288 0700
Fax 628 4417
The collection is housed in a Neoclassical building furnished as though it were a private house. In terms of size the collection is fairly modest, but in terms of quality it is superb, with a number of world-famous masterpieces, including Bellini's *Ecstasy of St Francis*, Holbein's *Sir Thomas More*, and Rembrandt's *The Polish Rider*.

Metropolitan Museum of Art
1000 Fifth Avenue
NY 10028
New York
Tel (001) (212) 879 5500
Fax 570 3879
Website http://www.metmuseum.org/
E-mail (for tours) tours@metmuseum.org;
(for information) education@metmuseum.org;
(for comments on website) webmaster@metmuseum.org
One of the world's largest and most varied collections of fine and applied art, it holds around 2 million works. It has one of the world's finest collections of Old Master paintings, with Rembrandt and Vermeer being among those best represented. American art, too, is present in force, with good examples of groups such as the Ash-can School and the Abstract Expressionists. The museum has an outpost at The Cloisters which presents a range of medieval art in a building made up of parts of medieval structures transported from Europe.

Museum of Modern Art
11 West 53rd Street
NY 10019
New York
Tel (001) (212) 708 9480
Fax 708 9889
Website http://www.moma.org/
E-mail (for website comments) comments@moma.org
MOMA boasts the world's largest holdings of modern painting and sculpture. The collection contains around 100,000 items and focuses on the development of modernism from the late 19th century (van Gogh's *Starry Night* is one of the most popular exhibits). The museum gives particular emphasis to the early and mid-part of the 20th century and Picasso's *Les Demoiselles d'Avignon* and *Night Fishing at Antibes* are among its most important paintings from this period. The Fauves, the German Expressionists, the Futurists, the Cubists, Dadaists, and Surrealists are among groups well represented. It also represents the American Abstract Expressionists and has a small range of Pop art. Its sculpture collection includes works by Brancusi, Giacometti, Moore, and Rodin.

Solomon R. Guggenheim Museum
1071 Fifth Avenue at 88th Street
NY 10128
New York
Tel (001) (212) 423 3600
Fax 423 3650
Website http://math240.lehman.cuny.edu/gugg/srgm.html
This museum is located in Frank Lloyd Wright's last important public building, an extraordinary, circular design with the works displayed on a continuous spiral ramp. The collection ranges from the late 19th century to contemporary art, and is particularly strong in works by leading avant-garde artists of the early 20th century (the collection of Kandinsky's works is the largest in the world). In 1992 the Guggenheim reopened after a two-year refurbishment programme. It celebrated its new, extended premises with a show of Russian avant-garde art from 1915 to 1932.

Philadelphia Museum of Art
26th Street at Benjamin Franklin Parkway
Philadelphia
PA 19130
Postal address: PO Box 7646, PA 19101
Tel (001) (215) 763 8100
Fax 236 4465
Website http://libertynet.org/pma/pmahome.html
E-mail (for comments on website) gregl@libertynet.org;
(education programme) pmaedud@libertynet.org
The collection is famous for its 20th-century holdings, particularly for its unrivalled collection of Marcel Duchamp's works and a superb representation of Brancusi's sculpture. There is much else besides, however, including a varied collection of Old Masters.

Freer Gallery of Art
Jefferson Drive at 12th Street SW
Washington DC 20560

Tel (001) (202) 357 4880
Fax 357 4911
Website (under Smithsonian Institution) http://www.si.edu/organiza/affil/
A collection which specializes in paintings by Whistler (who was a friend of the founder) as well as examples of Oriental art.

Hirshhorn Museum and Sculpture Garden
7th Street and Independence Avenue SW
Washington DC 20560
Tel (001) (202) 357 3091
Fax 786 2682
Website (under Smithsonian Institution)
http://ww.si.edu/organiza/museums/hirsh/homepage/hrsh.htm
Named after the Latvian-born financier and philanthropist Joseph H. Hirshhorn (1899–1981), the museum opened in 1974 with his personal collection of over 12,000 modern paintings and sculptures. The collection ranges from about 1880 to contemporary art, and is particularly strong in sculpture, in American painting, and in European painting since the Second World War. Willem de Kooning and Henry Moore are among the artists who are especially well represented.

National Gallery of Art
4th Street and Constitution Avenue NW
Washington DC 20565
Tel (001) (202) 737 4215
Fax 842 6176
Website (under Smithsonian Institution)
http://www.si.edu/organiza/affil/natgal/start.html
This is one of the most impressive collections of European and American painting in the world, and is especially rich in Italian Old Masters (one of the highlights is Leonardo da Vinci's portrait of Ginevra de' Benci, the only indisputed Leonardo painting in the USA). Other areas of strength include Dutch painting (with Rembrandt and Vermeer particularly well represented) and 18th-century British portraiture. Picasso's *Saltimbanques* is one of the outstanding works in the 20th-century collection.

A Selection of Christian and Classical themes in painting and sculpture

Christian themes

Tree of Jesse

Wells Cathedral (stained glass window, *c.*1340)
St Mary's, Abergavenny (late 14th-century wood carving)
Sucevita Monastery, Moldavia, Romania (external wall painting, 1596)

David

Donatello (1430s?, Bargello, Florence)
Andrea del Castagno (*c.*1450–7, National Gallery of Art, Washington, DC)
Verrocchio (*c.*1475, Bargello, Florence)
Michelangelo (1501–4, Accademia, Florence)
Bernini (1623, Galleria Borghese, Rome)

Herod's Feast (Salome's Dance)

Donatello, *The Feast of Herod* (font, *c.*1425, Siena Cathedral)
Beardsley (17 drawings published 1894, for Oscar Wilde's play *Salome*)

The Annunciation

Duccio (part of the *Maestà* altarpiece, 1308–11, National Gallery, London)
Master of the Flémalle?, *Mérode Altarpiece* (*c.*1425–30, Metropolitan Museum, New York)
Leonardo (*c.*1475, Uffizi, Florence)
Crivelli (1486, National Gallery, London)
Poussin (1657, National Gallery, London)
Dante Gabriel Rossetii (1850, Tate Gallery, London)

The Nativity

Giovanni Pisano, pulpit relief carving (1302–10, Pisa Cathedral, Italy)
Piero della Francesca (*c.*1470–5, National Gallery, London)
Geerten tot Sint Jans (*c.*1490, National Gallery, London)
Botticelli, *Mystic Nativity* (1500, National Gallery, London)

The Adoration of the Magi/Kings

Gentile da Fabriano (1423, Uffizi, Florence)
Botticelli, (*c.*1470, National Gallery, London)
Leonardo (*c.*1482, Uffizi, Florence)
Veronese (1573, National Gallery, London)
Velázquez (1619, Prado, Madrid)
Rubens (1623–4, King's College Chapel, Cambridge)

The Virgin/Madonna (and Child) (also includes scenes of the Holy Family and of the Virgin and Child with saints and/or donors)

First known representation of the Virgin (early 7th century, Monastery of St Catherine, Mt Sinai, Egypt)
Virgin and Child Enthroned (*c.*843–67, Hagia Sophia, Istanbul)
Notre Dame de la Belle Verrière (*c.*1150, stained-glass window, Chartres Cathedral, France)
Cimabue, *Madonna of Sta Trinità* (*c.*1280, Uffizi, Florence)
Duccio, *Maestà* (1308–11, Siena Cathedral Museum)
Giotto, *Ognissanti Madonna* (*c.*1310, Uffizi, Florence)
Giovanni Pisano (statue, *c.*1315, Prato Cathedral, Italy)
Jan van Eyck, *Madonna of Chancellor Rolin* (*c.*1435, Louvre, Paris)
Jean Fouquet, *Madonna and Child* (*c.*1450, Musée Royal des Beaux-Arts, Antwerp)
Piero della Francesca, *Madonna and Child with Federico da Montefeltro* (*c.*1475, Brera, Milan)
Crivelli, *Madonna and Child with Saints Francis and Sebastian* (1491, National Gallery, London)
Leonardo, *The Virgin and Child with St Anne and St John the Baptist* (cartoon, *c.*1500, National Gallery, London)
Giovanni Bellini, *The Madonna of the Meadow* (*c.*1510, National Gallery, London)
Raphael, *The Sistine Madonna* (1512–14, Gemäldegalerie, Dresden)
Parmigianino, *Madonna with the Long Neck* (*c.*1535, Uffizi, Florence)
Poussin, *The Holy Family* (*c.*1655, Hermitage, St Petersburg)
Maratta, *Virgin and Child* (*c.*1695, Pinacoteca, Vatican)
Henry Moore, *Madonna and Child* (1943–4, St Mathew's Church, Northampton)

The Flight into Egypt/Rest on the Flight into Egypt

Cranach (1504, Staatliche, Museen, Berlin)
Caravaggio (*c.*1595, Galleria Doria Pamphili, Rome)
Annibale Carracci (*c.*1604, Galleria Doria Pamphili, Rome)
Elsheimer (1609, Alte Pinakothek, Munich)

Rembrandt (1647, National Gallery, Dublin)
Giandomenico Tiepolo (22 etchings, published 1753)

The Baptism of Christ
Piero della Francesca (*c.*1450–60, National Gallery, London)
Verrocchio (*c.*1470, Uffizi, Florence)
Elsheimer (*c.*1598–1600, National Gallery, London)
Poussin (1649, National Gallery, Edinburgh, on loan from the Ellesmere Collection; part of the Seven Sacraments series)

The Transfiguration
Duccio (1306–11, panel from the *Maestà* altarpiece, National Gallery, London)
Giovanni Bellini (*c.*1490, Museo di Capodimonte, Naples)
Raphael (1517–20, Pinacoteca, Vatican)

The Last Supper
Andrea del Castagno (*c.*1445–50, Sta Apollonia, Florence)
Leonardo (*c.*1495, Sta Maria delle Grazie, Milan)
Veronese (renamed *The Feast in the House of Levi*, 1573, Accademia, Venice)
Tintoretto (1592–4, S. Giorgio Maggiore, Venice)
Emil Nolde (1909, Nolde Foundation, Seebüll, Germany)

The Agony in the Garden
Giovanni Bellini (*c.*1460, National Gallery, London)
Mantegna (*c.*1460, National Gallery, London)
Dürer (woodcut, 1511)
Correggio (*c.*1525, Wellington Museum, London)
El Greco (1597–1600, The Toldeo Museum of Art, Toledo, Ohio)

Ecce Homo
Corregio (*c.*1525–30, National Gallery, London)
Titian (1543, Kunsthistorisches Museum, Vienna)
Rembrandt (1634, National Gallery, London)
Cornelius (1839, Ludwigskirche, Munich)
Rouault (1952, Pinacoteca, Vatican)

The Way to Calvary
Dürer (woodcut, 1509)
Raphael (*c.*1516, Prado, Madrid)
Jacopo Bassano (*c.*1545, National Gallery, London)
Tintoretto (1536–87, Scuola di San Rocco, Venice)
Rubens (1636, Musées Royaux, Brussels)

The Crucifixion
Gero Crucifix (969–76, Cologne Cathedral)
Cimabue (*c.*1280, Sta Croce, Florence)
Giotto (*c.*1305, Arena Chapel, Padua)
Antonello da Messina (1475, National Gallery, London)
Master of the Virgo inter Virgines (*c.*1490, Bowes Museum, Barnard Castle, England)
Raphael, *The Mond Crucifixion* (*c.*1503, National Gallery, London)
Grünewald (*c.*1515, Musée d'Unterlinden, Colmar, Germany)
Tintoretto (1565–87, Scuola di San Rocca, Venice)
Ludovico Carracci, *Christ Crucified above Figures in Limbo* (1614, Sta Francesca Romana, Ferrara, Italy)
Velázquez (*c.*1632, Prado, Madrid)
Rembrandt, *The Three Crosses* (drypoint, 1653–*c.*1661)
Delacroix (1853, National Gallery, London)
Dali, *The Crucifixion of St John of the Cross* (1951, St Mungo Museum, Glasgow)
Bacon, *Three Studies for a Crucifixion* (1962, Solomon R. Guggenheim Museum, New York)

The Deposition
Rogier van der Weyden (*c.*1435, Prado, Madrid)
Giovanni Battista Rosso (1521, Galleria Pittorica, Volterra, Italy)
Rubens (1611–14, Antwerp Cathedral)
Rembrandt (*c.*1633, Alte Pinakothek, Munich)
Tiepolo (1750–60, National Gallery, London)

Pietà
Enguerrand Charonton?, *Avignon Pietà* (*c.*1460, Louvre, Paris)
Ercole de' Roberti (*c.*1485, Walker Art Gallery, Liverpool)
Michelangelo (1498–9, St Peter's Rome)
Titian (1573–6, Accademia, Venice)

The Stations of the Cross
Eric Gill (14 relief carvings, 1914–18, Westminster Cathedral, London)
Matisse (ceramic drawings, 1949–51, Chapel of the Rosary, Venice)

The Resurrection
Piero della Francesca (*c.*1460, Pinacoteca, Sansepolcro, Italy)
Gaudenzio Ferrari (*c.*1540, National Gallery, London)
El Greco (*c.*1600, Prado, Madrid)

The Assumption of the Virgin
Botticini (*c.*1475, National Gallery, London)
Titian (1516–18, Sta Maria Gloriosa dei Frari, Venice)
Correggio (*c.*1530, dome of Parma Cathedral, Italy)
Lanfranco (1625–7, dome of S. Andrea della Valle, Rome)

The Last Judgement
Giselbertus (*c.*1125–35, tympanum carving, Autun Cathedral)

Giotto (*c.*1305–6, Arena Chapel, Padua, Italy)
Rogier van der Weyden (*c.*1445–50, Hotel Dieu, Beaune, France)
Luca Signorelli (1499–1504, Orvieto Cathedral, Italy)
Michelangelo (1536–41, Sistine Chapel, Rome)
Rubens (*c.*1615, Alte Pinakothek, Munich)

Classical themes

The Judgement of Paris

Marcantonio Raimondi (*c.*1520, engraving from a design by Raphael)
Cranach (*c.*1550, Kunsthalle, Karlsruhe, Germany)
Balen (*c.*1600, Muzeu Brukenthal, Sibiu, Romania)
Wtewael (1615, National Gallery, London)
Rubens (*c.*1632–5, National Gallery, London)
Jan Both and Cornelis van Poelenburgh (*c.*1645–50, National Gallery, London)
Renoir (*c.*1913–14, Hiroshima Museum of Art)

Laocoön

Classical sculpture (*c.* AD 30, Vatican Galleries)
El Greco (*c.*1610, National Gallery, Washington, DC)

Diana and Actaeon

Titian (1556–9, National Gallery, Edinburgh, on loan from Ellesmere Collection)
Rottenhammer (1602, Alte Pinakothek, Munich)
Gainsborough (*c.*1785, British Royal Collection)

Apollo and Daphne

Antonio del Pollaiuolo (*c.*1470–80, National Gallery, London)
Bernini (1622–5, Galleria Borghese, Rome)
Poussin (*c.*1664, Louvre, Paris)

Bacchus and Ariadne

Titian (1522–3, National Gallery, London)
Annibale Carracci, *The Triumph of Bacchus and Ariadne* (1597–1600, central part of ceiling of Farnese Gallery, Rome)
Angelica Kauffmann (1794, Attingham Park, Shropshire, England)

Rape of Europa

Titian (1559–62, Isabella Stewart Gardner Museum, Boston)
Giambattista Tiepolo (*c.*1722, Accademia, Venice)
Serov (1910, Tretyakov Gallery, Moscow)

Venus

Botticelli, *The Birth of Venus* (*c.*1485, Uffizi, Florence)
Giorgione, *Venus Sleeping* (*c.*1510, Gemäldegalerie, Dresden)
Titian, *Venus of Urbino* (*c.*1538, Uffizi, Florence)
Boucher, *Triumph of Venus* (1740, Nationalmuseum, Stockholm)
Gibson, *Tinted Venus* (1851, Walker Art Gallery, Liverpool)

Hercules

Antonio del Pollaiuolo (bronze statuette, *c.*1480, Frick Collection, New York)
Bandinelli, *Hercules and Cacus* (1534, Piazza della Signoria, Florence)
Zurburán, *Hercules and Antaeus* (*c.*1634, Prado, Madrid)
Boucher, *Hercules and Omphale* (1730s, Pushkin Museum of Fine Arts, Moscow)
Thomas Hart Benson, *Achelous and Hercules* (1947, National Museum of American Art, Washington, DC)